S0-BDT-307

Applied Science

Applied Science

Engineering and Mathematics

Editor
Donald R. Franceschetti, Ph.D.
The University of Memphis

SALEM PRESS

A Division of EBSCO Information Services, Inc.
Ipswich, Massachusetts

GREY HOUSE PUBLISHING

∞ The paper used in these volumes conforms to the American National Standard for Permanence of Paper for Printed Library Materials, Z39.48-1992 (R1997)

Library of Congress Cataloging-in-Publication Data

Publisher's Cataloging-In-Publication Data
(Prepared by The Donohue Group, Inc.)

Applied science. Engineering & mathematics / editor, Donald R. Franceschetti.
-- [1st ed.].

p. : ill. ; cm.

Comprises articles on science and medicine extracted from the five-volume reference set Applied Science. Salem Press, 2012.
Includes bibliographical references and index.
ISBN: 978-1-61925-244-8

1. Engineering. 2. Mathematics. I. Franceschetti, Donald R., 1947- II. Title: Engineering & mathematics
III. Title: Engineering and mathematics

TA45 .A67 2013
600

Printed in the United States of America

CONTENTS

PUBLISHER'S NOTE

Applied Science: Engineering and Mathematics is a comprehensive volume that examines the relationship between engineering and mathematics, providing insight into the many ways in which both disciplines affect daily life. Understanding the interconnectedness of the different and varied branches of engineering and mathematics is important for anyone preparing for a career or endeavor in either field. Toward that end, essays look beyond basic principles to examine a wide range of topics, including industrial and business applications, historical and social contexts, and the impact a particular field of engineering or mathematics will have on future jobs and careers. A career-oriented feature details core jobs within the field, with a focus on fundamental and recommended coursework.

Applied Science: Engineering and Mathematics is specifically designed for a high-school audience and is edited to align with secondary or high-school curriculum standards. The content is readily accessible, as well, to patrons of both academic and university libraries. Librarians and general readers alike will also turn to this reference work for both basic information and current developments, presented in accessible language with copious reference aids. Pedagogical tools and elements, including a bibliographical directory of scientists, a timeline of major scientific milestones, and a glossary of key terms and concepts, round out this unprecedented and unique resource.

Scope of Coverage

Comprising 101 lengthy and alphabetically arranged essays on a broad range of subject fields, from long-established engineering fields to cutting edge fields such as spacecraft engineering, this excellent reference work addresses applied sciences in areas as diverse as aeronautics, energy, fluid dynamics, pneumatics, and climate modeling. *Applied Science: Engineering and Mathematics* also includes charts and tables, as well as "Fascinating Facts" related to each applied-science field..

Essay Length and Format

Essays in the encyclopedic set range in length from 3,000 to 4,000 words. All the entries begin with ready-reference top matter, including an indication of the relevant discipline or field of study and a summary statement in which the writer explains why the topic is important to the study of the applied sciences. A selection of key terms and concepts within that major field or technology is presented, with basic principles examined. Essays then place the subject field in historical or technological perspective and examine its development and implication. Also discussed are applications and applicable products and the field's impact on industry. Cross-references direct readers to related topics, and further reading suggestions accompany all articles.

Special Features

Several features continue to distinguish this reference series from other works on applied science. The front matter includes a table detailing common units of measure, with equivalent units of measure provided for the user's convenience. Additional features are provided to assist in the retrieval of information. An index illustrates the breadth of the reference work's coverage, directing the reader to appearances of important names, terms, and topics throughout the text, and is followed by the category list, which groups article titles by general area of interest.

The back matter includes several appendixes and indexes, including a biographical dictionary of scientists, a compendium of the people most influential in shaping the discoveries and technologies of applied science. A time line provides a chronology of all major scientific events across significant fields and technologies, from agriculture to computers to engineering to medical to space sciences. Additionally, a complete glossary covers all technical and other specialized terms used throughout the set, while a general bibliography offers a comprehensive list of works on applied science for students seeking out more information on a particular field or subject.

Acknowledgments

Many hands went into the creation of this work. Special mention must be made of editor Donald R.

Franceschetti, who played a principal role in shaping this volume and its contents. Thanks are also due to the many academicians and professionals who worked to communicate their expert understanding of the applied sciences to the general reader; a list of these individuals and their affiliations appears at the beginning of the volume. The contributions of all are gratefully acknowledged.

EDITOR'S INTRODUCTION: ENGINEERING AND MATHEMATICS

This volume presents the engineering and mathematics articles from the five-volume reference set *Applied Science*, for the benefit of students considering careers in engineering, mathematics, or computer science; their teachers; and their counselors. Other volumes cover technology, and science and medicine.

Something should be said at the outset about the relationships among science, technology, engineering, and mathematics. Terence Kealey states in his book *The Economic Laws of Scientific Research* (1996) that technology is the activity of manipulating nature, while science is the activity of learning about nature. To this we may add that engineering is another field devoted to manipulating nature. Leonard Mlodinow, in *Feynman's Rainbow* (2003) makes a distinction between the Greek way of approaching science and the Babylonian way. The Greeks, the great mathematicians of the ancient world, greatly admired pure thought. The Babylonians, technologists and engineers at heart, didn't care much about fine theoretical points but made important practical discoveries. Or, in the classical world, a distinction may be drawn between the Greeks and the Romans. The Greeks did beautiful mathematics and built beautiful small temples with roofs supported by many columns. The Romans realized that stone was much stronger under compression than under tension and so used arches to span much larger distances. Thus they could build much larger temples. They also used their arches to build aqueducts to carry water to populous cities.

Technology became far more scientific at the time of the Industrial Revolution. In England the process was tied to the emergence of the Royal Institution, founded in 1799, which made it possible for the general public to learn about technical matters in their spare time and without the training in classical languages and the social standing expected at the universities of Oxford and Cambridge. At about the same time these august institutions were joined by the "red brick" institutions, which emphasized the training of students for industrial leadership. In the United States, the venerable Ivy League schools—including Harvard, Yale, and Princeton—were joined by the newly established state universities. Some of these were designated as land-grant colleges under the Morrill Act (1862) and were set up to advance agriculture and the mechanical arts. This act provided that each state in the Union designate a parcel of land, income from the sale of which would be used to support public colleges and provide for agricultural experimental stations where new ideas in agriculture could be tested.

Educational practice is sometimes slow to recognize changing patterns in society. Colleges in the United States still award bachelor's, master's and doctoral degrees that have their roots in the Middle Ages. In fact the Bachelor of Arts, Master of Arts, and Doctor of Philosophy degrees are still awarded by the majority of American universities, although the coursework required and the major subjects offered differ greatly from those of a century ago. The bachelor's degree is still awarded to individuals who have completed a four-year course of study, although the emphasis may be on computer science or psycholinguistics instead of the liberal arts. Many schools now award the Bachelor of Science degree to graduates with majors in the sciences (although some, particularly in the Ivy League, award the traditional B.A. degree to all graduates). Additional study is required for the master's degree, and a period of intensive research and publication of a dissertation is required for the Doctor of Philosophy degree.

Despite the antiquity of the bachelor's, master's and doctoral designations, what the degrees actually meant was, for a time, quite flexible in early-twentieth-century America. Eventually a measure of quality control was achieved, with colleges being chartered by state governments and subject to review by regional accrediting agencies and, in some fields, by additional specialized agencies.

In modern usage an engineer is someone who has completed a program of study (usually four years in length) and been awarded a bachelor's degree in some field of engineering. Many states in the United States provide a route for engineers to be licensed as professional engineers (P.E.s) following college graduation, an apprenticeship period as an Engineer in Training (E.I.T.), and the passing of a rigorous examination. Some engineers regard the P.E. credential as equal to or more desirable than a Ph.D. or Doctor of Engineering (D.Eng.) degree. Engineering programs may be distinguished from technology programs by their mathematics requirement. Programs leading to a bachelor's degree in, say, electrical engineering usually include at least three semesters of calculus and one in differential equations before the Bachelor of Science in Electrical Engineering (B.S.E.E.) is awarded. In contrast, programs leading to a bachelor's degree in electrical engineering technology would require a single introductory course in the calculus. In the United States, both types of programs receive accreditation from the American Board for Engineering and Technology (ABET), which is responsible for maintaining standards of instruction in those fields.

Mathematics per se is one of the oldest of human pursuits, having grown out of the building crafts and the need to keep track of one's land, crops, and domestic animals. It reached one peak in classical Greece when the geometry of Euclid seemed to accurately describe the nature of physical space. (That is, the theorems of Euclid seemed to be valid for measured distances.) The Arabs developed algebra somewhat later, while Europe was stuck in the "Dark Ages." The calculus was developed in the seventeenth century, while statics came into its own in the nineteenth century. The twentieth century saw the extensive development of mathematical logic, leading to the design of digital computers. With the invention of the transistor in 1947 and the subsequent development of integrated circuits (computer "chips," which could be manufactured by photolithography), computer technology has grown explosively, while the physical size of computers has shrunk to the point that thousands of transistors can be contained in a cubic inch. Computer science and computer engineering now offer many opportunities for engineers and mathematicians.

To the Student

Your decision to study mathematics or engineering is one that, with persistence, will lead to a comfortable income and, more importantly, a variety of interesting work assignments. While you are in junior high and high school, be sure to practice writing and organizing presentations. If possible do not interrupt your study of mathematics, because mathematics courses build on each other. Do not postpone your first course in physics, on which all engineering is based. Resist the temptation to take an easy semester. In college join the student chapter of a professional society such as the Association for Computing Machinery or the Institute of Electrical and Electronic Engineers. Look for summer work with an engineering firm.

As a future mathematician or engineer, you are likely to be affected by two recent trends. One is the recent growth in on-line education. Although the widespread availability of computers and the Internet make it possible for students almost anywhere to study almost anything, there is still much to be said for the pre-professional socialization that goes on in college. For mathematics and engineering students, needing to keep to a schedule of assignments, work collaboratively with other students, develop presentation skills, and find information by research are all established aspects of college life. The second trend is the increasing importance of the adult student and the need for continuing education. While there are still individuals who work at the same job for 30 or more years, it is likely that you will change jobs several times over the course of your career, and you can count on needing periodic retraining.

A Brief Survey of Engineering and Mathematics

Some fields of engineering are of great antiquity. The seven wonders of the ancient world were basically works of civil engineering. The great Archimedes was both engineer and mathematician. The pyramids of Egypt and the aqueducts and roads of the Roman Empire are early examples of civil engineering. The design of fortifications and weapons systems may be taken as an example of military engineering. Mechanical engineering has about as long a history as civil and military engineering, but it really came into its own in the seventeenth century as it became apparent that the physics of Isaac Newton could be applied to mechanical devices. The invention of the steam engine greatly increased the demand for mechanical engineers and precision machinists: the Industrial Revolution in England was largely a consequence of the steam engine, which was first patented by James Watt. With the Watt steam engine as a "prime mover," factories were freed from their dependence on wind and moving water as power sources. Extensive coal mining was undertaken to provide fuel for the new source of mechanical power. Coal and later petroleum led to the fields of chemical and petroleum engineering. Nineteenth-century discoveries about electricity and magnetism would then lead to the field of electrical engineering, which grew explosively in the first half of the twentieth century.

At the close of World War II, there was considerable debate as to the proper relationship between science and government in peacetime. Some conservatives wanted to impose strong military control over scientific research. In 1945 Vannevar Bush, who had been director of the wartime Office of Scientific Research and Development, published the report *Science: The Endless Frontier*, which painted a very rosy picture of the gains to be derived by public investment in basic scientific research. Among other things, Bush emphasized the role that government could play in supporting research in universities, research to fill in the middle ground between purely academic research and research directed toward immediate objectives. Further, universities had a natural role to play in training the next generation of scientists and engineers. Bush's arguments led to establishment of the National Science Foundation and to expanded research funding within and by the National Institutes of Health in the United States. This trend was accelerated by the Soviet Union's launch of Sputnik 1, the first Earth satellite, in 1957, an event that shook the American public's faith in the inevitable superiority of American technology. In the twenty-first century, science and technology are supported by many sources, public and private, and the pace of development is perhaps even greater than at any time in the past.

The Preliterate World

According to archaeologists and anthropologists, the development of written language occurred relatively recently in human history, perhaps about 3000 B.C.E. Many of the basic components of technology date to those preliterate times, when humans struggled to secure the basic necessities—food, clothing, and shelter—against a background of growing population and changing climate. When food collection was limited to hunting and gathering, knowledge of the seasons and animal behavior was important for survival. The development of primitive stone tools and weapons greatly facilitated both hunting and obtaining meat from animal carcasses, as well as the preservation of the hides for clothing and shelter. Sometime in the middle Stone Age, humans obtained control over fire, making it possible to soften food by cooking and to separate some metals from their ores. Control over fire also made it possible to harden earthenware pottery and keep predatory animals away at night.

With gradually improved living conditions, human fertility and longevity both increased, as did competition for necessities of life. Spoken language, music, magical thinking, and myth developed as a means of coordinating activity. Warfare, along with more peaceful approaches, was adopted as a means of settling disputes, while society was reorganized to ensure access to the necessities of life, including protection from military attack.

The Ancient World

With the invention of written language, it became possible to enlarge and coordinate human activity on an unprecedented scale. Several new areas of technology and engineering were needed. Cities were established so that skilled workers could be freed from direct involvement in food production. Logistics and management became functions of the scribal class, the members of which could read and write. Libraries were built and manuscripts collected. The beginnings of mathematics may be seen in building, surveying, and wealth tabulation. Engineers built roads so that a ruler could oversee his enlarged domain and troops could move rapidly to where they were needed. Taxes were imposed to support the central government, and accounting methods were introduced. Aqueducts were needed to bring fresh water to the cities.

Astronomy, the Calendar, and Longitude

One might think that astronomy could serve as a paradigmatic example of fundamental science, but astronomy—and later space science—provides an example of a scientific activity often pursued for practical benefit. Efforts to fix the calendar, in particular, provide an interesting illustration of the interaction among pure astronomy, practicality, and religion.

The calendar provided a scheme of dates for planting and harvesting. Ancient stone monuments in Europe and the Americas may have functioned in part as astronomical observatories to keep track of the solstices and equinoxes. The calendar also provided a means for keeping religious feasts, such as Easter, in synchrony with celestial events, such as the vernal equinox. Roman emperor Julius Caesar introduced a calendar based on a year of 365 days. This calendar, called the Julian, would endure for more than 1,500 years. Eventually, however, it was realized that the Julian year was about six hours shorter than Earth's orbital period and so was out of sync with astronomical events. In 1583 Pope Gregory XIII introduced the system of leap-year days. The Gregorian calendar is out of sync with Earth's orbit by only one day in 3,300 years.

Modern astronomy begins with the work of Nicolaus Copernicus, Galileo Galilei, and Sir Isaac Newton. Copernicus was the first to advance the heliocentric model of the solar system, suggesting that it would be simpler to view Earth as a planet orbiting a stationary sun. Copernicus, however, was a churchman, and fear of opposition from church leaders (who were theologically invested in an Earth-centered hierarchy of the universe) led him to postpone publication of his ideas until the year of his death. Galileo was the first to use a technological advance, the invention of the telescope, to record numerous observations that called into question the geocentric model of the known universe. Among these was the discovery of four moons that orbit the planet Jupiter.

At the same time, advances in shipbuilding and navigation served to bring the problem of longitude to prominence. Out of sight of land, a sailing ship could easily determine its latitude on a clear night by noting the elevation of the pole star above the horizon. To determine longitude, however, required an accurate measurement of time, which was difficult to do onboard a moving ship at sea. Galileo was quick to propose that occultations by Jupiter of its moons could be used as a universal clock. The longitude problem also drew the attention of the great Newton. Longitude would eventually be solved by John Harrison's invention of a chronometer that could be used on ship, and the deciphering of it also fostered an improved understanding of celestial mechanics. John Harrison was a carpenter, and early technologist.

Since the launching of the first artificial satellites, astronomy—or, rather, space and planetary science—has assumed an even greater role in applied science. The safety of astronauts working in space requires understanding the dynamics of solar flares. A deeper understanding of the solar atmosphere and its dynamics could also have important consequences for long-range weather prediction.

The Scientific Revolution

The Renaissance and the Protestant Reformation marked something of a rebirth of scientific thinking. This "scientific revolution" would not have been possible without Gutenberg's printing press and the technology of printing with movable type. With wealthy patrons, natural philosophers felt secure in challenging the authority of Aristotle. Galileo published arguments in favor of the Copernican solar system. In the *Novum Organum* (1620; "New Instrument"), Sir Francis Bacon formalized the inductive method, by which generalizations could be made from observations, which then could be tested by further observation or experiment. In England in 1660, with the nominal support of the British Crown, the Royal Society was formed to serve as a forum for the exchange of scientific ideas and the support and publication of research results. Need for larger-scale studies brought craftsmen into the sciences, culminating in the recognition of the professional scientist. Earlier Bacon had proposed that the government undertake the support of scientific investigation for the common good. Bacon himself tried his hand at frozen-food technology. While on a coach trip, he conceived the idea that low temperatures could preserve meat. He stopped the coach, purchased a chicken from a farmer's wife, and stuffed it with snow. Unfortunately he contracted pneumonia while doing this experiment and died forthwith.

The Industrial Revolution followed on the heels of the scientific revolution in England. Key to the Industrial Revolution was the technology of the steam engine, the first portable source of motive power that was not dependent on human or animal muscle. The modern form of the steam engine owes much to James Watt, a self-taught technologist. The steam engine powered factories, ships, and, later, locomotives. In the case of the steam engine, technological advance preceded the development of the pertinent science—thermodynamics and the present-day understanding of heat as a random molecular form of energy.

It is not possible, of course, to do justice to the full scale of applied science and technology in this short space. In the remainder of this introduction, consideration will be given to only a few representative fields, highlighting the evolution of each area and its interconnectedness with fundamental and applied science as a whole.

Chemical Engineering

In 1792 the Scottish inventor William Murdock

discovered a way to produce illuminating gas by the destructive distillation of coal, producing a cleaner and more dependable source of light than previously was available and bringing about the gaslight era. The production of illuminating gas, however, left behind a nasty residue called coal tar. A search was launched to find an application for this major industrial waste. An early use, the waterproofing of cloth, was discovered by the Scottish chemist Charles Macintosh, resulting in the raincoat that now carries his name. In 1856 English chemist William Henry Perkin discovered the first of the coal-tar dyes, mauve. The color mauve, a deep lavender-lilac purple, had previously been obtained from plant sources and had become something of a fashion fad in Paris by 1857. The demand for mauve outstripped the supply of vegetable sources. The discovery of several other dyes followed.

The possibility of dyeing living tissue was rapidly seized on and applied to the tissues of the human body and the microorganisms that afflict it. German bacteriologist Paul Ehrlich proposed that the selective adsorption of dyes could serve as the basis for a chemically based therapy to kill infectious disease-bearing organisms.

Electrical Engineering

The history of electromagnetic devices provides an excellent example of the complex interplay of fundamental and applied science. The phenomena of static electricity and natural magnetism were described by Thales of Miletus in ancient times, but they remained curiosities through much of history. The magnetic compass was developed by Chinese explorers in about 1100 b.c.e., and the nature of Earth's magnetic field was explored by William Gilbert (physician to Queen Elizabeth I) around 1600. By the late eighteenth century, a number of devices for producing and storing static electricity were being used in popular demonstrations, and the lightning rod, invented by Benjamin Franklin, greatly reduced the damage due to lightning strikes on tall buildings. In 1800 Italian physicist Alessandro Volta developed the first electrical battery. Equipped with a source of continuous electric current, scientists made electrical and electromagnetic discoveries, practical and fundamental, at a breakneck pace.

The voltaic pile, or battery, was employed by British scientist Sir Humphry Davy to isolate a number of chemical elements for the first time. Davy also did demonstrations at the Royal Institution involving the effect of electricity on the bodies of recently hanged criminals. (He would later describe these to his friend Mary Shelley who went on to pen the novel *Frankenstein.*) In 1820 Danish physicist Hans Christian Ørsted discovered that any current-carrying wire is surrounded by an electric field. In 1831 English physicist Michael Faraday discovered that a changing magnetic field would induce an electric current in a loop of wire, thus paving the way for the electric generator and the transformer. In Albany, New York, schoolteacher Joseph Henry set his students the challenge of building the strongest possible electromagnet. Henry would move on to become professor of natural philosophy at Princeton University, where he invented a primitive telegraph.

The basic laws of electromagnetism were summarized in 1865 by Scottish physicist James Clerk Maxwell in a set of four differential equations that yielded a number of practical results almost immediately. These equations described the behavior of electric and magnetic fields in different media, including in empty space. In a vacuum it was possible to find wavelike solutions that appeared to move in time at the speed of light, which was immediately realized to be a form of electromagnetic radiation. Further, it turned out that visible light covered only a small frequency range. Applied scientists soon discovered how to transmit messages by radio waves, electromagnetic waves of much lower frequency.

The Computer

One of the most clearly useful of modern artifacts, the digital electronic computer, as it has come to be known, has a lineage that includes the most abstract of mathematics, the automated loom, the vacuum tubes of the early twentieth century, and the modern sciences of semiconductor physics and photochemistry. Although computing devices such as the abacus and slide rule themselves have a long history, the programmable digital computer has advanced computational power by many orders of magnitude. The basic logic of the computer and the computer program, however, arose from a mathematical logician's attempt to answer a problem arising in the foundations of mathematics.

From the time of the ancient Greeks to the end of the nineteenth century, mathematicians had assumed that their subject was essentially a study of the real world, the part amenable to purely deductive reasoning. This included the structure of space and the basic rules of counting, which led to the rules of arithmetic and algebra. With the discovery of non-Euclidean geometries and the paradoxes of set theory, mathematicians felt the need for a closer study of the foundations of mathematics, to make sure that the objects that might exist only in their minds could be studied and talked about without risking inconsistency.

David Hilbert, a professor of mathematics at the University of Göttingen, was the recognized leader of German mathematics. At a mathematics conference in 1928, Hilbert identified three questions about the foundations of mathematics that he hoped would be resolved in short order. The third of these was the so-called decidability problem: Was there was a foolproof procedure to determine whether a mathematical statement was true or false? Essentially, if one had the statement in symbolic form, was there a procedure for manipulating the symbols in such a way that one could determine whether the statement was true in a finite number of steps?

British mathematician Alan Turing presented an analysis of the problem by showing that any sort of mathematical symbol manipulation was in essence a computation and thus a manipulation of symbols not unlike the addition or multiplication one learns in elementary school. Any such symbolic manipulation could be emulated by an abstract machine that worked with a finite set of symbols that would store a simple set of instructions and process a one-dimensional array of symbols, replacing it with a second array of symbols. Turing showed that there was no solution in general to Hilbert's decision problem, but in the process he also showed how to construct a machine (now called a Turing machine) that could execute any possible calculation. The machine would operate on a string of symbols recorded on a tape and would output the result of the same calculation on the same tape. Further, Turing showed the existence of machines that could read instructions given in symbolic form and then perform any desired computation on a one-dimensional array of numbers that followed. The universal Turing machine was a programmable digital computer. The instructions could be read from a one-dimensional tape, a magnetically stored memory, or a card punched with holes, as was used for mechanized weaving of fabric.

The earliest electronic computers were developed at the time of World War II and involved numerous vacuum tubes. But vacuum tubes produce immense amounts of heat and involved the possibility that the heating element in one of the tubes might well burn out during the computation. In fact, it was standard procedure to run a program, one that required proper function of all the vacuum tubes, both before and after the program of interest. If the results of the first and last computations did not vary, one could assume that no tubes had burned out in the meantim-World War II ended in 1945. In addition to the critical role of computing machines in the design of the first atomic bombs, computational science had played an important role in predicting the behavior of targets. The capabilities of computing machines would grow rapidly following the invention of the transistor by John Bardeen, Walter Brattain and William Shockley in 1947. In this case, fundamental science led to tremendous advances in applied science.

The story of semiconductor science is worth telling. Silicon is unusual in displaying an increase in electrical conductivity as the temperature is raised. In general, when one finds an interesting property of a material, one tries to purify and refine the material. Purified silicon, however, lost most of its conductivity. On further investigation, it was found that tiny concentrations of impurities could vastly change both the amount of electrical conductivity and the mechanism by which it occurs. Because the useful properties of semiconductors depend critically on the impurities, or "dirt," in the material, solid-state (and other) physicists sometimes refer to the field as "dirt physics." Adding a small amount of phosphorus to pure silicon resulted in n-type conductivity, the type due to electrons moving in response to an electric field. Adding an impurity such as boron produced p-type conductivity, in which electron vacancies (in chemical bonds) moved through the material. Creating a p-type region next to an n-type produced a junction that let current flow in one direction and not the other, just as in a vacuum tube diode. Placing a p-type region between two n-types produced the equivalent of Lee de Forest's diode—a transistor. The transistor, however, did not require a heater and could be miniaturized.

The 1960s saw the production of integrated circuits—many transistors and other circuit elements on a single silicon wafer, or chip. Currently hundreds of thousands of circuit elements are available on a single chip, and anyone who buys a laptop computer commands more computational power than any government could control in 1950.

The computer has put new tools at the disposal of the engineer as well as scholars in other fields. Magnetic and optical methods of storage make it possible to record large amounts of information in a very small space. Storage of several trillion characters is now possible using far less than a cubic inch of space. Computer programs can now search through the millions of test items to find the one that best matches search criteria, in a second or so.

The notion that computers can actually "think" like people is still hotly debated. John McCarthy coined the term Artificial Intelligence in 1955. Since then computers have found new proofs of theorems in Euclidean geometry and have been used to discover trends in experimental data. Artificial Intelligence (AI) programs called "expert systems" have been used to diagnose diseases, devise new chemical syntheses and defeat chess masters at their own game.

Evolutionary computing is another recent development. Instead of trying to find the single best solution to certain types of problems, evolutionary computing forms a population of approximate solutions—sometimes called "chromosomes"—to the problem and then alters the population on the basis of "survival of the fittest," making more replicates of the better solution and discarding some of those solutions that are less fit. The solution process allows several biologically inspired mutations to occur between generations. Eventually a steady state is reached, with the optimum solution in the majority.

Donald R. Franceschetti

Further Reading

Bell Telephone Laboratories. A History of Engineering and Science in the Bell System: Electronics Technology, 1925–1975. Ed. by M. D. Fagen. 7 vols. New York: Bell Laboratories, 1975–1985. Provides detailed information on the development of the transistor and the integrated circuit.

Bodanis, David. Electric Universe: How Electricity Switched On the Modern World. New York: Three Rivers Press, 2005. Popular exposition of the applications of electronics and electromagnetism from the time of Joseph Henry to the microprocessor age.

Burke, James. Connections. Boston: Little Brown, 1978; reprint, New York: Simon & Schuster, 2007. Describes linkages among inventions throughout history.

Cobb, Cathy, and Harold Goldwhite. Creations of Fire: Chemistry's Lively History from Alchemy to the Atomic Age. New York: Plenum Press, 1995; reprint, Cambridge, Mass.: Perseus, 2001. History of pure and applied chemistry from the beginning through the late twentieth century.

Garfield, Simon. Mauve: How One Man Invented a Color That Changed the World. New York: W. W. Norton, 2000. Focuses on how the single and partly accidental discovery of coal tar dyes led to several new areas of chemical industry.

Kealy, Terence. The Economic Laws of Scientific Research. New York: St. Martin's Press, 1996. Makes the case that government funding of scientific research is relatively inefficient and emphasizes the role of private investment and hobbyist scientists.

Mlodinow, Leonard. Feynman's Rainbow: A Search for Beauty in Phycis and in Life. New York: Warner Books, 2003; reprint, New York: Vintage Books, 2011.

Schlader, Neil, ed. Science and Its Times: Understanding the Social Significance of Scientific Discovery. 8 vols. Detroit: Gale Group, 2000-2001. Massive reference work on the impactt of scientific and technological developments from the earliest times to the present.

Sobel, Dava. Longitude. New York: Walker & company, 2007. Story of the competition among scientists and inventors to develop a reliable means of determining longitude at sea.

Stokes, Donald E. Pasteur's Quadrant: Basic Science and Technological Innovation. Washington, D.C.: Brooking's Institution Press, 1997. Presents an extended argument that many fundamental scientific discoveries originate in application-driven research and that the distinction between pure and applied science is not, of itself, very useful.

CONTRIBUTORS

Ezinne Amaonwu
Rockville, Maryland

Michael Auerbach
Marblehead, Massachusetts

Craig Belanger
Journal of Advancing Technology

Raymond D. Benge
Tarrant County College

Harlan H. Bengtson
Southern Illinois University, Edwardsville

Victoria M. Breting-García
Houston, Texas

Michael A. Buratovich
Spring Arbor University

Byron D. Cannon
University of Utah

Christina Capriccioso
University of Michigan College of Engineering

Richard P. Capriccioso
University of Phoenix

Christine M. Carroll
American Medical Writers Association

Michael J. Caulfield
Gannon University

Martin Chetlen
Moorpark College

Thomas Drucker
University of Wisconsin-Whitewater

Jack Ewing
Boise, Idaho

June Gastón
Borough of Manhattan Community College, City University of New York

Gina Hagler
Washington, D.C.

Robert M. Hordon
Rutgers University

April D. Ingram
Kelowna, British Columbia

Micah L. Issitt
Philadelphia, Pennsylvania

Vincent Jorgensen
Sunnyvale, California

Bassam Kassab
Santa Clara Valley Water District

Narayanan M. Komerath
Georgia Institute of Technology

Jeanne L. Kuhler
Benedictine University

Dawn A. Laney
Atlanta, Georgia

M. Lee
Independent Scholar

Donald W. Lovejoy
Palm Beach Atlantic University

Marianne M. Madsen
University of Utah

Sergei A. Markov
Austin Peay State University

Roman Meinhold
Assumption University, Bangkok, Thailand

Randall L. Milstein
Oregon State University

David Olle
Eastshire Communications

Robert J. Paradowski
Rochester Institute of Technology

Ellen E. Anderson Penno
Western Laser Eye Associates

John R. Phillips
Purdue University Calumet

George R. Plitnik
Frostburg State University

Steven J. Ramold
Eastern Michigan University

Richard M. J. Renneboog
Independent Scholar

Charles W. Rogers
Southwestern Oklahoma State University

Lars Rose
Author, The Nature of Matter

Julia A. Rosenthal
Chicago, Illinois

Joseph R. Rudolph
Towson University

Billy R. Smith, Jr.
Anne Arundel Community College

Roger Smith
Portland, Oregon

Max Statman
Eastman Chemical Company

Polly D. Steenhagen
Delaware State University

Judith L. Steininger
Milwaukee School of Engineering

Rena Christina Tabata
University of British Columbia

John M. Theilmann
Converse College

Bethany Thivierge
Technicality Resources

Anh Tran
Wichita State University, Kansas

Christine Watts
University of Sydney

Shawncey Jay Webb
Taylor University

George M. Whitson III
University of Texas at Tyler

Edwin G. Wiggins
Webb Institute, Glen Cove, New York

Robin L. Wulffson
Faculty, American College of Obstetrics and Gynecology

Susan M. Zneimer
U.S. Labs, Irvine, California

COMMON UNITS OF MEASURE

Common prefixes for metric units—which may apply in more cases than shown below—include *giga-* (1 billion times the unit), *mega-* (one million times), *kilo-* (1,000 times), *hecto-* (100 times), *deka-* (10 times), *deci-* (0.1 times, or one tenth), *centi-* (0.01, or one hundredth), *milli-* (0.001, or one thousandth), and *micro-* (0.0001, or one millionth).

Unit	*Quantity*	*Symbol*	*Equivalents*
Acre	Area	ac	43,560 square feet 4,840 square yards 0.405 hectare
Ampere	Electric current	A *or* amp	1.00016502722949 international ampere 0.1 biot *or* abampere
Angstrom	Length	Å	0.1 nanometer 0.0000001 millimeter 0.000000004 inch
Astronomical unit	Length	AU	92,955,807 miles 149,597,871 kilometers (mean Earth-Sun distance)
Barn	Area	b	10^{-28} meters squared (approx. cross-sectional area of 1 uranium nucleus)
Barrel (dry, for most produce)	Volume/capacity	bbl	7,056 cubic inches; 105 dry quarts; 3.281 bushels, struck measure
Barrel (liquid)	Volume/capacity	bbl	31 to 42 gallons
British thermal unit	Energy	Btu	1055.05585262 joule
Bushel (U.S., heaped)	Volume/capacity	bsh *or* bu	2,747.715 cubic inches 1.278 bushels, struck measure
Bushel (U.S., struck measure)	Volume/capacity	bsh *or* bu	2,150.42 cubic inches 35.238 liters
Candela	Luminous intensity	cd	1.09 hefner candle
Celsius	Temperature	C	1° centigrade
Centigram	Mass/weight	cg	0.15 grain
Centimeter	Length	cm	0.3937 inch
Centimeter, cubic	Volume/capacity	cm^3	0.061 cubic inch
Centimeter, square	Area	cm^2	0.155 square inch
Coulomb	Electric charge	C	1 ampere second
Cup	Volume/capacity	C	250 milliliters 8 fluid ounces 0.5 liquid pint

Unit	*Quantity*	*Symbol*	*Equivalents*
Deciliter	Volume/capacity	dl	0.21 pint
Decimeter	Length	dm	3.937 inches
Decimeter, cubic	Volume/capacity	dm^3	61.024 cubic inches
Decimeter, square	Area	dm^2	15.5 square inches
Dekaliter	Volume/capacity	dal	2.642 gallons 1.135 pecks
Dekameter	Length	dam	32.808 feet
Dram	Mass/weight	dr *or* dr avdp	0.0625 ounce 27.344 grains 1.772 grams
Electron volt	Energy	eV	$1.5185847232839 \times 10^{-22}$ Btus $1.6021917 \times 10^{-19}$ joules
Fermi	Length	fm	1 femtometer 1.0×10^{-15} meters
Foot	Length	ft *or* ′	12 inches 0.3048 meter 30.48 centimeters
Foot, square	Area	ft^2	929.030 square centimeters
Foot, cubic	Volume/capacity	ft^3	0.028 cubic meter 0.0370 cubic yard 1,728 cubic inches
Gallon (British Imperial)	Volume/capacity	gal	277.42 cubic inches 1.201 U.S. gallons 4.546 liters 160 British fluid ounces
Gallon (U.S.)	Volume/capacity	gal	231 cubic inches 3.785 liters 0.833 British gallon 128 U.S. fluid ounces
Giga-electron volt	Energy	GeV	$1.6021917 \times 10^{-10}$ joule
Gigahertz	Frequency	GHz	—
Gill	Volume/capacity	gi	7.219 cubic inches 4 fluid ounces 0.118 liter
Grain	Mass/weight	gr	0.037 dram 0.002083 ounce 0.0648 gram
Gram	Mass/weight	g	15.432 grains 0.035 avoirdupois ounce

Unit	*Quantity*	*Symbol*	*Equivalents*
Hectare	Area	ha	2.471 acres
Hectoliter	Volume/capacity	hl	26.418 gallons 2.838 bushels
Hertz	Frequency	Hz	$1.08782775707767 \times 10^{-10}$ cesium atom frequency
Hour	Time	h	60 minutes 3,600 seconds
Inch	Length	in *or* "	2.54 centimeters
Inch, cubic	Volume/capacity	in^3	0.554 fluid ounce 4.433 fluid drams 16.387 cubic centimeters
Inch, square	Area	in^2	6.4516 square centimeters
Joule	Energy	J	$6.2414503832469 \times 10^{18}$ electron volt
Joule per kelvin	Heat capacity	J/K	$7.24311216248908 \times 10^{22}$ Boltzmann constant
Joule per second	Power	J/s	1 watt
Kelvin	Temperature	K	-272.15° Celsius
Kilo-electron volt	Energy	keV	$1.5185847232839 \times 10^{-19}$ joule
Kilogram	Mass/weight	kg	2.205 pounds
Kilogram per cubic meter	Mass/weight density	kg/m^3	$5.78036672001339 \times 10^{-4}$ ounces per cubic inch
Kilohertz	Frequency	kHz	—
Kiloliter	Volume/capacity	kl	—
Kilometer	Length	km	0.621 mile
Kilometer, square	Area	km^2	0.386 square mile 247.105 acres
Light-year (distance traveled by light in one Earth year)	Length/distance	lt-yr	5,878,499,814,275.88 miles 9.46×10^{12} kilometers
Liter	Volume/capacity	L	1.057 liquid quarts 0.908 dry quart 61.024 cubic inches
Mega-electron volt	Energy	MeV	—
Megahertz	Frequency	MHz	—
Meter	Length	m	39.37 inches
Meter, cubic	Volume/capacity	m^3	1.308 cubic yards

Unit	*Quantity*	*Symbol*	*Equivalents*
Meter per second	Velocity	m/s	2.24 miles per hour 3.60 kilometers per hour
Meter per second per second	Acceleration	m/s^2	12,960.00 kilometers per hour per hour 8,052.97 miles per hour per hour
Meter, square	Area	m^2	1.196 square yards 10.764 square feet
Metric. *See* unit name			
Microgram	Mass/weight	mcg *or* µg	0.000001 gram
Microliter	Volume/capacity	µl	0.00027 fluid ounce
Micrometer	Length	µm	0.001 millimeter 0.00003937 inch
Mile (nautical international)	Length	mi	1.852 kilometers 1.151 statute miles 0.999 U.S. nautical miles
Mile (statute or land)	Length	mi	5,280 feet 1.609 kilometers
Mile, square	Area	mi^2	258.999 hectares
Milligram	Mass/weight	mg	0.015 grain
Milliliter	Volume/capacity	ml	0.271 fluid dram 16.231 minims 0.061 cubic inch
Millimeter	Length	mm	0.03937 inch
Millimeter, square	Area	mm^2	0.002 square inch
Minute	Time	m	60 seconds
Mole	Amount of substance	mol	6.02×10^{23} atoms or molecules of a given substance
Nanometer	Length	nm	1,000,000 fermis 10 angstroms 0.001 micrometer 0.00000003937 inch
Newton	Force	N	0.224808943099711 pound force 0.101971621297793 kilogram force 100,000 dynes
Newton meter	Torque	N·m	0.7375621 foot-pound
Ounce (avoirdupois)	Mass/weight	oz	28.350 grams 437.5 grains 0.911 troy or apothecaries' ounce

AERONAUTICS AND AVIATION

FIELDS OF STUDY

Algebra; calculus; inorganic chemistry; organic chemistry; physical chemistry; optics; modern physics; statics; aerodynamics; thermodynamics; strength of materials; propulsion; propeller and rotor theory; vehicle performance; aircraft design; avionics; orbital mechanics; spacecraft design.

SUMMARY

Aeronautics is the science of atmospheric flight. Aviation is the design, development, production, and operation of flight vehicles. Aerospace engineering extends these fields to space vehicles. Transonic airliners, airships, space launch vehicles, satellites, helicopters, interplanetary probes, and fighter planes are all applications of aerospace engineering.

KEY TERMS AND CONCEPTS

- **Airfoil:** Structure, such as a wing or a propeller, designed to interact, in motion, with the surrounding airflow in a manner that optimizes the desired reaction, whether that be to minimize air resistance or to maximize lift.
- **Boundary Layer:** Thin region near a surface of an aircraft where the flow slows down because of viscous friction.
- **Bypass Ratio:** Ratio of turbofan engine mass flow rate bypassing the hot core, to that through the core.
- **Delta V:** Speed difference corresponding to the difference in energies between two orbital states.
- **Fuselage:** Body of an aircraft, other than engines, wings, tails, or control surfaces.
- **Lift To Drag Ratio:** Ratio of the lift to drag in cruise; the aerodynamic efficiency metric for transport aircraft and gliders.
- **Oblique Shock:** Thin wave in a supersonic flow through which flow turns and decelerates sharply.
- **Prandtl-Meyer Expansion:** Ideal model of a supersonic flow accelerating through a turn.
- **Stall:** Condition in which flow separates from most of a lifting surface, sharply lowering lift and raising drag.
- **Takeoff Gross Weight:** Mass or weight of an aircraft at takeoff with full payload and fuel load; the highest design weight for liftoff.
- **Wind Tunnel:** Facility where a smooth, uniform flow helps simulate flow around an object in flight.
- **Wing:** Object that generates lift with low drag and supports the weight of the aircraft in flight.

Definition and Basic Principles

Aeronautics is the science of atmospheric flight. The term ("aero" referring to flight and "nautics" referring to ships or sailing) originated from the activities of pioneers who aspired to navigate the sky. These early engineers designed, tested, and flew their own creations, many of which were lighter-than-air balloons. Modern aeronautics encompasses the science and engineering of designing and analyzing all areas associated with flying machines.

Aviation (based on the Latin word for "bird") originated with the idea of flying like the birds using heavier-than-air vehicles. "Aviation" refers to the field of operating aircraft, while the term "aeronautics" has been superseded by "aerospace engineering," which specifically includes the science and engineering of spacecraft in the design, development, production, and operation of flight vehicles.

A fundamental tenet of aerospace engineering is to deal with uncertainty by tying analyses closely to what is definitely known, for example, the laws of physics and mathematical proofs. Lighter-than-air airships are based on the principle of buoyancy, which derives from the law of gravitation. An object that weighs less than the equivalent volume of air experiences a net upward force as the air sinks around it.

Two basic principles that enable the design of heavier-than-air flight vehicles are those of aerodynamic lift and propulsion. Both arise from Sir Isaac Newton's second and third laws of motion. Aerodynamic lift is a force perpendicular to the direction of motion, generated from the turning of flowing air around an object. In propulsion, the reaction to the acceleration of a fluid generates a force that propels an object, whether in air or in the vacuum of space. Understanding these principles allowed aeronauts to design vehicles that could fly steadily despite being much heavier than the air they displaced and allowed rocket scientists to develop vehicles that could accelerate in space. Spaceflight occurs at speeds so high that the vehicle's kinetic energy is comparable to the potential energy due to gravitation. Here the principles of orbital mechanics derive from the laws of dynamics and gravitation and extend to the regime of relativistic phenomena. The engineering sciences of building vehicles that can fly, keeping them stable, controlling their flight, navigating, communicating, and ensuring the survival, health, and comfort of occupants, draw on every field of science.

Background and History

The intrepid balloonists of the nineteenth century were followed by aeronauts who used the principles of aerodynamics to fly unpowered gliders. The Wright brothers demonstrated sustained, controlled, powered aerodynamic flight of a heavier-than-air aircraft in 1903. The increasing altitude, payload, and speed capabilities of airplanes made them powerful weapons in World War I. Such advances improved flying skills, designs, and performance, though at a terrible cost in lives.

The monoplane design superseded the fabric-and-wire biplane and triplane designs of World War I. The helicopter was developed during World War II and quickly became an indispensable tool for medical evacuation and search and rescue. The jet engine, developed in the 1940's and used on the Messerschmitt 262 and Junkers aircraft by the Luftwaffe and the Gloster Meteor by the British, quickly enabled flight in the stratosphere at speeds sufficient to generate enough lift to climb in the thin air. Such innovations led to smooth, long-range flights in pressurized cabins and shirtsleeve comfort. Fatal crashes of the de Havilland Comet airliner in 1953 and 1954 focused attention on the science of metal fatigue.

The Boeing 707 opened up intercontinental air travel, followed by the Boeing 747, the supersonic Concorde, and the EADS Airbus A380. A series of manned research aircraft designated X-planes since the 1930's investigated various flight regimes and also drove the development of better wind tunnels and high-altitude simulation chambers. German ballistic missiles led to U.S. and Soviet missile programs that grew into a space race, culminating in the first humans landing on the Moon in 1969. Combat-aircraft development enabled advances that resulted in safer and more efficient airliners.

How It Works

Force Balance in Flight. Five basic forces acting on a flight vehicle are aerodynamic lift, gravity, thrust, drag, and centrifugal force. For a vehicle in steady level flight in the atmosphere, lift and thrust balance gravity (weight) and aerodynamic drag. Centrifugal force due to moving steadily around the Earth is too weak at most airplane flight speeds but is strong for a maneuvering aircraft. Aircraft turn by rolling the lift vector toward the center of curvature of the desired flight path, balancing the centrifugal reaction due to inertia. In the case of a vehicle in space beyond the atmosphere, centrifugal force and thrust counter gravitational force.

Aerodynamic Lift. Aerodynamics deals with the forces due to the motion of air and other gaseous fluids relative to bodies. Aerodynamic lift is generated perpendicular to the direction of the free stream as the reaction to the rate of change of momentum of air turning around an object, and, at high speeds, to compression of air by the object. Flow turning is accomplished by changing the angle of attack of the surface, by using the camber of the surface in subsonic flight, or by generating vortices along the leading edges of swept wings.

Propulsion. Propulsive force is generated as a reaction to the rate of change of momentum of a fluid moving through and out of the vehicle. Rockets carry all of the propellant onboard and accelerate it out through a nozzle using chemical heat release, other heat sources, or electromagnetic fields. Jet engines "breathe" air and accelerate it after reaction with fuel. Rotors, propellers, and fans exert lift force on the air and generate thrust from the reaction to this force. Solar sails use the pressure of solar radiation to push large, ultralight surfaces.

Static Stability. An aircraft is statically stable if a small perturbation in its attitude causes a restoring

Fascinating Facts About Aeronautics and Aviation

- The X-29 test vehicle demonstrated that an aircraft could be built to be statically unstable and yet maintain stable flight. The flight control computer reliably provides rapid updates of control surface actuators to compensate for disturbances before they amplify. Modern fighter aircraft are marginally unstable but use control systems to augment stability, thereby increasing maneuver performance.
- Future space travelers may fly up in hypersonic air-breathing vehicles that take off and land like airplanes. High-pressure intake air is liquefied in ducts cooled by the liquid-hydrogen fuel, enabling oxygen to be separated out from nitrogen and stored for use with hydrogen fuel beyond the atmosphere. This decreases takeoff weight.
- A long conductor trailed from a spacecraft generates an electric current when moving through the Earth's magnetic field. Conversely, a current passed through a tether between two objects in different orbits generates a force. This is a proposed electrodynamic space broom, which will drag debris into orbits where they will quickly burn up in the atmosphere.
- A conventional helicopter requires a tail rotor to provide anti-torque and keep the craft from spinning in reaction to the torque put in to turn the rotor. The rotor must also change its pitch angle through every cycle in order to keep the lift moment equal between the advancing side rotor blade and the retreating side blade. The counter-rotating rotor helicopter has two rotors on the same axis but rotating in opposite directions. This eliminates the need for a tail rotor and for large cyclic pitch variation. This has allowed helicopters to fly substantially faster. Counter-rotating compressor and turbine stages in jet engines eliminate the need for stator stages that do not contribute useful work, thereby reducing the mass and number of stages needed in modern aircraft engines.
- When aerodynamic lift is generated, the flow around the body is pushed down (against the direction of lift), while the flow outside the body's span gets lifted up. The energy in this updraft is wasted unless another aircraft benefits from following in close proximity off to one side. Birds use this feature routinely and so have air forces since World War I to save fuel. Swarms of small unmanned aerial vehicles, or micro spacecraft, can generate the same resolution and efficiency as a very large antenna by maintaining relative position in flight.

aerodynamic moment that erases the perturbation. Typically, the aircraft center of gravity must be ahead of the center of pressure for longitudinal stability. The tails or canards help provide stability about the different axes. Rocket engines are said to be stable if the rate of generation of gases in the combustion chamber does not depend on pressure stronger than by a direct proportionality, such as a pressure exponent of 1.

Flight Dynamics and Controls. Static stability is not the whole story, as every pilot discovers when the airplane drifts periodically up and down instead of holding a steady altitude and speed. Flight dynamics studies the phenomena associated with aerodynamic loads and the response of the vehicle to control surface deflections and engine-thrust changes. The study begins with writing the equations of motion of the aircraft resolved along the six degrees of freedom: linear movement along the longitudinal, vertical and sideways axes, and roll, yaw, and pitch rotations about them. Maneuvering aircraft must deal with coupling between the different degrees of freedom, so that roll accompanies yaw, and so on.

The autopilot system was an early flight-control achievement. Terrain-following systems combine information about the terrain with rapid updates, enabling military aircraft to fly close to the ground, much faster than a human pilot could do safely. Modern flight-control systems achieve such feats as reconfiguring control surfaces and fuel to compensate for damage and engine failures; or enabling autonomous helicopters to detect, hover over, and pick up small objects and return; or sending a space probe at thousands of kilometers per hour close to a planetary moon or landing it on an asteroid and returning it to Earth. This field makes heavy use of ordinary differential equations and transform techniques, along with simulation software.

Orbital Missions. The rocket equation attributed to Russian scientist Konstantin Tsiolkovsky related the speed that a rocket-powered vehicle gains to the amount and speed of the mass that it ejects. A vehicle launched from Earth's surface goes into a trajectory where its kinetic energy is exchanged for gravitational potential energy. At low speeds, the resulting trajectory intersects the Earth, so that the vehicle falls to the surface. At high enough speeds, the vehicle goes so far so fast that its trajectory remains in space and takes the shape of a continuous ellipse around Earth. At even higher kinetic energy levels, the vehicle goes into a hyperbolic trajectory, escaping Earth's orbit into the solar system. The key is thus to achieve enough tangential speed relative to Earth. Most rockets rise rapidly through the atmosphere so that the acceleration to high tangential speed occurs well above the atmosphere, thus minimizing air-drag losses.

Hohmann Transfer. Theoretically, the most efficient way to impart kinetic energy to a vehicle is impulsive launch, expending all the propellant instantly so that no energy is wasted lifting or accelerating propellant with the vehicle. Of course, this would destroy any vehicle other than a cannonball, so large rockets use gentle accelerations of no more than 1.4 to 3 times the acceleration due to gravity. The advantage of impulsive thrust is used in the Hohmann transfer maneuver between different orbits in space. A rocket is launched into a highly eccentric elliptical trajectory. At its highest point, more thrust is added quickly. This sends the vehicle into a circular orbit at the desired height or into a new orbit that takes it close to another heavenly body. Reaching the same final orbit using continuous, gradual thrust would require roughly twice as much expenditure of energy. However, continuous thrust is still an attractive option for long missions in space, because a small amount of thrust can be generated using electric propulsion engines that accelerate propellant to extremely high speeds compared with the chemical engines used for the initial ascent from Earth.

Applications and Products

Aerospace Structures. Aerospace engineers always seek to minimize the mass required to build the vehicle but still ensure its safety and durability. Unlike buildings, bridges, or even (to some degree) automobiles, aircraft cannot be made safer merely by making them more massive, because they must also be able to overcome Earth's gravity. This exigency has driven development of new materials and detailed, accurate methods of analysis, measurement, and construction. The first aircraft were built mostly from wood frames and fabric skins. These were superseded by all-metal craft, constructed using the monocoque concept (in which the outer skin bears most of the stresses). The Mosquito high-speed bomber in World War II reverted to wood construction for better performance. Woodworkers learned to align the grain (fiber direction) along the principal stress axes. Metal offers the same strength in all directions for the same thickness. Composite structures allow fibers with high tensile strength to be placed along the directions where strength is needed, bonding different layers together.

Aeroelasticity. Aeroelasticity is the study of the response of structurally elastic bodies to aerodynamic loads. Early in the history of aviation, several mysterious and fatal accidents occurred wherein pieces of wings or tails failed in flight, under conditions where the steady loads should have been well below the strength limits of the structure. The intense research to address these disasters showed that beyond some flight speed, small perturbations in lift, such as those due to a gust or a maneuver, would cause the structure to respond in a resonant bending-twisting oscillation, the perturbation amplitude rapidly rising in a "flutter" mode until structural failure occurred. Predicting such aeroelastic instabilities demanded a highly mathematical approach to understand and apply the theories of unsteady aerodynamics and structural dynamics. Modern aircraft are designed so that the flutter speed is well above any possible speed achieved. In the case of helicopter rotor blades and gas turbine engine blades, the problems of ensuring aeroelastic stability are still the focus of leading-edge research. Related advances in structural dynamics have enabled development of composite structures and of highly efficient turbo machines that use counter-rotating stages, such as those in the F135 engines used in the F-35 Joint Strike Fighter. Such advances also made it possible for earthquake-surviving high-rise buildings to be built in cities such as San Francisco, Tokyo, and Los Angeles, where a number of sensors, structural-dynamics-analysis software, and actuators allow the correct response to dampen the effects of earth movements even on the upper floors.

Smart Materials. Various composite materials such as carbon fiber and metal matrix composites have come to find application even in primary aircraft structures. The Boeing 787 is the first to use a composite main spar in its wings. Research on nano materials promises the development of materials with hundreds of times as much strength per unit mass as steel. Another leading edge of research in materials is in developing high-temperature or very low-temperature (cryogenic) materials for use inside jet and rocket engines, the spinning blades of turbines, and the impeller blades of liquid hydrogen pumps in rocket engines. Single crystal turbine blades enabled the development of jet engines with very high turbine inlet temperatures and, thus, high thermodynamic efficiency. Ceramic designs that are not brittle are pushing turbine inlet temperatures even higher. Other materials are "smart," meaning they respond actively in some way to inputs. Examples include piezoelectric materials.

Wind Tunnels and Other Physical Test Facilities. Wind tunnels, used by the Wright brothers to develop airfoil shapes with desirable characteristics, are still used heavily in developing concepts and proving the performance of new designs, investigating causes of problems, and developing solutions and data to validate computational prediction techniques. Generally, a wind tunnel has a fan or a high-pressure reservoir to add work to the air and raise its stagnation pressure. The air then flows through means of reducing turbulence and is accelerated to the maximum speed in the test section, where models and measurement systems operate.

The power required to operate a wind tunnel is proportional to the mass flow rate through the tunnel and to the cube of the flow speed achieved. Low-speed wind tunnels have relatively large test sections and can operate continuously for several minutes at a time. Supersonic tunnels generally operate with air blown from a high-pressure reservoir for short durations. Transonic tunnels are designed with ventilating slots to operate in the difficult regime where there may be both supersonic waves and subsonic flow over the test configuration. Hypersonic tunnels require heaters to avoid liquefying the air and to simulate the high stagnation temperatures of hypersonic flight and operate for millisecond durations. Shock tubes generate a shock from the rupture of a diaphragm, allowing high-energy air to expand into stationary air in the tube. They are used to simulate the extreme conditions across shocks in hypersonic flight. Many other specialized test facilities are used in structural and materials testing, developing jet and rocket engines, and designing control systems.

Avionics and Navigation. Condensed from the term "aviation electronics," the term "avionics" has come to include the generation of intelligent software systems and sensors to control unmanned aerial vehicles (UAVs), which may operate autonomously. Avionics also deals with various subsystems such as radar and communications, as well as navigation equipment, and is closely linked to the disciplines of flight dynamics, controls, and navigation.

During World War II, pilots on long-range night missions would navigate celestially. The gyroscopes in their aircrafts would spin at high speed so that their inertia allowed them to maintain a reference position as the aircraft changed altitude or accelerated. Most modern aircraft use the Global Positioning System (GPS), Galileo, or GLONASS satellite constellations to obtain accurate updates of position, altitude, and velocity. The ordinary GPS signal determines position and speed with fair accuracy. Much greater precision and higher rates of updates are available to authorized vehicle systems through the differential GPS signal and military frequencies.

Gravity Assist Maneuver. Yuri Kondratyuk, the Ukrainian scientist whose work paved the way for the first manned mission to the moon, suggested in 1918 that a spacecraft could use the gravitational attraction of the moons of planets to accelerate and decelerate at the two ends of a journey between planets. The Soviet Luna 3 probe used the gravity of the Moon when photographing the far side of it in 1959. American mathematician Michael Minovitch pointed out that the gravitational pull of planets along the trajectory of a spacecraft could be used to accelerate the craft toward other planets. The Mariner 10 probe used this "gravitational slingshot" maneuver around Venus to reach Mercury at a speed small enough to go into orbit around Mercury. The Voyager missions used the rare alignment of the outer planets to receive gravitational assists from Jupiter and Saturn to go on to Uranus and Neptune, before doing another slingshot around Jupiter and Saturn to escape the solar system. Gravity assist has become part of the mission planning for all exploration missions and even for missions near Earth, where the gravity of the Moon is used.

Impact on Industry

Aeronautics and aviation have had an immeasurable impact on industry and society. Millions of people fly long distances on aircraft every day, going about their business and visiting friends and relatives, at a cost that is far lower in relative terms than the cost of travel a century ago.

Every technical innovation developed for aeronautics and aviation finds its way into improved industrial products. Composite structural materials are found in everything from tennis rackets to industrial machinery. Bridges, stadium domes, and skyscrapers are designed with aerospace structural-element technology and structural-dynamics instrumentation and testing techniques. Electric power is generated in utility power plants using steam generators sharing jet engine turbo machine origins.

Satellite antennae are found everywhere. Much digital signal processing, central to digital music and cell phone communications, came from research projects driven by the need to extract low-level signatures buried in noise. Similarly, image-processing algorithms that enable computed tomography (CT) scans of the human body, eye and cardiac diagnostics, image and video compression, and laser printing came from aerospace image-processing projects. The field of geoinformatics has advanced immensely, with most mapping, navigation, and remote-sensing enterprises assuming the use of space satellites. The GPS has spawned numerous products for terrestrial drivers on land and navigators on the ocean. Aerospace medicine research has developed advances in diagnosing and monitoring the human body and its responses to acceleration, bone marrow loss, muscular degeneration, and their prevention through exercise, hypoxia, radiation protection, heart monitoring, isolation from microorganisms, and drug delivery. Teflon coatings developed for aerospace products are also used in cookware.

Careers and Course Work

Aerospace engineers work on problems that push the frontiers of technology. Typical employers in this industry are manufacturers of aircraft or their parts and subsystems, airlines, government agencies and laboratories, and the defense services. Many aerospace engineers are also sought by financial services and other industries seeking those with excellent quantitative (mathematical and scientific) skills and talents.

University curriculum generally starts with a year of mathematics, physics, chemistry, computer graphics, computer science, language courses, and an introduction to aerospace engineering, followed by sophomore-year courses in basic statics, dynamics, materials, and electrical engineering. Core courses include low-speed and high-speed aerodynamics, linear systems analysis, thermodynamics, propulsion, structural analysis, composite materials, vehicle performance, stability, control theory, avionics, orbital mechanics, aeroelasticity and structural dynamics, and a two-semester sequence on capstone design of flight vehicles. High school students aiming for such careers should take courses in mathematics, physics, chemistry and natural sciences, and computer graphics. Aerospace engineers are frequently required to write clear reports and present complex issues to skeptical audiences, which demands excellent communication skills. Taking flying lessons or getting a private pilot license is less important to aerospace engineering, as exhilarating as it is, and should be considered only if one desires a career as a pilot or astronaut.

The defense industry is moving toward using aircraft that do not need a human crew and can perform beyond the limits of what a human can survive, so the glamorous occupation of combat jet pilot may be heading for extinction. Airline pilot salaries are also coming down from levels that compared with surgeons toward those more comparable to bus drivers. Aircraft approach, landing, traffic management, emergency response, and collision avoidance systems may soon become fully automated and will require maneuvering responses that are beyond what a human pilot can provide in time and accuracy.

Opportunities for spaceflight may also be minimal unless commercial and military spaceflight picks up to fill the void left by the end of civilian programs discussed below. This is not a unique situation in aviation history. Early pilots, even much later than the intrepid "aeronauts," also worked much more for love of the unparalleled experience of flying, rather than for looming prospects of high-profile careers or the salaries paid by struggling startup airline companies. The only reliable prediction that can be made about aerospace careers is that they hold many surprises.

Social Context and Future Prospects

Airline travel is under severe stress in the first part of the twenty-first century. This is variously attributed to airport congestion, security issues, rising fuel prices, predatory competition, reduction of route monopolies, and leadership that appears to offer little vision beyond cost cutting. Meanwhile, the demand for air travel is rising all over the world. Global demand for commercial airliners is estimated at nearly 30,000 aircraft through 2030 and is valued at more than $3.2 trillion—in addition to 17,000 business jets valued at more than $300 billion.

Detailed design and manufacturing of individual aircraft are distributed between suppliers worldwide, with the wings, tails, and engines of a given aircraft often designed and built in different parts of the world. Japan and China are expected to increase their aircraft manufacturing, while major U.S. companies appear to be moving more toward becoming system integrators and away from manufacturing.

The human venture in space is also under stress as the U.S. space shuttle program ends without another human-carrying vehicle to replace it. The future of the one remaining space station is in doubt, and there are no plans to build another.

On the other hand, just over one century into powered flight, the human venture into the air and beyond is just beginning. Aircraft still depend on long runways and can fly only in a very limited range of conditions. Weather delays are still common because of uncertainty about how to deal with fluctuating winds or icing conditions. Most airplanes still consist of long tubes attached to thin wings, because designing blended wing bodies is difficult with the uncertainties in modeling composite structures. The aerospace and aviation industry is a major generator of atmospheric carbon releases. This will change only when the industry switches to renewable hydrogen fuel, which may occur faster than most people anticipate.

The human ability to access, live, and work in space or on extraterrestrial locations is extremely limited, and this prevents development of a large space-based economy. This situation may be expected to change over time, with the advent of commercial space launches. New infrastructure will encourage commercial enterprises beyond Earth.

The advancements in the past century are truly breathtaking and bode well for the breakthroughs that one may hope to see. Hurricanes and cyclonic storms are no longer surprise killers; they are tracked from formation in the far reaches of the oceans, and their paths are accurately predicted, giving people plenty of warning. Crop yields and other resources are accurately tracked by spacecraft, and ground-penetrating radar from Earth-sensing satellites has discovered much about humankind's buried ancient heritage and origins. Even in featureless oceans and deserts, GPS satellites provide accurate, reliable navigation information. The discovery of ever-smaller distant planets by orbiting space telescopes, and of unexpected forms of life on Earth, hint at the possible discovery of life beyond Earth.

Narayanan M. Komerath, Ph.D.

Further Reading

Anderson, John D., Jr. *Introduction to Flight.* 5th ed. New York: McGraw-Hill, 2005. This popular textbook, setting developments in a historical context, is derived from the author's tenure at the Smithsonian Air and Space Museum.

Bekey, Ivan. *Advanced Space System Concepts and Technologies, 2010-2030+.* Reston, Va.: American Institute of Aeronautics and Astronautics, 2003. Summaries of various advanced concepts and logical arguments used to explore their feasibility.

Design Engineering Technical Committee. *AIAA Aerospace Design Engineers Guide.* 5th ed. Reston, Va.: American Institute of Aeronautics and Astronautics, 2003. A concise book of formulae and numbers that aerospace engineers use frequently or need for reference.

Gann, Ernest K. *Fate Is the Hunter.* 1961. Reprint. New York: Simon and Schuster, 1986. Describes an incident that was the basis for a 1964 film of the same name. Autobiography of a pilot, describing the early days of commercial aviation and coming close to the age of jet travel.

Hill, Philip, and Carl Peterson. *Mechanics and Thermodynamics of Propulsion.* 2d ed. Upper Saddle River, N.J.: Prentice Hall, 1991. A classic textbook on propulsion that covers the basic science and engineering of jet and rocket engines and their components. Also gives excellent sets of problems with answers.

Jenkins, Dennis R. *X-15: Extending the Frontiers of Flight.* NASA SP-2007-562. Washington, D.C.: U.S. Government Printing Office, 2007. Contains various copies of original data sheets, memos, and pictures from the days when the X-15 research vehicle was developed and flown.

Lewis, John S. *Mining the Sky: Untold Riches from the Asteroids, Comets and Planets.* New York: Basic Books, 1997. The most readable answer to the question "What resources are there beyond Earth to make exploration worthwhile?" Written from a strong scientific background, it sets out the reasoning to estimate the presence and accessibility of extraterrestrial water, gases, minerals, and other resources that would enable an immense space-based economy.

Liepmann, H. W., and A. Roshko. *Elements of Gas Dynamics.* Mineola, N.Y.: Dover Publications, 2001. A textbook on the discipline of gas dynamics as applied to high-speed flow phenomena. Contains several photographs of shocks, expansions, and boundary layer phenomena.

O'Neill, Gerard K. *The High Frontier: Human Colonies in Space.* 3d ed. New York: William Morrow & Company, 1977. Reprint. Burlington, Ontario, Canada: Apogee Books, 2000. Sets out the logic, motivations, and general parameters for human settlements in space. This formed the basis for NASA/ASEE (American Society for Engineering Education) studies in 1977-1978 and beyond, to investigate the design of space stations for permanent habitation. Fascinating exposition of how ambitious concepts are systematically analyzed and engineering decisions are made on how to achieve them, or why they cannot yet be achieved.

Peebles, Curtis. *Road to Mach 10: Lessons Learned from the X-43A Flight Research Program.* Reston, Va.: American Institute of Aeronautics and Astronautics, 2008. A contemporary experimental flight-test program description.

Web Sites

Aerospace Digital Library
http://www.adl.gatech.edu

American Institute of Aeronautics and Astronautics
http://www.aiaa.org

Jet Propulsion Laboratory
A Gravity Assist Primer
http://www2.jpl.nasa.gov/basics/grav/primer.php

National Aeronautics and Space Administration
Born of Dreams, Inspired by Freedom. U.S. Centennial of Flight Commission
http://www.centennialofflight.gov

See also: Applied Mathematics; Applied Physics; Avionics and Aircraft Instrumentation; Communication; Computer Science; Spacecraft Engineering.

ALGEBRA

FIELDS OF STUDY

College algebra; precalculus; calculus; linear algebra; discrete mathematics; finite mathematics; computer science; science; engineering; finance.

SUMMARY

Algebra is a branch of applied mathematics that goes beyond the practical and theoretical applications of the numbers of arithmetic. Algebra has a definitive structure with specified elements, defined operations, and basic postulates. Such abstractions identify algebra as a system, so there are algebras of different types, such as the algebra of sets, the algebra of propositions, and Boolean algebra. Algebra has connections not only to other areas of mathematics but also to the sciences, engineering, technology, and other applied sciences. For example, Boolean algebra is used in electronic circuit design, programming, database relational structures, and complexity theory.

KEY TERMS AND CONCEPTS

- **Complement Of A Set:** Group of all elements that are not in a particular group but are in a larger group; the complement of a set *A* is the set of all elements that are not in *A* but are in the universal set and is written *A*r.
- **Conjunction:** Sentence formed by connecting two statements with "and"; the conjunction of statements *p* and *q* is written *p* Ù *q*.
- **Disjunction:** Sentence formed by connecting two statements with "or"; the disjunction of statements *p* or *q* is written *p* Ú *q*.
- **Equal Sets:** Sets whose members are identical; if every element in set *A* is in set *B*, and if every element in set *B* is in set *A*, *A* and *B are equal*.
- **Equivalent Sets:** Sets that have the same cardinality.
- **Negation:** Statement that changes the truth value of a given statement to its opposite truth value; negation of statement *p*, or not *p*, is written as ~*p*.
- **Set:** Well-defined collection of objects or elements.
- **Set Intersection:** Elements that two sets have in common; the intersection of set *A* and set *B* is the set of all elements in *A* and in *B* and is written *A* Ç *B*.
- **Set Union:** Set of elements that are in either or both of two sets; the union of set *A* and set *B* is the set of elements that are in either set *A* or set *B* or in both sets and is written *A* È *B*.
- **Subset:** Group of elements all of which are contained within a larger group of elements; if every element of set *A* is in set *B*, *A* is a subset of *B*.
- **Universal Set:** Universe of discourse or fixed set from which subsets are formed; written *U*.

Definition and Basic Principles

Algebra is a branch of mathematics. The word "algebra" is derived from an Arabic word that links the content of classical algebra to the theory of equations. Modern algebra includes a focus on laws of operations on symbolic forms and also provides a systematic way to examine relationships between such forms. The concept of a basic algebraic structure arises from understanding an important idea. That is, with the traditional definition of addition and multiplication, the identity, associative, commutative, and distributive properties characterize these operations with not only real numbers and complex numbers but also polynomials, certain functions, and other sets of elements. Even with modifications in the definitions of operations on other sets of elements, these properties continue to apply. Thus, the concept of algebra is extended beyond a mere symbolization of arithmetic. It becomes a definitive structure with specified elements, defined operations, and basic postulates. Such abstractions identify algebra as a system, and therefore, there are algebras of many different types, such as the algebra of sets, the algebra of propositions, and Boolean algebra.

The algebra of sets, or set theory, includes such fundamental mathematical concepts as set cardinality and subsets, which are a part of the study of various levels of mathematics from arithmetic to calculus and beyond. The algebra of propositions (logic or propositional calculus) was developed to facilitate the reasoning process by providing a way to symbolically represent statements and to perform calculations based on defined operations, properties, and truth tables. Logic is studied in philosophy, as well

Laws and Properties

Law or Property	Algebra of Sets (set theory)	Algebra of Propositions (logic or proposition calculus)	Boolean Algebra
	For nonempty sets *A*, *B*, and *C* that are subsets of a universal set *U* (∅ designates the empty set)	For propositions *p*, *q*, and *r* (T is a true proposition. F is a false proposition.)	For any elements *x*, *y*, and *z* in set *B* (The operation symbol × may be omitted.)
Identity property	$A \cup \varnothing = A$ $A \cap U = A$	$p \vee F \equiv p$ $p \wedge T \equiv p$	$x + 0 = x$ $x \times 1 = x$
Complement law	$A \cup A' = U$ $A \cap A' = \varnothing$	$p \vee {\sim}p \equiv T$ $p \wedge {\sim}p \equiv F$	$x + x' = 1$ $x \times x' = 0$
Involution law	$(A')' = A$	$\sim({\sim}p) \equiv p$	$(x')' = x$
Commutative property	$A \cup B = B \cup A$ $A \cap B = B \cap A$	$p \vee q \equiv q \vee p$ $p \wedge q \equiv q \wedge p$	$x + y = y + x$ $x \times y = y \times x$
Associative property	$(A \cup B) \cup C = A \cup (B \cup C)$ $(A \cap B) \cap C = A \cap (B \cap C)$	$(p \vee q) \vee r \equiv p \vee (q \vee r)$ $(p \wedge q) \wedge r \equiv p \wedge (q \wedge r)$	$(x + y) + z = x + (y + z)$ $(x \times y) \times z = x \times (y \times z)$
Distributive property	$A \cup (B \cap C) = (A \cup B) \cap (A \cup C)$ $A \cap (B \cup C) = (A \cap B) \cup (A \cap C)$	$(p \vee (q \vee r) \equiv (p \vee q) \wedge (p \vee r)$ $(p \wedge (q \wedge r) \equiv (p \wedge q) \vee (p \wedge r)$	$x + (y \times z) = (x + y) \times (x + z)$ $x \times (y + z) = (x \times y) + (x \times z)$
De Morgan's laws	$(A \cup B)' = A' \cap B'$ $(A \cap B)' = A' \cup B'$	$\sim(p \vee q) \equiv {\sim}p \wedge {\sim}q$ $\sim(p \wedge q) \equiv {\sim}p \vee {\sim}q$	$(x + y)' = x' \times y'$ $(x \times y)' = x' + y'$
Idempotent law	$(A \cup A) = A$ $(A \cap A) = A$	$p \vee p \equiv p$ $p \wedge p \equiv p$	$x + x = x$ $x \times x = x$
Absorption law			$x + (x \times y) = x$ $x \times (x + y) = x$
Domination law	$(A \cup U) = U$ $(A \cap \varnothing) = \varnothing$	$p \vee T \equiv T$ $p \wedge F \equiv F$	$x + 1 = 1$ $x \times 0 = 0$

as various areas of mathematics such as finite mathematics. Boolean algebra is the system of symbolic logic used primarily in computer science applications; it is studied in areas of applied mathematics such as discrete mathematics.

Boolean algebra can be considered a generalization of the algebra of sets and the algebra of propositions. Boolean algebra can be defined as a nonempty set *B* together with two binary operations, sum (symbol +) and product (symbol ×). There is also a unary operation, complement (symbol ¢). In set *B*, there are two distinct elements, a zero element (symbol 0) and a unit element (symbol 1), and certain laws or properties hold. The laws and properties table shows how laws and properties used in the algebra of sets and the algebra of propositions relate to those of Boolean algebra.

Background and History

The Algebra of Sets. In 1638, Italian scientist Galileo published *Discorsi e dimostrazioni matematiche: Intorno à due nuove scienze attenenti alla mecanica e i movimenti locali* (*Dialogues Concerning Two New Sciences*, 1900). In this work, Galileo recognized the basic concept of equivalent sets and distinguishing characteristics of infinite sets. During the nineteenth century, Bohemian mathematician Bernhard Bolzano studied infinite sets and their unique properties; English mathematician George Boole took an algebraic approach to the study of set theory. However, it was German mathematician Georg Cantor who developed a structure for set theory that later led to the modernization of the study of mathematical analysis.

Cantor had a strong interest in the arguments of medieval theologians concerning continuity and the infinite. With respect to mathematics, Cantor realized that not all infinite sets were the same. In 1874, his controversial work on infinite sets was published. After additional research, he established set theory as a mathematical discipline known as *Mengenlehre* (theory of assemblages) or *Mannigfaltigkeitslehre* (theory of manifolds).

The Algebra of Propositions and Boolean Algebra. During the nineteenth century, Boole, English mathematician Charles Babbage, German mathematician Gottlob Frege, and Italian mathematician Giuseppe Peano tried to formalize mathematical reasoning by an "algebraization" of logic.

Boole, who had clerical aspirations, regarded the human mind as God's greatest accomplishment. He wanted to mathematically represent how the brain processes information. In 1847, his first book, *The Mathematical Analysis of Logic: Being an Essay Towards a Calculus of Deductive Reasoning*, was published with limited circulation. He rewrote and expanded his ideas in an 1854 publication, *An Investigation of the Laws of Thought: On Which Are Founded the Mathematical Theories of Logic and Probabilities*. Boole introduced the algebra of logic and is considered the father of symbolic logic.

Boole's algebra was further developed between 1864 and 1895 through the contributions of British mathematician Augustus De Morgan, British economist William S. Jevons, American logician Charles Sanders Peirce, and German mathematician Ernst Schröder. In 1904, American mathematician Edward V. Huntington's *Sets of Independent Postulates for the Algebra of Logic* developed Boolean algebra into an abstract algebraic discipline with different interpretations. With the additional work of American mathematician Marshall Stone and Polish American logician Alfred Tarski in the 1930's, Boolean algebra became a modern mathematical discipline, with connections to several other branches of mathematics, including topology, probability, and statistics.

In his 1940 Massachusetts Institute of Technology master's thesis, Claude Elwood Shannon used symbolic Boolean algebra as a way to analyze relay and switching circuits. Boole's work thus became the foundation for the development of modern electronics and digital computer technology.

Outside the realm of mathematics and philosophy, Boolean algebra has found applications in such diverse areas as anthropology, biology, chemistry, ecology, economics, sociology, and especially computer science. For example, in computer science, Boolean algebra is used in electronic circuit design, programming, database relational structures, and complexity theory.

How It Works

Boolean algebra achieved a central role in computer science and information theory that began with its connection to set theory and logic. Set theory, propositional logic, and Boolean algebra all share a common mathematical structure that becomes apparent in the properties or laws that hold.

Set Theory. The language of set theory is used in the definitions of nearly all mathematical elements, and set theory concepts are integrated throughout the mathematics curriculum from the elementary to the college level. In primary school, basic set concepts may be introduced in sorting, combining, or classifying objects even before the counting process is introduced. Operations such as set complement, union, and intersection can be easily understood in this context.

For example, let the universal set *U* consist of six blocks, each of which is a different color. A block may be red, orange, yellow, violet, blue, or green. Using set notation, *U* = {red, orange, yellow, violet, blue, green}. Let four of the six blocks be sorted into two subsets, *A* and *B*, such that *A* = {red, yellow} and *B* = {blue, green}. The complement of set *A* is the set of blocks that are neither red nor yellow, *A*¢ = {orange, violet, blue, green}. The union of sets *A* and *B* is the set that contains all of the blocks in set *A* or set *B* or both, if there were any colors in common: *A* È *B* = {red, yellow, blue, green}. The intersection of sets *A* and *B* is the set of blocks that are in set *A* and in set *B*, any color that both sets have in common. Because the two sets of blocks have no color in common, *A* Ç *B* = Æ.

Above the primary level, the concepts of logic are

introduced. Daily life often requires that one construct valid arguments, apply persuasion, and make meaningful decisions. Thus, the development of the ability to organize thoughts and explain ideas in clear, precise terms makes the study of reasoning and the analysis of statements most appropriate.

Logic. In propositional algebra, statements are either true or false. A statement may be negated by using "not." Statements can be combined in a variety of ways by using connectives such as "and" and "or." The resulting compound statements are either true or false, based on given truth tables.

A compound statement such as "The First International Conference on Numerical Algebra and Scientific Computing was held in 2006 and took place at the Institute of Computational Mathematics of the Chinese Academy of Sciences in New York" can thus be easily analyzed, especially when written symbolically. The "and" connective indicates that the compound statement is a conjunction. Let *p* be "The First International Conference on Numerical Algebra and Scientific Computing was held in 2006," a true statement; let *q* be "(it) took place at the Institute of Computational Mathematics of the Chinese Academy of Sciences in New York," a false statement because the institute is in Beijing. The truth table for the conjunction indicates that the given compound statement is false: T Ù F º F.

Compound symbolic statements may require multistep analyses, but established properties and truth tables are still used in the process. For example, it is possible to analyze the two symbolic compound statements ~(*p* Ú *q*) and ~*p* Ù ~*q* and also to verify that they are logically equivalent. The truth tables for each compound statement can be combined in one large table to facilitate the process. The first two columns of the table show all possibilities for the truth values of two statements, *p* and *q*. The next three columns show the analysis of each of the parts of the two given compound statements, using the truth tables for negation, disjunction, and conjunction. The last two columns of the table have exactly the same corresponding T and F entries, showing that the truth value will be the same in all cases. This verifies that the two compound statements are logically equivalent. Note that the equivalence of these two propositions is one of De Morgan's laws: ~(*p* Ú *q*) º ~*p* Ù ~*q*.

Computer Circuits. Shannon showed how logic could be used to design and simplify electric circuits. For example, consider a circuit with switches *p* and *q* that can be open or closed, corresponding to the Boolean binary elements, 0 and 1. A series circuit corresponds to a conjunction because both switches must be closed for electric current to flow. A circuit where electricity flows whenever at least one of the switches is closed is a parallel circuit; this corresponds to a disjunction. Because the complement for a given switch is a switch in the opposite position, this corresponds to a negation table. When a circuit is represented in symbolic notation, its simplification may use the laws of logic, such as De Morgan's laws. The simplification may also use tables in the same way as the analysis of the equivalence of propositions, with 1 replacing T and 0 replacing F. Other methods may use Karnaugh maps, the Quine-McCluskey method, or appropriate software.

Computer logic circuits are used to make decisions based on the presence of multiple input signals. The signals may be generated by mechanical switches or by solid-state transducers. The various families of digital logic devices, usually integrated circuits, perform a variety of logic functions through logic gates. Logic gates are the basic building blocks for constructing digital systems. The gates implement the hardware logic function based on Boolean algebra. Two or more logic gates may be combined to provide the same function as a different type of logic gate. This process reduces the total number of integrated circuit packages used in a product.

Boolean expressions can direct computer hardware and also be used in software development by programmers managing loops, procedures, and blocks of statements.

Boolean Searches. Boolean algebra is used in information theory. Online queries are input in the form of logical expressions. The operator "and" is used to narrow a query and "or" is used to broaden it. The operator "not" is used to exclude specific words from a query.

For example, a search for information about "algebra freeware" may be input as "algebra or freeware," "algebra and freeware," or perhaps "algebra and freeware not games." The amount of information received from each query will be different. The first query will retrieve many documents because it will select those that contain "algebra," those that contain "freeware," and those that contain both terms. The second query will retrieve fewer documents because it will select only those documents that contain both

Truth Tables for Algebra of Propositions & Boolean Algebra

Algebra of Propositions			Boolean Algebra		
For propositions p and q (T is a true proposition; F is a false proposition)			For any elements x and y in set B		
Negation			Negation		
p	$\sim p$		x	x	
T	F		1	0	
F	T		0	1	
Disjunction			Sum		
p	q	$p \vee q$	x	y	$x + y$
T	T	T	1	1	1
T	F	T	1	0	1
F	T	T	0	1	1
F	F	F	0	0	0
Conjunction			Product		
p	q	$p \wedge q$	x	y	$x \times y$
T	T	T	1	1	1
T	F	F	1	0	0
F	T	F	0	1	0
F	F	F	0	0	0

terms. The last query will retrieve documents that contain both "algebra" and "freeware" but will exclude items containing the term "games."

Applications and Products

Logic Machines, Calculating Machines, and Computers. The "algebraization" of logic, primarily the work of De Morgan and Boole, was important to the transformation of Aristotelian logic into modern logic, and to the automation of logical reasoning. Several machines were built to solve logic problems, including the Stanhope demonstrator, Jevons's logic machine, and the Marquand machine. In the mid-nineteenth century, Jevons's logic machine, or logic piano, was among the most popular; it used Boolean algebra concepts. Harvard undergraduates William Burkhardt and Theodore Kalin built an electric version of the logic piano in 1947.

In the 1930's, Boolean algebra was used in wartime calculating machines. It was also used in the design of the first digital computer by John Atanasoff and his graduate student Clifford Berry. During 1944-1945, John von Neumann suggested using the binary mathematical system to store programs in computer memory. In the 1930's and 1940's, British mathematician Alan Turing and American mathematician Shannon recognized that binary logic was well suited to the development of digital computers. Just as Shannon's work served as the basis for the theory of switching and relay circuits, Turing's work became the basis for the field of automata theory, the theoretical study of information processing and computer design.

By the end of World War II, it was apparent that computers would soon replace logic machines. Later computer software and hardware developments confirmed that the logic process could be mechanized. Although research work continues to provide

Truth Table Verifying

$\sim(p \vee q) \equiv \sim p \wedge \sim q$

p	q	$\sim p$	$\sim q$	$p \vee q$	$\sim(p \vee q)$	$\sim p \wedge \sim q$
T	T	F	F	T	F	F
T	F	F	T	T	F	F
F	T	T	F	T	F	F
F	F	T	T	F	T	T

theoretical guidelines, automated reasoning programs such as those used in robotics development, are in demand by researchers to resolve questions in mathematics, science, engineering, and technology.

Integrated Circuit Design. Boolean algebra became indispensable in the design of computer microchips and integrated circuits. It is among the fundamental concepts of digital electronics that are essential to understanding the design and function of different types of equipment.

Many integrated circuit manufacturers produce complex logic systems that can be programmed to perform a variety of logical functions within a single integrated circuit. These integrated circuits include gate array logic (GAL), programmable array logic (PAL), the programmable logic device (PLD), and the complex programmable logic device (CPLD).

Engineering approaches to the design and analysis of digital logic circuits involves applications of advanced Boolean algebra concepts, including algorithmic state and machine design of sequential circuits, as well as digital logic simulation. The actual design and implementation of sizeable digital design problems involves the use of computer-aided design (CAD).

Computer Algebra Systems. During the 1960's and 1970's, the first computer algebra systems (CASs) emerged and evolved from the needs of researchers. Computer algebra systems are software that enable users to do tedious and sometimes difficult algebraic tasks, such as simplifying rational functions, factoring polynomials, finding solutions to a system of equations, and representing information graphically in two or three dimensions. The systems offer a programming language for user-defined procedures. Computer algebra systems have not only changed how algebra is taught but also provided a convenient tool for mathematicians, scientists, engineers, and technicians worldwide.

Among the first popular computer algebra systems were Reduce, Scratchpad, Macsyma (later Maxima), and Mu-Math. Later popular systems include MATLAB, Mathematica, Maple, and MathCAD.

In 1987, Hewlett-Packard introduced HP-28, the first handheld calculator series with the power of a computer algebra system. In 1995, Texas Instruments released the TI-92 calculator with advanced CAS capabilities based on Derive software. Manufacturers continue to offer devices such as these with increasingly

Fascinating Facts About Algebra

- Algebra has been studied since 2000 B.C.E., making it the oldest branch of written mathematics. Babylonian, Chinese, and Egyptian mathematicians proposed and solved problems in words, that is, using "rhetorical algebra."
- In 1869, British logician William S. Jevons, a student of the mathematician Augustus De Morgan, created a logic machine that used Boolean algebra. The popular machine was known as the logic piano because it had ivory keys and resembled a piano.
- English logician John Venn was heavily influenced by English mathematician George Boole, and his Venn diagrams, developed around 1880, facilitated conceptual and procedural understanding of Boolean algebra.
- In his 1936 paper, "On Computable Numbers, with an Application to the Entscheidungs problem," British mathematician Alan Turing characterized which numbers and functions in mathematics are effectively computable. His paper was an early contribution to recursive function theory, which was a topic of interest in several areas, including logic.
- With the publication of *A Symbolic Analysis of Relay and Switching Circuits* (1940) and "A Mathematical Theory of Communication" (1948), American mathematician Claude Elwood Shannon introduced a new area for the application of Boolean algebra. He showed that the basic properties of series and parallel combinations of electric devices such as relays could be adequately represented by this symbolic algebra. Since then, Boolean algebra has played a significant role in computer science and technology.
- In applied algebra, properties of groups can be used to analyze transformations and symmetry. The transformations include translating, rotating, reflecting, and dilating a pattern such as one in an M. C. Escher painting, or parts of an object such as a Rubik's cube.
- Applied algebra is used in cryptography to study codes and ciphers in problems involving data security and data integrity.
- Applied algebra is used in chemistry to study symmetry in molecular structure.

powerful functions; such devices tend to decrease in size and cost with advancements in technology.

Impact on Industry

Government and University Research. Boolean algebra has roots and applications in many areas, including topology, measure theory, functional analysis, and ring theory. Research and study of Boolean algebras therefore includes structure theory and model theory, as well as connections to other logics. Some of the techniques for analyzing Boolean functions have been used in such areas as computational learning theory, combinatorics, and game theory.

Computer algebra, originally known as algebraic computing, is concerned with the development, implementation, and application of algorithms that manipulate and analyze mathematical expressions. Practical and theoretical research includes the development of effective and efficient algorithms for use in computer algebra systems. Research includes engineering, scientific, and educational applications.

Linear algebra begins with the study of linear equations, matrices, determinants, function spaces, eigenvalues, and orthogonality. Research and development in applied linear algebra includes theoretical studies, algorithmic designs, and implementation of advanced computer architectures. Such research involves scientific, engineering, and industrial applications.

Engineering and Technology. The study of associative digital network theory comprises computer science, electrical engineering digital circuit design, and number theory. Such theory is of interest to researchers at industrial laboratories and instructors and students at technical institutions. The focus is on new research and developments in modeling and designing digital networks with respect to both mathematics and engineering disciplines. The unifying associative algebra of function composition (semigoup theory) is used in the study of the three main computer functions: sequential logic (state machines), arithmetic, and combinational (Boolean) logic.

Applied Science. There has been a dramatic rise in the power of computation and information technology. With it have come vast amounts of data in fields such as business, engineering, and science. The challenge of understanding the data has led to new tools and approaches, such as data mining. Data mining involves the use of algorithms to identify and verify structure from data analysis. Developments in the field of data mining have brought about increased focus on higher level mathematics. Such areas as topology, combinatorics, and algebraic structures (lattices and Boolean algebras) are often included in research.

Careers and Course Work

The applications of algebra are numerous, which means that those interested in algebra can pursue jobs and careers in a wide range of fields, including business, engineering, and science, particularly computer science.

Data Analyst or Data Miner. Data mining is a broad mathematical area that involves the discovery of patterns and hidden information in large databases, using algorithms. In applications of data mining, career opportunities emerge in e-commerce, security, forensics, medicine, bioinformatics and genomics, astrophysics, and chemical and electric power engineering. Course work should include a focus on higher level mathematics in such areas as combinatorics, topology, and algebraic structures.

Materials Engineer. Materials science is the study of the properties, processing, and production of such items as metallic alloys, liquid crystals, and biological materials. There are many career opportunities in research, manufacturing, and development in aerospace, electronics, biology, and nanotechnology. The design and analysis of materials depends on mathematical models and computational tools. Course work should include a focus on applied mathematics, including differential equations, linear algebra, numerical analysis, operations research, discrete mathematics, optimization, and probability.

Computer Animator or Digital Artist. Computer animation encompasses many areas, including mathematics, computer science, physics, biomechanics, and anatomy. Career opportunities arise in medical diagnostics, multimedia, entertainment, and fine arts. The algorithms for computer animation come from scientific relationships, statistics, signal processing, linear algebra, control theory, and computational geometry. Recommended mathematics course work includes statistics, discrete mathematics, linear algebra, geometry, and topology.

Financial Analyst. As quantitative methods transform the financial industry, banking, insurance, investment, and government regulatory institutions are among those relying on mathematical tools and computational models. Such tools and models are used to support investment decisions, to develop and price new securities, to manage risk, and to guide portfolio selection, management, and optimization. Course work should include a focus on the mathematics of finance, linear algebra, linear programming, probability, and descriptive statistics.

Social Context and Future Prospects

Algebra is part of two broad, rapidly growing fields, applied mathematics and computational science. Applied mathematics is the branch of mathematics that develops and provides mathematical methods to meet scientific, engineering, and technological needs. Applied mathematics includes not only discrete mathematics and linear algebra but also numerical analysis, operations research, and probability. Computational science integrates applied mathematics, science, engineering, and technology to create a multidisciplinary field developing and using innovative problem-solving strategies and methodologies.

Applied mathematics and computational science are used in almost every area of science, engineering, and technology. Business also relies on applied mathematics and computational science for research, design, and manufacture of products that include aircraft, automobiles, computers, communication systems, and pharmaceuticals. Research in applied mathematics therefore often leads to the development of new mathematical models, theories, and applications that contribute to diverse fields.

June Gastón, B.A., M.S.Ed., M.Ed., Ed.D.

Further Reading

Barnett, Raymond A., Michael R. Ziegler, and Karl E. Byleen. *Finite Mathematics for Business, Economics, Life Sciences, and Social Sciences.* 12th ed. Boston: Prentice Hall, 2011. Covers the mathematics of finance, linear algebra, linear programming, probability, and descriptive statistics, with an emphasis on cross-discipline principles and practices. Helps develop a functional understanding of mathematical concepts in preparation for application in other areas.

Cohen, Joel S. *Computer Algebra and Symbolic Computation: Elementary Algorithms.* Natick, Mass.: A. K. Peters, 2002. Examines mathematical fundamentals, practical challenges, formulaic solutions, suggested implementations, and examples in a few programming languages appropriate for building a computer algebra system. Further reading recommendations provided.

Cooke, Roger. *Classical Algebra: Its Nature, Origins, and Uses.* Hoboken, N.J.: Wiley-Interscience, 2008. Broad coverage of classical algebra that includes its history, pedagogy, and popularization. Each chapter contains thought-provoking problems and stimulating questions; answers are provided in the appendix.

Dunham, William. *Journey Through Genius: The Great Theorems of Mathematics.* New York: Penguin, 1991. Provides historical and technical information. Each chapter is devoted to a mathematical idea and the people behind it. Includes proofs.

Givant, Steven, and Paul Halmos. *Introduction to Boolean Algebras.* New York: Springer, 2009. An informal presentation of lectures given by the authors on Boolean algebras, intended for advanced undergraduates and beginning graduate students.

Van der Waerden, B. L. *Algebra.* New York: Springer, 2003. Reprint of the first volume of the 1970 translation of van der Waerden's *Moderne Algebra* (1930), designated one of the most influential mathematics textbooks of the twentieth century. Based in part on lectures by Emmy Noether and Emil Artin.

Web Sites

American Mathematical Society
http://www.ams.org

Mathematical Association of America
http://www.maa.org

Society for Industrial and Applied Mathematics
http://www.siam.org

See also: Applied Mathematics; Calculus; Computer Languages, Compilers, and Tools; Electronics and Electronic Engineering; Engineering Mathematics; Geometry; Numerical Analysis; Pattern Recognition; Topology.

ANTIBALLISTIC MISSILE DEFENSE SYSTEMS

FIELDS OF STUDY

Astronautical engineering; physics; electronics engineering; electrical engineering; mechanical engineering; optical engineering; software engineering; systems engineering.

SUMMARY

To protect people and their possessions from harm by incoming missiles launched by a potential enemy, antiballistic missile defense systems have been designed to detect, track, and destroy incoming missiles. These systems are designed to fire guided defensive missiles to hit the incoming missiles before they strike their targets.

KEY TERMS AND CONCEPTS

- **Ballistic Missile:** Missile that is powered only during the first part of its flight; includes missiles powered and steered during the end of their flight.
- **Boost Phase:** Initial part of a missile's flight, when it is under power.
- **Electromagnetic Pulse (EMP):** Burst of electromagnetic radiation given off by a nuclear explosion that sends a burst of energy through electronic devices and may severely damage them.
- **Exoatmospheric:** Outside or above the atmosphere.
- **Hit To Kill:** Destroying a warhead by hitting it precisely, often compared to hitting a bullet with a bullet.
- **Kill Vehicle:** Speeding mass that smashes into a warhead using its energy of motion (kinetic energy) to destroy the warhead.
- **Midcourse Phase:** Middle portion of a missile's flight, as it coasts through space.
- **Terminal Phase:** End of a missile's flight, as it reenters the atmosphere and nears its target.
- **Warhead:** Payload of the missile that does the desired damage; it can consist of high explosives, a nuclear bomb, or a kill vehicle.

Definition and Basic Principles

An intercontinental ballistic missile (ICBM) can deliver enough nuclear explosives to devastate a city. Because ICBMs plunge from the sky at very high speeds, unseen by the eye until they hit, they are difficult to defend against. That is the purpose of an antiballistic missile (ABM) system. Such systems have several essential tasks: to detect a missile launch or ascent, to track the missile during its midcourse and its terminal flight, and to calculate an intercept point for the defensive missile, fire a defensive missile, track the defensive missile, guide it to its target, and verify that the target has been destroyed.

Using multiple independently targeted reentry vehicle (MIRV) technology, ten to twelve independently targeted warheads and decoys can be delivered by a single ICBM (which is against the signed but unratified 1993 Strategic Arms Limitations Talks II treaty). Each deployed warhead requires a defensive missile to destroy it. Countries with MIRV technology could attack with enough warheads and decoys to saturate any ABM system, allowing at least some nuclear warheads to reach their targets. If the targeted country retaliates, the situation is likely to escalate into full-scale nuclear war.

Although ABM systems cannot prevent nuclear war, they do have some uses. For example, a limited system could defend against a few accidental launches of ICBMs or against a few missiles from rogue nations.

Background and History

The evolution of ABM systems has been driven by politics and perceived need. In the 1960's, the strategy of "mutual assured destruction," or MAD, was articulated by U.S. secretary of defense Robert McNamara. If either side launched a first strike against the other, the nonaggressor would still have enough warheads left to devastate the aggressor. An effective ABM system would have upset this balance.

The ABM treaty of 1972 was signed by U.S. president Richard Nixon and the Soviet Union's general secretary, Leonid Brezhnev, and was amended in 1974 to allow each side to have only one ABM site. The Soviet Union elected to place its single ABM system around Moscow, a system still maintained by Russia.

The United States built the Safeguard ABM system at the Grand Forks Air Force Base in North Dakota, where it guarded Minuteman III missiles. Safeguard operated for only a few months before it was closed down. The U.S. Missile Defense Agency was formed in 1983 (under a different name) and given the charge to develop, test, and prepare missile defense systems. Also in 1983, President Ronald Reagan announced a change: The offensive MAD would become the defensive building of an impenetrable shield under the Strategic Defense Initiative, popularly dubbed "Star Wars." Although Edward Teller, chief architect of the hydrogen bomb, assured President Reagan that the Strategic Defense Initiative could be implemented, it proved to be unfeasible.

After the Cold War ended with the dissolution of the Soviet Union in 1991, both the United States and Russia eventually concluded that the most likely use of their ABM systems would be against a limited strike by a country such as North Korea or Iran. To guard against Iranian and North Korean missiles, the United States has twenty-six ground-based interceptor (GBI) missiles at Fort Greeley, Alaska, and four at Vandenberg Air Force Base in California. (Two sites would not have been allowed under the ABM treaty, but the United States withdrew from the ABM treaty in June, 2002.) Warships equipped with the Aegis Combat System (a system that uses radar and computers to guide weapons aimed at enemy targets), with Standard Missile-3's (SM-3's), are stationed in the Mediterranean and the Black Sea to protect Europe.

Fascinating Facts About Antiballistic Missile Defense Systems

- The first ABM system deployed by the United States, the Safeguard system, had a fatal flaw that was known before it was built. Exploding warheads on the first salvo of Sprint missiles would blind Sprint's tracking radar.
- In 2008, the United States used a Standard Missile-3 to shoot down an orbiting satellite.
- American ABM systems can defend against an accidental launch or a limited attack by a rogue nation but not a full-scale attack by Russia or China.
- The United States can field a layered missile defense, where each successive layer deals with missiles that survived the previous layer. It would start with the long-range Ground-Based Interceptor, then the Aegis standard missiles, the THAAD missile, and finally the short-range Aegis missiles.
- The "sizzler" is a Russian-built cruise missile with a 300-kilometer maximum range. It flies at Mach 0.8 until it nears its target, when it accelerates to nearly Mach 3 and takes a zigzag path to its target—making it difficult to shoot it down. China, India, Algeria, and Vietnam have all purchased it.
- It may not be possible to defend against a full-scale attack by a determined aggressor who could place ten warheads and decoys in a single missile (banned by treaty). Each warhead and possibly each decoy would need to be targeted by a separate missile, unless several kill vehicles could fly on a single ABM (also banned by treaty).

How It Works

An ABM system must successfully perform several different functions to work properly. Initial detection may be done by a remote ground radar, by an airborne radar plane, or even by infrared sensors in space. Infrared sensors are particularly effective at spotting the hot rocket plume of a launching ICBM. Normal radar is line of sight and cannot detect targets beyond the curve of the Earth. Therefore, airborne radar is used; being higher, it can see farther.

To guard the United States from attack by ICBMs or submarine-launched missiles, three PAVE PAWS (Precision Acquisition Vehicle Entry Phased-Array Warning System) radars look outward from American borders. PAVE is a U.S. Air Force program, and the three radar systems are located on Air Force bases in Massachusetts, Alaska, and California. The systems track satellites and can spot a car-sized object in space 5,500 kilometers away. The initial detection of a long-range missile would probably come from a PAVE PAWS radar system. The radar and associated computers must also classify the objects they detect and determine if the objects are threatening.

If PAVE PAWS spots a suspicious object, a defensive missile site will be notified and will begin tracking the object with its on-site phased-array radar. If the object is still deemed a threat, permission to fire on it must be given either by direct command or by standing orders. Aided by a fire-control computer, the operator selects the target and fires one or two missiles at it. The missiles are guided using ground radar until they approach the target, when the missile's own radar or infrared sensors assume the tracking duties.

Modern missiles are usually hit-to-kill missiles, although some carry conventional explosives to ensure the kill. Next, the radar and computer must see if the target has been destroyed, and if not, defenses closer to the ICBM's target must be activated. These actions are all coordinated through the "command, control, and communications" resources of the on-site unit.

Initially, antiballistic missiles were designed to approach their targets and detonate a nuclear warhead, so great accuracy was not required. Close was good enough. The Safeguard ABM system had a long-range Spartan missile with a 5-megaton yield and the short-range Sprint missile, which carried a neutron bomb (an "enhanced radiation bomb"). The Spartan was to engage targets in space, and any surviving targets would be destroyed in the atmosphere by the Sprint. Unfortunately, the nuclear explosions would produce electromagnetic pulses, which would blind the Spartan/Sprint guiding radar. This problem encouraged designers to work toward a hit-to-kill technology. Russia has an ABM system around Moscow and has removed the nuclear warheads from these missiles and replaced them with conventional explosives.

Applications and Products

The hardware of an ABM system includes several key parts. Radars and infrared sensors are the eyes of an ABM system. Defensive missiles destroy the invading missiles, and computers calculate trajectories and direct the defensive missiles.

Radar. A phased-array radar antenna is a key component of any modern ABM system. It consists of an array of hundreds or thousands of small antennas mounted in a regular array of rows and columns on a wall. The radar can project a beam in a certain direction or receive a return echo from a particular direction by activating the small antennas in certain patterns. Because this is all done electronically without the radar antenna moving, many scan patterns can be run simultaneously. The Aegis SN/SPY-1 radar can simultaneously track more than one-hundred targets at a distance of up to 190 kilometers. Some multimission Navy destroyers have the AN/SPY-3 radar. It combines the functions of several radars into one, requires fewer operators, and is less visible to other radars. An advanced radar can also interrogate an IFF (identify friend or foe) device. Incoming missiles can also be identified by their flight paths and radar signatures.

Missiles. The Patriot system began as an antiaircraft system but was upgraded to defend against tactical missiles. It seemed to do well against scud missiles during the 1990-1991 Gulf War, but later analysis showed that most of the claimed kills were actually ripped apart by air resistance. The Patriot system is highly mobile because all its modules are truck or trailer mounted. One hour after the unit arrives on site, it can be up and running. The system uses the Patriot Advanced Capability-2 (PAC-2) missile with a range of 160 kilometers, a ceiling of 24 kilometers, a speed of Mach 5, and an explosive warhead. Its phased-array radar is difficult to jam.

The PAC-3 missile is smaller; four of them will fit in the space taken by one PAC-2 missile. This gives a missile battery more firepower. The missile travels at Mach 5, has a ceiling of 10 to 15 kilometers, and a maximum range of 45 kilometers. The warhead is hit-to-kill, backed up with an exploding fragmentation bomb on a proximity fuse. Two missiles are fired at a target, with the second missile firing a few seconds after the first. The second missile targets whatever might be a warhead left in the debris from the first warhead's impact. Under test conditions, the PAC-3 missile has scored twenty-one intercepts out of thirty-nine attempts.

The Theater High Altitude Area Defense missile system (THAAD) is designed to shoot down short, medium, and intermediate range missiles during their terminal phases. It uses the AN/TPY-2 radar. THAAD missiles also have some limited capability against ICBMs. Their effective range is about 200 kilometers with a peak altitude of 150 kilometers. Their warhead is a kinetic kill vehicle (KKV). When the missile nears its target, the KKV is explosively separated from its spent rocket. Guided by an advanced infrared sensor, steering rockets adjust the KKV's course so that it will hit dead on. In the six tests since 2006, the THAAD missile hit its target all six times.

The Ground-Based Midcourse Defense (GMD) system is designed to defend against a limited attack by intermediate- and long-ranged missiles. Its missile is the Ground-Based Interceptor (GBI), a three-stage missile with an exoatmospheric kill vehicle (EKV). The GBI is not mobile but is fired from an underground silo. Although there is on-site tracking radar, the GMD can receive early warnings from radars hundreds of kilometers away. The GBI travels at about 10 kilometers per second and has a ceiling of about

2,000 kilometers. Out of fourteen tests, GBIs have hit their targets eight times.

Aegis. Ticonderoga-class cruisers and Arleigh Burke-class destroyers all have Aegis Combat Systems (ACSs). Aegis was built to counter short- and medium-range ballistic missiles, aircraft, and other ships. Aegis combines several key parts: the AN/SPY-1 phased-array radar, the MK 99 Fire Control System, the Weapons Control System, the Command and Decision Suite, and the Standard Missile-2 (SM-2).

The SM-2's speed is Mach 3.5, and its range is up to 170 kilometers. The missile has radar and an infrared seeker for terminal guidance, and it has a blast fragmentation warhead. The SM-2 is being replaced by the SM-6, which has twice the range and better radar and is more agile so that it can better deal with the Russian "sizzler" cruise missile.

The third Aegis missile is the Standard Missile-3 (SM-3). It has four stages, a range of more than 500 kilometers, a ceiling of more than 160 kilometers, and a kinetic kill vehicle (KKV). It is guided by ground radar and by onboard infrared sensors. On February 21, 2008, an SM-3 missile was used to shoot down a failed U.S. satellite. The satellite was 240 kilometers above the ground, and the missile approached it at 36,667 kilometers per hour. The satellite had never reached its proper orbit and was coming down because of air resistance. The reason given for shooting it down was the large amount of toxic hydrazine fuel still aboard. Many viewed it as an excuse to test the antisatellite capability of the SM-3 because it was likely that the hydrazine would have been dispersed and destroyed when the satellite reentered the atmosphere. When equipped with the SM-3, Aegis can serve as an ABM system for assets within range. As has been noted, Aegis-equipped warships with the Standard Missile-3 (SM-3) are stationed in the Mediterranean and the Black Sea to protect Europe from Iranian missiles.

Lasers. The Airborne Laser Test Bed (ALTB) is mounted in a modified Boeing 747 designated the Boeing YAL-1. It has a megawatt-class chemical laser that gets its energy from the chemical reaction between oxygen and iodine compounds. It has successfully destroyed target missiles in flight, but they were not far away. It is unlikely that the laser's range will ever exceed 300 kilometers, and if the aircraft must loiter that close to the launch site, it is in danger of being shot down by the enemy nation's air defense. In 2009, Secretary of Defense Robert Gates recommended that the ALTB project be cut back to limited research.

Another laser project that showed promise was the Tactical High Energy Laser (THEL). The THEL is a deuterium fluoride chemical laser with a theoretical power of 100 kilowatts. It was a joint project with the United States and Israel and was able to shoot down Katyusha rockets but nothing larger. Although lasers show promise, it seems unlikely that they will be used in an ABM system anytime soon.

Impact on Governments and Industry

ABM systems have encouraged governments to cooperate. The Patriot system is to be replaced by the Medium Extended Air Defense System (MEADS), a joint project of the United States, Germany, and Italy. It is designed for quick setup so that it will be ready almost as quickly as it is unloaded. Although it is more capable than the Patriot system, MEADS has been streamlined so that it requires only one-fifth of the cargo flights to deliver it to its operation site. Its purpose is to protect against tactical ballistic missiles, unmanned aerial drones, cruise missiles, and aircraft. The ceiling of the PAC-3 MSE (missile segment enhancement) is 50 percent higher and its range is twice the range of the PAC-3. The United States together with Israel developed the Arrow missile to protect Israel from Iranian missiles. The United States, Russia, Israel, Japan, China, the Republic of China (Taiwan), India, North Korea, and Iran all have vigorous ABM programs and active defense industries. Other countries have smaller programs and use hardware manufactured elsewhere. For example, Patriot missile systems manufactured by Raytheon are in thirteen countries.

In the following list of the top defense companies involved in ballistic missile defense, no distinction has been made between companies that are the prime contractor or a subcontractor, and employee numbers and revenues (as opposed to profits) are for 2009. The numbers in parentheses are the ranks, by revenue, of these U.S. defense contractors.

1. Boeing Company is involved with the Airborne Warning and Control System (AWACS radar), Aegis SM-3, Arrow Interceptor, Ground-based Midcourse Defense (GMD) system, Patriot Advanced Capability-3 (PAC-3), and Strategic Missile and Defense Systems. Boeing has

about 160,000 employees and a revenue of $68.3 billion.

2. Lockheed Martin has 140,000 employees and revenue of $45 billion. It produces the Terminal High Altitude Area Defense (THHAD) weapon system, the Medium Extended Air Defense System (MEADS), Airborne Laser Test Bed, Space-Based Infrared System (SBIRS), and various other missiles and satellites.
3. Northrop Grumman has developed a Kinetic Energy Interceptor (KEI), is developing a satellite that will spot and track ICBMs, and is modernizing the Minuteman III missiles. The company has about 120,000 employees and revenue of about $32 billion.
4. General Dynamics is involved with Aegis and several satellite systems with infrared sensors that can track ballistic missiles. Its revenue is about $32 billion, and it has 91,200 employees.
5. Raytheon builds exoatmospheric kill vehicles, Standard Missiles-2, -3, and -6, airborne radar, the Patriot missile system, and other defense-related equipment. The company had more than 72,000 employees and earned $27 billion.
6. L-3 Communications is involved with the Ground-Based Midcourse Defense system, the Standard Missile-3, and the Aegis system. It has a revenue of $14 billion and a workforce of 64,000.
7. Orbital Sciences manufactures the GMD three-stage boost vehicle. Orbital has 3,100 employees and a revenue of $1.2 billion per year.

Altogether, these companies have around 650,000 employees and about $220 billion in revenue. Research shows that for every job in the defense industry, one to four more jobs are created in the community to meet the needs of the defense workers and their families. Reasonable guesses are that 20 percent of the employees work on antiballistic missile projects, and that each job draws one other job to the area. The effect on the economy of antiballistic missile programs is then 260,000 jobs and $88 billion each year.

Careers and Course Work

Many defense industry jobs in the United States require a security clearance; this means the applicant must be a U.S. citizen. A strong background in the physical sciences is necessary for the aerospace industry. High school students should take all the courses in physics, chemistry, computer science, and mathematics that they can. At least a bachelor's degree in science or engineering is required. Employees will need to write reports and make presentations, so students should take some classes in writing and speech. They may eventually become a team or unit leader; if so, they may wish they had taken a simple business management course. Those who are involved with research and development need a feel for how things work. It helps if they like to build or repair things. They should be creative and be able to think of new ways to do things.

Bachelor's degrees are sufficient for a number of aerospace positions: astronautical, computer, electrical, mechanical, optical engineer or physicist. Employees generally start as junior members of a team, but as time passes, they can become senior members and then perhaps team leaders with more responsibility. Astronautical engineers design, test, and supervise the construction of rockets, missiles, and satellites. Computer engineers interface hardware with computers, writing and debugging programs that instruct the hardware to do what is wished. Electrical engineers design, develop, and produce radio frequency data links for missile applications. Mechanical engineers design, analyze, and integrate cryogenic components and assemblies. Optical engineers develop solutions to routine technical problems and work with signal processing analysis and design as well as sensor modeling and simulation. Systems engineers design systems for missile guidance and control, computational fluid mechanics analysis, and wind tunnel testing. Physicists who work with ABM systems test electrical and mechanical components, measure radiation effects, and mitigate them if necessary.

Social Context and Future Prospects

People have always wondered whether money spent on an ABM system could be better spent elsewhere. Some maintain that even an excellent system would be unlikely to protect against some city destroyers. ICBMs with a dozen warheads and decoys could overwhelm any ABM system. Furthermore, if nation A thought that nation B was installing an effective ABM system, nation A might launch a preemptive strike before the ABM system became operational. At the very least, nation A would probably build more ICBMs and escalate the international arms race. Because of such considerations, it is generally conceded that a limited ABM system to deal with accidental launches or a few missiles from rogue nations makes sense.

The United States proposed defending Europe against Iranian missiles by placing defensive missiles and radar in Poland and the Czech Republic. Russia saw this move as a means to blunt a Russian attack on the United States. It threatened to respond with nuclear weapons if Poland or the Czech Republic allowed the installations. In 2009, President Barack Obama scrapped that plan and announced that Aegis-equipped warships would be stationed in the Mediterranean and Black Seas, where they could defend Europe. Russia welcomed this change, which makes it plain that simply building ABM installations may have serious political consequences.

Another consideration is that sooner or later another asteroid will hit Earth, and humans might want to do something about it. For example, asteroid (29075) 1950 DA has a 0.0033 percent (one-third of 1 percent) chance of hitting Earth on March 16, 2880. Experience and technology developed from the various ABM programs will most likely be of some use in dealing with errant asteroids, as their technology of guiding missiles toward a target may be adapted to deflect an incoming asteroid by just enough to prevent it from destroying the planet.

Charles W. Rogers, B.A., M.S., Ph.D.

Further Reading

Burns, Richard Dean. *The Missile Defense Systems of George W. Bush: A Critical Assessment.* Santa Barbara, Calif.: Praeger, 2010. A critical look at the fiscal and political costs to deploy a ground-based ABM system and the effects of trying to extend it to Europe.

Denoon, David. *Ballistic Missile Defense in the Post-Cold War Era.* Boulder, Colo.: Westview Press, 1995. An overview of various proposed ABM systems along with ways to judge if they would be worth the expense of constructing them.

Hey, Nigel. *The Star Wars Enigma: Behind the Scenes of the Cold War Race for Missile Defense.* Dulles, Va.: Potomac Books, 2006. An interesting story of who pushed for the Strategic Defense Initiative (SDI) and other ABM systems.

O'Rourke, Ronald. *Navy Aegis Ballistic Missile Defense (BMD) Program: Background and Issues for Congress.* Washington, D.C.: Congressional Research Service, 2010. O'Rourke, a specialist in naval affairs, discusses the past, present, and future of Aegis and especially the politics behind decisions affecting it.

Payne, Keith B. *Strategic Defense: "Star Wars" in Perspective.* Mansfield, Tex.: Hamilton Press, 1986. A wide-ranging examination of the various issues of SDI.

Sloan, Elinor C. "Space and Ballistic Missile Defense." In *Security and Defense in the Terrorist Era: Canada and the United States Homeland.* Montreal: McGill-Queen's University Press, 2010. How Canada's response to the threat of ballistic missiles is, and should be, different from that of the United States.

Web Sites

Boeing
Defense, Space, and Security
http://www.boeing.com/bds

Federation of American Scientists
Military Analysis Network
http://www.fas.org/programs/ssp/man/index.html

Lockheed Martin
Missiles and Fire Control
http://www.lockheedmartin.com/mfc/products.html

Northrop Grumman
Missile Defense
http://www.northropgrumman.com/missiledefense/index.html

Raytheon
Missile Systems
http://www.raytheon.com/businesses/rms

Union of Concerned Scientists
Nuclear Weapons and Global Security
http://www.ucsusa.org/nuclear_weapons_and_global_security

U.S. Department of State
Arms Control and International Security
http://www.state.gov

See also: Aeronautics and Aviation

APPLIED MATHEMATICS

FIELDS OF STUDY

Mechanical design; mechanical engineering; fluid dynamics; hydraulics; pneumatics; electronic engineering; physics; process modeling; physical chemistry; chemical kinetics; geologic engineering; geographic information systems; computer science; statistics; actuarial science; particle physics; epidemiology; investments; game theory; game design.

SUMMARY

Applied mathematics is the application of mathematical principles and theory in the real world. The practice of applied mathematics has two principal objectives: One is to find solutions to challenging problems by identifying the mathematical rules that describe the observed behavior or characteristic involved, and the other is to reduce real-world behaviors to a level of precise and accurate predictability. Mathematical rules and operations are devised to describe a behavior or property that may not yet have been observed, with the goal of being able to predict with certainty what the outcome of the behavior would be.

KEY TERMS AND CONCEPTS

- **Algorithm:** Effective method for solving problems that uses a finite series of specific instructions.
- **Boundary Conditions:** Values and properties that are required at the boundary limits of a mathematically defined behavior.
- **Markov Chain:** Probability model used in the study of processes that are considered to move through a finite sequence of steps, with repeats allowed.
- **Modeling:** Representation of real-world events and behaviors by mathematical means.
- **Particle-In-A-Box Model:** Classic mathematical analogy of the motion of an electron in an atomic orbital, subject to certain conditions, such as the requirement to have specific values at the limiting boundaries of the box.
- **Sparse System:** Any system whose parameter coefficients can be represented in a matrix that contains mostly zeros. Such systems are described as loosely coupled and typically represent members that are linked together linearly rather than as a network.
- **String Theory:** Mathematical representation of all subatomic particles as having the structures of strings rather than discrete spheres.
- **Wavelet Minimization:** Mathematical process whereby small random variations in frequency-dependent measurements (background noise) are removed to allow a clearer representation of the measurements.

Definition and Basic Principles

Applied mathematics focuses on the development and study of mathematical and computational tools. These tools are used to solve challenging problems primarily in science and engineering applications and in other fields that are amenable to mathematical procedures. The principal mathematical tool is calculus, often referred to as the mathematics of change. Calculus provides a means of quantitatively understanding how variables that cannot be controlled directly behave in response to changes in variables that can be controlled directly. Thus, applied mathematics makes it possible to make predictions about the behavior of an environment and thus gain some mastery over that environment.

For example, suppose a specific characteristic of the behavior of individuals within a society is determined by the combination of a large number of influencing forces, many of which are unknown and perhaps unknowable, and therefore not directly controllable. Study of the occurrence of that characteristic in a population, however, allows it to be described in mathematical terms. This, in turn, provides a valid means of predicting the future occurrence and behavior of that characteristic in other situations. Applied mathematics, therefore, uses mathematical techniques and the results of those techniques in the investigation or solving of problems that originate outside of the realm of mathematics.

The applications of mathematics to real-world phenomena rely on four essential structures: data structures, algorithms, theories and models, and computers and software. Data structures are ways of organizing information or data. Algorithms are specific methods of dealing with the data. Theories and models are used in the analysis of both data and ideas and represent the rules that describe either the way the data were formed or the behavior of the data. Computers and software are the physical devices that are used to manipulate the data for analysis and application. Algorithms are central to the development of software, which is computer specific, for the manipulation and analysis of data.

Background and History

Applied mathematics, as a field of study, is newer than the practices of engineering and building. The mathematical principles that are the central focus of applied mathematics were developed and devised from observation of physical constructs and behaviors and are therefore subsequent to the development of those activities. The foundations of applied mathematics can be found in the works of early Egyptian and Greek philosophers and engineers. Plane geometry is thought to have developed during the reign of the pharaoh Sesostris, as a result of agricultural land measurements necessitated by the annual inundation of the Nile River. The Greek engineer Thales of Miletus is credited with some of the earliest and most profound applications of mathematical and physical principles in the construction of some of his devices, although there is no evidence that he left a written record of those principles. The primary historical figures in the development of applied mathematics are Euclid and Archimedes. It is perhaps unfortunate that the Greek method of philosophy lacked physical experimentation and the testing of hypotheses but was instead a pure thought process. For this reason, there is a distinction between the fields of pure mathematics and applied mathematics, although the latter depends strictly on the former.

During the Middle Ages, significant mathematical development took place in Islamic nations, where *al geber*, which has come to be known as algebra, was developed, but the field of mathematics showed little progress in Europe. Even during the Renaissance period, mathematics was the realm almost exclusively of astronomers and astrologers. It is not certain that even Leonardo da Vinci, foremost of the Renaissance engineers and artists, was adept at mathematics despite the mathematical brilliance of his designs. The major historical development of applied mathematics began with the development of calculus by Sir Isaac Newton and Gottfried Wilhelm Leibniz in the seventeenth century. The applicability of mathematical principles in the development of scientific pursuits during the Industrial Revolution brought applied mathematics to the point where it has become essential for understanding the physical universe.

How It Works

Applied mathematics is the creation and study of mathematical and computational tools that can be broadly applied in science and engineering. Those tools are used to solve challenging problems in these and related fields of practice. In its simplest form, applied mathematics refers to the use of measurement and simple calculations to describe a physical condition or behavior.

A simple example might be the layout or design of a field or other area of land. Consider a need to lay out a rectangular field having an area of 2,000 square meters (m^2) with the shortest possible perimeter. The area (A) of any rectangular area is determined as the product of the length (l) and the width (w) of the area in question. The perimeter (P) of any rectangular area is determined as the sum of the lengths of all four sides, and in a rectangular area, the opposite sides are of equal length. Thus, $P = 2l + 2w$, and $A = l \times w = 2000\ m^2$. By trial and error, pairs of lengths and widths whose product is 2,000 may be tried out, and their corresponding perimeters determined. A field that is 2,000 meters long and 1 meter wide has an area of 2,000 square meters and a perimeter of 4,002 meters. Similarly, a field that is 200 meters long and 10 meters wide has the required area, and a perimeter of only 420 meters. It becomes apparent that the perimeter is minimized when the area is represented as a square, having four equal sides. Thus, the length of each side must be equal to the square root of 2,000 in magnitude. Having determined this, the same principles may be applied to the design of any rectangular area of any size.

The same essential procedures as those demonstrated in this simple example apply with equal validity to other physical situations and are, in fact, the very essence of scientific experimentation and research. The progression of the development of mathematical models and procedures in many different areas of application is remarkably similar. Development of a mathematical model begins with a simple expression to which refinements are made, and the

Fascinating Facts About Applied Mathematics

- Applied mathematics research toward the development of the quantum computer, which would operate on the subatomic scale, has led some researchers to postulate the simultaneous existence of multiple universes.
- Applied mathematics may have begun as geometry in ancient Egypt during the reign of the pharaoh Sesostris, who instituted the measurement of changes in land area for the fair assessment of taxes in response to the annual flooding of the Nile River.
- Ancient engineers apparently used certain esoteric principles of mathematics in the construction of the pyramids in Egypt and elsewhere, as the dimensions of these structures reflect the golden ratio (ϕ).
- The flow of traffic along a multilane highway can be described by the same mathematics that describes the flow of water in a system of pipes.
- The behavior of an electron bound to a single proton is described mathematically as a particle in a box, because it must conform to specific conditions determined by the confines of the box.
- The seemingly uniform movement of individual birds in flocks of flying birds demonstrates the chaos theory, in which discrete large-scale patterns arise from unique small-scale actions. This is also known as the butterfly effect.
- The growth rates of bananas in the tropics and other natural behaviors can be described exactly by differential equations.
- Three-dimensional MRI can be used to practice and perfect surgical techniques without ever touching an actual patient.

results of the calculation are compared with the actual behavior of the system under investigation. The changes in the difference between the real and calculated behaviors are the key to further refinements that, ideally, work to bring the two into ever closer agreement. When mathematical expressions have been developed that adequately describe the behavior of a system, those expressions can be used to describe the behaviors of other systems.

A key component to the successful application of mathematical descriptions or models is an understanding of the different variables that affect the behavior of the system being studied. In fluid dynamics, for example, obvious variables that affect a fluid are the temperature and density of the fluid. Less obvious perhaps are such variables as the viscosity of the fluid, the dipolar interactions of the fluid atoms or molecules, the adhesion between the fluid and the surface of the container through which it is flowing, whether the fluid is flowing smoothly (laminar flow) or turbulently (nonlaminar flow), and a number of other more obscure variables. A precise mathematical description of the behavior of such a system would include corrective terms for each and every variable affecting the system. However, a number of these corrective terms may be considered together in an approximation term and still produce an accurate mathematical description of the behavior.

An example of such an approximation may be found in the applied mathematical field of quantum mechanics, by which the behavior of electrons in molecules is modeled. The classic quantum mechanical model of the behavior of an electron bound to an atomic nucleus is the so-called particle-in-a-box model. In this model, the particle (the electron) can exist only within the confines of the box (the atomic orbital), and because the electron has the properties of an electromagnetic wave as well as those of a physical particle, there are certain restrictions placed on the behavior of the particle. For example, the value of the wave function describing the motion of the electron must be zero at the boundaries of the box. This requires that the motion of the particle can be described only by certain wave functions that, in turn, depend on the dimensions of the box. The problem can be solved mathematically with precision only for the case involving a single electron and a single nuclear proton that defines the box in which the electron is found. The calculated results agree extremely well with observed measurements of electron energy.

For systems involving more particles (more electrons and more nuclear protons and neutrons), the number of variables and other factors immediately exceeds any ability to be calculated precisely. A solution is found, however, in a method that uses an approximation of the orbital description, known as a Slater-type orbital approximation, rather than a precise mathematical description. A third-level Gaussian treatment of the Slater-type orbitals, or STO-3G analysis, yields calculated results for complex molecular structures that are in excellent agreement with the observed values measured in physical experiments. Although the level of mathematical technique is vastly more complex than in the simple area example, the basic method of finding an applicable method is almost exactly identical.

Applications and Products

Applied mathematics is essentially the application of mathematical principles and theories toward the resolution of physical problems and the description of behaviors. The range of disciplines in which applied mathematics is relevant is therefore very broad. The intangible nature of mathematics and mathematical theory tends to restrict active research and development to the academic environment and applied research departments of industry. In these environs, applied mathematics research tends to be focused rather than general in nature. Applied mathematics is generally divided into the major areas of computational mathematics: combinatorics and optimization, computer science, pure mathematics, and statistical and actuarial science. The breadth of the research field has grown dramatically, and the diversity of subject area applications is indicated by the applied mathematics research being conducted in atmospheric and biological systems applications, climate and weather, complexity theory, computational finance, control systems, cryptography, pattern recognition and data mining, multivariate data analysis and visualization, differential equation modeling, fluid dynamics, linear programming, medical imaging, and a host of other areas.

Computational Mathematics. Simply stated, computational mathematics is the process of modeling systems quantitatively on computers. This is often referred to in the literature as *in silico*, indicating that the operation or procedure being examined is carried out as a series of calculations within the silicon-based electronic circuitry of a computer chip and not in any tangible, physical manner. Research in computational mathematics is carried out in a wide range of subject areas.

The essence of computational mathematics is the development of algorithms and computer programs that produce accurate and reliable models or depictions of specific behaviors. In atmospheric systems, for example, one goal would be to produce mathematical programs that precisely depict the behavior of the ozone layer surrounding the planet. The objective of such a program would be to predict how the ozone layer would change as a result of alterations in atmospheric composition. It is not feasible to observe the effects directly and would ultimately be counterproductive if manifesting an atmospheric change resulted in the destruction of the ozone layer. Modeling the system *in silico* allows researchers to institute virtual changes and determine what the effect of each change would be. The reliability of the calculated effect depends directly on how accurately the model describes the existing behavior of the system.

Medical Imaging. An area in which applied mathematics has become fundamentally important is the field of medical imaging, especially as it applies to magnetic resonance imaging (MRI). The MRI technique developed from nuclear magnetic resonance (NMR) analysis commonly used in analytical chemistry to determine molecular structures. In NMR, measurements are obtained of the absorption of specific radio frequencies by molecules held within a magnetic field. The strength of each absorption and specific patterns of absorptions are characteristic of the structure of the particular molecule and so can be used to determine unequivocally the exact molecular structure of a material.

One aspect of NMR that has been greatly improved by applied mathematics is the elimination of background noise. A typical NMR spectrum consists of an essentially infinite series of small random signals that often hide the detailed patterns of actual absorption peaks and sometimes even the peaks themselves. The Fourier analysis methodology, in which such random signals can be treated as a combination of sine and cosine waves, also known as wavelet theory, eliminates a significant number of such signals from electromagnetic spectra. The result is a much more clear and precise record of the actual absorptions. Such basic NMR spectra are only one dimensional, however. The second generation modification of NMR systems was developed to produce a two-dimensional representation of the NMR absorption spectrum, and from this was developed the three-dimensional NMR imaging system that is known as MRI. Improvements that make the Fourier analysis technique ever more effective in accord with advances in the computational abilities of computer hardware are the focus of one area of ongoing applied mathematics research.

Population Dynamics and Epidemiology. Population dynamics and epidemiology are closely related fields of study. The former studies the growth and movements of populations, and the latter studies the growth and movements of diseases and medical conditions within populations. Both rely heavily for their mathematical descriptions on many areas of applied mathematics, including statistics, fluid dynamics, complexity theory, pattern recognition, data visualization, differential equation modeling, chaos theory, risk management, numerical algorithms and techniques, and statistical learning.

In a practical model, the movements of groups of individuals within a population are described by many of the same mathematical models of fluid dynamics that apply to moving streams of particles. The flow of traffic on a multilane highway or the movement of people along a busy sidewalk, for example, can be seen to exhibit the same gross behavior as that of a fluid flowing through a system of pipes. In fact, any population that can be described in terms of a flow of discrete particles, whether molecules of water or vehicles, can be described by the same mathematical principles, at least to the extent that the variables affecting their motion are known. Thus, the forces of friction and adhesion that affect the flow of a fluid within a tube are closely mimicked by the natural tendencies of drivers to drive at varying speeds in different lanes of a multilane highway. Window-shoppers and other slow-moving individuals tend to stay to the part of the sidewalk closest to the buildings, while those who walk faster or more purposefully tend to use the part of the sidewalk farthest away from the buildings, and this also follows the behavior of fluid flow.

The spread or movement of diseases through a population can also be described by many of the same mathematical principles that describe the movements of individuals within a population. This is especially true for diseases that are transmitted directly from person to person. For other disease vectors, such as animals, birds, and insects, a mathematical description must describe the movements of those particular populations, while at the same time reflecting the relationship between those populations and the human population of interest.

Statistical Analysis and Actuarial Science. Perhaps the simplest or most obvious use of applied mathematics can be found in statistical analysis. In this application, the common properties of a collection of data points, themselves measurements of some physical property, are enumerated and compared for consistency. The effectiveness of statistical methods depends on the appropriately random collection of representative data points and on the appropriate definition of a property to be analyzed.

Statistical analysis is used to assess the consistency of a common property and to identify patterns of occurrence of characteristics. This forms the basis of the practice of statistical process control (SPC) that has become the primary method of quality control in industry and other fields of practice. In statistical process control, random samples of an output stream are selected and compared to their design standard. Variations from the desired value are determined, and the data accumulated over time are analyzed to determine patterns of variation. In an injection-molding process, for example, a variation that occurs consistently in one location of the object being molded may indicate that a modification to the overall process must be made, such as adjusting the temperature of the liquid material being injected or an alteration to the die itself to improve the plastic flow pattern.

In another context, one that is tied to epidemiology, the insurance and investment industries make very detailed use of statistical analysis in the assessment of risk. Massive amounts of data describing various aspects of human existence in modern society are meticulously analyzed to identify patterns of effects that may indicate a causal relationship. An obvious example is the statistical relationship pattern between healthy lifestyle and mortality rates, in which obese people of all age groups have a higher mortality rate than their counterparts who maintain a leaner body mass. Similarly, automobile insurance rates are set much higher for male drivers between the ages of sixteen and twenty-five than for female drivers in that age group and older drivers because statistical analysis of data from accidents demonstrates that this particular group has the highest risk of involvement in a traffic accident. This type of data mining is a continual process as relationships are sought to describe every factor that plays a role in human society.

Impact on Industry

Statistical Process Control. The practice of statistical process control has had an unprecedented effect on modern society, especially in the manufacturing industry. Before the development of mass-production methods, parts and products were produced by skilled and semiskilled craftspeople. The higher the level of precision required for a production piece, the more skilled and experienced an artisan or craftsperson had to be. Manpower and cost of production quickly became the determining factors in the availability of goods.

The advent of World War II, however, ushered in an unprecedented demand for materials and products. The production of war materials required the development of methods to produce large numbers of products in a short period of time. Various accidents and mistakes resulting from the rapid pace of production also indicated the need to develop methods of quality control to ensure that goods were produced to the expected level of quality and dependability. The methods of quality assurance that were developed at that time were adopted by industries in postwar Japan and developed into a rigorous protocol of quality management based on the statistical analysis of key features of the goods being produced. The versatility of the system, when applied to mass-production methods, almost instantly eliminated the dependency on skilled tradespeople for the production of precision goods.

In the intervening half century, statistical process control became a required component of production techniques and an industry unto itself, spawning such rigorous programs as the International Organization for Standardization's ISO 9000, ISO 9001, and ISO 14000; the Motorola Corporation's Six Sigma; and Lean Manufacturing (derived from the Toyota Production System), which have become global standards.

Economics. The analysis of data by mathematical means, especially data mining and pattern recognition in population dynamics, has provided much of the basis of the economic theory that directs the conduct of business on a global scale. This has been especially applicable in regard to the operation of the stock market. In large part, the identification of patterns and trends in the historical data of trade and business provides the basic information for speculation on projected or future consumer trends. Although this is important for the operation of publicly traded businesses on the stock market, it is more important in the context of government. The same trends and patterns play a determining role in the design and establishment of government policies and regulations for both domestic and international affairs.

Medicine and Pharmaceuticals. Applied mathematics, particularly the use of Fourier analysis and wavelet theory, has sparked explosive growth in the fields of medicine and pharmaceutical development. The enhanced analytical methods available to chemists and bioresearchers through NMR and Fourier transform infrared (FTIR) spectroscopy facilitate the identification and investigation of new compounds that have potential pharmaceutical applications. In addition, advanced statistical methods, computer modeling, and epidemiological studies provide the foundation for unprecedented levels of research.

The science of medical diagnostic imaging has grown dramatically because of the enhancements made available through applied mathematics. Computer science, based on the capabilities of modern digital computer technology, has replaced the physical photographic method of X-ray diagnostics with a digital version that is faster and more efficient and has the additional capability of zooming in on any particular area of interest in an X-ray image. This allows for more accurate diagnostics and is especially important for situations in which a short response time is essential.

The methodology has enabled entirely new procedures, including MRI. Instead of relatively plain, two-dimensional images, real-time three-dimensional images can be produced. These images help surgeons plan and rehearse delicate surgical maneuvers to perfect a surgical technique before ever touching a patient with a scalpel. MRI data and modifications of MRI programming can be used to control the sculpting or fabrication of custom prosthetic devices. All depend extensively on the mathematical manipulation of data to achieve the desired outcome.

Careers and Course Work

The study of applied mathematics builds on a solid and in-depth comprehension of pure mathematics. The student begins by taking courses in mathematics throughout secondary school to acquire a solid foundational knowledge of basic mathematical principles before entering college or university studies. A specialization in mathematics at the college and university level is essential to practically all areas of study. As so many fields have been affected by applied mathematics, the methodologies being taught in undergraduate courses are a reflection of the accepted concepts of applied mathematics on which they are constructed. As the depth of the field of applied mathematics indicates, career options are, for all intents and purposes, unlimited. Every field of endeavor, from anthropology to zoology, has a component of applied mathematics, and the undergraduate must learn the mathematical methods corresponding to the chosen field.

Students who specialize in the study of applied mathematics as a career choice and proceed to postgraduate studies take courses in advanced mathematics and carry out research aimed at developing mathematics theory and advancing the application of those developments in other fields.

Particular programs of study in applied mathematics are included as part of the curriculum of other disciplines. The focus of such courses is on specific mathematical operations and techniques that are relevant to that particular field of study. All include a significant component of differential calculus, as appropriate to the dynamic nature of the subject matter.

Social Context and Future Prospects

Human behavior is well suited and highly amenable to description and analysis through mathematical principles. Modeling of population dynamics has become increasingly important in the contexts of service and regulation. As new diseases appear, accurate models to predict the manner in which they will spread play an ever more important role in determining the response to possible outbreaks. Similarly, accurate modeling of geological activities is increasingly valuable in determining the best ways to respond to such natural disasters as a tsunami or earthquake. The Earth itself is a dynamic system that is only marginally predictable. Applied mathematics is essential to developing models and theories that will lead to accurate prediction of the occurrence and ramifications of seismic events. It will also be absolutely necessary for understanding the effects that human activity may be having on the atmosphere and oceans, particularly in regard to the issues of global warming and the emission of greenhouse gases. Climate models that can accurately and precisely predict the effects of such activities will continue to be the object of a great deal of research in applied mathematics.

The development of new materials and engineering new applications with those materials is an ongoing human endeavor. The design of extremely small nanostructures that employ those materials is based on mathematical principles that are unique to the realm of the very small. Of particular interest is research toward the development of the quantum computer, a device that operates on the subatomic scale rather than on the existing scale of computer construction.

Richard M. J. Renneboog, M.Sc.

Further Reading

Anton, Howard. *Calculus: A New Horizon.* 6th ed. New York: John Wiley & Sons, 1999. Offers a clear and progressive introduction to many of the basic principles of applied mathematics.

De Haan, Lex, and Toon Koppelaars. *Applied Mathematics for Database Professionals.* New York: Springer, 2007. Focuses on the use of set theory and logic in the design and use of databases in business operations and communications.

Howison, Sam. *Practical Applied Mathematics: Modelling, Analysis, Approximation.* New York: Cambridge University Press, 2005. Provides a basic introduction to several practical aspects of applied mathematics, supported by in-depth case studies that demonstrate the applications of mathematics to actual physical phenomena and operations.

Kurzweil, Ray. *The Age of Spiritual Machines: When Computers Exceed Human Intelligence.* New York: Penguin Books, 2000. Presents a thought-provoking view of the nature of artificial intelligence that may arise as a direct result of applied mathematics.

Moore, David S. *The Basic Practice of Statistics.* 5th ed. New York: Freeman, 2010. Introduces statistical analysis at its most basic level and progresses through more complex concepts to an understanding of statistical process control.

Naps, Thomas L., and Douglas W. Nance. *Introduction to Computer Science: Programming, Problem Solving, and Data Structures.* 3d ed. St. Paul, Minn.: West Publishing, 1995. Presents a basic introduction to the principles of computer science and the development of algorithms and mathematical processes.

Rubin, Jean E. *Mathematical Logic: Applications and Theory.* Philadelphia: Saunders College Publishing,

1990. An introduction to the principles of mathematical logic that requires the reader to work closely through the material to obtain a good understanding of mathematical logic.

Washington, Allyn J. *Basic Technical Mathematics with Calculus.* 9th ed. Upper Saddle River, N.J.: Pearson Prentice Hall, 2009. Provides thorough and well-explained instruction in applied mathematics for technical pursuits in a manner that allows the reader to grasp many different mathematical concepts.

Web Sites

Mathematical Association of America
http://www.maa.org

Society for Industrial and Applied Mathematics
http://www.siam.org

See also: Algebra; Calculus; Computer Science; Engineering Mathematics; Game Theory; Geometry; Numerical Analysis; Pattern Recognition.

APPLIED PHYSICS

FIELDS OF STUDY

Physics; mathematics; applied mathematics; mechanical engineering; civil engineering; metrics; forensics; biological physics; photonics; microscopy and microanalysis; biomechanics; materials science; surface science; aeronautics; hydraulics; nanotechnology; robotics; medical imaging; radiology.

SUMMARY

Applied physics is the study and application of the behavior of condensed matter, such as solids, liquids, and gases, in bulk quantities. Applied physics is the basis of all engineering and design that requires the interaction of individual components and materials. The study of applied physics now extends into numerous other fields, including physics, chemistry, biology, engineering, medicine, geology, meteorology, and oceanography.

KEY TERMS AND CONCEPTS

- **Biomechanics:** The science of physical movement in living systems.
- **Condensed Matter:** Matter is a physical quantity that is normally recognized as having the form of a solid, a liquid, or a gas, also known as a bulk quantity.
- **Forensics:** The science of analysis to determine cause and effect after an event has occurred, highly dependent upon the application of laws of physics in the case of physical events.
- **Magnetic Levitation:** A technology that utilizes the mutual repulsion of like magnetic fields to maintain a physical separation between an object and a track (such as a high-speed train and its track bed), while keeping the two in virtual contact.
- **Magnetic Resonance Imaging (MRI):** A noninvasive diagnostic imaging technology based on nuclear magnetic resonance that provides detailed images of internal organs and the structures of living systems.
- **Nanoprobe:** A device designed to perform a specific function, constructed of nanometer (10-9 meter) scale components.
- **Newtonian Mechanics:** The behavior of condensed matter as described by Newton's laws of motion and inertia.
- **Nuclear Magnetic Resonance (NMR):** A property of certain atomic nuclei having unpaired electrons to absorb energy of a specific electromagnetic wavelength within a constant magnetic field.
- **Prosthetics:** The branch of applied physics that deals with the design of artificial components that replace or augment natural biological components; the artificial components that are designed and constructed to replace normal biological components.
- **Quantum Mechanics:** The physical laws describing the behavior of matter in quantities below the nanometer (10-9 meter) scale, especially of individual atoms and molecules.
- **Spectrometry:** An analytical technology that utilizes the specific electromagnetic energy of specific wavelengths to identify or quantify specific components of a material or solution.

Definition and Basic Principles

Applied physics is the study of the behavior and interaction of condensed matter. Condensed matter is commonly recognized as matter in the phase forms of solids, liquids, and gases. Each phase has its own unique physical characteristics, and each material has its own unique physical suite of properties that derive from its chemical identity. The combination of these traits determines how something consisting of condensed matter interacts with something else that consists of condensed matter.

The interactions of constructs of condensed matter are governed by the interaction of forces and other vector properties that are applied through each construct. A lever, for example, functions as the medium to transmit a force applied in one direction so that it operates in the opposite direction at another location. Material properties and strengths are also an intimate part of the interaction of condensed matter. A lever that cannot withstand the force applied to it laterally will bend or break rather than function as a lever. Similarly, two equally hard objects, such as a train car's wheels and the steel tracks upon which they roll, rebound elastically with little or no generation of friction. If, however, one of the materials is less hard than the other, as in the case of a steel rotor and a composition brake pad, a great deal of friction characterizes their interaction.

Background and History

Applied physics is of necessity the oldest of all practical sciences, dating back to the first artificial use of an object by an early hominid. The basic practices have been in use by builders and designers for many thousands of years. With the development of mathematics and measurement, the practice of applied physics has grown apace, relying as it still does upon the application of basic concepts of vector properties (force, momentum, velocity, weight, moment of inertia) and the principles of simple machines (lever, ramp, pulley).

The modern concept of applied physics can be traced back to the Greek philosopher Aristotle (c. 350 B.C.E.), who identified several separate fields for systematic study. The basic mathematics of physics was formulated by Pythagoras about two hundred years before, but was certainly known to the Babylonians as early as 1600 B.C.E. Perhaps the greatest single impetus for the advance of applied physics was the development of calculus by Sir Isaac Newton and the fundamental principles of Newtonian mechanics in the seventeenth century. Those principles describe well the behaviors of objects composed of condensed matter, only failing as the scale involved becomes very small (as below the scale of nanotechnology). Since the Industrial Revolution and into the twenty-first century, the advancing capabilities of technology and materials combine to enable even more advances and applications, yet all continue to follow the same basic rules of physics as the crudest of constructs.

How It Works

As the name implies, applied physics means nothing more or less than the application of the principles of physics to material objects. The most basic principles of applied physics are those that describe and quantify matter in motion (speed, velocity, acceleration, momentum, inertia, mass, force). These principles lead to the design and implementation of devices and structures, all with the purpose of performing a physical function or action, either individually or in concert.

A second, and equally important, aspect of applied physics is the knowledge of the physical properties and characteristics of the materials being used. This includes such physical properties as melting and boiling points, malleability, thermal conductivity, electrical resistivity, magnetic susceptibility, density, hardness, sheer strength, tensile strength, compressibility, granularity, absorptivity, and a great many other physical factors that determine the suitability of materials for given tasks. Understanding these factors also enables the identification of new applications for those materials.

Design. Applied physics is the essential basis for the design of each and every artificial object and construct, from the tiniest of nanoprobes and the simplest of levers to the largest and most complex of machines and devices. Given the idea of a task to be performed and a device to perform that task, the design process begins with an assessment of the physical environment in which the device must function, the forces that will be exerted against it, and the forces that it will be required to exert in order to accomplish the set task. The appropriate materials may then be selected, based on their physical properties and characteristics. Dimensional analyses determine the necessary size of the device, and also play a significant role in determining the materials that will be utilized.

All of these factors affect the cost of any device, an important aspect of design. Cost is an especially important consideration, as is the feasibility of replacing a component of the designed structure. For example, a device operating in a local environment in which replacement of a component, such as an actuating lever, is easily carried out may be constructed using a simple steel rod. If, however, the device is something like the Mars Rover, for which replacement of worn parts is not an option, it is far more reasonable and effective to use a less-corrosive, stronger, but more costly material, such as titanium, to make movable parts. The old design tenet that form follows function is an appropriate rule of thumb in the field of applied physics.

Applications and Products

In many ways, applied physics leads to the idea of creating devices that build other devices. At some point in time, an individual first used a handy rock to break up another rock or to pound a stick into the ground. This is a simple example of applied physics, when that individual then began to use a specific kind of rock for a specific purpose; this action advanced again when someone realized that attaching the rock to a stick provided a more effective tool. Applied physics has advanced far beyond the crude rock-on-a-stick hammer, yet the exact same physical principles apply now with the most elegant of impact devices. It would not be possible to itemize even a small percentage of the applications and products that have resulted from applied physics, and new physical devices are developed each day.

Civil Society. Applied physics underpins all physical aspects of human society. It is the basis for the design and construction of the human environment and its supporting infrastructure, the most obvious of which are roads and buildings, designed and engineered with consideration of the forces that they must withstand, both human-made and natural. In this, the highest consideration is given to the science and engineering of materials in regard to the desired end result, rather than to the machines and devices that are employed in the actual construction process. The principles of physics involved in the design and construction of supporting structures, such as bridges and high-rise towers, are of primary importance. No less important, however, are the physical systems that support the function of the structure, including such things as electrical systems, internal movement systems, environmental control systems, monitoring systems, and emergency response systems, to name a few. All are relevant to a specific aspect of the overall physical construct, and all must function in coordination with the other systems. Nevertheless, they all also are based on specific applications of the same applied physics principles.

Transportation. Transportation is perhaps the single greatest expenditure of human effort, and it has produced the modern automotive, air, and railway transportation industries. Applied physics in these areas focuses on the development of more effective and efficient means of controlling the movement of the machine while enhancing the safety of human and nonhuman occupants. While the physical principles by which the various forms of the internal combustion engine function are the same now as when the devices were first invented, the physical processes used in their operation have undergone a great deal of mechanical refinement.

A particular area of refinement is the manner in which fuel is delivered for combustion. In earlier gasoline-fueled piston engines, fuel was delivered via a mechanical fuel pump and carburetor system that functions on the Venturi principle. This has long since been replaced by constant-pressure fuel pumps and injector systems, and this continues to be enhanced as developers make more refinements based on the physical aspects of delivering a liquid fuel to a specific point at a specific time. Similarly, commercial jet engines, first developed for aircraft during World War II, now utilize the same basic physical principles as did their earliest counterparts. Alterations and developments that have been achieved in the interim have focused on improvement of the operational efficiency of the engines and on enhancement of the physical and combustion properties of the fuels themselves.

In rail transport, the basic structure of a railway train has not changed; it is still a heavy, massive tractor engine that tows a number of containers on very low friction wheel-systems. The engines have changed from steam-powered behemoths made up of as many as one-quarter-million parts to the modern diesel-electric traction engines of much simpler design.

Driven by the ever-increasing demand for limited resources, this process of refinement for physical efficiency progresses in every aspect of transportation on land, sea, and air. Paradoxically, it is in the area of railway transport that applied physics offers the greatest possibility of advancement with the development of nearly frictionless, magnetic levitation systems upon which modern high-speed bullet trains travel. Physicists continue to work toward the development of materials and systems that will be superconducting at ambient temperatures. It is believed that such materials will completely revolutionize not only the transportation sector but also the entire field of applied physics.

Medical. No other area of human endeavor demonstrates the effects of applied physics better than the medical field. The medical applications of applied physics are numerous, touching every aspect of medical diagnostics and treatment and crossing over into many other fields of physical science. The most obvious of these is medical imaging, in which X-ray diagnostics, magnetic resonance imaging (MRI), and other forms of spectroscopic analysis have become primary tools of medical diagnostics. Essentially, all of the devices that have become invaluable as medical tools began as devices designed to probe the physical nature of materials.

Spectrographs and spectrometers that are now routinely used to analyze the specific content of various human sera were developed initially to examine the specific wavelengths of light being absorbed or emitted by various materials. MRI developed from the standard technique of physical analytical chemistry called nuclear magnetic resonance spectrometry, or NMR. In this methodology, magnetic signals are recorded as they pass through a material sample, their patterns revealing intimate details about the three-dimensional molecular structure of a given compound. The process of MRI that has developed from this simpler application now permits diagnosticians to view the internal structures and functioning of living systems in real time. The diagnostic X ray that has become the single most common method of examining internal biological structures was first used as a way to physically probe the structure of crystals, even though the first practical demonstration of the existence of X rays by German physicist Wilhelm Röntgen in 1895 was as an image of the bones in his wife's hand.

Military. It is not possible to separate the field of applied physics from military applications and martial practice, and no other area of human endeavor so clearly illustrates the double-edged sword that is the nature of applied physics. The field of applied physics at once facilitates the most humanitarian of endeavors and the most violent of human aggressions. Ballistics, which gets its name from the Roman Latin word for "war," is the area of physics that applies to the motion of bodies moving through a gravitational field. The mathematical equations that describe the motion of a baseball being thrown from center field equally describe the motion of an arrow or a bullet in flight, of the motion of the International Space Station in its orbit, and of the trajectory of a warhead-equipped rocket. The physical principles that have permitted the use of nuclear fission as a source of reliable energy are the same principles that define the functioning of a nuclear warhead.

In the modern age it is desired that warfare be carried out as quickly and efficiently as possible as the means to restore order and peace to an embattled area, with as little loss of life and destruction of property as possible. To that end, military research relies heavily on the application of physics in the development of weapons, communications, and surveillance systems.

Digital Electronics. Applied physics has also led to the development of the transistor, commonly attributed to William Shockley, in 1951. Since that time, the development of ever-smaller and more efficient systems based on the semiconductor junction transistor has been rapid and continuous. This has inexorably produced the modern technology of digital electronics and the digital computer. These technologies combine remarkably well to permit the real-time capture and storage of data from the many and various devices employed in medical research and diagnostics, improve the fine control of transportation systems and infrastructure, and advance a veritable host of products that are more efficient in their use of energy than are their nondigital counterparts.

Impact on Industry

There is a common misconception that persons with degrees and expertise within the field of physics are not presented with sufficient opportunities for employment after graduating. This misconception reflects a difference in academic terminology versus business terminology only, and not reality. It is true that few physics graduates actually work as traditional physicists. Given the training and ability in mathematics and physical problem-solving that physicists (anyone with a degree in physics) have developed, they tend to acquire positions as industrial physicists instead, holding positions with industrial titles as managers, engineers, computer scientists, and technical staff.

Data from the American Institute of Physics indicate that physicists represent a significantly higher-than-average income group. About one-fifth of all physicists enter military careers, while about one-half of those with a doctorate remain in academia and teaching and about one-half of those with master's degrees make their careers in industry.

Given that applied physics is the fundamental basis of all human technology upon which modern society has been built, it is essentially impossible to assign a monetary value to the overall field. The direct economic value that derives from any individual aspect of applied physics is often measured in billions of U.S. dollars, and in some cases the effects are so far-reaching as to be immeasurable. The amount spent on space research, for example, by the U.S. government through the National Aeronautics and Space Administration (NASA), is less than 1 percent of the annual national budget of approximately $2.5 trillion. The return on this investment in terms of corporate and personal taxes, job creation, and economic growth, however, is conservatively estimated to be no less than seven-to-one. That is to say, every $1 million invested in space research generates a minimum of $7 million in tax revenues and economic growth.

Not the least significant of effects is the spin-off that enters the general economy as a direct result of applied physics research and development. Space research has provided a long list of consumer products and technical enhancements that touch everything from golf balls and enriched baby foods to industrial process modeling, virtual reality training, and education aids. The list of spin-off technologies and products from this source alone is very long, and it continues to grow as research advances. It is interesting to note that many of the products of space research that were developed with the intent of bettering a human presence on other planets have proved eminently applicable to bettering the human presence on Earth.

The wireless communications industry is one of the most obvious spin-offs from space research, and it now represents a significant segment of the overall economy of nations around the world. The value of this industry in Canada, for example, provides a good indication of the value of wireless communications to other nations. The wireless communications industry in Canada serves a country that has four persons per square kilometer, as compared with thirty persons per square kilometer in the United States. In 2010, the economic value of the wireless communications industry in Canada was approximately US$39 billion, of which $16.9 billion was a direct contribution to the gross domestic product (GDP) through the sale of goods and services. This contribution to the GDP compared favorably with those of the automotive manufacturing industry, the food manufacturing industry, and agricultural crop production.

Wireless communications produced up to 410,000 Canadian jobs, or about 2.5 percent of all jobs in Canada; this marks a significantly higher average salary than the national average. The value-added per employee in the wireless telecommunications industry was also deemed to be more than 2.5 times higher than the national average. While a direct extrapolation to the effect of this industry on the economy of the United States, and other nations, is not appropriate because of the differences in population distribution and geography, it is nevertheless easy to see that the wireless communications industry now represents a significant economic segment and has far-reaching effects.

Furthermore, wireless communication can greatly reduce unproductive travel time and the associated economic costs, while significantly improving the logistics of business operations. Wireless communications empowers small businesses, especially in rural

Fascinating Facts About Applied Physics

- When the semiconductor junction transistor was invented in 1951, it initiated the digital electronics revolution and formed the functional basis of the personal computer. In 1954, scientists at RAND envisioned a home computer in the year 2000, complete with a teletype Fortran-language programming station, a large CRT television monitor, rows of blinking lights and dials, and twin steering-wheels.
- Nuclear magnetic resonance (NMR) spectrometry is an analytical method that is sensitive enough to differentiate and measure the magnetic environments of individual atoms in molecules.
- Creating a Stone Age hammer by fastening a rock to a stick is as much a product of applied physics as is the creation of the International Space Station.
- A rock thrown through the air obeys exactly the same laws of ballistics as does a bullet fired from a high-powered rifle or a satellite orbiting a planet.
- The B2 Spirit stealth bomber, a U.S. military aircraft representing sophisticated applied physics, has a unit cost of more than US$1.15 billion (as of 2003).
- Spectrography, initially invented to identify specific wavelengths of emitted or absorbed light, is the basis for all analytical methods that rely on electromagnetic radiation.
- The formal identification of the science known as applied physics can be traced back to Aristotle, to circa 350 B.C.E.
- All human technology is the product of applied physics. Spin-off from such advanced activities as space research translates into a great number of everyday products, including cell phones, home computers, and digital egg timers, with an overall economic value of trillions of U.S. dollars annually.

areas. Medical devices represent another large economic segment that has grown out of applied physics research and development. Recent surveys (2008) suggest that it is one segment that is relatively unaffected by economic upswings and downswings. Given that medical methodology continues to advance on the strength of diagnostic devices, while the development of medical technology continues to advance in step with advancing medical methodology, it is logical that the effects of economic changes are mitigated in this field.

As of 1999, medical research in the United States represented an investment of more than $35 billion. The economic value of medical research was estimated in 1999 to be approximately $48 trillion for the elimination of death from heart disease and $47 trillion for a cure for cancer. Simply reducing cancer mortality by a mere 1 percent would generate an economic return of $500 billion. These estimates are all based on the assumed economic "value of life" as benefits to persons deriving from an extended ability to participate in society. The research and development of medical knowledge that would generate these economic returns cannot be called applied physics. Rather, the various technologies by which the overall medical research program is supported are the direct result of applied physics.

In the realm of military expenditure, it is estimated that approximately 10 percent of the U.S. GDP is directly related to military purposes. This includes research and development in applied physics technologies.

Careers and Coursework

Programs of study for careers in applied physics or in fields that are based on applied physics include advanced mathematics, physics, and materials science. Depending on the area of career specialization, coursework may also include electronics engineering, mechanical engineering, chemistry and chemical engineering, biology and biomechanics, instrumentation, and more advanced courses in mathematics and physics. The range of careers is broad, and the coursework required for each will vary accordingly.

Credentials and qualifications required for such careers range from a basic associate's degree from a community college to one or more postgraduate degrees (master's or doctorate) from a top-tier specialist university. An advanced degree is not a requirement for a career in an applied physics discipline, and the level of education needed for a successful career will depend on the nature of the discipline chosen.

Each discipline of applied physics has a core knowledge base of mathematics and physics that is common to all disciplines. Students should therefore expect to take several courses in these areas of study, including advanced calculus and a comprehensive program of physics. Other courses of study will depend on the particular discipline that a person chooses to pursue. For disciplines that focus on medical applications, students will also take courses in chemistry and biology, because understanding the principles upon which biological systems operate is fundamental to the effective application of physics and physical devices within those systems. Engineering disciplines, the most directly related fields of applied physics, require courses in materials science, mechanics, and electronics. As may be expected, there is a great deal of overlap of the study requirements of the different disciplines in the field of applied physics.

It is also worth noting that all programs require familiarity with the use of modern computers and general software applications, and that each discipline will make extensive use of specialized applications, which may be developed as proprietary software rather than being commercially available.

Social Context and Future Prospects

Applied physics has become such a fundamental underpinning of modern society that most people are unaware of its relevance. A case in point is the spin-off of technologies from the NASA space program into the realm of everyday life. Such common household devices as microwave ovens and cellular telephones, medical devices such as pacemakers and monitors, and the digital technology that permeates modern society were developed through the efforts of applied physics in exploring space. For most laypersons, the term "applied physics" conjures images of high-energy particle accelerators or interplanetary telescopes and probes, without the realization that the physical aspect of society is based entirely on principles of applied physics. As these cutting-edge fields of physical research continue, they will also continue to spawn new applications that are adapted continually into the overall fabric of modern society.

The areas of society in which applied physics has historically played a role, particularly those involving modern electronics, will continue to develop. Personal computers and embedded microcontrollers, for example, will continue to grow in power and application while decreasing in size, as research reveals new ways of constructing digital logic circuits through new materials and methods. The incorporation of the technology into consumer products, transportation, and other areas of the infrastructure of society is becoming more commonplace. As a result, the control of many human skills can be relegated to the autonomous functioning of the corresponding device. Consider, for example, the now-built-in ability of some automobiles to parallel park without human intervention.

The technologies that have been developed through applied physics permit the automation of many human actions, which, in turn, drives social change. Similarly, the ability for nearly instantaneous communication between persons in any part of the world, which has been provided through applied physics, has the potential to facilitate the peaceful and productive cooperation of large numbers of people toward resolving significant problems.

Richard M. Renneboog, M.Sc.

Further Reading

Askeland, Donald R. *The Science and Engineering of Materials.* 3d ed. London: Chapman & Hall, 1998. Includes chapters on the atomic nature of matter, the microstructure and mechanical properties of materials, the engineering of materials, the physical properties of materials, and the failure modes of materials.

The Britannica Guide to the One Hundred Most Influential Scientists. London: Constable & Robinson, 2008. Provides brief biographical and historical information about renowned people of science, from Thales of Miletus to Tim Berners-Lee. Includes chapters on Bacon, da Vinci, Kepler, Descartes, Watt, Galvani, Volta, Faraday, Kelvin, Maxwell, Tesla, Planck, Rutherford, Schrödinger, Fermi, Turing, Feynman, Hawking, and many other physical scientists.

Clegg, Brian. *Light Years: An Exploration of Mankind's Enduring Fascination with Light.* London: Judy Piatkus, 2001. Examines the history and relevance of the various concepts of philosophy and science in attempting to understand the physical nature of light.

Giancoli, Douglas C. *Physics for Scientists and Engineers with Modern Physics* Toronto, Ont.: Pearson Prentice Hall, 2008. Provides fundamental instruction in the basic physical laws, including Newtonian mechanics, thermodynamic properties, magnetism, wave mechanics, fluidics, electrical properties, optics, and quantum mechanics.

Kaku, Michio. *Visions: How Science Will Revolutionize the Twenty-first Century.* New York: Anchor Books, Doubleday, 1997. Examines the author's view of three great influences of the twenty-first century: computer science, biomolecular science, and quantum science.

Kean, Sam. *The Disappearing Spoon.* New York: Little, Brown, 2010. Provides a thoughtful and entertaining account of the persons and histories associated with each element on the periodic table, and how they have variously affected the course of society.

Sen, P. C. *Principles of Electric Machines and Power Electronics.* 2d ed. New York: John Wiley & Sons, 1997. Includes chapters on magnetic circuits, transformers, DC machines, asynchronous and synchronous machines, and special machines such as servomotors and stepper motors.

Valencia, Raymond P., ed. *Applied Physics in the Twenty-first Century.* New York: Nova Science, 2010. Includes chapters on nano metal oxide thin films, spectroscopic analysis, computational studies, magnetically modified biological materials, and plasma technology.

Web Sites

American Institute of Physics
www.aip.org

See also: Aeronautics and Aviation; Applied Mathematics; Bionics and Biomedical Engineering; Civil Engineering; Communication; Computer Science; Digital Logic; Transportation Engineering.

ARCHITECTURE AND ARCHITECTURAL ENGINEERING

FIELDS OF STUDY

Physics; geometry; calculus; chemistry; electric circuitry; lighting system design; urban planning and development; computer-assisted design; business and environmental law; accounting.

SUMMARY

Architecture and architectural engineering are fields involved in the design and construction of buildings. Working in close association with architects, architectural engineers translate people's needs and desires into physical space by applying a wide range of engineering and other technologies to provide building systems that are functional, safe, economical, environmentally healthy, and in harmony with the architect's aesthetic intent.

KEY TERMS AND CONCEPTS

- **Air Handler:** Device that blows air across a split-unit central air-conditioning evaporator coil into the air ducts of a building.
- **Building Envelope:** Technical term that refers to the components of a building, including the foundation, walls, roof, windows, and doors, and separating indoor and outdoor environments.
- **Building Regulations:** Building codes and permits developed in keeping with the best practices of the construction industry.
- **Computer-Aided Geometric Design (CAGD):** Use of computers to construct and represent free-form curves, surfaces, and volumes in the design process.
- **Concurrent Engineering:** Project design conducted by teams who work together rather than making serial contributions.
- **Design Charrette:** Intense multidisciplinary workshop lasting several days and held early in the planning process to engage the ideas and concerns of those responsible for the key components of a building project, including the architects, designers, engineers, contractors, managers, and consultants.
- **Fluid Mechanics:** Study of fluids and their properties.
- **Green Building Movement:** Organized effort toward sustainability in building design that emphasizes the life cycle of a building and its relationship to the natural landscape, energy efficiencies, indoor environmental quality, the reduction of toxins, and the use of recyclable, earth-friendly materials.
- **Heating, Ventilation, And Air-Conditioning Systems (HVAC):** Technologies providing heat, ventilation, and cooled air, essential components of built environments.
- **Life-Cycle Assessment:** Assessment of a product's life from raw materials to useful productivity to disposal.
- **Schematic Design:** Materials detailing the design structures and components of a project. It includes drawings, specifications, and cost projections.
- **Whole Building Design:** Design and building concept engaging architecture and engineering professionals in a team process to plan and implement the design and construction of highly integrated building and construction projects.

Definition and Basic Principles

Architecture and architectural engineering are complex and highly skilled fields. Architects develop the graphic design of buildings or dwellings and are often directly involved in their construction. Architectural engineers are certified professionals specializing in the application of engineering principles and systems technology to the design and function of a building. The term "architecture" sometimes includes the engineering technologies of building design. Architecture applies aesthetics, measurement, and design to the cooperative organization of human life. An essential dynamic of this process is the creation of technology. The engineering profession designs a magnitude of objects, structures, and environments to meet human needs and purposes. Building engineers specialize in the technologies that benefit human health and well-being within a built environment. These technologies are the direct result of exponential advances in the manipulation of chemical, hydraulic, thermal, electrical, acoustical, computational, and mechanical systems. Solar, wind, and nuclear power play an important role in the creation of sustainable buildings and environments, and the innovative use of new and recycled construction materials is an essential part of efficient planning and design.

Background and History

As early as about 10,000 B.C.E., small human settlements began to form along the fertile banks of the Nile River in Egypt, the Tigris and the Euphrates rivers in the Middle East, the Indus River in India, and the Yellow and Yangtze rivers in China. These settlements grew, becoming prosperous city-states that supported the erection of large buildings. Temples, pyramids, and public forums served as religious and political symbols of increasing private and public wealth and power. Elaborate structures of a grand scale required specialized instruments such as levers, rollers, inclined planes, saws, chisels, drills, cranes, and right-angle tools for surveying land surfaces and for masonry construction.

In ancient Egypt, individuals noted for their ability to provide shelter, public works, water supplies, and transportation infrastructure for their communities were known as "master builders." The ancient Greeks called these individuals *architektons*. In Roman armies, military engineers termed "architects" designed siege engines and artillery and built roads, bridges, baths, aqueducts, and military fortifications.

Developing Technologies. During the Middle Ages, craftsmen and artisans contributed considerable expertise and know-how to building construction and helped refine the tools of the trade. With the advent of the printing press, classical Greek, Roman, and Arabic treatises covering topics in architecture, mathematics, and the sciences were disseminated, and their content found its way into practical uses. Advances in graphic representation (such as the use of perspective) and the geometric manipulation of the Cartesian coordinate system enhanced the conceptual art of two- and three-dimensional spatial translations. Technologies flourished, creating the impetus that came to a head during the Industrial Revolution.

The Modern Era. Population pressures on urban centers during the eighteenth and nineteenth centuries accelerated the forward momentum of numerous technologies. The transition to fossil fuels, the development of steam power, and the distribution of electric power, as well as improvements in the production of iron and steel, transformed societies. These technologies had a profound impact on the principles of architectural design and building systems.

During the nineteenth century, the availability of energy and the technologies of mass production accelerated the standardization and distribution of construction materials worldwide. Heating flues, air ducts, elevator shafts, internal plumbing, and water and sewage conduits were developed in Great Britain and installed in hospitals, mills, factories, and public buildings. These innovative technologies were quickly adopted for private and commercial use in the United States, the British Commonwealth, and the continental nations.

New technologies for lighting, internal heating, and ventilation were first greeted with misgiving, but over time, building environments took on a life of their own in the minds of city health experts, architects, and engineers. A robust educational system developed to enable architects to master the increasing complexity and diversity of building design and construction technologies. Professional trade associations were formed to administer programs of certification and to determine public standards of health and safety that were codified and set within applicable statutory frameworks.

In the second half of the twentieth century, computers have enabled architects and architectural engineers to function with a greater degree of precision. Computer-aided drafting and design (CAD) software is an essential tool for architects. Computers also can be used to create simulations and animations, as well as detailed presentation graphics, tables, charts, and models, in a relatively short period of time.

How It Works

Architects and architectural engineers play prominent roles in the planning, design, construction, maintenance, and renovation of buildings. All processes of a project are initiated and directed to reflect the needs and desires of the project owner. These directives are modified by detailed sets of construction documents and best practices. Building codes are enforced to safeguard the health and safety of the owner, users, and residents; environmental standards are followed to protect property and to ensure the efficient use of natural resources; and attractive design features are chosen to enhance the community. The choice of construction materials, the visual and spatial relationship of the building to the environment and community, traffic safety, fire prevention, landscaping, and mechanical, electrical, plumbing, ventilation, lighting, acoustical, communication, and security systems are all essential features that determine the quality, functionality, and character of a building.

Articulating the Owner's Intent. The contract process begins with the owner's interest in a residential, public, or commercial building. The selection of an architect for the project will be based on the type of construction (light or heavy) and the particular purpose the building is intended to serve. The plans for simple one- and two-family dwellings frequently follow well-established homebuilder association guidelines and designs that can be modified to meet the future owner's budget and lifestyle. Larger, more complex designs (such as a church or a school) will be managed by a team of architects and engineers carefully chosen for their expertise in key aspects of the project. The selections of the architectural engineering team and the project manager are considered the most important steps in the design process.

The careful and deliberate coordination of effort and timely verbal and written communication are essential at all stages of the project. At the outset, the owner and those directly involved in the corporate administration of the project must work closely together to establish the values and priorities that will guide the design. Detailed feedback loops of self-assessment are conducted throughout the different phases of the project to ensure multiple levels of engagement and evaluation. In the introductory phase, questions are raised to clarify the owner's intentions and to begin addressing key issues such as the scope of the project, the projected use and life cycle of the building, the level of commitment to environmental quality issues, and health and energy requirements. Once these issues have been formulated, a site evaluation is conducted. This process further refines the relationship of the project to the natural surroundings and includes a thorough review of the environmental integrity of the land; local codes governing transportation, water supply, sewage, and electrical infrastructure; health and safety issues; and the relationship of the surroundings to the projected indoor environment.

Project Design. The architect, working with other professionals, uses the information obtained in the introductory phase to create a design that fits the owner's needs and desires, complies with all building codes, and suits the site.

In situations where project turnaround is rapid, a design charrette may be scheduled. This is an intense, rapid-fire meeting of key professionals to quickly develop a schematic design that meets the primary objectives of the project. Participants include administrative representatives from all the internal offices involved in the project. In addition, utilities managers, key community and industrial partners, technology experts, financial institutions, and other stakeholders in the project are invited to set goals, to discuss problems, and to resolve differences. The process is a demanding one, challenging participants to think beyond the boundaries of their professional biases to understand and bring to focus a building design that reflects the highest standards of their respective trades.

Project Documentation. Project documentation is a highly technical orchestration of graphics, symbols, written correspondence, schematic sequencing and scheduling, and audiovisual representation. Construction drawings are a precise, nonverbal map that translates physical spatial relationships into intelligible two- and three-dimensional projections and drawings. The sequence of presentation and the lexicographical notations used to represent and to communicate the dimensions, systems, materials, and objects within a constructed space are highly refined, requiring years of careful study and practice. Project documentation also includes all the legal and civic certifications, permits, and reviews required before a certificate of occupancy is finally issued.

The documents of the project must be recorded in a narrative that can be understood by the owner and the different teams of workers involved. Documentation is particularly important for those project teams working toward LEED (Leadership in Energy and Environmental Design) certification. The LEED Green Building Rating System is an internationally recognized, voluntary certification program that assists commercial and private property owners to build and maintain buildings according to the best practices of green building design. It is administered by the U.S. Green Building Council, a nonprofit organization whose members adhere to industry standards set to maximize cost savings, energy efficiencies, and the health of the indoor environment.

Systems Analyses and Energy Efficiencies. Whole-building performance is an essential concept for achieving energy efficiencies and sustainable mechanical systems. Critical decisions regarding the choice of materials, the orientation of the building and its effect on lighting, the dimensions of the space in question, and the requirements for heating, cooling, mechanical, electrical, and plumbing systems demand a high level of technical analysis and integration among a number of building professionals to create a viable base design amenable for human use and habitation. Other factors incorporated into quality building designs include accessibility for the disabled, the mitigation of environmental impacts, and legal compliance with building codes and requirements.

As a result of the tremendous advantages of computational technologies, systems engineers have an expanding pool of resources available to assist in the optimization of energy efficiencies. Simulation software packages such as EnergyPlus, DOE-2, and ENERGY-10 make it possible to evaluate the potential for using renewable energy sources, the effects of daylight, thermal comfort, ventilation rates, water conservation, potential emissions and contaminant containment strategies, and scenarios for recycling building materials. Landscape factors that affect a building's energy use include irrigation systems, the use of chemical insecticides and fertilizers, erosion control factors, the conservation of native vegetation, and adequate shade and wind protection. Engineers use simulations to evaluate a series of alternatives to provide mechanical, solar, electrical, hydraulic, and heating, ventilation, and air-conditioning systems that optimize critical energy features of the design.

Statement of Work and Contract Bid. The final construction documents include all the drawings, specifications, bidding information, contract forms, and cost features of the project. Certifications of compliance with all regulatory standards and applicable codes are included in the documentation and are verified and signed by the project architects and engineers. Once a bid is secured and a final contract is signed, the design team maintains a close relationship with the contractors and subcontractors to monitor the details of construction and systems installation. The final commissioning process serves to verify the realization of the owner's intent and to thoroughly test the systems installed in the building. The integrity of the building envelope, the function of all mechanical systems, and the stability of energy efficiency targets are validated by a third party.

Applications and Products

The primary applications of architecture and architectural engineering are residential, commercial, and public structures. The Bauhaus school, founded in 1919 in Weimar, Germany, expanded the traditional view of architecture by envisioning the construction of a building as a synthesis of art, technology, and craft. Walter Gropius, the founder of the school, wanted to create a new architectural style that reflected the fast-paced, technologically advanced modern world and was more functional and less ornate. Another architect associated with the school, Ludwig Mies van der Rohe developed an architectural style that is noted for its clarity and simplicity. The S. R. Crown Hall, which houses the College of Architecture at the Illinois Institute of Technology in Chicago, is considered to be the finest example of his work.

Modern architecture, characterized by simplicity of form and ornamentation that arises from the building's structure and theme, became the dominant style in the mid-1900's and has continued into the twenty-first century, despite the rise of postmodern architecture in the 1970's. Postmodernism incorporated elements such as columns strictly for aesthetic reasons. American architect Robert Venturi argued that ornamentation provided interest and variation. Venturi's architectural style is represented by the Guild House, housing for the elderly in Philadelphia, that originally featured a television antenna as a decorative element.

Architect Christopher Alexander has combined technology with architecture, incorporating inventions in concrete and shell design into aesthetically pleasing works, including the San Jose Shelter for the Homeless in California and the Athens Opera House in Greece. He argues for an organic approach to architecture, with people participating in the designs of the buildings that they will use and of the environments in which they will live.

New Paths. Alexander's work concerns itself with the environment in which buildings are constructed, which is a major part of architecture in the twenty-first century. Innovations in architecture and architectural engineering involve creating sustainable buildings or communities or creating buildings in unusual environments.

The Eden Project houses a series of eight large geodesic domes constructed in 2001 at St. Austell in Cornwall. These domes, or biomes, provide self-sustaining environments housing thousands of plants gathered from around the world. The tubular steel and thermoplastic structures use active and passive sources of heat, and innovative ventilation and water systems collect rainwater and groundwater for recirculation in the building envelope. The project's architects, Nicholas Grimshaw and Partners, have developed a system that helps other building professionals understand natural impacts on human environments.

Outside the city limits of Shanghai, Chinese architects, engineers, and planners are working with the London-based firm Arup to design and build the world's first eco-city on Chongming Island, the third largest island in China. Located on the edges of the Yangtze River, the new city, Dongtan, is unusual for the wetlands and the bird sanctuary on its periphery. The first phase of the project will provide a living community for 5,000 inhabitants on a plot that measures 1 square kilometer. By 2050, 500,000 people are to work and live in a city 30 square kilometers in size.

In the Netherlands, innovative designers are experimenting with amphibious housing, or homes that rise and fall with water levels. Along the banks of the Meuse River in Maasbommet, the internationally recognized construction company Dura Vermeer opened a small community of amphibious and floating homes in 2006. Designed by architect Ger Kengen, the amphibious models rest on a hollow foundation filled with foam and anchored on two mooring poles located at the front and back of the structure. The poles allow the house to float on rising water to a height of 18 feet. Floating houses remain on water year round.

Similar structures were designed in New Orleans, Louisiana, following Hurricane Katrina. In 2010, the Special No. 9 House, designed by John C. Williams Architects for the Brad Pitt Make It Right Foundation, won the American Institute of Architects Committee on the Environment (AIA/COTE) Top Ten Green Projects Award. The FLOAT house, also in New Orleans, is another Make It Right Foundation project developed by Morphosis Architects under the direction of architect and University of California, Los Angeles, professor Thom Mayne.

Habitats in Space. Perhaps the most spectacular applications of the methods of architectural engineering involve the manned space flight program. Many of the mechanical systems that make homes and public places so livable are essential for creating similar environments in space. The International Space Station of the National Aeronautics and Space Administration (NASA) is a monumental project involving the efforts of fifteen nations to develop sustainable human habitats in space. Orbiting nearly 250 miles above the Earth at a speed of 22,000 miles per hour, the space station uses eight massive solar and radiator panels to control heat and provide energy for the ship's modular systems. Heating and cooling systems are

Fascinating Facts About Architecture

- In 1972, shortly after the sixty-floor John Hancock Tower was built in Boston, its 4×11-foot windows began popping out and crashing to the sidewalk below. The 10,334 windows were replaced by 0.5-inch tempered glass, but the problem persisted.
- When 108-story, 1,451-foot-tall Sears Tower (renamed the Willis Tower in 2009) was built in 1974, it was the tallest building in the world. It held this position until 1998, when the 1,482.6-foot Petronas Twin Towers were built in Kuala Lumpur, Malaysia.
- Canadian-American architect Frank Gehry designed the Guggenheim Museum in Bilbao, Spain, completed in 1997. The museum features a curvaceous, free-form sculptural style and consists of a steel frame with titanium sheathing.
- In 2009, Custom-Bilt Metals developed a roofing product that combines traditional standing seam metal roofing with thin film solar technology. The lightweight system can be installed by a roofing contractor and an electrical contractor rather than a solar expert.
- Ancient Romans used underground heaters (hypocausts) to distribute warmth from fires through the double-layered floors and walls of public baths. This technology is considered one of the earliest heating, ventilation, and air-conditioning systems.
- The Pharos lighthouse of Alexandria, one of the seven wonders of the ancient world, was built in the third century B.C.E. by Sostratos of Cnidus. It stood about 400 feet tall, the height of a modern 40-story building, and its light was said to be visible from 100 miles away. It was destroyed in an earthquake in 1303.

essential for the maintenance of the ship, the internal environment, and the space suits worn during maneuvers outside the spaceship. Construction materials are in continuous development to find products that can withstand severe cold, radiation, and heat during reentry into Earth's atmosphere. In 2007, NASA developed a lightweight ceramic ablater material able to withstand temperatures of 5,000 degrees Fahrenheit. Other insulation materials are used for electric wires, paints, and protective cladding for the ship. Many of the thermal fabrics designed for space have been adapted by athletes and health providers worldwide.

Inside the ship, mechanical systems generate heat and other emissions. Air quality is regularly monitored for carbon dioxide and oxygen levels. Waste materials are carefully recycled or packaged for shuttle return to Earth for disposal. Transportation routes, cycles of delivery, points of entry and egress, fire safety, lighting and electrical systems, oxygen generators, efficient waste disposal systems, water distiller and filtration systems for the storage, treatment, and recycling of water—all the technologies that add so much to the quality of human life have been engineered to meet the needs of space travel. In turn, these novel space systems have enormous applicability in the design of products for use in homes and places of work on Earth.

Impact on Industry

Architecture and architectural engineering are inextricably linked to the trades and industries that support the manufacture and distribution of building materials. The choices of building materials and the systems employed to provide water, energy, lighting, and other elements vary according to architectural trends, regulations, and owners' preferences. The energy crisis of the 1970's deepened public concern for the conservation of the Earth's limited resources. Rising standards of health and consumer safety, stringent environmental regulations, booming urban and suburban populations, and an increased awareness of the potentially disastrous effects of climate change on built environments and communities had a cumulative effect on already struggling manufacturing sectors during the second half of the twentieth century. This intensified interest in sustainable products, systems, and architectural design.

Since the first global modeling studies of climate conditions were created for the 1972 United Nations Conference on the Human Environment, urban land managers, building architects, and engineers worldwide have responded by studying ecologically responsible building designs and renewable building technologies. The interest in sustainable architecture has resulted in the emergence of architectural firms specializing in green construction or in the preservation of historic structures, finding opportunities to adapt these buildings to modern uses. Building and construction supply companies also have developed ways to mitigate environmental concerns. For example, Swisstrax Corporation created a process that transforms rubber particles from old tires into modular flooring tiles.

Careers and Course Work

A career in architecture requires a five-year bachelor's degree in architecture or a four-year bachelor's degree plus a two-year master's degree in architecture. Those with undergraduate degrees in other areas may attend a three-year master's program in architecture. Usually students gain experience through an internship in an architectural firm. Architects with a degree from an accredited school and some practical experience must pass the Architect Registration Examination to obtain a license to practice.

The successful completion of a general program of study in architectural engineering requires demonstrated competencies in the mechanics of heating, ventilation, air-conditioning, plumbing, fire protection, electrical, lighting, transportation, and structural systems. Students may pursue a five-year program that terminates with a bachelor's degree in architectural engineering and generally leads to certification or a four-year program, followed by a master's degree. Students may also obtain a degree in architectural engineering technology, which specializes in the technology of building design.

Architectural engineering graduates must pass a series of exams administered by the National Council of Examiners for Engineering and Surveying to obtain a license to practice.

Social Context and Future Prospects

From the earliest records of human history, architects and engineers have advanced human civilization. The practical and aesthetic dimensions of architecture and engineering technologies have had profound social, political, economic, and religious impacts. The practices of architecture and engineering are just as much a product of cooperative human evolution as they are catalysts of change. Indeed, no thorough study of these fields is complete without a thoughtful and rigorous understanding of the environmental and cultural forces that stimulated their advances. Topographies, geographies, climates, political economies, known technologies, the commodities of exchange, and the natural resources available for human use stimulated human imagination and invention in novel and diverse ways in particular times and places. These innovations are reflected in the range of human responses to the need for food, shelter, safety, hygiene, and social interaction.

In modern society, these needs are modified by larger global concerns including the efficient use of energy and natural resources, pollution containment, and the need for sustainable life systems. Green architecture is likely to become an increasingly prominent part of the field. Architects will explore sustainable construction that looks at not only the building but also the environment and community, and they will pursue construction in unusual environments, such as housing on the water or in space. They will also look at ways to reuse existing buildings and structures, particularly those worthy of preservation. In choosing materials and systems, they will be investigating ways to recycle building materials and to incorporate systems that reduce pollution and minimize energy usage.

Victoria M. Breting-García, M.A.

Further Reading

Allison, Eric, and Lauren Peters. *Historic Preservation and the Livable City.* Hoboken, N.J.: John Wiley & Sons, 2011. Examines ways to preserve historical buildings and create livable city centers.

Farin, Gerald, Joseph Josef Hoschek, and Myung-Soo Kim. *Handbook of Computer-Aided Geometric Design.* Boston: Elsevier. 2002. A thorough introduction to computer-aided design and architectural drawing.

Fisanick, Christina, ed. *Eco-architecture.* Farmington Hills, Mich.: Greenhaven Press, 2008. A collection of essays written by advocates and opponents of green architecture.

Garrison, Ervan G. *A History of Engineering and Technology: Artful Methods.* 2d ed. Boca Raton, Fla.: CRC Press, 1999. An engaging historical survey documenting scientific, technological, architectural, and mechanical innovations that contributed to the fields of engineering and architectural design.

Goldberger, Paul. *Why Architecture Matters.* New Haven, Conn.: Yale University Press, 2009. Examines the psychological effects that buildings have on people.

Huth, Mark W. *Understanding Construction Drawings.* 5th ed. Clifton Park, N.Y.: Delmar Learning, 2010. Examines the construction process. Includes exercises for the design and interpretation of construction drawings for residential and commercial buildings.

Pohl, Jens G. *Building Science: Concepts and Applications.* Oxford, England: Wiley-Blackwell, 2011. Discusses environmental engineering and sustainable architecture.

Web Sites

American Institute of Architects
http://www.aia.org

American Institute of Building Design
http://www.aibd.org

National Institute of Building Sciences
http://www.nibs.org/index.php/nibshome.htm

Royal Institute of British Architects
http://www.architects.com

Society of American Registered Architects
http://www.sara-national.org

U.S. Department of Energy
Energy Efficiency and Renewable Energy, Buildings
http://www.eere.energy.gov/topics/buildings.html

See also: Civil Engineering; Environmental Engineering; Landscape Architecture and Engineering.

ARTIFICIAL INTELLIGENCE

FIELDS OF STUDY

Expert systems; knowledge engineering; intelligent systems; computer vision; robotics; computer-aided design and manufacturing; computer programming; computer science; cybernetics; parallel computing; electronic health record; information systems; mobile computing; networking; business; physics; mathematics; neural networks; software engineering.

SUMMARY

Artificial intelligence is the design, implementation, and use of programs, machines, and systems that exhibit human intelligence, with its most important activities being knowledge representation, reasoning, and learning. Artificial intelligence encompasses a number of important subareas, including voice recognition, image identification, natural language processing, expert systems, neural networks, planning, robotics, and intelligent agents. Several important programming techniques have been enhanced by artificial intelligence researchers, including classical search, probabilistic search, and logic programming.

KEY TERMS AND CONCEPTS

- **Automatic Theorem Proving:** Proving a theorem from axioms, using a mechanistic procedure, represented as well-formed formulas.
- **Computer Vision:** Technology that allows machines to recognize objects by characteristics, such as color, texture, and edges.
- **First-Order Predicate Calculus:** System of formal logic, including Boolean expressions and quantification, that is rich enough to be a language for mathematics and science.
- **Game Theory:** Technology that supports the development of computer programs or devices that simulate one or more players of a game.
- **Intelligent Agent:** System, often a computer program or Web application, that collects and processes information, using reasoning much like a human.
- **Logic Programming:** Programming methodology that uses logical expressions for data, axioms, and theorems and an inference engine to derive results.
- **Natural Language Processing:** How humans use language to represent ideas and to reason.
- **Neural Network:** Artificial intelligence system modeled after the human neural system.
- **Planning:** Set of processes, generally implemented as a program in artificial intelligence, that allows an organization to accomplish its objectives.
- **Robotics:** Science and technology used to design, manufacture, and maintain intelligent machines.

Definition and Basic Principles

Artificial intelligence is a broad field of study, and definitions of the field vary by discipline. For computer scientists, artificial intelligence refers to the development of programs that exhibit intelligent behavior. The programs can engage in intelligent planning (timing traffic lights), translate natural languages (converting a Chinese Web site into English), act like an expert (selecting the best wine for dinner), or perform many other tasks. For engineers, artificial intelligence refers to building machines that perform actions often done by humans. The machines can be simple, like a computer vision system embedded in an ATM (automated teller machine); more complex, like a robotic rover sent to Mars; or very complex, like an automated factory that builds an exercise machine with little human intervention. For cognitive scientists, artificial intelligence refers to building models of human intelligence to better understand human behavior. In the early days of artificial intelligence, most models of human intelligence were symbolic and closely related to cognitive psychology and philosophy, the basic idea being that regions of the brain perform complex reasoning by processing symbols. Later, many models of human cognition were developed to mirror the operation of the brain as an electrochemical computer, starting with the simple Perceptron, an artificial neural network described by Marvin Minsky in 1969, graduating to the backpropagation algorithm described by David E. Rumelhart and James L. McClelland in 1986, and culminating in a large number of supervised and nonsupervised learning algorithms.

When defining artificial intelligence, it is important to remember that the programs, machines, and models developed by computer scientists, engineers, and cognitive scientists do not actually have human intelligence; they only exhibit intelligent behavior. This can be difficult to remember because artificially intelligent systems often contain large numbers of facts, such as weather information for New York City; complex reasoning patterns, such as the reasoning needed to prove a geometric theorem from axioms; complex knowledge, such as an understanding of all the rules required to build an automobile; and the ability to learn, such as a neural network learning to recognize cancer cells. Scientists continue to look for better models of the brain and human intelligence.

Background and History

Although the concept of artificial intelligence probably has existed since antiquity, the term was first used by American scientist John McCarthy at a conference held at Dartmouth College in 1956. In 1955-1956, the first artificial intelligence program, Logic Theorist, had been written in IPL, a programming language, and in 1958, McCarthy invented Lisp, a programming language that improved on IPL. *Syntactic Structures* (1957), a book about the structure of natural language by American linguist Noam Chomsky, made natural language processing into an area of study within artificial intelligence. In the next few years, numerous researchers began to study artificial intelligence, laying the foundation for many later applications, such as general problem solvers, intelligent machines, and expert systems.

In the 1960's, Edward Feigenbaum and other scientists at Stanford University built two early expert systems: DENDRAL, which classified chemicals, and MYCIN, which identified diseases. These early expert systems were cumbersome to modify because they had hard-coded rules. By 1970, the OPS expert system shell, with variable rule sets, had been released by Digital Equipment Corporation as the first commercial expert system shell. In addition to expert systems, neural networks became an important area of artificial intelligence in the 1970's and 1980's. Frank Rosenblatt introduced the Perceptron in 1957, but it was *Perceptrons: An Introduction to Computational Geometry* (1969), by Minsky and Seymour Papert, and the two-volume *Parallel Distributed Processing: Explorations in the Microstructure of Cognition* (1986), by Rumelhart, McClelland, and the PDP Research Group, that really defined the field of neural networks. Development of artificial intelligence has continued, with game theory, speech recognition, robotics, and autonomous agents being some of the best-known examples.

How It Works

The first activity of artificial intelligence is to understand how multiple facts interconnect to form knowledge and to represent that knowledge in a machine-understandable form. The next task is to understand and document a reasoning process for arriving at a conclusion. The final component of artificial intelligence is to add, whenever possible, a learning process that enhances the knowledge of a system.

Knowledge Representation. Facts are simple pieces of information that can be seen as either true or false, although in fuzzy logic, there are levels of truth. When facts are organized, they become information, and when information is well understood, over time, it becomes knowledge. To use knowledge in artificial intelligence, especially when writing programs, it has to be represented in some concrete fashion. Initially, most of those developing artificial intelligence programs saw knowledge as represented symbolically, and their early knowledge representations were symbolic. Semantic nets, directed graphs of facts with added semantic content, were highly successful representations used in many of the early artificial intelligence programs. Later, the nodes of the semantic nets were expanded to contain more information, and the resulting knowledge representation was referred to as frames. Frame representation of knowledge was very similar to object-oriented data representation, including a theory of inheritance.

Another popular way to represent knowledge in artificial intelligence is as logical expressions. English mathematician George Boole represented knowledge as a Boolean expression in the 1800's. English mathematicians Bertrand Russell and Alfred Whitehead expanded this to quantified expressions in 1910, and French computer scientist Alain Colmerauer incorporated it into logic programming, with the programming language Prolog, in the 1970's. The knowledge of a rule-based expert system is embedded in the if-then rules of the system, and because each if-then rule has a Boolean representation, it can be seen as a form of relational knowledge representation.

Neural networks model the human neural system and use this model to represent knowledge. The brain is an electrochemical system that stores its knowledge in synapses. As electrochemical signals pass through a synapse, they modify it, resulting in the acquisition of knowledge. In the neural network model, synapses are represented by the weights of a weight matrix, and knowledge is added to the system by modifying the weights.

Reasoning. Reasoning is the process of determining new information from known information. Artificial intelligence systems add reasoning soon after they have developed a method of knowledge representation. If knowledge is represented in semantic nets, then most reasoning involves some type of tree search. One popular reasoning technique is to traverse a decision tree, in which the reasoning is represented by a path taken through the tree. Tree searches of general semantic nets can be very time-consuming and have led to many advancements in tree-search algorithms, such as placing bounds on the depth of search and backtracking.

Reasoning in logic programming usually follows an inference technique embodied in first-order predicate calculus. Some inference engines, such as that of Prolog, use a back-chaining technique to reason from a result, such as a geometry theorem, to its antecedents, the axioms, and also show how the reasoning process led to the conclusion. Other inference engines, such as that of the expert system shell CLIPS, use a forward-chaining inference engine to see what facts can be derived from a set of known facts.

Neural networks, such as backpropagation, have an especially simple reasoning algorithm. The knowledge of the neural network is represented as a matrix of synaptic connections, possibly quite sparse. The information to be evaluated by the neural network is represented as an input vector of the appropriate size, and the reasoning process is to multiply the connection matrix by the input vector to obtain the conclusion as an output vector.

Learning. Learning in an artificial intelligence system involves modifying or adding to its knowledge. For both semantic net and logic programming systems, learning is accomplished by adding or modifying the semantic nets or logic rules, respectively. Although much effort has gone into developing learning algorithms for these systems, all of them, to date, have used ad hoc methods and experienced limited success. Neural networks, on the other hand, have been very successful at developing learning algorithms. Backpropagation has a robust supervised learning algorithm in which the system learns from a set of training pairs, using gradient-descent optimization, and numerous unsupervised learning algorithms learn by studying the clustering of the input vectors.

Applications and Products

There are many important applications of artificial intelligence, ranging from computer games to programs designed to prove theorems in mathematics. This section contains a sample of both theoretical and practical applications.

Expert Systems. One of the most successful areas of artificial intelligence is expert systems. Literally thousands of expert systems are being used to help both experts and novices make decisions. For example, in the 1970's, Dell developed a simple expert system that allowed shoppers to configure a computer as they wished. In the 2010's, a visit to the Dell Web site offers a customer much more than a simple configuration program. Based on the customer's answers to some rather general questions, dozens of small expert systems suggest what computer to buy. The Dell site is not unique in its use of expert systems to guide customer's choices. Insurance companies, automobile companies, and many others use expert systems to assist customers in making decisions.

There are several categories of expert systems, but by far the most popular are the rule-based expert systems. Most rule-based expert systems are created with an expert system shell. The first successful rule-based expert system shell was the OPS 5 of Digital Equipment Corporation (DEC), and the most popular modern systems are CLIPS, developed by the National Aeronautics and Space Administration (NASA) in 1985, and its Java clone, Jess, developed at Sandia National Laboratories in 1995. All rule-based expert systems have a similar architecture, and the shells make it fairly easy to create an expert system as soon as a knowledge engineer gathers the knowledge from a domain expert. The most important component of a rule-based expert system is its knowledge base of rules. Each rule consists of an if-then statement with multiple antecedents, multiple consequences, and possibly a rule certainty factor. The antecedents of a rule are statements that can be true or false and that depend on facts that are either introduced into the system by a user or derived as the result of a rule being fired. For example, a fact could be red-wine and a simple rule could be if (red-wine) then (it-tastes-good). The expert system also has an inference engine that can apply multiple rules in an orderly fashion so that the expert system can draw conclusions by applying its rules to a set of facts introduced by a user. Although it is not absolutely required, most rule-based expert systems have a user-friendly interface and an explanation facility to justify its reasoning.

Theorem Provers. Most theorems in mathematics can be expressed in first-order predicate calculus. For any particular area, such as synthetic geometry or group theory, all provable theorems can be derived from a set of axioms. Mathematicians have written programs to automatically prove theorems since the 1950's. These theorem provers either start with the axioms and apply an inference technique, or start with the theorem and work backward to see how it can be derived from axioms. Resolution, developed in Prolog, is a well-known automated technique that can be used to prove theorems, but there are many others. For Resolution, the user starts with the theorem, converts it to a normal form, and then mechanically builds reverse decision trees to prove the theorem. If a reverse decision tree whose leaf nodes are all axioms is found, then a proof of the theorem has been discovered.

Gödel's incompleteness theorem (proved by Austrian-born American mathematician Kurt Gödel) shows that it may not be possible to automatically prove an arbitrary theorem in systems as complex as the natural numbers. For simpler systems, such as group theory, automated theorem proving works if the user's computer can generate all reverse trees or a suitable subset of trees that can yield a proof in a reasonable amount of time. Efforts have been made to develop theorem provers for higher order logics than first-order predicate calculus, but these have not been very successful.

Computer scientists have spent considerable time trying to develop an automated technique for proving the correctness of programs, that is showing that any valid input to a program produces a valid output. This is generally done by producing a consistent model and mapping the program to the model. The first example of this was given by English mathematician Alan Turing in 1931, by using a simple model now called a Turing machine. A formal system that is rich enough to serve as a model for a typical programming language, such as C++, must support higher order logic to capture the arguments and parameters of subprograms. Lambda calculus, denotational semantics, von Neuman geometries, finite state machines, and other systems have been proposed to provide a model onto which all programs of a language can be mapped. Some of these do capture many programs, but devising a practical automated method of verifying the correctness of programs has proven difficult.

Intelligent Tutor Systems. Almost every field of

Fascinating Facts About Artificial Intelligence

- In 1847, George Boole developed his algebra for reasoning that was the foundation for first-order predicate calculus, a logic rich enough to be a language for mathematics.
- In 1950, Alan Turing gave an operational definition of artificial intelligence. He said a machine exhibited artificial intelligence if its operational output was indistinguishable from that of a human.
- In 1956, John McCarthy and Marvin Minsky organized a two-month summer conference on intelligent machines at Dartmouth College. To advertise the conference, McCarthy coined the term "artificial intelligence."
- Digital Equipment Corporation's XCON, short for eXpert CONfigurer, was used in house in 1980 to configure VAX computers and later became the first commercial expert system.
- In 1989, international chess master David Levy was defeated by a computer program, Deep Thought, developed by IBM. Only ten years earlier, Levy had predicted that no computer program would ever beat a chess master.
- In 2010, the Haystack group at the Computer Science and Artificial Intelligence Laboratory at the Massachusetts Institute of Technology developed Soylent, a word-processing interface that lets users edit, proof, and shorten their documents using Mechanical Turk workers.

study has many intelligent tutor systems available to assist students in learning. Sometimes the tutor system is integrated into a package. For example, in Microsoft Office, an embedded intelligent helper provides popup help boxes to a user when it detects the need for assistance and full-length tutorials if it detects more help is needed. In addition to the intelligent tutors embedded in programs as part of a context-sensitive help system, there are a vast number of stand-alone tutoring systems in use.

The first stand-alone intelligent tutor was SCHOLAR, developed by J. R. Carbonell in 1970. It used semantic nets to represent knowledge about South American geography, provided a user interface to support asking questions, and was successful enough to demonstrate that it was possible for a computer program to tutor students. At about the same time, the University of Illinois developed its PLATO computer-aided instruction system, which provided a general language for developing intelligent tutors with touch-sensitive screens, one of the most famous of which was a biology tutorial on evolution. Of the thousands of modern intelligent tutors, SHERLOCK, a training environment for electronic troubleshooting, and PUMP, a system designed to help learn algebra, are typical.

Electronic Games. Electronic games have been played since the invention of the cathode-ray tube for television. In the 1980's, games such as Solitaire, Pac-Man, and Pong for personal computers became almost as popular as the stand-alone game platforms. In the 2010's, multiuser Internet games are enjoyed by young and old alike, and game playing on mobile devices is poised to become an important application. In all of these electronic games, the user competes with one or more intelligent agents embedded in the game, and the creation of these intelligent agents uses considerable artificial intelligence. When creating an intelligent agent that will compete with a user or, as in Solitaire, just react to the user, a programmer has to embed the game knowledge into the program. For example, in chess, the programmer would need to capture all possible configurations of a chess board. The programmer also would need to add reasoning procedures to the game; for example, there would have to be procedures to move each individual chess piece on the board. Finally, and most important for game programming, the programmer would need to add one or more strategic decision modules to the program to provide the intelligent agent with a strategy for winning. In many cases, the strategy for winning a game would be driven by probability; for example, the next move might be a pawn, one space forward, because that yields the best probability of winning, but a heuristic strategy is also possible; for example, the next move is a rook because it may trick the opponent into a bad series of moves.

Impact on Industry

United States government support has been essential in artificial intelligence research, including funding for the 1956 conference at which McCarthy introduced the term "artificial intelligence." The Defense Advanced Research Projects Agency (DARPA) was a strong early supporter of artificial intelligence, then reduced support for a number of years before again providing major support for research in basic and applied artificial intelligence. Industry support for development of artificial intelligence has generally emphasized short-range projects, and university research has developed both theory and applications. Although estimates of the total value of the goods and services produced by artificial intelligence technology in a year are impossible to determine, it is clear that it is in the range of billions of dollars a year.

Government, Industry, and University Research. Many government agencies have provided support for basic and applied research in artificial intelligence. In 1985, NASA released the CLIPS expert system shell, and it remains the most popular shell. The National Science Foundation (NSF) supports a wide range of basic research in artificial intelligence, and the National Institutes of Health (NIH) concentrates on applying artificial intelligence to health systems. Important examples of government support for artificial intelligence are the many NSF and NIH grants for developing intelligent agents that can be embedded in health software to help doctors identify at-risk patients, suggest best practices for these patients, manage their health care, and identify cost savings. DARPA also has a very active program in artificial intelligence, including the Deep Learning project, which supports basic research into "hierarchical machine perception and analysis, and applications in visual, acoustic and somatic sensor processing for detection and classification of objects and activities." The goal is to develop a better biological model of human intelligence than neural networks and to use this to provide machine support for high-level decisions made from sensory and learned information.

Industry and Business Sectors. Many, if not all, software companies use artificial intelligence in their software. For example, the Microsoft Office help system is based on artificial intelligence. Microsoft's Visual Studio programming environment, uses IntelliSense, an intelligent code completion system, and Microsoft's Xbox, like all game systems, uses many artificial intelligence techniques.

Medical technology uses many applications from artificial intelligence. For example, MEDai (an Elsevier company) offers business intelligence solutions that improve health care delivery for clinics and hospitals. Other businesses that make significant use of artificial intelligence are optical character recognition companies, such as Kurzweil Technologies; speech recognition companies, such as Dragon Speaking Naturally; and companies involved in robotics, such as iRobot.

Careers and Course Work

A major in computer science is the most common way to prepare for a career in artificial intelligence. One needs substantial course work in mathematics, philosophy, and psychology as a background for this degree. For many of the more interesting jobs in artificial intelligence, one needs a master's or doctoral degree. Most universities teach courses in artificial intelligence, neural networks, or expert systems, and many have courses in all three. Although artificial intelligence is usually taught in computer science, it is also taught in mathematics, philosophy, psychology, and electrical engineering. Taking a strong minor in any field is advisable for someone seeking a career in artificial intelligence because the discipline is often applied to another field.

Those seeking careers in artificial intelligence generally take a position as a systems analyst or programmer. They work for a wide range of companies, including those developing business, mathematics, medical, and voice recognition applications. Those obtaining an advanced degree often take jobs in industrial, government, or university laboratories developing new areas of artificial intelligence.

Social Context and Future Prospects

After artificial intelligence was defined by McCarthy in 1956, it has had a number of ups and downs as a discipline, but the future of artificial intelligence looks good. Almost every commercial program has a help system, and increasingly these help systems have a major artificial intelligence component. Health care is another area that is poised to make major use of artificial intelligence to improve the quality and reliability of the care provided, as well as to reduce its cost by providing expert advice on best practices in health care.

Ethical questions have been raised about trying to build a machine that exhibits human intelligence. Many of the early researchers in artificial intelligence were interested in cognitive psychology and built symbolic models of intelligence that were considered unethical by some. Later, many artificial intelligence researchers developed neural models of intelligence that were not always deemed ethical. The social and ethical issues of artificial intelligence are nicely represented by HAL, the Heuristically programmed ALgorithmic computer, in Stanley Kubrick's 1968 film *2001: A Space Odyssey,* which first works well with humans, then acts violently toward them, and is in the end deactivated.

Another important ethical question posed by artificial intelligence is the appropriateness of developing programs to collect information about users of a program. Intelligent agents are often embedded in Web sites to collect information about those using the site, generally without the permission of those using the Web site, and many question whether this should be done.

George M. Whitson III, B.S., M.S., Ph.D.

Further Reading

Giarratano, Joseph, and Peter Riley. *Expert Systems: Principles and Programming.* 4th ed. Boston: Thomson Course Technology, 2005. Provides an excellent overview of expert systems, including the CLIPS expert system shell.

Minsky, Marvin, and Seymour Papert. *Perceptrons: An Introduction to Computational Geometry.* Rev. ed. Boston: MIT Press, 1990. Originally printed in 1969, this work introduced many to neural networks and artificial intelligence.

Rumelhart, David E., James L. McClelland, and the PDP Research Group. *Parallel Distributed Processing: Explorations in the Microstructure of Cognition.* 1986. Reprint. 2 vols. Boston: MIT Press, 1989. Volume 1 gives an excellent introduction to neural networks, especially backpropagation. Volume 2 shows many biological and psychological relationships to neural nets.

Russell, Stuart, and Peter Norvig. *Artificial Intelligence: A Modern Approach.* 3d ed. Upper Saddle River, N.J.: Prentice Hall, 2010. The standard textbook on artificial intelligence, it provides a complete overview of the subject, integrating material from expert systems and neural networks.

Shapiro, Stewart, ed. *Encyclopedia of Artificial Intelligence.* 2d ed. New York: John Wiley & Sons, 1992. Contains articles covering the entire field of artificial intelligence.

WEB SITES

Association for the Advancement of Artificial Intelligence
http://www.aaai.org/home.html

Computer Society
http://www.computer.org

Defense Advanced Research Projects Agency
Deep Learning
http://www.darpa.mil/i2o/programs/deep/deep.asp

Institute of Electrical and Electronics Engineers
http://www.ieee.org

Massachusetts Institute of Technology
Computer Science and Artificial Intelligence Laboratory
http://www.csail.mit.edu

Society for the Study of Artificial Intelligence and Simulation of Behaviour
http://www.aisb.org.uk

See also: Computer Engineering; Computer Languages, Compilers, and Tools; Computer Science; Parallel Computing; Pattern Recognition; Robotics; Video Game Design and Programming.

ARTIFICIAL ORGANS

FIELDS OF STUDY

Biology; anatomy; biophysics; chemistry; physics; mathematics; physiology; genetics; immunology; molecular biology; organ transplantation; biomedical engineering.

SUMMARY

Artificial organs are complex systems of natural or manufactured materials used to supplement failing organs while they recover, sustain failing organs until transplantation, or replace failing organs that cannot recover. Some whole organs have artificial counterparts: heart, kidneys, liver, lungs, and pancreas. Smaller body parts also have artificial counterparts: blood, bones, heart valves, joints, skin, and teeth. In addition, there are mechanical support systems for circulation, hearing, and breathing. Artificial organs are composed of biomaterials, biological or synthetic materials that are adapted for use in medical applications.

KEY TERMS AND CONCEPTS

- **Biocompatibility:** Absence of immune reaction or rejection against biological or synthetic materials.
- **Biohybrid:** Interfacing of a biological material with a synthetic material.
- **Biomaterial:** Biological or synthetic material that is adapted for medical use.
- **Extracorporeal:** Outside the body; often used to describe large mechanical support systems.
- **Hemodialysis:** Removal of metabolic waste products and extra water from the blood in cases of kidney failure.
- **Hemoperfusion:** Removal of toxins from the blood in cases of liver failure.
- **Immunomodulation:** Exerting an affect on the immune system, either stimulation or suppression.
- **Organ Failure:** State in which an organ does not perform its natural functions.

Definition and Basic Principles

Artificial organs are complex systems that assist or replace failing organs. The human body is composed of ten major organ systems: nervous, circulatory, respiratory, digestive, excretory, reproductive, endocrine, integumentary (skin), muscular, and skeletal. The nervous system transmits signals between the brain and the body via the spinal cord and nerves. The circulatory system transports blood to deliver oxygen and nutrients to the body and to remove waste products. Its organs are the heart, blood, and blood vessels. It works closely with the respiratory system, in which the lungs and trachea perform oxygen exchange between the body and the environment. The digestive system breaks down food and absorbs its nutrients. Its organs include the esophagus, stomach, intestinal tract, and liver. The excretory system rids the body of metabolic waste in the forms of urine and feces. The reproductive system provides sex cells and in females the organs to develop and carry an embryo to term. The endocrine system consists of the pituitary, parathyroid, and thyroid glands, which secrete regulatory hormones. The integumentary system is the body's external protection system. Its organs include skin, hair, and nails. The muscular system recruits muscles, ligaments, and tendons to move the parts of the skeletal system, which consists of bones and cartilage.

Background and History

While he was still a medical student in 1932, renowned cardiac surgeon Michael E. DeBakey introduced a dual-roller pump for blood transfusion. It has since become the most widely used type of clinical pump for cardiopulmonary bypass and hemodialysis. Physician John H. Gibbon, Jr., of Philadelphia, developed the first clinically successful heart-lung pump. He initially demonstrated it in 1953, when he closed a hole between the atria of an eighteen-year-old girl.

In 1954, American physician Joseph Murray performed the first successful human kidney transplant from one identical twin to the other in Boston. In 1962, he performed the first kidney transplant in unrelated persons. In 1967, surgeon Christiaan Barnard performed the first successful human heart transplant in Cape Town, South Africa. The patient, a fifty-four-year-old man, lived another eighteen days.

Physician Willem J. Kolff is considered to be "the father of the artificial organ." In 1967, he emigrated from the Netherlands and spent a good deal of his career at the University of Utah, where he became a distinguished professor emeritus of internal medicine, surgery, and bioengineering. He led the designing of numerous inventions, including the modern kidney dialysis machine, the intra-aortic balloon pump, an artificial eye, an artificial ear, and an implantable mechanical heart.

American physician Robert K. Jarvik refined Kolff's design into the Jarvik-7 artificial heart, intended for permanent use. In 1982, at the University of Utah, American surgeon William C. DeVries implanted it into retired dentist Barney Clark, who survived 112 days.

How It Works

The existence and performance of artificial organs depend on the collaboration of scientists, engineers, physicians, manufacturers, and regulatory agencies. Each of these groups provides a different perspective of pumps, filters, size, packaging, and regulation.

Hemodynamics. The human heart acts as a muscular pump that beats an average of 72 times a minute. Each of the two ventricles pumps 70 milliliters of blood per beat or 5 liters per minute. Blood pressure is measured and reported as two numbers: the systolic pressure exerted by the heart during contraction and the diastolic pressure, when the heart is between contractions. Hemodynamics is the study of forces related to the circulation of the blood. The hemodynamic performance of artificial organs must match that of the natural body to operate efficiently without resulting in damage. Calculations may be made using computational fluid dynamics (CFD); relevant parameters include solute concentration, density, temperature, and water concentration. In addition to artificial hearts, which are intended to perform all cardiac functions, there is a mechanical circulatory implement called a ventricular assist device (VAD) that supports the function of the natural heart while it is recovering from a heart attack or surgery. Its pumping action may be pulsatile, in rhythmic waves matching those of the beating heart, or continuous.

Mass Transfer Efficiency. The human kidney acts as a filter to remove metabolic waste products from the blood. A person's kidneys process about 200 quarts of blood daily to remove two quarts of waste and extra water, which are converted into urine and excreted. Without filtration, the waste would build to a toxic level and cause death. Patients with kidney failure may undergo dialysis, in which blood is withdrawn, cleaned, and returned to the body in a periodic, continuous, and time-consuming process that requires the patient to remain relatively stationary. Portable artificial kidneys, which the patient wears, filter the blood while the patient enjoys the freedom of mobility. Filtration systems may involve membranes with a strict pore size to separate molecules based on size or columns of particle-based adsorbents to separate molecules by chemical characteristics. Mass transfer efficiency refers to the quality and quantity of molecular transport.

Scale. The development of artificial organs requires that biological processes that can be duplicated in the laboratory be scaled up to work within the human body without also magnifying the weaknesses. Biological functions occur at the organ, tissue, cellular, and molecular levels, which are on micro- and nanoscales. In addition, machines that work in the engineering laboratory must be scaled down to work within the human body without crowding the other organs. Novel power sources and electronic components have facilitated miniaturization. Size must also be balanced with efficiency and cost. Computer-aided design software is being used to create virtual three-dimensional models before fabrication.

Biomaterials. Artificial organs are made of natural and/or manufactured materials that have been adapted for medical use. The properties of these materials must be controlled down to the nanometer scale. The biological components may serve in gene therapy, tissue engineering, and the modification of physiological responses. The synthetic materials must be biocompatible, which means that they do not trigger an adverse physiological reaction such as blood clotting, inflammatory response, scar-tissue formation, or antibody production. The biomechanics of the artificial organ, such as friction and wear, must be known and parts must be sterile before use. Biomaterials have been developed for subspecialties such as orthopedics and ophthalmics.

Regulation. The body has natural feedback systems that allow the exchange of information with the brain for optimal regulation. Artificial organs that communicate directly with the brain are still in development. The present models require sensors

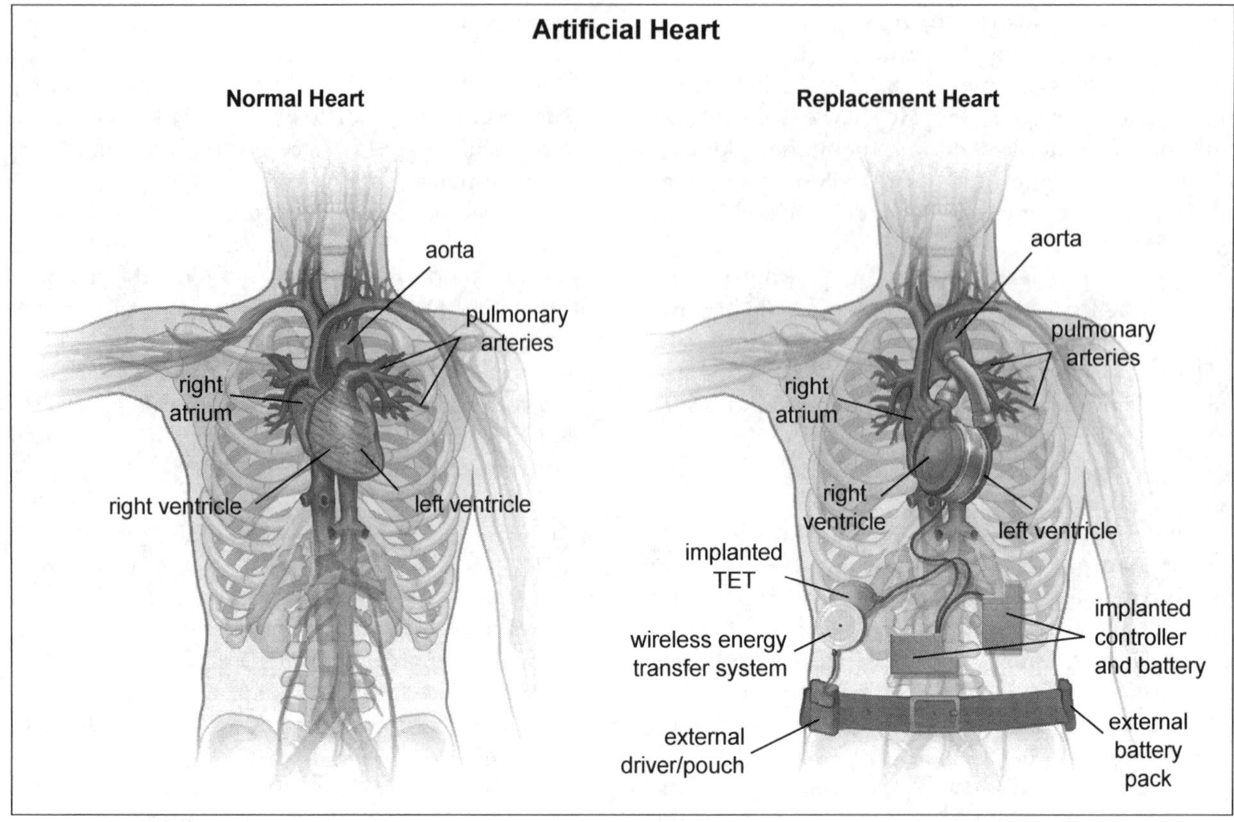

In an artificial heart, the lower two chambers of the heart are replaced.

and data systems that may be monitored by physicians. Implanted devices must be able to be inspected without direct observation. Another aspect of regulation is the uniform manufacturing of artificial organs in compliance with performance and patient safety specifications.

APPLICATIONS AND PRODUCTS

The collective knowledge of scientists, engineers, physicians, manufacturers, and regulatory agencies has produced the applications and products in the interdisciplinary realm of artificial organs.

Hemodynamics. Knowledge of hemodynamics, the study of blood-flow physics, has led to the development of artificial circulatory assistance. The ventricular assist device (VAD) supplements the contraction of the two lower chambers of the heart so the heart muscle does not have to work as hard while it is healing. The cardiopulmonary bypass pump, also known as a heart-lung machine, provides blood oxygenation and circulating pressure during open-heart surgery when the heart is stopped. A similar application called extracorporeal membrane oxygenation (ECMO) is used to assist neonates and infants in the intensive care unit and to maintain the viability of organs pending transplantation. The natural pressure generated by a healthy heart is used to send blood through versions of artificial lungs and kidneys without batteries.

Mass Transfer Efficiency. Information about molecular transport and delivery, known as mass transfer efficiency, has been applied to separation and secretion functions of artificial organs. In hemodialysis, toxins are removed from circulating blood that passes through a filter called a dialyzer. This process also removes excess salts and water to maintain a healthy blood pressure. The dialyzer is composed of a semipermeable membrane or cylinder of hollow synthetic fibers that separates out the metabolic-waste solutes in the incoming blood by diffusion into dialysate solution, leaving cleaner outgoing blood. Hemofiltration is a similar process; however, the filtration occurs without dialysate solution because instead of diffusion, the solutes are removed more quickly by hydrostatic pressure. Another separation technique in medical applications is apheresis, in which the constituents of blood are isolated. This may be achieved by gradient density centrifugation or absorption onto specifically coated beads. The therapeutic application is the absorptive removal of a specific blood component that is causing an adverse reaction in a patient, with the remaining components returned to the patient's circulatory system. The pathogenic blood component might be malignant white blood cells, excess platelets, low-density lipoprotein, autoantibodies, or plasma. The second application of apheresis is the separation of components following blood donation. Concentrated red blood cells are administered in the treatment of sickle-cell crisis or malaria. Plasmapheresis is used to collect fresh frozen plasma as well as rare antibodies and immunoglobulins.

Scale. Miniaturization of artificial organs has been facilitated by the application of smaller, more efficient batteries, transistors, and computer chips. For example, hearing aids once had to be worn with cumbersome amplifiers and batteries disguised in a purse or camera case with a carrying strap. Existing models fit completely in the ear canal and a computer chip facilitates digital rather than analogue processing for crisper sound. The artificial kidney has evolved into a wearable model that weighs 10 pounds and is seventeen times smaller than a conventional dialysis machine. Its hollow-fiber filter must be replaced once a week and its dialysate solution must be replenished daily. However, this maintenance is a trade-off that many patients are willing to make for freedom of movement. On the horizon is an artificial retina that depends on a miniature camera to transmit images. Conversely, research is under way to produce large-scale cultures of tissues on biohybrid matrices and scaffolding for transplantation.

Biomaterials. Synthetic materials are used in artificial organs. Dacron (polyethylene terephthalate) is a polyester fiber with high tensile strength and resistance to stretching whether wet or dry, chemical degradation, and abrasion. Patches of it are sewn to arteries to repair aneurysms. When tubing of it is used as an aortic valve bypass, the patient will not require subsequent blood-thinning medications. Gore-Tex (expanded polytetrafluoroethylene) is an especially strong microporous material that is waterproof. Vascular grafts made from it are supple and resist kinks and compression. It is also used for replacing torn anterior and posterior cruciate ligaments in the knee.

Perfluorocarbon fluids are synthetic liquids that carry dissolved oxygen and carbon dioxide with negligible toxicity, no biological activity, and a short retention time in the body. These features make them ideal for medical applications. One of these fluids, perfluorodecalin, is typically used as a blood substitute (also called a blood extender) because it mixes easily with blood without changing the hemodynamics. It increases the oxygen-carrying capacity of the blood and penetrates ischemic (oxygen-deprived) tissues especially easily because of its small particle size. This makes it particularly useful in the healing of ulcers and burns. It is also used in conjunction with ECMO in the life support of preterm infants to increase oxygenation and to keep the lungs inflated, reducing exertion. Furthermore, it is used in the preservation of harvested organs and cultured tissue for transplantation, extending their viable storage time.

Regulation. The application of regulatory systems has allowed artificial organs to be adjusted while they are in use. Artificial cardiac pacemakers, which supplement the natural electrical pacemaking capabilities of the heart to normalize a slow or irregular heartbeat, are externally programmable so that cardiologists are able to establish the optimal pacing parameters for each patient. Adjustments are made with radio frequency programming, so no further surgery is required. Contemporary hearing aids have volume controls that the wearer can adjust to suit changing surroundings. The inability to detect high- or low-pitch sounds is not a function of volume, yet pitch range can be adjusted in a hearing aid by an audiologist. Other parameters are also adjustable and the audiologist can reprogram the hearing aid as a person's hearing loss changes.

Impact on Industry

Professional Societies. The International Federation for Artificial Organs (IFAO) was founded in 1977 and serves three member societies: the American Society for Artificial Internal Organs (ASAIO) founded in 1955, the European Society for Artificial Organs (ESAO) founded in 1974, and the Japanese Society for Artificial Organs (JSAO) founded in 1962. The first president of the IFAO was "the father of the artificial organ," Willem J. Kolff.

Government Regulation. Artificial organs fall under the federal regulation of the United States Food and Drug Administration (FDA), Center for Devices and Radiological Health, Office of Science and Technology, Division of Physical Sciences. Artificial organs are considered to be medical devices, but they are some of the most complicated that the FDA evaluates. The Office of Science and Technology is charged with determining the scientific parameters that are most relevant to product evaluation and identifying valid techniques for measuring those parameters. The parameters for artificial organs include damage to blood components and neurological consequences. To establish quality-control standards for the production of artificial organs, the Division of Physical Sciences frequently consults national and international standards groups.

University Research. Several universities are frontrunners in the research and development of artificial organs. The University of California, San Francisco (UCSF) schools of pharmacy and medicine have a joint department, the Department of Bioengineering and Therapeutic Sciences, that has recently introduced a prototype of the first implantable artificial kidney. The University of Pittsburgh School of Medicine has partnered with the University of Pittsburgh Medical Center to establish the McGowan Institute for Regenerative Medicine, where they seek solutions to the repair or replacement of tissues and organs through the manipulation of cells, genes, other biomaterials, and bioengineering technologies.

Major Corporations. Corporations worldwide are involved with the production and marketing of artificial organs. Some specialize in a particular technology applicable to several organ systems and some specialize in a particular organ system and offer a variety of products. Some of these companies are: Abbott Laboratories; Abiomed Inc.; F. Hoffmann-La Roche Inc.; Alliqua, Inc.; Medtronic Inc.; SynCardia Systems, Inc.; Thoratec Corporation; Ventracor Ltd.; World-Heart Corporation; and Xenogenics Corporation.

Careers and Course Work

Universities offer various undergraduate and graduate programs related to artificial organs. The University of Pittsburgh's Swanson School of Engineering offers undergraduate degrees in bioengineering with concentration choices of cellular and medical product engineering, biomechanics, and biosignals and imaging. Brown University Department of Molecular Pharmacology, Physiology, and Biotechnology offers master's and doctoral degrees in artificial organs, biomaterials, and cellular technology.

Medical schools support researchers in the field of biotechnology development. Within the Michael E. DeBakey Department of Surgery at Baylor College of Medicine in Houston is the division of Transplant Surgery and Assist Devices. The University of Maryland School of Medicine's Department of Surgery has an Artificial Organs Laboratory as one of the surgical research laboratories. This is a collaborative field and scientists, biomedical engineers, physicians, and businesspeople work together in commercial ventures to design and fabricate functional artificial organs. Because artificial organs fall under the regulatory domain of the FDA as medical devices, manufacturers must undergo rigorous product development, clinical trials, and patent protection prior to FDA approval. Components are then made to custom specifications. Some companies produce a multitude of medical devices, while some specialize in specific technologies such as biotransport, dialysis, perfusion, and cell culture matrices.

Social Context and Future Prospects

The number of Americans older than sixty-five years of age is expected to double within the next twenty-five years. The fastest growing age group is people older than eighty-five years of age. Increasing life span of the general population is a direct result of improved health care. The shortage of donor organs is also increasing. As of 2010, more than 16,000 people were waiting for liver transplants, but each year, only about 6,500 kidneys become available. For the 93,000 people waiting for kidney transplants in 2010, only 17,000 donated kidneys became available for transplant. The need for artificial organs as a bridge to transplantation or even as a permanent substitute for failed organs is becoming increasingly urgent.

Once only made of synthetic components, artificial organs are becoming biohybrid organs: a combination of biological and synthetic components. Examples include functionally competent cells enveloped within immuno-protective artificial membranes and tissues cultured on chemically constructed matrices.

Fascinating Facts About Artificial Organs

- Aviator Charles Lindbergh worked with physician Alexis Carrel, a noted pioneer of vascular surgery, to find a means of oxygenating blood other than the lungs. They created a basic oxygen-exchange device that led to the development of the heart-lung bypass machine for artificial circulation.
- The first successful human organ transplant was performed in 1954. A kidney was transplanted from an identical twin donor.
- In the United States, the need for organ replacement therapies increases by 10 percent each year. In 2008, there were nearly 100,000 people on the national waiting list for organ transplants. Three of every four individuals on the list were waiting for a kidney, with an average wait exceeding five years. One patient waiting for an organ transplant dies every seventy-three minutes.
- About 350,000 Americans are reliant on hemodialysis because of kidney damage associated with diabetes and hypertension. Medicare spends $25 billion annually on treatments for kidney failure, which translates to spending 6 percent of its budget on 1 percent of its recipients.
- The National Organ Transplant Act of 1984 bans "valuable consideration" in exchange for providing an organ for transplantation. Thus, human organs cannot be bought or sold in the United States.
- In Spain, the law allows for the presumption of organ donation after death with consent unless the individual stated otherwise before death. As a result, the procurement rate of organ harvesting from cadavers is 35 percent higher than that in the United States. If the United States had the same presumed-consent law, nearly 14,000 more organs could be procured annually.

Experiments are underway to develop an antibacterial agent that can be incorporated into biomaterials to reduce the risk of infection from these organ surfaces. Emerging technologies also involve sensors and intelligent control systems, biological batteries and alternate power sources, and innovative delivery systems.

Other areas of research include the miniaturization of artificial organs for pediatric use and the development of smaller and more efficient batteries and sensors that will be capable of more accurate communication between the artificial organ and the brain. Another goal is to incorporate wireless capabilities so the artificial organ may be programmed, monitored, and recharged remotely so the patient has increased freedom of mobility.

Bethany Thivierge, B.S., M.P.H.

Further Reading

Fox, Renée C., and Judith P. Swazey. *Spare Parts: Organ Replacement in American Society*. New York: Oxford University Press, 1992. Discusses not only the progression of organ transplantation methods but also the emotional significance attached to the human body and its parts as well as the ethical concerns regarding organ replacement.

Hench, Larry L., and Julian R. Jones, eds. *Biomaterials, Artificial Organs, and Tissue Engineering*. Boca Raton, Fla.: CRC Press, 2005. Provides multiple essays and introductory topics on artificial organs and tissue engineering.

McClellan, Marilyn. *Organ and Tissue Transplants: Medical Miracles and Challenges*. Berkeley Heights, N.J.: Enslow, 2003. Explores the history of organ transplantation as well as the ensuing medical, ethical, and financial issues.

Sharp, Lesley A. *Bodies, Commodities, and Biotechnologies: Death, Mourning, and Scientific Desire in the Realm of Human Organ Transfer*. New York: Columbia University Press, 2008. Explores how organ transplantation and artificial organs have changed cultural attitudes toward the body.

Web Sites

American Society for Artificial Internal Organs
http://asaio.com

International Federation for Artificial Organs
http://www.ifao.org

Society for Biomaterials
http://www.biomaterials.org

See also: Biomechanical Engineering

AUDIO ENGINEERING

FIELDS OF STUDY

Acoustics; analogue signal processing; digital signal processing; electronic sound; electronics; live sound; microphone design and use; mixing and mastering; postproduction editing; recording software; sound compression; sound synchronization.

SUMMARY

Audio engineering is the capture, enhancement, and reproduction of sounds. It requires an aesthetic appreciation of music and sound quality, a scientific understanding of sound physics, and a technical familiarity with recording equipment and computer software. This applied science is essential to the music industry, film, television, and video game production, live television and radio broadcasting, and advertising. In addition, it contributes to educational services for the visually impaired and to forensic evidence analysis.

KEY TERMS AND CONCEPTS

- **Acoustic Sound:** Natural sound that is not produced or enhanced electronically.
- **Analogue Signal:** Information transmitted as a continuous signal that is responsive to change.
- **Compression:** Removing portions of a digital signal to store audio data in less space.
- **Digital Signal:** Information transmitted as a string of binary code without distortion.
- **Electronic Sound:** Sound that is produced or enhanced with technological manipulation, such as computer software.
- **Equalization:** Boosting or weakening bass or treble frequencies to achieve a desired sound.
- **Mastering:** Polishing the final audio product and compressing it for commercial reproduction.
- **Mixing:** Blending multiple recording tracks to achieve a desired effect.
- **Syncing:** Synchronizing an audio track with a video sequence.

Definition and Basic Principles

Audio engineering, also known as sound engineering and audio technology, is the recording, manipulation, and reproduction of sound, especially music. Audio engineers run recording sessions, work the equipment, and collaborate on the finished product. They are simultaneously technicians, scientists, and creative advisers. They work in recording studios, producing the following: instrumental and vocal music recordings; film and television soundtracks, syncing, and sound effects; music and voice-overs for radio and television commercials; and music and sound effects for video games.

A recording studio is a specialized environment designed to capture sounds accurately for enhancement and reproduction. The acoustic sounds produced by instruments and vocalists are picked up by strategically placed microphones and transmitted as analogue electrical signals to recording equipment, where they may be converted into digital data. Signals may be modified by the use of a mixing console, also called a mixing board or sound board, that changes the characteristics and balance of the input, which may be coming from multiple microphones or signals recorded in different sessions. The final product then undergoes mastering for commercial reproduction and compression for distribution in a digital format.

Audio engineers are not acoustic engineers: graduates of a formal university program in engineering who work with architects and interior designers to plan and install audio systems for large venues such as churches, school auditoriums, and concert halls. Audio engineers are engineers in the sense that they are needed to devise a creative solution to a complex sound challenge and oversee its implementation.

Background and History

Sound capture and reproduction, and thus audio engineering, began with the invention of the phonograph by Thomas Alva Edison in 1877. Sound was recorded on cylinders; the first were wrapped in tin foil and later ones in wax. By 1910, cylinders were replaced with disks, which held longer recordings, were somewhat louder, and could be more economically mass-produced. The disks were spun on a turntable at standard speeds—initially 78 revolutions per minute (rpm). Larger disks were played at 33 rpm and smaller disks were played at 45 rpm. Discs were originally made of shellac and later made of vinyl. They were played with needles (styli) made of industrial diamond, which held a point.

Concurrently, RCA was creating microphones that improved recorded sound quality. In the 1940's, sound began being recorded on magnetic tape and could be reproduced in stereo and as mixed multiple tracks. Digital technology appeared in the 1980's and by the turn of the century, digital recordings were produced with computer technology. Using data compression, digital audio recordings can produce quality replication of the original music in a format that requires less data storage (computer memory); MP3 is one such format and is popular for portable consumer music systems.

How It Works

Sound. Sound is the waves of pressure a vibrating object emits through air or water. The three most meaningful characteristics of sound waves are wavelength, amplitude, and frequency. The wavelength is the distance between equivalent points on consecutive waves, such as peak to peak. A short wavelength means that more waves are produced per second, resulting in a higher sound. The amplitude is the strength of the wave; the greater the amplitude, the greater the volume (loudness). The frequency is the number of wavelengths that occur in one second; the greater the frequency, the higher the pitch because the sound source is vibrating quickly.

Hearing. Hearing is the ability to receive, sense, and decipher sounds. To hear, the ear must direct the sound waves inside, sense the sound vibrations, and translate the sensations into neurological impulses that the brain can recognize. The outer ear funnels sound into the ear canal. It also helps the brain determine the direction from which the sound is coming.

When the sound waves reach the ear canal, they vibrate against the eardrum. These vibrations are amplified by the eardrum's movement against three tiny bones (the malleus, incus, and stapes) located behind it. The stapes rests against the cochlea, and when it transmits the sound, it creates waves in the fluid of the cochlea.

The cochlea is a coiled, fluid-filled organ that contains 30,000 hairs of different lengths that resonate at different frequencies. Vibrations of these hairs trigger complex electrical patterns that are transmitted along the auditory nerve to the brain, where they are interpreted.

The frequency of sound waves is measured in hertz (Hz). Humans have a hearing range from 20 to 20,000 Hz. Another name for the frequency of a sound wave is the musical pitch. Pitches are often referred to as musical notes, such as middle C.

The relative loudness of a sound compared with the threshold of human hearing is measured in decibels (dB). Conversation is usually conducted at 40 to 60 dB, while a car passing at 10 meters may be 80 to 90 dB, a jet engine 100 meters away may be 110 to 140 dB, and a rifle fired 1 meter away is 150 dB. Long-term (not necessarily continuous) exposure to sounds greater than 85 dB may cause hearing loss.

The human ear can discern between two musical instruments playing the same note at the same volume by the recognition of a sound characteristic called timbre. Often described by adjectives such as bright versus dark, smooth versus harsh, and regular versus random or erratic, timbre is often what distinguishes music from noise.

Sound Capture. Transducers are devices that change energy from one form into another. A microphone changes acoustical signals into electrical signals, while a speaker changes electrical signals into acoustical signals. The source of the incoming electrical signals may be immediate, such as a microphone or electrical musical instrument, or it may be a recording, such as a magnetic tape or compact disc.

Microphones come in many varieties, such as dynamic, ribbon, condenser, parabolic, and lavaliere. They also vary by their polar patterns, that is, their area of sensitivity to sounds coming in from different directions relative to the receiving membrane. They may be omnidirectional (sensitive to sounds coming from all directions), unidirectional (intended for directed sound reception), or cardioid (having a heart-shaped area of sensitivity). The choices of variety, polar pattern, and placement affect the quality and quantity of sound capture.

Signal Processing. The auditory electrical signal from a microphone is relatively weak, so it must be amplified before the sound can be deliberately modified through signal processing. Incoming sound may be modified in its analogue form or converted to digital data before alteration. Sound mixing is the process of blending sounds from multiple sources into a desired end product. It often starts with finding a balance between vocal and instrumental music or dissimilar instruments so that one does not overshadow the other. It involves the creation of stereo or surround sound from the placement of sound in the sound field to simulate directionality (left, center, or right). Equalizing adjusts the bass and treble frequency ranges. Effects such as reverberation may be added to create dimension. Signals may undergo gating and compression to remove unwanted noise and extraneous data selectively.

Sound Output. Auditory electrical signals may then be sent to speakers, where they are converted into acoustical signals to be heard by a live audience. Digital signals may be broadcast in real time over the Internet as streaming audio. Otherwise, the processed signals may be stored for future reproduction and distribution. Analogue signals may be stored on magnetic tape. Digital signals may be stored on a compact disc or subjected to MP3 encoding for storage on a computer or personal music player.

Applications and Products

Instrumental and Vocal Music. As specialists in the capture, enhancement, and reproduction of sound, audio engineers are crucial to successful recording sessions. They collaborate with producers and performers technically to generate the shared artistic vision. They determine the choice and placement of microphones and closely scrutinize the parameters of the incoming signals to collect sufficient data with which to work. They manage the scheduling of studio sessions to keep all participants working efficiently, especially when multiple tracks are being recorded and mixed at different times. They act professionally and deliver the finished product with the highest quality possible.

Audio engineers are also responsible for the restoration of classic recordings that would otherwise be lost. They rescue the raw data that was captured in the first recording, strengthen the sound while preserving the style of the original period, and return it to audiences in a contemporary format.

Because musicians go on concert tours, audio engineers accompany them to provide optimum live sound quality in each different venue. They conduct sound checks before performances and make adjustments for conditions such as wind on outdoor stages.

Film and Television. In the recording studio, audio engineers oversee the production of music soundtracks for films and television shows. Unlike songs that stand alone, the music must be carefully synchronized to the action of the film. It must also swell and ebb with precision to arouse audience emotion.

Foley recording is the production of sound effects that are inserted into videos after they are filmed to add realism and dramatic tension. Foley recording can be synchronized efficiently to video footage because the sound effects are produced in real time, not modified from stock recordings. In addition, sounds that do not exist in reality and so would not be catalogued in a prerecorded audio library must be created.

Live Broadcasting. Audio engineers may be seen sitting at mixing consoles or computers monitoring and adjusting the audio input and output quality at church services, lectures, theatrical performances, and events held in large auditoriums. They may similarly be found as part of a broadcasting team at live sporting events held outdoors, such as football or baseball games, golf tournaments, and the Olympic Games.

Radio and Television Commercials. Audio engineers are instrumental in the production of radio and television commercials, not only for their recording and sound-processing skills but also for their production skills. Because advertising time is sold in specific brief allotments, engineers must encourage the actors to perform at an accelerated pace and later edit the audio to fit within the time allowed. They may also be asked to recruit or audition competent musicians and voice actors to meet the client's needs.

Video Games. The skills of audio engineers enhance the production of popular video games. In addition to providing sound effects such as explosions and gunfire, engineers must create appropriate imaginary sounds such as spaceships landing, ambient sound effects such as slot machine bells and crowd murmurs, and realistic situational sounds, such as footsteps going from grass to gravel. In some cases, they may be called on to provide minor character voices or record spoken instructions.

Forensic Evidence Analysis. Police may seek the assistance of an experienced audio engineer to remove unnecessary background noise from covert recordings of suspected criminals and to make voiceprint comparisons with known exemplars. Voiceprints, vocal qualities that can be demonstrated on a sound spectrograph, are personal because each person's oral and pharyngeal anatomy is distinctive; however, they are not unique like fingerprints because children often sound like their parents and share similar voiceprints. Research has shown that the error rates of misidentifying suspects (false positives) and improperly eliminating suspects (false negatives) are respectably low.

Audio Books. Recordings of books originated in 1932 under the auspices of the American Foundation for the Blind as educational tools for the visually impaired. Books were recorded on shellac discs and played on a turntable. Books on audio cassettes came along twenty years later, and later audio books could be listened to on CDs or portable digital music devices. Audio engineers are responsible for processing the audio signal to optimize the clarity of human speech and editing numerous recitations into one continuous, flawless performance.

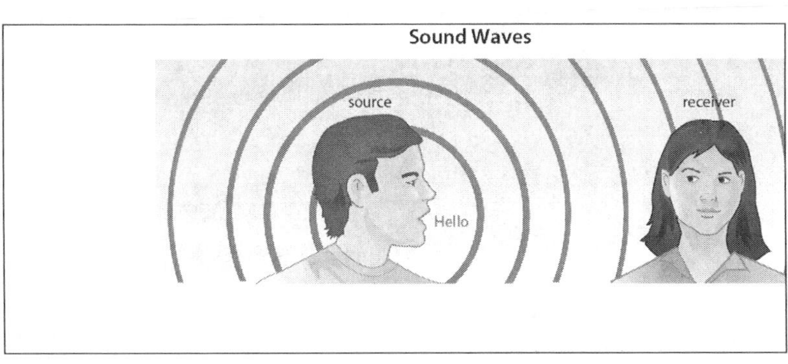

Sound is transmitted through the air as a sound wave.

Impact on Industry

Employment Statistics. According to the U.S. Bureau of Labor Statistics, nearly 200,000 people are presently employed in the audio engineering field. Further growth is expected, with a 17 percent increase predicted to occur between 2006 and 2016. Although the Consumer Electronics Association reported that from 2000 to 2009, American consumers spent 35 percent less on home stereo components (about $960 million), in that same time period, sales of portable digital music devices exploded to $5.4 billion. People are still willing to spend money on music, keeping audio engineers in business. As of May, 2006, the median annual income of audio engineers was $43,010; the top 10 percent were earning more than $90,770. Though audio engineers are needed in various industries (radio, television, film, and advertising), there is a great deal of competition for audio engineering jobs in large cities.

Professional Organizations. The more than 14,000 members of the Audio Engineering Society, an international group founded in 1948 in the United States, include recording engineers, broadcast technicians, acousticians, mixing engineers, equipment designers, and mastering engineers. They share creative and scientific information about audio standards and technologies. Sustaining members from manufacturing, research, and fields related to audio engineering include Bose, Dolby Laboratories, Klipsch Group, Motorola, and Sennheiser.

Careers and Course Work

Technical and vocational schools offer diploma and certificate programs in audio engineering along with internships for hands-on experience. The Musicians Institute in Los Angeles offers a certificate in audio engineering with subspecialties such as post-production or live-sound production. Other schools, such as the Conservatory of Recording Arts and Sciences in Gilbert, Arizona, provide certification for demonstrated proficiency in software and equipment related to audio engineering.

Some universities, such as Indiana University, offer a bachelor of arts in music degree with a concentration in the music industry and a track within that concentration in sound engineering. Texas State University School of Music offers a bachelor of science degree in sound recording technology. The Peabody Institute at Johns Hopkins University offers a bachelor of music degree in recording arts and sciences and a master of arts degree in audio sciences.

Many audio engineers are self-taught. Audio engineering is a multifaceted discipline, requiring a creative appreciation of music and sound quality, a scientific understanding of sound-wave physics, and precise technical familiarity with recording equipment and computer software.

Careers vary with expertise and experience. An assistant audio engineer is typically responsible for the setup and breakdown of a recording session, including the placement of the microphones. A staff engineer records the sound. A mixing engineer

Fascinating Facts About Audio Engineering

- Foley recording is the real-time recording of sound effects that are inserted into videos in post-production to add realism and dramatic tension. The field originated in 1927 and is named for Jack Foley, who worked on films at Universal Studios, adding sounds other than dialogue to "talkies."
- Late actor Don LaFontaine, best known for his commercial and film trailer voice-overs, was originally an audio engineer. He began his long, successful voice-over career in 1964, when he had to fill in at the microphone for a missing actor in an important presentation to MGM.
- The Moog synthesizer became popular in 1967 after it was played at the Monterey International Pop Festival. Its first commercial success as an electronic instrument was on the 1968 Wendy Carlos classical music album *Switched-On Bach.*
- In 1975, the University of Miami's Frost School of Music was the first American university to offer a bachelor of music degree in music engineering technology.
- Just as sound studio equipment can be used to clean up extraneous noise from old recordings for remastering, it can also be used in forensic voice analysis, also called voiceprinting. In 2002, former audio engineer Tom Owen enhanced a foreign radio broadcast of Osama bin Laden and compared it with a 1998 interview to determine that bin Laden was still alive.
- The proprietary computer software Auto-Tune, which adjusts recorded singing to the right note or pitch, has been described as "Photoshop for the human voice." It was creatively misused in the recording of Cher's 1998 hit "Believe."

coordinates multiple recording tracks to produce the desired effect. A mastering engineer adds the finishing touches to the final product and compresses it for mass duplication. A chief engineer works with the record producer, making the technical decisions that help achieve the artistic vision for the project.

Social Context and Future Prospects

Audio engineering is a combination of technology, science, and art. Advancements in audio engineering will come in all three areas. On the technical front, classic (especially pre-1920) recordings will continue to be found, researched, and digitally restored. Improved transducer materials are being sought and new computer software applications for signal processing are being developed. Surround sound is being refined to accompany three-dimensional and high-definition television programs and films as well as video games.

Scientific research into psychoacoustics, the study of sound perception, is expanding. The eventual understanding of how music affects a person's brain will advance the field of music therapy, which seems to touch every facet of a person's being to restore and maintain health. Researchers are also exploring the connections between sound characteristics and the perceptions of timbre and spatial placement and between these perceived attributes and listening preference.

The artistic manipulation of sound is broadening the definition of music and musical instruments. Computer-mediated music has inspired the creation of mobile phone and laptop orchestras. Music enhancement by selectively masking undesired frequencies of instruments and highlighting others is introducing new sound combinations previously not experienced.

Bethany Thivierge, B.S., M.P.H.

Further Reading

Friedman, Dan. *Sound Advice: Voiceover from an Audio Engineer's Perspective.* Bloomington, Ind.: AuthorHouse, 2010. An audio engineer who is passionate about his work presents basic yet often overlooked information, especially about recording equipment and studio etiquette.

Hampton, Dave. *The Business of Audio Engineering.* New York: Hal Leonard, 2008. Schools offer practical training, but this widely respected expert provides strategies for maneuvering within the industry of audio engineering, including professional presentation and customer relations.

_______. *So, You're an Audio Engineer: Well, Here's the Other Stuff You Need to Know.* Parker, Colo.: Outskirts Press, 2005. With years of experience working with musicians, singers, and other entertainment personalities, the author provides insights into the customer-service aspects of audio engineering.

Powell, John. *How Music Works: The Science and Psychology of Beautiful Sounds, from Beethoven to the Beatles and Beyond.* New York: Little, Brown, 2010. This book, which comes with a CD, presents the components of music and explains what makes music more than sound.

Talbot-Smith, Michael. *Sound Engineering Explained.* 2d ed. Woburn, Mass.: Focal Press, 2001. This book explains the basic principles of audio engineering.

_______, ed. *Audio Engineer's Reference Book.* 2d ed. Woburn Mass.: Focal Press, 1999. A complete resource that explains the scientific principles of sound and provides technical information about audio equipment.

Web Sites

Acoustical Society of America
http://acousticalsociety.org

Audio Engineering Society
http://www.aes.org

Society of Professional Audio Recording Services
http://spars.com

AVIONICS AND AIRCRAFT INSTRUMENTATION

FIELDS OF STUDY

Aerodynamics; aeronautical engineering; computer science; electrical engineering; electronics; hydraulics; mechanical engineering; meteorology; physics; pneumatics.

SUMMARY

Flight instrumentation refers to the indicators and instruments that inform a pilot of the position of the aircraft and give navigational information. Avionics comprises all the devices that allow a pilot to give and receive communications, such as air traffic control directions and navigational radio and satellite signals. Early in the history of flight, instrumentation and avionics were separate systems, but these systems have been vastly improved and integrated. These systems allow commercial airliners to fly efficiently and safely all around the world. Additionally, the integrated systems are being used in practically all types of vehicles—ships, trains, spacecraft, guided missiles, and unmanned aircraft—both civilian and military.

KEY TERMS AND CONCEPTS

- **Airspeed Indicator:** Aircraft's speedometer, giving speed based on the difference between ram and static air pressure.
- **Artificial Horizon:** Gyroscopic instrument that displays the airplane's attitude relative to the horizon.
- **Directional Gyro (Vertical Compass):** Gyroscopically controlled compass rose.
- **Global Positioning System (GPS):** Satellite-based navigational system.
- **Horizontal Situation Indicator (HSI):** Indicator that combines an artificial horizon with a directional gyro.
- **Instrument Flight:** Flight by reference to the flight instruments when the pilot cannot see outside the cockpit because of bad weather.
- **Instrument Landing System (ILS):** Method to guide an airplane to the runway using a sensitive localizer to align the aircraft with the runway and a glide slope to provide a descent path to the runway.
- **Radar Altimeter:** Instrument that uses radar to calculate the aircraft's height above the ground.
- **Ram:** Airflow that the aircraft generates as it moves through the air.
- **Turn And Bank Indicator:** Gyroscopic instrument that provides information on the angle of bank of the aircraft.

Definition and Basic Principles

Flight instrumentation refers to the instruments that provide information to a pilot about the position of the aircraft in relation to the Earth's horizon. The term "avionics" is a contraction of "aviation" and "electronics" and has come to refer to the combination of communication and navigational devices in an aircraft. This term was coined in the 1970's after the systems were becoming one integral system.

The components of basic flight instrumentation are the magnetic compass, the instruments that rely on air-pressure differentials, and those that are driven by gyroscopes and instruments. Air pressure decreases with an increase in altitude. The altimeter and vertical-speed indicator use this change in pressure to provide information about the height of the aircraft above sea level and the rate that the aircraft is climbing or descending. The airspeed indicator uses ram air pressure to give the speed that the aircraft is traveling through the air.

Other instruments use gyroscopes to detect changes in the position of the aircraft relative to the Earth's surface and horizon. An airplane can move around the three axes of flight. The first is pitch, or the upward and downward position of the nose of the airplane. The second is roll, the position of the wings. They can be level to the horizon or be in a bank position, where one wing is above horizon and the other below the horizon as the aircraft turns. Yaw is the third. When an airplane yaws, the nose of the airplane moves to the right or left while the airplane is in level flight. The instruments that use gyroscopes to show movement along the axes of flight are the turn and bank indicator, which shows the angle of the airplane's wings in a turn and the rate of turn in degrees per second; the artificial horizon, which indicates the airplane's pitch and bank; and the directional gyro, which is a compass card connected to a gyroscope. Output from the flight instruments can be used to operate autopilots. Modern inertial navigation systems (INS's) use gyroscopes, sometimes in conjunction with a Global Positioning System (GPS), as integrated flight instrumentation and avionics systems.

The radios that comprise the avionics of an aircraft include communications radios that pilots use to talk to air traffic control (ATC) and other aircraft and navigation radios. Early navigation radios relied on ground-based radio signals, but many aircraft have come to use GPS receivers that receive their information from satellites. Other components of an aircraft's avionics include a transponder, which sends a discrete code to ATC to identify the aircraft and is used in the military to discern friendly and enemy aircraft, and radar, which is used to locate rain and thunderstorms and to determine the aircraft's height above the ground.

Background and History

Flight instruments originally were separated from the avionics of an aircraft. The compass, perhaps the most basic of the flight instruments, was developed by the Chinese in the second century B.C.E. Chinese navy commander Zheng He's voyages from 1405 to 1433 included the first recorded use of a magnetic compass for navigation.

The gyroscope, a major component of many flight instruments, was named by French physicist Jean-Bernard-Léon Foucault in the nineteenth century. In 1909, American businessman Elmer Sperry invented the gyroscopic compass that was first used on U.S. naval ships in 1911. In 1916, the first artificial horizon using a gyroscope was invented. Gyroscopic flight instruments along with radio navigation signals guided American pilot Jimmy Doolittle to the first successful all-instrument flight and landing of an airplane in 1929. Robert Goddard, the father of rocketry, experimented with using gyroscopes in guidance systems for rockets. During World War II, German rocket scientist Wernher von Braun further developed Goddard's work to build a basic guidance system for Germany's V-2 rockets. After the war, von Braun and 118 of his engineers immigrated to the United States, where they worked for the U.S. Army on gyroscopic inertial navigation systems (INSs) for rockets. Massachusetts Institute of Technology engineers continued the development of the INS to use in Atlas rockets and eventually the space shuttle. Boeing was the first aircraft manufacturer to install INSs into its 747 jumbo jets. Later, the Air Force introduced the system to their C-141 aircraft.

Radios form the basis of modern avionics. Although there is some dispute over who actually invented the radio, Italian physicist Guglielmo Marconi first applied the technology to communication. During World War I, in 1916, the Naval Research Laboratory developed the first aircraft radio. In 1920, the first ground-based system for communication with aircraft was developed by General Electric. The earliest navigational system was a series of lights on the ground, and the pilot would fly from beacon to beacon. In the 1930's, the nondirectional radio beacon (NDB) became the major radio navigation system. This was replaced by the very high frequency omnidirectional range (VOR) system in the 1960's. In 1994, the GPS became operational and was quickly adapted to aircraft navigation. The great accuracy that GPS can supply for both location and time was adapted for use in INS.

How It Works

Flight Instruments. Flight instruments operate using either gyroscopes or air pressure. The instruments that use air pressure are the altimeter, the vertical speed indicator, and the airspeed indicator. Airplanes are fitted with two pressure sensors: the pitot tube, which is mounted under a wing or the front fuselage, its opening facing the oncoming air; and the static port, which is usually mounted on the side of the airplane out of the slipstream of air flowing past the plane. The pitot tube measures ram air; the faster the aircraft is moving through the air, the more air molecules enter the pitot tube. The static port measures the ambient air pressure, which decreases with increasing altitude. The airspeed indicator is driven by the force of the ram air calibrated to the ambient air pressure to give the speed that the airplane is moving through the sky. The static port's ambient pressure is translated into altitude above sea level by the altimeter. As air pressure can vary from location to location, the altimeter must be set to the local barometric setting in order to receive a correct altimeter reading. The vertical speed indicator also uses the ambient pressure from the static port. This instrument can sense changes in altitude and indicates feet per minute that the airplane is climbing or descending.

Other flight instruments operate with gyroscopes. These instruments are the gyroscopic compass, the turn and bank indicator, and the artificial horizon. The gyroscopic compass is a vertical compass card connected to a gyroscope. It is either set by the pilot or slaved to the heading indicated on the magnetic compass. The magnetic compass floats in a liquid that allows it to rotate freely but also causes it to jiggle in turbulence; the directional gyro is stabilized by its gyroscope. The magnetic compass will also show errors while turning or accelerating, which are eliminated by the gyroscope. The turn and bank indicator is connected to a gyroscope that remains stable when the plane is banking. The indicator shows the angle of bank of the airplane. The artificial horizon has a card attached to it that shows a horizon, sky, and ground and a small indicator in the center that is connected to the gyroscope. When the airplane pitches up or down or rolls, the card moves with the airplane, but the indicator is stable and shows the position of the aircraft relative to the horizon. Pilots use the artificial horizon to fly when they cannot see the natural horizon. The artificial horizon and directional gyro can be combined into one instrument, the horizontal situation indicator (HSI). These instruments can be used to supply information to an autopilot, which can be mechanically connected to the flight surfaces of the aircraft to fly it automatically.

Ground-Based Avionics. Ground-based avionics provide communications, navigational information, and collision avoidance. Communication radios operate on frequencies between 118 and 136.975 megahertz (MHz). Communication uses line of sight. Navigation uses VOR systems. The VOR gives a signal to the aircraft receiver that indicates the direction to or from the VOR station. A more sensitive type of VOR, a localizer, is combined with a glide slope indicator to provide runway direction and a glide path for the aircraft to follow when it is landing in poor weather conditions and the pilot does not have visual contact with the runway. Collision avoidance is provided by ATC using signals from each aircraft's transponders and radar. ATC can identify the aircrafts' positions and advise pilots of traffic in their vicinity.

Satellite-Based Systems. The limitation of line of sight for ground-based avionic transmitters is a major problem for navigation over large oceans or in areas of the world that have large mountain ranges or few transmitters. The U.S. military was very concerned about these limitations and the Department of Defense spearheaded the research and implementation of a system that addresses these problems. GPS is the United States's satellite system that provides navigational information. GPS can give location, movement, and time information. The system uses a minimum of twenty-four satellites orbiting the Earth that send signals to monitoring receivers on Earth. The receiver must be able to get signals from a minimum of four satellites in order to calculate an aircraft's position correctly. Although originally designed solely for military use, GPS is widely used by civilians.

Inertial Navigation Systems (INS). The INS is a self-contained system that does not rely on outside radio or satellite signals. INS is driven by accelerometers and gyroscopes. The accelerometer houses a small pendulum that will swing in relation to the aircraft's acceleration or deceleration and so can measure the aircraft's speed. The gyroscope provides information about the aircraft's movement about the three axes of flight. Instead of the gimbaled gyros, more precise strap-down laser gyroscopes have come to be used. The strap-down system is attached to the frame of the aircraft. Instead of the rotating wheel in the gimbaled gyroscopes, this system uses light beams that travel in opposite directions around a small, triangular path. When the aircraft rotates, the path traveled by the beam of light moving in the direction of rotation appears shorter than the path of the other beam of light moving in the opposite direction. The length of the path causes a frequency shift that is detected and interpreted as aircraft rotation. INS must be initialized: The system has to be able to detect its initial position or it must be programmed with its initial position before it is used or it will not have a reference point from which to work.

Applications and Products

Military INS and GPS Uses. Flight instrumentation and avionics are used by military aircraft as well as civilian aircraft, but the military have many other applications. INS is used in guided missiles and submarines. It can also be used as a stand-alone navigational system in vehicles that do not want to communicate with outside sources for security purposes. INS and GPS are used in bombs, rockets, and, with great success, unmanned aerial vehicles (UAVs) that are used for reconnaissance as well as delivering ordnance without placing a pilot in harm's way. GPS is used in almost all military vehicles such as tanks, ships, armored vehicles, and cars, but not in submarines as the satellite signals will not penetrate deep water. GPS is also used by the United States Nuclear Detonation Detection System as the satellites carry nuclear detonation detectors.

Navigation. Besides the use of flight instrumentation and avionics for aircraft navigation, the systems can also be used for almost all forms of navigation. The aerospace industry has used INS for guidance of spacecraft that cannot use earthbound navigation systems, including satellites that orbit the planet. INS systems can be initialized by manually inputting the craft's position using GPS or using celestial fixes to direct rockets, space shuttles, and long-distance satellites and space probes through the reaches of the solar system and beyond. These systems can be synchronized with computers and sensors to control the vehicles by moving flight controls or firing rockets. GPS can be used on Earth by cars, trucks, trains, ships, and handheld units for commercial, personal, and recreational uses. One limitation of GPS is that it cannot work where the signals could be blocked, such as under water or in caves.

Cellular Phones. GPS technology is critical for operating cellular phones. GPS provide accurate time that is used in synchronizing signals with base stations. If the phone has GPS capability built into it, as many smart phones do, it can be used to locate a mobile cell phone making an emergency call. The GPS system in cell phones can be used in cars for navigation as well for recreation such as guidance while hiking, biking, boating, or geocaching.

Tracking Systems. In the same manner that GPS can be used to locate a cell phone, GPS can be used to find downed aircraft or pilots. GPS can be used by biologists to track wildlife by placing collars on the animals, a major improvement over radio tracking that was line-of-sight and worked only over short ranges. Animals that migrate over great distances can be tracked by using only GPS. Lost pets can be tracked through GPS devices in their collars. Military and law enforcement use GPS to track vehicles.

Other Civilian Applications. Surveyors and mapmakers use GPS to mark boundaries and identify locations. GPS units installed at specific locations can detect movements of the Earth to study earthquakes, volcanoes, and plate tectonics.

Next Generation Air Transportation System (NextGen). While ground-based navigational systems such as the VOR are still used by pilots and radar is used by ATC to locate airplanes, the Federal Aviation Administration (FAA) is researching and designing NextGen, a new system for navigation and tracking aircraft that will be based on GPS in the National Airspace System (NAS). Using NextGen GPS navigation, aircraft will be able to fly shorter and more direct routes to their destinations, saving time and fuel. ATC will change to a satellite-based system of managing air traffic.

Impact on Industry

Throughout the history of aviation, the United States has been a leader in research and development of flight instrumentation and avionics. During the Cold War, the Soviet Union and China worked to develop their own systems but lagged behind the United States. China has produced its own avionics but has not found an international market for its products. China has partnered with U.S. companies such as Honeywell and Rockwell Collins to produce avionics.

Government and University Research. Most government research is spearheaded by the FAA and conducted in government facilities or in universities and businesses supported by grants. The FAA is concentrating on applying GPS technology to aviation. Ohio University's Avionics Engineering Center is researching differential GPS applications for precision runway approaches and landings, as GPS is not sensitive enough to provide the very precise positioning required. The University of North Dakota is researching GPS applications for NextGen. Embry-Riddle Aeronautical University is working with the FAA's Office of Commercial Space Transportation to develop a space-transportation information system that will tie in with air traffic control. Embry-Riddle is also partnering with Lockheed Martin, Boeing, and other companies on the NextGen system. The FAA has agreements with Honeywell and Aviation Communication & Surveillance Systems to develop and test airfield avionics systems to be used in NextGen. The military also does research on new products, but their efforts are usually classified.

Commercial and Industry Sectors. The major consumers of flight instrumentation and avionics are aircraft manufacturers and airlines. In the United States, Boeing is the foremost commercial manufacturer for civilian as well as military aircraft, while Lockheed Martin, Bell Helicopter, and Northrop Grumman produce military aircraft. In Europe, Airbus is the major aircraft manufacturer.

The GPS technology developed for aircraft has also been used by companies such as Verizon and AT&T for cellular phones, Garmin for GPS navigation devices, and by power plants to synchronize power grids.

Spacecraft and space travel sectors have several companies actively manufacturing rocket ships, spacecraft, and navigational systems. The British company Virgin Galactic and the American company Space Adventures operate spacecraft, and private citizens can book flights, although it is extremely expensive.

Flight instrumentation and avionics are being used to construct UAVs for the military. Northrop Grumman is working with the military to develop this technology further, and it has been used very successfully in Iraq and Afghanistan. Other companies and universities are researching the possibility of having unmanned commercial aircraft in the United States, although the problems with coordinating unmanned aircraft with traditional aircraft and the safety concerns of operating an aircraft without a pilot are daunting.

Major Corporations. Several companies manufacture avionics, most of which are in the United States. Some of these companies have joint ventures with other countries and have established factories in those countries. Major avionics and flight instrument manufacturers are Honeywell, Rockwell Collins, Bendix/King, ARINC, Kollsman, Narco Avionics, Sigtronics, and Thales.

Many companies continue to develop new flight and avionics systems for aircraft. These products often emphasize integrating existing systems and applying declassified military advances to civilian uses. For example, Rockwell Collins's Pro Line Fusion integrated avionics system uses the military head-up display (HUD) technology for civilian business jets.

Careers and Course Work

The possible careers associated with flight instruments and avionics include both civilian and military positions ranging from mechanics and technicians to designers and researchers. The education required for these occupations usually requires at least two years of college or technical training, but research and design may require a doctorate.

Maintenance and avionics technicians install and repair flight instruments and avionics. They may work on general aviation airplanes, commercial airliners, or military aircraft. With more and more modes of transportation using INS and GPS, mechanics and technicians may also be employed to install and repair these systems on other types of vehicles—ships, trains, guided missiles, tanks, or UAVs. NASA and private companies employ technicians to work with spacecraft. Most of these positions require an associate's degree with specialization as an aircraft or avionics technician, or the training may be acquired in the military.

As computers are becoming more and more important in these fields, the demand for computer

Fascinating Facts About Avionics and Aircraft Instrumentation

- During the Civil War, surveillance balloons communicated with ground crews using telegraph. The telegraph wires from the balloon could be connected with ground telegraph wires so that the observations could be relayed directly to President Lincoln in the White House.
- Before radios were installed in airplanes, inventors tried to communicate with people on the ground by dropping notes tied to rocks and by smoke signals.
- The technology that tells airplanes how high in the sky they are is used by skydivers to know when to open their parachutes and by scuba divers to know how far under the water they are.
- Some researchers are investigating using robotic dirigibles flying in the stratosphere to replace expensive communication satellites and unsightly cell phone towers.
- GPS, which is commonly used in the family car and is an "app" in smart phones, was originally developed by the military for defense.
- GPS can be used to help lost children and pets get home safely.
- Aircraft flight instrumentation is used to construct flight simulators, including flight simulators that can be used on a home computer.
- Someday unmanned airplanes may be used in commercial transportation, carrying freight and passengers.

technicians, designers, and programmers will increase. Jobs in these fields range from positions in government agencies such as the FAA or National Aeronautics and Space Administration (NASA) or the military to private-sector research and development. The education required for these occupations varies from high school or vocational computer training to doctorates in computer science or related fields.

Flight instrument and avionics systems are being designed and researched by persons who have been educated in mechanical, electrical, and aeronautical engineering, computer science, and related fields. Some of these occupations require the minimum of a bachelor's degree, but most require a master's or doctorate.

Social Context and Future Prospects

Aviation, made possible by flight instrumentation and avionics, has revolutionized how people travel and how freight is moved throughout the world. It has also dramatically changed how wars are fought and how countries defend themselves. In the future, flight instrumentation and avionics will continue to affect society not only through aviation but also through applications of the technology in daily life.

Military use of UAVs controlled by advances in flight instrumentation and avionics will continue to change how wars are fought. However, in the not-too-distant future this technology may be used in civilian aviation. UAVs could be used to inspect pipelines and perform surveys in unpopulated areas or rough terrain, but it is unsure whether they will be used for passenger flights. Many people will certainly be fearful of traveling in airplanes with no human operators. The FAA would have to develop systems that would incorporate unmanned aircraft into the airspace. However, the use of unmanned vehicles may be an important part of future space exploration.

Perhaps the avionics system that has had the most impact on society is GPS. As GPS devices are being made more compact and more inexpensively, they are being used more and more in daily life. GPS can permit underdeveloped countries to improve their own air-navigation systems more rapidly without the expensive of buying and installing expensive ground-based navigational equipment or radar systems used by air traffic control facilities.

Polly D. Steenhagen, B.S., M.S.

Further Reading

Collinson, R. P. G. *Introduction to Avionics Systems.* 2d ed. Boston: Kluwer Academic, 2003. A comprehensive and well-illustrated review of both civilian and military flight and avionics systems.

Dailey, Franklyn E., Jr. *The Triumph of Instrument Flight: A Retrospective in the Century of U.S. Aviation.* Wilbraham, Mass.: Dailey International, 2004. A history of instrument flight with factual information as well as the author's personal flying experiences.

El-Rabbany, Ahmed. *Introduction to GPS: The Global Positioning System.* 2d ed. Norwood, Mass.: Artech House, 2006. A thorough overview of GPS and a discussion of future applications for the technology.

Federal Aviation Administration. *Instrument Flying Handbook.* New York: Skyhorse, 2008. The FAA's official manual on instrument flying includes discussions of flight instruments and avionics and the two allow a pilot to fly by reference to the instruments.

Johnston, Joe. *Avionics for the Pilot: An Introduction to Navigational and Radio Systems for Aircraft.* Wiltshire, England: Airlife, 2007. A straightforward basic explanation of aircraft avionics, what they do, and how they operate.

Tooley, Mike. *Aircraft Digital Electronic and Computer Systems: Principles, Operation and Maintenance.* Burlington, Mass.: Butterworth-Heinemann, 2007. A more technical look at aircraft avionics and computer systems and how they work.

Web Sites

Aircraft Electronics Association
http://www.aea.net

Aviation Instrument Association
http://www.aia.net

Federal Aviation Administration
http://www.faa.gov

Official Government Information About the GPS
http://www.gps.gov

See also: Aeronautics and Aviation; Computer Science; Electrical Engineering; Mechanical Engineering; Pneumatics.

B

BIOMECHANICAL ENGINEERING

FIELDS OF STUDY

Biomedical engineering; biomechanics; physiology; nanotechnology; implanted devices; modeling; bioengineering; bioinstrumentation; computational biomechanics; cellular and molecular biomechanics; forensic biomechanics; tissue engineering; mechanobiology; micromechanics; anthropometics; imaging; biofluidics.

SUMMARY

Biomechanical engineering is a branch of science that applies mechanical engineering principles such as physics and mathematics to biology and medicine. It can be described as the connection between structure and function in living things. Researchers in this field investigate the mechanics and mechanobiology of cells and tissues, tissue engineering, and the physiological systems they comprise. The work also examines the pathogenesis and treatment of diseases using cells and cultures, tissue mechanics, imaging, microscale biosensor fabrication, biofluidics, human motion capture, and computational methods. Real-world applications include the design and evaluation of medical implants, instrumentation, devices, products, and procedures. Biomechanical engineering is a multidisciplinary science, often fostering collaborations and interactions with medical research, surgery, radiology, physics, computer modeling, and other areas of engineering.

KEY TERMS AND CONCEPTS

- **Angular Motion:** Motion involving rotation around a central line or point known as the axis of rotation.
- **Biofluidics:** Field of study that combines the characterization of fluids focused on flows in the body as well as environmental flows involved in disease process.
- **Dynamics:** Branch of mechanics that studies systems that are in motion, subject to acceleration or deceleration.
- **Kinematics:** Study of movement of segments of a body without regard for the forces causing the movement.
- **Kinesiology:** Study of human movement.
- **Kinetics:** Study of forces associated with motion.
- **Linear Motion:** Motion involving all the parts of a body or system moving in the same direction, at the same speed, following a straight (rectilinear) or curved (curvilinear) line.
- **Mechanics:** Branch of physics analyzing the resulting actions of forces on particles or systems.
- **Mechanobiology:** Emerging scientific field that studies the effect of physical force on tissue development, physiology, and disease.
- **Modeling:** Computerized analytical representation of a structure or process.
- **Statics:** Branch of mechanics that studies systems that are in a constant state of motion or constant state of rest.

DEFINITION AND BASIC PRINCIPLES

Biomechanical engineering applies mechanical engineering principles to biology and medicine. Elements from biology, physiology, chemistry, physics, anatomy, and mathematics are used to describe the impact of physical forces on living organisms. The forces studied can originate from the outside environment or generate within a body or single structure. Forces on a body or structure can influence how it grows, develops, or moves. Better understanding of how a biological organism copes with forces and stresses can lead to improved treatment, advanced diagnosis, and prevention of disease. This integration of multidisciplinary philosophies has lead to significant advances in clinical medicine and device design. Improved understanding guides the creation of artificial organs, joints, implants, and tissues. Biomechanical engineering also has a tremendous influence on the retail industry, as the results of laboratory research guide product design toward more comfortable and efficient merchandise.

Background and History

The history of biomechanical engineering, as a distinct and defined field of study, is relatively short. However, applying the principles of physics and engineering to biological systems has been developed over centuries. Many overlaps and parallels to complementary areas of biomedical engineering and biomechanics exist, and the terms are often used interchangeably with biomechanical engineering. The mechanical analysis of living organisms was not internationally accepted and recognized until the definition provided by Austrian mathematician Herbert Hatze in 1974: "Biomechanics is the study of the structure and function of biological systems by means of the methods of mechanics." Aristotle introduced the term "mechanics" and discussed the movement of living beings around 322 B.C.E in the first book about biomechanics, *On the Motion of Animals*. Leonardo da Vinci proposed that the human body is subject to the law of mechanics in the 1500's. Italian physicist and mathematician Giovanni Alfonso Borelli, a student of Galileo's, is considered the "father of biomechanics" and developed mathematical models to describe anatomy and human movement mechanically. In the 1890's German zoologist Wilhelm Roux and German surgeon Julius Wolff determined the effects of loading and stress on stem cells in the development of bone architecture and healing. British physiologist Archibald V. Hill and German physiologist Otto Fritz Meyerhof shared the 1922 Nobel Prize for Physiology or Medicine. The prize was divided between them: Hill won "for his discovery relating to the production of heat in the muscle"; Meyerhof won "for his discovery of the fixed relationship between the consumption of oxygen and the metabolism of lactic acid in the muscle."

The first joint replacement was performed on a hip in 1960 and a knee in 1968. The development of imaging, modeling, and computer simulation in the latter half of the 1900's provided insight into the smallest structures of the body. The relationships between these structures, functions, and the impact of internal and external forces accelerated new research opportunities into diagnostic procedures and effective solutions to disease. In the 1990's, biomechanical engineering programs began to emerge in academic and research institutions around the world.

How It Works

Biomechanical engineering science is extremely diverse. However, the basic principle of studying the relationship between biological structures and forces, as well as the important associated reactions of biological structures to technological and environmental materials, exists throughout all disciplines. The biological structures described include all life forms and may include an entire body or organism or even the microstructures of specific tissues or systems. Characterization and quantification of the response of these structures to forces can provide insight into disease process, resulting in better treatments and diagnoses. Research in this field extends beyond the laboratory and can involve observations of mechanics in nature, such as the aerodynamics of bird flight, hydrodynamics of fish, or strength of plant root systems, and how these findings can be modified and applied to human performance and interaction with external forces.

As in biomechanics, biomechanical engineering has basic principles. Equilibrium, as defined by British physicist Sir Isaac Newton, results when the sum of all forces is zero and no change occurs and energy cannot be created or destroyed, only converted from one form to another.

The seven basic principles of biomechanics can be applied or modified to describe the reaction of forces to any living organism.

1. The lower the center of mass, the larger the base of support; the closer the center of mass to the base of support, and the greater the mass, the more stability increases.
2. The production of maximum force requires the use of all possible joint movements that contribute to the task's objective.
3. The production of maximum velocity requires the use of joints in order—from largest to smallest.
4. The greater the applied impulse, the greater increase in velocity.
5. Movement usually occurs in the direction opposite that of the applied force.
6. Angular motion is produced by the application of force acting at some distance from an

axis, that is, by torque.

7. Angular momentum is constant when a body or object is free in the air.

The forces studied can be combinations of internal, external, static, or dynamic, and all are important in the analysis of complex biochemical and biophysical processes. Even the mechanics of a single cell, including growth, cell division, active motion, and contractile mechanisms, can provide insight into mechanisms of stress, damage of structures, and disease processes at the microscopic level. Imaging and computer simulation allow precise measurements and observations to be made of the forces impacting the smallest cells.

Applications and Products

Biomechanical engineering advances in modeling and simulation have tremendous potential research and application uses across many health care disciplines. Modeling has resulted in the development of designs for implantable devices to assist with organs or areas of the body that are malfunctioning. The biomechanical relationships between organs and supporting structures allow for improved device design and can assist with planning of surgical and treatment interventions. The materials used for medical and surgical procedures in humans and animals are being evaluated and some redesigned, as biomechanical science is showing that different materials, procedures, and techniques may be better for reducing complications and improving long-term patient health. Evaluating the physical relationship between the cells and structures of the body and foreign implements and interventions can quantify the stresses and forces on the system, which provides more accurate prediction of patient outcomes.

Biomechanical engineering professionals apply their knowledge to develop implantable medical devices that can diagnose, treat, or monitor disease and health conditions and improve the daily living of patients. Devices that are used within the human body are highly regulated by the U.S. Food and Drug Administration (FDA) and other agencies internationally. Pacemakers and defibrillators, also called cardiac resynchronization therapy (CRT) devices, can constantly evaluate a patient's heart and respond to changes in heart rate with electrical stimulation. These devices greatly improve therapeutic outcomes in patients afflicted with congestive heart failure.

Fascinating Facts About Biomechanical Engineering

- Biomechanical engineers design and develop many items used and worn by astronauts.
- Of all the engineering specialties, biomechanical engineering has one of the highest percentages of female students.
- Synovial fluid in joints has such low friction that engineers are trying to duplicate it synthetically to lubricate machines.
- Many biomechanical engineering graduates continue on to medical school.
- On October 8, 1958, in Sweden, forty-three-year-old Arne Larsson became the first person to receive an implanted cardiac pacemaker. He lived to the age of eighty-six.
- Many biomechanical engineers have conducted weightlessness experiments aboard the National Aeronautics and Space Administration's (NASA) C-9 microgravity aircraft.
- Cardiac muscles rest only between beats. Based on 72 beats a minute, they will pump an average of 3.0 trillion times during an eighty-year life span.

Patients with arrhythmias experience greater advantages with implantable devices than with pharmaceutical options. Cochlear implants have been designed to be attached to a patient's auditory nerve and can detect sound waves and process them in order to be interpreted by the brain as sound for deaf or hard-of-hearing patients. Patients who have had cataract surgery used to have to wear thick corrective lenses to restore any standard of vision but with the development of intraocular lenses that can be implanted into the eye, their vision can be restored, often to a better degree than before the cataract developed.

Artificial replacement joints comprise a large portion of medical-implant technology. Patients receive joint replacement when their existing joints no longer function properly or cause significant pain because of arthritis or degeneration. More than 220,000 total hip replacements were performed in the United States in 2003, and this number is expected to grow significantly as the baby boomer portion of the population ages. Artificial joints are normally fastened to the existing bone by cement, but advances in biomechanical engineering have lead to a new process called "bone ingrowth," in which the natural bone grows into the porous surface of the replacement joint. Biomechanical engineering contributes considerable knowledge to the design of the artificial joints, the materials from which they are made, the surgical procedure used, fixation techniques, failure mechanisms, and prediction of the lifetime of the replacement joints.

Computer-aided (CAD) design has allowed biomechanical engineers to create complex models of organs and systems that can provide advanced analysis and instant feedback. This information provides insight into the development of designs for artificial organs that align with or improve on the mechanical properties of biological organs.

Biomechanical engineering can provide predictive values to medical professionals, which can help them develop a profile that better forecasts patient outcomes and complications. An example of this is using finite element analysis in the evaluation of aortic-wall stress, which can remove some of the unpredictability of expansion and rupture of an abdominal aortic aneurysm. Biomechanical computational methodology and advances in imaging and processing technology have provided increased predictability for life- threatening events.

Nonmedical applications of biomechanical engineering also exist in any facet of industry that impacts human life. Corporations employ individuals or teams to use engineering principles to translate the scientifically proven principles into commercially viable products or new technological platforms. Biomechanical engineers also design and build experimental testing devices to evaluate a product's performance and safety before it reaches the marketplace, or they suggest more economically efficient design options. Biomechanical engineers also use ergonomic principles to develop new ideas and create new products, such as car seats, backpacks, or even equipment and clothing for elite athletes, military personnel, or astronauts.

Impact on Industry

Biomechanical engineering is a dynamic scientific field, and its vast range of applications is having a significant impact and influence on industry. Corporations realize the value of having their designs and products evaluated by biomechanical engineers to optimize the comfort and safety of consumers. Small modifications in the design of a product can influence consumers to select one product over several others, be it clothing, furniture, sporting equipment, or beverage and food containers. With so many products to choose from, it is important that one stand out as safer, more comfortable, or more efficient than another.

Biotechnology and health care are highly competitive, multibillion dollar industries. Biotechnology is an extremely research-intensive industry, so biotech companies employ teams of biomechanical engineers to research and develop devices, treatments, and diagnostic and monitoring devices to be used in health care. Implantable devices are used in the treatment and management of cardiovascular, orthopedic, neurological, ophthalmic, and various other chronic disorders. Orthopedic implants are the most common and include reconstructive joint replacements, spinal implants, orthobiologics, and trauma implants. In the United States, the demand for implantable medical devices is expected to reach a market value of $48 billion by 2014.

Surgeons, medical personnel, and health care administrators are always looking for new options that will provide their patients with optimal care, earlier diagnosis, pain reduction, and a decreased risk of complications. Aging and disabled persons have more effective options than ever before that allow them to continue vital, independent, and active lives. Industry is aligning with these developments, demands, and discoveries, and there is great competition among companies to be the first to capitalize. The FDA and other international regulatory bodies have strict standards and high levels of control for any device or implement that will be used on patients. The application and approval process can take many years. It is not uncommon for a biotechnology company to invest millions of dollars in the research and development of a medical device before it can be presented to the consumer.

Careers and Course Work

There are a variety of career choices in biomechanical engineering, and study in this field often evolves into specialized work in related areas. Students who earn a bachelor's degree from an accredited biomechanical engineering program may begin working in areas such as medical device, implant, or product design. Most teaching positions require a master's or doctoral degree. Some students continue to medical school.

Biomechanical engineering programs are composed of a cross section of course work from many disciplines. Students should have a strong aptitude for mathematics as well as biological sciences. Elements from engineering, physics, chemistry, anatomy, biology, and computer science provide core knowledge that is applied to mathematical modeling and computer simulation. Experimental work involving biological, mechanical, and clinical studies are performed to illustrate theoretical models and solve important research problems. The principles of biomechanical engineering can have vast applications, ranging from building artificial organs and tissues to designing products that are more comfortable for consumers.

Biomechanical engineering programs often are included as a subdiscipline of engineering or biomedicine. However, some schools, such as Stanford University, are creating interdisciplinary programs that offer undergraduate and graduate degrees in biomechanical engineering.

Social Context and Future Prospects

The diversity of studying the relationship between living structure and function has opened up vast opportunities in science, health care, and industry. In addition to conventional implant and replacement devices, the demand is growing for implantable tissues for cosmetic surgery, such as breast and tissue implants, as well as implantable devices to aid in weight loss, such as gastric banding.

Reports of biomechanical engineering triumphs and discoveries are appearing in the mainstream media, making the general public more aware of the scientific work being done and how it impacts daily life. Sports fans learn about the equipment, training, and rehabilitation techniques designed by biomechanical engineers that allow their favorite athletes to break performance records and return to work sooner after being injured or having surgery. The public is accessing more information about their own health options than ever before, and they are becoming knowledgeable about the range of treatments available to them and the pros and cons of each.

Biomechanical engineering and biotechnology is an area that is experiencing accelerated growth, and billions of dollars are being funneled into research and development annually. This growth is expected to continue.

April D. Ingram, B.Sc.

Further Reading

Ethier, C. Ross, and Craig A. Simmons. *Introductory Biomechanics: From Cells to Organisms.* Cambridge, England: Cambridge University Press, 2007. Provides an introduction to biomechanics and also discusses clinical specialties, such as cardiovascular, musculoskeletal, and ophthalmology.

Hall, Susan J. *Basic Biomechanics.* 5th ed. New York: McGraw-Hill, 2006. A good introduction to biomechanics, regardless of one's math skills.

Hamill, Joseph, and Kathleen M. Knutzen. *Biomechanical Basis of Human Movement.* 3d ed. Philadelphia: Lippincott, 2009. Integrates anatomy, physiology, calculus, and physics and provides the fundamental concepts of biomechanics.

Hay, James G., and J. Gavin Reid. *Anatomy, Mechanics, and Human Motion.* 2d ed. Englewood Cliffs, N.J.: Prentice Hall, 1988. A good resource for upper high school students, this text covers basic kinesiology.

Peterson, Donald R., and Joseph D. Bronzino, eds. *Biomechanics: Principles and Applications.* 2d ed. Boca Raton, Fla.: CRC Press, 2008. A collection of twenty articles on various aspects of research in biomechanics.

Prendergast, Patrick, ed. *Biomechanical Engineering: From Biosystems to Implant Technology.* London: Elsevier, 2007. One of the first comprehensive books for biomechanical engineers, written with the student in mind.

Web Sites

American Society of Biomechanics
http://www.asbweb.org

Biomedical Engineering Society
http://www.bmes.org

European Society of Biomechanics
http://www.esbiomech.org

International Society of Biomechanics
http://www.isbweb.org

World Commission of Science and Sports
http://www.wcss.org.uk

See also: Bioinformatics; Bionics and Biomedical Engineering; Calculus; Computer Science.

BIONICS AND BIOMEDICAL ENGINEERING

FIELDS OF STUDY

Biology; physiology; biochemistry; engineering; orthopedic bioengineering; physics; bionanotechnology; biomechanics; biomaterials; neural engineering; genetic engineering; tissue engineering; prosthetics.

SUMMARY

Bionics combines natural biologic systems with engineered devices and electrical mechanisms. An example of bionics is an artificial arm controlled by impulses from the human mind. Construction of bionic arms or similar devices requires the integrative use of medical equipment such as electroencephalograms (EEGs) and magnetic resonance imaging (MRI) machines with mechanically engineered prosthetic arms and legs. Biomedical engineering further melds biomedical and engineering sciences by producing medical equipment, tissue growth, and new pharmaceuticals. An example of biomedical engineering is human insulin production through genetic engineering to treat diabetes.

KEY TERMS AND CONCEPTS

- **Biologics:** Medicines produced from genes by manipulating genes and using genetic technology.
- **Biomaterials:** Substances, including metal alloys, plastic polymers, and living tissues, used to replace body tissues or as implants.
- **Bionanotechnology:** Construction of materials on a very small scale, enabling the use of microscopic machinery in living tissues.
- **Bionic:** Integrating biological function and mechanical devices.
- **Clone:** Genetically engineered organism with genetic composition identical to the original organism.
- **Human Genetic Engineering:** Genetic engineering focused on altering or changing visible human characteristics through gene manipulations.
- **Prosthesis:** Artificial or biomechanically engineered body part.
- **Recombinant DNA:** DNA created by the combination of two or more DNA sequences that do not normally occur together.

Definition and Basic Principles

The fields of biomedical engineering and bionics focus on improving health, particularly after injury or illness, with better rehabilitation, medications, innovative treatments, enhanced diagnostic tools, and preventive medicine.

Bionics has moved nineteenth-century prostheses, such as the wooden leg, into the twenty-first century by using plastic polymers and levers. Bionics integrates circuit boards and wires connecting the nervous system to the modular prosthetic limb. Controlling artificial limb movements with thoughts provides more lifelike function and ability. This mind and prosthetic limb integration is the "bio" portion of bionics; the "nic" portion, taken from the word "electronic," concerns the mechanical engineering that makes it possible for the person using a bionic limb to increase the number and range of limb activity, approaching the function of a real limb.

Biomedical engineering encompasses many medical fields. The principle of adapting engineering techniques and knowledge to human structure and function is a key unifying concept of biomedical engineering. Advances in genetic engineering have produced remarkable bioengineered medications. Recombinant DNA techniques (genetic engineering) have produced synthetic hormones, such as insulin. Bacteria are used as a host for this process; once human-insulin-producing genes are implanted in the bacteria, the bacteria's DNA produce human insulin, and the human insulin is harvested to treat diabetics. Before this genetic technique was developed in 1982 to produce human insulin, insulin-dependent diabetics relied on insulin from pigs or cows. Although this insulin was life saving for diabetics, diabetics often developed problems from the pig or cow insulin because they would produce antibodies against the foreign insulin. This problem disappeared with the ability to engineer human insulin using recombinant DNA technology.

Background and History

In the broad sense, biomedical engineering has existed for millennia. Human beings have always envisioned the integration of humans and technology to increase and enhance human abilities. Prosthetic devices go back many thousands of years: A three-thousand-year-old Egyptian mummy, for example, was found with a wooden big toe tied to its foot. In the fifteenth century, during the Italian Renaissance, Leonardo da Vinci's elegant drawings demonstrated some early ideas on bioengineering, including his helicopter and flying machines, which melded human and machine into one functional unit capable of flight. Other early examples of biomedical engineering include wooden teeth, crutches, and medical equipment, such as stethoscopes.

Electrophysiological studies in the early 1800's produced biomedical engineering information used to better understand human physiology. Engineering principles related to electricity combined with human physiology resulted in better knowledge of the electrical properties of nerves and muscles.

X rays, discovered by Wilhelm Conrad Röntgen in 1895, were an unknown type of radiation (thus the "X" name). When it was accidentally discovered that they could penetrate and destroy tissue, experiments were developed that led to a range of imaging technologies that evolved over the next century. The first formal biomedical engineering training program, established in 1921 at Germany's Oswalt Institute for Physics in Medicine, focused on three main areas: the effects of ionizing radiation, tissue electrical characteristics, and X-ray properties.

In 1948, the Institute of Radio Engineers (later the Institute of Electrical and Electronics Engineers), the American Institute for Electrical Engineering, and the Instrument Society of America held a conference on engineering in biology and medicine. The 1940's and 1950's saw the formation of professional societies related to biomedical engineering, such as the Biophysics Society, and of interest groups within engineering societies. However, research at the time focused on the study of radiation. Electronics and the budding computer era broadened interest and activities toward the end of the 1950's.

James D. Watson and Francis Crick identified the DNA double-helix structure in 1953. This important discovery fostered subsequent experimentation in molecular biology that yielded important information about how DNA and genes code for the expression of traits in all living organisms. The genetic code in DNA was deciphered in 1968, arming researchers with enough information to discover ways that DNA could be recombined to introduce genes from one organism into a different organism, thereby allowing the host to produce a variety of useful products. DNA recombination became one of the most important tools in the field of biomedical engineering, leading to tissue growth as well as new pharmaceuticals.

In 1962, the National Institutes of Health created the National Institute of General Medical Sciences, fostering the development of biomedical engineering programs. This institute funds research in the diagnosis, treatment, and prevention of disease.

Bionics and biomedical engineering span a wide variety of beneficial health-related fields. The common thread is the combination of technology with human applications. Dolly the sheep was cloned in 1996. Cloning produces a genetically identical copy of an existing life-form. Human embryonic cloning presents the potential of therapeutic reproduction of needed organs and tissues, such as kidney replacement for patients with renal failure.

In the twenty-first century, the linking of machines with the mind and sensory perception has provided hearing for deaf people, some sight for the blind, and willful control of prostheses for amputees.

How It Works

Restorative bionics integrates prosthetic limbs with electrical connections to neurons, allowing an individual's thoughts to control the artificial limb. Tiny arrays of electrodes attached to the eye's retina connect to the optic nerve, enabling some visual perception for previously blind people. Deaf people hear with electric devices that send signals to auditory nerves, using antennas, magnets, receivers, and electrodes. Researchers are considering bionic skin development using nanotechnology to connect with nerves, enabling skin sensations for burn victims requiring extensive grafting.

Many biomedical devices work inside the human body. Pacemakers, artificial heart valves, stents, and even artificial hearts are some of the bionic devices correcting problems with the cardiovascular system. Pacemakers generate electric signals that improve abnormal heart rates and abnormal heart rhythms. When pulse generators located in the pacemakers sense an abnormal heart rate or rhythm, they produce shocks to restore the normal rate. Stents are inserted into an artery to widen it and open clogged blood vessels. Stents and pacemakers are examples of specialized bionic devices made up of bionic materials compatible with human structure and function.

Cloning. Cloning is a significant area of genetic engineering that allows the replication of a complete living organism by manipulating genes. Dolly the sheep, an all-white Finn Dorset ewe, was cloned from a surrogate mother blackface ewe, which was used as an egg donor and carried the cloned Dolly during gestation (pregnancy). An egg cell from the surrogate was removed and its nucleus (which contains DNA) was replaced with one from a Finn Dorset ewe; the resulting new egg was placed in the blackface ewe's uterus after stimulation with an electric pulse. The electrical pulse stimulated growth and cell duplication. The blackface ewe subsequently gave birth to the all-white Dolly. The newborn all-white Finn Dorset ewe was an identical genetic twin of the Finn Dorset that contributed the new nucleus.

Recombinant DNA. Another significant genetic engineering technique involves recombinant DNA. Human genes transferred to host organisms, such as bacteria, produce products coded for by the transferred genes. Human insulin and human growth hormone can be produced using this technique. Desired genes are removed from human cells and placed in circular bacterial DNA strips called plasmids. Scientists use enzymes to prepare these DNA formulations, ultimately splicing human genes into bacterial plasmids. These plasmids are used as vectors, taken up and reproduced by bacteria. This type of genetic adaptation results in insulin production if the spliced genes were taken from the part of the human genome producing insulin; other cells and substances, coded for by different human genes, can be produced this way. Many biologic medicines are produced using recombinant DNA technology.

Applications and Products

Medical Devices. Biomedical engineers produce life-saving medical equipment, including pacemakers, kidney dialysis machines, and artificial hearts. Synthetic limbs, artificial cochleas, and bionic sight chips are among the prosthetic devices that biomedical engineers have developed to enhance mobility, hearing, and vision. Medical monitoring devices, developed by biomedical engineers for use in intensive care units and surgery or by space and deep-sea explorers, monitor vital signs such as heart rate and rhythm, body temperature, and breathing rate.

Equipment and Machinery. Biomedical engineers produce a wide variety of other medical machinery, including laboratory equipment and therapeutic equipment. Therapeutic equipment includes laser devices for eye surgery and insulin pumps (sometimes called artificial pancreases) that both monitor blood sugar levels and deliver the appropriate amount of insulin when it is needed.

Imaging Systems. Medical imaging provides important machinery devised by biomedical engineers. This specialty incorporates sophisticated computers and imaging systems to produce computed tomography (CT), magnetic resonance imaging (MRI), and positron emission tomography (PET) scans. In naming its National Institute of Biomedical Imaging and Bioengineering (NIBIB), the U.S. Department of Health and Human Services emphasized the equal importance and close relatedness of these subspecialties by using both terms in the department's name.

Computer programming provides important circuitry for many biomedical engineering applications, including systems for differential disease diagnosis. Advances in bionics, moreover, rely heavily on computer systems to enhance vision, hearing, and body movements.

Biomaterials. Biomaterials, such as artificial skin and other genetically engineered body tissues, are areas promising dramatic improvements in the treatment of burn victims and individuals needing organ transplants. Bionanotechnology, another subfield of biomedical engineering, promises to enhance the surface of artificial skin by creating microscopic messengers that can create the sensations of touch and pain. Bioengineers interface with the fields of physical therapy, orthopedic surgery, and rehabilitative medicine in the fields of splint development, biomechanics, and wound healing.

Medications. Medicines have long been synthesized artificially in laboratories, but chemically synthesized medicines do not use human genes in their production. Medicines produced by using human genes in recombinant DNA procedures are called biologics and include antibodies, hormones, and cell receptor proteins. Some of these products include human insulin, the hepatitis B vaccine, and human growth hormone.

Bacteria and viruses invading a body are attacked and sometimes neutralized by antibodies produced by the immune system. Diseases such as Crohn's disease, an inflammatory bowel condition, and psoriatic arthritis are conditions exacerbated by inflammatory antibody responses mounted by the affected person's immune system. Genetic antibody production in the form of biologic medications interferes with or attacks mediators associated with Crohn's and arthritis and improves these illnesses by decreasing the severity of attacks or decreasing the frequency of flare-ups.

Cloning and Stem Cells. Cloned human embryos could provide embryonic stem cells. Embryonic stem cells have the potential to grow into a variety of cells, tissues, and organs, such as skin, kidneys, livers, or heart cells. Organ transplantation from genetically identical clones would not encounter the recipient's natural rejection process, which transplantations must overcome. As a result, recipients of genetically identical cells, tissues, and organs would enjoy more successful replacements of key organs and a better quality of life. Human cloning is subject to future research and development, but the promise of genetically identical replacement organs for people with failed hearts, kidneys, livers, or other organs provides hope for enhanced future treatments.

Impact on Industry

Government and University Research. The National Institute of Biomedical Imaging and Bioengineering was established in 2001. It is dedicated to health improvement by developing and applying biomedical technologies. The institute supports research with grants, contracts, and career development awards. Many major universities around the world actively research and develop bionics and biomedical engineering. Top engineering schools, such as the University of Michigan, have undergraduate and graduate degree programs for biomedical engineering. Biomedical engineers work in industries producing medical equipment and pharmaceutical companies. They also conduct research and develop genetic therapies useful for treating a wide variety of illnesses.

Industry and Business Sectors. The pharmaceutical industry, along with the medical equipment industry, employs biomedical engineers. The medical industry, including the suppliers of laboratory equipment, imaging technology, bionics, and pharmaceuticals, drives much of the employment demand for biomedical engineers. Medical hardware, such as artificial hearts, pacemakers, and renal dialysis machines, remains an important part of biomedical engineering employment.

Careers and Course Work

According to the 2010-2011 edition of the *Occupational Outlook Handbook*, issued by the U.S. Department of Labor's Bureau of Labor Statistics, biomedical engineers are projected to have a 72 percent employment growth from 2010 to 2020, a rate of increase much greater than the average for all occupations. Biomedical engineers are employed as researchers and scientists, interfacing with a wide variety of disciplines and specialties.

Many career paths exist in biomedical engineering and bionics. Jobs exist for those holding bachelor's degrees through doctorates. Research scientists, usually with Ph.D. or M.D. degrees, can work in a variety of environments, from private companies to universities to government agencies, including the National Institutes of Health.

A typical undergraduate biomedical engineering curriculum for University of Michigan students, for example, includes three divisions of course study: subjects required by all engineering programs, advanced science and engineering mathematics, and biomedical engineering courses. All engineering students take courses in calculus, basic engineering concepts, computing, chemistry, physics, and humanities or social sciences. Required courses in advanced sciences in the second division include biology, chemistry, and biological chemistry, along with engineering mathematics courses. Biomedical engineering courses in the third division cover circuits and electrical systems, biomechanics, engineering materials, cell biology, physiology, and biomedical design. Students can concentrate in areas such as biomechanical, biochemical, and bioelectric engineering, with some modifications in course selection. The breadth and depth of the course work emphasize the link between engineering and biologic sciences—the essence of bionics and biomedical engineering.

A biomedical engineer is an engineer developing and advancing biomedical products and systems. This type of activity spans many specialties and presents many career opportunities. Biomedical engineering interfaces with almost every engineering discipline because bioengineered products require the breadth and depth of engineering knowledge. Major areas include the production of biomaterials, such as the type of plastic used for prosthetic devices.

Fascinating Facts About Bionics and Biomedical Engineering

- In 2010, scientists at the University of California, San Diego, developed biosensor cells that can be implanted in the brain to help monitor receptors and chemical signals that allow cells in the brain to communicate with one another. These cells may help scientists understand drug addiction.
- Vanderbilt engineers in 2010 began testing a knowledge repository and interactive software that will help surgeons more accurately and rapidly place electrodes in the brains of people with Parkinson's disease in a procedure called deep brain stimulation. The data allow for faster surgery and the implementation of best practices.
- In 2010, scientists at Vanderbilt University developed a robotic prosthesis for the lower leg that has powered knee and ankle joints. Intent recognizer software takes information from sensors, determines what the user wants to do, and provides power to the leg.
- The Wadsworth Center at the New York State Department of Health in 2009 developed a brain-computer interface that translates brain waves into action. It allowed a patient with amyotrophic lateral sclerosis who was no longer able to communicate with others because of failing muscles to write e-mails and convey his thoughts to others.
- In 2008, a research team at the Johns Hopkins University modified chondroitin sulfate, a natural sugar, so it could glue a hydrogel (like the material used in soft contact lenses) to cartilage tissue. It is hoped that this technique may help those experiencing joint pain from oseoarthritis, in which the natural cartilage in a joint disappears.

Bioinstrumentation involves computer integration in diagnostic machines and the control of devices. Genetic engineering presents many opportunities for life modifications. Clinical engineers help integrate new technologies, such as electronic medical records, into existing hospital systems. Medical imaging relieves the need for exploratory surgery and greatly enhances diagnostic capabilities. Orthopedic bioengineering plays important roles in prosthesis development and use, along with assisting rehabilitative medicine. Bionanotechnology offers the hope of using microscopic messengers to treat illness and advanced capabilities for artificial bionic devices. Systems physiology organizes the multidisciplinary approach necessary to complete complex bionic projects such as a functioning artificial eye.

Social Context and Future Prospects

Bionics technologies include artificial hearing, sight, and limbs that respond to nerve impulses. Bionics offers partial vision to the blind and prototype prosthetic arm devices that offer several movements through nerve impulses. The goal of bionics is to better integrate the materials in these artificial devices with human physiology to improve the lives of those with limb loss, blindness, or decreased hearing.

Cloned animals exist but cloning is not a yet a routine process. Technological advances offer rapid DNA analysis along with significantly lower cost genetic analysis. Genetic databases are filled with information on many life-forms, and new DNA sequencing information is added frequently. This basic information that has been collected is like a dictionary, full of words that can be used to form sentences, paragraphs, articles, and books, in that it can be used to create new or modified life-forms.

Biomedical engineering enables human genetic engineering. The stuff of life, genes, can be modified or manipulated with existing genetic techniques. The power to change life raises significant societal concerns and ethical issues. Beneficial results such as optimal organ transplantations and effective medications are the potential of human genetic engineering.

Richard P. Capriccioso, B.S., M.D., and Christina Capriccioso

Further Reading

Braga, Newton C. *Bionics for the Evil Genius: Twenty-five Build-It-Yourself Projects.* New York: McGraw-Hill, 2006. Step-by-step projects that introduce basic concepts in bionics.

Fischman, Josh. "Merging Man and Machine: The Bionic Age." *National Geographic* 217, no. 1 (January, 2010): 34-53. A well-illustrated consideration of the latest advances in bionics, with specific examples of people aided by the most modern prosthetic technologies.

Hung, George K. *Biomedical Engineering: Principles of the Bionic Man.* Hackensack, N.J.: World Scientific, 2010. Examines scientific bioengineering principles as they apply to humans.

Richards-Kortum, Rebecca. *Biomedical Engineering for Global Health.* New York: Cambridge University Press, 2010. Examines the potential of biomedical engineering to treat diseases and conditions throughout the world. Examines health care systems and social issues.

Smith, Marquard, and Joanne Morra, eds. *The Prosthetic Impulse: From a Posthuman Present to a Biocultural Future.* Cambridge, Mass.: MIT Press, 2007. Examines the developments in prosthetic devices and addresses the social aspects, including what it means to be human.

Web Sites

American Society for Artificial Internal Organs
http://www.asaio.com

Biomedical Engineering Society
http://www.bmes.org

National Institute of Biomedical Imaging and Bioengineering
http://www.nibib.nih.gov/HomePage

Rehabilitation Engineering and Assistive Technology Association of North America
http://resna.org

Society for Biomaterials
http://www.biomaterials.org

See also: Artificial Organs

BIOPROCESS ENGINEERING

FIELDS OF STUDY

Biology; engineering; bioengineering; medicine; genetic engineering; molecular biology.

SUMMARY

Bioprocess engineering is an interdisciplinary science that combines the disciplines of biology and engineering. It is associated primarily with the commercial exploitation of living things on a large scale. The objective of bioprocess engineering is to optimize either growth of organisms or the generation of target products. This is achieved mainly by the construction of controllable apparatuses. Both government agencies and private companies invest heavily in research within this area of applied science. Many traditional bioprocess engineering approaches (such as antibiotic production by microorganisms) have been advanced by techniques of genetic engineering and molecular biology.

KEY TERMS AND CONCEPTS

- **Biomass:** Mass of organisms or organic material; traditionally refers to the biomass of plants and microorganisms.
- **Bioreactor:** Apparatus for growing microbial, plant, or animal cells, with a practical purpose under controlled conditions; these closed systems range from small (5- to 10-milliliter), laboratory-scale devices to larger, industrial-scale devices of more than 500,000 liters.
- **Bioremediation:** Use of living organisms to clean up the environment.
- **Enzymes:** Biological catalysts made of proteins.
- **Fermentation:** Metabolic reaction that is necessary to generate energy in microbial cells; used to produce many important compounds, such as alcohol and acetone.
- **Fermenter:** Type of traditional bioreactor (involving either stirred or nonstirred tanks) in which cell fermentation takes place; in continuous-culture fermenters, nutrients are continuously fed into the fermentation vessel so that cells can ferment indefinitely, whereas in batch fermenters, nutrients are added in batches.

Definition and Basic Principles

Bioprocess engineering is the use of engineering devices (such as bioreactors) in biological processes carried out by microbial, plant, and animal cells in order to improve or analyze these processes. Large-scale manufacturing involving biological processes requires substantial engineering work. Throughout history, engineering has helped develop many bioprocesses, such as the production of antibiotics, biofuels, vaccines, and enzymes on an industrial scale. Bioprocess engineering plays a role in many industries, including the food, microbiological, pharmaceutical, biotechnological, and chemical industries.

Background and History

People have been using bioprocessing for making bread, cheese, beer, and wine—all fermented foods—for thousands of years. Brewing was one of the first applications of bioprocess engineering. However, it was not until the nineteenth century that the scientific basis of fermentation was established, with the studies of French scientist Louis Pasteur, who discovered the microbial nature of beer brewing and wine making.

During the early part of the twentieth century, large-scale methods for treating wastewater were developed. Considerable growth in this field occurred toward the middle of the century, when the bioprocess for large-scale production of the antibiotic penicillin was developed. The World War II goal of industrial-scale production of penicillin led to the development of fermenters by engineers working together with biologists from the pharmaceutical company Pfizer. The fungus *Penicillium* grows and produces antibiotics much more effectively under controlled conditions inside a fermenter.

Later progress in bioprocess engineering has followed the development of genetic engineering, which raises the possibility of making new products from genetically modified microorganisms and plants grown in bioreactors. Just as past developments in bioprocess engineering have required contributions from a wide range of disciplines, including microbiology, genetics, biochemistry, chemistry, engineering, mathematics, and computer science, future developments are likely to require cooperation among scientists in multiple specialties.

How It Works

Living cells may be used to generate a number of useful products: food and food ingredients (such as cheese, bread, and wine), antibiotics, biofuels, chemicals (enzymes), and human health care products such as insulin. Organisms are also used to destroy or break down harmful wastes, such as those created by the 2010 oil spill in the Gulf of Mexico, or to reduce pollution.

A good example of how bioprocess engineering works is the development of a bioprocess using bacteria for industrial production of the human hormone insulin. Without insulin, which regulates blood sugar levels, the body cannot use or store glucose properly. The inability of the body to make sufficient insulin causes diabetes. In the 1970's, the U.S. company Genentech developed a bioprocess for insulin production using genetically modified bacterial cells.

The initial stages involve genetic manipulation (in this case, transferring a human gene into bacterial DNA). Genetic manipulation is done in laboratories by scientists trained in molecular biology or biochemistry. After creating a genetically engineered bacterium, scientists grow it in a small tubes or flasks and study its growth characteristics and insulin production.

Once the bacterial growth and insulin production characteristics have been identified, scientists increase the scale of the bioprocess. They use or build small bioreactors (1-10 liters) that can monitor temperature, pH (acidity-alkalinity), oxygen concentration, and other process characteristics. The goal of this scale-up is to optimize bacterial growth and insulin production.

The next step is another scale-up, this time to a pilot-scale bioreactor. These bioreactors can be as large as 1,000 liters and are designed and built by engineers to study the response of bacterial cells to large-scale production. During a scale-up, decreased product yields are often experienced because the conditions in the large-scale bioreactors (temperature, pH, aeration, and nutrient supply) differ from those in small, laboratory-scale systems. If the pilot-scale bioreactors work efficiently, engineers will design industrial-scale bioreactors and supporting facilities (air supply, sterilization, and process-control equipment).

All these stages are part of upstream processing. An important part of bioprocess engineering is the product recovery process, or so-called downstream processing. Product recovery from cells often can be very difficult. It involves laboratory procedures such as mechanical breakage, centrifugation, filtration, chromatography, crystallization, and drying. The final step in bioprocess engineering is testing of the recovered product, in which animals are often used.

Applications and Products

A wide range of products and applications of bioprocess engineering are familiar, everyday items.

Foods, Beverages, Food Additives, and Supplements. Living organisms play a major role in the production of food. Foods, beverages, additives, and supplements traditionally made by bioprocess engineering include dairy products (cheeses, sour cream, yogurt, and kefir), alcoholic beverages (beer, wines, and distilled spirits), plant products (soy sauce, tofu, sauerkraut), and food additives and supplements (flavors, proteins, vitamins, and carotenoids).

Traditional fermenters with microorganisms are used to obtain products in most of these applications. A typical industrial fermenter is constructed from stainless steel. Mixing of the microbial culture in fermenters is achieved by mechanical stirring, often with baffles. Airlift bioreactors have also been applied in the manufacturing of food products such as crude proteins synthesized by microorganisms. Mixing and liquid circulation in these bioreactors are induced by movement of an injected gas (such as air).

Biofuels. Bioprocess engineering is used in the production of biofuels, including ethanol (bioethanol), oil (biodiesel), butanol, biohydrogen, and biogas (methane). These biofuels are produced by the action of microorganisms in bioreactors, some of which use attached (immobilized) microorganisms. Cells, when immobilized in matrices such as agar, polyurethane, or glass beads, stabilize their growth and increase their physiological functions. Many microorganisms exist naturally in a state similar to immobilization, either on the surface of soil particles or in symbiosis with other organisms.

Environmental Applications. Bioprocess engineering plays an important role in removing pollution from the environment. It is used in treatment of wastewater and solid wastes, soil bioremediation, and mineral recovery. Environmental applications are based on the ability of organisms to use pollutants or other compounds as their food sources. One of the most important and widely used environmental applications is the treatment of wastewater by microorganisms. Microbes eat organic and inorganic compounds in wastewater and clean it at the same time. In this application, microorganisms are placed inside bioreactors (known as digesters) specifically designed by engineers. Engineers have also developed biofilters, bioreactors for removing pollutants from the air. Biofilters are used to remove pollutants, odors, and dust from air by the action of microorganisms. In addition, the mining industry uses bioprocess engineering for extracting minerals such as copper and uranium through the use of bacteria. Microbial leaching uses leaching dumps or tank bioreactors designed by engineers.

Enzymes. Enzymes are used in the health, food, laundry, pulp and paper, and textile industries. They are produced mainly from fungi and bacteria using bioprocess engineering. One of these enzymes is glucose isomerase, important in the production of fructose syrup. Genetic manipulation provides the means to produce many different enzymes, including those not normally synthesized by microorganisms. Fermenters for enzyme production are usually up to 100,000 liters in volume, although very expensive enzymes may be produced in smaller bioreactors, usually with immobilized cells.

Antibiotics and Other Health Care Products. Most antibiotics are produced by fungi and bacteria. Industrial production of antibiotics usually occurs in fermenters (stirred tanks) of 40,000- to 200,000-liter capacity. The bioprocess for antibiotics was developed by engineers during World War II, although it has undergone some changes since the 1980's. Various food sources, including glucose and sucrose, have been adopted for antibiotic production by microorganisms. The modern bioprocess is highly efficient (90 percent). Process variables such as pH and aeration are controlled by computer, and nutrients are fed continuously to sustain maximum antibiotic production. Product recovery is also based on continuous extraction.

The other major health care products produced with the help of bioprocess engineering are steroids, bacterial vaccines, gene therapy vectors, and therapeutic proteins such as interferon, growth hormone, and insulin. Steroids are important hormones that are manufactured by the process of biotransformation, in which microorganisms are used to chemically modify an inexpensive material to create a desired product. Health care products are produced in traditional fermenters.

Biomass Production. Biomass is used as a fuel source, as a source of protein for human food or animal feed, and as a component in agricultural pesticides or fertilizer. Baker's yeast biomass is a major product of bioprocess engineering. It is required for making bread and other baked goods, beer, wine, and ethanol. Yeast is produced in large aerated fermenters of up to 200,000 liters. Molasses is used as a nutrient source for the cells. Yeast is recovered from the fermentation liquid by centrifugation and then is dried. People also use the biomass of algae. Algae are a source of animal feed, plant fertilizer, chemicals, and biofuels. Algal biomass is produced in open ponds, in tubular glass, or in plastic bioreactors.

Animal and Plant Cell Cultures. Bioprocess engineering incorporating animal cell culture is used primarily for the production of health care products such as viral vaccines or antibodies in traditional fermenters or bioreactors with immobilized cells. Antibodies, for example, are produced in bioreactors with hollow-fiber immobilized animal cells. Plant cell culture is also an important target of bioprocess engineering. However, only a few processes have been successfully developed. One successful process is the production of the pigment shikonin in Japan. Shikonin is used as a dye for coloring food and has applications as an anti-inflammatory agent.

Chemicals. There is an on-going trend in the chemical industry to use bioprocess engineering instead of pure chemistry for production of a variety of chemicals such as amino acids, polymers, and organic acids (citric, acetic, and lactic). Some of these chemicals (citric and lactic acids) are used as food preservatives. Many chemicals are produced in traditional fermenters by the action of microbes.

Impact on Industry

Bioprocess engineering plays a major role in many multibillion-dollar industries, including biotechnological, microbiological, food, chemical, and biofuel industries. Because bioprocess engineering is part of several industries, it is difficult to estimate its worldwide revenues. However, global revenues from enzyme production are more than $3 billion, and biofuel industry annual revenues are $46.5 billion.

The United States maintains a dominant global position in a number of industries because of advances created by bioprocess engineering research. The same is true for many European countries, as well as Japan, Israel, Canada, and Australia. Bioprocess engineering has affected industry in many developing countries as well. A good example is Brazil's use of bioprocessing in ethanol production. Brazil, the world's largest ethanol producer for 2010, ferments sugarcane in bioreactors to generate ethanol.

Much of the success of bioprocess engineering is because of the hard work of scientists, engineers, and technicians who have spent countless hours working to improve biological processes to increase the yields of desired products. Scientists are trying to create powerful microorganisms (superbugs) and also working to improve the efficiency of existing production processes and to make them more environmentally friendly. Scientists are studying biochemical processes and enzymes to develop new bioprocesses. In addition, engineers and scientists are designing and developing new types of bioreactors and fermenters.

Government and University Research. Many government agencies such as the U.S. Department of Energy, the National Science Foundation, and the U.S. Department of Agriculture provide funding for research in bioprocess engineering. The Department of Energy has several national laboratories that are involved in bioprocess engineering research. The vast majority of research is on biofuel generation by microorganisms and on environmental applications.

Industry and Business Sectors. Scientists employed by industry traditionally perform most of the research in bioprocess engineering. A significant proportion of the research in industry has been directed to health care or medical products (such as vaccines) and biofuels.

Fascinating Facts About Bioprocess Engineering

- Citric acid, a common supplement of soft drinks, is a major product of bioprocess engineering. It is produced in fermenters by the common mold *Aspergillus niger.*
- Bioprocess engineering is used to recover gold from gold ores. The bioprocess uses bacteria in bioreactors to attack ores, releasing the trapped gold.
- Most insecticides are produced by the genetically modified bacterium *Bacillus thuringiensis* in bioreactors.
- One of the bacteria commonly used to produce antibodies is *Escherichia coli,* largely because so much is known about *E. coli* protein expression and because gene manipulation is relatively easy in this bacterium.
- Lysine, an amino acid added to animal feed, is produced in fermenters using the bacterium *Corynebacterium glutamicum.* About 700,000 tons are produced this way each year.
- Global Cell Solutions and Hamilton has developed a benchtop incubator-bioreactor for high-density three-dimensional cell cultures. This type of bioprocess engineering may enable pharmaceutical companies to test the toxicity of their drugs without using animals.
- Metabolomics, a technique from functional genetics in which all the metabolities in a cell are analyzed and compared, allows scientists to optimize bioprocesses by improving the strains of bacteria and the medium used in fermentation.

Careers and Course Work

There is an increasing demand for students trained in bioprocess engineering who can convert new discoveries in biology into industrial applications. There are many career options for young specialists in bioprocess engineering. Their work may be in the areas of biological process development, manufacturing operations, environmental bioremediation, food technology, or therapeutic stem cell research. They may engage in the development and manufacture of gene therapy vectors, vaccines, or renewable biofuels.

Bioprocess engineering is widely used in industry. Many educational institutions offer bioprocess courses for undergraduates and degrees or concentrations in bioengineering or bioprocess engineering. Several community colleges offer associate degrees and certificate programs that typically prepare students to work in industry. Most of these programs are interdisciplinary. Graduates of these programs will have the knowledge and internship experience to enter directly into the bioprocess engineering workforce. Advanced degrees such as a master's degree or doctorate are necessary to obtain top positions in academia and industry in the bioprocess engineering area. Some universities such as Cornell University offer graduate programs in bioprocess engineering.

The basic courses for students interested in a career in bioprocess engineering are microbiology, plant biology, organic chemistry, biochemistry, agriculture, bioprocess engineering, and chemical engineering. Students must master basic engineering calculations and principles and understand physical and chemical processes including material and energy balances, reactor engineering, fluid flow and mixing, heat and mass transfer, filtration and centrifugation, and chromatography.

Careers in the bioprocess engineering field can take different paths. Biotechnological, microbiological, chemical, and biofuel companies are the biggest employers. People who are interested in research in bioprocess engineering can find jobs in government laboratories and universities. In universities, bioprocess engineers may divide their time between research and teaching.

Social Context and Future Prospects

The role of bioprocess engineering in industry is likely to expand because scientists are increasingly able to manipulate organisms to expand the range and yields of products and processes. Developments in this field continue rapidly.

Bioprocess engineering can potentially be the answer to several problems faced by humankind. One such problem is global warming, which is caused by rising levels of carbon dioxide and other greenhouse gases. A suggested method of addressing this issue is carbon dioxide removal, or sequestration, based on bioprocess engineering. This bioprocess uses microalgae (microscopic algae) in photobioreactors to capture the carbon dioxide that is discharged into the atmosphere by power plants and other industrial facilities. Photobioreactors are various types of closed systems made of an array of transparent tubes in which microalgae are cultivated and monitored under illumination.

The health care industry is another area where bioprocess engineers are likely to be active. For example, if pharmaceutical applications are found for stem cells, a bioprocess must be developed to produce a reliable, plentiful source of stem cells so that these drugs can be produced on a large scale. The process for growing and harvesting cells must be standardized so that the cells have the same characteristics and behave in a predictable manner. Bioprocess engineers must take these processes from laboratory procedures to industrial protocols.

In general, the future of bioprocess engineering is bright, although questions and concerns, primarily about using genetically modified organisms, have arisen. Public education in such a complex area of science is very important to avoid public mistrust of bioprocess engineering, which is very beneficial in most applications.

Sergei A. Markov, Ph.D.

Further Reading

Bailey, James E., and David F. Ollis. *Biochemical Engineering Fundamentals.* 2d ed. New York: McGraw-Hill, 2006. Covers all aspects of biochemical engineering in an understandable manner.

Bougaze, David, Thomas R. Jewell, and Rodolfo G. Buiser. *Biotechnology. Demystifying the Concepts.* San Francisco: Benjamin/Cummings, 2000. Classical book on biotechnology and bioprocessing.

Doran, Pauline M. *Bioprocess Engineering Principles.* London: Academic Press, 2009. A solid, basic textbook for students entering the field.

Glazer, Alexander N., and Hiroshi Nikaido. *Microbial Biotechnology: Fundamentals of Applied Microbiology.* New York: Cambridge University Press, 2007. In-depth analysis of the application of microorganisms in bioprocessing.

Heinzle, Elmar, Arno P. Biwer, and Charles L. Cooney. *Development of Sustainable Bioprocesses: Modeling and Assessment.* Hoboken, N.J.: John Wiley & Sons, 2007. Looks at making bioprocesses sustainable by improving them. Includes case studies on citric acid, biopolymers, antibiotics, and biopharmaceuticals.

Nebel, Bernard J., and Richard T. Wright. *Environmental Science: Towards a Sustainable Future.* 10th ed. Englewood Cliffs: Prentice Hall, 2008. Describes several bioprocesses used in waste treatment and pollution control.

Yang, Shang-Tian. *Bioprocessing for Value-Added Products from Renewable Resources: New Technologies and Applications.* Amsterdam: Elsevier, 2007. Reviews the techniques for producing products through bioprocesses and lists suitable organisms, including bacteria and algae, and describes their characteristics.

Web Sites

Biotechnology Industry Association
http://www.bio.org

International Society for BioProcess Technology
http://www.isbiotech.org

Society for Industrial Microbiology
http://www.simhq.org/index.aspx

U.S. Department of Agriculture
http://usda.gov

U.S. Department of Energy
Bioenergy
http://www.energy.gov/energysources/bioenergy.htm

See also: Proteomics and Protein Engineering

BRIDGE DESIGN AND BARODYNAMICS

FIELDS OF STUDY

Chemistry, civil engineering, construction, material engineering, mechanics, physics, structural engineering.

SUMMARY

Barodynamics is the study of the mechanics of heavy structures that may collapse under their own weight. In bridge building, barodynamics is the science of the support and mechanics of the methods and types of materials used in bridge design to ensure the stability of the structure. Concepts to consider in avoiding the collapse of a bridge are the materials available for use, what type of terrain will hold the bridge, the obstacle to be crossed (such as river or chasm), how long the bridge needs to be to cross the obstacle, what types of natural obstacles or disasters are likely to occur in the area (high winds, earthquakes), the purpose of the bridge (foot traffic, cars, railway), and what type of vehicles will need to cross the bridge.

KEY TERMS AND CONCEPTS

- **Arch Bridge:** Type of bridge where weight is distributed outward along two paths that curve toward the ground in the shape of an arch.
- **Beam Bridge:** Simplest type of bridge consisting of two or more supports holding up a beam; ranges from a complex structure to a plank of wood.
- **Cantilever Bridge:** Bridge type where two beams, well anchored, support another beam.
- **Cofferdam:** A temporary watertight structure that is pumped dry to enclose an area underwater and allow construction work on a bridge in the underwater area.
- **Keystone:** The wedge-shape stone at the highest point of an arch; function is to keep the other stones locked into place through gravity.
- **Pontoon Bridge:** A floating bridge supported by floating objects that contain buoyancy sufficient to support the bridge and any load it must carry; often a temporary structure.
- **Suspension Bridge:** A bridge hung by cables, which, in turn, hang from towers; weight is transferred from the cables to the towers to the ground.
- **Truss Bridge:** A type of bridge supported by a network of beams in triangular sections.

DEFINITION AND BASIC PRINCIPLES

Barodynamics is a key component of any bridge design. Bridges are made of heavy materials, and many concepts, such as tension and compression of building materials, and other factors, such as wind shear, torsion, and water pressure, come into play in bridge building.

Bridge designers and constructors must keep in mind the efficiency (the least amount of material for the highest-level performance) and economy (lowest possible costs while still retaining efficiency) of bridge building. In addition, some aesthetic principles must be followed; public outcry can occur when a bridge is thought to be "ugly." Conversely, a beautiful bridge can become a landmark symbol for an area, such as the Golden Gate Bridge has become for San Francisco.

The four main construction materials for bridges are wood, stone, concrete (including prestressed concrete), and iron (from which steel is made). Wood is nearly always available and inexpensive but is comparatively weak in compression and tension. Stone, another often available material, is strong in compression but weak in tension. Concrete, or "artificial stone," is, like stone, strong in tension and weak in compression.

The first type of iron used in bridges, cast iron, is strong in compression but weak in tension. The use of wrought iron helped in bridge building, as it is strong in compression but still has tensile strength. Steel, a further refinement of iron, is superior in compression and tensile strength, making it a preferred material for bridge building. Reinforced concrete or prestressed concrete (types of concrete with steel bars running through concrete beams) are also popular bridge-building materials because of their strength and lighter-weight design.

BACKGROUND AND HISTORY

From the beginning of their existence, humans have constructed bridges to cross obstacles such as rivers and chasms using the materials at hand, such as trees, stones, or vines. In China during the third century B.C.E., the emperor of the Qin Dynasty built canals to transport goods, but when these canals interfered with existing roads, his engineers built bridges of stone or wood over these canals.

However, the history of barodynamics in bridge building truly begins in Roman times, when Roman engineers perfected the keystone arch. In addition to the keystone and arch concepts, the Romans improved bridge-building materials such as cement and concrete and invented the cofferdam so that underwater pilings could be made for bridges. These engineers built a network of bridges throughout the Roman empire to keep communication with and transportation to and from Rome intact. The Romans made bridges of stone because of its durability, and many of these bridges are still intact today.

In the Middle Ages, bridges were an important part of travel and transportation of goods, and many bridges were constructed during this period to support heavy traffic. This is also the period when people began to live in houses built on bridges, in part because in walled cities, places to build homes were limited. Possibly the most famous inhabited bridge was London Bridge, the world's first stone bridge to be built over a tidal waterway where the water rose and fell considerably every twelve hours. However, in Paris in the sixteenth century, there were at least five inhabited bridges over the Seine to the Île de la Cité.

The first iron bridge was built in 1779. Using this material changed the entire bridge-building industry because of the size and strength of structure that became possible. In the 1870's, a fall in the price of steel made bridges made of this material even more popular, and in 1884, Alexandre Gustave Eiffel, of Eiffel Tower fame, designed a steel arch bridge that let wind pass through it, overcoming many of the structural problems with iron and steel that had previously existed. Iron and steel are still the most common materials to use in bridge building.

Suspension bridges began to be quite popular as they are the most inexpensive way to span a longer distance. In the early 1800's, American engineer John Roebling designed a new method of placing cables on suspension bridges. Famous examples of suspension bridges include the Golden Gate Bridge (completed in 1937) and Roebling's Brooklyn Bridge (completed in 1883).

Girder bridges were often built to carry trains in the early twentieth century. Though capable of carrying heavyweight railroad cars, this type of bridge is usually only built for short distances as is typical with beam-type bridges. In the 1950's, the box girder was designed, allowing air to pass through this type of bridge and making longer girder bridges possible.

How It Works

The engineering principles that must be used to construct even a simple beam bridge are staggering. Supports must be engineered to hold the weight of the entire structure correctly as well as any traffic that will cross the bridge. The bridge itself, or "span," must be strong enough to bear the weight of traffic and stable enough to keep that traffic safe. Spans must be kept as short as reasonably possible but must sometimes be built across long distances, for example, over deep water.

Arch. The Roman arch concept uses the pressure of gravity on the material forming the arch to hold the bridge together with the outward thrust contained by buttresses. It carries loads by compressing and exerting pressure on the foundation, which must be prevented from settling and sliding. This concept allowed bridges to be built that were longer than ever before. For example, a surviving bridge over the Tagus River in Spain has two central arches that are 110 feet wide and 210 feet above the water level. These arches are made of uncemented granite and each keystone weighs eight tons. This type of bridge is constructed by building a huge timber structure to support the bridge during the building phase, then winching blocks into place with a pulley system. After the keystone to the arch is put into place, the scaffolding is removed, leaving the bridge to stand alone.

Beam. This is the most common form of bridge and may be as simple as a log across a stream. This type of bridge carries a load by bending, which horizontally compresses the top of the beam and simultaneously causes horizontal tension on the bottom of the beam.

Truss. A truss bridge is popular because it requires a relatively small amount of construction material to carry a heavy load. It works like a beam bridge, carrying loads by bending and causing compression and tension in the vertical and diagonal supports.

Suspension. Suspension bridges are essentially steel-rope bridges: Thick steel cables are entwined like ropes into a larger and stronger steel cable or

Fascinating Facts About Bridge Design and Barodynamics

- The Tacoma Narrows Bridge across Puget Sound tore apart and fell into the water below in 1940 because of engineering miscalculations concerning vertical and torsional motion. Viewing motion pictures of the disaster helped engineers rethink the aerodynamics of building bridges.
- The I-35W bridge in Minneapolis collapsed in 2007. The deaths and injuries led to a renewed interest in maintenance and retrofitting of public bridges across the United States.
- London Bridge is possibly the most famous bridge in the world. It was the only bridge across the River Thames from the tenth century until the eighteenth century, when the original stone bridge was torn down and rebuilt with stone. It was rebuilt again in 1973 of steel and concrete to accommodate the heavy traffic using the bridge. Robert McCulloch bought the old stone bridge, numbering each stone, and had it rebuilt in Lake Havasu City, Arizona, where it remains a tourist attraction.
- The origin of the word "bridge" goes back to an Old English word "brycg," which is believed to be derived from the German word root "brugj." The name of the Belgian city Brugge can be translated as "a bridge or a place of bridges."
- The Ponte Vecchio in Florence has an added covered walk forming a top story above the shops that line the bridge. This addition was constructed in 1565 so members of the Medici family could walk across the river from the Uffizi to the Palazzo Pitti without descending to the street level and mingling with commoners.
- The world's first iron bridge was built in 1779 over the River Severn in an English industrial area called Coalbrookdale.
- In *Ramayana,* a mythological Indian epic, tales are told of bridges constructed by the army of Sri Rama, a mythological king of Ayodhya, from India to the island of Sri Lanka, a feat thought to have been impossible in that time period. However, space images taken by the National Aeronautics and Space Administration reveal an ancient bridge, called Adam's Bridge, in the Palk Strait with a unique curvature and composition that indicate that it is man-made.
- The Chinese made suspension bridges as early as 200 B.C.E.
- The oldest surviving stone bridge in China, the Zhaozhou Bridge, is thought to have been built during the Sui Dynasty, around the year 600.

rope. These thick, strong cables then suspend the bridge itself between pylons that support the weight of the bridge. A suspension bridge can be thought of as an upside-down arch, as the curved cables use tension and compression to support the load.

Cantilever. Cantilevered means something that projects outward and is supported at only one end (similar to a diving board). This type of bridge is generally made with three spans with the outside spans supported on the shore and the middle span supported by the outside spans. This is a type of beam bridge that uses tension in the lower spans and compression in the upper span to carry a load.

Pontoon. A pontoon bridge is built across water with materials that float. Each pontoon, or floating object, can support a maximum load equal to the amount of water it displaces. If the load placed on one pontoon-supported section exceeds the water displaced, the pontoon will submerge and cause the entire bridge to sink.

Applications and Products

Bridges are continuously being built to cross physical obstacles, and as the nature of materials changes, the ability to cross even larger obstacles becomes reality. Nature is the defining force on a bridge; most bridges fail because of flooding or other natural disasters.

Improvements in building materials are ongoing. For example, the Jakway Park Bridge in Buchanan County, Iowa, was the first bridge in North America to be built with ultrahigh performance concrete (UHPC) with pi-girders. This moldable material combines high compressive strength and flexibility and offers a wide range of design possibilities. It is very durable and has low permeability.

Bridge-building products may even be developed that help the environment. For example, the rebuilt I-35W bridge in Minnesota uses a concrete that is said to "eat smog." The concrete contains photo-catalytic titanium dioxide, which accelerates the decomposition of organic material. Other materials like this

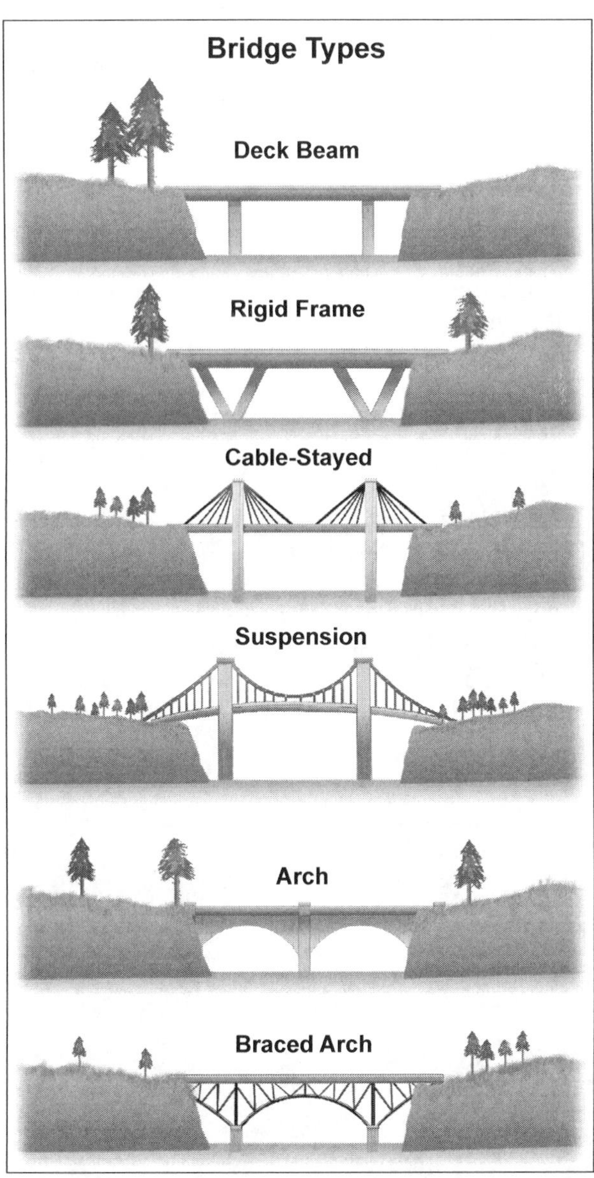

In describing the design of a bridge, there are four primary factors: span, placement, material, and form.

may change the future of bridge building.

Impact on Industry

Bridges have a significant impact on travel and transportation of goods. They are, generally, made for the public, for transportation or travel, and are often built with public funds. Therefore, bridge building has its greatest impact in the governmental or public-transportation areas. Bridge engineers often find jobs with government agencies, such as the U.S. Department of Transportation, or with private companies that are subcontractors on government projects.

Careers and Course Work

Those who engineer and design bridges may have backgrounds in a variety of fields, including architecture and design. However, those who are involved in the barodynamic aspects of bridge building are engineers, usually either civil, materials, or mechanical engineers. Earning a degree in one of these fields is required to get the training needed in geology, math, and physics to learn about the physical limitations and considerations in bridge building. Many bridge engineers have advanced degrees in a specific related field. After earning a degree, a candidate for this type of job usually works for a few years as an assistant engineer in a sort of apprenticeship, learning the specifics of bridge building such as drafting, blueprint reading, surveying, and stabilization of materials. To become a professional engineer (PE), one must then take a series of written exams to get his or her license.

Social Context and Future Prospects

Barodynamics is a rapidly changing field. As new materials are created and existing materials change, the possibilities for future improvements in this field increase. Development of future lightweight materials may change the way bridges are engineered and designed to make the structure stable and avoid collapse. Just as innovations in iron refinement changed the face of bridge building in the late 1800's, new materials may refine and improve bridge building even further, bringing more efficient and economical bridges.

Materials engineers are usually the people who provide the technological innovation to create these kinds of new materials. They examine materials on the molecular level to understand how materials can be improved and strengthened in order to provide better building materials for structures such as bridges. Possible future bridge-building materials include ceramics, polymers, and other composites. Two other rapidly growing materials fields that may affect bridge barodynamics are biomaterials and nanomaterials.

Marianne M. Madsen, M.S.

Further Reading

Blockley, David. *Bridges: The Science and Art of the World's Most Inspiring Structures.* New York: Oxford University Press, 2010. Written by a professor of engineering with a lay reader in mind; discusses basic forces such as tension, compression, and shear, and bridge failures. Includes a comprehensive history of bridge building with fifty illustrations.

Chen, Wai-Fah, and Lian Duan, eds. *Bridge Engineering Handbook.* Boca Raton, Fla.: CRC Press, 1999. Contains more than 1,600 tables, charts, and illustrations with step-by-step design procedures. Covers fundamentals, superstructure design, substructure design, seismic design, construction and maintenance, and worldwide practice; includes a special topics section.

Haw, Richard. *Art of the Brooklyn Bridge: A Visual History.* New York: Routledge, 2007. A visually interesting compilation of artists' renderings of the Brooklyn Bridge, contributing to the idea that bridges are artful as well as functional.

Tonias, Demetrios E., and Jim J. Zhao. *Bridge Engineering: Design, Rehabilitation, and Maintenance of Modern Highway Bridges.* 2d ed. New York: McGraw-Hill, 2007. Details the entire highway-bridge design process; includes information on design codes.

Unsworth, John F. *Design of Modern Steel Railway Bridges.* Boca Raton, Fla.: CRC Press, 2010. Focuses on new steel superstructures for railway bridges but also contains information on maintenance and rehabilitation and a history of existing steel railway bridges.

Van Uffelen, Chris. *Masterpieces: Bridge Architecture and Design.* Salenstein, Switzerland: Braun Publishing, 2010. Includes photos of sixty-nine bridges from around the world, displaying a variety of structures and materials.

Yanev, Bojidar. *Bridge Management.* Hoboken, N.J.: John Wiley & Sons, 2007. Contains case studies of bridge building and design; discusses bridge design, maintenance, and construction with topics such as objectives, tools, practices, and vulnerabilities.

Web Sites

American Society of Civil Engineers
http://www.asce.org

Design-Build Institute of America
http://www.dbia.org

National Society of Professional Engineers
http://www.nspe.org/index.html

See also: Civil Engineering; Earthquake Engineering; Structural Composites.

C

CALCULUS

FIELDS OF STUDY

Algebra; geometry; trigonometry.

SUMMARY

Calculus is the study of functions and change. It is the bridge between the elementary mathematics of algebra, geometry, and trigonometry, and advanced mathematics. Knowledge of calculus is essential for those pursuing study in fields such as chemistry, engineering, medicine, and physics. Calculus is employed to solve a large variety of optimization problems; one example is the so-called least squares solution method commonly used in statistics and elsewhere. The least squares function best fits a set of data points and can then be used to generalize or predict results based on that set.

KEY TERMS AND CONCEPTS

- **Antiderivative:** Function whose derivative is equal to that of a given function.
- **Continuity:** Characteristic manifested by a function when its output values are equal to the values of its limits.
- **Converge:** Action of an improper integral or series with a finite value.
- **Definite Integral:** Limit of a Riemann sum as the number of terms approaches infinity.
- **Derivative:** Function derived from a given function by means of the limit process, whose output equals the instantaneous rates of change of the given function.
- **Diverge:** Action of an improper integral or series with no finite value.
- **Gradient:** Vector whose components are each of the partial derivatives of a function of several variables.
- **Indefinite Integral:** Antiderivative of a given function.
- **Limit:** Number that the output values of a function approach the closer the input values of the function get to a specified target.
- **Riemann Sum:** Sum of the products of functional values and the lengths of the subintervals over which the function is defined.
- **Series:** Formal sum of an infinite number of terms; it may be convergent or divergent.
- **Taylor Series:** Series whose sums equal the output values of a given function, at least along some interval of input values.

DEFINITION AND BASIC PRINCIPLES

Calculus is the study of functions and their properties. Calculus takes a function and investigates it according to two essential ideas: rate of change and total change. These concepts are linked by their common use of calculus's most important tool, the limit. It is the use of this tool that distinguishes calculus from the branches of elementary mathematics: algebra, geometry, and trigonometry. In elementary mathematics, one studies problems such as "What is the slope of a line?" or "What is the area of a parallelogram?" or "What is the average speed of a trip that covers three hundred miles in five and a half hours?" Elementary mathematics provides methods or formulas that can be applied to find the answer to these and many other problems. However, if the line becomes a curve, how is the slope calculated? What if the parallelogram becomes a shape with an irregularly curved boundary? What if one needs to know the speed at an instant, and not as an average over a longer time period?

Calculus answers these harder questions by using the limit. The limit is found by making an approximation to the answer and then refining that approximation by improving it more and more. If there is a pattern leading to a single value in those improved approximations, the result of that pattern is called the limit. Note that the limit may not exist in some cases. The limit process is used throughout calculus to provide answers to questions that elementary mathematics cannot handle.

The derivative of a function is the limit of average slope values within an interval as the length of the interval approaches zero. The integral calculates the total change in a function based on its rate of change function.

BACKGROUND AND HISTORY

Calculus is usually considered to have come into being in the seventeenth century, but its roots were formed much earlier. In the sixteenth century, Pierre de Fermat did work that was very closely related to calculus's differentiation (the taking of derivatives) and integration. In the seventeenth century, René Descartes founded analytic geometry, a key tool for developing calculus.

However, it is Sir Isaac Newton and Gottfried Wilhelm Leibniz who share the credit as the (independent) creators of calculus. Newton's work came first but was not published until 1736, nine years after his death. Leibniz's work came second but was published first, in 1684. Some accused him of plagiarizing Newton's work, although Leibniz arrived at his results by using different, more formal methods than Newton employed.

Both men found common rules for differentiation, but Leibniz's notation for both the derivative and the integral are still in use. In the eighteenth century, the work of Jean le Rond d'Alembert and Leonhard Euler on functions and limits helped place the methods of Newton and Leibniz on a firm foundation. In the nineteenth century, Augustin-Louis Cauchy used a definition of limit to express calculus concepts in a form still familiar more than two hundred years later. German mathematician Georg Riemann defined the integral as a limit of a sum, the same definition learned by calculus students in the twenty-first century. At this point, calculus as it is taught in the first two years of college reached its finished form.

HOW IT WORKS

Calculus is used to solve a wide variety of problems using a common approach. First, one recognizes that the problem at hand is one that cannot be solved using elementary mathematics alone. This recognition is followed by an acknowledgment: There are some things known about this situation, even if they do not provide a complete basis for solution. Those known properties are then used to approximate a solution to the problem. This approximation may not be very good, so it is refined by taking a succession of better and better approximations. Finally, the limit is taken, and if the limit exists, it provides the exact answer to the original problem.

One speaks of taking a limit of a function $f(x)$ as x approaches a particular value, for example, $x = a$. This means that the function is examined on an interval around, but not including $x = a$. Values of $f(x)$ are taken on that interval as the varying x values get closer and closer to the target value of $x = a$. There is no requirement that $f(a)$ exists, and many times it does not. Instead the pattern of functional values is examined as x approaches a. If those values continue to approach a single target value, it is that value that is said to be equal to the limit of $f(x)$ as x approaches a. Otherwise, the limit is said not to exist. This method is used in both differential calculus and integral calculus.

Differential Calculus. Differentiation is a term used to mean the process of finding the derivative of a function $f(x)$. This new function, denoted *f*¢(*x*), is said to be "derived" from $f(x)$. If it exists, *f*¢(*x*) provides the instantaneous rate of change of $f(x)$ at x. For curves (any line other than a straight line), the calculation of this rate of change is not possible with elementary mathematics. Algebra is used to calculate that rate between two points on the graph, then those two points are brought closer and closer together until the limit determines the final value.

Shortcut methods were discovered that could speed up this limit process for functions of certain types, including products, quotients, powers, and trigonometric functions. Many of these methods go back as far as Newton and Leibniz. Using these formulas allows one to avoid the more tedious limit calculations. For example, the derivative function of sine x is proven to be cosine x. If the slope of sine x is needed at $x = 4$, the answer is known to be cosine 4, and much time is saved.

Integral Calculus. A natural question arises: If *f*¢(*x*) can be derived from $f(x)$, can this process be reversed? In other words, suppose an $f(x)$ is given. Can an $F(x)$ be determined whose derivative is equal to $f(x)$? If so, the $F(x)$ is called an antiderivative of $f(x)$; the process of finding $F(x)$ is called integration. In general, finding antiderivatives is a harder task than finding derivatives. One difficulty is that constant functions all have derivatives equal to zero, which means that without further information, it is impossible to determine which constant is the correct one. A bigger problem is that there are functions, such as sine (x^2), whose derivatives are reasonably easy to calculate but for which no elementary function serves as an antiderivative.

The definite integral can be thought of as an attempt to determine the amount of area between the graph of $f(x)$ and the x-axis, usually between a left and right endpoint. This cannot typically be answered using elementary mathematics because the shape of the graph can vary widely. Riemann proposed approximating the area with rectangles and then improving the approximation by having the width of the rectangles used in the approximation get smaller and smaller. The limit of the total area of all rectangles would equal the area being sought. It is this notion that gives integral calculus its name: By summing the areas of many rectangles, the many small areas are integrated into one whole area.

As with derivatives, these limit calculations can be quite tedious. Methods have been discovered and proven that allow the limit process to be bypassed. The crowning achievement of the development of calculus is its fundamental theorem: The derivative of a definite integral with respect to its upper limit is the integrand evaluated at the upper limit; the value of a definite integral is the difference between the values of an antiderivative evaluated at the limits. If one is looking for the definite integral of a continuous $f(x)$ between $x = a$ and $x = b$, one need only find any antiderivative $F(x)$ and calculate $F(b) - F(a)$.

APPLICATIONS AND PRODUCTS

Optimization. A prominent application of the field of differential calculus is in the area of optimization, either maximization or minimization. Examples of optimization problems include What is the surface area of a can that minimizes cost while containing a specified volume? What is the closest that a passing asteroid will come to Earth? What is the optimal height at which paintings should be hung in an art gallery? (This corresponds to maximizing the viewing angle of the patrons.) How shall a business minimize its costs or maximize its profits?

All of these can be answered by means of the derivative of the function in question. Fermat proved that if $f(x)$ has a maximum or minimum value within some interval, and if the derivative function exists on that interval, then the derivative value must be zero. This is because the graph must be hitting either a peak or the bottom of a valley and has a slope of zero at its highest or lowest points. The search for optimal values then becomes the process of finding the correct function modeling the situation in question, finding its derivative, setting that derivative equal to zero, and solving. Those solutions are the only candidates for optimal values. However, they are only candidates because derivatives can sometimes equal zero even if no optimal value exists. What is certain is that if the derivative value is not zero, the value is not optimal.

The procedure discussed here can be applied in two dimensions (where there is one input variable) or three dimensions (where there are two input variables).

Surface Area and Volume. If a three-dimensional object can be expressed as a curve that has been rotated about an axis, then the surface area and volume of the object can be calculated using integrals. For example, both Newton and Johannes Kepler studied the problem of calculating the volume of a wine barrel. If a function can be found that represents the curvature of the outside of the barrel, that curve can be rotated about an axis and pi (p) times the function squared can be integrated over the length of the barrel to find its volume.

Hydrostatic Pressure and Force. The pressure exerted on, for example, the bottom of a swimming pool of uniform depth is easily calculated. The force on a dam due to hydrostatic pressure is not so easily computed because the water pushes against it at varying depths. Calculus discovers the answer by integrating a function found as a product of the pressure at any depth of the water and the area of the dam at that depth. Because the depth varies, this function involves a variable representing that depth.

Arc Length. Algebra is able to determine the length of a line segment. If that path is curved, whether in two or three dimensions, calculus is applied to determine its length. This is typically done by expressing the path in parametric form and integrating the function representing the length of the vector that is tangent to the path. The length of a path winding through three-dimensional space, for example, can be determined by first expressing the path in the parametric form $x = f(t)$, $y = g(t)$, and $z = h(t)$, in which f, g, and h are continuous functions defined for some interval of values of t. Then the square root of the sum of the squares of the three derivatives is integrated to find the length.

Kepler's Laws. In the early seventeenth century,

Kepler formulated his three laws of planetary motion based on his analysis of the observations kept by Tycho Brahe. Later, calculus was used to prove that these laws are correct. Kepler's laws state that any planet's orbit around the Sun is elliptical, with the Sun at one focus of the ellipse; that the line joining the Sun to the planet sweeps out equal areas in equal times; and that the square of the period of revolution is proportional to the cube of the length of the major axis of the orbit.

Probability. Accurate counting methods can be sufficient to determine many probabilities of a discrete random variable. This would be a variable whose values could be, for example, counting numbers such as 1, 2, 3, and so on, but not numbers inbetween, such as 2.4571. If the random variable is continuous, so that it can take on any real number within an interval, then its probability density function must be integrated over the relevant interval to determine the probability. This can occur in two or three dimensions.

One common example is determining the likelihood that a customer's wait time is longer than a specified target, such as ten minutes. If the manager knows the average wait time that a customer experiences at an establishment is, for example, six minutes, then this time can be used to determine a probability density function. This function is integrated to determine the probability that a person's wait time will be longer than ten minutes, less than three minutes, between five and thirteen minutes, or within any range of times that is desired.

Impact on Industry

The study of calculus is the foundation of more advanced work in mathematics, engineering, economics, and many areas of science. In some cases, it is these other disciplines that affect industry, but much of the time, the effect of calculus can be seen directly. Businesses, government, and industry throughout the developed world continue to apply calculus in a wide variety of settings.

Government and University Research. Knowing the importance of calculus in many fields, the National Science Foundation funded many projects in the 1990's that were designed to renew and refresh calculus education. The foundation's Division of Mathematical Sciences continues to fund projects related to both education and research. Projects funded in 2009 included research on the suspension of aerosol particles in the atmosphere, including relating these concepts to the teaching of calculus, and research in the area of partial differential equations and their applications.

Industry and Business. Many branches of engineering use calculus methods and results. In chemical engineering, knowledge of vector calculus and Taylor polynomials is particularly important. In electrical engineering, these same topics, together with an understanding of integration techniques, are emphasized. Mechanical engineering makes significant use of vector calculus and the solving of differential

Fascinating Facts About Calculus

- Archimedes, who lived in the third century B.C.E., derived the formula for the volume of a sphere by using a method that foreshadowed integration. This was understood only when Archimedes's explanation of his method was discovered on a palimpsest in 1906.
- The controversies as to whether Sir Isaac Newton or Gottfried Wilhelm Leibniz (or both) should be credited as the founders of calculus and whether Leibniz stole Newton's ideas led to Leibniz's disgrace. In fact, his secretary was the only mourner to attend Leibniz's funeral.
- The geometric figure known as Gabriel's horn or Toricelli's trumpet is found by revolving the curve $1/x$ about the x-axis, beginning at $x = 1$. Calculus shows that this object, an infinitely long horn shape, has an infinite surface area but only finite volume.
- The Bernoulli brothers, Jakob and Johann, proved that a chain hanging from two points has the shape of a catenary, not a parabola, and that of all possible shapes, the catenary has the lowest center of gravity and thus the minimal potential energy.
- The logical foundations of calculus were not established until well after the methods themselves had been employed. One of the critics who spurred this development was George Berkeley, a bishop of the Church of Ireland, who accused mathematicians of accepting calculus as a matter of faith, not of science.
- Calculus, combined with probability, is used in the pricing, construction, and hedging of derivative securities for the financial market.

equations. The latter are often solved by numerical procedures such as Euler's method or the Runge-Kutta method when exact solutions are either impossible or impractical to obtain.

In finance, series are used to find the present value of revenue streams with an unlimited number of perpetuities. Calculus is also used to model and calculate levels of risk and benefit for investment schemes. In economics, calculus methods find optimal levels of production based on cost and revenue functions.

Careers and Course Work

A person preparing for a career involving the use of calculus will most likely graduate from a university with a degree in mathematics, physics, actuarial science, statistics, or engineering. In most cases, engineers and actuaries are able to join the profession after earning their bachelor's degree. For actuaries, passing one or more of the exams given by the Society of Actuaries or the Casualty Actuarial Society is also expected, which requires a thorough understanding of calculus.

In statistics, a master's degree is typically preferred, and to work as a physicist or mathematician, a doctorate is the standard. In terms of calculus-related course work, in addition to the calculus sequence, students will almost always take a course in differential equations and perhaps one or two in advanced calculus or mathematical analysis.

In 2008, about 428,000 people were working as either chemical, electrical, or mechanical engineers in the United States, most in either the manufacturing or service industries. Private industry and the government employed 23,000 statisticians. Insurance carriers, brokers, agents, or other offices provided jobs for 20,000 actuaries. About 15,000 physicists worked as researchers for private industry or the government. The Bureau of Labor Statistics counted only 3,000 individuals as simply mathematicians but tallied 55,000 mathematicians who teach at the college or university level. As of 2010, all these careers were projected to have a job growth rate at or above the national average.

Social Context and Future Prospects

Calculus itself is not an industry, but it forms the foundation of other industries. In this role, it continues to power research and development in diverse fields, including those that depend on physics. Physics derives its results by way of calculus techniques. These results in turn enable developments in small- and large-scale areas. An example of a small-scale application is the ongoing development of semiconductor chips in the field of electronics. Large-scale applications are in the solar and space physics critical for ongoing efforts to explore the solar system. These are just two examples of calculus-based fields that will continue to have significant impact in the twenty-first century.

Michael J. Caulfield, B.S., M.S., Ph.D.

Further Reading

Bardi, Jason Socrates. *The Calculus Wars: Newton, Leibniz, and the Greatest Mathematical Clash of All Time.* New York: Thunder's Mouth Press, 2006. Examines the controversy over who should be considered the originator of calculus.

Dunham, William. *The Calculus Gallery: Masterpieces from Newton to Lebesgue.* Princeton, N.J.: Princeton University Press, 2005. Focuses on thirteen individuals and their notable contributions to the development of calculus.

Kelley, W. Michael. *The Complete Idiot's Guide to Calculus.* 2d ed. Indianapolis: Alpha, 2006. Begins by examining what calculus is, then leads the reader through calculus basics.

Simmons, George F. *Calculus Gems: Brief Lives and Memorable Mathematics.* 1992. Reprint. Washington, D.C.: The Mathematical Association of America, 2007. Includes dozens of biographies from ancient times to the nineteenth century, together with twenty-six examples of the remarkable achievements of these people.

Stewart, James. *Essential Calculus.* Belmont, Calif.: Thomson Brooks/Cole, 2007. A standard text that relates all of the concepts and methods of calculus, including examples and applications.

Web Sites

American Mathematical Society
http://www.ams.org

Mathematical Association of America
http://www.maa.org

Society for Industrial and Applied Mathematics
http://www.siam.org

See also: Applied Mathematics; Engineering Mathematics; Numerical Analysis.

CHAOTIC SYSTEMS

FIELDS OF STUDY

Mathematics; chemistry; physics; ecology; systems biology; engineering; theoretical physics; theoretical biology; quantum mechanics; fluid mechanics; astronomy; sociology; psychology; behavioral science.

SUMMARY

Chaotic systems theory is a scientific field blending mathematics, statistics, and philosophy that was developed to study systems that are highly complex, unstable, and resistant to exact prediction. Chaotic systems include weather patterns, neural networks, some social systems, and a variety of chemical and quantum phenomena. The study of chaotic systems began in the nineteenth century and developed into a distinct field during the 1980's.

Chaotic systems analysis has allowed scientists to develop better prediction tools to evaluate evolutionary systems, weather patterns, neural function and development, and economic systems. Applications from the field include a variety of highly complex evaluation tools, new models from computer and electrical engineering, and a range of consumer products.

KEY TERMS AND CONCEPTS

- **Aperiodic:** Of irregular occurrence, not following a regular cycle of behavior.
- **Attractor:** State to which a dynamic system will eventually gravitate—a state that is not dependent on the starting qualities of the system.
- **Butterfly Effect:** Theory that small changes in the initial state of a system can lead to pronounced changes that affect the entire system.
- **Complex System:** System in which the behavior of the entire system exhibits one or more properties that are not predictable, given a knowledge of the individual components and their behavior.
- **Dynamical System:** Complex system that changes over time according to a set of rules governing the transformation from one state to the next.
- **Nonlinear:** Of a system or phenomenon that does not follow an ordered, linear set of cause-and-effect relationships.
- **Randomness:** State marked by no specific or predictable pattern or having a quality in which all possible behaviors are equally likely to occur.
- **Static System:** System that is without movement or activity and in which all parts are organized in an ordered series of relationships.
- **System:** Group interrelated by independent units that together make up a whole.
- **Unstable:** State of a system marked by sensitivity to minor changes in a variety of variables that cause major changes in system behavior.

Definition and Basic Principles

Chaotic systems analysis is a way of evaluating the behavior of complex systems. A system can be anything from a hurricane or a computer network to a single atom with its associated particles. Chaotic systems are systems that display complex nonlinear relationships between components and whose ultimate behavior is aperiodic and unstable.

Chaotic systems are not random but rather deterministic, meaning that they are governed by some overall equation or principle that determines the behavior of the system. Because chaotic systems are determined, it is theoretically possible to predict their behavior. However, because chaotic systems are unstable and have so many contributing factors, it is nearly impossible to predict the system's behavior. The ability to predict the long-term behavior of complex systems, such as the human body, the stock market, or weather patterns, has many potential applications and benefits for society.

The butterfly effect is a metaphor describing a system in which a minor change in the starting conditions can lead to major changes across the whole system. The beat of a butterfly's wing could therefore set in motion a chain of events that could change the universe in ways that have no seeming connection to a flying insect. This kind of sensitivity to initial conditions is one of the basic characteristics of chaotic systems.

Background and History

French mathematician Henri Poincaré and Scottish theoretical physicist James Clerk Maxwell are considered two of the founders of chaos theory and complexity studies. Maxwell and Poincaré both worked on problems that illustrated the sensitivity of complex systems in the late nineteenth century. In 1960, meteorologist Edward Lorenz coined the term "butterfly effect" to describe the unstable nature of weather patterns.

By the late 1960's, theoreticians in other disciplines began to investigate the potential of chaotic dynamics. Ecologist Robert May was among the first to apply chaotic dynamics to population ecology models, to considerable success. Mathematician Benoit Mandelbrot also made a significant contribution in the mid-1970's with his discovery and investigation of fractal geometry.

In 1975, University of Maryland scientist James A. Yorke coined the term "chaos" for this new branch of systems theory. The first conference on chaos theory was held in 1977 in Italy, by which time the potential of chaotic dynamics had gained adherents and followers from many different areas of science.

Over the next two decades, chaos theory continued to gain respect within the scientific community and the first practical applications of chaotic systems analysis began to appear. In the twenty-first century, chaotic systems have become a respected and popular branch of mathematics and system analysis.

How It Works

There are many kinds of chaotic systems. However, all chaotic systems share certain qualities: They are unstable, dynamic, and complex. Scientists studying complex systems generally focus on one of these qualities in detail. There are two ultimate goals for complex systems research: first, to predict the evolution of chaotic systems and second, to learn how to manipulate complex systems to achieve some desired result.

Instability. Chaotic systems are extremely sensitive to changes in their environment. Like the example of the butterfly, a seemingly insignificant change can become magnified into systemwide transformations. Because chaotic systems are unstable, they do not settle to equilibrium and can therefore continue developing and leading to unexpected changes. In the chaotic system of evolution, minor changes can lead to novel mutations within a species, which may eventually give rise to a new species. This kind of innovative transformation is what gives chaotic systems a reputation for being creative.

Scientists often study complexity and chaos by creating simulations of chaotic systems and studying the way the systems react when perturbed by minor stimuli. For example, scientists may create a computer model of a hurricane and alter small variables, such as temperature, wind speed and other factors, to study the ultimate effect on the entire storm system.

Strange Attractors. An attractor is a state toward which a system moves. The attractor is a point of equilibrium, where all the forces acting on a system have reached a balance and the system is in a state of rest.

Because of their instability, chaotic systems are less likely to reach a stable equilibrium and instead proceed through a series of states, which scientists sometimes call a dynamic equilibrium. In a dynamic equilibrium, the system is constantly moving toward some attractor, but as it moves toward the first attractor, forces begin building that create a second attractor. The system then begins to shift from the first to the second attractor, which in turn leads to the formation of a new attractor, and the system shifts again. The forces pulling the system from one attractor to the next never balance each other completely, so the system never reaches absolute rest but continues changing.

Visual models of complex systems and their strange attractors display what mathematicians call fractal geometry. Fractals are patterns that, like chaotic systems, are nonlinear and dynamic. Fractal geometry occurs throughout nature, including in the formation of ice crystals and the branching of the circulatory system in the human body. Scientists study the mathematics behind fractals and strange attractors to find patterns that can be applied to complex systems in nature. The study of fractals has yielded applications for medicine, economics, and psychology.

Emergent Properties. Chaotic systems also display emergent properties, which are system-wide behaviors that are not predictable from knowledge of the individual components. This occurs when simple behaviors among the components combine to create more complex behaviors. Common examples in nature include the behavior of ant colonies and other social insects.

Mathematic models of complex systems also yield emergent properties, indicating that this property is driven by the underlying principles that lead to the creation of complexity. Using these principals, scientists can create systems, such as computer networks, that also display emergent properties.

Applications and Products

Medical Technology. The human heart is a chaotic system, and the heartbeat is controlled by a set of nonlinear, complex variables. The heartbeat appears periodic only when examined through an imprecise measure such as a stethoscope. The actual signals that compose a heartbeat occur within a dynamic equilibrium.

Applying chaos dynamics to the study of the heart is allowing physicians to gain more accuracy in determining when a heart attack is imminent, by detecting minute changes in rhythm that signify the potential for the heart to veer away from its relative rhythm. When the rhythm fluctuates too far, physicians use a defibrillator to shock the heart back to its rhythm. A new defibrillator model developed by scientists in 2006 and 2007 uses a chaotic electric signal to more effectively force the heart back to its normal rhythmic pattern.

Consumer Products. In the mid-1990's, the Korean company Goldstar manufactured a chaotic dishwasher that used two spinning arms operated with an element of randomness to their pattern. The chaotic jet patterns were intended to provide greater cleaning power while using less energy.

Chaotic engineering has also been used in the manufacture of washing machines, kitchen mixers, and a variety of other simple machines. Although most chaotic appliances have been more novelty than innovation, the principles behind chaotic mechanics have become common in the engineering of computers and other electrical networks.

Business and Industry. The global financial market is a complex, chaotic system. When examined mathematically, fluctuations in the market appear aperiodic and have nonlinear qualities. Although the market seems to behave in random ways, many believe that applying methods created for the study of chaotic systems will allow economists to elucidate hidden developmental patterns. Knowledge of these patterns might help predict and control recessions and other major changes well before they occur.

The application of chaos theory to market analysis produced a subfield of economics known as fractal market analysis, wherein researchers conduct economic analyses by using models with fractal geometry. By looking for fractal patterns and assuming that the market, like other chaotic systems, is highly sensitive to small changes, economists have been able to build more accurate models of market evolution.

Fascinating Facts About Chaotic Systems

- Chaos theory gained popular attention after the 1993 film adaptation of Michael Crichton's novel *Jurassic Park* (1990). In the film, a scientist played by Jeff Goldblum evokes chaos theory to warn that complex systems are impossible to control.
- Fractal image compression uses the fractal property of self similarity to compress images.
- Chaotic systems can be found at both the largest and the smallest levels of the universe, from the evolution of galaxies to molecular activity.
- Fractal geometry is commonly found in the development of organisms. Among plants with fractal organization are broccoli, ferns, and many kinds of flowers.
- Some philosophers and neuroscientists have suggested that the creative properties of complex systems are what allow for free will in the human mind.
- The Korean company Daewoo marketed a washing machine known as the "bubble machine" in 1990, which the company claimed was the first consumer device to use chaos theory in its operation.

Impact on Industry

The fact that chaos theory applies to so many fields means that there are numerous funding opportunities available. Many countries offer some government funding for studies into chaos theory and related fields.

In the United States, the National Science Foundation offers a grant for researchers studying theoretical biology, which has been awarded to several research teams studying chaotic systems. The National Science Foundation, National Institute of Standards and Technology, National Institutes of Health, and the Department of Defense have also provided some funding for research into chaotic dynamics.

Any product created using chaotic principles may be subject to regulation governing the industry in question. For instance, the chaotic defibrillator is subject to regulation from the Food and Drug Administration, which has oversight over any new medical technology. Other equipment, such as the chaotic dishwasher, will be subject to consumer safety regulations at the federal and regional levels.

Careers and Course Work

Those seeking careers in chaotic systems analysis might begin by studying mathematics or physics at a university. Those with backgrounds in other fields such as biology, ecology, economics, sociology, and computer science might also choose to focus on chaotic systems during their graduate training.

There are a number of graduate programs offering training in chaotic dynamics. For instance, the Center for Interdisciplinary Research on Complex Systems at Northeastern University in Boston, Massachusetts, offers programs training students in many types of complex system analyses. The Center for Nonlinear Phenomena and Complex Systems at the Université Libre de Bruxelles offers programs in thermodynamics and statistical mechanics.

Careers in chaotic systems span a range of fields from engineering and computer network design to theoretical physics. Trained researchers can choose to contribute to academic analyses and other pertinent laboratory work or focus on creating applications for immediate consumer use.

Social Context and Future Prospects

As chaotic systems analysis spread in the 1980's and 1990's, some scientists began theorizing that chaotic dynamics might be an essential part of the search for a grand unifying theory or theory of everything. The grand unifying theory is a concept that emerged in the early nineteenth century, when scientists began theorizing that there might be a single set of rules and patterns underpinning all phenomena in the universe.

The idea of a unified theory is controversial, but the search has attracted numerous theoreticians from mathematics, physics, theoretical biology, and philosophy. As research began to show that the basic principles of chaos theory could apply to a vast array of fields, some began theorizing that chaos theory was part of the emerging unifying theory.

Because many systems meet the basic requirements to be considered complex chaotic systems, the study of chaos theory and complexity has room to expand. Scientists and engineers have only begun to explore the practical applications of chaotic systems, and theoreticians are still attempting to evaluate and study the basic principles behind chaotic system behavior.

Micah L. Issitt, B.S.

Further Reading

Ford, Kenneth W. *The Quantum World: Quantum Physics for Everyone.* Cambridge, Mass.: Harvard University Press, 2005. An exploration of quantum theory and astrophysics written for the general reader. Contains coverage of complexity and chaotic systems applied to quantum phenomena.

Gleick, James L. *Chaos: Making a New Science.* Rev. ed. New York: Penguin Books, 2008. Technical popular account of the development of chaos theory and complexity studies. Topics covered include applications of chaos systems analysis to ecology, mathematics, physics, and psychology.

Gribbin, John. *Deep Simplicity: Bringing Order to Chaos and Complexity.* New York: Random House, 2005. An examination of how chaos theory and related fields have changed scientific understanding of the universe. Provides many examples of complex systems found in nature and human culture.

Mandelbrot, Benoit, and Richard L. Hudson. *The Misbehavior of Markets: A Fractal View of Financial Turbulence.* New York: Basic Books, 2006. Provides a detailed examination of fractal patterns and chaotic systems analysis in the theory of financial markets. Provides examples of how chaotic analysis can be used in economics and other areas of human social behavior.

Stewart, Ian. *Does God Play Dice? The New Mathematics of Chaos.* 2d ed. 1997. Reprint. New York: Hyperion, 2005. An evaluation of the role that order and chaos play in the universe through popular explanations of mathematic problems. Includes accessible descriptions of complex mathematical ideas underpinning chaos theory.

Strogatz, Peter H. *Sync: How Order Emerges from Chaos in the Universe, Nature and Daily Life.* 2003. Reprint. New York: Hyperion Books, 2008. Strogatz provides an accessible account of many aspects of complex systems, including chaos theory, fractal organization, and strange attractors.

Web Sites

Society for Chaos Theory in Psychology and Life Sciences
http://www.societyforchaostheory.org

University of Maryland
Chaos Group at Maryland
http://www-chaos.umd.edu

See also: Applied Mathematics; Applied Physics; Climate Modeling; Electrical Engineering; Fractal Geometry; Risk Analysis and Management.

CHARGE-COUPLED DEVICES

FIELDS OF STUDY

Physics; electrical engineering; electronics; optics; mathematics; semiconductor manufacturing; solid-state physics; chemistry; mechanical engineering; photography; computer science; computer programming; signal processing.

SUMMARY

Originally conceptualized as a form of optical volatile memory for computers, charge-coupled devices were also seen to be useful for capturing light and images. The ability to capture images that can be read by a computer in digital form was profoundly useful, and charge-coupled devices quickly become known for their imaging abilities. Besides their utility in producing digital images that could be analyzed and manipulated by computers, charge-coupled devices also proved to be more sensitive to light than film. As technology advanced, charge-coupled devices became less expensive and more capable, and eventually they replaced film as the primary means of taking photographs.

KEY TERMS AND CONCEPTS

- **Bias Frame:** Image frame of zero length exposure, designed to measure pixel-to-pixel variation across the charge-coupled device array.
- **Blooming:** Bleeding of charge from a full charge, often appearing as lines extending from overexposed pixels.
- **Dark Frame:** Image taken with the shutter closed, designed to measure noise and defects affecting images in the charge-coupled device.
- **Flat Field:** Image taken of uniform illumination across the charge-coupled device chip, designed to measure defects in the chip and optical system affecting images.
- **Photoelectric Effect:** Physical process whereby photons of light knock electrons off of atoms.
- **Pixel:** Individual charge-collecting region and ultimately a single element of a final image.
- **Quantum Efficiency:** Measure of the percentage of photons hitting a device that will produce electrons.
- **Radiation Noise:** Spurious charges produced in an image due to ionizing radiation, frequently produced by cosmic rays.
- **Readout Noise:** Electronic noise affecting image quality due to shifting and reading the charge in the pixels.

Definition and Basic Principles

A charge-coupled device (CCD) is an electronic array of devices fabricated at the surface of a semiconductor chip. The typical operation of a CCD is to collect electrical charge in specific locations laid out as an array on the surface of the chip. This electrical charge is normally produced by light shining onto the chip, producing an electrical charge. Each collecting area is called a pixel and consists of an electrical potential well that traps charge carriers. The charge in one area of the array is read, and then the charges on other parts of the chip are shifted from potential well to potential well until all have been read. The brighter the light shining on a pixel, the more charge it will have. Thus, an image can be constructed from the data collected by each pixel. When the CCD is placed at the focal plane of an optical system, such as a lens, then the image on the chip will be the same as the image seen by the lens; thus CCD chips can be used as the heart of a camera system. Because the data is collected electronically in digital form for each pixel, the image from a CCD is inherently a digital image and can be processed by a computer. For this reason, CCDs (and digital cameras) have become more popular than film as a way to take photographs.

Background and History

The charge-coupled device was invented based on an idea by Willard S. Boyle and George E. Smith of AT&T Bell Laboratories for volatile computer memory. They were seeking to develop an electrical analogue to magnetic bubble memory. Boyle and Smith postulated that electric charge could be used to store data in a matrix on a silicon chip. The charge could be moved from location to location within that array, shifting from one holding cell to another by applying appropriate voltages. The name "charge-coupled device" stemmed from this shifting of electrical charge around on the device. Initial research aimed to perfect the CCD as computer memory, but the inventors soon saw that it may be even more useful in converting light values into electrical charge that could be read. This property led to the development of the CCD as an imaging array. Subsequent developments made pixel sizes smaller, the arrays larger, and the price lower. The CCD became the detector at the heart of digital cameras, and their ease of use quickly made digital cameras popular. By the early 2000's, digital cameras and their CCDs had become more popular than film cameras.

How It Works

At the heart of a charge-coupled device is a semiconductor chip with electrical potential wells created by doping select areas in an array on the surface of the device. Doping of a semiconductor involves fabricating it with a select impurity that would tend to have either an extra electron or one too few electrons for the normal lattice structure of the semiconductor. Fabricating the CCD with select areas so doped would tend to make any electrical charge created at the surface of the semiconductor chip want to stay in the region of the potential well. By applying the proper electrical voltage between adjacent wells, however, charge is permitted to flow from one well to another. In this way, charge anywhere in the array can be moved from well to well to a location where it can be read.

The Photoelectric Effect. The key to using a charge-coupled device is to put charge on it. This is done by shining light onto the surface of the semiconductor chip. When photons of sufficient energy shine on a material, the absorbed light can knock electrons from atoms. This is known as the photoelectric effect, a physical phenomenon first observed by German physicist Heinrich Hertz in 1887 and explained by German physicist Albert Einstein in 1905. While the photoelectric electric effect works in many materials, in a semiconductor it produces a free electron and a missing electron, called a hole. The hole is free to move just like an electron. Normally, the electron and hole recombine (the electron going back to the atom it came from in the semiconductor lattice) after the electron-hole pair are created. However, if an electric field is present, the electron may go one way in the semiconductor and the hole another. Such an electric field is present in an operating CCD, and so the electrons will congregate in regions of lower electrical potential (electrical potential wells), building up an electrical charge proportional to the intensity of the light shining on the chip. The potential wells are arranged in an array across the surface of the CCD. The individual wells serve as the individual picture elements in a digital image and are called pixels. A nearly identical technology to the CCD uses different materials and is called the complementary metal oxide semiconductor (CMOS).

Shifting Charges. Once charge is produced and captured on the CCD chip, it must be measured in order for the device to be of any use. Typically, charge is read only from one location on the CCD. After that charge is read, all of the charges in all the pixels on a row are shifted over to the adjacent pixel to allow the next charge to be read from the readout position. This process repeats until all charges in the row have been read. All columns of pixels are then shifted down one row, shifting the charges in the second row into the row that has just been read. The process of reading that row continues until all charges are read and all the columns shift down again, repopulating the empty row of charge with new charges. All the charges are shifted around until they have been read. If a pixel has too much charge (being overexposed), then the charge can bleed between adjacent pixels. This creates spurious surplus charge on charges throughout a column and appears as vertical lines running through the image. This image defect is known as blooming.

Color Images. Charge-coupled devices respond to light intensity, not color. Therefore, all images are inherently black and white. There are several techniques to get color images from CCDs. The simplest and most cost-effective method is to take pictures using different color filters and then to create a final image by adding those images in the colors of the filters. This is the technique usually used in astronomy. The disadvantage of this technique, though, is that it requires at least three images taken in succession. This does not permit "live" or action photography. A separate technique is to have filters constructed over each pixel on the CCD array. This permits one chip to take multiple images simultaneously. The disadvantage of this technique, besides the cost of constructing such an array of filters, is that each image uses only about one-third of the pixels on the chip, thus losing image quality and detail. Most color digital cameras use this technique. Though other strategies exist for making color images with CCDs, these two are the most common.

Applications and Products

CCDs and CMOSs are the detectors in virtually every digital imaging system. CCDs have also displaced the imaging tubes in television cameras.

Cameras. Because CCDs produce digital images, they can be directly viewed using computers, sent by e-mail, coded into Web pages, and stored in electronic media. Digital images can also be viewed almost immediately after taking them, unlike film, which must first be developed. This has made digital cameras very popular. In addition to providing nearly instant pictures, CCDs can be manufactured quite small, permitting cameras that are much smaller than used to exist. This has allowed cameras to be placed in cell phones, computers, tablet computers, and many other devices. Traditionally, the pixels in CCDs have been larger than the grains in film, so digital image quality has suffered. However, technological advantages have permitted CCD pixels to be made much smaller, and by 2010, most commercial CCD camera systems compared very favorably with film systems in terms of image quality, with high-end CCD cameras often performing better than most film cameras.

Scientific Applications. Some of the first applications of CCDs were in the scientific community. CCDs permit very accurate images, but these images also contain precise data about light intensity. This is particularly important in chemistry, where spectral lines can be studied using CCD detectors. It is also important in astronomy, where digital images can be studied and measured using computers. Astronomical observatories were among the first to adopt CCD imaging systems, and they are still leaders in working with companies to develop new more powerful and larger CCD arrays.

Satellite Images. Early satellite cameras used film that had to be dropped from orbit to be collected below. Soon, television cameras displaced film, but as soon as the first reliable charge-coupled devices capable of imaging became available, that technology was far better than any other. As of 2011, nearly all satellite imaging systems, both civilian and military, rely on a type of CCD technology.

Impact on Industry

Government and University Research. The digital nature of CCDs makes them of great use in scientific images, and since they are more sensitive than film, they can record dimmer light sources. This makes CCDs particularly useful for astronomy. By the 1990's, practically all professional astronomical research was done using CCDs. By 2010, nearly all amateur astronomers were using CCDs in astrophotography.

Industry and Business. CCDs are used in fax machines, scanners, and even many small copiers. The ubiquity of these devices changes business communications. Web cameras attached to computers make videos easy to produce and have led to an explosion of individual amateur video clips being posted online in such places as YouTube.

Major Corporations. The widespread use of CCDs in digital cameras has displaced film in most photographic activities. Many companies that used to make film have had to adjust their business model to accommodate the change in customer activities. Companies that built their reputations on fine film products have had to shift their product lines to stay competitive. Most of these companies now sell more digital-related products than film-related products.

Careers and Course Work

Charge-coupled devices have become the most common imaging system in use. Thus, any career that uses images will come into contact with CCDs. For most people using CCDs, there is no different training or course work than would be needed to use any camera.

Manufacturing CCDs or developing new CCD technology, however, requires specialized training. CCDs are semiconductor devices, so their manufacture requires a background in semiconductor-manufacturing technology. Both two-year and four-year degrees in semiconductor manufacturing technology exist, and these degrees require courses in electronics, semiconductor physics, semiconductor-manufacturing technology, mathematics, and related disciplines. Manufacturing CCDs is not much different from manufacturing any other semiconductor device.

Construction of equipment, such as cameras, to hold CCDs involves both optics and electronics. Manufacture of such equipment requires little detailed knowledge of these areas other than what is required for any other camera or electronic device. Most of the components are manufactured ready to be assembled. Some specialized electronics knowledge is needed in order to design the circuit boards

Fascinating Facts About Charge-Coupled Devices

- A charge-coupled device (CCD) is the imaging system at the heart of every digital camera, Web cam, and cell phone camera. CCDs are also found in scanners, fax machines, and many other places where digital images are needed.
- By 2005, digital-related sales revenue exceeded film-related sales revenue for Kodak, a company known for its high-quality film and film photography products.
- Astronomical and military applications of CCDs led the way before widespread commercial applications became available.
- In order to reduce thermal noise, astronomical charge-coupled devices are often cooled with cryogenic fluids, such as liquid nitrogen.
- The first experimental CCD imaging system, constructed in 1970 at AT&T Bell Laboratories, had only eight pixels.
- The first commercially produced CCD imaging system, made available by Fairchild Electronics in 1974, had only 10,000 pixels, arranged in a 100-by-100 array.
- Because charge-coupled devices are sensitive to infrared light, they can be used in some solid-state night-vision devices—and some commercial cameras and video cameras have night-vision capability.

to operate the CCDs, however.

Developing improved CCD technology or research into better CCDs or technologies to replace CCDs requires much more advanced training, generally a post-graduate degree in physics or electrical engineering. These degrees require extensive physics and mathematics courses, along with electrical engineering course work and courses in chemistry, materials science, electronics, and semiconductor technology. Most jobs related to research and development are in university or corporate laboratory settings.

Social Context and Future Prospects

CCDs were once very esoteric devices. Early digital cameras had CCDs with large pixels, producing pictures of inferior quality to even modest film cameras. However, CCDs quickly became less expensive and of higher quality and soon became ubiquitous. Most cameras sold by 2005 were digital cameras. Most cell phones now come equipped with cameras having CCD technology. Since most people own a cell phone, this puts a camera in the hands of almost everyone all the time. The number of photographs and videos being made has skyrocketed far beyond what has ever existed since the invention of the camera. Since these images are digital, they can easily be shared by e-mail or on Web pages. Social networking on the Internet has permitted these pictures to be more widely distributed than ever. There is no indication that this will change in the near future. The possibility exists that in the near future two-way video phone calls may become commonplace.

Though charge-coupled devices are far superior to film, other competing technologies that can do the same thing are being developed. A very similar technology (so similar that it is often grouped with CCD technology) is the complementary metal oxide semiconductor (CMOS). CMOS technology works the same as CCD technology for the end user, but the details of the chip operation and manufacture are different. However, with CCD technology becoming so commonplace in society, it is likely that other imaging technological developments will follow.

Raymond D. Benge, Jr., B.S., M.S.

Further Reading

Berry, Richard, and James Burnell. *The Handbook of Astronomical Image Processing.* 2d ed. Richmond, Va.: Willmann-Bell, 2005. Though the book discusses charge-coupled devices somewhat, it mainly focuses on how to use the images taken by such devices to produce pictures and do scientific research. Comes with a CD-ROM.

Holst, Gerald C., and Terrence S. Lomheim. *CMOS/CCD Sensors and Camera Systems.* 2d ed. Bellingham, Wash.: SPIE, 2011. A technical review of CCD and CMOS imaging systems and cameras with specifics useful for engineering applications and for professionals needing to know the specific capabilities of different systems.

Howell, Steve B. *Handbook of CCD Astronomy.* 2d ed. New York: Cambridge University Press, 2006. This excellent and easy to understand handbook covers how to use charge-coupled device cameras in astronomy. The book is written for the amateur astronomer but contains good information on how CCDs work.

Janesick, James R. *Scientific Charge-Coupled Devices.* Bellingham, Wash.: SPIE, 2001. A very technical, thorough overview of how charge-coupled devices work.

Jorden, P. R. "Review of CCD Technologies." *EAS Publications Series* 37 (2009): 239-245. A good overview of charge-coupled devices, particularly as they relate to astronomy and satellite imaging systems.

Nakamura, Junichi, ed. *Image Sensors and Signal Processing for Digital Still Cameras.* Boca Raton, Fla.: CRC Press, 2006. A thorough and somewhat technical overview of CCD and CMOS technologies. Includes a good history of digital photography and an outlook for future technological developments.

Williamson, Mark. "The Latent Imager." *Engineering & Technology* 4, no. 14 (August 8, 2009): 36-39. A very good history of the development of charge-coupled device imaging systems and some of their uses.

Web Sites

American Astronomical Society
http://aas.org

Institute of Electrical and Electronics Engineers
http://www.ieee.org

Professional Photographers of America
http://www.ppa.com

See also: Computer Engineering; Computer Science; Electrical Engineering; Mechanical Engineering.

CHEMICAL ENGINEERING

FIELDS OF STUDY

Chemistry; mathematics; physics; electrical engineering; fluid dynamics; electronics; mechanical engineering; process control; material engineering; safety engineering; biology; communications; economics; critical path scheduling.

SUMMARY

Chemical engineering, sometimes also called processing engineering, is the field of engineering that studies the conversion of raw chemicals into useful products by means of chemical transformation. Chemical engineering applies engineering concepts to design, construction, operation, and improvement of processes that create products from chemicals. For example, chemical engineering converts petroleum into products such as gasoline, lubricants, petrochemicals, solvents, plastics, processed food, electronic components, pharmaceuticals, agricultural chemicals, paints, and inks. Chemical engineering relies on all the technologies used in chemical and related industries, including distillation, chemical kinetics, mass transport and transfer, heat transfer, control instrumentation, and other unit operations, as well as economics and communications.

KEY TERMS AND CONCEPTS

- **Activation Energy:** Amount of energy required to make a chemical reaction occur.
- **Azeotrope:** Mixture of two or more liquids in such a ratio that its composition cannot be changed by simple distillation; also known as constant-boiling mixture.
- **Catalyst:** Substance that changes the rate of a chemical reaction but is not consumed by it.
- **Chemical Decomposition:** Chemical reaction in which a chemical is decomposed into elements or smaller compounds; also known as analysis.
- **Chemical Reaction:** Interaction between chemicals that changes one or both of them in some way.
- **Distillation:** Process in which more volatile components are separated from materials that have a higher boiling point.
- **Intermediate:** Short-lived, unstable chemical in a chemical reaction.
- **Kinetics:** Study of chemical reaction rates and how these rates are affected by temperature, pressure, and catalysts.
- **Petroleum Fraction:** End product from refining petroleum.
- **Process Control:** Maintaining proper operations by controlling temperature, flows, levels, and other factors.
- **Product:** Result of a chemical reaction.
- **Reactant:** Substance initially involved in a chemical reaction and that is consumed in the process.
- **Reagent:** Substance added to a system to produce a chemical reaction; usually used to test for the presence of a specific substance.
- **Synthesis:** Chemical reaction that ends with the combination of two or more chemicals.
- **Unit Operations:** Individual steps that in combination with other steps are used to convert raw materials into finished products.

Definition and Basic Principles

Chemical engineering is the discipline that studies chemical reactions used to manufacture useful products, which appear in almost everything people have and use. Chemical engineering also is involved with reduction and removal of waste, improvement of air and water quality, and production of new sources of energy. It is also the responsibility of chemical engineering to ensure the safety of all involved through design, training, and operating procedures.

The dividing line between chemistry and chemical engineering is hazy and there is much overlap. Primarily chemistry discovers and develops new reactions, new chemicals, and new analytical tools. It is the role of chemical engineering to take these discoveries and use them to evaluate the economic possibilities of a new product or a new process for making an existing product. Chemical engineers determine what processes or unit operation are needed to carry out the reactions required to produce, recover, refine, and store a particular chemical. In the case of a new process, chemical engineers consider what other products could be made using the process. They also examine what products newly discovered chemicals can be used to make.

The largest field of chemical engineering applications is the petroleum and petrochemical industries. Crude oil, if merely distilled, would yield less than 35 percent gasoline. Through reactions such a catalytic cracking and platforming, these yields have come to approach 90 percent. Other treatments produce other fuels, lubricants, and waxes. The lighter components of crude oil became the raw materials used to make most of the plastics, fibers, solvents, synthetic rubber, paper, antifreeze, pharmaceutical drugs, agricultural chemicals, and paints that are seen in people's daily lives.

Background and History

Chemical engineering started in Germany with the availability of coal tar, a by-product of metallurgical coke. German chemists in the late nineteenth century used these tars to make aniline dyes and pharmaceuticals, most famously aspirin. Large-scale production meant that organic chemists needed help from engineers, usually mechanical, who designed and built the larger units needed. It became a tradition in Germany to pair a chemist with a mechanical engineer who would work together on this type of operation.

Americans first trained engineers in the chemical sciences to improve their understanding of chemical processes. In 1891, the Massachusetts Institute of Technology became the first school to offer a degree in chemical engineering. By 1904, the first handbook of chemical engineering, an encyclopedia of useful physical, chemical, and thermodynamic properties of chemicals and other information needed for design work, was published. A desire among professionals to share ideas, discoveries, and other useful information led to the founding in 1908 of the American Institute of Chemical Engineering.

World War I increased demand for military chemicals and the fuel needed for a mechanized war. Previously, chemicals had been developed in batches, but the petroleum industry converted to using continuous operations to raise the production rate. Soon the industry was developing systems to control production, first by empirical methods and then by scientific and mathematical techniques.

How It Works

The design, construction, and operation of chemical engineering projects are commonly divided into various unit operations. These unit operations, singly or in combination, require a basic knowledge and understanding of many scientific, mathematical, and economic principles.

Mathematics drives all aspects of chemical engineering. Calculations of material and energy balances are needed to deal with any operation in which chemical reactions are carried out. Kinetics, the study dealing with reaction rates, involves calculus, differential equations, and matrix algebra, which is needed to determine how chemical reactions proceed and what products are made and in what ratios. Control system design additionally requires the understanding of statistics and vector and non-linear system analysis. Computer mathematics including numerical analysis is also needed for control and other applications.

Chemistry, especially organic and physical, is the basis of all chemical processes. A full understanding of organic chemistry is essential in the fields of petroleum, petrochemicals, pharmaceuticals, and agricultural chemicals. A great deal of progress was made in the field of organic chemistry beginning in the 1880's. Physical chemistry is the foundation of understanding how materials behave with respect to motion and heat flow. The study of how gases and liquids are affected by heat, pressure, and flows is needed for the unit operations of mass transfer, heat transfer, and distillation.

Distillation, in which more volatile components are separated from less volatile materials, requires knowledge of individual physical and thermodynamic properties and how these interact with one another in mixtures. For example, materials with different boiling points sometimes form a constant-boiling mixture, or azeotrope, which cannot be further refined by simple distillation. A well-known example of this is ethyl alcohol and water. Although ethyl alcohol boils at 78 degrees Celsius as compared with 100 degree Celsius for water, a mixture of ethyl alcohol and water that contains 96 percent (by weight) or 190-proof ethyl alcohol is a constant-boiling mixture. Further concentration is not possible without extraordinary means.

Inorganic chemistry deals with noncarbon chemistry and is often considered basic as it is taught in high school and the first year of college. Many of the chemical industries do make inorganic chemicals such as ammonia, caustic, chlorine, oxygen, salts, cement, glass, sulfur, carbon black, pigments, fertilizers, and sulfuric, hydrochloric, and nitric acids. Catalysts are inorganic materials that influence organic reactions, and understanding them is very important. The effectiveness of any catalyst is determined by its chemical composition and physical properties such as surface area, pore size, and hardness.

Analytical chemistry was once strictly a batch process, in which a sample would be collected and taken to a laboratory for analysis. However, modern processes continually analyze material during various stages of manufacture, using chromatography, mass spectroscopy, color, index of refraction, and other techniques.

The disciplines of mechanical, electrical, and electronic engineering are needed by the chemical engineer to be able to consider process design, materials of construction, corrosion, and electrical systems for the motors and heaters. Electronic engineering is basic knowledge needed for control systems and computer uses.

Safety engineering is the study of all aspects of design and operation to find potential hazards to health and physical damage and how to correct them. This work begins at the start of any project. Not only must each and every part of a process be examined but also each chemical involved must be checked for hazards, either by itself or in combinations with other chemicals and materials in which it will come in contact. Safety engineering also is a part of operator training and the writing of operating procedures used for the proposed operation.

Communication is key to all progress. Chemical engineers must be able to interact with others; no idea—whether for a new process or product or an improvement in an existing operation—can be implemented unless others are convinced of its value and are willing to invest in it. Coherent and easily understood reports and presentations are as important as any other part of a project. Operating procedures must be reviewed with the operation personnel to ensure that they are understood and can be followed.

Economics is the driving force behind all design, construction, and operations. The chemical engineer must be conversant in finance, banking, accounting, and worldwide business practices. Cost estimates, operating balances to determine actual costs, market forces, and financing are a necessary part of the work of any chemical engineer.

Applications and Products

Chemical engineering is involved in every step in bringing a process from the laboratory to full-scale production. This involves determining methods to make the process continuous, safe, environmentally compatible, and economically sound. During these steps, chemical engineers determine the methods and procedures needed for a full-scale plant. Mathematical modeling is used to test various steps in the process, controls, waste treatment, environmental concerns, and economic feasibility.

Acetonitrile Process. The acetonitrile process demonstrates how these disciplines combine to produce a process. Acetonitrile is a chemical used as a solvent and an intermediate for agricultural chemicals. The chemistry of this process involves reacting acetic acid and ammonia to make acetonitrile and water. The reaction between these raw materials is carried out over a catalyst at 400 degrees Celsius. The reaction takes place in tubes loaded with a catalyst of phosphoric acid deposited on an alumina ceramic support, which allows the reaction to take place at high rates. The reaction tubes are located in a gas-fired furnace designed to provide even temperatures throughout the length of the tubes. The exiting gases are cooled and condensed by scrubbing with water. This mixture enters a train of distillation columns, which first removes a constant-boiling mixture of the acetonitrile along with some water. In another distillation column, an azeotroping agent is used to produce an overhead mixture that when condensed, produces two layers with all the water in one layer. The water layer is removed, and the other layer is recycled. The base material leaving the distillation column contains the water-free acetonitrile, which is then redistilled to produce the finished product, which is ready to package and ship.

Such a process involves chemical engineers in the design and assembly of all the equipment needed for the process: the furnace, reactor tubes, distillation columns, tanks, pumps, heat exchangers, and piping. The chemical engineers also create the controls, operating procedures, hazardous material data sheets, and startup and shut-down instructions, and provide operator training. Once the plant is running, the role of the chemical engineer becomes operating and improving the unit.

Petroleum and Petrochemical Applications. Chemical engineering is used in the petroleum and petrochemical industries. At first, all petroleum products were produced by simple batch distillations of crude oil. Chemical engineering developed continuous distillation processes that permitted marked increases in refinery production rates. Then high-temperature cracking methods permitted the conversion of high-boiling petroleum fractions (end products of refining) to useful products such as more gasoline. This was followed by the use of catalytic cracking that improved the gasoline output even more.

Distillation can take many forms in addition to simple atmospheric distillation. Chemical engineers can determine the need to use pressure distillation for the purification of components that are normally gases. Vacuum distillation may be useful if there are high-boiling components, which are sensitive to the elevated temperatures required for normal distillation. With the need for aviation gasoline and other high-octane fuels, chemical engineering developed such methods as platforming, which uses the addition of a platinum catalyst to speed up certain reactions, to convert lower-boiling petroleum components into high-octane additives.

Petrochemical industries convert surplus liquefiable gases into solvents, plastics, synthetic rubber, adhesives, coatings, paints, inks, intermediates for agricultural chemicals, food additives, and many more items.

Inorganics. Sulfuric acid is typical of a major inorganic product that requires the type of process improvement that is provided by chemical engineering. Sulfuric acid is a high-volume, low-profit-margin material that has been in production for more than a hundred years. The chemistry is well known. Sulfur is vaporized, mixed with air, and passed over a catalyst to make sulfur trioxide, which is then adsorbed into water to make sulfuric acid. Chemical engineers look for better catalysts and seek to improve the purity of raw materials and increase control over temperature, flows, safety, and environmental concerns.

Biological Applications. The ancient biological process of fermentation is best known for its role in producing alcoholic beverages and breads. Although many people do not realize that fermentation is a chemical reaction, it is much the same as other organic reactions. The biological component is a microorganism that acts as the catalyst. Biochemical engineering products created using fermentation include acetone and butyl acetate, chemicals that were in critical demand by the aviation industry in World War I.

Biochemical engineering also made antibiotics available on a large scale. In the 1920's, Sir Alexander Fleming discovered the antibiotic properties of penicillin, which was produced in laboratory flasks, a few grams at a time. During World War II, the need for penicillin increased, and chemical engineers developed a large-scale process for producing penicillin from corn. This process was adapted to produce other antibiotics, and chemical engineering processes were developed for the manufacture of many synthetic drugs. Genetic modifications are being developed that are expected to produce the drugs of the future.

Fibers. In 1905, the first synthetic fiber, reconstituted cellulose, commonly known as rayon, was developed. The second synthetic fiber, cellulose acetate, which was developed in 1924, is still manufactured in large amounts for use in cigarette filters. The first fully synthetic fiber was nylon, a commonly used polyamide, followed by polyesters. Chemically treated cotton, known as permanent press, is another product of chemical engineering. In addition to fibers, the textile industry uses lubricants, dyes, pigments, coatings, and inks, all derived by chemical processes.

Plastics. The first thermoplastic, a material that could be reheated and molded, were the cellulosics that are used to make large signs, toys, automobile parts, and other objects that can be easily fabricated. Acrylic polymers such as methyl methacrylate are formed into optically clear sheeting and used as a nonshattering replacement for plate glass and for eyeglass and camera lenses. Polyethylene, developed during World War II as a superior coating for the wiring needed in radar, is still used as an insulator. The most common use for polyethylene and polypropylene is as the thin film used in plastic bags. Molded items such as bottles, containers, kitchen items, and packaging are manufactured from these polymers as well as polyesters, nylon, polyvinyl chloride (PVC), polystyrene, Teflon, and polycarbonates. The manufacture of all these plastics was developed through chemical engineering and depends on it for operation and improvement.

Refrigerants. The first refrigerants were sulfur dioxide and ammonia. These substances were rather hazardous, so a new range of refrigerants was developed. These refrigerants, commonly known as Freon, are halogenated hydrocarbons that can be tailored to the needs of the application. Home air-conditioning, automotive air-conditioning, and industrial applications are examples of these uses.

Nuclear Energy. The nuclear energy industry requires many solvents and reaction agents for the separation and purification of the fuel used in the reactors. Special coatings and other materials used in the vicinity of intense radiation were developed by the chemical industry. Recovery of the spent nuclear fuel requires specially developed techniques, again requiring solvents and reactions.

Coatings. A typical example of corrosion is the rusting of iron. Corrosion causes the loss of equipment as well as physical and health hazards not only in the chemical industry but also in almost every aspect of modern life. Corrosion is avoided through metal alloys such as stainless steel and protective coatings. These coatings can be tailored to protective needs inside and outside of a product.

Water Treatment. Raw water, especially for industrial purposes, requires processing to remove suspended solids and soluble organics as well as inorganic ions such as sodium, calcium, iron, chlorides, sulfates and nitrates. Chemical engineering is used to design and manufacture the ion-exchange resins, coagulants, and adsorbents needed, as well as the procedures for use and regeneration of ion-removal systems.

Waste Treatment. Many chemical processes produce by-products or waste that may be hazardous to the environment and represent uncaptured value. The role of process improvement engineers is to first find ways to reduce waste and failing that, to develop methods to convert these wastes into nonhazardous materials that will not harm the environment.

Agricultural Applications. Agricultural chemicals such as insecticides, herbicides, fungicides, fertilizers, seed coatings, animal feed additives, and medicines such as hormones and antibiotics all are produced by chemical means. Other chemicals are used in the preparation of products for harvesting and transporting to market.

Food Processing. The manufacture of food stuffs such as dairy products, breakfast foods, soups, bread, canned goods, frozen foods, and other processed foods, as well as meats, fruits, and vegetables, use engineering operations such as heat transfer to heat or cool.

Fascinating Facts About Chemical Engineering

- Chaim Azreil Weizmann developed a process to produce acetone and butyl acetate, chemicals in critical demand during World War I. In return, the British government supported the establishment of a Jewish state, Israel, which named Weizmann its first president in 1948.
- The pure organic chemical produced in the greatest quantity per year is sucrose sugar. The process of converting raw sugarcane or beets into uniform-size sucrose crystals involves extraction, heat transfer, multieffect evaporation, adsorption, crystallization, filtration, and drying—not to mention mass transfer and instrumentation.
- One of the chemical by-products of paper manufacture is synthetic vanilla. One paper mill could produce the world's annual demand in one day.
- The first synthetic plastic, Bakelite, was developed to be a substitute for ivory in the manufacture of billiard balls.
- Aluminum, once more costly than gold or platinum, was so expensive that Russia minted aluminum coins to demonstrate its wealth and power. The Hall electrochemical engineering process for producing aluminum has lowered the cost of aluminum to where it has become a very low cost material.
- Latex paints were developed as a substitute for oil-based paints, not only to produce a water-based paint that would not contribute to air pollution but also to produce a drip-proof product.
- DuPont employee Roy Plunkett stored some tetrafluoroethylene gas in a container. The next day, he found it had polymerized, leaving a waxy substance called polytetrafluoroethylene, commonly known as Teflon.
- Water and the element gallium, a material used in semiconductor manufacture, are the only two significant substances known to expand when they are cooled and frozen.

Impact on Industry

Chemical engineering has an impact on almost every aspect of people's lives. The chemical engineering profession is international in its scope. Chemical plants exist in almost every country in the world; however, almost all petrochemical industries are located in countries with or near petroleum refineries. The United States dominates this industry, followed by Europe and the Middle East. Industrial processes are developed throughout the world and are made available to others through licensing. Engineering research is carried out by government agencies, universities, and major industrial concerns.

Government and University Research. The U.S. government has many agencies and programs that promote the development of new energy products or the improvement of existing ones, the reduction of chemical by-products, and the recovery of energy and useful products from existing waste sites. These agencies include the Bureau of Mines, the Environmental Protection Agency, the Department of Defense, the Department of Agriculture, and even the National Aeronautics and Space Administration. Government grants fund many academic research programs designed to develop new materials and processes that use less energy and produce fewer and less hazardous by-products.

One example of government-funded academic research is the recovery of liquid and gaseous fuels from coal. During World II, the Germans developed a process to make aviation gasoline and other fuels from low-grade coal. At the end of the war, the U.S. government confiscated the German data and gave it to Texas A&M University. In the 1970's, the Bureau of Mines and Texas A&M began converting this mass of literature into a program to design a research-scale plant to develop methods for the economic production of fuels from coal and oil shale. As naturally occurring petroleum becomes more expensive and harder to find, the recovery of fuel products from coal is likely to become practical and to lead to the creation of new industries.

Another area of government-funded academic research is the desalinization of sea water. Fresh water is becoming scarcer, and multiple studies are seeking to find ways to produce large quantities of fresh water.

Industry and Business Sectors. Chemical engineering affects almost every industry and business that manufactures, handles, transports, and sells chemicals. Chemical engineers are called on to help evaluate the practicality and earning potential of investment proposals in which the manufacture or use of chemicals is to be considered.

The major corporations in the chemical industry include DuPont, Union Carbide, the Dow Chemical Company, Celanese Chemicals, Monsanto, Eastman Chemical, ExxonMobil, Texaco, BP, and Royal Dutch Shell. Other major producers of chemicals include Procter & Gamble, Kraft Foods, Quaker Oats, Sherman-Williams, and PPG Pittsburgh Paints. Major pharmaceutical companies such as Pfizer and Ely Lily require biochemical engineering to produce the medicines they sell.

Careers and Course Work

A bachelor's degree in chemical engineering takes four to five years of study. Course work includes a great deal of chemistry, including organic, physical, and analytic chemistry; chemical engineering courses, and mechanical, electrical, and civil engineering courses. Advanced mathematics courses are also required. Courses in English, economics, history, and public speaking will also help further the career of a chemical engineer.

A bachelor's degree is generally sufficient for an entry-level position at most industrial companies. Many universities offer engineering degree programs in which students take classes one semester and work at a chemical plant the next semester, repeating this pattern until graduation. This enables students to pay for their education and to gain relevant work experience, as well as to make contacts within the industry.

A master's degree in chemical or computer engineering or a master's of business administration degree will help advance a career in chemical engineering. A doctorate is required for those seeking jobs in colleges and universities.

Social Context and Future Prospects

Many of the major issues that face society, particularly those concerning the supply of energy and water, the environment, and global warming, require immediate and continuing action from scientists such as chemists and chemical engineers.

Energy. The predominant sources of energy are coal, liquid petroleum, and natural gas. Coal use produces air pollutants such as sulfur dioxide, nitrogen oxides, mercury, and more carbon dioxide per British thermal unit of energy generated than any of the other energy sources. Liquid petroleum and natural gas also produce carbon dioxide and other air pollutants. Problems regarding the limited supply of all three sources of energy are causing worldwide economic and political disruptions. Chemical engineering may be able to provide economical answers in that it can create clean oil and gas from coal and oil shale, produce better biofuels from nonfood agricultural crops, and develop materials to make solar energy and hydrogen fuel cells practical.

The Environment. Both air and water pollution are affected by chemicals that are the result of manufacturing, handling, and disposing of materials such as solvents, insecticides, herbicides, and fertilizers. Chemical engineering has reduced factory emissions through the development of water-based coatings. In addition, new methods of converting solid wastes into usable fuels will help the environment and provide new fuel sources.

Water Access. The supply of fresh water for personal and agricultural use is already limited and will become more so as the world population grows. Two methods for recovering fresh water are distillation and reverse osmosis. To become practical, distillation systems must use better materials to prevent corrosion. Chemical engineers are likely to develop new alloys and ways to manufacture them, as well as better heat-recovery methods to reduce cost and lower carbon dioxide emissions. Although the long-range effect of increased concentrations of carbon dioxide on the climate may not be known for sure, chemical engineers are striving to develop methods to control and reduce these emissions.

Reverse osmosis uses membranes that allow water to pass through but not soluble salts. Chemical engineers are working on improved membrane materials that will allow higher pressures to improve efficiency and membrane life.

Max Statman, B.S., M.S.

Further Reading

Dethloff, Henry C. *A Unit Operation: A History of Chemical Engineering at Texas A&M University.* College Station: Texas A&M University Press, 1988. Describes how chemical engineering helped create the petrochemical industry that dominates Gulf Coast industry.

Dobre, Tanase, and José G. Sanchez Marcano. *Chemical Engineering: Modeling, Simulation, and Similitude.* Weinheim, Germany: Wiley-VCH, 2007. Looks at how computer-aided modeling is used to develop, implement, and improve industrial processes. Covers the entire process, including mathematical modeling, results analysis, and performance evaluation.

Lide, David R., ed. *CRC Handbook of Chemistry and Physics: A Ready-Reference Book of Chemical and Physical Data.* 90th ed. Boca Raton, Fla.: CRC Press, 2009. A vital source of information for designing chemical processes, analyzing results, and estimating costs.

Perry, R. H., and D. W. Green, eds. *Perry's Chemical Engineers Handbook.* 8th ed. New York: McGraw-Hill, 2007. First published in 1934, this handbook provides information about the processes, operations, and equipment involved in chemical engineering, as well as chemical and physical data, and conversion factors. More than seven hundred illustrations.

Towler, Gavin P., and R. K. Sinnott. *Chemical Engineering Design.* Oxford, England: Butterworth-Heinemann, 2009. Examines how chemical engineers design chemical processes and discusses all the elements and factors involved.

Web Sites

American Chemical Society
http://portal.acs.org

American Institute of Chemical Engineering
http://www.aiche.org

Society of Chemical Industry
http://www.soci.org

See also: Coal Gasification; Coal Liquefaction; Petroleum Extraction and Processing.

CIVIL ENGINEERING

FIELDS OF STUDY

Mathematics; chemistry; physics; engineering mechanics; strength of materials; fluid mechanics; soil mechanics; hydrology; surveying; engineering graphics; environmental engineering; structural engineering; transportation engineering.

SUMMARY

Civil engineering is the branch of engineering concerned with the design, construction, and maintenance of fixed structures and systems, such as large buildings, bridges, roads, and other transportation systems, and water supply and wastewater-treatment systems. Civil engineering is the second oldest field of engineering, with the term "civil" initially used to differentiate it from the oldest field of engineering, military engineering. The major subdisciplines within civil engineering are structural, transportation, and environmental engineering. Other possible areas of specialization within civil engineering are geotechnical, hydraulic, construction, and coastal engineering.

KEY TERMS AND CONCEPTS

- **Abutment:** Part of a structure designed to withstand thrust, such as the end supports of a bridge or an arch.
- **Aqueduct:** Large pipe or conduit used to transport water a long distance.
- **Backfill:** Material used in refilling an excavated area.
- **Cofferdam:** Temporary structure built to keep water out of a construction zone in a river.
- **Design Storm:** Storm of specified return period and duration at a specified location, typically used for storm water management design.
- **Foundation:** Ground that is used to support a structure.
- **Freeboard:** Difference in height between the water level and the top of a tank, dam, or channel.
- **Girder:** Large horizontal structural member, supporting vertical loads.
- **Invert:** Curved, inside, bottom surface of a pipe.
- **Percolation Test:** Test to determine the drainage capability of soil.
- **Rapid Sand Filter:** System for water treatment by gravity filtration through a sand bed.
- **Reinforced Concrete:** Concrete that contains wire mesh or steel reinforcing rods to give it greater strength.
- **Sharp-Crested Weir:** Obstruction with a thin, sharp upper edge, used to measure flow rate in an open channel.
- **Tension Member:** Structural member that is subject to tensile stress.
- **Ultimate Bearing Capacity:** Theoretical maximum pressure that a soil can support from a load without failure

Definition and Basic Principles

Civil engineering is a very broad field of engineering, encompassing subdisciplines ranging from structural engineering to environmental engineering, some of which have also become recognized as separate fields of engineering. For example, environmental engineering is included as an area of specialization within most civil engineering programs, many colleges offer separate environmental engineering degree programs.

Civil engineering, like engineering in general, is a profession with a practical orientation, having an emphasis on building things and making things work. Civil engineers use their knowledge of the physical sciences, mathematics, and engineering sciences, along with empirical engineering correlations to design, construct, manage, and maintain structures, transportation infrastructure, and environmental treatment equipment and facilities.

Empirical engineering correlations are important in civil engineering because useable theoretical equations are not available for all the necessary engineering calculations. These empirical correlations are equations, graphs, or nomographs, based on experimental measurements, that give relationships among variables of interest for a particular engineering application. For example the Manning equation gives an experimental relationship among the flow rate in an open channel, the slope of the channel, the depth of water, and the size, shape, and material of the bottom and sides of the channel. Rivers, irrigation ditches, and concrete channels used to transport wastewater in a treatment plant are examples of open channels. Similar empirical relationships are used in transportation, structural, and other specialties within civil engineering.

Background and History

Civil engineering is the second oldest field of engineering. The term "civil engineering" came into use in the mid-eighteenth century and initially referred to any practice of engineering by civilians for nonmilitary purposes. Before this time, most large-scale construction projects, such as roads and bridges, were done by military engineers. Early civil engineering projects were in areas such as water supply, roads, bridges, and other large structures, the same type of engineering work that exemplifies civil engineering in modern times.

Although the terminology did not yet exist, civil engineering projects were carried out in early times. Examples include the Egyptian pyramids (about 2700-2500 B.C.E.), well-known Greek structures such as the Parthenon (447-438 B.C.E.), the Great Wall of China (220 B.C.E.), and the many roads, bridges, dams, and aqueducts built throughout the Roman Empire.

Most of the existing fields of engineering split off from civil engineering or one of its offshoots, as new fields emerged. For example, with increased use of machines and mechanisms, the field of mechanical engineering emerged in the early nineteenth century.

How It Works

In addition to mathematics, chemistry, and physics, civil engineering makes extensive use of principles from several engineering science subjects: engineering mechanics (statics and strength of materials), soil mechanics, and fluid mechanics.

Engineering Mechanics—Statics. As implied by the term "statics," this area of engineering concerns objects that are not moving. The fundamental principle of statics is that any stationary object must be in static equilibrium. That is, any force on the object must be cancelled out by another force that is equal in magnitude and acting in the opposite direction. There can be no net force in any direction on a stationary object, because if there were, it would be moving in that direction. The object considered to be in static equilibrium could be an entire structure or it could be any part of a structure down to an individual member in a truss. Calculations for an object in static equilibrium are often done through the use of a free body diagram, that is a sketch of the object, showing all the forces external to that object that are acting on it. The principle then used for calculations is that the sum of all the horizontal forces acting on the object must be zero and the sum of all the vertical forces acting on the object must be zero. Working with the forces as vectors helps to find the horizontal and vertical components of forces that are acting on the object from some direction other than horizontal or vertical.

Engineering Mechanics—Strength of Materials. This subject is sometimes called mechanics of materials. Whereas statics works only with forces external to the body that is in equilibrium, strength of materials uses the same principles and also considers internal forces in a structural member. This is done to determine the required material properties to ensure that the member can withstand the internal stresses that will be placed on it.

Soil Mechanics. Knowledge of soil mechanics is needed to design the foundations for structures. Any structure resting on the Earth will be supported in some way by the soil beneath it. A properly designed foundation will provide adequate long-term support for the structure above it. Inadequate knowledge of soil mechanics or inadequate foundation design may lead to something such as the Leaning Tower of Pisa. Soil mechanics topics include physical properties of soil, compaction, distribution of stress within soil, and flow of water through soil.

Fluid Mechanics. Fundamental principles of physics are used for some fluid mechanics calculations. Examples are conservation of mass (called the continuity equation in fluid mechanics) and conservation of energy (also called the energy equation or the first law of thermodynamics). Some fluid mechanics applications, however, make use of empirical (experimental) equations or relationships. Calculations for flow through pipes or flow in open channels, for example, use empirical constants and equations.

Knowledge from Engineering Fields of Practice. In addition to these engineering sciences, a civil engineer uses accumulated knowledge from the civil engineering areas of specialization. Some of the important fields of practice are hydrology, geotechnical engineering, structural engineering, transportation engineering, and environmental engineering. In each of these fields of practice, there are theoretical equations, empirical equations, graphs or nomographs, guidelines, and rules of thumb that civil engineers use for design and construction of projects related to structures, roads, storm water management, or wastewater-treatment projects, for example.

Civil Engineering Tools. Several tools available for civil engineers to use in practice are engineering graphics, computer-aided drafting (CAD), surveying, and geographic information systems (GIS). Engineering graphics (engineering drawing) has been a mainstay in civil engineering since its inception, for preparation of and interpretation of plans and drawings. Most of this work has come to be done using computer-aided drafting. Surveying is a tool that has also long been a part of civil engineering. From laying out roads to laying out a building foundation or measuring the slope of a river or of a sewer line, surveying is a useful tool for many of the civil engineering fields. Civil engineers often work with maps, and geographic information systems, a much newer tool than engineering graphics or surveying, make this type of work more efficient.

Codes and Design Criteria. Much of the work done by civil engineers is either directly or indirectly for the public. Therefore, in most of civil engineering fields, work is governed by codes or design criteria specified by some state, local, or federal agency. For example, federal, state, and local governments have building codes, state departments of transportation specify design criteria for roads and highways, and wastewater-treatment processes and sewers must meet federal, state, and local design criteria.

Applications and Products

Structural Engineering. Civil engineers design, build, and maintain many and varied structures. These include bridges, towers, large buildings (skyscrapers), tunnels, and sports arenas. Some of the civil engineering areas of knowledge needed for structural engineering are soil mechanics/geotechnical engineering, foundation engineering, engineering mechanics (statics and dynamics), and strength of materials.

When the Brooklyn Bridge was built over the East River in New York City (1870-1883), its suspension span of 1,595 feet was the longest in the world. It remained the longest suspension bridge in North America until the Williamsburg Bridge was completed in New York City in 1903. The Brooklyn Bridge joined Brooklyn and Manhattan, and helped establish the New York City Metropolitan Area.

The Golden Gate Bridge, which crosses the mouth of San Francisco Bay with a main span of 4,200 feet, had nearly triple the central span of the Brooklyn Bridge. It was the world's longest suspension bridge from its date of completion in 1937 until 1964, when the Verrazano-Narrows Bridge opened in New York City with a central span that was 60 feet longer than that of the Golden Gate Bridge. The Humber Bridge, which crosses the Humber estuary in England and was completed in 1981, has a single suspended span of 4,625 feet and is the longest suspension bridge in the world.

One of the most well-known early towers illustrates the importance of good geotechnical engineering and foundation design. The Tower of Pisa, commonly known as the Leaning Tower of Pisa, in Italy started to lean to one side very noticeably, even during its construction (1173-1399). Its height of about 185 feet is not extremely tall in comparison with towers built later, but it was impressive when it was built. The reason for its extreme tilt (more than 5 meters off perpendicular) is that it was built on rather soft, sandy soil with a foundation that was not deep enough or spread out enough to support the structure. In spite of this, the Tower of Pisa has remained standing for more than six hundred years.

Another well-known tower, the Washington Monument, was completed in 1884. At 555 feet in height, it was the world's tallest tower until the Eiffel Tower, nearly 1,000 feet tall, was completed in 1889. The Washington Monument remains the world's tallest masonry structure. The Gateway Arch in St. Louis, Missouri, is the tallest monument in the United States, at 630 feet.

The 21-story Flatiron Building, which opened in New York City in 1903, was one of the first skyscrapers. It is 285 feet tall and its most unusual feature is its triangular shape, which was well suited to the wedge-shaped piece of land on which it was built. The 102-floor Empire State Building, completed in 1931 in New York City with a height of 1,250 feet, outdid the Chrysler Building that was under construction at the same time by 204 feet, to earn the title of the world's tallest building at that time. The Sears Tower (now the Willis Tower) in Chicago is 1,450 feet tall and was the tallest building in the world when it was completed in 1974. Several taller buildings have been constructed since that time in Asia.

Some of the more interesting examples of tunnels go through mountains and under the sea. The Hoosac Tunnel, built from 1851 to 1874, connected New York State to New England with a 4.75-mile railway tunnel through the Hoosac Mountain in northwestern Massachusetts. It was the longest railroad tunnel in the United States for more than fifty years. Mount Blanc Tunnel, built from 1957 to 1965, is a 7.25-mile long highway tunnel under Mount Blanc in the Alps to connect Italy and France. The Channel Tunnel, one of the most publicized modern tunnel projects, is a rather dramatic and symbolic tunnel. It goes a distance of 31 miles beneath the English Channel to connect Dover, England, and Calais, France.

Transportation Engineering. Civil engineers also design, build, and maintain a wide variety of projects related to transportation, such as roads, railroads, and pipelines.

Many long, dramatic roads and highways have been built by civil engineers, ever since the Romans became the first builders of an extensive network of roads. The Appian Way is the most well known of the many long, straight roads built by the Romans. The Appian Way project was started in 312 B.C.E. by the Roman censor, Appius Claudius. By 244 B.C.E., it extended about 360 miles from Rome to the port of Brundisium in southeastern Italy. The Pan-American Highway, often billed as the world's longest road, connects North America and South America. The original Pan-American Highway ran from Texas to Argentina with a length of more than 15,500 miles. It has since been extended to go from Prudhoe Bay, Alaska, to the southern tip of South America, with a total length of nearly 30,000 miles. The U.S. Interstate Highway system has been the world's biggest earthmoving project. Started in 1956 by the Federal Highway Act, it contains sixty-two highways covering a total distance of 42,795 miles. This massive highway construction project transformed the American system of highways and had major cultural impacts.

The building of the U.S. Transcontinental Railroad was a major engineering feat when the western portion of the 2,000-mile railroad across the United States was built in the 1860's. Logistics was a major part of the project, with the need to transport steel rails and wooden ties great distances. An even more formidable task was construction of the Trans-Siberian Railway, the world's longest railway. It was built from 1891 to 1904 and covers 5,900 miles across Russia, from Moscow in the west to Vladivostok in the east.

The Denver International Airport, which opened in 1993, was a very large civil engineering project. This airport covers more than double the area of all of Manhattan Island.

The first oil pipeline in the United States was a 5-mile-long, 2-inch-diameter pipe that carried 800 barrels of petroleum per day. Pipelines have become much larger and longer since then. The Trans-Alaska Pipeline, with 800 miles of 48-inch diameter pipe, can carry 2.14 million barrels per day. At the peak of construction, 20,000 people worked 12-hour days, seven days a week.

Water Resources Engineering. Another area of civil engineering practice is water resources engineering, with projects like canals, dams, dikes, and seawater barriers.

The oldest known canal, one that is still in operation, is the Grand Canal in China, which was constructed between 485 B.C.E. and 283 C.E. The length of the Grand Canal is more than 1,000 miles, although its route has varied because of several instances of rerouting, remodeling, and rebuilding over the years. The 363-mile-long Erie Canal was built from 1817 to 1825, across the state of New York from Albany to Buffalo, thus overcoming the Appalachian Mountains as a barrier to trade between the eastern United States and the newly opened western United States. The economic impact of the Erie Canal was tremendous. It reduced the cost of shipping a ton of cargo between Buffalo and New York City from about $100 per ton (over the Appalachians) to $4 per ton (through the canal).

The Panama Canal, constructed from 1881 to 1914 to connect the Atlantic and Pacific oceans through the Isthmus of Panama, is only about 50 miles long, but its construction presented tremendous challenges because of the soil, the terrain, and the tropical illnesses that killed many workers. Upon its completion, however, the Panama Canal reduced the travel distance from New York City to San Francisco by about 9,000 miles.

When the Hoover Dam was built from 1931 to

Fascinating Facts About Civil Engineering

- The Johnstown flood in Pennsylvania, on May 31, 1889, which killed more than 2,200 people, was the result of the catastrophic failure of the South Fork Dam. The dam, built in 1852, held back Lake Conemaugh and was made of clay, boulders, and dirt. An improperly maintained spillway combined with heavy rains caused the collapse.
- The I-35W bridge over the Mississippi River in Minneapolis, Minnesota, collapsed during rush-hour traffic on August 1, 2007, causing 13 deaths and 145 injuries. The collapse was blamed on undersized gusset plates, an increase in the concrete surface load, and the weight of construction supplies and equipment on the bridge.
- On November 7, 1940, 42-mile-per-hour winds twisted the Tacoma Narrows Bridge and caused its collapse. The bridge, with a suspension span of 2,800 feet, had been completed just four months earlier. Steel girders meant to support the bridge were blocking the wind, causing it to sway and eventually collapse.
- Low-quality concrete and incorrectly placed rebar led to shear failure, collapsing the Highway 19 overpass in Laval, Quebec, on September 30, 2006.
- The first design for the Gateway Arch in St. Louis, Missouri, had a fatal flaw that made it unstable at the required height. The final design used 886 tons of stainless steel, making it a very expensive structure.
- On January 9, 1999, just three years after it was built, the Rainbow Bridge, a pedestrian bridge across the Qi River in Sichuan Province, collapsed, killing 40 people and injuring 14. Concrete used in the bridge was weak, parts of it were rusty, and parts had been improperly welded.
- On December 7, 1982, an antenna tower in Missouri City, Texas, collapsed, killing 2 riggers on the tower and 3 who were in an antenna section that was being lifted. U-bolts holding the antenna failed, and as it fell, it hit a guy wire on the tower, collapsing the tower. The engineers on the project declined to evaluate the rigger's plans.
- On February 26, 1972, coal slurry impoundment dam 3 of the Pittston Coal Company in Logan County, West Virginia, failed, four days after it passed inspection by a federal mine inspector. In the Buffalo Creek flood that resulted, 125 people were killed. The dam, which was above dams 1 and 2, had been built on coal slurry sediment rather than on bedrock.

1936 on the Colorado River at the Colorado-Arizona border, it was the world's largest dam, at a height of 726 feet and crest length of 1,224 feet. The technique of passing chilled water through pipes enclosed in the concrete to cool the newly poured concrete and speed its curing was developed for the construction of the Hoover Dam and is still in use. The Grand Coulee Dam, in the state of Washington, was the largest hydrolectric project in the world when it was built in the 1930's. It has an output of 10,080 megawatts. The Itaipu Dam, on the Parana River, along the border of Brazil and Paraguay, is also one of the largest hydroelectric dams in the world. It began operation in 1984 and is capable of producing 13,320 megawatts.

Dikes, dams and similar structures have been used for centuries around the world for protection against flooding. The largest sea barrier in the world is a 2-mile-long surge barrier in the Oosterschelde estuary of the Netherlands, constructed from 1958 to 1986. Called the Dutch Delta Plan, the purpose of this project was to reduce the danger of catastrophic flooding. The impetus that brought this project to fruition was a catastrophic flood in the area in 1953. A major part of the barrier design consists of sixty-five huge concrete piers, weighing in at 18,000 tons each. These piers support tremendous 400-ton steel gates to create the sea barrier. The lifting and placement of these huge concrete piers exceeded the capabilities of any existing cranes, so a special U-shaped ship was built and equipped with gantry cranes. The project used computers to help in guidance and placement of the piers. A stabilizing foundation used for the concrete piers consists of foundation mattresses made up of layers of sand, fine gravel, and coarse gravel. Each foundation mattress is more than 1 foot thick and more than 650 feet by 140 feet, with a smaller mattress placed on top.

IMPACT ON INDUSTRY

In view of its status as the second oldest engineering discipline and the essential nature of the type of work done by civil engineers, it seems reasonable that civil engineering is well established as an important field of engineering around the world. Civil engineering is the largest field of engineering in the United States. The U.S. Bureau of Labor Statistics estimates that 278,400 civil engineers were employed in the United States in 2008. The bureau also projected that civil engineering employment would grow at the rate of 24 percent rate until 2018, which is much faster than average for all occupations. Civil engineers are employed by a wide variety of government agencies, by universities for research and teaching, by consulting engineering firms, and by industry.

Consulting Engineering Firms. This is the largest sector of employment for civil engineers. There are many consulting engineering firms around the world, ranging in size from small firms with a few employees to very large firms with thousands of employees. In 2010, the *Engineering News Record* identified the top six U.S. design firms: AECOM Technology, Los Angeles; URS, San Francisco; Jacobs, Pasadena, California; Fluor, Irving, Texas; CH2M Hill, Englewood, Colorado; and Bechtel, San Francisco. Many consulting engineering firms have some electrical engineers and mechanical engineers, and some even specialize in those areas; however, a large proportion of engineering consulting firms are made up predominantly of civil engineers. About 60 percent of American civil engineers are employed by consulting engineering firms.

Construction Firms. Although some consulting engineering firms design and construct their own projects, some companies specialize in constructing projects designed by another firm. These companies also use civil engineers. About 8 percent of American civil engineers are employed in the nonresident building construction sector.

Other Industries. Some civil engineers are employed in industry, but less than 1 percent of American civil engineers are employed in an industry other than consulting firms and construction firms. The industry sectors that hire the most civil engineers are oil and gas extraction and pipeline companies.

Government Agencies. Civil engineers work for many federal, state, and local government agencies. For example, the U.S. Department of Transportation uses civil engineers to handle its many highway and other transportation projects. Many road or highway projects are handled at the state level, and each state department of transportation employs many civil engineers. The U.S. Corps of Engineers and the Department of the Interior's Bureau of Reclamation employ many civil engineers for their many water resources projects. Many cities and counties have one or more civil engineers as city or county engineers, and many have civil engineers in their public works departments. About 15 percent of American civil engineers are employed by state governments, about 13 percent by local government, and about 4 percent by the federal government.

University Research and Teaching. Because civil engineering is the largest field of engineering, almost every college of engineering has a civil engineering department, leading to a continuing demand for civil engineering faculty members to teach the next generation of civil engineers and to conduct sponsored research projects. This applies to universities not only in the United States but also around the world.

Careers and Course Work

A bachelor's degree in civil engineering is the requirement for entry into this field. Registration as a professional engineer is required for many civil engineering positions. In the United States, a graduate from a bachelor's degree program accredited by the Accreditation Board for Engineering and Technology is eligible to take the Fundamentals of Engineering exam to become an engineer in training. After four years of professional experience under the supervision of a professional engineer, one is eligible to take the Professional Engineer exam to become a registered professional engineer.

A typical program of study for a bachelor's degree in civil engineering includes chemistry, calculus and differential equations, calculus-based physics, engineering graphics/AutoCAD, surveying, engineering mechanics, strength of materials, and perhaps engineering geology, as well as general education courses during the first two years. This is followed by fluid mechanics, hydrology or water resources, soil mechanics, engineering economics, and introductory courses for transportation engineering, structural engineering, and environmental engineering, as well as civil engineering electives to allow specialization in one of the areas of civil engineering during the last two years.

A master's degree in civil engineering that provides additional advanced courses in one of the areas of specialization, an M.B.A., or engineering management master's degree complement a bachelor's of science degree and enable their holder to advance more rapidly. A master's of science degree would typically lead to more advanced technical positions, while an M.B.A. or engineering management degree would typically lead to management positions.

Anyone aspiring to a civil engineering faculty or research position must obtain a doctoral degree. In that case, to provide proper preparation for doctoral level study, any master's-level study should be in pursuit of a research-oriented master of science degree rather than a master's degree in engineering or a practice-oriented master of science degree.

Social Context and Future Prospects

Civil engineering projects typically involve basic infrastructure needs such as roads and highways, water supply, wastewater treatment, bridges, and public buildings. These projects may be new construction or repair, maintenance or upgrading of existing highways, structures, and treatment facilities. The buildup of such infrastructure since the beginning of the twentieth century has been extensive, leading to a continuing need for the repair, maintenance, and upgrading of existing structures. Also, governments tend to devote funding to infrastructure improvements to generate jobs and create economic activity during economic downturns. All of this leads to the projection for a continuing strong need for civil engineers.

Harlan H. Bengtson, B.S., M.S., Ph.D.

Further Reading

Arteaga, Robert R. *The Building of the Arch.* 10th ed. St. Louis, Mo.: Jefferson National Parks Association, 2002. Describes how the Gateway Arch in St. Louis, Missouri, was built up from both sides and came together at the top. Contains excellent illustrations.

Davidson, Frank Paul, and Kathleen Lusk-Brooke, comps. *Building the World: An Encyclopedia of the Great Engineering Projects in History.* Westport, Conn.: Greenwood Press, 2006. Examines more than forty major engineering projects from the Roman aqueducts to the tunnel under the English Channel.

Hawkes, Nigel. *Structures: The Way Things Are Built.* 1990. Reprint. New York: Macmillan, 1993. Discusses many well-known civil engineering projects. Chapter 4 contains information about seventeen projects, including the Great Wall of China, the Panama Canal, and the Dutch Delta Plan. Contains illustrations and discussion of the effect of the projects.

National Geographic Society. *The Builders: Marvels of Engineering.* Washington, D.C.: National Geographic Society, 1992. Documents some of the most ambitious civil engineering projects, including roads, canals, bridges, railroads, skyscrapers, sports arenas, and exposition halls. Discussion and excellent illustrations are included for each project.

Weingardt, Richard G. *Engineering Legends: Great American Civil Engineers—Thirty-two Profiles of Inspiration and Achievement.* Reston, Va.: American Society of Civil Engineers, 2005. Looks at the lives of civil engineers who were environmental experts, transportation trendsetters, builders of bridges, structural trailblazers, and daring innovators.

Web Sites

American Institute of Architects
http://www.aia.org

American Society of Civil Engineers
http://www.asce.org

Institution of Civil Engineers
http://www.ice.org.uk

WBGH Educational Foundation and PBS
http://www.pbs.org/wgbh/buildingbig/index.html

See also: Architecture; Bridge Design and Barodynamics; Hydraulic Engineering; Transportation Engineering.

CLIMATE ENGINEERING

FIELDS OF STUDY

Atmospheric chemistry; ecology; meteorology; plant biology; ecosystem management; marine systems and chemistry; geoengineering; environmental engineering; aeronautical engineering; naval architecture; engineering geology; applied geophysics; Earth system modeling.

SUMMARY

Climate engineering (more commonly known as geoengineering) is a field of science that aims to deliberately control both micro (local) and macro (global) climates—by actions such as seeding clouds or shooting pollution particles into the upper atmosphere to reflect the Sun's rays—with the intention of reversing the effects of global warming.

KEY TERMS AND CONCEPTS

- **Albedo:** Measure of the reflectivity of the Earth surface (that is, the proportion of solar radiation or energy that is reflected back into space).
- **Carbon Dioxide Removal (CDR):** Geoengineering techniques and technologies that remove carbon dioxide from the atmosphere in an attempt to combat climate change and global warming.
- **Climate Change:** Alterations over time in global temperatures and rainfall because of human-caused increases in greenhouse gases, such as carbon dioxide and methane.
- **Fossil Fuel:** Deposit of either solid (coal), liquid (oil), or gaseous (natural gas) hydrocarbon derived from the decomposition of organic plant and animal matter over many million of years within the Earth's crust.
- **Ocean Acidification:** Process in which the pH (acidity) level of the Earth's oceans becomes more acidic because of the increase in atmospheric carbon dioxide.
- **Phytoplankton:** Microscopic photosynthesizing aquatic plants that are responsible for absorbing carbon dioxide and producing more than half of the world's oxygen.
- **Radiative Forcing:** Measure of the effect that climatic factors have in modifying the energy balance (incoming and outgoing energy) of the Earth's atmosphere.
- **Solar Radiation Management (SRM):** Geoengineering techniques and technologies that reflect solar radiation from the Earth's atmosphere (or surface) back into space in an attempt to combat climate change and global warming.
- **Stratosphere:** The region of the Earth's atmosphere (approximate altitude of 10 to 50 kilometers) below the mesosphere and above the troposphere.

Definition and Basic Principles

The term "geoengineering" comes from the Greek word "geo," meaning earth, and the word "engineering," a field of applied science that incorporates and uses data from scientific, technical, and mathematical sources in the invention and execution of specific structures, apparatuses, and systems.

Not to be confused with geotechnical engineering, which is related to the engineering behavior of earth materials, geoengineering is a field of science that aims to manipulate both micro (local) and macro (global) climates with the intention of reversing the effects of climate change and global warming. Although geoengineering, or climate engineering, can be undertaken on the local level, such as cloud seeding, most of the proposed technologies are based on a worldwide scale for the wholesale mediation of the global climate and the effects of climate change.

Background and History

Human society and ancient cultures had long believed power over the weather to be the province of gods, but the advent of human-made climate control began with the concept of rainmaking (pluviculture) in the mid-nineteenth century. James Pollard Espy, an American meteorologist, was instrumental in developing the thermal theory of storms. Although such a discovery placed him in the annals of scientific history, he also became known (and disparaged) for his ideas regarding artificial rainmaking. According to Espy, the burning of huge areas of forest would create sufficient hot air and updraft to create clouds and precipitation.

In 1946, Vincent Schaefer, a laboratory technician at General Electric Research Laboratory in New York, generated a cloud of ice from water droplets that had been supercooled by dry ice. That same year, Bernard Vonnegut discovered that silver iodide smoke produced the same result. These processes came to be known as cloud seeding. Cloud seeding attempts to encourage precipitation to fall in arid or drought-stricken agricultural areas by scattering silver iodide in rain clouds. Traditional climate modification has generally been limited to cloud seeding programs on a regional or local level.

In the following years, particularly during the Cold War, researchers in countries including the United States and the Soviet Union investigated climate control for its potential as a weapon. The Soviets also looked at climate control to warm the frozen Siberian tundra. However, in the 1970's, as concern regarding the greenhouse effect began to be expressed within conventional scientific circles, the concept of climate engineering took on new global significance.

Cesare Marchetti, an Italian physicist, first coined the term "geoengineering" in 1977. The word initially described the specific process of carbon dioxide capture and storage in the ocean depths as a means of climate change abatement. Since then, however, the term "geoengineering" has been used to cover all engineering work performed to manipulate the global and local climate. In later years, many researchers have begun using the term "climate engineering," which is a more accurate way to describe this type of applied science.

How It Works

Significant scientific evidence points to human activities, particularly the burning of fossil fuels, as playing a role in global climate change. An increase in the level of greenhouse gases, in particular carbon dioxide and methane, has been identified as the main culprit in relation to global climate change. The majority of scientists agree that the safest and best way to tackle climate change is through a reduction in fossil fuel consumption via the implementation of green technology, societal change, and industry regulation. However, although carbon reduction technology is already available and affordable, many scientists are increasingly concerned that carbon reduction schemes will not be introduced in time to stop the possible and probable effects of climate change. As a result, interest in geoengineering technology as a means to provide rapid solutions to climate change has increased.

There are two main areas of geoengineering research—carbon dioxide removal (CDR) and solar radiation management (SRM). Techniques based on CDR propose to remove carbon dioxide from the atmosphere and store it, while SRM techniques seek to reflect solar radiation from the Earth's atmosphere (or surface) back into space. Despite having the same objective of combating climate change and reducing global temperatures, these two approaches differ significantly in regard to their implementations, time scales, temperature effects, and possible consequences. According to the Royal Society of London, CDR techniques "address the root cause of climate change by removing greenhouse gases from the atmosphere," whereas SRM techniques "attempt to offset effects of increased greenhouse gas concentrations by causing the Earth to absorb less solar radiation."

The CDR approach removes carbon dioxide from the atmosphere and sequesters it underground or in the ocean. Many consider this approach to be the more attractive of the two as it not only helps reduce global temperatures but also works to combat issues such as ocean acidification caused by escalating carbon dioxide levels. Conversely, SRM techniques have no effect on atmospheric carbon dioxide levels, instead using reflected sunlight to reduce global temperatures.

Applications and Products

Traditionally, climate modification took the form of regional or local cloud seeding programs. In cloud seeding, a substance, usually silver iodide, is scattered in rain clouds to cause precipitation in arid or drought-stricken agricultural areas. By the start of the twenty-first century, however, deleterious global climate change and increasing levels of atmospheric carbon dioxide had pushed climate engineering to the forefront of science. Geoengineering technology, once considered as more fringe then functional, is being investigated as a possible weapon in the fight against global warming.

Despite the surge in research and interest, no longitudinal or large-scale geoengineering projects have been conducted. Geoengineering applications and products are generally regarded as highly speculative and environmentally untested, with significant ambiguity in regard to global and institutional regulation. Although climate engineering theories abound, only a limited few have captured global attention for their feasibility and applicability.

Iron Fertilization of the Oceans. The intentional introduction of iron into the upper layers of certain areas of the ocean to encourage phytoplankton blooms is a form of CDR. The concept relies on the fact that increasing certain nutrients—such as iron—in nutrient-poor areas stimulates phytoplankton growth. Carbon dioxide is absorbed from the surface of the ocean during the processes of photosynthesis; when the phytoplankton, marine animals, and plankton die and sink in the natural cycle, that carbon is removed from the atmosphere and sequestered in the ocean's depths.

Scrubbers and Artificial Trees. Both scrubbers and artificial trees aim to remove and store carbon dioxide from the Earth's atmosphere and assist in reducing the effect of climate change. Scrubbing towers involve the use of large wind turbines funneling air into specially designed structures, where the air reacts with a number of chemicals to form water and carbonate precipitates, essentially capturing the carbon. These carbon precipitates can then be stored. The use of artificial trees seeks the same result but by a different method. Large artificial trees or structures act as filters to capture and convert atmospheric carbon dioxide into a carbonate, which is then removed and stored.

Biochar. Biochar is a form of charcoal created from the pyrolysis (chemical decomposition by heating) of plant and animal waste. It captures and stores carbon by sequestering it in biomass. The biochar can be returned to the soil as fertilizer, as it helps the soil retain water and necessary nutrients, as well as to store carbon.

Stratospheric Sulfur Aerosols. Stratospheric sulfur aerosols are minute sulfur-rich particles that are found in the Earth's stratosphere and are often observed following significant volcanic activity (such as after the 1991 Mount Pinatubo eruption). The presence of these aerosols in the stratosphere results in a cooling effect. The SRM geoengineering technique of intentionally releasing sulfur aerosols into the stratosphere is based on the concept that they would produce a cooling or dimming effect by reflecting solar radiation. A workable delivery system has not yet been developed, but proposals include using high-altitude aircraft, balloons, and rockets.

Orbital Mirrors and Space Sunshades. The SRM technique of orbital mirrors and space sunshades entails the release of many billions (possibly trillions) of small reflective objects at a Lanrangian point in space to partially reflect solar radiation or impede it from entering the Earth's atmosphere. The theory is that the decrease in sunlight hitting the Earth's surface would help decrease average global temperatures.

Marine Cloud Whitening. Marine cloud whitening involves increasing the reflective properties of cloud cover so that solar radiation entering the Earth's atmosphere is reflected back into space. Proposed methods for achieving this include mounting large-scale mist-producing structures on seafaring vessels. The theory is that the spray of minute water droplets released by these structures would increase cloud cover and whitening, which would in turn increase sunlight reflection.

Reflective Roofs. Reflective or white roofs are often considered to be the most cost effective and easily implemented SRM method of reducing global temperatures. The concept relies on reflecting solar radiation back into space by using white materials (or paint) on the surface of building roofs.

Impact on Industry

Geoengineering has been experiencing a revolution. For the most part, many scientists and governments have not been interested in climate engineering because of questions about its feasibility or have been opposed to research and implementation of such technology. Because of the global consequences of climate change, however, many government agencies, universities, industries, corporations, and businesses worldwide are becoming interested in researching and working in this field of applied science.

Government and University Research. The governments of the United States and the United Kingdom have increasingly expressed interest in researching geoengineering technology and determining its applicability. Although the U.S. government has traditionally shied away from the climate engineering controversy, the Congress heard testimony related to geoengineering in November, 2009. This follows on the July, 2009, joint seminar, "Geoengineering: Challenges and Global Impacts," held by the United

Kingdom's House of Commons, the Institute of Physics, the Royal Society of Chemistry, and the Royal Academy of Engineering. The seminar explored and discussed the possibility of applying geoengineering technology to help mitigate the effects of climate change.

Although a growing number of governments are studying the possible use of geoengineering technology to help alleviate the effects of climate change, many scientists and other experts hesitate to embrace and apply the technologies. They cite the unanticipated consequences of these yet untested theories and technologies.

An increasing number of universities are exploring the potential of climate engineering. Some of the more well-known institutions that conduct geoengineering research include the Department of Atmospheric Chemistry at the Max Planck Institute for Chemistry, the Carnegie Institution for Science's Department of Global Ecology at Stanford University, the Earth Institute at Columbia University, the Energy and Environmental Systems Group at the University of Calgary, and the Oxford Geoengineering Institute.

Major Organizations. Many world-renowned organizations have entered into the geoengineering debate. Although there is growing support for increasing research into such technology, many organizations are hesitant to fully endorse its actual implementation and use. Such organizations include the Intergovernmental Panel on Climate Change, which published information about climate engineering in its fourth assessment report, "Climate Change 2007," and the American Meteorological Society, which released a guarded geoengineering policy statement in 2009. One of the most comprehensive reports, *Geoengineering the Climate: Science, Governance, and Uncertainty* (2009), was issued by the Royal Society of London, the world's oldest scientific academy. The society detailed the potential applications and costs of geoengineering and stated that although some climate engineering schemes are utterly implausible or are being endorsed without due consideration, some are more realistic. Its report, the first by any national science academy, concludes that much greater research is required before the implementation of any such technology is undertaken.

Careers and Course Work

Most commonly, students who wish to pursue careers in climate engineering begin by majoring in scientific fields such as atmospheric science, marine systems, and civil engineering. However, given the multitude of fields covered in geoengineering, a career in this applied science could follow many different paths, and students should have a solid understanding of subjects such as atmospheric chemistry, ecology, meteorology, plant biology, ecosystem management, marine systems and chemistry, and engineering. The majority of graduate programs in this area are open to students who have backgrounds in engineering, applied sciences, or closely related disciplines. University research covers many areas, and students who obtain a doctorate or master's degree in climate engineering can expect to have careers in geoengineering research and design, atmospheric sciences, aeronautical and nautical engineering, and environmental management consulting.

Social Context and Future Prospects

In the past, global geoengineering was regarded as more science fiction than fact. With greenhouse gas emissions continuing to increase from the burning of fossil fuel, however, the concept of deliberately engineering the Earth's climate is garnering interest and gaining credibility.

Engineering the climate could assist in lowering atmospheric carbon dioxide and reducing the impact of climate change. Although the majority of geoengineering scientists stress that such technology should be used only for emergency quick fixes or as a last resort, many are also stressing that research into such technology is imperative. The 2009 Royal Society report strongly advocated increased research and recommended the world's governments allocate some £100 million ($165 million), collectively, per year to examine geoengineering options. Given the probable economic and environmental costs associated with climate change, many researchers have claimed that implementing geoengineering technology may be cheaper and more viable than doing nothing. Many researchers are also quick to state, however, that while dollar costs may be affordable, the possible environmental and economic costs if the technology fails to work or has unexpected deleterious effects may be immeasurable.

Despite the move into more mainstream science, climate engineering is still controversial and full of

Fascinating Facts About Climate Engineering

- Human activity, particularly the burning of fossil fuels such as oil and coal, is responsible for the release of some 30 billion metric tons of carbon dioxide into the Earth's atmosphere every year.
- Most climate engineering proposals fall into one of two fundamentally different approaches–the removal of carbon dioxide from the atmosphere or the reflection of solar radiation from Earth's atmosphere or surface back into space.
- Climate engineers claim they can cool the planet and reverse ice-sheet melting by mimicking a natural volcanic eruption through the release of sulfate aerosols into the Earth's stratosphere, where they will reflect sunlight back into space.
- Giant artificial trees, ocean fertilization, and trillions of space mirrors are some of the most popular proposals being considered and researched in the fight against global warming.
- Reflecting the Sun's rays back into the atmosphere could quickly affect global temperatures but would not have an impact on the levels of carbon dioxide in the atmosphere.
- Biochar, the end result of the burning of animal and agricultural waste, can help clean the air by storing carbon dioxide that would have been released into the atmosphere during decomposition of the waste, and it assists plants by aiding in the storage of carbon from photosynthesis.

significant technological, social, ethical, legal, diplomatic, and safety challenges. The concept of purposely modifying the climate to correct climate change caused by humans has been labeled as, at best, ironic and, at worst, catastrophic. Significant concern has been raised about encouraging geoengineering research and technology. For example, many conservation organizations are concerned that access to such technology not only will promote the "if we build it, they will use it" mentality but also will lessen the resolve of governments and people to tackle climate change by reducing people's ecological footprint and consumption of fossil fuels.

In addition, geoengineering technology may ignite tensions between nations The ethical ramifications of climate engineering are unclear, and significant confusion and uncertainty exists regarding who should implement and control the global thermostat. If a country implements technology to fix one area and inadvertently adversely alters the climatic patterns in another country, who pays for the mishap? The consequences of climate manipulation on a global scale will almost certainly be unequal across nations. Such concerns stress the importance of conducting further research and of using caution regarding any technological advance in geoengineering.

Christine Watts, Ph.D., B.App.Sc., B.Sc.

FURTHER READING

Flannery, Tim F. *The Weather Makers: How Man Is Changing the Climate and What It Means for Life on Earth.* New York: Grove Press, 2006. A comprehensive look at how human activity has influenced the global climate and its impact on life on the Earth.

Keith, David. "Geoengineering the Climate: History and Prospect." *Annual Review of Energy and the Environment* 25 (November, 2000): 245-284. Provides an interesting review of the history of climate control technology and future directions.

Launder, Brian, and J. Michael T. Thompson, eds. *Geoengineering Climate Change: Environmental Necessity or Pandora's Box?* New York: Cambridge University Press, 2010. Presents a comprehensive examination of the problems of climate change and the potential of different geoengineering technologies.

The Royal Society. *Geoengineering the Climate: Science, Governance, and Uncertainty.* London: The Royal Society, 2009. Twelve experts in the fields of science, economics, law, and social science conducted this study of the main techniques of climate engineering, focusing on how well they might work and examining possible consequences.

WEB SITES

American Geophysical Union
Geoengineering the Climate
http://www.agu.org/sci_pol/positions/geoengineering.shtml

Intergovernmental Panel on Climate Change
http://www.ipcc.ch

See also: Climate Modeling; Climatology; Environmental Engineering.

CLIMATE MODELING

FIELDS OF STUDY

Physics; environmental science; oceanography; meteorology; climatology; atmospheric physics; earth sciences; computer science; computer programming; advanced mathematics; biology; sociology; chemistry; game theory; statistics; agriculture; forestry; anthropology.

SUMMARY

The goal of climate models is to provide insight into the Earth's climate system and the interactions among the atmosphere, oceans, and landmasses. Computer climate modeling provides the most effective means of predicting possible future changes and the potential effect such changes might have on societies and ecosystems. Shifting multiple variables and the unpredictability of climate model components have produced findings that fuel intense scientific and political debate. Climate models will continue to be refined as computing power increases, allowing models to be run with more integrative components and using wider, more refined space and time scales.

KEY TERMS AND CONCEPTS

- **Climate:** Regional weather conditions averaged over an extended time period.
- **Climate Modification:** Changes to the climate due to either natural processes or human activities.
- **Climate System:** System consisting of the atmosphere, hydrosphere, cryospher, land surface, and biosphere and their interactions.
- **Cryosphere:** Portion of Earth covered by sea ice, mountain glaciers, snow cover, ice sheets, or permafrost.
- **Feedback:** Process changing the relationship between a forcing agent acting on climate and the climate's response to that agent.
- **General Circulation Model:** Model simulating the state of the entire atmosphere, ocean, or both at chosen locations over a discrete, selected incremental time scale.
- **Global Warming:** Event characterized by measurable increases in annual mean surface temperature of the Earth, most noticeably occurring since about 1860.
- **Greenhouse Gases:** Atmospheric gases that absorb infrared radiation; the most important are carbon dioxide, methane, water vapor, and chlorofluorocarbons.
- **Infrared Absorbers:** Molecules such as carbon dioxide and water vapor that absorb electromagnetic radiation at infrared wavelengths.
- **Infrared Radiation:** Electromagnetic radiation with wavelengths between microwaves and visible red light.
- **Microclimate:** Local climate regime distinguished by physical characteristics and processes associated with local geography.
- **Radiative Cooling:** Process by which the Earth's surface cools by emitting longwave radiation.
- **Regime:** Preferred or dominant state of the climate system.

Definition and Basic Principles

All climate models are mathematical models derived from differential equations defining the principles of physics, chemistry, and fluid dynamics driving the observable processes of the Earth's climate. Climate models can be highly complex three-dimensional computer simulations using multiple variables and tens of thousands of differential equations requiring trillions of computations or fairly simple two-dimensional projections with a single equation defining a sole observable process. In all climate models, each additional physical process incorporated into the model increases the level of complexity and escalates the mathematical parameters needed to define the process's potential effects.

Background and History

The ability to predict weather and climate patterns could have a major impact on the health and well-being of societies. The possibility to warn of impending harm, prepare for changes, estimate outcomes, and chose more opportune times for essential human activities has been a dream of cultures for thousands of years. Although observational records of climate phenomena have been kept and interpreted for generations, the advent of computer models and advanced procedures for gathering and assimilating global data allows for quantitative assessments of the complex interactions between climate processes.

Climate models are governed by the laws of physics, which produce climate system components affecting the atmosphere, cryosphere, oceans, and land. No climate model would exist without an understanding of Newtonian mechanics and the fundamental laws of thermodynamics. By the late eighteenth century, scientific understanding of these physical laws allowed for numerical calculations to define observable climate processes. By the early twentieth century, scientists understood atmospheric phenomena as the product of preceding phenomena defined by physical laws. If one has accurate observable data of the atmosphere at a particular time and understands the physical laws under which atmospheric phenomena take place, it is possible to predict the outcome of future atmospheric changes. Mathematical equations were subsequently developed to define atmospheric motions and simulate observable processes and features of climate system components.

By the late 1940's, electronic computing provided a means by which greater numbers of equations could be assigned to data and calculated to define observable climate phenomena. Throughout the 1950's, predictive climate models were refined through changes to equations, resulting in modifications to models of atmospheric circulation. During the 1960's, similar equations were used to define ocean, land, and ice processes and their role in climate. Once this was done, the development of coupled models began. Coupled models involve combining observational processes to achieve more realistic outcomes. For each new process, however, multiple observations must be added and defined by additional equations. As the quality of observational data increases, the complexity of the climate model grows. Each observation of sea ice movement, cloud density, atmospheric chemistry, land cover, water temperature, precipitation, and human activities can be defined mathematically and added to a climate model's program. By defining the multiple variables of climate processes and adding these to model programs, scientists became able to produce predictive simulations of climate ranging from small regional models to global general circulation models.

How It Works

The purpose of all climate models is to simulate, over a given time period, regional or global patterns of climatic processes: wind, temperature, precipitation, ocean currents, sea levels, and sea ice. By imposing changes to the physical process within a model simulation—such as altering the amount of solar input, carbon dioxide, or ice cover—predictions can be made concerning possible future climate scenarios. Although models constructed from purely observational evidence result in less accurate predictions, the use of computers to process numerical models representing multiple levels of climate system feedbacks improves the accuracy of predictions for multiple components of the climate model.

When formulating a truly practical climate model, all components of Earth's climate system must be taken into consideration. The model must represent the atmosphere, the oceans, ice and snow cover on both land and sea, and landmasses, including their biomass. The interactions between these climate components occur in multiple ways, at differing intensities, and over varied time scales: Practical climate models must be able to reflect these dynamics.

The key to understanding climate and climatic change is energy balance. Multiple feedback processes occur within Earth's climate system, and these interactions constantly fluctuate to maintain an efficient energy balance. Positive feedback processes amplify variations to the climate system; negative feedbacks dampen them. Examples of feedbacks include radiation feedback from clouds and water vapor; changes to reflective power from ice, snow, and deserts; and changes in ocean temperatures. These and many other forms of feedback must be considered when preparing a climate model. Because the physics of energy transfer between climate system processes is so varied, resolution scale becomes a major problem in preparing data for modeling. Flux compensations must be made in models to account for differences in energy transfers among atmosphere, oceans, and land. Heat, momentum, water densities, size of phenomena such as currents or eddies, energy radiation, rainfall, wind, humidity, barometric pressure, atmospheric chemistry, and natural phenomena such as forest fires and volcanic eruptions all must be considered.

The purpose of preparing climate models is to answer the fundamental questions of what can be reliably predicted about climate and at what time scales these predictions can be made. The difficulty with all climate modeling is that the computers used must be powerful enough to deal with the complexity of observable climate system data; that the data used must be as free as possible of errors in interpretation; and that all observational data have limitations and these limit interpretive predictability. These factors are the limitations confining all existing climate-modeling capabilities and set the boundary ranges for predicting climate phenomena in space and time.

Variability of climate over space and time scales is a key component of Earth's history. Throughout Earth's history, climate has naturally fluctuated, often to extremes in comparison to present-day climatic conditions. Being able to distinguish between naturally occurring and human-induced climate changes is an important aspect of climate modeling. The implications of climate models suggesting that human behaviors are altering the atmosphere—and in turn disrupting the planet's energy balance—drives vigorous debates and conflict.

Applications and Products

All climate models are simulations that make predictions about climate processes. Climate change skeptics and deniers make that case that model predictions are merely simulations and not "real" data produced by "real" science. Without climate models, however, there would be no climate data. No climate observation, no satellite imagery or observation, no meteorological data, no atmospheric sampling, no remote sensing, and no chemical analysis exists without passing through a series of computer data models. Almost everything known about Earth's climatic processes exists because of models. At present, almost all knowledge and understanding of climate change comes from three kinds of computer models: simulation models of weather and climate; reanalysis models re-creating climate history from historical weather data, including worldwide weather stations and remote data platforms; and data models combining and adjusting measurements from many different sources, including ice cores, gas analysis, tree rings, observation data, and archaeological and geological data. Because the amount of weather and climate data available are so vast and diverse and from so many different sources and at differing time scales, it can be managed and organized only by computer analysis. The result of modern computer-simulated climate models has been to create a stable and reliable basis for making predictions about climate change and the byproducts of that change.

Impact on Industry

Understanding how climate works and having the ability to predict possible changes to climate over certain regions and at specific time scales can have substantial impacts on the social, economic, and cultural well-being of any nation. More accurate prediction of cycles of increased precipitation or drought, rises in temperature, or shifts in oceanic currents allows people to determine more opportune times to plant and harvest crops, to change patterns of land use, to predict marine harvests, and to better plan other human activities. The ability to provide general predictions about these economically important aspects of society help control product costs and plan for social needs. High-quality climate models can also provide warnings that specific human behaviors may be affecting the environment and, as a result, people's quality of life. Climate models reflecting processes of deforestation, desertification, industrial insertion of greenhouse gases into the atmosphere, and ozone depletion all help establish grounds to alter human behaviors to increase survivability. Because of these factors and how they may affect the well-being and security of nations, governments throughout the world have created programs, earmarked research funds, and mandated agencies to study and model climate. A number of high-profile international treaties, most notably the Kyoto Protocol, have been initiated to address the observable and predicted effects of climate change.

Most major universities have climate change programs or research centers. A number of host government climate research centers and many universities have formed research consortiums to engage in regional climate studies and to share resources and supercomputing time. The result of high-profile interest in climate modeling related to issues of climate change provides large sums of research monies to qualified researchers and research facilities.

Although climate models are certainly helpful in predicting marine harvests, crop futures, energy needs, potential droughts, food shortages, and water resources, the biggest effect on industry from climate modeling is the concern they have generated

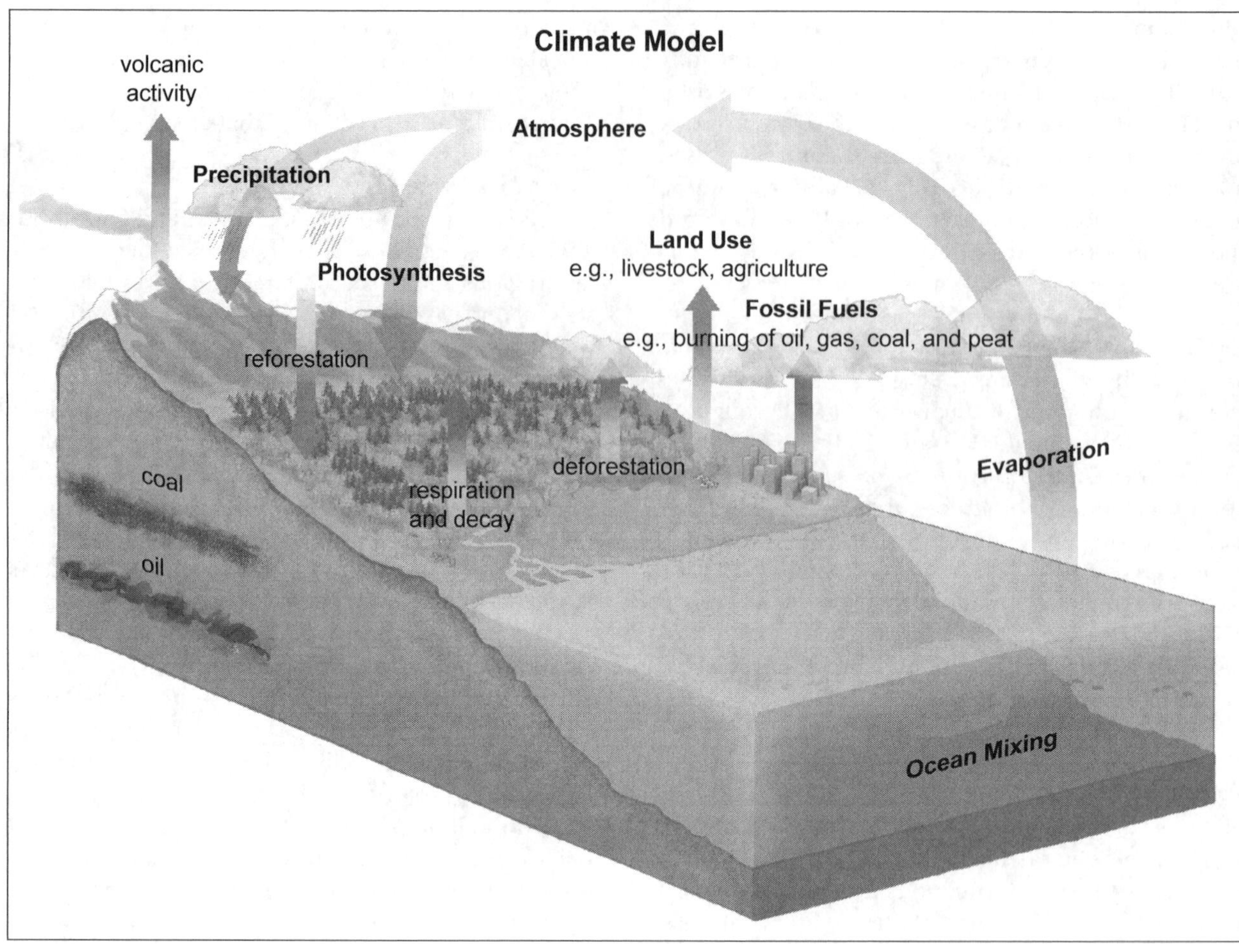

regarding certain human behaviors. The use of environmentally detrimental technologies and the release of large amounts of greenhouse gases such as carbon dioxide, methane, sulfur dioxide, and chlorofluorocarbons into the atmosphere has begun to significantly alter Earth's future livability. Individuals do not wish to hear that driving their automobiles or heating their homes is damaging the planet. Companies do not want to be told that their methods of doing business are producing more harm than good and that changing these methods will require large financial investments. Additionally, governments do not wish to alter their strategies for national growth because their attempts to modernize are environmentally detrimental.

Since the mid-nineteenth century, many of humankind's economic, technological, and cultural advancements have been achieved through the exploitation of fossil fuels and other resources. Certain methods of forest harvesting have been shown to promote desertification, disrupt water resources, change the amount of radiation from the sun reflected in a region, and alter regional climate patterns. Because trees absorb carbon dioxide, the loss of large tracts of forest lessens Earth's ability to remove and recycle carbon dioxide from the atmosphere. Burning petroleum and coal results in the discharge of greenhouse gases and the release of harmful particulates into the atmosphere. Climate models examining historical and observational climate data and human behaviors have demonstrated a direct link between the use of fossil fuels and changes in the atmosphere. Climate observations and models show that increased greenhouse gas levels have produced real, noticeable, historically unprecedented, human-induced changes in the atmosphere and to climate patterns. In addition, the insertion of industrial chemicals—such as chlorofluorocarbons—into the atmosphere is known to deplete ozone and have a serious effect on both the climate and human health.

Laws and Regulations. In attempts to control industrial and commercial release of greenhouse gases and other toxic chemicals into the atmosphere, governments have imposed regulations on consumers and companies. Laws and regulations have been enacted that mandate limits on automobile exhaust, industrial air pollution, aerosol propellants, power plant emissions, field burning, biofuels, and certain agricultural practices. The difficulty with these attempts to alter industrial, personal, and governmental behaviors is that their reach is limited. Although one nation, state, or individual seeks to reduce harmful emissions, not all nations, states, or individuals share the same goals. Earth's atmosphere is an open system of circulation without human-imposed boundaries; therefore, pollutants created in one region affect all others.

Financial Impact. The impact of climate modeling on industry is often financial, decreasing corporations' profits. As climate models become more accurate in their predictions, certain industrial practices are likely to be found clearly responsible for some negative atmospheric effects. Altering business practices often requires large financial investments in new technologies; these investments cut into short-term profits. Although such investments are likely to result in long-term growth and allow the business to survive, unquestionably, they negatively affect short-term profits. For many nations and all nonprofit industries, the bottom line is of utmost importance. Changing the way one does business, retooling, reequipping, changing priorities, changing business models, and altering national and institutional identities is unthinkable for many. Some nations and many businesses have reacted to regulations and attempts to formulate new laws restricting atmospheric pollution by denying the existence of climate change. They seek to maintain the status quo for as long as possible.

Fascinating Facts About Climate Modeling

- Some of the earliest known long-term observations leading to predictive climate models are those of ancient native peoples of South and Central America who noted regional climate changes related to what are now known as El Niño and La Niña ocean currents.
- In the early 1960's, Warren M. Washington and Akira Kasahara, scientists with the National Center for Atmospheric Research, developed a computer model of atmospheric circulation. Data were input to a CDC computer using punch cards and seven-channel digital magnetic tape, and data were output through two line printers, a card punch, a photographic plotter, and standard magnetic tape.
- In December, 2009, the Copenhagen United Nations Climate Conference was the most-searched topic on the Google Internet search engine. In December, 2010, "climate change" was one of the top phrases searched daily on the Internet.
- In 2010, the National Center for Atmospheric Research released the Community Earth System Model, which creates computer simulations of Earth's past, present, and future climates. Experiments using this model will be part of the Intergovernmental Panel on Climate Change's 2013-1014 assessments.
- The complexity of climate models requires that they be run on supercomputers. The fastest supercomputer in the United States as of November, 2010, was the Cray XT5, at the U.S. Department of Energy's Oak Ridge Leadership Computing Facility. It can perform at 1.75 petaflops (quadrillions of calculations per second).
- Pioneering climate modeler Warren M. Washington was awarded the National Medal of Science by President Barack Obama on November 17, 2010. Washington was also a member of the Intergovernmental Panel on Climate Change, which received the 2007 Nobel Peace Prize.

Careers and Course Work

Careers in climate modeling are limited. Most positions involving climate modeling are in academic institutions, government agencies, and private research organizations. The majority of climate modelers are employed within university settings, which usually means that their time must be divided between active research and teaching responsibilities. In some university consortiums, such as the University

Corporation for Atmospheric Research, climate modelers work in teams. Climate modelers working for government agencies focus on predictive applications to help fulfill their governmental branch mandates. At the international level, a number of climate modelers work under contract or are funded by the United Nations and may do research for any of a number of United Nations agencies or advisory boards. Climate modelers at the federal level may find work with the U.S. Geological Survey, the Department of Agriculture, all branches of the military and intelligence communities, the National Oceanic and Atmospheric Agency, the National Aeronautics and Space Administration, the National Center for Atmospheric Research, and the Department of Energy. A limited number of climate modeling positions are available in private research organizations that offer services to businesses, lobbying organizations, agricultural interests, commodities traders, maritime shipping and fishing concerns, and politicians.

Students interested in careers involving climate modeling need to take classes in chemistry, physics, atmospheric sciences, geosciences, meteorology, oceanography, biology, mathematics, statistics, computer science, and environmental science, and obtain a bachelor of science degree. Almost all careers in climate modeling require a graduate-level education. Although obtaining a master's degree may allow for an entry-level position or journeyman status within certain agencies, for nearly all climate-modeling opportunities in academia, civil service, or private research, a doctorate is a necessity. In graduate school, studies in advanced mathematics and computer programming combined with intensely focused studies in climatology, oceanography, and physics are the norm. Individual areas of research interest require students to narrow their courses to reflect the direction of their research and make course and seminar selections accordingly.

As interest in climate change issues continues to grow, the need for qualified climate modeling scientists will increase. The rate at which observable changes to the climate begin to reflect predictions made by climate models will most likely raise the value and significance of trained climatologists and their predictive modeling skills.

Social Context and Future Prospects

Using climate models to accurately predict future climate trends is the ultimate goal. If climate models can help define the difference between natural climate fluctuations and human-induced climate change, positive human activities can be developed to counter the impact of environmentally unsustainable behaviors. Accurate predictive climate models can also increase economic outcomes by indicating more opportune times for planting, harvesting, and fishing, and by predicting droughts and temperature shifts. Climate shifts are known to be associated with pandemic outbreaks of disease: Climate models may allow people to prepare in advance for disease-formulating conditions.

Existing climate modeling is limited by computational power and the nonlinear nature of climate system phenomena. As computing technology advances and observational data of climate processes over longer time scales becomes available, flux compensations will be more accurately defined and the accuracy of future climate models will increase. Public acceptance of climate model predictions will remain complicated as long as politics and economics, rather than observational facts, are allowed to drive the climate change debate.

Randall L. Milstein, B.S., M.S., Ph.D.

Further Reading

Edwards, Paul N. *A Vast Machine: Computer Models, Climate Data, and the Politics of Global Warming.* Cambridge, Mass.: MIT Press, 2010. Tells the history of how scientists learned to understand the atmosphere.

Kiehl, J. T., and V. Ramanathan. *Frontiers of Climate Modeling.* New York: Cambridge University Press, 2006. A good general overview of climate modeling, with an emphasis on how greenhouse gases are altering the climate system.

McGuffie, Kendall, and Ann Henderson-Sellers. *A Climate Modelling Primer.* West Sussex, England: John Wiley & Sons, 2005. Explains the basis and mechanisms of existing physical-observation-based climate models.

Mote, Philip, and Alan O'Neill, eds. *Numerical Modeling of the Global Atmosphere in the Climate System.* Boston: Kluwer Academic, 2000. Meant for those actively creating climate models, this work explains the uses of numerical constraints to define climate processes.

Robinson, Walter A. *Modeling Dynamic Climate Systems.* New York: Springer, 2001. A basic book for understanding climate modeling that includes a good description of how climate systems function and interact with each other and vary over time and space.

Trenberth, Kevin E., ed. *Climate System Modeling.* New York: Cambridge University Press, 2010. A comprehensive textbook covering the most important topics for developing a climate model.

Washington, Warren M., and Claire L. Parkinson. *An Introduction to Three-Dimensional Climate Modeling.* Sausalito, Calif.: University Science Books, 2005. An introduction to the use of three-dimensional climate models. Includes a history of climate modeling.

WEB SITES

American Meteorological Society
http://www.ametsoc.org

Intergovernmental Panel on Climate Change
http://www.ipcc.ch

National Center for Atmospheric Research
http://ncar.ucar.edu

National Oceanic and Atmospheric Administration
National Climatic Data Center
http://www.Ncdc.noaa.gov

National Weather Association
http://www.nwas.org

University Corporation for Atmospheric Research
http://www2.ucar.edu

U.S. Geological Survey
Climate and Land Use Change
http://www.usgs.gov/climate_landuse

World Meteorological Organization
Climate
http://www.wmo.int/pages/themes/climate/index_en.php

See also: Climate Engineering; Climatology; Industrial Pollution Control.

CLIMATOLOGY

FIELDS OF STUDY

Atmospheric sciences; meteorology; physical geography; climate change; climate classification; climate zones; tree-ring analysis; climate modeling; bioclimatology; climate comfort indices; hydroclimatology.

SUMMARY

Climatology deals with the science of climate, which includes the huge variety of weather events. These events change at periods of time that range from months to millennia. Climate has such a profound influence on all forms of life, including human life, that people have made numerous attempts to predict future climatic conditions. These attempts resulted in research efforts to try to understand future changes in the climate as a consequence of anthropogenic and naturally caused activity.

KEY TERMS AND CONCEPTS

- **Climate:** Average weather for a particular area.
- **Climograph:** Line graph that shows monthly mean temperatures and precipitation.
- **Coriolis Effect:** Rghtward (Northern Hemisphere) and leftward (Southern Hemisphere) deflection of air due to the Earth's rotation.
- **Energy Balance:** Balance between energy received from the Sun (shortwave electromagnetic radiation) and energy returned from the Earth (longwave electromagnetic radiation).
- **Front:** Boundary between air masses that differ in temperature, moisture, and pressure.
- **Greenhouse Effect:** Process in which longwave radiation is trapped in the atmosphere and then radiated back to the Earth's surface.
- **Occluded Front:** Front that occurs when a cold front overtakes a warm front and forces the air upward.
- **Semipermanent Low-Pressure Centers:** Semipermanent patterns of low pressure that occur in northern waters off the Alaskan Aleutian Islands and Iceland.
- **Sublimation:** Change from ice in a frozen solid state to water vapor in a gaseous state without going through a liquid state.
- **Troposphere:** Lowest level of the atmosphere that contains water vapor that can condense into clouds.

Definition and Basic Principles

"Weather" pertains to atmospheric conditions that constantly change, hourly and daily. In contrast, "climate" refers to the long-term composite of weather conditions at a particular location, such as a city or a state. Climate at a location is based on daily mean conditions that have been aggregated over periods of time that range from months and years to decades and centuries. Both weather and climate involve measurements of the same conditions: air temperature, water vapor in the air (humidity), atmospheric pressure, wind direction and speed, cloud types and extent, and the amount and kind of precipitation.

Estimates of ancient climates, going back several thousand years or more, are produced in various ways. For example, the vast amount of groundwater discovered in southern Libya indicates that during some period in the past, that part of the Sahara Desert was much wetter. In ancient Egypt, Nilometers, stone markers built along the banks of the Nile, were used to gauge the height of the river from year to year. They are similar to the staff gauges that are used by the U.S. Geological Survey to indicate stream or canal elevation. The height of the Nilometer reflects the extent of precipitation and associated runoff in the headwaters of the Nile in east central Africa.

Background and History

Measurements of precipitation were being made and recorded in India during the fourth century B.C.E. Precipitation records were kept in Palestine about 100 C.E., Korea in the 1440's, and in England during the late seventeenth century. Galileo invented the thermometer in the early 1600's. Physicist Daniel Fahrenheit created a measuring scale for a liquid-in-glass thermometer in 1714, and Swedish astronomer Anders Celsius developed the centigrade scale in 1742. Italian physicist Evangelista Torricelli, who worked with Galileo, invented the barometer in 1643.

The first attempt to explain the circulation of the atmosphere around the Earth was made by English astronomer Edmond Halley, who published a paper charting the trade winds in 1686. In 1735, English meteorologist George Hadley further explained the movement of the trade winds, describing what became known as a Hadley cell, and in 1831, Gustave-Gaspard Coriolis developed equations to describe the movement of air on a rotating Earth. In 1856, American meteorologist William Ferrel developed a theory that described the mid-latitude atmospheric circulation cell (Ferrel cell). In 1860, Dutch meteorologist Christophorus Buys Ballot demonstrated the relationship between pressure, wind speed, and direction (which became known as Buys Ballot's law).

The first map of average annual isotherms (lines connecting points having the same temperature) for the northern hemisphere was created by German naturalist Alexander von Humboldt in 1817. In 1848, German meteorologist Heinrich Wilhelm Dove created a world map of monthly mean temperatures. In 1845, German geographer Heinrich Berghaus prepared a global map of precipitation. In 1882, the first world map of precipitation using mean annual isohyets (lines connecting points having the same precipitation) appeared.

How It Works

Earth's Global Energy Balance. The Earth's elliptical orbit about the Sun ranges from 91.5 million miles at perihelion (closest to the Sun) on January 3, to 94.5 million miles at aphelion (furthest from the Sun) on July 4, averaging 93 million miles. The Earth intercepts about two-billionth of the total energy output of the Sun. Upon reaching the Earth, a portion of the incoming radiation is reflected back into space, while another portion is absorbed by the atmosphere, land, or oceans. Over time, the incoming shortwave solar radiation is balanced by a return to outer space of longwave radiation.

The Earth's atmosphere extends to an estimated height of about 6,000 miles. Most of it is made up of nitrogen (78 percent by volume) and oxygen (about 21 percent). Of the remaining 1 percent, carbon dioxide (CO_2) accounts for about 0.0385 percent of the atmosphere. This is a minute amount, but carbon dioxide can absorb both incoming shortwave radiation from the sun and outgoing longwave radiation from the Earth. The measured increase in carbon dioxide since the early 1900's is a major cause for concern as it is a very good absorber of heat radiation, which adds to the greenhouse effect.

Air Temperature. Air temperature is a fundamental constituent of climatic variation on the Earth. The amount of solar energy that the Earth receives is governed by the latitude (from the equator to the poles) and the season. The amount of solar energy reaching low-latitude locations is greater than that reaching higher-latitude sites closer to the poles. Another factor pertaining to air temperature is the fivefold difference between the specific heat of water (1.0) and dry land (0.2). Accordingly, areas near the water have more moderate temperatures on an annual basis than inland continental locations, which have much greater seasonal differences.

Anthropogenic (human-induced) changes in land cover in addition to aerosols and cloud changes can result in some degree of global cooling, but this is much less than the combined effect of greenhouse gases in global warming. The gases include carbon dioxide from the burning of fossil fuels (coal, oil, and natural gas), which has been increasing since the second half of the twentieth century. Other gases such as methane (CH_4), chlorofluorocarbons (CFCs), ozone (O_3), and nitrous oxide (NO_3) also create additional warming effects.

Air temperature is measured at 5 feet above the ground surface and generally includes the maximum and minimum observation for a twenty-four-hour period. The average of the maximum and minimum temperature is the mean daily temperature for that particular location.

Earth's Available Water. Water is a tasteless, transparent, and odorless compound that is essential to all biological, chemical, and physical processes. Almost all the water on the Earth is in the oceans, seas, and bays (96.5 percent) and another 1.74 percent is frozen in ice caps and glaciers. Accordingly, 98.24 percent of the total amount of water on this planet is either frozen or too salty and must be thawed or desalinated. About 0.76 percent of the world's water is fresh (not saline) groundwater, but a large portion of this is found at depths too great to be reached by drilling. Freshwater lakes make up 0.007 percent, and atmospheric water is about 0.001 percent of the total. The combined average flows of all the streams on Earth—from tiny brooks to the mighty Amazon River—account for 0.0002 percent of the total.

Air Masses. The lowest layer of the atmosphere is the troposphere, which varies in height from 10 miles at the equator and lower latitudes to 4 miles at

Fascinating Facts About Climatology

- Mean sea levels for the world have increased an average of 0.07 inches per year from 1904 to 2003, an amount that is much larger than the average rate in the past.
- Serbian astronomer Milutin Milankovitch developed the astronomical hypothesis in 1938 by observing that glacial and interglacial periods were related to insolation variations caused by small cycles in the Earth's axial rotation and orbit about the Sun.
- The larger input of solar energy received at and near the equator creates a very large circulation path known as the Hadley cell. The air converges at a narrow zone known as the intertropical convergence zone, which seasonally varies from 15 degrees south in northern Australia to 25 degrees north in northern India, representing a latitudinal shift of 40 degrees.
- The coldest temperatures in the world were recorded at the Russian weather station at Vostok, Antarctica, at 78 degrees south: –127 degrees Fahrenheit in 1958 and –128.5 degrees Fahrenheit on July 21, 1983, at an ice sheet elevation of 11,220 feet.
- There is a fivefold difference between the specific heat of water (1.0) and dry land (0.2). This results in a one- to two-month lag in average maximum and minimum temperatures after the summer and winter solstices.
- Atlantic storm names can be repeated after six years unless the hurricane was particularly severe. The names of noteworthy storms–such as Camille (1969), Hugo (1989), Andrew (1992), Floyd (1999), and Katrina (2005)–are not used again, so as to avoid confusion in later years.

the poles. Different types of air masses within the troposphere can be delineated on the basis of their similarity in temperature, moisture, and to a certain extent, air pressure. Air masses develop over continental and maritime locations that strongly determine their physical characteristics. For example, an air mass starting in a cold, dry interior portion of a continent would develop thermal, moisture, and pressure characteristics that would be substantially different from those of an air mass that developed over water. Atmospheric dynamics also allow air masses to modify their characteristics as they move from land to water and vice versa.

Air mass and weather front terminology were developed in Norway during World War I. The Norwegian meteorologists were unable to get weather reports from the Atlantic theater of operations; consequently, they developed a dense network of weather stations that led to impressive advances in atmospheric modeling.

Greenhouse Effect. Selected gases in the lower parts of the atmosphere trap heat and radiate some of that heat back to Earth. If there was no natural greenhouse effect, the Earth's overall average temperature would be close to 0 degrees Fahrenheit rather than 57 degrees Fahrenheit.

The burning of coal, oil, and gas makes carbon dioxide the major greenhouse gas. Carbon dioxide accounts for nearly half of the total amount of heat-producing gases in the atmosphere. In mid-eighteenth century Great Britain, before the Industrial Revolution, the estimated level of carbon dioxide was about 280 parts per million (ppm). Estimates for the natural range of carbon dioxide for the past 650,000 years are 180-300 ppm. All of these values are less than the October, 2010, estimate of 387 ppm. Since 2000, atmospheric carbon dioxide has been increasing at a rate of 1.9 ppm per year. The radiative effect of carbon dioxide accounts for about one-half of all the factors that affect global warming. Estimates of carbon dioxide values at the end of the twenty-first century range from 490 to 1,260 ppm.

The second most important greenhouse gas is methane (CH_4), which accounts for about 14 percent of all of the global warming factors. This gas originates from the natural decay of organic matter in wetlands, but anthropogenic activity in the form of rice paddies, manure from farm animals, the decay of bacteria in sewage and landfills, and biomass burning (both natural and human-induced) doubles the amount produced.

Chlorofluorocarbons (CFCs) absorb longwave energy (warming effect) but also have the ability to destroy stratospheric ozone (cooling effect). The warming radiative effect is three times greater than the cooling effect. CFCs account for about 10 percent of all of the global warming factors. Tropospheric ozone (O_3) from air pollution and nitrous oxide (N_2O) from motor vehicle exhaust and bacterial emissions from nitrogen fertilizers account for about

10 percent and 5 percent, respectively, of all the global warming factors.

Several human actions lead to a cooling of the Earth's climate. For example, the burning of fossil fuels results in the release of tropospheric aerosols, which acts to scatter incoming solar radiation back into space, thereby lowering the amount of solar energy that can reach the Earth's surface. These aerosols also lead to the development of low and bright clouds that are quite effective in reflecting solar radiation back into space.

Applications and Products

Climatology involves the measurement and recording of many physical characteristics of the Earth. Therefore, numerous instruments and methods have been devised to perform these tasks and obtain accurate measurements.

Measuring Temperature. At first glance, it would appear that obtaining air temperatures would be relatively simple. After all, thermometers have been around since 1714 (Fahrenheit scale) and 1742 (Celsius scale). However, accurate temperature measurements require a white (high-reflectivity) instrument shelter with louvered sides for ventilation, placed where it will not receive direct sunlight. The standard height for the thermometer is 5 feet above the ground.

Remote-Sensing Techniques. Oceans cover about 71 percent of the Earth's surface, which means that large portions of the world do not have weather stations and places where precipitation can be measured with standard rain gauges. To provide more information about precipitation in the equatorial and tropical parts of the world, the National Aeronautics and Space Administration (NASA) and the Japanese Aerospace Exploration Agency began a program called the Tropical Rainfall Monitoring Mission (TRMM) in 1997. The TRMM satellite monitors the area of the world between 35 degrees north and 35 degrees south latitude. The goal of the study is to obtain information about the extent of precipitation, its intensity, and length of occurrence. The major instruments on the satellite include radar to detect rainfall, a passive microwave imager that can acquire data about precipitation intensity and the extent of water vapor, and a scanner that can examine objects in the visible and infrared portions of the electromagnetic spectrum. The goal of collecting this data is to obtain the necessary climatological information about atmospheric circulation in this portion of the Earth so as to develop better mathematical models for determining large-scale energy movement and precipitation.

Geostationary Satellites. Geostationary orbiting earth satellites (GOES) enable researchers to view images of the planet from what appears to be a fixed position. To achieve this, these satellites circle the globe at a speed that is in step with the Earth's rotation. This means that the satellite, at an altitude of 22,200 miles, will make one complete revolution in the same twenty-four hours and direction that the Earth is turning above the Equator. At this height, the satellite is in a position to view nearly half the planet at any time. On-board instruments can be activated to look for special weather conditions such as hurricanes, flash floods, and tornadoes. The instruments can also be used to make estimates of precipitation during storm events.

Rain Gauges. The accurate measurement of precipitation is not as simple as it may seem. Collecting rainfall and measuring it is complicated by the possibility of debris, dead insects, leaves, and animal intrusions occurring. Standards were established, although the various national climatological offices use more than fifty types of rain gauges. The location of the gauge, its height above the ground, the possibility for splash and evaporation, its distance from trees, and turbulence all affect the results. Accordingly, all gauge records are really estimates. Precipitation estimates are also affected by the number of gauges per unit area. The number of gauges in a sample area of 3,860 square miles for Britain, the United States, and Canada is 245, 10, and 3, respectively. Although the records are reported to the nearest 0.01 inch, discrepancies occur in the official records. It is important to have a sufficiently dense network of rain gauges in urban areas. Some experts think that 5 to 10 gauges per 100 square miles is necessary to obtain an accurate measure of rainfall.

Doppler Radar. Doppler radar was first used in England in 1953 to pick up the movement of small storms. The basic principle behind Doppler radar is that the back-scattered radiation frequency detected at a certain location changes over time as the target, such as a storm, moves. The mode of operation requires a transmitter that is used to send short but powerful microwave pulses. When a foreign object (or target) is intercepted, some of the outgoing energy is returned to the transmitter, where a receiver can pick up the signal. An image (or echo) from the target can then be enlarged and shown on a screen. The target's distance is revealed by the time between transmission and return. The radar screen can indicate not only where the precipitation is taking place but also its intensity by the amount of the echo's brightness. Doppler radar has developed into a very useful device for determining the location of storms and the intensity of the precipitation and for obtaining good estimates of the total amount of precipitation.

Impact on Industry

Global Perspective. The World Meteorological Organization, headquartered in Geneva, Switzerland, was established to encourage weather station networks that would facilitate the acquisition of climatic data. In 2007, the organization decided to expand the global observing system and other related observing systems, including the global ocean observing system, global terrestrial observing system, and the global climate observing system. Data are being collected by 10,000 manned and automatic surface weather stations, 1,000 upper-air stations, more than 7,000 ships, 100 moored and 1,000 floating buoys that can drift with the currents, several hundred radars, and more than 3,000 commercial airplanes, which record key aspects of the atmosphere, land, and ocean surfaces on a daily basis.

Government Research. About 183 countries have meteorological departments. Although many are small, the larger countries have well-established organizations. In the United States, meteorology and climatology are handled by the National Oceanic and Atmospheric Administration. The agency's National Climatic Data Center has records that date back to 1880 and provide invaluable information about previous periods for the United States and the rest of the world. For example, sea ice in the Arctic Ocean typically reaches its maximum extent in March. The coverage at the end of March, 2010, was 5.8 million square miles. This was the seventeenth consecutive March with below-average coverage. The center issues many specialized climate data publications as well as monthly temperature and precipitation summaries for all fifty states.

University Research. In the United States, forty-eight states (except Rhode Island and Tennessee) have either a state climatologist or someone who has comparable responsibility. Most of the state climatologists are connected with state universities, particularly the colleges that started as land-grant institutions.

The number of cooperative weather stations established to take daily readings of temperature and precipitation in each of these states has varied in number since the late nineteenth century.

Industry and Business Sectors. The number, size, and capability of private consulting firms has increased over the years. Some of the first of these firms were frost-warning services that served the citrus and vegetable growers in Arizona, Florida, and California. These private companies expanded considerably as better forecasting and warning techniques were developed. For example, AccuWeather.com has seven global forecast models and fourteen regional forecast models for the United States and North America and also prepares a daily weather report for *The New York Times.*

Careers and Course Work

Although many consider meteorologists to be people who forecast weather, the better title for such a person is atmospheric scientist. For example, climatologists focus on climate change, and environmental meteorologists are interested in air quality. Broadcast meteorologists work for television stations. The largest number of jobs in the field are with the National Weather Service, which employs about one-third of the atmospheric scientists who work for the federal government.

Meteorologists are predicted to have above-average employment growth in the 2010's. Employment at the National Weather Station generally requires a bachelor's degree in meteorology, or at least twenty-four credits in meteorology courses along with college physics and physical science classes. Anyone who wants to work in applied research and development needs a master's degree. Research positions require a doctorate.

The median annual average salary in 2008 was $81,290. Entry-level positions at the National Weather Service earn about $35,000, and those in the highest 10 percent bracket earn more than $127,000.

Social Context and Future Prospects

Climate change may be caused by both natural

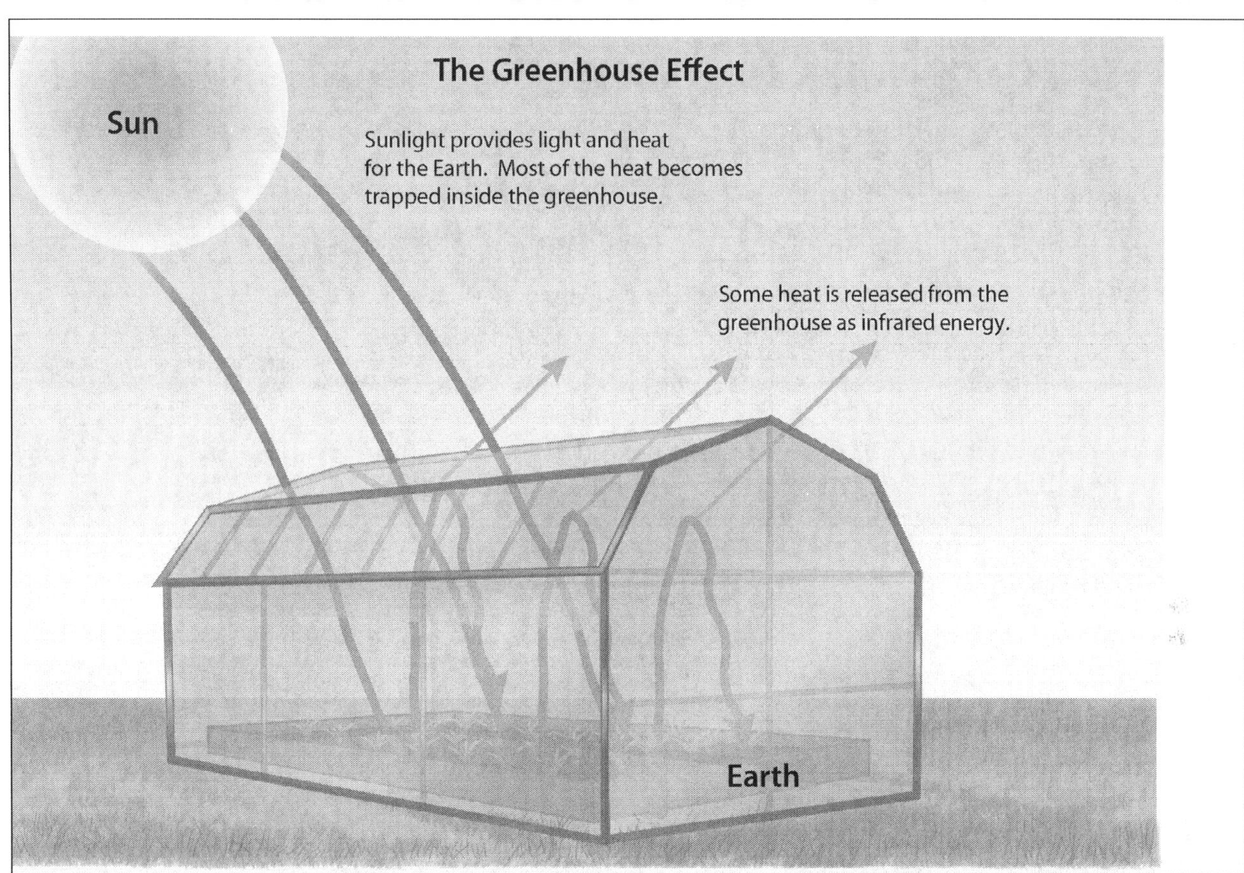
The Greenhouse Effect
Sun
Sunlight provides light and heat
for the Earth. Most of the heat becomes
trapped inside the greenhouse.
Some heat is released from the
greenhouse as infrared energy.
Earth

The greenhouse effect refers to the process in which longwave radiation is trapped in the atmosphere and then radiated back to the Earth's surface.

internal and external processes in the Earth-Sun system and human-induced changes in land use and the atmosphere. The United Nations Framework Convention on Climate Change states that the term "climate change" should refer to anthropogenic changes that affect the composition of the atmosphere as distinguished from natural causes, which should be referred to "climate variability." An example of natural climate variability is the global cooling of about 0.5 degrees Fahrenheit in 1992-1993 that was related to the 1991 Mount Pinatubo volcanic eruption in the Philippines. The 15 million to 20 million tons of sulfuric acid aerosols that were released into the stratosphere reflected incoming radiation from the sun and created a cooling effect. Many experts suggest that climate change is caused by human activity, as evidenced by the above-normal temperatures in the 2000's. Based on a variety of techniques that estimate temperatures in previous centuries, the year 2005 was the warmest in the last thousand years.

Numerous observations strongly suggest a continuing warming trend. Snow and ice have retreated from areas such as Mount Kilimanjaro, which at 19,340 feet is the highest mountain in Africa, and glaciated areas in Switzerland. In the Special Report on Emission Scenarios (2000), the Intergovernmental Panel on Climate Change examined the broad spectrum of possible concentrations of greenhouse gases by examining the growth of population and industry along with the efficiency of energy use. The panel estimated future trends using computer climate models. For example, it estimated that the global temperature would increase 35.2-39.2 degrees Fahrenheit by the year 2100.

Given the effect that climate change will have on humanity, many agencies and organizations will be doing research in the area, and climatologists are likely to be needed by a variety of governmental and private entities.

Robert M. Hordon, B.A., M.A., Ph.D.

FURTHER READING

Coley, David A. *Energy and Climate Change: Creating a Sustainable Future.* Hoboken, N.J.: John Wiley & Sons, 2008. A detailed review of energy topics and their relationship to climate change and energy technologies.

Gautier, Catherine. *Oil, Water, and Climate: An Introduction.* New York: Cambridge University Press, 2008. A good discussion of the impact of fossil fuel burning on climate change.

Lutgens, Frederick K., and Edward J. Tarbuck. *The Atmosphere: An Introduction to Meteorology.* 9th ed. Upper Saddle River, N.J.: Prentice Hall, 2004. A useful and standard text that is written with considerable clarity.

Strahler, Alan. *Introducing Physical Geography.* 5th ed. Hoboken, N.J.: John Wiley & Sons, 2011. An excellent text covering weather and climate with superlative illustrations, clear maps, and lucid discussions.

Wolfson, Richard. *Energy, Environment, and Climate.* New York: W. W. Norton, 2008. Provides an extensive discussion of the relationship between energy and climate change.

WEB SITES

American Association of State Climatologists
http://www.stateclimate.org

American Meteorological Society
http://www.ametsoc.org

Intergovernmental Panel on Climate Change
http://www.ipcc.ch

International Association of Meteorology and Atmospheric Sciences
http://www.iamas.org

National Oceanic and Atmospheric Administration
National Climatic Data Center
http://www.Ncdc.noaa.gov

National Weather Association
http://www.nwas.org

U.S. Geological Survey
Climate and Land Use Change
http://www.usgs.gov/climate_landuse

World Climate Research Programme
http://www.wcrp-climate.org

World Meteorological Organization
http://www.wmo.int/pages/themes/climate/index_en.php

See also: Climate Engineering; Climate Modeling.

COAL GASIFICATION

FIELDS OF STUDY

Process engineering; chemical engineering; mechanical engineering; electrical engineering; systems engineering; mining engineering; chemistry; physics; earth science; statistics.

SUMMARY

Coal gasification is the chemical and physical process of converting coal into coal gas, a type of synthesis gas (syngas) that is composed of varying amounts of carbon monoxide and hydrogen gas. The syngas is subsequently used as fuel for power generation or as feedstock in chemical processes such as the production of synthetic fuels and production of fertilizers.

KEY TERMS AND CONCEPTS

- **Biomass:** Renewable energy source derived from a living or recently living organism such as wood, alcohol, or oil.
- **Coal:** Combustible, sedimentary rock that is primarily composed of carbon and smaller amounts of sulfur, hydrogen, oxygen, and nitrogen and used as the largest source of energy for the generation of electricity worldwide and as one of the main feedstocks to create steel.
- **Coal Gas:** Type of synthesis gas composed of a mixture of carbon monoxide and hydrogen gas.
- **Feedstock:** Raw material that is used for processing or manufacturing.
- **Fischer-Tropsch Gasification Process:** Catalyzed chemical reaction in which carbon monoxide and hydrogen are converted into liquid hydrocarbons of various forms, typically used to produce synthetic petroleum substitute for use as lubrication oil or as fuel. Named for German chemists Hans Fischer and Franz Tropsch.
- **Flue Gas:** Exhaust gas that exits into the atmosphere from a device such as a boiler, furnace, or steam generator through a pipe-like channel called a flue.
- **Fuel Cell:** Electrochemical cell that uses an external source fuel to generate an electric current.
- **Gasification:** Process of converting carbonaceous material such as coal, petroleum coke, refuse-derived fuel (RDF), or biomass into gas that can be used as an energy source.
- **Integrated Gasification Combined Cycle (IGCC):** High-efficiency gasification system that uses two types of turbines, combustion and steam.
- **Synthesis Gas:** Gas mixture that contains carbon monoxide and hydrogen; also known as syngas.
- **Synthetic Fuel:** Liquid fuel created from carbon-rich materials such as coal, natural gas, oil shale, biomass, or industrial waste.
- **Underground Coal Gasification (UCG):** Industrial method of converting coal that is underground, or in situ, within natural coal seams into a combustible gas to be used for heating, power generation, or other industrial uses; also called in situ coal gasification (ISCG).

Definition and Basic Principles

Coal gasification is a method of converting coal into a combustible gas that can be used for heating, power generation, and manufacture of hydrogen, synthetic natural gas, or diesel fuel. Coal gasification plants are in operation throughout the world. Unlike conventional coal-fired power plants, gasification involves a thermochemical process that breaks down coal into its basic chemical constituents. This is accomplished in modern-day gasifiers by reacting coal with a mixture of steam and air or oxygen under high temperature and pressure. The end product is a gaseous mixture of carbon monoxide, hydrogen, and other gas compounds. Many experts predict that coal gasification will lead the way to a clean-energy future: Whereas burning coal can contribute to global warming by increasing the concentration of carbon dioxide in the atmosphere, coal gasification produces lean gas because pollutants or impurities such as sulfur and mercury are removed in the system.

Background and History

Coal gasification dates back to about 1780, when it was first developed, but the technology and its applications have evolved significantly over the years. When coal gasification was first implemented, the carbon monoxide that was produced was used as a source of energy for municipal lighting and heating because industrial-scale production of natural gas was not yet available. This coal gas was referred to as blue gas, producer gas, water gas, town gas, or fuel gas. Natural gas became widely available in the 1940's and by the 1950's, coal gas was nearly replaced with natural gas because it burned cleaner, had a greater heating value, and was safer to use. Later, when the price of alternative fuels such as oil and natural gas were low, interest in coal gasification fell further; however, renewed interest has arisen for coal gasification solutions because of the high cost of oil and gas, and energy security and environmental considerations.

How It Works

Coal Selection and Preparation. The first step in successful coal gasification is the careful selection, analysis, and preparation of coal. Coal is first analyzed for its percentage of sulfur, fixed carbon, oxygen, ash, and other volatile content. In general, use of a coal feedstock with low sulfur, low moisture, high fixed-carbon content, and low ash content will result in low oxygen consumption, a high volume of syngas, and a small volume of waste-product generation. Some adjustments to gasification systems can be made to accommodate for different coal qualities.

Coal Gasifier. Central to coal gasification is the gasifier. A gasifier converts hydrocarbon feedstock into gaseous components by providing heat under pressure along with steam. A gasifier, unlike a combustor, relies on the careful regulation of the quantity of air or oxygen that is permitted to enter the reaction so that only a small portion of the fuel, the coal, burns completely. At a high temperature (about 900 degrees Celsius) and high pressure, the oxygen and water in the gasifier partially oxidizes the coal. During this stage, called the steam-forming reaction, rather than burning, the coal feedstock is converted into a syngas mixture of carbon dioxide, carbon monoxide, molecular hydrogen, and water in the form of vapor. To get the syngas out of the gasifier, the syngas is cooled to room temperature by using exhausts and filters that remove solid particles.

Types of Gasifiers. There are three main types of gasification technologies: entrained flow, moving bed, and fluid bed. Of the three, the entrained-flow gasifier possesses the greatest efficiency. Because it operates at a high temperature, nearly 99 percent of the coal is converted into high-purity syngas, as the majority of the tar and oil contained in the coal is destroyed with the high heat. The one drawback, however, is that the entrained-flow gasifier has a high oxygen demand, and this is easily exacerbated through use of a coal feedstock with high ash content. The moving-bed gasifier has the lowest oxygen demand. In this system, coal moves slowly in a downward fashion and is gasified by a counter-current blast. The low operational temperature tends to be inhibitory to the reaction rate, resulting in lower syngas purity and volume generation. The fluid-bed gasifier, named so because of the fluidlike manner in which coal particles behave in the gasifier, facilitates good interaction between the coal feedstock and oxygen without requiring a membrane, leading to lower overall cost to implement and operate. The fluid-bed gasifier is the best option for low-rank (low-quality) coals; however, overall, it possesses the lowest conversion rate of carbon, resulting in a low-purity syngas.

Integrated Gasification Combined Cycle. The technology that combines coal gasification and the subsequent burning of the gas is called integrated gasification combined cycle (IGCC). IGCC combines a coal gasifier with a gas turbine and a steam turbine to produce electric power. The hydrogen-rich syngas from the gasifier is purified to remove acidic compounds and particulates. It then enters the first turbine (the gas turbine) to generate electricity, and the waste heat from the gas turbine works to power the second turbine (the steam turbine), which in turn produces additional electricity. Because steps are taken to remove the majority of acidic compounds and particulates, the resulting combustion-exhaust gas, or flue gas, that leaves the gas turbines has minimal effect on the environment. The combined use of a gas turbine and steam turbine, which both work to produce electricity, makes IGCC the preferred technology in this carbon-constrained world. It is more energy-efficient and generates less carbon dioxide per ton of coal used compared with conventional coal-fired power plants.

Applications and Products

Underground Coal Gasification. Coal gasification can be applied to coal in situ— in underground coal seams. Gasification of coal contained underground is called underground coal gasification (UCG). It is particularly well suited for technologically or economically unmineable coal deposits. The traditional method of extracting coal is through the excavation of open pits to expose coal-containing seams or through the excavation of underground mines. The design, construction, equipment, and labor to build such mines are a costly undertaking. UCG-candidate coal seams possess one or more features that render it unsuitable, from a technical and economic standpoint, for conventional coal mining: faulting, volcanic intrusions, complex depositional and tectonic features, and environmental constraints.

The UCG process uses injection and production wells drilled from the surface to access the underground coal, and the coal is not mined to the surface. A horizontal connection underground between the injector and extractor is made normally by hydrofracturing, which is a process that uses high-pressure water to break up the rock or coal. Through the injection well, oxidants (such as water and air or a water and oxygen mixture) are sent down into the coal seam. As in the case of conventional coal gasification, the coal is heated to temperatures that would normally cause the coal to burn, but through careful control of the oxidant flow, the coal is separated into syngas. The product syngas is drawn out through the second well.

Gasification By-Products. Some by-products of coal gasification have commercial value, so they are isolated and set aside to be sold or used on-site for industrial use. Mineral components in the coal that are unable to gasify leave the gasifier or fall to the bottom of the gasifier as an inert, glasslike material, known as slag, which can be used in cement or road construction.

Mercury that is isolated by passing the syngas through a bed of charcoal has no commercial value, but the final cleaning step that follows in the acid gas removal units handles sulfur impurities, which are converted into valuable by-products. Sulfur impurities are converted to hydrogen sulfide (H_2S) and carbonyl sulfide (COS), which are isolated in the form of elemental sulfur or sulfuric acid, which can be used for industrial processes such as incorporation into fertilizer.

Nitrogen, in the form of ammonia, is also extracted out of the product gas stream for industrial use. Other usable by-products include tar and phenols.

Fascinating Facts About Coal Gasification

- The benefits of coal gasification are based on the capability to cleanse or capture as much as 99 percent of the impurities from coal-derived gases that have been associated with global warming.
- Gasification plants typically consume 30 percent less water than conventional coal-fired power plants. Gasification plants can also be designed so that water is never discharged from the system but instead circulates to be used over and over again.
- The fuel efficiency of a coal gasification power plant can exceed 50 percent. On the other hand, in a conventional coal combustion plant, only one-third of the energy value of coal is converted into usable energy. The remainder is lost as heat and is usually not recovered.
- Integrated gasification combined cycle (IGCC) may help people realize a clean-energy future. Hydrogen is rapidly gaining popularity as a clean-burning fuel source for vehicles, and IGCC is an economic solution to generating hydrogen.
- Syngas produced from coal gasification can be used for power generation in IGCC or for making fertilizers, methanol, synthetic natural gas, hydrogen, or carbon dioxide. Alternatively, syngas put through the Fischer-Tropsch process can produce diesel fuel, kerosene, jet fuel, gasoline, detergents, waxes, and lubricants.
- Underground coal gasification is not yet commercial. There are several pilot projects in North America.
- Coal is used to generate about 50 percent of the electricity in North America.
- Each person in North America uses just under four tons of coal per year.
- Gasifiers can use a dry or a wet coal feed. A dry coal feed requires about 25 percent less oxygen.
- Gasification is thought to become the driving force for the development and widespread acceptance of hydrogen-powered automobiles and fuel cells.

Impact on Industry

Impact on Seaborne Coal Trade. There is significant merit to the deployment of coal gasification technology, particularly for emerging countries such as China, India, and Africa, as low plant development and operational costs are involved. As the price for thermal coal used for conventional coal-fired plants and the price for metallurgical coal used for making steel (to build rail systems to transport coal) continues its upward trend in trading price, more countries will see that embracing implementation of UCG for their coal resources on hand is the most economical choice. This may lead to a relative reduction in the volume of the seaborne coal trade market.

Alternative Feedstocks. In addition to the research to advance coal gasification plant efficiency and environmental performance, the application of other feedstocks such as municipal waste and biomass are currently being explored. If a significant volume of conventionally unrecyclable material such as municipal waste or solid human waste can be converted into useful energy, there will be further development in the area of isolating certain materials.

Enhanced Oil Recovery. Another area that could flourish because of advancement in coal gasification is oil extraction. An optional addition to coal gasification is the isolation of carbon dioxide (CO_2) and subsequent storage underground or in depleted oil wells. This process is called CO_2 sequestration or carbon capture. The collected carbon dioxide may be transported and used for other industrial processes such as enhanced oil recovery (EOR), which is an oil-extraction method that uses carbon dioxide to help extraction of oil that is difficult to recover by conventional means. When CO_2 is injected in underground oil fields, it increases the pressure and enhances the recovery of oil remaining underground. In EOR, once the oil field is depleted, the oil field is sealed to trap CO_2 permanently underground, instead of allowing it to release into the atmosphere. Coal gasification systems can also be designed subsequently to route the product gas to another processing device called a Fischer-Tropsch reactor. Fischer-Tropsch reactors are used to produce alkanes, which are hydrocarbons often added to natural gas, gasoline, and diesel fuel.

CAREERS AND COURSE WORK

Courses in advanced engineering such as process engineering, chemical engineering, petroleum engineering, civil engineering, mechanical engineering, electrical engineering, as well as mathematics and physics will form the foundational requirements for students interested in pursuing careers as gasification or process engineers. Software programs specific to engineering such as CAD and AutoCAD are also essential areas of mastery. Earning a bachelor of science or bachelor of applied science degree in any of the aforementioned fields would prepare a student for graduate studies in a similar field. A professional engineer (PE) license, earned usually through fulfillment of work in a field of engineering for a prescribed number of years and successful completion of an engineering ethics examination and a comprehensive engineering examination, will facilitate career advancement. A master's degree or doctorate would equip students to pursue advanced career opportunities in industry. To obtain an upper management or executive role in industry, one would likely additionally be required to hold a management or finance degree, such as a master of business administration (M.B.A.). An advanced degree is typically not a requirement for a technician or administrator position in this field. A coal gasification engineer or manager in the field would work closely with various other engineering specialties to build, monitor, and maintain coal gasification plants. They may work with vendors and customers for procurement of equipment, feedstock, or other industrial materials and supplies.

SOCIAL CONTEXT AND FUTURE PROSPECTS

Fuel flexibility is important in an increasingly carbon-constrained world. Future technology conceivably could accommodate economical use of a wide variety of feedstocks such as municipal waste, biomass, and recycled materials. Existing coal gasification technologies perform best on high rank, costly coal or petroleum refinery products but are inefficient and expensive to operate when using poorer-quality coal, despite being available worldwide in abundance. Advancements to bring down the cost of construction and operation of gasification plants will be important to ensure widespread use of the technology. One way to bring down the operational cost of gasification plants is to reduce the cost of oxygen used in the gasification process. Oxygen is made by an expensive cryogenic process. Research involving the use of ceramic membranes is demonstrating promising results for the separation of oxygen from air at higher temperatures. Innovations in membrane technology may enhance the utility of coal gasifiers. The development of inexpensive membranes that can readily separate hydrogen from syngas would be useful.

Economical sequestration of hydrogen will help drive the use and advancement of hydrogen fuel cells and hydrogen-powered vehicles.

Rena Christina Tabata, B.Sc., M.Sc.

Further Reading

Asplund, Richard W. *Profiting from Clean Energy: A Complete Guide to Trading Green in Solar, Wind, Ethanol, Fuel Cell, Carbon Credit Industries, and More.* Hoboken, N.J.: John Wiley & Sons, 2008. This guidebook provides summaries on clean energy topics and informs on how to invest financially in the clean-energy market.

Bell, David A., Brian F. Towler, and Maohong Fan. *Coal Gasification and Its Applications.* Burlington, Mass.: William Andrew, 2011. Includes chapters on the fundamentals of gasification, gasification technologies, gas cleaning processes, gasification kinetics, conversion of syngas to electricity, and process economics.

De Souza-Santos, Marcio L. *Solid Fuels Combustion and Gasification: Modeling, Simulation, and Equipment Operations.* 2d ed. Boca Raton, Fla.: CRC Press, 2010. The book covers operational features of equipment dealing with combustion and gasification of solid fuels, such as coal and biomass, presents basic concepts of solid and gas combustion mechanism, introduces fundamental approaches to formulate mathematical models for gasification systems, and case studies demonstrating computer simulation.

Fehl, Pamela. *Green Careers: Energy.* New York: Ferguson, 2010. The handbook provides comprehensive coverage on the range of green careers both traditional and new age, including those that focus on conservation of resources and the development and application of alternative energy sources.

Girard, James. *Principles of Environmental Chemistry.* 2d ed. Sudbury, Mass.: Jones and Bartlett, 2010. This textbook covers the principles of environmental chemistry and is suitable for students who have a basic fluency of general chemistry.

Schaeffer, Peter. *Commodity Modeling and Pricing: Methods for Analyzing Resource Market Behavior.* Hoboken, N.J.: John Wiley & Sons, 2008. This textbook introduces readers to the economic analysis and modeling of commodity markets, including those pertaining to coal and clean-energy trade.

Taylor, Allan T., and James Robert Parish. *Career Opportunities in the Energy Industry.* New York: Ferguson, 2008. Includes career profiles of a variety of technological specialties including petroleum engineering, energy processing, and mining.

Williams, A., et al. *Combustion and Gasification of Coal.* New York: Taylor & Francis, 2000. Provides an overview of coal combustion and gasification with special attention to the properties of coal, combustion mechanism of coal, combustion in fluidized beds, the gasification process, industrial applications, and environmental impact.

Web Sites

Coal Gasification News
http://www.coalgasificationnews.com

Underground Coal Gasification Association
http://www.ucgp.com

United States Department of Energy
Fossil Energy: Coal Gasification R&D
http://www.fossil.energy.gov

World Coal Association
http://www.worldcoal.org

See also: Chemical Engineering; Electrical Engineering; Mechanical Engineering.

COAL LIQUEFACTION

FIELDS OF STUDY

Energy; chemistry; environmental science; political economy; physics.

SUMMARY

Coal liquefaction is a catalytic process that converts different varieties of coal into synthetic petroleum by reacting coal with hydrogen gas at high temperature and under high pressure. The resultant synoil product has the potential to contribute significantly to the U.S. petroleum supply and reduce the need for the United States and other coal-producing countries to rely on imported oil.

KEY TERMS AND CONCEPTS

- **Anthracite Coal:** Coal composed of 90 percent carbon and prized for coal-fired electrification because of its low level of impurities.
- **Bituminous Coal:** Most common form of coal, which has anthracite's high heating value but usually contains a high level of sulfur.
- **Commercialization:** Idea of developing only those technologies that can compete in the marketplace and leaving that development to the private sector, which can profit from those technologies.
- **Direct Liquefaction:** Process in which coal is subjected to hydrogenation under high pressure and at a high temperature, thereby converting it directly into a synthetic liquid fuel; also known as the Bergius process.
- **Fossil Fuel:** General term applied to any fuel created by the fossilization of plant and animal matter dating back millions of years; these fuels are coal, petroleum, and natural gas.
- **Hydrogenation:** The chemical process resulting from the addition of hydrogen to an element at a very high temperature or in the presence of a catalyst.
- **Indirect Liquefaction:** Process for converting coal into oil or synfuel by first gasifying it; also know as the Fischer-Tropsch method.
- **Pyrolysis:** Process of chemically decomposing organic materials by incineration in an oxygen-free environment.

DEFINITION AND BASIC PRINCIPLES

Coal liquefaction involves converting bituminous or (more rarely) anthracite coal into a liquid that can be refined into the end product. The focus is on transportation fuels to lower the direct and indirect costs that oil-importing states incur in purchasing large volumes of petroleum from abroad.

The chemical processes being used in the twenty-first century favor the indirect Fischer-Tropsch method of coal liquefaction. In this process, coal is initially subjected to very high heat to create a charred substance that can be combined with carbon dioxide and steam to produce a synthesis gas composed of hydrogen and carbon monoxide. The gas is then chemically subjected to a metallic catalyst, which transforms it into a synthetic crude oil. The resultant synthetic oil can then be refined into the desired fuel.

Whatever the method, the amount of coal necessary to produce oil is about the same: Nearly three-quarters of a ton of coal is required to produce slightly over one barrel of oil. Therefore, large-scale production of oil from coal would be a massive undertaking.

BACKGROUND AND HISTORY

Neither the idea nor the practice of producing oil from coal is new. As early as 1819, Charles Macintosh distilled naphtha from coal for the purpose of waterproofing textiles. However, the major breakthroughs in coal liquefaction did not occur until the period between the early 1900's, when the two processes long used as the starting points for producing oil from coal—the Bergius and Fischer-Tropsch methods—were developed.

Since then, political considerations have trumped economics in pushing countries into developing oil from coal on only two occasions: when Germany experienced an oil shortage during World War II and when South Africa feared the imposition of an oil embargo against its white minority government during apartheid. However, the high cost of imported oil or concern with its inadequate or insecure supply have sparked interest in coal liquefaction on four separate occasions. In 1928, Standard Oil of New Jersey and I. G. Farben joined forces to pursue a significant liquefaction project based on the Bergius process, but their venture was short-lived. No sooner was the

Standard Oil-Farben deal made than the market conditions that had brought it about disappeared. The Great Depression dampened the Western demand for oil, and the discovery of rich oil fields in Texas in 1930 meant an increase in the supply of cheap crude oil in the United States.

During World War II and shortly after, concerns over the security of the oil supply and fears that the supply would not be adequate to meet the rapid postwar expansion drove a short period of government-sponsored research into creating oil from coal in the United States. However, soon after two pilot plants were established to test the Bergius and Fischer-Tropsch processes in the postwar period, imported oil from the vast, newly developed fields of the Middle East began to arrive in such abundance that an oil glut appeared likely. In 1949, the National Petroleum Council began questioning the commercial feasibility of coal-based oil, and four years later, citing the changed economic climate, the administration of President Dwight D. Eisenhower ordered the government-owned research facilities closed.

During the next twenty years, private industry continued to conduct coal liquefaction experiments, and major petroleum corporations acquired ownership of several coal companies with coal liquefaction pilot plants. Nevertheless, with the cost of oil remaining low throughout this period, research projects tended to remain very small in scale. It was not until the oil crises of the 1970's, which drove the price of imported oil from under $3 per barrel in mid-1973 to over $35 per barrel in 1979, that a third round of interest in a commercial synthetic fuels industry began. This time, major research and development programs emerged across five continents, and both public agencies and private companies initiated multibillion-dollar projects. The outcome, nonetheless, was again the same. Within a decade of the 1973 oil crisis, the rising price of oil had generated a massive recession in the oil-importing world. Demand for oil dropped precipitously, and when the price of oil began to drop, Organization of the Petroleum Exporting Countries (OPEC) members began trying to undersell one another to gain a wider share of the shrinking market, further accelerating the downward spiral in the price of oil. By then, President Ronald Reagan had drained the Synthetic Fuel Corporation created by the U.S. Congress in 1980 of the $88 billion allocated to it for the development of an American coal liquefaction and gasification industry, and around the world, others were following his lead.

Except for a momentary spike in the price of oil when Iraq invaded Kuwait in 1990, which generated fears that the supply of oil would be disrupted, low oil prices remained the norm from the mid-1980's until the U.S. invasion of Iraq in 2003. The growing demand for oil in China and India had already pushed the price toward $30 per barrel, but that figure was too low for investors to revisit the synthetic fuels option. However, in 2003, when the United States invaded Iraq, the gap between OPEC's production capacity and global demand for oil was closing. By the end of 2003, imported petroleum had reached more than $30 per barrel; two years later, it had doubled to $60 per barrel. Costs eventually peaked at more than $147 per barrel in July, 2008, before the high price of oil dampened demand and oil prices began to drop sharply. By then, interest in coal liquefaction as a commercial alternative to OPEC oil had been resuscitated, and several pilot projects based on the Fischer-Tropsch process were under way around the world.

How It Works

Although there are variations in terms of catalytic agents, the degree of heat, and the level of pressure employed, there are three broad methods for liquefying coal into oil: hydrogenation, catalytic synthesis, and pyrolysis.

Hydrogenation. The first hydrogenation process to develop was the Bergius process, which produces heavy crude oil directly from coal by adding hydrogen to unprocessed coal and recycled heavy oil. Hydrogenation is the central element in the Begrius process; however, unlike the Fischer-Tropsch approach, which also uses hydrogenation, the Bergius system employs a combination of various catalysts (typically tungsten, tin, or nickel oleate) to convert the carbon directly, under very high pressure and at much higher temperatures (from 700 to more than 925 degrees Fahrenheit), into oil liquids capable of being refined into industrial and heating oil and transportation fuels.

Developed in Germany by Friedrich Bergius in 1913 and sold to the German chemical conglomerate I. G. Ferben in 1924, this process was the more popular means of coal liquefaction in Germany from 1925 until the end of World War II. It was resurrected there again during the oil crises of the 1970's with the creation of a demonstration plant in Bottrop, a coal-mining center in the Ruhr region; however, that plant closed in the early 1990's.

Catalytic Synthesis. The liquefaction method more commonly used in the twenty-first century was also developed in Germany by Franz Fischer and Hans Tropsch in 1923. This indirect method produces oil from coal by first gasifying the hydrocarbons through a combination of complex chemical reactions involving catalytic synthesis. At the heart of the process, typically at temperatures ranging from 300 to 575 degrees Fahrenheit, carbon monoxide is mixed with hydrogen in a series of catalytic chemical reactions.

As of 2011, the Republic of South Africa's synthetic fuels industry and all confirmed liquefaction projects under way were employing some variant of this process and for understandable reasons. In a world in which liquid fuels and gas are important sources of energy, the Fischer-Tropsch process has the advantage of producing both petroleum liquids and a high volume of natural gas liquids and ethane.

Pyrolysis. In the pyrolysis process for converting coal into oil, high heat is applied to coal in an oxygen-free environment, decomposing it into oil and coal tar, then the oil is subjected to hydrogenation to remove sulfur and other extraneous content before it is processed into fuel sources. Substantially fewer demonstration efforts have been devoted to this process because the amount of oil produced per ton of coal is less than that yielded by the Fischer-Tropsch or Bergius method, and no major liquefaction operation at even the pilot plant level has been based on it.

Applications and Products

Because the Earth's coal reserves are substantially greater than its oil reserves, a general consensus exists within the energy industry that a liquefied coal industry will eventually emerge around the globe. That day, however, has yet to come. Hence, any discussion of the products that can emerge from such an industry must necessarily be divided into two parts: the likely applications of a future liquefied coal industry and what has occurred within the framework of the Fischer-Tropsch method in the Republic of South Africa, the one country where a significant oil-from-coal industry exists.

The development of a liquefied coal industry in the oil-importing world is likely. The United States, for example, throughout the first decade of the twenty-first century, was persistently importing 60 percent or more of its petroleum needs. Mid-decade testimony before the U.S. Congress placed the cost of these imports—a direct transfer of wealth from the United States to petroleum-exporting states—at a minimum of $320 billion per year, with oil in the $80 per barrel range. Indirect costs, measured in terms of lost jobs, inflationary erosion of purchasing power, and other consequences of this transfer of wealth, were estimated at an additional $800 billion per year.

Fascinating Facts About Coal Liquefaction

- For six years during World War II, Nazi Germany used a network of twenty-five synthetic fuel plants, manned mostly by slave labor, that, at peak production, converted coal into the 124,000 barrels of oil per day that Germany used to power most of its war machine before the U.S. Army Air Force closed those I. G. Farben plants in 1944.
- After World War II, the United States employed German scientists to explore the Fischer-Tropsch method as a possible way of creating fuel in the United States. The project was code-named Operation Paperclip, but it collapsed when the discovery of oil in Kuwait and new pools in Venezuela ended postwar fears of an oil shortage.
- Representative Carl Perkins from Kentucky's coal-mining district, the same congressman who encouraged the federal government to explore coal liquefaction in 1946, also spearheaded the quest for synthetic fuels in the 1970's that led to the creation of the Synthetic Fuels Corporation.
- In campaigning during his failed 1980 run for the Senate from oil-producing Oklahoma, Edward Noble–who was appointed by President Ronald Reagan to head the Synthetic Fuels Corporation—pledged that if elected, he would shut down the corporation.
- In 2005, the U.S. Air Force launched a coal-to-fuel project designed to produce aviation fuel but abandoned it in 2009, when the drop in oil prices made it unlikely that the fuel could be produced at competitive prices.
- Audi has competed in the Le Mans races with R15 TDI cars powered by synthetic diesel fuel produced using the Fischer-Tropsch method.

Moreover, because more than half of the petroleum used in the United States goes into the transportation sector, it is assumed that the majority of any synthetic oil produced in the United States or elsewhere in the oil-importing world will be used in that sector to achieve the target goal of reducing the amount of oil imported as much as possible.

As for South Africa, liquefied coal has figured into its overall energy equation since 1955, when the South Africa Coal Oil and Gas Corporation (SASOL) began liquefying cheap South African coal mined by inexpensive local labor into state-subsidized gasoline. That first Fischer-Tropsch-based plant was subsequently devoted entirely to producing gas from coal, but in the interim, SASOL opened two additional coal-into-oil plants that together produce nearly 200,000 barrels of oil per day. SASOL has more than half a century of experience in converting coal to liquids, and in addition to transportation fuel, it produces a variety of chemical solvents and waxes using the Fischer-Tropsch method. However, even the increase in petroleum prices after 2003 did not encourage it to expand its liquefaction operations, and South Africa still imports about two-thirds of its petroleum needs.

Impact on Industry

The U.S. coal industry is marred by collaboration between mine owners and workers in evading safety standards that, if implemented, could close mines and eliminate jobs because of their cost. Most coal goes to the production of electricity, and the cleaner coal deposits of the West, largely harvested by deep-shovel strip mining, is preferred for environmental reasons over the deep-shaft coal of the East and Midwest. A vibrant coal liquefaction industry, whatever the process used, would revive the economically depressed coal-mining areas in the East and Midwest. Consequently, it is not hard to find enthusiastic supporters for a coal liquefaction industry among those members of Congress who represent these regions, as occurred in 1979-1980 when Congress created the quasi-public Synthetic Fuels Corporation. Japan, Germany, Britain, Australia, and China's coal regions would likewise profit from a major commitment to an coal liquefaction industry.

Careers and Course Work

Careers involving the development of alternative energy sources will almost certainly expand in the twenty-first century as the demand for oil inevitably continues to rise even if Western countries are able to reduce the number of internal combustion engines on their highways. With that demand, a long-term rise in the price of oil is also probable. Consequently, scientists trained in the applied fields of chemistry, earth sciences, geology, environmental science, and physics, with specialization in alternative fuel development, can expect to be in demand. Whether those specializing in coal liquefaction technologies will be swept up in that demand and find more employment opportunities, however, is more debatable. The case for pursuing coal liquefaction has altered little in a century.

On the plus side, coal remains far more abundantly available than petroleum and is found in significant deposits on every continent. Moreover, new liquefaction processes based on the Fischer-Tropsch method have reportedly reduced the cost of coal liquefaction to levels that should be competitive at the likely cost of future OPEC oil imports. On the negative side, the déjà vu quality of these arguments has prevented investors from diving heavily into coal liquefaction projects as has the developed world's swing toward more environmentally clean energy sources. By its nature, coal is a dirtier fossil fuel than oil or natural gas, and increased coal production for and processing in large volumes in synthetic oil industries carries a risk of greater air pollution and groundwater contamination. On balance, then, in at least the near future, those inclined to pursue courses in chemistry and thermal physics for future careers involving alternative energy fuels would probably be better off specializing in coal gasification or the production of oil from tar sands than coal liquefaction.

Social Context and Future Prospects

The economic assumptions driving early twenty-first century interest in coal liquefaction are that coal can be converted into oil for about $40 per barrel and that it is highly unlikely that OPEC oil will ever again drop below that figure because the production costs have become higher for some OPEC members. To this may be added a third assumption, that Western societies will understand that the social costs of importing so much oil are greater than the environmental risks involved in reducing those imports.

Moving to a large-scale synthetic oil industry, however, is not blocked by just the uncertainties surrounding the likelihood of greater environmental restrictions being placed on the industry. In addition, some experts fear that given the low production costs of OPEC producers in the Arabian peninsula (where Saudi Arabia can still produce oil at $2 per barrel), these oil producers might drop the price of oil to the point where synthetic fuels cannot compete, and coal liquefaction investors will lose—as they did in the 1980's—billions of dollars. Therefore, although several synthetic fuel projects were begun or announced between 2004 and 2010, including a Shell project in Malaysia and Shell and Marathon Oil projects in Qatar, the size of those ventures has generally been that of pilot plants. As a global commercial undertaking, coal liquefaction remains a technology in search of its time.

Joseph R. Rudolph, Jr., Ph.D.

FURTHER READING

Bartis, James T., Frank Camm, and David S. Ortiz. *Producing Liquid Fuels from Coal: Prospects and Policy Issues.* Santa Monica, Calif.: RAND Corporation, 2008. Good description of the issues, written by a cautious advocate of oil-from-coal projects.

Crow, Michael, et al. *Synthetic Fuel Technology Development in the United States: A Retrospective Assessment.* Westport, Conn.: Greenwood Press, 1988. Examines the economic and political factors affecting synthetic fuel projects during the Carter and Reagan administrations.

International Energy Agency. *Coal Liquefaction: A Technology Review.* Paris: Organization for Cooperation and Development, 1982. An excellent, easy-to-understand explanation of coal liquefaction technologies and the oil-from-coal projects being pursued in Europe in the early 1980's.

Speight, J. G. *Synthetic Fuels Handbook: Properties, Process, and Performance.* New York: McGraw-Hill, 2008. An easy-to-follow explanation of the costs and benefits of various synthetic fuel technologies, including coal liquefaction.

Toman, Michael, et al. *Unconventional Fossil-based Fuels: Economic and Environmental Tradeoffs.* Santa Monica Calif.: RAND Corporation, 2008. A rather upbeat view of the potential cost competitiveness of oil derived from coal and oil sands, combined with a hard look at the economic benefits versus environmental costs of developing these sources.

Yanarella, Ernest J., and William C. Green, eds. *The Unfulfilled Promise of Synthetic Fuels: Technological Failure, Policy Immobilism, or Commercial Illusion.* New York: Greenwood Press, 1987. A one-stop reader on the politics and technology of synthetic fuels at the time of the last big push to develop liquefied coal in the United States and abroad.

Yeng, Chi-Jen. *Belief-Based Energy Technology in the United States: A Comparative Study of Nuclear Power and Synthetic Fuel Policies.* Amherst, N.Y.: Cambria Press, 2009. An excellent study of the importance of political clout as well as technological feasibility in explaining energy choices in the American political process.

WEB SITES

Coal Utilization Research Center
Clean Coal 101
http://www.coal.org/clean_coal_101/index.asp

U.S. Department of Energy
National Energy Technology Laboratory, Clean Coal Demonstrations
http://www.netl.doe.gov/technologies/coalpower/cctc/cctdp/bibliography/misc/bibm_cl.html

See also: Chemical Engineering; Coal Gasification.

COMMUNICATION

FIELDS OF STUDY

Interpersonal communication; conflict management; group dynamics; intercultural communication; linguistics; nonverbal communication; semiotics; communication technology; negotiation and mediation; organizational behavior; public speaking; speech pathology; rhetorical theory; diction; mass communication; advertising; broadcasting and telecommunications; journalism; media ethics; public relations.

SUMMARY

Communication is the complex, continuous, two-way process of sending and receiving information in the form of messages. Communication engages all the senses and involves speech, writing, and myriad nonverbal methods of data exchange. It is a vital component in the everyday existence of all living creatures. An interdisciplinary, multidimensional field that incorporates language, linguistic structure, symbols, interpretation, and meaning in both personal and professional life, communication is an essential part of any division of scientific study.

KEY TERMS AND CONCEPTS

- **Etymology:** Study of the origins of words and their changes in meaning and form over time.
- **Feedback:** Response of a recipient to a message received.
- **Haptics:** Study of how touch and bodily contact affects communication.
- **Kinesics:** Study of how nonverbal forms of communication such as posture, gestures, and facial expressions affect meaning.
- **Paralinguistics:** Study of the nonverbal elements that individualize speech to help convey meaning, including intonation, pitch, word stress, volume, and tempo.
- **Phonetics:** Study of how speech is physically produced and received.
- **Proxemics:** Study of the use of personal space in face-to-face communication.
- **Semantics:** Interpretation of meaning in language.
- **Semiotics:** Study of how signs, symbols, and signifiers create meaning.
- **Syntax:** Study of the grammatical rules of a language that aid effective communication.

Definition and Basic Principles

Communication is the science and art of transmitting information. Communication science is objective, involving research, data, methodologies, and technological approaches to procedures of information exchange. Communication art is subjective, concentrating on the aesthetics and effectiveness with which messages are composed, sent, received, interpreted, understood, and acted on. Art and science are equally important in understanding how communication is supposed to work, why it succeeds or fails, and how best to use that knowledge to improve the dissemination of information. Communication is the oldest, broadest, most complicated, and most versatile of all scientific disciplines, with attitude-changing, life-influencing applications in every field of human endeavor.

The process of communication requires three basic components: a sender, a message, and a receiver. The sender encodes information into a message to send to the recipient for decoding. The message usually has an overt purpose, based on the sender's desire: to inform, educate, persuade, or entertain. Some messages can also be covert: to attract attention, make connections, gain support, or sell something.

In theory, the communication ideal is to match sender intent to receiver interpretation of a message closely enough to generate a favorable response. In practice, communication is often unsuccessful in achieving such objectives, considering the quirky, complicated nature of human beings and the variety and intricacy of languages used to transmit messages.

There are two fundamental forms of communication: verbal and nonverbal. Verbal communication includes interpersonal or electronically transmitted conversations and chats, lectures, audiovisual mass media such as television and radio, and similar forms of oral speech. Verbal communication also includes written communication: books, letters, magazines, signs, Web sites, and e-mails. Nonverbal communication incorporates an infinite range of facial expressions, gestures and sign language, body language, and paralanguage. In face-to-face interactions, nonverbal messages frequently convey more meaning than verbal content.

Background and History

Human communication has been a vital part of evolution. Protohumans communicated through gestures before developing the ability to speak, perhaps as far back as 150,000 years ago. An opposable thumb allowed early humans to make tools and to leave long-lasting marks, such as cave paintings. As communities formed, spoken language evolved. Verbal language spawned written language; pictographs appeared 10,000 years ago. Cuneiform originated several thousand years later in Mesopotamia about the time hieroglyphics were born in Egypt. By 3000 B.C.E., writing had developed independently in China, India, Crete, and Central America. Written symbols became alphabets, which were organized into words and structured linguistic systems.

From the dawn of civilization, humans have reached out to one another, devising ingenious solutions to expand range, enhance exactitude, and ensure the permanence of messages. A library was begun in Greece in the sixth century B.C.E., and paper was invented soon afterward. By 1455, Johannes Gutenberg had devised a printing press with movable type and printed his first communication: the Gutenberg Bible. Each innovation represented a leap forward in the spread of information.

Communication has greatly accelerated and expanded during the last two hundred years as it became a scientific study. The nineteenth century witnessed the development of inventions such as railroads, the telegraph, postal systems, the typewriter and phonograph, automobiles, and telephones. The twentieth century ushered in movies, airplanes, radio and television, audio and video recorders, communication satellites, photocopiers, and facsimile machines. Since the 1980's personal computers, cellular phones, fiber optics, the Internet, and a variety of mobile, handheld devices have widened and sped human integration into the information age.

How It Works

Though necessary to interaction throughout human history, communication is still an imperfect science. As in the past, modern senders have vastly different abilities in composing comprehensible messages. Contemporary receivers, for a variety of reasons, may have an equally difficult time interpreting messages.

Channeling. How a message is sent—the medium of transmission, called a channel in communications science—is key to the communication process. An oral message too faint to be heard, a written message too garbled to be understood, or a missed nonverbal signal can interrupt, divert, delay, or derail the passage of information from one source to another. In the modern world, there are dozens of methods of sending messages: telephone, text messaging, ground mail, mass media, e-mail or social networking via the Internet, and face-to-face contact are most common. Deciding the best means of information transmission depends on a number of factors, including the purpose of the message, the amount and profundity of data to be sent, the audience for the message, and the desired outcome. All media have distinct advantages and disadvantages. Each type of communication requires a particular skill set from both sender and receiver.

A vital consideration is the relationship between sender and receiver: They must have something in common. A message written in Chinese and sent to someone who reads only English is ineffective. Likewise, shouting at the hearing impaired or sending text messages to newborn infants are counterproductive activities. Even among individuals with a great degree of similarity in language and culture there can be gaping discrepancies between sender intent and receiver interpretation. Ultimately, communication is a result-oriented discipline. Whatever the purpose of a message, a response of some kind from the recipient (feedback) is expected. If there is no reply, or an inappropriate response is generated, the process is incomplete and communication fails.

Context. Communication does not occur in a vacuum. Many factors determine and define how the process of information transfer will unfold and the likelihood of successful transmission, reception, and response to a message.

Psychological context, for example, entails the emotional makeup of individuals who originate and receive information: the desires, needs, values, and personalities of the participants, which may be in harmony or at variance. Environmental context deals with the physical setting of an interaction: weather, surroundings, noise level, or other elements with the potential to impair communication. Situational context is specific to the relationship between the participants in communication: Senders tailor messages differently to receivers who are friends, relatives, coworkers, or strangers. Linguistic context concerns the relationship among words, sentences, and paragraphs used throughout a speech or written work that help to clarify meaning. Interpretation of any part of a message is relative to the preceding and following parts of the message.

Particularly relevant to effective communication in the ever-widening global community is cultural context. Every culture has particular, ingrained rules governing verbal and nonverbal behavior among its members. What might be acceptable within one cultural group—such as extended eye contact, the use of profanity, or frequent touching during conversation—might be offensive to a different cultural group. High-context cultures are homogeneous communities with long-established, well-defined traditions that help preserve cultural values; such cultures are found throughout Asia, Africa, the Middle East, and South America. Low-context cultures, like the United States, are heterogeneous, a blend of many traditions that has produced a less rigid, broader-based, more open-ended set of behavioral rules. Though adhering to certain patterns learned from diverse national and ethnic heritages, Americans as a group are more geared toward individual values. When high- and low-context cultures collide in communication, such differences in attitude can wreak havoc on the implied intent of a message and the interpretation of meaning.

Communication Barriers. There are universal emotions (such as fear, surprise, happiness, sadness, and anger), common to all peoples in all cultures at all times, that can serve as building blocks to understanding. However, numerous obstacles interfere with the clear, unambiguous transmission of messages at one end, and the full grasp of meaning at the other end. Some of the more prevalent physical, social, and psychological impediments to be overcome include racial or ethnic prejudice, human ego, noise (in transmission equipment or surroundings), and distractions. Gender issues are also a primary concern. For a number of reasons, men and women think, speak, and act quite differently in interpersonal relations. There are generational issues, too: Children, parents, and grandparents can speak different languages. It is the task of communication to identify and find methods of circumventing or accommodating such information blockers.

Applications and Products

Linguistics. The scientific study of the structure and function of spoken, written, and nonverbal languages, linguistics has many academic, educational, social, and professional applications. A foundation in linguistics is essential in understanding how words are formed and fitted together to create meaning and establish connections among people who are simultaneously senders and receivers of billions of messages over a lifetime. The stronger a linguistic base, the easier it is to discriminate among the plethora of messages received daily from disparate sources, to interpret meaning and respond accordingly, to prioritize, and to bring personal order to a disordered world of information overload.

Linguistics incorporates numerous threads, each worthy of close examination in its own right—word origins, vocabulary, grammar, phonetics, symbols, gestures, dialects, colloquialisms, or slang—that contribute to the rich tapestry of communication. Linguistics can be approached from several basic, broadly overlapping directions.

Structural linguistics deals with how languages are built and ordered into rules for communication. This subdiscipline, a basis for scientific research, clinical studies, sociological explorations, or scholarly pursuits, focuses on how speech is physically produced. Structural linguistics involves acoustic features, the comparison of different languages, grammatical concepts, sentence construction, vocabulary function, and other analytical aspects of language.

Historical linguistics concentrates on the development of written, oral, and nonverbal languages over time. An appropriate field for historians, educators, and comparative linguists, the discipline is concerned with why, where, when, and how words change in pronunciation, spelling, usage, and meaning.

Geolinguistics, a newer discipline with sociological, political, and professional implications in the modern global marketplace, focuses on living languages. The study is primarily concerned with historical and contemporary causes and effects of language distribution and interaction as a means to improve international and intercultural communication.

Interpersonal Communication. Interpersonal communication is that which takes place between two or more people. An understanding of the process is useful in grasping how message transfer, receipt, interpretation, and response succeed or fail in the real world.

A broad discipline relevant to all applied science—since it involves everyday relationships and interactions among people—interpersonal communication serves as the basis for a considerable amount of scientific and popular study. Thousands of books and articles are published annually on various aspects of the subject: how to talk to a spouse, how to deal with a child, how to succeed in business, how to communicate with pets.

Interpersonal communication can be subdivided by specific fields of study or professional emphasis.

- Conflict management and problem solving are communication-based specialties invaluable for students of family therapy, sociology, psychology, psychiatry, law, criminology, business management, and political science.
- Gender, racial, sexual, intergenerational, and cultural communication issues are the concerns of many disciplines, particularly sociologists, psychologists, intercultural specialists, and international business students.
- Health communication, a subdiscipline that involves translation of jargon to plain language and the development of empathy, is of particular relevance to medical and psychology students, and to intercultural therapists and clinicians.
- Organizational communication, including departmental interaction, group dynamics, and behavior, is indispensable to the successful function of businesses, professional associations, government, and the military.

Mass Communication. This is the study of the dissemination of information through various media (newspapers, magazines, billboards, books, radio, television, Web sites, blogs, and a host of other methods) with the potential to influence large audiences. Modern mass media is the culmination of centuries of technology enhancing the power and glory of communication. For better or worse, messages sent and received are the manifestation of qualities and possibilities that have elevated humans to a predominant position on earth. As languages have been defined and refined, as methods of transmission have improved, the range of communication has expanded. The distribution of data, once confined to the limitations of the human voice, through a series of quantum leaps is now a worldwide phenomenon. There are multiple outlets (Internet, satellite television, international phone lines) available to send messages, solicit responses, and record aftereffects on a global scale. Mass communication offers numerous opportunities for specialization.

Advertising, marketing, and public relations represent the persuasive power of communication. The modern world is a global marketplace. There are billions of potential customers for every conceivable product, service, or cause, and there are thousands of businesses and agencies whose task it is to create consumer demand. Marketing is sales strategy: planning, researching, testing, budgeting, setting objectives, and measuring results. Advertising is message strategy: what to say; how, when, where, and how often to say it to achieve desired aesthetic and marketing goals. Public relations is image strategy: the manipulation of words and pictures that establish and preserve public perception of corporations, institutions, organizations, governmental entities, or individuals.

Performance or public communication often involves a particular motivation: to generate immediate response from an audience. Motivational speakers, lecturers, debaters, politicians, actors, and other public performers all have the common goal of eliciting instant emotion in listeners or viewers.

Telecommunications refers to electronic means of sending information, primarily via radio or television broadcast, but the field also incorporates broadband, mobile wireless, information technology, networks, cable, satellite, unified communication, and emergency communication. A burgeoning, far-ranging industry, telecommunications offers global possibilities in business management, on-air and on-screen performance, research, marketing, journalism, programming, advertising, editing, sales, information science, and technology.

Creative communication concerns the composition of messages for a variety of informational purposes and is a component found to some degree throughout all facets of mass media. Writers of all kinds, graphic artists, photographers, designers, illustrators, and critics have the ability to influence behavior through the use of words or pictures.

Related fields of study fall under the umbrella of mass communication. Demographics, the collection of data that quantify an intended audience according to economic, political, ethnic, religious, professional, or educational factors, is an important consideration for many segments of mass media. There are also numerous legal, ethical, environmental, political, and regulatory issues to be dealt with that require specialized training.

Impact on Industry

A few facts demonstrate the scope and trend of contemporary communication, a field that continues to grow exponentially in all directions. More than a million new books were released in 2009, three-quarters of which were self-published; digital book sales are expected to increase 400 percent by 2015. In 2010, the leading one hundred worldwide corporations—led by Proctor & Gamble's $8.68 billion expenditures—spent an average of $1 billion apiece in a $900 billion advertising industry. In 2011, the telecommunications industry will comprise $4 trillion, about 3 percent of the entire global economy. Internet advertising, expected to top $105 billion, will soon surpass newspaper advertising. Though the United States leads the way in online spending, with Japan in second place, China has supplanted Germany for third place, with Central and Eastern Europe, the Middle East, and Africa all fast-growing regions for such spending. Some 1.2 billion people, more than 20 percent of Earth's inhabitants, were connected to the Internet in 2010, and the penetration of the medium (about 75 percent in the United States) to the farthest corners of the world is expected to continue apace throughout the century. There is a constant influx of new individuals and groups establishing a presence through Web sites, blogs, and social networks. Mind-boggling numbers of messages about a bewildering array of subjects are sent and received daily, making effective communication more important than ever.

Academics. Though rhetoric and oratory have been studied since the days of Aristotle, communication science as a formal discipline dates from the 1940's, when research facilities were established at Columbia University, the University of Chicago, and the University of Illinois, Urbana-Champaign. In the twenty-first century, virtually all universities and colleges worldwide offer communication studies, most have degree programs, and many offer postgraduate work in various areas of concentration. The Annenberg School for Communication at the University of Pennsylvania, for example, has a long tradition in examining the effects of mass media on culture. Rensselaer Polytechnic Institute specializes in technical communications. New York University has particular interest in exploring media ecology, the study of media environments. At many institutions, original research in a wide variety of communication-related studies is encouraged and supported via grants and scholarships.

Government. The U.S. Government Printing Office (GPO) was established 150 years ago to keep track of mountains of information generated in the course of governing behavior across a sprawling, complicated society. The agency physically or electronically prints and disseminates—to officials, libraries, and

Fascinating Facts About Communication

- Humans the world over typically express basic emotions–happiness, sadness, surprise, anger, fear, disgust, and embarrassment–through virtually identical facial expressions, though such emotions can be hidden or controlled with training.
- Throughout his or her lifetime, the average American will spend an entire year (365 days × 24 hours per day = 8,760 hours) talking on the telephone.
- By first grade, average children understand about 10,000 words; by fifth grade, they understand about 40,000 words.
- Radio took thirty-eight years to accumulate 50 million listeners; television took thirteen years to acquire 50 million viewers and the Internet only five years for 50 million users.
- Humans typically speak at a rate of 100 to 175 words per minute but are capable of listening intently to information transmitted at 600 to 800 words per minute.
- More than a dozen languages are spoken by at least 100 million people. Mandarin Chinese is the widest-spoken language, with about 900 million speakers. English, in third place behind Mandarin and Spanish, is the first language of nearly 500 million people.

the public— Supreme Court documents, the text of all congressional proceedings and legislation, presidential and executive department communications, and the records or bulletins of many independent agencies. The Federal Communications Commission (FCC), founded in 1934 and segmented to deal with various issues, regulates wired and wireless interstate and international transmissions via radio, television, satellite, cable, and fiber optics. Likewise, such bureaus as the Federal Trade Commission (FTC) oversee the commerce generated from correspondence between businesses and citizens. Millions of dollars in research funding is available at federal and state government levels across a spectrum of concerns related to communication. Finally, many government agencies, bureaus, and administrations—from the Census Bureau to the National Aeronautics and Space Administration to the U.S. Geological Survey and the White House—disperse aggregated data, images, and other information to the tax-paying public.

Corporations. While universities study causes and effects and governments regulate legal form and use, the business world presents numerous opportunities for practical application of communication principles. A broad-based, interdisciplinary field, communication allows many points of corporate entry, such as technology, creative disciplines, public relations, management, or sales. A computer maven can find a niche in many modern industries; prowess in journalism can be adapted to other forms of writing; performance skills can translate from television to film to stage; artistic talent can be suited to graphic design or commercial illustration; an individual who can successfully sell one product can easily learn to sell another product. Corporations reward innovation.

Many professional organizations exist to provide detailed, updated data on issues that affect both for-profit and not-for-profit industries. The National Communication Association, the International Communication Association, and a United Nations agency, the International Telecommunications Union, are leaders in the exchange of ideas, knowledge, research, technology, and regulatory information pertinent to communication.

Careers and Course Work

Communication is an all-purpose discipline that offers a multitude of career paths to personally and professionally rewarding occupations in a booming, always relevant field. Core courses—English language and grammar, sociology, psychology, history and popular culture, speech and business—provide a firm foundation on which a successful specialization can be built in any of several broad, compatible areas always in demand.

In academics, communication offers careers in teaching, counseling, research, and technology. Students should plan to pursue degrees past the bachelor's level, adding courses in education, communication theory, media ethics and history, information technology, and telecommunications in the undergraduate program.

Many creative niches are available in corporate communications, journalism, consumer advertising, public relations, freelance writing and art, and the media. Verbal artists need course work in rhetoric, literature, composition, linguistics, persuasion, and technical writing. Visual artists require graphic design, illustration, semiotics, and computer-aided design. On-screen or on-air performers would benefit from courses in organization, diction, public speaking, nonverbal communication, and mass media.

There are dozens of jobs and professions (positions including editor, agent, producer, director, publicist, and journalist) in media, government, politics, corporations, nonprofits, and other organizations for those with skills and credentials in communication and its various subfields. Courses in interpersonal and intercultural communication, business management, group dynamics, interviewing, problem solving, negotiation, and motivation are particularly useful.

Communication also has many applications in the social and health sciences. These include family therapy, psychiatry, marriage counseling, speech pathology, gender and sexuality services, legal providers, ethical concerns, international relations, and intercultural, intergenerational, and interspecies specialties that often require advanced degrees and original research. Course work can include concentrated studies in sociology and psychology, cultural history and geolinguistics, law, and foreign languages.

Social Context and Future Prospects

From the beginning of civilization, communication has been the glue that binds together all elements of human society. In the modern global community—with billions of information exchanges passing rapidly among members of a vast, diverse, largely receptive audience eager to connect—competence in oral and written skills is mandatory for personal and professional success. More than ever, messages must be concise, unambiguous, accurate, and compelling to cut through clutter streaming from dozens of sources.

Several issues are of particular concern in contemporary communication studies. Publishing is changing from print to electronic. There is high interest across the multitrillion-dollar telecommunications industry in green technology that reduces air and noise pollution. A new field, **unified communications**, which promotes system interoperability, is expected to reach worldwide revenues of $14.5 billion by 2015. With fresh markets of information exchangers emerging around the world, there is great demand for intercultural expertise.

All disciplines are subject to economic climate. If employment slumps in one field, such as journalism, skill sets can often be easily and profitably applied to a related field, such as advertising or public relations. People and management and verbal skills translate well across industries, geography, and time.

One of the greatest challenges facing communication in the age of the Internet is the educated assessment and consumption of the vast array of information available to receivers at all levels. The need for courses and instruction in critical thinking, always important, has risen exponentially as various electronic sources of information have proliferated—especially when those sources are not well understood, their authoritativeness is open to question, and their existence is ephemeral.

For the foreseeable future, there will always be news to report, information to supply, products and services to sell, causes to promote, politicians to elect, legislation to be enacted, and government actions to document and disseminate. There will always be ethical debates about what can be done versus what should be done. There will always be a place for those who advance the art and science of communication through performance, education, therapy, research, or innovation. And there will always be a need for people who can consistently connect successfully with fellow humans through the evocative use of words and images.

Jack Ewing

Further Reading

Baran, Stanley J., and Dennis K. Davis. *Mass Communication Theory: Foundations, Ferment, and Future.* 6th ed. Boston: Wadsworth, 2011. This textbook examines the field of mass communication, exploring theories and providing examples that aid in understanding the roles and ethics of various media.

Belch, George, and Michael Belch. *Advertising and Promotion: An Integrated Marketing Communications Perspective.* 9th ed. New York: McGraw-Hill/Irwin, 2008. This work spotlights the variety of strategies and methods used to build relationships between advertisers and customers in highly competitive, in-demand communication specialties.

Knapp, Mark L., and Judith A. Hall. *Nonverbal Communication in Human Interaction.* 7th ed. Boston: Wadsworth, 2009. A thorough examination of the theory, research, and psychology behind one of the most significant aspects of interpersonal communication.

Seiler, William J., and Melissa L. Beall. *Communication: Making Connections.* 8th ed. Needham Heights, Mass.: Allyn & Bacon, 2010. This book concentrates on the theory and practice of speech communication as applied to public discourse, interpersonal relationships, and group dynamics in a variety of settings.

Tomasello, Michael. *Origins of Human Communication.* Cambridge, Mass.: MIT Press, 2008. An award-winning study of how human communication evolved from gestures, founded on a fundamental need to cooperate for survival.

Varner, Iris, and Linda Beamer. *Intercultural Communication in the Global Workplace.* 5th ed. New York: McGraw-Hill/Irwin, 2010. This work addresses the impact of the Internet in particular on communication, especially as it affects relationships between businesses and governments in a shrinking world.

Web Sites

International Communication Association
http://www.icahdq.org

International Telecommunications Union
http://www.itu.int/en/pages/default.aspx

Linguistic Society of America
http://www.lsadc.org

National Communication Association
http://www.natcom.org

See also: Grammatology; Information Technology.

COMMUNICATIONS SATELLITE TECHNOLOGY

FIELDS OF STUDY

Avionics; computer engineering; computer information systems; computer network technology; computer numeric control technology; computer science; electrical engineering; electronics engineering; mechatronics; physics; telecommunications technology; television broadcast technology.

SUMMARY

Communications satellite technology has evolved from its first applications in the 1950's to become a part of most people's daily lives and thereby producing billions of dollars in yearly sales. Communications satellites were initially used to help relay television and radio signals to remote areas of the world and to aid navigation. Weather forecasts routinely make use of images transmitted from communications satellites. Telephone transmissions over long distances, including fax, cellular phones, pagers, and wireless technology, are all examples of the increasingly large impact that communications satellite technology continues to have on daily, routine communications.

KEY TERMS AND CONCEPTS

- **Baseband:** Transmission method for communications signals that uses the entire range of frequencies (bandwidth) available. It differs from broadband transmission, which is divided into different frequency ranges to allow multiple signals to travel simultaneously.
- **Bit:** Binary digit, of which there are only two, the numbers zero and one. Computer technology is based on the binary number system because early computers contained switches that could be only on or off.
- **Browser:** Software program that is used to view Web pages on the Internet.
- **Downlinking:** Transmitting data from a communications satellite or spacecraft to a receiver on Earth.
- **Downloading:** Process of accessing information on the Internet and then allowing a browser to make a copy to save on a personal computer.
- **Gravity:** Force of attraction between two objects, expressed as a function of their masses and the distance between them. Typically, the Earth's gravity is the most important consideration for satellites, and it has the constant value of 9.8 meters per second squared (m/s^2).
- **Hyperlinks:** Clickable pointers to online content in Web pages other than the page one is reading.
- **Internet Service Provider (ISP):** Organization that provides access to the Internet for a fee.
- **Ionosphere:** Part of the upper atmosphere that is ionized because of radiation from the sun, and therefore it affects the propagation of the radio waves within the electromagnetic spectrum.
- **Kilobit:** Quantity equal to 1,000 bits.
- **Orbit:** Curved path that an object travels in space because of gravity.
- **Period:** Time required to complete one revolution around the Earth.
- **Satellite:** Object that travels around the Earth.
- **Transponder:** The electronic component of a communications satellite that automatically amplifies and retransmits signals that are received.
- **Uplinking:** Transmitting data from a station on Earth up to a communications satellite or spacecraft.

Definition and Basic Principles

Sputnik 1, launched on October 4, 1957, by the Soviet Union, was the first artificial satellite. It used radio transmission to collect data regarding the distribution of radio signals within the ionosphere in order to measure density in the atmosphere. In addition to space satellites, the most common artificial satellites are the satellites used for communication, weather, navigation, and research. These artificial satellites travel around the Earth because of human action, and they depend on computer systems to function. A rocket is used to launch these artificial satellites so that they will have enough speed to be accelerated into the most common types of circular orbits, which require speeds of about 27,000 kilometers per hour. Some satellites, especially those that are to be used at locations far removed from the Earth's equator, require elliptical-shaped orbits instead, and their acceleration speeds are 30,000 kilometers per hour. If a launching rocket applies too much energy to an artificial satellite, the satellite may acquire enough energy to reach its escape velocity of 40,000 kilometers per hour and break free from the Earth's gravity. It is important that the satellite be able to maintain a constant high speed. If the speed is too low, gravity may cause the satellite to fall back down to the Earth's surface. There are also natural satellites that travel without human intervention, such as the Moon.

Background and History

In 1945, science fiction writer Arthur C. Clarke first described the concept of satellites being used for the mass distribution of television programs in his article "Extra-Terrestrial Relays," published in *Wireless World.* John Pierce, who worked at Bell Telephone Laboratories, further expanded on the idea of using satellites to repeat and relay television channels, radio signals, and telephone calls in his article "Orbital Radio Relays," published in the April, 1955, issue of *Jet Propulsion.* The first transatlantic telephone cable was opened by AT&T in 1956. The first transatlantic call was made in 1927, but it traveled via radio waves. The cable vastly improved the signal quality. The Soviet Union launched Sputnik 1, the first satellite, in 1957, which began the Space Race between the Soviet Union and the United States. The Communications Satellite Act of 1962 was passed by the United States Congress to regulate and assist the developing communications satellite industry. The first American television satellite transmission was made on July 10, 1962, five years into the Space Race, with the National Space and Aeronautics Administration's (NASA) launch of the world's first communications satellite, AT&T's Telstar.

The many new communications satellites followed, with names such as Relay, Syncom, Early Bird, Anik F2, Westar, Satcom, and Marisat. Since the 1970's, communications satellites have allowed remote parts of the world to receive television and radio, primarily for entertainment purposes. Technology advances have continued to evolve and now use these satellites to facilitate mobile phone communication and high-speed Internet applications.

How It Works

Communications satellites orbit the Earth and use microwave radio relay technology to facilitate communication for television, radio, mobile phones, weather forecasting, and navigation applications by receiving signals within the six-gigahertz (GHz) frequency range and then relaying these signals at frequencies within the four-GHz range. Generally there are two components required for a communications satellite. One is the satellite itself, sometimes called the space segment, which consists of the satellite and its telemetry controls, the fuel system, and the transponder. The other key component is the ground station, which transmits baseband signals to the satellite via uplinking and receives signals from the satellite via downlinking.

These communications satellites are suspended around the Earth in different types of orbits, depending on the communication requirements.

Geostationary Orbits. Geostationary orbits are most often used for communications and weather satellites because this type of orbit has permanent latitude at zero degrees, which is above the Earth's equator, and only longitudinal values vary. The result is that satellites within this type of orbital can use a fixed antenna that is pointed toward one location in the sky. Observers on the ground view these types of satellites as motionless because their orbit exactly matches the Earth's rotational period. Numerically, this movement equates to an orbital velocity of 1.91 miles per second, or a period of 23.9 hours. Because this type of orbit was first publicized by the science fiction writer Arthur C. Clarke in the 1940's, it is sometimes called a Clarke orbit. Systems that use geostationary satellites to provide images for meteorological applications include the Indian National Satellite System (INSAT), the European Organisation for the Exploitation of Meteorological Satellites' (EUMETSAT) Meteosat, and the United States' Geostationary Operational Environmental Satellites (GOES). These geostationary meteorological satellites provide the images for daily weather forecasts.

Molniya Orbits. Molniya orbits have been important primarily in the Soviet Union because they require less energy to maintain in the area's high latitudes. These high latitudes cause low grazing angles, which indicate angles of incidence for a beam of electromagnetic energy as it approaches the surface of the Earth. The angle of incidence specifically measures the deviation of this approach of energy from a straight line. As a result, geostationary satellites would orbit too low to the Earth's surface and their signals would have significant interference. Because of Russia's high latitudes, Molniya orbits are more energy efficient than geostationary orbits. The word Molniya comes from the Russian word for "lightning," and these orbits have a period of twelve hours, instead of the twenty-four hours characteristic of geostationary orbits. Molniya orbits have a large amount of incline, with an angle of incidence of about 63 degrees.

Low Earth Orbits. Low earth orbit (LEO) refers to a satellite orbiting between 140 and 970 kilometers above the Earth's surface. The periods are short, only about ninety minutes, which means that several of them are necessary to provide the type of uninterrupted communication characteristic of geostationary orbits, which have twenty-four-hour periods. Although a larger number of low earth orbits are needed, they have lower launching costs and require less energy for signal transmission because of how close to the Earth they orbit.

Applications and Products

DISH Network and Direct Broadcast Satellites. DISH Network is a type of direct broadcast satellite (DBS) network that communicates using small dishes that have a diameter of only 18 to 24 inches to provide access to television channels. This DBS service is available in several countries through many commercial direct-to-home providers, including DIRECTV in the United States, Freesat and Sky Digital in the United Kingdom, and Bell TV in Canada. These satellites transmit using the upper portion of the microwave Kμ band, which has a range of between 10.95 and 14.5 GHz. This range is divided based on thegeographic regions requiring transmissions. Law enforcement also uses the frequencies of the electromagnetic spectrum to detect traffic-speed violators.

Fixed Service Satellites. Besides the DBS services, the other type of communication satellite is called a fixed service satellite (FSS), which is useful for cable television channel reception, distance learning applications for universities, videoconferencing applications for businesses, and local television stations for live shots during the news broadcasts. Fixed service satellites use the lower frequencies of the Kμ bands and the C band for transmission. All of these frequencies are within the microwave region of the electromagnetic spectrum. The frequency range for the C band is about 4 to 8 GHz, and generally the C band functions better when moisture is present, making it especially useful for weather communication.

Intercontinental Telephone Service. Traditional landline telephone calls are relayed to an Earth station via the public switched telephone network (PSTN). Calls are then forwarded to a geostationary satellite to allow intercontinental phone communication. Fiber-optic technology is decreasing the dependence on satellites for this type of communication.

Iridium Satellite Phones. Iridium is the world's largest mobile satellite communications company. Satellites are useful for mobile phones when regular mobile phones have poor reception. These phones depend only on open sky for access to an orbiting satellite, making them very useful for ships on the open ocean for navigational purposes. Iridium manufactures several types of satellite phones, including the Iridium 9555 and 9505A, and models that have water resistance, e-mail, and USB data ports. Although it has the largest market share for satellite phones in Australia, Iridium does face competition from two other satellite phone companies: Immarsat and Globalstar.

Satellite Trucks and Portable Satellites. Trucks equipped with electrical generators to provide the power for an attached satellite have found applications for mobile transmission of news, especially after natural disasters. Some of these portable satellites use the C-band frequency for the transmission of information via the uplink process, which requires rather large antennas, whereas other portable satellites were developed in the 1980's to use the K_u band for transmission of information.

Global Positioning System (GPS). GPS makes use of communications satellite technology for navigational purposes. The GPS was first developed by the government for military applications but has become widely used in civilian applications in products such as cars and mobile phones.

Impact on Industry

Cable Television and Satellite Radio. Ted Turner led the way in the 1980's with the application of communications satellite technology for distributing cable television news and entertainment channels (CNN, TBS, and TNT, to name a few). Since the 1980's, access to additional cable television stations has continued to grow and has evolved to include access to radio stations as well, such as SiriusXM radio. Communications satellite technology is transforming the radio industry by allowing listeners to continue to listen to the radio stations of their choice no matter where they are. Specifically, many new cars incorporate the Sirius S50 Satellite Radio Receiver and MP3/ WMA player. However, the overall effect of communications satellites on the radio industry has not been as significant as it has on the television industry.

Internet Access and Cisco. In addition to providing access to additional television stations, the satellite dish can be used for Internet access as well. Downloading data occurs at speeds faster than 513 kilobits per second and uploading is faster than 33 kilobits, which is more than ten times faster than the dial-up modems that use the plain old telephone service. Cisco has been a leader in developing the tools to connect the world via the Internet for more than twenty years. High-speed connections to the Internet have been replacing the slower dial-up modems that connect via telephone lines. High-speed connections use fiber distributed data interface (FDDI), which is composed of two concentric rings of fiber-optic cable that allow it to transmit over a longer distance. This is expensive, so Cisco developed internet routing in space (IRIS), which will have a huge impact on the industry as it places routers on the communications satellites that are in orbit. These routers will provide more effective transmission of data, video, and voice without requiring any transmission hubs on the ground at all. IRIS technology will radically transform Internet connections in remote, rural areas. An Internet server connects to the antenna of a small satellite dish about the same size of the dishes used to enhance television reception. This connection speed can be as fast as 1.5 megabits per second and is a viable alternative to wireless networks. Wireless networks have been transforming the way that people communicate on a daily basis, by using radio and infrared signals for a variety of handheld devices (iPhones, iPods, iPads). Huge economic growth potential exists for Cisco and other companies that can effectively incorporate more reliable satellite technology, such as IRIS.

EchoStar and High-Speed Internet Access. Further indication of the large economic effect of communications satellites is from the February, 2011, report that EchoStar Corporation is going to pay more than $1 billion to purchase Hughes Communications. Hughes Communications uses satellites to provide high-speed, broadband access to the Internet. EchoStar, located in Englewood, Colorado, is the major provider of satellites and fiber networking technology to Dish Network Corporation (the second largest television provider in the United States), as well as additional military, commercial, and government customers that involve various data, video, and voice communications. The combined impact of merging Hughes Communications and EchoStar will be to continue the expansion of communications satellite technology.

Careers and Course Work

Careers working with communications satellite technology can be found primarily in the radio, television, and mobile phone industries. Specifically, these careers involve working with the wired telecommunications services that often include direct-to-home satellite television distributors as well as the newer wireless telecommunications carriers that provide mobile telephone, Internet, satellite radio, and navigational services. Government organizations also need employees who are trained in working with communications satellite technology for weather forecasting and other environmental applications as well as communication of data between public-safety officials.

The highest salaries are earned by those with a bachelor's degree in avionics technology, computer engineering, computer science, computer information systems, electrical engineering, electronics engineering, physics, or telecommunications technology, although a degree in television broadcast technology can also lead to lucrative career after obtaining several years of on-the-job-training. The work environments for those with these types of degrees are primarily office and technology, with more than 14 percent of the workers working more than 40 hours per week. Although the telecommunications and communications technology industries are expected to continue to grow faster than many other industries, the actual job growth is expected to be less than other high-growth industries because of computer optimization. Those without a bachelor's degree are the most at-risk, as their jobs also can involve lifting and climbing around electrical wires outdoors in a variety of weather and locations. The National Coalition for Telecommunications Education and Learning (NACTEL), the Communications Workers of America (CWA), and the Society of Cable Telecommunications Engineers (SCTE) are sources of detailed career information for anyone interested in communications satellite technology.

Social Context and Future Prospects

Advances in satellite technology have accompanied the rapid evolution of computer technology to such an extent that some experts describe this media revolution as an actual convergence of all media (television, motion pictures, printed news, Internet, and mobile phone communications). In 1979, Nicholas Negroponte of the Massachusetts Institute of Technology began giving lectures describing this future convergence of all forms of media. As of the twenty-first century, this convergence seems to be nearly complete. Television shows can be viewed on the Internet, as can news from cable television news stations such as CNN, Fox, and MSNBC. Hyperlinks provide digital connections between information that can be accessed from almost anywhere in the world instantly because of communication satellite technology. The result is that there is a twenty-four-hour news cycle, and the effects are sometimes positive but can also be negative if the wrong information is broadcast. The instantaneous transmission of political and social unrest by communications satellite technology can lead to further actions, as shown by the 2011 protests in Egypt, Iran, Yemen, Libya, and Bahrain.

Jeanne L. Kuhler, B.S., M.S., Ph.D.

Fascinating Facts About Communications Satellite Technology

- In 2011, there are more than 2,000 communications satellites orbiting the Earth.
- There are at least nineteen Orbiting Satellite Carrying Amateur Radio (OSCAR). These artificial communication satellites have been launched by individuals.
- The Communications Satellite Act was passed in 1962 with the intention of making the United States a worldwide leader in communications satellite technology so that it could encourage peaceful relations with other nations. This act also established the Communications Satellite Corporation (Comsat), with headquarters in Washington, D.C.
- In 2003, Optus and Defence C1 carried sixteen antennas to provide eighteen satellite beams across Australia and New Zealand, making it one of the most advanced communications satellites ever launched.
- Sputnik, the first artificial satellite, was only 58 centimeters in diameter.
- The first known ideas for communications satellites were actually made public not by scientists or engineers but by the writers Arthur C. Clarke and Edward Everett Hale.

Further Reading

Baran, Stanley J., and Dennis K. Davis. *Mass Communication Theory: Foundations, Ferment, and Future.* 5th ed. Boston: Wadsworth Cengage Learning, 2008. Provides a detailed historical discussion of the communications technologies and their impacts on society.

Bucy, Erik P. *Living in the Information Age: A New Media Reader.* 2d ed. Belmont, Calif.: Wadsworth Thomson, 2005. Describes the societal implications of the evolution of media technology and the convergence of communications media made possible by new technologies.

Giancoli, Douglas C. *Physics for Scientists and Engineers with Modern Physics.* 4th ed. Upper Saddle River, N.J.: Pearson Education, 2008. This introductory physics textbook provides mathematical information regarding satellites and their orbits.

Grant, August E., and Jennifer Meadows. *Communication Technology Update and Fundamentals.* 12th ed. Burlington, Mass.: Focal Press, 2010. This introductory textbook provides technical information regarding communications satellites in terms of historical development, detailed applications, and existing uses.

Hesmondhalgh, David. *The Cultural Industries.* 2d ed. Thousand Oaks, Calif.: Sage, 2007. Describes the worldwide cultural effects of communications technologies in the past and present.

Mattelart, Armand. *Networking the World: 1794-2000.* Translated by Liz Carey-Libbrecht and James A. Cohen. Minneapolis: University of Minnesota Press, 2000. Describes the impact of communications satellites and other technologies on political, economic, and cultural phenomena, including nationalism, liberalism, and universalism.

Parks, Lisa, and Shanti Kumar, eds. *Planet TV:, A Global Television Reader.* New York: New York University Press, 2003. This textbook focuses on the application of communication satellites to television programming and discusses its historical evolution as well as its impact on the societies of various nations, especially India.

Web Sites

Communications Workers of America
http://www.cwa-union.org

National Coalition for Telecommunications Education and Learning
http://www.nactel.org

Society of Cable Telecommunications Engineers
http://www.scte.org

See also: Avionics and Aircraft Instrumentation; Computer Engineering; Computer Networks.

COMPUTER ENGINEERING

FIELDS OF STUDY

Computer engineering; computer science; computer programming; computer information systems; electrical engineering; information systems; computer information technology; software engineering.

SUMMARY

Computer engineering refers to the field of designing hardware and software components that interact to maximize the speed and processing capabilities of the central processing unit (CPU), memory, and the peripheral devices, which include the keyboard, monitor, disk drives, mouse, and printer. Because the first computers were based on the use of on-and-off mechanical switches to control electrical circuits, computer hardware is still based on the binary number system. Computer engineering involves the development of operating systems that are able to interact with compilers that translate the software programs written by humans into the machine instructions that depend on the binary number system to control electrical logic circuits and communication ports to access the Internet.

KEY TERMS AND CONCEPTS

- **Basic Input-Output System (BIOS):** Computer program that allows the central processing unit (CPU) of a computer to communicate with other computer hardware.
- **Browser:** Software program that is used to view Web pages on the Internet.
- **Icon:** Small picture that represents a file, program, disk, menu, or option.
- **Internet Service Provider (ISP):** Organization that provides paid access to the Internet.
- **Logical Topology:** Pathway within the physical network devices that directs the flow of data. The bus and ring are the only two types of logical topology.
- **Mainframe:** Large, stand-alone computer that completes batch processing (groups of computer instructions completed at once).
- **Protocol:** Set of rules to be followed that allows for communication.
- **Server:** Computer that is dedicated to managing resources shared by users (clients).

Definition and Basic Principles

Much of the work within the field of computer engineering focuses on the optimization of computer hardware, which is the general term that describes the electronic and mechanical devices that make it possible for a computer user (client) to utilize the power of a computer. These physical devices are based on binary logic. Humans use the decimal system for numbers, instead of the base two-number system of binary logic, and naturally humans communicate with words. A great deal of interface activity is necessary to bridge this communication gap, and computer engineering involves additional types of software (programs) that function as intermediate interfaces to translate human instructions into hardware activity. Examples of these types of software include operating systems, drivers, browsers, compilers, and linkers.

Computer hardware and software generally can be arranged in a series of hierarchical levels, with the lowest level of software being the machine language, consisting of numbers and operands that the processor executes. Assembly language is the next level, and it uses instruction mnemonics, which are machine-specific instructions used to communicate with the operating system and hardware. Each instruction written in assembly language corresponds to one instruction written in machine code, and these instructions are used directly by the processor. Assembly language is also used to optimize the runtime execution of application programs. At the next level is the operating system, which is a computer program written so that it can manage resources, such as disk drives and printers, and can also function as an interface between a computer user and the various pieces of hardware. The highest level includes applications that humans use on a daily basis. These are considered the highest level because they consist of statements written in English and are very close to human language.

Background and History

The first computers used vacuum tubes and mechanical relays to indicate the switch positions of *on* or *off* as the logic units corresponding to the binary digits of 0 or 1, and it was necessary to reconfigure them each time a new task was approached. They were large enough to occupy entire rooms and required huge amounts of electricity and cooling. In the 1930's the Atanasoff-Berry Computer (ABC) was created at Iowa State University to solve simultaneous numerical equations, and it was followed by the electronic numerical integrator and computer (ENIAC), developed by the military for mathematical operations.

The transistor was invented in 1947 by John Bardeen, Walter Brattain, and William Shockley, which led to the use of large transistors as the logic units in the 1950's. The integrated circuit chip was invented by Jack St. Clair Kilby and Robert Norton Noyce in 1958 and caused integrated circuits to come into usage in the 1960's. These early integrated circuits were still quite large and included transistors, diodes, capacitors, and transistors. Modern silicon chips can hold these components and as many as 55 million transistors. Silicon chips are called microprocessors, because each microprocessor can hold these logic units within just over a square inch of space.

How It Works

Hardware. The hardware, or physical components, of a computer can be classified according to their general uses of input, output, processing, and storage. Typical input devices include the mouse, keyboard, scanner, and microphone that facilitate communication of information between the human user and the computer. The operation of each of these peripheral devices requires special software called a driver, which is a type of controller, that is able to translate the input data into a form that can be communicated to the operating system and controls input and output peripheral devices.

A read-only memory (ROM) chip contains instructions for the basic input-output system (BIOS) that all the peripheral devices use to interact with the CPU. This process is especially important when a user first turns on a computer for the boot process.

When a computer is turned on, it first activates the BIOS, which is software that facilitates the interactions between the operating system, hardware, and peripherals. The BIOS accomplishes this interaction by first running the power-on self test (POST), which is a set of routines that are always available at a specific memory address in the read-only memory. These routines communicate with the keyboard, monitor, disk drives, printer, and communication ports to access the Internet. The BIOS also controls the time-of-day clock. These tasks completed by the BIOS are sometimes referred to as booting up (from the old expression "lift itself up by its own bootstraps"). The last instruction within the BIOS is to start reading the operating system from either a boot disk in a diskette drive or the hard drive. When shutting down a computer there are also steps that are followed to allow for settings to be stored and network connections to be terminated.

The CPU allows instructions and data to be stored in memory locations, called registers, which facilitate the processing of information as it is exchanged between the control unit, arithmetic-logic unit, and any peripheral devices. The processor interacts continuously with storage locations, which can be classified as either volatile or nonvolatile types of memory. Volatile memory is erased when the computer is turned off and consists of main memory, called random-access memory (RAM) and cache memory. The fundamental unit of volatile memory is the flip-flop, which can store a value of 0 or 1 when the computer is on. This value can be flipped when the computer needs to change it. If a series of 8 flip-flops is hooked together, an 8-bit number can be stored in a register. Registers can store only a small amount of data on a temporary basis while the computer is actually on. Therefore, the RAM is needed for larger amounts of information. However, it takes longer to access data stored in the RAM because it is outside the processor and needs to be retrieved, causing a lag time. Another type of memory, called cache, is located in the processor and can be considered an intermediate type of memory between registers and main memory.

Nonvolatile memory is not erased when a computer is turned off. It consists of hard disks that make up the hard drive or flash memory. Although additional nonvolatile memory can be purchased for less money than volatile memory, it is slower. The hard drive consists of several circular discs called platters that are made from aluminum, glass, or ceramics and covered by a magnetic material so that they can develop a magnetic charge. There are read and write heads made of copper so that a magnetic field develops that is able to read or write data when interacting with the platters. A spindle motor causes the platters to spin at a constant rate, and either a stepper motor or voice coil is used as the head actuator to initiate interaction with the platters.

The control unit of the CPU manages the circuits for completing operations of the arithmetic-logic unit. The arithmetic-logic unit of the CPU also contains circuits for completing the logical operations, in addition to data operations, causing the CPU essentially to function as the brain of the computer. The CPU is located physically on the motherboard. The motherboard is a flat board that contains all the chips needed to run a computer, including the CPU, BIOS, and RAM, as well as expansion slots and power- supply connectors. A set of wires, called a bus, etched into the motherboard connects these components. Expansion slots are empty places on the motherboard that allow upgrades or the insertion of expansion cards for various video and voice controllers, memory expansion cards, fax boards, and modems without having to reconfigure the entire computer. The motherboard is the main circuit board for the entire computer.

Circuit Design and Connectivity. Most makers of processor chips use the transistor-transistor logic (TTL) because this type of logic gate allows for the output from one gate to be used directly as the input for another gate without additional electronic input, which maximizes possible data transmission while minimizing electronic complications. The TTL makes this possible because any value less than 0.5 volt is recognized as the logic value of 0, while any value that exceeds 2.7 volts indicates the logic value of 1. The processor chips interact with external computer devices via connectivity locations called ports. One of the most important of these ports is called the Universal Serial Bus (USB) port, which is a high-speed, serial, daisy-chainable port in newer computers used to connect keyboards, printers, mice, external disk drives, and additional input and output devices.

Software. The operating system consists of software programs that function as an interface between the user and the hardware components. The operating system also assists the output devices of printers and monitors. Most of the operating systems being used also have an application programming interface (API), which includes graphics and facilitates use. APIs are written in high-level languages (using statements approximating human language), such as C++, Java, and Visual Basic.

Applications and Products

Stand-Alone Computers. Most computer users rely on relatively small computers such as laptops and personal computers (microcomputers). Companies that manufacture these relatively inexpensive computers have come into existence only since the early 1980's and have transformed the lives of average Americans by making computer usage a part of everyday life. Before microcomputers came into such wide usage, the workstation was the most accessible smaller-size computer. It is still used primarily by small and medium-size businesses that need the additional memory and speed capabilities. Larger organizations such as universities use mainframe computers to handle their larger power requirements. Mainframes generally occupy an entire room. The most powerful computers are referred to as supercomputers, and they are so expensive that often several universities will share them for scientific and computational activities. The military uses them as well. They often require the space of several rooms.

Inter-Network Service Architecture, Interfaces, and Inter-Network Interfaces. A network consists of two or more computers connected together in order to share resources. The first networks used coaxial cable, but now wireless technologies allow computer devices to communicate without the need to be physically connected by a coaxial cable. The Internet has been a computer-engineering application that has transformed the way people live. Connecting to the Internet first involved the same analogue transmission used by the plain old telephone service (POTS), but connections have evolved to the use of fiber-optic technology and wireless connections. Laptop computers, personal digital assistants (PDAs), cell phones, smart phones, RFID (radio frequency identification), iPods, iPads, and Global Positioning Systems (GPS) are able to communicate, and their developments have been made possible by the implementation of the fundamental architectural model for inter-network service connections called the Open Systems Interconnection (OSI) model. OSI is the layered architectural model for connecting networks. It was developed in 1977 and is used to make troubleshooting easier so that if a component fails on one computer, a new, similar component can be used to fix the problem, even if the component was manufactured by a different company.

OSI's seven layers are the application, presentation, session, transport, network, data link, and physical layers.

The application and presentation layers work together. The application layer synchronizes applications and services in use by a person on an individual computer with the applications and services shared with others via a server. The services include e-mail, the World Wide Web, and financial transactions. One of the primary functions of the presentation layer is the conversion of data in a native format such as extended binary coded decimal interchange code (EBCDIC) into a standard format such as American Standard Code for Information Interchange (ASCII).

As its name implies, the primary purpose of the session layer is to control the dialog sessions between two devices. The network file system (NFS) and structured query language (SQL) are examples of tools used in this layer.

The transport layer controls the connection-oriented flow of data by sending acknowledgments to data senders once the recipient has received data and also makes sure that segments of data are retransmitted if they do not reach the intended recipient. A router is one of the fundamental devices that works in this layer, and it is used to connect two or more networks together physically by providing a high degree of security and traffic control.

The data link layer translates the data transmitted by the components of the physical layer, and it is within the physical layer that the most dramatic technological advances made possible by computer engineering have had the greatest impact. The coaxial cable originally used has been replaced by fiber-optic technology and wireless connections.

Fiber Distributed Data Interface (FDDI) is the fiber-optic technology that is a high-speed method of networking, composed of two concentric rings of fiber-optic cable, allowing it to transmit over a longer distance but at greater expense. Fiber-optic cable uses glass threads or plastic fibers to transmit data. Each cable contains a bundle of threads that work by reflecting light waves. They have much greater bandwidth than the traditional coaxial cables and can carry data at a much faster rate of 100 gigabits per second because light travels faster than electrical signals. Fiber-optic cables are also not susceptible to electromagnetic interference and weigh less than coaxial cables.

Wireless networks use radio or infrared signals for cell phones and a rapidly growing variety of handheld devices, including iPhones, iPods, and tablets. New technologies using Bluetooth and Wi-Fi for mobile connections are leading to the next phase of inter-network communications with the Internet called cloud computing, which is the direction for the most economic growth. Cloud computing is basically a wireless Internet application where servers supply resources, software, and other information to users on demand and for a fee.

Smart phones that use Google's new Android operating system for mobile phones have a PlayStation emulator called PSX4Droid that allows PlayStation games to be played on these phones. Besides making games easily accessible, cloud computing is also making it easier and cheaper to do business all around the world with applications such as Go To Meeting, which is one example of video conferencing technology used by businesses.

Impact on Industry

The first developments in technology that made personal computers, the Internet, and cell phones more easily accessible to the average consumer have primarily been made by American companies. Computer manufacturers include IBM and Microsoft, which were followed by Dell, Compaq, Gateway, Intel, Hewlett-Packard, Sun Microsystems, and Cisco. Cisco continues to be the primary supplier of the various hardware devices necessary for connecting networks, such as routers, switches, hubs, and bridges. Mobile computing is being led by American companies Apple, with its iPhone, and Research in Motion (RIM), with its Blackberry. These mobile devices require collaboration with phone companies such as AT&T and Verizon, causing them to grow in worldwide dominance as well. Most U.S. technology companies are international in scope, with expansion still expected in the less-developed countries of Southeast Asia and Eastern Europe.

Microsoft was founded in 1975 by Bill Gates and Paul Allen. All the personal computers (PCs) that became more accessible to consumers depended on Microsoft's operating system, Windows. Over the years, Microsoft has made numerous improvements with each new version of its operating system, and it has continued to dominate the PC market to such an extent that the U.S. government sued Microsoft, charging it with violating antitrust laws. As of 2011, Microsoft remains the largest software company in the world, although it has been losing some of its dominance as an innovative technology company because of the growth of Google, with its Internet search engine software, and Apple, with its new, cutting-edge consumer electronics, including the iPod, iPad, and iPhone. In 2010, Microsoft started selling the Kinect for Xbox 360, which is a motion-capture device used for games that easily connects to a television and the Internet. More than 2.5 million of the Kinect for Xbox 360 units have been sold in the United States since November, 2010. To try to compete with Apple's dominance in the mobile-technology market, Microsoft announced in January, 2011, that it would collaborate with chip manufacturer Micron, to incorporate Micron's flash memory into mobile devices. A new version of Microsoft's Windows operating system was to use processor chips manufactured by ARM Holdings rather than those by Intel, because the ARM chips contain lower power, which facilitates mobile communication. Microsoft continues to dominate the PC operating-system market worldwide.

Apple is one of the most successful U.S. companies because of its Macintosh computer and iPod. Introduced in 2001, the iPod uses various audio file formats, such as MP3 and WAV, to function as a portable media player. To further expand on the popularity of the iPod, Apple opened its own online media and applications stores to allow consumers to access music and video applications via the Internet. In 2007, Apple expanded its market share by introducing the iPhone, and adding visual text messaging, emailing of digital photos using its own camera capabilities, and a multi-touch screen for enhanced access to the Internet. The Apple iPhone is essentially a miniature handheld computer that also functions as a phone. The apps that are be downloaded from the Apple Store online are computer programs that can be designed to perform just about any specific task. There are more than 70,000 games available for purchase for the iPhone, and more than 7 million apps have been sold as of January, 2011.

Google was started by Larry Page and Sergey Brin, who met as graduate students in Stanford University's computer science department, in 1996. Ten years later, the Oxford dictionary added the term "google," a verb meaning to do an Internet search on a topic using the Google search engine. Despite attempts by several competitors, including Microsoft, to develop search engine software, Google has dominated the market. In addition to its search engine, Google has developed many new innovative software applications, including Gmail, Google Maps, and its own operating system called Android, designed specifically to assist Google with its move into the highly lucrative mobile-technology market.

Fascinating Facts About Computer Engineering

- It is possible to warm up the grill before arriving home by connecting to the Internet and attaching an iGrill to a USB port. Recipes can also be downloaded directly from the Internet for the iGrill.
- The average laptop computer has more processing power than the ENIAC, which had more than 20,000 vacuum tubes and was large enough to fill entire rooms.
- By about 2015, batteries in digital devices will consume oxygen from the air in order to maintain their power.
- By 2015, it is expected that some mobile devices will be powered by kinetic energy and will no longer depend on batteries.
- A job growth rate of 31 percent is projected for computer engineers by 2018.

Careers and Course Work

Knowledge of the design of electrical devices and logic circuits is important to computer engineers, so a strong background in mathematics and physics is helpful. Since there is a great deal of overlap between computer engineering and electrical engineering, any graduate with a degree in electrical engineering may also find work within the computer field.

One of the most rapidly expanding occupations within the computer engineering field is that of network systems and data communication analyst. According to the Bureau of Labor Statistics, this occupation is projected to grow by 53.4 percent to almost 448,000 workers between by 2018, making it the second-fastest growing occupation in the United States. The related occupation of computer systems analyst is also expected to grow faster than the average for all jobs through 2014. Related jobs include network security specialists and software engineers. All of these occupations require at least a bachelor's degree, typically in computer engineering, electrical engineering, software engineering, computer science, information systems, or network security.

Social Context and Future Prospects

The use of cookies, a file stored on an Internet user's computer that contains information about that user, an identification code or customized preferences, that is recognized by the servers of the Web sites the user visits, as a tool to increase online sales by allowing e-businesses to monitor the preferences of customers as they access different Web pages is becoming more prevalent. However, this constant online monitoring also raises privacy issues. In the future there is the chance that one's privacy could be invaded with regard to medical or criminal records and all manner of financial information. In addition, wireless technologies are projected to increase the usage of smart phones, the iPad, the iPod, and many new consumer gadgets, which can access countless e-business and social networking Web sites. Thus, the rapid growth of Internet applications that facilitate communication and financial transactions will continue to be accompanied by increasing rates of identity theft and other cyber crimes, ensuring that network security will continue to be an important application of computer engineering.

Jeanne L. Kuhler, B.S., M.S., Ph.D.

Further Reading

Cassel, Lillian N., and Richard H. Austing. *Computer Networks and Open Systems: An Application Development Perspective.* Sudbury, Mass.: Jones & Bartlett Learning, 2000. This textbook describes the OSI architecture model.

Das, Sumitabha. *Your UNIX: The Ultimate Guide.* New York: McGraw-Hill, 2005. This introductory textbook provides background and instructions for using the UNIX operating system.

Dhillon, Gurphreet. "Dimensions of Power and IS Implementation." *Information and Management* 41, no. 5 (2004): 635-644. Describes some of the work done by managers when choosing which computer languages and tools to implement.

Irvine, Kip R. *Assembly Language for Intel-Based Computers.* 5th ed. Upper Saddle River, N.J.: Prentice Hall, 2006. This introductory textbook provides instruction for using assembly language for Intel processors.

Kerns, David V., Jr., and J. David Irwin. *Essentials of Electrical and Computer Engineering* 2d ed. Upper Saddle River, N.J.: Prentice Hall, 2004. A solid introductory text that integrates conceptual discussions with modern, relevant technological applications.

Magee, Jeff, and Jeff Kramer. *Concurrency: State Models and Java Programming.* Hoboken, N.J.: John Wiley & Sons, 2006. This intermediate-level textbook describes the software engineering techniques involving control of the timing of different processes.

Silvester, P. P., and D. A. Lowther. *Computer Engineering Circuits, Programs, and Data.* New York: Oxford University Press, 1989. This is an introductory text for engineering students.

Sommerville, Ian. *Software Engineering.* 9th ed. Boston: Addison-Wesley, 2010. Text presents a broad, up-to-date perspective of software engineering that focuses on fundamental processes and techniques for creating software systems. Includes case studies and extensive Web resources.

Web Sites

Computer Society
http://www.computer.org

Institute of Electrical and Electronics Engineers
http://www.ieee.org

Software Engineering Institute
http://www.sei.cmu.edu

See also: Computer Graphics; Computer Languages, Compilers, and Tools; Computer Networks; Computer Science; Electrical Engineering; Internet and Web Engineering; Pattern Recognition; Software Engineering.

COMPUTER GRAPHICS

FIELDS OF STUDY

Three-dimensional (3-D) design; calculus; computer programming; computer animation; digital modeling; graphic design; multimedia applications; software engineering; Web development; vector graphics and design; drawing; animation.

SUMMARY

Computer graphics involves the creation, display, and storage of images on a computer with the use of specialized software. Computer graphics fills an essential role in many everyday applications. 3-D animation has revolutionized video games and has resulted in box-office hits in theaters. Virtual images of people's bodies and tissues are used in medicine for teaching, surgery simulations, and diagnoses. Educators and scientists are able to develop 3-D models that illustrate principles in a more comprehensible manner than a two-dimensional (2-D) image can. Through the use of such imagery, architects and engineers can prepare virtual buildings and models to test options prior to construction. Finally, businesses use computer graphics to prepare charts and graphs for better comprehension during presentations.

KEY TERMS AND CONCEPTS

- **Application Programming Interface (API):** Set of functions built into a software program that allows communication with another software program.
- **Computer-Aided Design (CAD):** Software used by architects, engineers, and artists to create drawings and plans.
- **Computer Animation:** Art of creating moving images by means of computer technology.
- **Graphical User Interface (GUI):** Program by which a user interacts with a computer by controlling visual symbols on the screen.
- **Graphics Processing Unit (GPU):** Specialized microprocessor typically found on a video card that accelerates graphics processing.
- **Pixels:** Abbreviation for "picture elements," the smallest discrete components of a graphic image appearing on a computer screen or other graphics output device.
- **Raster Graphics:** Digital images that use pixels arranged in a grid formation to represent an image.
- **Rendering:** Process of generating an image from a model by means of computer programs.
- **Three-Dimensional (3-D) model:** Representation of any 3-D surface of an object using graphics software.
- **Vector Graphics:** Field of computer graphics that uses mathematical relationships between points and the paths connecting them to describe an image.

DEFINITION AND BASIC PRINCIPLES

The field of computer graphics uses computers to create digital images or to modify and use images obtained from the real world. The images are created from internal models by means of computer programs. Two types of graphics data can be stored in a computer. Vector graphics are based on mathematical formulas that generate geometric images by joining straight lines. Raster graphics are based on a grid of dots known as pixels, or picture elements. Computer graphics can be expressed as 2-D, 3-D, or animated images.

The graphic data must be processed in order to render the image and display on a computer movie screen. The work of computer-graphics programmers has been facilitated by the development of application programming interfaces (APIs), notably the Open Graphics Library (OpenGL). OpenGL provides a set of standard operations to render images across a wide variety of platforms (operating systems).The graphics processing unit (GPU) found in a video card facilitates presentation of the image.

There are subtle differences between the responsibilities of the computer-graphics specialist and graphic or Web designers. Computer-graphics specialists develop programs to display visual images or models, while designers are creative artists who use programs to communicate a specific message effectively. The end products of graphic designers are seen in various print media, while Web designers produce digital media.

Background and History

The beginning of computer graphics has been largely attributed to Ivan Sutherland, who developed a computer drawing program in 1961 for his dissertation work at the Massachusetts Institute of Technology. This program, Sketchpad, was a seminal event in the area of human-computer interaction, as it was one of the first to use graphical user interfaces (GUIs). Sutherland used a light pen containing a photoelectric cell that interacted with elements on the monitor. His method was based on vector graphics. Sketchpad provided vastly greater possibilities for the designer or engineer over previous methods based on pen and paper.

Other researchers further developed vector-graphics capabilities. Raster-based graphics using pixels was later developed and is the primary technology being used. The mouse was invented and proved more convenient than a light pen for selecting icons and other elements on a computer screen. By the early 1980's, Microsoft's personal computer (PC) and Apple's Macintosh were marketed using operating systems that incorporated GUIs and input devices that included the mouse as well as the standard keyboard.

Major corporations developed an early interest in computer graphics. Engineers at Bell Telephone Laboratories, Lawrence Berkeley National Laboratory, and Boeing developed films to illustrate satellite orbits, aircraft vibrations, and other physics principles. Flight simulators were developed by Evans & Sutherland and General Electric.

The invention of video graphics cards in the late 1980's followed by continual improvements gave rise to advances in animation. Video games and full-length animated motion pictures have become large components of popular culture.

How It Works

Types of Images. Vector graphics uses mathematical formulas to generate lines or paths that are connected at points called vertices to form geometric shapes (usually triangles). These shapes are joined in a meshwork on the surfaces of figures. Surfaces on one plane are two-dimensional, while connecting vertices in three dimensions will produce 3-D images. Two-dimensional images are more useful for applications such as advertising, technical drawing, and cartography. Raster images, on the other hand, develop images based on pixels, or picture elements. Pixels can be thought of as tiny dots or cells that contain minute portions of the image and together compose the image on the computer screen. The bits of information that the pixels are able to process determine the resolution or sharpness of the image. Raster images are much more commonly used in computer graphics and are essential for 3-D and animation work.

Graphics Pipeline. The process of creating an image from data is known as the graphics pipeline. The pipeline consists of three main stages: modeling, rendering, and display.

Modeling begins with a specification of the objects, or components of a scene, in terms of shape, size, color, texture, and other parameters. These objects then undergo a transformation involving their correct placement in a scene.

Rendering is the process of creating the actual image or animation from the scene. Rendering is analogous to a photograph or an artist's drawing of a scene. Aspects such as proper illumination and the visibility of objects are important at this stage.

The final image is displayed on a computer monitor with the use of advanced software, as well as computer hardware that includes the motherboard and graphics card. The designer must keep in mind that the image may appear differently on different computers or different printers.

OpenGL. OpenGL is an API that facilitates writing programs across a wide variety of computer languages and hardware and software platforms. OpenGL consists of graphics libraries that provide the programmer with a basic set of commands to render images. OpenGL is implemented through a rendering pipeline (graphics pipeline). Both vector and raster data are accepted for processing but follow different steps. At the rasterization stage, all data are converted to pixels. At the final step, the pixels are written into a 2-D grid known as a framebuffer. A framebuffer is the memory portion of the computer allocated to hold the graphics information for a single frame or picture.

Maya. Maya (trade name Autodesk Maya) is computer-graphics software that has become the industry standard for generating 3-D models for game development and film. Maya is particularly effective in producing dazzling animation effects. Maya is imported into OpenGL or another API to display the models on the screen.

Video Games. Video-game development takes a specialized direction. The term "game engine" was coined to refer to software used to create and render video games on a computer screen. Many features need to work together to create a successful game, including 2-D and 3-D graphics, a "physics engine" to prepare realistic collision effects, sound, animation, and many other functions. Because of the highly competitive nature of the video-game industry, it is necessary to develop new games rapidly. This has led to reusing or adapting the same game engine to create new games. Some companies specialize in developing so-called middleware software suites that are conceived to contain basic elements on which the game programmer can build to create the complete game.

Television. To create 3-D images for television, individual objects are created in individual layers in computer memory. This way, the individual objects can move independently without affecting the other objects. TV graphics are normally produced a screen at a time and can be layered with different images, text, backgrounds, and other elements to produce rich graphic images. Editing of digital graphics is much faster and efficient than the traditional method of cutting and pasting film strips.

Film. The film *Avatar* (2009) illustrated how far 3-D animation had been developed. The production used a technique called performance or motion tracking. Video cameras were attached to computers and focused on the faces of human actors as they performed their parts. In this manner, subtle facial expressions could be transferred to their animated avatars. Backgrounds, props, and associated scenery moved in relation to the actors.

Applications and Products

Game Development. Game development has become a major consumer industry, grossing $20 billion in 2010. Ninety-two percent of juveniles play video or computer games. Major players in the field include Sony PlayStation, Nintendo, and Microsoft Xbox. The demographics of video-game players are changing, resulting in a greater number of female and adult players. Whereas previously video games focused on a niche market (juvenile boys), in the future game developers will be increasingly directing their attention to the mass market. These market changes will result in games that are easier to play and broader in subject matter.

Film. The influence of computer graphics in the film industry is largely related to animation. Of course, animation predates computer graphics by many years, but it is realism that gives animation its force. Entire films can be animated, or animation can play a supplemental role. The fantasy of Hollywood was previously based on constructing models and miniatures, but now computer-generated imagery can be integrated into live action.

Television. The conversion of the broadcast signal for television from analogue to digital, and later to high definition, has made the role of the computer-graphics designer even more important. It is common for many shows to have a computer-graphics background instead of a natural background, such as when a weatherperson stands in front of a weather map. This development results in more economical productions, since there are no labor costs involved in preparing sets or a need to store them.

Computer graphics was used in television advertising before film, since it is more economical to produce. A combination of dazzling, animated graphics with a product or brand-name can leave a lasting impression on the viewer.

Medicine. Computer graphics typically works in concert with other advanced technologies, such as computed tomography, magnetic resonance, and positron emission tomography, to aid in diagnosis and treatment. The images obtained by these technologies are reconstructed by computer-graphics techniques to form 3-D models, which can then be visualized. The development of virtual human bodies has proven invaluable to illustrate anatomical structures in healthy and diseased subjects. These virtual images have found application in surgery simulations and medical education and training. The use of patient-specific imaging data guides surgeons to conduct minimally invasive interventions that are less traumatic for the patient and lead to faster healing. Augmented reality provides a larger view of the surgical field and allows the surgeon to view structures that are below the observed surface.

Science. Computer graphics has proven valuable to illustrate scientific principles and concepts that are not easily visible in the natural environment. By preparing virtual 3-D models of molecular structures or viruses moving through tissues, a student or scientist who is a visual learner is able to grasp these concepts.

Architecture and Engineering. Computer-aided design has greatly helped the fields of architecture and engineering. Computer graphics was initially used only as a replacement for drafting with pencil and paper. However, the profession has come to recognize its value in the early stages of a project to help designers check and reevaluate their proposed designs. Multimedia designs such as animations and panoramas are very useful in client presentations. The designs allow clients to walk through a building virtually and interactively look around interior spaces. Engineers can also test the effect of various inputs to a system, model, or circuit.

Business. Presentation of numeric data in graphs and charts is an important application of computer graphics in business. Market trends, production data, and other business information presented in graphic form are often more understandable to an audience or reader.

Education. Computer graphics has proven very useful in education because of the power of visualization in the learning process. There are many benefits to using computer graphics in education: Students learn at their own pace at their own time, the instruction is interactive rather than passive, the student is engaged in the learning process, and textual and graphic objects can be shared among applications via tools such as cutting and pasting, clipboards, and scrapbooks.

Impact on Industry

The worldwide market for hardware, software, and services for 3-D visualization is expected to grow rapidly. Spending in the defense and government markets reached $16.5 billion in 2010 and is expected to increase to $20 billion by 2015. The popularity of games and the film industry has resulted in huge investments in developing 3-D technology. With the technology already in place, world governments can acquire 3-D visualization at a much more reasonable price than previously. The military sector is the largest market, followed by civil-aviation and civil-service sectors, including law-enforcement and emergency-response agencies.

Government and University Research. Although the U.S. government played a leading role in the early development of computer graphics, it has come to play a more collaborative role with universities. A typical example is the Graphics and Visualization Center, which was founded by the National Science Foundation to pursue interdisciplinary research at five universities (Brown, California Institute of Technology, Cornell, University of North Carolina, and University of Utah). The center pursues research in four main areas of computer graphics: modeling, rendering, user interfaces, and high-performance architectures. The Department of Defense also conducts research involving computer graphics.

On the local level, computer graphics has proven useful in preparing presentation material, such as visually appealing graphs, charts, and diagrams. Complex projects and issues can be presented in a manner that is more understandable than traditional speeches or handouts. County and community

Fascinating Facts About Computer Graphics

- *Avatar* (2009) became the highest-grossing film of all time, earning $2.7 billion worldwide and demonstrating how compelling stories told using computer graphics had become.
- Creation of virtual persons allows surgeons to perform a trial run of their procedures before the actual surgery.
- Real-time visualization models prepared by architects allow buyers to walk through virtual homes and change the views they see.
- Scientists can study chemical reactions with molecules in real time to gain insight into the structural changes taking place among the molecules.
- Engineers can prepare a 3-D model of an engine under development and study the effects of various changes.
- The features visible in 3-D through Google Earth—topography, cities, street views, trees, and other details–dramatically demonstrate the uses of computer graphics when wedded to satellite technologies.
- 3-D visualization of atmosphere and terrain has resulted in improved weather forecasting and stunning presentations by news meteorologists.
- 3-D augmented reality is being studied to animate traffic-simulation models to aid in the planning and design of highway construction projects. In the 3-D simulation tests, the user has the opportunity to drive a virtual or real car.

planners can explore "what if" scenarios for land-use projects and investigate the potential for change. Computers can enhance information used in the planning process or explain the scope of a project. The presentation material can be modified to incorporate community suggestions over a series of meetings.

Universities tend to offer courses and degree programs in the types of research they specialize in. Brown, Penn State, and several University of California campuses are working on scientific visualization that has applications in designing virtual humans and medical illustrations. Universities in Canada and Europe are also active in these areas. The University of Utah, a longtime leader in the field, conducts research in geometric design and scientific computing.

Industry and Business. The computer-graphics industry will continue to grow, resulting in an increased demand for programmers, artists, scientists, and designers. Pixar, DreamWorks, Disney, Warner Bros., Square Enix, Sony, and Nintendo are considered top animation producers in the film and video-game industries. Architectural and engineering consulting firms have been contracted to use virtual images in planning public buildings such as stadiums. Virtual people help to determine traffic flow through buildings. If congestion points are observed as virtual crowds travel to concession stands or restrooms, for example, changes can be made to provide more efficient flow. Virtual models were also used to study traffic flow in Hong Kong harbor.

Computer graphics are a big improvement over architectural scale models. They can readily portray a variety of alternative plans without incurring large costs.

Careers and Course Work

Computer-graphics specialists must have a unique combination of artistic and computer skills. They must have good math, programming, and design skills, and be able to visualize 3-D objects. The specialist must be creative, detail oriented, and able to work well individually and as part of a team.

Course offerings can vary considerably among universities, so the student must consider the specialties of the prospective schools in relation to the field in which he or she is most interested. For example, the student may want to focus more on software design than on graphics design. Essential courses can include advanced mathematics, programming, computer animation, geometric design and modeling, multimedia systems and applications, and software engineering.

Computer graphics can be applied to a vast number of fields, and its influence can only increase. In addition to animation in video games and film, computer graphics has proven valuable in architectural and engineering design, education, medicine, business, and cartography. Typical positions for a computer-graphics specialist include 3-D animator or modeler, special effects designer, video game designer, and Web designer.

Computer-graphics specialists can work in the public or private sector; they can also work independently. Although companies prefer to hire candidates with a bachelor's degree, many workers in the field have only an associate's degree or vocational certificate.

In the private sector, computer-graphics specialists can be employed by architectural, construction, and engineering consulting firms, electronics and software manufacturing companies, and petrochemical, food processing, and energy industries. In the public sector, they can work at all levels of government and in hospitals and universities.

Social Context and Future Prospects

Computer graphics will continue to have a profound effect on the visual arts, freeing the artist from the need to master technical skills in order to focus on creativity and imagination in his or her work. The artist can experiment with unlimited variations in structures and designs in a single work to see which produces the desired effect. Continuing advances in producing virtual images and 3-D animation will enhance understanding of scientific principles and processes in education, medicine, and science.

David Olle, B.S., M.S.

Further Reading

McConnell, Jeffrey. *Computer Graphics: Theory into Practice.* Boston: Jones & Bartlett, 2006. The basic principles of graphic design are amply presented with reference to the human visual system. OpenGL is integrated with the material, and examples of 3-D graphics are illustrated.

Shirley, Peter, and Steve Marschner. *Fundamentals of Computer Graphics.* 3d ed. Natick, Mass.: A. K. Peters, 2009. This book emphasizes the underlying mathematical fundamentals of computer graphics rather than learning particular graphics programs.

Vidal, F. P., et al. "Principles and Applications of Computer Graphics in Medicine." *Computer Graphics Forum* 25, no. 1 (2006): 113-137. Excellent review of the state of the art. Discusses software development, diagnostic aids, educational tools, and computer-augmented reality.

Web Sites

ACMSIGGRAPH (Association for Computing Machinery's Special Interest Group on Graphics and Interactive Techniques)
http://www.siggraph.org

Computer Society
http://www.computer.org

Institute of Electrical and Electronics Engineers
http://www.ieee.org

OpenGL Overview
http://www.opengl.org/about/overview

See also: Calculus; Computer Engineering; Computer Networks; Computer Science; Software Engineering; Video Game Design and Programming.

COMPUTER LANGUAGES, COMPILERS, AND TOOLS

FIELDS OF STUDY

Computer science; computer programming; information systems; information technology; software engineering.

SUMMARY

Computer languages are used to provide the instructions for computers and other digital devices based on formal protocols. Low-level languages, or machine code, were initially written using the binary digits needed by the computer hardware, but since the 1960's, languages have evolved from early procedural languages to object-oriented high-level languages, which are more similar to English. There are many of these high-level languages, with their own unique capabilities and limitations, and most require some type of compiler or other intermediate translator to communicate with the computer hardware. The popularity of the Internet has created the need to develop numerous applications and tools designed to share data across the Internet.

KEY TERMS AND CONCEPTS

- **Application:** Computer program that completes a specific task.
- **Basic Input-Output System (BIOS):** Computer program that allows the central processing unit of a computer to communicate with other computer hardware.
- **Browser:** Software program used to view Web pages on the Internet.
- **Compiler:** Program that converts source code in a text file into a format that can be executed by a computer.
- **Graphical User Interface (GUI):** Visual interface that allows a user to position a cursor over a displayed object or icon and click on that object or icon to make a selection.
- **Interpreter:** Computer tool that is much faster and more efficient than a compiler.
- **Mainframe:** Large stand-alone computer that completes batch processing (groups of computer instructions completed at once).
- **Operating System:** Computer program written in a language so that it can function as an interface between a computer user and the hardware that runs the computer by managing resources.
- **Portability:** Ability of a program to be downloaded from a remote location and executed on a variety of computers with different operating systems.
- **Protocol:** Set of rules to be followed to allow for communication.
- **Server:** Computer dedicated to managing network activities, such as printers and email, which are shared by many users (clients).
- **Structured Query Language (SQL):** Computer language used to access databases.

Definition and Basic Principles

The traditional process of using a computer language to write a program has generally involved the initial design of the program using a flowchart based on the purpose and desired output of the program, followed by typing the actual instructions for the computer (the code) into a file using a text editor, and then saving this code in a file (the source code file). A text editor is used because it does not have the formatting features of a word processor. An intermediate tool called a compiler then has been used to convert this source code into a format that can be run (executed) by a computer.

However, as of 2011, there are new tools that are much faster and more efficient than compilers. Therefore, many compilers have been replaced by these new tools, called interpreters. Larger, more complex programs have evolved that have required an additional step to link external files. This process is called linking and it joins the main, executable program created by the compiler to other necessary programs. Finally, the executable program is run and its output is displayed on the computer monitor, printed, or saved to another digital file. If errors are found, the process of debugging is followed to go back through the code to make corrections.

Background and History

Early computers such as ENIAC (Electronic Numerical Integrator and Computer), the first general-purpose computer, were based on the use of switches that could be turned on or off. Thus, the binary digits of 0 and 1 were used to write machine code. In addition to being tedious for a programmer, the code had to be rewritten if used on a different type of machine, and it certainly could not be used to transmit data across the Internet, where different computers all over the world require access to the same code.

Assembly language evolved by using mnemonics (alphabetic abbreviations) for code instead of the binary digits. Because these alphabetic abbreviations of assembly language no longer used the binary digits, additional programs were developed to act as intermediaries between the human programmers writing the code and the computer itself. These additional programs were called compilers, and this process was initially known as compiling the code. This compilation process was still machine and vendor dependent, however, meaning, for example, that there were several types of compilers that were used to compile code written in one language. This was expensive and made communication of computer applications difficult.

The evolution of computer languages from the 1950's has accompanied technological advances that have allowed languages to become increasingly powerful, yet easier for programmers to use. FORTRAN and COBOL languages led the way for programmers to develop scientific and business application programs, respectively, and were dependent on a command-line user interface, which required a user to type in a command to complete a specific task. Several other languages were developed, including Basic, Pascal, PL/I, Ada, Lisp, Prolog, and Smalltalk, but each of these had limited versatility and various problems. The C and C++ languages of the 1970's and 1980's, respectively, have emerged as being the most useful and powerful languages, are still in use, and have been followed by development tools written in the Java and Visual Basic languages, including integrated development environments with editors, designers, debuggers, and compilers all built into a single software package.

How It Works

BIOS and Operating System. The programs within the BIOS are the first and last programs to execute whenever a computer device is turned on or off. These programs interact directly with the operating system (OS). The early mainframe computers that were used in the 1960's and 1970's depended on several different operating systems, most of which are no longer in usage, except for UNIX and DOS. DOS (Disk Operating System) was used on the initial microcomputers of the 1980's and early 1990's, and it is still used for certain command-line specific instructions.

Graphical User Interfaces (GUIs). Microsoft dominates the PC market with its many updated operating systems, which are very user-friendly with GUIs. These operating systems consist of computer programs and software that act as the management system for all of the computer's resources, including the various application programs most taken for granted, such as Word (for documents), Excel (for mathematical and spreadsheet operations), and Access (for database functions). Each of these applications is actually a program itself, and there are many more that are also available.

Since the 1980's, many programming innovations increasingly have been built to involve the client-server model, with less emphasis on large main frames and more emphasis on the GUIs for smaller microcomputers and handheld devices that allow consumers to have deep color displays with high resolution and voice and sound capabilities. However, these initial GUIs on client computers required additional upgrades and maintenance to be able to interact effectively with servers.

World Wide Web. The creation of the World Wide Web provided an improvement for clients to be able to access information, and this involvement of the Internet led to the creation of new programming languages and tools. The browser was developed to allow an end user (client) to be able to access Web information, and hypertext markup language (HTML) was developed to display Web pages. Because the client computer was manufactured by many different companies, the Java language was developed to include applets, which are mini-programs embedded into Web pages that could be displayed on any type of client computer. This was made possible by a special type of compiler-like tool called the Java Virtual Machine, which translated byte code. Java remains the primary computer language of the Internet.

Applications and Products

FORTRAN and COBOL. FORTRAN was developed by a team of programmers at IBM and was first released in 1957 to be used primarily for highly numerical and scientific applications. It derived its name from formula translation. Initially, it used punched cards for input, because the text editors were not available in the 1950's. It has evolved but still continues to be used primarily in many engineering and scientific programs, including almost all programs written for geology research. Several updated versions have been released. FORTRAN77, released in 1980, had the most significant language improvements, and FORTRAN2000, released in 2002, is the most recent. COBOL (Common Business-Oriented Language) was released in 1959 with the goal of being used primarily for tracking retail sales, payroll, inventory control, and many other accounting-related activities. It is still used for most of these business-oriented tasks.

C and C++. The C computer language was the predecessor to the C++ language. Programs written in C were procedural and based on the usage of functions, which are small programming units. As programs grew in complexity, more functions were added to a C program. The problem was that eventually it became necessary to redesign the entire program, because trying to connect all of the functions, which added one right after the other, in a procedural way, was too difficult. C++ was created in the 1980's based on the idea of objects grouped into classes as the building blocks of the programs, which meant that the order did not have to be procedural anymore. Object-oriented programming made developing complex programs much more efficient and is the current standard.

Microsoft.NET. In June, 2000, Microsoft introduced a suite of languages and tools named Microsoft.NET along with its new language called Visual C#. Microsoft.NET is a software infrastructure that consists of many programs that allow a user to write programs for a range of new applications such as server components and Web applications by using new tools. Although programs written in Java can be run on any machine, as long as the entire program is written in Java, Microsoft.NET allows various programs to be run on the Windows OS. Additional advantages of Microsoft.NET involve its new language of Visual C#. Visual C# provides services to help Web pages already in existence, and C# can be integrated with the Visual Basic and Visual C++ languages, which facilitate the work of Web programmers by allowing them to update existing Web applications, rather than having to rewrite them.

The Microsoft.NET framework uses a common type system (CTS) tool to compile programs written in Cobol.NET, PerlNET, Visual Basic.NET, Jscript, Visual C++, Fortran.NET, and Visual C# into an intermediate language. This common intermediate language (CIL) can then be compiled to a common language runtime (CLR). The result is that the .NET programming environment promotes interoperability to allow programs originally written in different languages to be executed on a variety of operating systems and computer devices. This interoperability is crucial for sharing data and communication across the Internet.

Fascinating Facts About Computer Languages, Compilers, and Tools

- Job prospects for software engineers are expected to be excellent, with salaries typically in the range of $70,000 to $80,000 per year.
- As of October, 2010, Java is the most widely used computer language, followed by C++.
- Using computer languages to write programs for video games is expected to be one of the most in-demand skills in the gaming area.
- Computer languages are necessary for a wide variety of everyday applications, ranging from cell phones, microwaves, and security systems to banking, online purchases, Web sites, auction sites (such as eBay), social-networking sites (such as Facebook), and Internet search engines (such as Google).
- Businesses, including banks, almost exclusively use COBOL, which uses two-digit rather than four-digit years. The use of two-digit years was the basis for the Y2K fear, in which it was predicted that on January 1, 2000, all the banking and other business-related applications would default to the year 1900, documents would be lost, and chaos would ensue. January 1, 2000, passed without incident.
- In 2010, Mattel introduced a Barbie doll called Computer Software Engineer Barbie. She has her own laptop with the Linux operating system and an iPhone.

Impact on Industry

Microsoft was founded in 1975 by Bill Gates, who dropped out of Harvard, and Paul Allen. Their first software sold was in the BASIC language, which was the first language to be used on personal computers (PCs). As PC prices decreased, the number of consumers able to purchase a PC increased, as did the revenue and market dominance of Microsoft. Microsoft is the largest software company in the world, with annual revenues of more than $32 billion. Because it has gained such widespread dominance in the technology field, it offers its own certifications in order to maintain quality standards in its current and prospective employees. Among the most commonly held and useful certifications offered by Microsoft are the Microsoft Certified Applications Developer (MCAD), Microsoft Certified Solution Developer (MCSD), Microsoft Certified Professional Developer (MCPD), Microsoft Certified Systems Engineer (MCSE), Microsoft Certified Systems Administrator (MCSA), Microsoft Certified Database Administrator (MCDBA), and Microsoft Certified Trainer (MCT).

Sun Microsystems began with four employees in 1982 as a computer hardware manufacturer. With the rise of the Internet, the company seized the opportunity to create its computer language with its special tool, the Java Virtual Machine, which could be downloaded to any type of machine. This Java Virtual Machine contained a program called an interpreter, rather than a compiler, to act as an interface between the specific machine (platform) and the user. The result was a great enhancement in the interoperability of data on the Internet, and Java became the primary programming language on the Web.

Sun Microsystems also created Java servlets to allow more interaction with dynamic, graphical Web pages written in Java. However, although these servlets work very well with Java, they also must be compiled as classes before execution, which takes extra time.

Apple, one of the most successful American companies, has experienced phenomenal growth in the first decade of the twenty-first century. Its position as a market leader in consumer electronics is because of the success of its Macintosh computer accompanied by several Internet-enabled gadgets, such as the iPod, introduced in 2001, which uses various audio file formats, such as MP3 and WAV, to function as a portable media player. To further expand on the popularity of the iPod, Apple opened its own online media and applications stores, iTunes in 2003 and the App Store in 2008, to allow consumers to access music, video, and numerous applications via the Internet. In 2007, Apple expanded its market share by releasing the iPhone, which added to the phone the functions of text messaging, emailing of digital photos using its own camera capabilities, and a multi-touch screen for enhanced Internet access. In 2010, Apple further expanded with the release of the iPad, an electronic tablet that allows users access to the Internet to read, write, listen to, and view almost anything, including e-mail, movies, books, music, and magazines. The iPad also has a Global Positioning System (GPS) and a camera.

Careers and Course Work

Although it is becoming more common for entry-level job seekers to have a bachelor's degree in a computer-related field, either through a university computer-science or business-school department, it is possible to find rewarding employment as a programmer, software engineer, application developer, Web programmer or developer, database administrator, or software support specialist with just an associate's degree or relevant experience. This career field is unique in the large number of certifications that can be obtained to enhance job skills and increase the likelihood of finding employment, especially with just an associate's degree or related experience. Databases that can be accessed from within programming languages have also created the need for database administrators. There are many vendor-specific certifications for numerous job openings as database administrators who know computer languages such as SQL.

The future job prospects for software engineers are projected by the Bureau of Labor Statistics to be better than the job prospects for programmers for the period from 2008 to 2018. There is a great deal of similarity between these two careers, but generally programmers spend more time writing code in various languages, while the software engineers develop the overall design for the interaction of several programs to meet the needs of a customer. Software engineers are required more often to have a bachelor's or master's degree, usually in software engineering. The job growth for software engineers is expected to grow by 32 percent over the time period of 2008 to 2018, which is much faster than most other occupations.

Social Context and Future Prospects

The Internet continues to bring the world together at a rapid pace, which has both positive and negative ramifications. Clearly, consumers have much easier access to many services, such as online education, telemedicine, and free search tools to locate doctors, learn more about any topic, comparison shop and purchase, and immediately access software, movies, pictures, and music. However, along with this increase in electronic commerce involving credit card purchases, bank accounts, and additional financial transactions has been the increase of cybercrime. Thousands of dollars are lost each year to various Internet scams and hackers being able to access private financial information. Some programmers even use computer languages to produce viruses and other destructive programs for purely malicious purposes, which have a negative impact on computer security.

Modern computer programs follow standard engineering principles to solve problems involving detail-driven applications ranging from radiation therapy and many medical devices to online banking, auctions, and stock trading. These online application programs require the special characteristics of Web-enabled software such as security and portability, which has given rise to the development of additional tools and will continue to produce the need for increased security features within computer languages and tools.

Jeanne L. Kuhler, B.S., M.S., Ph.D.

Further Reading

Das, Sumitabha. *Your UNIX: The Ultimate Guide.* 2d ed. Boston: McGraw-Hill, 2006. This textbook provides background and instructions for using the UNIX operating system.

Dhillon, Gupreet. "Dimensions of Power and IS Implementation." *Information and Management* 41, no. 5 (May, 2004): 635-644. This article describes some of the work done by managers when choosing which computer languages and tools to implement.

Guelich, Scott, Shishir Gundavaram, and Gunther Birznieks. *CGI Programming with Perl.* 2d ed. Cambridge, Mass.: O'Reilly, 2000. This introductory overview text describes the use of the common gateway interface programming and its interactions with the Perl programming language.

Horstmann, Cay. *Big Java.* 4th ed. Hoboken, N.J.: John Wiley & Sons, 2010. This book describes the most recent version of the Java language, along with its history and special features of the Java interpreter.

Snow, Colin. "Embrace the Role and Value of Master Data Management," *Manufacturing Business Technology*, 26, no. 2 (February, 2008): 92-95. This article, written by the VP and research director of Ventana Research, covers how to implement data management and avoid the errors that cost businesses millions of dollars each year.

Web Sites

Association of Information Technology Professionals
http://www.aitp.org

The Computer Society
http://www.computer.org

Computing Technology Industry Association
http://www.comptia.org/home.aspx

Microsoft
Microsoft Certification Overview
http://www.microsoft.com/learning/en/us/certification/cert-overview.aspx

See also: Computer Engineering; Computer Networks; Computer Science; Internet and Web Engineering; Software Engineering.

COMPUTER NETWORKS

FIELDS OF STUDY

Computer science; computer engineering; information technology; telecommunications; engineering; communications; mathematics; physics.

SUMMARY

Computer networks consist of the hardware and software needed to support communications and the exchange of data between computing devices. The computing devices connected by computer networks include large servers, business workstations, home computers, and a wide array of smart mobile devices. The most popular computer network application is e-mail, followed by exchanging audio and video files. Computer networks provide an infrastructure for the Internet, which in turn provides support for the World Wide Web.

KEY TERMS AND CONCEPTS

- **Cloud Computing:** Software developed to execute on multiple servers, and delivered by a Web service.
- **Internet:** Another name for a set of networked computers.
- **Local Area Network (LAN):** Network that logically uses a shared media to connect its computers.
- **Media:** Term used to describe the physical components used to actually connect computers in a network.
- **Network Protocol:** Description of the rules and syntax used for computer networking software.
- **Packet:** Description of a block of data exchanged by two computers. For some protocols, the packet is the entire message, while for others, a message consists of many packets.
- **Routers:** Computing devices that receive and retransmit packets of information between networks.
- **Transmission Control Protocol/Internet Protocol (TCP/IP):** Primary network protocol used for wide area networks.

Definition and Basic Principles

A computer network is a collection of computer devices that are connected in order to facilitate the sharing of resources and information. Underlying the computer network is a communications network that establishes the basic connectivity of the computing devices. This communications network is often a wired system but can include radio and satellite paths as well. Devices used on the network include large computers, used to store files and execute applications; workstations, used to execute small applications and interact with the servers; home computers, connected to the network through an Internet service provider (ISP); and mobile devices, connected to the network by radio wave transmissions. Middleware is the software that operates on top of the communications network to provide a software layer for developers to add high-level applications, such as a search engine, to the basic network.

The high-level applications are what most people using the network see as the network. Some of the most important computer network applications provide communications for users. Older examples of this are e-mail, instant messaging, chat rooms, and videoconferencing. Newer examples of communications software are the multitude of social networks, such as Facebook. Other high-level applications allow users to share files. One of the oldest and still quite popular file-sharing programs is the file transfer protocol (FTP) program; the newer Flickr makes it easy to share photographs. Another way to use computer networks is to share computing power. The Telnet program (terminal emulation program for TCP/IP networks) allowed one to use an application on a remote mainframe in the early days of networking; and modern Web services allow one to run an application on a mobile device while getting most of the functionality from a remote server.

Background and History

The scientists who developed the early computers in the 1950's recognized the advantage of connecting computing devices. Teletype machines were in common use at that time, and many of the early computers were "networked" with these teletype machines over wired networks. By 1960, American Telephone and Telegraph (AT&T) had developed the first modem to allow terminal access to mainframes, and in 1964, IBM and American Airlines introduced the SABRE (Semi-Automated Business Research Environment) networked airline reservation system.

The Defense Department created ARPANET (Advanced Research Projects Agency Network) in 1966 to connect its research laboratories with college researchers. The early experience of ARPANET led the government to recognize the importance of being able to connect different networks so they could interoperate. One of the first efforts to promote interoperability was the addition of packet switching to ARPANET in 1969. In 1974, Robert Kahn and Vinton Cerf published a paper on packet-switching networks that defined the TCP/IP protocol, and in 1980, the U.S. government required all computers it purchased to support TCP/IP. When Microsoft added a TCP/IP stack to its Windows 95 operating system, TCP/IP became the standard wide area network in the world.

The development of the microcomputer led to the need to connect these devices to themselves and the wide area networks. In 1980, the Institute of Electrical and Electronics Engineers (IEEE) 802 standard was announced. It has provided most of the connectivity to networks since that time, although many wireless computing devices can connect with Bluetooth.

How It Works

Computer networks consist of the hardware needed for networking, such as computers and routers; the software that provides the basic connectivity, such as the operating system; and middleware and applications, the programs that allow users to use the network. In understanding how these components work together, it is useful to look at the basic connectivity of the wide area network and contrast that to the way computers access the wide area network.

Wide Area Networks. A wide area network is one that generally covers a large geographic area and in which each pair of computers connect via a single line. The first wide area networks consisted of a number of connected mainframes with attached terminals. By connecting the mainframes, a user on one system could access applications on every networked computer. IBM developed the Systems Network Architecture (SNA), which was a popular early wide area network in the United States. The X.25 packet switching protocol standard provided a common architecture for the early wide area networks in Europe. Later, as all computers provided support for the TCP/IP protocol, it became possible for different computer networks to work together as a single network. In the twenty-first century, any device on a single network, whether attached as a terminal by X.25 or as a part of a local area network (LAN), can access applications and files on any other network on the Internet. This complete connectivity allows browsers on any type of mobile device to access content over the World Wide Web because it runs transparently over the Internet.

Routing. A key to making the Internet operate efficiently is the large number of intermediate devices that route IP packets from the computer making a connection to the computer receiving the data. These devices are called routers, and they are the heart of the Internet. When a message is sent from a computer, it is decomposed into IP packets, each having the IP address of the receiver. Then each packet is forwarded to a border router of the sender. From the border router, the packet is routed through a set of intermediate routers, using an algorithm-like best path, and delivered to the border router of the receiver, and then the receiver. Once all the packets have arrived, the message is reassembled and used by the receiver.

Local Area Networks. A local area network (LAN) is a collection of computers, servers, printers, and the like that are logically connected over a shared media. A media access control protocol operates over the LAN to allow any two devices to communicate at the link level as if they were directly connected. There are a number of LAN architectures, but Ethernet is the most popular LAN protocol for connecting workplace computers on a LAN to other networks. Ethernet is also usually the final step in connecting home computers to other networks using regular, cable, and ADSL (Asymmetric Digital Subscriber Line) modems. The original coaxial cable Ethernet, developed by Robert Metcalfe in 1973, has been largely supplanted by the less expensive and easier to use twisted-pair networks. The IEEE 802.11 wireless LAN is a wireless protocol that is almost as popular as Ethernet.

Wireless Networks. A laptop computer's initial connection to a network is often through a wireless device. The most popular wireless network is the IEEE 802.11, which provides a reliable and secure connection to other networks by using an access point, a radio transmitter/receiver that usually includes border-routing capabilities. The Bluetooth network is another wireless network that is often used for peer-to-peer connection of cameras and cell phones to a computer and then through the computer to other networks. Cell phones also provide network connectivity of computing devices to other networks. They use a variety of techniques, but most of them include using a cell phone communications protocol, such as code division multiple access (CDMA) or Global System for Mobile Communications (GSM), and then a modem.

Applications and Products

Computer networks have many applications. Some improve the operation of networks, while others provide services to people who use the network. The World Wide Web supplies many exciting new applications, and technologies such as cloud computing appear to be ready to revolutionize computing in the future.

Network Infrastructure. Computer networks are complex, and many applications have been developed to improve their operation and management. A typical example of the software developed to manage networks is Cisco's Internetwork Operating System (IOS) software, an operating system for routers that also provides full support for the routing functions. Another example of software developed for managing computer networks is the suite of software management programs provided by IBM's Tivoli division.

Communications and Social Networking. Communications programs are among the most popular computer applications in use. Early file-sharing applications, such as FTP, retain their popularity, and later applications such as Flickr, which shares photographs, and eDonkey2000, which shares all types of media files, are used by all. Teleconferencing is used to create virtual workplaces; and Voice over Internet Protocol (VoIP) allows businesses and home users to use the Internet for digital telephone service with services provided by companies such as Vonage.

The earliest computer network communications application, e-mail, is still the largest and most successful network application. One accesses e-mail services through an e-mail client. The client can be a stand-alone program, such as Outlook Express or Eudora, or it can be browser based, using Internet Explorer or Foxfire. The e-mail server can be a proprietary system such as Lotus Notes or one of the many Web-based e-mail programs, such as Hotmail, Gmail or Yahoo! Mail.

One of the most important types of applications being developed for computer networks is social networking sites. Facebook is probably the most widely used social network, although sites such as MySpace and Twitter are also very popular. Facebook was conceived by Mark Zuckerberg in 2004 while he was still a student at Harvard University. Facebook members can set up a profile that they can share with other members. The site supports a message service, event scheduling service, and news feed. By setting up a friend relationship with other members, one can quickly rekindle old friendships and acquaintances. Although Zuckerberg's initial goal was to create a friendly and interactive environment for individuals, it has fast become a great way to promote businesses, organizations, and even politicians. As of 2010, Facebook did not charge a subscription fee and had a reasonably successful business model selling customer information and advertising as well as working with partners.

Cloud Computing. Cloud computing is a generic name for applications on the Web that are implemented as a Web service and used by desktop computers, wireless-connected laptops, and smartphones. A Web service application exposes its functionality by placing a description of itself on a Universal Description, Discovery, and Integration (UDDI) server. When a Web application, for example, a Web page in a browser, requests data from the Web service, the service sends the data to the user over the Internet, just as if the service was on the user's computer. The real appeal of the Web service is the promise of true interoperability of applications running on a wide variety of computing devices and requesting services from an equally wide variety of Web servers.

As an example of a useful Web service, assume that an individual has written a Web application to select the best automobile insurance to purchase. The application would return key information about each insurance company and its rates. Other programmers could develop similar applications, but there would be differences in some of the details and the cost for accessing the application. An enterprising smartphone developer could then create an application to assist people in buying a new car, and as a part of the interface, display the automobile insurance information. The smartphone developer could insert the individual's automobile insurance Web service information, or that of a competitor, in the automobile purchase interface. The smartphone user would see the auto insurance data displayed, with no idea that the information was coming from the first individual's application.

Next to e-mail, the most common application is word processing, and Microsoft Word is the dominant word processor. Microsoft has announced Azure Word, part of Azure Office, as a version of Word that will execute in the cloud. Because most of the work of Azure Word will be done on Microsoft servers and customers will pay for use at the time of use, Azure Word should run on smartphones almost as well as on a desktop. Google has already released a word processor of its own to run in the cloud, and as more and more applications are produced to run on the cloud, it is predicted that this will become the dominant form of computing, replacing the desktops of the early twenty-first century.

Impact on Industry

Although computer networks are important for the communications infrastructure in each country, internetworking (networking between countries) is almost as important. For example, when someone does a search using Google for the best air cleaner to purchase, results are returned from the United States, Sweden, Switzerland, and China. This would not be possible unless the underlying networks were able to work together. Although early networking hardware and software were largely produced in the United States, there were significant contributions from Europe as well. For example, the token-ring LAN was developed in England and the X.25 standard was developed in Europe. By the twenty-first century, computer network hardware and software were being produced all over the world.

Government, Industry, and University Research. Most governments recognize the importance of having a good national computer network and support research to develop networks and standards for networking. One of the best examples of this is ARPANET, a network developed by the United States government to connect the Department of Defense laboratories with university scientists. ARPANET started in 1969 with a few nodes; by 1970, it was growing at the rate of one node a month; and by 1975, it included nodes from nondefense organizations. In 1980, Larry Landweber produced a plan to expand the ARPANET to a network for everyone, including some individual access from telephone networks. When the U.S. government decided to become the world leader in supercomputers in 1985, it also supported the creation of the National Science Foundation Network (NSFNET) as a high-speed backbone for ARPANET and a number of feeder networks to complement ARPANET. By 1998, the government was allowing considerable commercial traffic on NSFNET, and the passage of the National Research and Education Network Act in 1991 led to the creation of the Internet.

The International Telecommunications Union (ITU), located in Geneva, Switzerland and founded

Fascinating Facts About Computer Networks

- Two students, Richard Bartle and Roy Trubshaw, went online in 1978 with Multi-User Dungeons (MUD), a game that allowed two users to compete against each other. Extensions of MUD in the 2010's have many of the features of social networks.
- Robert Metcalfe's doctoral thesis at Harvard University defined Ethernet, but it was initially rejected. In 1973, an extended version was accepted.
- In 1971, Ray Tomlinson sent the first e-mail over ARPANET, demonstrating that it was possible to exchange e-mail between different computer systems.
- Microsoft embedded a TCP/IP stack as a part of its Windows 95 operating system. When combined with an earlier government requirement that all government computers support TCP/IP, this made TCP/IP the only wide-area network.
- In 1990, Tim Berners-Lee demonstrated that a browser could access content on a Web server, thus creating the World Wide Web.
- In 1991, James Gosling created Java as a computer language to support embedded systems such as a home security system. Later, it was adapted to Web programming and became a big success.
- In the early 1980's, court decisions and changes in telecommunications laws eliminated the high tariffs charged on telecommunications, which facilitated the development of wide-area networking.

in 1965, is an international organization that supports developing international standards for networks on behalf of all countries. It actually guides the development of networks in Europe. The X.25 protocol for wide area networks was approved by the ITU in 1976 and was used for commercial networks in Europe and the United States after its release. The ITU remains active in standards development for networks, including a green networking initiative.

In the early days of computer networking in the United States, most research done at universities and companies was government related. In spite of this, considerable computer network research was done, especially for local area networks. European universities did considerable research in basic network theory and applications, with the development of token-ring networks in English universities being a prime example.

Computer Network Companies. A great deal of the network research is done by industries. In 1979, Robert Metcalfe cofounded 3Com, a company to implement the Ethernet system he invented, and it remains active. In 1990, a twisted-pair implementation of Ethernet, called 10BaseT, was produced and soon became the most widely deployed implementation of Ethernet. Novell, founded in Utah in 1979, developed a network operating system, Netware, that ran on top of Ethernet (and other LAN protocols as well) and was the dominant LAN during the 1980's and 1990's. IBM developed the Token-Ring network in the mid-1980's, and although it never achieved commercial acceptance, it remains in use. Digital Equipment Corporation's DecNet and IBM's SNA were mainstays for mainframe networks in the 1980's, but these companies, and many more, have come to produce a wide variety of TCP/IP networks.

Microsoft entered the LAN market in 1984 with its production of a LAN to support IBM personal computer networks, but Microsoft and IBM decided to develop separate products for networks with the introduction of the Personal System/2 (PS/2). Still, Microsoft's highly successful Window's New Technology (NT) network operating system and IBM's successful LAN Manager network operating system for the PS/2 had much in common. Both Microsoft and IBM continue to be leaders in research in computer networking.

Considerable research is done in computer networking by smaller companies. Cisco leads in router research and many other areas, such as wireless access points, video-conferencing equipment, and network management software. Symantec and MacAfee develop a great deal of security software for networks. Google, Yahoo!, and Microsoft spend millions every year on research and development to develop a better search engine.

Careers and Course Work

A major in computer science is the traditional way to prepare for a computer networking job. Students first need courses in ethics, mathematics, and physics to form the basis for a computer science degree. Then they take about thirty-six hours of computer hardware and software courses. Those getting a computer science degree often take jobs developing network software or managing a network.

A major in information systems is another way to prepare for a computer networking job. Students must take courses in mathematics and business as a background for this degree. Then they take about thirty hours of courses on information systems development. Those getting degrees in computer information systems often take jobs as network managers, especially in small businesses.

In addition to the traditional academic programs that prepare someone for a computer networking job, a large number of professional training programs result in certification. Novell was the first company to develop a certification program for its NetWare network operating system. Microsoft followed with its Microsoft Certified Professional program for Windows Server operating system, which includes networking. The Cisco Certified Network Professional program is a large program that covers all networking, not just routers.

Social Context and Future Prospects

The development of computer networks and network applications has often resulted in some sticky legal issues. File exchange programs, such as Flickr, are very popular but can have real copyright issues. Napster developed a very successful peer-to-peer file exchange program but was forced to close after only two years of operation by a court decision in 2001. Legislation has played an important role in the development of computer networks in the United States. The initial deregulation of the communications system and the breakup of AT&T in 1982 reduced the cost of the base communications system for computer networks and thus greatly increased the number of networks in operation. Opening up NSFNET for commercial use in 1991 made the Internet possible.

Networking and access to the Web have been increasingly important to business development and improved social networking. The emergence of cloud computing as an easy way to do computing promises to be just as transformative. Although there are still security and privacy issues to be solved, most believe that in the future, people will be using mobile devices to access a wide variety of Web services through the cloud. For example, using a smartphone, people will be able to communicate with friends and associates, transact business, compose letters, pay bills, read books, and watch films.

George M. Whitson III, B.S., M.S., Ph.D.

Further Reading

Cerf, V. G., and R. E. Kahn. "A Protocol for Packet Network Interconnection." *IEEE Transactions on Communication Technology* 22 (May, 1974): 627-641. The article that defined the Internet and a TCP/IP network.

Dumas, M. Barry, and Morris Schwartz. *Principles of Computer Networks and Communications.* Upper Saddle River, N.J.: Pearson/Prentice Hall, 2009. Covers all aspects of computer networks, including signal carriers and fundamentals, local area networks, wide area networks, protocols, wireless networks, security, and network design and implementation.

Forouzan, Behrouz. *Data Communications and Networking.* 4th ed. New York: McGraw-Hill, 2007. A complete, readable text that provides excellent coverage of the physical implementation of networking components.

Metcalfe, Robert, and David Boggs. "Ethernet: Distributed Packet Switching for Local Computer Networks." *Communications of the ACM* 19, no. 7 (July, 1976): 395-404. Described the fundamentals of Ethernet and provided the foundation for local area networks.

Stallings, William. *Data and Computer Communications.* 9th ed. Upper Saddle River, N.J.: Prentice Hall, 2011. An excellent overview of the entire field of computer networks. Includes many diagrams illustrating network architectures.

Tanenbaum, Andrew. *Computer Networks.* 5th ed. Upper Saddle River, N.J.: Prentice Hall, 2010. An excellent theoretical coverage of computer networks.

Web Sites

Computer Society
http://www.computer.org

Institute of Electrical and Electronics Engineers
http://www.ieee.org

Internet Society
http://www.isoc.org/internet/history/brief.shtml

See also: Communications; Computer Engineering; Computer Science; Information Technology; Internet and Web Engineering; Software Engineering.

COMPUTER SCIENCE

FIELDS OF STUDY

Computer science; computer engineering; electrical engineering; mathematics; artificial intelligence; computer programming; human-computer interaction; software engineering; databases and information management; bioinformatics; game theory; scheduling theory; computer networking; computer and data security; computer forensics; computer simulation and modeling; computation methods (including algorithms); ethics; computer graphics; multimedia.

SUMMARY

Computer science is the study of real and imagined computers, their hardware and software components, and their theoretical basis and application. Almost every aspect of modern life involves computers in some way. Computers allow people to communicate with people almost anywhere in the world. They control machines, from industrial assembly equipment to children's toys. Computers control worldwide stock market transactions and analyze and regulate those markets. They allow physicians to treat patients even when the physicians and patients cannot meet in the same location. Researchers endeavor to make the computers of science fiction everyday realities.

KEY TERMS AND CONCEPTS

- **Algorithm:** Step-by-step procedure for solving a problem; a program can be thought of as a set of algorithms written for a computer.
- **Artificial Intelligence:** Study of how machines can learn and mimic natural human and animal abilities; it can involve enhancing or going beyond human abilities.
- **Binary System:** Number system that has only two digits, 0 and 1; forms the basis of computer data representation.
- **Computer:** Real or theoretical device that accepts data input, has a running program, stores and manipulates data, and outputs results.
- **Computer Network:** Group of computers and devices linked together to allow users to communicate and share resources.
- **Database:** System that allows for the fast and efficient storage and retrieval of vast amounts of data.
- **Hardware:** Tangible part of a computer, consisting mostly of electronic and electrical components and their accessories.
- **Programming:** Using an artificial programming language to instruct a computer to perform tasks.
- **Software:** Programs running within a computer.
- **Theoretical Computer:** Imaginary computer made up by researchers to gain insights into computer science theory.

Definition and Basic Principles

Computer science is the study of all aspects of computers, applied and theoretical. However, considerable disagreement exists over the definition of such basic terms as computer science, computer, hardware, and software. This disagreement can be seen as a testament to the vitality and relative youth of this field. The Association for Computing Machinery's computing classification system, developed in 1998, is an attempt to define computer science.

The science part of computer science refers to the underlying theoretical basis of computers. Broadly speaking, computation theory, part of mathematics, looks at what mathematical problems are solvable. In computer science, it focuses on which problems can be solved using a computer. In 1936, English mathematician Alan Turing attempted to determine the limits of mechanical computation using a theoretical device, now called a Turing machine. Mathematics also forms the basis for research in programming languages, artificial languages developed for computers. Because computers do not have the ability to think like humans, these languages are very formal with strict rules on how programs using these languages can be written and used.

Another part of the underlying structure of computer science is engineering. The physical design of computers involves a number of disciplines, including electrical engineering and physics. The quest for smaller, faster, more powerful devices has led to research in fields such as quantum physics and nanotechnology.

Computer science is not just programming. Computer scientists view programming languages as tools to research such issues as how to create better programs, how information is represented and used by computers, and how to do away with programming languages altogether and instead use natural languages (those used by people).

Background and History

Computer science can be seen as the coming together of two separate strands of development. The older strand is technology and machines, and the newer is the theoretical one.

Technology. The twentieth century saw the explosive rise of computers. Computers began to incorporate electronic components instead of the mechanical components that had gone into earlier machines, and they made use of the binary system rather than the decimal system. During World II, major developments in computer science arose from the attempt to build a computer to control antiaircraft artillery. For this project, the Hungarian mathematician John von Neumann wrote a highly influential draft paper on the design of computers. He described a computer architecture in which a program was stored along with data. The decades after World War II saw the development of programming languages and more sophisticated systems, including networks, which allow computers to communicate with one another.

Theory. The theoretical development of computer science has been primarily through mathematics. One early issue was how to solve various mathematical equations. In the eighteenth and nineteenth centuries, this blossomed into research on computation theory. At a conference in Germany in 1936 to investigate these issues, Turing presented the Turing machine concept.

The World War II antiaircraft project resulted not only in the development of hardware but also in extensive research on the theory behind what was being done as well as what else could be done. Out of this ferment eventually came such work as Norbert Wiener's cybernetics and Claude Shannon's information theory. Modern computer science is an outgrowth of all this work, which continues all around the world in industry, government, academia, and various organizations.

How It Works

Computer Organization and Architecture. The most familiar computers are stand-alone devices based on the architecture that von Neumann sketched out in his draft report. A main processor in the computer contains the central processing unit, which controls the device. The processor also has arithmetic-processing capabilities.

Electronic memory is used to store the operating system that controls the computer, numerous other computer programs, and the data needed for running programs. Although electronic memory takes several forms, the most common is random access memory (RAM), which, in terms of size, typically makes up most of a computer's electronic memory. Because electronic memory is cleared when the computer is turned off, some permanent storage devices, such as hard drives and flash drives, were developed to retain these data.

Computers have an input/output (I/O) system, which communicates with humans and with other devices, including other computers. I/O devices include keyboards, monitors, printers, and speakers.

The instructions that computers follow are programs written in an artificial programming language. Different kinds of programming languages are used in a computer. Machine language, the only language that computers understand, is used in the processor and is composed solely of the binary digits 0 and 1.

Computers often have subsidiary processors that take some of the processing burden away from the main processor. For example, when directed by the main processor, a video processor can handle the actual placing of images on a monitor. This is an example of parallel processing, in which more than one action is performed simultaneously by a computer. Parallel processing allows the main processor to do one task while the subsidiary processors handle others. A number of computers use more than one main processor. Although one might think that doubling the number of processors doubles the processing speed, various problems, such as contention for the same memory, considerably reduce this advantage. Individual processors also use parallel processing to speed up processing; for example, some processors use multiple cores that together act much like individual processors.

Modern computers do not actually think. They are best at performing simple, repetitious operations incredibly fast and accurately. Humans can get bored and make mistakes, but not computers (usually).

Mathematics. Mathematics underlies a number of areas of computer science, including programming languages. Researchers have long believed that programming languages incorporating rules that are mathematically based will lead to programs that contain fewer errors. Further, if programs are mathematically based, then it should be possible to test programs without running them. Logic could be used to deduce whether the program works. This process is referred to as program proving.

Mathematics also underlies a number of algorithms that are used in computers. For example, computer games have a large math and physics component. Game action is often expressed in mathematical formulas that must be calculated as the game progresses. Different algorithms that perform the same task can be evaluated through an analysis of the relative efficiencies of different approaches to a problem solution. This analysis is mathematically based and independent of an actual computer.

Software. Computer applications are typically large software systems composed of a number of different programs. For example, a word-processing program might have a core set of programs along with programs for dictionaries, formatting, and specialized tasks. Applications depend on other software for a number of tasks. Printing, for example, is usually handled by the operating system. This way, all applications do not have to create their own basic printing programs. The application notifies the operating system (OS), which can respond, for example, with a list of all available printers. If an application user wishes to modify printer settings, the OS contacts the printer software, which can then display the settings. This way, the same interface is always used. To print the application, the user sends the work to the OS, which sends the work to a program that acts as a translator between the OS and the printer. This translator program is called a driver. Drivers work with the OS but are not part of it, allowing for new drivers to be developed and installed after an OS is released.

Machine language, since it consists only of 0's and 1's, is difficult for people to readily interpret. For human convenience, assembly language, which uses mnemonics rather than digits, was developed. Because computers cannot understand this language, a program called an assembler is used to translate assembly language to machine language. For higher-level languages such as C++ and Java, a compiler program is used to translate the language statements first to assembly language and then to machine language.

Applications and Products

Computers have penetrated into nearly every area of modern life, and it is hard to think of an area in which computer technology is not used. Some of the most common areas in which computers are used include communications, education, digitalization, and security.

Communication. In the early twentieth century, when someone immigrated to the United States, they knew that communication with those whom they were leaving behind would probably be limited. In the modern world, people across the world can almost instantly communicate with anyone else as long as the infrastructure is available. Products such as Skype and Vonage allow people to both see and talk to other people throughout the world. Instead of traveling to the other person's location, people can hold meetings through software such as Cisco's WebEx or its TelePresence, which allows for face-to-face meetings.

One of the most far-reaching computer applications is the Internet, a computer network that includes the World Wide Web. People are increasingly relying on the Internet to provide them with news and information, and traditional sources, such as printed newspapers and magazines are declining in popularity. Newer technologies such as radio and television have also been affected. The Internet seems to be able to provide everything: entertainment, information, telephone service, mail, business services, and shopping.

Telephones and the way people think of telephones have also been revolutionized. Telephones have become portable computing devices. Apple iPhones and Research in Motion (RIM) Blackberry phones are smart phones, which provide access to e-mail and the Internet; offer Global Positioning Systems (GPS) that guide the user to a selected destination; download motion pictures, songs, and other entertainment; and shoot videos and take photographs. Applications (apps) are a burgeoning industry for the iPhone. Apps range from purely entertaining applications to those that provide useful information, such as weather and medical data. Some people no longer have traditional land-line telephones and rely instead on their cell phones.

Smart phones demonstrate convergence, a trend toward combining devices used for individual activities into a single device that performs various activities and services. For example, increasingly high-definition televisions are connected to the Internet. Devices such as Slingbox allow users, wherever they are, to control and watch television on their desktop computer, laptop computer, or mobile phone. Digital video recorders (DVRs) allow users to record a television program and watch it at a later time or to pause live broadcasts. DVRs can be programmed to record shows based on the owner's specifications. Televisions are becoming two-way interactive devices, and viewers are no longer passive observers. Although televisions and DVRs are not themselves computers, they contain microprocessors, which are miniature computers.

Networking is possible through a vast network infrastructure that is still being extended and improved. Companies such as Belkin provide the cable, Cisco the equipment, and Microsoft the software that supports this infrastructure for providers and users. It is common for homes to have wireless networks with a number of different devices connected through a modem, which pulls out the Internet signal from the internet service provider (ISP), and a router, which manages the network devices.

Education. Distance education through the Internet (online education) is becoming more and more commonplace. The 2010 Horizon Report noted that people expect anywhere, anytime learning. Learning (or course) management systems (LMS or CMS) are the means by which the courses are delivered. These can be proprietary products such as Blackboard or Desire2Learn, or nonproprietary applications such as Moodle. Through these management systems, students can take entire courses without ever meeting their instructor in person. Applications such as Wimba and Elluminate allow classes to be almost as interactive as traditional classes and record those sessions for future viewing. Those participating can be anywhere as long as they can connect to the Internet. Through such software, students can ask questions of an instructor, student groups can meet online, and an instructor can hold review sessions.

Digitalization. This area is concerned with how data are translated for computer usage. Real-world experience is usually thought of as continuous phenomena. Long-play vinyl records (LPs) and magnetic tapes captured a direct analogy for the original sound; therefore, any change to the LP or tape meant a change to the sound quality, usually introducing distortions. Each copy (generation) that was made of that analogue signal introduced further distortions. LPs were several generations removed from the original recording, and additional distortions were added to accommodate the sound to the LP medium.

Rather than record an analogue of the original sound, digitalization translates the sound into discrete samples that can be thought of as snapshots of the original. The more samples (the higher the sampling rate) and the more information captured by each sample (the greater the bit depth), the better the result. Digitalization enables the storage of data of all types. The samples are encoded into one of the many binary file formats, some of which are open and some proprietary. MPEG-1 Audio Layer 3 (MP3) is an open encoding standard that is used in a number of different devices. It is a lossy standard, which means that the result is of a lower quality than the original. This tradeoff is made to gain a smaller file size, enabling more files to be stored. The multimedia equivalent is MPEG-4 Part 14 (MP4), also a lossy format. Lossless formats such as Free Lossless Audio Codec (FLAC) and Apple Lossless have better sound quality but require more storage space.

These formats and others have resulted in a proliferation of devices that have become commonplace. iPods and MP3 players provide portable sound in a relatively small container and, in some cases, play videos. Some iPods and computers can become media centers, allowing their users to watch television shows and films.

In a relatively short span of time, digital cameras have made traditional film cameras obsolete. Camera users have choices as to whether they want better quality (higher resolution) photographs or lower resolution photos but the capacity to store more of them in the same memory space. Similarly, DVDs have made tape videos obsolete. Blu-ray, developed by Sony, allows for higher quality images and better sound by using a purple laser rather than the standard red laser. Purple light has a higher frequency than red light, which means that the size (wavelength) of purple light is smaller than that of red light. This can be visualized as allowing the purple

Fascinating Facts About Computer Science

- Konrad Zuse, working in Germany in the 1930's and 1940's, developed what many consider the first electronic computer. He also invented a programming language, Plankalkül, which is considered the first high-level computer language. His work was not generally known until after World War II.
- In 1973, a U.S. Federal District Court ruled that John Atanasoff, a professor at Iowa State University, was the inventor of the first electronic digital computer, the Atanasoff-Berry Computer, designed in 1939 and successfully tested in 1942.
- The 1952 U.S. presidential election was the first time that a computer was used to predict an outcome. The CBS News computer predicted that Dwight Eisenhower would defeat Adlai Stevenson, even though a number of polls predicted Stevenson's victory. Because of this discrepancy, the projection was not announced for some time.
- The UNIVAC, used to predict the outcome of the 1952 U.S. presidential election, was a warehouse-sized computer. Most modern laptop computers exceed the UNIVAC's computing power and capabilities.
- A Global Positioning System device with augmented reality capabilities can not only give directions to a restaurant but also provide its menu. If the device can access personal information from such sources as Facebook, it can be tailored to make restaurant recommendations based on an individual's preferences.
- The RoboBees project of the School of Engineering and Applied Sciences at Harvard University is attempting to create robotic bees. These bees are designed not only to fly but also to reproduce other bee activities, including colony behaviors. The project is funded by the National Science Foundation.
- Users of mobile devices use touch-screen technology, which requires a relatively large display for finger access. Patrick Baudisch's nanoTouch device has a touchpad on the back of the device, allowing it to be much smaller.

laser to get into smaller spaces than the red laser can; thus the 0's and 1's on a Blu-ray disc can be smaller than those on a standard DVD disc, so more data can be stored on the same size disc. Because the recordings are encoded in binary, they can be easily and exactly copied and manipulated without any distortion, unless that distortion is deliberately introduced, as with lossy file formats.

Security. The explosion of digital communication and products has caused some individuals to illegally, or at least unethically, exploit people's dependence on these technologies. These people write malware, programs designed to harm or compromise devices and send spam (digital junk mail). Malware includes viruses (programs that embed themselves in other programs and then make copies of themselves and spread), worms (programs that act like viruses but do not have to be embedded in another program), Trojan horses (malicious programs, often unknowingly downloaded with something else), and key loggers (programs that record all keystrokes and mouse movements, which might include user names and passwords). The key-stroke recorder is a type of spyware, programs that collect information about the device that they are on and those using it without their knowledge. These are often associated with Internet programs that keep track of people's browsing habits.

Spam can include phishing schemes, in which a person sends an e-mail that appears to come from a well-known organization such as a bank. The e-mail typically states that there is a problem with the recipient's account and to rectify it, the recipient must verify his or her identity by providing sensitive information such as his or her user name, password, and account number. The information, if given, is used to commit fraud using the recipient's identity.

Another type of spam is an advance-fee fraud, in which the sender of an e-mail asks the recipient to help him or her claim a substantial sum of money that is due the sender. The catch is that the recipient must first send money to "show good faith." The money that the recipient is to get in return never arrives. Another asks the recipient to cash a check and send the scammer part of the money. By the time the check is determined to be worthless, the recipient has sent the scammer a considerable sum of money. These scams are a form of social engineering, in which the scammer first obtains the trust of the e-mail recipient so that he or she will help the scammer.

These problems are countered by companies such as Symantec and McAfee, which produce security suites that contain programs to manage these threats. These companies keep databases of threats and require that their programs be regularly updated to be able to combat the latest threats. Part of these suites is a firewall, which is designed to keep the malware and spam from reaching the protected network.

Impact on Industry

It is hard to think of any industry that has not felt the impact of computers in a major, perhaps transforming, way. The United States, the United Kingdom, and Japan led research and development in computer science, but Israel, the European Union, China, and Taiwan are increasingly engaging in research. Although India is not yet a major center of research, it has been very successful in setting up areas in Mumbai and elsewhere that support technology companies from around the world. Other countries, such as Pakistan, have tried to duplicate India's success. Indian and Chinese computer science graduates have been gaining a reputation worldwide as being very well prepared.

Industry and Business Sectors. Computer science has led to the creation of electronic retailing (e-tailing). Online sales rose 15.5 percent in 2009, while brick and mortar (traditional) sales were up 3.6 percent from the previous year. Online-only outlets such as Amazon.com have been carving out a significant part of the retail market. Traditional businesses, such as Macy's and Home Depot, are finding that they must have an online presence. Online sales for the 2009 Christmas season were estimated at $27 billion.

Some traditional industries are finding that survival has become difficult. Newspapers and magazines have been experiencing a circulation decline, as more and more people read the free online versions of these publications or get their news and information from other Internet sites. Some newspapers, such as *The Wall Street Journal*, sell subscriptions to their Web sites or offer subscriptions to enhanced Web sites. Travel agencies are losing business, as their clients instead use online travel sites such as Expedia.com or the Web sites of airlines and hotels to make their own travel arrangements. DVD rental outlets such as Blockbuster (which declared Chapter 11 bankruptcy in September, 2010) have been declining in the face of competition from Netflix, which initially offered a subscription service and sent DVDs through the U.S. Postal Service. As consumers have turned to streaming television programs and films, Netflix has entered this business, but faces competition from Amazon.com, Google, Apple, and cable and broadband television providers.

CD and DVD sales have been affected not only by online competition but also by piracy. Since digital sources can be copied without error, copying of CDs and DVDs has become rife. Online file sharing through such applications as BitTorrent has caused the industry to fight back through highly publicized lawsuits and legislation. The Digital Millennium Copyright Act of 1998 authorized, among many other measures, the use of Digital Rights Management (DRM) to restrict copying. Many DVDs and downloads use DRM protection. However, a thriving industry has sprung up to block the DRM features and allow copying. This is a highly contentious area, which engendered more controversy with the passage of the 2010 Digital Economy Act in the United Kingdom.

A thriving industry has been developed using virtual reality. Linden Research's Second Life is one of the most popular virtual reality offerings. Members immerse themselves in a world where they have the opportunity to be someone that they might never be in real life. A number of organizations use Second Life for various purposes, including education.

Professional Organizations. Probably the two largest computer organizations are the Association for Computer Machinery (ACM), based in New York, and Computer Society of the Institute of Electrical and Electronic Engineers (IEEE), based in Washington, D.C. The ACM Special Interest Groups explore all areas of computer science. When computer science higher education was a new frontier and the course requirements for a degree varied widely, it was the ACM that put forward a curriculum that has become the standard for computer science education in colleges and universities in the United States. Its annual Turing Award is one of the most prestigious awards for computing professionals.

The IEEE partners with the American National Standards Institute (ANSI) and the International Organization for Standardization on many of its wireless specifications, such as 1394 (FireWire) and Ethernet (802.3). Both computer organizations sponsor various contests for students from middle school to college.

Research. The National Science Foundation, a federal agency created in 1950, is responsible for about 20 percent of all federally funded basic research at colleges and universities in the United States. An example project is the California Regional Consortium for Engineering Advances in Technological Excellence (CREATE), a consortium of seven California community colleges, which works toward innovative technical education in community colleges. The consortium has funded a number of computer-networking initiatives.

The Defense Advanced Research Projects Agency (DARPA), part of the U.S. Department of Defense, funds research into a number of areas, including networks, cognitive systems, robotics, and high-priority computing. This agency initially funded the project that became the basis of the Internet.

Careers and Course Work

Computer science degrees require courses in mathematics (including calculus, differential equations, and discrete mathematics), physics, and chemistry as well as the usual set of liberal arts courses. Lower-division computer science courses are usually heavily weighted toward programming. These might include basic programming courses in C++ or Java, data structures, and assembly language. Other courses usually include computer architecture, networks, operating systems, and software engineering. A number of specialty areas, such as computer engineering, affect the exact mix of course work.

Careers in software engineering typically require a bachelor's degree in computer science. The U.S. Bureau of Labor Statistics sees probable growth in the need for software engineers. Computer skills are not all that employers want, however. They also want engineers to have an understanding of how technology fits in with their organization and its mission.

Those who wish to do research as computer scientists will generally require a doctorate in computer science or some branch of computer science. Computer scientists are often employed by universities, government agencies, and private industries such as IBM. AT&T Labs also has a tradition of Nobel Prize-winning research and was where the UNIX operating system and the C and C++ programming languages were developed.

Social Context and Future Prospects

Computers are considered an essential part of modern life, and their influence continues to revolutionize society. The advantages and disadvantages of an always-wired society are being keenly debated. People not only are connected but also can be located by computer devices with great accuracy. Because people are doing more online, they are increasingly creating an online record of what they have done at various times in their lives. Many employers routinely search online for information on job candidates, and people are finding that comments they made online and the pictures they posted some years ago showing them partying are coming back to haunt them. People are starting to become aware of these pitfalls and are taking actions such as deleting these possibly embarrassing items and changing their settings on social network sites such as Facebook to limit access.

These issues bring up privacy concerns, including what an individual's privacy rights are on the Internet and who owns the data that are being produced. For example, in 1976, the U.S. Supreme Court ruled that financial records are the property of the financial institution not the customer. This would seem to suggest that a company that collects information about an individual's online browsing habits—not the individual whose habits are being recorded—owns that information. Most of these companies state that individuals are not associated with the data that they sell and all data are used in aggregate, but the capacity to link data with specific individuals and sell the information exists. Other questions include whether an employer has a right to know whether an employee visits a questionable Internet site. Certainly the technology to accomplish this is available.

These concerns lead to visions of an all-powerful and all-knowing government such as that portrayed in George Orwell's *Nineteen Eighty-Four* (1949). With the Internet a worldwide phenomenon, perhaps no single government can dictate regulations governing the World Wide Web and enforce them. The power of technology will only grow, and theses issues and many others will only become more pressing.

Martin Chetlen, B.S.E.E, M.C.S.

Further Reading

Brooks, Frederick P., Jr. *The Mythical Man-Month: Essays on Software Engineering.* Anniversary ed.

Boston: Addison-Wesley, 2008. Brooks, who was the leader of the IBM 360 project, discusses computer and software development. The first edition of this book was published in 1975, and this 2008 edition adds four chapters.

Gaddis, Tony. *Starting Out With C++*. 6th ed. Boston: Addison-Wesley, 2009. A good introduction to a popular programming language.

Goldberg, Jan, and Mark Rowh. *Great Jobs for Computer Science Majors*. Chicago: VGM, 2003. Designed for a young and general audience, this career guide includes tools for self-assessment, researching computer careers, networking, choosing schools, and understanding a variety of career paths.

Henderson, Harry. *Encyclopedia of Computer Science and Technology*. Rev. ed. New York: Facts On File, 2009. An alphabetical collection of information about computer science. Contains bibliography, chronology, list of awards, and a list of computer organizations.

Kidder, Tracy. *The Soul of the New Machine*. 1981. Reprint. New York: Back Bay Books, 2000. Chronicles the development of a new computer and the very human drama and comedy that surrounded it.

Schneider, G. Michael, and Judith L. Gersting. *Invitation to Computer Science*. 5th ed. Boston: Course Technology, 2010. Gives a view of the breadth of computer science, including social and ethical issues.

Web Sites

Association for Computer Machinery
http://www.acm.org

Computer History Museum
http://www.computerhistory.org

Computer Society
http://www.computer.org

Institute of Electrical and Electronic Engineers
http://www.ieee.org

See also: Communications; Computer Engineering; Computer Languages, Compilers, and Tools; Computer Networks; Information Technology; Internet and Web Engineering; Parallel Computing; Robotics; Software Engineering; Virtual Reality.

CRYPTOLOGY AND CRYPTOGRAPHY

FIELDS OF STUDY

History; computer science; mathematics; systems engineering; computer programming; communications; cryptographic engineering; security engineering; statistics.

SUMMARY

Cryptography is the use of a cipher to hide a message by replacing it with letters, numbers, or characters. Traditionally, cryptography has been a tool for hiding communications from the enemy during times of war. Although it is still used for this purpose, it is more often used to encrypt confidential data, messages, passwords, and digital signatures on computers. Although computer ciphers are based on manually applied ciphers, they are programmed into the computer using complex algorithms that include algebraic equations to encrypt the information.

KEY TERMS AND CONCEPTS

- **Algorithm:** Mathematical steps to solve a problem with a computer.
- **Ciphertext:** Meaningless text produced by applying a cipher to a message.
- **Cleartext:** Original message; also known as plaintext.
- **Cryptosystem:** Computer program or group of programs that encrypt and decrypt a data file, message, password, or digital signature.
- **Key:** Number that is used both in accessing encrypted data and in encrypting it.
- **Polyalphabetic Cipher:** Cipher that uses multiple alphabets or alphabetic arrangements.
- **Prime Number:** Number that can be divided only by itself and the numeral one.
- **Proprietary Information:** Information that is the property of a specific company.

Definition and Basic Principles

Cryptography is the use of a cipher to represent hidden letters, words, or messages. Cryptology is the study of ciphers. Ciphers are not codes. Codes are used to represent a word or concept, and they do not have a hidden meaning. An example is the international maritime signal flags, part of the International Code of Signals, known by all sailors and available to anyone else. Ciphers are schemes for changing the letters and numbers of a message with the intention of hiding the message from others. An example is a substitution cipher, in which the order of letters of the alphabet is rearranged and then used to represent other letters of the alphabet. Cryptography has been used since ancient times for communicating military plans or information about the enemy. In modern times, it is most commonly thought of in regard to computer security. Cryptography is critical to storing and sharing computer data files and using passwords to access information either on a computer or on the Internet.

Background and History

The most common type of cipher in ancient times was the substitution cipher. This was the type of cipher employed by Julius Caesar during the Gallic Wars, by the Italian Leon Battista Alberti in a device called the Alberti Cipher Disk described in a treatise in 1467, and by Sir Francis Bacon of Great Britain. In the 1400's, the Egyptians discovered a way to decrypt substitution ciphers by analyzing the frequency of the letters of the alphabet. Knowing the frequency of a letter made it easy to decipher a message.

German abbot Johannes Trithemius devised polyalphabetic ciphers in 1499, and French diplomat Blaise de Vigenère did the same in 1586. Both used the tableau, which consists of a series of alphabets written one below the other. Each alphabet is shifted one position to the left. De Vigenère added a key to his tableau. The key was used to determine the order in which the alphabets were used.

The Greeks developed the first encryption device, the scytale, which consisted of a wooden staff and a strip of parchment or leather. The strip was wrapped around the staff and the message was written with one letter on each wrap. When wrapped around another staff of the same size, the message appeared. In the 1780's, Thomas Jefferson invented a wheel cipher that used wooden disks with the alphabet printed around the outside. They were arranged side by side on a spindle and were turned to create a huge number of ciphers. In 1922, Jefferson's wheel cipher was adopted by the U.S. Army. It used metal disks and was named the M-94.

Four inventors—Edward H. Hebern, United States, in 1917; Arthur Scherbius, Germany, in 1918; Hugo Alexander Koch, Netherlands, in 1919; and Arvid Gerhard Damm, Sweden, in 1919—independently developed cipher machines that scrambled letters using a rotor or a wired code wheel. Scherbius, an electrical engineer, called his machine the Enigma, which looked like a small, narrow typewriter. The device changed the ciphertext with each letter that was input. The German navy and other armed forces tried the machine, and the German army redesigned the machine, developing the Enigma I in 1930 and using various models during World War II. The Japanese also developed an Enigma-like device for use during the war. The Allies were unable to decrypt Enigma ciphers, until the Polish built an Enigma and sold it to the British. The German military became careless in key choice and about the order of words in sentences, so the cipher was cracked. This was a factor in the defeat of Germany in World War II.

In the 1940's, British intelligence created the first computer, Colossus. After World War II, the British destroyed Colossus. Two Americans at the University of Pennsylvania are credited with creating the first American computer in 1945. It was named the Electronic Numerical Integrator and Calculator (ENIAC). It was able to easily decrypt manual and Enigma ciphers.

How It Works

The data, message, or password starts out as cleartext or plaintext. Once it is input into the computer, a computer algebra system (CAS) performs the actual encryption, using the key and selected algebraic equations, based on the cryptographic algorithm. It stores the data, message, or password as ciphertext.

Most encryption systems require a key. The length of the key increases the complexity of the cryptographic algorithm and decreases the likelihood that the cipher will be cracked. Modern key lengths range from 128 to 2,048 bits. There are two key types. The first is a symmetric or secret key that is used both to encrypt and to decrypt the data. A different key generates different ciphertext. It is critical to keep the key secret and to use a sufficiently complex cryptographic algorithm. The algebraic equations used with a symmetric key are two-way equations.

Symmetric key cryptography uses a polyalphabetic encryption algorithm, such as a block cipher, a stream cipher, or a hash function to convert the data into binary code. A block cipher applies the same key to each block of data. Stream ciphers encrypt the data bit by bit. They can operate in several ways, but two common methods are self-synchronizing and synchronous. The self-synchronizing cipher encrypts the data bit by bit, using an algebraic function applied to the previous bit. Synchronous ciphers apply a succession of functions independent of the data.

Public key or asymmetric cryptography uses two different keys: a public key and a private key. The public key can be distributed to all users, whereas the private key is unknown. The private key cannot be calculated from the public key, although there is a relationship between the two numbers. The public key is used for encryption and the private key is used for decryption. The algebraic equations used in public key cryptography can be calculated only one way, and different equations are used to encrypt and decrypt. The calculations are complex and often involve factorization of a large number or determining the specific logarithm of a large number. Public key algorithms use block ciphers and the blocks may be variable in size.

Digital signatures can be linked to the public key. A digital signature provides a way to identify the creator of the data and to authenticate the source. A digital signature is difficult to replicate. The typical components of a digital signature are the public key, the user's name and e-mail address, the expiration date of the public key, the name of the company that owns the signature, the serial number of the digital identification number (a random number), and the digital signature of the certification authority.

A hash function, or message digest, is an encryption algorithm that uses no key. A hash function takes a variable length record and uses a one-way function to calculate a shorter, fixed-length record. Hash functions are difficult to reverse, and so they are used to verify a digital signature or password. If the new hash is the same as the encrypted version, then the password, or digital signature, is accepted. There are a number of hash algorithms. Typically, they break the file into even-sized blocks and apply either a random number or a prime number to each block. Some hash functions act on all the values in the block, and others work on selected values. Sometimes they generate

Fascinating Facts About Cryptology and Cryptography

- Codes and ciphers have been used since the beginning of recorded time, when humans drew figures on a cave wall to communicate ideas.
- During the Gallic Wars, Julius Caesar devised one of the first substitution ciphers by printing the ciphertext alphabet with the letters moved four places to the left under the plaintext alphabet.
- In 1795 Thomas Jefferson created the wheel cipher using thirty-six removable and numbered alphabetic disks on a spindle. The sender and receiver must arrange the disks in the same order to communicate a message.
- Hash means to cut up and mix, and hash programs do just that. They break files into pieces and mix them with random numbers.
- The scytale, which was invented by the Spartans during the Greek and Persian wars, was used in the 2006 film the *Da Vinci Code*.
- Random numbers generated by a computer program are not truly random. Computer programs use either a physical process such as clock drift to generate numbers and then try to compensate for any biases in the numbers, or they use a smaller number, or seed, to calculate a random number.

duplicate hashes, which are called collisions. If there is any change in the data, the hash changes.

There are specific security standards that cryptographic data must meet. They are authentication, privacy/confidentiality, integrity, and nonrepudiation. These standards not only protect the security of the data but also verify the identity of the user, the validity of the data, and that the sender provided the data.

Applications and Products

Cryptographic software is created so that it can interact with a variety of computer systems. Some of this software is integrated into other computer programs, and some interfaces with other computer systems. Most cryptographic programs are written in the computer languages of Java, JavaScript, C, C+, or C++. The types of software used for cryptography include computer algebra systems, symmetric key algorithms, public key algorithms, hash functions schemes, and digital signature algorithms.

CAS Software. A computer algebra system (CAS) is a software package that performs mathematical functions. Basic CAS software supports functions such as linear algebra, algebra of polynomials, calculus, nonlinear equations, functions, and graphics. More complex CAS software also supports command lines, animation, statistics, number theory, differential equations, networking, geometric functions, graphing, mathematical maximization, and a programming language. Some of the CAS software that is used for encryption of data includes FriCAS 1.0.3, Maple 12, Mathematica 6.0, Matlab 7.2.0.283, Maxima 5.19.1, and Sage 3.4.

Symmetric Key Algorithms. Symmetric key algorithms work best for storing data that will not be shared with others. This is because of the need to communicate the secret key to the receiver, which can compromise its secrecy. There are two types of secret key encryption algorithms: stream ciphers and block ciphers. The standard for secret key encryption is the Data Encryption Standard (DES) software that the National Bureau of Standards has adopted. Other secret key encryption algorithms include: IDEA (International Data Encryption Algorithm), Rivest Ciphers (RC1 through RC6), Blowfish, Twofish, Camellia, MISTY1, SAFER (Secure and Fast Encryption Routine), KASUMI, SEED, ARIA, and Skipjack. They are block ciphers, except for RC4, which is a stream cipher.

Public Key Algorithms. Asymmetric or public key algorithms support the review of digital signatures and variable length keys. They are used for data that are sent to another businesses or accessed by users, because there is no need to keep the public key secure. The U.S. National Institute of Standards and Technology (NIST) has adopted a new encryption cipher, called Advanced Encryption Standard (AES). AES uses a public key and a block algorithm and is considered to be more secure than DES. Some examples of public key algorithms are: RSA1, Diffie-Hellman, DSA, ElGamal, PGP (Pretty Good Privacy), ECC (Elliptic Curve Cryptography), Public-Key Cryptography Standards (PKCS 1 through 15), Cramer-Shoup, and LUC.

Hash Function Algorithms. Hash function algorithms are not considered actual encryption, although they are often used with encryption algorithms. Hash functions are used to verify passwords or digital signatures, to look up a file or a table of data, to store data, to verify e-mail users, to verify data integrity, and to verify the parties in electronic funds transfer (EFT). Some examples of actual hash algorithms are: SHA-1, SHA-2, MD2, MD4, MD5, RIPEMD, Haval, Whirlpool, and Tiger.

Digital Signatures. A digital signature scheme is used with a public key system and may be verified by a hash. It can be incorporated into the algorithms or just interface with them. If a digital signature algorithm is used, it requires that the user know both the public and the private keys. For hash functions, there may be no key. The digital key also verifies that the message or data were not altered during transmission. Digital signature schemes are used to verify a student's identity for access to an academic record, to verify the identity of a credit card user or a banking account owner, to verify the identity of an e-mail user, to verify the company identities in electronic funds transfers, to verify the source of data that is being transferred, and to verify the identity of the user who is storing data. Some digital signature algorithms include RSA, DSA, the elliptical curve variant ElGamel, variants of Schnorr and Pointcheval-Stern, Rabin, pairing-based schemes, undeniable signature, and aggregate signature.

Impact on Industry

As society becomes more dependent on computers, data security by encryption becomes more important. Data need to be available to those who use them to perform their jobs, but they must be safe from hacking and unregistered users. Encryption of computerized data and passwords affects all types of businesses, including health care providers, the government, and schools.

Government. The U.S. government has standards for computer cryptography that are used to encrypt some of its own data. IBM designed Data Encryption Standard (DES) in the 1970's, and it has become the most commonly used secret key algorithm. DES has some shortcomings, but modifications have been made to strengthen it. In 2001, the National Institute of Standards and Technology adopted Advanced Encryption Standard, a more secure cryptosystem, for its applications. The U.S. government has to ensure the security of its own data, such as Internal Revenue records, personnel data, and confidential military data. In addition, it provides computers for its personnel. These require security of work data, messages, passwords, and e-mail. Some government agencies are responsible for gathering intelligence about foreign nations. This may involve attempting to decrypt messages sent by these countries.

Universities and Schools. Universities require a secure computer system to store student demographic and health data; student grades and course work; computer access information; personnel information, including salaries; research data; and Internet use. Most students have their own personal computers and are able to access some areas of the university's computer system. Their digital signatures must be linked to a table that reflects their limited access. Faculty members, both on and off campus, are able to access the university's system, and their access also has limits. Researchers use the university system to store data for their projects and to perform analyses. Much of the research on encryption of data has been performed by university researchers. University staff have access to additional data based on their responsibilities. Even elementary and secondary schools have computerized data that include student records and grades, as well as personnel information. Some school systems permit parents to access the grades of their children online. Security is important to protect this data.

Industry and Business. All big businesses and most small businesses have computers and use encryption algorithms. Computers have made it possible to store business data in a small space, to aggregate and analyze information, and to protect proprietary data through encryption. Banking has been changed to a primarily electronic business with the use of automated teller machines (ATMs), debit cards, electronic funds transfer, and credit cards. A key or password within the debit and credit cards is verified by a hash function. Access to an ATM is granted by a public key or credit card number, along with the PIN (personal identification number). Some banking transactions can even be done at home on a personal computer. Encryption keeps this information confidential. Health insurance companies record customer transactions on a computer, process claims, list participating health care providers, link employees to their employer's individual health insurance plans, perform aggregate reporting for income taxes (1099s), and report on employee claims. Actuaries use aggregate claims data to determine insurance premiums. All these functions require that only employees access computerized data. Similar types of data use are performed by most businesses.

Careers and Course Work

Most cryptographers have at least a bachelor's degree in computer science, mathematics, or engineering. Often they have either a master's degree or a doctorate. A number of universities and information technology institutes offer nondegree programs in cryptography. Cryptographers are persons who are knowledgeable in encryption programming, computer security software, data compression, and number theory, as well as firewalls and Windows and Internet security. People with this background may be employed in computer software firms, criminology, universities, or information technology.

There is a voluntary accreditation examination for cryptographers, which is administered by the International Information Systems Security Certification Consortium. The examination is called the Certified Information Systems Security Professional (CISSP) examination. To be certified, a cryptographer must have five years of relevant job experience and pass the CISSP examination. Recertification must be done every three years, either by retaking the examination or by earning 120 continuing professional education credits.

Position titles include cryptoanalysts, cryptosystem designers, cryptographic engineers, and digital rights professionals. A cryptoanalyst is involved in the examination and testing of cryptographic security systems. Cryptosystem designers develop the complex cryptographic algorithms. Cryptographic engineers work with the hardware of cryptographic computer systems. Digital rights professionals are responsible for the security of encrypted data, passwords, and digital signatures. They may be certificate administrators and be responsible for accepting new users and approving their digital signatures.

Social Context and Future Prospects

Computer technology is an important part of contemporary life. Encryption of data, messages, passwords and digital signatures makes this possible. Without data and password security, both personal and professional users would find that anything that they loaded onto their computers would be available to hackers and spies, and they would be vulnerable to computer viruses that could damage their computers and corrupt their files. It would be difficult to use a computer under these circumstances because of the lack of dependability of the data and of the computer system. Many workers perform their jobs on a computer, and some are able to work from home. Wireless systems are increasingly being used. These capabilities require adequate security.

Despite the complexity of modern cryptography, there still are risks of attacks on computer data, messages, and passwords. No cryptography program is without its weak point. The likelihood of breaking a cryptographic algorithm is assessed by how long it would take to break the cipher with a high-speed computer. A common attack on key cryptography is brute force. This involves trying all the possible number combinations in order to crack the key. The longer the key is, the longer this will take. The only way to create an unbreakable cipher is to use a one-time pad, in which the secret key is encrypted using an input number that is used only once. Each time the data are accessed, another random number of the same length is used as the secret key.

Cryptosystems are not totally secure. As computer technology and cryptography knowledge advances, any particular encryption algorithm is increasingly likely to be broken. Cryptoanalysts must constantly evaluate and modify or abandon encryption algorithms as needed. A future risk to cryptographic systems is the development of a quantum computer, based on quantum theory. It is thought that a quantum computer could crack all the cryptographic ciphers in existence; however, no quantum computer has been created yet.

Christine M. Carroll, B.S.N., R.N., M.B.A.

Further Reading

Blackwood, Gary. *Mysterious Messages: A History of Codes and Ciphers.* New York: Penguin Books, 2009. Provides a comprehensive history of the development of cryptography from ancient history through modern times.

Kahate, Atul. *Cryptography and Network Security.* 2d ed. Boston: McGraw-Hill Higher Education, 2009. Describes the concepts involved with the use of cryptography for network security.

Katz, Jonathan, and Yehuda Lindell. *Introduction to Modern Cryptography.* Boca Raton, Fla.: Chapman & Hall/CRC, 2008. Explores both public and private key cryptography. Looks at various models and systems.

Lunde, Paul, ed. *The Book of Codes: Understanding the World of Hidden Messages.* Los Angeles: University of California Press, 2009. Covers codes and ciphers used in society and the history of cryptography. Chapter 13 discusses the development of the computer and computer cryptography.

Seife, Charles. *Decoding the Universe.* New York: Penguin Group, 2006. Discusses information theory, which is the basis of computer cryptography.

Web Sites

American Cryptogram Association
http://cryptogram.org

International Association for Cryptographic Research
http://www.iacr.org

International Financial Cryptography Association
http://ifca.ai

International Information Systems Security Certification Consortium
https://www.isc2.org

See also: Computer Networks; Computer Science; Information Technology; Internet and Web Engineering; Software Engineering.

D

DIGITAL LOGIC

FIELDS OF STUDY

Mathematics; electronics; physics; analytical chemistry; chemical engineering; mechanical engineering; computer science; biomedical technology; cryptography; communications; information systems technology; integrated-circuit design; nanotechnology; electronic materials

SUMMARY

Digital logic is electronic technology constructed using the discrete mathematical principles of Boolean algebra, which is based on binary calculation, or the "base 2" counting system. The underlying principle is the relationship between two opposite states, represented by the numerals 0 and 1. The various combinations of inputs utilizing these states in integrated circuits permit the construction and operation of many devices, from simple on-off switches to the most advanced computers.

KEY TERMS AND CONCEPTS

- **Boolean Algebra:** A branch of mathematics used to represent basic logic statements.
- **Clock Speed:** A specific frequency that controls the rate at which data bits are changed in a digital logic circuit.
- **Gate:** A transistor assembly that combines input signals according to Boolean logic to produce a specific output signal.
- **Interleaf:** To incorporate sections of different processes within the same body in such a way that they do not interfere with each other.
- **Karnaugh Map:** A tabular representation of the possible states resulting from combinations of a specific number of binary inputs.
- **Sample:** To take discrete measurements of a specific quantity or property at a specified rate.

DEFINITION AND BASIC PRINCIPLES

Digital logic is built upon the result of combining two signals that can have either the same or opposite states, according to the principles of Boolean algebra. The mathematical logic is based on binary calculation, or the base 2 counting system. The underlying principle is the relationship between two opposite states, represented by the numerals 0 and 1.

The states are defined in modern electronic devices as the presence or absence of an electrical signal, such as a voltage or a current. In modern computers and other devices, digital logic is used to control the flow of electrical current in an assembly of transistor structures called gates. These gates accept the input signals and transform them into an output signal. An inverter transforms the input signal into an output signal of exactly opposite value.

An AND gate transforms two or more input signals to produce a corresponding output signal only when all input signals are present. An OR gate transforms two or more input signals to produce an output signal if any of the input signals are present. Combinations of these three basic gate structures in integrated circuits are used to construct NAND (or not-AND) and NOR (or not-OR) gates, accumulators, flip-flops, and numerous other digital devices that make up the functioning structures of integrated circuits and computer chips.

BACKGROUND AND HISTORY

Boolean algebra is named for George Boole (1815-1864), a self-taught English scientist. This form of algebra was developed from Boole's desire to express concrete logic in mathematical terms; it is based entirely on the concepts of true and false. The intrinsically opposite nature of these concepts allows the logic to be applied to any pair of conditions that are related as opposites.

The modern idea of computing engines began with the work of Charles Babbage (1791-1871), who envisioned a mechanical "difference engine" that would calculate results from starting values. Babbage did not see his idea materialize, though others using his ideas were able to construct mechanical difference engines.

The development of the semiconductor junction transistor in 1947, attributed to William Shockley, John Bardeen, and Walter Brattain, provided the means to produce extremely small on-off switches that could be used to build complex Boolean circuits. This permitted electrical signals to carry out Boolean algebraic calculations and marked the beginning of what has come to be known as the digital revolution. These circuits helped produce the many modern-day devices that employ digital technology.

How It Works

Boolean Algebra. The principles of Boolean algebra apply to the combination of input signals rather than to the input signals themselves. If one associates one line of a conducting circuit with each digit in a binary number, it becomes easy to see how the presence or absence of a signal in that line can be combined to produce cumulative results. The series of signals in a set of lines provides ever larger numerical representations, according to the number of lines in the series. Because the representation is binary, each additional line in the series doubles the amount of information that can be carried in the series.

Bits and Bytes. Digital logic circuits are controlled by a clock signal that turns on and off at a specific frequency. A computer operating with a CPU (central processing unit) speed of 1 gigahertz (10^9 cycles per second) is using a clock control that turns on and off 1 billion times each second. Each clock cycle transmits a new set of signals to the CPU in accord with the digital logic circuitry. Each individual signal is called a bit (plural byte) of data, and a series of 8 bits is termed one byte of data. CPUs operating on a 16-bit system pass two bytes of data with each cycle, 32-bit systems pass four bytes, 64-bit systems pass eight bytes, and 128-bit systems pass sixteen bytes with each clock cycle.

Because the system is binary, each bit represents two different states (system high or system low). Thus, two bits can represent four (or 2^2) different states, three bits represents eight (or 2^3) different states, four bits represents sixteen (or 2^4) different states, and so on. A 128-bit system can therefore represent 2^{128} or more than 3.40×10^{38} different system states.

Digital Devices. All digital devices are constructed from semiconductor junction transistor circuits. This technology has progressed from individual transistors to the present technology in which millions of transistors can be etched onto a small silicon chip. Digital electronic circuits are produced in "packages" called integrated circuit, or IC, chips. The simplest digital logic device is the inverter, which converts an input signal to an output signal of the opposite value. The AND gate accepts two or more input signals such that the output signal will be high only if all of the input signals are high. The OR gate produces an output high signal if any one or the other of the input signals is high.

All other digital logic devices are constructed from these basic components. They include NAND gates, NOR gates, X-OR gates, flip-flops that produce two simultaneous outputs of opposite value, and counters and shift registers, which are constructed from series of flip-flops. Combinations of these devices are used to assemble accumulators, adders, and other components of digital logic circuits.

One other important set of digital devices is the converters that convert a digital or analogue input signal to an analogue or digital output signal, respectively. These find extensive use in equipment that relies on analogue, electromagnetic, signal processing.

Karnaugh Maps. Karnaugh maps are essential tools in designing and constructing digital logic circuits. A Karnaugh map is a tabular representation of all possible states of the system according to Boolean algebra, given the desired operating characteristics of the system. By using a Karnaugh map to identify the allowed system states, the circuit designer can select the proper combination of logic gates that will then produce those desired output states.

Applications and Products

Digital logic has become the standard structural operating principle of most modern electronic devices, from the cheapest wristwatch to the most advanced supercomputer. Applications can be identified as programmable and nonprogrammable.

Nonprogrammable applications are those in which the device is designed to carry out a specific set of operations as automated processes. Common examples include timepieces, CD and DVD players, cellular telephones, and various household appliances. Programmable applications are those in which an operator can alter existing instruction sets or provide new ones for the particular device to carry out. Typical examples include programmable logic controllers, computers, and other hybrid devices into which some degree of programmability has been incorporated, such as gaming consoles, GPS (global positioning system) devices, and even some modern automobiles.

Digital logic is utilized for several reasons. First, the technology provides precise control over the processes to which it is applied. Digital logic circuits function on a very precise clock frequency and with a rigorously defined data set in which each individual bit of information represents a different system state that can be precisely defined millions of times per second, depending on the clock speed of the system. Second, compared with their analogue counterparts, which are constructed of physical switches and relays, digital circuits require a much lower amount of energy to function. A third reason is the reduced costs of materials and components in digital technology. In a typical household appliance, all of the individual switches and additional wiring that would be required of an analogue device are replaced by a single small printed circuit containing a small number of IC chips, typically connected to a touchpad and LCD (liquid crystal display) screen.

Programmable Logic Controller. One of the most essential devices associated with modern production methods is the programmable logic controller, or PLC. These devices contain the instruction set for the automated operation of various machines, such as CNC (computer numerical control) lathes and milling machines, and all industrial robotics. In operation, the PLC replaces a human machinist or operator, eliminating the effects of human error and fatigue that result in undesirable output variability. As industrial technology has developed, the precision with which automated machinery can meet demand has far exceeded the ability of human operators.

PLCs, first specified by General Motors Corporation (now Company) in 1968, are small computer systems programmed using a reduced-instruction-set programming language. The languages often use a ladder-like structure in which specific modules of instructions are stacked into memory. Each module consists of the instructions for the performance of a specific machine function. More recent developments of PLC systems utilize the same processors and digital logic peripherals as personal computers, and they can be programmed using advanced computer programming languages.

Digital Communications. Digital signal processing is essential to the function of digital communications. As telecommunication devices work through the use of various wavelengths of the electromagnetic spectrum, the process is analogue in nature. Transmission of an analogue signal requires the continuous, uninterrupted occupation of the specific carrier frequency by that signal, for which analogue radio and television transmission frequencies are strictly regulated.

A digital transmission, however, is not continuous, being transmitted as discrete bits or packets of bits of data rather than as a continuous signal. When the signal is received, the bits are reassembled for audio or visual display. Encryption codes can be included in the data structure so that multiple signals can utilize the same frequency simultaneously without interfering with each other. Data reconstruction occurs at a rate that exceeds human perception so that the displayed signal is perceived as a continuous image or sound. The ability to interleaf signals in this way increases both the amount of data that can be transmitted in a limited frequency range and the efficiency with which the data is transmitted.

A longstanding application of digital logic in telecommunications is the conversion of analogue source signals into digital signals for transmission, and then the conversion of the digital signal back into an analogue signal. This is the function of digital-to-analogue converters (DACs), and analogue-to-digital converters (ACDs). A DAC uses sampling to measure the magnitude of the analogue signal, perhaps many millions of times per second depending upon the clock speed of the system. The greater the sampling rate, the closer its digital representation will be to the real nature of the analogue signal. The ACD accepts the digital representation and uses it to reconstruct the analogue signal as its output.

One important problem that exists with this method, however, is what is called aliasing, in which the DAC analogue output correctly matches the digital representation, but at the wrong frequencies. Present-day telecommunications technology is eliminating these steps by switching to an all-digital format that does not use analogue signals.

Servomechanisms. Automated processes controlled by digital logic require other devices by which the function of the machine can be measured. Typically, a measurement of some output property is automatically fed back into the controlling system and used to adjust functions so that desired output parameters are maintained. The adjustment is carried out through the action of a servomechanism, a mechanical device that performs a specific action in the operation of the machine. Positions and rotational speeds are the principal properties used to gauge machine function.

In both cases, it is common to link the output property to a digital representation such as Gray code, which is then interpreted by the logic controller of the machine. The specific code value read precisely describes either the position or the rotational speed of the component, and is compared by the controller to the parameters specified in its operating program. Any variance can then be corrected and proper operation maintained.

IMPACT ON INDUSTRY

Digital logic devices in industry, often referred to as embedded technology, represent an incalculable value, particularly in industrial production and fabrication, where automated processes are now the rule rather than the exception.

Machine shops once required the services of highly skilled machinists to control the operation of machinery for the production of quality parts. Typically, each machine required one master machinist per working shift, and while the work pieces produced were generally of excellent quality, dimensional tolerance was variable. In addition, human error and inconsistency caused by fatigue and other factors resulted in significant waste, especially given that master machinists represented the high end of the wage scale for skilled tradespersons.

Compare this earlier shop scenario to a modern machine shop, in which a single millwright ensures the proper operating condition of several machines, each of which is operated by a digital logic controller. The machines are monitored by unskilled laborers who must ensure that raw parts are supplied to the machine as needed and who must report any discrepancies to the millwright for needed maintenance. The actual function of the machine is automated through a PLC or similar device, providing output products that are dimensionally consistent, with minimal waste, and are produced at the same rate throughout the entire operating period.

It is easy to realize the economic value to industry of even just this one example of change in production methods. Given that almost all industrial production methods that can be automated with digital logic have been, the effects on productivity and profitability are fairly obvious. Automation, however, reduces the requirement for skilled tradespersons. Machine operators now represent a much lower position on the wage scale than did their skilled predecessors. The reduced production costs in this area have led to the relocation of many company facilities to parts of the world where daily wages are typically much lower and willing workers are plentiful.

PLCs and other digital logic controls automate almost every industrial process. In many cases the function is entirely automatic, not even requiring the presence of a human overseer. Sophisticated machine design in these cases replaces each action that a human would otherwise be required to carry out. In other cases, human intervention is required, as some aspect of the process cannot be left to automatic control. For example, die-casting of parts from magnesium alloys involves a level of art over science that calls for continuous human attention to produce an acceptable cast and prevent a piece from jammingin the die. Between these two more extreme situations are those common ones in which digital logic replaces the need for an operator to carry out several minor actions with the press of a single button. Once that button is pressed, the operation proceeds automatically, and the operator can only stand and watch while preparing for the next cycle of the machinery.

CAREERS AND COURSEWORK

An understanding of digital logic and its applications is obtained through courses of study in electronics engineering technology. This will necessarily include the study of mathematics and physics as foundation courses. Because digital logic is used to control physical devices, students can also expect to study mechanics and mechanical engineering, control systems and feedback, and perhaps hydraulic and pneumatic systems technology.

The development of new electronic materials and production methods will draw students to the study of materials science and chemistry, while others will specialize in circuit design and layout. Graphic methods are extremely important in this

Fascinating Facts About Digital Logic

- Graphene, a one-atom-thick form of carbon that may allow transistors the size of single molecules, is one of the most exotic materials ever discovered, even though people have unknowingly been writing and drawing with it for centuries.
- The first functional transistor used two pieces of gold leaf that had been split in half with a razor blade.
- The term "transistor" was devised to indicate a resistor that could transfer electrical signals.
- Digital logic devices have automated nearly all modern production processes.
- An X-OR (Exclusive-OR) gate outputs a positive signal when just one or the other of its input signals is positive.
- Quantum dots can be thought of as artificial atoms containing a single electron with a precisely defined energy.
- Moore's law, stated by Gordon Moore of Intel Corporation in a 1965 paper, predicts that the number of transistors that can be put on a computer chip doubles about every eighteen months.
- The most sophisticated of digital logic devices, the CPU chips of advanced computers, are constructed from only three basic transistor gate designs, the NOT, AND, and OR gates. All other logic gates are constructed from these three.

area, and specialized design programs are essential components of careers involving logic circuit design.

Introductory and college-level courses will focus on providing a comprehensive understanding of logic gates and their functions in the design of relatively simple logic circuits. Special attention may be given to machine language programming, because this is the level of computer programming that works directly with the logical hardware. Typically, study progresses in these two areas, as students will design and build more complex logic circuits with the goal of interfacing a functional device to a controlling computer.

Postgraduate electronics engineering programs build on the foundation material covered in college-level programs and are highly specialized fields of study. This level will take the student into quantum mechanics and nanotechnology, as research continues to extend the capabilities of digital logic in the form of computer chips and other integrated circuits. This involves the development and study of new materials and methods, such as graphenes and fullerenes, superconducting organic polymers, and quantum dots, with exotic properties.

Social Context and Future Prospects

At the heart of every consumer electronic device is an embedded digital logic device. The transistor quickly became the single most important feature of electronic technology, and it has facilitated the rapid development of everything from the transistor radio to space travel. Digital logic, as embedded devices, is becoming an ever more pervasive feature of modern technology; an entire generation has now grown up not knowing anything but digital computers and technology. Even the accoutrements of this generation are rapidly being displaced by newer versions, as tablet computers and smartphones displace more traditional desktop personal computers, laptops, and cell phones. The telecommunications industry in North America is in the process of making a government-regulated switch-over to digital format, making analogue transmissions a relic of the past.

Research to produce new materials for digital logic circuits and practical quantum computers is ongoing. The eventual successful result of these efforts, especially in conjunction with the development of nanotechnology, will represent an unparalleled advance in technology that may usher in an entirely new age for human society.

Richard M. Renneboog, M.Sc.

Further Reading

Brown, Julian. *Minds, Machines, and the Multiverse: The Quest for the Quantum Computer.* New York: Simon & Schuster, 2000. Discusses much of the history of digital logic systems, in the context of ongoing research toward the development of a true quantum computer.

Bryan, L. A., and E. A. Bryan. *Programmable Controllers: Theory and Application.* Atlanta: Industrial Text Company, 1988. Provides a detailed review of basic digital logic theory and circuits, and discusses the theory and applications of PLCs.

Holdsworth, Brian, and R. Clive Woods. *Digital Logic Design.* 4th ed. Woburn, Mass.: Newnes/Elsevier Science, 2002. A complete text providing a thorough treatment of the principles of digital logic and digital logic devices.

Jonscher, Charles. *Wired Life: Who Are We in the Digital Age?* London: Bantam Press, 1999. Discusses the real and imagined effects that the digital revolution will have on society in the twenty-first century.

Marks, Myles H. *Basic Integrated Circuits.* Blue Ridge Summit, Pa.: TAB Books, 1986. A very readable book that provides a concise overview of digital logic gates and circuits, then discusses their application in the construction and use of integrated circuits.

Miczo, Alexander. *Digital Logic Testing and Simulation.* 2d ed. Hoboken, N.J.: John Wiley & Sons, 2003. An advanced text that reviews basic principles of digital logic circuits, then guides the reader through various methods of testing and fault simulations.

Web Sites

ASIC World
http://www.asic-world.com/

NobelPrize.org
http://www.nobelprize.org

See also: Algebra; Applied Mathematics; Applied Physics; Communication; Computer Engineering; Computer Science; Electronics and Electronic Engineering; Engineering Mathematics; Information Technology; Integrated-Circuit Design; Mechanical Engineering; Nanotechnology.

EARTHQUAKE ENGINEERING

FIELDS OF STUDY

Civil engineering; earthquake engineering; engineering seismology; geology; geotechnical engineering.

SUMMARY

Earthquake engineering is a branch of civil engineering that deals with designing and constructing buildings, bridges, highways, railways, and dams to be more resistant to damage by earthquakes. It also includes retrofitting existing structures to make them more earthquake resistant.

KEY TERMS AND CONCEPTS

- **Asperity:** Surface roughness that projects outward from the surface.
- **Epicenter:** Point on the Earth's surface directly above the hypocenter.
- **Hypocenter:** Point beneath the Earth's surface where an earthquake originates.
- **Love Wave:** Wave formed by the combination of secondary waves and primary waves on the surface; causes the ground to oscillate from side to side perpendicular to the propagation direction of the wave and is the most destructive.
- **P Wave:** First wave from an earthquake to reach a seismograph; travels through the body of the Earth, including through a liquid; also called a primary wave.
- **Rammed Earth:** Mixture of damp clay, sand, and a binder such as crushed limestone that is poured into a form and then rammed down by thrusting with wooden posts; after it dries the forms are removed.
- **Rayleigh Wave:** Wave formed by the combination of secondary waves and primary waves on the surface; causes the ground to oscillate in a rolling motion parallel with the direction of the wave.
- **S Wave:** Wave that reaches a seismic station after a primary wave; travels through the body of the Earth, but not through a liquid; also called secondary wave.

Definition and Basic Principles

Worldwide, each year there are about eighteen major earthquakes (magnitude 7.0 to 7.9) strong enough to cause considerable damage and one great earthquake (magnitude 8.0 or greater) strong enough to destroy a city. The outermost layer of the Earth is the rocky crust where humans live. The continental crust of the Earth is 30 to 50 kilometers thick, while the oceanic crust is 5 to 15 kilometers thick. The crust is broken up into about two dozen plates that fit together like pieces of a jigsaw puzzle, with the larger plates being hundreds to thousands of kilometers across. As these plates move about on the underlying mantle at rates of a few to several centimeters per year, they rub against neighboring plates. Asperities (irregularities) from one plate lock with those of an adjacent plate and halt the motion. While the plate boundary is held motionless, the rest of the plate continues in motion in response to the forces on it, and this action builds up stress in the boundary rocks.

Finally, when the stress on the boundary rocks is too great the asperities are sheared off as the boundary rock surges ahead several centimeters to several meters. It is this sudden lurching of the rock that produces earthquake waves. The point of initial rupture produces the most waves and is called the hypocenter, while the point on the surface directly above the hypocenter is called the epicenter. The epicenter is usually the site of the worst damage on the surface. Earthquake engineers can design structures to reduce the damage and the number of deaths, but the limited resources available means that not everything that might be done is done. The philosophy generally adopted is that while a strong earthquake may damage most structures they should remain standing at least long enough for the people in them to evacuate safely. Essential structures such as hospitals should not only remain standing but should be still usable after the quake.

Background and History

Earthquakes have plagued mankind throughout history. The Antioch (now in Turkey) earthquake in 526 killed an estimated 250,000 people. A thousand years later in 1556, an estimated 830,000 died in Shaanxi, China. In ancient times, earthquakes were ascribed to various fanciful causes such as air rushing out of deep caverns. Chinese mathematician Zhang Heng (78-139) is credited with the invention of a Chinese earthquake detector consisting of a large, nearly spherical vessel with eight dragon heads projecting outward from its circumference. A brass ball is loosely held in each dragon's mouth. A pendulum is suspended inside the vessel so that if an earthquake sets it swinging, it will strike a dragon causing the ball to fall from its mouth and into the waiting mouth of a toad figure. The sound of the ball striking the metal toad alerts the operator that an earthquake has occurred, and whichever toad has the ball indicates the direction to the epicenter.

The scientific study of earthquakes did not blossom until the twentieth century. In 1935, American seismologist Charles Richter with German seismologist Beno Gutenberg developed the Richter scale for measuring the intensity of an earthquake based on the amplitude of the swinging motion of the needle on a seismometer. The Richter scale was superseded in 1979 by the moment magnitude scale based on the energy released by the quake. This scale was developed by Canadian seismologist Thomas C. Hanks and Japanese American seismologist and Kyoto Prize winner Hiroo Kanamori and is the same as the Richter scale for quakes of magnitude 3 through 7.

Following the 1880 Yokohama earthquake, the Seismological Society of Japan was founded to see what might be done to reduce the consequences of earthquakes. It was the world's first such society and marks the beginning of earthquake engineering. The Japanese were forced into such a leadership role by being an industrial society sitting on plate boundaries and therefore subject to frequent earthquakes. In 1893, Japanese seismologist Fusakichi Omori and British geologist John Milne studied the behavior of brick columns on a shake table (to simulate earthquakes). Toshikata Sano, a professor of structural engineering at the Imperial University of Tokyo, published "Earthquake Resistance of Buildings" in 1914, and by the 1930's, several nations had adopted seismic building codes. Knowledge from earthquake engineering was beginning to be put into practice.

The early twenty-first century has seen several devastating quakes. On January, 12, 2010, a magnitude-7 earthquake struck Haiti, killing an estimated 316,000 people, injuring 300,000, and leaving more than one million people homeless. On March 11, 2011, a magnitude-8.9 quake occurred off the east coast of Honshu, Japan, causing a massive tsunami that destroyed entire villages and also affected places as far away as Australia and the West Coast of the United States. This was the strongest quake in Japanese history. In addition to an estimated 10,000 deaths and almost 8,000 people reported missing, Japanese officials had to deal with subsequent leaks at three nuclear reactors in the affected region.

How It Works

Earthquakes may be classified by their depth. Shallow-focus quakes have a hypocenter less than 70 kilometers deep and are the most destructive. Mid-focus quakes originate between 70 kilometers and 300 kilometers deep, and deep-focus quakes originate deeper than 300 kilometers and are the least destructive.

P Waves and S Waves. Underground quakes emit two kinds of waves, P waves (primary waves) and S waves (secondary waves). P waves are longitudinal or compression waves just like sound waves. The rock atoms vibrate along the direction in which the wave moves. P waves travel about 5,000 meters per second in granite. S waves are transverse waves, so atoms vibrate up and down perpendicular to the direction of wave travel. The speed of an S wave is about 60 percent that of a P wave. The difference in arrival times of P waves and S waves at a seismic station provides an estimate of the distance to the hypocenter. Therefore during a site evaluation, earthquake engineers must locate any nearby faults and the location of past hypocenters. Then they must try to determine the most likely hypocenters of future earthquakes, which they hope will be deep and far away.

Love Waves and Rayleigh Waves. S waves and P waves interact at the Earth's surface to produce two new types of surface waves: Love waves and Rayleigh waves. These are the waves that destroy buildings and knock people off their feet. Love waves are named for British mathematician Augustus Edward Hough Love, who developed the theory about the waves in his book *Some Problems About Geodynamics* (1911). They cause the ground to oscillate from side to side perpendicular to the propagation direction of the wave. Love waves are the greatest source of destruction outside the epicenter. Rayleigh waves are named for British physicist and Nobel Prize winner John William Strutt (Lord Rayleigh). They cause the ground to oscillate in a rolling motion parallel with the direction of the wave. The greater the wave amplitude, the more violent the shaking. A careful examination of the rocks and soil underlying a site should give information on propagation, damping, and direction of likely earthquake waves.

Core Sampling. To complete the site investigation, core samples may need to be taken to look for damp and insufficiently compacted soil. The shaking of an earthquake can turn damp sand into jelly, a process called liquefaction. Sand grains are surrounded by liquid and cannot cling together. Buildings sink into liquified soil. If it looks like liquefaction may be a problem, support piers must go from the building's foundation down into bedrock. If that is not possible, densification and grout injection may stabilize the soil. Liquefaction caused a segment of roadway to drop 2.4 meters in the 1964 Alaska earthquake. It caused great destruction in the Marina District of the 1989 Loma Prieta earthquake, and contributed to the destruction of Christchurch, New Zealand, in the February 22, 2011, magnitude-6.3 earthquake.

If the site involves a slope, or if the site is a railroad embankment, an earthquake might cause the slope to collapse. In this case the angle of repose of the slope could be reduced. If that is not possible, a retaining wall may be required, or a geotextile (made from polyester or polypropylene) covered with sand could be used to stabilize the slope.

Applications and Products

Geologists and geophysicists have learned enough about earthquakes to be able to identify ways in which the damage they cause can be minimized by appropriate human behavior. In general, there are two main approaches to mitigating the effects of earthquakes: Build the structure to withstand a quake, and make the structure invisible to the earthquake waves, but if that cannot be done, dampen the waves as quickly as possible.

Designing for Earthquake Safety. Earthquake engineers can test design ideas by carefully modeling a structure on a computer, inputting the location and strength of the various materials that will be used to construct the building. They can then use the computer to predict the results of various stresses. They can also build a scaled-down model, or even a full-scale model, on a shake table, a platform that can be shaken to simulate an earthquake. When using a scale model, care must be taken so that it is not only geometrically similar to the full-size structure but that other factors such as the velocity of waves moving through the structure are also scaled down. The ultimate test is to build the structure and see how it fares.

Retrofitting Old Structures. The single act that has the potential to save the most lives is to fortify adobe houses against earthquakes. About 50 percent of the population in developing countries live in houses made of adobe or rammed earth, since dirt is cheap and is the ubiquitous building material. Adobe bricks are made by mixing water into 50 percent sand, 35 percent clay, and 15 percent binder-either straw or manure (said to repel insects). The mixture is poured and patted into a mold and then the mold is turned upside down so that the new brick will fall onto the ground to dry in the sun. The bricks may be assembled into a wall using a wet mixture of sand and clay as mortar. Mortar joints should be no more than 20 millimeters thick to avoid cracking. Rammed earth uses a similar mixture of sand and clay but uses lime, cement, or asphalt as a stabilizer. The wet mixture is poured into a form and then tamped down, or rammed, by workers with thick poles. After it sets, the forms are removed.

In an earthquake, both rammed earth and adobe crack and shatter. Walls tumble down, and roofs that had rested on the walls collapse onto people. The magnitude-7 quake that struck Haiti in 2011 flattened a large part of its capital, Port-Au-Prince. It was so devastating because its hypocenter was shallow (only 13 kilometers deep) and only 25 kilometers away, and the poorly constructed adobe houses fell on people and buried them.

Earthquake engineers have figured out relatively inexpensive ways to strengthen adobe houses. Laying bamboo lengthwise in the mortar strengthens the wall as does drilling holes in the bricks and inserting vertical bamboo sticks so that they tie rows of blocks together. A continuous wooden or cement ring should go all around the top of the walls to tie the walls together and to provide a way to fasten ceiling and roof joists to the walls. If an existing house is being retrofitted, vertical and horizontal strips of wire mesh should be nailed onto the walls both inside and outside at corners and around windows and doors. In practice, the mesh strips used range from seven centimeters to sixty centimeters in width. The strips are attached by driving nails through metal bottle caps and into the adobe.

Bridges, Dams, and Isolation Bearings. Structures can be protected by strengthening them to withstand an earthquake or by isolating them from the ground so that earthquake energy does not enter the structure. Dams are built to the first plan, and bridges follow the second plan. To ensure that a concrete dam survives an earthquake, extra reinforcing steel bars (rebars) would be used. The dam must either rest directly on bedrock or on massive pillars that extend downward to bedrock. If there are known fault lines in the area so that it is known in which direction land will move, it may be possible to construct a slip joint. A slip joint works like sliding closet doors, where one door can slide in front of the other. The Clyde Dam in Central Otago, New Zealand, has a slip joint that will allow the land on either side to slip up to two meters horizontally and one meter vertically. Finally, the dam should be built several meters higher than originally planned so that if an earthquake causes the impounded water to slosh back and forth it will not overtop the dam.

Bridges have long used bearings in the form of several hardened steel cylinders between a flat bridge plate and a flat supporting plate. These bearings allow the bridge to move as it expands or contracts with temperature. The same method can be used to allow motion during an earthquake without damaging the bridge, but now the bridge is isolated from the pier since the pier can move back and forth while the bridge remains stationary.

Waves that shake the foundation of a building send some of that vibrational energy up into the building. Placing a bridge or a building on lead and rubber bearings lessens the energy transmitted to the bridge or building. A typical bearing consists of a large block of alternating steel and rubber layers surrounding a vertical lead cylinder. It is quite rigid in the vertical direction but allows considerable motion in the horizontal direction. Since the rubber heats up as it is deformed, it converts the horizontal motion into heat and thereby damps this motion. The Museum of New Zealand and the New Zealand Parliament buildings stand on lead-rubber bearings and are thereby partially isolated from ground motion. Ironically, Christchurch, New Zealand, was considering reinforcing a large number of buildings prior to the February, 2011, quake, but had not proceeded very far because of the cost. That cost was a pittance compared with the rebuilding cost.

Mass Dampers. A building of a certain height, mass, and stiffness will tend to oscillate or resonate at a given frequency. Small oscillations can quickly build up into large oscillations, just as repeatedly giving a small push at the right frequency to a child in a playground swing will cause the swing to move in a large arc. If the upper floors of a tall building sway too much, people get motion sickness, and the structure gets weakened and may eventually fall. Mass dampers are usually huge concrete blocks mounted on tracks on an upper floor. When sensors detect lateral motion of the building's upper floors, motors drive the block in the opposite direction to the building's motion. This pushes the building in the opposite direction from its motion and causes that motion to die out. A tuned mass damper oscillates at the natural frequency of the building. This technique also works with motion caused by earthquake waves.

Taipei 101 in Taiwan is the world's second-tallest building. It is 101 stories (508 meters) tall, and its mass wind damper is a 660-metric ton metal sphere suspended like a pendulum between the eighty-seventh and ninety-second floors. If the building sways, the pendulum is driven in the opposite direction. This passively tuned mass wind damper reduces the building's lateral motion by more than half. Two six-metric-ton pendula are positioned in the tower to control its motion.

X-Braces and Pneumatic Dampers. The vertical and horizontal beams of a building's steel framework form rectangles. Consider a vertical rectangle in the wall at the bottom of the building. The bottom of the rectangle is fastened to the foundation and will move with a seismic wave, but the top of the rectangle is fastened to the building above it so that inertia will tend to hold it fixed. As the foundation moves laterally, the rectangle will be deformed into a parallelogram.

Two diagonal beams making an "X" in the rectangle will keep it from deforming very much. The beautiful seventy-one story Pearl River Tower in Guangzhou, China, uses massive X-braces to keep the tower from swaying in the wind or an earthquake. These beams can clearly be seen in construction photographs of the tower. The tower uses integral wind turbines and solar cells to be largely energy self-sufficient. Rather than using the X-braces to stiffen the tower, the centers of the diagonal beams could have been clamped together with break-lining material between them, then they would damp the horizontal motion. Diagonally mounted massive hydraulic pistons can also be used to strengthen a structure and simultaneously to damp out the earthquake energy in a building.

Pyramids. Large amplitude horizontal motion can also be avoided if earthquake motion is not concentrated at the building's natural frequency, but is spread over many frequencies. A building can be designed not to amplify waves of certain frequencies and to deflect some waves and absorb others. The speed of a wave traveling up a building depends on the amount of stress present, the amount of mass per meter of height, and the frequency of the wave. A bullwhip made of woven leather thongs makes a good analogue. Near the handle, the whip is as thick as the handle, but it tapers to a single thin thong at the far end. When the handle is given a quick backward and forward jerk, a wave speeds down the length of the whip. If the momentum (mass times velocity) of a whip segment is to remain constant, as the mass of a segment decreases (because of the taper), the speed must increase. The distinctive whip crack occurs as the whip end exceeds the speed of sound. In a similar fashion, the speed of a wave traveling up a pyramid-shaped building changes, and as the speed changes, the frequency changes. If a pyramidal building is properly designed, earthquake waves will attenuate as they try to pass upward. This is the idea behind the design of the forty-eight story Transamerica Pyramid building in San Francisco. It is not essential that the shape of the building be a pyramid since changing the mass density or the tension in the steel structure can have a similar effect.

Impact on Industry

Government and University Research. Most earthquake engineering work is done by consulting companies, while research is carried out by universities and research institutes. The United States Geological Survey (USGS) is the federal agency tasked with recording and reporting earthquake activity nationwide. Data is provided by the Advanced National Seismic System (ANSS), a nationwide array of seismic stations. USGS maintains several active research projects. The Borehole Geophysics and Rock Mechanics project drills deep into fault zones to measure heat flow, stress, fluid pressure, and the mechanical properties of the rocks. The Earthquake Geology and Paleoseismology project seeks out and analyzes the rocks pertaining to historic and prehistoric earthquakes.

Earthquake Engineering Research Institute (EERI) of Oakland, California, carries out various research projects. One project involves surveying concrete buildings that failed during an earthquake in an effort to discover the top ten reasons for the failure of these buildings. The Pacific Earthquake Engineering Research Center (PEER) at the University of California, Berkeley, has a 6-meter-by-6 meter shaking table. It can move horizontally and vertically and can rotate about three different axes. It can carry structures weighing up to 45 tons and subject them to horizontal accelerations of 1.5 times gravity. Recent projects include the seismic-qualification testing of three types of 245-kilovolt disconnect switches, testing a friction pendulum system (for damping vibrations), testing a two-story wood-frame house, and testing a reinforced concrete frame.

The John A. Blume Earthquake Engineering Center at Stanford University pursues the advancement of research, education, and practice of earthquake engineering. Scientists there did the research into the earthquake risk for the 2-mile-long Stanford Linear Accelerator, the Diablo Canyon Nuclear Power Plant, and many other sites. The founder, John Blume, is quoted as reminding a reporter that the center designed "earthquake-resistant" buildings not "earthquake-proof" buildings, and added, "Don't say 'proof' unless you're talking about whiskey."

Feeling that more coordinated efforts were needed, the Japanese formed the Japan Association for Earthquake Engineering (JAEE) in January, 2001. The association was to be involved with the evaluation of seismic ground motion and active faults, resistance measures before an earthquake, education on earthquake disaster reduction, and sponsoring meetings and seminars where new techniques

Fascinating Facts About Earthquake Engineering

- San Francisco is moving toward Los Angeles at about 5 centimeters per year. They should be across from each other in about 12 million years.
- There are about 500,000 earthquakes each year, but only about 100,000 are strong enough to be felt by people.
- Only about 100 earthquakes each year are strong enough to do any damage.
- The largest recorded earthquake was a magnitude 9.5 in Chile on May 22, 1960.
- On March 28, 1964, a woman in Anchorage, Alaska, was trying to remove a stuck lid from a jar of fruit. At the instant she tapped the lid against the corner of the kitchen counter, a 9.2 magnitude earthquake struck. For a few moments she feared she had caused the quake.
- Without a tuned mass damper, the top floors of very tall buildings can sway back-and-forth 30 centimeters or more in a strong wind, let alone an earthquake.
- The 2004 Indian Ocean 9.0-magnitude earthquake released energy equivalent to 9,560,000 megatons of TNT, about 1,000 times the world supply of strategic nuclear warheads.

could be shared and analyzed. After an earthquake, they hope to aid in damage assessment, emergency rescue and medical care, and in evaluating what building techniques worked and what did not work.

Major Corporations. ABS Consulting, with headquarters in Houston, Texas, is a worldwide risk-management company. It has 1,400 employees and uses earthquake engineers when it evaluates the earthquake risk for a site.

Air Worldwide provides risk analysis and catastrophe modeling software and consulting services. It has offices in Boston, San Francisco, and several major cities in other countries. It hires civil engineers to perform seismic design studies and to prepare plans for structural engineering.

ARUP, an engineering consulting company headquartered in London, employs 10,000 people worldwide. To reduce the lateral movement of the upper stories of buildings, it employs its damped outrigger system. It uses large hydraulic cylinders to tie the central pillar of the building to the outer walls. The alternative is to make the building stiffer with more concrete and steel (which costs several million dollars) and then add a tuned mass damper (which ties up a great deal of space). The company used a few dozen of these dampers in the beautiful twin towers of the St. Francis Shangri-La Place in Manila, Philippines.

International Seismic Application Technology (ISAT), with headquarters in La Mirada, California, uses earthquake engineers to do site studies and to design seismic-restraint systems for plumbing, air-conditioning ducts, and electrical systems in buildings.

Miyamoto International is a global earthquake and structural engineering firm that specializes in designing earthquake engineering solutions. It has offices throughout California, and in Portland-Vancouver and Tokyo, and specializes in viscous and friction cross-bracing dampers.

The Halcrow Group based in London does seismic-hazard analysis, design, and remediation. It also does earthquake site response analysis and liquefaction assessment and remediation. It has done site evaluations for nuclear power plants, dams, and intermediate level nuclear waste storage. It is a large company with 8,000 employees and many interests.

Careers and Course Work

Earthquake engineering is a subset of geotechnical engineering, which itself is a branch of civil engineering. Perhaps the most direct route would be to attend a university such as the University of California, Los Angeles, which offers both graduate and undergraduate degrees in engineering, and get an undergraduate degree in civil engineering. An undergraduate should take principles of soil mechanics, design of foundations and earth structures, advanced geotechnical design, fundamentals of earthquake engineering, soil mechanics laboratory, and engineering geomatics. Graduate courses should include advanced soil mechanics, advanced foundation engineering, soil dynamics, earth retaining structures, advanced cyclic and monotonic soil behavior, geotechnical earthquake engineering, geoenvironmental engineering, numerical methods in geotechnical engineering, and advanced soil mechanics laboratory.

Other schools will have their own version of the program. For example, Stanford University offers a

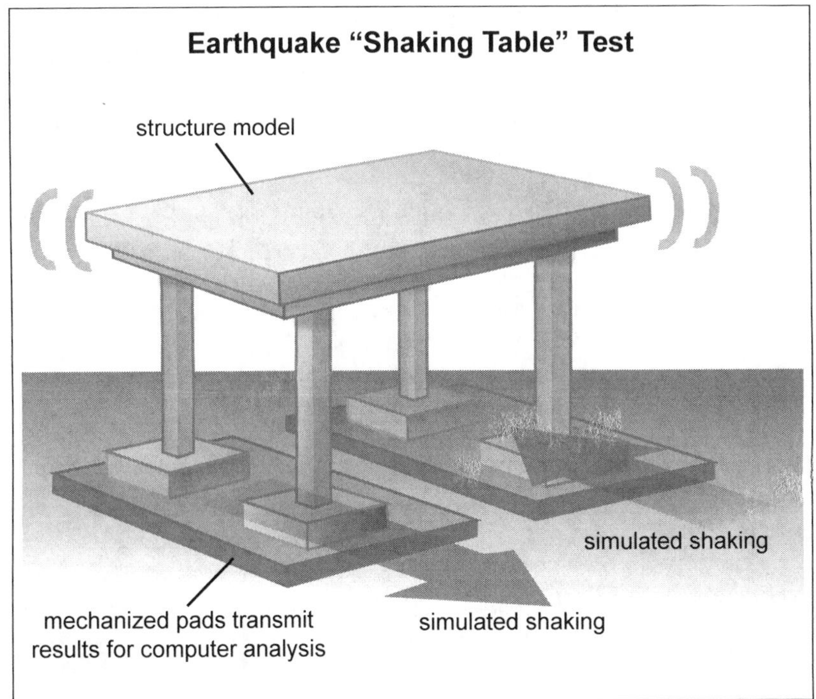

An earthquake shake table is used to test the resistance of certain components or structures to seismic activity.

master's degree in structural engineering and geomechanics. The program requires a bachelor's degree in civil engineering including courses in mechanics of materials, geotechnical engineering, structural analysis, design of steel structures, design of reinforced concrete structures, and programming methodology. The University of Southern California; the University of California, San Diego; the University of California, Berkeley; and the University of Alaska at Anchorage all have earthquake engineering programs. On the east coast, the Multidisciplinary Center for Earthquake Engineering Research at the State University of New York at Buffalo and the Center for Earthquake Engineering Simulation at Rensselaer Polytechnic Institute in Troy, New York, are centers for earthquake engineering.

Social Context and Future Prospects

Although earthquake engineering has made a lot of progress, some areas of society have been surprisingly slow to implement proven measures. Many of the buildings that collapsed in the New Zealand earthquake in February, 2011, would have remained standing had they been reinforced to the recommended standard. They were not reinforced because of cost, but that cost was a small fraction of what it will now cost to rebuild. The January, 2010, Haitian earthquake was so deadly because there are no national building standards. The December, 2003, earthquake in Bam, Iran, was so devastating because building codes were not followed. In particular, enough money was budgeted to build the new hospitals to standards that would have kept them standing. The hospitals collapsed into piles of rubble while corrupt officials (according to expatriates) enriched themselves. Building codes will do no good until they are enforced, and they will not be enforced without honest officials.

On a more positive note, an exciting, recent proposal is to make a building invisible to earthquake waves. Earthquake surface waves cause the damage. The speed of such waves depends upon the density and rigidity of the rock and soil they traverse. Consider a wave coming toward a building almost along a radius, and suppose that the building's foundation is surrounded by a doughnut-shaped zone in which the speed of an earthquake wave is increased above that of the surrounding terrain. The incoming wave will necessarily bend away from the radius. One or more properly constructed doughnuts, or rings, should steer the earthquake waves around the building. No doubt there will be problems in implementing this method, but it seems promising. It may even be possible to surround a town with such rings and thereby protect the whole town.

Charles W. Rogers, B.A., M.S., Ph.D.

Further Reading

Bozorgnia, Yousef, and Vitelmo V. Bertero, eds. *Earthquake Engineering: From Engineering Seismology to Performance-Based Engineering.* Boca Raton, Fla.: CRC Press, 2004. Provides a good overview of the problems encountered in earthquake engineering and ways to solve them. Requires a good science and math background.

Building Seismic Safety Council for the Federal Emergency Management Agency of the Department of Homeland Security. *Homebuilder's Guide to Earthquake-Resistant Design and Construction.* Washington, D.C.: National Institute of Building Sciences, 2006. A gold mine for the non-engineer or the prospective engineer. Introduces the terms and techniques of earthquake-resistant structures in an understandable fashion.

Kumar, Kamalesh. *Basic Geotechnical Earthquake Engineering.* New Delhi, India: New Age International, 2008. Emphasizes site properties and preparation, when to expect liquefaction and what to do about it. Easily read by the science-savvy layperson.

Stein, Ross S. "Earthquake Conversation." *Scientific American* 288 (January, 2003): 72-79. Active faults are responsive to even a small increase in stress that they acquire when there is a quake in a nearby fault. This may make earthquake prediction more accurate.

Yanev, Peter I., and Andrew C. T. Thompson. *Peace of Mind in Earthquake Country: How to Save Your Home, Business, and Life.* 3d ed. San Francisco: Chronicle Books, 2008. An excellent introductory treatment of earthquakes, how they damage structures, and what may be done beforehand to reduce damage. Discusses building sites and possible problems such as liquefaction.

Web Sites

ArchitectJaved.com
Earthquake Resistant Structures
http://articles.architectjaved.com

Earthquake Engineering Research Institute
http://www.eeri.org/site

Seismological Society of America
http://www.seismosoc.org

See also: Bridge Design and Barodynamics; Civil Engineering.

EARTHQUAKE PREDICTION

FIELDS OF STUDY

Geology; oceanography; geophysics; seismology; structural engineering; materials science; soil science.

SUMMARY

Earthquakes are among the most potentially devastating catastrophes human beings can face, and ways of predicting their occurrence in advance are urgently needed. Short-range predictions are needed so that people can evacuate dangerous locations or shutdown critical services, such as nuclear power plants and transportation systems. Long-range predictions are needed to help communities upgrade their building codes and identify at-risk areas where construction or other land uses should not be permitted.

KEY TERMS AND CONCEPTS

- **Elastic Rebound:** Sudden release of strain built up by fault creep over a long period of time.
- **Fault:** Crack in the Earth's crust along which differential movement has occurred.
- **Foreshock:** Preliminary vibration before a major earthquake.
- **Locked Fault:** Fault along which anticipated movement is overdue.
- **Moment Magnitude:** Scale for measuring earthquake size based on the amount of energy released.
- **Primary Wave:** First group of vibrations to arrive from an earthquake.
- **Richter Scale:** Scale for measuring earthquake size based on the amount of shaking.
- **Secondary Wave:** Second group of vibrations to arrive from an earthquake.

Definition and Basic Principles

Earthquake prediction is the act of determining in advance that an earthquake is likely to take place. The prediction must address four factors: the time period during which the earthquake is expected to occur, the area that the earthquake is expected to affect, the type of shaking that will occur, and the likelihood that the earthquake will actually take place. Short-range predictions estimate that an earthquake will occur within the next few hours, or at most, the next few days. These are based on premonitions of an impending earthquake, such as changes in groundwater levels in wells, an increased number of foreshocks, or unusual animal behavior. Long-range predictions, which estimate that an earthquake may take place within a certain number of years, are generally based on the past history of the area, augmented by Global Positioning System (GPS) data or the digging of trenches in critical areas. Short-range warnings are valuable if the earthquake occurs, but if the anticipated earthquake does not materialize, they have disrupted people's lives unnecessarily. Long-range predictions are useful primarily because they often result in reviews of building codes and reconsideration of the suitability of some land areas for construction and development.

Background and History

The scientific understanding of earthquakes and the development of methods for predicting them date only to the time of the great San Francisco earthquake in 1906. Before that time, the relationship between faulting and earthquakes was not clearly understood. H. F. Reid, a professor at The Johns Hopkins University, was appointed to California's state-funded commission to study the great earthquake. Based on his examination of the terrain surrounding the city, he was able to show that points on opposite sides of the San Andreas fault had moved differentially as much as 6 meters before breakage. Reid coined the term "elastic rebound" to describe what then happened. Strain had built up in the rocks by fault creep, as the two sides of the fault drifted in opposite directions. When breakage finally occurred, the rocks returned to their unstrained positions just as a rubber band returns to its original shape after the band snaps, and this snapping back of the rocks was what had caused the earthquake. Reid led the way to an understanding of how the breakage of faults causes earthquakes, thereby enabling scientists to turn their attention to ways of predicting them in advance.

How It Works

Short-Range Predictions. A variety of methods has been used to predict earthquakes on a short-term basis, and several of them are related to changes in underground water. Levels in wells may change, turbidity may appear, or water temperatures may rise. Chemical changes have also been noted. Kobe, Japan, experienced a devastating earthquake in 1995. Subsequent to the earthquake, an analysis of the water produced at the city's bottling plant revealed that the chemistry of the water had been steadily changing for two weeks before the earthquake. The water at the bottling plant is being studied daily for warnings of the next earthquake.

Mexico City uses a different method of predicting earthquakes. Fault lines are located along the Pacific coast, 300 kilometers away from Mexico City. A network of seismographs is used to determine that an earthquake has taken place, and that information is transmitted by satellite radio to the city, warning residents 50 to 80 seconds before the damaging waves arrive. Fault lines lie near and under the Los Angeles and Tokyo metropolitan areas, so there is less time to prepare for an earthquake, but seismographs around the city can detect the less destructive primary waves 10 to 15 seconds before the destructive secondary waves arrive. This gives computerized control systems enough time to shut off the power to specified industries and close down the mass transit systems.

Changes in the behavior of animals before an earthquake has been noted in some instances, most notably before the 1906 San Francisco earthquake. The earthquake occurred around five o'clock in the morning, and afterward residents reported that dogs at the pound had howled incessantly from midnight on and that horses had stamped and neighed in their livery stables during that same period. Presumably the animals heard high-pitched sounds, not audible to humans, caused by dilation of the rock preliminary to the final breaking.

Dilatancy, the opening of small cracks in the rocks of a fault zone before breakage, may cause other signals that an earthquake is about to occur. These signals include the tremors known as foreshocks, which often precede major earthquakes, as well as changes in the bedrock's magnetism, electrical resistance, or ability to transmit seismic waves. Radon gas may also be released into wells, and in some instances, changes in the land surface will also be noted. These changes can be monitored using sensitive instruments known as tiltmeters. Because earthquakes often accompany volcanic eruptions, any volcano overlooking a major city—such as Vesuvius beside Naples, Italy, or Mount Rainier next to Seattle, Washington—is monitored daily by seismographs, watching for the slightest change in vibrations that might indicate that an eruption is imminent. Other warning signs from a volcano include steam clouds coming from the crater, ash falling in the surrounding area, tilting of the volcano's flanks as magma swells into the crater, and changes in water features around the mountain, such as springs and wells.

Long-Range Predictions. The historical record of an area's previous earthquakes is useful in making long-range predictions of future shocks and is helpful in the preparation of seismic risk maps. A seismic risk map of the United States would show the West Coast, Alaska, and Hawaii as the highest risk zones. A repeated pattern of earthquakes is also helpful in making predictions. Parkfield, California, has experienced a magnitude 6 or 8 earthquake about every twenty-two years since 1857. Parkfield experienced an earthquake on September 28, 2004, so the town might expect its next earthquake in 2026. Not all segments of a fault break at the same time, and the inactive segments are known as seismic gaps. People living in areas where the seismic risk is high need to remember that an earthquake is possible at any time.

Careful fieldwork by geologists is another way to predict future shocks. The study of a fault trace across the landscape may reveal offset features that provide clues as to how often and when earthquake movements have taken place. Trenches may be dug for an examination of the soil layers, and a pilot hole is being drilled at Parkfield for a study of the San Andreas fault at depth.

Seismologist have begun using GPS measurements to analyze the differential creep on two sides of a fault. Using such measurements, in April, 2008, scientists predicted that the Enriquillo fault, near Haiti's capital city of Port-au-Prince, was capable of a magnitude 7.2 earthquake. Just twenty months later, in January, 2010, a devastating 7.0 earthquake did occur.

Applications and Products

Construction in earthquake-prone areas requires special precautions, and attention must be given to the type of structure planned. Some structures, such as nuclear power plants and large dams, should never be built in areas with active faults. Engineers and developers must determine the type of earthquakes affecting the area, how their vibrations might affect the intended structure, and how long the shaking could last. Knowledge of the area's geology is critical because soil and bedrock characteristics can influence how an earthquake will affect structures. For earthquakes with magnitudes up to 5, damage should be slight; for earthquakes of magnitudes 5 to 7, damage should be easily repaired; and for magnitude 7 or higher earthquakes, the structure must not collapse although significant repairs, or even demolition, may be required.

Nature of the Substratum. Structures built on solid bedrock invariably suffer less damage in an earthquake than structures built on filled land or water-saturated sediments. Filled land or water-saturated sediments lose their cohesiveness because their loose structure amplifies the vibrations from the earthquake. They can begin behaving like a liquid, in a process known as liquefaction, causing severe damage to the structures on the surface. Liquefaction occurred during a 7.5 magnitude earthquake in Niigata, Japan, on June 16, 1964, and during a magnitude 9.2 earthquake in Anchorage, Alaska, on March 18, 1964. In Niigata, part of which was built on sandy soil, hundreds of buildings, including multistory concrete-reinforced apartment buildings, tilted and fell over. In Anchorage, clay beds beneath the Turnagain Heights subdivision liquefied, and more than 200 acres covered with homes slid toward the sea. The damage to homes, streets, sidewalks, water and sewer lines, and electrical systems was massive.

Proper Building Materials. Adobe and mud-walled structures are the weakest buildings and invariably collapse during severe earthquakes. Stucco, unreinforced brick, and concrete-block structures also suffer severe damage; their walls prove inflexible to shaking once the mortar that binds them is broken. Surprisingly, wooden structures survive earthquakes quite well. Wood is lighter weight than metal or concrete, and if the frame of the building is properly secured to the ground, the nails and screws in the wood provide enough flexibility to resist shaking. Wood is preferable for smaller private homes in earthquake-prone areas, although concrete can be used for larger structures, provided it is heavily reinforced with steel bars.

Construction of Larger Buildings. The most important consideration in earthquake-proof construction is that the foundation, floors, walls, and roof of a building be tied together to withstand both horizontal and vertical stresses. Open lower floors are hazardous and should be reinforced with added pillars or diagonal bracing. Tall buildings and skyscrapers need to sway to prevent cracking, and the use of diagonal steel beams will permit this. Such buildings may even be placed on a layered steel or rubber structure that acts as a shock absorber to decrease such swaying. A good example of this is the pyramid-shaped Transamerica building in San Francisco. In addition to its tapering shape, which reduces the amount of mass in the higher floors, the skyscraper sits on a two-story-high cushion of massive diagonal trusses, designed to help the building absorb earthquake vibrations.

Retrofitting Older Structures. Many older structures are being reinforced to provide protection during earthquakes using a process known as retrofitting. The object is to make the existing structure more resistant to stresses, ground motion, and possible soil failure. High-strength steel, fiber-reinforced polymers, or fiber-reinforced concrete can be used to build new structural walls, and columns can be jacketed with steel. Stronger columns can be added and diagonal bracing placed on all floors. Foundation work might include jacking up the building to place springs or other flexible materials between it and the ground.

Quake Preparedness. People in earthquake-prone zones need to be ready to live for long periods without electricity and other services following a severe shock. Battery-powered radios, flashlights, and first-aid kits are essential, and homeowners need to know the location of the fuse box and shut-off valves for water and gas. Heavy appliances must be securely anchored to the floor, and shelves, bookcases, and cupboards firmly attached to walls. Heavy objects should not be placed on top shelves. Families need to have a plan for reuniting in the event that members become separated during an earthquake. During the earthquake itself, people who are outside should remain outside, away from possible falling objects, and people who are inside should seek shelter under a heavy piece of furniture. Those driving cars should come to a stop, away from overhead objects.

Impact on Industry

Most research related to earthquake prediction is centered in the United States, Russia, Japan, and China because these countries have experienced serious earthquakes in the past. Research includes laboratory and field studies of rock behavior before, during, and after earthquakes, as well as the monitoring of activity along major faults. Both government agencies and universities conduct this research.

Government Agencies. The United States Geological Survey (USGS) plays a major role in earthquake studies and earthquake prediction for the United States. In addition to providing general information regarding earthquakes, it makes available data on the latest earthquakes in the United States and around the world. For selected areas, such as California and Nevada where the potential for earthquakes is high, the USGS provides an index map, updated constantly, showing where the latest earthquake activity has been observed. The agency also makes short-range predictions of future earthquake activity, such as the probability of strong shaking in critical areas within the next twenty-four hours, and estimates of the probability of aftershocks in areas where earthquakes have recently occurred. The USGS also makes long-range predictions of future earthquake activity. It forecasts a 62 percent probability of a magnitude 6.7 or greater earthquake, capable of causing widespread damage, striking the San Francisco Bay Area before 2032.

The Alaska Earthquake Information Center (AEIC) provides services similar to those provided by the USGS, which serves primarily the lower forty-eight states. The center is important for Alaska residents because the state is just as prone to earthquakes as California, having had the world's second largest earthquake in 1964. The AEIC receives daily reports of earthquakes from a network of more than four hundred seismograph stations and makes these available on its Web site. The center makes predictions of future earthquakes, although shocks of magnitude 8 and lower are so common that no effort is made to predict them. For earthquakes of magnitude 8 and greater, predictions are made when possible, giving the expected location, magnitude, and time.

A third government agency involved in earthquake monitoring and prediction is the National Oceanographic and Atmospheric Administration (NOAA), which operates tsunami warning centers in Hawaii and Palmer, Alaska. Because of the great loss of life during the Indian Ocean tsunami of 2004, the Hawaii warning center has expanded its coverage to the Indian Ocean and the Caribbean until centers for those two areas can be created. The Alaska center serves the coastal areas of Canada and the United States, but not Hawaii.

Academic Research. The California Institute of Technology (Caltech) is a leading institution involved in earthquake research and prediction on the West Coast of the United States. Charles Richter and Beno Gutenberg developed the Richter scale for measuring the magnitude of earthquakes at Caltech in the 1930's. Geophysicists in the division

Fascinating Facts About Earthquake Prediction

- According to an ancient piece of Chinese folklore, before an earthquake, hibernating snakes come out of their holes, sometimes freezing to death in the cold.
- During the 1906 San Francisco earthquake, a cow fell into a small, shallow crack that opened up in the ground and was buried with only her tail protruding.
- A series of strong earthquakes occurred near New Madrid, Missouri, in 1811 and 1812. They were so severe that they made church bells ring in Boston and caused the Mississippi River to flow in the opposite direction during the shaking.
- During the massive 8.8 magnitude earthquake in Chile on February 27, 2010, the city of Concepción moved 3 meters to the west.
- The famed tenor, Enrico Caruso, was in San Francisco for a performance at the time of the 1906 earthquake. Terrified, he ordered his trunks packed, and at great expense, he managed to flee the city, carrying a framed photograph of President Theodore Roosevelt autographed to himself as his identification. He vowed never to sing in the city again and never did.
- The city of Los Angeles is moving north toward San Francisco along the San Andreas fault at the rate of about 1 centimeter a year. If movement continues at this rate, in millions of years, the two cities will meet.

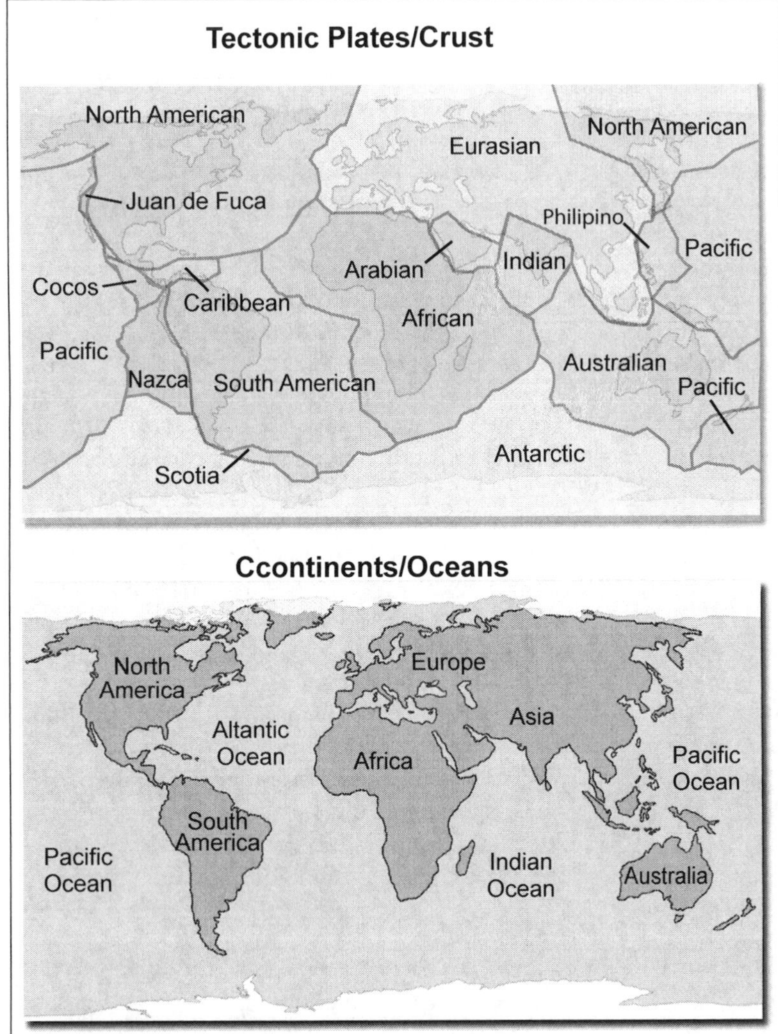

Earthquakes occur along the boundaries of tectonic plates, which span both continents and oceans.

of geological and planetary sciences supervise Caltech's seismological laboratory and also help maintain the Southern California Seismic Network and the Southern California Earthquake Data Center.

On the East Coast of the United States, Columbia University's Lamont-Doherty Earth Observatory also has a long tradition of earthquake research. Besides performing observational studies in the field, scientists in the seismology, geology, and tectonophysics division operate the Lamont-Doherty Cooperative Seismograph Network and also contribute to research for the Center for Hazards and Risk at Columbia's Earth Institute.

Careers and Course Work

A basic course in general geology is the logical starting point for a student interested in pursuing a career related to earthquake prediction. Most colleges and universities include such a course among their undergraduate offerings. This course provides the student with basic factual information related to faulting, which is the cause of most earthquakes, and also provides an introduction to the seismograph, which is the instrument scientists use to measure the strength of earthquakes and determine where they have taken place. Graduate study is probably a must for students who wish to obtain research positions in the field of earthquake prediction. Useful courses that the student can take in graduate school include structural geology, which deals with the types of faulting and their causes, and geophysics, which details the types of earthquake waves and how to interpret them using the seismograph. One course in a related field that is useful in earthquake prediction is engineering and materials science. This course provides important information regarding building codes for earthquake-prone areas and appropriate land-use planning for them. Soil science is another useful course because information about the past history of faults is often obtained by digging trenches through them. For students seeking employment opportunities in fields related to earthquake prediction, government agencies and the research departments of colleges and universities would be the logical place to start.

Social Context and Future Prospects

The importance of predicting earthquakes in advance of their occurrence cannot be stressed too greatly. Many of the world's best known cities lie near, or directly on top of, one or more major through-going faults, and the destruction resulting from movement on such faults can be catastrophic. San Francisco and Oakland, California, for example, lie in a narrow sliver of crust, bordered on one side by the San Andreas fault and on the other by the equally dangerous Hayward fault. Destructive earthquakes are likely to take place in Mexico City, Los Angeles, and Tokyo, which have had major earthquakes in the past and need adequate warning systems. Cities that have not had severe shocks in the recent past but are thought to be at high risk are Seattle and Istanbul. Seattle has not had a major earthquake in several hundred years, but geologists doing fieldwork around the city have found evidence that a massive earthquake took place there in the 1700's. Seattle residents need to prepare for a repeat of such an earthquake. Istanbul has not had a major earthquake in many years, but geologists point out that it is at the west end of a fault similar to the San Andreas. Fourteen earthquakes with magnitudes of 6 or higher have taken place along this fault since 1936, with each earthquake moving closer to the city. Despite all the advances that have been made in long-range earthquake prediction, scientists can still give people only a vague idea of when a large earthquake will hit their area.

Donald W. Lovejoy, B.S., M.S., Ph.D.

Further Reading

Berke, Philip R., and Timothy Beatley. *Planning for Earthquakes: Risk, Politics, and Policy.* Baltimore: The Johns Hopkins University Press, 1992. Excellent recommendations for land-use planning and construction techniques in earthquake-prone areas. Contains many helpful photographs illustrating the types of hazardous situations discussed.

Bolt, Bruce A. *Earthquakes.* 5th ed. New York: W. H. Freeman, 2006. Written for the lay reader. The chapter "Events that Precede Earthquakes" is a thorough summary of ways to predict earthquakes. Useful diagrams and photographs are included.

Chester, Ray. *Furnace of Creation Cradle of Destruction: A Journey to the Birthplace of Earthquakes, Volcanoes, and Tsunamis.* New York: AMACM, 2008. Explains plate tectonics and faulting, and how they are capable of causing earthquakes, volcanoes, and tsunamis. The section on "Mitigating Against Earthquakes" is of particular interest.

Hough, Susan Elizabeth. *Predicting the Unpredictable: The Tumultuous Science of Earthquake Prediction.* Princeton, N.J.: Princeton University Press, 2010. Examines the science of earthquake prediction, including early attempts, the Chinese prediction efforts, spectacular failures, and social and political issues.

_______. *Richter Scale: Measure of an Earthquake, Measure of a Man.* Princeton, N.J.: Princeton University Press, 2007. The life of the man who invented the Richter scale. Contains a chapter on the problems relating to predicting California earthquakes.

Winchester, Simon. *A Crack in the Edge of the World: America and the Great California Earthquake of 1906.* New York: HarperCollins, 2006. A highly readable account of the most famous earthquake in the United States. Excellent eyewitness accounts and information regarding the behavior of animals just before the earthquake.

Yeats, Robert S., Kerry Sieh, and Clarence R. Allen. *The Geology of Earthquakes.* New York: Oxford University Press, 1997. A comprehensive analysis of the types of earthquakes, with a chapter on assessing the risk of future movement on a fault. Useful diagrams and photographs.

Web Sites

Alaska Earthquake Information Center
http://www.aeic.alaska.edu

National Oceanic and Atmospheric Administration
Earthquakes
http://www.noaawatch.gov/themes/quake.php

U.S. Geological Survey
http://earthquakes.usgs.gov

See also: Earthquake Engineering; Engineering Seismology; Risk Analysis and Management.

ECOLOGICAL ENGINEERING

FIELDS OF STUDY

Chemistry; physics; calculus; biology; agriculture; physiology; genetics; statistics; electronics; GIS systems; geography; geology; hydrology; marine science; biotechnology; ethics; public policy.

SUMMARY

Ecological engineers design sustainable systems using ecological principles to create, restore, and conserve natural systems. The multidisciplinary synergy of the engineering sciences and biological knowledge makes it possible to design and maintain strong, self-evolving ecosystems to support a variety of life forms in a wide range of habitats. Human communities are planned so that they contribute to the balanced flow of energy and products to the surrounding environment. Plant and animal communities are created or restored, often in keeping with regulatory guidelines that mandate the restoration of particular locales to their original, prehuman conditions. Scales of applicability range from microscopic exchanges to the processes of much larger ecosystems.

KEY TERMS AND CONCEPTS

- **Abiotic Environment:** Nonliving and chemical factors that influence the organisms in a biotic environment.
- **Agroecosystem:** Multilevel assemblage of ecological processes, including microbial interactions within soils; the relationships of plants, crops, and herd animals; and the study of farming landscapes and communities.
- **Biotic Environment:** Living organisms within a particular environment and their relationships with one another to form a particular ecosystem.
- **Drilosphere:** All soils affected by earthworm activities, including external structures (middens, burrows, diapause chambers, surface and below-ground casts) and the earthworm's gut.
- **Horizontal Gene Transfer:** Any process that allows an organism to incorporate genetic material from an organism other than a parent or outside its own species; also known as lateral gene transfer.
- **Keystone and Foundation Species:** Species that define the structure and function of an ecosystem; for example, many forests are endangered as pathogens and insects destroy particular foundation tree species on which the forest ecosystem depends for coherence.
- **Laser Scanning:** Laser-based instrumentation used to measure complete environmental systems, including fixed beam, rotation beam, and distance measurers.
- **Light Detection and Ranging (Lidar):** Type of laser scanning used as airborne laser systems to take measurements of landscapes for the collection of detailed spatial data sets.
- **Macroecology:** Study of large-scale patterns and processes and how they relate to local assemblage structures; principles of evolution, ecology, and historical contingency are referenced to explain species richness, range size, abundance, and body size of particular life species.
- **Metal Toxicity:** Inhibition of microbial activity by certain metals—including arsenic, barium, cadmium, chromium, lead, mercury, and nickel—that interact with cellular proteins in complex ways.
- **Microclimate:** Small but distinct climate zone within a larger climate area; slight differences in landscape climates (frequently due to shade trees or mountains and valleys) allow for variegated and unusual species to occupy a common terrain.
- **Micrometeorology:** Study of the atmosphere just above the ground.
- **Mobile Genetic Element:** Small piece of DNA that can replicate and insert copies at random locations within cells.
- **Rhizosphere:** Area of soil that surrounds the roots of plants.
- **Soil Organic Matter (SOM) Dynamics:** Response of soil to its environment; the decomposition of organic residue by microorganisms and its alteration in agroecosystems are essential indicators of soil quality.
- **Soil Properties:** Characteristics that define the soil of a locale, including measurements of color, texture, and acidity; soil properties influence nutrient and water uptake and the disposition of organic materials.

- **Terrestrial Carbon Sequestration:** Photosynthetic process of removing carbon from the atmosphere and storing it in biomass.
- **Urban Ecology:** Impact of urban development on native ecosystems.

Definition and Basic Principles

The field of ecological engineering represents the synthesis of a variety of knowledge sets coordinated to address the challenges of holistic ecosystem design. Ecological engineering seeks to redress the fragmentation of scientific thought and activity, particularly in reference to the natural world and its processes. The term was first coined by Howard T. Odum, a prolific writer and researcher noted for founding the Center for Wetlands at the University of Florida. He was the son of Howard W. Odum, a noted sociologist and director of the Institute for Research in the Social Sciences at the University of North Carolina. The young Odum and his older brother, Eugene, helped establish the intellectual and scientific foundations of ecology in the twentieth century. The Odum brothers coauthored *Fundamentals of Ecology*, a successful textbook published in 1953 in which they state that mathematics and statistics were essential to the development of a unified system of ecosystem dynamics.

During the nineteenth and twentieth centuries, ecologists made substantial contributions to the knowledge of undisturbed ecosystems. Therefore, it is only within the past century that human relationships have been taken into account in the study of the evolution of natural systems. In 2004, the Ecological Society of America published a report to its governing board outlining a set of responses to the growing need for including sustainable ecological precepts in all phases of land use and development. Plans for a sustainable future are based on four basic principles:

- The rise of human-dominated ecosystems is inevitable.
- Ecosystems of the future will include a variety of conserved, created, and restored landscapes.
- Ecological science must become a core body of knowledge influencing global economics and policies.
- New and unprecedented regional and global partnerships among private and public entities are necessary to promote sustainable life systems.

Practitioners are careful to point out that ecological engineering is not synonymous with environmental engineering or biotechnology. Ecological engineers rely on the self-organizing capacities of ecosystem dynamics to create, restore, or conserve particular communities and habitats. Ecological-systems designers have a wealth of research materials available for consultation and comparison. These include efforts in habitat reconstruction, stream and wetland restoration, and a variety of wastewater, reclamation, and conservation projects. The scientific practices of observation, measurement, and documentation of the complex procedures and results of an ecosystem project are essential. The results of many restoration projects provide valuable information for understanding the processes of succession, the formation of biological macrocommunities in particular locales, and their relationships within broader ecosystems. Time is a fundamental component of ecosystem design. Many projects are undertaken with the realization that their fulfillment will exceed the lifetimes of their founders. Biodiversity, adaptation systems, and succession patterns are important dimensions of ecosystem design that require time.

Background and History

The use of myriad tools and technologies to construct and engineer life systems is an ancient human practice and a key factor in the evolution of the human species. Many tools and technologies are artifacts of cultures rooted in particular biological communities. Some sustainable traditions have been practiced for centuries and have made important contributions to the design and maintenance of contemporary ecosystems.

In the nineteenth century, the Industrial Revolution accelerated the rate of technological innovation in profound ways. Awareness of the aesthetic value of the natural landscape deepened as whole cultures and biological species disappeared in the wake of rapid urbanization, progressive conservation, and the colonization of the wilderness. Advances in urbanization; the consumption of fossil fuels; the mass fabrication of commodities; the mechanization of agriculture; the transformation of wetlands and deserts into vast farmlands; the unregulated extraction of minerals, metals, and inorganic materials for industrial expansion; increased demands for hydroelectric and nuclear power; a rapidly developing transportation and communications network; widespread deforestation; the unprotected destruction of wildlife; the release of radioactive isotopes, airborne pesticides, and vehicle emissions; and the creation of synthetic chemicals and materials in the wake of World War I and World War II ravaged European and American landscapes. Accelerated industrial production in the postwar period lifted economies while creating vulgar panoramas of urban blight: air and waterways darkened with toxic industrial, agricultural, and residential wastes; scarred mountains and forests; and a host of health and sanitation issues directly related to the toxic environment of the modern world.

The natural world is the subject of many of the supreme artistic and literary expressions of human culture. It is the basis of unique knowledge systems, the value of which is of particular importance to local communities. Since antiquity, classical treatises on the subject of natural history have included detailed studies that have formed the foundation for continuing work in the fields of medicine, chemistry, physics, biology, botany, mineralogy, geography, astronomy, and geology. Natural historians made remarkable contributions to the study of natural systems in the eighteenth and nineteenth centuries. The concept of evolution is an ever-present thread in the dialogue of natural philosophers. That tendril of thought took root in the discourse that followed the publication of *The Origin of Species by Means of Natural Selection* by Charles Darwin in 1859. As of 2011, evolution is understood as a complex process of genetic variation. That reality drives early twenty-first century research in ecology.

The environmental movement of the twentieth century created the grassroots momentum for a global reassessment of the controversial roles of science and technology in designing structures and processes that are functional, aesthetically pleasing, and sustainable. Environmental justice was an important topic of sociopolitical inquiry. Following the enactment of the National Environmental Policy Act in 1970, tens of thousands of regulations were enforced. Under the Toxic Substances Control Act of 1976, more than 80,000 toxic substances were registered by the Environmental Protection Agency (EPA). Similar standards have been adopted worldwide. Extensive research in pollution control provided new sources of expertise needed to remediate critical environmental conditions. Advances in computational technologies greatly facilitated the creation of detailed data sets, microassays, simulations, and measurement technologies. Extensive land surveys helped to substantiate the terms and application of soil and marine biotechnologies. Thermophiles and acidophils, xenobiotics, endocrine disrupters, bioaccumulation, biofilters, bioscrubbers, activated sludge, wastewater stabilization ponds; biosubstitutions, terrestrial phyto systems, and genetic manipulation: These and other terms form a common vocabulary and are found in all substantive research and textbooks addressing topics of environmental engineering and biotechnology.

How It Works

Ecological engineering is know-how applied to the design of a variety of environments. The mechanics of ecological self-maintenance are the foundations of healthy ecosystems design. Practitioners adopt a wide variety of natural techniques and materials to create, manage, or remediate a range of ecosystems.

Technologies rely on the chemical processes of sunlight, soil, temperature, and water in the design of living landscapes that assist in ecosystem management. For instance, constructed wetlands are designed so that natural biofilms and open spaces are formed to break down pollutants into usable bioproducts. Plant evaporation and transpiration, composting, and biogas production are other natural processes with a wide range of applications worldwide. The maintenance of designed ecosystems is similar to that of horticulture. This includes the routine maintenance of beds and screens and sampling and documentation of fluids and solids. The treatment of wastewater effluents in natural systems includes the removal of large solids, the aerobic and anaerobic treatment of organic materials, and regular sampling of safely reusable organic by-products.

Applications and Products

The ecological engineering literature embraces ecological technologies compatible with cultures and practices around the globe. Biological processes are the basis for these technologies. The complexity of biological relationships is the essential reality that all ecological engineers strive to replicate and support.

Phytoengineering. Phytoengineering technologies rely on plants as primary ecosystem providers. Designs and processes are coordinated for wastewater treatment, environmental remediation, wetlands remediation, and sustainable processes for industry, agriculture, and urban communities. Traditional biotechnologies favored the removal or destruction of contaminated soils. Phytoremediation is a successful in situ alternative. Plants work with contaminated soils to decompose toxins, accumulate chemical wastes, and create nutrient-rich composting materials.

Natural Wastewater Treatment Systems. Wastewater and storm-water effluents can be treated and reused in natural systems of plants, algae, and other living organisms. Sequenced constructed ecosystems mimic natural processes of land and water habitats. Applications include greenhouses, rain gardens, aquatic systems, and wetlands remediation.

Wetlands Remediation. For some localities, wetlands are an essential component of wastewater treatment. In the Northern European countries of Denmark, Norway, Finland, and Sweden, full-scale filter-bed systems are used to recapture phosphorus from septic effluents. In the European Union, groundwater is treated as a living ecosystem. Microorganisms and subterranean fauna are valuable indicators of the health of a particular source. The mapping of aquifers and other groundwater structures is an essential prerequisite for the accurate assessment and classification of these habitats. Similar assessments of groundwater resources and dependent wetlands, legislated under the Water Framework Directive, set in motion a systematic testing of all groundwater-dependent water resources in Europe. In April, 2009, the International Commission on Groundwater of the International Association of Hydrological Sciences and the UNESCO Division of Water Sciences convened a special session to evaluate the results of this testing program. Sweden, Denmark, Norway, Finland, the Netherlands, England, Wales, Scotland, and Austria participated in the consortium to assess risks of damage to related wetlands. Groundwater salinity is of particular concern to residents of Ravenna, Italy, where subsidence has introduced saltwater into groundwaters that feed pine forests. The forests are dying as a consequence of these shifts in water quality.

Green Roof Technologies. Roofs constructed of living plants and grasses are recognized for their ability to absorb sunlight and serve as effective solar wastewater evaporation systems. Sod roofs were ubiquitous features of Northern European landscapes, where birch bark and turf grasses were plentiful. Many plant, flower, bird, and insect species thrived on established terraces. As of 2011, some species are found only in green roof habitats. Nearly 10 percent of all German homes have green roofs, and 50 percent of all new construction in Germany must include vegetative covering. Green roofs are an important component of that compliance. In Switzerland flat green roof vegetation is required for renovated structures. Similar innovations are occurring in the United States.

Ecological Sanitation. Composting toilets provide a safe and healthy option for sanitation systems worldwide. It relies on "dry" biochemical processes that do not require water. Aerobic decomposition of human feces is a viable technology particularly in rural and desert communities where water is unavailable for flushing toilets.

Agroecology. Self-sustaining agricultural systems engineered to accommodate climate change are of great interest to nations throughout the world. Agroecology addresses these systemic concerns with the intent of balancing crops and animal stocks to maximize their relationships to their environment. This includes impacts on soil quality, effective pest control, discharges into water supplies, and the release of noxious gases and particulate matter into the atmosphere. For example, in southern Mexico, researchers are working with the Lacandon Maya to recover indigenous agroforestry systems of soil fertility and rainforest conservation. These centuries-old practices included the collection of plant species and management of the succession processes of the forest. In Europe and the Americas, similar efforts are under way to restore vast tracts of grasslands and their species and populations.

Light Detection and Ranging (Lidar) Systems. Laser-generated light beams are unique in that they are emitted in a single direction with a constant wavelength and amplitude and total phase correspondence. These characteristics allow for very detailed mappings of surfaces, topical variations, the height of surface objects, and the physical characteristics of different terrains. Low-level and high-level airborne laser scanning provide data of varying resolution. Digital terrain models (DTMs) provide valuable information about forest roads and vegetation, changes in beach topography, channel flows of river floodplains, as well as monitor flood events.

Wireless Sensor Networks (WSNs). Computation and the use of wireless sensors have greatly enhanced scientific knowledge about soil microenvironments. Light, humidity, and temperature in a particular locale can be measured using high-resolution grids of autonomous sensors organized to provide minute-by-minute descriptions of the soil's condition.

Atlantic Rainforest Sensor Net Research. In 2009, Microsoft Research collaborated with researchers from The Johns Hopkins University, the Universidade de São Paolo, the Fundação de Amparo à Pesquisa do Estatdo de São Paulo (FAPESP), and the Instituto Nacional de Pesquisas Espaciais (INPE, the Brazilian National Institute for Space Research) to collect data about the microecology of the Serra do Mar rainforest located in the southeast corner of Brazil. To collect the data, a grid of hundreds of wireless sensors collected 18 million data points over a four-week period. It is hoped that similar data sets will help researchers better understand the relationship of the rainforest to its environment.

Ecosystem Bioinformatics (Ecoinformatics). Undergraduate courses in ecological or ecosystem bioinformatics teach the analysis of gene sequencing and expression within microbial communities. Increasing knowledge in molecular biology and the capacities of computer data systems make it possible to construct algorithms to calculate the mechanics of gene expression within biological systems.

Impact on Industry

There is an urgent need for of eco-friendly partnerships between private industry and public welfare institutions. Effective communications systems and critical-thinking skills are needed for the broad dissemination of ecological principles across traditional disciplines. The formation of public values that take into consideration the effect of human activity on other life systems should be part of the essential core curriculum at all levels of education. Such partnerships go beyond the local community to include international consortiums concerned with the full range of effects of commodities exchange, the genetic transfer of plants and animals, effective sanitation and waste-disposal practices, and the disciplined use and distribution of chemical substances.

Corporate leadership is an important driver for sustainable global development. Some leading chemical producers are developing novel production processes that reduce hazardous materials and effluents. These include innovative manufacturing systems, solventless processes, and new separation processes. The investment in "green chemistry" technologies is predicted to reap healthy savings, reducing the costs of hazardous-materials containment and on-site contaminant remediation.

Careers and Course Work

In the fall semester of 2007, Oregon State University offered the nation's first accredited ecological engineering degree program. Since then, other universities have begun offering similar accredited programs of study. These include Ohio State University; Purdue University; University of California, Berkeley; University of Maryland; Texas A&M University; and Washington State University.

The bachelor's degree in ecological engineering is remarkable for its breadth and multidisciplinary program of study. Typical requirements include core engineering courses in fluid mechanics, thermodynamics, circuits, vector calculus, differential equations, and mechanics. It will also include course work in ecology, organic chemistry, biology, microbiology, hydrology, biosystems modeling, statistical methods, and environmental technology and design. Graduate and doctoral programs in ecological engineering focus on issues, systems, and technologies related to ecological design. These include water-resource management and pollution control, biotechnology, environmental sensing, systems modeling, informatics, and soil and wetland treatment programs. Ecological engineering graduates pursue careers as environmental engineering project managers and risk advisers, researchers, statisticians, and land use and public policy consultants. Specific design projects include riparian restoration, ecological monitoring with sensor arrays, and the development of sustainable urban, industrial, and agricultural systems. Professionals play important roles in both the private and public sector. As environmental consultants they work in multidisciplinary teams to design and engineer systems for the remediation of wastewaters and the recycling of soil nutrients and contaminants. They work in federal agencies to plan systems for natural resource protection and utilization. Bioremediation, water conservation, and the protection of endangered species are important dimensions of healthy ecosystem design.

Fascinating Facts About Ecological Engineering

- Earthworms are among the most celebrated of all ecosystem engineers. They are noted for their amazing capacity to consume litter over large environments of forest and savannah, often consuming an entire seasonal leaf fall. Leaving behind tons of castings, earthworms have profound impacts on local soils and the microorganisms that inhabit them.
- Hydroponically grown sunflowers are used to absorb radioactive metals. Their hyper-accumulation capacities were an ideal solution for soil bioremediation at the Chernobyl nuclear site in Ukraine.
- Salt-tolerant plants, known as halophytes, can be used to restore damaged soils. Once toxic salt levels have been reduced, the plants die and other species of plants and grasses can be planted or restored to previously unusable lands. These technologies are particularly valuable in areas subject to urine accumulation.
- Macrophytes are large aquatic plants found in standing waters common in wetland habitats. Examples include aquatic angiosperms (flowering plants), pteridophytes (ferns), and bryophytes (mosses and liverworts). They are important sources of nutrient uptake. Harvesting the plants and removing decaying organic debris reduces the re-pollution of wetland resources.

SOCIAL CONTEXT AND FUTURE PROSPECTS

There is a universal recognition that environmental protection is not enough to address the continuing degradation of the natural environment. The pace of chemical production and distribution continues to accelerate to meet global demand for industrial and agricultural products. Natural resources continue to be depleted to meet the needs of expanding communities worldwide. Ecological engineering represents a paradigm shift that favors ecological principles as the foundation of global economic and cultural prosperity. Sustainable design is an essential feature of that paradigm and it is the basis of the practice of ecological engineering. Traditional engineering practices and technologies are adapted for use in a variety of self-sustaining ecosystem designs. Many systems accommodate human lifestyle patterns while protecting biodiversity. Other systems are designed to restore or remediate particular biological systems such as a forest, a river, a farmland or pasture, or an industrial site. Sustainable cities can be found in locations around the globe.

Victoria M. Breting-García, B.A., M.A.

FURTHER READING

Allen, T. F. H., M. Giampietro, and A. M. Little. "Distinguishing Ecological Engineering from Environmental Engineering." *Ecological Engineering* 20, no. 5 (2003): 389-407. Helps refine the definition of ecological engineering as a field the practices of which are distinct from those of the environmental engineer.

Cuddington, Kim, et al., eds. *Ecosystem Engineers: Plants to Protists.* Burlington, Mass.: Academic Press, 2007. Presents studies of nonhuman ecosystem engineers; provides a good background for understanding how ecosystems change over time.

Daily, Gretchen C., and Paul R. Ehrlich. "Managing Earth's Ecosystems: An Interdisciplinary Challenge." *Ecosystems* 2, no. 4 (July/August, 1999): 277-280. Provides a substantive framework for understanding the emerging links in science and technology that support ecosystem engineering efforts.

Kingsland, Sharon E. *The Evolution of American Ecology, 1890-2000.* Baltimore: The Johns Hopkins University Press, 2005. A good chronology for understanding the progression of study and observation that strengthened the field of ecology in the twentieth century.

Palmer, Margaret A., et al. "Ecological Science and Sustainability for the Twenty-first Century." *Frontiers in Ecology and the Environment* 3, no. 1 (February, 2005): 4-11. Presents a good conceptual understanding of the core assumptions that motivate the professional concerns of ecological engineers.

Spellman, Frank R., and Nancy E. Whiting. *Environmental Science and Technology: Concepts and Applications.* 2d ed. Lanham, Md.: Government Institutes, 2006. An excellent introduction to the definitions, concepts, and practices of environmental science and technology. These are the foundations for understanding the emergence of ecological engineering as a distinct profession located at the juncture of the biological and applied sciences.

Taylor, Walter P. "What Is Ecology and What Good Is

It?" *Ecology* 17, no. 3 (July, 1936): 333-346. Provides a valuable understanding of the history of ecological thought in the twentieth century.

Wargo, John. *Green Intelligence: Creating Environments That Protect Human Health.* New Haven, Conn.: Yale University Press, 2009. A biting update on research in environmental protection, emphasizing the hazards of synthetic chemicals.

Worster, Donald. *Nature's Economy: A History of Ecological Ideas.* 2d ed. New York: Cambridge University Press, 1994. This collection of classic essays in the field of ecology provides a strong foundation for understanding contemporary ecosystem technologies as a part of a flow of ideas and convictions that have flowered since the eighteenth century. The seminal contributions of Carolus Linnaeus, Gilbert White, Charles Darwin, Henry David Thoreau, Rachel Carson, Aldo Leopold, Eugene Odum, and others are presented in the light of twentieth-century research in environmental history.

Web Sites

Ecological Engineering Group
Ecological Engineers and Designers
http://www.ecological-engineering.com

Ecological Society of America
http://www.esa.org

Michigan State University
Green Roof Research Program
http://www.hrt.msu.edu/greenroof

Oregon State University
Biological and Ecological Engineering (BEE)
http://bee.oregonstate.edu

See also: Calculus

ELECTRICAL ENGINEERING

FIELDS OF STUDY

Physics; quantum physics; thermodynamics; chemistry; calculus; multivariable calculus; linear algebra; differential equations; statistics; electricity; electronics; computer science; computer programming; computer engineering; digital signal processing; materials science; magnetism; integrated circuit design engineering; biology; mechanical engineering; robotics; optics.

SUMMARY

Electrical engineering is a broad field ranging from the most elemental electrical devices to high-level electronic systems design. An electrical engineer is expected to have fundamental understanding of electricity and electrical devices as well as be a versatile computer programmer. All of the electronic devices that permeate modern living originate with an electrical engineer. Items such as garage-door openers and smart phones are based on the application of electrical theory. Even the computer tools, fabrication facilities, and math to describe it all is the purview of the electrical engineer. Within the field there are many specializations. Some focus on high-power analogue devices, while others focus on integrated circuit design or computer systems.

KEY TERMS AND CONCEPTS

- **Alternating Current (AC):** Current that alternates its potential difference, changes its rate of flow, and switches direction periodically.
- **Analogue:** Representation of signals as a continuous set of numbers such as reals.
- **Binary:** Counting system where there are only two digits, 0 and 1, which are best suited for numerical representations in digital applications.
- **Capacitance:** Measure of potential electrical charge in a device.
- **Charge:** Electrical property carried by all atomic particles (protons, neutrons, and electrons).
- **Current:** Flow of electrical charge from one region to another.
- **Digital:** Representation of signals as a discrete number, such as an integer.
- **Digital Signal Processing (DSP):** Mathematics that describes the processing of digital signals.
- **Direct Current (DC):** Current that flows in one direction only and does not change its potential difference.
- **Inductance:** Measure of a device's ability to store magnetic flux.
- **Integrated Circuit (IC):** Microscopic device where many transistors have been etched into the surface and then connected with wire.
- **Resistance:** Measure of how easily current can flow through a material.
- **Transistor:** Three-terminal device where one terminal controls the rate of flow between the other two.
- **Voltage:** Measure of electrical potential energy between two regions.

Definition and Basic Principles

Electrical engineering is the application of multiple disciplines converging to create simple or complex electrical systems. An electrical system can be as simple as a lightbulb, power supply, or switch and as complicated as the Internet, including all its hardware and software subcomponents. The spectrum and scale of electrical engineering is extremely diverse. At the atomic scale, electrical engineers can be found studying the electrical properties of electrons through materials. For example, silicon is an extremely important semiconductive material found in all integrated circuit (IC) devices, and knowing how to manipulate it is extremely important to those who work in microelectronics.

While electrical engineers need a fundamental background in basic electricity, many (if not most) electrical engineers do not deal directly with wires and devices, at least on a daily basis. An important subdiscipline in electrical engineering includes IC design engineering: A team of engineers are tasked with using computer software to design IC circuit schematics. These schematics are then passed through a series of verification steps (also done by electrical engineers) before being assembled. Because computers are ubiquitous, and the reliance on good computer programs to perform complicated operations is so important, electrical engineers are adept computer programmers as well. The steps would be the same in any of the subdisciplines of the field.

Background and History

Electrical engineering has its roots in the pioneering work of early experimenters in electricity in the eighteenth and nineteenth centuries, who lent their names to much of the nomenclature, such as French physicist André-Marie Ampère and Italian physicist Alessandro Volta. The title electrical engineer began appearing in the late nineteenth century, although to become an electrical engineer did not entail any special education or training, just ambition. After American inventor Thomas Edison's direct current (DC) lost the standards war to Croatian-born inventor Nicola Tesla's alternating current (AC), it was only a matter of time before AC power became standard in every household.

Vacuum tubes were used in electrical devices such as radios in the early twentieth century. The first computers were built using warehouses full of vacuum tubes. They required multiple technicians and programmers to operate because when one tube burst, computation could not begin until it had been identified and replaced.

The transistor was invented in 1947 by John Bardeen, Walter Brattain, and William Shockley, employees of Bell Laboratories. By soldering together boards of transistors, electrical engineers created the first modern computers in the 1960's. By the 1970's, integrated circuits were shrinking the size of computers and the purely electrical focus of the field.

As of 2011, electrical engineers dominate IC design and systems engineering, which include mainframes, personal computers, and cloud computing. There is, of course, still a demand for high-energy electrical devices, such as airplanes, tanks, and power plants, but because electricity has so many diverse uses, the field will continue to diversify as well.

How It Works

In a typical scenario, an electrical engineer, or a team of electrical engineers, will be tasked with designing an electrical device or system. It could be a computer, the component inside a computer, such as a central processing unit (CPU), a national power grid, an office intranet, a power supply for a jet, or an automobile ignition system. In each case, however, the electrical engineer's grasp on the fundamentals of the field are crucial.

Electricity. For any electrical application to work, it needs electricity. Once a device or system has been identified for assembly, the electrical engineer must know how it uses electricity. A computer will use low voltages for sensitive IC devices and higher ones for fans and disks. Inside the IC, electricity will be used as the edges of clock cycles that determine what its logical values are. A power grid will generate the electricity itself at a power plant, then transmit it at high voltage over a grid of transmission lines.

Electric Power. When it is determined how the device or application will use electricity, the source of that power must also be understood. Will it be a standard AC power outlet? Or a DC battery? To power a computer, the voltage must be stepped down to a lower voltage and converted to DC. To power a jet, the spinning turbines (which run on jet fuel) generate electricity, which can then be converted to DC and will then power the onboard electrical systems. In some cases, it's possible to design for what happens in the absence of power, such as the battery backup on an alarm clock or an office's backup generator. An interesting case is the hybrid motor of certain cars such as the Toyota Prius. It has both an electromechanical motor and an electric one. Switching the drivetrain seamlessly between the two is quite a feat of electrical and mechanical engineering.

Circuits. If the application under consideration has circuit components, then its circuitry must be designed and tested. To test the design, mock-ups are often built onto breadboards (plastic rows of contacts that allow wiring up a circuit to be done easily and quickly). An oscilloscope and voltmeter can be used to measure the signal and its strength at various nodes. Once the design is verified, if necessary the schematic can be sent to a fabricator and mass manufactured onto a circuit board.

Digital Logic. Often, an electrical engineer will not need to build the circuits themselves. Using computer design tools and tailored programming languages, an electrical engineer can create a system using logic blocks, then synthesize the design into a circuit. This is the method used for designing and fabricating application-specific integrated circuits (ASICs) and field-programmable gate arrays (FPGAs).

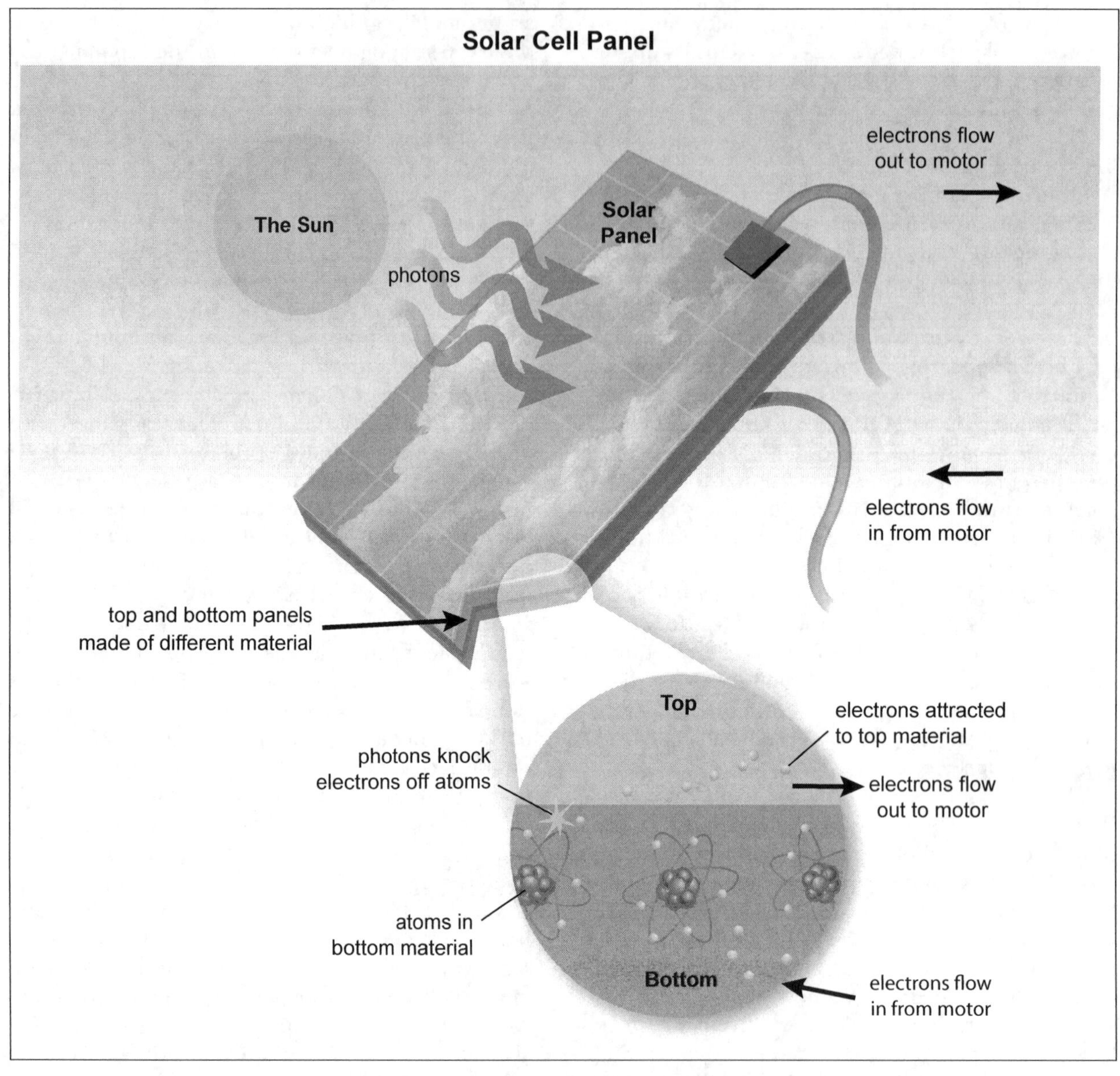

Solar panels are a collection of solar cells that convert light into electricity.

Digital Signal Processing (DSP). Since digital devices require digital signals, it is up to the electrical engineer to ensure that the correct signal is coming in and going out of the digital circuit block. If the incoming signal is analogue, it must be converted to digital via an analogue-to-digital converter, or if the circuit block can only process so much data at a time, the circuit block must be able to time slice the data into manageable chunks. A good example is an MP3 player: The data must be read from the disk while it is moving, converted to sound at a frequency humans can hear, played back at a normal rate, then converted to an analogue sound signal in the headphones. Each one of those steps involves DSP.

Computer Programming. Many of the steps above can be abstracted out to a computer programming language. For example, in a logical programming language such as Verilog, an electrical engineer can write lines of code that represent the logic. Another program can then convert it into the schematics of an IC block. A popular programming language called SPICE can simulate how a circuit will behave, saving the designer time by verifying the circuit works as expected before it is ever assembled.

Applications and Products

The products of electrical engineering are an integral part of our everyday life. Everything from cell phones and computers to stereos and electric lighting encompass the purview of the field.

For example, a cell phone has at every layer the mark of electrical engineering. An electrical engineer designed the hardware that runs the device. That hardware must be able to interface with the established communication channels designated for use. Thus, a firm knowledge of DSP and radio waves went into its design. The base stations with which the cell phone communicates were designed by electrical engineers. The network that allows them to work in concert is the latest incarnation of a century of study in electromagnetism. The digital logic that allows multiple phone conversations to occur at the same time on the same frequency was crafted by electrical engineers. The whole mobile experience integrates seamlessly into the existing landline grid. Even the preexisting technology (low voltage wire to every home) is an electrical engineering accomplishment—not to mention the power cable that charges it from a standard AC outlet.

One finds the handiwork of electrical engineers in such mundane devices as thermostats to the ubiquitous Internet, where everything from the network cards to the keyboards, screens, and software are crafted by electrical engineers. Electrical engineers are historically involved with electromagnetic devices as well, such as the electrical starter of a car or the turbines of a hydroelectric plant. Many devices that aid artists, such as sound recording and electronic musical instruments, are also the inspiration of electrical engineers.

Below is a sampling of the myriad electrical devices that are designed by electrical engineers.

Computers. Computer hardware and often computer software are designed by electrical engineers. The CPU and other ICs of the computer are the product of hundreds of electrical engineers working together to create ever-faster and more miniature devices. Many products can rightfully be considered computers, though they are not often thought of as such. Smart phones, video-game consoles, and even the controllers in modern automobiles are computers, as they all employ a microprocessor. Additionally, the peripherals that are required to interface with a computer have to be designed to work with the computer as well, such as printers, copiers, scanners, and specialty industrial and medical equipment.

Test Equipment. Although these devices are seldom seen by the general public, they are essential to keeping all the other electrical devices in the world

Fascinating Facts About Electrical Engineering

- The first transistor, invented in 1947 at the famous Bell Laboratories, fit in the palm of a hand.
- As of 2011, the smallest transistor is the size of a molecule, about a nanometer across, or 1/100,000 the thickness of a hair.
- In the 1890's Thomas Edison partnered with Harold Brown, a self-described electrical engineer, to shock dogs with either AC or DC current to see which was more fatal. Not surprisingly, Edison, a huge backer of DC, found AC to be deadly and DC not so.
- A Tesla coil, which is a high-voltage device, can shoot an arc of electricity in the air. The larger the coil, the bigger the arc. A small coil might spark only a few centimeters, but a large one can spark for a meter or more.
- The Great Northeast Blackout, which occurred on November 9, 1965, was the largest in U.S. history. It affected 30 million people, including 800,000 riders trapped in New York City's subways. It was caused by the failure of a single transmission line relay.
- In the lower 48 states, there are about 300,000 kilometers of interconnected transmission lines, operated by about 500 different companies.
- The term "debug" comes from the days of large vacuum-tube computers when bugs could quite literally get into the circuitry and cause shorts. To debug was to remove all the bugs.

working. For example, an oscilloscope can help an electrical engineer test and debug a failing circuit because it can show how various nodes are behaving relative to each other over time. A carpenter might use a wall scanner to find electrical wire, pipes, and studs enclosed behind a wall. A multimeter, which measures voltage, resistance, and current, is handy not just for electrical engineers but also for electricians and hobbyists.

Sound Amplifiers. Car stereos, home theaters, and electric guitars all have one thing in common: They all contain an amplifier. In the past, these have been purely analogue devices, but since the late twentieth century, digital amplifiers have supplanted their analogue brethren due to their ease of operation and size. Audiophiles, however, claim that analogue amplifies sound better.

Power Supplies. These can come in many sizes, both physically and in terms of power. Most people encounter a power supply as a black box plugged into an AC outlet with a cord that powers electrical devices such as a laptop, radio, or television. Inside each is a specially designed power inverter that converts AC power to the required volts and amperes of DC power.

Batteries. Thomas Edison is credited with creating the first portable battery, a rechargeable box that required only water once a week. Batteries are an electrochemical reaction, that is the realm of chemistry, and demonstrate how far afield electrical engineering can seem to go while remaining firmly grounded in its fundamentals. Battery technology is entering a new renaissance as the charge life is extending and the size is shrinking. Edison first marketed his "A" battery for use in electric cars before they went out of fashion. Electric cars that run on batteries may be making a comeback, and their cousin, the hybrid, runs on both batteries and combustion.

The Power Grid. This is one of the oldest accomplishments of electrical engineering. A massive nationwide interdependent network of transmission lines delivers power to every corner of the country. The power is generated at hydroelectric plants, coal plants, nuclear plants, and wind and solar plants. The whole thing works such that if any one section fails, the others can pick up the slack. Wind and solar pose particular challenges to the field, as wind and sunshine do not flow at a constant rate, but the power grid must deliver the same current and voltage at all times of day.

Electric Trains and Buses. Many major cities have some kind of public transportation that involves either an electrified rail, or bus wires, or both. These subways, light rails, and trolleys are an important part of municipal infrastructure, built on many of the same principles as the power grid, except that it is localized.

Automobiles. There are many electronic parts in a car. The first to emerge historically is the electric starter, obviating the hand crank. Once there was a battery in the car to power the starter, engineers came up with all sorts of other uses for it: headlamps, windshield wipers, interior lighting, a radio (and later tape and CD players), and the dubious car alarm, to name a few. The most important electrical component of modern automobiles is the computer-controlled fuel injector. This allows for the right amount of oxygen and fuel to be present in the engine for maximum fuel efficiency (or for maximum horsepower). The recent success of hybrids, and the potentially emerging market of all electric vehicles, means that there is still more electrical innovation to be had inside a century-old technology.

Medical Devices. Though specifically the domain of biomedical engineering, many, if not most, medical devices are designed by electrical engineers who have entered this subdiscipline. Computed axial tomography (CAT) scanners, X rays, ultrasound, and magnetic resonance imaging (MRI) machines all rely on electromagnetic and nuclear physics applied in an electrical setting (and controlled by electronics). These devices can be used to look into things other than human bodies as well. Researchers demonstrated that an MRI could determine if a block of cheese had properly aged.

Telecommunications. This used to be an international grid of telephone wires and cables connecting as many corners of the globe where wire could be strung. As of 2011, even the most remote outposts can communicate voice, data, and video thanks to advances in radio technology. The major innovation in this field has been the ability for multiple connections to ride the same signal. The original cell phone technology picked a tiny frequency for each of its users, thus limiting the number of total users to a fixed division in that band. Mobile communication has multiple users on the same frequency, which opens up the band to more users.

Broadcast Television and Radio. These

technologies are older but still relevant to the electrical engineer. Radio is as vibrant as ever, and ham radio is even experiencing a mini renaissance. While there may not be much room for innovation, electrical engineers must understand them to maintain them, as well as understand their derivative technologies.

Lighting. Light-emitting diodes (LEDs), are low-power alternatives to incandescent bulbs (the lightbulb that Thomas Edison invented). They are just transistors, but as they have grown smaller and more colors have been added to their spectrum, they have found their way into interior lighting, computer monitors, flashlights, indicator displays, and control panels.

Impact on Industry

New electrical devices are being introduced every day in a quantity too numerous to document. In 2006, consumer electronics alone generated $169 trillion in revenue. Nonetheless, electrical engineering has a strong public research and development component. Increasingly, businesses are partnering with universities to better capitalize on independent research. As of 2011, most IC design is done in the United States, Japan, and Europe, and most of the manufacturing is outsourced to Taiwan and Singapore. As wind and solar power become more popular, so does the global need for electrical engineers. Spain, Portugal, and Germany lead the European Union in solar panel use and production. In the United States, sunny states such as California give financial incentives to homes and businesses that incorporate solar.

University Research. University research is funded by the United States government in the forms of grants from organizations such as the Defense Advanced Research Projects Agency (DARPA) and the National Aeronautics and Space Administration (NASA). The National Science Foundation (NSF) indirectly supports research through fellowships. The rest of the funding comes from industry. Research can be directed at any of the subdisciplines of the field: different material for transistors or new mathematics for better DSP or unique circuit configuration that optimizes a function. Often, the research is directed at combining disciplines, such as circuits that perform faster DSP. DARPA is interested in security and weapons, such as better digital encryption and spy satellites. Solar power is also a popular area of research. There is a race to increase the performance of photovoltaic devices so that solar power can compete with gas, coal, and petroleum in terms of price per kilowatt hour. Universities that are heavily dependent on industry funding tend to research in areas that are of concern to their donors.

For example, Intel, the largest manufacturer of microprocessors, is a sponsor of various University of California electrical engineering departments and in 2011 announced their new three-dimensional transistor. The technology is based on original research first described by the University of California, Berkeley, in 2000 and funded by DARPA.

The Internet was originated from a Cold War era DARPA project. The United States wanted a communications network that would survive a first-strike nuclear assault. When the Cold War ended, the technology found widespread use in the civilian sphere. The Internet enabled universities to share research data and libraries a decade before it became a household commodity.

Business Sector. More than half of the electrical engineers employed in the United States are working in electronics. Consumer electronics include companies such as Apple, Sony, and Toshiba that make DVD players, video-game consoles, MP3 players, laptops, and computers. But the majority of the engineering takes place at the constituent component level. Chip manufacturers such as Intel and Advanced Micro Devices (AMD) are household names, but there are countless other companies producing all the other kinds of microchips that find their way into refrigerators, cars, network storage devices, cameras, and home lighting. The FPGA market alone was $2.75 billion, though few end consumers will ever know that they are in everything from photocopiers to cell phone base stations.

In the non-chip sector, there are behemoths such as General Electric, which make everything from lightbulbs to household appliances to jet engines. There are about 500 electric power companies in the contiguous forty-eight states of the United States. Because they are all connected to each other and the grid is aging, smart engineering is required to bring new sources online such as solar and wind. In 2009, the Federal Communications Commission (FCC) issued the National Broadband Plan, the goal of which is to bring broadband Internet access to every United

States citizen. Telecommunications companies such as AT&T and cable providers are competing fiercely to deliver ever-faster speeds at lower prices to fulfill this mandate.

Careers and Course Work

Electrical engineering requires a diverse breadth of background course work—math, physics, computer science, and electrical theory—and a desire to specialize while at the same time being flexible to work with other electrical engineers in their own areas of expertise. A bachelor of science degree in electrical engineering usually entails specialization after the general course work is completed. Specializations include circuit design, communications and networks, power systems, and computer science. A master's degree is generally not required for an electrical engineer to work in the industry, though it would be required to enter academia or to gain a deeper understanding in the specialization. An electrical engineer wishing to work as an electrical systems contractor will probably require professional engineer (PE) certification, which is issued by the state after one has several years of work experience and has passed the certification exam.

Careers in the field of electrical engineering are as diverse as its applications. Manufacturing uses electrical engineers to design and program industrial equipment. Telecommunications employs electrical engineers because of their understanding of DSP. More than half of all electrical engineers work in the microchip sector, which uses legions of electrical engineers to design, test, and fabricate ICs on a continually shrinking scale. Though these companies seem dissimilar—medical devices, smart phones, computers (any device that uses an IC)—they have their own staffs of electrical engineers that design, test, fabricate, and retest the devices.

Electrical engineers are being seen more and more in the role of computer scientist. The course work has been converging since the twentieth century. University electrical engineering and computer science departments may share lecturers between the two disciplines. Companies may use electrical engineers to solve a computer-programming problem in the hopes that the electrical engineer can debug both the hardware and software.

Social Context and Future Prospects

Electrical engineering may be the most underrecognized driving force behind modern living. Everything from the electrical revolution to the rise of the personal computer to the Internet and social networking has been initiated by electrical engineers. This field first brought electricity into our homes and then ushered in the age of transistors. Much of the new technology being developed is consumed as software and requires computer programmers. But the power grid, hardware, and Internet that powers it were designed by electrical engineers and maintained by electrical engineers.

As the field continues to diversify and the uses for electricity expands, the need for electrical engineers will expand, as will the demands placed on the knowledge base required to enter the field. Electrical engineers have been working in the biological sciences, a field rarely explored by the electrical engineer. The neurons that comprise the human brain are an electrical system, and it makes sense for both fields to embrace the knowledge acquired in the other.

Other disciplines rely on electrical engineering as the foundation. Robotics, for example, merge mechanical and electrical engineering. As robots move out of manufacturing plants and into our offices and homes, engineers with a strong understanding of the underlying physics are essential. Another related field, biomedical engineering, combines medicine and electrical engineering to produce lifesaving devices such pacemakers, defibrillators, and CAT scanners. As the population ages, the need for more advanced medical treatments and early detection devices becomes paramount. Green power initiatives will require electrical engineers with strong mechanical engineering and chemistry knowledge. If recent and past history are our guides, the next scientific revolution will likely come from electrical engineering.

Vincent Jorgensen, B.S.

Further Reading

Adhami, Reza, Peter M. Meenen III, and Dennis Hite. *Fundamental Concepts in Electrical and Computer Engineering with Practical Design Problems.* 2d ed. Boca Raton, Fla.: Universal Publishers, 2005. A well-illustrated guide to the kind of math required to analyze electrical circuits, followed by sections on circuits, digital logic, and DSP.

Davis, L. J. *Fleet Fire: Thomas Edison and the Pioneers of the Electric Revolution.* New York: Arcade, 2003. The stunning story of the pioneer electrical engineers, many self-taught, who ushered in the electric revolution.

Gibilisco, Stan. *Electricity Demystified.* New York: McGraw-Hill, 2005. A primer on electrical circuits and magnetism.

Mayergoyz, I. D., and W. Lawson. *Basic Electric Circuit Theory: A One-Semester Text.* San Diego: Academic Press, 1997. Introductory textbook to the fundamental concepts in electrical engineering. Includes examples and problems.

McNichol, Tom. *AC/DC: The Savage Tale of the First Standards War.* San Francisco: Jossy-Bass, 2006. The riveting story of the personalities in the AC/DC battle of the late nineteenth century, focusing on Thomas Edison and Nicola Tesla.

Shurkin, Joel N. *Broken Genius: The Rise and Fall of William Shockley, Creator of the Electronic Age.* New York: Macmillian, 2006. Biography of the Nobel Prize-winning electrical engineer and father of the Silicon Valley, who had the foresight to capitalize on invention of the transistor but ultimately went down in infamy and ruin.

Web Sites

Association for Computing Machinery
http://www.acm.org

Computer History Museum
http://www.computerhistory.org

Institute of Electrical and Electronics Engineers
http://www.ieee.org

National Society of Professional Engineers
http://www.nspe.org/index.html

See also: Bionics and Biomedical Engineering; Computer Engineering; Computer Networks; Computer Science; Mechanical Engineering; Robotics.

ELECTRICAL MEASUREMENT

FIELDS OF STUDY

Electronics; electronics technology; instrumentation; industrial machine maintenance; electrical engineering; metrology; avionics; physics; electrochemistry; robotics; electric power transmission and distribution services.

SUMMARY

Electrical measurement has three primary aspects: the definition of units to describe the electrical properties being measured; the modeling, design, and construction of instrumentation by which those units may be applied in the measurement process; and the use of measurement data to analyze the functioning of electric circuits. The measurement of any electrical property depends on the flow of electric current through a circuit. A circuit can exist under any conditions that permit the movement of electric charge, normally as electrons, from one point to another. In a controlled or constructed circuit, electrons move only in specific paths, and their movement serves useful functions.

KEY TERMS AND CONCEPTS

- **Ampere:** Measure of the rate at which electrons are passing through an electric circuit or some component of that circuit.
- **Analogue:** Electric current that is continuous and continuously variable in nature
- **Capacitance:** Measure of the ability of a nonconducting discontinuous structure to store electric charge in an active electric circuit.
- **Digital:** Electric current that flows in packets or bits, with each bit being of a specific magnitude and duration.
- **Electromagnetism:** Property of generating a magnetic field by the flow of electricity through a conductor.
- **Hertz:** Unit of measurement defined as exactly one complete cycle of any process over a time span of exactly one second.
- **Inductance:** Measure of the work performed by an electric current in generating a magnetic field as it passes through an induction coil.
- **Left-Hand Rule:** Rule of thumb in determining the direction of the north pole of the magnetic field in a helically wound electromagnetic coil.
- **Phase Difference:** Measure of the extent to which a cyclic waveform such as an alternating voltage follows or precedes another.
- **Resistance:** Measure of the ability of electric current to flow through a circuit or some component part of a circuit or device; the inverse of conductance.

Definition and Basic Principles

Electrical measurement refers to the quantification of electrical properties. As with all forms of measurement, these procedures provide values relative to defined standards. The basic electrical measurements are voltage, resistance, current, capacitance, and waveform analysis. Other electrical quantities such as inductance and power are generally not measured directly but are determined from the mathematical relationships that exist among actual measured properties of an electric circuit.

An electric circuit exists whenever physical conditions permit the movement of electrons from one location to another. It is important to note that the formation of a viable electric circuit can be entirely accidental and unexpected. For example, a bolt of lightning follows a viable electric circuit in the same way that electricity powering the lights in one's home follows a viable electric circuit.

Electrons flow in an electric circuit because of differences in electrical potential between one end of the circuit and the other. The flow of electrons in the circuit is called the current. When the current flows continuously in just one direction, it is called direct current, or DC. In direct current flow, the potential difference between ends of the circuit remains the same, in that one end is always relatively positive and the other is relatively negative. A second type of electric current is called alternating current, or AC. In alternating current, the potential difference between ends of the circuit alternates signs, switching back and forth from negative to positive and from positive to negative. The electrons in alternating current do not flow from one end of the circuit to the other but instead oscillate back and forth between the ends at a specific frequency.

The movement of electrons through an electric circuit is subject to friction at the atomic level and to other effects that make it more or less difficult for the electrons to move about. These effects combine to restrict the flow of electrons in an overall effect called the resistance of the circuit. The current, or the rate at which electrons can flow through a circuit, is directly proportional to the potential difference (or applied voltage) and inversely proportional to the resistance. This basic relationship is the foundation of all electrical measurements and is known as Ohm's law.

Another basic and equally important principle is Kirchoff's current law, which states that the electric current entering any point in a circuit must always be equal to the current leaving that point in the circuit. In the light of the definition of electric current as the movement of electrons from point to point through a conductive pathway, this law seems concrete and obvious. It is interesting to note, however, that it was devised in 1845, well before the identification of electrons as discrete particles and the discovery of their role in electric current.

Electrical measurement, like all measurement, is a comparative process. The unit of potential difference—called the volt—defines the electrical force required to move a current of one ampere through a resistance of one ohm. Devices that measure voltage are calibrated against this standard definition. This definition similarly defines the ohm but not the ampere. The ampere is defined in terms of electron flow, such that a current of one ampere represents the movement of one coulomb of charge (equivalent to 6.24×10^{18} electrons) past a point in a period of one second.

The capacitance of a device in an electric circuit is defined as the amount of charge stored in the device relative to the voltage applied across the device. The inductance of a device in an electric circuit is more difficult to define but may be thought of as the amount of current stored in the device relative to the voltage applied across the device. In both cases, the current flow is restricted as an accumulation of charge within the device but through different methods. Whereas a capacitor restricts the current flow by presenting a physical barrier to the movement of electrons, an inductor restricts current flow by effectively trapping a certain amount of flowing current within the device.

Background and History

Wild electricity—lightning and other natural phenomena that result from differences in the oxidation potentials of different materials—has been observed and known for ages. Artificially produced electricity may have been known thousands of years ago, although this has not been proven conclusively. For example, artifacts recovered from some ancient Parthian tombs near Baghdad, Iraq, bear intriguing similarities in construction to those of more modern electrochemical cells or batteries. Reconstructions of the ancient device have produced electric current at about 0.87 volts, and other observations indicate that the devices may have been used to electroplate metal objects with gold or silver.

The modern battery, or voltaic pile, began with the work of Alessandro Volta in 1800. During the nineteenth century, a number of other scientists investigated electricity and electrical properties. Many of the internationally accepted units of electrical measurement were named in their honor.

How It Works

Ohm's Law. The basis of electrical measurement is found in Ohm's law, derived by Georg Simon Ohm in the nineteenth century. According to Ohm's law, the current flowing in an electric circuit is directly proportional to the applied voltage and inversely proportional to the resistance of the circuit. In other words, the greater the voltage applied to the circuit, the more current will flow. Conversely, the greater the resistance of the circuit, the less current will flow. This relationship can be stated mathematically as $E = I \times R$ (where E = voltage, I = current, and R = resistance), in which voltage is represented as the product of current and resistance.

Given this relationship, it is a relatively simple matter to design a device that uses two specific properties to determine the third. By constructing a device that employs set values of voltage and resistance, one can measure current. Similarly, by constructing a device that describes a system in which current and resistance are constant, one can measure voltage, and by devising a system in which current and voltage are regulated, one can measure resistance.

If the three primary properties of a circuit are known, then all other properties can be determined by arithmetic calculations. The capacitance of a circuit or circuit component, for example, is the amount of charge stored in the device at a given applied voltage, and the amount of charge is proportional to the current in the device. Similarly, the inductance in a circuit or circuit component depends on the current passing through the device at a given voltage.

Units of Measurement. All electrical properties must have an associated defined standard unit to be measurable. To that end, current is measured in amperes, named after Louis Ampere. Potential difference, sometimes called electromotive force, is measured in volts, named after Volta. Resistance is measured in ohms, named after Ohm. Power is measured in watts, named after James Watt. Capacitance is measured in farads, named after Michael Faraday. Inductance is measured in henrys, named after Joseph Henry. Conductance, the reciprocal of resistance, is measured in siemens, named after Ernst W. von Siemens. Frequencies are measured in hertz, or cycles per second, named after Gerhard Hertz.

Basic Electricity Concepts. Electricity can be produced in a continuous stream known as direct current (DC), in which electrons flow continuously from a negative source to a positive sink. The potential difference between the source and the sink is the applied voltage of the circuit, and it does not change. Electricity can also be produced in a varying manner called alternating current (AC), in which electron flow oscillates back and forth within the circuit. The applied voltage in such a system varies periodically between positive and negative values that are equal in magnitude. It is important to understand that circuits designed to operate with one type of applied voltage do not function when the other type of voltage is applied. In other words, a circuit designed to perform certain functions when supplied with a constant voltage and direct current will not perform those functions when supplied with a varying voltage and alternating current. The fundamental concept of Ohm's law applies equally to both cases, but other characteristics such as phase and frequency differences and voltage waveform make the relationships more complex in alternating current applications. Electrical measurement devices are designed to accommodate these characteristics and are capable of extremely fine differentiation and precision.

Applications and Products

The easiest electrical properties to measure accurately are voltage and resistance. Thus, the primary application tool of electrical measurement is the common volt-ohm meter (VOM), either as an analogue device or as its digital counterpart, the digital volt-ohm meter (DVOM).

Basic Analogue Measuring Devices. Two systems are required for any measuring device. One system is the structure by which the unknown value of a property is measured, and the other is the method of indicating that value to the user of the device. This latter feature was satisfied through the application of electromagnetic induction in the moving coil to produce a D'Arsonval movement. The strength of a magnetic field produced by current flowing through a coil is directly proportional to the magnitude of that current. In a basic analogue measuring device, the small coil is allowed to pivot freely within a permanent magnetic field. The amount by which the coil pivots is determined by the amount of current flowing through it; a needle attached to the coil indicates the appropriate value being measured on a calibrated scale. Analogue meters used to measure most electrical properties employ the moving-coil system to indicate the value of the property being measured.

Basic Digital Measuring Devices. The advent of digital electronics brought about a revolution in electrical measurement devices. Digital meters use a number of different systems to produce the desired information. The basic operation of digital devices is controlled by a central clock cycle in such a way that the value of the inputs are effectively measured thousands of separate times per second rather than continuously. This is known as sampling. Because the flow of electricity is a continuous or analogue process, sampling converts the analogue input value to a digital data stream. The data values can then be manipulated to be displayed directly as a numerical readout, eliminating the guesswork factor involved in reading a needle scale indicator. Another advantage of digital measurement devices is their inherent sensitivity. Through the use of an operational amplifier, or op-amp integrated circuits (IC), input signals can be amplified by factors of hundreds of thousands. This allows extremely small electrical values to be measured with great accuracy.

Other Measuring Devices. One of the most valuable devices in the arsenal of electrical measurement is the oscilloscope, which is available in both analogue and digital models. Like a typical meter, the oscilloscope measures electrical inputs, but it also has the capability to display the reading as a dynamic trace on a display screen. The screen is typically a cathode-ray tube (CRT) display, but later versions use liquid crystal display (LCD) screens and even can be used with desktop and laptop computers.

Another highly useful device for electrical measurement is the simple logic probe used with digital circuitry. In digital electronics, the application of voltages, and therefore the flow of current, is not continuous but appears in discrete bits as either an applied voltage or no applied voltage. These states are known as logic high and logic low, corresponding to on and off. The manipulation of these bits of data is governed by strict Boolean logic (system for logical operations on which digital electronics is based). Accordingly, when a digital device is operating properly, certain pins (or leads) of an integrated-circuit connection must be in the logic high state while others are in the logic low state. The logic probe is used to indicate which state any particular pin is in, generally by whether an indicator light in the device is on or off. Unlike a meter, the logic probe does not provide quantitative data or measurements, but it is no less invaluable as a diagnostic tool for troubleshooting digital circuitry.

Ancillary Devices. A more specific analytical device is the logic analyzer. Designed to read the input or pin signals of specific central processing unit (CPU) chips in operation, the logic analyzer can provide a running record of the actual programming used in the function of a digital electronic circuit. Another device that is used in electrical measurement, more typically as an electrical source than as a measuring device, is the waveform generator. This device is used to provide a specific shape of input voltage for an electric circuit to verify or test the function of the circuit.

Indirect Applications. Because of the high sensitivity that can be achieved in electrical measurement, particularly in the application of digital electronic devices, the measurement of certain electrical properties is widely used in analytical devices. The most significant electrical property employed in this way is resistance measurement. Often, this is measured as its converse property—conductivity. Gas-phase detectors on analytical devices such as gas chromatographs, high-performance liquid chromatographs, and combination devices with mass spectrometers are designed to measure the resistance, or conductivity, of the output stream from the device. For example, in gas chromatography, a carrier gas is passed through a heated column packed with a chromatography medium. When a mixture of compounds is injected into the gas stream, the various components become separated as they pass through the column. As the materials exit the column, they pass through a detector that measures changes in conductivity that occur with each different material. The changes are recorded either as a strip chart or as a collection of data.

The sensitive measurement of electrical resistance is made possible by the use of specific electric circuits, most notably the Wheatstone bridge circuit. In a Wheatstone bridge circuit, four resistance values are used, connected in a specific order. Two sets of resistance values are connected parallel to each other, with each set containing two resistance values connected in series with each other. One of the resistance values is the unknown quantity to be measured, while the other three are precisely known values. The voltage between the two midpoints of the series circuits of the bridge changes very precisely with any change in one of the resistance values. In addition, the output voltage signal can be amplified by several orders of magnitude, making even very small changes in output voltage meaningful. The role of digital electronics and op-amps cannot be overstated in this area.

Impact on Industry

Electrical measurement has had a tremendous impact on the manner in which industry is carried out, particularly in regard to automation and robotic control, and in process control. The electrical measurement of process variables in manufacturing and processing can be applied to control circuitry that automatically adjusts and maintains the correct and proper functioning of the particular system. This both renders the process more precise and eliminates problems that are caused by human error, as well as reduces the number of personnel required for hands-on operational checking and process maintenance.

Electrical measurement is central to essentially

all modern methods of testing and analysis and can be attributed to several factors. The technology is superbly adaptable to such functions and is fundamental to the operations of detection and electronic control. It is also extremely sensitive and precise, with the capability of measuring extremely small values of voltage, resistance, and current with precision. A thriving industry has subsequently developed to provide the various control mechanisms needed for the operation of modern methods of production and analysis. The procedures of electrical measurement also have a central role in medical analysis and other high-tech sciences such as physics and chemistry. In these fields, the devices used to carry out everything from the routine analysis of large groups of urine samples in a biomedical analytical laboratory to single-run experiments in the newest, largest subatomic particle colliders depend on electrical measurement for their functioning.

Nanoscience and Nanotechnology. Scientists continuously discover new materials and develop them for applications that capitalize on their specific electrical properties. They also modify existing measurement devices and design completely new ones. This often requires the redesign of electric circuitry, particularly the miniaturization of the devices. In the field of nanotechnology, for example, control circuitry for such small devices may consist of no more than a network of metallic tracings on a surface that measures correspondingly small electrical values.

Basic Electric Service. The mainstay of electrical measurement in industry is the service industry that maintains the operation of the continental electric grid and the power supplied to individual locations. Distribution of electric power calls for close monitoring of the network by which electric power is carried to individual residences, factories, and other locations. It also requires the installation and operation of new sources of electric power. This requires a skilled workforce that is capable of using basic electrical measurement practices. Accordingly, a significant component of training programs provided by colleges, universities, and vocational schools focuses on the practical aspects of the physical delivery of electric power, where measurement is a foundation skill. Simple economics demands that the amount of electricity being produced and the amount being consumed must be known. The effective distribution of that electricity, both generally and within specific installations, is vitally important to its continued functioning.

Fascinating Facts About Electrical Measurement

- Devices nearly 2,000 years old that were recovered from tombs near Baghdad, Iraq, were almost certainly batteries used to gold plate other metal objects. The technology of the battery was lost and not rediscovered until 1800.
- Henry Cavendish discovered what would be known as Ohm's law about fifty years earlier than Georg Simon Ohm did. However, as Cavendish did not publish his observations, the law has been attributed to Ohm.
- The most commonly and easily measured electrical quantity is voltage, or potential difference.
- Essentially all electrical measurements are determined by the relationship between voltage, resistance, and current (Ohm's law).
- Kirchoff's current law was stated before electrons were discovered and identified, and electricity was still regarded as some kind of mysterious fluid that certain materials contained.
- A bolt of lightning results when electrons flow through a regular electric circuit formed by the presence of charged particles (ions) in the atmosphere between clouds and the surface of the ground.
- An average lightning bolt transfers only about 5 coulombs of charge, but the transfer takes place so quickly that the electric current is about 30,000 amperes.
- Power, as measured in watts, is used to train elite cyclists. Over the course of an hour, professional cyclist Lance Armstrong can consistently generate 350 watts of electricity; the average in-shape cyclist, only about 100 watts.
- A static shock that a person can hear, feel, and see is about 250 volts of electricity.
- Conductance is the reciprocal of resistance; the unit of conductance was originally called the mho, as the reciprocal of the ohm.

Careers and Course Work

Studies related to the use of electrical devices

provide a good, basic knowledge of electrical measurement. A sound basis in physics and mathematics will also be required. College and university level course work will depend largely on the chosen area of specialization. Options at this level range from basic electrical service technician to fundamental physics. At a minimum, students will pursue studies in mathematics, physical sciences, industrial technologies (chemical, electrical, and mechanical), and business at the undergraduate level or as trade students. More advanced studies usually consist of specialized courses in a chosen field. The study of applied mathematics will be particularly appropriate in advanced studies, as this branch of mathematics provides the mathematical basis for phase relationships, quantum boundaries, and electron behavior that are the central electrical measurement features of advanced practices.

In addition, as technologies of electrical measurement and application and regulations governing the distribution of electric energy change, those working in the field can expect to be required to upgrade their working knowledge on an almost continual basis to keep abreast of changes.

Social Context and Future Prospects

Economics drives the production of electricity for consumption, and in many ways, the developed world is very much an electrical world. A considerable amount of research and planning has been devoted to the concept of the smart grid, in which the grid itself would be capable of controlling the distribution of electricity according to demand. Effective electrical measurement is absolutely necessary for this concept to become a working reality.

At a very basic level, the service industry for the maintenance of electrically powered devices will continue to provide employment for many, particularly as the green movement tends to shift society toward recycling and recovering materials rather than merely disposing of them. Accordingly, repair and refurbishment of existing units and the maintenance of residential wiring systems would focus more heavily on effective troubleshooting methods to determine the nature of flaws and correct them before any untoward incident should occur.

The increased focus on alternative energy sources, particularly on the development of solar energy and fuel cells, will also place higher demands on the effectiveness of electrical measurement technology. This will be required to ensure that the maximum amount of usable electric energy is being produced and that the electric energy being produced is also being used effectively.

Richard M. J. Renneboog, M.Sc.

Further Reading

Clark, Latimer. *An Elementary Treatise on Electrical Measurement for the Use of Telegraph Inspectors and Operators*. London: E. and F. N. Spon, 1868. An interesting historical artifact that predates the identification of electrons and their role in electric current and measurement.

Herman, Stephen L. *Delmar's Standard Textbook of Electricity*. 5th ed. Clifton Park, N.Y.: Delmar Cengage Learning, 2009. A good basic presentation of the fundamental principles of electricity and electrical measurement.

Herman, Stephen L., and Orla E. Loper. *Direct Current Fundamentals*. 7th ed. Clifton Park, N.Y.: Thomson Delmar Learning, 2007. An introductory level textbook offering basic information regarding direct current electricity and electrical measurement.

Lenk, John D. *Handbook of Controls and Instrumentation*. Englewood Cliffs, N.J.: Prentice Hall, 1980. Describes general electrical measurement applications in principle and practice, as used for system and process control purposes.

Malvino, Albert Paul. *Malvino Electronic Principles*. 6th ed. New York: Glencoe/McGraw-Hill Books, 1999. Presents a detailed analysis of the electronic principles and semiconductor devices behind digital electronic applications in electrical measurement.

Strobel, Howard A., and William R. Heineman. *Chemical Instrumentation: A Systematic Approach*. 3d ed. New York: John Wiley & Sons, 1989. Describes the fundamentals of design and the operation of instrumentation used in chemistry and chemical applications, all of which ultimately use and depend on electrical measurement.

Tumanski, Slawomir. *Principles of Electrical Measurement*. New York: Taylor and Francis, 2006. Provides a thorough, descriptive introduction to the principles of electrical measurement and many of the devices that are used to manipulate and measure electric signals.

Web Sites
U.S. Department of Energy
Electric Power
http://www.energy.gov/energysources/electricpower.htm

U.S. Energy Information Administration
Electricity
http://www.eia.doe.gov/fuelelectric.html

See also: Electrical Engineering; Electrometallurgy; Electronics and Electronic Engineering; Fossil Fuel Power Plants; Integrated-Circuit Design.

ELECTRIC AUTOMOBILE TECHNOLOGY

FIELDS OF STUDY

Automotive engineering; chemical engineering; clean-energy technologies; electrical engineering; informatics; information technologies; materials engineering; mathematics; physics; resource management.

SUMMARY

Electric vehicles have been around even longer than internal combustion engine cars. With health issues resulting from the modern use of internal combustion engines, the automotive industry is intensifying its efforts to produce novel machines that run on electricity. Many cars come with drivetrains that can accept electric propulsion, offering quieter, healthier transportation options. Although consumers still seem to shy away from completely electric vehicles, hybrid vehicles that use both internal combustion engines and electric power are in use in many cities around the world.

KEY TERMS AND CONCEPTS

- **Aromatic Compound:** Carbon-based chemical such as benzene or toluene; sometimes harmful.
- **Battery:** Combination of one or more electrochemical cells used to convert stored chemical energy into an electric current.
- **Combustion:** Process by which a substance (fuel) reacts with oxygen to form carbon dioxide, water, heat, and light.
- **Current:** Flow of electricity through a material.
- **Drivetrain:** Mechanical system that transmits power or torque from one place (such as the motor) to another (such as the wheels).
- **Electrochemical Cell:** Device that can derive electrical energy from chemical reactions or facilitate chemical reactions by applying electricity.
- **Fuel:** Chemical compound that can provide energy by its conversion to water or carbon dioxide, either by combustion or by electrochemical conversion.
- **Fuel Cell:** Device that combines oxygen from the air with a fuel such as hydrogen to form water, heat, and electricity by splitting off protons from the fuel and allowing only those protons to pass through a dense membrane.
- **Hybrid Vehicle:** Car using a combination of several essentially different drive mechanisms, usually an internal combustion engine in combination with an electric motor, which in turn can derive its power from a battery or a fuel cell.
- **Lithium-Ion Battery:** Modern battery that involves lithium in storing and converting energy, has a high energy density, but degrades fast at elevated temperatures.
- **Membrane:** Dense dividing wall between two compartments.
- **Polarization:** Magnetization of a material with a defined polarity.
- **Proton Exchange Membrane Fuel Cell:** Polymer-based electrochemical device that converts various fuels and air into water, heat, and electric energy at low operating temperatures (70 to 140 degrees Celsius).
- **Solid Oxide Fuel Cell:** Ceramic electrochemical device that converts various fuels and air into water, heat, and electric energy at high operating temperatures (600 to 1,000 degrees Celsius).
- **Torque:** Tendency of a force to rotate an object around a defined axis or pivot; also known as moment force.
- **Voltage:** Electromotive force of electricity; also defined as the work required to move a charged object through an electric field.

Definition and Basic Principles

Electric vehicles are driven by an electric motor. The electricity for this motor can come from different sources. In vehicle technology, electrical power is usually provided by batteries or fuel cells. The main advantages of these devices are that they are silent, operate with a high efficiency, and do not have tailpipe emissions harmful to humans and the environment. In hybrid vehicles, two or more motors coexist in the vehicle. When large quantities of power are required rapidly, the power is provided by combusting fuels in the internal combustion engine; when driving is steady, or the car is idling at a traffic light, the car is entirely driven by the electric motor, thereby cutting emissions while providing the consumer with the normal range typically associated with traditional cars that would rely entirely on internal combustion engines. Electric vehicles make it possible for drivers to avoid having to recharge at a station. Recharging can occur at home, at work, and in parking structures, quietly, cleanly, and without involving potentially carcinogenic petroleum products.

Background and History

Electric vehicles have been around since the early 1890's. Early electric vehicles had many advantages over their combustion-engine competitors. They had no smell, emitted no vibration, and were quiet. They also did not need gear changes, a mechanical problem that made early combustion-engine cars cumbersome. In addition, the torque exhibited by an electric engine is superior to the torque generated by an equivalent internal combustion engine. However, battery technology was not yet sufficiently developed, and consequently, as a result of low charge, storage capacity in the batteries, and rapid developments in internal combustion engine vehicle technology, electric vehicles declined on the international markets in the early 1900's.

At the heart of electric vehicles is the electric motor, a relatively simple device that converts electric energy into motion by using magnets. This technology is typically credited to the English chemist and physicist Michael Faraday, who discovered electromagnetic induction in 1831. These motors require some electrical power, which is typically provided by batteries, as it was done in the early cars, or by fuel cells. Batteries were described as early as 1748 by one of the founding fathers of the United States, Benjamin Franklin. These devices can convert chemically stored energy into a flow of electrons by converting the chemicals present in the battery into different chemicals. Depending on the materials used, some of these reactions are reversible, which means that by applying electricity, the initial chemical can be re-created and the battery reused. The development of fuel cells is typically credited to the English lawyer and physicist Sir William Grove in 1839, who discovered that flowing hydrogen and air over the surfaces of platinum rods in sulfuric acid creates a current of electricity and leads to the formation of water. The devices necessary to develop an electric car had been around for many decades before they were first assembled into a vehicle.

A major circumstance that led to the commercial success of combustion-engine vehicles over electric vehicles was the discovery and mining of cheap and easily available oil. Marginal improvements in battery technology compared with internal combustion engine technology occurred during the twentieth century. As a result of stricter emissions standards near the end of the twentieth century, global battery research began to reemerge and has significantly accelerated in the 2010's. Some early results of this research are on the market.

During the 1990's, oil was still very cheap, and consumers, especially in North America, demanded heavier and larger cars with stronger motors. During this decade, General Motors (GM) developed an electric vehicle called the EV1, which gained significant, though brief, international positive attention before it was taken off the market shortly after its introduction. All produced new cars were destroyed, and the electric-vehicle program was shut down. The development of electric vehicles was then left to other companies.

Barely twenty years later, following a significant negative impact on the car manufacturing companies in North America from the 2009 financial crisis and their earlier abandonment of research and development of electric car technology, North American companies tried to catch up with the electric vehicle technology of the global vehicle manufacturing industry. During the hiatus, global competitors surpassed North American companies by creating modern, fast, useful electrical vehicles such as the Nissan Cube from Japan and the BMW ActiveE models from Germany. GM, at least, has made a full turnaround after receiving government financial incentives to develop battery-run vehicles and has actively pursued a new electric concept called Chevrolet Volt. Similar to hybrid cars, the Volt has a standard battery, but because early twenty-first century batteries are not yet meeting desired performance levels, the Volt also has a small engine to extend its range. Installing two different power sources in a vehicle, one electric and one combustion based, makes sense in order to develop a product that has lower emissions but the same range as combustion-engine-based vehicles.

How It Works

Power Source. Gasoline, which is mainly a

chemical called octane, is a geologic product of animals and plants that lived many millions of years ago. They stored the energy of the Sun either directly from photosynthesis or through digestion of plant matter. The solar energy that is chemically stored in gasoline is released during combustion.

The storage of energy in batteries occurs through different chemicals, depending on the type of battery. For example, typical car-starter lead-acid batteries use the metal lead and the ceramic lead oxide to store energy. During discharge, both these materials convert into yet another chemical called lead sulfate. When a charge is applied to the battery, the original lead and lead oxide are re-created from the lead sulfate. Over time, some of the lead and lead oxide are lost in the battery, as they separate from the main material. This can be seen as a black dust swimming in the sulfuric acid of a long-used battery and indicates that the battery can no longer be recharged to its full initial storage capacity. This happens to all types of batteries. Modern lithium-ion batteries used in anything from vehicles to mobile phones use lithium-cobalt/nickel/manganese oxide, and lithium graphite. These batteries use lithium ions to transport the charges around, while allowing the liberated electrons to be used in an electric motor. Other batteries use zinc to store energy—for example, in small button cells. Toxic materials such as mercury and cadmium have for some years been used in specific types of batteries, but have mostly been phased out because of the potential leaching of these materials into groundwater after the batteries' disposal.

Fuel cells do not use a solid material to store their charge. Instead, low-temperature proton exchange membrane fuel cells use gases such as hydrogen and liquid ethanol (the same form of alcohol found in vodka) or methanol as fuels. These materials are pumped over the surface of the fuel cells, and in the presence of noble-metal catalysts, the protons in these fuels are broken away from the fuel molecule and transported through the electrolyte membrane to form water and heat in the presence of air. The liberated electrons can, just as in the case of batteries, be used to drive an electric motor. Other types of fuel cells, such as molten carbonate fuel cells and solid oxide fuel cells, can use fuels such as carbon in the form of coal, soot, or old rubber tires and operate at 800 degrees Celsius with a very high efficiency.

Converting Electricity into Motion. Most electric motors use a rotatable magnet the polarity of which can be reversed inside a permanent magnet. Once electricity is available to an electric motor, electrons, traveling through an electric wire and coiled around a shaft that can be magnetized, generate an electrical field that polarizes the shaft. As a result, the shaft is aligned within the external permanent magnet, since reverse polarities in magnets are attracted to each other. If the polarity of the rotatable shaft is now reversed by changing the electron flow, the magnet reverses polarity and rotates 180 degrees. If the switching of the magnetic polarity is precisely timed, constant motion will be created. Changes in rotational speed can be achieved by changing the frequency of the change in polarization. The rotation generated by an electric motor can then be used like the rotation generated by an internal combustion engine by transferring it to the wheels of the vehicle.

Research and Development. Many components of electric vehicles can be improved by research and development. In the electric motor, special magnetic glasses can be used that magnetize rapidly with few losses to heat, and the magnet rotation can occur in a vacuum and by using low-friction bearings. Materials research of batteries has resulted in higher storage capacities, lower overall mass, faster recharge cycles, and low degradation over time. However, significant further improvements can still be expected from this type of research as the fundamental understanding of the processes occurring in batteries become better understood.

Novel fuel cells are being developed with the goal of making them cheaper by using non-precious-metal catalysts that degrade slowly with time and are reliable throughout the lifetime of the electric motor. The U.S. Department of Energy has set specific lifetime and performance targets to which all these devices have to adhere to be useful on the commercial market. In fact, electric car prototypes are already available and comparable to vehicles using internal combustion engines. As of 2011, the main factor preventing deep mass-market penetration is cost, but that is continuously addressed by research and development of novel batteries and fuel cells that are lighter, use less expensive precious-metal catalysts, last longer, and are more reliable than previous devices. Since the 2010's, the development of these devices has significantly accelerated, especially because of international funding that is being poured into clean-energy technologies.

There are, however, disadvantages to all energy-conversion technologies. Internal combustion engines require large amounts of metals, including iron, chromium, nickel, manganese, and other alloying elements. They also require very high temperatures in forges during production. Additionally, the petroleum-based fuels contain carcinogenic chemicals, and the exhausts are potentially dangerous to humans and the environment, even when catalytic converters are used; to function well, these devices require large amounts of expensive and rare noble metals such as palladium and platinum. The highest concentrations of oil deposits have been found in politically volatile regions, and oil developments in those regions have been shown to increase local poverty and to cause severe local environmental problems.

Batteries require large quantities of rare-earth elements such as lanthanum. Most of these elements are almost exclusively mined in China, which holds a monopoly on the pricing and availability of these elements. Some batteries use toxic materials such as lead, mercury, or cadmium, although the use of these elements is being phased out in Europe. Lithium-ion batteries can rapidly and explosively discharge when short-circuited and are also considered a health risk. Electricity is required to recharge batteries, and it is often produced off-site in reactors whose emissions and other waste can be detrimental to human health and the environment.

Fuel cells require catalysts that are mostly made from expensive noble metals. Severe price fluctuations make it difficult to identify a stable or predictable cost for these devices. The fuels used in fuel cells, mostly hydrogen and methanol or ethanol, have to be produced, stored, and distributed. As of 2011, the majority of the hydrogen used is derived via a water-gas shift reaction, where oxygen is stripped off the water molecules and binds with carbon molecules from methane gas, producing hydrogen with carbon dioxide as a by-product; the process requires large quantities of natural gas. Methanol or ethanol can be derived from plant matter, but if it is derived from plants originally intended as food, food prices may increase, and arable land once used for food production then produces fuels instead.

Nevertheless, while the advantages and disadvantages of cleaner energy technologies such as fuel cells and batteries must be weighed against their ecological and economic impacts, it is important to remember that they are significantly cleaner than current internal combustion engine technologies.

Applications and Products

Batteries. Battery technology still needs to be developed to be lighter without reducing the available charge. This means that the energy density of the battery (both by mass and by volume) needs to increase in order to improve a vehicle's range. Furthermore, faster recharge cycles have to be developed that will not negatively impact the degradation of the batteries. Overnight recharge cycles are possible, and good for home use, but a quick recharge during a shopping trip should allow the car to regain a significant proportion of its original charge. Repeated recharge cycles at different charge levels as well as long-time operation with large temperature fluctuations should not detrimentally affect the microstructure of the batteries, so the power density of the batteries will remain intact. Furthermore, operation in very cold environments, in which the charge carriers inside the battery are less mobile, should be realized for a good market penetration. The introduction of the Chevrolet Volt into the North American market in March, 2011, resulted in a disappointment, as customers appeared unwilling to pay a premium for battery-operated cars. Stricter policies enforcing the conversion of a more significant proportion of cars into electric vehicles are necessary to change the market, especially in North America.

Personal Vehicles. GM's EV1 was an attempt to market electric vehicles in North America in the 1990's. It was fast and lightweight and had all the amenities required by consumers but was discontinued by the manufacturer because it was not commercially viable. The 2011 edition of GM's Chevrolet Volt was almost indistinguishable from other GM station wagons, but the cost for the battery-powered car proved too high for a market that was used to very cheap vehicles with internal combustion engines. Electric vehicle technology is arguably much more advanced in Asia. Asian vehicle manufacturers were up to ten years ahead of the rest of the world in producing hybrid-electric vehicles, and they are set up to be ahead in the manufacturing of completely electric vehicles as well. For example, battery-only vehicles such as the Toyota iQ and the Nissan Leaf can drive up to 100 miles on a single battery charge with performance similar to that of an internal combustion engine car. European manufacturers, such as Renault, have teamed up with Nissan so as to not be left behind in the electric vehicle business and offered their first electric vehicle lineup to the European markets in 2011. Volkswagen has produced several studies including a concept called SpaceUp, but Volkswagen CEO Martin Winterkorn said in 2011 that battery technology was not mature enough for vehicles, and that Volkswagen would not produce any mass-market devices before 2013. Car manufacturer Fisker developed two plug-in hybrid electric vehicles, Nina and Karma, and planned to sell 1,000 units in the United States in 2011. As of 2011, Tesla Motors had two electric vehicles ready for the North American market: Model S and Roadster. Think City also manufactures small electric vehicles. However, while demand outside North America is large, demand in the United States is low. With major international governmental tax incentives in place, all vehicle manufacturers are developing at least some studies of electric vehicles for auto shows. Whether these models will actually go on to be developed for commercial markets remains to be seen.

Utility Vehicles and Trucks. To develop a green image, some municipalities considered switching their fleets to electric vehicles based on fuel cells or batteries. Ford has created a model called Transit Connect, which is an electrified version of its Ford Transporter. Navistar has developed an electric truck, the eStar, and expects to sell a maximum of 1,000 units each year until 2015 in the United States. Smith Electric Vehicles has developed several models, such as the Newton, for the expected demand in electric-utility vehicles. Additionally, there are many small companies producing small utility trucks, such as the Italian manufacturer Alkè. The products of these companies are small, practical multi-purpose vehicles for cities and municipalities.

Bicycles and Scooters. Small electric motor-assisted bicycles and electric scooters have been in use since the early twentieth century. Other small electric vehicles include wheelchairs, skateboards, golf carts, lawnmowers, and other equipment that typically does not require much power. For customers looking for faster vehicles, Oregon-based Brammo produces electric motorcycles that have a range of 100 miles at a speed of 100 miles per hour. These electric vehicles have comparable performance to any internal combustion motorcycle but lack any tailpipe or noise emissions.

Mass Transit. In North America, many cities had electric public transit similar to the San Francisco cable cars until they were sold to car manufacturers who decommissioned them. As a result, most public transit systems rely heavily on diesel engine buses. Some of the public transit companies have considered testing fuel-cell- or battery-powered electric buses, but all these efforts have remained very small, with a handful of buses running at any time throughout Europe and North America. For example, during the 2010 Olympics, the Canadian Hydrogen and Fuel Cell Association ran fuel-cell electric vehicle busses between the cities of Vancouver and Whistler. United Kingdom-based manufacturer Optare has produced battery-powered buses since 2009. Cities, such as Seattle, that use electric overhead lines to power trolley buses, trams, and trains have had much higher impact in terms of actual transported passengers. All these systems constitute electric vehicles, but all of them are dependent on having electric wires in place before they can operate. On the other hand, once the wires are in place, the public transit systems can operate silently and cleanly, using electricity provided through an electric grid instead of a battery or a fuel cell.

Forklifts. In spaces with little ventilation, the exhaust of internal combustion engines can be harmful and potentially toxic to humans, which is why warehouse forklifts are typically powered by electric engines. Traditionally, these engines are powered by batteries, but the recharging time of several hours often requires the purchase of at least twice as many batteries as forklifts—or twice as many forklifts as drivers—to be able to work around the clock. Using fuel cells as a power source, forklifts such as the ones produced by Vancouver-based company Cellex require only a short time at a hydrogen refueling station before being ready for use. Such a short downtime of fuel-cell powered electric forklifts compared with battery-powered forklifts allows warehouses to operate with less machinery, cutting back on the initial capital cost of operation.

Impact on Industry

Government Research. Globally, most governments have introduced specific targets for the electrification of vehicles to reduce pollution. In Germany, for example, the federal government plans to have at least one million electric vehicles on the streets by 2020. This is an ambitious target that is backed by significant funding for government, university, and private-sector research. However, in 2009-2010, the German government offered car wreckage flat-rate premium to all car owners, many of whom traded their used but still very usable cars for newer cars with larger motors and increased emissions. Although the funding of this incentive program significantly benefited automakers worldwide, incentives to improve the performance of electric motors and the market penetration of electric vehicles were not included. As a consequence, no net reduction in vehicle-emission pollution was achieved, and the number of electric cars actually on the road remained negligible. About 1,500 electric vehicles were registered in Germany in 2010 out of a total of about 41 million cars. With no incentives in place as of 2011, the number of registered electric vehicles has not significantly changed either. Asian countries appear to invest more into an electric-vehicle infrastructure, with China set to place a half million electric vehicles on its roads in 2012, although it remains doubtful whether this target can be met completely. The U.S. Department of Energy's February, 2011, Status Report stated the ambitious target of having one million electric vehicles on the road by 2015. This is based on 2010 statistics in which 97 percent of cars sold had conventional internal combustion engines and 3 percent had hybrid electric engines. While one million cars sounds like a lot, in the actual car market, this amount will not represent any significant step in reducing pollution from internal combustion engines. Significantly higher numbers of electric vehicle usage are required to make cleaner inner-city air a reality.

University Research. Universities have always been involved in battery research, but only since the 2010's has there been a significant increase in attention to clean-energy storage technologies from super capacitors to flywheels, hydrogen, and battery technologies. Although university projects suffered some detrimental impact from the 2009 world economic crisis, the research, especially in batteries, was minimally affected and has since seen significant increases in funding, scope, and technology-focused applications. One example is the Institute for Electrochemical Energy Storage and Conversion at the University of Ulm in Germany, which has significantly expanded its activities in battery research. Similar expansions in battery-technology research can be seen around the globe.

Industry and Business. One of the main business sectors for electronic vehicles includes forklifts and other indoor equipment for confined or explosive environments, such as mines. Other sectors that have made profits over the past years include electric scooters and small electric bicycles. And while there is significant media attention on electric cars, both fuel cell and battery based, as well as hybrid-electric vehicles, their contribution to the global automobile market is negligibly small and requires significant governmental incentives to become a larger part of the global automotive economy.

Major Corporations. Although North America was a global leader in electric vehicles and battery research, the car industry mostly dismissed

Fascinating Facts About Electric Automobile Technology

- The average power capacity of rechargeable batteries has tripled since the 1970's.
- Fuel cells produce electricity with clean water as the only exhaust. During space missions, this water has been used for drinking.
- The clean and quiet Brammo Enertia electric motorcycle can go 100 miles at a velocity of 100 miles per hour before requiring a recharge.
- Hitting the accelerator in a car with an electric motor results in a much faster acceleration than in a comparable gasoline-run car, since the torque of the electric motor far exceeds that of the internal combustion engine.
- Modern electric vehicles from major carmakers are indistinguishable from their standard line of models, as this reduces potential customer barriers to purchasing electric or electric-hybrid vehicles and lowers manufacturing cost.
- Non-rechargeable batteries that are not properly disposed of may leach toxic chemicals into the ground over time.
- It takes significantly more energy to create a new car than it does to keep an old car in good shape and run it—especially if the new car uses an internal combustion engine.

electric vehicle developments in the 1990's, as North American consumers demanded larger, heavier, and more powerful vehicles with internal combustion engines. As of 2011, North American car manufacturers lag behind the rest of the world in developing mature vehicles that will be accepted in the market. Both Ford, with its Focus electric vehicle, and GM, with its Chevrolet Volt, have produced early studies of electric vehicles that they intend to develop for the North American market. They compete with all major international car manufacturers that are introducing the electric vehicles worldwide, as well as domestic corporations that focus on electric vehicles, such as the California-based luxury-car manufacturer Tesla Motors and Missouri-based electric truck and commercial vehicle manufacturer Smith Electric Vehicles. Internationally, there is a large number of electric car manufacturing businesses starting up, reviving a business that had become the monopoly of very few corporations. Because of more stringent emissions targets and significant tax incentives, North American car manufacturers have returned to some electric car developments in the 2010's. Whether these are just greening initiatives and convenient tax-reduction programs or whether these cars can become a major business in the North American market remains to be seen.

Most car manufacturers have created spin-off companies that develop customer-specific batteries for modern vehicles. For example, the European car manufacturer Daimler and the international chemical corporation Evonik and its subsidiary Li-Tec formed a new company called Deutsche Accumotive, with the sole aim of producing better batteries for the international vehicle market. This comes as a result of the dominance of lithium-ion batteries in the low-end markets by Chinese manufacturers and high-end markets by Japan and South Korea. Most batteries are used in power tools and small handheld devices such as mobile phones and laptop computers. The clean technology consulting firm Pike Research estimates that the global annual market for lithium-ion batteries for vehicles is $8 billion.

Careers and Course Work

As gasoline prices increase, it becomes more important to have lighter vehicles that require less material during manufacturing, as these have to be mined and transported around the world and machined using energy coming primarily from fossil fuels. Additionally, vehicles should become more efficient, to reduce the operating cost for vehicle owners. All these issues are addressed by selecting and designing better, novel materials. Those interested in a career in electric vehicle manufacturing or design would do well studying materials, mechanical, chemical, mining, or environmental engineering for designing novel cars, highly efficient motors, better batteries, and cheaper, more durable fuel cells. The mathematical modeling of the electrochemistry involved in electric motors is also very important to understand how to improve electric devices, and studies in chemistry and physics may lead to improvements in the efficiency of vehicles.

After earning a bachelor's degree in one of the above-mentioned areas, an internship would be ideal. After an internship, one's career path can be extremely varied. In the research sector, for example, working on catalysts for batteries and fuel cells in a chemical company could include developing new materials that involve inexpensive, nontoxic, durable noble metals that are at least as efficient as traditional catalysts. This is only one example of many potential careers in the global electric vehicle market.

Social Context and Future Prospects

Energy consumption per capita is increasing continuously. The majority of power production uses the combustion of fossil fuels with additional contributions from hydroelectric and nuclear energy conversion. These energy-conversion methods create varying kinds of pollution and dangers to the environment such as habitat destruction, toxic-waste production, or radiation, as seen in nuclear reactors hit by earthquakes, equipment malfunction, or operator errors. The increasing demand for a finite quantity of fossil fuels has the potential to increase the cost of these resources significantly. Another undesirable consequence of the thermochemical conversion of fossil fuels by combustion is environmental contamination. The reaction products from combustion can be harmful to humans on a local scale and have been cited as contributing to global climate change. The remaining ash of coal combustion contains heavy metals and radioactive isotopes that can be severely damaging to health, as seen in the 2008 Kingston Fossil Plant coal ash slurry spill in Tennessee.

Furthermore, fossil fuel resources are unevenly distributed over the globe, leading to geopolitical unrest as a result of the competition for resource access. As a consequence of upheavals in the Middle East and North Africa in 2011, oil and food prices have soared, and may continue doing so.

Clearly, the energy demands of society need to be satisfied in a more appropriate, sustainable, and efficient way. Cleaner devices for energy conversion are batteries and fuel cells. They operate more efficiently, produce less pollution, are modular, and are less likely to fail mechanically since they have fewer moving parts than energy conversion based on combustion.

The advantages of electric vehicles are clear: a world in which all or most vehicles are quiet, with no truck engine brakes to rattle windows from a mile away and no lawnmowers disturbing the quiet or fresh air of a neighborhood; a society with no harmful local emissions from any of the machines being used, allowing people to walk by a leaf blower without having to hold their breath and to live next to major roads without risking chronic diseases from continually breathing in harmful emissions. All this could already be humanity's present-day reality if people were willing to change their habits and simply use electric motors instead of combustion engines.

Lars Rose, M.Sc., Ph.D.

Further Reading

Cancilla, Riccardo, and Monte Gargano, eds. *Global Environmental Policies: Impact, Management and Effects.* Hauppauge, N.Y.: Nova Science Publishers, 2010. Features multifaceted chapters on various international aspects of environmental policies, laying the groundwork for a change in consumer attitude toward new clean energy and infrastructure.

Hoel, Michael, and Snorre Kverndokk. "Depletion of Fossil Fuels and the Impact of Global Warming." *Resource and Energy Economics* 18, no. 2 (June, 1996): 115-136. Gives an explanation of the calculation of global oil depletion.

Husain, Iqbal. *Electric and Hybrid Vehicles: Design Fundamentals.* 2d ed. Boca Raton, Fla.: CRC Press, 2011. A very technical but comprehensive book that provides an overview of modern electric and electric-hybrid vehicle technologies.

Root, Michael. *The TAB Battery Book: An In-Depth Guide to Construction, Design, and Use.* New York: McGraw-Hill, 2011. Provides a good, readable background for all different types of batteries and the challenges in research and development.

Taylor, Peter J., and Frederick H. Buttel. "How Do We Know We Have Global Environmental Problems? Science and the Globalization of Environmental Discourse." *Geoforum* 23, no. 3 (1992): 405-416. Outlines the science behind global environmental issues.

U.S. Army Center for Health Promotion and Preventive Medicine. "Engine Emissions—Health and Medical Effects." http://phc.amedd.army.mil/PHC%20Resource%20Library/FS65-039-1205.pdf. Outlines the detrimental acute and chronic effects of engine exhaust intake via the lungs and the skin.

U.S. Department of Energy. "One Million Electric Vehicles By 2015." http://www.energy.gov/media/1_Million_Electric_Vehicle_Report_Final.pdf. Details the U.S. plan to achieve a sale rate of at least 1.7 percent electric vehicles every year until 2015, also indicating preferred companies involved for the North American electric vehicle market.

Web Sites

American Society for Engineering Education
http://www.asee.org

Electric Auto Association
http://www.electricauto.org

Electric Drive Transportation Organization
http://www.electricdrive.org

European Association for Battery, Hybrid and Fuel Cell Electric Vehicles
http://www.avere.org/www/index.php

See also: Chemical Engineering; Electrical Engineering.

ELECTROMETALLURGY

FIELDS OF STUDY

Chemistry; physics; engineering; materials science.

SUMMARY

Electrometallurgy includes electrowinning, electroforming, electrorefining, and electroplating. The electrowinning of aluminum from its oxide (alumina) accounts for essentially all the world's supply of this metal. Electrorefining of copper is used to achieve levels of purity needed for its use as an electrical conductor. Electroplating is used to protect base metals from corrosion, to increase hardness and wear resistance, or to create a decorative surface. Electroforming is used to produce small, intricately shaped metal objects.

KEY TERMS AND CONCEPTS

- **Anode:** Electrode where electrons return to the external circuit.
- **Cathode:** Electrode where electrons react with the electrolyte.
- **Electrode:** Interface between an electrolyte and an external circuit; usually a solid rod.
- **Electroforming:** Creation of metal objects of a desired shape by electrolysis.
- **Electrolysis:** Chemical change produced by passage of an electric current through a substance.
- **Electrolyte:** Water solution or molten salt that conducts electricity.
- **Electroplating:** Creation of an adherent metal layer on some substrate by electrolysis.
- **Electrorefining:** Increasing the purity of a metal by electrolysis.
- **Electrowinning:** Obtaining a metal from its compounds by the use of electric current.
- **Equivalent Weight:** Atomic weight of a metal divided by the integer number of electron charges on each of its ions.
- **Faraday:** Least amount of electric charge needed to plate out one equivalent weight of a metal.
- **Pyrometallurgy:** Reduction of metal ores by heat and chemical reducing agents such as carbon.

Definition and Basic Principles

Electrometallurgy is that part of metallurgy that involves the use of electric current to reduce compounds of metals to free metals. It includes uses of electrolysis such as metal plating, metal refining. and electroforming. Electrolysis in this context does not include the cosmetic use of electrolysis in hair removal.

Background and History

The field of electrometallurgy started in the late eighteenth century and was the direct result of the development of a source of electric current, the cell invented by Alessandro Volta of Pavia, Italy. Volta placed alternating disks of zinc and copper separated by brine-soaked felt disks into a long vertical stack. In March, 1800, Volta described his "pile" in a letter to the Royal Society in London. His primitive device could produce only a small electric current, but it stimulated scientists all over Europe to construct bigger and better batteries and to explore their uses. Luigi Valentino Brugnatelli, an acquaintance of Volta, used a voltaic pile in 1802 to do electroplating. British surgeon Alfred Smee published *Elements of Electrometallurgy* in 1840, using the term "electrometallurgy" for the first time.

Michael Faraday discovered the basic laws of electrolysis early in the nineteenth century and was the first to use the word "electrolysis." Both Sir Humphry Davy and Robert Bunsen used electrolysis to prepare chemical elements. The development of dynamos based on Faraday's ideas made possible larger currents and opened up industrial uses of electrolysis, most notably the Hall-Héroult process for the production of aluminum around 1886.

The Aluminum Company of America (known as Alcoa since 1999) was founded in 1909 and became a major producer of aluminum by the Hall-Héroult process. Canada-based company Rio Tinto Alcan and Norway-based Norsk Hydro are also major producers of aluminum. Magnesium was first made commercially in Germany by I. G. Farben in the 1890's, and later in the United States by Dow Chemical. In both instances, electrolysis of molten magnesium chloride was used to produce the magnesium.

How It Works

Electrolysis involves passage of an electric current through a circuit containing a liquid electrolyte, which may be a water solution or a molten solid. The current is carried through the electrolyte by migration of ions (positively or negatively charged particles). At the electrodes, the ions in the electrolyte undergo chemical reactions. For example, in the electrolysis of molten sodium chloride, positively charged sodium ions migrate to one electrode (the cathode) and negatively charged chloride ions to the other electrode (the anode). As the sodium ions interact with the cathode, they absorb electrons from the external circuit, forming sodium metal. This absorption of electrons is known as reduction of the sodium ions. At the anode, chloride ions lose electrons to the external circuit and are converted to chlorine gas—an oxidation reaction.

The chemical reaction occurring in electrolysis requires a minimum voltage. Sufficient voltage also must be supplied to overcome internal resistance in the electrolyte. Because an electrolyte may contain a variety of ions, alternative reactions are possible. In aqueous electrolytes, positively charged hydrogen ions are always present and can be reduced at the cathode. In a water solution of sodium chloride, hydrogen ions are reduced to the exclusion of sodium ions, and no sodium metal forms. The only metals that can be liberated by electrolysis from aqueous solution are those such as copper, silver, cadmium, and zinc, whose ions are more easily reduced than hydrogen ions.

An electrochemical method of liberating active metals depends on finding a liquid medium more resistant to reduction than water. The use of molten salts as electrolytes often solves this problem. Salts usually have high melting points, and it can be advantageous to choose mixtures of salts with lower melting temperatures, as long as the new cations introduced do not interfere with the desired reduction reaction. Choice of electrode materials is important because it can affect the purity of the metal being liberated. Usually inert electrodes are preferable.

In electroplating, it is necessary to produce a uniform adherent coating on the object that forms the cathode. Successful plating requires careful control of several factors such as temperature, concentration of the metal ion, current density at the cathode, and cleanliness of the surface to be plated. The metal ions in the electrolyte may need to be replenished as they are being depleted. Current density is usually kept low and plating done slowly. Sometimes additives that react with the metal ions or modify the viscosity or surface tension of the medium are used in the electrolyte. Successful plating conditions are often discovered by experiment and may not be understood in a fundamental sense.

Electroforming consists of depositing a metal plate on an object with the purpose of preparing a duplicate of the object in all its detail. Nonconducting objects can be rendered conductive by coating them with a conductive layer. The plated metal forms on a part called a mandrel, which forms a template and from which the new metal part is separated after the operation. The mandrel may be saved and used again or dissolved away with chemicals to separate it from the new object.

In electrorefining, anodes of impure metals are immersed in an electrolyte that contains a salt of the metal to be refined—often a sulfate—and possibly sulfuric acid. When current is passed through, metal dissolves from the anode and is redeposited on the cathode in purer form. Control of the applied voltage is necessary to prevent dissolution of less easily oxidized metal impurities. Material that is not oxidized at the anode falls to the bottom of the electrolysis cell as anode slime. This slime can be further processed for any valuable materials it may contain.

Applications and Products

Electrowinning of Aluminum. Aluminum, the most abundant metal in the Earth's crust, did not become readily available commercially until the development of the Hall-Héroult process. This process involves electrolysis of dry aluminum oxide (alumina) dissolved in cryolite (sodium aluminum hexafluoride). Additional calcium fluoride is used to lower the melting point of the cryolite. The process runs at about 960 degrees Celsius and uses carbon electrodes. The alumina for the Hall-Héroult process is obtained from an ore called bauxite, an impure aluminum oxide with varying amounts of compounds such as iron oxide and silica. The preparation of pure alumina follows the Bayer process: The alumina is extracted from the bauxite as a solution in sodium hydroxide (caustic soda), reprecipitated by acidification, filtered, and dried. The electrolysis cell has a carbon coating at the bottom that forms a cathode, while the anodes are graphite rods extending down into the molten salt electrolyte. As aluminum forms, it forms a pool at the bottom of the cell and can be removed periodically

The anodes are consumed as the carbon reacts with the oxygen liberated by the electrolysis. The alumina needs to be replenished from time to time in the melt. The applied voltage is about 4.5 volts, but it rises sharply if the alumina concentration is too low. The electrolysis cells are connected in series, and there may be several hundred cells in an aluminum plant. The consumption of electric power amounts to about 15,000 kilowatt-hours per ton of aluminum. (A family in a home might consume 150 kilowatt-hours of power in a month.) Much of the power goes for heating and melting the electrolyte. The cost of electric power is a significant factor in aluminum manufacture and makes it advantageous to locate plants where power is relatively inexpensive, for example, where hydropower is available. Although some other metals are produced by electrolysis, aluminum is the metal produced in the greatest amount, tens of millions of metric tons per year.

Aluminum is the most commonly used structural metal after iron. As a low-density strong metal, aluminum tends to find uses where weight saving is important, such as in aircraft. When automobile manufacturers try to increase gas mileage, they replace the steel in vehicles with aluminum to save weight. Aluminum containers are commonly used for foods and beverages and aluminum foil for packaging.

Alcoa has developed a second electrolytic aluminum process that involves electrolysis of aluminum chloride. The aluminum chloride is obtained by chlorinating aluminum oxide. The chlorine liberated at the anode can be recycled. Also the electric power requirements of this process are less than for the Hall-Héroult process. This aluminum chloride reduction has not been used as much as the oxide reduction.

Electrowinning of Other Metals. Magnesium, like aluminum, is a low-density metal and is also manufactured by electrolysis. The electrolysis of molten magnesium chloride (in the presence of other metal chlorides to lower the melting point) yields magnesium at the cathode. The scale of magnesium production is not as large as that of aluminum, amounting to several hundred thousand metric tons per year. Magnesium is used in aircraft alloys, flares, and chemical syntheses.

Sodium metal comes from the electrolysis of molten sodium chloride in an apparatus called the Downs cell after its inventor J. C. Downs. Molten sodium forms at the cathode. Sodium metal was formerly very important in the process for making the gasoline additive tetraethyl lead. As leaded fuel is no longer sold in the United States, this use has declined, but sodium continues to be used in organic syntheses, the manufacture of titanium, and as a component (with potassium) in high-temperature heat exchange media for nuclear reactors.

Lithium and calcium are obtained in relatively small quantities by the electrolysis of their chlorides. Lithium is assuming great importance for its use in high-performance batteries for all types of applications but particularly for powering electric automobiles. A lightweight lithium-aluminum alloy is used in the National Aeronautics and Space Administration's Ares rocket and the external tank of the space shuttle.

Electrorefining Applications. Metals are often obtained from their ores by pyrometallurgy (sequences of heating) and the use of reducing agents such as carbon. Metals obtained this way include iron and copper. Iron can be refined electrolytically, but much iron is used for steel production without refining.

Copper, however, is used in applications where purity is important. Copper, when pure, is ductile and an excellent electrical conductor, so it needs to be refined to be used in electrical wiring. Copper anodes (blister copper) are suspended in a water solution containing sulfuric acid and copper sulfate with steel cathodes. Electrolysis results in dissolution of copper from the anode and migration of copper ions to the cathode, where purified metal is deposited. The result is copper of 99.9 percent purity. A similar procedure may be used in recycling copper. Other metals that are electrorefined include aluminum, zinc, nickel, cobalt, tin, lead, silver, and gold. Materials are added to the electrolyte to make the metal deposit more uniform: glue, metal chlorides, levelers, and brighteners may be included. The details of the conditions for electrorefining vary depending on the metal. The electrorefining done industrially worldwide ranges from 1,000 to 100,000 metric tons of metals per year.

Electroforming Applications. The possibility of reproducing complicated shapes on both large and small scale is an advantage of electroforming. Objects that would be impossible to produce because of their intricate shapes or small sizes are made by electroforming. The metal used may be a single metal or an alloy. The metals used most often are nickel and copper. The manufacture of compact discs for recording sound makes use of electroforming in the reproduction of the bumps and grooves in a studio disk of glass, which is a negative copy, and is then used to make a mold from which plastic discs can be cast. Metal foil can be electroformed by using a rotating mandrel surrounded by a cylindrical anode. The foil is peeled off the mandrel in a continuous sheet. The electroforming of copper foil for electronic applications is the largest application of electroforming. About $2 billion per year is spent on electroforming.

Electroplating Applications. Plating is done to protect metal surfaces from corrosion, to enhance the appearance of a surface, or to modify its properties in other ways such as to increase hardness or reflectivity. Familiar applications include chromium plating of automobile parts, silver plating of tableware and jewelry, and gold plating of medals and computer parts. Plating can also be done on nonmetallic surfaces such as plastic or ceramics. The manufacture of circuit boards requires a number of steps, some of which involve the electroplating of copper, lead, and tin. Many switches and other electrical contacts are plated with gold to prevent corrosion. The metals involved in commercial electroplating are mostly deposited from an aqueous electrolyte. This excludes metals such as aluminum or magnesium, which cannot be liberated in water solution. If a molten salt electrolyte is used, aluminum plating is possible, but it is seldom done.

Fascinating Facts About Electrometallurgy

- The engine block of the airplane built by Orville and Wilbur Wright in 1903 was made of aluminum obtained by electrolysis.
- In the early nineteenth century, aluminum was so expensive that it was used in jewelry.
- Roughly 5 percent of the electricity in the United States is used to make aluminum.
- The price of aluminum fell to its lowest recorded level in 1993; $0.53 per pound.
- The U.S. industry that uses the most aluminum is the beverage industry. Americans use 80 billion aluminum cans per year.
- When Prince Charles of England was invested as Prince of Wales in 1969, the coronet he received was made partly by electroforming.
- The Voyager space vehicle launched in 1977 contained a gold-plated copper disk electroplated with a patch of pure uranium-238 isotope. If this disk were recovered by some advanced alien civilization, the disk's age could be determined from the radioactive decay rate of the isotope.

Impact on Industry

In 2006, 33.4 million tons of primary aluminum was produced worldwide, with the majority of it being made in China, the United States, Russia, Canada, Brazil, Australia, Norway, and India. Magnesium production in 2008 was 719,000 tons. These two metals account for most of the electrochemical metal production. Because of the high economic value of these industries, the processes involved are the subjects of research at government institutions and universities and at major corporations. Continuing research is done to improve the energy efficiency of processes and to mitigate pollution problems. A continuing problem with the Hall-Héroult process is emission of fluorine-containing compounds that are either toxic or are greenhouse gases. The high temperature of the process leads to corrosion problems in the cells, but attempts to lower the temperature can lead to reduced solubility of alumina in the electrolyte.

Alcoa and Rio Tinto Alcan are two of the largest aluminum producers in the world. Each has tens of thousands of employees spread over forty countries. Alcoa maintains a technical research center near Pittsburgh. In 2007, the company signed an agreement with a consortium of universities in Russia to sponsor research and development. Similar agreements were made in both India and China. Rio Tinto Alcan makes grants to several Canadian universities for research on aluminum production.

The National Research Council of Canada has an aluminum technical center in Saguenay, Quebec, and the E. B. Yeager Center for Electrochemical Science at Case Western University is in Cleveland, Ohio. The center sponsors workshops, lectures, and research in all areas of electrochemistry, including the use of computer-controlled processes to achieve optimum conditions in manufacturers. U.S. government laboratories such as Argonne National Laboratory in Illinois also do research on electrochemistry. The U.S. Department of Energy supports university research at various institutions, including Michigan Technological University, which received a $2 million grant for research on uses for aluminum smelting wastes.

Careers and Course Work

The path to a career in electrometallurgy is through a bachelor's or master's degree in chemistry or chemical engineering, although careers in research require a doctorate. The course work involves a thorough grounding in physical science (chemistry, physics), two years of calculus, and courses in computer science and engineering principles. Most large state universities offer programs in chemistry and chemical engineering. A few universities, including the University of Utah and the University of Nevada at Reno, offer specialized work in electrometallurgy. Rio Tinto Alcan and Alcoa offer summer internships for students and Rio Tinto Alcan also offers scholarships.

Multiday, personal development courses in electroplating are widely available through groups such as the National Society for Surface Finishing.

Social Context and Future Prospects

The electometallurgy industry, like many industries, poses challenges for society. Metals have great value and many uses, which are an essential part of modern life. Unfortunately, electrometallurgy consumes huge amounts of energy and uses tons of unpleasant chemicals. In addition, aluminum plants emit carbon dioxide and fluorine compounds. However, the use of electricity to produce metals probably remains the cleanest and most efficient method.

In the future, the process of electrometallurgy probably will become more efficient and less polluting. Also, new techniques are likely to be developed to permit additional metals to be obtained by electrometallurgy. Titanium metal continues to be made by reduction of its chloride by sodium metal, but the Fray Farthing Chen (FFC) Cambridge process announced in 2000 shows promise as an electrolytic method for obtaining titanium. The process involves reduction of titanium oxide by electrically generated calcium in a molten calcium chloride medium. Titanium is valued for its strength, light weight, high temperature performance, and corrosion resistance. These qualities make it essential in jet engine turbines. Titanium is stable in the human body and can be used for making artificial knees and hips for implantation.

John R. Phillips, B.S., Ph.D.

Further Reading

Chen, G. Z., D. J. Fray, and T. W. Farthing. "Direct Electrochemical Reduction of Titanium Dioxide to Titanium in Molten Calcium Chloride." *Nature* 407 (2000): 361-364. First description of the FFC-Cambridge method—a possibly game-changing procedure for reducing metal oxides.

Curtis, Leslie. *Electroforming*. London: A & C Black, 2004. A simple description of electroforming with practical directions for applications.

Geller, Tom. "Common Metal, Uncommon Past." *Chemical Heritage* 25 no. 4 (Winter, 2007): 32-36. Historical details on the discovery and manufacture of aluminum.

Graham, Margaret B. W., and Bettye H. Pruitt. *R&D for Industry: A Century of Technical Innovation at Alcoa*. New York: Cambridge University Press, 1990. An extensive history that is mostly nontechnical. Much discussion of organizational and management matters.

Kanani, Nasser. *Electroplating: Basic Principles, Processes and Practice*. New York: Elsevier, 2005. A thorough discussion of electroplating processes including measurements of thickness and adherence using modern instruments.

Mertyns, Joost. "From the Lecture Room to the Workshop: John Frederic Daniell, the Constant Battery and Electrometallurgy Around 1840." *Annals of Science* 55, no. 3 (July, 1998): 241-261. Early developments in batteries, electroplating, and electroforming are described.

Pletcher, Derek, and Frank C. Walsh. *Industrial Electrochemistry*. 2d ed. New York: Blackie Academic & Professional, 1992. Particularly good discussion of metal plating and electroforming with both theoretical and practical aspects treated. List of definitions and units.

Popov, Konstantin I., Stojan S. Djokić, and Branimir N. Grgur, eds. *Fundamental Aspects of Metallurgy*. New York: Kluwer Academic/Plenum, 2002.

Provides background on metallurgy, then the specifics of metal disposition.

Web Sites

The Electrochemical Society
http://www.electrochem.org

International Society of Electrochemistry
http://www.ise-online.org

See also: Aeronautics and Aviation; Chemical Engineering; Metallurgy.

ELECTRONIC MATERIALS PRODUCTION

FIELDS OF STUDY

Mathematics; physics; chemistry; crystallography; quantum theory; thermodynamics

SUMMARY

While the term "electronic materials" commonly refers to the silicon-based materials from which computer chips and integrated circuits are constructed, it technically includes any and all materials upon which the function of electronic devices depends. This includes the plain glass and plastics used to house the devices to the exotic alloys and compounds that make it possible for the devices to function. Production of many of these materials requires not only rigorous methods and specific techniques but also requires the use of high-precision analytical methods to ensure the structure and quality of the devices.

KEY TERMS AND CONCEPTS

- **Biasing:** The application of a voltage to a semiconductor structure (transistor) to induce a directional current flow in the structure.
- **Czochralski Method:** A method of pulling material from a molten mass to produce a single large crystal.
- **Denuded Zone:** Depth and area of a silicon wafer that contains no oxygen precipitates or interstitial oxygen.
- **Epi Reactor:** A thermally programmable chamber in which epitaxial growth of silicon chips is carried out.
- **Gettering:** A method of lowering the potential for precipitation from solution of metal contaminants in silicon, achieved by controlling the locations at which precipitation can occur.
- **Polysilicon (Metallurgical Grade Silicon):** A form of silicon that is 99 percent pure, produced by the reaction of silicon dioxide (SiO_2) and carbon (C) to produce silicon (Si) and carbon monoxide (CO) at a temperature of 2000 degrees Celsius.

DEFINITION AND BASIC PRINCIPLES

Electronic materials are those materials used in the construction of electronic devices. The major electronic material today is the silicon wafer, from which computer chips and integrated circuits (ICs) are made. Silicon is one of a class of elements known as semiconductors. These are materials that do not conduct electrical currents appreciably unless acted upon, or "biased," by an external voltage. Another such element is germanium.

The construction of silicon chips requires materials of high purity and consistent internal structure. This, in turn, requires precisely controlled methods in the production of both the materials and the structures for which they are used. Large crystals of ultrapure silicon are grown from molten silicon under strictly controlled environmental conditions. Thin wafers are sliced from the crystals and then polished to achieve the desired thickness and mirror-smooth surface necessary for their purpose. Each wafer is then subjected to a series of up to five hundred, and sometimes more, separate operations by which extremely thin layers of different materials are added in precise patterns to form millions of transistor structures. Modern CPU (central processing unit) chips have between 10^7 and 10^9 separate transistors per square centimeter etched on their surfaces in this way.

One of the materials added by the thin-layer deposition process is silicon, to fill in spaces between other materials in the structures. These layers must be added epitaxially, in a way that maintains the base crystal structure of the silicon wafer.

Other materials used in electronic devices are also formed under strictly controlled environmental conditions. Computers could not function without some of these materials, especially indium tin oxide (ITO) for what are called transparent contacts and indium nitride for light-emitting diodes in a full spectrum range of colors.

BACKGROUND AND HISTORY

The production of modern electronic materials began with the invention of the semiconductor bridge transistor in 1947. This invention, in turn, was made possible by the development of quantum theory and the vacuum tube technology with which electronic devices functioned until that time.

The invention of the transistor began the development of electronic devices based on the semiconducting character of the element silicon. Under the influence of an applied voltage, silicon can be induced to conduct an electrical current. This feature allows silicon-based transistors to function somewhat like an on-off switch according to the nature of the applied voltage.

In 1960, the construction of the functional laser by American physicist and Nobel laureate Arthur Schawlow began the next phase in the development of semiconductor electronics, as the assembly of transistors on silicon substrates was still a tedious endeavor that greatly limited the size of transistor structures that could be constructed. As lasers became more powerful and more easily controlled, they were applied to the task of surface etching, an advance that has produced ever smaller transistor structures. This development has required ever more refined methods of producing silicon crystals from which thin wafers can be cut for the production of silicon semiconductor chips, the primary effort of electronic materials production (though by no means the most important).

How It Works

Melting and Crystallization. Chemists have long known how to grow large crystals of specific materials from melts. In this process, a material is heated past its melting point to become liquid. Then, as the molten material is allowed to cool slowly under controlled conditions, the material will solidify in a crystalline form with a highly regular atomic distribution.

Now, molten silicon is produced from a material called polysilicon, which has been stacked in a closed oven. Specific quantities of doping materials such as arsenic, phosphorus, boron, and antimony are added to the mixture, according to the conducting properties desired for the silicon chips that will be produced. The polysilicon melt is rotated in one direction (clockwise); then, a seed crystal of silicon, rotating in the opposite direction (counterclockwise), is introduced. The melt is carefully cooled to a specific temperature as the seed crystal structure is drawn out of the molten mass at a rate that determines the diameter of the resulting crystal.

To maintain the integrity of the single crystal that results, the shape is allowed to taper off into the form of a cone, and the crystal is then allowed to cool completely before further processing. The care with which this procedure is carried out produces a single crystal of the silicon alloy as a uniform cylinder, whose ends vary in diameter first as the desired extraction rate was achieved and then due to the formation of the terminal cone shape.

Wafers. In the next stage of production, the nonuniform ends of the crystal are removed using an inner diameter saw. The remaining cylinder of crystal is called an ingot, and is then examined by X ray to determine the consistency and integrity of the crystal structure. The ingot then will normally be cut into smaller sections for processing and quality control.

To produce the rough wafers that will become the substrates for chips, the ingot pieces are mounted on a solid base and fed into a large wire saw. The wire saw uses a single long moving wire to form a thick network of cutting edges. A continuous stream of slurry containing an extremely fine abrasive provides the cutting capability of the wire saw, allowing the production of many rough wafers at one time. The rough wafers are then thoroughly cleaned to remove any residue from the cutting stage.

Another procedure rounds and smooths the edges of each wafer, enhancing its structural strength and resistance to chipping. Each wafer is also laser-etched with identifying data. They then go on to a flat lapping procedure that removes most of the machining marks left by the wire saw, and then to a chemical etching process that eliminates the marking that the lapping process has left. Both the lapping process and the chemical etching stage are used to reduce the thickness of the wafers.

Polishing. Following lapping and rigorous cleaning, the wafers move into an automated chemical-mechanical polishing process that gives each wafer an extremely smooth mirror-like and flat surface. They are then again subjected to a series of rigorous chemical cleaning baths, and are then either packaged for sale to end users or moved directly into the epitaxial enhancement process.

Epitaxial Enhancement. Epitaxial enhancement is used to deposit a layer of ultrapure silicon on the surface of the wafer. This provides a layer with different properties from those of the underlying wafer material, an essential feature for the proper functioning of the MOS (metal-oxide-semiconductor) transistors that are used in modern chips. In this process, polished wafers are placed into a programmable oven and spun in an atmosphere of trichlorosilane gas. Decomposition of the trichlorosilane

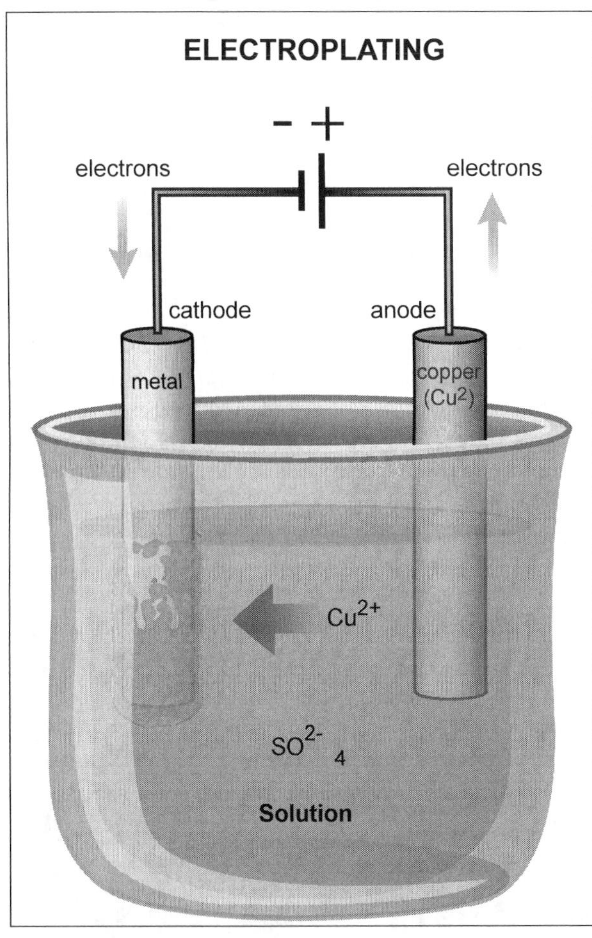

Electroplating is the creation of an adherent metal layer on some substrate by electrolysis.

deposits silicon atoms on the surface of the wafers. While this produces an identifiable layer of silicon with different properties, it also maintains the crystal structure of the silicon in the wafer. The epitaxial layer contains no imperfections that may exist in the wafer and that could lead to failure of the chips in use.

From this point on, the wafers are submitted to hundreds more individual processes. These processes build up the transistor structures that form the functional chips of a variety of integrated circuit devices and components that operate on the principles of digital logic.

Applications and Products

Microelectronics. The largest single use of silicon chips is in the microelectronics industry. Every digital device functions through the intermediacy of a silicon chip of some kind. This is as true of the control pad on a household washing machine as it is of the most sophisticated and complex CPU in an ultramodern computer.

Digital devices are controlled through the operation of digital logic circuits constructed of transistors built onto the surface of a silicon chip. The chips can be exceedingly small. In the case of integrated circuit chips, commonly called ICs, only a few transistors may be required to achieve the desired function.

The simplest of these ICs is called an inverter, and a standard inverter IC provides six separate inverter circuits in a dual inline package (DIP) that looks like a small rectangular block of black plastic about 1 centimeter wide, 2 centimeters long, and 0.5 centimeters thick, with fourteen legs, seven on each side. The actual silicon chip contained within the body of the plastic block is approximately 5 millimeters square and no more than 0.5 millimeters thick. Thousands of such chips are cut from a single silicon wafer that has been processed specifically for that application.

Inverters require only a single input lead and a single output lead, and so facilitate six functionalities on the DIP described. However, other devices typically use two input leads to supply one output lead. In those devices, the same DIP structure provides only four functionalities. The transistor structures are correspondingly more complex, but the actual chip size is about the same. Package sizes increase according to the complexity of the actual chip and the number of leads that it requires for its function and application to physical considerations such as dissipation of the heat that the device will generate in operation.

In the case of a modern laptop or desktop computer, the CPU chip package may have two hundred leads on a square package that is approximately 4 centimeters on a side and less than 0.5 centimeters in thickness. The actual chip inside the package is a very thin sheet of silicon about 1 square centimeter in size, but covered with several million transistor structures that have been built up through photo-etching and chemical vapor deposition methods, as described above. Examination of any service listing of silicon chip ICs produced by any particular manufacturer will quickly reveal that a vast number of different ICs and functionalities are available.

Solar Technology. There are several other current uses for silicon wafer technology, and new uses are yet to be realized. Large quantities of electronic-grade silicon wafers are used in the production of functional solar cells, an area of application that is experiencing high growth, as nonrenewable energy resources become more and more expensive. Utilizing the photoelectron effect first described by Albert Einstein in 1905, solar cells convert light energy into an electrical current. Three types are made, utilizing both thick (> 300 micrometers [μm]) and thin (a few μm) layers of silicon. Thick-layer solar cells are constructed from single crystal silicon and from large-grain polycrystalline silicon, while thin-layer solar cells are constructed by using vapor deposition to deposit a layer of silicon onto a glass substrate.

Microelectronic and Mechanical Systems. Silicon chips are also used in the construction of microelectronic and mechanical systems (MEMS). Exceedingly tiny mechanical devices such as gears and single-pixel mirrors can be constructed using the technology developed for the production of silicon chips. Devices produced in this way are by nature highly sensitive and dependable in their operation, and so the majority of MEMS development is for the production of specialized sensors, such as the accelerometers used to initiate the deployment of airbag restraint systems in automobiles. A variety of other products are also available using MEMS technology, including biosensors, the micronozzles of inkjet printer cartridges, microfluidic test devices, microlenses and arrays of microlenses, and microscopic versions of tunable capacitors and resonators.

Other Applications. Other uses of silicon chip technology, some of which is in development, include mirrors for X-ray beams; mirrors and prisms for application in infrared spectroscopy, as silicon is entirely transparent to infrared radiation; and the material called porous silicon, which is made electrochemically from single-crystal silicon and has itself presented an exceptionally varied field of opportunity in materials science.

As mentioned, there are also many other materials that fall into the category of electronic materials. Some, such as copper, gold, and other pure elements, are produced in normal ways and then subjected to methods such as zone refining and vapor deposition techniques to achieve high purity and thin layers in the construction of electronic devices. Many exotic elements and metallic alloys, as well as specialized plastics, have been developed for use in electronic devices. Organic compounds known as liquid crystals, requiring no extraordinary synthetic measures, are normally semisolid materials that have properties of both a liquid and a solid. They are extensively used as the visual medium of thin liquid crystal display (LCD) screens, such as would be found in wristwatches, clocks, calculators, laptop and tablet computers, almost all desktop monitors, and flat-screen televisions.

Another example is the group of compounds made up of indium nitride, gallium nitride, and aluminum nitride. These are used to produce light-emitting diodes (LEDs) that provide light across the full visible spectrum. The ability to grow these LEDs on the same chip now offers a technology that could completely replace existing CRT (cathode ray tube) and LCD technologies for visual displays.

Impact on Industry

Electronic materials production is an entire industry unto itself. While the products of this industry are widely used throughout society, they are not used in the form in which they are produced. Rather, the products of the electronic materials industry become input supplies for further manufacturing processes. Silicon chips, for example, produced by any individual manufacturer, are used for in-house manufacturing or are marketed to other manufacturers, who, in turn, use the chips to produce their own particular products, such as ICs, solar cells, and microdevices.

This intramural or business-to-business market aspect of the electronic materials production industry, with its novel research and development efforts and especially given the extent to which society now relies on information transfer and storage, makes ascribing an overall economic value to the industry impossible. One has only to consider the number of computing devices produced and sold each year around the world to get a sense of the potential value of the electronic materials production industry.

Ancillary industries provide other materials used by the electronic materials production industry, many of which must themselves be classified as electronic materials. An electric materials company, for example, may provide polishing and surfacing materials, photovoltaic materials, specialty glasses, electronic packaging materials, and many others.

Given both the extremely small size and sensitivity

of the structures created on the surface of silicon chips and the number of steps required to produce those structures, quality control procedures are stringent. These steps may be treated as part of a multistep synthetic procedure, with each step producing a yield (as the percentage of structures that meet functional requirements). In silicon-chip production, it is important to understand that only the chips that are produced as functional units at the end of the process are marketable. If a process requires two hundred individual construction steps, even a 99 percent success rate for each step translates into a final yield of functional chips of only 0.99^{200}, or 13.4 percent. The majority of chip structures fail during construction, either through damage or through a step failure. It is therefore imperative that each step in the construction of silicon chips be precisely carried out.

To that end, procedures and quality control methods have been developed that are applicable in other situations too. Clean room technology that is essential for maximizing usable chip production is equally valuable in biological research and medical treatment facilities, applied physics laboratories, space exploration, aeronautics repair and maintenance facilities, and any other situations in which steps to protect either the environment or personnel from contamination must be taken.

Careers and Coursework

Electronic materials production is a specialist field that requires interested students to take specialist training in many subject areas. For many such careers, a university degree in solid state physics or electronic engineering is required. For those who will specialize in the more general field of materials science, these subject areas will be included in the overall curriculum. Silicon technology and semiconductors are also primary subject areas. The fields of study listed here are considered prerequisites for specialist study in the field of electronic materials production, and students can expect to continue studies in these subjects as new aspects of the field develop.

Researchers are now looking into the development of transistor structures based on graphene. This represents an entirely new field of study and application, and the technologies that develop from it will also set new requirements for study. High-end spectrometric methodologies are essential tools in the study and development of this field, and students can expect to take advanced study and training in the use of techniques such as scanning probe microscopy.

Fascinating Facts About Electronic Materials Production

- Large single crystals of silicon are grown from a molten state in a process that literally pulls the molten mass out into a cylindrical shape.
- About 70 percent of silicon chips fail during the manufacturing process, leaving only a small percentage of chips that are usable.
- Silicon is invisible to infrared light, making it exceptionally useful for infrared spectroscopy and as mirrors for X rays.
- Quantum dot and graphene-based transistors will produce computers that are orders of magnitude more powerful than those used today.
- The photoelectric effect operating in silicon allows solar cells to convert light energy into an electrical current.
- Semiconductor transistors were invented in 1947 and integrated circuits in 1970, and the complexity of electronic components has increased by about 40 percent each year.
- Copper and other metals dissolve very quickly in liquid silicon, but precipitate out as the molten material cools, often with catastrophic results for the silicon crystal.
- Porous silicon is produced electrochemically from single crystals of silicon. Among its other properties, porous silicon is highly explosive.

Social Context and Future Prospects

Moore's law has successfully predicted the progression of transistor density that can be inscribed onto a silicon chip. There is a finite limit to that density, however, and the existing technology is very near or at that limit. Electronic materials research continues to improve methods and products in an effort to push the Moore limit.

New technologies must be developed to make the use of transistor logic as effective and as economic as possible. To that end, there exists a great deal of research into the application of new materials. Foremost is the development of graphene-based transistors and quantum dot technology, which will drive the level of technology into the molecular and atomic scales.

Richard M. Renneboog, M.Sc.

Further Reading

Akimov, Yuriy A., and Wee Shing Koh. "Design of Plasmonic Nanoparticles for Efficient Subwavelength Trapping in Thin-Film Solar Cells." *Plasmonics* 6 (2010): 155-161. This paper describes how solar cells may be made thinner and lighter by the addition of aluminum nanoparticles on a surface layer of indium tin oxide to enhance light absorption.

Askeland, Donald R. *The Science and Engineering of Materials.* London: Chapman & Hall, 1998. A recommended resource, this book provides a great deal of fundamental background regarding the physical behavior of a wide variety of materials and processes that are relevant to electronic materials production.

Falster, Robert. "Gettering in Silicon: Fundamentals and Recent Advances." *Semiconductor Fabtech* 13 (2001). This article provides a thorough description of the effects of metal contamination in silicon and the process of gettering to avoid the damage that results from such contamination.

Zhang, Q., et al. "A Two-Wafer Approach for Integration of Optical MEMS and Photonics on Silicon Substrate." *IEEE Photonics Technology Letters* 22 (2010): 269-271. This paper examines how photonic and micro-electromechanical systems on two different silicon chips can be precisely aligned.

Zheng, Y., et al. "Graphene Field Effect Transistors with Ferroelectric Gating." *Physical Review Letters* 105 (2010). This paper discusses the experimental development and successful testing of a graphene-based field-effect transistor system using gold and graphene electrodes with SiO_2 gate structures on a silicon substrate.

Web Sites

SCP Symposium (June 2005) "Silicon Starting Materials for Sub-65nm Technology Nodes."
http://www.memc.com/assets/file/technology/papers/SCP-Symposium-Seacrist.pdf

University of Kiel "Electronic Materials Course."
http://www.tf.uni-kiel.de/matwis/amat/elmat_en/index.html

See also: Applied Physics; Computer Engineering; Computer Science; Electronics and Electronic Engineering; Integrated-Circuit Design.

ELECTRONICS AND ELECTRONIC ENGINEERING

FIELDS OF STUDY

Mathematics; physics; electronics; electrical engineering; automotive mechanics; analytical technology; chemical engineering; aeronautics; avionics; robotics; computer programming; audio/video technology; metrology; audio engineering; telecommunications; broadcast technology; computer technology; computer engineering; instrumentation

SUMMARY

A workable understanding of the phenomenon of electricity originated with proof that atoms were composed of smaller particles bearing positive and negative electrical charges. The modern field of electronics is essentially the science and technology of devices designed to control the movement of electricity to achieve some useful purpose. Initially, electronic technology consisted of devices that worked with continuously flowing electricity, whether direct or alternating current. Since the development of the transistor in 1947 and the integrated circuit in 1970, electronic technology has become digital, concurrent with the ability to assemble millions of transistor structures on the surface of a single silicon chip.

KEY TERMS AND CONCEPTS

- **Cathode Rays:** Descriptive term for energetic beams emitted from electrically stimulated materials inside of a vacuum tube, identified by J. J. Thomson in 1897 as streams of electrons.
- **Channel Rays:** Descriptive term for energetic beams having the opposite electrical charge of cathode rays, emitted from electrically stimulated materials inside a vacuum tube, also identified by J. J. Thomson in 1897.
- **Gate:** A transistor structure that performs a specific function on input electrical signals to produce specific output signals.
- **Operational Amplifier (Op-Amp):** An integrated circuit device that produces almost perfect signal reproduction with high gains of amplification and precise, stable voltages and currents.
- **Sampling:** Measurement of a specific parameter such as voltage, pressure, current, and loudness at a set frequency determined by a clock cycle such as 1 MHz.
- **Semiconductor:** An element that conducts electricity effectively only when subjected to an applied voltage.
- **Zener Voltage:** The voltage at which a Zener diode is designed to operate at maximum efficiency to produce a constant voltage, also called the breakdown voltage.

Definition and Basic Principles

The term "electronics" has acquired different meanings in different contexts. Fundamentally, "electronics" refers to the behavior of matter as affected by the properties and movement of electrons. More generally, electronics has come to mean the technology that has been developed to function according to electronic principles, especially pertaining to basic digital devices and the systems that they operate. The term "electronic engineering" refers to the practice of designing and building circuitry and devices that function on electronic principles.

The underlying principle of electronics derives from the basic structure of matter: that matter is composed of atoms composed of smaller particles. The mass of atoms exists in the atomic nucleus, which is a structure composed of electrically neutral particles called neutrons and positively charged particles called protons. Isolated from the nuclear structure by a relatively immense distance is an equal number of negatively charged particles called electrons. Electrons are easily removed from atoms, and when a difference in electrical potential (voltage) exists between two points, electrons can move from the area of higher potential toward that of lower potential. This defines an electrical current.

Devices that control the presence and magnitude of both voltages and currents are used to bring about changes to the intrinsic form of the electrical signals so generated. These devices also produce physical changes in materials that make comprehensible the information carried by the electronic signal.

Background and History

Archaeologists have found well-preserved Parthian relics that are now believed to have been rudimentary, but functional, batteries. It is believed that these ancient devices were used by the Parthians to plate objects with gold. The knowledge was lost until 1800, when Italian physicist Alessandro Volta reinvented the voltaic pile. Danish physicist and chemist Hans Christian Oersted demonstrated the relationship between electricity and magnetism in 1820, and in 1821, British physicist and chemist Michael Faraday used that relationship to demonstrate the electromagnetic principle on which all electric motors work. In 1831, he demonstrated the reverse relationship, inventing the electrical generator in the process.

Electricity was thought, by American statesman and scientist Benjamin Franklin and many other scientists of the eighteenth and nineteenth centuries, to be some mysterious kind of fluid that might be captured and stored. A workable concept of electricity was not developed until 1897, when J. J. Thomson identified cathode rays as streams of light electrical particles that must have come from within the atoms of their source materials. He arbitrarily ascribed their electrical charge as negative. Thomson also identified channel rays as streams of massive particles from within the atoms of their source materials that are endowed with the opposite electrical charge of the electrons that made up cathode rays. These observations essentially proved that atoms have substructures. They also provided a means of explaining electricity as the movement of charged particles from one location to another.

With the establishment of an electrical grid, based on the advocacy of alternating current by Serbian American engineer and inventor Nikola Tesla (1856-1943) , a vast assortment of analogue electrical devices were soon developed for consumer use, though initially these devices were no more than electric lights and electromechanical applications based on electric motors and generators.

As the quantum theory of atomic structure came to be better understood and electricity better controlled, electronic theory became much more important. Spurred by the success of the electromagnetic telegraph of American inventor Samuel Morse (1791-1872), scientists sought other applications. The first major electronic application of worldwide importance was wireless radio, first demonstrated by Italian inventor Guglielmo Marconi (1874-1937). Radio depended on electronic devices known as vacuum tubes, in which structures capable of controlling currents and voltages could operate at high temperatures in an evacuated tube with external contacts. In 1947, American physicist William Shockley and colleagues invented the semiconductor-based transistor, which could be made to function in the same manner as vacuum tube devices, but without the high temperatures, electrical power consumption, and vacuum construction of those analogue devices.

In 1970, the first integrated circuit "chips" were made by constructing very small transistor structures on the surface of a silicon chip. This gave rise to the entire digital technology that powers the modern world.

Applications and Products

Electronics are applied in practically every conceivable manner today, based on their utility in converting easily-produced electrical current into mechanical movement, sound, light, and information signals.

Basic Electronic Devices. Transistor-based digital technology has replaced older vacuum tube technology, except in rare instances in which a transistorized device cannot perform the same function. Electronic circuits based on vacuum tubes could carry out essentially the same individual operations as transistors, but they were severely limited by physical size, heat production, energy consumption, and mechanical failure. Nevertheless, vacuum tube technology was the basic technology that produced radio, television, radar, X-ray machines, and a broad variety of other electronic applications.

Electronic devices that did not use vacuum tube technology, but which operated on electronic and electromagnetic principles, were, and still are, numerous. These devices include electromagnets and all electric motors and generators. The control systems for many such devices generally consisted of nothing more than switching circuits and indicator lights. More advanced and highly sensitive devices required control systems that utilized the more refined and correspondingly sensitive capabilities available with vacuum tube technology.

Circuit Boards. The basic principles of electricity, such as Ohm's resistance law and Kirchoff's current law and capacitance and inductance, are key features in the functional design and engineering of analogue electronic systems, especially for vacuum-tube control systems. An important application that facilitated the general use and development of electronic systems of all kinds is printed circuit board technology.

A printed circuit board accepts standardized components onto a nonconducting platform made initially of compressed fiber board, which was eventually replaced by a resin-based composite board. A circuit design is photo-etched onto a copper sheet that makes up one face of the circuit board, and all nonetched copper is chemically removed from the surface of the board, leaving the circuit pattern. The leads of circuit components such as resistors, capacitors, and inductors are inserted into the circuit pattern and secured with solder connections.

Mass production requirements developed the flotation soldering process, whereby preassembled circuit boards are floated on a bed of molten solder, which automatically completes all solder connections at once with a high degree of consistency. This has become the most important means of circuit board production since the development of transistor technology, being highly compatible with mechanization and automation and with the physical shapes and dimensions of integrated circuit (IC) chips and other components.

Digital Devices. Semiconductor-based transistors comprise the heart of modern electronics and electronic engineering. Unlike vacuum tubes, transistors do not work on a continuous electrical signal. Instead, they function exceedingly well as simple on-off switches that are easily controlled. This makes them well adapted to functions based on Boolean algebra. All transistor structures consist of a series of "gates" that perform a specific function on the electronic signals that are delivered to them.

Digital devices now represent the most common (and rapidly growing) application of electronics and electronic engineering, including relatively simple consumer electronic devices such as compact fluorescent light bulbs and motion-detecting air fresheners to the most advanced computers and analytical instrumentation. All applications, however, utilize an extensive, but limited, assortment of digital components in the form of IC chips that have been designed to carry out specific actions with electrical or electromagnetic input signals.

Input signals are defined by the presence or absence of a voltage or a current, depending upon the nature of the device. Inverter gates reverse the sense of the input signal, converting an input voltage (high input) into an output signal of no voltage (low output), and vice versa. Other transistor structures (gates) called AND, NAND, OR, NOR and X-OR function to combine input signals in different ways to produce corresponding output signals. More advanced devices (for example, counters and shift registers) use combinations of the different gates to construct various functional circuits that accumulate signal information or that manipulate signal information in various ways.

One of the most useful of digital IC components is the operational amplifier, or Op-Amp. Op-Amps contain transistor-based circuitry that boosts the magnitude of an input signal, either voltage or current, by five orders of magnitude (100,000 times) or more, and are the basis of the exceptional sensitivity of the modern analytical instruments used in all fields of science and technology.

Electrical engineers are involved in all aspects of the design and development of electronic equipment. Engineers act first as the inventors and designers of electronic systems, conceptualizing the specific functions a potential system will be required to carry out. This process moves through the specification of the components required for the system's functionality to the design of new system devices. The design parameters extend to the infrastructure that must support the system in operation. Engineers determine the standards of safety, integrity, and operation that must be met for electronic systems.

Consumer Electronics. For the most part, the term "electronics" is commonly used to refer to the electronic devices developed for retail sale to consumers. These devices include radios, television sets, DVD and CD players, cell phones and messaging devices, cameras and camcorders, laptops, tablets, printers, computers, fax and copy machines, cash registers, and scanners. Millions such devices are sold around the world each day, and numerous other businesses have formed to support their operation.

Impact on Industry

With electrical and electronic technology now intimately associated with all aspects of society, the impact of electronics and electronic engineering on industry is immeasurable. It would be entirely fair to say that modern industry could not exist without electronics. Automated processes, which are ubiquitous, are not possible without the electronic systems that control them.

The transportation industry, particularly the

automotive industry, is perhaps the most extensive user of electronics and electronic engineering. Modern automobiles incorporate an extensive electronic network in their construction to provide the ignition and monitoring systems for the operation of their internal combustion engines and for the many monitoring and control systems for the general safe operation of the vehicle; an electronic network also informs and entertains the driver and passengers. In some cases, electronic systems can completely take control of the vehicle to carry out such specific programmable actions as speed control and parallel parking. Every automated process in the manufacture of automobiles and other vehicles serves to reduce the labor required to carry out the corresponding tasks, while increasing the efficiency and precision of the process steps. Added electronic features also increase the marketability of the vehicles and, hence, manufacturer profits.

Processes that have been automated electronically also have become core components of general manufacturing, especially in the control of production machinery. For example, shapes formed from bent tubing are structural components in a wide variety of applications. While the process of bending the tubing can be carried out by the manual operation of a suitably equipped press, an automated process will produce tube structures that are bent to exact angles and radii in a consistent manner. Typically, a human operator places a straight tube into the press, which then positions and repositions the tube for bending over its length, according to the program that has been entered into the manufacturing system's electronic controller. Essentially, all continuous manufacturing operations are electronically controlled, providing consistent output.

Electronics and electronic engineering make up the essence of the computer industry; indeed, electronics is an industry worth billions of dollars annually. Electronics affects not only the material side of industry but also the theoretical and actuarial side. Business management, accounting, customer relations, inventory and sales data, and human resources management all depend on the rapid information-handling that is possible through electronics.

XML (extensible markup language) methods and applications are being used (and are in development) to interface electronic data collection directly to physical processes. This demands the use of specialized electronic sensing and sampling devices to convert measured parameters into data points within the corresponding applications and databases. XML is an application that promises to facilitate information exchange and to promote research using large-scale databases. The outcome of this effort is expected to enhance productivity and to expand knowledge in ways that will greatly increase the efficiency and effectiveness of many different fields.

An area of electronics that has become of great economic importance in recent years is that of electronic commerce: the exclusive use of electronic communication technology for the conduct of business between suppliers and consumers. Electronic communications encompasses interoffice faxing, e-mail exchanges, and the Web commerce of companies

Fascinating Facts About Electronics and Electronic Engineering

- In 1847, George Boole developed his algebra for reasoning that was the foundation for first-order predicate calculus, a logic rich enough to be a language for mathematics.
- In 1950, Alan Turing gave an operational definition of artificial intelligence. He said a machine exhibited artificial intelligence if its operational output was indistinguishable from that of a human.
- In 1956, John McCarthy and Marvin Minsky organized a two-month summer conference on intelligent machines at Dartmouth College. To advertise the conference, McCarthy coined the term "artificial intelligence."
- Digital Equipment Corporation's XCON, short for eXpert CONfigurer, was used in-house in 1980 to configure VAX computers and later became the first commercial expert system.
- In 1989, international chess master David Levy was defeated by a computer program, Deep Thought, developed by IBM. Only ten years earlier, Levy had predicted that no computer program would ever beat a chess master.
- In 2010, the Haystack group at the Computer Science and Artificial Intelligence Laboratory at the Massachusetts Institute of Technology developed Soylent, a word-processing interface that lets users edit, proof, and shorten their documents using Mechanical Turk workers.

such as eBay, Amazon, Google, and of the New York and other stock exchanges. The commercial value of these undertakings is measured in billions of dollars annually, and it is expected to continue to increase as new applications and markets are developed.

The fundamental feature here is that these enterprises exist because of the electronic technology that enables them to communicate with consumers and with other businesses. The electronics technology and electronics engineering fields have thus generated entirely new and different daughter industries, with the potential to generate many others, all of which will depend on persons who are knowledgeable in the application and maintenance of electronics and electronic systems.

Careers and Coursework

Many careers depend on knowledge of electronics and electronic engineering because almost all machines and devices used in modern society either function electronically or utilize some kind of electronic control system. The automobile industry is a prime example, as it depends on electronic systems at all stages of production and in the normal operation of a vehicle. Students pursuing a career in automotive mechanics can therefore be expected to study electronic principles and applications as a significant part of their training. The same reasoning applies in all other fields that have a physical reliance on electronic technology.

Knowledge of electronics has become so essential that atomic structure and basic electronic principles, for example, have been incorporated into the elementary school curriculum. Courses of study in basic electronics in the secondary school curriculum are geared to provide a more detailed and practicable knowledge to students.

Specialization in electronics and other fields in which electronics play a significant role is the province of a college education. Interested students can expect to take courses in advanced mathematics, physics, chemistry, and electronics technology as part of the curriculum of their specialty programs. Normally, a technical career, or a skilled trade, requires a college-level certification and continuing education. In some cases, recertification on a regular schedule is also required to maintain specialist standing in that trade.

Students who plan to pursue a career in electronic engineering at a more theoretical level will require, at minimum, a bachelor's degree. A master's degree can prepare a student for a career in forensics, law, and other professions in which an intimate or specialized knowledge of the theoretical side of electronics can be advantageous. (The Vocational Information Center provides an extensive list of careers involving electronics at http://www.khake.com/page19.html.)

Social Context and Future Prospects

It is difficult, if not impossible, to imagine modern society without electronic technology. Electronics has enabled the world of instant communication, wherein a person on one side of the world can communicate directly and almost instantaneously with someone on the other side of the world. As a social tool, such facile communication has the potential to bring about understanding between peoples in a way that has until now been imagined only in science fiction.

Consequently, this facility has also resulted in harm. While social networking sites, for example, bring people from widely varied backgrounds together peacefully to a common forum, network hackers and so-called cyber criminals use electronic technology to steal personal data and disrupt financial markets.

Electronics itself is not the problem, for it is only a tool. Electronic technology, though built on a foundation that is unlikely to change in any significant way, will nevertheless be transformed into newer and better applications. New electronic principles will come to the fore. Materials such as graphene and quantum dots, for example, are expected to provide entirely new means of constructing transistor structures at the atomic and molecular levels. Compared with the 50 to 100 nanometer size of current transistor technology, these new levels would represent a difference of several orders of magnitude. Researchers suggest that this sort of refinement in scale could produce magnetic memory devices that can store as much as ten terabits of information in one square centimeter of disk surface. Although the technological advances seem inevitable, realizing such a scale will require a great deal of research and development.

Richard M. Renneboog, M.Sc.

Further Reading

Gates, Earl D. *Introduction to Electronics.* 5th ed. Clifton Park, N.Y.: Cengage Learning, 2006. This book presents a serious approach to practical electronic theory beginning with atomic structure and progressing through various basic circuit types to modern digital electronic devices. Also discusses various career opportunities for students of electronics.

Mughal, Ghulam Rasool. "Impact of Semiconductors in Electronics Industry." *PAF-KIET Journal of Engineering and Sciences* 1, no. 2 (July-December, 2007): 91-98. This article provides a learned review of the basic building blocks of semiconductor devices and assesses the effect those devices have had on the electronics industry.

Petruzella, Frank D. *Introduction to Electricity and Electronics 1.* Toronto: McGraw-Hill Ryerson, 1986. A high-school level electronics textbook that provides a beginning-level introduction to electronic principles and practices.

Platt, Charles. *Make: Electronics.* Sebastopol, Calif.: O'Reilly Media, 2009. This book promotes learning about electronics through a hands-on experimental approach, encouraging students to take things apart and see what makes those things work.

Robbins, Allen H., and Wilhelm C. Miller. *Circuit Analysis Theory and Practice.* Albany, N.Y.: Delmar, 1995. This textbook provides a thorough exposition and training in the basic principles of electronics, from fundamental mathematical principles through the various characteristic behaviors of complex circuits and multiphase electrical currents.

Segura, Jaume, and Charles F. Hawkins. *CMOS Electronics: How It Works, How It Fails.* Hoboken, N.J.: John Wiley & Sons, 2004. The introduction to basic electronic principles in this book leads into detailed discussion of MOSFET and CMOS electronics, followed by discussions of common failure modes of CMOS electronic devices.

Singmin, Andrew. *Beginning Digital Electronics Through Projects.* Woburn, Mass.: Butterworth-Heinemann, 2001. This book presents a basic introduction to electrical properties and circuit theory and guides readers through the construction of some simple devices.

Strobel, Howard A., and William R. Heineman. *Chemical Instrumentation: A Systematic Approach.* 3d ed. New York: John Wiley & Sons, 1989. This book provides an exhaustive overview of the application of electronics in the technology of chemical instrumentation, applicable in many other fields as well.

Web Sites

Institute of Electrical and Electronics Engineers
http://www.ieee.org

See also: Audio Engineering; Communication; Computer Engineering; Electrical Engineering; Electrical Measurement; Electronic Materials Production; Electronics and Electronic Engineering; Information Technology; Integrated-Circuit Design; Radio.

ENERGY-EFFICIENT BUILDING

FIELDS OF STUDY

Physics; thermodynamics; architecture; materials science; construction methods; electronics; plumbing; carpentry; building technology; fluid dynamics

SUMMARY

Energy-efficient building practices involve the construction of buildings using as little energy as possible and with minimal environmental impact, beginning with construction techniques and materials and continuing with the ongoing, energy-efficient operation and maintenance of a building.

KEY TERMS AND CONCEPTS

- **Convective Air Current:** A circulatory movement of air caused by the tendency of warm air to rise and cool air to descend.
- **Mass Effect:** A physical characteristic or behavior of a material that is related to the mass of that material present.
- **Thermal Mass:** The thermal energy that can be stored within the material of a body such as the inner section of a Trombe wall or the interior of a water-filled thermal window.
- **Trombe Wall:** A two-part wall structure consisting of a transparent thermal capture outer wall and a dense, massive thermal storage inner wall.

Definition and Basic Principles

Energy-efficient building is a construction methodology designed and regulated to minimize the energy consumed in building, operating, and maintaining a given structure. The basic principle of minimizing the energy required for the operation of the building is simplicity itself. The practice of energy-efficient building is at the beginning of its development, however, and a great deal of future research will conceptualize and test materials and structures to minimize the cost and maximize the effectiveness of the technology.

The most basic principle of energy-efficient building is the capture of solar thermal or geothermal energy (or both) that can be stored in some mass of material, such as water or concrete. The energy is then released into the building in a controlled manner to maintain environmental conditions within the building.

A second main principle of energy-efficient building is the prevention of thermal transfer through the use of insulation and insulating structures. This works to prevent heat loss in winter and heat gain in summer. The third principle ensures that the daily operation and maintenance of a building is to consume as little energy as possible.

Background and History

Builders have long known how to construct buildings to make full use of the natural environment for heating, cooling, and ventilation. With the Industrial Revolution of the eighteenth century, and particularly since the establishment of electrical grids, artificial heating, cooling, ventilating, and lighting systems have rapidly developed. They now represent one of the largest general uses of energy throughout the world. As the human population continues to grow, traditional methods of energy production have not been able to meet demand, leading to the necessary development of new materials and more efficient methods of using energy.

Central to this movement are methods that utilize natural heat and light sources to reduce the demand for electrical energy and combustible fuels, and methods to reduce or eliminate energy losses from structures. Initially a fringe philosophy of the 1950's and 1960's, passive solar heating and other methods, for example, have now become a rapidly developing part of mainstream building techniques.

French engineer Felix Trombe (1906-1985) invented what came to be called the Trombe wall structure in the late 1950's. The photoelectric effect that makes solar cells function was first explained by Albert Einstein in 1905, but it was not until the transistor revolutionized electronics in the mid-twentieth century that silicon became a sufficiently viable commodity to produce solar cells on a large scale. The physical principles that drive these different applications now form the foundation of the energy-efficiency movement.

How It Works

The basic concept of energy-efficient building is to construct buildings with as little energy consumption and environmental impact as possible, beginning with construction techniques and materials to the ongoing operation and maintenance of the completed building. Traditional building techniques and materials remain fundamental components of construction practice, although a significant outcome of the move to higher energy efficiency is the development and application of new materials and methods.

Applied physics provides the theoretical framework of energy-efficient building. Convection currents and how they arise have long been known, but it is only relatively recently that this knowledge has been applied to modern building practices. The principle of convection states that warm air, having a lower density than cooler air, rises to float on top of the cooler air, setting up a circulatory movement of air within a room. Passive solar heating makes extensive use of this principle, and building designers consider the fluid dynamics of air movement to maximize solar-heating efficiency. Mechanical devices such as fans to assist convection are often integral to building design. Structures such as the Trombe wall capture and store heat energy. Openings at the top and bottom of the inner wall facilitate convection, while the massive structure itself radiates heat into the interior of the building.

Foundations represent a significant heat exchange mechanism between the interior and exterior of a building, often producing undesirable moisture-related side effects. Traditional methods rely on sealing the exterior of the foundation wall and either lining the interior of the wall with Styrofoam or creating a dead space with furring strips and paneling. Concrete blocks containing a dead air-space help in insulating foundation walls, but their strength is limited. Poured concrete is a far more effective and structurally versatile construction method, but concrete is a poor thermal barrier. For energy-efficient building, poured concrete has been rethought so that insulated concrete forms are now used in many applications to provide a more efficient thermal barrier.

Above-ground construction, particularly of residential buildings, traditionally utilizes stud-frame construction, in which an insulating material is inserted between the vertical studs of wall structures. The studs themselves serve as thermal conductors connecting the interior and exterior of the building, especially if steel studs are used in place of wooden studs. The use of structural insulated panels, or SIPs, in exterior wall construction provides an uninterrupted layer of insulation throughout the entire wall, with wall strength that meets or exceeds the requirements of a standard stud wall, and an insulating factor between R-45 and R-48 compared with the typical R-20 to R-25 rating for standard stud walls.

Standard glazed windows have very low insulation ratings, generally no more than R-10. A thermal mass window system, however, functions as a solar energy collector by using the mass of a 7.5-centimeter thick layer of water sandwiched between inner and outer glass plates to capture passive solar heat. Such windows are able to maintain a temperature of between 65 degrees and 115 degrees Fahrenheit, radiating heat into a building, even when the exterior temperature is as low as 34 degrees Fahrenheit.

The heating and cooling of buildings traditionally represents high energy consumption, either through the combustion of fuels or through the use of electricity to provide heat or to power cooling systems. Concerns about the long-term availability of fuels and electrical energy demand that more efficient means be utilized. Geothermal heating and cooling systems, coupled with specialized materials that permit facile heat exchange and capture at relatively low temperatures, have thus become more and more important and are rapidly becoming the standard systems for new constructions.

An energy-efficient building quickly ceases to be so if its ongoing operation and maintenance turn to the use of materials and mechanisms that consume more energy. For example, replacing compact fluorescent light bulbs with standard incandescent light bulbs increases energy consumption. Building maintenance may also specify the use of certain cleaning materials based on the energy consumed in their manufacture or transportation requirements.

Applications and Products

LEED Certification. LEED (Leadership in Energy and Environmental Design) is an accreditation program of the Green Building Certification Institute, with several designated areas of specialization and three levels of achievement for accreditation. The basic level of accreditation, LEED green associate, demonstrates fundamental knowledge of green construction practices, design principles, and operations. In this context, "green" indicates that the process is effective at reducing energy consumption and at conserving resources. Essentially, it serves to assure others that the completed building is in accord with green construction policies and standards. To qualify for green associate accreditation, a candidate must be able to provide evidence of having been associated with a LEED-registered project, show evidence of employment in a sustainable field, or prove completion of a program of education incorporating green construction principles. The candidate must also successfully complete an examination within specified constraints.

The second tier of LEED certification is accredited professional and the third is accredited professional fellow. Each level represents a significantly higher degree of experience and training in LEED practices. LEED currently provides accreditations in the areas of new construction, existing buildings' operation and maintenance, commercial interiors, core and shell, schools, retail, health care, homes, and neighborhood development. Operations and maintenance provides standards by which services can be measured to maximize operational efficiency and minimize environmental impact and energy consumption. LEED for homes provides a consensus-based, voluntary rating system that is also third-party verified to ensure that homes meet the applicable standards. The new construction protocol ensures that LEED standards and green principles are included in every aspect of building design.

LEED for commercial interiors is geared to ensuring that the interior environments of buildings are healthy and productive workspaces with reduced operating and maintenance costs and reduced environmental impacts. It also provides the direction for tenants and interior designers to make sustainable choices for their own inhabited spaces. LEED certification enables a better understanding of decisions made in project design and construction. Certification better ensures that the work conforms to the spirit of the design and to the design drawings.

Building Materials. Reconsideration of the performance of materials has resulted in numerous modifications to those materials and the structures in which they are used. Energy-efficient modifications are incorporated into the building structure even before the foundation is formed. Foundations themselves are historically responsible for unregulated heat transfer into and out of buildings in the underground levels. Heated basements tend to lose much of their heat to conductive basement walls, while the coolness of those same walls in warm weather contributes to moisture, mildew, and mold problems.

Energy-efficient building design using insulated concrete forms minimizes unregulated heat transfer across basement walls. Primarily made from reinforced Styrofoam or a proprietary wood fiber and concrete mixture called Durisol, insulated forms are stacked like bricks inside a retaining form and then filled with concrete in the usual way. Once the concrete has set, a well-sealed, insulated wall remains.

Structural insulated panels (SIPs) are thick foam panels that have been bonded on both sides to oriented strand board or to plywood. SIP construction replaces the standard stud-wall structure of a building, locking together in such a way that studs are not required for the formation of a straight, strong wall and an air-tight building envelope. The energy requirements of a house built using SIP construction can be as much as 50 percent lower with no further modifications.

Thermal mass (TM) windows are designed to use the heat-storing capacity of a mass of material to absorb solar thermal energy and radiate the stored heat into the interior. Their construction is no more complex than two sheets of thick glass enclosing a mass of water. A TM window system may contain nearly 1,500 kilograms of water and have the ability to capture and store tens of thousands of British thermal units (BTUs) of energy. Other designs incorporate aerogel insulation with water or use phase-changing materials to capture thermal energy.

Trombe walls also use the principle of thermal mass. A Trombe wall consists of a thick solid wall of concrete or other dense material, open at the top and bottom to allow convective air circulation. A second outer wall of glass windows adjacent to the solid wall acts as the solar thermal collector.

Geothermal and In-Floor Heating Systems. The constant underground temperature provides a ready means of heating and cooling through the use of geothermal loop systems. A circulating carrier fluid can be used to transfer heat into the building in winter or to remove excess heat from the building in summer. Though geothermal systems can be used to retrofit an existing building, it is more commonly used in new construction to incorporate an in-floor radiant heating system. The circulating fluid, usually just water, moves through an array of piping built into the floor of the structure to heat the floor itself, which then radiates heat upward to warm the interior of the building. The system also can be used to cool floors in the heat of summer.

Passive Solar Heating. This is the original method of solar heating used to reduce the energy requirements of a building. It is based on the capture of solar thermal energy to directly heat the interior of a building, just as a glass greenhouse captures thermal energy. The method heats air that then circulates by convection, requiring that the building be oriented appropriately when built to maximize the effect.

Other Methods of Energy Consumption. Energy-efficient buildings by definition consume less energy than do other buildings. Part of the reduction in energy consumption, defined in standards, includes the use of such low-energy lighting as compact fluorescent light bulbs, the use of insulated flooring, and the use of windows that reflect away excess thermal energy to reduce or eliminate the need for air conditioning.

Impact on Industry

Energy-efficient building represents a paradigm shift in the ways industry approaches infrastructure and facilities. Historically, industry constructed commercial and residential buildings to be simply functional. In this paradigm, the construction of the building and its ongoing operation and maintenance were not regarded as integral features, but as tasks to be accomplished separately. While the functional stability and integrity of buildings have always been considerations, this construction focused on the use of standard building materials to produce a structure that needed only to withstand the external environment while providing the desired interior capacity.

Integral to design in the new paradigm of energy-efficient building, however, is the ongoing nature, functionality, and maintenance of the building after it has been constructed. Plain poured concrete foundations shaped within bare wooden forms become poured concrete foundations formed with and poured within precision structures of insulated concrete forms. Energy consuming forced-air heating and central air conditioning is replaced, or, at minimum, augmented by passive solar heating, geothermal ground-loop heating and cooling, and Trombe walls. Plain glazed, double-glazed, and even triple-glazed windows are supplanted by water-filled thermal mass windows. To meet energy-efficiency certification standards, the building industry now considers the energy costs associated with products used in the upkeep of energy-efficient buildings.

As with all such shifts in industry practices, associated costs are high at the outset, as the new methods and technologies begin to supplant the older ones. In this period of adjustment, the primary and secondary industries needed to provide materials and education are in short supply and, therefore, carry premium values. In contrast, the secondary industries that support the older established methodologies are functioning to economy of scale and, therefore, represent considerably lower costs. Inevitably, however, as new methods prove their worth, they attain the desirability that leads to established secondary industries, and their associated costs decrease as those new secondary industries move toward economy-of-scale operation to match the increasing demand for their products.

At the same time, new industries are produced that provide specialized expertise with the new methods of energy-efficient building. A concrete-forming company, for example, will add proficiency with insulated concrete forms to the traditional services it provides and so continue as a viable business, else it may find that the demands for its services steadily decrease while those for a newly established competitor that specializes in energy-efficient methods rapidly increase.

One of the fastest growing demands for energy-efficient methodologies in recent years, particularly in areas where seasonal temperatures vary widely from winter to summer, has been for geothermal heating and cooling systems. Because they are amenable to augmentation via heat exchangers by a variety of heating and cooling technologies, such as heat pumps and boiler systems, and because they work well with in-floor systems, geothermal loop systems have become an attractive means of energy-efficient heating and cooling in new structures. They also have proved valuable in reducing the energy consumption of existing heating and cooling systems as a retrofitted augmentation method, particularly for residential purposes.

Retrofitting of existing structures for energy

Fascinating Facts About Energy-Efficient Building

- Thermal mass windows can capture and store as much as 100,000 BTUs of thermal energy on a sunny winter day, which will hold a building's temperature between 60 degrees and 115 degrees Fahrenheit.
- A Trombe wall captures solar energy to heat a massive inner wall that then radiates that heat into the interior of a building through air convection.
- Energy-efficient houses can be built from discarded tires that have been packed full of earth.
- A thermal mass window may hold 1,500 kilograms or more of water.
- Geothermal ground-loop systems for heating and cooling make use of the constant temperature of the earth that exists a short distance below ground level.
- Insulated foundation walls in a building can reduce energy consumption for heating and cooling by as much as 20 percent.
- Structural insulated panel (SIP) construction can reduce a building's energy requirements by 50 percent.
- It is possible to construct modern houses with all the conveniences entirely self-sufficient in energy needs and therefore requiring no additional heating or cooling.

efficiency represents both a significant cost and, over time, a substantial return on investment. While a small retrofitting project may be carried out in a matter of months, large projects may require a number of years. During the retrofit period, funding must be made available, which could negatively affect cash flow. However, as savings accrue from the functioning retrofit, those funds and the overall economic value generated by the building as additional cash flow eventually surpass the initial costs.

Careers and Coursework

Standardized techniques and theoretical principles of energy-efficient building practices are in their infancy as subject matter in academic curricula. Persons interested in pursuing a career in the building trades as specialists in energy-efficient building can expect to take courses and trades-training based on established and traditional building methods. Specialization will be acquired through upgrading programs and on-the-job experience. This training focus can be expected to change as the demand for expertise in energy-efficient building increases and the methodology becomes the accepted norm of building practices.

Certification programs such as LEED are only beginning to become incorporated into trades-training. The certification procedure is offered only outside standard training curricula, although this is often provided through the facilities of a local trade school. Students can therefore expect to take basic courses in mathematics, general physics and chemistry, architectural drafting and design, and the strength of materials as part of the training curriculum for basic trades. Specialized trades such as electrician, plumber, framer, and general construction worker will have more advanced hands-on requirements to ensure knowledge of standard, safe working methods according to appropriate regulations. With such training in hand, or with documented equivalent experience, a person is then able to pursue certification. This hands-on experience is considered an essential prerequisite for a career in energy-efficient building.

Social Context and Future Prospects

Access to the energy required to power an ever-growing society has become an issue of major concern, especially in recent years. Accordingly, developing the means to minimize the use of energy and to maximize the production of energy has also become important.

Electrical energy is the mainstay of modern industry, while both electricity and combustible fuels are the primary means of providing temperature control in buildings. Traditional methods of producing electrical energy have been hydroelectric generation, coal and gas-fired generating plants, and nuclear-powered generating stations. These methods are aging and the technologies are becoming outdated, but significant efforts are being made to augment or replace their capabilities with renewable sources such as solar- and wind-generated electricity. At the same time, efforts are expanding to reduce the energy requirements of buildings and infrastructure. At the forefront of this movement are energy-efficient building design and construction methods.

It is now possible to construct residential

buildings that require no additional input of energy to maintain and operate. Such zero-energy housing, however, is a specialty or niche market. The cost of constructing such a building is somewhat higher than the cost of a traditionally built home. This factor alone is sufficient to hold the demand for zero-energy housing to a minimum. That said, as secure access to energy becomes increasingly difficult, the demand and market for zero-energy buildings and construction methods is likely to follow an exponential growth curve.

Richard M. Renneboog, M.Sc.

Further Reading

Gulli, Cathy. "The Big Chill." *Maclean's* 123, no. 13 (June, 2010): 63-64. This article explores the energy drain associated with the wide-scale use of air conditioners.

Hordeski, Michael F. *Megatrends for Energy Efficiency and Renewable Energy.* Lilburn, Ga.: Fairmont Press, 2011. This book examines the entwined relationships between energy and industries concerned with building, power, fuels, conservation, and automation.

Jefferey, Yvonne, Liz Barclay, and Michael Grosvenor. *Green Living for Dummies.* Hoboken, N.J.: John Wiley & Sons, 2008. This brief introductory book provides many insights into the philosophy of energy-efficient building.

Johnston, David and Scott Gibson. *Green From the Ground Up: Sustainable, Healthy, and Energy-efficicient Home Construction.* Newtown, Conn.: Taunton Press, 2008. A builder's guide that presents the rationale of green construction and focuses on applying those principles with standard construction practices.

Web Sites

Energy Efficient Building Technologies
"Affordable Zero Energy Homes"
http://www.eebt.org

Fraser Basin
"Energy Efficiency and Buildings"
http://www.fraserbasin.bc.ca/publications/documents/caee_manual_2007.pdf

International Energy Agency
"Technology Roadmap: Energy-efficient Buildings—Heating and Cooling Equipment"
http://www.iea.org/papers/2011/buildings_roadmap.pdf

United States Department of Energy
"Elements of an Energy-efficient House"
http://www.nrel.gov/docs/fy00osti/27835.pdf

See also: Applied Physics; Architecture and Architectural Engineering; Civil Engineering; Electronics and Electronic Engineering; Environmental Engineering; Fluid Dynamics; Structural Composites; Urban Planning and Engineering.

ENERGY STORAGE TECHNOLOGIES

FIELDS OF STUDY

Biochemistry; chemistry; electrical engineering; electronics; hydraulics; materials science; mechanical engineering; physics; quantum mechanics; superconductivity; thermodynamics.

SUMMARY

Energy storage technologies provide primary power sources for portable devices and vehicles and are employed in electrical grids to act as backups for generators in order to ensure a stable, steady energy supply. Energy storage is particularly needed for grids that rely on renewable energy sources, such as solar and wind power, so that during periods without sunlight or wind when generators are not operating, electricity can still be sent to consumers. Storage technologies fall into three broad categories: mechanical energy (kinetic or potential) and thermal energy systems; electrochemical systems; and electrical storage systems.

KEY TERMS AND CONCEPTS

- **Electrolyte:** Solution, or the molten state of a substance, that conducts electricity through the presence of ions.
- **Energy Density:** Amount of energy stored in a system per unit of volume.
- **Frequency Regulation:** Continuous and instantaneous balancing of supply and demand on an electrical grid.
- **Peak Load:** Period when there is high demand on an electricity grid.
- **Power:** Rate at which energy is transferred or work done, measured in watts.
- **Power Density:** Amount of energy a system transfers (power) per unit of volume.
- **Storage Time:** Discharge time of stored energy at a given power.
- **Watt Hour (Wh):** Energy delivery of one watt for one hour, a basic unit in electrical generation and transfer.

Definition and Basic Principles

Energy storage is the artificial containment of energy for controlled release. It is different in kind from the various forms of natural energy storage, such as the storage of radiant energy from the sun in wood, oil, or coal, which frequently supplies the energy for the generation of the electrical power that supports modern technological civilization. Energy storage technologies provide the physical means for energy containment and are based on a variety of mechanical, electrochemical, or electromagnetic principles. As such, they are rated for usefulness and efficiency based upon how much energy they can contain (energy density) and how they release that energy (power density).

Energy storage devices can supply either primary power or secondary power. Devices such as batteries, capacitors, and fuel cells, for instance, may provide primary power, usually for portable electronics or vehicles—anything that must be used apart from the steady supply of an electrical power grid. Portable radios, smoke detectors, watches, electric cars, and emergency lighting are examples. As secondary, or supplemental, power sources, energy storage technologies are charged by a power grid and then return the energy back to the grid as needed to manage peak electrical loads, improve power quality, ensure frequency regulation, or make up for failing production, as when a generator must be taken off-line or when absence of sunlight or wind idles a solar- or wind-powered generator.

Background and History

Archaeologists found evidence of what might have been a battery dating from the third century B.C.E. in the form of a clay pot containing iron and copper rods; the pots could have produced electrical discharges if filled with vinegar, although scholars disagree about this interpretation. It is even less well known when another major energy storage device first saw use, the reservoir or millpond. It consisted of a dam or diverted stream that filled a pond that supplied water on demand to turn a waterwheel.

The first modern artificial energy storage systems were the Leyden jar and the electrochemical battery. German physicist Ewald Georg von Kleist invented the Leyden jar in 1744. A capacitor, it stored static electricity in a glass jar coated inside and out with metal foil and had a conducting rod stuck through an insulating stopper. In a 1800 letter to England's Royal Society, Italian physicist Alessandro Volta described the first electrochemical battery, an arrangement of stacked disks of tin or zinc paired with disks of copper, brass, or silver, each pair separated by a disk of paper or leather moistened in an electrolyte solution. Whereas the Leyden jar delivered electricity only in a burst, the battery gave a steady current, which proved invaluable to experiments in physics and chemistry.

Other devices for storing energy followed steadily: the fuel cell in 1839, compressed air energy storage in 1870, the rechargeable flow battery in 1884, thermal energy storage in 1886, pumped hydro storage in 1908, flywheel energy storage in 1950, super capacitors in 1966, and superconducting magnetic energy storage in 1986.

How It Works

Mechanical Systems. Pumped water (pumped hydro) is the most common system worldwide for storing energy to serve electrical grids. During low demand on the grid, usually at night, water is pumped from a lower to an upper reservoir. When electrical demand increases and load leveling or supplemental electricity is needed, water in the upper reservoir is released to flow downhill and turn generators. It is also possible to use underground cavities or the open sea for storage. Compressed air energy storage (CAES) also takes advantage of off-peak electricity. Air is compressed, stored underground, and used when needed to turn a gas turbine. The main difficulty with pumped hydro and CAES is finding suitable terrain for reservoirs or underground cavities to store air.

Flywheels store kinetic energy in a rotating mass in the form of a disk or rotor. The flywheel is accelerated during off-peak load by an electrical motor. Later this stored energy can be used to turn an electrical generator and return energy to a grid. Flywheels are categorized as low speed if the rotor spins at fewer than 6,000 revolutions per minute (rpm); these usually have steel rotors and turn on metal ball bearings. High-speed flywheels turn at up to 50,000 rpm and have rotors of composite materials that turn on magnetic bearings. The drawback of flywheels is primarily the degree of standby energy loss. Springs, long used to store mechanical energy in clocks, for instance, are also proposed to store energy for electrical generation.

Thermal energy storage (TES) involves heat stored in saltwater ponds, molten salt, bricks, water reservoirs, pressurized water, or steam by converting electrical energy or by collecting heat directly, as through geothermal or solar heating. Conversely, during low demand in a cooling system, electricity can run refrigerators to make ice, which during high demand can be returned to the system to enhance cooling. Waste electricity during off-peak demand can also be used to produce liquid oxygen and nitrogen that can later be boiled to operate a turbine. In all cases, mechanical and thermal systems are in theory continually renewable as long as machinery and materials remain sound.

Electrochemical Systems. Large or small, batteries combine cells in a sequence. A cell comprises an electron source, the anode, and an electron acceptor, the cathode, that, immersed in an electrolyte, directly converts chemical energy to electricity. Batteries come in two major types: primary, or disposable (alkaline, lithium cells, and zinc-carbon batteries), and secondary, or rechargeable (lead-acid, nickel-cadmium, nickel-iron, nickel-metal hydride, and lithium-ion batteries). The number of charge-discharge cycles that secondary batteries can go through before wearing out, their charging and storage times, and their energy density and power density vary and determine their range of applications.

Flow cells (also redox flow cells, flow batteries) are similar to batteries, except that the electrodes are catalysts for the chemical reaction, which occurs as a microporous membrane allows ions to pass from one electrolyte solution to another. Among flow cells are types that use zinc and bromine, vanadium in two types with different states, or polysulfide and bromine as the pairs of electrolytes. The advantages of flow cells are that they are capable of a large number of cycles, and the electrolytes can be replenished.

Hydrogen fuel cells store energy by employing an electrolyzer to produce hydrogen. It is stored until a fuel cell splits the hydrogen into ions and electrons. The electrons flow through a wire, producing an electric current, while the ions, after passing through an electrolyte between the anode and cathode, are reattached to the electrons reentering the cell from the wire and in the presence of oxygen-form wastewater.

Biological storage is yet another means of storing energy chemically through biological conversion, as is done with adenosine triphosphate (ATP) in the mitochondria of cells. Although interest in artificial biological systems is likely to increase, it is a nascent technology.

Electricity Storage Systems. Superconducting magnetic energy storage (SMES) entails storing energy in a magnetic field produced by passing direct current through a coil. This technique does not work with normal electrical wire because resistance and heat loss dissipates energy from the coil. Superconducting wire, having almost no resistance, is required. The superconductors, however, must be kept at very low temperatures to function. Niobium-based wires, for instance, require a temperature of just more than 9 Kelvin. Ceramic materials that superconduct at much higher temperatures may be used, and room-temperature superconductors are theoretically possible, obviating the need for complex cooling systems for the coil.

Super capacitors are also known as double-layer capacitors, electrochemical capacitors, ultra capacitors, and pseudo capacitors. They store energy by separating electrostatic charges in electrode plates on either side of a liquid or organic electrolyte. One plate attracts positive ions from the electrolyte, the other negative ions, thereby storing electrical energy in two layers. Super capacitors are reversible and capable of 10,000 charge-discharge cycles.

Applications and Products

Batteries and Cells. Batteries are ubiquitous in technological society, needed to power virtually all devices that are portable and many vehicles, but also used in utility-scale energy storage. Disposable batteries have long powered household devices, such as the D-cell 1.5 volt batteries in flashlights and toys, but rechargeable batteries are the power source in developing technologies, such as laptop computers, cell phones, and MP3 players. Lithium-ion batteries quickly dominated the portable-devices market because they have high-energy density (300 to 400 kilowatt hours, or kWh, per cubic meter), high efficiency, and long recharging life (about 3,000 cycles). Arrays of batteries are also used for large-scale energy storage. Zinc-bromide battery systems, sometimes mounted on trailers for transportation, have capacities of 1 megawatt (MW) for 3 megawatt hours (MWh); units can be linked for further capacity. Sodium-sulfur batteries are widely used in Japan, storing up to 34 MW per 245 MWh for frequency regulation and to receive energy from wind generators. The lead-acid battery, the most developed battery

Fascinating Facts About Energy Storage Technologies

- The first modern battery was described in 1800 by Alessandro Volta, a professor at the University of Pavia in Italy. His stacked zinc-silver cells, moistened with an electrolyte solution, became known as a voltaic pile.
- The world market for batteries in 2010 was estimated at about $15 billion and is expected to grow steadily.
- In 2010, the U.S. Information Administration predicted that total world consumption of marketed energy would rise from 495 quadrillion British Thermal Units (Btu) in 2007 to 739 quadrillion Btu's in 2035, an increase of 49 percent, or an average of 1.4 percent a year. To meet the demand, energy storage systems will be needed to make energy available efficiently.
- Three advantages of flywheels for energy storage are that they are low maintenance, environmentally benign, and last long—up to twenty years.
- Global productions from solar cells reached 21 gigawatts in 2009, enough to supply electricity to 5.5 million houses yearly. Because solar cells do not generate without sunlight, batteries are required to supply a continuous current.
- Pumped hydro is the most commonly employed non-electrochemical energy storage technology and has been used in the United States since 1929.
- In the United States, interest in energy storage technologies soared during the energy crisis of the mid-1970's.
- A laptop computer battery and a full Thermos of hot coffee store about the same amount of energy, but the Thermos does it for less than 1 percent of the battery's cost, which is why thermal energy storage systems are attractive.
- The lithium-ion batteries storing solar-generated power on the Mars rovers Spirit and Opportunity were designed to provide about 900 watt hours of power per day for at least 90 days. They were still working six years later.

technology, provided the storage core for vehicles for generations. Nickel-cadmium and nickel-metal hydride batteries are competing systems, especially for hybrid and electric vehicles, because they are more efficient and powerful. Home-energy storage from solar photovoltaic cell arrays include lead-acid, lithium, and metal-hydride batteries connected in banks. Nickel-hydrogen, nickel-cadmium, and lithium-ion batteries power space exploration, variously serving on satellites, the Hubble Space Telescope, interplanetary probes, and Opportunity and Spirit, the two rovers that landed on Mars in 2004.

Flow cells are ideal for storage systems in remote locations. Vanadium redox systems, for instance, deliver up to 500 kW for up to ten hours. Zinc-bromine systems have been produced for 50-kWh and 500-kWh systems to reinforce weak distribution networks or prevent power fluctuations. Hydrogen fuel cells can potentially do almost anything a battery can do: provide backup power, perform power leveling, run handheld devices, and supply primary or auxiliary power to cars, trucks, buses, and boats. In many cases they are more efficient than petrochemical fuels. A hydrogen fuel cell in a vehicle that uses an electric motor, for example, can be 40 to 60 percent efficient, compared with the 35 percent peak efficiency of the internal combustion engine.

Mechanical and Thermal Systems. Although flywheels have been tested as power storage in cars and have aerospace applications, mechanical and thermal systems are primarily backups for large-scale power generation or cooling systems. Pumped hydro serves electrical grids worldwide: more than 90 gigawatts (GW) of storage and about 3 percent of global generation capacity in 2010. Systems have efficiencies of 70 to 85 percent. Some flywheels can convert 90 percent of their kinetic energy to electricity that will last for several hours. Telecommunications systems employ flywheels delivering 2 kW for 6 kWh, and megawatts can be stored in linked arrays of flywheels. CAES is used to store energy for electrical grids. The first large-scale commercial system, capable of 290 MW, came online in Huntorf, Germany, in 1978. Later systems varied in output from 110 to 300 MW, and as of 2011, a 2.7 GW system was planned for Norton, Ohio, using an abandoned limestone mine some 670 meters deep.

TES systems have somewhat more variety of use. Passive solar power systems used in houses and commercial buildings soak up energy from the sun that can be used to heat interior water or air. Likewise, systems that use off-peak energy to make ice that later contributes to refrigeration operate in large buildings to save electricity costs. For instance, the Dallas Veterans Affairs Medical Center incorporated a chilled-water TES that reduced electricity consumption nearly 3,000 kW and saved more $200,000 in its first year of operation.

Superconducting Magnetic Energy Storage (SMES). SMES is used by some electrical utilities to improve the system reliability and transfer capacity so that grid operators can compensate for damage from storms, voltage variation, and increasing consumer demand. The first distributed SMES system began operation in Wisconsin in 2000; it employs SMES mounted on 48-foot trailers that can be deployed to support the grid in rural areas. The largest users from the 1990's on, however, were industries making plastics, paper, aluminum, and chemicals.

Super Capacitors. Super capacitors are widely used in consumer electronics of all kinds for power backup and flash charging. They also figure in braking systems, such as energy buffers for elevators, and in motor ignition, especially for the big engines in buses, trucks, tanks, and submarines. However, they also have application in larger systems to compensate for voltage drops and can be used in hybrid systems with rechargeable batteries, such as electric and hybrid vehicles.

Impact on Industry

Energy storage is a rapidly growing field worldwide. Developed nations, especially those with a heavy industry or high tech base, look to energy storage to help solve problems in maintaining high-quality electrical transmission on their grids, but many nations also use energy storage in remote areas and for conservation. As grids rely more and more on renewable-energy generation—wind, wave, and solar power—energy storage becomes increasingly important to provide current throughout the day. Among leaders in energy storage technologies are the European Union, Japan, the United States, Great Britain, Australia, and Brazil. Several international and regional organizations foster research and development, such as Instituto Para o Desenvolvimento de Energias Alternativas na América Latina (IDEAL); the International Energy Agency (IEA), an autonomous organization based in Paris; and the Electricity

Storage Association (ESA) in the United States.

Government and University Research. The Department of Energy (DOE) started its Energy Storage Systems Program at Sandia National Laboratories in New Mexico in the mid-1970's as a response to the first oil crisis. It concentrates on developing technologies in batteries, CAES, and flywheels for use in electrical grids. The DOE also fosters research at university and industry laboratories and holds an annual conference. The Northeastern Center for Chemical Energy Storage, based at the State University of New York at Stony Brook, brings together researchers from American universities and the Lawrence Berkeley National Laboratory, Argonne National Laboratory, and Brookhaven National Laboratory. The Center for Advanced Energy Studies at Idaho National Laboratory in Idaho Falls is another leading research facility.

Industry and Business. Many utilities and industries are expanding both research facilities and development budgets for energy storage systems. ESA and the U.S. Advanced Battery Consortium promote long-range research and development in common with national laboratories and universities, and industry newsletters, such as the *Clean Technology Business Review*, keep potential users apprised of new products. In the car industry alone, for example, battery development to support electric or hybrid cars became a major research focus in the 2000's. The leaders include Chrysler, Daimler, Ford, General Motors, Honda, and Toyota.

Energy Storage Suppliers. The number of established and start-up companies offering energy storage systems grew rapidly beginning in the 1990's. Among them are: A123 Systems, Cobasys, Compact Power, EVionyx, Exide Technologies, NGK Insulators, Optima, Saft (batteries), Deeya Energy, Premium Power Corporation, Prudent Energy, RedFlow, ZBB Energy Corporation (flow cells), GE Energy Industrial Solutions, Saft (capacitors), Active Power, Beacon Power, Piller Power Systems, Urenco Power Technologies (flywheels), CAES Development Company, Dresser-Rand, Energy Storage and Power, Ridge Energy Storage (CAES), American Superconductor, Bruker Energy and Supercon Technologies, and SuperPower (SMES).

Careers and Course Work

The substantial increase in research and development since the 1990's, as well as proliferating venture capital investments, in energy storage technologies means that they will continue to offer career opportunities for scientists, engineers, and businesspeople worldwide through the twenty-first century.

College students interested in pursing a career in energy storage can begin by taking science, mathematics, and engineering courses. At some point selecting a major will prepare one for a particular type of energy storage technology: For example, physics for those interested in SMES and flywheels; electrical engineering, chemistry, or automotive engineering for those interest in batteries, capacitors, and fuel cells and their applications; and mechanical and civil engineering for CAES and pumped hydro. Refining the specialty in graduate school to obtain a master's degree or doctorate would position a person to work at the forefront of a crucial technology. However, someone with a science background would be prepared for a marketing or management position within an energy storage company or utility after obtaining a master of business administration (M.B.A.).

For instance, a physicist who wrote a doctoral dissertation on high-temperature superconducting materials in energy storage systems might find employment at a company such as American Superconductor, a utility, a university or government research laboratory, or a university faculty. An undergraduate degree in mechanical engineering could lead a person, after obtaining an M.B.A., to joining junior management for a corporation like General Electric, working for a utility or construction company, or starting an energy storage company. The prospect of a system based upon biological storage of energy for utilities, consumer devices, or industry was, as of 2011, promising. Because research is not far advanced, experts in biochemistry have the opportunity to be pioneers.

Social Context and Future Prospects

Energy storage helps solve several pressing problems in costs and availability of power. Studies repeatedly argue that electrical grids cannot be expanded sufficiently or fast enough to meet projected consumer demand during the twenty-first century without incorporating energy storage systems. Moreover, in many cases although the initial construction or purchasing and installation costs are high, energy storage systems save money in the long run by obviating investment in expensive new generators, load management systems, and transmission lines. The savings can help keep the price of electricity low, especially for manufacturers, and so reduce the power industry's contribution to inflation. Energy storage will also prove crucial to ensure steady supply from solar-, wind-, or wave-powered energy generation for large grids. This is true on a smaller scale for houses or buildings that rely on autonomous renewable energy systems—most likely solar or wind—for their electricity and heating. As consumers demand better batteries for their portable devices and vehicles, research in batteries and fuel cells will concentrate on reducing recharging time and increasing cycling life, storage time, power density, and energy density.

Energy storage systems will also affect the environment, although not always beneficially. In operation most do not pollute or disrupt an ecosystem and so are environmentally benign. On the other hand, construction of CAES or pumped hydro is disruptive, and manufacture of the other systems frequently entails waste products that are toxic. In balance, however, energy storage systems will prove to have less environmental impact than electricity produced from coal-, nuclear-, or gas-powered or hydroelectric generators.

Roger Smith, Ph.D.

Further Reading

Baxter, Richard. *Energy Storage: A Nontechnical Guide.* Tulsa, Okla.: PennWell, 2006. Intended for policy makers and business analysts, this guide argues for a distributed energy supply strategy and in that light surveys several storage technologies, such as pumped-hydrogen, compressed air, batteries, capacitors, and magnetic energy.

Brunet, Yves, ed. *Energy Storage.* Hoboken, N.J.: John Wiley & Sons, 2010. The author discusses hydrogen, super capacitors, flywheels, thermal, and gravitational storage systems as he considers applications in electric power generation, transmission, and distribution systems and in vehicles and buildings.

Dinçer, Ibrahim, and Marc A. Rosen. *Thermal Energy Storage: Systems and Applications.* 2d ed. Hoboken, N.J.: John Wiley & Sons, 2011. Although concentrating on the nature and applications of thermal energy storage, this technical introduction also describes mechanical, chemical, biological, magnetic, and hydrogen systems

Huggins, Robert A. *Energy Storage.* New York: Springer, 2010. For engineering students, this book presents the fundamentals of energy storage, covering various fuels, phase transitions, heat capacity, and mechanical, hydrogen, and electromagnetic storage systems.

Press, Roman J., et al. *Introduction to Hydrogen Technology.* Hoboken, N.J.: John Wiley & Sons, 2009. Addresses available energy resources, the chemical background of hydrogen reactions and properties, hydrogen technology, and the science and application of fuel cells for an audience who understands basic chemistry and mathematics.

Root, Michael. *The TAB Battery Book: An In-Depth Guide to Construction, Design, and Use.* New York: McGraw-Hill, 2011. Although this books requires some knowledge of chemistry and mathematics to appreciate fully, the introductory chapters about the history and basics of batteries provides a clear overview.

Zito, Ralph. *Energy Storage: A New Approach.* Hoboken, N.J.: Wiley-Scrivener, 2010. Discusses storage of massive energy and explains the basic chemical processes that underlie storage no matter how the energy is generated.

Web Sites

Electricity Storage Association
http://www.electricitystorage.org

Sandia National Laboratory
Energy Storage Systems Program
http://www.sandia.gov/ess/index.html

World Energy Council
http://www.electricitystorage.org

See also: Electrical Engineering; Mechanical Engineering.

ENGINEERING

FIELDS OF STUDY

Physics; chemistry; computer science; mathematics; calculus; design; systems; processes; materials; circuitry; electronics; environmental science; miniaturization; biology; aeronautics; fluids; gases; technical communication.

SUMMARY

Engineering is the application of scientific and mathematical principles for practical purposes. Engineering is subdivided into many disciplines; all create new products and make existing products or systems work more efficiently, faster, safer, or at less cost. The products of engineering are ubiquitous and range from the familiar, such as microwave ovens and sound systems in movie theaters, to the complex, such as rocket propulsion systems and genetic engineering.

KEY TERMS AND CONCEPTS

- **Analogue:** Technology for recording a wave in its original form.
- **Design:** Series of scientifically rigorous steps engineers use to create a product or system.
- **Digital:** Technology for sampling analogue waves and turning them into numbers; these numbers are turned into voltage.
- **Energy:** Capacity to do work; can be chemical, electrical, heat, kinetic, nuclear, potential, radiant, radiation, or thermal in nature.
- **Feasibility Study:** Process of evaluating a proposed design to determine its production difficulty in terms of personnel, materials, and cost.
- **Force:** Anything that produces or prevents motion; a force can be precisely measured.
- **Matter:** Anything that occupies space and has weight.
- **Power:** Time rate of doing work.
- **Prototype:** Original full-scale working model of a product or system.
- **Quantum Mechanics:** Science of understanding how a particle can act like both a particle and a wave.
- **Specifications:** Exact requirements engineers must comply with to create products or services.
- **Work:** Product of a displacement and the component of the force in the direction of the displacement.

Definition and Basic Principles

Engineering is a broad field in which practitioners attempt to solve problems. Engineers work within strict parameters set by the physical universe. Engineers first observe and experiment with various phenomena, then express their findings in mathematical and chemical formulas. The generalizations that describe the physical universe are called laws or principles and include gravity, the speed of light, the speed of sound, the basic building or subatomic particles of matter, the chemical construction of compounds, and the thermodynamic relationship that to produce energy requires energy. The fundamental composition of the universe is divided into matter and energy. The potential exists to convert matter into energy and vice versa. The physical universe sets the rules for engineers, whether the project is designing a booster rocket to lift thousands of tons into outer space or creating a probe for surgery on an infant's heart.

Engineering is a rigorous, demanding discipline because all work must be done with regard to the laws of the physical universe. Products and systems must withstand rigorous independent trials. A team in Utah, for example, must be able to replicate the work of a team in the Ukraine. Engineers develop projects using the scientific method, which has four parts: observing, generalizing, theorizing, and testing.

Background and History

The first prehistoric man to use a branch as a lever might be called an engineer although he never knew about fulcrums. The people who designed and built the pyramids of Giza (2500 B.C.E.) were engineers. The term "engineer" derives from the medieval Latin word *ingeniator*, a person with "ingenium," connoting curiosity and brilliance. Leonardo da Vinci, who used mathematics and scientific principles in everything from his paintings to his designs for military fortifications, was called the Ingegnere

Generale (general engineer). Galileo is credited with seeking a systematic explanation for phenomena and adopting a scientific approach to problem solving. In 1600, William Gilbert, considered the first electrical engineer, published *De magnete, magneticisque corporibus et de magno magnete tellure* (*A New Natural Philosophy of the Magnet, Magnetic Bodies, and the Great Terrestrial Magnet*, 1893; better known as *De magnete*) and coined the term "electricity." Until the Industrial Revolution of the eighteenth and nineteenth centuries, engineering was done using trial and error. The British are credited with developing mechanical engineering, including the first steam engine prototype developed by Thomas Savery in 1698 and first practical steam engine developed by James Watt in the 1760's.

Military situations often propel civilian advancements, as illustrated by World War II. The need for advances in flight, transportation, communication, mass production, and distribution fostered growth in the fields of aerospace, telecommunication, computers, automation, artificial intelligence, and robotics. In the twenty-first century, biomedical engineering spurred advances in medicine with developments such as synthetic body parts and genetic testing.

How It Works

Engineering is made up of specialties and subspecialties. Scientific discoveries and new problems constantly create opportunities for additional subspecialties. Nevertheless, all engineers work the same way. When presented with a problem to solve, they research the issue, design and develop a solution, and test and evaluate it.

For example, to create tiles for the underbelly of the space shuttle, engineers begin by researching the conditions under which the tiles must function. They examine the total area covered by the tiles, their individual size and weight, and temperature and frictional variations that affect the stability and longevity of the tiles. They decide how the tiles will be secured and interact with the materials adjacent to them. They also must consider budgets and deadlines.

Collaboration. Engineering is collaborative. For example, if a laboratory requires a better centrifuge, the laboratory needs designers with knowledge in materials, wiring, and metal casting. If the metal used is unusual or scarce, mining engineers need to determine the feasibility of providing the metal. At the assembly factory, an industrial engineer alters the assembly line to create the centrifuge. Through this collaborative process, the improved centrifuge enables a biomedical engineer to produce a life-saving drug.

Communication. The collaborative nature of engineering means everyone relies on proven scientific knowledge and symbols clearly communicated among engineers and customers. The increasingly complex group activity of engineering and the need to communicate it to a variety of audiences has resulted in the emergence of the field of technical communications, which specializes in the creation of written, spoken, and graphic materials that are clear, unambiguous, and technically accurate.

Design and Development. Design and development are often initially at odds with each other. For example, in an architectural team assigned with creating the tallest building in the world, the design engineer is likely to be very concerned with the aesthetics of the building in a desire to please the client and the city's urban planners. However, the development engineer may not approve the design, no matter how beautiful, because the forces of nature (such as wind shear on a mile-high building) might not allow for facets of the design. The aesthetics of design and the practical concerns of development typically generate a certain level of tension. The ultimate engineering challenge is to develop materials or methods that withstand these forces of nature or otherwise circumvent them, allowing designs, products, and processes that previously were impossible.

Testing. With computers, designs that at one time took days to draw can be created in hours. Similarly, computers allow a prototype (or trial product) to be quickly produced. Advances in computer simulation make it easier to conduct tests. Testing can be done multiple times and under a broad range of harsh conditions. For example, computer simulation is used to test the composite materials that are increasingly used in place of wood in building infrastructures. These composites are useful for a variety of reasons, including fire retardation. If used as beams in a multistory building, they must be able to withstand tremendous bending and heat forces. Testing also examines the materials' compatibility with the ground conditions at the building site, including the potential for earthquakes or other disasters.

Financial Considerations. Financial parameters often vie with human cost, as in biomedical advancements. If a new drug or stent material is rushed into production without proper testing to maximize the profit of the developing company, patients may suffer. Experimenting with new concrete materials without determining the proper drying time might lower the cost of their development, but buildings or bridges could collapse. Dollars and humanity are always in the forefront of any engineering project.

Applications and Products

The collaborative nature of engineering requires the cooperation of engineers with various types of knowledge to solve any single problem. Each branch of engineering has specialized knowledge and expertise.

Aerospace. The field of aerospace engineering is divided into aeronautical engineering, which deals with aircraft that remain in the Earth's atmosphere, and astronautical engineering, which deals with spacecraft. Aircraft and spacecraft must endure extreme changes in temperature and atmospheric pressure and withstand massive structural loads. Weight and cost considerations are paramount, as is reliability. Engineers have developed new composite materials to reduce the weight of aircraft and enhance fuel efficiency and have altered spacecraft design to help control the friction generated when spacecraft leave and reenter the Earth's atmosphere. These developments have influenced earthbound transportation from cars to bullet trains.

Architectural. The field of architectural engineering applies the principles of engineering to the design and construction of buildings. Architectural engineers address the electrical, mechanical, and structural aspects of a building's design as well as its appearance and how it fits in its environment. Areas of concern to architectural engineers include plumbing, lighting, acoustics, energy conservation, and heating, ventilation, and air conditioning (HVAC). Architectural engineers must also make sure that buildings they design meet all regulations regarding accessibility and safety in addition to being fully functional.

Bioengineering. The field of bioengineering involves using the principles of engineering in biology, medicine, environmental studies, and agriculture. Bioengineering is often used to refer to biomedical engineering, which involves the development of artificial limbs and organs, including ceramic knees and hips, pacemakers, stents, artificial eye lenses, skin grafts, cochlear implants, and artificial hands. However, bioengineering also has many other applications, including the creation of genetically modified plants that are resistant to pests, drugs that prevent organ rejection after a transplant operation, and chemical coatings for a stent placed in a heart blood vessel that will make the implantation less stressful for the body. Bioengineers must concern themselves with not only the biological and mechanical functionality of their creations but also financial and social issues such as ethical concerns.

Chemical. Everything in the universe is made up of chemicals. Engineers in the field of chemical engineering develop a wide range of materials, including fertilizers to increase crop production, the building materials for a submarine, and fabric for everything from clothing to tents. They may also be involved in finding, mining, processing, and distributing fuels and other materials. Chemical engineers also work on processes, such as improving water quality or developing less-polluting, readily available, inexpensive fuels.

Civil. Some of the largest engineering projects are in the field of civil engineering, which involves the design, construction, and maintenance of infrastructure such as roads, tunnels, bridges, canals, dams, airports, and sewage and water systems. Examples include the interstate highway system, the Hoover Dam, and the Brooklyn Bridge. Completion of civil engineering projects often results in major shifts in population distribution and changes in how people live. For example, the highway system allowed fresh produce to be shipped to northern states in the wintertime, improving the diets of those who lived there. Originally, the term "civil engineer" was used to distinguish between engineers who worked on public projects and "military engineers" who worked on military projects such as topographical maps and the building of forts. The subspecialties of civil engineering include construction engineering, irrigation engineering, transportation engineering, soils and foundation engineering, geodetic engineering, hydraulic engineering, and coastal and ocean engineering

Computer. The field of computer engineering has two main focuses, the design and development of hardware and of the accompanying software. Computer hardware refers to the circuits and architecture of the computer, and software refers to the computer programs that run the computer. The hardware does only what the software instructs it to do, and the software is limited by the hardware. Computer engineers may research, design, develop, test, and install hardware such as computer chips, circuit boards, systems, modems, keyboards, printers, or computers embedded in various electronic products, such as the tracking devices used to monitor parolees. They may also create, maintain, test, and install software for mainframes, personal computers, electronic devices, and smartphones. Computer programs range from simple to complex and from familiar to unfamiliar. Smartphone applications are extremely numerous, as are applications for personal computers. Software is used to track airplanes and other transportation, to browse the Web, to provide security for financial transactions and corporations, and to direct unmanned missiles to a precisely defined target. Computers can operate from a remote location. For example, anaerobic manure digesters are used to convert cattle manure to biogas that can be converted to energy, a biosolid that can be used as bedding or soil amendment, and a nonodorous liquid stream that can be used as fertilizer. These digesters can be placed on numerous cattle farms in different states and operated and controlled by computers miles away.

Electrical. Electrical engineering studies the uses of electricity and the equipment to generate and distribute electricity to homes and businesses. Without electrical engineering, digital video disc (DVD) players, cell phones, televisions, home appliances, and many life-saving medical devices would not exist. Computers could not turn on. The Global Positioning System (GPS) in cars would be useless, and starting a car would require using a hand crank. This field of engineering is increasingly is involved in investigating different ways to produce electricity, including alternative fuels such as biomass and solar and wind power.

Environmental. The growth in the population of the world has been accompanied by increases in consumption and the production of waste. Environmental engineering is concerned with the reduction of existing pollution in the air, in the water, and on land, and the prevention of future harm to the environment. Issues addressed include pollution from manufacturing and other sources, the transportation of clean water, and the disposal of nonbiodegradable materials and hazardous and nuclear waste. Because pollution of the air, land, and water crosses national borders, environmental engineers need a broad, global perspective.

Fascinating Facts About Engineering

- The Bering Strait is 53 miles of open water between Alaska and Russia. In the 2009 Bering Strait Project Competition, people submitted plans for bridging the strait, including a combination bridge-tunnel with passageways for migrating whales and the capability of circulating the frigid arctic waters of the north to help fend off global warming.
- The Gotthard Base Tunnel network is being built under the Alps from Switzerland to Milan, Italy. At about 35 miles, the two-way tunnel will be the longest in the world and will reduce the time to make the trip by automobile from 3.5 hours to 1 hour. High-speed trains traveling at speeds of 155 miles per hour will make more than two hundred trips through the tunnel per day.
- A typical desktop computer can handle 100 million instructions per second. As of June, 2010, the fastest supercomputer was the Cray Jaguar at Oak Ridge National Laboratory. Its top speed is 1.75 petaflops (1 quadrillion floating point operations) per second.
- The world's smallest microscope weighs about as much as an egg. Instead of using a lens to magnify, it generates holographic images of microscopic particles or cells using a light emitting diode (LED) to illuminate and a digital sensor to capture the image.
- The National Institutes of Health has developed an implant made of silk and metal that when placed in the brain can detect impending seizures and send out electric pulses to halt them. It can also send out electric signals to prostheses used by people with spinal cord injuries.
- Stanford University is developing the computers and technology for a driverless car. The 2010 autonomous race car, dubbed Shelley, is an Audi TTS equipped with a differential Global Positioning System accurate to an inch. It calculates the right times to brake and accelerate while turning.

Industrial. Managing production and delivery of any product is the expertise of industrial engineers. They observe the people, machines, information, and technology involved in the process from start to finish, looking for any areas that can be improved. Increasingly, they use computer simulations and robotics. Their goals are to increase efficiency, reduce costs, and ensure worker safety. For example, worker safety can be improved through ergonomics and the use of less-stressful, easier-to-manipulate tools. The expertise of industrial engineers can have a major impact on the profitability of companies.

Manufacturing. Manufacturing engineering examines the equipment, tools, machines, and processes involved in manufacturing. It also examines how manufacturing systems are integrated. Its goals are to increase product quality, safety, output, and profitability by making sure that materials and labor are used optimally and waste—whether of time, labor, or materials—is minimized. For example, engineers may improve machinery that folds disposable diapers or that machines the gears for a truck, or they may reconfigure the product's packaging to better protect it or facilitate shipping. Increasingly, robots are used to do hazardous, messy, or highly repetitive work, such as painting or capping bottles.

Mechanical. The field of mechanical engineering is the oldest and largest specialty. Mechanical engineers create the machines that drive technology and industry and design tools used by other engineers. These machines and tools must be built to specifications regarding usage, maintenance, cost, and delivery. Mechanical engineers create both power-generating machinery such as turbines and power-using machinery such as elevators by taking advantage of the compressibility properties of fluids and gases.

Nuclear. Nuclear engineering requires expertise in the production, handling, utilization, and disposal of nuclear materials, which have inherent dangers as well as extensive potential. Nuclear materials are used in medicine for radiation treatments and diagnostic testing. They also function as a source of energy in nuclear power plants. Because of the danger of nuclear materials being used for weapons, nuclear engineering is subject to many governmental regulations designed to improve security.

Impact on Industry

The U.S. economy and national security are closely linked to engineering. For example, engineering is vital for developing ways to reduce the cost of energy, decrease American reliance on foreign sources of energy, and conserve existing natural resources. Because of its importance, engineering is supported and highly regulated by government agencies, universities, and corporations. However, some experts question whether the United States is educating enough engineers, especially in comparison with China and India. The number of engineering degrees awarded each year in the United States is not believed to be keeping pace with the demand for new engineers.

Government Research. The U.S. government has a vested interest in the commerce, safety, and military preparedness of the nation. It both funds and regulates development through subcontractors, laws, guidelines, and educational initiatives. For example, the Americans with Disabilities Act of 1990 made it mandatory that public buildings be accessible for all American citizens. Its passage spurred innovations in engineering such as kneeling buses, which make it possible for those in wheelchairs to board a bus. The U.S. Department of Defense (DOD) is the largest contractor and the largest provider of funds for engineering research. For example, it issued 80,000 specifications for the creation of a synthetic jet fuel. The Federal Drug Administration (FDA) concentrates its efforts on supplier control and testing of products and materials. Funding for research in engineering is also provided by the National Science Foundation. The government has also sponsored educational initiatives in science, technology, engineering, and mathematics, an example of which is the America Competes Act of 2007.

Academic Research. Universities, often in collaboration with governmental agencies, conduct research in engineering. Also, universities have entered into partnerships with private industry as investors have sought to capitalize on commercial possibilities presented by research, as in the case of stem cell research in medicine. Universities are also charged with providing rigorous, up-to-date education for engineers. Numerous accrediting agencies, including the Accreditation Board for Engineering and Technology, ensure that graduates from engineering programs have received an adequate and appropriate education. Attending an institution without accreditation is not advisable.

Industry and Business. Engineering has a role in virtually every company in every industry, including nonprofits in the arts, if only because these companies use computers in their offices. Consequently, some fields of engineering are sensitive to swings in the economy. In a financial downturn, no one develops office buildings, so engineers working in architecture and construction are downsized. When towns and cities experience a drop in tax income, projects involving roads, sewers, and environmental cleanup are delayed or canceled, and civil, mechanical, and environmental engineers lose their jobs. However, some economic problems can actually spur developments in engineering. For example, higher energy costs have led engineers to create sod roofs for factories, which keep the building warmer in winter and cooler in summer, and to develop lighter, stronger materials to use in a airplanes.

Careers and Course Work

To pursue a career in engineering, one must obtain a degree from an accredited college in any of the major fields of engineering. A bachelor's degree is sufficient for some positions, but by law, each engineering project must be approved by a licensed professional engineer (P.E.). To gain P.E. registration, an engineer must pass the comprehensive test administered by the National Society of Professional Engineers and work for a specified time period. In addition, each state has its own requirements for being licensed, including an exam specific to the state. An engineer with a bachelor's degree may work as an engineer with or without P.E. registration, obtain a master's degree or doctorate to work in a specialized area of engineering or pursue an academic career, or obtain an M.B.A. in order to work as a manager of engineers and products.

Social Context and Future Prospects

Engineering can both prolong life through biomedical advances such as neonatal machinery and destroy life through unmanned military equipment and nuclear weaponry. An ever-increasing number of people and their concentration in urban areas means that ways must be sought to provide more food safely and to ensure an adequate supply of clean, safe drinking water. These needs will create projects involving genetically engineered crops, urban agriculture, desalination facilities, and the restoration of contaminated rivers and streams. The never-ending quest for energy will remain a fertile area for research and development. Heated political debates about taxing certain fuels and subsidizing others are part of the impetus behind solar, wind, and biomass development and renewed discussions about nuclear power.

The lack of minorities, including women, Hispanics, African Americans, and Native Americans, in engineering is being addressed through education initiatives. Women's enrollment in engineering schools has hovered around 20 percent since about 2000. African Americans make up about 13 percent of the U.S. population yet only about 3,000 blacks earn bachelor's degrees in engineering each year. About 4,500 Hispanics, who represent about 15 percent of the U.S. population, earn bachelor's degrees in engineering each year.

Judith L. Steininger, B.A., M.A.

Further Reading

Addis, Bill. *Building: Three Thousand Years of Design Engineering and Construction.* New York: Phaidon Press, 2007. Traces the history of building engineering in the Western world, covering the people, buildings, classic texts, and theories. Heavily illustrated.

Baura, Gail D. *Engineering Ethics: An Industrial Perspective.* Boston: Elsevier Academic Press, 2006. Thirteen case studies examine problems with products, structures, and systems and the role of engineers in each. Chapters cover the Ford Explorer rollovers, the San Francisco-Oakland Bay Bridge earthquake collapse, the Columbia space shuttle explosion, and the 2003 Northeast blackout.

Dieiter, George E., and Linda C. Schmidt. *Engineering Design.* 4th ed. Boston: McGraw-Hill Higher Education, 2009. Looks at design as the central activity of engineering. Provides a broad overview of basic topics and guidance on the design process, including materials selection and design implementation.

Nemerow, Nelson L., et al., eds. *Environmental Engineering: Environmental Health and Safety for Municipal Infrastructure, Land Use and Planning, and Industry.* Hoboken, N.J.: John Wiley & Sons, 2009. Covers environmental issues such as waste disposal for industry, the residential and institutional environment, air pollution, and surveying and mapping for environmental engineering.

Petroski, Henry. *Success Through Failure: The Paradox of Design.* 2006. Reprint. Princeton, N.J.: Princeton University Press, 2008. Examines failure as a motivator for engineering and defines success as "anticipating and obviating failure." Chapters deal with bridges, buildings, and colossal failures.

Yount, Lisa. *Biotechnology and Genetic Engineering.* 3d ed. New York: Facts On File, 2008. Covers the history of genetic engineering and biotechnology, including the important figures. Contains a bibliography and index.

Web Sites

American Association of Engineering Societies
http://www.aaes.org

American Engineering Association
http://www.aea.org

American Society for Engineering Education
http://www.asee.org

Institute of Electrical and Electronics Engineers (IEEE)
http://www.ieee.org

National Society of Professional Engineers
http://www.nspe.org

See also: Biomechanical Engineering; Chemical Engineering; Civil Engineering; Computer Engineering; Electrical Engineering; Electronics and Electronic Engineering; Engineering Mathematics; Environmental Engineering; Mechanical Engineering; Software Engineering; Spacecraft Engineering.

ENGINEERING MATHEMATICS

FIELDS OF STUDY

Algebra; geometry; trigonometry; calculus (including vector calculus); differential equations; statistics; numerical analysis; algorithmic science; computational methods; circuits; statics; dynamics; fluids; materials; thermodynamics; continuum mechanics; stability theory; wave propagation; diffusion; heat and mass transfer; fluid mechanics; atmospheric engineering; solid mechanics.

SUMMARY

Engineering mathematics focuses on the use of mathematics as a tool within the engineering design process. Such use includes the development and application of mathematical models, simulations, computer systems, and software to solve complex engineering problems. Thus, the solution of the engineering problem might be a component, system, or process.

KEY TERMS AND CONCEPTS

- **Algorithmic Science:** Study, implementation, and application of real-number algorithms for solving problems of continuous mathematics that arise in the realms of optimization and numerical analysis.
- **Applied Mathematics:** Branch of mathematics devoted to developing and applying mathematical methods, including math-modeling techniques, to solve scientific, engineering, industrial, and social problems. Areas of focus include ordinary and partial differential equations, statistics, probability, operational analysis, optimization theory, solid mechanics, fluid mechanics, numerical analysis, and scientific computing.
- **Bioinformatics:** Field that uses sophisticated mathematical and computational tools for problem solving in diverse biological disciplines.
- **Computational Science:** Broad field blending applied mathematics, computer science, engineering, and other sciences that uses computational methods in problem solving.

Definition and Basic Principles

Engineering mathematics entails the development and application of mathematics (such as algorithms, models, computer systems, and software) within the engineering design process. In engineering mathematics course work and professional research, a variety of tools may be used in the collection, analysis, and display of data. Standard tools of measurement include rules, spirit levels, micrometers, calipers, and gauges. Software tools include Maplesoft, Mathematica, MATLAB, and Excel. Engineering mathematics has roots and applications in many areas, including algorithmic science, applied mathematics, computational science, and bioinformatics.

Background and History

Since engineering is such a broad area, engineering mathematics includes a variety of applications. In general, a link between engineering and mathematics is established when mathematical descriptions of physical systems are formulated.

Links may involve precise mathematical relationships or formulas. For example, Galileo Galilei's pioneering work on the study of the motion of physical objects led to the equations of accelerated motion, $v = at$ and $d = ½ at^2$, in which velocity is v, acceleration a, time t, and distance d. His work paved the way for Newtonian physics.

Other links are established through empirical relationships that have the status of laws. French physicist Charles-Augustin de Coulomb discovered that the force between two electrical charges is proportional to the product of the charges and inversely proportional to the square of the distance between them: $F = kq1q2/d2$ (force is F, constant of variation k, charges $q1$ and $q2$, and distance d). The coulomb, a measure of electrical charge was named for him. Coulomb's work was the first in a sequence of related discoveries by other notable scientists, many of whose findings led to additional laws. The list includes Danish physicist Hans Christian Ørsted, French physicists André-Marie Ampère, Jean-Baptiste Biot, and Félix Savart, British physicist and chemist Michael Faraday, and Russian physicist Heinrich Lenz.

Sometimes mathematical expressions of principles

apply almost universally. In physics, for example, the conservation laws indicate that in a closed system certain measurable quantities remain constant: mass, momentum, energy, and mass-energy. Lastly, systems of equations are required to describe physical phenomena of various levels of complexity. Examples include English astronomer and mathematician Sir Isaac Newton's equations of motion, Scottish physicist and mathematician James Clerk Maxwell's equations for electromagnetic fields, and Swiss mathematician and physicist Leonhard Euler's and French engineer Claude-Louis Navier and British mathematician and physicist George Gabriel Stokes's (Navier-Stokes) equations in fluid mechanics.

Further links between engineering and mathematics are discovered through the ongoing development, extension, modification, and generalization of equations and models in broader physical systems. For example, the Euler equations used in fluid mechanics can be connected to the conservation laws of mass and momentum.

How It Works

Many problems in engineering mathematics lead to the construction of models that can be used to describe physical systems. Because of the power of technology, a model may be derived from a system of a few equations that may be linear, quadratic, exponential, or trigonometric—or a system of many equations of even greater complexity. In engineering, such equations include ordinary differential equations, differential algebraic equations, and partial differential equations.

As the system of equations is solved, the mathematical model is formulated. Models are expressed in terms of mathematical symbols and notation that represent objects or systems and the relationships between them. Computer software, such as Maplesoft, Mathematica, MATLAB, and Excel, facilitates the process.

Engineers have available many models of physical systems. The development, extension, and modification of existing models, and the development of new models, are the subject of ongoing research. In this way, engineering mathematics continues to advance.

The ultimate test of a mathematical model is whether it truly reflects the behavior of the physical system under study. Computational experiments can be run to test the model for unexpected characteristics of the system and possibly optimize its design. However, models are approximations, and the accuracy of computed results must be evaluated through some form of error analysis.

The level of complexity of the construction and use of models depends on the engineering application. Further appreciation of the utility of a model may be gained by examining the use of a new model in engineering mathematics that impacts several scientific and technological areas. A mathematical model can now be used to investigate how materials break: One led to a new law of physics that depicts fracturing before it happens, or even as it occurs. In addition to the breakage of materials such as glass and concrete used in construction, the model enables better examination of bone breakage in patients with pathologies such as osteoporosis.

Applications and Products

Cell Biology. Advances in research in cell growth and division have proved helpful in disease detection, pharmaceutical research, and tissue engineering. Biologists have extensively explored cell growth and mass and the relationship between them. Using microsensors, bioengineers can now delineate colon cancer cell masses and divisions over given time periods. They have found that such cells grow faster as they grow heavier. With additional cell measurements and mathematical modeling, the scientists examined other properties such as stiffness. They also performed simulations to study the relationship between cell stiffness, contact area, and mass measurement.

Genetics. New genes involved in stem-cell development can be found, quickly and inexpensively, along the same pathway as genes already known. When searching for genes involved in a particular biological process, scientists try to find genes with a symmetrical correlation. However, many biological relationships are asymmetric and can now be found using Boolean logic in data-mining techniques. Engineering and medical researchers can then examine whether such genes become active, such as those in developing cancers. This research is expected to lead to advances in disease diagnosis and cancer therapy.

Energy. A new equation could help to further the use of organic semiconductors. The equation represents the relationship between current and voltage at the junctions of the organic semiconductors. Research in the use of organic semiconductors may

Fascinating Facts About Engineering Mathematics

- In aircraft design, computational simulations have been used extensively in the analysis of lift and drag. Advanced computation and simulation are now essential tools in the design and manufacture of aircraft.
- Auto-engineering researchers have developed a simulation model that can significantly decrease the time and cost of calibrating a new engine. Unlike statistics-based models, the new physics-based model can generate data for transient behavior (acceleration or deceleration between different speeds).
- Although the number of nuclear weapons held by each country in the world is securely guarded information, the rising demands of regulatory oversight require computing technology that exceeds current levels. A knowledge of the mathematical development and use of uncertainty models is critical for this application.
- Researchers have used nontraditional mathematical analyses to identify evolving drug resistance in strains of malaria. Their goal is to enable the medical community to react quickly to inevitable drug resistance and save lives. They also want to increase the life span of drugs used against malaria. The researchers used mathematical methods that involved graphing their data in polar coordinates and in rectangular coordinates. The results of the study are of particular interest to biomedical engineers focusing on genetics research.
- Bioengineers studying Alzheimer's disease found that amyloid beta peptides generate calcium waves. Following these waves in brain-cell networks, the scientists discovered voltage changes that signify intracellular communication. This research involves mathematical modeling of brain networks.
- Medical and engineering researchers have developed a mathematical model reflecting how red blood cells change in size and hemoglobin content during their four-month life span. This model may provide valuable clinical information that can be used to predict who is likely to become anemic.
- Mathematically gifted philosophers Sir Isaac Newton and Gottfried Leibniz invented the calculus that is now part of engineering programs.
- French mathematician Joseph Fourier introduced a mathematical series using sines and cosines to model heat flow.
- Albert Einstein proposed his general theory of relativity, incorporating the Riemannian or elliptical geometry concept that space can be unbounded without being infinite. The theory of relativity, along with later nuclear-energy research, led to development of the atomic and hydrogen bombs and theoretical understanding of thermonuclear fusion as the energy source powering the Sun and stars.
- Among the prominent figures involved in the design and development of electronic computers is American mathematician Norbert Wiener. Wiener is the founder of the science of cybernetics, the mathematical study of the structure of control systems and communications systems in living organisms and machines.

lead to advances in solar cells, displays, and lighting. Engineers have been studying organic semiconductors for about 75 years but have only recently begun to discover innovative applications.

Impact on Industry

Government and University Research. Engineering mathematics has roots and applications in many research areas, including algorithmic science, applied mathematics, computational science, and bioinformatics. Theoretical, empirical, and computational research includes the development of effective and efficient mathematical models and algorithms. Professional research includes diverse medical, scientific, and industrial applications. Even research in university-level engineering mathematics includes work that challenges students to integrate and apply course-work knowledge from many disciplines as they examine cracks in solids, mixing in small geometries, the crumpling of paper, lattice packings in curved geometries, materials processes, and predict optimal mechanics for device applications.

Engineering and Technology. Engineering mathematics is among the tools used to create, develop, and maintain products, systems, and processes. Applications are found in many areas, including nanotechnology. For example, radio-frequency designs are approaching higher frequency ranges and higher levels of complexity. The underlying mathematics of such designs is also complex. It requires new modeling techniques, mathematical methods, and simulations with mixed analogue and digital signals. Ordinary differential equations, differential algebraic equations, and partial differential algebraic equations are used in the analyses. The goal is to predict the circuit behavior before costly production begins. In such research, algorithms have been modified and new algorithms created to meet simulation demands.

Other Applied Sciences. There has been a dramatic rise in the power of computation and information technology. With it have come vast amounts of data in various fields of applied science and engineering. The challenge of understanding the data has led to new tools and approaches, such as data mining. Applied mathematics includes the use of mathematical models and control theory that facilitate the study of epidemics, pharmacokinetics, and physiologic systems in the medical industry. In telematics, models are developed and used in the enhancement of wireless mobile communications.

Careers and Course Work

Data Analyst or Data Miner. Data mining involves the discovery of hidden but useful information in large databases. In applications of data mining, career opportunities emerge in medicine, science, and engineering. Data mining involves the use of algorithms to identify and verify previously undiscovered relationships and structure from rigorous data analysis. Course work should include a focus on higher-level mathematics in such areas as topology, combinatorics, and algebraic structures.

Materials Science. Materials science is the research, development, and manufacture of such items as metallic alloys, liquid crystals, and biological materials. There are many career opportunities in aerospace, electronics, biology, and nanotechnology. Research and development uses mathematical models and computational tools. Course work should include a focus on applied mathematics, including differential equations, linear algebra, numerical analysis, operations research, discrete mathematics, optimization, and probability.

Ecological and Environmental Engineering. The work of professionals in these fields covers many areas that transcend pollution control, public health, and waste management. It might, for example, involve the design, construction, and management of an aquatic ecosystem or the research and development of appropriate sustainable technologies. Course work should include a focus on higher-level mathematics in such areas as calculus, linear algebra, differential equations, and statistics.

Meteorology and Climatology. These career areas incorporate not only atmospheric, hydrologic, and oceanographic sciences but modeling, forecasting, geoengineering, and geophysics. In general, historical weather data and current data from satellites, radar, and monitoring equipment are combined with other measurements to develop, process, and analyze complex models using high-performance computers. Current research areas include global warming and the impact of atmospheric radiation and industrial pollutants. Mathematics courses in meteorology and atmospheric science programs include calculus, differential equations, linear algebra, statistics, computer science, numerical analysis, and matrix algebra or computer systems.

Social Context and Future Prospects

Within engineering mathematics, an interdisciplinary specialty has emerged, computational engineering. Computational engineering employs mathematical models, numerical methods, science, engineering, and computational processes that connect various fields of engineering science. Computational engineering emerged from the impact of supercomputing on engineering analysis and design. Computational modeling and simulation are vitally important for the development of high-technology products in a globally competitive marketplace. Computational engineers develop and use advanced software for real-world engineering analysis and design problems. The research work of engineering professionals and academics has potential for applications in several engineering disciplines.

June Gastón, B.A., M.S.Ed., M.Ed., Ed.D.

Further Reading

Gribbin, John. *The Scientists: A History of Science Told Through the Lives of Its Greatest Inventors.* New York: Random House, 2003. This compelling text on the history of modern science marks the subject's discoveries and milestones through the great thinkers who were integral to its creation, both well known and not as well known.

Merzbach, Uta C., and Carl B. Boyer. *A History of Mathematics.* 3d ed. Hoboken, N.J.: John Wiley &

Sons, 2011. An excellent and highly readable history of the subject that chronicles the earliest principles as well as the latest computer-aided proofs.

Schäfer, M. *Computational Engineering, Introduction to Numerical Methods.* New York: Springer, 2006. Schäfer includes applications in fluid mechanics, structural mechanics, and heat transfer for newer fields such as computational engineering and scientific computing, as well as traditional engineering areas.

Shiflet, Angela B., and George W. Shiflet. *Introduction to Computational Science: Modeling and Simulation for the Sciences.* Princeton, N.J.: Princeton University Press, 2006. Two approaches to computational science problems receive focus: system dynamics models and cellular automaton simulations. Other topics include rate of change, errors, simulation techniques, empirical modeling, and an introduction to high-performance computing. Numerous examples, exercises, and projects explore applications.

Stroud, K.A. *Engineering Mathematics.* 6th ed. New York: Industrial Press, 2007. Providing a broad mathematical survey, this innovative volume covers a full range of topics from basic arithmetic and algebra to challenging differential equations, Laplace transforms, and statistics and probability.

Velten, Kai. *Mathematical Modeling and Simulation: Introduction for Scientists and Engineers.* Weinheim, Germany: Wiley-VCH, 2009. Velten explains the principles of mathematical modeling and simulation. After treatment of phenomenological or data-based models, the remainder of the book focuses on mechanistic or process-oriented models and models that require the use of differential equations.

Web Sites

American Mathematical Society
http://www.ams.org/home/page

National Society of Professional Engineers
http://www.nspe.org

Society for Industrial and Applied Mathematics
http://www.siam.org

See also: Algebra; Applied Mathematics; Calculus; Geometry; Numerical Analysis; Trigonometry.

ENGINEERING SEISMOLOGY

FIELDS OF STUDY

Engineering; geology; physics; geophysics; earth science; electrical engineering; volcanology; volcanic seismology; plate tectonics; geodynamics; mineral physics; tectonic geodesy; mantle dynamics; seismic modeling; seismic stratigraphy; statistical seismology; computer science; mathematics; earthquake engineering; paleoseismology; archeoseismology; historical seismology; structural engineering; geography

SUMMARY

Engineering seismology is a scientific field focused on studying the likelihood of future earthquakes and the potential damage such seismic activity can cause to buildings and other structures. Engineering seismology utilizes computer modeling, geological surveys, existing data from historical earthquakes, and other scientific tools and concepts. Engineering seismology is particularly useful for the establishment of building codes and for land-use planning.

KEY TERMS AND CONCEPTS:

- **Duration:** The length of time ground motion occurs during an earthquake.
- **Epicenter:** The surface-level geographic point located directly above an earthquake's hypocenter.
- **Focal Depth:** The depth of an earthquake's hypocenter.
- **Ground Motion:** The ground-level shaking that occurs during an earthquake.
- **Hypocenter:** The point of origin of an earthquake.
- **Love Waves:** Seismic waves that occur in a side-to-side motion.
- **Magnitude:** An earthquake's size and relative strength.
- **Rayleigh Waves:** Seismic waves that occur in a circular, rolling fashion.
- **Richter Scale:** Logarithmic scale used to assign a numerical value to the magnitude of an earthquake.
- **Source Parameters:** A series of earthquake characteristics, including distance, duration, energy, and the types of waves that occur.
- **Stress Drop:** The amount of energy released when locked tectonic plates separate, causing an earthquake.
- **Wave Propagation Path:** The directions in which seismic waves travel in an earthquake.

DEFINITION AND BASIC PRINCIPLES

Engineering seismology (also known as earthquake engineering) is a multidisciplinary field that assesses the effects of earthquakes on buildings, bridges, roads, and other structures. Seismology engineers work in the design and construction of structures that can withstand seismic activity. They also assess the damages and effects of seismic activity on existing structures. Engineering seismologists analyze such factors as quake duration, ground motion, and focal depth in assessing the severity of seismic events and how those events affect fabricated structures. They also consider source parameters, which help seismologists zero in on a seismic event's location and the speed and trajectory at which the quake's resulting waves are traveling.

Earthquake engineers also study theoretical concepts and models related to potential earthquakes and historical seismic events. Such knowledge can help engineers and architects design structures that can withstand as powerful an earthquake as the geographic region has produced (or possibly will produce). Mapping systems and programs and mathematical and computer-based models are essential to engineering seismologists' work. Such techniques are also useful for archeologists and paleontologists, both of whom may use engineering seismology concepts to understand how the earth has evolved over millions of years and how ancient civilizations were affected by seismic events.

BACKGROUND AND HISTORY

Throughout human history, people have struggled to understand the nature of earthquakes and, as a result, have faced the challenges of preparing for these seismic events. Some ancient civilizations attributed earthquakes to giant snakes, turtles, and other creatures living and moving beneath the earth's surface. In the fourth century B.C.E., Aristotle was the first to speculate that earthquakes were not caused by supernatural forces but rather were natural events. However, little scientific study on earthquakes took place for hundreds of years, despite the occurrence of many major seismic events (including the eruption of Mount Vesuvius in Italy in 79 C.E., which was preceded by a series of earthquakes).

In the mid-eighteenth century, however, the British Isles experienced a series of severe earthquakes, which created a tsunami that destroyed Lisbon, Portugal, killing tens of thousands of people. Scientists quickly developed an interest in cataloging and understanding seismic events. In the early nineteenth century, Scottish physicist and glaciologist James D. Forbes invented the inverted pendulum seismometer, which gauged not only the severity of an earthquake but also its duration.

Throughout history, seismology has seen advances that immediately followed significant seismic events. Engineering seismology, which is proactive, represents a departure from reactionary approaches to the study of earthquakes. Today, engineering seismology uses seismometers, computer modeling, and other advanced technology and couples it with historical data for a given site. The resulting information helps civil engineers and architects construct durable buildings, bridges, and other structures and assess the risks to existing structures posed by an area's seismic potential.

How it Works

To understand engineering seismology, one must understand the phenomenon of earthquakes. Earthquakes may be defined as the sudden shaking of the earth's surface as caused by the movement of subterranean rock. These massive rock formations (plates), resting on the earth's superheated core, experience constant movement caused predominantly by gravity. While some plates move above and below one another, others come into contact with one another as they pass. The boundaries formed by these passing plates are known as faults. When passing plates lock together, stored energy builds up gradually. The plates eventually give, causing that energy to be released and sent from the quake's point of origin (the hypocenter) outward to the surface in the form of seismic (or surface) waves. Such waves occur either in a circular, rolling fashion (Rayleigh waves) or in a twisting, side-to-side motion (Love waves).

The field of seismology has developed only over the last few centuries, largely because of major, devastating seismic events. The practice of engineering seismology has grown in demand in recent years, mainly because of the modern world's dependency on major cities, infrastructure (such as bridges, roadways, and rail systems), and energy resources (including nuclear power plants and offshore oil rigs). Earthquake engineers, therefore, have two main areas of focus: studying seismology and developing structures that can withstand the force of an earthquake.

To study seismic activity and earthquakes, engineering seismologists may use surface-based detection systems, such as seismometers, to monitor and catalog tremors. They also employ equipment—including calibrators and accelerometers—that is lowered into deep holes. Such careful monitoring practices help seismologists and engineering seismologists better understand a region's potential for seismic activity.

When earthquakes occur, engineering seismologists quickly attempt to locate the hypocenter and the epicenter (the surface point that lies directly above the hypocenter). They are able to do so by monitoring two types of waves—P and S waves—that move much quicker than surface waves and, therefore, act as precursors to surface waves. These engineers also work to determine the magnitude (a measurement of an earthquake's size) of the event.

Magnitude may be based on a number of key factors (or source parameters), including duration, distance to the epicenter and hypocenter, the size and speed of the surface waves, the amount of energy (known as the stress drop) that is released from the hypocenter, P and S waves, and the directions in which surface waves move (the wave propagation path). Analyzing an earthquake's magnitude provides an accurate profile of the quake and the conditions that caused it.

In addition to developing a profile of a region's past seismic activity, earthquake engineers use such information to ascertain the type of activity a geographic region may experience in the future. For example, scientific evidence suggests that the level of stress drop is a major contributor to the severity of seismic activity that can cause massive destruction in major urban centers. Similarly, studies show that the duration of ground motion (the "shaking" effects of an earthquake) may be more of a factor in the

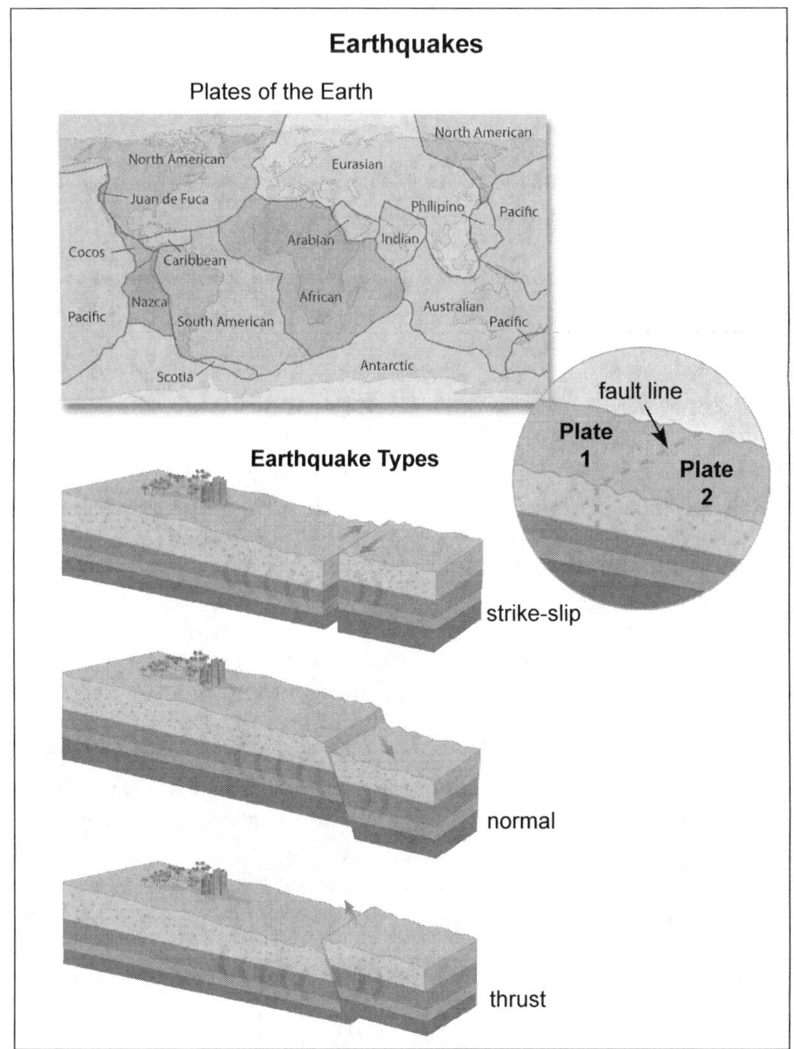

Earthquakes mostly occur along fault lines, which are cracks in the Earth's crust, formed from the movement of two tectonic plates.

amount of damage to buildings and other structures than stress drop.

The field of engineering seismology is less than one century old, but in the twenty-first century, it plays an important role in urban development and disaster prevention. Earthquake seismologists work with civil engineers and architects to design buildings, roads, bridges, and tunnels that may withstand the type of seismic activity that has occurred in the past.

Applications and Products

Engineering seismology applies knowledge of seismic conditions, events, and potential to the design and development of new and existing fabricated structures. Among the methods and applications employed by earthquake seismologists are the following:

Experimentation. Engineering seismologists may construct physical scale models of existing structures or proposed structures. Using data from a region's known seismic history, the engineers attempt to recreate an earthquake by placing these models on so-called shake tables, large mechanical platforms that simulate a wide range of earthquake types. After the "event," engineering seismologists examine the simulation's effects on the model structure, including its foundations, support beams, and walls. This approach enables the engineers and architects to directly examine the pre- and post-simulation structure and determine what sort of modifications may be warranted.

Computer Models. One of the most effective tools utilized by engineering seismologists is computer modeling. Through the application of software, engineering seismologists can input a wide range of source parameters, ground motion velocities, wave types, and other key variables. They also can view how different structural components withstand (or fail to withstand) varying degrees of seismic activities without the expense and construction time of a shake table. Computer modeling has become increasingly useful when attempting to safeguard against earthquake damage to dams, nuclear power plants, and densely developed urban centers. Computer modeling is also used by engineering seismologists to predict the path of destruction that often occurs after an earthquake, destruction such as that caused by fires or flooding.

Seismic Design Software. Earthquake engineers

study seismic activity in terms of how it affects structures. To this end, engineers must attempt to predict how earthquakes will strike an area. Seismic design software is used to create a map of a region's seismic activity and how those conditions will potentially cause structural damage. The software enables government officials to establish formalized building codes for buildings, bridges, power plants, and other structures. This software is easily obtained on the Web and through the U.S. Geological Survey (USGA) and other organizations.

Mathematics. Engineering seismology is an interdisciplinary field that relies heavily on an understanding of physics and mathematics. To quantify the severity of earthquakes, to calculate the scope of seismic activity, and, in general, to create a profile of a region's seismic environment, engineering seismologists utilize a number of mathematical formulae. One of the most-recognized of these formulae is the Richter scale, which was developed in 1935 by American seismologist and physicist Charles Richter. The Richter scale uses a logarithm to assign a numerical value (with no theoretical limit) to establish the magnitude of an earthquake. The Richter scale takes into account the amplitude (the degree of change) between seismic waves and the distance between the equipment that detects the quake and the quake's epicenter.

Earthquake engineers use such mathematical data as part of their analyses when working with civil engineers on construction projects. Earthquake engineers also are increasingly called upon by government officials to use this data to assess individual structure and citywide structural deficiencies that resulted in earthquake destruction. Forensic engineering was called into service in 2009, when Australian emergency officials intervened in Padang, Indonesia, after a magnitude 7.6 quake devastated that city. Engineers used mathematical formulae and statistical data to assess system-wide structural deficiencies in Padang rather than analyzing damage on a structure-by-structure basis. In light of the countless variables involved with studying earthquakes and their effects on fabricated structures, the use of established logarithms, data sets, and mathematical formulae is a time-honored practice of engineering seismologists.

Sensors. Not all earthquakes cause immediate and significant damage to affected structures. According to the USGA, the Greater San Francisco area experienced more than eighty earthquakes in 2011 alone, with none of those quakes registering higher than a 2.3 on the Richter scale. However, seismic activity on a small but frequent basis can cause long-term damage to structures. For example, seismic events can shift soil pressure on underground structures (such as pipes and foundations). Earthquake engineers are therefore highly reliant on sensor equipment, which enables them to gauge the effects of frequent seismic activity not only on above-ground structures but also on the ground itself.

To examine shifts in soil pressure caused by seismic activity, seismology engineers used an array of tactile pressure sensors, which were originally designed for artificial intelligence systems but were later utilized for the purposes of designing car seats and brake pad systems.. The use of such equipment helps engineers study the long-term effects of seismic activity on water pipes, underground cables, and underground storage tanks.

Arguably one of the best-known types of seismic detector systems is the seismograph. The seismograph uses a pendulum-based system to detect ground motion from seismic activity. Originally, the modern seismograph was designed to detect only significant earthquakes and tremors, and it could be found only in stable environments (such as a laboratory). Today, however, there are many different types of seismographs; some may be placed underground, others can be used in the field, while others are so sensitive that they can detect distant explosions or minute tremors.

Impact on Industry

Engineering seismology is a relatively new combination of civil engineering and seismology, along with other fields (such as geology, emergency management, and risk management). Its uses have proven invaluable, however, as many urban centers in earthquake-prone regions, such as Tokyo and San Francisco, have benefited from the careful application of safe building practices and disaster mitigation programs that are borne of engineering seismology. Japan, the United States, and Switzerland are among the leaders in earthquake engineering, along with Australia and New Zealand. Engineering seismology also involves a range of public and private organizations, such as the following:

Governments. National and regional governments

Fascinating Facts About Engineering Seismology

- There are approximately 500,000 detectable earthquakes in the world annually, 100,000 of which may be felt and only 100 that cause damage.
- Between 1975 and 1995, only the U.S. states of Florida, North Dakota, Iowa, and Wisconsin did not experience any earthquakes.
- The earliest known "seismograph" was introduced in 132 C.E. in China. It took the shape of a hollow urn (with a hidden pendulum inside) with dragon heads adorning the sides and frogs at the base directly under each dragon (which held a single ball). During an earthquake, a dragon would drop its ball into the mouth of the frog below, revealing the direction of the waves. The device once detected an earthquake four hundred miles away.
- Scientists cannot presently predict earthquakes; they can only calculate the potential for a quake to strike.
- The largest recorded earthquake in the world was a magnitude 9.5 quake in Chile in 1960.
- In 1931, there were 350 operating seismic stations in the world. In 2011, there existed more than 4,000 .
- The largest recorded earthquake in the United States was a magnitude 9.2 quake in Alaska in 1964.
- A magnitude 8.8 earthquake in Chile in 2010 shifted that country's coastline approximately 1,640 feet toward the Pacific Ocean.

play an important role in the application of the findings of engineering seismologists. The USGA, for example, offers a wide range of resources and services for studying seismic activity and the dangers it poses. Additionally, the Federal Emergency Management Agency (FEMA), the National Science Foundation, the National Institute of Standards, and the USGA combine their resources to operate the National Earthquake Hazards Reduction Program, which seeks to reduce property losses and human casualties caused by earthquakes through careful engineering seismology practices (including mapping seismically active areas and generating building codes).

Engineering Seismology Societies. Engineering seismologists share information and theories with their peers through professional associations and societies, many of which are global in nature. The International Association for Earthquake Engineering is one such network, holding worldwide conferences on engineering seismology every four years. This organization has branch societies in Japan, Europe, and the United States, each working locally but also contributing to the larger association.

Universities. Because of the contributions engineering seismology provides to the field of civil engineering, earthquake engineering continues to evolve within this educational discipline. Universities such as Stanford and the University of California, Berkeley, offer such programs. Many other universities feature coursework in environmental engineering, which includes earthquake engineering and seismology studies. Furthermore, a large number of universities house research laboratories, shake tables, and seismograph stations covering seismic activity throughout a given geographic area.

Consulting Firms. There is a considerable financial benefit to constructing a building, bridge, or other structure that will survive in a seismically active environment. Oil companies and other energy corporations, mining operations, and other businesses frequently seek the advice of private engineering consultants who offer seismic monitoring services. Construction companies also look to these consultants, seeking structural analyses and other services. The person who introduced the Richter scale, Charles Richter, founded a consulting firm upon his retirement from the California Institute of Technology in 1970.

Careers and Coursework

Students interested in engineering seismology should pursue and complete a bachelor's degree program in a related field, such as geology or civil engineering. They should also obtain a master's degree, preferably in a field of relevance to earthquake engineering, such as environmental engineering, structural engineering, geology, or seismology. Engineering seismologists' competitiveness as job candidates is improved greatly when they also earn a doctorate.

Engineering seismologists must receive training in the geosciences, including seismology, geology, and physics. These fields include courses in geodynamics, plate tectonics, statistical seismology, and mineral dynamics. They must also study civil engineering, structural design, and computer science (which must include training in computer modeling, digital mapping systems, and design software, which are essential in this arena). Furthermore, engineering seismologists must demonstrate excellent mathematical skills, particularly in geometry, algebra, and calculus. Finally, earthquake engineers must be trained in the use of many of the technical systems and devices that seismologists must use to monitor earthquakes.

Social Context and Future Prospects

The study of earthquakes is a practice that dates back hundreds of years. Earthquake engineering specifically, however, represents an evolution toward a practical application of the study of seismic activity to the design and construction of large buildings, power plants, and other structures.

Engineering seismologists work closely with seismologists and civil engineers. On the former front, these engineers help design and operate detection equipment and systems to help explain seismic activity. This collaboration is critical, as improved knowledge of seismic activity can save lives and property.

For example, Japan has long utilized engineering seismology practices in its urban centers. The magnitude 8.9 earthquake in that country in March, 2011, did not devastate Tokyo because of strong building codes that, among other things, cause skyscrapers to sway with the region's seismic waves rather than stand in rigid fashion. Comparatively, the magnitude 7.0 Haiti earthquake in 2010 virtually flattened the country's capital, Port-au-Prince, and outlying areas, largely because Haiti did not have earthquake-safe building codes. Its buildings were built using insufficient steel and on slopes with no reinforcing foundation or support systems. One observer in Haiti reported that Port-au-Prince would likely not have survived even a magnitude 2.0, much less the 7.0 quake it did have.

The significance of the 2011 Japan disaster and of the rare 5.8 Virginia earthquake that struck the East Coast of the United States in August, 2011, continues to cast light on the need for earthquake engineering in structural design and construction. As regions with a history of major seismic activity (and those regions with the potential for such activity) continue to grow in size and population, engineering seismologists are likely to remain in high demand.

Michael P. Auerbach, B.A., M.A.

Further Reading

Chopra, Anil K. *Dynamics of Structures: Theory and Applications to Earthquake Engineering.* 3d ed. Upper Saddle River, N.J.: Prentice Hall, 2007. This book analyzes the theories of structural dynamics and applies the effects of earthquakes and seismology to them, including structural design and energy dissipation models.

Griffith, M. C., et al. "Earthquake Reconnaissance: Forensic Engineering on an Urban Scale." *Australian Journal of Structural Engineering* 11, no. 1 (2010): 63-74. This article describes the observations of a team of Australian aid workers who were dispatched to Indonesia to assess structural damages caused by a magnitude 7.6 earthquake there in 2009.

Saragoni, G. Rudolfo. "The Challenge of Centennial Earthquakes to Improve Modern Earthquake Engineering." *AIP Conference Proceedings,* 1020, no. 1 (July 8, 2008): 1113-1120. This article uses the 1906 San Francisco earthquake as a point of reference for reviewing the evolution of modern engineering seismology.

Slak, Tomaz, and Vojko Kilar. "Development of Earthquake Resistance in Architecture from an Intuitive to an Engineering Approach." *Prostor* 19, no. 1 (2011): 252-263. This article reviews the application of earthquake engineering concepts to the design of modern structures.

Stark, Andreas. *Seismic Methods and Applications.* Boca Raton, Fla.: BrownWalker Press, 2008. This book first reviews the principles of seismology and relevant geosciences and then proceeds to the practical application of these principles to structural design and engineering.

Villaverde, Roberto. *Fundamental Concepts of Earthquake Engineering.* New York: CRC Press, 2009. This book features a review of the history of the field of engineering seismology and includes some examples of how certain seismic wave types and other conditions are taken into account in building design.

See also: Applied Physics; Architecture and Architectural Engineering; Civil Engineering; Earthquake Engineering; Earthquake Prediction; Engineering; Environmental Engineering; Urban Planning and Engineering.

ENVIRONMENTAL BIOTECHNOLOGY

FIELDS OF STUDY

Biology; microbiology; biochemistry; genetics; virology; agronomy; chemistry; toxicology; environmental monitoring; microbiology, earth and planetary sciences; ecology and evolutionary biology; environmental engineering; chemical and biomolecular engineering; atmospheric science.

SUMMARY

Environmental biotechnology, also known as biotechnical pollution control, is a rapidly developing science that uses biological resources to protect and restore the environment. It has significant implications and applications in both the prevention of air, soil, and water pollution and the restoration of contaminated environments.

KEY TERMS AND CONCEPTS

- **Bioaugmentation:** Form of bioremediation that involves increasing the number or activity level of microorganisms that assist in pollutant reduction.
- **Bioenrichment:** Form of bioremediation that involves adding nutrients or oxygen to environments to increase the breakdown of contaminants.
- **Biofilms:** An aggregated layer of microbial populations formed on aqueous (water) environments; sometimes referred to as microbial mats and known colloquially as slime.
- **Bioremediation:** Use of microorganisms, such as fungi and bacteria, and their enzymes to return a contaminated environment to its original condition.
- **Biostimulation:** Form of bioremediation in which the natural processes of degradation are encouraged through the introduction of certain stimuli, such as nutrients and additional substrates.
- **Biotechnology:** Exploitation and manipulation of living organisms to produce valuable and useful products or results.
- **End-of-Pipe Technology:** Treatment of waste and pollution that has already contaminated the air, soil, or water.

Definition and Basic Principles

Environmental biotechnology is a multidisciplinary science, a synthesis of environmental engineering, biochemistry, and microbiology. Those involved in environmental biotechnology generally concentrate on the development, application, adaptation, and management of biological systems and organisms to repair and prevent environmental damage caused by pollution and on the advancement of green technologies and sustainable development. Fundamentally, environmental biotechnology is the study of how microorganisms, plants, and their enzymes can assist in the restoration, remediation, preservation, and sustainable use of the world's natural environment. The primary role of environmental biotechnologists is to create a better balance between human development and the natural environment.

All fields of biotechnology have seen rapid growth and progress in the 1990's and 2000's. The field of environmental biotechnology, in particular, has benefited from advancements in genetic engineering and modern microbiological concepts, which offer both traditional and innovative solutions to different forms of contamination occurring in different mediums (soil, water, and air).

Background and History

The term "biotechnology" was being used as early as 1919 but did not occur frequently in scientific literature until the 1960's and 1970's, with the publication of the *Journal of Biotechnology*. Although the use of biotechnology within the medical and agricultural industry can be traced back many years, the purposeful use of biotechnology to mitigate environmental issues has a much shorter history.

Although people were familiar with the concept of biotechnology, for the most part, efforts were focused on the medical and agricultural industries. The arrival of the Industrial Revolution, however, rapidly altered and influenced the environment through the release of toxic pollutants into the waterways and soil. As people became wealthier, the demand for goods grew, and with the rise in industrial and agricultural production came the increase in the impact on the environment. Although the concept of using natural degradation processes was not new—for many years, communities had composted and relied on natural processes and microbes in the breakdown and treatment of sewage—environmental biotechnology was not yet considered a science.

In the 1960's, however, chemical pollution and its adverse effects came under significant public scrutiny. One of the landmark cases involved the chemical dichloro-diphenyl-trichloroethane (DDT), which was widely used as a pesticide but later was found to seriously affect bird populations and to cause cancer in humans. Other cases of chemical pollution and its adverse effects on people and the environment were becoming known, including mercury poisoning in Japan, Agent Orange in Vietnam, industrial sludge in the United States, and the devastating effects of oil spills. By the 1970's, it had become clear that the environment was becoming sullied and that contaminants were adversely affecting people. Concern regarding the environment led to the development of many laws and regulations in both developed and developing countries regarding the proper management of waste and pollution control. It was in the light of this political and social awareness that the field of environmental biotechnology emerged. Although environmental biotechnology originally focused on the treatment of wastewater, the field expanded to include areas of study such as soil contamination, solid-waste treatment, and air purification methods.

In 1992, during the United Nations Conference on Environment and Development in Rio de Janeiro, environmental biotechnology was recognized and embraced as a crucial tool for both repairing and preventing environmental and health issues caused by humans. Since this conference, the field of environmental biotechnology has advanced at a rapid rate and has grown to provide innovative approaches to the sustainable development and protection of the world's ecosystems.

How It Works

Traditional methods of waste removal, such as landfill and incineration, cannot cope with the sheer volume of waste created by human populations. This situation has increased the need to develop alternative environmentally sound treatments and techniques. Environmental biotechnology seeks to positively affect pollution control and waste management.

A rise in consumption since the 1990's has been accompanied by a corresponding increase in the release of pollutants into the environment. Some of these contaminants, particularly those that also are naturally occurring, can be digested, degraded, or removed from the soil and water through the action of microorganisms. However, some of these human-created pollutants rarely occur naturally, and the accumulation of such substances can have a serious ecological impact.

The use of biotechnology in the treatment of waste and pollution is not a new idea. For more than a century, many communities have relied on natural processes and microbes to break down and treat sewage. The fundamental aim of environmental biotechnology is to use organisms to control contamination and treat waste. In the process called bioremediation, microorganisms, including fungi and bacteria and their enzymes, are used to return a contaminated environment to its original condition. Naturally occurring biological degradation processes are purposely employed to remove contaminants from areas where they have been released. The use of such processes requires a solid scientific understanding of the contaminant, its impact, and the affected ecosystem.

The concept of environmental biotechnology depends on the notion that all living organisms, such as flora, fauna, bacteria, and fungi, consume nutrients for their survival and, in doing so, produce waste byproducts. Not all organisms require the same nutrients nor react in the same manner, however. Some organisms, such as certain bacteria and microorganisms, flourish on chemicals and toxins that are actually poisonous or harmful to other organisms or ecosystems. The fact that some microorganisms and various strains of microbial species react differently to chemical toxins and environmental pollutants has advanced the concept of using genetic manipulation techniques.

Environmental biotechnology aims to provide a natural approach to tackling environmental issues, from identification of biohazards to restoration of industrial, agricultural, and natural areas affected by contamination. Central to the concept of environmental biotechnology is the ability to determine which contaminants are present, for how long, and in what quantity, and what recovery method is applicable. There are four basic concepts and approaches in the field of environmental biotechnology:

bioremediation, prevention, detection and monitoring, and genetic engineering.

Applications and Products

Environmental pollution can be a legacy of former industrial practices or a product of unsustainable present-day practices. One of the most serious environmental issues facing the world in the twenty-first century is the production of very large quantities of waste, the majority of which becomes landfill. Industrialized nations also produce significant quantities of chemicals that often end up in the soil and water. Environmental biotechnology seeks ways to combat the escalating ecological problems associated with such pollution and waste. Technologies that have been developed and implemented include bioremediation of water and soil, biomonitoring using biosensors and bioassays, and bioprocessing.

Bioremediation. Bioremediation is usually classified as either in situ or ex situ. In situ bioremediation entails treating the contamination in place and relies on the ability of the microorganisms to metabolize or remove the contaminants inside the naturally occurring system; ex situ remediation entails the polluted material being removed from the contaminated site and treated elsewhere and relies on some form of artificial engineering and input. The process of bioremediation can occur naturally or be encouraged through artificial stimulus. The process of attenuation occurs under natural conditions and incorporates the normal chemical, biological, and physical processes, such as aerobic and anaerobic degradation, that eliminate or reduce soil and water contaminants.

Biostimulation and Bioaugmentation. Biostimulation is a form of bioremediation in which the natural processes of degradation are encouraged through the introduction of certain stimuli, such as nutrients and additional substrates. Bioaugmentation is a form of bioremediation that involves increasing the activity of the microorganisms that assist in pollutant reduction or augmenting existing yet insufficient populations of microorganisms. Bioenrichment is another form of bioremediation that involves adding nutrients or oxygen to environments to increase the breakdown of contaminants.

Biodetectors. Biodetectors such as bioassays and biosensors are used to monitor, assess, and analyze biological material, and provide important information and data about the effects and concentration of pollutants in the natural environment. Bioassays are procedures or experiments in which the quantity of a contaminant is estimated by measuring its effect on living organisms. Although they can be relatively slow and expensive, they are essential for the assessment and prediction of real and potential effects of pollution on the natural environment.

Biosensors are devices that detect and measure minute amounts of or changes in concentration of chemical substances within an environmental area and translate that information into data. Because of their ability to detect even tiny quantities of targeted chemicals with greater speed and at less cost than bioassays, biosensors have become important tools for the monitoring and control of pollution levels, both before and after bioremediation measures are implemented.

Bioprocessing. Bioprocessing is a process that uses living cells or organisms to produce specific outcomes. Communities of microbial organisms perform a comprehensive range of bioprocesses within the natural environment, which can be exploited to benefit both the environment and industry. Bioprocesses used in environmental biotechnology include microbial enhanced oil recovery (MEOR), biological treatments of polluted air, biodesulfurization, conversion of pollutants into useful products such as fertilizers and green energy, and microbial exploration technology.

Biofilm Control. Biofilms, often referred to as slime, occur on the surface of aqueous environments and are caused by a complex accumulation of microorganisms. Biofilms are usually considered undesirable, as they are frequently associated with odors, infections, fouling, and corrosion, but they can be beneficial under some circumstances such as the treatment of wastewater.

Genetic Engineering and Manipulation. The advancement of genetics and genetic manipulation has had significant impact on environmental biotechnology. Research in molecular genetics has provided novel techniques for the detection and degradation of contaminants through the manipulation and enhancement of the microbes' ability to adapt themselves genetically to different pollutants. The ecologically useful and improved organisms are classified as genetically engineered microbes (GEM).

IMPACT ON INDUSTRY

Major Organizations. Many organizations worldwide are focusing on environmental biotechnology research. The International Society of Environmental Biotechnology is an umbrella organization that aims to enhance the progress and promotion of environmental biotechnology within industry and governments on an international scale. In Australia, the Australian Environmental Biotechnology Cooperative Research Centre focuses its research on biofilm control, microbial detection and control, and bioprocesses, and the Commonwealth Scientific and Industrial Research Organisation (CSIRO) focuses its research on the use of microorganisms and biotechnological processes in the treatment of mining contamination and wastewater. In the United States, significant research on environmental biotechnology is undertaken at the Center for Environmental Biotechnology at the Lawrence Berkeley National Laboratory. Additionally, as world leaders in biotechnology and research, both the United States and the European Union (EU) have been instrumental in the evolution of environmental biotechnology, culminating in the establishment of the EU-U.S. Task Force on Biotechnology Research.

Government and University Research. Environmental biotechnology is an essential tool for the sustainability of the Earth's natural ecosystems and the balance between people and nature. Environmental pollution is a global issue, particularly the contamination of the world's oceans and the atmosphere. As such, many governments have initiated policies and regulations for waste removal and pollution control.

Although pollution is a global problem and requires international cooperation and control, the United States, China, India, and the European Union are home to a large proportion of the world's population and are the greatest contributors (by country) of pollution. Therefore, it is crucial that the governments of these countries research and implement environmental biotechnology, pollution control, and ecological restoration. The Chinese government has introduced programs to develop environmental technology, including the National Medium- and Long-Term Science and Technology Development Plan in 2006. The Indian government's Environmental Biotechnology Division was introduced to provide solutions for the abatement of contaminants, with the additional objectives of conserving endangered species and developing useful products from waste.

The U.S. government has also introduced initiatives such as the Environmental Remediation Sciences Program, formed in 2005 through the consolidation of the Natural and Accelerated Bioremediation Research (NABIR) program and the Environmental Management Science Program (EMSP), and administered by the U.S. Department of Energy's Office of Science. This program aims to address some of the United States' most pressing environmental recovery and contamination issues.

A growing number of international universities and institutes contribute significantly to environmental biotechnology. These include the Center

Fascinating Facts About Environmental Biotechnology

- Environmental biotechnology ranges from basic natural organic processes, such as fermentation and biodegradation, to enhanced and human-manipulated procedures, such as genetic engineering.
- Oil spills can be cleaned up using a bioremediation accelerator, which when sprayed on an oil sheen, creates a hydrocarbon-eating bacterium that breaks down hydrocarbons into water and carbon dioxide.
- Bioremediation can occur within soils, on and below the surface. Water and nutrients are mixed into the soil to facilitate bacterial growth. The bacteria used in bioremediation break down the carbon chains of harmful organic contaminants, leaving water and carbon dioxide.
- Scientists at the University of California, San Diego, have discovered an enzyme that, when manipulated, causes plants to absorb more carbon dioxide while retaining more water. This enzyme may address the dual problem of increased carbon levels and decreased water resources.
- Biotechnology is a $30 billion-a-year industry. There are more than 1,450 biotechnology companies in the United States alone.
- In the gold-mining industry in South Africa, bacteria are used to isolate gold from gold ore, thus producing less waste and avoiding the expenditure of a large amount of smelting energy.

for Environmental Biotechnology at Arizona State University, the School of Biotechnology and Biomolecular Sciences and the Centre for Marine Biofouling and Bio-Innovation at the University of New South Wales, the Center for Environmental Biotechnology at the University of Tennessee, the Centre on Environmental Biotechnology at the University of Kalyani (India), and the University of Idaho's Environmental Biotechnology Institute.

Careers and Course Work

Students who wish to pursue a career in environmental biotechnology can have a degree in a number of fields, including environmental science, chemistry, microbiology, and biomolecular engineering. Many universities provide undergraduate and postgraduate degrees in environmental biotechnology. Upon course completion, students should have a solid understanding of environmental and contamination processes, and the theories and technologies used in environmental biotechnology to mitigate environmental damage.

Environmental biotechnology graduates can pursue such careers as environmental and bioengineering consultants and water-recycling and water resource managers in the private sector, nongovernmental organizations (NGOs), and specialized government organizations and agencies, and as researchers and professors at universities.

Social Context and Future Prospects

In the early twenty-first century, the population of the world edged toward 7 billion, increasing the amount of pollution that reaches the land, water, and atmosphere. Many natural ecosystems are struggling to cope with the remnants of old toxic contamination and the influx of new contamination. The ecological, social, and economic costs of pollution are immeasurable, and environmental recovery is one of the most important problems facing the global community.

Conventional biotechnology processes and techniques have relied on end-of-pipe technologies, that is treatment of waste and pollution that has already contaminated the air, soil, or water. Although such methods are necessary, particularly in remediation of existing pollution, many environmental biotechnologists think that end-of-pipe methods should be regarded as last-resort efforts, rather than preferred methods. As such, environmental biotechnology is moving from first-generation technology based on naturally occurring processes and microorganisms to second-generation technology based on high-tech anthropomorphic enhancement and manipulation of natural processes and microorganisms.

The future of environmental biotechnology lies in following an integrated environmental protection approach, with the fundamental goal being to control pollution before it enters the natural ecosystem and to recover already polluted areas. An essential step in this goal is to create controls and pass legislation to reduce the incidence of contamination. However, many researchers believe that the future of environmental engineering also will be closely aligned with the advancement and application of molecular and genetic methods. Decreasing or mitigating greenhouse gas pollution in the Earth's atmosphere is of vital importance to global health, so research in environmental biotechnology is focusing on biological organisms and processes that may help.

Christine Watts, Ph.D., B.App.Sc., B.Sc.

Further Reading

Cummings, Stephen, ed. *Bioremediation: Methods and Protocols.* New York: Humana Press, 2010. Experts in the field of environmental biotechnology present innovative and imaginative bioremediation techniques in pollution removal.

Evans, Gareth, and Judith Furlong. *Environmental Biotechnology: Theory and Application.* New York: John Wiley & Sons, 2003. A detailed examination of environmental biotechnology, focusing on present-day practices, the potential for biotechnological interventions, and microbial techniques and methods.

Illman, Walter, and Pedro Alvarez. "Performance Assessment of Bioremediation and Natural Attenuation." *Critical Reviews in Environmental Science and Technology,* 39, no. 4 (April, 2009): 209-270. A critical review of the state-of-the-art in performance assessment methods. Discusses future research directions in bioremediation, natural attenuation, chemical fingerprinting, and molecular biological tools.

Jördening, Hans-Joachim, and Josef Winter. *Environmental Biotechnology: Concepts and Applications.* Weinheim, Germany: Wiley-VCH, 2005. A solid foundation for students wishing to study environmental biotechnology. Examines in detail the microbiological treatment of waste and pollution in water, soil, and air.

Scragg, Alan. *Environmental Biotechnology.* 2d ed. New York: Oxford University Press, 2005. Examines the multitude of ways in which environmental biotechnology is applied in pollution control, environmental management, and removal of oil and minerals.

Thakur, Indu Shekhar. *Environmental Biotechnology: Basic Concepts and Applications.* New Delhi: I. K. International, 2006. A comprehensive examination of environmental processes and the many possible applications of environmental biotechnology, such as bioremediation, bioprocessing, and bioleaching.

Web Sites

International Society for Biotechnology
http://www3.inecol.edu.mx/iseb

Lawrence Berkeley National Laboratory, Earth Sciences Division
Center for Environmental Biotechnology
http://esd.lbl.gov/CEB

U.S. Department of Energy, Office of Biological and Environmental Research
Environmental Remediation Sciences Program
http://www.er.doe.gov/ober/ERSD/ersphome.html

See also: Ecological Engineering; Environmental Engineering; Industrial Pollution Control; Sanitary Engineering; Sewage Engineering.

ENVIRONMENTAL ENGINEERING

FIELDS OF STUDY

Mathematics; chemistry; physics; biology; geology; engineering mechanics; fluid mechanics; soil mechanics; hydrology.

SUMMARY

Environmental engineering is a field of engineering involving the planning, design, construction, and operation of equipment, systems, and structures to protect and enhance the environment. Major areas of application within the field of environmental engineering are wastewater treatment, water-pollution control, water treatment, air-pollution control, solid-waste management, and hazardous-waste management. Water-pollution control deals with physical, chemical, biological, radioactive, and thermal contaminants. Water treatment may be for the drinking water supply or for industrial water use. Air-pollution control is needed for stationary and moving sources. The management of solid and hazardous wastes includes landfill and incinerators for disposal of solid waste and identification and management of hazardous wastes.

KEY TERMS AND CONCEPTS

- **Activated Sludge Process:** Biological wastewater-treatment system for removing waste organic matter from wastewater.
- **Baghouse:** Air-pollution control device that filters particulates from an exhaust stream; also called a bag filter.
- **Biochemical Oxygen Demand (BOD):** Amount of oxygen needed to oxidize the organic matter in a water sample.
- **Catch Basin:** Chamber to retain matter flowing from a street gutter that might otherwise obstruct a sewer.
- **Digested Sludge:** Wastewater biosolids (sludge) that have been stabilized by an anaerobic or aerobic biological process.
- **Effluent:** Liquid flowing out from a wastewater-treatment process or treatment plant.
- **Electrostatic Precipitator:** Air-pollution control device to remove particulates from an exhaust stream by giving the particles a charge.
- **Nonpoint Source:** Pollution source that cannot be traced back to a single emission source, such as storm-water runoff.
- **Oxidation Pond:** Large shallow basin used to treat wastewater, using sunlight, bacteria, and algae.
- **Photochemical Smog:** Form of air pollution caused by nitrogen oxides and hydrocarbons in the air that react to form other pollutants because of catalysis by sunlight.
- **Pollution Prevention:** Use of conscious practices or processes to reduce or eliminate the creation of wastes at the source.
- **Primary Treatment:** Wastewater treatment to remove suspended and floating matter that will settle from incoming wastewater.
- **Sanitary Landfill:** Site used for disposal of solid waste that uses liners to prevent groundwater contamination and is covered daily with a layer of earth.
- **Secondary Treatment:** Wastewater treatment to remove dissolved and fine suspended organic matter that would exert an oxygen demand on a receiving stream.
- **Trickling Filter:** Biological wastewater-treatment process in which wastewater trickles through a bed of rocks with a coating containing microorganisms.

Definition and Basic Principles

Environmental engineering is a field of engineering that split off from civil engineering as the importance of the treatment of drinking water and wastewater was recognized. This field of engineering was first known as sanitary engineering and dealt almost exclusively with the treatment of water and wastewater. As awareness of other environmental concerns and the need to do something about them grew, this field of engineering became known as environmental engineering, with the expanded scope of dealing with air pollution, solid wastes, and hazardous wastes, in addition to water and wastewater treatment.

Environmental engineering is an interdisciplinary field that makes use of principles of chemistry, biology, mathematics, and physics, along with engineering sciences (such as soil mechanics, fluid mechanics, and hydrology) and empirical engineering correlations and knowledge to plan for, design, construct, maintain, and operate facilities for treatment of liquid and gaseous waste streams, for prevention of air pollution, and for management of solid and hazardous wastes.

The field also includes investigation of sites with contaminated soil and/or groundwater and the planning and design of remediation strategies. Environmental engineers also provide environmental impact analyses, in which they assess how a proposed project will affect the environment.

Background and History

When environmental engineering, once a branch of civil engineering, first became a separate field in the mid-1800's, it was known as sanitary engineering. Initially, the field involved the water supply, water treatment, and wastewater collection and treatment.

In the middle of the twentieth century, people began to become concerned about environmental quality issues such as water and air pollution. As a consequence, the field of sanitary engineering began to change to environmental engineering, expanding its scope to include air pollution, solid- and hazardous-waste management, and industrial hygiene.

Several pieces of legislation have affected and helped define the work of environmental engineers. Some of the major laws include the Clean Air Act of 1970, the Safe Drinking Water Act of 1974, the Toxic Substances Control Act of 1976, the Resource Recovery and Conservation Act (RCRA) of 1976, and the Clean Water Act of 1977.

How It Works

Environmental engineering uses chemical, physical, and biological processes for the treatment of water, wastewater, and air, as well as in-site remediation processes. Therefore, knowledge of the basic sciences—chemistry, biology, and physics—is important along with knowledge of engineering sciences and applied engineering.

Chemistry. Chemical processes are used to treat water and wastewater, to control air pollution, and for site remediation. These chemical treatments include chlorination for disinfection of both water and wastewater, chemical oxidation for iron and manganese removal in water-treatment plants, chemical oxidation for odor control, chemical precipitation for removal of metals or phosphorus from wastewater, water softening by the lime-soda process, and chemical neutralization for pH (acidity) control and for scaling control.

The chemistry principles and knowledge that are needed for these treatment processes include the ability to understand and work with chemical equations, to make stoiciometric calculations for dosages, and to determine size and configuration requirements for chemical reactors to carry out the various processes.

Biology. The major biological treatment processes used in wastewater treatment are biological oxidation of dissolved and fine suspended organic matter in wastewater (secondary treatment) and stabilization of biological wastewater biosolids (sludge) by anaerobic digestion or aerobic digestion.

Biological principles and knowledge that are useful in designing and operating biological wastewater treatment and biosolids digestion processes include the kinetics of the biological reactions and knowledge of the environmental conditions required for the microorganisms. The required environmental conditions include the presence or absence of oxygen and the appropriate temperature and pH.

Physics. Physical treatment processes used in environmental engineering include screening, grinding, comminuting, mixing, flow equalization, flocculation, sedimentation, flotation, and granular filtration. These processes are used to remove materials that can be screened, settled, or filtered out of water or wastewater and to assist in managing some of the processes. Many of these physical treatment processes are designed on the basis of empirical loading factors, although some use theoretical relationships such as the use of estimated particle settling velocities for design of sedimentation equipment.

Soil Mechanics. Topics covered in soil mechanics include the physical properties of soil, the distribution of stress within the soil, soil compaction, and water flow through soil. Knowledge of soil mechanics is used by environmental engineers in connection with design and operation of sanitary landfills for solid waste, in storm water management, and in the investigation and remediation of contaminated soil and groundwater.

Fluid Mechanics. Principles of fluid mechanics are used by environmental engineers in connection with the transport of water and wastewater through pipes and open channels. Such transport takes place in water distribution systems, in sanitary sewer collection systems, in storm water sewers, and in wastewater-treatment and water-treatment plants. Design and sizing of the pipes and open channels make use of empirical relationships such as the Manning equation for open channel flow and the Darcy-Weisbach equation for frictional head loss in pipe flow. Environmental engineers also design and select pumps and flow measuring devices.

Hydrology. The principles of hydrology (the science of water) are used to determine flow rates for storm water management when designing storm sewers or storm water detention or retention facilities. Knowledge of hydrology is also helpful in planning and developing surface water or groundwater as sources of water.

Practical Knowledge. Environmental engineers make use of accumulated knowledge from their work in the field. Theoretical equations, empirical equations, graphs, nomographs, guidelines, and rules of thumb have been developed based on experience. Empirical loading factors are used to size and design many treatment processes for water and wastewater. For example, the design of rapid sand filters to treat drinking water was based on a specified loading rate in gallons per minute of water per square foot of sand filter. Also the size required for a rotating biological contactor to provide secondary treatment of wastewater was determined based on a loading rate in pounds of biochemical oxygen demand (BOD) per day per 1,000 square feet of contactor area.

Engineering Tools. Tools such as engineering graphics, computer-aided drafting (CAD), geographic information systems (GIS), and surveying are available for use by environmental engineers. These tools are used for working with plans and drawings and for laying out treatment facilities or landfills.

Codes and Design Criteria. Much environmental engineering work makes use of codes or design criteria specified by local, state, or federal government agencies. Examples of such design criteria are the storm return period to be used in designing storm sewers or storm water detention facilities and the loading factor for rapid sand filters. Design and operation of treatment facilities for water and wastewater are also based on mandated requirements for the finished water or the treated effluent.

Applications and Products

Environmental engineers design, build, operate, and maintain treatment facilities and equipment for the treatment of drinking water and wastewater, air-pollution control, and the management of solid and hazardous wastes.

Air-Pollution Control. Increasing air pollution from industries and power plants as well as automobiles led to passage of the Clean Air Act of 1970. This law led to greater efforts to control air pollution. The two major ways to control air pollution are the treatment of emissions from fixed sources and from moving sources (primarily automobiles).

The fixed sources of air pollution are mainly the smokestacks of industrial facilities and power plants. Devices used to reduce the number of particulates emitted include settling chambers, baghouses, cyclones, wet scrubbers, and electrostatic precipitators. Electrostatic precipitators impart the particles with an electric charge to aid in their removal. They are often used in power plants, at least in part because of the readily available electric power to run them. Water-soluble gaseous pollutants can be removed by wet scrubbers. Other options for gaseous pollutants are adsorption on activated carbon or incineration of combustible pollutants. Because sulfur is contained in the coal used as fuel, coal-fired power plants produce sulfur oxides, particularly troublesome pollutants. The main options for reducing these sulfur oxides are desulfurizing the coal or desulfurizing the flue gas, most typically with a wet scrubber using lime to preciptitate the sulfur oxides.

Legislation has greatly reduced the amount of automobile emissions, the main moving source of air pollution. The reduction in emissions has been accomplished through catalytic converters to treat exhaust gases and improvements in the efficiency of automobile engines.

Water Treatment. The two main sources for the water supply are surface water (river, lake, or reservoir) and groundwater. The treatment requirements for these two sources are somewhat different.

For surface water, treatment is aimed primarily at removal of turbidity (fine suspended matter) and perhaps softening the water. The typical treatment processes for removal of turbidity involve the addition of chemicals such as alum or ferric chloride. The chemicals are rapidly mixed into the water so that they react with alkalinity in the water, then slowly mixed (flocculation) to form a settleable precipitate. After sedimentation, the water passes through a sand filter and finally is disinfected with chlorine. If the water is to be softened as part of the treatment, lime, $Ca(OH)_2$, and soda ash, Na_2CO_3, are used in place of alum or ferric chloride, and the water hardness (calcium and magnesium ions) is removed along with its turbidity.

Groundwater is typically not turbid (cloudy), so it does not require the type of treatment used for surface water. At minimum, it requires disinfection. Removal of iron and manganese by aeration may be needed, and if the water is very hard, it may be softened by ion exchange.

Wastewater Treatment. The Clean Water Act of 1977 brought wastewater treatment to a new level by requiring that all wastewater discharged from municipal treatment plants must first undergo at least secondary treatment. Before the passage of the legislation, many large cities located on a river or along the ocean provided only primary treatment in their wastewater-treatment plants and discharged effluent with only settleable solids removed. All dissolved and fine suspended organic matter remained in the effluent. Upgrading treatment plants involved added a biological treatment to remove dissolved and fine suspended organic matter that would otherwise exert an oxygen demand on the receiving stream, perhaps depleting the oxygen enough to cause problems for fish and other aquatic life.

Solid-Waste Management. The main options for solid-waste management are incineration, which reduces the volume for disposal to that of the ash that is produced, and disposal in a sanitary landfill. Some efforts have been made to reuse and recycling materials to reduce the amount of waste sent to incinerators or landfills. A sanitary landfill is a big improvement over the traditional garbage dump, which was simply an open dumping ground. A sanitary landfill uses liners to prevent groundwater contamination, and each day, the solid waste is covered with soil.

Hazardous-Waste Management. The Resource Conservation and Recovery Act (RCRA) of 1976 provides the framework for regulating hazardous-waste handling and disposal in the United States. One very useful component of RCRA is that it specifies a very clear and organized procedure for determining if a particular material is a hazardous waste and therefore subject to RCRA regulations. If the material of interest is indeed a waste, then it is defined to be a hazardous waste if it appears on one of RCRA's lists of hazardous wastes, if it contains one or more hazardous chemicals that appear on an RCRA list, or if it has one or more of the four RCRA hazardous waste characteristics as defined by laboratory tests. The four RCRA hazardous waste characteristics are flammability, reactivity, corrosivity, and toxicity. The RCRA regulations set standards for secure landfills and treatment processes for disposal of hazardous waste.

Much work has been done in investigating and cleaning up sites that have been contaminated

Fascinating Facts About Environmental Engineering

- In March, 1987, a barge loaded with municipal solid waste departed from Islip, New York, headed to a facility in North Carolina, where state officials turned it away. It traveled to six states and three countries over seven months trying to find a place to unload its cargo before it was allowed to return to New York.
- The activated sludge process for treating wastewater was invented in England in 1914. Interest in the activated sludge process spread rapidly, and it soon became the most widely used biological wastewater-treatment process in the world.
- The use of chlorine to disinfect drinking water supplies began in the late 1800's and early 1900's. It dramatically reduced the incidence of waterborne diseases such as cholera and typhoid fever.
- In the book *Silent Spring* (1962), Rachel Carson described the negative effect of the pesticide dichloro-diphenyl-trichloroethane (DDT) on birds. This book increased environmental awareness and is often cited as the beginning of the environmental movement.
- The first comprehensive sewer system in the United States was built in Chicago in 1850. The city level was raised 10 to 15 feet so that gravity would drain the sewers into the Chicago River, which emptied into Lake Michigan.
- A "solid waste" as defined by the U.S. Resource Conservation and Recovery Act may be solid, liquid, or semi-solid in form.

by hazardous wastes in the past. In some cases, funding is available for cleanup of such sites through the Comprehensive Environmental Response, Compensation, and Liability Act of 1980 (known as CERCLA or Superfund) or its amendment, the Superfund Amendments and Reauthorization Act (SARA) of 1986.

Impact on Industry

Increased interest in environmental issues in the last quarter of the twentieth century has made environmental engineering more prominent. The U.S. Bureau of Labor Statistics shows environmental engineering as the eighth largest field of engineering, with an estimated 54,300 environmental engineers employed in the United States in 2008. The bureau projects a 31 percent rate of growth in environmental engineering employment through much of the 2010's, which is much higher than the average for all occupations. Environmental engineers are employed by consulting engineering firms, industry, universities, and federal, state, and local government agencies.

Consulting Engineering Firms. Slightly more than half of all environmental engineers in the United States are employed by firms that engage in consultng in architecture, engineering, management, and scientific and technical issues. Some engineering consulting companies specialize in environmental projects, while others have an environmental division or simply have some environmental engineers on staff. The U.S. environmental consulting industry is made up of about 8,000 companies, ranging in size from one-person shops to huge multinational corporations. Some of the largest environmental consulting firms are CH2M Hill, Veolia Environmental Services North America, and Tetra Tech. Two engineering and construction firms with large environmental divisions are Bechtel and URS.

Government Agencies. Environmental engineers are employed by government agencies at the local, state, and federal levels. The U.S. Environmental Protection Agency and state environmental agencies employ the most environmental engineers, but many other government agencies, such as the Army Corps of Engineers, the Bureau of Reclamation, the Department of Agriculture, the Department of Defense, the Federal Emergency Management Agency, and the Natural Resources Conservation Service also have environmental engineers on their staffs. At the local government level, environmental engineers are used by city and county governments, in city or county engineering offices, and in public works departments.

University Research and Teaching. Colleges and universities employ environmental engineers to teach environmental engineering and the environmental component of civil engineering programs. Civil engineering is one of the largest engineering specialties, and its former subspecialty, environmental engineering, is taught at numerous colleges and universities around the world.

Industry. A small percentage of environmental engineers work in industry, at companies such as 3M, Abbott Laboratories, BASF, Bristol-Myers Squibb, Chevron, the Dow Chemical Company, DuPont, and IBM.

Careers and Course Work

An entry-level environmental engineering position can be obtained with a bachelor's degree in environmental engineering or in civil or chemical engineering with an environmental specialization. However, because many positions require registration as an engineer in training or as a professional engineer, it is important that the bachelor's degree program is accredited by the Accreditation Board for Engineering and Technology (ABET). Students must first graduate from an accredited program before taking the exam to become a registered engineer in training. After four years of experience, the engineer in training can take another exam for registration as a professional engineer.

A typical program of study for an environmental engineering degree at the undergraduate level includes the chemistry, calculus-based physics, and mathematics that is typical of almost all engineering programs in the first two years of study. It also may include biology, additional chemistry, and engineering geology. The last two years of study will typically include hydrology, soil mechanics, an introductory course in environmental engineering, and courses in specialized areas of environmental engineering such as water treatment, wastewater treatment, air-pollution control, and solid- and hazardous-waste management.

Master's degree programs in environmental engineering fall into two categories: those designed primarily for people with an undergraduate degree in environmental engineering and those for people with an undergraduate degree in another type of engineering. Some environmental engineering positions require a master's degree. A doctoral degree in environmental engineering is necessary for a position in research or teaching at a college or university.

Social Context and Future Prospects

Many major areas of concern in the United States and around the world are related to the environment. Issues such as water-pollution control, air-pollution control, global warming, and climate change all need the work of environmental engineers. These issues, as well as the need for environmental engineers, are likely to remain concerns for much of the twenty-first century.

Water supply, wastewater treatment, and solid-waste management all involve infrastructure, needing repair, maintenance, and upgrading, which are all likely to need the help of environmental engineers.

Harlan H. Bengtson, B.S., M.S., Ph.D.

Further Reading

Anderson, William C. "A History of Environmental Engineering in the United States." In *Environmental and Water Resources History*, edited by Jerry R. Rogers and Augustine J. Fredrich. Reston, Va.: American Society of Civil Engineers, 2003. Describes the development of environmental engineering in the United States, starting in the 1830's, through the growth of environmental awareness in the 1970's, and into the twenty-first century. Identifies and discusses significant pioneers in the field.

Davis, Mackenzie L., and Susan J. Masten. *Principles of Environmental Engineering and Science.* 2d ed. Boston: McGraw-Hill Higher Education, 2009. An introduction to the field of environmental engineering. Includes illustrations and maps.

Juuti, Petri S., Tapio S. Katko, and Heikki Vuorinen. *Environmental History of Water: Global Views on Community Water Supply and Sanitation.* London: IWA, 2007. Provides information on the history of the water supply and sanitation around the world.

Leonard, Kathleen M. "Brief History of Environmental Engineering: 'The World's Second Oldest Profession.'" *ASCE Conference Proceedings* 265, no. 47 (2001): 389-393. Describes the evolution of environmental engineering from its earliest beginnings.

Spellman, Frank R., and Nancy E. Whiting. *Environmental Science and Technology: Concepts and Applications.* 2d ed. Lanham, Md.: Government Institutes, 2006. Provides background basic science and engineering science information as well as an introduction to the different areas of environmental engineering.

Vesilind, P. Aarne, Susan M. Morgan, and Lauren G. Heine. *Introduction to Environmental Engineering.* 3d ed. Stamford, Conn.: Cengage Learning, 2010. A holistic approach to solving environmental problems with two unifying themes—material balances and environmental ethics.

Web Sites

American Academy of Environmental Engineers
http://www.aaee.net

American Society of Civil Engineers
http://www.asce.org

U.S. Environmental Protection Agency
http://www.epa.gov

See also: Civil Engineering; Industrial Pollution Control; Landscape Architecture and Engineering; Sanitary Engineering; Sewage Engineering.

ENZYME ENGINEERING

FIELDS OF STUDY

Biology; biochemistry; biotechnology; chemistry; food science; genetics; medicine; microbiology; pharmacology.

SUMMARY

Catalysts accelerate the rate of chemical reactions without being essentially changed, and enzymes are biological catalysts that accelerate the rate of reactions that occur in living systems. Enzyme engineering identifies enzymes that have potentially useful catalytic activities and chemically or structurally modifies them to increase their activity, change their substrate specificity, change the types of reactions they catalyze, or change the properties of enzymes and the manner in which they are regulated. Engineered enzymes can generate completely novel molecules or new, improved ways to synthesize useful molecules.

KEY TERMS AND CONCEPTS

- **Active Site:** Pocket or cleft in an enzyme surrounded by amino acids that specifically bind the substrate and catalyze its conversion to the product.
- **Amino Acids:** Chemical building blocks of proteins that contain a nitrogen-containing amino group ($-NH_2$), a carboxylic acid ($-COOH$), and a functional group that varies between amino acids and determines their chemical properties.
- **Catalyst:** Substance that increases the rate of a reaction without being consumed or permanently changed in the process.
- **Cofactors:** Non-amino acid molecules that are associated with enzymes and necessary for enzymatic function.
- **Enzyme:** Proteins, groups of proteins, or ribonucleic acids (RNAs) that are produced by living organisms and catalyze biochemical reactions.
- **Protease:** Enzyme that degrades proteins.
- **Substrate:** Molecule or molecules physically engaged by an enzyme to accelerate the chemical reaction that consists of the conversion of the substrate or substrates to the product or products.
- **Transition State:** Intermediate chemical structure formed during the process of a chemical reaction that represents the highest energy state of the reaction and usually degenerates to form the product.

Definition and Basic Principles

Enzymes are widely used as catalysts in several industrial ventures, ranging from food to synthetic chemistry to many other industrial processes. However, enzymes often show insufficient substrate selectivity, poor stability, and catalytic activities that are not robust enough for industrial use. To remedy this shortcoming, enzyme engineering builds new enzymes or modifies existing enzymes to give them novel, useful properties, or the ability to catalyze valuable chemical reactions. Engineered enzymes come in several forms. Semisynthetic enzymes (synzymes) have specific amino acids that have been chemically modified. These modifications can significantly alter the activity, specificity, or properties of the enzyme. Directed evolution subjects the gene that encodes the enzyme to multiple rounds of mutation. The variant enzymes generated by these mutagenic genes are then sieved by some kind of selection scheme that identifies those mutant forms that display the desired characteristics or activities.

A cheaper strategy is rational design. Rational design uses detailed knowledge of the structure of proteins to identify those regions that are essential for its function and properties. By changing only those amino acids thought to be necessary for the modification of that function, the enzyme is potentially tailored for a new function with little investment.

Enzyme engineers use catalytic antibodies or abzymes. Antibodies are Y-shaped proteins made by vertebrate immune systems that bind to specific chemicals. Abzymes bind to chemicals and force them into the transition state of a chemical reaction, which accelerates the formation of the product from the reactants.

Background and History

Enzyme engineering arose only after advances in several other fields made it possible to determine the primary amino acid sequences and three-dimensional structure of enzymes and directly manipulate them at the molecular level. Swedish biochemist Pehr Victor Edman gave birth to protein sequencing in 1950, when he designed the Edman degradation reactions that can determine the primary amino acid sequence of proteins. In 1958, English biochemist John Cowdery Kendrew used X-ray crystallography to solve the three-dimensional structure of the muscle oxygen-storing protein myoglobin. In the 1970's, American biochemist Herbert Wayne Boyer and American geneticist Stanley Norman Cohen pioneered molecular cloning techniques that gave scientists the means to clone genes and insert them into bacteria for propagation.

The first studies in enzyme engineering examined the effects of mutations on enzyme active sites. Beta-lactamase, the enzyme used by bacteria to degrade beta-lactam antibiotic (penicillin, ampicillin, and amoxicillin) was one of the first enzymes examined by enzyme engineering. In 1978, Canadian chemist Michael Smith and his colleagues invented site-directed mutagenesis, which gave biochemists a much better way to place targeted mutations into the genes that encode enzymes and thereby change their primary amino acid sequence. In 1986, the laboratories of Peter Schultz (University of California, Berkeley) and Richard Lerner (Research Institute of Scripps Clinic) made the first catalytic antibodies that could split ester bonds.

How It Works

Semisynthetic Enzymes. Enzymes that are modified by chemical means are known as semisynthetic enzymes. There are two main ways to produce semisynthetic enzymes: atom replacement or group attachment.

Atom replacement exchanges one atom within an enzyme for a different atom. Such replacements can modify enzyme activity or change the substrate specificity of the enzyme. Group attachment involves the use of particular chemical reagents to attach particular molecules to enzymes. Attaching additional molecules to enzymes can also markedly change enzyme activity and substrate specificity.

Directed Evolution. Directed evolution randomly changes amino acids in a protein without prior knowledge of the exact function of each amino acid. The first step, diversification, takes the gene that encodes the enzyme of interest and replicates it many times while using a copying machinery that is inherently error-prone. This introduces random mutations into the gene and creates a large collect of gene variants that are usually grown in bacteria. The second step, selection, tests or screens these enzyme variants for a desired property. Once the desired variants are identified, they undergo the third step, amplification, which replicates the identified variants and sequences them in order to determine which mutations produced the desired properties. Collectively, these three steps constitute one round of directed evolution, and the vast majority of such experiments require multiple rounds. The goal is to find those variant enzymes that show the most desired characteristics to the greatest extent. Directed-evolution studies suffer from the need to make huge numbers of mutants that produce no discernable effect, since up to 90 percent of all mutants made are uninformative.

Semirational Design. This enzyme engineering strategy employs sophisticated computer programs that assemble all the available structural information of the enzyme under study and predict how the mutations introduced into different locations within the enzyme might affect its activity. The enzyme engineer then notes the predicted changes that will potentially generate the desired property changes and uses this information to conduct targeted mutagenesis experiments. Targeted mutagenesis experiments introduce mutations into specific locations of a protein. Once these mutations are made, the variant enzyme with the engineered changes is tested to determine if it has the specific properties the enzyme engineer was hoping to produce in the enzyme. These approaches combine structural information with rational design. Two computer programs that make such predictions include Protein Sequence-Activity Relationship or ProSAR and Combinatorial Active-Site Saturation Test, otherwise known as CASTing.

Rational Design. If a great deal of structural information about the enzyme in question is available, then that structural information informs which amino acids should be changed. Many rational design attempts have not succeeded because of uncertainties regarding protein structure.

De Novo Design. A computer builds an enzyme around the transition state of a reaction from scratch. The computer begins by designing the active site by placing specific amino acids in strategic positions so that they efficiently bind the transition state of the chemical reaction and stabilize it. The program then constructs a protein backbone that supports and properly positions the active-site amino acids but still provides a coherent protein structure that is predictably stable under the desired conditions.

This particular strategy suffers from gaps in the ability to predict protein structure accurately and correlate this ideal structure with enzymatic activity. For example, two enzymes (retro-aldol enzyme and a Kemp elimination catalyst) were built completely from scratch by using computer programs. However, both enzymes required further optimization by directed evolution to achieve maximum activity.

Catalytic Antibodies. The immune system of some vertebrates makes Y-shaped proteins that specifically bind to and neutralize foreign substances that invade the body. Immunizing laboratory animals with stable analogues of the transition states of various reactions directs the immune systems of those animals to synthesize antibodies that cannot only bind particular chemical reactants but force them into the transition state of the reaction, which subsequently forms the product.

Applications and Products

Pharmaceutical Production. Beta-lactam and cephalosporin antibiotics are commonly prescribed to combat various illnesses. Both of these drugs kill bacteria by inhibiting the synthesis of the bacterial cell wall. Beta-lactam antibiotics include such widely recognized drugs as penicillin, ampicillin, and amoxicillin, whereas cephalosporin antibiotics include such popularly used antibiotics as Ceftin (cefuroxime), Kephlex (cephalexin), and Ceclor (cefaclor). Unfortunately, with repeated use, bacteria can become resistant to commonly used antibiotics, and making new, improved antibiotics is essential to treat some of the more recent and aggressive infectious diseases. To make new cephalosporin antibiotics, enzyme engineers have used enzymes called acylases to convert simple starting chemicals into various versions of these drugs. By engineering these acylase enzymes, pharmaceutical companies have been able to make new cephalosporin and beta-lactam antibiotics that have novel properties and can kill bacteria that are resistant to older drugs.

Enzymes as Medicines. When a person is cut, blood oozes from the damaged tissue. Fortunately, blood clotting (also known as coagulation) eventually stanches this blood flow. Blood clotting is an essential part of wound healing, but it is also a very highly regulated event. The formation of blood clots inside undamaged blood vessels clogs those vessels and leads to heart attacks if clots form inside the vessels that surround the heart, or a stroke, if they occur within vessels that surround the brain. The human body has ways to destroy unnecessary clots. An enzyme called tissue plasminogen activating factor (TPA) activates other enzymes in the body that degrade harmful clots. Commercially available, native TPA is called Alteplase, which has a half-life in the bloodstream of four to six minutes. Engineered forms of TPA are also clinically available. Reteplase, a shortened version of TPA (consists of 357 of the 527 amino acids of Alteplase), has a longer half-life (thirteen to sixteen minutes). Tenecteplase, which has two amino acid changes (substitutes asparagine^{103} with a threonine and asparagine^{114} with glutamine), has an even longer half-life of twenty to twenty-four minutes.

Engineered enzymes are also used in enzyme-replacement therapies. Several genetic diseases, known as lysosomal storage diseases, result from the inability to make functional versions of enzymes that degrade various biological molecules. The accumulation of these molecules kills brain cells and causes the death of the patient. Engineered enzymes used in enzyme-replacement therapies include Cerezyme (imiglucerase, used to treat Gaucher's disease), Naglazyme (galsulfase, used to treat mucopolysaccharidosis VI), Myozyme (alglucosidase alfa, used to treat Pompe disease), and Aldurazyme (laronidase, used to treat mucopolysaccharidosis I).

Enzyme Immobilization. By attaching enzymes to surfaces, embedding them in gel matrices, hollow fibers, or cross-linking them to each other, enzymes are immobilized on insoluble surfaces. This increases their stability, simplifies their recycling, and increases the tolerance of enzymes to high levels of substrate and products. Detergent enzyme preparations, such as Alcalase, immobilize the protease subtilisin by attaching it to insoluble particles. Attaching the enzyme to inert material increases its reuse as it degrades proteinaceous matter.

Making Enzymes Soluble in Organic Solvents. Enzymes usually work in water, but many reactions between organic chemicals occur in organic solvents. Although Russian chemist Alexander Klibanov showed that several enzymes are active in organic

Fascinating Facts About Enzyme Engineering

- Hemophilia is a genetic disease characterized by an inability to form blood clots. Because blood clotting requires the coordinated and sequential activity of a host of blood-based enzymes called clotting factors, hemophilia patients are treated with infusions of purified clotting enzymes. Unfortunately, some patients form antibodies against clotting factors, which abrogates the efficacy of such treatments. However, enzyme engineers have made a truncated version of clotting factor VIII (B-domain deleted recombinant factor VIII or BDDrFVIII) that is smaller than the native enzyme but just as active. This protein is not recognized by the immune system nearly as often as the native clotting factor VIII.
- Another clotting factor called factor VII (FVII) can bypass the need for other clotting factors if it is present in high enough concentrations. FVII is made in an inactive form that is only activated when the front tip of the protein is removed. Enzyme engineers have made a form of FVII that does not require activation called recombinant FVIIa (rFVIIa; the "a" stands for active). Infusions of rFVIIa can help clot the blood of hemophilia patients who have antibodies against other clotting factors and can also successfully treat patients who suffer from uncontrolled bleeding from trauma, surgery, anticoagulant drugs, or pregnancy.
- Metabolic engineering manipulates enzymes that direct metabolic pathways. Such manipulation can generate organisms that synthesize industrially useful compounds or degrade environmental pollutants. Animal feed production uses a microorganism called *Methylophilus methylotrophus* to convert methanol to animal protein, but by inserting genes from the common intestinal bacterial *Escherichia coli* into *M. methylotrophus*, this metabolically engineered organism can make much more protein from the same initial mass of methanol, thus increasing the overall efficiency of animal protein production and decreasing production costs.
- Antibody-directed enzyme prodrug therapy (ADEPT) uses an enzyme to a human antibody that tightly binds to a tumor-specific surface protein. When these antibodies bind to cancer cells, the enzyme converts an anticancer drug that is present in an inactive form (a prodrug) into a cancer-killing chemical. Because the prodrugs are activated only at the surface of the tumors, they cause few side effects and effectively kill tumors. As of 2011, ADEPT is in clinical trials.

solvents, many enzymes are neither soluble in organic solvents nor work properly in such environments. Attaching a molecule called polyethylene glycol (PEG) to some enzymes makes them soluble and active in organic solvents and allows them to make things such as polyester, peptides (small proteins), esters (sweet-smelling things found in foods), and amides (nitrogen-containing compounds). Such modified enzymes also have clinical uses. For example, the enzyme asparaginase can kill cancer cells but is toxic, unstable, and some patients have severe allergies to it. PEG-treated asparaginase is not as toxic as the native enzyme, is much more stable, and does not cause allergy. PEG-asparaginase is used to treat tumors in humans.

Abzymes. A notable variety of reactions are catalyzed by catalytic antibodies that range from forming or breaking carbon-carbon bonds, rearrangements, hydrolysis of various bonds, transfer of chemical groups, and even an industrial reaction called the Diels-Alder reaction. However, abzymes are very expensive and tedious to make, and their catalytic activity is well below that of enzymes. Yet they do provide tailor-made catalysts when no other such reagent exists.

Impact on Industry

Government and University Research. Regulation of research in academic venues that use genetically engineered organisms is specified by the National Institutes of Health (NIH) Guidelines for Research Involving Recombinant DNA Molecule. The Recombinant DNA Advisory Committee (RAC) is responsible for executing the regulations in the NIH guidelines. These guidelines typically regulate all work on pathogenic (disease-causing) organisms, prohibit genetic engineering of human beings without specific approvals for gene-therapy trials, and also prohibit the release of genetically engineered organisms into the environment without the express approval of government agencies such as the Environmental Protection Agency (EPA). However, research on specific enzymes receives little mention in these guidelines. Only the organisms that house the recombinant enzymes are regulated. Because the NIH is the largest funding agency in the United States, it sets the rules that are followed by other research-funding organizations.

Industry and Business. Any enzyme used to produce products that are sold as food or drugs comes under the regulatory auspices of the Food and Drug Administration (FDA). However, since enzymes do not usually end up in the finished product, manufacturers are not required to list them on product labels. Furthermore, if the enzyme is made by a genetically engineered organism, or if the enzyme itself is the result of enzyme engineering, the company is not required by law to notify the FDA. Engineered enzymes are subject to little regulatory oversight, and companies are free to experiment at will, provided that such enzymes do not become a part of the finished product.

Major Companies. Enzyme engineering has yet to make a significant mark on industry. Nevertheless, there are some companies that have made and marketed engineered enzymes.

The largest maker of enzymes for the food industry is Novo Nordisk, which makes its enzymes from genetically engineered microorganisms. Few of the enzymes made by this company are engineered, although Novo Nordisk also produces several different types of insulin, some of which are engineered. Enzyme engineering research is a part of the research and development at Novo Nordisk.

Pharmaceutical companies that make engineered enzymes usually have individual enzymes that they produce and market. As of 2011, there is no clear industrial leader in the production of engineered enzymes for clinical use. Enzon Pharmaceuticals makes PEG-conjugated asparaginase (Oncaspar), which was approved by the FDA in 2006 as a treatment for tumors. This is a prime example of an engineered enzyme that is being used in a clinical setting. Reteplase (retavase), an engineered version of TPA, is made by Hospira Incorporated but is marketed by EKR Therapeutics. Other companies, such as Genentech and Pfizer, have conducted extensive research on enzymes in order to make drugs that are inhibitors of important enzymes for clinical use.

Because enzyme engineering is rather labor intensive and requires a great deal of time to discover something commercially useful, there are few engineered enzymes on the market. However, given the interest in this field it is only a matter of time before engineered enzymes begin to appear in larger numbers and companies begin to appear that specialize in enzyme engineering.

Careers and Course Work

Anyone who wishes to study enzyme engineering must have a good understanding of general, organic, and physical chemistry and biochemistry. Knowledge of calculus is also essential as is mastery of computers, since many structural studies of enzymes use somewhat sophisticated computer programs. Enzyme engineers also must have some mastery of the tools of molecular biology, gene cloning, and microbiology. Enzyme engineers will also need graduate training, since many of the tools used in enzyme engineering are simply beyond the typical undergraduate laboratory curriculum. A bachelor's degree in chemistry or biochemistry is necessary to work as a technician in the enzyme engineering field. A master's degree will also be sufficient for a technician, but to run an enzyme engineering lab, a Ph.D. in chemistry or biochemistry is required.

Enzyme engineers work either in academia or industry. In academia, they will run their own lab and train graduate students. Most of the research in academic settings is not applied, but theoretical. Academic enzyme engineers usually try to develop new technologies for enzyme engineering, or use enzyme engineering techniques to study various enzymes. In industry, enzyme engineering research is much more applied, since the goal is to optimize an enzyme for a specific synthetic process that saves time, money, and resources.

Enzyme engineering is almost certainly the next frontier in biochemistry. Many industries make copious use of enzymes already, and the need to tailor these enzymes to fit the needs of industrial uses is pressing. Enzyme engineers need to be good collaborators, visionary, and very patient, since most experiments require extensive trials before something interesting is discovered.

Social Context and Future Prospects

Because modified enzymes can make certain products more cheaply, the public response to modified enzymes is generally positive. However, the genetically modified organisms (GMOs) that are used to produce these enzymes give many people pause, since the introduction of GMOs into the environment may have long-term consequences that are presently unrecognized. Strict government regulation that forbids the release of GMOs into the environment without approval allays most of these concerns, but some people are still troubled by the use of GMOs to make products that they eventually end up eating or using in some other manner.

Enzyme engineering is one of the up-and-coming fields in chemistry and biochemistry. Since the 1990's, the use of enzymes in industrial and academic chemistry has greatly increased. There are many advantages to using enzymes in that they can act outside cells and under mild conditions that minimize troublesome side effects, are environmentally innocuous, compatible with other enzymes, and are very efficient, though highly selective catalysts. The largest drawback of using enzymes is that the right enzyme is sometimes not available to catalyze the desired reaction. Enzyme engineering can eliminate this significant drawback.

Furthermore, as biochemists achieve a more profound understanding of protein structure, cheaper and faster ways of doing enzyme engineering, such as rational design, become more successful and practical. This will shorten the time required for enzyme engineering experiments and reduce its cost. Companies are already looking intently at enzyme engineering as a significant investment for their research and development departments.

Michael A. Buratovich, B.S., M.A., Ph.D.

Further Reading

Arnold, Frances H., and George Georgiou, eds. *Directed Enzyme Evolution: Screening and Selection Methods.* Totowa, N.J.: Humana Press, 2010. Laboratory protocol book that describes, in great detail with figures and graphs, some rather ingenious techniques for screening mutant clones of enzyme genes.

_______. *Directed Evolution Library Creation: Methods and Protocols.* Totowa, N.J.: Humana Press, 2010. Encyclopedic collection of protocols for generating libraries of randomly mutagenic enzyme genes in bacteria, with tables, graphs, and some figures.

Faber, Kurt. *Biotransformations in Organic Chemistry: A Textbook.* 5th ed. New York: Springer-Verlag, 2004. A very clear, useful textbook on the uses of enzymes in chemistry that includes a chapter on engineered enzymes.

Park, Sheldon J., and Jennifer R. Cochran, eds. *Protein Engineering and Design.* Boca Raton, Fla.: CRC Press, 2010. Covers the broader field of protein engineering—methods of developing altered proteins for novel applications—in two sections: one on experimental protein engineering and the other on computational design. Includes discussion of enzyme engineering using both rational and combinatorial approaches.

Scheindlin, Stanley. "Clinical Enzymology: Enzymes As Medicine." *Molecular Interventions* 7, no. 1 (February, 2007): 4-8. An absorbing and readable summary of the use of engineered enzymes in clinical diagnoses and treatments.

Web Sites

International Enzyme Engineering Symposium
http://www.enzymeengineering.ege.edu.tr

National Institutes of Health
Recombinant DNA Advisory Committee
http://oba.od.nih.gov/rdna_rac/rac_about.html

See also: Chemical Engineering

FIBER-OPTIC COMMUNICATIONS

FIELDS OF STUDY

Physics; chemistry; solid-state physics; electrical engineering; electromechanical engineering; mechanical engineering; materials science and engineering; telecommunications; computer programming; broadcast technology; information technology; electronics; computer networking; mathematics; network security.

SUMMARY

The field of fiber optics focuses on the transmission of signals made of light through fibers made of glass, plastic, or other transparent materials. The field includes the technology used to create optic fibers as well as modern applications such as telephone networks, computer networks, and cable television. Fiber optics are used in almost every part of daily life in technologies such as fax machines, cell phones, television, computers, and the Internet.

KEY TERMS AND CONCEPTS

- **Attenuation:** Loss of light power as the signal travels through fiber-optic cable.
- **Bandwidth:** Range of frequencies within which a fiber-optic transmitting device can transmit data or information.
- **Broadband:** Telecommunications signal with a larger-than-usual bandwidth.
- **Dispersion:** Spreading of light-signal pulses as they travel through fiber-optic cable.
- **Endoscope:** Fiber optic medical device that is used to see inside the human body without surgery.
- **Fiber-Optic Cable:** Cable consisting of numerous fiber-optic fibers lined with a reflective core medium to direct light.
- **Fiberscope:** First device that was able to transmit images over a glass fiber.
- **Light-Emitting Diodes (LED):** Light source sometimes used in fiber-optic systems to transmit data.
- **Receiver:** In fiber-optics systems, a device that captures the signals transmitted through the fiber-optic cable and then translates the light back into electronic data.
- **Semiconductor Laser:** Laser sometimes used in fiber-optic systems to transmit data.
- **Transmitter:** In fiber optic systems, a device that codes electronic data into light signals.

Definition and Basic Principles

The field of fiber optics focuses on the transmission of signals made of light through fibers made of glass, plastic, or other transparent media. Using the principles of reflection, optical fibers transmit images, data, or voices and provide communications links for a variety of applications such as telephone networks, computer networks, and cable television.

Background and History

The modern field of fiber optics developed from a series of important scientific discoveries, principles, technologies, and applications. Early work in use of light as a signal by French engineer Claude Chappe, British physicist John Tyndall, Scottish physicist Alexander Graham Bell, and American engineer William Wheeler in the eighteenth and nineteenth centuries laid the foundation for harnessing light through conductible materials such as glass. These experiments also served as proof of the concept that sound could be transmitted as light. The failures in the inventions indicated further areas of work before use in practical applications: The main available light source was the Sun, and the light signal was reduced by travel through the conductible substance. For example, in 1880 Bell created a light-based system of sound transmission or photophone that was abandoned for being too affected by the interruption of the light transmission beam. In the 1920's, the transmission of facsimiles (faxes) or television images through light signals via glass or plastic rods or pipes was patented by Scottish inventor John Logie Baird and American engineer Clarence Hansell. The fiberscope, developed in the 1950's, was able to transmit low- resolution images of metal welds over a glass fiber. In the mid-1950's, Dutch scientist Abraham Van Heel reported a method of gathering fibers into bundles and coating them in a clear coating or cladding that decreased interference between the fibers and reduced distortion effects from the outside. In 1966, English engineer George Hockham and Chinese physicist Charles Kao published a theoretical method designed to dramatically decrease the amount of light lost as it traveled through glass fibers. By 1970, scientists at Corning Glass Works created fibers that actualized Hockham and Kao's theoretical method. In the mid-1970's, the first telephone systems using fiber optics were piloted in Atlanta and Chicago. By 1984, other major cities on the Eastern seaboard were connected by AT&T's fiber-optic systems. In 1988, the first transatlantic fiber-optic cable connected the United States to England and France.

By the late 1980's, fiber-optic technology was in use for such medical applications as the gastroscope, which allowed doctors to look inside a patient and see the image transmitted along the fibers. However, more work was still needed to allow effective and accurate transmission of electronic data for computer work.

In the mid-twentieth century, the use of fiber optics accelerated in number of applications and technological advances. Scientists found a way to create a glass fiber coated in such a way that the light transmitted moved forward at full strength and signal. Coupled with the development of the semiconductor laser, which could emit a high-powered, yet cool and energy-efficient, targeted stream of light, fiber optics quickly became integrated into existing and new technology associated with computer networking, cable television, telephone networks, and other industry applications that benefited from high-speed and long-distance data transfer.

How It Works

The major elements required for fiber-optics transmission include: long flexible fibers made of transparent materials such as glass, plastic, or plastic-clad silica; a light-transmittal source such as a laser of light-emitting diode (LED); cables or rods lined with a reflective core medium to direct light; and a receiver to capture the signal. Many systems also include a signal amplifier or optoelectronic repeater to increase the transmission distance of a signal. Electronic data is coded into light signals using the transmitter. The light signals then move down the fibers bouncing off the reflective core of the fibers to the receiver. The receiver captures the signals and then translates the light back into electronic data. This process is used to transmit data in the form of images, sound, or other signals down the rods at the speed of light.

Applications and Products

Information Transmittal. Fiber-optics technology revolutionized the ability to transfer data between computers. Networked computers share and distribute information via a main computer (a server) and its connected computers (nodes). The use of fiber optics exponentially increases the data-transmission speed and ability of computers to communicate. In addition, fiber-optics data transfer is more secure than lines affected by magnetic interference. Industries that use information and data transmission through networks include banking, communications, cable television, and telecommunications. Fiber-optic information transmission has advantages over copper-cable transmission in that it is relatively easy to install, is lighter weight, very durable, can transmit for long distances at a higher bandwidth, and is not influenced by electromagnetic disruptions such as lightning or fluorescent lighting fixture transformers.

Modern Communications. The use of fiber-optics technology in telephone communication has increased the capacity, ease, and speed of standard copper-wired phones. The quality of voices over the phone is improved, as the sound signal is no longer distorted by distance or is subject to time delay. Fiber-optic lines are not affected by electromagnetic interference and are less subject to security breaches related to unauthorized access to phone calls and data transfer over phone lines. Additionally, fiber-optic cables are less expensive and easier to install than copper wire or coaxial cables, and since the 1980's they have been installed in many areas. Fiber-optic cabling can be used to provide high-speed Internet access, cable television, and regular telephone service over one line. In addition to traditional phone lines and home-based services, fiber-optic links between mobile towers and networks also allow the use of smart phones, which can be used to send and receive e-mails, surf the Internet, and have device-specific applications such as Global Positioning Systems.

Manufacturing. The increased globalization of the manufacturing of goods requires information, images, and data to be transmitted quickly from one location to another (known as point-to-point connections). For example, a car may be assembled in Detroit, but one part may be made in Mexico, another in Taiwan, and a third in Alabama. The logistics to make sure all the parts are of appropriate quality and quantity to be shipped to Detroit for assembly are coordinated through networked computer systems and fiber-optic telephone lines. In addition, the ability to use fiber optics to capture and transmit images down a very small cable allows quality-control personnel to "see" inside areas that the human eye cannot. As an example, a fiberscope can be used to inspect a jet engine's welding work within combustion chambers and reactor vessels.

The Internet. According to the United Nations' International Telecommunication Union (ITU), the number of Internet users across the world met the two billion mark in January 2011. Much like standard telephone service, the capacity, ease, and speed to the Internet has been greatly increased by the replacement of phone-based modem systems, cable modems, and digital subscriber line (DSL) by fiber-optics wired systems. Although fiber-optic connections directly to homes in the United States are not available in all areas, some companies use fiber-optic systems down major networking lines and then split to traditional copper wiring for houses.

Medicine. Fiber optics have significantly altered medical practice by allowing physicians to see and work within the human body using natural or small surgical openings. The fiberscopes or endoscopes are fiber-optics-based instruments that can image and illuminate internal organs and tissues deep within the human body. A surgeon is able to visualize an area of concern without performing large-scale exploratory surgery. In addition to viewing internal body surfaces, laproscopic surgery using fiber-optic visualization allows the creation of very small cuts to target and perform surgery reducing overall surgical risks and recovery time in many cases. Beyond the use of endoscopes, fiber-optic technology has been used to update standard medical equipment so that it may be used in devices that emit electromagnetic fields. As an example, companies have developed a fiber-optic pulse oximeter to be used to measure heart rate and oxygen saturation during magnetic resonance imaging (MRI).

Broadcast Industry. The broadcast industry has moved much of its infrastructure to fiber-optics technology. This change has also allowed the creation and transmission of television signals with increased clarity and picture definition known as high-definition television (HDTV). The use of fiber optics and its increased data-transmission ability was key in 2009, when all television stations changed from analogue to digital signals for their content broadcasts.

Military. The military began using fiber optics as a reliable method of communications early in the development of the technology. This quick implementation was due to recognition that fiber optics cables were able to withstand demanding conditions and temperature extremes while still transferring information accurately and quickly. Programs such as the Air Force's Airborne Light Optical Fiber Technology (ALOFT) program helped move fiber-optic technology along even as it served as proof of concept: fiber-optic signal transmission could transmit data reliably even in outer space. Beyond communications, the military uses fiber-optic gyroscopes (FOGs) in navigation systems to direct guided missiles accurately. Additionally, fiber optics have been used to increase the accuracy of rifle-bullet targeting by using sensitive laser-based fiber-optic sensors that adjust crosshairs on the scope based on the precise measurement of the barrel's deflection.

Traffic Control. According to the United States Department of Transportation, traffic signals that are not synchronized result in nearly 10 percent of all traffic delays and waste nearly 300 million vehicle-hours nationwide each year. Fiber optics has been used as part of intelligent transport systems to help coordinate traffic signals and improve the flow of cars via real-time monitoring of congestion, accidents, and traffic flow. Beyond traffic congestion, some cities capture data on cars running red lights, paying tolls, and the license plates moving through toll roads, tunnels, and bridges.

Impact on Industry

The total value of the fiber-optic communications industry is difficult to estimate as different aspects of the industry are divided and captured under different financial projections. For example, according to market-research firm BCC Research, the total estimated global market value for fiber-optic connectors in 2010 was an estimated $1.9 billion with an annual average growth rate of 9.6 percent. These projections are different from the total estimated global market value for fiber-optic circulators, which, as of 2011, is forecast to increase annually at 14.29 percent according to technology-forecasting firm ElectroniCast. Further projections are divided into categories such as medical fiber optics, glass and fiber manufacturing, and fiber optic components.

The fiber-optics industry has a global presence; however, the United States is a major player, dominating the optical cable and fiber markets internationally. According to market-research firm First Research, the U.S. glass and fiber-optic manufacturing industry includes about 2,000 companies with combined annual revenue of $20 billion. The majority of the market is concentrated on companies such as Corning and PPG Industries, and 80 percent of the market is captured by the fifty largest companies. Japan and the countries of the European Union are also instrumental in the fiber-optic communications industries as both manufacturers and consumers.

Government and University Research. The United States federal government funds fiber-optics research through branches such as the Department of Defense (DOD). One main category of basic research funded through the DOD includes lasers and fiber optics in communications and medicine. This research funding encourages joint ventures between universities and corporations such as the Lockheed Martin-Michigan Technological University project to develop a fiber-optic-based circuit board manufacturing process.

Industry and Business. Research and development programs in the industry and business sector continue to search for the next technological innovation in fiber-optics access and technology. Upcoming updates in speed and improved processing from industry will soon be seen in the next generation of fiber-optic cables. In addition, businesses are working to connect phone services, the Internet, and cable directly to an increasing number of private homes via fiber-optic cables.

Military. The military continues to fund research and implement new fiber-optics technologies related to several applications. Monitoring systems based on fiber optics are being developed to detect chemical weapons, explosives, or biohazardous substances based on a specific wavelength emitted by the substance or device. The military has also integrated use of smart phones into its operation activities with applications such as BulletFlight, which helps snipers determine the most effective angle from which to fire and input data to account for changes based on altitude and weather conditions.

Fascinating Facts About Fiber-Optic Communications

- As of 2011, all new undersea cables are made of optical fibers.
- Airplanes use fiber-optic cabling in order to keep overall weight down and increase available capacity.
- Fiber optics were an integral part of the television cameras sent to film the first Moon walk in 1969.
- Industry analysts predict that sometime in the early twenty-first century 98 percent of copper wire will have been replaced by fiber-optic cable.
- A fiber-optic fiber is thinner than a human hair.
- A fiber-optics system is capable of transmitting more than the equivalent of a twenty-four-volume encyclopedia worth of information in one second.

Careers and Course Work

There are many careers in the fiber-optics industry and entry-level requirements vary significantly by position. Given the wide spectrum of difference between the careers, a sampling of careers and course work follows.

Professional, management, and sales occupations generally require a bachelor's degree. Technical occupations often require specific course work but not necessarily a bachelor's degree. However, it is easier to obtain employment and gain promotions with a degree, especially in larger, more competitive markets. Advanced schooling usually is required for supervisory positions—including technical occupations—which have greater responsibility and higher salaries. These positions comprise about 19 percent of the fiber-optics communications industry careers.

Engineering roles in the fiber-optics industry range from cable logistics and installation planning to research and development positions in fiber optics and lasers. Positions may be found in universities, corporations, and the military. Engineers may specialize in a particular area of fiber optics such as communication systems, telecommunications design, or computer network integration with fiber-optic technology. Education requirements for entry-level positions begin with a bachelor's degree in engineering, computer science, or a related field.

Telecommunications equipment installers and repairers usually acquire their skills through formal training at technical schools or college, where they major in electronics, communications technology, or computer science. Military experience in the field, on-the-job training with a software manufacturer, or prior work as a telecommunications line installer may also provide entry into more complicated or complex positions.

Optics physicists work in the fiber-optics industry in research and development. The role of the optics physicist is to develop solutions to fiber-optics communications quandaries using the laws of physics. Most optics physicists have a doctorate in physics, usually with a specialization in optics. They also tend to spend several years after obtaining their doctorate performing academic research before moving to industry positions.

Specialized roles in computer software engineering and networking in the fiber-optic telecommunications industry also exist. Much like the engineering roles, individuals may specialize in a particular area of fiber optics such as networking, communication systems, telecommunications design, data communications, or computer software. Education requirements for entry-level positions begin with a bachelor's degree with a major in engineering, computer science, or a related field.

Social Context and Future Prospects

Fiber-optic communications technologies are constantly changing and integrating new innovations and applications. Some countries, such as Japan, have fully embraced use of fiber optics in the home as well as in business; however, the investment in infrastructure is not as fully actualized in other areas. The consumer demand for faster, better access to the Internet and related data-transmittal applications is driving the move from standard copper wiring to fiber optics. New types of fibers will increase fiber-optic application beyond telecommunications into more medical, military, and industrial uses. Though wireless technology use could negatively check industry growth, the strong consumer demand and increasing number of fiber-optics applications suggest that the fiber-optic industry will continue to grow. However, the industry may have more moderate growth as the telecommunication industry experiences decreased growth. This was seen during the economic recession of 2008 to 2010, as consumers held off upgrading from copper cabling to fiber optics. According to a report by Global Industry Analysts, the recession's impact on fiber-optic cabling ended in 2011. Overall, the report anticipates significant growth in the industry as more fiber-optic cable networks are installed and businesses, consumers, and telecom providers invest in advanced tools to facilitate the new networks. Employment in the wired telecommunications industry is expected to decline by 11 percent during the period from 2008 to 2018; however, telecommunications jobs focused on fiber optics are expected to rise.

Dawn A. Laney, B.A., M.S., C.G.C., C.C.R.C.

Further Reading

Allen, Thomas B. "The Future Is Calling," *National Geographic* 200, Issue 6 (December 2001): 76. An interesting and well-written description of the growth of the fiber-optics industry.

Belson, Ken. "Unlike U.S., Japanese Push Fiber Over Profit." *The New York Times.* October 3, 2007. Compares the United States' and Japan's approach to updating infrastructure with fiber-optic cabling.

Crisp, John, and Barry Elliott. *Introduction to Fiber Optics.* 3d ed. Burlington, Mass.: Elsevier, 2005. An excellent text for anyone, of any skill level, who wants to learn more about fiber optics from the ground up. Each chapter ends with review questions.

Goff, David R. *Fiber Optic Reference Guide: A Practical Guide to Communications Technology.* 3d ed. Burlington, Mass.: Focal Press, 2002. An excellent review of the history of fiber optics, the basic principles of the technology, and information on practical applications particularly in communications.

Hayes, Jim. *FOA Reference Guide to Fiber Optics: Study Guide to FOA Certification.* Fallbrook, Calif.: The Fiber Optic Association, 2009. A useful guide that details the design and installation of fiber optic cabling networks including expansive coverage of the components and processes of fiber optics.

Hecht, Jeff. *City of Light: The Story of Fiber Optics.* Rev. ed. New York: Oxford University Press, 1999. A readable history of the development of fiber optics.

Web Sites

Fiber Optic Association
http://www.thefoa.org

International Telecommunication Union (ITU)
http://www.itu.int

U.S. Bureau of Labor Statistics
Career Guide to Industries: Telecommunications
http://www.bls.gov/oco/cg/cgs020.htm

See also: Computer Engineering; Computer Networks; Electrical Engineering; Mechanical Engineering.

FLUID DYNAMICS

FIELDS OF STUDY

Physics; mathematics; engineering; chemistry; suspension mechanics; hydrodynamics; computational fluid dynamics; microfluidic systems; coating flows; multiphase flows; viscous flows.

SUMMARY

Fluid dynamics is an interdisciplinary field concerned with the behavior of gases, air, and water in motion. An understanding of fluid dynamic principles is essential to the work done in aerodynamics. It informs the design of air and spacecraft. An understanding of fluid dynamic principles is also essential to the field of hydromechanics and the design of oceangoing vessels. Any system with air, gases, or water in motion incorporates the principles of fluid dynamics.

KEY TERMS AND CONCEPTS

- **Aerodynamics:** Study of air in motion.
- **Boundary Layer:** Region between the wall of a flowing fluid and the point where the flow speed is nearly equal to that of the fluid.
- **Continuum:** Continuous flow of fluid.
- **Fluid:** State of matter in which a substance cannot maintain a shape on its own.
- **Hydrodynamics:** Study of water in motion.
- **Ideal Fluids:** Fluids without any internal friction (viscosity).
- **Incompressible Flows:** Those in which density does not change when pressure is applied.
- **Inviscid Fluid:** Fluid without viscosity.
- **Newtonian Fluids:** Fluids that quickly correct for shear strain.
- **Shear Strain:** Stress in a fluid that is parallel to the fluid motion velocity or streamline.
- **Streamline:** Manner in which a fluid flows in a continuum with unbroken continuity.
- **Viscosity:** Internal friction in a fluid.

Definition and Basic Principles

Fluid dynamics is the study of fluids in motion. Air, gases, and water are all considered to be fluids. When the fluid is air, this branch of science is called aerodynamics. When the fluid is water, it is called hydrodynamics.

The basic principles of fluid dynamics state that fluids are a state of matter in which a substance cannot maintain an independent shape. The fluid will take the shape of its container, forming an observable surface at the highest level of the fluid when it does not completely fill the container. Fluids flow in a continuum, with no breaks or gaps in the flow. They are said to flow in a streamline, with a series of particles following one another in an orderly fashion in parallel with other streamlines. Real fluids have some amount of internal friction, known as viscosity. Viscosity is the phenomenon that causes some fluids to flow more readily than others. It is the reason that molasses flows more slowly than water at room temperature.

Fluids are said to be compressible or incompressible. Water is an incompressible fluid because its density does not change when pressure is applied. Incompressible fluids are subject to the law of continuity, which states that fluid flows in a pipe are constant. This theory explains why the rate of flow increases when the area of the pipe is reduced and vice versa. The viscosity of a fluid is an important consideration when calculating the total resistance on an object.

The point where the fluid flows at the surface of an object is called the boundary layer. The fluid "sticks" to the object, not moving at all at the point of contact. The streamlines further from the surface are moving, but each is impeded by the streamline between it and the wall until the effect of the streamline closest to the wall is no longer a factor. The boundary layer is not obvious to the casual observer, but it is an important consideration in any calculations of fluid dynamics.

Most fluids are Newtonian fluids. Newtonian fluids have a stress-strain relationship that is linear. This means that a fluid will flow around an object in its path and "come together" on the other side without a delay in time. Non-Newtonian fluids do not have a linear stress-strain relationship. When they encounter shear stress their recovery varies with the type of non-Newtonian fluid.

A main consideration in fluid dynamics is the amount of resistance encountered by an object moving through a fluid. Resistance, also known as drag, is made up of several components but all have in common that they occur at the point where the object meets the fluid. The area can be quite large as in the wetted surface of a ship, the portion of a ship that is below the waterline. For an airplane, the equivalent is the body of the plane as it moves through the air. The goal for those who work in the field of fluid dynamics is to understand the effects of fluid flows and minimize their effect on the object in question.

Background and History

Swiss mathematician Daniel Bernoulli introduced the term "hydrodynamics" with the publication of his book *Hydrodynamica* in 1738. The name referred to water in motion and gave the field of fluid dynamics its first name, but it was not the first time water in action had been noted and studied. Leonardo da Vinci made observations of water flows in a river and was the one who realized that water was an incompressible flow and that for an incompressible flow, $V =$ constant. This law of continuity states that fluid flow in a pipe is constant. In the late 1600's, French physicist Edme Mariotte and Dutch mathematician Christiaan Huygens contributed the velocity-squared law to the science of fluid dynamics. They did not work together but they both reached the conclusion that resistance is not proportional to velocity; it is instead the square of the velocity.

Sir Isaac Newton put forth his three laws in the 1700's. These laws play a fundamental part in many branches of science, including fluid dynamics. In addition to the term hydrodynamics, Bernoulli's contribution to fluid dynamics was the realization that pressure decreases as velocity increases. This understanding is essential to the understanding of lift. Leonhard Euler, the father of fluid dynamics, is considered by many to be the preeminent mathematician of the eighteenth century. He is the one who derived what is today known as the Bernoulli equation from the work of Daniel Bernoulli. Euler also developed equations for inviscid flows. These equations were based on his own work and are still used for compressible and incompressible fluids.

The Navier-Stokes equations result from the work of French engineer Claude-Louis Navier and British physicist George Gabriel Stokes in the mid-nineteenth century. They did not work together, but their equations apply to incompressible flows. The Navier-Stokes equations are still used. At the end of the nineteenth century, Scottish engineer William John Macquorn Rankine changed the understanding of the way fluids flow with his streamline theory, which states that water flows in a steady current of parallel flows unless disrupted. This theory caused a fundamental shift in the field of ship design because it changed the popular understanding of resistance in oceangoing vessels. Laminar flow is measured today by use of the Reynolds number, developed by British engineer and physicist Osborne Reynolds in 1883. When the number is low, viscous forces dominate. When the number is high, turbulent flows are dominant.

American naval architect David Watson Taylor designed and operated the first experimental model basin in the United States at the start of the twentieth century. His seminal work, *The Speed and Power of Ships*, first published in 1910, is still read. Taylor played a role in the use of bulbous bows on vessels of the navy. He also championed the use of airplanes that would be launched from naval craft underway in the ocean.

The principles of fluid dynamics took to the air in the eighteenth century with the work done by aviators such has the Montgolfier brothers and their hot-air balloons and French physicist Louis-Sébastien Lenormand's parachute. It was not until 1799 when English inventor Sir George Cayley designed the first airplane with an understanding of the roles of lift, drag, and propulsion, that aerodynamics came under scrutiny. Cayley's work was soon followed by the work of American engineer Octave Chanute. In 1875, he designed several biplane gliders, and with the publication of his book *Progress in Flying Machines* in 1894, he became internationally recognized as an aeronautics expert.

The Wright brothers are rightfully called the first aeronautical engineers because of the testing they did in their wind tunnel. By using balances to test a variety of different airfoil shapes, they were able to correctly predict the lift and drag of different wing shapes. This work enabled them to fly successfully at Kitty Hawk, North Carolina, on December 17, 1903.

German physicist Ludwig Prandtl identified the boundary layer in 1904. His work led him to be known as the father of modern aerodynamics. Russian scientist Konstantin Tsiolkovsky and American physicist Robert Goddard followed, and Goddard's first successful liquid propellant rocket launch in 1926 earned him the title of the father of modern rocketry.

All of the principles that applied to hydrodynamics- —

Fascinating Facts About Fluid Dynamics

- The pitot tube is a simple device invented by French hydraulic engineer Henri Pitot in the 1700's. It is used to measure air speed in wind tunnels and on aircraft.
- The Bernoulli principle, that pressure decreases as velocity increases, explains why an airplane wing produces lift.
- English engineer William Froude was the first to prove the validity of scale-model tests in the design of full-size vessels. He did this in the 1870's, building and operating a model basin for this purpose.
- The Froude number is a dimensionless number that measures resistance. The greater the Froude number, the greater the resistance.
- Froude performed the seminal work on the rolling of ships.
- Alfred Thayer Mahan's book *The Influence of Sea Power Upon History:1660-1783*, published in 1890, made such a powerful case for the importance of a strong navy that it caused the major powers of that time to invest heavily in new technology for their fleets.
- American naval architect David Watson Taylor was a rear admiral in the U.S. Navy. He was meticulous in his work and developed procedures still in use in model basins. The Taylor Standard Series was a series of trials run with specific models. The results could be used to estimate the resistance of a ship effectively before it was built.
- The bulbous bow is a torpedo-shaped area of the bow of a ship. It is below the waterline. It reduces the resistance of a ship by reducing the impact of the waves on the front bow of a ship underway.
- David Watson Taylor studied the phenomenon of suction between two vessels moving close together in a narrow channel. He was called as an expert witness for the Olympic-Hawke trial in 1911.
- Bioengineers examine fluid flows when designing pacemakers and other medical equipment that will be implanted in the human body.

the study of water in motion—applied to aerodynamics: the study of air in motion. Together these principles comprise the field of fluid dynamics.

How It Works

When an object moves through a fluid such as gas or water, it encounters resistance. How much resistance depends upon the amount of internal friction in the fluid (the viscosity) as well as the shape of the object. A torpedo, with its streamlined shape, will encounter less resistance than a two-by-four that is neither sanded nor varnished. A ship with a square bow will encounter more resistance than one with a bulbous bow and V shape. All of this is important because with greater resistance comes the need for greater power to cover a given distance. Since power requires a fuel source and a way to carry that fuel, a vessel that can travel with a lighter fuel load will be more efficient. Whether the design under consideration is for a tractor trailer, an automobile, an ocean liner, an airplane, a rocket, or a space shuttle, these basic considerations are of paramount importance in their design.

Applications and Products

Fluid dynamics plays a part in the design of everything from automobiles to the space shuttle. Fluid dynamic principles are also used in medical research by bioengineers who want to know how a pacemaker will perform or what effect an implant or shunt will have on blood flow. Fire flows are also being studied to aid in the science of wildfire management. Until now the models have focused on heat transfer but new studies are looking at fire systems and their fluid dynamic properties. Sophisticated models are used to predict fluid flows before model testing is done. This lowers the cost of new designs and allows the people involved to gain a thorough understanding of the trade-off between size and power given a certain design and level of resistance.

Impact on Industry

Before English engineer William Froude, ships were built based on what had worked and what should work. Once Froude proved that scale-model testing could reliably predict the performance of full-scale vessels, the entire process of vessel design was forever altered. The experience with scale models in a model basin transferred to the testing of scale models of airplanes and automobiles in wind tunnels. Computational fluid dynamic models are used for testing key elements of everything from ships to skyscrapers. The use of fluid dynamic theories to predict performance is ongoing and continues to be vital to engineers.

Careers and Course Work

Fluid dynamics plays a part in a host of careers. Naval architects use fluid dynamic principles to design vessels. Aeronautical engineers use the principles to design aircraft. Astronautical engineers use fluid dynamic principles to design spacecraft. Weapons are constructed with and understanding of fluids in motion. Automotive engineers must understand fluid dynamics to design fuel-efficient cars. Architects must take the motion of air into their design of skyscrapers and other large buildings. Bioengineers use fluid dynamic principles to their advantage in the design of components that will interact with blood flow in the human body. Land-management professionals can use their understanding of fluid flows to develop plans for protecting the areas under their care from catastrophic loss due to fires. Civil engineers take the principles of fluid dynamics into consideration when designing bridges. Fluid dynamics also plays a role in sports: from pitchers who want to improve their curveballs to quarterbacks who are determined to increase the accuracy of their passes.

Students should take substantial course work in more than one of the primary fields of study related to fluid dynamics (physics, mathematics, computer science, and engineering), because the fields that depend upon knowledge of fluid dynamic principles draw from multiple disciplines. In addition, anyone desiring to work in fluid dynamics should possess skills that go beyond the academic, including an aptitude for mechanical details and the ability to envision a problem in more than one dimension. A collaborative mind-set is also an asset, as fluid dynamic applications tend to be created by teams.

Social Context and Future Prospects

The science of fluid dynamics touches upon a number of career fields that range from sports to bioengineering. Anything that moves through liquids such as air, water, or gases is subject to the principles of fluid dynamics. The more thorough the understanding, the more efficient vessel and other designs will be. This will result in the use of fewer resources in the form of power for inefficient designs and help create more efficient aircraft and launch vehicles as well as medical breakthroughs.

Gina Hagler, B.A., M.B.A.

Further Reading

Anderson, John D., Jr. *A History of Aerodynamics and Its Impact on Flying Machines.* New York: Cambridge University Press, 1997. Includes several chapters that deal with the theories of fluid dynamics and their application to flight.

Carlisle, Rodney P. *Where the Fleet Begins: A History of the David Taylor Research Center.* Washington, D.C.: Naval Historical Center, 1998. A detailed history of the work done at the David Taylor Research Center.

Çengel, Yunus A., and John M. Cimbala. *Fluid Mechanics: Fundamentals and Applications.* Boston: McGraw-Hill, 2010. An essential text for those seeking familiarity with the principles of fluid dynamics.

Darrigol, Olivier. *Worlds of Flow: A History of Hydrodynamics from the Bernoullis to Prandtl.* New York: Oxford University Press, 2005. A thorough account of the progress in hydrodynamic and fluid dynamic theory.

Eckert, Michael. *The Dawn of Fluid Dynamics: A Discipline Between Science and Technology.* Weinheim, Germany: Wiley-VCH, 2006. An introduction to fluid dynamics and its applications.

Ferreiro, Larrie D. *Ships and Science: The Birth of Naval Architecture in the Scientific Revolution, 1600-1800.* Cambridge, Mass.: MIT Press, 2007. A fully documented account of the transition from art to science in the field of naval architecture.

Johnson, Richard W., ed. *The Handbook of Fluid Dynamics.* Boca Raton, Fla.: CRC Press, 1998. A definitive text on the principles of fluid dynamics.

Mahan, A. T. *The Influence of Sea Power Upon History, 1660-1783.* 1890. Reprint. New York, Barnes & Noble Books, 2004. Mahan wrote the book that changed the way nations viewed the function of their navies.

Web Sites

American Physical Society
http://www.aps.org

National Agency for Finite Element Methods and Standards
http://www.nafems.org

Society of Naval Architects and Marine Engineers
http://www.sname.org

Von Karman Institute for Fluid Dynamics
https://www.vki.ac.be

See also: Civil Engineering; Computer Science; Engineering; Naval Architecture and Marine Engineering; Spacecraft Engineering.

FOSSIL FUEL POWER PLANTS

FIELDS OF STUDY

Electrical engineering; mechanical engineering; pressure vessel engineering; industrial electronics; pipe fitting; pressure welding; power plant operations; industrial maintenance; millwrighting; geology; mining; deep borehole technologies; power distribution technology; fluid mechanics; hydraulics; environmental chemistry; combustion chemistry; heat transfer technology; mathematical modeling; computer simulation; controls and switching technology; transportation; management; project coordination; human resources.

SUMMARY

Fossil fuels are the organic residues of geological processes and include the various grades of coal, natural gas, petroleum, and crude oil. By definition, all fossil fuels are nonrenewable resources. Fossil fuel power plants all function in fundamentally the same manner and rely on a principle that has not changed significantly since the earliest applications of steam power. In short, the fuel is combusted to generate heat, producing pressurized steam that in turn drives the turbines of large electric generators.

Coal was the first major fossil fuel to be exploited as an energy source for the steam-powered plants that drove the Industrial Revolution of the eighteenth and nineteenth centuries. When steam technology was applied to the large-scale generation of electricity, coal became the fuel of choice because of its ready availability. Coal remains the major fuel source for fossil fuel power plants, although natural gas and, to a much lesser extent, petroleum-based fuels have been considered.

Significant costs, both economic and environmental, have been identified in the use of fossil fuels for the production of electricity. Economically, the prices of fossil fuels have been driven upward by the market economy, and environmentally, the combustion of large quantities of carbon-based fossil fuels releases a great deal of carbon dioxide and other gases into the atmosphere.

KEY TERMS AND CONCEPTS

- **Carbon Dioxide:** Molecule formed by the combination of one atom of carbon with two atoms of oxygen.
- **Combustion:** Chemical process of oxidation of a fuel material, generally a carbon-based fuel, by molecular oxygen that results in the release of thermal energy in the form of flames.
- **Cooling Tower:** Structure within a power plant that serves to reduce the temperature of the exhaust gas stream from the combustion stage, allowing the recovery of quantities of fly ash, sulfur, and other pollutants.
- **Energy Density:** Amount of usable energy that can be extracted from the unit mass or volume of a material.
- **Fluidization:** Treatment that causes a nonfluid material to behave in a fluidlike manner; typically achieved by passing a strong current of air uniformly through a mass of solid particles so that the particles are suspended in the moving air stream.
- **Fluidized Bed Combustion:** Combustion process carried out by injecting the fuel into a bed of solid particulate matter fluidized with a stream of air.
- **Fossil:** Geologic remnant of a once-living organism; includes mineralized body structures and the carbonaceous materials, such as coal, crude oil, and natural gas, resulting from geologic processes.
- **Fossil Fuel:** Carbonaceous material resulting from geologic processes that is used as a fuel for combustion.
- **Generator:** Device that produces electric current by the interaction of electrical conductors with the moving magnetic field produced by spinning magnets within the device.
- **Heat Exchange:** Transfer of thermal energy from a material at one temperature to a material at a lower temperature.
- **Pulverization:** Process of reducing the physical structure of a material to extremely fine particles.
- **Turbine:** Device that converts linear motion of a pressurized fluid into rotational motion that may then be used to power other devices such as generators.
- **Vaporization:** Conversion of a material from a liquid state to a gas state.

Definition and Basic Principles

Fossil fuel power plants are generating stations that rely on the combustion of fossil fuels to produce electricity. Only three fossil fuels—coal, petroleum, and natural gas—are used for this purpose. The term "power plant" does not refer only to facilities that generate electricity but rather to any facility whose function is to produce usable power, whether electrical, mechanical, hydraulic, pneumatic, or another type. In common usage, however, power plant generally refers to those facilities that are used to generate electricity.

All fossil fuels are the remnants of organisms that existed, in most cases, many millions of years ago. Time and geologic processes involving heat and pressure chemically and physically altered the form of these organisms, turning them into mineralogical fossils (such as mineralized bones found in sedimentary rock formations) and the carbonaceous forms of coal, crude oil, and natural gas. When these carbonaceous materials are refined, they can be used as combustion fuels in fossil fuel power plants.

The combustion process is carried out in a variety of ways, from standard internal combustion engines using natural gas, gasoline, or diesel oil, to fluidized bed combusters using pulverized coal powder. Internal combustion engines are used to drive a generator directly, while other combustion methods are used to heat water and produce pressurized steam through heat exchange. The steam is then used to drive a turbine that in turn drives electric generators. The spent steam is generally recycled through the system. The exhaust steam from combustion is passed through a treatment process to reduce or eliminate contaminants formed from materials that were in the fuel.

Ideally, the combustion process would produce only carbon dioxide—from the combustion of coal—or carbon dioxide and water—from the combustion of hydrocarbon fuels such as natural gas and refined petroleum fuels. In practice, however, fossil fuels contain a percentage of materials other than carbon and hydrogen, such as sulfur, metals (including iron, mercury, and lead), and nonmetals (including phosphorus, silicon, and arsenic). In addition, air used to supply oxygen for combustion also contains about 78 percent nitrogen and about 1 percent of other gases. At the temperatures of combustion, these impurities can react with oxygen to produce a variety of pollutant by-products such as sulfur dioxide, nitrogen oxides, and fly ash.

The combustion of fossil fuels results in a very large quantity of carbon dioxide being released into the atmosphere, where it can act as a greenhouse gas. A greenhouse gas has the effect of trapping heat that normally would be radiated out of the atmosphere and into space. Many experts believe that the carbon dioxide released by the burning of fossil fuels has been a primary factor in global warming, which refers to an increase in the mean annual temperature of the planet.

Background and History

Coal has been used as a fuel for combustion for thousands of years. It was reportedly used by native North Americans when the first European settlers arrived, and it was undoubtedly used by other peoples throughout the world because of the ease with which it could be extracted from the ground in certain areas. It became the fuel of choice beginning in the eighteenth century and had almost completely replaced wood as the dominant fuel of industry by the early 1900's because of its more favorable energy density. Coal's increasing popularity as a fuel also drove the growth of the coal-mining industry, in turn increasing its availability.

With the development of the large-scale generation of electricity and its many applications, coal-fired power plants were used to drive electric generators where suitable water power, such as at voluminous waterfalls, was not available. The convenience and versatility of a common electric grid resulted in the growth of the electric generation industry. Small and localized generation systems ranging from low-output gasoline-powered home generators to large diesel-powered industrial generating stations can provide emergency and local service if the common grid is not available. Large generating stations using fossil fuels have been built and continue to be built in areas where coal or other fossil fuels are readily available.

How It Works

Fossil Fuels. Coal and petroleum are the remnants of plants and animals that lived millions of years ago. Over the years, geologic processes compressed and chemically altered the plants and animals in such a way that coal consists almost entirely of pure carbon, crude petroleum consists almost entirely of a vast assortment of hydrocarbons, and natural gas consists almost entirely of methane, ethane, and propane, which are simple hydrocarbon gases. Coal can be found at varioius depths in the Earth's crust, in veins ranging from only a few centimeters to hundreds of meters thick. It is mined out as a solid, rocky, and relatively lightweight material and used in forms ranging from crude lump coal to a fine powder that is fluidlike in its behavior.

Crude oil and natural gas are found only at depths of hundreds and thousands of meters. As liquids or fluids, these materials have migrated downward through porous rock over a long period of time, until further progress is prevented by an impervious rock layer. There, they collect, often in large pools of oil and gas that can be recovered only after being found through careful exploration and deep borehole drilling. Natural gas requires no further processing before being used as a combustion fuel, unless it is classed as sour gas, meaning it contains an unacceptably large proportion of foul-smelling hydrogen sulfide and other poisonous gases that must be removed. Petroleum, or crude oil, must be heavily refined before it can be used as a fuel. The crude oil is subjected to thermal cracking, which breaks down and separates the various hydrocarbon components into usable portions from light petroleum ethers such as pentanes and hexanes to heavy tars such as asphalt. The most well-known fractions refined from petroleum are gasoline, kerosene, diesel fuel, and waxes, and various grades of lubricating oils and greases.

Combustion. Combustion is a chemical reaction between a material and oxygen. The reaction is an oxidation-reduction process in which one material becomes chemically oxidized and the other becomes chemically reduced. In the context of fossil fuels, the material that becomes oxidized is coal, fuel liquids, or natural gas. Combustion of these carbonaceous materials converts each atom of carbon in the fuel molecules to a molecule of carbon dioxide, according to the general equation:

$$C + O_2 \rightarrow CO_2$$

This conversion (greatly simplified here) releases an amount of energy that can then be transferred to and captured by a moderator—typically water—through the use of heat exchangers (devices that facilitate the transfer of heat from one material to another material). Combustion of hydrocarbons also produces water as an output of the reaction, in which two atoms of hydrogen combine with one atom of oxygen to produce one molecule of water. Other reactions corresponding to the combustion of impurities in the fuel stream also take place, and their products are ejected in the exhaust flow from the combustion process.

Turbines and Generators. Steam under pressure, produced by heating water via the combustion of fuels, is directed into a mechanical device called a turbine. Turbines can basically be described as high-tech versions of the ancient water wheel. The pressure

Fascinating Facts About Fossil Fuel Power Plants

- Some 2 billion metric tons of carbon dioxide have been added to the atmosphere from the burning of fossil fuels in the 1800's and 1900's.
- About 70 percent of the world's power plants use fossil fuels. In 1984, fossil fuels accounted for 82 percent of all commercial energy production, and 91 percent of the U.S. energy supply. According to the U.S. Department of Energy, fossil fuels (coal, natural gas, and oil) account for more than 85 percent of the United States' energy consumption, including nearly two-thirds of electricity and nearly all fuels used for transportation.
- Coal was used as a heating fuel by North American native peoples, and presumably other peoples around the world dating back hundreds of years ago.
- Fossil fuel power plants come in all sizes, from little 750-watt gasoline-powered portable generators to large coal-fired or natural gas-fired facilities producing hundreds of megawatts of electricity.
- Carbon dioxide capture and storage technology would take the carbon dioxide produced from fossil fuel combustion and store it a kilometer or more underground. There, slow geologic processes might one day convert it back to free oxygen and carbonaceous deposits that future generations could use as fuel.
- Some fossil fuels are more than 650 million years old.
- In ancient Greece, Archimedes invented a device that directed steam outward in opposing directions, causing the device to rotate. It would be known later as an external combustion steam engine.

of the flowing gas (steam) pushes against a series of vanes attached to an armature (electric component) in the structure of the turbine, driving them to spin the armature with force. This converts the linear fluid motion of the steam into the rotary mechanical motion of the turbine. Turbines are coupled to an electric generator so that their rotation results in the generation of electricity.

A generator is another rotary device that, in its most basic concept, consists of a magnet spinning inside a cage of conducting wires. The magnetic field of the magnet also spins at the same rate that the magnet spins. The movement of the magnetic field through the conductors in the surrounding cage produces an electromotive force (EMF) in the conductors, which is measured in volts. If the generator is connected into a circuit, this EMF causes current to flow in the circuit. Strictly speaking, generators produce direct current (DC) electricity, while alternators produce alternating current (AC) electricity. AC electricity is the standard form of electricity used in national power grids around the world.

Both generators and alternators are available in various output capacities and are driven by many types of rotary engines, from small internal combustion engines to large industrial steam turbines.

Applications and Products

Fossil fuel power plants, in the context of electric generating stations, produce only one product: electricity. Any and all other materials that come from them are considered ancillary or waste by-products. The ultimate goal of operating a fossil fuel power plant is to maximize the output of electricity from each unit of fuel consumed, while also minimizing any and all undesirable outputs. To that end, the efficiencies of control design, the data feedback process, economics, fuel processing, and a host of other aspects of the electric power generation industrial complex are examined each year. Not the least of these considerations is the placement and construction of new facilities and the maintenance of older facilities.

At one time, the competitive cost and the availability of natural gas and crude oil nearly spelled the demise of the coal-fired power plant. However, the prices of natural gas and crude oil were driven upward, both artificially and naturally, making these choices less attractive. Nuclear power plants were initially welcomed by the public, but their popularity declined. Because of these circumstances, the continuing demand for electricity caused fossil fuel power plants—especially coal-fired plants—to regain their position of prominence in electric power generation. The operation of fossil fuel plants spawned related industries: the development of coal-mining methods and machinery, oil and gas exploration and recovery, fossil fuel transportation and preprocessing, specialized construction and trades, environmental assessment and maintenance operations, financial and administrative companies, plant operations and control technology, industrial maintenance, and grid supply and service.

Concern about greenhouse gases resulted in the birth of an industry aimed at capturing and storing the carbon dioxide produced by the power plants. Different approaches are being developed, but possibly the most promising technology is the sequestering of carbon dioxide in deep underground water formations. This has engendered a whole new area of research and development in regard to compression and recovery technology.

The operation of fuel fossil power plants generates chemicals, some of which have been recovered during preprocessing and exhaust gas scrubbing procedures. Sulfur recovered from preprocessing and from entrapment of sulfur dioxide in the exhaust gas stream is used to produce sulfuric acid, an important industrial chemical, as well as numerous other sulfur-containing compounds. Similarly, nitrogen oxides recovered from the combustion process provide nitric acid and other nitrogen-containing compounds. Interestingly, the entire plastics industry grew out of research to find uses for compounds recovered from coal tar, a by-product of the coal processing industry in the nineteenth century. In modern times, however, essentially all plastics are derived from petroleum.

Impact on Industry

Several countries rely almost exclusively on electricity produced by fossil fuel power plants, especially coal-fired plants. According to the International Energy Agency, in 2009, South Africa and Poland both obtained almost 95 percent of their electric supply from coal-fired power plants, while the United States obtained almost exactly half of its electric supply from coal-fired power plants. Although several

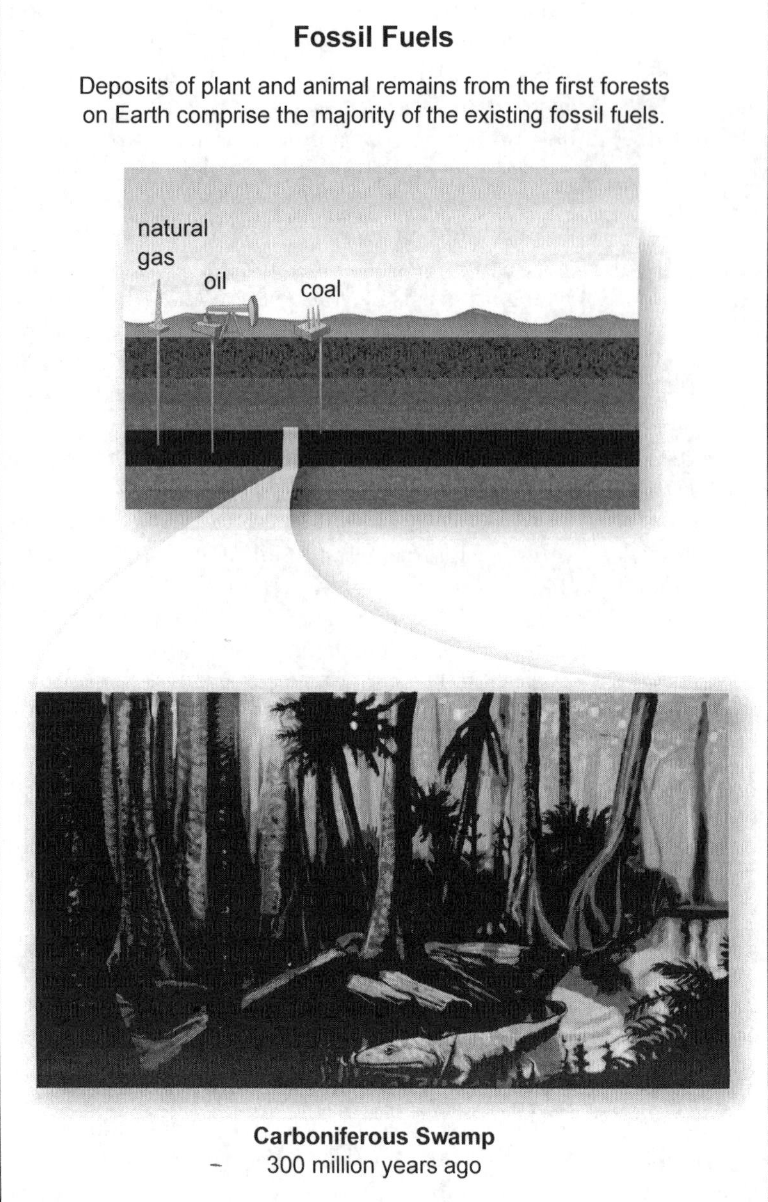

major hydroelectric projects are under way in China, coal-fired power plants are also being built at a great rate to help establish an adequate electrical infrastructure. Fossil-fuel power plants have a reputation for high reliability because of their well-established technology, and they are less costly than nuclear and other alternative electricity-generating methods. Additionally, they can be constructed in much less time than would be needed for other types of generating stations of the same capacity.

Fossil Fuels
Deposits of plant and animal remains from the first forests on Earth comprise the majority of the existing fossil fuels.
natural gas
oil
coal
Carboniferous Swamp
300 million years ago

Modern industry depends on a constant, adequate supply of electric energy for its very existence. The application of electric motors and the ubiquitous nature of electric arc welding in fabrication and manufacturing processes have defined modern industry and created power demands that are hard to match through any other means of power delivery. Stricter regulations regarding the handling of waste materials produced during the combustion of fossil fuels have also had a significant impact on the industry as a whole. Private and governmental power-generating organizations have been required both to reduce the production of waste materials and to assume responsibility for the handling of wastes that were produced in past years. In addition, international agreements for the reduction of greenhouse gas emissions have placed more pressure on the industry.

Careers and Course Work

Students undertaking any program that will lead them into a career related to fossil fuel power plants, whether directly or indirectly, will be required to exit high school with an understanding of physics, chemistry, mathematics, and business and technology. Biology will also be required if the career path chosen is directed toward environmental studies. College and university level course work will depend greatly on the area of specialization chosen, as the range of options at this level is immense. At a minimum, students will continue studies in mathematics, physical sciences, industrial technologies (chemical, electrical, and mechanical), and business as undergraduates or as trade students. More advanced studies will take the form of specialist courses in a chosen field. In addition, as technologies and regulations change, those working in the field of fossil fuel power plants can expect to be required to upgrade their working knowledge on an almost continual basis to keep abreast of changes.

Social Context and Future Prospects

The world's reliance on the ready availability of coal and the industrial convenience of fossil fuel power plants, especially in developing countries, essentially guarantees that those facilities will be part of the landscape for many years to come, despite intensive efforts to develop alternative power sources. New technologies being developed, such as carbon dioxide capture and storage (CCS), combined geothermal-fossil fuel cogeneration, and biomass cogeneration, will require generations of power plant workers who understand both the older and the new technologies as fossil fuel power plants are managed toward a zero emissions platform (ZEP). Emission-abatement plans also include elimination of the heavy metals component of fossil fuels, as the combustion of those materials has the effect of greatly concentrating any heavy metals, including radioactive trace elements such as uranium and thorium, in the ash residues. It is therefore not inconceivable that, in the future, work in a fossil fuel power plant will also require training in working with radioactive materials.

Richard M. J. Renneboog, M.Sc.

Further Reading

Borowitz, Sidney. *Farewell Fossil Fuels: Reviewing America's Energy Policy.* New York: Plenum Trade, 1999. Examines the role of fossil fuels in energy production in the United States and other Western nations, and the technologies that can be and are being developed to replace fossil fuels.

Breeze, Paul. *Power Generation Technologies.* Boston: Elsevier, 2005. A comprehensive treatment of the generation of electric energy by different power plant technologies and other methods, with consideration of their economic and environmental impacts

Droege, Peter, ed. *Urban Energy Transition: From Fossil Fuels to Renewable Power.* New York: Elsevier, 2008. Looks at principles, policies, and practices that affect energy use, the technology involved, ways to alter the urban environment, and programs being employed in London and elsewhere.

Evans, Robert L. *Fueling Our Future: An Introduction to Sustainable Energy.* New York: Cambridge University Press, 2007. Looks at energy supply and demand and how to reduce reliance on fossil fuels.

Leyzerovich, Alexander S. *Steam Turbines for Modern Fossil-fuel Power Plants.* Lilburn, Ga.: Fairmont Press, 2008. Closely examines the engineering and operating properties of modern steam turbines in fossil fuel power plants.

Web Sites

Tennessee Valley Authority
Fossil-fuel Generation
http://www.tva.gov/power/fossil.htm

U.S. Department of Energy
Fossil Energy
http://fossil.energy.gov/index.html

World Coal Institute
Coal and Electricity
http://www.worldcoal.org/coal/uses-of-coal/coal-electricity

Zero Emissions Platform
European Technology Platform for Zero Emission Fossil Fuel Power Plants
http://www.zeroemissionsplatform.eu

See also: Coal Gasification; Industrial Pollution Control; Petroleum Extraction and Processing.

FRACTAL GEOMETRY

FIELDS OF STUDY

Calculus; computer science; geometry; measure theory; metric topology; probability; set theory.

SUMMARY

Fractal geometry is a branch of mathematics that is used to study irregular or fragmented shapes, such as clouds, trees, mountains, and coastlines. Applications of fractal geometry are widespread, from the sciences and medicine to social sciences and the arts, and especially in human anatomy, ecology, physics, geology, economics, and computer graphics. Related areas include chaotic systems and turbulent systems.

KEY TERMS AND CONCEPTS

- **Affine Transformations:** Transformations on the Euclidean plane include translation, scaling, rotation, and reflection. Affine transformations can be used to convert one geometric shape into another.
- **Brownian Motion:** Result of thermal molecular motion in a liquid environment.
- **Chaos Theory:** Behavior of nonlinear dynamic systems, the processes of which are governed by an underlying order and the outcomes of which demonstrate sensitivity to initial conditions.
- **Complex System:** Refers to the properties of a nonlinear system, in which the behavior of the system as a whole is different from those of its parts.
- **Diffusion-Limited Aggregation (DLA):** Clustering of particles in random walk due to Brownian motion. DLA is prevalent in diffusion systems. Lichtenberg figures (or electric trees) are a common example of DLA.
- **Dimension:** Space a set occupies near its points.
- **Dynamical System:** Set of possible states with a rule that defines the present in terms of past conditions.
- **Feedback Loop:** Process of a transformational system where the input of a cycle gives a different output, which becomes the input of a new transformation. Feedback processes are fundamental tools for studying natural systems.
- **Fractal:** With reference to a mapped set, the set can be considered fractal if it is detailed at smaller scales, has an irregular structure, demonstrates self-similarity, has a Hausdorff-Besicovitch dimension greater than its topological dimension, and can be defined by a simple statement that is often reiterated. Fractals can be geometric (nonrandom) or random.
- **Fractal Dimension:** Measure of a geometric object that can have fractional values. It refers to the measure of how fast the length, area, or volume of an object increases with a decrease in scale. Fractal dimension can be calculated by box counting or by evaluating the information dimension of an object.
- **Generator:** Collection of scaled copies of an initiator.
- **Hausdorff-Besicovitch Dimension:** Mathematical statement used to obtain a dimension that is not a whole number, commonly written as $d = log (N)/ log (r)$.
- **Initial Condition:** Starting point of a dynamic system.
- **Initiator:** Starting shape of a reiterated self-similar object.
- **Julia Set:** Boundary of the set of points of a function z that escape into infinity under repeated iteration by $f(z)$.
- **Lacuna:** Gap or space within a structure.
- **Lacunarity:** In reference to fractal objects, lacunarity describes the weave and texture of the fractal object, with particular reference to the open and filled spaces of that object.
- **Lindenmayer (L-) System:** Named for Hungarian biologist Aristid Lindenmayer, it is a system used to model the development of plants. It is defined as a parallel rewriting system, one that takes a simple form and then reiteratively replaces its parts by a set of rules. It can generate a fractal with a dimension between one and two.
- **Mandelbrot Set:** Set of all complex numbers c such that iteration of the function $f(x) = x^2 + c$, starting at $x = 0$, does not go into infinity.
- **Noise:** Signal with fractal properties.
- **Rule:** Principle that describes how scaled copies of a generator replace copies of an initiator.
- **Self-Similarity:** Structure that can be broken down into smaller pieces, all of which are individual replicas of the main object under repeated magnification. There is a relationship between

the scaling factor and the number of pieces to which the structure can be reduced.

- **Symmetry:** Universal scientific and philosophical category used in reference to the structure of matter. It is defined by the invariance of particular features under different types of transformations.

Definition and Basic Principles

Fractal geometry is a workable geometric middle ground between the excessive geometric order of Euclid and the geometric chaos of general mathematics. According to French American mathematician Benoit Mandelbrot in *Proceedings of the Royal Society of London* (May 8,1989), it is based on a form of symmetry previously underused, namely, "invariance under contraction or dilation":

> Fractal geometry is conveniently viewed as a [language that has proven] its value by its uses. Its uses in art and pure mathematics, being without "practical" application, can be said to be poetic. Its uses in various areas of the study of materials and [of] other areas of [engineering] are examples of practical prose. Its uses in [physical] theory, especially in conjunction with the basic equations of mathematical physics, combine poetry and high prose. Several of the problems that fractal [geometry] tackles involve old mysteries, some of them already known to primitive man, others mentioned in the Bible, and others familiar to every landscape artist.

Background and History

Fractal geometry was first defined in 1975 by Benoit Mandelbrot, a creative and prolific twentieth-century polymath who pioneered a synthesis of mathematical paradigms for interpreting rational surface roughness. *Les Objets Fractals,* translated into English in 1977 under the title *Fractals: Form, Chance, and Dimension,* explored Mandelbrot's innovative approach to understanding the properties of mathematical and natural forms. The manuscript was revised and republished in 1983 under the title *The Fractal Geometry of Nature.* This book illustrated principles that reordered understanding of dimension, symmetry, and scalar transformations. It is considered one of the classic works of twentieth-century mathematics. As a result of rational inquiry into shapes and nonlinear physical phenomena, fractal analysis can be applied with exceptional effectiveness in both pure and applied mathematics. Advances in computation during the late twentieth century made it possible to evaluate unique objects constructed after millions of iterations of a function. Repetitive calculations that are not feasible in human time were performed with novel speed and duration via the computer, producing extraordinary mappings of functions, the visual beauty of which astounded its practitioners. The results of fractal analysis transformed and aligned knowledge in many domains, including physics, religion, the mechanical and biological sciences, economics, statistics, music, and many fields of art.

Mandelbrot's investigations regarding the nature of roughness or nonlinearity in the structural composition and generation of mathematical and physical objects are products of the twentieth century. However, fractal analysis is synthetic and embraces systems of thought originating in the preceding centuries. In retrospect, the principles described by fractal theory provide a platform for integrating a variety of human expressions regarding infinity. Throughout history, mathematicians intuitively understood the implicit beauty and symmetry of mathematical systems. Glimpses of the infinite were magnified and demonstrated in fractal forms with particular appeal to the human sense of vision. Moreover, the language of fractals can be grasped and used in a variety of fields requiring varying levels of mathematical and computational ability. Finally, fractal symmetry provides an accessible framework for understanding universal structures and dynamics.

The antecedents to fractal geometry can be seen in the following areas.

Projective Geometry. Projective geometry refers to a class of geometric properties describing invariance under projection. Scholarship suggests that this phenomenon was understood by early geometers—perhaps by Euclid himself. Mathematicians of the seventeenth century are credited for realizing these properties in an attempt to make geometry more practical. French mathematicians Gérard Desargues, Blaise Pascal, and Philippe de la Hire provided mathematical foundations for new applications in the nineteenth century. Subsequent formulations of the works of Desargues are fundamental to novel descriptions of space and time. French mathematicians Jean-Victor Poncelet and Michel Chasles continued the work of Desargues. Similar innovations were made in the fields of analytic geometry, non-Euclidean geometry, and Riemannian geometry.

Geometers have relied on the definitions and assumptions described in Euclid's *Elements of Geometry* (1893), a textbook that established time-honored principles for studying geometric properties and their mathematical relationships. These fundamental truths formed the basis of rational descriptions of shapes and measurements to the present day. During the nineteenth century, scholars found the ideal shapes of Euclidean geometry insufficient to analyze the natural phenomena described by science. In particular, Euclid's Postulate 5 defied proof by the brightest minds. The investigations of German mathematician Carl Friedrich Gauss, Hungarian mathematician János Bolyai, and Russian mathematician Nikolai Ivanovich Lobachevsky set the groundwork for the descriptions of the hyperbolic plane, a concept that would have stunning implications in descriptions of the mechanics of space and time, a field pioneered by Albert Einstein.

Cartography. In his 2002 essay "A Maverick's Apprenticeship," Mandelbrot mentioned his familiarity with maps as a result of an uncle's tutoring. It is of note that well-established cartographic publishing houses and government agencies supported the works of many nineteenth- and twentieth-century mathematicians who made magnificent contributions to the field of cartography. Expanding centers of global trade, competition for world hegemony, and the remarkable advances of science and technology created new demands for accurate systems of measurement. Novel uses of geometric projections stimulated inventive minds engaged in the proliferation of maps representing expanding enterprises in topographical land surveillance, marine navigation, meteorology, oceanography, climatology, geology, mineralogy, biogeography, and demography. Many two- and three-dimensional projections were developed to represent different sets of data visually. Accurate scalar measurements of coastlines and elevated land features were absolutely essential to mariners and land developers, who relied on visual data to guide their enterprises. Mathematics was an indispensable tool of the trade.

Symmetry. The adoption of the concept of symmetry was a very gradual process. It was an accepted condition of mathematical equations in the nineteenth century, but its laws and functions in relation to physical phenomena were poorly understood. In the field of physics, French mineralogist René Just Haüy's analysis of the geometric forms of crystals and subsequent studies of the symmetry of crystalline structure and properties were eventually applied to the study of other natural and dynamic systems. Mandelbrot's introduction of fractal theory helped to refine core concepts of symmetry, particularly those of proportion and scale. The concept of symmetry is considered a universal property that unifies all natural and aesthetic phenomena.

Artists were among the first groups to embrace the fractal as an exquisite audiovisual medium for creative expression. The processes of fractal symmetry have stimulated profound intellectual and technical affinities among mathematicians, scientists, theologians, software designers, composers, architects, graphic artists, and writers who understand the universal principles of harmony that fractals represent.

Monsters. German mathematician Karl Weierstrass's proof of a non-differential function replicated earlier work by Bernhardt Riemann and opened the door for further analysis of non-differentiable curves. Functional analysis and formal logic were essential practices of the mathematician's craft. Graphic representations of complex functions were not in practice and, in the case of some functions, their representations simply could not be mapped in human time. Subsequently, many leading mathematicians rebuffed the anomalous solutions of nondifferential functions, banishing them to the periphery of mathematics. As Mandelbrot explained in his introduction to *The Fractal Geometry of Nature*, many symbolic solutions of non-differential functions were anomalous to the tradition of differential curves and thus dismissed as "pathological" and belonging to a "gallery of monsters" and relegated to the periphery of mathematics.

Mandelbrot also pointed out that these unique mathematical concepts are now the fundamental tools defining natural phenomena in the world around us. He referenced particular mathematicians who made singular contributions to the theory of fractals; their theories and the objects named after them are standards of the canon of fractal geometry.

In 1883, German mathematician Georg Cantor, one of the founders of point-set topology, introduced the Cantor set (also known as the triadic Cantor dust), a self-similar disconnected function. In 1890, Italian mathematician Giuseppe Peano defined the function of a space-filling curve, and in 1891, German mathematician David Hilbert provided variations of the Peano curve and included graphic representations of the models described. In 1915, Polish mathematician Waclaw Sierpinski introduced the Sierpinski gasket (also called a triangle or sieve).

In 1919, German mathematician Felix Hausdorff developed the concept of fractional dimension, a measure theory originated by Greek mathematician Constantin Carathéodory in 1914. Russian mathematician Abram Samoilovitch Besicovitch developed the idea of fractional dimension between 1929 and 1934. Taken together, the concepts defined by Hausdorff and Besicovitch were used by Mandelbrot to define what he termed the "fractal dimension" of a surface.

In 1938, French mathematician Paul Lévy introduced another geometric object, known as Lévys Dragon, which demonstrated a triangular set that could be configured (or tiled) to fill a curved space. Other contributors to fractal theory mentioned by Mandelbrot are Bohemian mathematician Bernard Bolzano, French mathematician Henri-Leon Lebesgue, American mathematician William Fogg Osgood, and Russian mathematician Pavel Samuilovich Urysohn. Finally, the works of French mathematicians Gaston Maurice Julia, Henrí Poincaré, and Pierre Joseph Louis Fatou are memorialized in Mandelbrot's graphic representations of the Julia set and the Mandelbrot set described below.

Unusual Data Sets and Nonlinear Phenomena. During the late 1950's and 1960's Mandelbrot was an employee at IBM, and computer technologies were in their infancy. One of the first problems Mandelbrot was asked to resolve was that of "noise" during data transmission across telephone wires. Occasional errors were problematic. Mandelbrot graphed the sequence of erratic transmissions and noticed a regular, self-similar series that reminded him of the phenomena described by Georg Cantor many years before. Mandelbrot realized that the binary nature of electronic data transmission permitted intermittent switching of signals that would prove disastrous to the system IBM was interested in marketing. The recognition of these signal patterns confirmed his deep-seated hunches about the nature of the "monsters" that lurked in the works of several turn-of-the-century mathematicians. These transmission data sets illustrated the powerful concepts of reiteration and self-similarity.

However, not all natural phenomena are self-identical, nor are they fixed or static. At the time, Mandelbrot was aware that nonlinear mechanics were an important component of dynamic systems, particularly in the study of turbulence and galaxy formation. Edward Lorenz, a meteorologist at the Massachusetts Institute of Technology, was pioneering new investigations in chaos theory relative to the unreliable prediction of weather patterns. Better working concepts were needed for explaining the mechanics of difficult nonlinear phenomena such as turbulence and clustering. Mandelbrot recognized the potential of the Hausdorff-Besicovitch dimension to describe irregular phenomena and so chose the coastline of Great Britain as a teaching rubric for working through the concept of fractional dimension. In subsequent studies of scalar proportions, he recognized that the shape of the coastline was similar to the form of the Koch curve, which was developed by Swedish mathematician Helge von Koch in 1904. Manipulating the parameters of the curve, Mandelbrot was able to duplicate mathematically, with exceptional accuracy, the shapes of various landforms. These and other mathematical data sets are the basis of subsequent innovations in image data compression, a technology on which the magnificent computer-generated landscapes of movies such as *Star Wars: Episode VI—Return of the Jedi* and *Star Trek II: The Wrath of Khan* are based.

Self-Similar Systems: The Julia Set and the Mandelbrot Set. In 1918, Gaston Maurice Julia and Pierre Joseph Louis Fatou published independent works describing the process of iterating rational functions. Mandelbrot's uncle gave him original reprints of the articles, suggesting they were worth pursuing, but he did not follow up on them until much later, when employed at IBM.

Access to computers gave him an exceptional opportunity to test the reiterative models described by Julia and Fatou. Instead of simply solving equations, he ran them as a reiterative feedback loop, using the results of one calculation to serve as the inputs of the next run. After millions of iterations, he mapped the results on a grid and was astounded by the results, now known as a Julia set. Working with the graphics of different functions, Mandelbrot devised his own complex quadratic polynomial, such that $f(x) = x^2 +$

c, starting at $x = 0$, does not go into infinity, and then ran it. The result was a point set now known as the Mandelbrot set. The Mandelbrot set is unique in its graphic mapping of recursive self-similarity applicable at all magnitudes examined. Its manifestation had a profound and immediate effect on scholars and artists worldwide, who found in the magnificent representations of nonlinear functions a new and incontrovertible paradigm for understanding the universal shapes and forms of the natural world.

Describing Chaos. The dynamics of complex systems are notoriously difficult to characterize, much less to control. Nevertheless, Mandelbrot's application of the concepts of the random walk, Brownian motion, diffusion-limited aggregation, galaxy clusters, fractal attractors, percolation, fractal nets and lattices, L-systems, box counting, and multi-fractal surface dimensions created new conceptual models for rationalizing chaotic systems. Novel cluster patterns are used in the research design of global feedback systems to regulate chaotic systems. In May, 2001, researchers at the Fritz Haber Institute of the Max Planck Society in Germany studying catalytic reactions announced the controlled design of repeating chemical clusters with fractal-like patterns. These patterns and formations anticipated new methods of controlling chemical systems turbulence.

Fractals and the Natural Sciences. The first conference on Fractals in Natural Sciences was held in Budapest, Hungary, from August 30 to September 2, 1993. The major subjects covered included topics related to DNA sequencing and biorhythms; complex bacterial colony formation; correlations of galaxy distribution; fractal tectonics, erosion, and river networks; the geometry of large diffusion-limited aggregates; the use of wavelets to characterize fractals beyond fractal dimension; fixed-scale transformations; diffusion properties of dynamical systems; transfers across fractal electrodes; crack branching; the fractal structure of electrodeposits; the vibrations of fractals; experiments with surface and kinetic roughness; self-organized criticality; granular segregation and pattern formation; and nonlinear ocean waves. The proceeds of the conference were published by the World Scientific Publishing Company in 1994 and provide an excellent starting point for understanding the immediate reception of fractal theory in the biological sciences, chemistry, earth science, and physics.

How It Works

The Feedback Loop. The iterated feedback loop is a fundamental concept of fractal geometry. A feedback loop is a transformational system where the input of a cycle (or mathematical function) gives a different output, which becomes the input of a new transformation. Feedback processes are fundamental tools for studying natural systems. Examples of iterated mathematical systems can be found in history over thousands of years. However, the capability of the feedback loop to generate complex systems was not fully understood until electronic and computational technologies were developed. The iterated feedback system is a standard process for generating geometric fractals. The function of an iterated object can be linear or nonlinear and can include affine, projective, and Möbius transformations. It is an essential component of the description of chaotic systems.

Fractal Dimension. Describing the dimensions of a given space was a topic that reordered the landscape of mathematics and physics in the nineteenth and twentieth centuries. Euclidean space is three-dimensional, a definition that refers to the range of motion of objects in a physical space. D = 1 refers to points on a straight line, D = 2 refers to points located on a plane, and D = 3 refers to points located within a cube.

Smoothness is an essential feature of Euclidean geometry. Idealized figures such as the cube, the sphere, and the triangle are abstractions that facilitate human understanding of fundamental spatial relationships. These concepts had enormous use in the applied arts and sciences and are the basic tools of technology. However, natural forms such as a flower, a mountain, or a riverbed defied classic mathematical descriptions. Mandelbrot's breakthrough was the realization that Euclidean parameters could not be used to define natural forms because they are irregular. Roughness is an essential characteristic of the natural world, and it requires a different set of mathematical tools for analysis.

Mandelbrot used the fraction as a description of the non-integer dimensions that pertain to natural surfaces. The dimensions of natural objects are fractional composites of the straight line, the plane, and the cube. That fractal dimension is summarized in the equation $d = \log(N)/\log(r)$, where r refers to a scaling factor that indicates the roughness of an object. When these values are coordinated on a log-log plot, the steepness of the slope of the line indicates the fractal dimension or roughness of the object.

Technologists and statisticians working with graphic variables describing biological and physical phenomena immediately recognized similarities in the results of time-series representations and the graphic plots of fractal analytics. Heartbeats, physical motions, geological processes, and the biological progression of species in a particular ecosystem could be defined using fractal nomenclature. Institutions all over the world use fractal analysis as a standard tool for interpreting physical systems. The BENOIT fractal analysis software, patented by TruSoft International, provides methods for measuring data sets that are chaotic and not amenable to traditional analytics. These include methods for plotting self-similar and self-affine fractal characteristics. Self-similar methods apply calculations of box dimension, perimeter-area dimension, information dimension, mass dimension, and the ruler method. Self-affine analytics require tools that calculate the scalar properties of the data aligned on the horizontal and vertical axes at constant sampling intervals. These include measurements of fractal Brownian motion, fractional Gaussian noise, power-spectral analysis, variogram analysis, and wavelet analysis.

Iterated Function Systems. Iterated function systems make it possible to re-create natural objects using mathematical descriptions collected into data sets with accompanying rules for computation. This process is also known as fractal image compression, and it is the technology that makes computer-generated landscapes and visual effects in film possible. Very simply, a table of numbers (or matrix formulation) is created describing the affine transformations desired. These follow the conventional order of scalings, reflections, rotations, and translations.

Applications and Products

In his introduction to *The Fractal Geometry of Nature*, Mandelbrot wrote: "More generally, I claim that many patterns of Nature are so irregular and fragmented, that, compared with *Euclid*—a term used in this work to denote all of standard geometry—Nature exhibits not simply a higher degree but an altogether different level of complexity. The number of distinct scales of length of natural patterns is for all practical purposes infinite."

Biological Diversity. Understanding biodiversity and the power laws that govern its rich complexity is one of the great challenges of science in the twenty-first century. Scaling relationships in nature provide a powerful paradigm for understanding how biological systems evolve and their relationship to the physical landscape. Environmental factors have a measurable influence on diffusion distributions of species.

The Earth Sciences. Computational molecular modeling is slowly transforming the metrics of geochemistry, making it possible to analyze the geometric structures and properties of materials at microscopic levels. Materials such as clays, composed of fine-grain silicates, are considered complex systems, the dynamics of which are still poorly understood. Water, too, is a substance the surface of which is difficult to characterize. Surface chemistries at the edge of a molecule are often different from that of a flat surface, a reality that challenges accurate manipulations of surface dynamics.

Chemical and Biochemical Processes. In physical chemistry and biochemistry, the fractal dimension of surface porosities is a remarkable statistic for the evaluation of processes affecting adsorption rates, chemical clustering, dispersion and uptake algorithms, spectroscopy, photochemistry, estimates of agglomeration dynamics and electrolyte deposition, the behavior of electrolytic dendrites, and the study of disordered systems and catalytic rates. Controlling for nanoscale surface roughness is an important part of developing effective thermal conductors for use in microelectronics. Similar systems analyses of the properties of granular surfaces are of particular importance to the understanding of the chemistry of interstellar dust particles and how ice crystals adhere to them, essential processes in the evolution of stars and galaxies.

Self-Organized Nanostructures. One of the remarkable properties of nanoparticles (NPs) is their capacity for self-organization. In the early twenty-first century, researchers noted that nanoparticles interact to form rings, linear chains, and hyperbranched (dentritic) structures. Of particular interest is the step-growth organization of NPs into polymer-like aggregates with varying types of isomers. These properties suggest the potential use of nanostructures in a variety of products. In 2008, researchers at the University of Wisconsin-Madison reported the creation of nanotrees, spiraling branched objects. Scientists are developing nanoforests with an array of structures. These have great promise for

Fascinating Facts About Fractal Geometry

- In February, 2011, researchers at the department of gerontology at the National Institute for Longevity Science in Aichi, Japan, and the Thayer School of Engineering in New Hampshire reported the use of a portable device that can model the gait of patients with Parkinson's disease, a progressive disorder of the central nervous system. A sensor attached to the device measures the patient's movements in three dimensions and then uses fractal analysis to assign a measure to the movement. This provides a parameter for understanding the progression of the disease. The higher the fractal measure, the more complex are the movements of the patient. A fractal dimension of 1.3 is given to walking patterns of people not afflicted with the disease, while Parkinson's patients have a value of 1.48 or higher.
- Researchers studying the power relationships of the metabolic rate and body mass of different animals were surprised to find a proportional increase with a fractal surface value of 2.25. This value reflects the remarkable efficiency of convoluted, space-filling physical systems with fractal properties on a three-dimensional plane that allows for the compact organization of transport services. Like the coastline of Great Britain, physiognomic surfaces have an infinite measure.
- In the laboratories of professor Eshel Ben-Jacob of Tel-Aviv University, in collaboration with professor Herbert Levine of University of California, San Diego's Center for Theoretical Biological Physics, bacterial forms grown in petri dishes display remarkable fractal organization, demonstrating principles of intelligent cooperation in response to a variety of environmental stressors.
- In July, 2010, physicists at Rice University in Houston, Texas; the Max Planck Institute for Chemical Physics of Solids and the Max Planck Institute for the Physics of Complex Systems, both in Dresden, Germany; and the Vienna University of Technology have reported that after seven years of research on high-temperature superconductors they have established quantum-critical scaling properties at work during the transition from one quantum phase to another. In experiments with a heavy-fermion metal containing ytterbium, rhodium, and silicon, researchers identified thermodynamic scaling properties as a result of a fermi-volume collapse.

revolutionizing the production of high-performance integrated circuits, biosensors, solar cells, light-emitting diodes (LEDs), and lasers.

Epigenetics. Evidence of self-organization at the nanoparticle level has provided new insights aligning the structural self-organization of life-forms from the symmetries of atomic structure through the allometric scalings of organisms to the forces arranging the patterns of stars. Nevertheless, continuing research into the dynamics of the epigenome reinforce the enormous complexity of the pathways of genetic inheritance, suggesting that genetic expression is influenced by a recursive and scaling array of multidimensional patterns affecting phenotype expression. Fractional dimension analytics are a valuable tool of bioinformatics and are used to measure the complexity of self-similar biological organisms and their relationship to other organisms in their environment. Time is an essential dimension of epigenetic models.

Software. TruSoft International's BENOIT fractal analysis system enables the user to measure the fractal dimension and the Hurst exponent of data sets using eleven different methods.

Impact on Industry

Properties of self-similarity and proportion are ubiquitous in the world around us and are deeply imprinted in society's cultures and practices. Understanding those principles and deliberately applying them to the design and production of human systems and product life cycles offer a compelling model for the development of thriving world cultures in balance with particular geographies and climates. Self-similarity, reiteration of motifs, and interesting transformations of properties are novel landscape tools for developing the natural aesthetics. Multidimensional fractal models of urban landforms and the structures and activities that occupy them both in the past and the present can be used to make critical macro-level evaluations of how human structures and production systems harmonize with the wider environment over time. Similar studies of fractal dimensionality can be used to evaluate agroforestry systems design and biochemical efficiencies. These virtual types of macro-level evaluations are essential for modeling a thriving world eco-economy based on sustainable design in harmony with particular locales and cultures.

The beauty of a fractal is its conceptual simplicity, one that even a child can understand. In the article "Science Starts Early" published in the February 25, 2011, issue of *Science*, author Frank C. Keil commented on the ability of infants and young children to make sophisticated judgments about causal relationships and patterns. The fractal is a magnificent tool for encouraging a child's thoughtful and informed understanding of the effectiveness of simple processes to create a range of effects. Persistence is an essential part of the effort to impress upon future generations the importance of understanding and respecting the complex dynamics of life systems. Industrial leaders are an important part of that process.

Careers and Course Work

Fractal geometry is a subject applicable to a broad range of academic subjects and careers. Dynamic geometry software programs such as Geometer's Sketchpad provide interactive tools for learning fractal structures at the high school level. At the university level, introductory fractals and chaos theory courses have been designed for and taught to undergraduate liberal arts majors. Chaos games, iterated function systems, fractional dimension, cellular automata, and artificial life are basic topics addressed. Familiarity with fractal software programs is essential for careers in which the product is the visual image, including photography, fashion design, graphic design, urban landscape design, the technologies of film and sound, medical imaging, scientific illustration, laser optics, and land topography mapping technologies.

Upper-level undergraduate and graduate courses in fractal mathematics require training in writing mathematical proofs, advanced calculus, metric topology, and measure theory. Fractal mathematics is an essential feature of Internet engineering design, economics, mechanical and electrical engineering, the physical and life sciences, computer-game design, electronics, telecommunications, hydrology, geography, demography, and other statistical disciplines.

Social Context and Future Prospects

In 1993, Benoit Mandelbrot was awarded the Wolf Foundation Prize for Physics. On April 25, 2003, he shared the Japan Prize for Science and Technology of Complexity with James A. Yorke, professor of mathematics and physics at the University of Maryland. Both men were cited for the creation of the science and technology of universal concepts in complex systems, namely fractals and chaos. These structures underlie complex phenomena in a wide range of fields, demonstrating the importance of understanding the behavior of systems as a whole as opposed to the reduction of phenomena into discrete elements for observation. Both the Wolf Foundation Prize and the Japan Prize were given in recognition of unique contributions to the progress of science and technology and the promotion of peace and prosperity to humankind.

Fractal theory continues to integrate human knowledge, providing a workable interface that links very different cognitive domains to examine complex physical phenomena in unusual ways. Computer technologies and interactive software programs search for ways to engage the evolutionary capacity of the human eye to assist in the processing of multiple and continuous streams of data. Fractal mappings and time series analyses are important new components of human logic. These are built on new realms of thinking about and modeling geometric shapes, setting the stage for the creation and manipulation of workable multidimensional geometries and design technologies.

Victoria M. Breting-García, B.A., M.A.

Further Reading

Brown, James H., et al. "The Fractal Nature of Nature: Power Laws, Ecological Complexity and Biodiversity." *Philosophical Transactions of the Royal Society B: Biological Sciences* 357, no. 1421 (May 29, 2002): 619-626. This essay provides a good working example of how fractal metrics are applied in primary field research.

Carter, Paul. "Dark With Excess of Bright: Mapping the Coastlines of Knowledge" in *Mappings*. Edited by Denis Cosgrove. London: Reaktion Books, 1999. This essay explores human interactions with coastlines and how that engagement stimulated a formal analysis of the processes governing the natural world. It is an excellent prequel to Mandelbrot's essay "How Long Is the Coast of Great Britain? Statistical Self-Similarity and Fractional Dimension."

Edgar, Gerald A., ed. *Classics on Fractals*. Boulder, Colo.: Westview Press, 2004. This highly regarded textbook is designed as an introduction to the theory and practice of fractal mathematics. It is a unique collection of primary mathematical documents describing non-linear mathematical concepts as they were understood in the nineteenth and early twentieth centuries.

Falconer, Kenneth. *Fractal Geometry: Mathematical Foundations and Applications.* 2d ed. Chichester, England: John Wiley & Sons, 2003. Introduces researchers and graduate-level students to the current mathematical models of fractal geometry. Includes substantial sets of notes and references to other foundational texts and essays.

Frame, Michael, and Benoit B. Mandelbrot, eds. *Fractals, Graphics, and Mathematical Education.* Washington, D.C.: Mathematical Association of America, 2002. An excellent description of how teachers at different levels of mathematical education are incorporating fractal theory into their mathematics curriculum.

Mandelbrot, Benoit B. *The Fractal Geometry of Nature.* San Francisco: W. H. Freeman, 1983. This is considered the foundational text in the development of fractal geometry.

_______. "How Long Is the Coast of Britain? Statistical Self-Similarity and Fractional Dimension." *Science* 156, no. 3775 (May 5, 1967): 636-638. This is one of several classic introductory essays written by Mandelbrot on metrics and fractional dimensions.

Peitgen, Heinz-Otto, Hartmut Jürgens, and Dietmar Saupe. *Chaos and Fractals: New Frontiers of Science.* 2d ed. New York: Springer-Verlag, 2004. This well-organized textbook explains fractal geometry and how it is used to describe chaotic or erratic systems.

Siegmund-Schultze, Reinhard. *Mathematicians Fleeing From Nazi Germany: Individual Fates and Global Impact.* Princeton, N.J.: Princeton University Press, 2009. Describes the magnitude of the intellectual and cultural transformations that occurred in academia in the wake of World War II.

Web Sites

American Mathematical Society
http://www.ams.org

Fractal Foundation
http://fractalfoundation.org

Mathematical Association of America
http://www.maa.org

See also: Calculus; Chaotic Systems; Computer Graphics; Computer Science; Geometry.

G

GAME THEORY

FIELDS OF STUDY

Economics; finance; political science; biology; psychology; decision science; mathematics.

SUMMARY

Game theory is a tool that has come to be used to explain and predict decision making in a variety of fields. It is used to explain and predict both human and nonhuman behavior as well as the behavior of larger entities such as nation states. Game theory is often quite complex mathematically, especially in new fields such as evolutionary game theory.

KEY TERMS AND CONCEPTS

- **Cooperative Game:** Game in which rational actors find a means to cooperate so as to enhance the utility of each.
- **Dominant Strategy:** Strategy that gives a player a higher payoff than other strategies no matter what the opponent does.
- **Evolutionary Game:** Game that changes over time as players' strategies change and as payoffs change. These games have multiple equilibria.
- **Multiple Equilibria:** Many potential equilibria that have no clearly superior strategies.
- **Nash Equilibrium:** Position in which both players have selected a strategy and neither player can change his or her strategy independently without ending up in a worse position.
- **Noncooperative Game:** Game in which rational actors will not cooperate even though doing so may be to their advantage.
- **Repeated Game:** Game that occurs through iterations, either indefinitely or with a finite ending point.
- **Utility:** Benefit that a player achieves from following a strategy.

Definition and Basic Principles

Game theory is a means of modeling individual decision making when the decisions are interdependent and when the actors are aware of this interdependence. Individual actors (which may be defined as people, firms, and nations) are assumed to maximize their own utility—that is, to act in ways that will provide the greatest benefits to them. Games take a variety of forms such as cooperative or noncooperative; they may be played one time only or repeated either indefinitely or with a finite ending point, and they may involve two or more players. Game theory is used by strategists in a variety of settings.

The underlying mathematical assumptions found in much of game theory are often difficult to understand. However, the explanations drawn from game theory are often intuitive, and people can engage in decision making based on game theory without understanding the underlying mathematics. The formal approach to decision making that is part of game theory enables decision makers to clarify their options and enhance their ability to maximize their utility.

Classical game theory has quite rigid assumptions that govern the decision process. Some of these assumptions are so rigid that critics have argued that game theory has few applications beyond controlled settings.

Later game theorists developed approaches to strategy that have made game theory applicable in many decision-making applications. Advances such as evolutionary game theory take human learning into account, and behavioral game theory factors emotion into decision making in such a way that it can be modeled and predictions can be made.

Background and History

The mathematical theory of games was developed by the mathematician John von Neumann and the economist Oskar Morgenstern, who published *The Theory of Games and Economic Behavior* in 1944. Various scholars have contributed to the development of game theory, and in the process, it has become useful for scholars and practitioners in a variety of fields, although economics and finance are the disciplines most commonly associated with game theory.

Some of the seminal work in game theory was done by the mathematician John Nash in the early 1950's, with later scholars building on his work. Nash along with John Harsanyi and Reinhard Selten received a Nobel Prize in Economic Sciences in 1994 for their work in game theory. The political scientist Thomas Schelling received a Nobel Prize in Economic Sciences in 2005 for his work in predicting the outcomes of international conflicts.

How It Works

Game theory is the application of mathematical reasoning to decision making to provide quantitative estimates of the utilities of game players. Implicit in the decision-making process is the assumption that all actors act in a self-interested fashion so as to maximize their own utilities. Rationality, rather than altruism or cooperation, is a governing principle in much of game theory. Actors are expected to cooperate only when doing so benefits their own self-interest.

Noncooperative Games. Most games are viewed as noncooperative in that the players act only in their own self-interest and will not cooperate, even when doing so might lead to a superior outcome. A well-known noncooperative game is the prisoners' dilemma. In this game, the police have arrested Moe and Joe, two small-time criminals, for burglary. The police have enough information to convict the two men for possession of burglar's tools, a crime that carries a sentence of five years in prison. They want to convict the two men for burglary, which carries a ten-year sentence, but this requires at least one man to confess and testify against the other. It would be in each man's best interest to remain silent, as this will result in only a five-year sentence for each. The police separate Moe and Joe. Officers tell Moe that if he confesses, he will receive a reduced sentence (three years), but if Joe confesses first, Moe will receive a harsher sentence (twelve years). At the same time, other officers give Joe the same options. The police rely on the self-interest of each criminal to lead him to confess, which will result in ten-year sentences for both men. If the various payoffs are examined, cooperation (mutual silence), which means two five-year sentences, is the most rewarding overall. However, if the two criminals act rationally, each will assume that the other will confess and therefore each will confess. Because each man cannot make sure that the other will cooperate, each will act in a self-interested fashion and end up worse off than if he had cooperated with the other by remaining silent.

As some game theorists such as Robert Axelrod have demonstrated, it is often advantageous for players to cooperate so that both can achieve higher payoffs. In reality, as the prisoners' dilemma demonstrates, players who act rationally often achieve an undesirable outcome.

Sequential Games. Many games follow a sequence in which one actor takes an action and the other reacts. The first actor then reacts, and so the game proceeds. A good way to think of this process is to consider a chess game in which each move is countered by the other player, but each player is trying to think ahead so as to anticipate his or her opponent's future moves. Sequential games are often used to describe the decision process of nations, which may lead to war if wrong reactions occur.

Sequential games can be diagramed using a tree that lists the payoffs at each step, or a computer program can be used to describe the moves. Some sequential games are multiplayer games that can become quite complicated to sort out. Work in game theory suggests that equilibrium points change at each stage of the game, creating what are called sequential equilibria.

Simultaneous Games. In some games, players make their decisions at the same time instead of reacting to the actions of the other players. In this case, they may be trying to anticipate the actions of the other player so as to achieve an advantage. The prisoners' dilemma is an example of a simultaneous game. Although rational players may sometimes cooperate in a sequential game, they will not cooperate in a simultaneous game.

Applications and Products

Game theory is most commonly used in economics, finance, business, and politics, although its applications have spread to biology and other fields.

Economics, Finance, and Business. Game theory was first developed to explain economic decision

Game Theory

Strategy for the Prisoners' Dilemma

	Prisoner B stays silent (cooperates)	Prisoner B confesses (defects)
Prisoner A stays silent (cooperates)	Each serves 1 month	Prisoner A: 1 year Prisoner B: goes free
Prisoner A confesses (defects)	Prisoner A: goes free Prisoner B: 1 year	Each serves 3 months

making, and it is widely used by economists, financial analysts, and individuals. For example, a firm may want to analyze the impact of various options for responding to the introduction of a new product by a competitor. Its strategists might devise a payoff matrix that encompasses market responses to the competitor's product and to the product that the company introduces to counter its competitor. Alternatively, a firm might prepare a payoff matrix as part of decision process concerning entry into a new market.

Game theory cannot be used to predict the stock market. However, some game theorists have devised models to explain investor response to such events as an increase in the interest rate by the Federal Reserve Board. At the international level, there are various games that can be used to explain and predict the responses of governments to the imposition of various regulations such as tariffs on imports. Because of the large number of variables involved, this sort of modeling is quite complex and still in its infancy.

Some businesses might be tempted to develop a game theoretic response (perhaps using the prisoners' dilemma) to the actions of workers. For example, a company can develop a game theoretic response to worker demands for increased wages that will enable the company to maximize its utility in the negotiating process. Even if a game does not play out exactly as modeled, a company gains by clarifying its objectives in the development of a formal payoff matrix. Companies can also develop an approach to hiring new employees or dealing with suppliers that draws on game theory to specify goals and strategies to be adopted.

Politics. Some of the most interesting work in game theory has occurred in explaining developments in international affairs, enabling countries to make better decisions in the future. Game theory is often used to explain the decision process of the administration of President John F. Kennedy during the Cuban Missile Crisis. Other game theoretic explanations have been developed to examine a country's decision to start a war, as in the Arab-Israeli conflicts. Game theorists are able to test their formal game results against what actually occurred in these cases so as to enhance their models' ability to predict.

Other game theorists have devised game theoretic models that explain legislative decision making. Much of the legislative process can be captured by game theoretic models that take into account the step-by-step process of legislation. In this process, members from one party propose legislation and the other party responds, then the first part often responds, and the responses continue until the legislation is passed or defeated. Most of these models are academic and do not seem to govern legislative decision making, at least not explicitly.

Biology. Some evolutionary biologists have used game theory to describe the evolutionary pattern of some species. One relationship that is often described in game theoretic terms is the coevolution of predators and prey in a particular area. In this case, biologists do not describe conscious responses but rather situations in which a decline in a prey species affects predators or an increase in predators leads to a decline in prey and a subsequent decline in predators. Biologists use this relationship (called the hawk-dove game by the biologist John Maynard Smith) to show how species will evolve so as to better fit an evolutionary niche, such as the development of coloration that enables prey to better conceal itself from predators.

Impact on Industry

Much initial work in game theory was done by academics and was not readily adopted by practitioners. Over time, business and political leaders have come to understand that the application of game theoretic principles could possibly enhance their decision making. For example, in the 1950's, organizations such as the RAND Corporation developed game theoretic models that created different scenarios for how the United States should respond to the potential of nuclear war with the Soviet Union. These models were quite cumbersome and their predictability was questioned because of the rigidity of the models and the difficulties of running extensive computer simulations. More powerful computers and the evolution of game theory (including behavioral game theory and the recognition of sequential equilibria) have enabled game theorists to develop quite sophisticated simulation models, some of which can be run on laptop computers, although some international relations and financial models still require supercomputers.

Simple Interpersonal Models. Some businesses have applied the formal approach to decision making used in game theory to labor relations or in developing pay scales. Many of these simple models can be worked out quickly so that executives can make quick decisions. Executives trained in game theory can follow a somewhat formal approach that enables them to evaluate the impact of their decisions in a clear fashion. Of course, employees can also counter using game theoretic logic. The use of ad hoc game theory is becoming commonplace in many businesses.

Game Theoretic Simulations. Assisted by powerful computers that can combine massive amounts of data in multitudes of scenarios, decision makers use game theory in making major decisions that affect market share, profit, and the long-term success of firms. Some financial institutions use game theory to consider the impact of financial instruments such as derivatives. Analysts in such governmental agencies as the Congressional Budget Office use game theoretic simulations to study the potential impact of legislation such as quotas or tariffs or of providing assistance to various domestic industries. In these cases, game theoretic approaches are combined with econometric models to examine the effect of various possible courses of action. National security analysts use various games to predict the responses of foreign countries to moves made by the United States. In one case, analysts used theory to examine responses that U.S. policy makers might follow in responding to an attack with a biological weapon.

Improved Decision Making. Because nations and businesses do not act in isolation, they are starting to use game theory to predict the responses of other nations and businesses. These models are complex, but they still follow the same principles found in simple games such as the prisoners' dilemma. The development of evolutionary game theory has enabled game theorists to better predict the outcomes of games with repeated rounds that have multiple equilibria, as have the use of concepts from psychology.

Fascinating Facts About Game Theory

- Although the motion picture *A Beautiful Mind* (2001) is a fascinating account of John Nash's life, including his work in game theory and his struggles with mental illness, it is not always an accurate introduction to game theory.
- Most game theory texts have at least one example drawn from domestic relations, often concerning how a husband and wife might go about reaching a decision regarding a night out so as to maximize their utilities and reduce domestic discord.
- The common sense nature of game theory (in this case, a cooperative game) can be used to explain why soldiers in the trenches on both sides of the front in World War I refrained from firing at each other at certain times of the day such as mealtimes.
- Most games are not true zero-sum games in which one player wins and the other loses but instead have mixed payoffs for players.
- Game theory is governed by the garbage-in, garbage-out dictum in that poor initial assumptions lead to flawed predicted outcomes.

Careers and Course Work

Game theory can be used by individuals to help them to make better decisions, but it is also used to develop business or political strategy, to predict aspects of financial markets, and to describe evolutionary processes in biology. Many people practice game theory, often without knowing it. Formal game theory, driven by extensive mathematical modeling and computer applications, is used in industry and government to help guide decision making. Only a few mathematicians are full-time game theorists; most game theorists are people who use game theory to enhance decisions or to describe decision processes.

Although game theory is a field of applied mathematics, many graduate programs in social sciences require students to familiarize themselves with at least the basics. Students planning on advanced work in game theory can benefit from courses in statistics and formal logic, as well as any computer programming and any course work that enhances their ability to deal with quantitative material. As behavioral game theory develops, gaining knowledge of psychology (and possibly neuroscience) will become important for some applications. Many game theorists, however, are self-taught.

Social Context and Future Prospects

Game theory is an evolving field that can become esoteric and divorced from the realities of practical decision making. It can also be an intuitive, essentially nonmathematical approach to enhancing decision making. Both formal and informal game theoretic approaches are likely to be used in business and government in the future. However, most observers agree that sophisticated actors who are aware of the principles of game theory are likely to prevail over those who follow an ad hoc approach to decision making.

Game theory is not a perfect guide to decision making. At times, it has led to overly simplistic approaches that are derived from a narrow view of utility. With the introduction of newer conceptual frameworks, game theory has become less rigid and better able to model human decision making. In the future, sophisticated computer simulations based on game theory are likely to be used for corporate and governmental decision making.

John M. Theilmann, Ph.D.

Further Reading

Binmore, K. G. *Game Theory: A Very Short Introduction.* New York: Oxford University Press, 2007. A renowned game theorist explains game theory in a nontechnical, entertaining manner, looking at famous figures such as John Nash and applications in many areas, including drivers negotiating heavy traffic.

Bueno de Mesquita, Bruce. *The Predictioneer's Game: Using the Logic of Brazen Self-Interest to See and Shape the Future.* New York: Random House, 2009. Describes various predictive applications of game theory without delving deeply into mathematics. Focuses on rationality and people's interests and beliefs.

Fisher, Len. *Rock, Paper, Scissors: Game Theory in Everyday Life.* New York: Basic Books, 2008. A scientist examines game theory by using its strategies in everyday life, such as a crowded supermarket and congested Indian roads. His emphasis is on cooperation.

Gintis, Herbert. *Game Theory Evolving: A Problem-Centered Introduction to Modeling Strategic Behavior.* 2d ed. Princeton, N.J.: Princeton University Press, 2009. Looks at how game theory is moving from a formal study of rational behavior to a way to study behavior (not always rational) in social situations such as teams.

Miller, James. *Game Theory at Work: How to Use Game Theory to Outthink and Outmaneuver Your Competition.* New York: McGraw-Hill, 2003. Basic work with business applications, based on a strong perspective from neoclassical economics.

Web Sites

Economic and Social Research Council
Centre for Economic Learning and Social Evolution
http://else.econ.ucl.ac.uk/newweb/index.php

Game Theory Society
http://www.gametheorysociety.org

See also: Applied Mathematics; Risk Analysis and Management.

GEOINFORMATICS

FIELDS OF STUDY

Calculus, analytic geometry, trigonometry, computer science, earth science, statistics, geography, mapmaking, digital imaging and graphics design, database management systems.

SUMMARY

The term "geoinformatics" refers to a collection of information systems and technologies used to create, collect, organize, analyze, display, and store geographic information for specific end-user applications. The field represents a paradigm shift from traditional discipline-based systems such as cartography, geodesy, surveying, photogrammetry, and remote sensing to a data systems management protocol that includes all earlier technologies and combines them to create new models of spatial information. Computation is an essential foundation of all geoinformatics systems.

KEY TERMS AND CONCEPTS

- **Datum:** Starting point from which a system of measurements of the Earth's size is computed.
- **Geodatabase:** Essential structure of a geographic information system (GIS) including selected geographic data sets, particular object definitions, and descriptions of relationships. Information is stored in a variety of formats and includes nominal, ordinal, interval, ratio, and scalar data.
- **Geographic Data Model:** Core component of a GIS system. The data model provides a geographic template for a particular industry application.
- **Geographic Information System (GIS):** Collection of computer hardware, software, and geographic data used to reference spatial geographic information. It is an analytic system to input, edit, manipulate, study, compare, and output geographic data collected from a variety of sources.
- **Geography:** Study of the Earth's surface and climate and the interrelations between them.
- **Geoid:** Physical model of the Earth. It is recognized by its irregular shape.
- **Geoprocessing:** Refers to a set of operations that can be selected to create particular geographic data outputs. These include software applications for file conversions, geographic data overlays, intersection and union of particular features, data extraction, proximity analytics, and user-facilitated data management.
- **Map:** Visual representation of the surface of the Earth or other celestial body.
- **Map Projection:** Projection of a three-dimensional figure onto a flat two-dimensional plane.
- **Query:** Carefully structured data search format designed to collect particular information from a database set.
- **Raster Data:** Digital information stored on a grid of pixels assigned a particular value. These data sets create numeric data well suited for the storage of continuous data collected over large regions.
- **Scale:** Numerical statement of the mathematical relationship between a unit of distance on the ground and its cartographic representation.
- **Spatial Cognition:** Cognitive ability to make judgments about spatial patterns and relationships.
- **Topology:** Mathematical study of objects the shapes of which are distorted under manipulation. For example, a circle can be stretched into the shape of an ellipse.
- **Vector Model:** Geographic data with lines and shapes plotted on an x-y coordinate system.

Definition and Basic Principles

Geoinformatics is a complex, multidisciplinary field of knowledge specializing in the creation, collection, storage, classification, manipulation, comparison, and evaluation of spatially referenced information for use in a variety of public and private practices. Its technologies are rooted in mapping, land surveying, and communication technologies that are thousands of years old. The exponential growth of science-based knowledge and mathematical expertise during the nineteenth and twentieth centuries greatly assisted the accumulation of verifiable geographic information describing the Earth and its position among the myriad celestial entities occupying the known universe. Detailed geometric descriptions and photographic materials make it possible to translate, measure, and order the surface of the Earth into multidimensional coordinate systems that provide a rich and detailed visual language for understanding the relationships and features of locations too vast to be easily comprehended by the human senses. Geographic communication systems are greatly enhanced by the proliferation of computation technologies including database management systems, laser-based surveys, digital satellite photo technologies, and computer-aided design (CAD).

Background and History

The components and design structures of maps form the fundamental language of geographic information systems (GIS). Maps illustrate a variety of environments, both real and imagined, and the structures and life forms that fill them. The geographies of the Earth's land masses and the relationships of land and water to the Moon and stars are the foundations of advanced mathematics and the physical sciences, subjects that are continually modified by new measurements of spatial-temporal coordinates. Aerial and nautical photography introduced a new and vital reality to cartographic representation in the twentieth century. Advances in sonar and radar technologies, the telephone and telegraph, radio broadcast, mass transportation, manned space flight, spectrometry, the telescope, nuclear physics, biomedical engineering, and cosmology all rely on the power of the map to convey important information about the position, structure, and movement of key variables.

Cartography. Mapping is an innate cognitive ability; it is common in a variety of animate species. In human practice, mapping represents an evolutionary process of symbolic human communication. Its grammar, syntax, and elements of style are composed of highly refined systems of notation, projections, grids, symbols, aesthetics, and scales. These time-honored features and practices of cartographic representation demand careful study and practice. Maps are the most ancient documents of human culture and civilization and, as such, many are carefully preserved by private and public institutions all over the world. The World Wide Web has made it possible to share, via the Internet, facsimiles of these precious cultural artifacts; the originals are protected for posterity. Scholars study the intellectual and technical processes used to create and document maps. These provide valuable clues about the beliefs and assumptions of the cultures, groups, and individuals that contract and prepare them.

Geodesy. The cartographer artfully translates the spatial features of conceptual landscapes into two- and three-dimensional documents using data and symbols selected to illustrate specific relationships or physical features of a particular place for the benefit or education of a group or individual enterprise. Geodesy is an earth science the practitioners of which provide timely geographical measurements used by cartographers to create accurate maps. Since earliest recorded times, geodesists have utilized the most current astronomical and geographical knowledge to measure the surface of the Earth and its geometric relationship to the Sun and Moon. The roots of geodetic measurement systems are buried in the ancient cultures and civilizations of Egypt, China, Mesopotamia, India, and the Mediterranean. Enlightened scholars and astute merchants of land and sea traveled the known world and shared manuscripts, instruments, personal observation, and practical know-how for comprehending the natural world. Hellenic scholars brilliantly advanced the study of astronomy and the Earth's geography. Their works formed the canon of cartographical and astronomical theory in the Western world from the time of the great voyages of discovery in the fifteenth and sixteenth centuries.

Alexandria was the intellectual center of the lives of the Greek polymath Eratosthenes of Cyrene and the Roman astronomer Ptolemy. Both wrote geographical and astronomical treatises that were honored and studied for centuries. Eratosthenes wrote *Peri tes avametreoeos tes ges* (On the Measurement of the Earth) and the three-volume *Geographika* (Geography), establishing mathematics and precise linear measurements as necessary prerequisites for the accurate geographical modeling of the known world. His works are considered singular among the achievements of Greek civilization. He is particularly noted for his calculations of the earth's circumference, measurements that were not disputed until the seventeenth century.

Ptolemy, following Eratosthenes, served as the librarian at Alexandria. He made astronomical observations from Alexandria between the years 127 and 141, and he was firm in his belief that accurate geographical maps were derived from the teachings of astronomy and mathematics. The *Almagest* is a treatise devoted to a scientific, methodical description of the movement of the stars and planets. The beauty and integrity of his geocentric model set a standard for inquiry that dominated astronomy for centuries. He also wrote the eight-volume *Geographia* (Geography), in which he established latitude and longitude coordinates for the major landmasses and mapped the world using a conic projection of prime and linear meridians and curved parallels. His instructions for creating a coordinate system of mapping, including sectional and regional maps to highlight individual countries, are still in practice.

At the end of the nineteenth century, geophysics became a distinct science. Its intellectual foundations provided rising petroleum corporations new technologies for identifying and classifying important land features and resources. In the United States, extensive surveys were conducted for administrative and military purposes. The U.S. Coast Survey was established by President Thomas Jefferson in 1807. The Army Corps of Engineers, the Army Corps of Topographical Engineers, and the United States Naval Observatory supported intensive geophysical research including studies of harbors and rivers, oceans and land topographies. In 1878, the U.S. Coast Survey became the U.S. Coast and Geodetic Survey. In 1965, the U.S. Coast and Geodetic Survey was reincorporated under the Environmental Sciences Services Administration, and in 1970 it was reorganized as the National Oceanic and Atmospheric Administration (NOAA).

Underwater acoustics and geomagnetic topographies were critical to the success of naval engagements during World War I and World War II. The International Union of Geodesy and Geophysics (IUGG) was founded in 1919. As of 2011, it is one of thirty scientific unions participating in the International Council for Science (ICSU). That same year the American National Committee of the International Union of Geodesy and Geophysics combined with the Committee on Geophysics of the National Research Council; in 1972 the committee was independently organized as the American Geophysical Union. Graduate geophysics programs became prominent in major universities after World War II.

Geodesy is a multidisciplinary effort to calculate and document precisely the measurements of the Earth's shape and gravitational field in order to define accurate spatial-temporal locations for points on its surface. Land surveys and geomensuration, or the measure of the Earth as a whole, are essential practices. Geodetic data is used GIS—these include materials from field surveys, satellites, and digital maps. The Earth's shape is represented as an ellipsoid in current mathematical models. Three-dimensional descriptors are applied to one quadrant of the whole in a series of calculations. All cartographic grid systems and subsequent measurements of the Earth begin with a starting point called a datum. The first datum was established in North America in 1866 and its calculations were used for the 1927 North American Datum (NAD 27). In 1983, a new datum was established (NAD 83) and it is the basis of the standard geodetic reference system (GRS 80). It was again modified and became the World Geodetic System in 1984 (WGS 84). WGS 84 utilizes constant parameters for the definition of the Earth's major and minor axes, semimajor and semiminor axes, and various ratios for calculating the flattening at the poles. The geoid is another geodetic representation of the Earth, as is the sphere.

Photogrammetry. The word photogrammetry refers to the use of photographic techniques to produce accurate three-dimensional information about the topology of a given area. Accurate measurements of spaces and structures are obtained through various applications including aerial photography, aerotriangulation, digital mapping, topographic surveys, and database and GIS management. Precise, detailed photographs make it possible to compare and analyze particular features of a given environment. Aerial photographs are particularly useful to engineers, designers, and planners who need visual information about the site of a project or habitat not easily accessible by other means. The International Society for Photogrammetry and Remote Sensing (ISPRS) advances the knowledge of these technologies in more than 100 countries worldwide.

Remote Sensing. Like photogrammetry, remote sensing is an art, science, and technology specializing in the noncontact representation of the Earth and the environment by measuring the wavelengths of different types of radiation. These include passive and active technologies, both of which collect data from natural or emitted sources of electromagnetic energy. These include cosmic rays, X rays, ultraviolet light, visible light, infrared and thermal radiation, microwaves, and radio waves. Aerial photography and digital imaging are traditional passive remote

Fascinating Facts About Geoinformatics

- The 2011 State of the Birds report is the first of its kind to document the value of public lands to the stability of migratory birds in North America. The report was prepared by combining data sets provided by the National Gap Analysis Program (GAP) at the University of Idaho, the U.S. Geological Survey, and the Cornell Lab of Ornithology. The Cornell Lab Bird Database and Species Distribution Modeling programs provided geographic data representing 600,000 bird checklists collected from 107,000 locations. This information was used as an overlay correlated to the GAP Protected Area Database of the United States. (PAD-US). Combined, these data sets allowed researchers to use statistics to evaluate the relationship of public land management to the stability of bird populations in those areas.
- Peak Freaks has been providing guided expeditions of high-altitude mountains for twenty years. This year's Everest 2011 team is collaborating with 3D Reality Maps to create highly detailed three-dimensional mappings of Mount Everest. The technology used was created by DigitalGlobe, a global provider of high-resolution imagery, and the German Aerospace Center (DLR).
- GIS technologies can be used to create simulations of crowds and their movements in particular locations during different types of community events. These simulations include geometric descriptions and sets of rules of engagement. Such descriptions are particularly useful for planners of street festivals and carnivals, shopping malls, art galleries, and coordinators of large religious pilgrimages and ceremonies.
- The Magellan space probe was launched from the space shuttle in 1989. It conducted six 243-day cycles around the planet Venus, using synthetic aperture radar (SAR) image-mapping technologies to create detailed maps of the planet's surface. The fourth, fifth, and sixth orbits created mappings of the planet's gravitational field using radio signals to Earth. These images have provided detailed visual records of the geological dynamics of the planet including studies of volcanoes, lava flows, and terrestrial tectonic movements.
- The Landsat Program is jointly administered by the National Aeronautics and Space Administration (NASA) and the U.S. Geological Survey. Its satellites are used to provide digital photographs of the Earth's surface. The first Landsat satellite was launched on July 23, 1972. The Landsat Data Continuity Mission (LDCM) is scheduled to launch in December, 2012, and will provide essential data for the study of the Earth's climate and ecosystems.

sensing technologies based on photographic techniques and applications. Later applications include manned space and space shuttle photography and Landsat satellite imagery. Radar interferometry and laser scanning are common examples of active remote sensing technologies. These products are used for the documentation of inaccessible or dangerous spaces. Examples include studies of particular environments at risk or in danger.

Global Navigation Satellite Systems (GNSS). Satellites and rapidly advancing computer technologies have transformed GIS products. Satellite systems of known distance from the Earth receive radio transmissions, which are translated and sent back to Earth as a signal giving coordinates for the position and elevation of a location. Navstar is an American Global Positioning System (GPS) originating from a World War II radio transmission system known as loran, which is an acronym for long-range navigation. It is the only GNSS system in operation. On April 27, 2008, the European Space Agency announced the successful launch of the second Galileo In-Orbit Validation Element (GIOVE-B), one of thirty identical satellites planned to complete a European satellite navigation system similar to the Global Positioning System in use in the United States and the Global Navigation Satellite System (GLONASS) in Russia. Students can monitor the activities of remote sensing satellite systems such as the Landsat, Seasat, and Terra satellites used to create new map profiles of terrestrial and extraterrestrial landscapes, many of which are available online for study. These satellite systems are used to collect data including measurements of the Earth's ozone shield, cloud mappings, rainfall distributions, wind patterns, studies of ocean phenomena, vegetation, and land-use patterns.

How It Works

Geographic information systems are built on geometric coordinate systems representing particular locations and terrestrial characteristics of the Earth or other celestial bodies. The place and position of particular land features form the elementary and most regular forms of geographic data. Maps provide visual information about key places of interest and structural features that assist or impede their access. Land-survey technologies and instrumentation are ancient and are found in human artifacts and public records that are thousands of years old. They are essential documents of trade, land development, and warfare. Spatial coordinates describing the Moon and stars and essential information about particular human communities are chinked into rocks, painted on the walls of caves, and hand-printed on graphic media all over the world. Digital photographs, satellite data, old maps, field data, and measurements provide new contexts for sharing geographic information and knowledge about the natural world and human networks of exchange.

GIS devices include high-resolution Landsat satellite imagery, light detecting and ranging (lidar) profiles, computer-aided design (CAD) data, and database management systems. Cross-referenced materials and intricately detailed maps and overlays create opportunities for custom-designed geographic materials with specific applications. Data that is created and stored in a GIS device provide a usable base for building complex multi-relational visualizations of landscapes, regions, and environments.

How a GIS component is used depends on the particular applications required. Service providers will first conduct a needs assessment to understand what information will be collected, by whom, and how it will be used by an individual or organization. The careful design of relationships connecting data sets to one another is an essential process of building an effective GIS system. Many data sets use different sets of coordinates to describe a geographical location and so algorithms need to be developed to adjust values that can have significant effects on the results of a study.

Depending on the needs of an organization, the rights to use some already-established data sets can be acquired. Other data can be collected by the user and stored in appropriate data files. This includes records already accumulated by an individual or organization. Converting and storing records in GIS format is particularly useful for creating the documents needed for a time-series analysis of particular land features and regional infrastructures. Digital records protect against the loss of valuable information and create flexible mapping models for communicating with internal and external parties.

Applications and Products

Satellite and computer technologies have transformed the way spatial information is collected and communicated worldwide. With exponential increases in speed and detail, satellites provide continuous streams of information about Earth's life systems. More than fifty satellites provide free access to GPS coordinates. These are used on a daily basis by people from all walks of life, providing exceptional mapping applications for determining the location and elevation of roadways and waterways.

The Environmental Systems Research Institute (ESRI) is a pioneer in the development of GIS landscape-analysis systems. This includes the development of automated software products with Web applications. Users can choose from a menu of services including community Web-mapping tools; ArcGIS Desktop with ArcView, ArcEditor, and ArcInfo applications; and two applications for use in educational settings, ArcExplorer Java Edition for Education (AEJEE) and Digital Worlds (DW). ESRI software is an industry standard used worldwide by governments and industries. The ESRI International User Conference is attended by thousands of users and is one of the largest events of its kind.

The Global Geodetic Observing System (GGOS) is another application developed by the International Association of Geodesy. It provides continuing observations of the Earth's shape, gravity field, and rotational motion. These measurements are integrated into the Global Earth Observing System of Systems (GEOSS), an application that provides high-quality reference materials for use by groups such as the Group on Earth Observations and the United Nations.

Impact on Industry

The Open Geospatial Consortium (OGC) comprises 417 corporate, government, and university entities working to establish freely available standards for use by private and commercial institutions. Electronic location resources and technologies continue to be integrated into market-driven applications in industry and government. These include three-dimensional information systems; information processing standards for architecture, engineering, and construction; building information models for life-cycle manufacturing, defense and intelligence systems, homeland security, disaster management, and emergency services; sustainable natural resource and ecosystem management; mass market Web services; and university research. The OGC Web site lists all of its members with e-mail and Web links to their resources.

Careers and Course Work

Course work and certification in geoinformatics at the undergraduate and graduate levels can be completed in tandem with course work in other earth science and engineering fields. Courses include selections of introductory materials, computer programming, database management and design, statistics, bioinformatics, geostatistics, remote sensing, various foundational courses in mapping techniques and spatial analysis, and computer laboratory exercises to gain familiarity with a number of GIS software programs. Some course work will emphasize applications in land development, transportation analysis, environmental science, public health, and a variety of engineering and architectural design schematics.

Careers specializing in geoinformatics technologies are directly involved in data-driven hardware and software applications. Practitioners must master the organizational and design skills necessary to produce highly detailed, error-free maps and related documents for use in private industry and government-related agencies. Familiarity with field research and land-survey technologies are also desirable.

Social Context and Future Prospects

As a result of the World Wide Web, computer and telecommunications technologies continue to provide novel platforms for connecting individuals, groups, and communities worldwide. Location is an essential feature of such networks, and geographic information systems are needed to provide timely spatial information for individual and cooperative ventures. The safety and integrity of global communications systems and data systems continue to challenge political values in free and democratic societies. Nevertheless, geographic information systems will continue to be integrated into a variety of applications, contributing to the safety of individuals, the integrity of the world's natural resources, and the profitability of enterprises around the globe.

Victoria M. Breting-García, B.A., M.A.

Further Reading

Bender, Oliver, et al., eds. *Geoinformation Technologies for Geocultural Landscapes: European Perspectives.* Leiden, the Netherlands: CRC Press, 2009. These essays explain the many ways geoinformation technologies are used to model and simulate the changing reality of landscapes and environments.

DeMers, Michael N. *Fundamentals of Geographic Information Systems.* 4th ed. Hoboken, N.J.: John Wiley & Sons, 2009. Provides useful historical background for the technologies and systems presented. Each chapter ends with a review of key terms, review questions, and a helpful list of references for further study.

Galati, Steven R. *Geographic Information Systems Demystified.* Boston: Artech House, 2006. Introductory textbook provides graphic examples of the concepts presented and includes a very useful glossary of terms.

Harvey, Francis. *A Primer of GIS: Fundamental Geographic and Cartographic Concepts.* New York: Guildford Press. 2008. This well-organized and clearly written textbook presents a thorough review of GIS with questions for analysis and vocabulary lists at the end of every chapter.

Kolata, Gina Bari. "Geodesy: Dealing with an Enormous Computer Task." *Science* vol. 200, no. 4340 (April 28, 1978): pp. 421-466. This article explains the enormous task of computing the 1983 North American Datum.

Konecny, Gottfried. *Geoinformation: Remote Sensing, Photogrammetry and Geographic Information Systems.* London: Taylor & Francis, 2003. This detailed introduction to geoinformation includes hundreds of illustrations and tables, with a chapter-by-chapter bibliography.

Rana, Sanjay, and Jayant Sharma, eds. *Frontiers of Geographic Information Technology.* New York: Springer, 2006. This collection of essays covers rising geographic technologies. Chapter 1: "Geographic Information Technologies—An Overview" and Chapter 4: "Agent-Based Technologies and GIS:

Simulating Crowding, Panic, and Disaster Management" are of note.

Scholten, Henk J., Rob van de Velde, and Niels van Manen, eds. *Geospatial Technology and the Role of Location in Science.* New York: Springer, 2009. The essays collected in this volume represent a series of presentations made at the September, 2007, conference held by the Spatial Information Laboratory of the VU University Amsterdam. They address the use of geospatial technologies at the junction of historical research and the sciences.

Thrower, Norman J. W. *Maps and Civilization: Cartography in Culture and Society.* 3d ed. Chicago: University of Chicago Press. 2007. This book is an excellent introduction to the field of cartography.

Web Sites

American Geophysical Union
http://www.agu.org

Geodetic Systems
Photogrammetry Information
http://www.geodetic.com/photogrammetry.htm

International Union of Geodesy and Geophysics
http://www.iugg.org

Library of Congress
Map Collections
http://memory.loc.gov/ammem/gmdhtml/gmdhome.html

University of California, Berkeley, Library
Earth Sciences & Map Library
http://www.lib.berkeley.edu/EART/browse.html

See also: Calculus; Computer Science; Geometry; Plane Surveying; Trigonometry.

GEOMETRY

FIELDS OF STUDY

Civil engineering; architecture; surveying; agriculture; environmental and conservation sciences; mechanical engineering; computer-aided design; molecular design; nanotechnology; physical chemistry; biology; physics; graphics; computer game programming; textile and fabric arts; fine arts; cartography; geographic information systems; Global Positioning Systems; medical imaging; astronomy; robotics.

SUMMARY

Geometry, which literally means "earth measurement," is absolutely critical to most fields of physical science and technology and especially to any application that involves surfaces and surface measurement. The term applies on all scales, from nanotechnology to deep space science, where it describes the relative physical arrangement of things in space. Although the first organized description of the principles of geometry is ascribed to the ancient Greek philosopher Euclid, those principles were known by others before him, and certainly had been used by the Egyptians and the Babylonians. Euclidean, or plane, geometry deals with lines and angles on flat surfaces (planes), while non-Euclidean geometry applies to nonplanar surfaces and relationships.

KEY TERMS AND CONCEPTS

- **Conic Section:** Geometric form that can be defined by the intersection of a plane and a cone; examples are a circle, an ellipse, a parabola, a hyperbola, a line, and a point.
- **Euclidean:** Describing geometric principles included in the range of principles described by Euclid.
- **Fibonacci Series:** Geometric sequence described by Italian mathematician Leonardo of Pisa (known as Fibonacci) in about 1200, beginning with zero and in which each subsequent number is the sum of the previous two: 0, 1, 1, 2, 3, 5, 8, 13, 21, 34, 55, . . .
- **Geometric Isomers:** Chemical term used to denote molecular structures that differ only by the relative arrangement of atoms in different molecules.
- **Golden Ratio:** Represented as φ or Φ, a seemingly ubiquitous naturally occurring ratio or proportion having the approximate value 1.618018513; also known as the golden mean, the golden section, and the divine proportion.
- **Pi:** Represented as π, an irrational number that is equivalent to the ratio of the circumference of a circle to its diameter, equal to about 3.14159265358979323846264 (the decimal part is believed to be nonrepeating and indeterminate, or unending).
- **Plane:** Flat surface that would be formed by the translation of a line in one direction.
- **Polygon:** Two-dimensional, or plane, geometric shape having a finite number of sides; designated as a regular polygon if all sides are of equal length.
- **Polyhedron:** Three-dimensional geometric shape having a finite number of planar faces; designated as a regular polyhedron if all faces are equivalent.
- **Postulate:** Rule that is accepted as true without proof.
- **Pythagorean:** Describing a relation to the theorem and principles ascribed to Pythagoras, especially the properties of right triangles.
- **Theorem:** Rule that is accepted as true but requires a rigorous proof.
- **Torus:** Structure formed by the translation of a circle through space along a path defined by a second circle whose plane is orthogonal to that of the first circle.

Definition and Basic Principles

Geometry is the branch of mathematics concerned with the properties and relationships of points, lines, and surfaces, and the space contained by those entities. A point translated in a single direction describes a line, while a line translated in a single direction describes a plane. The intersections and rotations of various identities describe corresponding structures that have specific mathematical relationships and properties. These include angles and numerous two- and three-dimensional forms. Geometric principles can also be extended into realms encompassing more than three dimensions, such as with the incorporation of time as a fourth dimension in Albert Einstein's space-time continuum. Higher dimensional analysis is also possible and is the subject of theoretical studies.

The basic principles of geometry are subject to various applications within different frames of reference. Plane geometry, called Euclidean geometry after the ancient Greek mathematician Euclid, deals with the properties of two-dimensional constructs such as lines, planes, and polygons. The five basic principles of plane geometry, called postulates, were described by Euclid. The first four are accepted as they are stated and require no proof, although they can be proven. The fifth postulate differs considerably from the first four in nature, and attempts to prove or disprove it have consistently failed. However, it gave rise to other branches of geometry known as non-Euclidean. The first four postulates apply equally to all branches of Euclidean and non-Euclidean geometry, while each of the non-Euclidean branches uses its own interpretation of the fifth postulate. These are known as hyperbolic, elliptic, and spherical geometry.

The point, defined as a specific location within the frame of reference being used, is the common foundation of all branches of geometry. Any point can be uniquely and unequivocally defined by a set of coordinates relative to the central point or origin of the frame of reference. Any two points within the frame of reference can be joined with a single line segment, which can be extended indefinitely in that direction to produce a line. Alternatively, the movement of a point in a single direction within the frame of reference describes a line. Any line can be translated in any orthogonal direction to produce a plane. Rotation of a line segment in a plane about one of its end points describes a circle whose radius is equal to the length of that line segment. In any plane, two lines that intersect orthogonally produce a right angle (an angle of 90 degrees). These are the essential elements of the first four Euclidean postulates. The fifth, which states that nonparallel lines must intersect, could not be proven within Euclidean geometry, and its interpretation under specific conditions gives rise to other frames of reference.

Background and History

The Greek historian Herodotus soundly argued that geometry originated in ancient Egypt. During the time of the legendary pharaoh Sesostris, the farmland of the empire was apportioned equally among the people, and taxes were levied accordingly. However, the annual inundation of the Nile River tended to wash away portions of farmland, and farmers who lost land in this way complained that it was unfair for them to pay taxes equal to those whose farms were complete. Sesostris is said to have sent agents to measure the loss of land so that taxation could be made fair again. The agents' observations of the relationships that existed gave rise to an understanding of the principles of geometry.

It is well documented that the principles of geometry were known to people long before the ancient Greeks described them. The Rhind papyrus, an Egyptian document dating from 2000 B.C.E., contains a valid geometric approximation of the value of pi, the ratio of the circumference of a circle to its diameter. The ancient Babylonians were also aware of the principles of geometry, as is evidenced by the inscription on a clay tablet that is housed in Berlin. The inscription has been translated as an explanation of the relationship of the sides and hypotenuse of a right triangle in what is known as the Pythagorean theorem, although the tablet predates Pythagoras by several hundred years.

From these early beginnings to modern times, studies in geometry have evolved from being simply a means of describing geometric relationships toward encompassing a complete description of the workings of the universe. Such studies allow the behavior of materials, structures, and numerous processes to be predicted in a quantifiable way.

How It Works

Geometry is concerned with the relationship between points, lines, surfaces, and the spaces enclosed or bounded by those entities.

Points. A point is any unique and particular location within a frame of reference. It exists in one dimension only, having neither length nor width nor breadth but only location. The location of any point can be uniquely and unequivocally defined by a set of coordinate values relative to the central point of the particular reference system being used. For example, in a Cartesian coordinate system—named after French mathematician René Descartes although the method was described at the same time by Pierre de Fermat—the location of any point in a two-dimensional plane is described completely by an *x*-coordinate and a *y*-coordinate, as (x, y), relative to the origin point at (0, 0). Thus, a point located at (3, 6) is 3 units away from the origin in the direction corresponding to positive values of *x*, and 6 units away from the origin in the direction corresponding to positive values of *y*. Similarly, a point in a three-dimensional Cartesian system is identified by three coordinates, as (x, y, z). In Cartesian coordinate systems, each axis is orthogonal to the others. Because orthogonality is a mathematical property, it can also be ascribed to other dimensions as well, allowing the identification of points in theoretical terms in *n*-space, where *n* is the number of distinct dimensions assigned to the system. For the purposes of all but the most theoretical of applications, however, three dimensions are sufficient for normal physical representations.

Points can also be identified as corresponding to specific distances and angles, also relative to a coordinate system origin. Thus, a point located in a spherical coordinate system is defined by the radius (the straight-line distance from the origin to the point), the angle that the radius is swept through a plane, and the angle that the radius is swept through an orthogonal plane to achieve the location of the point in space.

Lines. A line can be formed by the translation of a point in a single direction within the reference system. In its simplest designation, a line is described in a two-dimensional system when one of the coordinate values remains constant. In a three-dimensional system, two of the three coordinate values must remain constant. The lines so described are parallel to the reference coordinate axis. For example, the set of points (x, y) in a two-dimensional Cartesian system that corresponds to the form $(x, 3)$—so that the value of the *y*-coordinate is 3 no matter what the value of *x*—defines a line that is parallel to the *x*-axis and always separated from it by 3 units in the positive *y* direction.

Lines can also be defined by an algebraic relationship between the coordinate axes. In a two-dimensional Cartesian system, a line has the general algebraic form $y = mx + b$. In three-dimensional systems, the relationship is more complex but can be broken down into the sum of two such algebraic equations involving only two of the three coordinate axes.

Planes and Surfaces. Planes are described by the translation of a line through the reference system, or by the designation of two of the three coordinates having constant values while the third varies. A plane can be thought of as a flat surface. A curved surface can be formed in an analogous manner by translating a curved line through the reference system, or by definition, as the result of a specific algebraic relationship between the coordinate axes.

Angles. Intersecting lines have the property of defining an angle that exists between them. The angle can be thought of as the amount by which one line must be rotated about the intersection point in order to coincide with the other line. The magnitude, or value, of angles rigidly determines the shape of structures, especially when the structures are formed by the intersection of planes.

Conic Sections. A cone is formed by the rotation of a line at an angle about a point. Conic sections are described by the intersection of a plane with the cone structure. As an example, consider a cone formed in the three-dimensional Cartesian system by rotating the line $x = y$ about the *y*-axis, forming both a positive and a negative cone shape that meet at the origin point. If this is intersected by a plane parallel to the *x-z* plane, the intersection describes a circle. If the plane is canted so that it is not parallel to the *x-z* plane and intersects only one of the cone ends, the result is an ellipse. If the plane is canted further and positioned so that it intersects the positive cone on one side of the *y*-axis and the negative cone on the other side of the *y*-axis, the intersection defines a hyperbola. If the plane is canted still further so that it intersects both cones on the same side of the *y*-axis, then a parabola is described.

Applications and Products

It is impossible to describe even briefly more than a small portion of the applications of geometry because geometry is so intimately bound to the structures and properties of the physical universe. Every physical structure, no matter its scale, must and does adhere to the principles of geometry, since these are the properties of the physical universe. The application of geometry is fundamental to essentially every field, from agriculture to zymurgy.

GIS and GPS. Geographical information systems (GIS) and Global Positioning Systems (GPS) have been developed as a universal means of location identification. GPS is based on a number of satellites orbiting the planet and using the principles of geometry to define the position of each point on the Earth's surface. Electronic signals from the various satellites triangulate to define the coordinates of each point. Triangulation uses the geometry of triangles and the strict mathematical relationships that exist between the angles and sides of a triangle, particularly the sine law, the cosine law, and the Pythagorean theorem.

GIS combines GPS data with the geographic surface features of the planet to provide an accurate "living" map of the world. These two systems have revolutionized how people plan and coordinate their movements and the movement of materials all over the world. Applications range from the relatively simple GPS devices found in many modern vehicles to precise tracking of weather systems and seismic activity. An application that is familiar to many through the Internet is GoogleEarth, which presents a satellite view of essentially any place on the planet at a level of detail that once was available from only the most top secret of military reconnaissance satellites. The system also allows a user to view traditional map images in a way that allows them to be scaled as needed and to add overlays of specific buildings, structures, and street views. Anyone with access to the Internet can quickly and easily call up an accurate map of almost any desired location on the planet.

GIS and GPS have provided a whole new level of security for travelers. They have also enabled the development of transportation security features such as General Motors' OnStar system, the European Space Agency's Satellite Based Alarm and Surveillance System (SASS), and several other satellite-based security applications. They invariably use the GPS and GIS networks to provide the almost instantaneous location of individuals and events as needed.

CAD. Computer-aided design (CAD) is a system in which computers are used to generate the design of a physical object and then to control the mechanical reproduction of the design as an actual physical object. A computer drafting application such as AutoCAD is used to produce a drawing of an object in electronic format. The data stored in the drawing file include all the dimensions and tolerances that define the object's size and shape. At this point, the object itself does not exist; only the concept of it exists as a collection of electronic data. The CAD application can calculate the movements of ancillary robotic machines that will then use the program of instructions to produce a finished object from a piece of raw material. The operations, depending on the complexity and capabilities of the machinery being directed, can include shaping, milling, lathework, boring or drilling, threading, and several other procedures. The nature of the machinery ranges from basic mechanical shaping devices to advanced tooling devices employing lasers, high-pressure jets, and plasma- and electron-beam cutting tools.

The advantages provided by CAD systems are numerous. Using the system, it is possible to design and produce single units, or one offs, quickly and precisely to test a physical design. Adjustments to production steps are made very quickly and simply by adjusting the object data in the program file rather than through repeated physical processing steps with their concomitant waste of time and materials. Once perfected, the production of multiple pieces becomes automatic, with little or no variation from piece to piece.

Metrology. Closely related to CAD is the application of geometry in metrology, particularly through the use of precision measuring devices such as the measuring machine. This is an automated device that uses the electronic drawing file of an object, as was produced in a CAD procedure, as the reference standard for objects as they are made in a production facility. Typically, this is an integral component of a statistical process control and quality assurance program. In practice, parts are selected at random from a production line and submitted to testing for the accuracy of their construction during the production process. A production piece is placed in a custom jig or fixture, and the calibrated measuring machine then goes through a series of test measurements to determine the correlation between the features of the actual piece and the features of the piece as they are designated in the drawing file. The measuring machine is precisely controlled by electronic mechanisms and is capable of highly accurate measurement.

Game Programming and Animation. Basic and integral parts of both the video game and the motion-picture industry are described by the terms "polygon," "wire frame," "motion capture," and "computer-generated imagery" (CGI). Motion capture uses a series of reference points attached to an actor's body. The reference points become data points in a computer file, and the motions of the actor are recorded as the geometric translation of one set of data points into another in a series. The data points can then be used

Fascinating Facts About Geometry

- The translation of an ancient Babylonian clay tablet shows that the relationship between the sides of a right triangle, commonly called the Pythagorean theorem, was known at least 2,000 years before Pythagoras lived.
- The ancient Greek mathematician Thales of Miletus is credited with devising the principles of geometry known as Euclid's postulates and is also heralded as the father of modern engineering.
- The fifth postulate of Euclidean geometry has never been proven or disproven.
- The golden ratio, or ϕ, is found almost everywhere in nature and even describes processes and trends in such nonphysical fields as economics and sociology.
- The value of pi, the ratio of the circumference of a circle to its diameter, is a finite value whose nonrepeating decimal portion is infinitely long; it has been calculated to more than a trillion (10^{12}) decimal places.

to generate a wire-frame drawing of a figure that corresponds to the character whose actions have been imitated by the actor during the motion-capture process. The finished appearance of the character is achieved by using polygon constructions to provide an outward texture to the image. The texture can be anything from a plain, smooth, and shiny surface to a complex arrangement of individually colored hairs. Perhaps the most immediately recognizable application of the polygon process is the generation of dinosaur skin and aliens in video games and films.

The movements of the characters in both games and films are choreographed and controlled through strict geometric relationships, even to the play of light over the character's surface from a single light source. This is commonly known as ray tracing and is used to produce photorealistic images.

Impact on Industry

One cannot conceive of modern agriculture without the economic assessment scales of yield, fertilizer rates, pesticide application rates, and seed rates, all on a per hectare or per acre basis. Similarly, one cannot envisage scientific applications and research programs that do not rely intimately on the principles of geometry in both theory and practice. Processes such as robotics and CAD have become fundamental aspects of modern industry, representing entirely different paradigms of efficiency, precision, and economics than existed before. The economic value of computer-generated imagery and other computer graphics applications is similarly inestimable. For example, James Cameron's film *Avatar* (2009), which featured computer-generated imagery, returned more than $728 million within two weeks of its release date. Indeed, many features simply would not exist without computer-generated imagery, which in turn would not exist without the application of geometry. Similarly, computer gaming, from which computer-aided imagery developed, is a billion-dollar industry that is relies entirely on the mathematics of geometry.

Academic Research. All universities and colleges maintain a mathematics department in which practical training and theoretical studies in the applications of geometric principles are carried out. Similarly, programs of research and study in computer science departments examine new and more effective ways to apply geometry and geometric principles in computing algorithms. Graphic arts programs in particular specialize in the development and application of computer-generated imagery and other graphics techniques relying on geometry. Training in plane surveying and mechanical design principles are integral components of essentially all programs of training in civil and mechanical engineering.

Industry and Business. A wide variety of businesses are based almost entirely on the provision of goods and services that are based on geometry. These include contract land-surveying operations that serve agricultural needs; environmental conservation and management bodies; forestry and mining companies; municipal planning bodies; transportation infrastructure and the construction industries; data analysis; graphics programming; advertising; and cinematographic adjuncts, to name but a few. The opportunity exists for essentially anyone with the requisite knowledge and some resources to establish a business that caters to a specific service need in these and other areas.

Another aspect of these endeavors is the provision of materials and devices to accommodate the services. Surveyors, for example, require surveying equipment such as transit levels and laser source-detectors in order to function. Similarly, construction contractors typically require custom-built roof trusses and other structures. Graphics and game design companies often subcontract needed programming expertise. Numerous businesses and industries exist, or can be initiated, to provide needed goods.

Careers and Course Work

Geometry is an essential and absolutely critical component of practically all fields of applied science. A solid grounding in mathematics and basic geometry is required during secondary school studies. In more advanced or applied studies at the post-secondary level, in any applied field, mathematical training in geometrical principles will focus more closely on the applications that are specific to that field of study. Any program of study that integrates design concepts will include subject-specific applications of geometric principles. Applications of geometric principles are used in mechanical engineering, manufacturing, civil engineering, industrial plant operations, agricultural development, forestry management, environmental management, mining, project management and logistics, transportation, aeronautical engineering, hydraulics and fluid dynamics, physical chemistry, crystallography, graphic design, and game programming.

The principles involved in any particular field of study can often be applied to other fields as well. In economics, for example, data mining uses many of the same ideological principles that are used in the mining of mineral resources. Similarly, the generation of figures in electronic game design uses the same geometric principles as land surveying and topographical mapping. Thus, a good grasp of geometry and its applications can be considered a transferable skill usable in many different professions.

Social Context and Future Prospects

Geometry is set to play a central role in many fields. Geometry is often the foundation on which decisions affecting individuals and society are made. This is perhaps most evident in the establishment and construction of the most basic infrastructure in every country in the world, and in the most high-tech advances represented by the satellite networks for GPS and GIS. It is easy to imagine the establishment of similar networks around the Moon, Mars, and other planets, providing an unprecedented geological and geographical understanding of those bodies. In between these extremes that indicate the most basic and the most advanced applications of geometry and geometrical principles are the typical everyday applications that serve to maintain practically all aspects of human endeavor. The scales at which geometry is applied cover an extremely broad range, from the ultra-small constructs of molecular structures and nanotechnological devices to the construction of islands and buildings of novel design and the ultra-large expanses of interplanetary and even interstellar space.

Richard M. J. Renneboog, M.Sc.

Further Reading

Bar-Lev, Adi. "Big Waves: Creating Swells, Wakes and Everything In-Between." *Game Developer* 15, no. 2, (February, 2008): 14-24. Describes the application of geometry in the modeling of liquid water actions in the computer graphics of gaming and simulations.

Bonola, Roberto. *Non-Euclidean Geometry: A Critical and Historical Study of Its Development.* 1916. Reprint. Whitefish, Mont.: Kessinger, 2007. A facsimile reproduction of a classic work on the history of alternate geometries.

Boyer, Carl B. *History of Analytic Geometry.* Mineola, N.Y.: Dover Publications, 2004. Traces the history of analytic geometry from ancient Mesopotamia, Egypt, China, and India to 1850.

Darling, David. *The Universal Book of Mathematics: From Abracadabra to Zeno's Paradoxes.* 2004. Reprint. Edison, N.J.: Castle Books, 2007. An encyclopedic account of things mathematical.

Heilbron, J. L. *Geometry Civilized: History, Culture, and Technique.* Reprint. New York: Oxford University Press, 2003. A very readable presentation of the history of geometry, including geometry in cultures other than the Greek, such as the Babylonian, Indian, Chinese, and Islamic cultures.

Herz-Fischler, Roger. *A Mathematical History of the Golden Number.* Mineola, N.Y.: Dover Publications, 1998. A well-structured exploration of the golden number and its discovery and rediscovery throughout history.

Holme, Audun. *Geometry: Our Cultural Heritage.* New York: Springer, 2002. An extensive discussion of the historical development of geometry from prehistoric to modern times, focusing on many major figures in its development.

Livio, Mario. *The Golden Ratio: The Story of Phi, the World's Most Astonishing Number.* New York: Broadway Books, 2002. Describes how the occurrence of the golden ratio, also known as the divine ratio, defined by the ancient mathematician Euclid, seems to be a fundamental constant of the physical world.

Szecsei, Denise. *The Complete Idiot's Guide to Geometry.* 2d ed. Indianapolis: Alpha Books, 2007. A fun, step-by-step presentation that walks the reader through the mathematics of geometry and assumes absolutely no prior knowledge beyond basic arithmetic.

West, Nick. "Practical Fluid Dynamics: Part 1." *Game Developer* 14, no. 3 (March, 2007): 43-47. Introduces the application of geometric principles in the modeling of smoke, steam, and swirling liquids in the graphics of computer games and simulations.

Web Sites

American Mathematical Society
http://www.ams.org

Mathematical Association of America
http://www.maa.org

See also: Applied Mathematics; Computer Graphics; Engineering; Engineering Mathematics; Fractal Geometry; Topology; Trigonometry; Video Game Design and Programming.

GRAMMATOLOGY

FIELDS OF STUDY

Linguistics; anthropology; archaeology; history; literature psychology; classics; English; foreign languages; computer science.

SUMMARY

Grammatology is the science of writing. It is closely related to the field of linguistics but concentrates on written expression whereas linguistics is primarily oriented toward oral expression. Grammatology is an interdisciplinary field that is applied to research in a variety of fields studying human interaction and communication and the various cultures and civilizations. Anthropologists, psychologists, literary critics, and language scholars are the main users of grammatology. Electracy, digital literacy, is a concept that applies the principles of grammatology to electronic media.

KEY TERMS AND CONCEPTS

- **Abjad:** Script using consonants only, no vowels.
- **Abugida:** Script using consonants that contain inherent vowels, modifications to the character to denote other vowels; also known as alphasyllabary.
- **Alphabetic:** Script using both consonants and vowels.
- **Cuneiform Writing:** System of wedge-shaped marks used for communication.
- **Logography:** System of writing using a sign to indicate a word or a combination of words; also known as word writing.
- **Script:** Set of basic symbols or characters.
- **Syllabary:** Writing system using a combination of symbols for a word.
- **Text:** Object produced by writing that disseminates information.
- **Writing:** Set of marks or characters used for communication.

Definition and Basic Principles

Grammatology is the scientific examination of writing. Oral communication is the natural means of communication for human beings. Oral language uses neither instruments nor objects that are not part of the human body. Writing, by contrast, is not natural language; it requires an object (paper, stone, wood, or even sand) on which marks can be made and some kind of tool with which to make the marks. Although it is possible to use one's fingers to make marks in a soft medium such as sand, such "writing" lacks an essential feature of writing as defined by grammatology: Writing lasts beyond the moment of its creation. It preserves thoughts, ideas, stories, and accounts of various human activities. In addition, writing, in contrast to pictography (drawn pictures), can be understood only with knowledge of the language that it represents, because writing uses symbols to create a communication system. The discipline defines what writing is by using a system of typologies, or types. An example of this is the classification of languages by script (the kind of basic symbols used). Another way of classifying languages is by the method used to write them; linear writing systems use lines to compose the characters.

Grammatology also investigates the importance of writing in the development of human thought and reasoning. Grammatologists address the role of writing in the development of cultures and its importance in the changes that occur in cultures. Grammatology proposes the basic principle that changes in ways of communication have profound effects on both the individual and the societal life of people. It addresses the technology of writing and its influence on the use and development of language. It deals extensively with the impact of the change from oral to handwritten communication, from handwritten texts to printed texts, and from printed texts to electronic media texts.

Background and History

Polish American historian Ignace Gelb, the founder of grammatology, was the first linguist to propose a science of writing independent of linguistics. In 1952, he published *A Study of Writing: The Foundations of Grammatology*, in which he classified writing systems according to their typologies and set forth general principles for the classification of writing systems. As part of his science of writing systems, Gelb traced the evolution of writing from pictures that he referred to as "no writing" through semasiography to full writing or scripts that designate sounds by characters. Gelb divided languages into three typologies: logo-syllabic systems, syllabic systems, and alphabetic systems. Peter T. Daniels, one of Gelb's students, added two additional types to this typology: abugida and abjad. Gelb also included as part of the science of writing an investigation of the interactions of writing and culture and of writing and religion as well as an analysis of the relationship of writing to speech or oral language. Although Gelb did not elaborate extensively on the concept of writing as a technology, he did address the importance of the invention of printing and the changes that it brought about in dissemination of information. This aspect of writing (by hand or device) as technology, as an apparatus of communication, has become one of the most significant areas of the application of grammatology to other disciplines.

Gelb's work laid the foundation for significant research and development in several fields involved in the study of language. Eric Havelock, a classicist working in Greek literature, elaborated on the idea of how the means of communication affects thought. He developed the theory that Greek thought underwent a major change or shift when the oral tradition was replaced by written texts. In Havelock's *Preface to Plato* (1963), he presented his argument that the transition to writing at the time of Plato altered Greek thought and consequently all Western thought by its effect on the kind of thinking available through written expression. Havelock stated that what could be expressed and how it could be expressed in written Greek was significantly different from what could be expressed orally. He devoted the rest of his scholarship to elaboration of this premise and investigation of the difference between the oral tradition and writing. Havelock's theory met with criticism from many of his colleagues in classics, but it received a favorable reception in many fields, including literature and anthropology.

Starting from Havelock's theory, Walter J. Ong, a literary scholar and philosopher, further investigated the relationship between written and spoken language and how each affects culture and thought. He proposed that writing is a technology and therefore has the potential to affect virtually all aspects of an oral culture. Addressing the issue of writing's effect on ways of thinking, Ong stated that training in writing shifted the mental focus of an individual. Ong reasoned that oral cultures are group directed, and those using the technology of writing are directed toward the individual. His most influential work is *Orality and Literacy: The Technologizing of the Word* (1982). The British anthropologist Sir John "Jack" Rankine Goody applied the basic premise of grammatology (that the technology of writing influences and changes society) in his work in comparative anthropology. For Goody, writing was a major cause of change in a society, affecting both its social interaction and its psychological orientation.

In *De la grammatologie* (1967; *Of Grammatology*, 1976), French philosopher and literary critic Jacques Derrida examined the relationship of spoken language and writing. By using Gelb's term "grammatology" as a term of literary criticism, Derrida narrowed the significance of the term. Grammatology became associated with the literary theory of deconstruction. However, Derrida's work had more far-reaching implications, and for Derrida, as for Gelb, grammatology meant the science of writing. Derrida was attempting to free writing from being merely a representation of speech.

In *Applied Grammatology: Post(e)-pedagogy from Jacques Derrida to Joseph Beuys* (1985), Gregory Ulmer moved from deconstructing writing to a grammatology that he defined as inventive writing. Although Ulmer's work moves away from the areas of systems and classifications of writing as established by Gelb, it does elaborate on Gelb's notion of how literacy or the technology of writing changes human thought and interaction. Ulmer's work brings Gelb's theories and work into the electronic age. Gelb and his followers investigated the development of writing systems and the technologies of handwritten and printed texts and their influences on human beings, both in their social interactions and in their thought processes. Just as print influenced the use, role, and significance of writing, electronic media is significantly affecting writing, both in what it is and what it does.

How It Works

Classification by Typologies. Written languages are classified based on particular features of the language. Grammatologists analyze written languages to determine whether the basic element of the language is a syllable, a letter, a word, or some other element. They also look for similarities among languages that they compare. From these analyses, they classify languages in existing typologies or create new ones. The established typologies are logographic, syllabic, alphabetic, abjad, and abugida.

Mechanisms and Apparatuses. Grammatology identifies the mechanisms or the apparatuses of writing. It examines how writing is produced, what type of tools are used to produce it, and the kind of texts that result from writing. It also considers the speed and breadth of dissemination of information through writing. Grammatology looks at writing systems from a practical viewpoint, considering questions of accuracy and ease of usage.

Relationship of Oral and Written Language. Grammatologists investigate the relationship of oral and written language by comparing a language in its oral and written form. They look for differences in sentence structure, vocabulary, and complexity of usage.

Writing and Social Structure. Grammatologists note the time that writing appears in a culture and analyze the social, political, economic, and family structures to ascertain any and all changes that appear after the introduction of writing. They also attempt to identify the type of thinking, ideas, and beliefs prevalent in the society before and after the introduction of writing and determine if writing brought about changes in these areas. Once they have established evidence of change, they address the ways in which writing causes these modifications. In addition, they investigate these same areas when the mechanism of writing in a culture changes, such as when print replaces handwritten texts and when electronic media replaces print.

Fascinating Facts About Grammatology

- The Roman alphabet reached its definitive form about seven centuries after the first writing was done in Latin. During this time, the alphabet underwent considerable graphic changes.
- Languages that use systems of writing or scripts that contain no vowels are classified as abjad. Two of these languages are Arabic and Hebrew.
- Chinese is a language whose basic element is the logogram. Japanese contains many of the same logograms. However, an individual capable of reading a Japanese text is not able to read a Chinese text because of the disparity in the grammar of the two languages.
- Throughout history, human beings have believed that writing possessed magical powers. The writing of poetry has been traditionally associated with divine inspiration. The poet and the poet's muse play an important role in concepts of artistic creation in writing.
- The writing and the architecture of a specific period often share certain characteristics. Handwriting from the Carolingian period resembles Romanesque architecture by its use of roundness, and handwriting from the later Gothic period of European history reflects Gothic architecture by its angularity.
- After the arrival of Europeans, especially missionaries, in North America, members of certain Indian tribes such as the Cherokee and Algonquin attempted to invent writing systems for their spoken languages.
- The Egyptian civilization used a system of writing known as hieroglyphics, which survived for more than three thousand years.

Applications and Products

Pedagogy. The science of grammatology is important in the development of pedagogies that adequately address the needs and orientations of students. The way information is presented and taught varies with the medium by which students are accustomed to receiving knowledge. In an oral culture, stories, poems, and proverbs that use repetitive signifiers are employed to enable students to memorize information. Handwritten texts are suited to cultures in which a small quantity of information is disseminated among a limited number of students. Print text cultures rely heavily on the dissemination of information through books and other printed sources. Instruction is classroom based with face-to-face contact between the teacher and student. Assignments are written out and turned in to the professor to be corrected and returned. Cultures that rely on electronic media call for different pedagogical structures. Grammatology provides a means of analyzing and addressing this need.

Anthropology. Anthropologists use grammatology as an integral part of their research techniques to ascertain the development of civilizations and cultures. The time when writing appeared in a culture (or the lack of writing in a culture), the type of writing, and the classes of society that use writing are all important to the study and understanding of a culture or civilization.

Literary Criticism. Grammatology enables literary critics to dissect or deconstruct texts and perform close analysis of both the structure and the meaning of a work. It has also been the basis for the creation of new types of literary criticism, including deconstruction and postdeconstruction.

Linguistics. Grammatology, which deals with the written form of language, plays an important role in linguistics, the study of languages. It classifies languages, addresses how graphemes (units of a writing system) and allographs (letters) are used, compares written to oral forms of languages, and identifies archaic forms and vocabularies. The study of writing forms is necessary to linguists tracing the development of a particular language. Texts written in the language during different periods permit linguists to compare vocabulary, syntax, and prosaic and poetic forms of the language throughout its development. Written texts also provide evidence of borrowings from other languages.

Archaeology. The ability to analyze and classify written language is an essential part of the work of the archaeologist. Knowledge regarding when particular written forms were used and what cultures used them plays an important role in the dating of artifacts.

Electronic Media. Grammatology is highly applicable to cyberlanguage and electronic media as these phenomena affect writing. Texting and e-mail have created what may be viewed as a subtext or alternative to the traditional print culture. The principles and techniques of analysis of grammatology provide a means for classifying the language of texting and e-mail and for evaluating the effects of these new forms of language on thought and on societal interaction.

Impact on Industry

The academic community is the sector in which grammatology is primarily used; therefore, its impact on industry is indirect. However, because grammatologists describe the changes that occur in human thought processes as the means and mechanisms of written communication change, they contribute significantly to the development of strategies of teaching and presentation of material in ways appropriate to the needs of those receiving the information.

Government, University, and Private Foundation Grants. Many sources of funding are available to researchers using grammatological principles and techniques of investigation in their research projects. The United States government funds many projects through the National Endowment for the Humanities. Individual universities such as those in the University of California system provide research grants and fellowships for graduate students, postdoctoral candidates, and faculty involved in research into writing and its importance and implications. Private foundations such as the Guggenheim Foundation also provide grants and residencies to encourage research in areas involving grammatology.

Cyberlanguage and Electronic Communication. Universities and graduate schools such as the European Graduate School are developing departments addressing cyberlanguage and electronic communication studies in which grammatologists who are interested in applying their skills to electronic communication can find interesting and challenging positions.

Careers and Course Work

Courses in linguistics, the history of language, and the composition and grammar of the language or languages with which an individual wishes to work form the basis of preparation for a career in grammatology as a subdiscipline of linguistics. However, grammatology is also relevant to a wide variety of careers. Majors in the history of language and in various disciplines such as anthropology, pedagogy, history, literature, psychology, electronic media, and cyberlanguage can be combined with the study of grammatology. Grammatologists, for the most part, work in one of these fields and apply their knowledge of writing systems and their importance to that field. They usually have a doctorate and are employed as university professors or in scientific research centers or consulting firms.

Social Context and Future Prospects

With the changes in the mechanisms and apparatus of writing that have taken place in the second half of the twentieth century and continue to take place in the twenty-first, grammatology continues to be an essential science. Research in electronic media and its relation to human thought processes will be an area in which grammatology will play an important role. Because grammatology is a science used in conjunction with a considerable number of areas of scientific research, it should continue to play a significant role in human intellectual life.

Shawncey Jay Webb, Ph.D.

Further Reading

Daniels, Peter T., and William Bright, eds. *The World's Writing Systems*. New York: Oxford University Press, 1996. These essays continue and refine the work done by Gelb.

Gelb, Ignace. *A Study of Writing: The Foundations of Grammatology*. 1952. Reprint. Chicago: University of Chicago Press, 1989. The foundational text for grammatology.

Olson, David R., and Michael Cole, eds. *Technology, Literacy and the Evolution of Society: Implications of the Work of Jack Goody*. Mahwah, N.J.: Lawrence Erlbaum Associates, 2006. An interdisciplinary look at Goody's work on modes of communication in societies and their influence.

Ong, Walter J. *Orality and Literacy: The Technologizing of the Word*. 2d ed. Reprint. New York: Routledge, 2009. Ong elaborates on the work of Havelock concerning the impact of writing on oral culture. Identifies writing as a learned technology.

Rogers, Henry. *Writing Systems: A Linguistic Approach*. Malden, Mass.: Wiley-Blackwell, 2005. A very detailed presentation of world languages and their writing systems that describes the classification of writing systems.

Ulmer, Gregory. *Applied Grammatology: Post(e)-pedagogy from Jacques Derrida to Joseph Beuys*. 1985. Reprint. Baltimore: The Johns Hopkins University Press, 1992. Presents Ulmer's early thoughts on how electronic media was changing the technology of writing. Also provides an excellent explanation of Derrida's views.

_______. *Electronic Monuments*. Minneapolis: University of Minnesota Press, 2005. Ulmer's view of the effects of electracy, or digital literacy, on society and human interaction.

Web Sites

American Association for Applied Linguistics
http://www.aaal.org

International Cognitive Linguistics Association
http://www.cognitivelinguistics.org

Linguistic Society of America
http://www.lsadc.org

See also: Communications; Computer Languages, Compilers, and Tools; Typography.

HYDRAULIC ENGINEERING

FIELDS OF STUDY

Physics; mathematics; computer programming; numerical analysis and modeling; material science; civil engineering; water resources engineering; fluid mechanics; hydrostatics; fluid kinematics; hydrodynamics; hydraulic structures; reservoir operations; dam design; open-channel flow; channel design; bridge design; river navigation; coastal engineering; water supply; hydraulic transients; pipeline design; storm drainage; irrigation; water reclamation and recycling; sanitary engineering; environmental engineering; hydraulic machinery; hydroelectric power; pump design.

SUMMARY

Hydraulic engineering is a branch of civil engineering concerned with the properties, flow, control, and uses of water. Its applications are in the fields of water supply, sewerage evacuation, water recycling, flood management, irrigation, and the generation of electricity. Hydraulic engineering is an essential element in the design of many civil and environmental engineering projects and structures, such as water distribution systems, wastewater management systems, drainage systems, dams, hydraulic turbines, channels, canals, bridges, dikes, levees, weirs, tanks, pumps, and valves.

KEY TERMS AND CONCEPTS

- **Froude Number:** Dimensionless number for open-channel flow, defined as the ratio of the velocity to the square root of the product of the hydraulic depth and the gravitational acceleration. The Froude number equals 1 for critical flow.
- **Gravity Flow:** Flow of water with a free surface, as in open channels or partially full pipes.
- **Ideal Fluid Flow:** Hypothetical flow that assumes a frictionless flow and no fluid viscosity; also known as inviscid flow.
- **Incompressible Fluid:** Fluid that assumes constant fluid density; applies to fluids such as water and oil, but not gases, which are compressible.
- **Laminar Flow:** Streamlined flow in a pipe in which a fluid particle follows an observable path. It occurs at low velocities, in high-viscosity fluids, and at low Reynolds numbers, typically less than 2,000.
- **Pressure Flow:** Flow under pressure, such as water in a pipe flowing at capacity.
- **Real Fluid Flow:** Flow that assumes frictional (viscous) effects; also known as viscid flow.
- **Reynolds Number:** Dimensionless number defined by the product of the pipe diameter and the flow velocity, divided by the fluid's kinematic viscosity.
- **Steady Flow:** Flow that remains constant over time at every point.
- **Subcritical Flow:** Relatively low-velocity and high-depth flow in an open channel. Its Froude number is smaller than 1.
- **Supercritical Flow:** Relatively high-velocity and small-depth flow in an open channel. Its Froude number is larger than unity.
- **Turbulent Flow:** Flow characterized by the irregular motion of particles following erratic paths. It may be found in streams that appear to be flowing very smoothly and in swirls and large eddies caused by a disturbance source.
- **Uniform Flow:** Flow that occurs when the cross section and the velocity remain constant along the channel or pipe.
- **Unsteady Flow:** Flow that changes with time; also known as hydraulic transient flow.
- **Varied Flow:** Flow that occurs when the free water surface and the velocity vary along the flow path; also known as nonuniform flow.

Definition and Basic Principles

Hydraulic engineering is a branch of civil engineering that focuses on the flow of water and its role in civil engineering projects. The principles of hydraulic engineering are rooted in fluid mechanics. The conservation of mass principle (or the continuity principle) is the cornerstone of hydraulic analysis and design. It states that the mass going into a control volume within fixed boundaries is equal to the rate of increase of mass within the same control volume. For an incompressible fluid with fixed boundaries, such as water flowing through a pipe, the continuity equation is simplified to state that the inflow rate is equal to the outflow rate. For unsteady flow in a channel or a reservoir, the continuity principle states that the flow rate into a control volume minus the outflow rate is equal to the time rate of change of storage within the control volume.

Energy is always conserved, according to the first law of thermodynamics, which states that energy can neither be created nor be destroyed. Also, all forms of energy are equivalent. In fluid mechanics, there are mainly three forms of head (energy expressed in unit of length). First, the potential head is equal to the elevation of the water particle above an arbitrary datum. Second, the pressure head is proportional to the water pressure. Third, the kinetic head is proportional to the square of the velocity. Therefore, the conservation of energy principle states that the potential, pressure, and kinetic heads of water entering a control volume, plus the head gained from any pumps in the control volume, are equal to the potential, pressure, and kinetic heads of water exiting the control volume, plus the friction loss head and any head lost in the system, such as the head lost in a turbine to generate electricity.

Hydraulic engineering deals with water quantity (flow, velocity, and volume) and not water quality, which falls under sanitary and environmental engineering. However, hydraulic engineering is an essential element in designing sanitary engineering facilities such as wastewater-treatment plants.

Hydraulic engineering is often mistakenly thought to be petroleum engineering, which deals with the flow of natural gas and oil in pipelines, or the branch of mechanical engineering that deals with a vehicle's engine, gas pump, and hydraulic breaking system. The only machines that are of concern to hydraulic engineers are hydraulic turbines and water pumps.

Background and History

Irrigation and water supply projects were built by ancient civilizations long before mathematicians defined the governing principles of fluid mechanics. In the Andes Mountains in Peru, remains of irrigation canals were found, radiocarbon dating from the fourth millennium B.C.E. The first dam for which there are reliable records was built before 4000 B.C.E. on the Nile River in Memphis in ancient Egypt. Egyptians built dams and dikes to divert the Nile's floodwaters into irrigation canals. Mesopotamia (now Iraq and western Iran) has low rainfall and is supplied with surface water by two major rivers, the Tigris and the Euphrates, which are much smaller than the Nile but have more dramatic floods in the spring. Mesopotamian engineers, concerned about water storage and flood control as well as irrigation, built diversion dams and large weirs to create reservoirs and to supply canals that carried water for long distances. In the Indus Valley civilization (now Pakistan and northwestern India), sophisticated irrigation and storage systems were developed.

One of the most impressive dams of ancient times is near Marib, an ancient city in Yemen. The 1,600-foot-long dam was built of masonry strengthened by copper around 600 B.C.E. It holds back some of the annual floodwaters coming down the valley and diverts the rest of that water out of sluice gates and into a canal system. The same sort of diversion dam system was independently built in Arizona by the Hohokam civilization around the second or third century C.E.

In the Szechwan region of ancient China, the Dujiangyan irrigation system was built around 250 B.C.E. and still supplies water in modern times. By the second century C.E., the Chinese used chain pumps, which lifted water from lower to higher elevations, powered by hydraulic waterwheels, manual foot pedals, or rotating mechanical wheels pulled by oxen.

The Minoan civilization developed an aqueduct system in 1500 B.C.E. to convey water in tubular conduits in the city of Knossos in Crete. Roman aqueducts were built to carry water from large distances to Rome and other cities in the empire. Of the 800 miles of aqueducts in Rome, only 29 miles were above ground. The Romans kept most of their aqueducts underground to protect their water from enemies and diseases spread by animals.

The Muslim agricultural revolution flourished during the Islamic golden age in various parts of Asia and Africa, as well as in Europe. Islamic hydraulic engineers built water management technological complexes, consisting of dams, canals, screw pumps, and *norias*, which are wheels that lift water from a river into an aqueduct.

The Swiss mathematician Daniel Bernoulli published *Hydrodynamica* (1738; *Hydrodynamics by Daniel Bernoulli*, 1968), applying the discoveries of Sir Isaac Newton and Gottfried Wilhelm Leibniz in mathematics and physics to fluid systems. In 1752, Leonhard Euler, Bernoulli's colleague, developed the more generalized form of the energy equation.

In 1843, Adhémar-Jean-Claude Barré de Saint Venant developed the most general form of the differential equations describing the motion of fluids, known as the Saint Venant equations. They are sometimes called Navier-Stokes equations after Claude-Louis Navier and Sir George Gabriel Stokes, who were working on them around the same time.

The German scientist Ludwig Prandtl and his students studied the interactions between fluids and solids between 1900 and 1930, thus developing the boundary layer theory, which theoretically explains the drag or friction between pipe walls and a fluid.

How It Works

Properties of Water. Density and viscosity are important properties in fluid mechanics. The density of a fluid is its mass per unit volume. When the temperature or pressure of water changes significantly, its density variation remains negligible. Therefore, water is assumed to be incompressible. Viscosity, on the other hand, is the measure of a fluid's resistance to shear or deformation. Heavy oil is more viscous than water, whereas air is less viscous than water. The viscosity of water increases with reduced temperatures. For instance, the viscosity of water at its freezing point is six times its viscosity at its boiling temperature. Therefore, a flow of colder water assumes higher friction.

Hydrostatics. Hydrostatics is a subdiscipline of fluid mechanics that examines the pressures in water at rest and the forces on floating bodies or bodies submerged in water. When water is at rest, as in a tank or a large reservoir, it does not experience shear stresses; therefore, only normal pressure is present. When the pressure is uniform over the surface of a body in water, the total force applied on the body is a product of its surface area times the pressure. The direction of the force is perpendicular (normal) to the surface. Hydrostatic pressure forces can be mathematically determined on any shape. Buoyancy, for instance, is the upward vertical force applied on floating bodies (such as boats) or submerged ones (such as submarines). Hydraulic engineers use hydrostatics to compute the forces on submerged gates in reservoirs and detention basins.

Fluid Kinematics. Water flowing at a steady rate in a constant-diameter pipe has a constant average velocity. The viscosity of water introduces shear stresses between particles that move at different velocities. The velocity of the particle adjacent to the wall of the pipe is zero. The velocity increases for particles away from the wall, and it reaches its maximum at the center of the pipe for a particular flow rate or pipe discharge. The velocity profile in a pipe has a parabolic shape. Hydraulic engineers use the average velocity of the velocity profile distribution, which is the flow rate over the cross-sectional area of the pipe.

Bernoulli's Theorem. When friction is negligible and there are no hydraulic machines, the conservation of energy principle is reduced to Bernoulli's equation, which has many applications in pressurized flow and open-channel flow when it is safe to neglect the losses.

Applications and Products

Water Distribution Systems. A water distribution network consists of pipes and several of the following components: reservoirs, pumps, elevated storage tanks, valves, and other appurtenances such as surge tanks or standpipes. Regardless of its size and complexity, a water distribution system serves the purpose of transferring water from one or more sources to customers. There are raw and treated water systems. A raw water network transmits water from a storage reservoir to treatment plants via large pipes, also called transmission mains. The purpose of a treated water network is to move water from a water-treatment plant and distribute it to water retailers through transmission mains or directly to municipal and industrial customers through smaller distribution mains.

Some water distribution systems are branched, whereas others are looped. The latter type offers more reliability in case of a pipe failure. The hydraulic engineering problem is to compute the steady velocity or flow rate in each pipe and the pressure at each junction node by solving a large set of continuity equations and nonlinear energy equations that characterize the network. The steady solution of a branched network is easily obtained mathematically; however, the looped network initially offered challenges to engineers. In 1936, American structural engineer Hardy Cross developed a simplified method that tackled networks formed of only pipes. In the 1970's and 1980's, three other categories of numerical methods were developed to provide solutions for complex networks with pumps and valves. In 1996, engineer Habib A. Basha and his colleagues offered a perturbation solution to the nonlinear set of equations in a direct, mathematical fashion, thus eliminating the risk of divergent numerical solutions.

Hydraulic Transients in Pipes. Unsteady flow in pipe networks can be gradual; therefore, it can be modeled as a series of steady solutions in an extended period simulation, mostly useful for water-quality analysis. However, abrupt changes in a valve position, a sudden shutoff of a pump because of power failure, or a rapid change in demand could cause a hydraulic transient or a water hammer that travels back and forth in the system at high speed, causing large pressure fluctuations that could cause pipe rupture or collapse.

The solution of the quasi-linear partial differential equations that govern the hydraulic transient problem is more challenging than the steady network solution. The Russian scientist Nikolai Zhukovsky offered a simplified arithmetic solution in 1904. Many other methods–graphical, algebraic, wave-plane analysis, implicit, and linear methods, as well as the method of characteristics–were introduced between the 1950's and 1990's. In 1996, Basha and his colleagues published another paper solving the hydraulic transient problem in a direct, noniterative fashion, using the mathematical concept of perturbation.

Open-Channel Flow. Unlike pressure flow in full pipes, which is typical for water distribution systems, flow in channels, rivers, and partially full pipes is called gravity flow. Pipes in wastewater evacuation and drainage systems usually flow partially full with a free water surface that is subject to atmospheric pressure. This is the case for human-built canals and channels (earth or concrete lined) and natural creeks and rivers.

The velocity in an open channel depends on the area of the cross section, the length of the wetted perimeter, the bed slope, and the roughness of the channel bed and sides. A roughness factor is estimated empirically and usually accounts for the material, the vegetation, and the meandering in the channel.

Open-channel flow can be characterized as steady or unsteady. It also can be uniform or varied flow, which could be gradually or rapidly varied flow. A famous example of rapidly varied flow is the hydraulic jump.

When high-energy water, gushing at a high velocity and a shallow depth, encounters a hump, an obstruction, or a channel with a milder slope, it cannot sustain its supercritical flow (characterized by a Froude number larger than 1). It dissipates most of its energy through a hydraulic jump, which is a highly turbulent transition to a calmer flow (subcritical flow with a Froude number less than 1) at a higher depth and a much lower velocity. One way to solve for the depths and velocities upstream and downstream of the hydraulic jump is by applying the conservation of momentum principle, the third principle of fluid mechanics and hydraulic engineering. The hydraulic jump is a very effective energy dissipater that is used in the designs of spillways.

Hydraulic Structures. Many types of hydraulic structures are built in small or large civil engineering projects. The most notable by its size and cost is the dam. A dam is built over a creek or a river, forming a reservoir in a canyon. Water is released through an outlet structure into a pipeline for water supply or into the river or creek for groundwater recharge and environmental reasons (sustainability of the biological life in the river downstream). During a large flood, the reservoir fills up and water can flow into a side overflow spillway–which protects the integrity of the face of the dam from overtopping–and into the river.

The four major types of dams are gravity, arch, buttress, and earth. Dams are designed to hold the immense water pressure applied on their upstream face. The pressure increases as the water elevation in the reservoir rises.

Hydraulic Machinery. Hydraulic turbines transform the drop in pressure (head) into electric power. Also, pumps take electric power and transform it into water head, thereby moving the flow in a pipe to a higher elevation.

Fascinating Facts About Hydraulic Engineering

- An interesting application of Bernoulli's theorem outside the field of hydraulic engineering is the lift force on an airplane wing. The air velocity over the longer, top side of the wing is faster than the velocity along the shorter underside. Bernoulli's theorem proves that the pressure on the top of the wing is lower than the pressure on the bottom, which results in a net upward force that lifts the plane in the air.
- Hoover Dam, built between 1931 and 1936, is a 726-foot-tall concrete arch-gravity dam on the border between Nevada and Arizona. It is the second tallest dam in the United States after the 770-foot-tall Oroville Dam in California, built between 1961 and 1968. The world's tallest dam is the 984-foot-tall Nurek Dam in Tajikistan, an earth-fill embankment dam constructed between 1961 and 1980.
- One of the first recorded dam failures was in Grenoble, France, in 1219. Since 1865, twenty-nine dam failures have been recorded worldwide, ten of which occurred in the United States.
- The 305-foot-tall Teton Dam in Idaho became the highest dam to fail in June, 1976, a few months after construction was complete. Spring runoff had filled the reservoir with 80 million gallons of water, which gushed through the mostly evacuated downstream towns, killing eleven people and causing $1 billion in damages.
- Hurricane Katrina hit New Orleans in August, 2005, causing the failure of levees and flood walls at about fifty locations. Millions of gallons of water flooded 85 percent of the coastal city, and more than 1,800 people died.
- A tidal bore is a moving hydraulic jump that occurs when the incoming high tide forms a wave that travels up a river against the direction of the flow. Depending on the water level in the river, the tidal bore can vary from an undular wave front to a shock wave that resembles a wall of water.
- A water hammer is hydraulic transient in a pipe, characterized by dangerously large pressure fluctuations caused by the sudden shutdown of a pump or the rapid opening or closing of a valve. The high-velocity wave travels back and forth in the pipe, causing the pressure to fluctuate from large positive values that could burst the pipe to very low negative values that could cause the walls of the pipe to collapse.

There are two types of turbines, impulse and reaction. The reaction turbine is based on the steam-powered device that was developed in Egypt in the first century C.E. by Hero of Alexandria. A simple example of a reaction turbine is the rotating lawn sprinkler.

Pumps are classified into two main categories, centrifugal and axial flow. Pumps have many industrial, municipal, and household uses, such as boosting the flow in a water distribution system or pumping water from a groundwater well.

Impact on Industry

The vast field of hydraulics has applications ranging from household plumbing to the largest civil engineering projects. The field has been integral to the history of humankind. The development of irrigation, water supply systems, and flood protection shaped the evolution of societies.

Hydraulic engineering is not a new field. Its governing principles were established starting in the eighteenth century and were refined through the twentieth century. Modern advances in the industry have been mainly the development of commercial software that supports designers in their modeling.

Since the 1980's, water distribution systems software has been evolving. Software can handle steady flow, extended period simulations, and hydraulic transients. Wastewater system and storm drainage software are also being developed.

One-dimensional open-channel software can be used for modeling flow in channels and rivers and even to simulate flooding of the banks. However, two-dimensional software is better for modeling flow in floodplains and estuaries, although most software still has convergence problems. Three-dimensional software is used for modeling the flow in bays and lakes.

Careers and Course Work

Undergraduate students majoring in civil or environmental engineering usually take several core courses in hydraulic engineering, including fluid mechanics, water resources, and fluid mechanics laboratory. Advanced studies in hydraulic engineering lead to a master of science or a doctoral degree. Students with a bachelor's degree in a science or another engineering specialty could pursue an advanced degree in hydraulic engineering, but they may need to take several undergraduate level courses before starting the graduate program.

Graduates with a bachelor's degree in civil engineering or advanced degrees in hydraulics can work for private design firms that compete to be chosen to work on the planning and design phases of large governmental hydraulic engineering projects. They can work for construction companies that bid on governmental projects to build structures and facilities that include hydraulic elements, or for water utility companies, whether private or public.

To teach or conduct research at a university or a research laboratory requires a doctoral degree in one of the branches of hydraulic engineering.

Social Context and Future Prospects

In the twenty-first century, hydraulic engineering has become closely tied to environmental engineering. Reservoir operators plan and vary water releases to keep downstream creeks wet, thus protecting the biological life in the ecosystem.

Clean energy is the way to ensure sustainability of the planet's resources. Hydroelectric power generation is a form of clean energy. Energy generated by ocean waves is a developing and promising field, although wave power technologies still face technical challenges.

Bassam Kassab, B.Eng., M.Eng., M.Sc.

Further Reading

Basha, Habib A. "Nonlinear Reservoir Routing: Particular Analytical Solution." *Journal of Hydraulic Engineering* 120, no. 5 (May, 1994): 624-632. Presents a mathematical solution for the flood routing equations in reservoirs.

Basha, Habib A., and W. El-Asmar. "The Fracture Flow Equation and Its Perturbation Solution." *Water Resources Research* 39, no. 12 (December, 2003): 1365. Shows that the perturbation method could be used not only on steady and transient water distribution problems but also on any nonlinear problem.

Boulos, Paul F. "H2ONET Hydraulic Modeling." *Journal of Water Supply Quarterly, Water Works Association of the Republic of China (Taiwan)* 16, no. 1 (February, 1997): 17-29. Introduces the use of modeling software (the H2ONET) in water distribution networks.

Boulos, Paul F., Kevin E. Lansey, and Bryan W. Karney. *Comprehensive Water Distribution Systems Analysis Handbook for Engineers and Planners.* 2d ed. Pasadena, Calif.: MWH Soft Press, 2006. Includes chapters on master planning and water-quality simulation.

Chow, Ven Te. *Open-Channel Hydraulics.* 1959. Reprint. Caldwell, N.J.: Blackburn, 2008. Contains chapters on uniform flow, varied flow, and unsteady flow.

Finnemore, E. John, and Joseph B. Franzini. *Fluid Mechanics With Engineering Applications.* 10th international ed. Boston: McGraw-Hill, 2009. Features chapters on kinematics, energy principles, hydrodynamics, and forces on immersed bodies.

Walski, Thomas M. *Advanced Water Distribution Modeling and Management.* Waterbury, Conn.: Haestad Methods, 2003. Examines water distribution modeling, has a chapter on hydraulic transients.

Web Sites

American Society of Civil Engineers
http://www.asce.org

International Association of Hydro-Environment Engineering and Research
http://www.iahr.net

International Association of Hydrological Sciences
http://iahs.info

United States Society on Dams
http://www.ussdams.org

U.S. Bureau of Reclamation
Waterways and Concrete Dams Group
http://www.usbr.gov/pmts/waterways

See also: Civil Engineering; Fluid Dynamics.

I

INDUSTRIAL POLLUTION CONTROL

FIELDS OF STUDY

Chemistry; meteorology; engineering; environmental studies; marine biology.

SUMMARY

Industries that have contributed historically to problems of air, water, and soil pollution are too many to name. Some obvious examples are mining and metallurgy, pharmaceuticals, and all industries processing petroleum. Because different industries use different methods to produce their products, different approaches must be adopted and adapted for controlling pollutants that result.

KEY TERMS AND CONCEPTS

- **Acid Rain:** Occurs when air pollutants rise and collect in clouds, falling later to Earth in the form of chemically charged rain.
- **Adsorption:** Chemical process in which gases, liquids, or any dissolved substances adhere to a solid body next to it or in which it is contained.
- **Digestion:** Biochemical decomposition of organic materials to obtain different, less polluting, organic compounds.
- **Effluent:** Any liquid waste product leaving a source of pollution.
- **Entrainment:** Droplets of mist entering the air when bubbles form in boiling liquids; such droplets can carry polluting particles contained in the liquid source.
- **Sludge Seeding:** Injection of chemically active additives to sewage (or other biodegradable waste) to speed up the process of decomposition.

Definition and Basic Principles

Because various industries use chemicals or fossil fuels, by-products of such processes may occur as waste materials that are either emitted into the air or evacuated by means of polluted water drainage, sometimes into streams, lakes, or groundwater. Control of such pollutants requires both general legal guidelines, provided by the Environmental Protection Agency (EPA), and control or recycling procedures responding to the specific nature of the environmental pollution problem identified.

Background and History

Ultimately, industrial pollution goes back to the origins of the Industrial Revolution in Europe and its migration to the United States in the nineteenth century. In the early Industrial Revolution, methods of production were considerably closer to nature than they would become when heavy industries, most driven by fossil fuel energy, became predominant. Although water-driven industrial machines, particularly in the developing textile industry, might have wasted or diverted water, truly harmful effects (mainly from increases in chemical additives) were slow in coming.

Legislative actions in the United States aiming at control of industrial pollution— starting with concerns about water quality—have been numerous and diverse. Early measures, the Rivers and Harbors Appropriation Act of 1899 and organization of the Smoke Prevention Association in 1907, were followed by a very broad but important precedent: the Public Health Service Act of 1912. Over ensuing decades (except during the heavy emergency production period of World War II), particular industrial sectors became targets of pollution control laws, beginning with the Oil Pollution Act of 1924. Following World War II a major turning point came with the Water Pollution Control Act (WPCA) of 1948. In 1972, after multiple revisions beginning in 1961, when responsibility for the WPCA shifted to the new Department of Health, Education and Welfare, it became known as the Clean Water Act (CWA). This legislation established federally supervised procedures (originally to be overseen by the U.S. Surgeon General) for safeguarding the quality of water in public reservoirs, but also "interstate waters and [their] tributaries" as well as the "sanitary condition of surface and underground waters."

The CWA came in the wake of President Kennedy's "Special Message to the Congress on Natural Resources" on February 23, 1961. After this date it was clear that questions of water pollution needed to be considered along with industrial pollution in the air and in surface and subsurface soils. In 1970, Congress created the EPA. At the same time, President Nixon signed an executive order establishing the National Industrial Pollution Control Council (NIPCC). This special council, which included several representatives of private industry, was to help coordinate responsibilities of the new EPA, with specific emphasis on matters of industrial pollution.

The Pollution Prevention Act of 1990 represented a landmark in industrial pollution control. It shifted control emphasis from efforts to reduce the quantity of pollutants released by manufacturers to efforts to reduce pollution in earlier stages of industrial processing. This goal appeared in Section 6604, which created a new EPA Office of Pollution Prevention. Proponents of the 1990 laws argued that—in the long run—manufacturers stood to save money through "source reduction," both by reducing quantities of expensive (and polluting) materials and replacing them by less costly nonpolluting substitute materials. Manufacturers could also avoid looming liability costs (in the form of fines). The 1990 law also encouraged finding ways to treat industrial waste products so that they might be recycled, either in internal production operations or for use by other manufacturers.

How It Works

Methods of pollution control differ from industry to industry. The selected industries below demonstrate several approaches to pollution problems in different key sectors.

Mining and Metallurgy. Issues of pollution occur in all mining and metallurgical industries. Iron mining and steel production operations face problems widely shared with many other industrial subsectors. Pollution in the steel industry begins at mining sites. Water that has been used to treat extracted iron ore— containing iron sulfide and iron pyrite—has to be disposed of somehow. If oxidation of such materials occurs, harmful secondary chemicals, including sulfuric acid and carbon dioxide, result. Acid water requires chemical neutralization, usually with limestone additives, followed by careful disposal of the resulting sludge.

The later stage leading to steel—when iron ore is reduced in blast furnaces—presents a different set of pollution control problems. Three pollutants are produced at this point. Air and water is polluted when a mixture of coke, iron ore, and calcium carbonate is blasted in extremely hot furnaces. Once the process yields a percentage of molten metal, masses of solid waste need to be disposed of.

It may be possible to extract some useful substances, mainly hydrocarbons, through condensation, and some coke oven gases can be recycled for fuel needs elsewhere. The net effect of blast furnace operations, however, remains a menace to the environment. Efforts have been made to treat iron ore without coke-burning blast furnaces. These include direct reduction of crushed iron ore using a hydrogen additive, which produces pig iron that is transformable into steel in less polluting electric furnaces. Another method—called the HIsmelt process— uses force injected coal and oxygen to pre-reduce iron ore at an earlier stage.

Other problems arise when open hearth (or electric) furnaces and rolling mills are used to produce steel itself. Dangerous substances come from chemical by-products (called mill scale—a layer of metal oxides on new steel) in the rolling process. The steel industry has increasingly converted mill scale sludge into a concentrate that is about 70 percent iron, at the same time removing oil contaminants that—in earlier mill scale disposal processes—escaped into the air.

Control of pollution in the giant copper industry involves similar chemical by-products. Methods of control, however, particularly avoiding the need for high- temperature smelting, are different. Copper can be obtained by solvent-extraction means: Weak acid is percolated, both through ore and copper-bearing waste materials that would otherwise enter slag heaps. Copper ores containing sulfides, however, require new forms of treatment that use environmentally friendly bacterial additions.

Cement Industry. The cement industry ranks ninth in the list of major polluters in the United States; it is also a major (sixth-ranking) consumer of energy. After crushing limestone, shale, and sand components, mixed materials pass through high-temperature kilns. Here, a major pollutant, kiln dust, composed of at least eight polluting chemicals, enters the air.

Developing higher-efficiency machinery to filter kiln dust can help reduce its harmful chemical content. The cement industry does, however, have other commercially attractive options for disposal of this potential pollutant. Certain (not all) agricultural soils, for example, can be fertilized with kiln dust and it can also be used as a cattle grain supplement.

Pharmaceutical Industry. As a subbranch of the total chemical industry, the pharmaceutical industry must be as closely supervised as a potential polluter as it is for potentially dangerous contributions to commercial medicine. Some key processes, particularly fermentation of organic substances that yield antibiotics, produce harmful waste products. Some of these can be neutralized by careful calculation of biological additives to what are essentially pharmaceutical septic tanks.

In cases where wastes contain highly concentrated chemicals, neutralization (or adjustment of base-acid pH ratios) can be obtained via inorganic chemical additives. It should be noted, however, that such procedures—unless they are carefully managed by chemical experts—can produce other forms of risk, in some cases leading to fires or explosions.

Applications and Products

Because polluting chemicals are frequently combatted by using other chemicals that neutralize their noxious effects, a number of potentially beneficial products may eventually come out of pollution control research. Any ongoing results of such specialized research are known to specialized chemists and as such remain beyond the ken of the general public.

By contrast, the public is becoming increasingly aware of much less high-tech applications that can make recycling not only ecologically beneficial but a significant contribution to middle levels of the economy. Most obvious are efforts to recycle paper products. Even though waste paper is biodegradable and therefore not highly polluting itself, industrial processes for reusing wastepaper not only save trees, they produce less pollution.

A similar observation applies to the recycling of plastic materials, all of which are in one way or another by-products of petroleum. Even the simplest methods of recycling plastics, therefore, promise reduced levels of industrial pollution linked to extraction and processing of crude oil.

In terms of higher levels of technology, each time a major pollution problem occurs, scientists and engineers find themselves recruited to develop effective ways to deal with the specifics of each occurrence. The variety of submarine applications that were tried (some for the first time) during the months-long British Petroleum (BP) *Deepwater Horizon* oil spill in the Gulf of Mexico in 2010 is a prime example of this challenge. The oil spill, which began after a violent explosion and fire destabilized the drilling platform itself, illustrated another characteristic of pollution control demands: Highly specialized technical applications are used to prevent fire-producing disasters, but, if such fires occur, they must be controlled and extinguished with a minimum of pollution.

Finally, the BP oil spill, and even more dramatically, the 2011 nuclear reactor catastrophe in Fukushima, Japan, that followed the earthquake provided the public with unprecedented exposure to diverse scientific applications, some of which admittedly fell short of their stated goals.

Impact on Industry

Pollution control obviously comes with a price tag. At each stage of new legislation the question of how costs can be reduced is raised. In recent years, major efforts have been made to recycle a number of industrially produced metals, particularly scrap copper. In the United States, more than half the copper used each year comes from recycling. Although reuse means less pollution, any eventual reduction in demand for products made from new raw materials can have economic repercussions for key segments of the industrial sector.

If one adopts a longer-term perspective, there is a real possibility that a number of key U.S. industries—faced with the necessity of revamping their mode of production to meet stricter pollution standards—may either cease operations altogether, or restructure their operations by resorting to a form of outsourcing to industrial sites in other countries where antipollution legislation is less stringent.

One of the biggest questions that loomed as the twenty-first century began was how to balance the risk of widespread ecological damage with what appears to be a binding necessity on a global level: expanding activities of the oil drilling and oil refining industry. The impact of the 2010 BP oil spill in the Gulf of

Mexico is a major example of this quandary.

On the one hand, the enormous economic costs incurred to control a spill of this magnitude is in most cases billed to the private industrial concerns responsible for the pollution. On the other hand, if and when such procedures achieve the longer-term goal of taking measures to prevent similar troubles on the same or similar sites, the petroleum industry in particular seems to hold a position of bargaining with various agencies over terms governing their future operations. Such terms might include resumption of operations in the same location with similar equipment, even if public opinion seems skeptical of pampering the petroleum industry.

Needless to say, it is unlikely that large-scale petroleum companies would willingly accept government-imposed cutbacks in the scale of their operations, especially in areas where known oil reserves promise continuing profits.

Increasingly stringent demands by governments for costly preventive measures may, however, have the effect of raising production costs. In such cases, the petroleum industry's decision to pass such costs on to consumers (not only private consumers, but other fossil-fuel-dependent industries) could produce a spiraling impact, with consequences that would have to be measured carefully— if only to maintain necessary levels of stability in all sectors of the economy.

Careers and Course Work

Anyone seeking a career in industrial pollution control should plan his or her course work so that essential scientific fields, especially chemistry and biology, are covered at the undergraduate level and narrowed to specific subareas in graduate course work. Students nearing the end of their university training may qualify for a number of federally funded short- or medium-term or summer internships, especially with the EPA. Although many openings are for clerical work, these temporary posts can provide valuable practical training in areas such as pollution policy analysis. Where applicants possess fairly advanced technical skills, work on EPA in-field projects is a possibility.

The EPA is the most important governmental agency involved in industrial pollution matters. It is headquartered in Washington, D.C., and has ten regional offices, all of which are equipped with laboratories staffed with trained experts, many of whom specialize in industrial pollution treatment.

One branch of the EPA, the Office of Chemical Safety and Pollution Prevention, offers an example of dovetailing between different specialized fields of training and employment. Its functions involve toxic substances (including pesticide manufacture), food and drug safety, and pollution prevention in general.

Needless to say, almost every industry, heavy and light, hires personnel trained to trace sources of pollution in their production process and to devise methods to control it. In the automobile industry, such control concerns are present at two levels: factory hazardous waste and lowering exhaust emissions from new models.

Most local governments staff offices responsible

Fascinating Facts About Industrial Pollution Control

- So-called carbon taxes require industries that release CO_2 into the air to pay a supplemental tax to help fund vital environmental-preservation measures. A carbon tax was introduced in 2010 on the coal industry in India. The first European country to impose a carbon tax, in 1990, was Finland. At least ten countries in Europe had some form of carbon tax by 2010.
- In unusual cases, such as in a court case involving Iowa farmers, local private interests may prefer continuation of what amounts to industrial pollution. In the case noted above, local farmers successfully challenged EPA sanctions against a cement factory the dust of which was polluting the air—but depositing free fertilizing chemicals on surrounding farmlands.
- Despite continuing concerns about carbon dioxide emissions from automobile exhaust, much higher totals of carbon dioxide gas enter the atmosphere from industrial pollution.
- Industrial pollution affects not only the local regions in which industries are found. Pollutants spread rapidly in very wide patterns, either in the atmosphere or in underground water systems.
- Increasingly, methods are being found to recycle industrial waste products that can be useful and even sold commercially. This is sometimes done by means of additional chemical treatment. Plastic and paper recycling have become part of daily consumer habits all over the globe.

for pollution control. Among many possible examples, one can cite local air pollution control districts, such as the one created as early as 1971 in California's Glenn County (near Sacramento), which maintainsclose association with the state's department of agriculture. One of its key concerns is to maintain informational networks with local industries that may need technical consultation based on specific local resources of mixed private and governmental agencies.

Social Context and Future Prospects

In many formerly dominant industrial countries, especially the United States and European countries such as England, France, and Germany, economic and technological changes have made certain industrial subsectors much less important than they were up to the second half of the twentieth century. Metallurgy in general is an example of such changing conditions.

Even though some formerly very polluting heavy industries have reduced operations (or closed down entirely), lingering residue from earlier years of maximum production poses major health and safety problems for locally surrounding communities. Public reactions to such ecological dangers have in some cases led to citizen-based movements to oblige responsible powers, whether owners of industries themselves or governmental agencies, to clean up residual pollution.

A number of such movements could be cited, some in distant countries such as India (the Vasundhara research and policy advocacy group, for example, which supports initiatives to protect local communities from pollution and natural resource depletion) and Vietnam (see "Further Reading"). More and more cases are surfacing in the United States. Some more or less spontaneous local organizations depend on the media to gain support. An example appeared in *The New York Times* of April 4, 2011: An interfaith local movement pit itself against a defunct chromium- producing plant in Jersey City, New Jersey. Other more structured bodies, such Nexleaf Analytics (a private group working with the Center for Embedded Networked Sensing at University of California, Los Angeles), seek to use advanced technology to calculate levels of danger from various pollutants.

In most developed and developing societies, emphasis on recycling programs has begun to have an effect, not only on basic consumer patterns (either by reducing or eliminating entirely the use of certain industrial products that may end up as harmful additions to waste-disposal programs), but also on industries themselves. Examples of the latter may involve increased attention to ways in which polluting industrial by-products can be used in followup procedures that actually convert them to useful (and, in some cases, commercially valuable) by-products.

As technology advances, it is to be hoped that increasingly effective methods will be found to control (or even better, prevent) industrial pollution. For example, research should yield wider use of good chemicals to reduce the polluting effects of bad chemicals that must be used by some industrial manufacturers. Although most contributors to a special 1994-1995 issue of the *Georgia Law Review* dedicated to technical aspects of prevention (see "Further Reading"), a main suggestion by all was that future cooperation between scientists, lawmakers, and industrial manufacturers is becoming increasingly critical.

Such cooperation depends, however, on legislators' willingness to remain firm in the face of pressures to water down control regulations. An example of this problem appeared in *The New York Times* of April 16, 2011, citing Republican Party efforts to cut back both the EPA's budget and governmental regulations controlling industrial wastewater disposal. Opponents of EPA controls argued that the extra costs involved represented an additional burden to industries and might force them to cut back economically critical operations.

Byron D. Cannon, M.A., Ph.D.

Further Reading

Bishop, Paul L. *Pollution Prevention: Fundamentals and Practice.* Boston: McGraw-Hill, 2000. A study emphasizing prevention before pollution becomes a major problem involving more costly controls.

Eckenfelder, W. Wesley. *Industrial Water Pollution Control.* 3d ed. Boston: McGraw-Hill, 2000. Covers diverse aspects of the most common and, along with air pollution, potentially disastrous aspects of industrial waste control, including not only pollution of waterways but also groundwater reserves.

Hirschorn, Joel S. "Pollution Prevention Comes of Age." *Georgia Law Review* 29 (1994-1995): 325-347. This is a general review of what was, after the Pollution Prevention Act of 1990, a relatively new emphasis on methods to stop pollution at early to midpoint stages of industrial production, as well as finding ways to produce reusable products from industrial waste.

Lund, Herbert F., ed. *Industrial Pollution Control Handbook.* New York: McGraw-Hill, 1971. Although somewhat dated, this volume contains very useful technical information on different control processes in different industrial sectors.

Rodgers, William H., Jr. "The National Industrial Pollution Control Council: Advise or Collude?" *Boston College Law Review* vol. 13, issue 4 (1972): 719-747. A highly qualified professional familiar with the workings of government and industry wrote this criticism of the administrative steps to coordinate pollution control agencies.

Sell, Nancy J. *Industrial Pollution Control: Issues and Techniques.* 2d ed. New York: Van Nostrand Reinhold, 1992. This general study of different industrial pollution sectors continues to be widely cited in more recent literature for its comprehensive and accurate analysis.

WEB SITES

Air and Waste Management Association
http://www.awma.org

National Pollution Prevention Roundtable
http://www.p2.org

Worldwide Pollution Control Association
http://wpca.info

See also: Engineering; Meteorology; Thermal Pollution Control.

INFORMATION TECHNOLOGY

FIELDS OF STUDY

Information systems; computer science; software engineering; systems engineering; networking; computer security; Internet and Web engineering; computer engineering; computer programming; usability engineering; business; mobile computing; project management; information assurance; computer-aided design and manufacturing; artificial intelligence; knowledge engineering; mathematics; robotics.

SUMMARY

Information technology is a new discipline with ties to computer science, information systems, and software engineering. In general, information technology includes any expertise that helps create, modify, store, manage, or communicate information. It encompasses networking, systems management, program development, computer hardware, interface design, information assurance, systems integration, database management, and Web technologies. Information technology places a special emphasis on solving user issues, including helping everyone learn to use computers in a secure and socially responsible way.

KEY TERMS AND CONCEPTS

- **Application Software:** Software, such as word processors, used by individuals at home and at work.
- **Cloud Computing:** Software developed to execute on multiple servers and delivered by a Web service.
- **Computer Hardware:** Computing devices, networking equipment, and peripherals, including smartphones, digital cameras, and laser printers.
- **Human-Computer Interaction:** Ways in which humans use or interact with computer systems.
- **Information Assurance:** Discipline that not only secures information but also guarantees its safety and accuracy.
- **Information Management:** Storage, processing, and display of information that supports its use by individuals and businesses.
- **Networking:** Sharing of computer data and processing through hardware and software.
- **Programming:** Development of customized applications, using a special language such as Java.
- **Software Integration:** Process of building complex application programs from components developed by others.
- **Systems Integration:** Technique used to manage the computing systems of companies that use multiple types of software, such as an Oracle database and Microsoft Office.
- **Web Systems:** Database-driven Web applications, such as an online catalog, as well as simple Web sites.

Definition and Basic Principles

Information technology is a discipline that stresses systems management, use of computer applications, and end-user services. Although information technology professionals need a good understanding of networking, program development, computer hardware, systems development, software engineering, Web site design, and database management, they do not need as complete an understanding of the theory of computer science as computer scientists do. Information technology professionals need a solid understanding of systems development and project management but do not need as extensive a background in this as information systems professionals. In contrast, information technology professionals need better interpersonal skills than computer science and information systems workers, as they often do extensive work with end-users.

During the second half of the twentieth century, the world moved from the industrial age to the computer age, culminating in the development of the World Wide Web in 1990. The huge success of the Web as a means of communications marked a transition from the computer age, with an emphasis on technology, to the information age, with an emphasis on how technology enhances the use of information. The Web is used in many ways to enhance the use and transfer of information, including telephone service, social networking, e-mail, teleconferencing, and even radio and television programs. Information technology contains a set of tools that make it easier to create, organize, manage, exchange, and use information.

Background and History

The first computers were developed during World War II as an extension of programmable calculating machines. John von Neumann added the stored program concept in 1944, and this set off the explosive growth in computer hardware and software during the remainder of the twentieth century. As computing power increased, those using the computers began to think less about the underlying technology and more about what the computers allowed them to do with data. In this way, information, the organization of data in a way that facilitates decision making, became what was important, not the technology that was used to obtain the data. By 1984, organizations such as the Data Processing Management Association introduced a model curriculum defining the training required to produce professionals in information systems management, and information systems professionals began to manage the information of government and business.

In 1990, Computer scientist Tim Berners-Lee developed a Web browser, and the Web soon became the pervasive method of sharing information. In addition to businesses and governmental agencies, individuals became extensive users and organizers of information, using applications such as Google (search engine) and Facebook (social networking). In the early 2000's, it became clear that a new computer professional, specializing in information management, was needed to complement the existing information systems professionals. By 2008, the Association of Computing Machinery and the IEEE Computer Society released their recommendations for educating information technology professionals, authenticating the existence of this new computing field. The field of information technology has become one of the most active areas in computer science.

How It Works

The principal activity of information technology is the creation, storage, manipulation, and communication of information. To accomplish these activities, information technology professionals must have a background from a number of fields and be able to use a wide variety of techniques.

Networking and Web Systems. Information is stored and processed on computers. In addition, information is shared by computers over networks. Information technology professionals need to have a good working knowledge of computer and network systems to assist them in acquiring, maintaining, and managing these systems. Of the many tasks performed by information technology professionals, none is more important than installing, updating, and training others to use applications software. The Web manages information by storing it on a server and distributing it to individuals using a browser, such as Internet Explorer or Foxfire. Building Web sites and applications is a big part of information technology and promises to increase in importance as more mobile devices access information provided by Web services.

Component Integration and Programming. Writing programs in a traditional language such as Java will continue to be an important task for information technology professionals, as it had been for other computer professionals in the past. However, building new applications from components using several Web services appears to be poised to replace traditional programming. For all types of custom applications, the creation of a user-friendly interface is important. This includes the careful design of the screen elements so that applications are easy to use and have well thought-out input techniques, such as allowing digital camera and scanner input at the appropriate place in a program, as well as making sure the application is accessible to visually and hearing impaired people.

Databases. Data storage is an important component of information technology. In the early days of computers, most information was stored in files. The difficulty of updating information derived from a file led to the first database systems, such as the information management system created by IBM in the 1960's. In the 1970's, relational databases became the dominant method of storing data, although a number of competing technologies are also in use. For example, many corporate data are stored in large spreadsheets, many personal data are kept in word-processing documents, and many Web data are stored directly in Web pages.

Information Security, Privacy, and Assurance. Regardless of how information is collected, stored, processed, or transferred, it needs to be secure. Information security techniques include developing security plans and policies, encrypting sensitive data during storage and transit, having good incident response and recovery teams, installing adequate

Fascinating Facts About Information Technology

- In 1993, Intel introduced the Pentium processor as a Cray (supercomputer) on a chip. It contained 3.1 million transistors and was capable of performing 112 million instructions per second. The Pentium was the first of a number of low-cost but powerful processors that changed information technology.
- The iPhone, introduced in 2007, is a powerful and popular Internet-enabled smartphone. It allows users to access e-mail, the Web, and many cloud applications in a touch-friendly environment and promises to change the way people access Internet content.
- In 1981, IBM introduced a personal computer, which was copied by many other companies. Soon businesses, government agencies, and individuals were able to use computers to improve their productivity.
- Microsoft introduced Windows 3.1 in 1992, and its immediate popularity made Microsoft Windows the most popular operating system for personal computers.
- In 1980, Edgar Codd published "A Relational Model of Data for Large Shared Data Banks," a paper that defined the relational data model and led to the development of relational databases such as Oracle and DB2, which became the dominant storage technology for information technology.
- In 1989, Tim Berners-Lee introduced HTML (HyperText Markup Language) as a way to represent data on the Web. Later, Java and JavaScript were added to make programming on the Web almost as rich as programming on the desktop.
- Cloud computing, or providing software as a set of services to a computing device, is predicted to be the primary software model for the future. The origins of cloud computing go back to the early days of distributed computing, but exactly who should get credit for this exciting new area is not clear.

security controls, and providing security training for all levels of an organization. In addition to making sure that an organization's data are secure, it is also important to operate the data management functions of an organization in such a way that each individual's personal data are released only to authorized parties. Increasingly, organizations want information handled in such a way that the organization can be assured that the data are accurate and have not been compromised.

Professionalism. Information technology professionals need to inform themselves of the ethical principles of their field (such as the Association of Computing Machinery' s code of ethical and professional conduct) and conscientiously follow this code. Increasingly, laws are being passed about how to handle information, and information technology professionals need to be aware of these laws and to follow them.

Applications and Products

There are more useful applications of information technology than can be enumerated, but they all involve making information accessible and usable.

End-User Support. One of the most important aspects of information technology is its emphasis on providing support for a wide variety of computer users. In industrial, business, and educational environments, user support often starts immediately after someone is given access to a computer. An information technology professional assists the user with the login procedure, shows him or her how to use e-mail and various applications. After this initial introduction, technology professionals at a help desk answer the user's questions. Some companies also provide training courses. When hardware and software problems develop, users contact the information technology professionals in the computer support department for assistance in correcting their problems.

Information technology educational programs usually cover the theory and practice of end-user support in their general courses, and some offer courses dedicated to end-user support. This attention to end-user support is quite different from most programs in computer science, information systems, or software engineering and is one of the most important differences between information technology programs and other areas of computer science.

The Electronic Medical Record. A major use of information technology is in improving the operation of hospitals in providing medical services, maintaining flexible schedules, ensuring more accurate billing, and reducing the overall cost of operation. In much the same way, information technology helps doctors in clinics and individual practices improve the quality of their care, scheduling, billing, and operations. One of the early successes in using computers in medicine was the implementation of e-prescribing. Doctors can easily use the Internet to determine the availability of a drug needed by a patient, electronically send the prescription to the correct pharmacy, and bill the patient, or their insurance company, for the prescription. Another success story for medical information technology is in the area of digital imaging. Most medical images are created digitally, and virtually all images are stored in a digital format. The Digital Imaging and Communications in Medicine (DICOM) standard for storage and exchange of digital images makes it possible for medical images to be routinely exchanged between different medical facilities. Although the process of filling prescriptions and storing images is largely automated, the rest of medicine is still in the process of becoming fully automated. However, computers are likely to become as much a part of medical facilities as they are of financial institutions.

One of the major goals of information technology is the development of a true electronic medical record (EMR). The plan is to digitize and store all the medical information created for an individual by such activities as filling out forms, taking tests, and receiving care. This electronic medical record would be given to individuals in its entirety and to hospital and insurance companies as needed. Electronic medical records would be available to the government for data mining to determine health policies. Many professionals in information technology are likely to be involved in the collection, storage, and securitization of electronic medical records.

Geographic Information Management. Maps have been used by governments since the beginning of recorded history to help with the process of governing. Geographic information systems (GIS) are computerized systems for the storage, retrieval, manipulation, analysis, and display of geographic data and maps. In the 1950's, the first GIS were used by government agencies to assist in activities such as planning, zoning, real estate sales, and highway construction. They were developed on mainframes and accessed by government employees on behalf of those desiring the information contained in the systems. The early GIS required substantial computing resources because they used large numbers of binary maps but were relatively simple as information retrieval systems.

Geographic information systems still serve the government, but they are also used by industry and educational institutions and have gained many applications, including the study of disease, flood control, census estimates, and oil discovery. GIS information is easily accessed over the Internet. For example, zoning information about Dallas, Texas, is readily available from the city's Web site. Many general portals also provide GIS information to the public. For example, Google Maps allows travelers to print out their route and often provides a curbside view of the destination. Modern geographic information systems are complex, layered software systems that require expertise to create and maintain.

Network Management. Computer networking developed almost as quickly as the computing field itself. With the rapid acceptance of the Internet in the 1990's, computer connectivity through networking became as common as the telephone system. As computer networks developed, a need also developed for professionals to manage these networks. The first network managers were generally hardware specialists who were good at running cable through a building, adding hardware to a computer, and configuring network software so that it would work. Modern network specialists need all these early capabilities but also must be network designers, managers, and security experts.

Database Management. One of the major functions of information technology is the storage of information. Information consists of data organized for meaningful usage. Both data and information regarding how the data are related can be stored in many ways. For example, many corporate data are stored in word-processing documents, spreadsheets, and e-mails. Even more corporate data are stored in relational databases such as Oracle. Most businesses, educational institutions, and government agencies have database management specialists who spend most of their time determining the best logical and physical models for the storage of data.

Mobile Computing and the Cloud. Mobile phones have become so powerful that they rival small computers, and Web services and applications for computers or mobile devices are being developed at a breathtaking rate. These two technological developments are working together to provide many Web applications for smartphones. For example, any cell phone can play MP3 music. Literally dozens of Web services have been developed that will automatically download songs to cell phones and bill the owner a small fee. In addition, many smartphone applications will download and play songs from Web services.

Cloud computing and storage is beginning to change how people use computers in their homes. Rather than purchasing computers and software, some home users are paying for computing as a service and storing their data in an online repository. Microsoft's Live Office and Google's Apps are two of the early entrants in the software and storage as a service market.

Computer Integrated Manufacturing. Computer integrated manufacturing (CIM) provides complete support for the business analysis, engineering, manufacturing, and marketing of a manufactured item, such as a car. CIM has a number of key areas including computer-aided design (CAD), supply chain management, inventory control, and robotics.

Many of the areas of CIM require considerable use of information technology. For example, CAD programs require a good data management program to keep track of the design changes, complex algorithms to display the complicated graphical images, fast computer hardware to display the images quickly, and good project management tools to assist in completing the project on time. Robots, intelligent machines used to build products, require much innovative computer and machine hardware, very complex artificial intelligence algorithms simulating human operation of machine tools, and sophisticated computer networks for connecting the robotic components on the factory floor.

Computer Security Management. One of the most active areas of information technology is that of computer security management. As theoretical computer scientists develop new techniques to protect computers, such as new encryption algorithms, information technology specialists work on better ways to implement these techniques to protect computers and computer networks. Learning how to acquire the proper hardware and software, to do a complete risk analysis, to write a good security policy, to test a computer network for vulnerabilities, and to accredit a business's securing of its computers requires all the talents of an information technologist.

Web Site Development. The World Wide Web was first introduced in 1990, and since that time, it has become one of the most important information distribution technologies available. Information technology professionals are the backbone of most Web site development teams, providing support for the setup and maintenance of Web servers, developing the HTML pages and graphics that provide the Web content, and assisting others in a company in getting their content on the Web.

Impact on Industry

The use of computers became pervasive in companies, government agencies, and universities in the second half of the twentieth century. Information technology gradually developed as a separate discipline in the late 1970's. By 1980, companies, government agencies, and universities recognized that it was the information provided by the computer that was important, not the computers themselves. Although access to and the ability to use the information was what people wanted, getting the information was difficult and required a number of skills. These included the building and maintenance of computers and networks, the development of computer software, providing end-user support, and using management information and decision support systems to add value to the information. Government and industry made their information needs known; universities and some computer manufacturers did basic and applied research developing hardware and software to meet these needs, and information technology professionals provided support to allow those in government and industry to actually use the information.

Although estimates of the total value of the goods and services provided by information technology in a year are impossible to determine, it is estimated to be close to $1 trillion per year.

Government, Industry, and University Research. The United States government exerts a great influence on the development of information technology because it purchases a large amount of computer hardware and software and makes substantial use of the information derived from its computers and networks. Companies such as IBM, SAP, and Oracle have geared much of their research and development toward meeting the needs of their government customers. For example, IBM's emphasis on COBOL for its mainframes was a result of a government requirement that all computers purchased by the government support COBOL compilers. SAP's largest customer base in the United States is federal, state, and local government agencies. One of the main reasons for Oracle's acquisition of PeopleSoft was to enhance its position in the educational enterprise resource planning market.

The National Science Foundation, the National Institutes of Health, and the U.S. Department of Defense have always provided support for university research for information technology. For example, government agencies supporting cancer research gave a great deal of support to Oracle and to university researchers in the development of a clinical research data management system to better track research on cancer. The government also has been a leader in developing educational standards for government information technology professionals, and these standards have greatly influenced university programs in information technology. For example, after the September 11, 2001, terrorist attacks on the United States, the National Security Agency was charged with coming up with security standards that all government employees needed to meet. The agency created an educational branch, the Committee on National Security Systems, which defined the standards and developed a set of security certifications to implement the standards. Its certifications have had a significant influence on the teaching of computer security.

The computer industry has been very active in developing better hardware and software to support information technology. For example, one of the original relational database implementations, System R and SQL/Data System, was developed at the IBM Research Laboratory in San Jose, California, in the 1970's, and this marked the beginning of the dominance of relational databases for data storage. There have also been a number of successful collaborations between industry and universities to improve information technology. For example, the Xerox PARC (later PARC) laboratories, established as a cooperative effort between Xerox and several universities, developed the Ethernet, the graphical user interface, and the laser printer. Microsoft has one of the largest industry research programs, which supports information technology development centers in Redmond, Washington; Cambridge, England; and six other locations. In 2010, Microsoft announced Azure, its operating system to support cloud computing. Azure will allow individuals to access information from the cloud from almost anywhere on a wide variety of devices, providing information on demand.

Information Technology Companies. Many successful large information technology companies are operating in the United States. Dell and HP (Hewlett Packard) concentrate on developing the computing hardware needed to store and process information. IBM produces both the hardware and software needed to store and process information. In the early 2010's, Microsoft produced more information technology software than any other company, but SAP was reported to have the greatest worldwide sales volume. Oracle, the largest and most successful database company, was the leader in providing on-site storage capacity for information, but Microsoft and Google had begun offering storage as a service over the Internet, a trend predicted to continue. Smaller but important information technology companies include Adobe Systems, which developed portable document format (PDF) files for transporting information and several programs to create PDF files.

Careers and Course Work

The most common way to prepare for a career in information technology is to obtain a degree in information systems. Students begin with courses in mathematics and business and then take about thirty hours of courses on information systems development. Those getting degrees in information systems often take information technology jobs as systems analysts, data modelers, or system managers, performing tasks such as helping implement a new database management system for a local bank.

A degree in computer science or software engineering is another way to prepare for an information technology career. The courses in ethics, mathematics, programming, and software management that are part of these majors provide background for becoming an information technology professional. Those getting degrees in these areas often take information technology jobs as programmers or system managers, charged with tasks such as helping write a new program to calculate the ratings of a stock fund.

Degree programs in information technology are available at some schools. Students begin with courses in problem solving, ethics, communications, and management, then take about thirty-six hours of courses in programming, networking, human-computer communications, databases, and Web systems.

Those getting an information technology degree often take positions in network management, end-user support, database management, or data modeling. A possible position would be a network manager for a regional real estate brokerage.

In addition to obtaining a degree, many information technology professionals attend one of the many professional training and certification programs. One of the original certification programs was the Novell Certified Engineer, followed almost immediately by a number of Microsoft certification programs. These programs produced many information technology professionals for network management. Cisco has created a very successful certification program for preparing network specialists in internetworking (connecting individual local-area networks to create wide-area networks).

Social Context and Future Prospects

The future of information technology is bright, with good jobs available in end-user support, network management, programming, and database management. These areas, and the other traditional areas of information technology, are likely to remain important. Network management and end-user support both appear to be poised for tremendous growth over the next few years. The growth in the use of the Internet and mobile devices requires the support of robust networks, and this, in turn, will require a large number of information technology professionals to install, repair, update, and manage networks. The greater use of the Internet and mobile devices also means that there will be a large number of new, less technically aware people trying to use the Web and needing help provided by information technology end-user specialists.

A large number of new applications for information technology are being developed. One of the areas of development is in medical informatics. This includes the fine-tuning of existing hospital software systems, development of better clinical systems, and the integration of all of these systems. The United States and many other nations are committed to the development of a portable electronic health record for each person. The creation of these electronic health records is a massive information technology project and will require a large workforce of highly specialized information technology professionals. For example, to classify all of the world's medical information, a new language, Health Level 7, has been created, and literally thousands of specialists have been encoding medical information into this language.

The use of mobile devices to access computing from Web services is another important area of information technology development. Many experts believe the use of Web-based software and storage, or cloud computing, may be the dominant form of computing in the future, and it is likely to employ many information technology professionals.

Another important area of information technology is managing information in an ethical, legal, and secure way, while ensuring the privacy of the owners of the information. Security management specialists must work with network and database managers to ensure that the information being processed, transferred, and stored by organizations is properly handled.

George M. Whitson III, B.S., M.S., Ph.D.

Further Reading

Miller, Michael. *Cloud Computing: Web-Based Applications That Change the Way You Work and Collaborate.* Indianapolis, Ind.: Que, 2009. Provides an excellent overview of the new method of using computing and storage on demand over the cloud.

Reynolds, George. *Information Technology for Managers.* Boston: Course Technology, 2010. A short but very complete introduction to information technology by one of the leading authors of information technology books.

Senn, James. *Information Technology: Principles, Practices, and Opportunities.* 3d ed. Upper Saddle River, N.J.: Prentice Hall, 2004. An early information technology book that emphasizes the way information technology is used and applied to problem solving.

Shneiderman, Ben, and Catherine Plaisant. *Designing the User Interface.* Boston: Addison-Wesley, 2010. An excellent book about developing good user interfaces. In addition to good coverage on developing interfaces for desktop applications, it covers traditional Web applications and applications for mobile devices.

Stair, Ralph, and George Reynolds. *Principles of Information Systems.* 9th ed. Boston: Course Technology, 2010. Contains not only a complete introduction to systems analysis and design but also a very good description of security, professionalism, Web technologies, and management information systems.

Web Sites

Association for Computing Machinery
http://www.acm.org

CompTIA
http://www.comptia.org/home.aspx

IEEE Computer Society
http://sites.computer.org/ccse

Information Technology Industry Council
http://www.itic.org

TechAmerica
http://www.itaa.org

See also: Communication; Computer Networks; Computer Science; Internet and Web Engineering; Software Engineering.

INTEGRATED-CIRCUIT DESIGN

FIELDS OF STUDY

Mathematics; physics; electronics; quantum mechanics; mechanical engineering; electronic engineering; graphics; electronic materials science.

SUMMARY

The integrated circuit (IC) is the essential building block of modern electronics. Each IC consists of a chip of silicon upon which has been constructed a series of transistor structures, typically MOSFETs, or metal oxide semiconductor field effect transistors. The chip is encased in a protective outer package whose size facilitates use by humans and by automated machinery. Each chip is designed to perform specific electronic functions, and the package design allows electronics designers to work at the system level rather than with each individual circuit. The manufacture of silicon chips is a specialized industry, requiring the utmost care and quality control.

KEY TERMS AND CONCEPTS

- **Clock:** A multitransistor structure that generates square-wave pulses at a fixed frequency; used to regulate the switching cycles of electronic circuitry.
- **Fan-Out:** The number of transistor inputs that can be served by the output from a single transistor.
- **Flip-Flop:** A multitransistor structure that stores and maintains a single logic state while power is applied or until it is instructed to change.
- **Vibrator:** A multitransistor structure that outputs a constant sinusoidal pulse by switching between two unstable states, rather than by modifying an input signal.

Definition and Basic Principles

An integrated circuit, or IC, is an interconnected series of transistor structures assembled on the surface of a silicon chip. The purpose of the transistor assemblages is to perform specific operations on electrical signals that are provided as inputs. All IC devices can be produced from a structure of four transistors that function to invert the value of the input signal, called NOT gates or inverters. All digital electronic circuits are constructed from a small number of different transistor assemblages, called gates, that are built into the circuitry of particular ICs. The individual ICs are used as the building blocks of the digital electronic circuitry that is the functional heart of modern digital electronic technology.

The earliest ICs were constructed from bipolar transistors that function as two-state systems, a system high and a system low. This is not the same as "on" and "off," but is subject to the same logic. All digital systems function according to Boolean logic and binary mathematics.

Modern ICs are constructed using metal oxide semiconductor field effect transistor (MOSFET) technology, which has allowed for a reduction in size of the transistor structures to the point in which, at about 65 nanometers (nm) in size, literally millions of them can be constructed per square centimeter (cm^2) of silicon chip surface. The transistors function by conducting electrical current when they are biased by an applied voltage. Current ICs, such as the central processing units (CPUs) of personal computers, can operate at gigahertz frequencies, changing that state of the transistors on the IC billions of times per second.

Background and History

Digital electronics got its start in 1906, when American inventor Lee de Forest constructed the triode vacuum tube. Large, slow, and power-hungry as they were, vacuum tubes were nevertheless used to construct the first analogue and digital electronic computers. In 1947, American physicist William Shockley and colleagues constructed the first semiconductor transistor junction, which quickly developed into more advanced silicon-germanium junction transistors.

Through various chemical and physical processes, methods were developed to construct small transistor structures on a substrate of pure crystalline silicon. In 1958, American physicist and Nobel laureate Jack Kilby first demonstrated the method by constructing germanium-based transistors as an IC chip, and American physicist Robert Noyce constructed the first silicon-based transistors as an IC chip. The transistor structures were planar bipolar in nature, until the CMOS (complementary metal oxide semiconductor) transistor was invented in 1963. Methods for producing CMOS chips efficiently were not developed for another twenty years.

Transistor structure took another developmental leap with the invention of the field effect transistor (FET), which was both a more efficient design than that of semiconductor junction transistors and easier to effectively manufacture. The MOSFET structure also is amenable to miniaturization and has allowed designers to engineer ICs that have one million or more transistor structures per centimeters squared.

How It Works

The electronic material silicon is the basis of all transistor structures. It is classed as a pure semiconductor. It is not a good conductor of electrical current or insulate well against electrical current. By adding a small amount of some impurity to the silicon, its electrical properties can be manipulated such that the application of a biasing voltage to the material allows it to conduct electrical current. When the biasing voltage is removed, the electrical properties of the material revert to their normal semiconductive state.

Silicon Manufacture. Integrated-circuit design begins with the growth of single crystals of pure silicon. A high-purity form of the material, known as polysilicon, is loaded into a furnace and heated to melt the material. At the proper stage of melt, a seed crystal is attached to a slowly turning rod and introduced to the melt. The single crystal begins to form around the seed crystal and the rotating crystal is "pulled" from the melt as it grows. This produces a relatively long, cylindrical, single crystal that is then allowed to cool and set.

Wafers. From this cylinder, thin wafers or slices are cut using a continuous wire saw that produces several uniform slices at the same time. The slices are subjected to numerous stages of polishing, cleaning, and quality checking, the end result of which is a consistent set of silicon wafers suitable for use as substrates for integrated circuits.

Circuitry. The integrated circuit itself begins as a complex design of electronic circuitry to be constructed from transistor structures and "wiring" on the surface of the silicon wafer. The circuits can be no more than a series of simple transistor gates (such as invertors, AND-gates, and OR-gates), up to and including the extremely complex transistor circuitry of advanced CPUs for computers.

Graphics technology is extremely important in this stage of the design and production of integrated circuits, because the entire layout of the required transistor circuitry must be imaged. The design software also is used to conduct virtual tests of the circuitry before any ICs are made. When the theoretical design is complete and imaged, the process of constructing ICs can begin.

Because the circuitry is so small, a great many copies can be produced on a single silicon wafer. The actual chips that are housed within the final polymer or ceramic package range in size from two to five cm^2. The actual dimensions of the circa 1986 Samsung KS74AHCT240 chip, for example, are just 1 cm x 2 cm. The transistor gate sizes used in this chip are 2 micrometers (um) (2×10^{-6} meters [m]), and each chip contains the circuitry for two octal buffers, constructed from hundreds of transistor gate structures. Transistor construction methods have become much more efficient, and transistor gate sizes are now measured in nanometers (10^{-9} m) rather than um, so that actual chip sizes have also become much smaller, in accord with Moore's law. The gate structures are connected through the formation of aluminum "wires" using the same chemical vapor deposition methodology used to form the silicon oxide and other layers needed.

Photochemical Etching. The transistor structures of the chip are built up on the silicon wafer substrate through a series of steps in which the substrate is photochemically etched and layers of the necessary materials are deposited. Absolute precision and an ultraclean environment are required at each step. The processes are so sensitive that any errant speck of dust or other contaminant that finds its way to the wafer's surface renders that part of the structure useless.

Accounting for losses of functional chips at each stage of the multistep process, it is commonly the case that as little as 10 percent of the chips produced from any particular wafer will prove viable at the end of their construction. If, for example, the procedure requires one hundred individual steps, not including quality-testing steps, to produce the final product, and a mere 2 percent of the chips are lost at each step, then the number of viable chips at the end of the procedure will be 0.98^{100}, or 13.26 percent of the number of chips that could ideally have been produced.

Each step in the formation of the IC chip must be tested to identify the functionality of the circuitry as it is formed. Each such step and test procedure adds significantly to the final cost of an IC. When the ICs are completed, the viable ones are identified, cut from the wafer, and enclosed within a protective casing of resin or ceramic material. A series of leads are also built into this "package" so that the circuitry of the IC chip can be connected into an electronic system.

Applications and Products

Bipolar Transistors and MOSFETS. Transistors are commonly pictured as functioning as electronic on-off switches. This view is not entirely correct. Transistors function by switching between states according to the biasing voltages that are applied. Bipolar switching transistors have a cut-off state in which the applied biasing voltage is too low to make the transistor function. The normal operating condition of the transistor is called the linear state. The saturation state is achieved when the biasing voltage is applied to both poles of the transistor, preventing them from functioning. MOSFET transistors use a somewhat different means, relying on the extent of an electric field within the transistor substrate, but the resulting functions are essentially the same.

The transistor structures that form the electronic circuitry of an IC chip are designed to perform specific functions when an electrical signal is introduced. For simple IC circuits, each chip is packaged to perform just one function. An inverter chip, for example, contains only transistor circuitry that inverts the input signal from high to low or from low to high. Typically, six inverter circuits are provided in each package through twelve contact points. Two more contact points are provided for connection to the biasing voltage and ground of the external circuit. It is possible to construct all other transistor logic gates using just inverter gates. All ICs use this same general package format, varying only in their size and the number of contact points that must be provided.

MOSFETS have typical state switching times of something less than 100 nanoseconds, and are the transistor structures of choice in designing ICs, even though bipolar transistors can switch states faster. Unlike bipolars, however, MOSFETS can be constructed and wired to function as resistors and can be made to a much smaller scale than true resistors in the normal production process. MOSEFTS are easier to manufacture than are bipolars as they can be made much smaller in VLSI (very large scale integration) designs. MOSFETS also cost much less to produce.

NOTs, ANDs, ORs, and Other Gates. All digital electronic devices comprise just a few basic types of circuitry—called logic elements—of which there are two basic types: decision-making elements and storage elements. All logic elements function according to the Boolean logic of a two-state (binary) system. The only two states that are allowed are "high" and "low," representing an applied biasing voltage that either does or does not drive the transistor circuitry to function. All input to the circuitry is binary, as is all output from the circuitry.

Decision-making functions are carried out by logic gates (the AND, OR, NOT gates and their constructs) and memory functions are carried out by combination circuitry (flip-flops) that maintains certain input and output states until it is required to change states. All gates are made up of a circuit of interconnected transistors that produces a specific output according to the input that it receives.

A NOT gate, or inverter, outputs a signal that is the opposite of the input signal. A high input produces a low output, and vice versa. The AND gate outputs a high signal only when all input signals are high, and a low output signal only if any of the input signals are low. The OR gate functions in the opposite sense, producing a high output signal if any of the input signals are high and a low output signal only when all of the input signals are low. The NAND gate, which can be constructed from either four transistors and one diode or five diodes, is a combination of the AND and NOT gates. The NOR gate is a combination of the OR and NOT gates. These, and other gates, can have any number of inputs, limited only by the fan-out limits of the transistor structures.

Sequential logic circuits are used for timing, sequencing, and storage functions. The flip-flops are the main elements of these circuits, and memory functions are their primary uses. Counters consist of a series of flip-flops and are used to count the number of applied input pulses. They can be constructed to count up or down, as adders, subtracters, or both. Another set of devices called shift registers maintains a memory of the order of the applied input pulses, shifting over one place with each input pulse. These can be constructed to function as series devices, accepting one pulse (one data bit) at a time in a linear fashion, or as parallel devices, accepting several data bits, as bytes and words, with each input pulse. The devices also provide the corresponding output pulses.

Operational amplifiers, or OP-AMPs, represent a special class of ICs. Each OP-AMP IC contains a self-contained transistor-based amplifying circuit that provides a high voltage gain (typically 100,000 times or more), a very high input impedance and low output impedance, and good rejection of common-mode signals (that is, the presence of the same signal on both input leads of the OP-AMP).

Combinations of all of the gates and other devices constructed from them provide all of the computing power of all ICs, up to and including the most cutting-edge CPU chips. Their manufacturing process begins with the theoretical design of the desired functions of the circuitry. When this has been determined, the designer minimizes the detailed transistor circuitry that will be required and develops the corresponding mask and circuitry images that will be required for the IC production process. The resulting ICs can then be used to build the electronic circuitry of all electronic devices.

Impact on Industry

Digital-electronic control mechanisms and appliances are the mainstays of modern industry, mainly because of the precision with which processes and quality can be monitored through digital means. Virtually all electronic systems now operate using digital electronics constructed from ICs.

Digital electronics has far-reaching consequences. For example, in September, 2011, the telecommunications industry in North America ceased using analogue systems for broadcasting and receiving commercial signals. This development can be considered a continuation of a process that began with the invention of the Internet, with its entirely digital protocols and access methods. As digital transmission methods and infrastructure have developed to facilitate the transmission of digital communications between computer systems, the feasibility of supplanting traditional analogue communications systems with their digital counterparts became apparent. Electronic technology and the digital format enable the transmission of much greater amounts of information within the same bandwidth, with much lower energy requirements.

Modern industry without the electronic capabilities provided by IC design is almost inconceivable, from both the producer and the consumer points of view. The production of ICs themselves is an expensive proposition because of the demands for ultraclean facilities, special production methods, and the precision of the tools required for producing the transistor structures.

The transistors are produced on the surface of silicon wafers through photoetching and chemical-vapor-deposition methodologies. Photoetching requires the use of specially designed masks to provide the shape and dimensions of the transistors' component structures. As transistor dimensions have decreased from 35 um to 65 nm, the cost of the masks has risen exponentially, from about $100,000 to more than $3 million. At the same time, however, semiconductor industry revenues have increased from about $20 million in 1955 to more than $200 billion in the early twenty-first century, and they continue to increase at a rate of about 6 percent per year. This is mirrored by the amount of investment in research and development. In 2006, for example, such investment amounted to some $19.9 billion.

The consumer aspect of IC design is demonstrated by the market proliferation of consumer electronic devices and appliances. It seems as though everything, from compact fluorescent light bulbs and motion-sensing air fresheners to high-definition television monitors, personal computers, cellular telephones, and even automobiles depends on IC technology. This is a quantity whose value rivals the gross domestic product of the United States.

Careers and Coursework

IC design and manufacturing is a high-precision field. Students who are interested in pursuing a career working with devices constructed with ICs will require a sound basic education in mathematics and physics, as the basis for the study of electronics technology and engineering. As specialists in these fields, the focus will be on using ICs as the building blocks of electronic circuits for numerous devices and appliances. A recognized college degree or the equivalent is the minimum requirement for qualification as an electronics technician or technologist. The minimum

Fascinating Facts About Integrated-Circuit Design

- The ENIAC computer, built in 1947, weighed more than 30 tons; used 19,000 vacuum tubes, 1,500 relays, about one-half-million resistors, capacitors, and inductors; and consumed almost 200 kilowatts of electricity per hour.
- In 1997, the ENIAC computer was reproduced on a single triple-layer CMOS chip that was 7.44 millimeters (mm) x 5.29 mm in size and contained 174,569 transistors.
- In 1954, RAND scientists predicted that by the year 2000, some homes in the United States could have their own computer room, including a teletype terminal for FORTRAN programming, a large television set for graphic display, rows of blinking lights, and a double "steering wheel."
- The point transistor was invented in 1947, and the first integrated circuits were demonstrated in 1950 by two different researchers, leading directly to the first patent infringement litigation involving transistor technology.
- The flow of electricity in a bipolar transistor, and hence its function, is controlled by a bias voltage applied to one or the other pole.
- The flow of electricity, and hence its function, of a MOSFET is controlled by the extent of the electrical field formed within the substrate materials.
- Silicon wafers sliced from cylindrical single crystals of silicon undergo dozens of production steps before they are suitable for use as substrates for ICs.
- The production of ICs from polished silicon wafers can take hundreds of steps, after which as few as 10 percent of the ICs will be usable.

qualification for a career in electronics engineering is a bachelor's degree in that field from a recognized university. In addition, membership in professional associations following graduation, if maintained, will require regular upgrading of qualifications to meet standards and keep abreast of changing technology.

Electronics technology is an integral component of many technical fields, especially for automotive mechanics and transportation technicians, as electronic systems have been integrated into transportation and vehicle designs of all kinds.

For those pursuing a career in designing integrated circuitry and the manufacture of IC chips, a significantly more advanced level of training is required. Studies at this level include advanced mathematics and physics, quantum mechanics, electronic theory, design principles, and graphics. Integrated circuitry is cutting-edge technology, and the nature of ICs is likely to undergo rapid changes that will, in turn, require the IC designer to acquire knowledge of entirely new concepts of transistors and electronic materials. A graduate degree (master's or doctorate) from a recognized university will be the minimum qualification in this field.

Social Context and Future Prospects

IC technology is on the verge of extreme change, as new electronic materials such as graphene and carbon nanotubes are developed. Research with these materials indicates that they will be the basic materials of molecular-scale transistor structures, which will be thousands of times smaller and more energy-efficient than VLSI technology based on MOSFETs. The computational capabilities of computers and other electronic devices are expected to become correspondingly greater as well.

Such devices will utilize what will be an entirely new type of IC technology, in which the structural features are actual molecules and atoms, rather than what are by comparison mass quantities of semiconductor materials and metals. As such, future IC designers will require a comprehensive understanding of both the chemical nature of the materials and of quantum physics to make the most effective use of the new concepts.

The scale of the material structures as well will have extraordinary application in society. It is possible, given the molecular scale of the components, that the technology could even be used to print ultrahigh resolution displays and computer circuitry that would make even the lightest and thinnest of present-day appliances look like the ENIAC (Electronic Numerical Integrator and Computer) of 1947, which was the first electronic computer.

The social implications of such miniaturized technology are far-reaching. The RFID (radio-frequency identification) tag is now becoming an important means of embedding identification markers directly into materials. RFID tags are tiny enough to be included as a component of paints, fuels, explosives, and other materials, allowing identification of the exact source of the material, a useful implication for forensic investigations and other purposes. Even the RFID tag, however, would be immense compared with the molecular scale of graphene and nanotube-based devices that could carry much more information on each tiny particle.

The ultimate goal of electronic development, in current thought, is the quantum computer, a device that would use single electrons, or their absence, as data bits. The speed of such a computer would be unfathomable, taking seconds to carry out calculations that would take present-day supercomputers billions of years to complete. The ICs used for such a device would bear little resemblance, if any, to the MOSFET-based ICs used now.

Richard M. Renneboog, M.Sc.

Further Reading

Brown, Julian R. *Minds, Machines, and the Multiverse: The Quest for the Quantum Computer.* New York: Simon & Schuster, 2000. Provides historical insight and speculates about the future of computer science and technology.

Kurzweil, Ray. *The Age of Spiritual Machine: When Computers Exceed Human Intelligence.* New York: Penguin, 2000. Offers some speculations about computers and their uses through the twenty-first century.

Marks, Myles H. *Basic Integrated Circuits.* Blue Ridge Summit, Pa.: Tab Books, 1986. An introduction to functions and uses of integrated circuits.

Zheng, Y., et al. "Graphene Field Effect Transistors with Ferroelectric Gating." *Physical Review Letters* 105 (2010). Reports on the experimental development and successful testing of a graphene-based field-effect transistor system using gold and graphene electrodes with SiO_2 gate structures on a silicon substrate.

Web Sites

U.S. Department of Commerce
http://www.bis.doc.gov/defenseindustrialbaseprograms

See also: Applied Mathematics; Applied Physics; Computer Engineering; Computer Science; Electronic Materials Production; Electronics and Electronic Engineering; Quality Control.

INTERNET AND WEB ENGINEERING

FIELDS OF STUDY

Computer science; computer programming; computer information systems; computer information technology; software engineering; Web design.

SUMMARY

The term "Internet" is often used to describe the web of computer networks connected to each other digitally that is accessible to the public around the world. The World Wide Web was developed by Oxford University graduate Tim Berners-Lee in 1989 while working at the European Council for Nuclear Research. The World Wide Web makes it possible for users to browse through various documents on different Web sites by clicking on hyperlinks located on Web pages. Numerous hardware and software advances have made the Internet an indispensable tool for business transactions and personal communication.

KEY TERMS AND CONCEPTS

- **Browser:** Software program that is used to view Web pages on the Internet.
- **Firewall:** Hardware device placed between the private network and Internet connection that prevents unauthorized users from gaining access to data on the network.
- **Hacker:** Person who tries to breach the security of other computers by using programming skills.
- **Hypertext Transfer Protocol (HTTP):** Set of standard rules that allow HTML data to be transferred between computers.
- **Operating System:** Computer program designed to manage the resources (including input and output devices) used by the central processing unit and that functions as an interface between a computer user and the hardware that runs the computer.
- **Portability:** Ability of a program to be downloaded from a remote location and executed on a variety of computers with different operating systems.
- **Static Web Page:** Web page that contains content that cannot be changed by the user.
- **Web Application:** Collection of Web pages that work together.
- **Wide Area Network (WAN):** Computer network that connects central processing units (CPUs), servers, and other computing devices at more than one physical site.

Definition and Basic Principles

The Internet is a type of wide area network (WAN) because digital devices of many different types, such as a personal computer (PC), the Apple Macintosh computer, or even cell phones, can all connect to the Internet from locations all over the world. Although sharing information between computers via the network connections that comprise the Internet is free, connecting to the Internet itself is not free in North America and requires the use of a private company that functions as an Internet service provider (ISP). Many of these ISPs were phone companies because during the 1990's most connections to the Internet were made via analogue dial-up connections through the phone lines made possible because of a device called a modem. A modem functions by translating the digital signal from a sending computer to the analogue signals transmitted by phone lines and then back into the digital signal required by the receiving computer.

New hardware devices and software have continuously been developed to make sharing information via the Internet easier and faster. A language called hypertext markup language (HTML) was created to design and format the user interface to Web page content for transmission over the Internet, based on hypertext transfer protocol (HTTP). As of 2011, there are many Web applications available to computer and cell phone users all over the world.

Background and History

The precursor to the Internet was created in 1969 by the Advanced Research Projects Agency (now the Defense Advanced Research Projects Agency) of the United States Department of Defense. This project proposed a method to link together the computers at several universities (University of California, Los Angeles and Santa Barbara, Stanford, and the University of Utah) to share computational data via networks. This network became known as ARPANET. In the 1980's the International Organization for Standardization (ISO) implemented its Open Systems Interconnection—Reference Model (OSI—RM) to facilitate the interoperability of different hardware components. The creation in the 1990's of new software and programming languages, such as browsers and HTML, made the explosive growth of Internet activity and Web sites possible.

Along with the increasing use of microcomputers, there has also been an increase in the need for trained workers because companies need to be able to share information. This need for sharing information was one of the driving forces in the development of the Internet and the client-server model common to Web architecture that now allows workers to share files and access centralized databases.

How It Works

To transmit data across the Internet, there must be a level of communication possible between devices, analogous to two people shaking hands. For computers, this ability to communicate is called interoperability, and it can be classified as connection-oriented or connectionless. The connection-oriented mode of communication is somewhat similar to the process of a phone call because it requires that the sender of a message wait for the intended recipient to answer before any data is actually sent. It is more secure, but slower, for transmission of data over the Internet than the connectionless mode of data transmission. The connectionless mode has more of a broadcast nature, with data being transmitted before any secure connection is established. The connectionless mode is analogous to sending a letter or postcard through the regular mail. Different protocols, developed according to the Institute of Electrical and Electronics Engineers (IEEE), are used for communication across the Internet, depending on whether the connection or connectionless mode is best for a given situation.

Logical and Physical Topology Classifications. The different protocols function via alternative topologies that can be classified as logical or physical. Logical topologies describe the theoretical, but not visible, pathways for transmission of data signals throughout the physical topologies, which describe the tangible hardware connections of the actual pieces of equipment on a network. The two logical topologies are bus and ring, and the five physical topologies are bus, star, ring, mesh, and cellular.

The Seven Layers of the OSI Web Architecture Model. The OSI model was developed to be the Web architectural model for designing components that allow communication across the Internet. It continues to be useful as a theoretical construct to facilitate the development of different pieces of software and hardware from different manufacturers according widely accepted standards. If one single piece of hardware equipment within the network for one home or business fails, then that one piece can be purchased, even from a different manufacturer, and successfully replace the faulty equipment. This network architecture model was developed by the ISO in 1977, primarily to provide standardization among different manufacturers and vendors to allow data communication to flow uninterrupted across the Internet through different nations with complete interoperability of the software and hardware components. Therefore, the OSI model also helps with the diagnosis of connection problems by dividing the communication between two computers connected by a network into seven different layers.

The lowest level is called the physical layer because it consists of the most fundamental hardware devices that connect computers and transmit bits, including coaxial, twisted-pair, and fiber-optic cabling and connectors. Next is the data link Layer, which manages timing, flow control, error control, and encryption. It also recognizes the physical device addresses, which consist of 48 bits provided by manufacturers. Most bridges and switches are used to repeat data across different networks according to protocols such as Ethernet through this layer. The network layer uses the Internet protocol for five classes of addresses for devices, based on a 32-bit logical (not physical) address. These addresses are typically assigned by a network administrator for an organization. Routers consist of both hardware and software programs that can make decisions regarding the choice of routes for data transmission across different networks. The transport layer uses the transmission control protocol, which is a reliable, connection-oriented protocol that transmits data between two points at the same time. The session layer coordinates individual connection sessions and uses Structured Query Language (SQL) to retrieve information from databases. The presentation layer uses image, sound, and video formats such as graphics interchange format (GIF), musical instrument data interface (MIDI), and motion picture experts group (MPEG). It also encrypts data and compresses data for secure transmission. The top layer, the application layer, has the most direct interaction with a computer user. It contains the protocols for e-mail, Web access, operating systems, and file transfer protocol. These seven layers work together through encapsulation to make the transmission of data from one device to another device seamless.

Applications and Products

In 1990, the World Wide Web began as a public web of networks linking digital content described using HTML at different digital locations defined by a uniform resource locator (URL). Initially the digital content was static and slow to display. To display forms that could interact with the user, Brendan Eich of Netscape developed the language initially called LiveScript in 1995, which later became widely known as JavaScript.

Web Applications. HTML was useful for allowing static Web pages to be displayed but not dynamic Web pages. Technologies were developed to provide dynamic Web page access, including ASP, Perl/CGI, JavaScript, and Cold Fusion. ColdFusion and ASP are examples of server-side technologies, which means that the computer of an individual user (client) is not required to have any special software or hardware to access the information from a remote geographical location other than any type of browser that allows access to the Internet. JavaScript, on the other hand is a client-side technology.

ColdFusion MX 6.1 is one example of a Web application server. Other examples are CGI and ASP.NET. These examples of server-side technologies are more secure because individual clients do not have any direct access to the code. Java libraries can be imported into Web pages by using ColdFusion. ColdFusion was created by Macromedia in 2003, and is supported by the operating systems Windows Server 2003, Red Hat Linux 8 and 9, and Solaris 9. ColdFusion's compiler can take Java byte code and directly compile it into ColdFusion code, which speeds up the Web page interactions.

Linus Torvalds created Linux in 1991 while he was a student at the University of Helsinki. Linux has continued to grow in popularity as an operating system ideal for Internet applications because it is free, open source (without any hidden or proprietary interfaces), and compatible with many devices and Web applications, including Oracle and IBM databases, and many browsers. Moreover, it is powerful enough to be able to handle many of the applications that were originally handled by the UNIX operating system. Some of the applications that use Linux include digital camera programs, sound cards, printers, modems, and electronic pocket organizers such as the PalmPilot.

To decrease the time for Web pages to be displayed, frames were introduced in the middle 1990's to divide a document into pieces and display them gradually. Navigator 2.0 was the browser created by Netscape and the first to use both frames and JavaScript. Microsoft's Explorer was the next browser to implement frames, and the creation of Dynamic HTML (DHTML) along with Cascading Style Sheets (CSS) and the Document Object Model (DOM) in the late 1990's revolutionized Web pages by allowing parts of Web pages to be modified and also provided structures to Web pages. Hidden frames, Extensible Markup Language (XML) and ActiveX controls such as XMLHttp became widespread after 2001.

These technologies were taken to the next level to make Web pages more dynamic in 2005 by Jesse James Garrett, cofounder of Adaptive Path. He called this new technology Ajax, an acronym for asynchronous JavaScript and XML. The XML parser is software that allows one to use XML to create one's own markup that can read and interpret XML data. Google was one of the first companies to incorporate this technology into Google Maps, Google Suggest, and Gmail Web applications still in use. Many additional companies have adopted Ajax as well, including Amazon, Microsoft, and Yahoo. Ajax increases interactivity with Web pages for users over the traditional "click and wait" process. JavaScript embedded within a Web page sends a message to a Web server that allows data to be retrieved and transmitted to the user without having to wait for the entire Web page to be displayed.

Flash is a programming language introduced by Macromedia to create dynamic images on a Web page by using vector graphic images instead of the bitmap images (common in digital cameras) that are smaller and slow down the dynamic animations and graphics of a Web page. This language has the by-product stand-alone applications of Flash Player and Shockwave Flash files used to display movies on the Internet. It uses a JavaScript application programming interface and CSS. CSS are used by HTML and Flash as a template to keep consistent appearance among Web pages located on different Web sites.

Common Gateway Interface. Common Gateway Interface (CGI) is used to write programs that interact with Web pages to process information a person types into a form for registration, credit card purchase, or search on a Web site. CGI is a tool that functions as an interface between code written in Perl (Practical Extraction and Report Language) and a Web server. These CGI applications are referred to as scripts instead of programs. Because CGI is able to interact dynamically to Web users, it is designed to locate URLs instantly using HTTP and HTML. Embedded with CGI scripts there are tags to create cache information, so that a user can quickly return to Web sites. Because the general use of CGI is to process dynamic data for forms and to assist with searches through forms, it also interacts with Perl.

Active Server Pages, ColdFusion, and PHP. The primary problem with CGI is that it requires a new and separate execution of the CGI script each time updated input is necessary. Therefore, several alternatives have been developed. One is called the Active Server Pages (ASP) engine, which is included as part of a Web server, so the extra step of accessing the Web server is not needed. In addition, ASP can be used with several programming languages, including JavaScript and Visual Basic, making ASP more versatile than CGI. ColdFusion allows functions to be programmed and then called from within a Web page, allowing more flexibility. Both ColdFusion and PHP are integrated into Web servers. PHP was developed in 1994 as the acronym for personal home page because it was used as a scripting language to facilitate the development of Web pages. PHP can be embedded within the HTML code of a Web page to make it more dynamic because the PHP code is interpreted rather than compiled.

Impact on Industry

The growth of the Internet has led to the creation of many successful companies that are constantly leading the way in terms of innovative technologies and therefore new jobs. Among these are many companies originally founded in the United States that have since expanded worldwide, including Apple, Cisco, Google, Microsoft, Oracle, Sun Microsystems, Sybase, and Yahoo.

Cisco began in 1984 in San Francisco, primarily as a producer of some of the physical hardware devices such as routers and switches needed for the joining together of networks. It has since emerged as one of the largest technology companies in the world and has evolved into a company producing innovative software for Internet communication, such as its WebEx for video conferencing. Because it has become such an industry leader in the manufacture of networking devices, it has its own certifications including Cisco Certified Network Associate (CCNA) and Cisco Certified Design Associate (CCDA). These certifications are just a few of the most common certifications designed to enhance the skills for current and future employees.

Google was started by Larry Page and Sergey Brin, Stanford doctoral students, in 1996. The company has developed many new innovative software applications to accompany its search engine, including Gmail and Google Maps. As a result, its stock price has been more than $500 per share for many years.

Research in Motion (RIM) introduced its first smart phone to customers in 2002. These early smart phones allowed customers to send e-mail and text messages and access Web pages. RIM even developed its own software, including a new operating system, Internet service, database management system, and the innovative concept of push e-mail, whereby messages are sent instantly. Like Cisco, RIM has developed several certifications designed to help current and future employees attain the necessary job skills. Among the RIM certifications are BlackBerry Certified Enterprise Server Consultant and BlackBerry Certified Server Support Specialist. Because these certifications, like those offered by Cisco, depend primarily on successfully passing various tests, there have been many opportunities for well-paying jobs within the technology sector for individuals possessing enough knowledge to pass these exams, but who may not necessarily have spent years to obtain a college degree.

Careers and Course Work

According to the Bureau of Labor Statistics, this career field is expected to grow by 23 percent during the period from 2008 to 2018, which is much faster

Fascinating Facts About Internet and Web Engineering

- Employment of network systems and data communications analysts is estimated to increase by 53 percent from 2008 to 2018, according to the Bureau of Labor Statistics.
- Employment of database administrators, who are needed by all types of private companies and public organizations to manage online database access, is projected to grow by 20 percent from 2008 to 2018, according to the Bureau of Labor Statistics.
- Google currently has several thousand job openings at almost a dozen sites located throughout the United States. It has several more thousand job openings worldwide.
- Facebook was created in 2004 by Mark Zuckerberg, who was a college student at Harvard. By the end of 2010 it had become the largest social networking site in the world.
- At least 68 percent of the victims of cyber crime in 2005 suffered monetary losses of at least $10,000.

than average. This career field refers to network administrators, network engineers, systems engineers, database administrators, help desk engineers and technicians, Web designers, security analysts, programmers, and project managers. There are ample opportunities for employment as help desk engineers and technicians, Web designers, database administrators, and network administrators for individuals who have an associate's degree in an appropriate field or work experience and certifications. In addition to some of the vendor-specific certifications offered by Cisco, RIM, Oracle, Sun Microsystems, andMicrosoft, there are several certifications that are vendor neutral such as CompTIA's Network+, Security+, and iNet+. Network administrators design networks for organizations and maintain the security of these networks on a daily basis, as well as install shared software, assign logical Internet protocol (IP) addresses, and back up data. Web designers use Web page languages such as HTML, Perl, Java, and VB.NET. Certifications in these languages indicate useful skills as well. The bachelor's degree in computer science or a computer-related field is especially helpful to secure employment as a security analyst or project manager.

Social Context and Future Prospects

The economy can be described as a digital one because the Internet is used as a tool to facilitate all kinds of online financial transactions involving the common use of bank accounts, credit cards, airline and hotel reservations, government benefits, and the stock market. These financial transactions can be completed using both wired and wireless computer internetworking. The rapid growth of Internet applications that facilitate communication and financial transactions has also been accompanied by a growth in identity theft and other cyber crimes. Web architecture has developed both in terms of providing convenience for consumer transactions and social networking sites, such as Facebook, and also in terms of providing security through the use of firewalls and new security software to try to prevent cyber crimes. Despite the security risks, the usage of the Internet for personal, financial, educational, and business-related activities is expected to increase, especially in many of the newly developing countries that have been lagging behind the United States and other developed nations.

Jeanne L. Kuhler, B.S., M.S., Ph.D.

Further Reading

Castro, Elizabeth. *HTML, XHTML, and CSS.* 6th ed. Berkeley, Calif.: Peachpit Press, 2007. This introductory text describes how a novice can set up an individual Web site and provides ample helpful screen shots.

Longino, Carlo. "Your Wireless Future." *Business 2.0.* May 22, 2006. http://money.cnn.com/2006/05/18/technology/business2_wirelessfuture_intro This article discusses the future of wireless technology in all sectors—business, entertainment, and communications.

Love, Chris. *ASP.NET 3.5 Website Programming: Problem-Design-Solution.* Indianapolis: Wiley Publishing, 2010. Describes the process for the design and construction of a Web site using ASP.NET 3.5.

Prayaga, Lakshmi, and Hamsa Suri. *Programming the Web with ColdFusion MX 6.1 Using XHTML.* Boston: McGraw-Hill, 2004. Explains the use of the ColdFusion language, with representative code examples and screen shots.

Spaanjaars, Imar. *Beginning ASP.NET 3.5 in C# and VB.* Indianapolis: Wiley Publishing, 2008. This text, although for beginners, assumes that readers have had some type of introduction to Web technology and programming.

Umar, Amjad. "IT Infrastructure to Enable Next Generation Enterprise." *Information Systems Frontiers* 7, no. 3 (July, 2005): 1201-1205. Looks at advances in network protocols and design.

Underdahl, Brian. *Macromedia Flash MX: The Complete Reference.* Berkeley, Calif.: McGraw-Hill/Osborne, 2003. This comprehensive volume covers all aspects of Flash.

Web Sites

Association of Internet Researchers
http://aoir.org

Institute of Electronics and Electrical Engineers
http://www.ieee.org

WebProfessionals
http://webprofessionals.org

World Wide Web Consortium (W3C)
http://www.w3.org

See also: Computer Graphics; Computer Networks; Computer Science; Software Engineering.

L

LANDSCAPE ARCHITECTURE AND ENGINEERING

FIELDS OF STUDY

Architecture; ecology; forestry; landscape engineering; industrial design; landscape design; environmental science; ecological restoration; natural resource management; physical, natural, and social sciences; land-use planning; environmental psychology; botany; landscape ecology; geographic information systems (GIS); computer modeling.

SUMMARY

Landscape architecture is the development and design of landscapes that take into account environmental and aesthetic concerns while addressing practical human use of the space. Landscape architecture aims to incorporate the natural environment into the design and construction of a space in a minimally disruptive way that also conserves natural resources in the short and long term. This can pose challenges and create limitations in terms of techniques, applications, and judgments used in the design and construction processes. Landscape architects design and plan spaces such as gardens, parks, campuses, and commercial centers, and restore historic sites and degraded environments, such as landfills and mined land.

KEY TERMS AND CONCEPTS

- **Degraded Environment and Ecosystem:** Space that has been negatively affected by humans (through events such as oil spills or the clear cutting of forests) or natural disasters (such as earthquakes or tsunamis).
- **Greenfield Site:** Forested area, agricultural land, or any area outside of a city where development can occur without hindrance from previous structures and that has been identified for industrial or commercial development.
- **Hardscape:** Inanimate or nonliving elements of landscaping, including features such as water fountains, masonry, and woodwork.
- **Landscape Design:** Art of designing and altering landscape features for function or aesthetics; frequently divided into hardscape and softscape design.
- **Landscape Ecology:** Study of the relationship between spatial patterns and ecological processes on a large number of diverse landscapes.
- **Native Vegetation:** Plant or tree that is indigenous to the area.
- **Softscape:** Animate or living elements of landscaping, such as plants and trees.
- **Sustainability:** Degree of success in integration of a site's natural land, water, and energy resources with human needs and structures in development of an area.

DEFINITION AND BASIC PRINCIPLES

Landscape architecture attempts to unify the human-constructed environment with the existing natural environment to create a practical, sustainable, and aesthetically pleasing space. Landscape architecture tends to incorporate urban planning, ecological restoration, and green infrastructure in the design of open-air public spaces.

The title of landscape architect is often used interchangeably with landscape designer, and although both do undertake landscape design, the American Society of Landscape Architects states that the two professions are different because a landscape architect must have a higher level of training and skill and usually needs post-secondary education. Fundamentally, it takes a greater investment of time to study landscape architecture.

All visual, ecological, and cultural issues involved in the development of the natural landscape must be considered in landscape architecture. Essentially, landscape architecture involves designing open-air public spaces that meet certain environmental, social, and aesthetic goals while making prudent use of finite land and natural resources. It follows the theories, principles, and practices of general architecture and also considers and assesses the social, ecological, and geological processes occurring within and affecting the landscape.

Background and History

Human society has long attempted to blend natural and artificial environments, as evidenced by the English theater gardens of the seventeenth and eighteenth centuries, the Zen gardens in China and Japan, and large urban parks in the United States. Although the design philosophies and approaches toward unifying architecture with the natural environment differ, the concept is shared by Western and Eastern cultures.

Landscape architecture, particularly as it relates to gardening and building design, is a discipline as old as human society and the human desire to control and modify the natural landscape. The term "landscape architecture" was first coined by the Scottish gentleman Gilbert Laing Meason in 1828. The concept of landscape architecture reached the mainstream about thirty years later, when Frederick Law Olmsted and Calvert Vaux were commissioned to design Central Park in New York City. Olmsted is considered the father of American landscape architecture. The profession continued to gain significant credibility and growth at the beginning of the twentieth century with the founding of the American Society of Landscape Architects in 1899 and the introduction of a landscape architecture program at Harvard University in 1900. This program was instrumental in shifting the focus of landscape architecture from private gardens to open-air public spaces.

Landscape architecture has had a long history in the West, but countries in other parts of the world have also been active in the field. Since the mid-twentieth century, a significant amount of research in and application of landscape architecture has taken place in China. There, landscape architecture has advanced significantly, focusing on incorporating gardens into urban landscapes.

How It Works

Landscape architecture involves the design and development of outdoor public areas that are not only functional but also aesthetically pleasing and environmentally friendly. It is the job of a landscape architect to blend design, function, beauty, and the natural environment to achieve desired social, environmental, and aesthetic outcomes.

Typically, landscape architecture involves overseeing project design, often of urban regeneration programs; site inspections of both ecological aspects and preexisting human-made structures; and environmental impact assessments. Landscape architects also design projects and present their proposals before government bodies charged with planning and conservation responsibilities. The design process usually involves a number of steps, including preplanning, project planning, preliminary and final design, report writing and approvals, and implementation and development. Landscape architects work in conjunction with architects, surveyors, hydrologists, environmental scientists, foresters, and engineers to determine the design and arrangement of structures and infrastructure and the most appropriate way to protect and use natural areas and resources. Landscape architects analyze the natural elements present in the area and produce detailed designs that incorporate environmental aspects of construction such as climate, soil, topography, vegetation, and habitat, with structures for human use, including walkways, fountains, and decorative features.

Applications and Products

Landscape architecture is a multidisciplinary field of applied science and engineering, and its applications are vast and varied. Wherever humans and nature meet, landscape architecture theory and application is necessary. The applications of landscape architecture are diverse and range from small to large. They include the design and management of vast wilderness areas and national parks, small-scale urban gardens and public parks, resorts, residential estates, and open areas in business parks. Landscape architecture also deals with the reclamation of environmentally degraded areas, such as landfills, and the preservation of historic sites and districts. Landscape architects must be versatile, as their work can encompass urban and rural areas and involve both terrestrial and aquatic environments.

For the most part, however, landscape architecture is employed in urban design and planning, environmental restoration, ecotourism, parks, gardens, recreational area design and development, and green infrastructure implementation.

Urban Design. Urban design generally refers to the organizational and developmental stage of landscape architecture in towns and cities and often involves development of public spaces in urban areas. Landscape architects use the knowledge gained through preliminary assessments to determine the ideal site for development within a specific urban environmental space. For instance, when choosing a site for what is to be a widely used structure or area, a landscape architect would want the site to be easily accessible, near public transportation or a densely populated residential area. This would minimize travel time and expense, reduce car emissions and pollution, and conserve fossils fuels.

Environmental Restoration. The growing human population is placing significant pressure on the natural landscape. The impact that humans have on nature can be seen in polluted bodies of water, abandoned industrial areas, dilapidated waterfronts, neglected railroad yards, and landfills. As humans continue to encroach into natural habitats and true greenfield sites become rarer, landscape architects and designers look toward degraded areas, either to restore them to their natural state or to permit renewed human use. The reclamation of derelict urban and rural areas for public use is vital for the sustainable development of human communities. It not only demonstrates a commitment to sustainable land use and protection of the environment but also improves the quality of life for members of the community.

Ecotourism, Parks, Gardens, and Recreational Areas. Many people living within urban areas use and visit parks and gardens. Landscape architects design and develop these areas for both recreational and aesthetic purposes, while minimizing their impact on the natural environment. Landscape architects also design resorts and ecotourism infrastructures. Tourism based on ecological principles has become increasingly popular. One of the applications of landscape architecture is the development of green infrastructure in wilderness areas that will accommodate tourists without adversely affecting the surrounding landscape and the native flora and fauna.

Fascinating Facts About Landscape Architecture and Engineering

- Frederick Law Olmstead, the father of American landscape architecture, was involved in the design of Central Park in New York City and the grounds surrounding the U.S. Capitol in Washington, D.C. He also designed the campuses of Yale University in New Haven, Connecticut, and Iowa State University in Ames, Iowa.
- In 2008, for the first time in the history of human society, more than half the world's human population lived in urban areas. This urbanization increased the recreational and aesthetic importance of park and garden areas.
- Landscape architecture was once viewed as not more than a way to embellish the space around a building, but it is increasingly viewed as a central aspect of urban space design.
- American landscape architect Thomas Church designed more than 2,000 residential gardens over the course of his career and public gardens at Stanford University and the University of California, Berkeley. He focused on principles of unity, scale, function, and simplicity in his designs.
- The U.S. National Parks Service announced plans in 2009 to restructure and "rehabilitate" the National Mall in Washington, D.C., so that it would be more sustainable and environmentally friendly. Plans include soil restoration and replacement, as well as the planting of more trees and the development of a sustainable irrigation system.

Impact on Industry

Landscape architects, designers, and engineers have moved away from the manipulation of landscapes to the design of landscapes that are aligned with the environmental elements within a given area. Instead of fighting environmental conditions, modern landscape design works with existing conditions to provide for human use. This shift in approach requires that landscape architects not only know the visible conditions but also understand subsurface conditions (such as groundwater and soils), conservation and larger planning requirements and directives, and the laws that guide their efforts.

Major Organizations. The profession of landscape architecture is globally represented by the International Federation of Landscape Architects, founded at Cambridge, England, in 1948. The federation aims to cultivate a public understanding of landscape architecture as being environmentally, culturally, and socially responsible. Another major organization is the American Society of Landscape Architects, a professional association for landscape architects founded in 1899, with members in the United States and in more than sixty countries around the world. Despite the significant number of organizations involved in landscape architecture, the level of government legislation and regulation within the field varies significantly among countries.

Many countries possess professional landscape architecture institutes and agencies that are responsible for the promotion and regulation of the profession. Such institutes include the Council of Landscape Architectural Registration Boards in the United States and Canada, the Landscape Institute in Great Britain, the Canadian Society of Landscape Architects, and the Australian Institute of Landscape Architects.

Government Research and Regulation. Most governments apply landscape architecture principles and practices to urban design and town planning. Evaluating the environmental impact of development within urban areas is becoming ever more important. In particular, many governments are investigating the reclamation of degraded sites, including their restoration and potential use.

In 2009, Barack Obama's administration created the Partnership for Sustainable Communities to encourage the sustainable development and restoration of urban areas. In addition, the U.S. Congress has considered bills that relate to numerous aspects of landscape architecture, such as green infrastructure, historic restoration, and environmental reclamation. In other countries, including China, concepts such as underground landscape design and architecture are being considered in an effort to maximize use of the available space.

Careers and Course Work

For entry into the landscape architecture profession, a person should obtain a bachelor of landscape architecture or a bachelor of science in landscape architecture from an accredited university. Programs in landscape architecture have been offered at more than sixty universities in the United States, including Harvard University, Pennsylvania State University, and the University of Wisconsin, Madison. Students who seek a career in landscape architecture will usually have a talent for design and possess strong analytical skills. They should also have a strong interest in the environment. Landscape architecture courses include the history of landscape architecture, surveying, landscape design and construction, landscape ecology, plant and soil science, site design, and urban and regional planning.

Upon completion of their undergraduate studies, students will have gained knowledge and awareness of urban and landscape design, ecological planning, heritage conservation, regional landscape planning, land development, park and recreation planning, and environmental reclamation and restoration. In addition, students will be familiar with computer-aided design, model building, geographic information systems (GIS), and three-dimensional visualization and simulation.

In many countries, including the United States, the United Kingdom, Canada, and Australia, graduates must register to obtain a license to practice landscape architecture, and they may have to continue their education to maintain registration. Registration processes and procedures, however, differ from country to country. Many landscape architects are self-employed, but graduates can also embark on careers in the private sector, in specialized government organizations and agencies, and in universities undertaking teaching and research.

Social Context and Future Prospects

As the public's environmental consciousness develops, greater opportunities in landscape architecture will emerge, as will the necessity for lasting and significant decisions regarding landscape planning and design. Novel projects such as green roofing to reduce the environmental and economic costs of heating and cooling and water conservation and pollution reduction through design and management of storm water are being investigated and implemented by landscape architects in the early twenty-first century.

In addition, as GIS and computer graphics technologies continue to advance, so will the ability of landscape architects to design projects using simulated three-dimensional visualizations and modeling. In particular, GIS-based data can assist in creating realistic representations of softscape design elements such as vegetation and improve terrain and landscape imagery. Advances in technology will allow landscape architects to produce highly detailed simulations of natural and urban environments, enabling them to provide much more accurate predictions of possible environmental impacts.

Christine Watts, Ph.D., B.App.Sc., B.Sc.

Further Reading

Benedict Mark A., and Edward T. McMahon. *Green Infrastructure: Linking Landscapes and Communities.* Washington, D.C.: Island Press, 2006. Examines the large-scale design and integrated action involved in the advancement of green infrastructure and focuses on smart land development.

Downton, Paul. *Ecopolis: Architecture and Cities for a Changing Climate.* Collingwood, Australia: CSIRO, 2009. Discusses architectural theories and practices, planning, and ecology, with a focus on the development of sustainable future cities through the integration of academic and practical knowledge.

Dramstad, Wenche, James D. Olson, and Richard T. T. Forman. *Landscape Ecology Principles in Landscape Architecture and Land-Use Planning.* Washington, D.C.: Island Press, 1996. An overview of landscape ecology theory and principles with a focus on how they are applied in design and planning.

Newman, Peter, and Isabella Jennings. *Cities as Sustainable Ecosystems: Principles and Practices.* Washington, D.C.: Island Press, 2008. Examines how living systems can be applied to sustainable urban design strategies through the exploration of flows of energy, materials, and information. Considers the relationship between human and nonhuman elements.

Register, Richard. *EcoCities: Rebuilding Cities in Balance with Nature.* Gabriola, B.C.: New Society Publishers, 2006. This book examines the concepts and ideas behind rebuilding urbanized areas based on the principles of ecology, with the aim of long-term sustainability, cultural vitality, and environmental health.

Swaffield, Simon. *Theory in Landscape Architecture: A Reader.* Philadelphia: University of Pennsylvania Press, 2002. A resource for seminal theory in the field, covering fifty years of study on the relationship between landscape and social, cultural, and political structures.

Web Sites

American Society of Landscape Architects
http://www.asla.org

International Federation of Landscape Architects
http://www.iflaonline.org

See also: Civil Engineering; Ecological Engineering; Environmental Engineering.

M

MECHANICAL ENGINEERING

FIELDS OF STUDY

Acoustics; algebra; applied mathematics; calculus; chemistry; control theory; drafting; dynamics; economics; electronics; fluid dynamics; graphics; heat transfer; kinematics; materials science; mechanics; optics; physics; product design; robotics; statics; structural analysis; system design; thermodynamics.

SUMMARY

Mechanical engineering is the field of technology that deals with engines, machines, tools, and other mechanical devices and systems. This broad field of innovation, design, and production deals with machines that generate and use power, such as electric generators, motors, internal combustion engines, and turbines for power plants, as well as heating, ventilation, air-conditioning, and refrigeration systems. In many universities, mechanical engineering is integrated with nuclear, materials, aerospace, and biomedical engineering. The tools used by scientists, engineers, and technicians in other disciplines are usually designed by mechanical engineers. Robotics, microelectromechanical systems, and the development of nanotechnology and bioengineering technology constitute a major part of modern research in mechanical engineering.

KEY TERMS AND CONCEPTS

- **Computer-Aided Design (CAD):** Using computer software to design objects and systems and develop and check the graphical representation of a design.
- **Computer-Aided Manufacturing (CAM):** Use of computer software and software-guided machines to perform manufacturing operations starting with computer-aided design.
- **Dynamics:** Application of the laws of physics to determine the acceleration and velocity of bodies and systems.
- **Engineering Economics:** Science focused on determining the best course of action in designing and manufacturing a given system to reduce uncertainty and maximize the return on the investment.
- **Fluid Mechanics:** Science describing the behavior of fluids.
- **Heat Transfer:** Science relating the flow of thermal energy to temperature differences and material properties.
- **Kinematics:** Science of determining the relationship between the movement of different elements of a system, such as a machine.
- **Machine Design:** Science of designing the most suitable individual elements and their integration into machines, relating the stresses and loads that the machine must handle, its lifetime, the properties of the materials used, and the cost of manufacturing and using the machine.
- **Manufacturing:** Generation of copies of a product, based on good design, in sufficient numbers to satisfy demand at the best return on investment.
- **Materials Science:** Study of the detailed structure and properties of various materials used in engineering.
- **Mechanical System:** Grouping of elements that interact according to mechanical principles.
- **Metrology:** Science of measurement dealing with the design, calibration, sensitivity, accuracy, and precision of measuring instruments.
- **Robotics:** Science of designing machines that can replace human beings in the execution of specific tasks, such as physical activities and decision making.
- **Statics:** Science of how forces are transmitted to, from, and within a structure, determining its stability.
- **Strength Of Materials:** Science of determining the deflection of objects of different shapes and sizes under various loading conditions.
- **Thermodynamics:** Science dealing with the relationships between energy, work, and the properties of matter. Thermodynamics defines the best performance that can be achieved with power conversion, generation, and heat transfer systems.

Definition and Basic Principles

Mechanical engineering is the field dealing with the development and detailed design of systems to perform desired tasks. Developed from the discipline of designing the engines, power generators, tools, and mechanisms needed for mass manufacturing, it has grown into the broadest field of engineering, encompassing or touching most of the disciplines of science and engineering. Mechanical engineers take the laws of nature and apply them using rigorous mathematical principles to design mechanisms. The process of design implies innovation, implementation, and optimization to develop the most suitable solution to the specified problem, given its constraints and requirements. The field also includes studies of the various factors affecting the design and use of the mechanisms being considered.

At the root of mechanical engineering are the laws of physics and thermodynamics. Sir Isaac Newton's laws of motion and gravitation, the three laws of thermodynamics, and the laws of electromagnetism are fundamental to much of mechanical design.

Starting with the Industrial Revolution in the nineteenth century and going through the 1970's, mechanical engineering was generally focused on designing large machines and systems and automating production lines. Ever-stronger materials and larger structures were sought. In the 1990's and first part of the twenty-first century, mechanical engineering saw rapid expansion into the world of ever-smaller machines, first in the field of micro and then nano materials, probes and machines, down to manipulating individual atoms. In this regime, short-range forces assume a completely different relationship to mass. This led to a new science integrating electromagnetics and quantum physics with the laws of motion and thermodynamics. Mechanical engineering also expanded to include the field of system design, developing tools to reduce the uncertainties in designing increasingly more complex systems composed of larger numbers of interacting elements.

Background and History

The engineering of tools and machines has been associated with systematic processes since humans first learned to select sticks or stones to swing and throw. The associations with mathematics, scientific prediction, and optimization are clear from the many contraptions that humans developed to help them get work done. In the third century B.C.E., for example, the mathematician Archimedes of Syracuse was associated with the construction of catapults to hurl projectiles at invading armies, who must themselves have had some engineering skills, as they eventually invaded his city and murdered him. Tools and weapons designed in the Middle Ages, from Asia to Europe and Africa, show amazing sophistication. In the thirteenth century, Mesopotamian engineer Al-Jazari invented the camshaft and the cam-slider mechanism and used them in water clocks and water-raising machines. In Italy, Leonardo da Vinci designed many devices, from portable bridges to water-powered engines.

The invention of the steam engine at the start of the Industrial Revolution is credited with the scientific development of the field that is now called mechanical engineering. In 1847, the Institution of Mechanical Engineers was founded in Birmingham, England. In North America, the American Society of Civil Engineers was founded in 1852, followed by the American Society of Mechanical Engineers in 1880. Most developments came through hard trial and error. However, the parallel efforts to develop retrospective and introspective summaries of these trials resulted in a growing body of scientific knowledge to guide further development.

Nevertheless, until the late nineteenth century, engineering was considered to be a second-rate profession and was segregated from the "pure" sciences. Innovations were published through societies such as England's Royal Society only if the author was introduced and accepted by its prominent members, who were usually from rich landed nobility. Publications came from deep intellectual thinking by amateurs who supposedly did it for the pleasure and amusement; actual hands-on work and details were left to paid professionals, who were deemed to be of a lower class. Even in America, engineering schools were called trade schools and were separate from the universities that catered to those desiring liberal arts educations focused on the classics and languages from the Eurocentric point of view.

Rigorous logical thinking based on the experience of hands-on applications, which characterizes mechanical engineering, started gaining currency with the rise of a culture that elevated the dignity of labor in North America. It gained a major boost with the urgency brought about by several wars. From the time of the American Civil War to World War I, weapons such as firearms, tanks, and armored ships saw significant advancements and were joined by airplanes and motorized vehicles that functioned as ambulances. During these conflicts, the individual heroism that had marked earlier wars was eclipsed by the technological superiority and scientific organization delivered by mechanical engineers.

Concomitantly, principles of mass production were applied intensively and generated immense wealth in Europe and America. Great universities were established by people who rose from the working classes and made money through technological enterprises. The Great Depression collapsed the established manufacturing entities and forced a sharp rise in innovation as a means of survival. New engineering products developed rapidly, showing the value of mechanical engineering. World War II and the subsequent Cold War integrated science and engineering inseparably. The space race of the 1960's through the 1980's brought large government investments in both military and civilian aerospace engineering projects. These spun off commercial revolutions in computers, computer networks, materials science, and robotics. Engineering disciplines and knowledge exploded worldwide, and as of 2011 there is little superficial difference between engineering curricula in most countries of the world.

The advent of the Internet accelerated and completed this leveling of the knowledge field, setting up sharper impetus for innovation based on science and engineering. Competition in manufacturing advanced the field of robotics, so that cars made by robots in automated plants achieve superior quality more consistently than those built by skilled master craftsmen. Manufacturing based on robotics can respond more quickly to changing specifications and demand than human workers can.

Beginning in the 1990's, micro machines began to take on growing significance. Integrated microelectromechanical systems were developed using the techniques used in computer production. One by one, technology products once considered highly glamorous and hard to obtain—from calculators to smart phones—have been turned into mass-produced commodities available to most at an affordable cost. Other products—from personal computers and cameras to cars, rifles, music and television systems, and even jet airliners—are also heading for commoditization as a result of the integration of mechanical engineering with computers, robotics, and micro electromechanics.

How It Works

The most common idea of a mechanical engineer is one who designs machines that serve new and useful functions in an innovative manner. Often these machines appear to be incredibly complex inside or extremely simple outside. The process of accomplishing these miraculous designs is systematic, and good mechanical engineers make it look easy.

System Design. At the top level, system design starts with a rigorous analysis of the needs to be satisfied, the market for a product that satisfies those needs, the time available to do the design and manufacturing, and the resources that must be devoted. This step also includes an in-depth study of what has been done before. This leads to "requirements definition," where the actual requirements of the design are carefully specified. Experienced designers believe that this step already determines more than 80 percent of the eventual cost of the product.

Next comes an initial estimate of the eventual system characteristics, performed using simple, commonsense logic, applying the laws of nature and observations of human behavior. This step uses results from benchmarking what has been achieved before and extrapolating some technologies to the time when they must be used in the manufacturing of the design. Once these rudimentary concept parameters and their relationships are established, various analyses of more detailed implications become possible. A performance estimation then identifies basic limits and determines whether the design "closes," meeting all the needs and constraints specified at the beginning. Iterations on this process develop the best design. Innovations may be totally radical, which is relatively rare, or incremental in individual steps or aspects of the design based on new information, or on linking developments in different fields. In either case, extensive analysis is required before an innovation is built into a design. The design is then analyzed for ease and cost of manufacture. The "tooling," or specific setups and machines required for mass manufacture, are considered.

A cost evaluation includes the costs of maintenance through the life cycle of the product. The entire process is then iterated on to minimize this cost. The design is then passed on to build prototypes, thereby gaining more experience on the manufacturing techniques needed. The prototypes are tested extensively to see if they meet the performance required and predicted by the design.

When these improvements are completed and the manufacturing line is set up, the product goes into mass manufacture. The engineers must stay engaged in the actual performance of the product through its delivery to the end user, the customer, and in learning from the customer's experience in order to design improvements to the product as quickly as possible. In modern concurrent engineering practice, designers attempt to achieve as much as possible of the manufacturing process design and economic optimization during the actual product design cycle in order to shorten the time to reach market and the cost of the design cycle. The successful implementation of these processes requires both technical knowledge and experience on the part of the mechanical engineers. These come from individual rigorous fields of knowledge, some of which are listed below.

Engineering Mechanics. The field of engineering mechanics integrates knowledge of statics, dynamics, elasticity, and strength of materials. These fields rigorously link mathematics, the laws of motion andgravitation, and material property relationships to derive general relations and analysis methods. Fundamental to all of engineering, these subfields are typically covered at the beginning of any course of study.

In statics, the concept of equilibrium from Newton's first law of motion is used to develop free-body diagrams showing various forces and reactions. These establish the conditions necessary for a structure to remain stable and describe relations between the loads in various elements.

In dynamics, Newton's second law of motion is used to obtain relations for the velocity and acceleration vectors for isolated bodies and systems of bodies and to develop the notions of angular momentum and moment of inertia.

Strength of materials is a general subject that derives relationships between material properties and loads using the concepts of elasticity and plasticity and the deflections of bodies under various types of loading. These analyses help the engineer predict the yield strength and the breaking strength of various structures if the material properties are known. Metals were the preferred choice of material for engineering for many decades, and methods to analyze structures made of them were highly refined, exploiting the isotropy of metal properties. Modern mechanical engineering requires materials the properties of which are much less uniform or exotic in other ways.

Graphics and Kinematics. Engineers and architects use graphics to communicate their designs precisely and unambiguously. Initially, learning to draw on paper was a major part of learning engineering skills. As of 2011, students learn the principles of graphics using computer-aided design (CAD) software and computer graphics concepts. The drawing files can also be transferred quickly into machines that fabricate a part in computer-aided manufacturing (CAM). Rapid prototyping methods such as stereo lithography construct an object from digital data generated by computer graphics.

The other use of graphics is to visualize and perfect a mechanism. Kinematics develops a systematic method to calculate the motions of elements, including their dependence on the motion of other elements. This field is crucial to developing, for instance, gears, cams, pistons, levers, manipulator arms, and robots. Machines that achieve very complex motions are designed using the field of kinematics.

Robotics and Control. The study of robotics starts with the complex equations that describe how the different parts satisfy the equations of motion with multiple degrees of freedom. Methods of solving large sets of algebraic equations quickly are critical in robotics. Robots are distinguished from mere manipulator arms by their ability to make decisions based on the input, rather than depend on a telepresence operator for commands. For instance, telepresence is adequate to operate a machine on the surface of the Moon, which is only a few seconds of round-trip signal travel time from Earth using electromagnetic signals. However, the round-trip time for a signal to Mars is several minutes, so a rover operating there cannot wait for commands from Earth regarding how to negotiate around an obstacle. A fully robotic rover is needed that can make decisions based on what its sensors tell it, just as a human present on the scene might do.

Entire manufacturing plants are operated using robotics and telepresence supervision. Complex maneuvers such as the rendezvous between two spacecraft, one of which may be spinning out of control, have been achieved in orbits in space, where the dynamics are difficult for a human to visualize. Flight control systems for aircraft have been implemented using robotics, including algorithms to land the aircraft safely and more precisely than human pilots can. These systems are developed using mathematical methods for solving differential equations rapidly, along with software to adjust parameters based on feedback.

Materials. The science of materials has advanced rapidly since the late twentieth century. Wood was once a material of choice for many engineering products, including bridges, aircraft wings, propellers, and train carriages. The fibrous nature of wood required considerable expertise from those choosing how to cut and lay sections of wood; being a natural product, its properties varied considerably from one specimen to another. Metals became much more convenient to use in design and fabrication because energy to melt and shape metals cheaply became available. Various alloys were developed to tailor machinery for strength, flexibility, elasticity, corrosion resistance, and other desirable characteristics. Detailed tables of properties for these alloys were included in mechanical engineering handbooks.

Materials used to manufacture mass-produced items have migrated to molded plastics made of hydrocarbons derived from petroleum. The molds are shaped using such techniques as rapid prototyping and computer-generated data files from design software. Composite materials are tailored with fiber bundles arrayed along directions where high-tensile strength is needed and much less strength along directions where high loads are not likely, thus achieving large savings in mass and weight.

Fluid Mechanics. The science of fluid mechanics is important to any machine or system that either contains or must move through water, air, or other gases or liquids (fluids). Fluid mechanics employs the laws of physics to derive conservation equations for specific packets of fluid (the Lagrangian approach) or for the flow through specified control volumes (Eulerian approach). These equations describe the physical laws of conservation of mass, momentum, and energy, relating forces and work to changes in flow properties. The properties of specific fluids are related through the thermal and caloric equations expressing their thermodynamic states. The speed of propagation of small disturbances, known as the speed of sound, is related to the dependence of pressure on density and hence on temperature. Various nondimensional groupings of flow and fluid properties—such as the Reynolds number, Mach number, and Froude number—are used to classify flow behavior. Increasingly, for many problems involving fluid flow through or around solid objects, calculations starting from the conservation equations are able to predict the loads and flow behavior reliably using the methods of computational fluid mechanics (CFD). However, the detailed prediction of turbulent flows remains beyond reach and is approximated through various turbulence models. Fluid-mechanic drag and the movements due to flow-induced pressure remain very difficult to calculate to the accuracy needed to improve vehicle designs.

Methods for measuring the properties of fluids and flows in their different states are important tools for mechanical engineers. Typically, measurements and experimental data are used at the design stage, well before the computational predictions become reliable for refined versions of the product.

Thermodynamics. Thermodynamics is the science behind converting heat to work and estimating the best theoretical performance that a system can achieve under given constraints. The three basic laws of temperature are the zeroth law, which defines temperature and thermal equilibrium; the first law, which describes the exchange between heat, work, and internal energy; and the second law, which defines the concept of entropy. Although these laws were empirically derived and have no closed-form proof, they give results identical to those that come from the law of conservation of energy and to notions of entropy derived from statistical mechanics of elementary particles traced to quantum theory. No one has yet been able to demonstrate a true perpetual-motion machine, and it does not appear likely that anyone will. From the first law, various heat-engine cycles have been invented to obtain better performance suited to various constraints. Engineers working on power-generating engines, propulsion systems, heating systems, and air-conditioning and refrigeration systems try to select and optimize thermodynamic cycles and then use a figure of merit—a means of evaluating the performance of a device or system against the best theoretical performance that could be achieved—as a measure of the effectiveness of their design.

Heat Transfer. Heat can be transferred through conduction, convection, or radiation, and all three modes are used in heat exchangers and insulators. Cooling towers for nuclear plants, heat exchangers for nuclear reactors, automobile and home air-conditioners, and the radiators for the International Space Station are all designed from basic principles of these modes of heat transfer. Some space vehicles are designed with heat shields that are ablative. The Thermos flask (which uses an evacuated space between two silvered glass walls) and windows with double and triple panes with coatings are examples of widely used products designed specifically to control heat transfer.

Machine Design. Machine design is at the core of mechanical engineering, bringing together the various disciplines of graphics, solid and fluid mechanics, heat transfer, kinematics, and system design in an organized approach to designing devices to perform specific functions. This field teaches engineers how to translate the requirements for a machine into a design. It includes procedures for choosing materials and processes, determining loads and deflections, failure theories, finite element analysis, and the basics of how to use various machine elements such as shafts, keys, couplings, bearings, fasteners, gears, clutches, and brakes.

Metrology. The science of metrology concerns measuring systems. Engineers deal with improving the accuracy, precision, linearity, sensitivity, signal-to-noise ratio, and frequency response of measuring systems. The precision with which dimensions are measured has a huge impact on the quality of engineering products. Large systems such as airliners are assembled from components built on different continents. For these to fit together at final assembly, each component must be manufactured to exacting tolerances, yet requiring too much accuracy sharply increases the cost of production. Metrology helps in specifying the tolerances required and ensuring that products are made to such tolerances.

Acoustics and Vibrations. These fields are similar in much of their terminology and analysis methods. They deal with wavelike motions in matter, their effects, and their control. Vibrations are rarely desirable, and their minimization is a goal of engineers in perfecting systems. Acoustics is important not only because minimizing noise is usually important, but also because engineers must be able to build machines to generate specific sounds, and because the audio signature is an important tool in diagnosing system status and behavior.

Production Engineering. Production engineering deals with improving the planning and implementation of the production process, designing efficient and precise tools to produce goods, laying out efficient assembly sequences and facilities, and setting up the flow of materials and supplies into the production line, and the control of quality and throughput rate. Production engineering is key to implementing the manufacturing step that translates engineering designs into competitive products.

APPLICATIONS AND PRODUCTS

Conventional Applications. Mechanical engineering is applied to the design, manufacture, and testing of almost every product used by humans and to the machines that help humans build those products. The products most commonly associated with mechanical engineering include all vehicles such as railway trains, buses, ships, cars, airplanes and spacecraft, cranes, engines, and electric or hydraulic motors of all kinds, heating, ventilation and air-conditioning systems, the machine tools used in mass manufacture, robots, agricultural tools, and the machinery in power plants. Several other fields of engineering such as aerospace, materials, nuclear, industrial, systems, naval architecture, computer, and biomedical developed and spun off at the interfaces of mechanical engineering with specialized applications. Although these fields have developed specialized theory and knowledge bases of their own, mechanical engineering continues to find application in the design and manufacture of their products.

Innovations in Materials. Carbon nanotubes have been heralded as a future super-material with strength hundreds of times that of steel for the same mass. As of the first decade of the twenty-first century, the longest strands of carbon nanotubes developed are still on the order of a few centimeters. This is a very impressive length-to-diameter ratio. Composite materials incorporating carbon already find wide use in various applications where high temperatures must be encountered. Metal matrix composites find use in primary structures even for commercial aircraft. Several "smart structures" have been developed, where sensors and actuators are incorporated into a material that has special properties to respond to stress and strain. These enable structures that will twist in a desired direction when bent or become stiffer or more flexible as desired, depending on electrical signals sent through the material. Materials capable of handling very low (cryogenic) temperatures are at the leading edge of research applications. Magnetic materials with highly organized structure have been developed, promising permanent magnets with many times the attraction of natural magnets.

Sustainable Systems. One very important growth area in mechanical engineering is in designing replacements for existing heating, ventilation, and air-conditioning systems, as well as power generators, that use environmentally benign materials and yet achieve high thermodynamic efficiencies, minimizing heat emission into the atmosphere. This effort demands a great deal of innovation and is at the leading edge of research, both in new ways of generating power and in reducing the need for power.

Impact on Industry

Having developed as a discipline to formalize knowledge on the design of machines for industry, mechanical engineering is at the core of most industries. The formal knowledge and skills imparted by schools of mechanical engineering have revolutionized human industry, bringing about a huge improvement in quality and effectiveness. The disciplined practice of mechanical engineering is responsible for taking innumerable innovations to market success. In the seventeenth through twentieth centuries, rampant industrialization destroyed many long-lasting community skills and occupations, replacing them with mass manufacturing concentrated and collocated with water resources, power sources, and transportation hubs. This has led to many problems as rural communities atrophied and their young people migrated to the unfamiliar and crowded environment of cities in search of well-paying jobs.

The effects on the environment and climate have also been severe. Heightened global concerns about the environment and climate change and new technological innovation may find mechanical engineers again at the head of a new revolution. This may start a drive to decentralize energy resources and production functions, permitting small communities and enterprises to flourish again.

Fascinating Facts About Mechanical Engineering

- Robotic surgery enables surgeons to conduct very precise operations by eliminating the problems of hand vibrations and by using smaller steps than a human can take.
- The REpower 5M wind turbine in Germany, rated at 5 megawatts, is 120 meters high and has 61.5-meter radius blades, more than one and one-half times as long as each wing of an Airbus A380 jetliner.
- The General Electric H System integrates a gas turbine, steam turbine, generator, and heat-recovery steam generator to achieve 60 percent efficiency.
- The nanomotor built in 2003 by Alex Zettl, a physics professor at the University of California, Berkeley, and his research group is 500 nanometers in diameter with a carbon nanotube shaft 5 to 10 nanometers in diameter.
- The crawler-transporter used to move space shuttles on to the launch pad weighs 2,721 metric tons and has eight tracks, two on each corner. Its platform stays level when the crawler moves up a five-degree incline.
- Solar thermo-acoustic cooker-refrigerators use the heat from the Sun to drive acoustic waves in a tube, which convects heat and creates a low temperature on one side.
- Much of the world's telecommunications are carried by undersea fiber-optic cables that connect all continents except Antarctica. The first telegraph cable across the English Channel was laid in 1850.
- In 1650, Thomas Savery invented a steam engine to pump water out of coal mines.

Careers and Course Work

Mechanical engineers work in nearly every industry, in an innumerable variety of functions. The curriculum in engineering school accordingly focuses on giving the student a firm foundation in the basic knowledge that enables problem solving and continued learning through life. The core curriculum starts with basic mathematics, science, graphics and an introduction to design and goes on to engineering mechanics and the core subjects and specialized electives. In modern engineering schools, students have the opportunity to work on individual research and design projects that are invaluable in providing the student with perspective and integrating their problem-solving skills.

After obtaining a bachelor's degree, the mechanical engineer has a broad range of choices for a career. Traditional occupations include designing systems for energy, heating, ventilation, air-conditioning, pressure vessels and piping, automobiles, and railway equipment. Newer options include the design of bioengineering production systems, microelectromechanical systems, optical instrumentation, telecommunications equipment, and software. Many mechanical engineers also go on to management positions.

Social Context and Future Prospects

Mechanical engineering attracts large numbers of students and offers a broad array of career opportunities. Students in mechanical engineering schools have the opportunity to range across numerous disciplines and create their own specialties. With nano machines and biologically inspired self-assembling robots becoming realities, mechanical engineering has transformed from a field that generally focused on big industry to one that also emphasizes tiny and efficient machines. Energy-related studies are likely to become a major thrust of mechanical engineering curricula. It is possible that the future will unfold a post-industrial age where the mass-manufacture paradigm of the Industrial Revolution that forced the overcrowding of cities and caused extensive damage to the environment is replaced by a widely distributed industrial economy that enables small communities to be self-reliant for essential services and yet be useful contributors to the global economy. This will create innumerable opportunities for innovation and design.

Narayanan M. Komerath, Ph.D.

Further Reading

Avallone, Eugene A., Theodore Baumeister III, and Ali M. Sadegh. *Marks' Standard Handbook for Mechanical Engineers.* 11th ed. New York: McGraw-Hill, 2006. Authoritative reference for solving mechanical engineering problems. Discusses pressure sensors and measurement techniques and their applications in various parts of mechanical engineering.

Calvert, Monte A. *The Mechanical Engineer in America, 1830-1910: Professional Cultures in Conflict.* Baltimore, Md.: Johns Hopkins University Press, 1967. Discusses the life of the mechanical engineer in nineteenth-century America. The author describes the conflict between the shop culture originating in the procedures of the machine shop and the school culture of the engineering colleges that imparted formal education.

Freitas, Robert A., Jr., and Ralph C. Merkle. *Kinematic Self-Replicating Machines.* Georgetown, Tex.: Landes Bioscience, 2004. A review of the theoretical and experimental literature on the subject of self-replicating machines. Discusses the prospects for laboratory demonstrations of such machines.

Hill, Philip G., and Carl R. Peterson. *Mechanics and Thermodynamics of Propulsion.* 2d ed. Reading, Mass.: Addison-Wesley, 1992. This textbook on propulsion covers the basic science and engineering of jet and rocket engines and their components. Also gives excellent sets of problems with answers.

Lienhard, John H., IV, and John H. Lienhard V. *A Heat Transfer Textbook.* 4th ed. Mineola, N.Y.: Dover Publications, 2011. An excellent undergraduate text on the subject.

Liepmann, H. W., and A. Roshko. *Elements of Gas Dynamics.* Reprint. Mineola, N.Y.: Dover Publications, 2001. Classic textbook on the discipline of gas dynamics as applied to high-speed flow phenomena. Contains several photographs of shocks, expansions, and boundary layer phenomena.

Pelesko, John A. *Self Assembly: The Science of Things That Put Themselves Together.* Boca Raton, Fla.: Chapman and Hall/CRC, 2007. Discusses natural self-assembling systems such as crystals and soap films and goes on to discuss viruses and self-assembly of DNA cubes and electronic circuits. Excellent introduction to a field of growing importance.

Shames, Irving H. *Engineering Mechanics: Statics and Dynamics.* 4th ed. Upper Saddle River, N.J.: Prentice Hall, 1997. A classic textbook that integrates both statics and dynamics and uses a vector approach to dynamics. Used by undergraduates and professionals all over the world since the 1970's in its various editions. Extensive work examples.

Shigley, Joseph E., Charles R. Mischke, and Richard G. Budynas. *Mechanical Engineering Design.* 7th ed. New York: McGraw-Hill, 2004. Classic undergraduate textbook showing students how to apply mathematics, physics, thermal sciences, and computer-based analysis to solve problems in mechanical engineering. Includes sections on quality control, and the computer programming sections provide an insight into the logic used with the high-level languages of the 1980's.

Siciliano, Bruno, et al. *Robotics: Modelling, Planning and Control.* London: Springer-Verlag, 2010. Rigorous textbook on the theory of manipulators and wheeled robots, based on kinematics, dynamics, motion control, and interaction with the environment. Useful for industry practitioners as well as graduate students.

Web Sites

American Society of Heating, Refrigerating and Air-Conditioning Engineers
http://www.ashrae.org

American Society of Mechanical Engineers
http://www.asme.org

National Society of Professional Engineers
http://www.nspe.org/index.html

Society of Automotive Engineers International
http://www.sae.org

Society of Manufacturing Engineers
http://www.sme.org

See also: Algebra; Applied Mathematics; Calculus; Engineering; Robotics.

METALLURGY

FIELDS OF STUDY

Materials science; materials engineering; mechanical engineering; physical engineering; mining; chemical engineering; electrical engineering; environmental engineering.

SUMMARY

Starting as an art and a craft thousands of years ago, metallurgy has evolved into a science concerned with processing and converting metals into usable forms. The conversion of rocky ores into finished metal products involves a variety of activities. After the ores have been mined and the metals extracted from them, the metals need to be refined into purer forms and fashioned into usable shapes such as rolls, slabs, ingots, or tubing.

Another part of metallurgy is developing new types of alloys and adapting existing materials to new uses. The atomic and molecular structure of materials are manipulated in controlled manufacturing environments to create materials with desirable mechanical, electrical, magnetic, chemical, and heat-transfer properties that meet specific performance requirements.

KEY TERMS AND CONCEPTS

- **Bessemer Process:** Steelmaking process in which air is blown through molten pig iron contained in a furnace so that impurities can be removed by oxidation.
- **Blast Furnace:** Smelting furnace for the production of pig iron in which hot air is blown upward into the furnace as iron ore, coke, and limestone are supplied through the top, producing chemical reactions; the molten iron and slag are collected at the bottom.
- **Carburizing:** Process of adding carbon to the surface by exposing a metal to a carbon-rich atmosphere under high temperatures, allowing carbon to diffuse into the surface, making the surface more wear resistant.
- **Ductility:** Characteristic of metal that enables it to be easily molded or shaped, without fracturing.
- **Extrusion:** Process in which a softened metal is forced through a shaped metal piece or die, creating an unbroken ribbon of product.
- **Flux:** Substance added to molten metals to eliminate impurities or encourage fusing.
- **Forging:** Process of shaping metal by heating it in a forge, then beating or hammering it.
- **Galvanizing:** Process of coating steel with zinc to prevent rust.
- **Metal Alloy:** Homogeneous mixture of two or more metals.
- **Ore:** Mineral from which metal is extracted.
- **Plastic Deformation:** Permanent distortion of a metal under the action of applied stresses.
- **Recrystallization:** Process by which deformed grains in a metal or alloy are replaced by undeformed grains; reduces the strength and hardness of a material.
- **Recrystallization Temperature:** Approximate minimum temperature at which a cold-worked metal becomes completely recrystallized within a specified time.
- **Sintering:** Process of turning a metal powder into a solid by pressure and heating it to a temperature below its boiling point.
- **Slag:** By-product consisting of impurities produced during the refining of ore or melting of metal.

Definition and Basic Principles

Metallurgy is the science of extracting metals and intermetallic compounds from their ores and working and applying them based on their physical, chemical, and atomic properties. It is divided into two main areas: extractive metallurgy and physical metallurgy.

Extractive Metallurgy. Extractive metallurgy, also known as process or chemical metallurgy, deals with mineral dressing, the converting of metal compounds to more treatable forms and refining them. Mineral dressing involves separating valuable minerals of an ore from other raw materials. The ore is crushed to below a certain size and ground to powder. The mineral and waste rock are separated, using a method based on the mineral's properties. After that, water is removed from the metallic concentrate or compound. Because metallic compounds are often complex mixtures (carbonates, sulfides, and oxides), they need to be converted to other forms for easier processing and refining. Carbonates are converted to oxides; sulfides to oxides, sulfates, and chlorides; and oxides to sulfates, and chlorides. Depending on the type of metallic compound, either pyrometallurgy or hydrometallurgy is used for conversion. Both processes involve oxidation and reduction reactions. In oxidation, the metallic element is combined with oxygen, and in reduction, a reducing agent is used to remove the oxygen from the metallic element. The difference between pyrometallurgy and hydrometallurgy, as their names imply, is that the former uses heat while the latter uses chemicals. These two processes also include refining the metallic element in the final stage of extractive metallurgy when heat and chemicals are used. Electrometallurgy refers to the use of the electrolytic process for refining metal elements, precipitating dissolved metal values, and recovering them in solid form.

Physical Metallurgy. Physical metallurgy deals with making metal products based on knowledge of the crystal structures and properties (chemical, electrical, magnetic, and mechanical) of metals. Metals are mixed together to make alloys. Heat is used to harden metals, and their surfaces can be protected with metallic coating. Through a process called powder metallurgy, metals are turned into powders, compressed, and heat-treated to produce a desired product. Metals can be formed into their final shapes by such operations as casting, forging, or plastic deformation. Metallography is a subfield of metallurgy that studies the microstructure of metals and alloys by various methods, especially by light and electron microscopes.

Background and History

Metallurgy came into being in the Middle East around 5000 B.C.E. with the extraction of copper from its ore. The discovery of the first alloy, bronze, the result of melting copper and tin ores together, initiated the Bronze Age (4000-3000 B.C.E.). Melting iron ore with charcoal to obtain iron marked the beginning of the Iron Age in Anatolia (2000-1000 B.C.E.). Gold, silver, and lead were separated from lead-bearing silver in Greece about 500 B.C.E. Mercury was produced from cinnabar around 100 B.C.E., and it was later used to recover and refine various metals. Around 30 B.C.E., brass, the second alloy, was made from copper and zinc in Egypt, and another alloy, steel, was produced in India.

From the sixth to the nineteenth centuries, metallurgy focused on the development and improvement of the processes involved in obtaining iron, making steel, and extracting aluminum and magnesium from their ores. The blast furnace was developed in the eighth century and spread throughout Europe. During the sixteenth century, the first two books on metallurgy were written by Vannoccio Biringuccio, an Italian metalworker, and by Georgius Agricola, a German metallurgist.

Modern metallurgy began during the eighteenth century. Abraham Darby, an English engineer, developed a new furnace fueled by coke. Another English engineer, Sir Henry Bessemer developed a steelmaking process in 1856. Great Britain became the greatest iron producer in the world, and Spain and France also produced large amounts of iron. About 1886, American chemist Charles Martin Hall and a French metallurgist Paul-Louis-Toussaint Héroult independently developed a way to extract aluminum from its ore, which became known as the Hall-Héroult process. Aluminum soon became an important metal in manufactured goods.

Metallurgy did not emerge as a modern science with two branches, extractive and physical, until the twentieth century. The development and improvement of metallurgy were made possible by the application of knowledge of the chemical and physical principles of minerals.

How It Works

Crushing and Grinding of Ores. In the first step of mineral dressing, two kinds of mechanized crushers are used to reduce ores. Jaw crushes reduce ores to less than 150 millimeters (mm) and cone crushers to less than 10-15 mm. Different kinds of grinding mills are used to reduce crushed ores to powder: cylinder mills filled with grinding bodies (stones or metal balls), autogenous mills (coarse crushed ores grinding themselves), semiautogenous mills using some grinding bodies, and roll crushers, which combine crushing and grinding.

Separating Valuable and Waste Minerals. The process used in the next step of mineral dressing depends on the properties of the minerals. Magnetic separation is used for strongly magnetic minerals such as iron ore and iron-bearing ore. Gold, tin, and tungsten ores require gravity separation. A process called flotation separation is widely used for hydrophilic (water-attracting) intergrown ores containing copper, lead, and zinc. Electrostatic separation work best with particles of different electric charges such as mineral sands bearing zircon, rutile, and monazite.

Pyrometallurgy. Pyrometallurgy is a method of converting metallic compounds to different forms for easier processing and refining by using oxidation and reduction reactions.

The first conversion process, roasting, has two main types: One type changes sulfide compounds to oxides, and the other reduces an oxide to a metal. Other types of roasts convert sulfides to sulfates or change oxides to chlorides. These processes are carried out in different kinds of steel roasters.

The second conversion process, smelting, separates a metallic compound into two parts: an impure molten metal and a molten slag. The two types of smelting are reduction and matte, and the processes are done in many kinds of blast furnaces. Coke is used for fuel and limestone as a flux for making slag. Reduction smelting converts an oxide feed material to a metal and an oxide slag. Matte smelting converts a sulfide feed material to a mixture of nickel, copper, cobalt, and iron sulfides as well as an iron and silicon oxide slag.

Refining, a process of removing any impurities left after roasting or smelting, also can be done in a blast furnace. Iron, copper, and lead can be refined in oxidation reaction that removes impurities as an oxide slag or an oxide gas. Fire refining can separate copper from its impurities of zinc, tin, iron, lead, arsenic, and antimony. Similarly, lead can be separated from such impurities as tin, antimony, and arsenic, and zinc from impurities of cadmium and lead.

Hydrometallurgy. Another method of converting metallic compounds to different forms is hydrometallurgy. It uses several types of leach solvents: ammonium hydroxide for sulfides and carbonates; sulfuric acid, sodium carbonate, or sodium hydroxide for oxides; and sulfuric acid or water for sulfates. The dissolved metal values are then recovered from the leaching solution in solid form. Although numerous recovery processes exist, they usually involve electrolysis. By a process called precipitation, gold that has been dissolved in sodium cyanide and placed in contact with zinc is separated from the solution and gathers on zinc. In another process called electrolytic deposition, or electrowinning, an electric current is passed through the leach solution with dissolved metals, causing metal ions to deposit at the cathode. Copper, zinc, nickel, and cobalt can be obtained this way.

Electrometallurgy. Electrolysis can be used to refine metallic elements as well as to recover them after hydrometallurgical treatment. Copper, nickel, lead, gold, and silver can be refined this way. In this method, for example, impure copper is used as the anode. When the electric current passes through the solution, atoms of pure gold travel to the cathode, acquire electrons, and become neutral copper atoms. Electrolysis is also the process for recovering copper, aluminum, and magnesium in hydrometallurgy.

Alloys. Alloys are made by mixing pure metals together to obtain a substance with increased strength, increased corrosion resistance, lower cost, lower melting points, or desirable magnetic, thermal, or electrical properties. They are usually made by melting the base metals and adding alloying agents. Stainless steel, a mixture of steel, nickel, and chromium, is stronger and more chemically resistant than the base metals from which it was formed.

Powder Metallurgy. Powder metallurgy is the process of reducing metals and nonmetals to powder and shaping the powder into a solid form under great heat and pressure. Metal powders are usually produced by atomization of streams of molten metal with a spinning disk or with a jet of water, air, or inert gas. After the powders are cold pressed for initial adhesion, they are heated to temperatures about 80 percent below the melting point of the major component. Friction between powders and pressing dies is reduced by adding lubricants, and porosity in the final product is eliminated by applying pressure.

Metal Forming. Metals are usually cast into ingots in iron molds. Casting is also carried out in molds made of sand, plaster of Paris, or glue. Permanent casting uses pressure or centrifugal action. Plastic deformation is performed on metals to change their properties and dimensions. If done below the recrystalization temperature, the process is called cold working; above this temperature but below the melting or burning point, it is called hot working. Techniques involved include rolling, pressing, extrusion, stamping, forging, and drawing. Surface treatments of metals include protective coating and hardening. In metallic coating, zinc and other metals such as chromium, cadmium, lead, and silver are often used. Surface hardening of metals is usually done with heat in a gas rich in carbon or in ammonia and hydrogen.

Applications and Products

The most important applications of metallurgy involve common metals and alloys and powder metallurgy technology.

Copper. Copper is ductile and malleable, and it resists corrosion and conducts heat and electricity. Copper and its alloy brass (copper plus zinc) are used to make coins, household fixtures (doorknobs, bolts), and decorative-art objects (statues, sculptures, imitation-gold jewelry). It is also used in transportation vehicles and has many electrical applications (transformers, motors, generators, wiring harnesses). Its alloy bronze (copper plus tin) is used in plumbing and heating applications (water pipes, cooking utensils). Aluminum-bronze is used to make tools and parts for aircraft and automobiles. Manganese-bronze is used to make household fixtures and ship propellers.

Iron. Iron is ductile, malleable, and one of the three magnetic elements (the others are cobalt and nickel). Cast iron is resistant to corrosion and used to make manhole covers and engine blocks for gasoline and diesel engines. Wrought iron is used to make cooking utensils and outdoor household items such as fencing and furniture. Most iron is used to make steel. Steel is used as a structural material in the construction of large, heavy projects (bridges, ships, buildings) and automobile parts (body frames, radial-ply tires). When chromium and nickel are combined with steel, steel becomes stainless, and it is used to make flatware and surgical tools. Steel combined with cobalt is used to make jet engines and gas turbines.

Gold. Applications of gold are based on such properties as its electrical and thermal conductivity, ductility, malleability, resistance to corrosion, and infrared reflectivity. Gold serves as a medium of exchange and money. Its decorative applications include jewelry, golf leaf on the surfaces of buildings, and flourishes on ceramics or glassware. More practical applications include components for electronic devices (telephones, computers), parts for space vehicles, and dental fillings, crowns, and bridges.

Silver. Silver is ductile and very malleable, conducts heat, and has the highest electrical conductivity of all metals. It is used to make cutlery, jewelry, coins, long-life batteries, photographical films, and electronic components (circuits, contacts), and in dentistry. Its alloy, sterling silver (silver plus copper) is also used to make jewelry and tableware. German silver (silver plus nickel) is another alloy used for silverware.

Platinum. This ductile and malleable material is one of the densest and heaviest metals. It is resistant to corrosion and conducts electricity well. It is used to make jewelry, electronic components (hard disk

Fascinating Facts About Metallurgy

- Iron is the most abundant element, making up 34.6 percent of Earth, and the most used of all metals.
- The magnetic property of steel allows recyclers to reclaim millions of tons of iron and steel from garbage.
- Meteors were the source of much of the wrought iron used in early human history.
- Because gold is so malleable, 1 gram of gold can be hammered into a sheet 1 square meter in size. It can also be made so thin that it appears transparent.
- Nitric acid can be used to determine if gold is present in ore. This "acid test," proving an ore's value, has come to mean a decisive test proving an item's worth or quality.
- Half of the world's gold is held by the Republic of South Africa. The second gold-producing nation, in terms of volume, is the United States.
- A lead pencil might more accurately be called a graphite pencil, as it contains a shaft of graphite, not lead.
- An alloy of equal parts of silver and aluminum is as hard as bronze.
- The amount of platinum mined each year is 133 tons, less than one-tenth of the 1,782 tons of gold mined each year.
- Uranium is very dense. A one-gallon container filled with uranium would weigh about 150 pounds, while that same container filled with milk would weigh around 8 pounds.
- To produce about 40 million kilowatt hours of electricity, it would take 16,000 tons of coal or 1 ton of natural uranium.
- In Japan, all government-subsidized dental alloys must have a palladium content of at least 20 percent.

drive coatings, fiber-optics cables), and spark plug components. It is important in making the glass for liquid crystal displays (LCDs), in the petrol industry as an additive and refining catalyst, in medicine (anticancer drugs, implants), and in dentistry. Its alloys (platinum plus cobalt or metals in the platinum groups) are mostly used to make jewelry.

Mercury. Sometimes called quicksilver, mercury is the only common metal that is liquid at ordinary temperatures. It is a fair conductor of electricity and of high density. It is used in barometers and thermometers, to recover gold from its ore, and to manufacture chlorine and sodium hydroxide. Its vapor is used in street lights, fluorescent lamps, and advertizing signs. Mercury compounds have various uses, such as insecticides, rat poisons, disinfectants, paint pigments, and detonators. Mercury easily is alloyed with silver, gold, and cadmium.

Lead. Lead is malleable, ductile, resistant to corrosion, and of high density. Its softness is compensated for by alloying it with such metals as calcium, antimony, tin, and arsenic. Lead is a component in lead-acid batteries, television and computer screens, ammunition, cables, solders, and water drains, and is used as a coloring element in ceramic glazes.

Magnesium. Magnesium is the lightest structural metal, with low density (two-thirds that of aluminum), superior corrosion performance, and good mechanical properties. Because it is more expensive than aluminum, its applications are somewhat limited. Magnesium and its alloys are used in the bicycle industry, racing car industry (gearbox casings, engine parts), and aerospace industry (engines, gearbox casings, generator housings, wheels).

Manganese. Manganese is a hard but very brittle, paramagnetic metal. Mostly it is used in steel alloys to increase strength, hardness, and abrasion resistance. It can be combined with aluminum and antimony to form ferromagnetic compounds. It is used to give glass an amethyst color, in fertilizers, and in water purification.

Cobalt. Cobalt has a high melting point and retains its strength at high temperatures. It is used as a pigment for glass, ceramics, and paints. When alloyed with chromium and tungsten, it is used to make high-speed cutting tools. It is also alloyed to make magnets, jet engines, and gas turbine engines.

Tungsten. Tungsten has the highest melting point and the lowest thermal expansion of all metals, high electrical conductivity, and excellent corrosion resistance. It is used to make lightbulb filaments, electric contacts, and heating elements; as an additive for strengthening steel; and in the production of tungsten carbide. Tungsten carbide is used to make dies and punches, machine tools, abrasive products, and mining equipment.

Chromium. Chromium is a hard but brittle metal of good corrosion resistance. It is mostly alloyed with other metals, especially steel, to make final products harder and more resistant to corrosion. It is also used in electroplating, leather tanning, and refractory brick making, and as glass pigments.

Cadmium. Cadmium is resistant to corrosion, malleable, ductile, and of high electrical and thermal conductivity. It is mostly used in rechargeable nickel-cadmium batteries. It is also used to make electronic components and pigments for plastics, glasses, ceramics, enamels, and artists' colors.

Nickel. Nickel is a hard, malleable, and ductile metal that is highly resistant to corrosion. Like chromium, it is used to make stainless steel. Alloyed with copper, it is used for ship propellers and chemical industry plumbing. Other uses include rechargeable batteries, coinage, foundry products, plating, burglar-proof vaults, armor plates, and crucibles.

Aluminum. Aluminum has a density about one-third that of steel, high resistance to corrosion, and excellent electrical and thermal conductivity. Moreover, this nontoxic metal reflects light and heat well. This versatile metal can be used to replace other materials depending on the application. It is widely used in such areas as food packaging and protection (foils, beverage cans), transportation (vehicles, trains, aircraft), marine applications (ships, support structures for oil and gas rigs), and buildings and architecture (roofing, gutters, architectural hardware). Other applications of aluminum include sporting goods, furniture, road signs, ladders, machined components, and lithographic printing plates.

Special Alloys. Fusible alloys are mixtures of cadmium, bismuth, lead, tin, antimony, and indium. They are used in automatic sprinklers and in forming and stretching dies and punches. Superalloys are developed for aerospace and nuclear applications: columbium for reactors, tantalum for rocket nozzles and heat exchangers in nuclear reactors, and a nickel-based alloy for jet and rocket engines and electric heating furnaces. The alloy of tin and niobium has superconductivity. It is used in constructing superconductive magnets that generate high field strengths without consuming much power.

Powder Metallurgy Applications and Products. Powder metallurgy was developed in the late 1920's primarily to make tungsten-carbide cutting tools and self-lubricating electric motor bearings. The technique was then applied in the automobile industry, where it is used to make precision-finished machine parts, permanent metal filters, bearing materials, and self-lubricating bearings. It is useful in fabricating products that are difficult to make by other methods, such as tungsten-carbide cutting tools, super magnets of aluminum-nickel alloy, jet and missile applications of metals and ceramics, and wrought powder metallurgy tool steel.

Impact on Industry

Metallurgy is used in many industries, and the challenges associated with it vary by industry. All industries, for example, seek ways to achieve fuel savings, and where metallurgy is concerned, they want high metallic yields and high-quality products.

Aluminum production requires large amounts of electricity for the electrolytic smelting process. Electricity accounts for 25 percent of the cost of producing aluminum. The techniques of ladle metallurgy (degassing, desulfurization) require high temperatures for preheating, also resulting in high energy costs. Hydrogen and oxides can cause porosity in solidified aluminum, which decreases the mechanical properties of the final product. High oxygen levels ranging from 25 percent in converters to 90 percent in flash-smelting furnaces are used for primary smelting, resulting in low productivity as well as high operating costs.

New technologies in metallurgy not only solve operational problems but also try to offer such benefits as high productivity, availability and reliability of power, safety, and environmental acceptance. For example, Alstom, a company in transport and energy infrastructure, provides air-quality control systems for the aluminum industry. To abate emissions from pot lines, anodes, and green anodes, Alstom systems take the gas from the pots and absorb hydrofluorides on alumina. Additional sulfur is removed from the gas before it is released to the atmosphere. Alstom has also developed an energy recovery system to decrease pot amperage in smelters, thereby ensuring high availability and reliability of power.

Air Liquide provides solutions to the problems caused by the need to preheat the ladle by using oxy-fuel burners, based on the evolution in refractory materials. The necessary calories needed to keep a metal at the suitable pouring temperature during the metal transfer into the ladle are stocked in the refractory. Air Liquide also provides solution to the porosity problem in aluminum. Its technique involves injecting hydrogen through porous plugs. The same company also developed an innovative scheme to lower the use of oxygen in smelting furnaces. It includes a process that allows the direct production of oxygen at 95 percent purity under medium pressure without an oxygen compressor and a balancing system for liquid and gaseous oxygen for handling fast-changing regimes and optimizing working conditions at low operating costs.

Problems in metallurgy are also solved by simply replacing the old technologies. For example, operating problems in the copper and nickel converter include tuyere blockage and refractory erosion at the tuyere line. Using an oxygen injector to replace conventional tuyeres not only permits enrichment levels of up to 60 percent of oxygen and increases converter productivity but also reduces volumes of toxic off-gases.

The metallurgy industry, like many other industries, must constantly deal with new laws and regulations and expends considerable time and money to meet environmental requirements. The industry must adapt its extraction processes and invest in emission-reducing and energy-saving technologies. For example, the steel industry has replaced all its open-hearth furnaces. The aluminum industry has used prebaked instead of Soderberg electrodes in the electrolytic cells. The copper industry has replaced the reverberatory furnace with the flash-smelting furnace. The common reason behind these changes was to reduce toxic off-gases as much as possible. Stacks have become much taller to dispose of sulfur dioxide.

A process based on the combination of biology and hydrometallurgy—also called bioleaching, biohydrometallurgy, microbial leaching, and biomining—has gained some interest from those who want to replace some of the traditional metallurgical processes. Although conventional metallurgy uses smelting of ores at high temperatures, bioleaching involves dissolving metals from ores using microorganisms. Copper, for example, can be leached by the activity of the bacterium *Acidithiobacillus ferrooxidans*. Canada used this process to extract uranium in 1970, and South Africa experimented it with gold during the 1980's. Several countries have used bioleaching for a number of metals (copper, silver, zinc, nickel). The process is helpful in the recovery of low-grade ores, which cannot be economically processed with chemical methods; however, it creates a problem because it results in excessive disposal and handling of a mixture of ferrous hydroxide and gypsum.

Careers and Course Work

High school students who wish to work as metallurgical technicians must take at least two years of mathematics and two years of science, including a physical science. Shop courses of any kind are also helpful. Positions in the metallurgical industry are typically in the areas of production, quality control, and research and development, which share many concerns and often require similar skills from prospective metallurgical technicians. Two years of study in metallurgy or materials science at a community college or technical college is therefore strongly recommended. Metallurgical technicians occupy a middle ground between engineers and skilled trade workers. Representative entry-level jobs include metallurgical laboratory technicians, metallographers, metallurgical observers, metallurgical research technicians, and metallurgical sales technicians. Students who are interested in these kinds of jobs should have an interest in science and average mathematical ability. Prospective technicians must be willing to participate in a wide variety of work and must be able to communicate well. Companies employing metallurgical technicians can be found in a wide variety of industries. Working environments vary depending on the area of activities.

A number of colleges and universities offer four-year programs in metallurgy. If students wish to become metallurgical engineers, they will need a bachelor's degree in materials science or metallurgical engineering. The first two years of college focus on subjects such as chemistry, physics, mathematics, and introductory engineering. In the following years, courses will focus on metallurgy and related engineering areas. Students who wish to become metallurgical engineers should be interested in nature and enjoy problem solving. They also need to have good communication skills. There are basically three areas in which metallurgical engineers work: extractive, physical, and mechanical metallurgy. Their work environment varies depending on their area of specialty. Companies employing metallurgical engineers include metal-producing and processing companies, aircraft companies, machinery and electrical equipment manufacturers, the federal government, engineering consulting firms, research institutes, and universities.

Social Context and Future Prospects

Metallurgy faces the challenges of reducing the effect of its processes on the air and land, making more efficient use of energy, and increasing the amount of recycling. To these ends, industries are increasingly using clean technologies and developing methods of oxygen combustion that drastically reduce emissions of carbon dioxide and other pollutants. For example, zinc, copper, and nickel are being recovered from their ores through a technique in which pressure leaching is performed in an acid medium, followed by electrolysis in a conventional sulfuric acid medium. The technique produces no dust, no slag, and no sulfur dioxide and therefore is environmentally acceptable. It has been applied for zinc sulfide concentrates in 1980's, for nickel sulfides in Canada, and for copper sulfide concentrates in the United States. In addition, between 1994 and 2003, the steel industry reduced its releases of chemicals to the air and water by 69 percent. However, it still releases manganese, chromium, and lead to the air, and efforts are concentrating on this problem.

Becoming more energy efficient has long been a goal of the metallurgy industry, especially as it benefits the bottom line. A wide range of approaches has been employed, including lowering generation costs (for example, by generating energy rather than purchasing it), capturing energy (such as gases) produced during various metallurgy processes, making use of energy-efficient equipment and techniques, and monitoring the production process. The steel industry in North America has reduced its consumption of energy by 60 percent since World War II. One way that the industry reduced energy consumption was by using scrap steel instead of natural resources to produce steel.

Metallurgy companies have made efforts to increase recycling. Each year, more steel is recycled in the United States than paper, aluminum, plastic, and glass combined. In 2008, more than 75 million tons of steel were either recycled or exported for recycling. All new steel produced in the United States in 2008 contained at least 25 percent steel scrap on average.

Despite these environmental challenges, metallurgy is an important area and will continue to grow and develop. In addition to developing ways to lessen metallurgy's impact on the environment, engineers are likely to develop improved processes for extraction of ores and processing materials into products. New techniques are likely to develop in response to the need to impart metals with additional qualities and to conserve natural resources by reusing metals and finding uses for by-products. Waste disposal and reduction also are likely to remain areas of research.

Anh Tran, Ph.D.

Further Reading

Abbaschian, Reza, Lara Abbaschian, and Robert E. Reed-Hill. *Physical Metallurgical Principles.* Stamford, Conn.: Cengage Learning, 2009. A comprehensive introduction to physical metallurgy for engineering students.

Boljanovic, Vukota. *Metal Shaping Processes: Casting and Molding, Particulate Processing, Deformation Processes, and Metal Removal.* New York: Industrial Press, 2010. Describes the fundamentals of how metal is shaped into products.

Bouvard, Didier. *Powder Metallurgy.* London: ISTE, 2008. Looks at the thermo-mechanical processes used to turn powdered metal into metal parts and also describes applications.

Brandt, Daniel A., and J. C. Warner. *Metallurgy Fundamentals.* 5th ed. Tinley Park, Ill.: Goodheart-Willcox, 2009. Examines metallurgy, focusing on iron and steel but also covering nonferrous metals.

Pease, Leander F., and William G. West. *Fundamentals of Powder Metallurgy.* Princeton, N.J.: Metal Powder Industries Federation, 2002. A primer on powder metallurgy.

Popov, K. I., Stojan D. Djokić, and Branimir N. Grgur. *Fundamental Aspects of Electrometallurgy.* New York: Plenum, 2002. Examines the theory and mechanisms of electrometallurgy.

Vignes, Alain. *Handbook of Extractive Metallurgy.* Hoboken, N.J.: John Wiley & Sons, 2010. Examines how metals are transformed from ore into liquids ready for pouring.

Web Sites

American Institute of Mining, Metallurgical, and Petroleum Engineers
http://www.aimeny.org

The Minerals, Metals, and Materials Society
http://www.tms.org

Mining and Metallurgical Society of America
http://www.mmsa.net

See also: Chemical Engineering; Electrometallurgy; Environmental Engineering.

N

NAVAL ARCHITECTURE AND MARINE ENGINEERING

FIELDS OF STUDY

Thermodynamics; fluid mechanics; mechanics of materials; machine design; engineering; hydrodynamics; ship structures; propulsion technologies.

SUMMARY

The process of designing a ship involves two complementary disciplines: naval architecture and marine engineering. Naval architecture has two subdivisions: hydrodynamics and ship structures. Hydrodynamics is concerned with the interaction between the moving ship hull and the water in which it floats. Ship structures is concerned with building a hull that has the strength needed to withstand the forces to which it is subject. Marine engineering is concerned with all the machinery that goes into the ship. The machinery must perform the following tasks: propulsion and steering, electric power generation and distribution, and cargo handling.

KEY TERMS AND CONCEPTS

- **Bridge:** Place from which the ship's movement and navigation is controlled.
- **Deck:** Horizontal surface in a ship, like the floors in a building.
- **Deckhouse:** Upper part of the ship that sits on the deck; also known as a ship's superstructure.
- **Hull:** Lower part of the ship. Part of it is submerged in the water.
- **Knot:** Unit of measure of ship speed. It represents one nautical mile per hour or about 1.15 land miles per hour.
- **Long Ton:** Unit of weight used for ships and their cargoes. It amounts to 2,240 pounds.
- **Rudder:** Movable device that projects below the ship's hull and is used to steer the ship.
- **Waterline:** Highest point on the hull that is submerged in the water when the ship is stationary.

Definition and Basic Principles

A ship is a very complex object. Some ships must be able to load and unload themselves, while others depend on facilities in ports of call. A ship must propel itself from place to place and control its direction of motion. It must produce its own electricity and freshwater. In addition to the space devoted to cargo, a ship must provide space for fuel, freshwater, living accommodations for crew and perhaps passengers, propulsion, and related machinery. A ship must float, and it must remain upright in all sorts of sea conditions. There are many different kinds of ships: aircraft carriers, submarines, containerships, tankers, passenger ships, ferries, and ships that transport liquefied natural gas, to name a few. The size of a ship is chosen based on cargo-carrying capacity, but it is limited by channel depth, pier length, and other characteristics of the ports it must enter. Most ships are part of a transportation business. To maximize profit, the costs of construction and operation must be minimized, while the income from transporting cargo must be maximized. Working together, naval architects and marine engineers design each ship to satisfy the requirements mentioned above. A design team is assembled that represents the needed areas of expertise. The team may work for a year or more to produce the complete design. When the design work is finished, one or more shipyards may be invited to submit bids for construction.

Background and History

One of the earliest well-documented ships is the funeral ship of Pharaoh Cheops that was built in about 2600 B.C.E. The Greeks and the Romans built large oar-powered ships called biremes and triremes. Bartholomew Diaz sailed three caravels around the tip of Africa in 1488, and Christopher Columbus reached the New World in 1492. All these ships were built by artisans who knew no theory of ship design. Scientific ship design began in France and Spain in the late 1600's. Steam was first used to propel watercraft in the late 1700's, and the first successful steam powered vessel, *Charlotte Dundas*, made its first voyage on the River Clyde in Scotland in 1802. By 1816, steam-powered passenger ships sailed regularly between Brighton, England, and Le Havre, France. William H. Webb designed and built clipper ships along the East River in New York City in the years before the Civil War. Webb and his contemporaries and predecessors called themselves shipbuilders, but they designed their ships too. During the twentieth century, the functions of designing ships and building them became separate.

How It Works

The two principal issues in designing the hull of a ship are its structural strength and the force required to push it through the water at the desired speed. Other issues include stability and the way the ship moves up and down and side to side in waves.

Designing the Hull. Forces on a ship's hull may be divided into static (constant) forces and dynamic (variable) forces. There are two static forces: the force of gravity pulling down on the ship and its contents and the upward force exerted by the water on the hull.

Wave action is responsible for the dynamic forces, but some of these forces are direct results of the waves and some are indirect. When the ship's bow meets a wave, the wave exerts an upward force, lifting the bow. As the ship passes through the wave, the upward force moves along the ship from bow to stern. When the bow reaches the low point between two wave crests, the bow drops and the stern rises. Wave action causes the ship to flex, like a board that is supported at its ends does when someone jumps up and down at the middle.

Wave action also causes the ship to roll from side to side. As the ship rolls, liquids in its tanks slosh from side to side. This causes dynamic forces on the walls of the tanks. Waves may slap against the sides of the ship, and sometimes waves put water onto the decks.

Resistance and Stability. As a ship moves through the water, it must push aside the water ahead of it. This water moves away from the ship as waves. Water flowing along the sides of the ship exerts a friction force on the hull. The combination of these forces is called the resistance. The propulsion force provided by a ship's propeller must overcome the resistance. Naval architects can accurately predict the friction resistance, but the resistance associated with pushing the water aside is usually determined by testing a scale model.

A ship must be stable. This means it must float upright in still water, and it must return to an upright position after waves cause it to tilt. A naval architect must perform detailed calculations to ensure that the ship being designed meets requirements for stability.

Propulsion and Auxiliary Machinery. It is the responsibility of the marine engineer to select the machinery that will provide the required power. There are ships powered by steam turbines, gas turbines, and diesel engines. The propeller may be driven directly by the engine, it may be driven through a speed-reducing gear, or the engine may drive an electric generator that provides power to a motor that drives the propeller.

The marine engineer must also select pumps, piping, oil purifiers, speed-reducing gears, heat exchangers, and many other pieces of auxiliary machinery. The marine engineer designs the systems that connect all of the components and allow them to work together. A ship must generate its own electricity and produce freshwater from seawater. Machinery must be provided to control the direction of motion of the ship.

Applications and Products

Naval architects and marine engineers are called on to design many different types of ships. Each ship type has its own unique design requirements. The designers must consider many factors, including how deep the channels are in the intended ports of call, what type of cargo will be carried, and how time sensitive it is. Several examples are discussed below.

Passenger Ships. Passenger ships may be liners or cruise ships. Liners are used to transport people from one place to another. Before air travel became common, this is how people traveled across oceans. In the early twenty-first century, most passenger ships are cruise ships. They embark passengers at a port, take them to visit interesting places, and return them to the same port where they embarked. Cruise ships have extensive passenger-entertainment facilities, which may range from rock-climbing walls to casinos. Most cruise ships are propelled by diesel engines. These engines may drive the propellers directly, or they may drive electric generators. In the latter case, the propellers are driven by electric motors.

Containerships. Shipping of cargo in standard

rectangular containers has revolutionized the shipping business. At specialized containership ports, a ship can be loaded or unloaded in twelve hours or less by large container cranes mounted on the dock. The largest containerships can carry more than 10,000 twenty-foot long containers. Containers can hold all manner of cargo from cameras and flat-screen TVs to food and cut flowers. Containerships are typically the fastest category of merchant ships. Speeds between 20 and 25 knots are common because these ships carry time-sensitive cargo.

Tankers. Tankers range from ultra-large crude-oil carriers that carry more than 300,000 long tons of crude to small coastal tankers that carry 10,000 long tons or less. Ultra-large crude-oil ships transport large amounts of crude oil from sources in the Persian Gulf to refineries in Europe and North America. There are many other categories of tankers. The most specialized are chemical tankers, which are equipped to carry corrosive cargo such as sulfuric acid and highly flammable cargo such as gasoline.

LNG Ships. These tankers are designed to carry liquefied natural gas (LNG). Natural gas, which is mainly methane, changes from gas to liquid at −160 degrees Celsius. Although the tanks are heavily insulated, some heat does leak into the LNG. This causes a small amount of LNG to vaporize. One of the major decisions in designing an LNG ship is how to handle this gas. Many LNG ships are propelled by steam turbines, and the gas that boils off from the tanks is burned in the boilers. Other LNG ships are propelled by diesel engines. In some cases, the boil-off gas (BOG) is burned in the engines. On other diesel-powered ships the BOG is condensed and returned to the tanks. Many LNG ships operate at about 19 knots. Depending on their size, they may require as much as 50,000 horsepower for propulsion.

Naval Surface Ships. Naval surface ships range from huge aircraft carriers, which may weigh 100,000 long tons or more, to much smaller destroyers and frigates, which may weigh 4,000 long tons. Modern aircraft carriers and cruisers are often nuclear powered. Destroyers and frigates are often powered by gas turbines. On an aircraft carrier the main weapons are the aircraft. Other warships are armed with guns, missiles, and torpedoes. The ships must have the capability to locate and track enemy targets and to launch weapons at them. They must also be able to defend themselves against weapons launched by enemy ships and aircraft.

Fascinating Facts About Naval Architecture and Marine Engineering

- Modern containerships can carry more than 10,000 containers of twenty feet in length. Container shipping has drastically reduced the cost of moving cargo by sea.
- Containerships can be loaded and unloaded in twenty-four hours or less. Containers are quickly loaded onto trucks or railcars for land transportation.
- Nuclear submarines can remain underwater for several months. They produce their own oxygen by splitting water molecules, and nuclear reactors require no air to produce power.
- A crude oil tanker can carry 300,000 long tons of crude halfway around the world.
- An aircraft carrier may have more than 5,000 people aboard. It is like a small city.
- LNG ships carry liquefied natural gas at a temperature of −160 degrees Celsius. As small amounts boil off from the tanks, it may be re-liquefied or burned in the ship's engines.
- Until the middle of the twentieth century most ships were steam powered, but in the early twenty-first century most are powered by diesel engines.

Submarines. Submarines range from one-person research vessels powered by batteries to the U.S. Navy's large, nuclear-powered missile submarines. Two things make submarines different from other ships: First, they must be able to operate without access to the atmosphere, and second, their hulls must withstand the pressure of the sea at the depths where they operate. Navy submarines are powered by nuclear reactors so that they do not need air for the combustion of fuel. The oxygen required to support human life aboard a submarine is produced by using electricity to split water molecules into hydrogen and oxygen, a process called electrolysis.

There are many other types of ships: ferries, roll-on/roll-off ships, heavy lift ships, fishing vessels, dredges, tugboats, and yachts of all shapes and sizes. Two very interesting specialized ships are hovercraft and hydrofoils. Hovercraft ride on a cushion of air that is trapped between the hull and the surface of the sea. Hydrofoils are supported by "wings" that are submerged in the water. Both hovercraft and hydrofoils are capable of much higher speeds than conventional ships.

Impact on Industry

Ship-design companies range in size from a single individual to companies employing hundreds of naval architects, marine engineers, and support personnel. A team of experts may spend a year or more designing a large, complex ship. Detailed drawings and specifications are the final product of a ship-design company. Major ship-design companies in the United States include Gibbs & Cox, and the naval architecture and marine engineering division of Alion Science and Technology. Smaller companies include Herbert Engineering in California and Robert Allan in Canada.

There is a modest amount of ship building in the United States, much of it devoted to the U.S. Navy and Coast Guard. Northrop Grumman Shipbuilding operates shipyards on the Gulf Coast and in Newport News, Virginia. At Newport News, it builds nuclear-powered aircraft carriers and submarines as well as surface ships for the Navy and Coast Guard. General Dynamics has shipyards in Connecticut and San Diego, California. General Dynamics Electric Boat specializes in building nuclear submarines for the Navy, while its San Diego facility builds surface ships for the Navy and for commercial service. By law, ships that carry cargo from one American port to another must be registered in the United States. With some exceptions, such ships must be built in the United States as well. These are called cabotage laws, and the specific U.S. law is called the Jones Act.

In Asia, there is an extensive shipbuilding industry in Japan, Korea, and China. These three countries are the top shipbuilding nations in the world. China ranks third, but its industry is growing very rapidly. Europe has large shipyards in France, Italy, and Denmark. Drydocks World in Dubai provides ship repair and new construction, mainly in support of the oil business.

Standards for ship construction are issued by governments and by the insurance industry. In the United States, the Coast Guard issues such regulations. Companies that write insurance on ships manage their risk through organizations called classification societies. The American Bureau of Shipping is one such society. International organizations such as the International Maritime Organization, part of the United Nations, are concerned with safe operation of ships and the prevention of pollution.

Technical societies such as the Society of Naval Architects and Marine Engineers and the American Society of Naval Engineers provide a forum for the exchange of information among professionals in this field. These organizations hold annual meetings at the national level, and they have local groups, called sections, that meet more frequently in cities around the country. Such organizations also publish books and technical journals that cover the latest developments in the field. The Royal Institution of Naval Architects and the Institute of Marine Engineering, Science and Technology are organizations with a similar purpose based in the United Kingdom.

Careers and Course Work

A small number of colleges and universities in the United States offer degrees in naval architecture, marine engineering, or a combination of the two. Similar programs are offered in Canada, Europe, and Asia. Students of naval architecture take courses in strength of materials, ship structures, hydrodynamics, ship resistance, and propeller design. Marine engineering programs include thermodynamics, heat transfer, and machine design. Advanced mathematics is a part of both programs. A bachelor of science degree is the minimum requirement, and many working professionals in these fields have master of science degrees.

Most ship design is performed by companies that produce only the plans and specifications for a ship. The would-be shipowner takes these documents to one or more shipyards and invites bids for the actual construction. Shipyards also employ naval architects and marine engineers who deal with design issues that arise during the construction of a ship.

Social Context and Future Prospects

When cargo must be transported across large bodies of water, such as the Atlantic and Pacific oceans, there are only two possible ways to do it—ships and airplanes. Lightweight, high-value, time-sensitive cargo goes by air, but more mundane cargo goes by ship. Ships are slow, but in many cases, time is not of the essence. Crude oil and natural gas are abundant in a part of the world where demand for these materials is low. A steady stream of large tankers carries crude oil nearly halfway around the world. LNG ships do the same. It is hard to imagine an alternative. Although it is possible to transport cargo from France to Algeria by road, it is not practical to do so.

Transportation by ship is far less expensive than by air. Among other factors, the fuel consumed to move a given amount of cargo a given distance is much less. Shipping products in standard rectangular containers has drastically reduced the cost and time required for handling cargo at both ends of its travel by sea.

It appears that cargo transportation by ship will remain an important business for the foreseeable future.

Edwin G. Wiggins, B.S., M.S., Ph.D.

FURTHER READING

Benford, Harry. *Naval Architecture for Non-Naval Architects.* Jersey City, N.J.: Society of Naval Architects and Marine Engineers, 1991. The author states that one of his intended audiences is "high school students contemplating a career in the marine field." Includes many good black-and-white photos and line drawings.

Ferreiro, Larrie D. *Ships and Science: The Birth of Naval Architecture in the Scientific Revolution, 1600-1800.* Cambridge, Mass.: MIT Press, 2007. Although technical in some chapters, most of the book is readable for the layperson. This is a very thorough history of naval architecture.

Gardiner, Robert, ed. *The Earliest Ships: The Evolution of Boats Into Ships.* London: Conway Maritime Press, 2004. Easy to read with many excellent black-and-white illustrations. Traces shipbuilding back to the Greeks, the Vikings, and the Celts.

Kemp, John F., and Peter Young. *Ship Construction Sketches and Notes.* 2d ed. Boston: Butterworth-Heinemann, 1997. Contains line drawings on half the pages with explanations on the facing pages.

Rowen, Alan L., et al. *Introduction to Practical Marine Engineering.* New York: Society of Naval Architects and Marine Engineers, 2005. Comprehensive, easily readable coverage of marine engineering with extensive illustrations.

Van Dokkum, Klass. *Ship Knowledge: Ship Design, Construction and Operation.* 3d ed. Enkhuizen, the Netherlands: Dokmar Maritime, 2008. Profusely illustrated in color. Very comprehensive nonmathematical coverage.

Zubaly, Robert B. *Applied Naval Architecture.* Centreville, Md.: Cornell Maritime Press, 1996. Written as a textbook for students who aim to become deck officers in the merchant marine. It contains some high-school-level math, and it is easy to read.

WEB SITES

American Society of Naval Engineers
http://www.navalengineers.org/Pages/default.aspx

Royal Institution of Naval Architects
http://www.rina.org.uk

Society of Naval Architects and Marine Engineers
http://www.sname.org

See also: Engineering; Engineering Mathematics; Submarine Engineering.

NUMERICAL ANALYSIS

FIELDS OF STUDY

Chemistry; computer science; mathematics; engineering; physics.

SUMMARY

Numerical analysis is the study of how to design, implement, and optimize algorithms that provide approximate values to variables in mathematical expressions. Numerical analysis has two broad subareas: first, finding roots of equations, solving systems of linear equations, and finding eigenvalues; and second, finding solutions to ordinary and partial differential equations (PDEs). Much of this field involves using numerical methods (such as finite differences) to solve sets of differential equations. Examples are Brownian motion of polymers in solution, the kinetics of phase transition, the prediction of material microstructures, and the development of novel methods for simulating earthquake mechanics.

KEY TERMS AND CONCEPTS

- **Algorithm:** Finite set of steps that provide a definite solution to a mathematical problem.
- **Derivative:** Instantaneous rate of change with respect to time of a dependent variable.
- **Differential Equation:** Equation including a derivative; if the equation has at least one partial derivate, it is called a partial differential equation.
- **Eigenvalue:** Solution λ of the linear equation $T(x) = \lambda x$ for a linear operator T.
- **Error Analysis:** Attaching a value to the difference between the actual solution of a system and a numerical approximation to the solution.
- **Finite Difference:** Approximating a derivative, such as y', by a difference, such as y/x.
- **Finite Difference Method:** Replacing derivatives in an equation by their finite difference approximation and solving for one of the resulting variables.
- **Partial Derivative:** Instantaneous rate of change of a dependent variable of at least two independent variables with respect to one of the independent variables.
- **Solution:** Value, or set of values that, when substituted into an equation, or system of equations, satisfies the equation or system.

Definition and Basic Principles

Most of the phenomena of science have discrete or continuous models that use a set of mathematical equations to represent the phenomena. Some of the equations have exact solutions as a number or set of numbers, but many do not. Numerical analysis provides algorithms that, when run a finite number of times, produce a number or set of numbers that approximate the actual solution of the equation or set of equations. For example, since π is transcendental, it has no finite decimal representation. Using English mathematician Brook Taylor's series for the arctangent, however, one can easily find an approximation of π to any number of digits. One can also do an error analysis of this approximation by looking at the tail of the series and see how closely the approximation came to the exact solution.

Finding roots of polynomial equations of a single variable is an important part of numerical analysis, as is solving systems of linear equations using Gauss-Jordan elimination (named for German mathematician Carl Friedrich Gauss) and finding eigenvalues of matrices (using triangulation techniques). Numeric solution of ordinary differential equations (using simple finite difference methods like Swiss mathematician Leonhard Euler's formula, or more complex methods like the German mathematicians C. Runge and J. W. Kutta's Runge-Kutta algorithm and partial differential equations (using finite element or grid methods) are the most active areas in numerical analysis.

Background and History

Numerical analysis existed as a discipline long before the development of computers. By 1800, Lagrangian polynomials, named for Italy-born mathematician Joseph-Louis Lagrange, were being used for general approximation, and by 1900 the Gauss-Jordan technique for solving systems of equations was in common use. Ordinary differential equations with boundary conditions were being solved by Gauss's method in 1810, by English mathematician John

Couch Adams's difference methods in 1890, and by Runge-Kutta in 1900. Analytic solutions of partial differential equations (PDEs) were being developed by 1850, finite difference solutions by 1930, and finite element solutions by 1956.

The classic numerical analysis textbook, *Introduction to Numerical Analysis*, written by American mathematician Francis Begnaud Hildebrand in 1956, had substantial sections on numeric linear algebra and ordinary differential equations, but the algorithms were computed with desktop calculators. In these early days, much time was spent finding multiple representations of a problem in order to get a representation that worked best with desktop calculators. For example, a great deal of effort was spent on deriving Lagrange polynomials to be used for approximating curves. The early computers, like the Electronic Numerical Integrator and Computer (ENIAC), built by John Mauchly and John Presper Eckert in 1946, were immediately applied to the existing numerical analysis algorithms and stimulated many new algorithms as well.

Modern computer-based numerical analysis really got started with the 1947 paper by John von Neumann and Herman Goldstine, "Numerical Inverting of Matrices of High Order," which appeared in the *Bulletin of the American Mathematical Society*. Following that, many new and improved techniques were developed for numerical analysis including cubic-spline approximation, sparse-matrix packages, and the finite element method for elliptic PDEs with boundary condition.

How It Works

Numerical analysis has many "fundamental" techniques. Below are some of the best known and most useful.

Approximation and Error. It is believed that the earliest examples of numerical analysis were developed by the Greeks and others as methods of finding numerical quantities that were approximations of values of variables in simple equations. For example, the length of the hypotenuse of a 45-degree right triangle with side of length 1, is $\sqrt{2}$, which has no exact decimal representation but has rational approximations to any degree of accuracy. For example, 1.4, 1.414, and 1.4142 are the first three elements of the standard sequence of approximations to the $\sqrt{2}$. Another fundamental idea is error (a bound of the absolute difference between a value and its approximation), for example, the error in the approximation of $\sqrt{2}$ by 1.4242, to ten digits, is 0.000013563.

There are many examples of approximation in numerical analysis. Newton-Raphson method, named for English physicist Sir Isaac Newton and English mathematician Joseph Raphson, is an iterative method used to find a real root of a differentiable function and is included in most desktop math packages. Function can also be approximated by a combination of simpler functions, for example, representing a periodic function as (French mathematician Jean Baptiste Joseph) Fourier series of trigonometric functions; representing a piecewise smooth function as a sum of cubic splines (a special polynomial of degree 3), and representing any function as a (French mathematician Edmond) Laguerre series (a sum of special polynomials of various degrees).

Solution of Systems of Linear Equations. Systems of linear equations were studied shortly after the introduction of variables (examples existed in early Babylonia). Solving a system of linear equations involves determining whether a solution exists and then using either a direct or an iterative method to find the solution. The earliest iterative solutions of linear system were developed by Gauss, and newer iterative algorithms are still published. Many of the problems of science are expressed as systems of linear equations, such as balancing chemical equations, and are solved when the linear system is solved.

Finite Differences. Many of the equations used in numerical analysis contain ordinary or partial derivatives. One of the most important techniques used in numerical analysis is to replace the derivatives in an equation, or system of equations, with equivalent finite differences of the same order, and then develop an iterative formula from the equation. For example, in the case of a first-order differential equation with an initial value condition, such as $f(x) = F[x, f(x)]$, $y_0 = f(x_0)$, one can replace the derivative by using equivalent differences, $f'(x_1) = [f(x_1) - f(x_0)]/(y_1 - x_0)$ and solve the resulting equation for y_1. After getting y_1, one can use the same technique to find approximations for *yi* given x*i* for *i* greater than 2. There are many examples of using finite differences in solving differential equations. Some use forward differences, such as the example above; others use backward differences, much like antiderivatives; and still others use higher-order differences.

Grid Methods. Grid methods provide a popular technique for solving a partial differential equation with n independent variables that satisfies a set of boundary conditions. A grid vector is a vector of the independent variables of a partial differential equation, often formed by adding increments to a boundary point. For example, from (t, x) one could create a grid by adding t to t and x to x systematically. The basic assumption of the grid method is that the partial differential equation is solved when a solution has been found at one or more of the grid points. While grid methods can be very complex, most of them follow a fairly simple pattern. First, one or more initial vectors are generated using the boundary information. For example, if the boundary is a rectangle, one might choose a corner point of the rectangle as an initial value; otherwise one might have to interpolate from the boundary information to select an initial point. Once the initial vectors are selected, one can develop recursive formulas (using techniques like Taylor expansions or finite differences) and from these generate recursive equations over the grid vectors. Adding in the information for the initial values yields sufficient information to solve the recursive equations and thus yields a solution to the partial differential equation. For example, given a partial differential equation and a rectangle the lower-left corner point of which satisfies the boundary condition, one often generates a rectangular grid and set of recursive equations that, when solved, yield a solution of the partial differential equation. Many grid methods support existence and uniqueness proofs for the PDE, as well as error analysis at each grid vector.

Applications and Products

The applications of numerical analysis, including the development of new algorithms and packages within the field itself, are numerous. Some broad categories are listed below.

Packages to Support Numerical Analysis. One of the main applications of numerical analysis is developing computer software packages implementing sets of algorithms. The best known and most widely distributed package is LINPACK, a software library for performing numeric linear algebra developed at Stanford University by students of American mathematician George Forsythe. Originally developed in FORTRAN for supercomputers, it is now available in many languages and can be run on large and small computers alike. Another software success story is the development of microcomputer numerical analysis packages. In 1970, a few numerical analysis algorithms existed for muMath (the first math package for microcomputers), but by 2011, MATLAB, Mathematica, and Mathcad all incorporated many numerical analysis algorithms. MATLAB actually has a version of LINPACK.

Astronomy, Biology, Chemistry, Geology, and Physics. Those in the natural sciences often express phenomena as variables in systems of equations, differential equations or PDEs. Sometimes a symbolic solution is all that is necessary, but numeric answers are also sought. Astronomers use numeric integration to estimate the volume of Saturn a million years ago; entomologists can use numeric integration to predict the size of the fire ant population in Texas twenty years from the present. In physics, the solutions of differential equations associated with dynamic light scattering can be used to determine the size of polymers in a solution. In geology, some of the Earth's characteristics, such as fault lines, can be used as variables in models of the Earth, and scientists have predicted when earthquakes will occur by solving these differential equations.

Medicine. Many of the phenomena of medicine are represented by ordinary or partial differential equations. Some typical applications of numerical analysis to medicine include estimating blood flow for stents of different sizes (using fluid flow equations), doing a study across many physicians of diaphragmatic hernias (using statistical packages), and determining the optimal artificial limb for a patient (solving some differential equations of dynamics).

Engineering. Engineers apply the natural sciences to real-world problems and often make heavy use of numerical analysis. For example, civil and mechanical engineering have structural finite element simulations and are among the biggest users of computer time at universities. In industry, numerical analysis is used to design aerodynamic car doors and high-efficiency air-conditioner compressors. Electrical engineers, at universities and in industry, always build a computer model of their circuits and run simulations before they build the real circuits, and these simulations use numeric linear algebra, numeric solution of ordinary differential equations, and numeric solution of PDEs.

Fascinating Facts About Numerical Analysis

- In 1923, New Zealand-born astronomer Leslie Comrie taught the first numerical analysis course at Swarthmore College in Pennsylvania.
- In 1970, the first IMSL was released; its subroutines were called from FORTRAN programs.
- American electrical engineer Seymour Cray used an approximation to division, rather than the actual floating point division, for the CDC 6600 to get a faster arithmetic logic unit (ALU).
- The simplex method for optimization of well-formed problems was introduced by American mathematician George Dantzig and demonstrates the value of numerical analysis in business.
- The finite element method was introduced in 1942 by German mathematician Richard Courant.

Finite Element Packages. For some problems in numerical analysis, a useful technique for solving the problem's differential equation is first to convert it to a new problem, the solution of which agrees with that of the original equation. The most popular use of the finite element, to date, has been to solve elliptic partial PDEs. In most versions of finite element packages, the original equations are replaced by a new system of equations that agree at the boundary points. The new system of differential equations is easier to solve "inside" the boundary than is the original system and is proved to be close to the solution of the original system. Much care is taken in selecting the original finite element grid and approximating functions. Examples of finite element abound, including modeling car body parts, calculating the stiffness of a beam that needs to hold a number of different weights and simulating the icing of an airplane wing.

Many packages have been developed for finite element algorithms. In fact, a Google search of "best finite element package" revealed more than fifty packages available with names ranging from ALGOR to Zebulon.

Impact on Industry

Research and development in numerical analysis is done in industry, academia, and government. Numerical approximation was closely related to the development of the real number system until the work of Sir Issac Newton, then with astronomy and physics, and after the 1950's with computers.

Government and University Research. Most research done on numerical analysis was done at universities until the 1900's. For example, Swiss mathematician Leonhard Euler was an early leader in numeric solution of linear and differential equations at a number of Russian schools of higher learning. Typical of the early leaders in numerical analysis in the 1800's was Newton, who taught at several universities in England as he developed calculus, including the beginnings of approximation theory and root determination. In the 1900's, many governments provided support in numerical analysis research at labs for building and testing better weapons systems. With the development of the digital computer in the 1940's, there was an explosion of research and development of numeric methods for digital computers. George Forsythe, who founded Stanford University's computer science department, also founded a Stanford-based numerical analysis group, which did much to develop numeric linear algebra, and American mathematician Alston Householder was a leader in numerical analysis at Oak Ridge National Laboratory in Oak Ridge, Tennessee, in the 1960's, supporting work on the first atomic bomb.

Numerical Analysis Corporations. Many of the early mechanical calculator companies, such as NCR, Monroe, and Friden, employed mathematicians to develop formulas to improve the speed of their business calculations. Computer companies—including IBM, CDC, and DEC—realized the importance of having good numerical analysis libraries to support scientific computing. In addition, a number of spin-off companies appeared that produced numerical analysis software, such as IMSL (an old, but still widely used library), SAS (a library that has many statistical uses), and MATLAB (a numerical analysis package designed for microcomputers).

Academic, Business, and Industry Uses of Numerical Analysis. Since engineering is essential for most manufacturing, and numerical analysis is a central tool in engineering, numerical analysis was crucial to the Industrial Revolution of the 1900's. Creating a car with a computer-aided design (CAD) system uses finite element techniques to get optimal airflow over the car: Calculating the wind direction for the nightly weather report requires finding a numeric solution to weather-related partial differential equations; and finding numeric solutions of kinetic energy differential equations allows chemists to model what happens at a molecular level when a liquid freezes. Much academic research uses simulation techniques, and numerical analysis provides considerable support for this. For example, many circuits are being designed for nontraditional media, such as cloth and thin films. Before such a circuit is physically built, it is modeled with a hybrid simulation model (a model with discrete and continuous parts). Finding the optimal layout for the circuit generally involves finding numeric solutions to systems of ordinary and partial differential equations.

Careers and Course Work

Students who enter one of the careers that use numerical analysis typically major in mathematics or physics. One needs substantial course work in computer science, mathematics, and physics. A bachelor's degree is sufficient for those seeking positions as programmers, although a master's degree is helpful. For a position involving development of new algorithms, one generally needs a master's or doctoral degree. A formal minor in an area that regularly uses numerical analysis, such as biology, is recommended.

Those seeking careers involving numerical analysis take a wide variety of positions. These include programmers, algorithm designers, managers of scientific labs, and computer science instructors.

Social Context and Future Prospects

Advances in numerical algorithms have made a number of advances in science possible, such as improved weather forecasting, and have improved life for everyone. If scientific theories are to live up to their full potential in the future, ways of finding approximations to values of the variables used in these theories are needed. The increased complexity of these scientific models is forcing programmers to design, implement, and test new and more sophisticated numerical analysis algorithms.

George M. Whitson III, B.S., M.S., Ph.D.

Further Reading

Burden, Richard L., and J. Douglas Faires. *Numerical Analysis.* 8th ed. Boston: Brooks/Cole, 2011. The first edition was written in the 1980's, and this is still one of the most popular textbooks in use.

Hildebrand, F. B. *Introduction to Numerical Analysis.* 2d ed. New York: McGraw-Hill, 1974. An early numerical analysis text, still widely available in libraries and online, that does an exceptional job describing the basic concepts.

Iserles, Arieh. *A First Course in the Numerical Analysis of Differential Equations.* 2d ed. New York: Cambridge University Press, 2009. This text gives an excellent overview of the various techniques used to solve ordinary differential equations.

Moler, Cleve B. *Numerical Computing with MATLAB.* Philadelphia: Society for Industrial and Applied Mathematics, 2008. A good description of one of the most popular desktop packages for numerical analysis, written by its main architect.

Overton, Michael L. *Numerical Computing with IEEE Floating Point Arithmetic.* Philadelphia: Society for Industrial and Applied Mathematics, 2001. A complete discussion of the standard representation of the real numbers as used on a computer.

Ralston, Anthony, and Philip Rabinowitz. *A First Course in Numerical Analysis.* 2d ed. Mineola, N.Y.: Dover Publications, 2001. A very good introductory book on numeric solution of differential equations.

Strauss, Walter A. *Partial Differential Equations: An Introduction.* 2d ed. Hoboken, N.J.: John Wiley & Sons, 2008. An excellent introduction to the numeric solution of partial differential equations.

Web Sites

American Mathematical Society
http://www.ams.org

Mathematical Association of America
http://www.maa.org

Society for Industrial and Applied Mathematics
http://www.siam.org

See also: Applied Mathematics; Applied Physics; Calculus; Computer Science; Parallel Computing.

P

PAPER ENGINEERING

FIELDS OF STUDY

Mathematics; physics; chemistry; chemical engineering; pulp and paper engineering; forestry management; silviculture; mechanical engineering; nanotechnology

SUMMARY

Paper is an important technological development. Although paper-like materials had already been in use for thousands of years, true paper from separated cellulose fibers was invented in China in the year 105. The manner of its production was a closely guarded secret. Since then, paper has become one of the most versatile and useful materials known, and it is used for a multitude of purposes.

KEY TERMS AND CONCEPTS

- **Calendaring:** Compression of a random fiber sheet between heated steel rollers to compact the fibers and harden the paper.
- **Cellulose:** A complex polymer that forms the cell walls of plants.
- **Cockling:** A waviness acquired by paper that has dried without tension.
- **Fourdrinier:** A machine that uses a wire mesh belt to deliver a continuous sheet of raw paper from a slurry of fibers to the calendaring rollers.
- **Hemicellulose:** Plant molecules that are less complex than cellulose.
- **Lignin:** A polymer that, together with cellulose, forms the cell walls of plants.
- **Silviculture:** The growth and management of woodlands and forests as a commercial crop.
- **Snailing:** Streaks or marks on the surface of paper, as though a snail had crawled across it.
- **Vellum:** A highly durable, paper-like material made from processed sheep or calf skin.

Definition and Basic Principles

At its most basic, paper is a random collection of cellulose fibers that have been manipulated and compressed to form a flat sheet. The variety of fiber-refining and fiber-processing methods that are used make possible an infinite variety of paper types.

True paper is a semisynthetic material made by physically and chemically treating a cellulose source material to isolate and separate the individual cellulose fibers. Through the pulping process, different source materials provide cellulose fibers of different molecular lengths and properties. The cellulose fibers are made into a slurry, which can then be drawn into a sheet of a desired thickness, typically in a Fourdrinier machine. The sheet of loose cellulose fibers then goes through a calendaring process, in which heated rollers express the residual water from the slurry and compress the loose fibers into a thin, hardened sheet. This dry finished sheet is then made into the final form of the paper by trimming and rolling processes.

Background and History

The word "paper" is taken from the ancient material called papyrus, a paper-like writing material developed and used approximately five thousand years ago in ancient Egypt. Papyrus was a constructed material, made by weaving together the fine fibers of the papyrus bush and compressing the resulting weave into a hardened, paper-like sheet.

True paper was invented in China by a court official, Cai Lun, in the year 105. Cai Lun found that by treating cotton rags in a certain way, the material would break down into a mushy mass of fibers. When more water was added, the mixture could be pressed and dried into a thin, hard sheet that could be used for writing. This true paper became popular because it was much lighter than bamboo mats and less expensive than silk, both of which were used as writing materials at the time. Cai's method of making paper eventually made its way to Europe, where it became the material of choice for use with the printing press in the fifteenth century. Industrialization and mechanization since that time have led to the modern paper industry and to the thousands of types and forms of paper that are a basic commodity of modern society.

How It Works

Paper is produced by processing different materials to isolate cellulose fibers, which are then formed into a sheet structure. Essentially, any cellulosic material can be used as the source of cellulose fibers, but the exact nature of the fibers and the type of processing required to obtain them depend a great deal on the source.

Most paper is made by chemically pulping wood from softwood trees, yet the finest papers are made from cotton. The difference is in the type of fibers each material contains and the processing that is required. Cellulose molecules in wood are combined with shorter molecules called hemicelluloses and with a hard substance called lignin, which gives trees and other plants their structural strength. The cellulose fibers must be separated from these components before they can be used for making paper. This requires a detailed procedure of chemical and physical treatment to obtain usable wood cellulose fibers. Compared with wood cellulose, cotton fiber provides long, pure cellulose fibers that require little or no additional treatment. The length of the cotton fibers also ensures that the intertwining of the fibers in the slurry stage is more extensive, which in turn produces a much tougher final paper structure. The difference is readily seen by comparing a sheet of cotton-based paper with a sheet of ordinary bond paper.

The formation of paper begins when the cellulose fibers, from whatever source, are blended with water to form a slurry. The density of the slurry can be adjusted to any consistency, according to the type of paper being produced. Colorants and other additives are introduced to the slurry at this point. A Fourdrinier machine, named for its English inventor, is commonly used to produce continuous sheet paper. In a Fourdrinier machine, a continuous mesh belt passes through the slurry, drawing with it a uniform coating of the mixture. Most of the liquid in the slurry passes through the mesh structure of the belt, leaving behind the randomly mixed cellulose fiber.

The fiber sheet then goes into calendaring machinery, in which heated rollers express the remaining liquid from the sheet and dry the cellulose fiber mass while compressing the fibers together into the harder and more rigid structure of paper. After calendaring, the paper sheet is trimmed and taken up on rolls to be used as feedstock for final production. Rolled stock is subjected to further trimming, shaping, and finishing treatments, based on the ultimate use of the particular type of paper.

Applications and Products

Cellulose fiber is one of the most versatile materials known. The range of properties that can be imparted to paper produced from cellulose fiber is essentially infinite. The intrinsic properties of the paper depend on the average length and purity of the cellulose fibers themselves. Other properties are imparted during the physical processing of the paper, and still others are provided by combining the paper with other materials during formation and finishing procedures.

Papers are typically named according to their intended use, their method of production, or the source of the cellulose fibers that were used. A conservative listing of paper types contains approximately two hundred individual references in several general categories. A list of the various pulps and pulping procedures used is similarly long.

Abrasive Papers. Paper is used as the backing or support material for a number of products designed for abrasive purposes, the most familiar being common sandpaper. Sandpaper is produced in several standard grit designations, ranging from the extremely coarse 30 grit to the extremely fine polishing paper called crocus cloth. Sandpapers are normally used dry, but they are also formulated to be used wet with water or other solvents.

The production of sandpaper consists of coating the backing paper with an appropriate adhesive to which is then applied an even and consistent distribution of the desired abrasive grit material. Once the adhesive has cured, the sandpaper is cut to a standard size.

Absorbent Papers. A variety of papers are produced without extensive calendaring or sizing, leaving a relatively thick, fiber-like paper that readily absorbs liquids. Most commonly used in the production of various sanitary paper products, such as toilet paper and paper towels, this class of papers also includes blotting paper, which was essential for absorbing excess ink from written documents.

The bulky nature of absorbent papers also makes them permeable to liquids. Based on this property, absorbent papers are used as filter media in coffee makers, in laboratory procedures for the recovery of crystalline solids, in fuel and oil filters for all types of machinery, in air filters for computer hard drives, and in many other similar applications.

Barrier Papers. Barrier papers are produced with treatments to make them impermeable and nonabsorbent. The treatment may be incorporated into the paper during its formation, added after as a finishing procedure, or through lamination of the paper to an impermeable material. Barrier papers are used to resist or prevent the passage of fluids such as air, water, and oils. This class includes papers for several different applications, including aluminum-foil-laminated paper (for chocolate bar wrappers), bacon paper and blood-proof or butcher's paper (for wrapping fat- and blood-containing meats), Bakelite paper, waxed paper (such as candy-twist tissue papers, butter wrapping-paper, common household waxed paper), and all manner of cellophane papers. Polyethylene plastic sheeting has been substituted for some of the barrier papers, especially in food wrapping applications.

Cellophanes. Cellophane looks and acts like a hydrocarbon-based plastic material, but is not one. The viscose process is used to extract cellulose as cellulose xanthate, which is then treated and processed to regenerate the cellulose as Rayon, a material used extensively as fibers. The same material, however, can be made in sheet form, in which it is essentially a modified cellulose paper.

Industrial Papers. The industrial papers comprise one of the largest uses of paper products worldwide. Industrial papers are not subject to the refining, bleaching, and other finishing treatments required for finer papers. Instead, some industrial grade papers may actually contain small wood fibers among the cellulose. This class of papers is normally used for packaging material and utility applications that do not call for a more refined paper.

One application is in making low-grade newsprint-type paper used as separators for sheets of glass, plastics, and other materials. Another application is heavy brown manila paper, used for wrapping bulk matter such as raw, cured tobacco and for the manufacture of heavy-duty envelopes and corrugated cardboard boxes. The industrial papers include numerous construction-grade cardstocks and fiberboard materials such as tar-coated roofing paper and exterior wall sheathing, as well as low-, medium-, and high-density construction fiberboard. Seemingly more like wood products than paper in their applications, they are nevertheless manufactured from cellulosic pulps using papermaking methods.

Laminated Papers. This is a designation of paper types that crosses into many other classes, including barrier and industrial papers. Laminated papers are produced by laminating sheets of the particular type of paper to either other sheets or to other sheet materials. Cardstocks and fiberboards are typically made by laminating together several sheets of a thinner construction or industrial paper, while barrier papers are often made by laminating a base paper to aluminum or other metal foil or to some kind of plastic sheet.

Supercalendared Papers. This class of papers includes all of the various types of papers that are processed with a high degree of calendaring to provide an extremely smooth, high-gloss finish. Papers used for printing magazines are typically supercalendared. The paper sheet known as glassine is a special case of supercalendared paper because of the extra processing required for the preparation of the pulp from which it is made and because of the higher calendaring pressures used in its formation. The result is an almost impermeable, translucent paper that has a texture reminiscent of very thin glass, hence the name "glassine."

Impact on Industry

The paper industry has helped develop civilization over the past several thousand years. Today it provides so many products of both common and specialized usage that life without paper is almost inconceivable. Three major industrial sectors are driven by paper engineering: the paper industry itself, forestry, and machine design and manufacturing. Several ancillary industries also play a role in the overall paper industry.

Paper Industry. Cellulose, from which all paper products are formed, is a member of the class of chemical compounds known as natural products. Cellulose is a very large and complex biopolymer whose individual molecules are made up of between 1,000 and 3,500 (on average) molecules of glucose chemically linked in a head-to-tail order. The number of glucose molecules in cellulose varies from source to source, with the longest coming from cotton and the shorter being typical of wood. In addition, the actual amount in any given molecule from any given source also varies.

Industrial and University Research. Cellulose chemistry is, accordingly, not necessarily an exact science. Because of this, a great deal of chemical research is conducted around cellulose and the other components, such as lignins and hemicelluloses, which the pulping process must also accommodate. Not the least of these is the application of nanotechnology to the chemical and physical aspects of cellulose preparation and paper formation. A certain amount of genetic and silvicultural research is also carried out to seek ways to improve both the productivity and the cellulose content of various source materials.

Data Storage. The demand for paper and paper-based products continues to grow, despite the move to electronic data storage and communications and despite other strategies to reduce paper use. Understandably, society demands a more concrete record of its transactions than mere electronic and magnetic records, which can disappear in an instant.

The increased demand for papers also comes from world population growth and from the need to protect products and materials consumed by that population. As a result, the production capacity of the paper industry's resource base has been stressed, raising demands for conservation and recycling. The recycling industry now recovers billions of tons of recyclable paper products annually, to serve as feedstock for remanufacturing.

Forestry and Manufacturing. The forestry industry, specifically in softwoods for source materials for paper products, is a multibillion-dollar industry and the subject of contentious international agreements and of treaties in North America. The production side of the industry affects different sectors of employment and manufacturing, such as the logging industry and industrial equipment required for its support. Manufacturers produce all types of forestry-related machinery, from chainsaws to automatic tree harvesters, transport trucks, loaders, and other devices designed to work with trees and tree logs that can each weigh thousands of kilograms. Paradoxically, reforestation is typically carried out by planting seedling trees, but with little mechanization. Trees are a renewable resource, of course. However, trees of any useful size, especially slow-growing softwoods, require decades to grow to a useful size for industry, while it takes as little as five minutes to fell a tree that has stood for one hundred years or more. The rate at which the trees are harvested can quickly exceed the available supply. Overharvesting, which harms the environment, also can lead to disastrous economic effects, so severe controls over the forestry industry are necessary.

Fascinating Facts About Paper Engineering

- Papyrus, from which paper gets its name, was never used to make paper.
- Paper made from cellulose fibers was invented in China in the year 105. The method of making paper was a closely guarded secret for centuries.
- Control of paper manufacturing in Renaissance Italy was a source of great economic and political power, much like paper money is in modern times.
- Civilization likely would not exist in its present form without the invention of paper and the printing press, both of which enabled mass communication and literacy.
- The variety of paper types and formulations is essentially infinite.
- All true paper is made from cellulose fibers that are biopolymeric molecules having molecular weights between 100,000 and 600,000 Daltons (atomic mass).
- The qualities of a paper depend on the source of the cellulose fibers, the chemical and physical treatment of the fibers, the nature of any other materials added during processing, and the nature of the finishing process.
- Paper can be thick, soft, and bulky like paper towels and bathroom tissue, or it can be hard and glassy, like glassine, a type of translucent paper that has a texture similar to a very thin glass sheet.
- Paper can be as clear as glass. Cellophane and rayon, for example, are forms of cellulose that have been recovered using the viscose process.
- Papermaking methods can be used to make papers from other fiber-like materials, including carbon nanotubes.

CAREERS AND COURSEWORK

Paper engineering is a technology-intensive field.

Students interested in a career in this field will require a sound basic education in mathematics and physics. Completion of a secondary school program is sufficient for work as a forestry or paper-manufacturing laborer, while occupations at a more technical level will require a higher education. A college degree is the minimum requirement for a career in supervisory and management positions, with higher business management roles calling for a master's degree in business administration (M.B.A.).

Other careers within the paper engineering field demand specialist training with a university graduate degree (master's or doctorate). These careers tend toward the research and development aspects of paper engineering. Careers involving machine design will require specialization in mechanical and electronic engineering. Pulp and paper chemistry is a specialized field unto itself and requires that interested students have a strong background in organic chemistry, physical chemistry, and analytical chemistry before specializing in carbohydrate and cellulose chemistry.

The application of nanotechnology to the chemical and physical aspects of pulp and paper engineering is a new field of endeavor. Careers in this area will require specialist training in nanotechnology and in chemistry and physics.

Social Context and Future Prospects

Paper engineering is such an intimate part of the historical development of human civilization that society without it is unimaginable. For that reason alone, it can be expected that paper will continue in that role as new forms and applications of the material are developed. Paper engineering serves another purpose, however, related to the most advanced materials known. The long history of paper made from cellulosic fibers has provided a wealth of information and understanding regarding the behavior of fibers in general; this historical fact also is applicable to knowledge of fiber-like molecules of other materials.

Graphene, a natural material, has only recently been recognized and characterized as perhaps the most significant material found. It has a rather paper-like molecular structure and unique chemical and physical properties that could revolutionize electronics and electronic technology. Related compounds called fullerenes are essentially small graphene molecules that have folded over and "zipped together" through chemical bonds. They range in size from the relatively small buckminsterfullerene (C_{60}) molecule to the macromolecules known as carbon nanotubes.

The physical behavior of large carbon nanotubes closely mimics the behavior of cellulose fibers, enabling the material to be formed into a carbon nanotube paper. This represents an entirely new area of research and development that likely would not have been discovered without the scientific understanding of paper and paper engineering. It is perhaps in this role that paper engineering will have its greatest effect in future applications and technologies.

Richard M. Renneboog, M.Sc.

Further Reading

Bajpai, Pratima. *Environmentally Friendly Production of Pulp and Paper.* Hoboken, N.J.: John Wiley & Sons, 2010. This book provides a thorough, yet concise, overview of many aspects of the pulp and paper production process.

Ek, Monica, Göran Gellerstedt, and Gunnar Henrikson, eds. *Pulp and Paper Chemistry and Technology, Volume 4: Paper Products, Physics, and Technology.* Berlin: Walter de Gruyter, 2009. Although intended for specialists, the many illustrations and electronmicrographs in this book provide a great deal of basic visual information.

Web Sites

PaperOnWeb.org
http://www.paperonweb.com

University of Minnesota. Forest Products Management Development Institute
http://www.forestprod.org/cdromdemo/pf/pf8.html

See also: Chemical Engineering; Engineering.

PARALLEL COMPUTING

FIELDS OF STUDY

Computer engineering; computer programming; computer science; engineering; mathematics; networking; physics; information systems; information technology.

SUMMARY

Parallel computing involves the execution of two or more instruction streams at the same time. Parallel computing takes place at the processor level when multiple threads are used in a multi-core processor or when a pipelined processor computes a stream of numbers, at the node level when multiple processors are used in a single node, and at the computer level when multiple nodes are used in a single computer. There are many memory models used in parallel computing, including shared cache memory for threads, shared main memory for symmetric multiprocessing (SMP) and distributed memory for grid computing. Parallel computing is often done on supercomputers that are capable of achieving processing speeds of more than 10^6 Gflop/s.

KEY TERMS AND CONCEPTS

- **Computer:** Set of nodes functioning as a unit that uses a network or high-speed bus for communication.
- **Flop:** Floating point operation; generally a central processing unit addition, subtraction, multiplication, or division.
- **Grid Computing:** Distributing a single computer job over multiple independent computers.
- **MIMD:** Stands for multiple instruction multiple data. Computer with multiple processors each of which executes an instruction stream using a local data set.
- **Node:** Set of processors functioning as a unit that is tightly coupled by a switch or high-speed network.
- **Processor:** Unit of a computer that performs arithmetic functions and exercises control of the instruction stream.
- **SIMD:** Stands for single instruction multiple data. Computer with multiple processors each of which executes the same instructions on their local data set.
- **Symmetric Multiprocessing:** Use of shared memory for processing by two or more independent processors of a node of a computer.
- **Vector Processor:** Central processing unit that executes an array of data, processing the data sequentially by decomposing the arithmetic units into components that execute multiple numbers in parallel.

Definition and Basic Principles

Several types of parallel computing architectures have been used over the years, including pipelined processors, specialized SIMD machines and general MIMD computers. Most parallel computing is done on hybrid supercomputers that combine features from several basic architectures. The best way to define parallel processing in detail is to explain how a program executes on a typical hybrid parallel computer.

To perform parallel computing one has to write a program that executes in parallel; it executes more than one instruction simultaneously. At the highest level this is accomplished by decomposing the program into parts that execute on separate computers and exchanging data over a network. For grid computing, the control program distributes separate programs to individual computers, each using its own data, and collects the result of the computation. For in-house computing, the control program distributes separate programs to each node (the node is actually a full computer) of the same parallel computer and, again, collects the results. Some of the control programs manage parallelization themselves, while others let the operating system automatically manage parallelization.

The program on each computer of a grid, or node of a parallel computer, is itself a parallel program, using the processors of the grid computer, or node, for parallel processing over a high-speed network or bus. As with the main control program, parallelization at the node can be under programmer or operating system control. The main difference between control program parallelization at the top level and the node level is that data can be moved more quickly around one node than between nodes.

The finest level of parallelization in parallel programming comes at the processor level (for both grid and in-house parallel programming). If the processor uses pipelining, then streams of numbers are processed in parallel, using components of the arithmetic units, while if the processor is multi-core, the processor program decomposes into pieces that run on the different threads.

Background and History

In 1958, computer scientists John Cocke and Daniel Slotnick of IBM described one of the first uses of parallel computing in a memo about numerical analysis. A number of early computer systems supported parallel computing, including the IBM MVS series (1964-1995), which used threadlike tasks; the GE Multics system (1969), a symmetric multiprocessor; the ILLIAC IV (1964-1985), the most famous array processor; and the Control Data Corporation (CDC) 7600 (1971-1983), a supercomputer that used several types of parallelism.

American computer engineer Seymour Cray left CDC and founded Cray Research in 1972. The first Cray 1 was delivered in 1977 and marked the beginning of Cray's dominance of the supercomputer industry in the 1980's. Cray used pipelined processing as a way to increase the flops of a single processor, a technique also used in many of the Reduced Instruction Set Computing (RISC), such as the i860 used in the Intel Hypercube. Cray developed several MIMD computers, connected by a high-speed bus, but these were not as successful as the MIMD Intel Hypercube series (1984-2005) and the Thinking Machines Corporation SIMD Connection Machine (1986-1994). In 2004, multi-core processors were introduced as the latest way to do parallel computing, running a different thread on each core, and by 2011, many supercomputers were based on multi-core processors.

Many companies and parallel processing architectures have come and gone since 1958, but the most popular parallel computer in the twenty-first century consists of multiple nodes connected by a high-speed bus or network, where each node contains many processors connected by shared memory or a high-speed bus or network, and each processor is either pipelined or multi-core.

How It Works

The history of parallel processing shows that during its short lifetime (1958 through the present), this technology has taken some very sharp turns resulting in several distinct technologies, including early supercomputers that built super central processing units (CPUs), SIMD supercomputers that used many processors, and modern hybrid parallel computers that distribute processing at several levels. What makes this more interesting is that all of these technologies remain active, so a full explanation of parallel computing must include several rather different technologies.

Pipelined and Super Processors. The early computers had a single CPU, so it was natural to improve the CPU to provide an increase in speed. The earliest computers had CPUs that provided control and did integer arithmetic. One of the first improvements to the CPU was to add floating point arithmetic to the CPU (or an attached coprocessor). In the 1970's, several people, most notably Seymour Cray, developed pipelined processors that could process arrays of floating point numbers in the CPU by having CPU processor components, such as part of the floating point multiplier, operate on numbers in parallel with the other CPU components. The Cray 1, X-MP and Y-MP were the leaders in supercomputers for the next few years. Cray, and others, considered using gallium arsenide rather than silicon for the next speed improvement for the CPU, but this technology never worked. A number of companies attempted to increase the clock speed (the number of instructions per second executed by the CPU) using other techniques such as pipelining, and this worked reasonably well until the 2000's.

A number of companies were able to build a CPU chip that supported pipelining, with Intel's i860 being one of the first. While chip density increased as predicted by Moore's law (density doubles about every two years), signal speed on the chip limited the size of pipelined CPU that could be put on a chip. In 2005, Intel introduced its first multi-core processor that had multiple CPUs on a chip. Applications software was then developed that decomposed a program into components that could execute a different thread on each processor, thus achieving a new type of single CPU parallelism.

Another idea has developed for doing parallel processing at the processor level. Some supercomputers are being built with multiple graphics processing units (GPUs) on a circuit board, and some have been built with a mix of CPUs and GPUs on a board. A variety of techniques is being used to combine these processors into nodes and computers, but they are similar to the existing hybrid parallel computers.

Data Transfer. Increasing processor speed is an important part of supercomputing, but increasing the speed of data transfers between the various components of a supercomputer is just as important. A processor is housed on a board that also contains local memory and a network or bus connection module. In some cases, a board houses several processors that communicate via a bus, shared memory, or a board-level network. Multiple boards are combined to create a node. In the node, processors exchange data via a (backplane) bus or network. Nodes generally exchange data using a network, and if multiple computers are involved in a parallel computing program, the computers exchange data via a transmission-control protocol/Internet protocol (TCP/IP) network.

Flynn's Taxonomy. In 1972, American computer scientist Michael J. Flynn described a classification of parallel computers that has proved useful. He divided the instruction types as single instruction (SI) and multiple instruction (MI); and divided the data types as single data (SD) and multiple data (MD). This led to four computer types: SISD, SIMD, MISD and MIMD. SISD is an ordinary computer, and MISD can be viewed as a pipelined processor. The other classifications described architectures that were extremely popular in the 1980's and 1990's and are still in use. SIMD computers are generally applied to special problems, such as numeric solution of partial differential equations, and are capable of very high performance on these problems. Examples of SIMD computers include the ILLIAC IV of the University of Illinois (1974), the Connection Machine (1980's), and several supercomputers from China. MIMD computers are the most general type of supercomputer, and the new hybrid parallel processors can be seen as a generalization of these computers. While there have been many successful MIMD computers, the Intel Hypercube (1985) popularized the architecture, and some are still in use.

Software. Most supercomputers use some form of Linux as their operating system, and the most popular languages for developing applications are FORTRAN and C++. Support for parallel processing at the operating-system level is provided by code unit. At the program level, one can use blocking/unblocking message passing, access a thread library, or use a section of shared memory with the OpenMP API (application program interface).

Fascinating Facts About Parallel Computing

- In 1974, Seymour Cray founded Cray Research, which produced the world's first supercomputer.
- Grid computing developed during the 1990's as a way to harness the combined computing power of the multitude of computers on the Internet. There is no agreement as to who first came up with the concept of grid computing, but by 2002, there was Open Grid Services Architecture (OGSA).
- Intel announced its support for multi-core central processing units (CPUs) in 2004, thus establishing this architecture as key to parallel processing.
- In 1997, the Deep Blue IBM supercomputer defeated Garry Kasparov, a chess master, two wins to one, establishing that supercomputers could exhibit human intelligence.
- The Tianhe-1A, a massively parallel processor at the National Supercomputing Center in Tianjin, China, was officially timed by the TOP500 as the world's fastest computer.

Applications and Products

Weather Prediction and Climate Change. One of the first successful uses of parallel computing was in predicting weather. Information, like temperature, humidity, and rainfall, has been collected and used to predict the weather for more than 500 years. Many early computers were used to process weather data, and as the first supercomputers were deployed in the 1970's, some of them were used to provide faster and more accurate weather forecasts. In 1904, Norwegian physicist and meteorologist Vilhelm Bjerknes proposed a differential-equation model for weather forecasting that included seven variables, including temperature, rainfall, and humidity. Many have added to this initial model since its introduction, producing complex weather models with many variables that are ideal for supercomputers, and this has led to many government agencies involved in weather prediction using supercomputers. The European Centre for Medium-Range Weather Forecasts (ECMWF) was using a CDC 6600 by 1976; the National Center for Atmospheric Research (NCAR) used an early model by Cray; and in 2009, the National Oceanic and Atmospheric Administration (NOAA) announced the purchase of two IBM 575s to run complex weather models, which improved forecasting of hurricanes and tornadoes. Supercomputer modeling of the weather has also been used to determine previous events in weather, such as the worldwide temperature of the Earth during the age of the dinosaurs, as well as to predict future phenomena such as global warming.

Efficient Wind Turbines. Mathematical models describing the characteristics of airflow over a surface, such as an airplane wing, using partial differential equations (Claude-Louis Navier, George Gabriel Stokes, Leonhard Euler, and others) have existed for more than a hundred years. The solution of these equations for a variable, such as the lift applied to an airplane wing for a given airflow, has used computers since their invention, and as soon as supercomputers appeared in the 1970's, they were used to solve these types of problems. There also has been great interest in using wind turbines to generate electricity. Researchers are interested in designing the best blade to generate thrust without any loss due to vortices, rather than developing a wing with the maximum lift. The EOLOS Wind Energy Research Consortium at the University of Minnesota used the Minnesota Supercomputer Institute's Itasca supercomputer to perform simulations of airflow over a turbine (three blades and their containing device) and as a result of these simulations was able to develop more efficient turbines.

Multispectral Image Analysis. Satellite images consist of large amounts of data. For example, Landsat 7 images consists of seven tables of data, where each entry in a table represents a different magnetic wavelength (blue, green, red, or thermal-infrared) for a 30-meter-square pixel of the Earth's surface. A popular use of Landsat data is to determine what a particular set of pixels represents, such as a submerged submarine. One approach used to classify Landsat pixels is to build a backpropagation neural network and train it to recognize pixels (determining the difference between water over the submarine and water that is not). A number of neural networks has been implemented on supercomputers over the years to identify multispectral images.

Biological Modeling. Many applications of supercomputers involve the modeling of biological processes on a supercomputer. Both continuous modeling, involving the solution of differential equations, and discrete modeling, finding selected values of a large set of values, has made use of supercomputers., Dr. Dan Siegal-Gaskins of the Ohio State University built a continuous model of cress cell growth, consisting of seven differential equations and twelve unknown factors, to study why some cress cells divide into trichomes, a cell that assists in growth, as opposed to an ordinary cress cell. The model was run on the Ohio Supercomputer Center's IBM 1350 Cluster, which has 9,500 core CPUs and a peak computational capability of 75 teraflops. After running a large number of models, and comparing the results of the models to the literature, Siegal-Gaskins decided that three proteins were actively involved in determining whether a cell divided into a trichome or ordinary cells. Many examples of discrete modeling in biology using supercomputers are also available. For example, many DNA and protein-recognition programs can only be run on supercomputers because of their computational requirements.

Astronomy. There are many applications of supercomputers to astronomy, including using a supercomputer to simulate events in the future, or past, to test astronomical theories. The University of Minnesota Supercomputer Center simulated what a supernova explosion originating on the edge of a giant interstellar molecular gas cloud would look like 650 years after the explosion.

Impact on Industry

In the United States, most of the parallel computing development has been done by industry, but often with considerable government support.

Government, Industry, and University Research. Early research and development on supercomputing was done by IBM, its task processing was an early form of thread processing, and GE, with the Multics multiprocessing. Seymour Cray's development at CDC and Cray of pipelining was done by private industry, but government purchase of these computers was essential for their success. Government support was essential for the early SIMD computers like the ILLIAC series, with the ILLIAC IV being housed at a military base after its development. Government grants have been supportive of supercomputer research during the entire history of parallel computing. For example, there was considerable government support for the Connection Machine. IBM has done a lot of research on supercomputing over the years, with its chess-playing computer, Deep Blue, being the best-known example.

An interesting development is the substantial government support for research for quantum computing. Most scientists agree that quantum computing should be classified as a form of parallel computing, and there is an active community of researchers developing quantum algorithms, but no one is sure if a quantum computer has been constructed yet. If, or when, a quantum computer is developed, it will probably redefine parallel computing.

Major Corporations. The supercomputer industry is relatively small and concentrated in a few large companies. A lot of the early leaders in this field are no longer in the supercomputer business, including CDC, the Thinking Machine (Connection Machine), and even Intel. Cray Research is still the leader in supercomputer production in the United States, but it uses a hybrid architecture on most of its computers that has little in common with the Cray 1. IBM has a strong presence in the supercomputer field with two computers in the top ten of the TOP500, a list of the 500 fastest supercomputers in the world, according to a standard LINPACK benchmark. Dell has developed a strong parallel processing presence that emphasizes clustering, similar to the old virtual memory system (VMS) clustering, as well as other parallelization technologies, and also has a number of computers in the TOP500.

An interesting development in the twenty-first century has been the rise of China as a major player in the parallel computing field. In fact, in the 2010 TOP500 list, the number one and three supercomputers are Chinese, with the second position belonging to Cray. There are a number of other supercomputer vendors in the United States including Oracle (through its purchase of Sun), SGI, and HP. Japan, France, and Russia also have at least one super computer in the top ten, demonstrating the interest across the world in supercomputer research, development, and manufacture.

Careers and Course Work

A major in computer science, engineering, mathematics, or physics is most often the course selected to prepare for a career in parallel computing. It is advisable for those going into parallel processing to have a strong minor in an application field such as biology, medicine, or meteorology. One needs substantial course work in mathematics and computer science, especially scientific programming, for a career in parallel computing. There are a number of positions available for those interested in extending operating systems and building packages to be used by application programmers, and for these positions, one generally needs a master's or doctoral degree. Most universities teach courses in computer science and mathematics. Those parts of parallel computing related to the construction of devices are generally taught in computer science, electrical engineering, or computer engineering, while those involved in developing systems and application software are usually taught in computer science and mathematics. Taking a number of courses in programming and mathematics is advisable for anyone seeking a career in parallel computing.

Those seeking careers in parallel computing take a wide variety of positions. A few go to work for companies that develop hardware such as Intel, AMD, and Cray. Others go to work for companies that build system software such as Cray or IBM, or develop parallel computing applications for a wide range of organizations.

Social Context and Future Prospects

Parallel processing can be used to solve some of societies' most difficult problems, such as determining when, and where, a hurricane is going to hit, thus improving people's standard of living. An interesting phenomenon is the rapid development of supercomputers for parallel computing in Europe and Asia. While this will result in more competition for the United States' supercomputer companies, it should also result in a wider use in the world of supercomputers, and improve the worldwide standard of living.

Supercomputers have always provided technology for tomorrow's computers, and technology developed for today's supercomputers will provide faster processors, system buses, and memory for future home and office computers. Looking at the growth of the supercomputer industry in the beginning of the twenty-first century and the state of computer technology, there is good reason to believe that the supercomputer industry will experience just as much growth in the 2010's and beyond. If quantum computers become a reality in the future, that will only increase the power and variety of future supercomputers.

George M. Whitson III, B.S., M.S., Ph.D.

Further Reading

Culler, David, Jaswinder Pal Singh, and Anoot Gupta. *Parallel Computer Architecture: A Hardware/Software Approach.* San Francisco: Morgan Kaufmann, 1999. Has a good survey on modern architectures, including an interesting discussion about the convergence of the modern parallel computing architectures.

Hwang, Kai, and Doug Degroot, eds. *Parallel Processing for Supercomputers and Artificial Intelligence.* New York: McGraw-Hill, 1989. An older text that provides an outstanding overview of the architectures and applications in the early years of parallel processing.

Kirk, David, and Wen-mei W. Hwu. *Programming Massively Parallel Processors: A Hands-On Approach.* Burlington, Mass.: Morgan Kaufmann, 2010. A useful text describing the methods used for CPU and GPU programming used in modern massively parallel processors.

Rauber, Thomas, and Gudula Rünger. *Parallel Programming: For Multicore and Cluster Systems.* Berlin: Springer-Verlag, 2010. An interesting text that compares cluster programming to thread programming.

Varoglu, Sevin, and Stephen Jenks. "Architectural support for thread communications in multi-core processors." *Parallel Computing* 37, issue 1 (January 2011): 26-41. A very good article on writing threaded programs.

Web Sites

Association for Computing Machinery
http://www.acm.org

EOLOS
http://www.eolos.umn.edu

TOP500
http://www.top500.org

See also: Computer Engineering; Computer Science.

PATTERN RECOGNITION

FIELDS OF STUDY

Statistics; mathematics; physics; chemistry; engineering; computer science; artificial intelligence; machine learning; psychology; marketing and advertising; economics; forensics; physiology; complexity science; biology.

SUMMARY

Pattern recognition is a branch of science concerned with identifying patterns within any type of data, from mathematical models to visual and auditory information. Applied pattern recognition aims to create machines capable of independently identifying patterns and using patterns to perform tasks or make decisions. The field involves cooperation between statistical analysis, mechanical and electrical engineering, and applied mathematics. Pattern recognition technology emerged in the 1970's from work in advanced theoretical mathematics. The development of the first pattern recognition systems for computers occurred in the 1980's and early 1990's. Pattern recognition technology is found both in complex analyses of systems such as economic markets and physiology and in everyday electronics such as personal computers.

KEY TERMS AND CONCEPTS

- **Algorithm:** Rule or set of rules that describe the steps taken to solve a certain problem.
- **Artificial Intelligence:** Field of computer science and engineering seeking to design computers that have the capability to use creative problem-solving strategies.
- **Character Recognition:** Programs and devices used to allow machines and computers to recognize and interpret data symbolized with numbers, letters, and other symbols.
- **Computer-Aided Diagnosis (CAD):** Field of research concerned with using computer analysis to more accurately diagnose diseases.
- **Machine Learning:** Field of mathematics and computer science concerned with building machines and computers capable of learning and developing systems of computing and analysis.
- **Machine Vision:** Field concerned with developing systems that allow computers and other machines to recognize and interpret visual cues.
- **Neural Network:** Patterns of neural connections and signals occurring within a brain or neurological system.
- **Pattern:** Series of distinguishable parts that has a recognizable association, which in turn forms a relationship among the parts.

Definition and Basic Principles

Pattern recognition is a field of science concerned with creating machines and programs used to categorize objects or bits of data into various classes. A pattern is broadly defined as a set of objects or parts that have a relationship. Recognizing patterns therefore requires the ability to distinguish individual parts, identify the relationship between the parts, and remember the pattern for future applications. Pattern recognition is closely linked to machine learning and artificial intelligence, which are fields concerned with creating machines capable of learning new information and using it to make decisions. Pattern recognition is one of the tools used by learning machines to solve problems.

One example of a pattern recognition machine is a computer capable of facial recognition. The computer first evaluates the face using a visual sensor and then divides the face into parts, including the eyes, lips, and nose. Next, the system assesses the relationship between individual parts, such as the distance between the eyes, and notes features such as the length of the lips. Once the machine has evaluated an individual's face, it can store the data in memory and later compare the information against other facial scans. Facial recognition machines are most often used to confirm identity in security applications.

Background and History

The earliest work on pattern recognition came from theoretical statistics. Early research concentrated on creating algorithms that would later be used to control pattern recognition machines. Pattern recognition became a distinct branch of mathematics and statistics in the early 1970's. Engineer Jean-Claude Simon, one of the pioneers of the field, began publishing papers on optical pattern recognition in the early 1970's. His work was followed by a number of researchers, both in the United States and abroad. The first international conference on pattern recognition was held in 1974, followed by the creation of the International Association for Pattern Recognition in 1978.

In 1985, the United States spent $80 million on the development of visual and speech recognition systems. During the 1990's, pattern recognition systems became common in household electronics and were also used for industrial, military, and economic applications. By the twenty-first century, pattern recognition had become a robust field with applications ranging from consumer electronics to neuroscience. The field continues to evolve along with developments in artificial intelligence and computer engineering.

How It Works

Pattern recognition systems are based on algorithms, or sets of equations that govern the way a machine performs certain tasks. There are two branches of pattern recognition research: developing algorithms for pattern recognition programs and designing machines that use pattern recognition to perform a function.

Obtaining Data. The first step in pattern recognition is to obtain data from the environment. Data can be provided by an operator or obtained through a variety of sensory systems. Some machines use optical sensors to evaluate visual data, and others use chemical receptors to detect molecules or auditory sensors to evaluate sound waves.

Most pattern recognition computers focus on evaluating data according to a few simple rules. For example, a visual computer may be programmed to recognize only red objects and to ignore objects of any other color, or it may be programmed to look only at objects larger than a target length.

Translating Data. Once a machine has intercepted data, the information must be translated into a digital format so that it can be manipulated by the computer's processor. Any type of data can be encoded as digital information, from spatial relationships and geometric patterns to musical notes.

A character recognition program is programmed to recognize symbols according to their spatial geometry. In other words, such a program can distinguish the letter *A* from the letter *F* based on the unique organization of lines and spaces. As the machine identifies characters, these characters are encoded as digital signals. A user can then manipulate the digital signals to create new patterns, using a computer interface such as a keyboard.

A character recognition system that is familiar to many computer users is the spelling assistant found on most word-processing programs. The spelling assistant recognizes patterns of letters as words and can therefore compare each word against a preprogrammed list of words. If a word is not recognized, the program looks for a word that is similar to the one typed by the user and suggests a replacement.

Memory and Repeated Patterns. In addition to recognizing patterns, machines must also be able to use patterns in problem solving. For example, the spelling assistant program allows the computer to compare patterns programmed into its memory against input given by a user. Word processors can also learn to identify new words, which become part of the machine's permanent memory.

Advanced pattern recognition machines must be able to learn without direct input from a user or engineer. Certain learning robots, for instance, are programmed with the capability to change their own programming according to experience. With repeated exposure to similar patterns, the machines can become faster and more accurate.

Applications and Products

Computer-Aided Diagnosis. Pattern recognition technology is used by hospitals around the world in the development of computer-aided diagnosis (CAD) systems. CAD is a field of research concerned with using computer analysis to more accurately diagnose disease. CAD research is usually conducted by radiologists and also involves participation from computer and electrical engineers. CAD systems can be used to evaluate the results taken from a variety of imaging techniques, including radiography, computed tomography (CT), magnetic resonance imaging (MRI), and ultrasound. Radiologists can use CAD systems to evaluate disorders affecting any body system, including the pulmonary, cardiac, neurologic, and gastrointestinal systems.

At the University of Chicago Medical Center, the radiology department has obtained more than seventy patents for CAD systems and related technology. Among other projects, specialists have been working on systems to use CAD to evaluate potential tumors.

Speech Recognition Technology. Some pattern recognition systems allow machines to recognize and respond to speech patterns. Speech recognition computers function by recording and analyzing sound waves; individual speech patterns are unique, and the computer can use a recorded speech pattern for comparison. Speech recognition technology can be used to create security systems in which an individual's speech pattern is used as a passkey to gain access to private information. Speech recognition programs are also used to create dictation machines that translate speech into written documents.

Neural Networks. Many animals, including humans, use pattern recognition to navigate their environments. By examining the way that the brain behaves when confronted with pattern recognition problems, engineers are attempting to design artificial neural networks, which are machines that emulate the behavior of the brain.

Biological neural networks have a complex, nonlinear structure, which makes them unpredictable and adaptable to new problems. Artificial neural networks are designed to mimic this nonlinear function by using sets of algorithms organized into artificial neurons that imitate biological neurons. The artificial network is designed to be adaptive, so that repeated exposure to similar problems creates strong connections among the artificial neurons—similar to memory in the human brain.

Although artificial neural networks have just begun to emerge, they could potentially be used for any pattern recognition application from economic analysis to fingerprint identification. Many engineers have come to believe that artificial neural networks are the future of pattern recognition technology and will eventually replace the linear algorithms that have been used most frequently.

Military Applications. Pattern recognition is one of the most powerful tools in the development of military technology, including both surveillance and offensive equipment. For example, the Tomahawk cruise missile, sometimes called a smart bomb, is an application of pattern recognition used for offensive military applications. The missile uses a digital scene area matching correlation (DSAMC) system to guide the missile toward a specific target identified by the pilot. The missile is equipped with sensors, an onboard computer, and flight fins that can be used to adjust its trajectory. After the missile is fired, the DSAMC adjusts the missile's flight pattern by matching images from its visual sensors with the target image.

Fascinating Facts About Pattern Recognition

- Pattern recognition is the basis of several popular games, including Set, a game in which players compete to identify matching series of cards from a deck.
- Digital cameras often use pattern recognition to identify the location of the most prominent features of a photograph to determine the correct focus and other variables.
- Banking and financial analysis groups use pattern recognition to identify potential patterns in financial data, especially when making predictions about future developments in the economic market.
- In 2009, popular Web browser Google revealed a new program using visual pattern recognition to allow Web browsers to more accurately search and identify search terms within images.
- Researchers in Australia have begun using pattern recognition to create a series of sleeves that help athletes train by detecting muscle movements and producing auditory cues to let athletes know when they are moving their muscles in the correct sequence.
- In 2007, Australian scientists revealed a system using pattern recognition to read brain waves and to predict a person's actions. The technology could potentially be used to predict the onset of seizures and other physical issues.

Impact on Industry

Academic projects in pattern recognition receive funding from a variety of public and private granting agencies. In the United States, grants for pattern recognition technology have come from the National Institutes of Health, the National Science Foundation, and a variety of other sources. The CEDAR research group at the University of Buffalo has received funding from the U.S. Postal Service to develop technology using pattern recognition for document analysis. In October, 2009, Con Edison received $136 million from the U.S. Department of Energy for research into the development of an automated pattern recognition program that would be used to prevent blackouts and power failures.

In addition, numerous private and public companies have made investments in pattern recognition technology because of its numerous marketable applications. As the use of pattern recognition systems expands globally, researchers and engineers will have increasing opportunities to find funding from foreign investors.

Careers and Course Work

The most direct route to achieve a career in pattern recognition would be to receive advanced training in electrical engineering. Professional statisticians, mathematicians, neurobiologists, and physicists also participate in pattern research. Alternatively, some medical professionals work with pattern recognition, most notably radiologists who participate in CAD research and development.

Texts and training materials in pattern recognition generally require a strong background in mathematics and statistics. Basic knowledge of statistical analysis is a prerequisite for the most basic college courses in statistical engineering. Those hoping to work at the forefront of pattern recognition research will also need experience with machine learning, artificial intelligence, artificial neural network design, and related areas.

Social Context and Future Prospects

Pattern recognition technology has become familiar to many consumers. Voice-activated telephones, fingerprint security for personal computers, and character analysis in word-processing software are just a few of the many applications that affect daily life. Advances in medicine, military technology, and economic analysis are further examples of how pattern recognition has come to shape the development of society.

Projects that may represent the future of pattern recognition technology include the development of autonomous robots, space exploration, and evaluating complex dynamics. In the field of robotics, pattern recognition is being used in an attempt to create robots with the ability to locate objects in their environment. The National Aeronautics and Space Administration has begun research to create probes capable of using pattern recognition to find objects or sites of interest on alien landscapes. Combined with research on artificial neural networks, automated systems may soon be capable of making complex decisions based on the recognition of patterns.

Micah L. Issitt, B.S.

Further Reading

Brighton, Henry, and Howard Selina. *Introducing Artificial Intelligence.* Edited by Richard Appignanesi. 2003. Reprint. Cambridge, England: Totem Books, 2007. Introduces many aspects of artificial intelligence, including machine learning, geometric algorithms, and pattern recognition, and provides examples of how specialists in artificial intelligence and pattern recognition create applications from their research.

Frenay, Robert. *Pulse: The Coming Age of Systems and Machines Inspired by Living Things.* 2006. Reprint. Lincoln: University of Nebraska Press, 2008. An introduction to bioengineering, artificial intelligence, and other organism-inspired machines. Provides an introduction to the philosophy and politics of artificial intelligence research.

Marques de Sá, J. P. *Pattern Recognition: Concepts, Methods and Applications.* New York: Springer Books, 2001. A nonspecialist treatment of issues surrounding pattern recognition. Looks at many of the principles underlying the discipline as well as discussions of applications. Although full understanding requires knowledge of advanced statistics and mathematics, the book can be understood by students with a basic science background.

McCorduck, Pamela. *Machines Who Think: A Personal Inquiry into the History and Prospects of Artificial Intelligence.* 2d ed. Natick, Mass.: A. K. Peters, 2004. An introduction to the history and development of artificial intelligence written for the general reader. Contains information about pattern recognition systems and applications.

Singer, Peter Warren. *Wired for War: The Robotics Revolution and Conflict in the Twenty-first Century.* New York: Penguin Press, 2009. An exploration of the use of intelligent machines in warfare and defense. Contains information on a variety of military

applications for pattern recognition systems.
Theodoridis, Sergio, and Konstantinos Koutroumbas. *Pattern Recognition.* 4th ed. Boston: Academic Press, 2009. A complex introduction to the field, with information about pattern recognition algorithms, neural networks, logical systems, and statistical analysis.

Web Sites

IEEE Computer Society
http://www.computer.org

International Association for Pattern Recognition
http://www.iapr.org

See also: Artificial Intelligence; Computer Engineering; Computer Science; Robotics.

PETROLEUM EXTRACTION AND PROCESSING

FIELDS OF STUDY

Engineering; petroleum engineering; chemical engineering; mechanical engineering; environmental engineering; control engineering; geosciences; geology; geophysics; hydrogeology; mining; seismology; chemistry; physics; mathematics; business management.

SUMMARY

Petroleum extraction and processing is the human exploitation of fossil fuels consisting of hydrocarbons in the form of crude oil and natural gas. The primary use of processed petroleum products is as powerful fuels, particularly for transportation in the form of gasoline, diesel fuel, or jet fuel, and for heating purposes, which account for 84 percent of petroleum use. The remainder of processed petroleum products is used in the petrochemical industry as fuel additives and to create applications such as plastics, specialty chemicals, solvents, fertilizer, pesticides, and pharmaceuticals. Worldwide, about 37 billion barrels of crude oil and natural gas are extracted each year.

KEY TERMS AND CONCEPTS

- **Barrel:** Standard unit of measuring the volume of oil as well as oil and gas products; equals 42 gallons or 159 liters.
- **Crude Oil:** Liquid part of petroleum, a mix of different hydrocarbon chains.
- **Distillation:** Physical separation by heating of mixed components with different volatilities.
- **Downstream:** Processing of crude oil and natural gas, transporting and marketing the processed products to consumers.
- **Exploration:** Searching for geological formations likely to hold petroleum reservoirs and accessing them.
- **Feedstock:** Chemical compound to be processed into a higher value chemical in a petrochemical plant.
- **Field:** Area covered by an underground petroleum reservoir.
- **Fraction:** End product of refining petroleum.
- **Natural Gas:** Volatile part of petroleum.
- **Refining:** Processing crude oil through distillation, cracking of fractions, reforming, and blending.
- **Reservoirs:** Large underground deposits of fossilized hydrocarbons.
- **Rig:** Mechanism to drill an oil or gas well.
- **Upstream:** Exploring and producing petroleum.

Definition and Basic Principles

Petroleum extraction and processing encompass the activities through which crude oil and natural gas are taken from their natural reservoirs below the surface of the earth on land and sea and treated so they can be used as fuels and materials for the petrochemical industry. Reservoirs can be detected by applying the results of petroleum geology, the science describing under what circumstances oil and gas reservoirs were created during ancient times and how they can be found.

Once a likely reservoir is identified, exploratory drilling through the ground begins, either on land or on sea. When a well strikes a reservoir, the size of its area is estimated. Depending on whether crude oil or natural gas is dominant, the reservoir is called an oil or gas field. Many wells are drilled to extract oil and gas from fields that can be quite vast, extending for hundreds or thousands of square kilometers (or miles) under the surface. As of 2010, there were about 40,000 producing oil and gas fields in the world.

Usually, natural pressure forces gas and oil to the surface. In mature wells, however, pressure has to be added by technical means. After gas and oil are extracted, both are sent by pipelines for further processing.

Processing natural gas separates almost all other components from its key ingredient, methane. Processing crude oil occurs in a refinery. Because the different components of crude oil vaporize at different temperatures, they can be separated by distillation. To obtain the most desirable end products, the components are further treated through chemical processes called cracking and reforming.

Background and History

Modern petroleum use was based on demand for kerosene for lighting. Polish pharmacist Ignacy

Łukasiewicz discovered how to distill kerosene from petroleum in 1853. He dug the first modern oil well in 1846 and the world's first refinery in 1856.

Retired railway conductor Edwin Drake dug the first oil well in the United States at Titusville, Pennsylvania, on August 28, 1859. His drilling method, using a pipe to cover a drill suspended from a rig and leading the pipe into the earth to prevent collapse of the borehole, was widely copied as he did not patent it. It is still the basic drilling mechanism. Horizontal rather than vertical drilling, which allows easier access to the reservoir of an oil field, revolutionized crude oil extraction in the 1990's.

The appearance and spread of the internal combustion engine by the late nineteenth century led to crude oil being refined to produce gasoline. The quest to extract more gasoline led to the application of thermal cracking in 1913, and by 1975, many subsequent new cracking techniques had culminated in residual hydrocracking. Beginning in 1916, processes were invented to decrease the unwanted sulfur content of crude oil. Since 1939, visbreaking has been used to make smoother flowing products, and isomerization and alkylation have increased gasoline yield and quality since 1940. Modern refineries try to make these processes as efficient and clean as possible.

How It Works

Exploration. Almost all exploration of new oil and gas fields is done by geophysical means such as seismic reflection, or by other seismic, gravity, or magnetic surveys. These surveys use the different densities of subterranean rock formations to identify source rock formations likely to hold hydrocarbons. Based on this information, three-dimensional computer models are generated and analyzed. Once petroleum engineers and geologists have decided on a promising prospect, exploratory drilling begins.

Extraction. Oil and gas deposits are accessed by drilling a well from a rig, either on land or in the sea. Rigs are adapted to their particular environment but share some common elements. Suspended from a metal tower called a derrick, the drilling bit hangs from its drilling string over the well bore, the borehole where the well is dug. As the drilling bit goes deeper, the borehole is fitted with a pipe to prevent its collapse. The pipe is placed in a larger steel casing, leaving space between the pipe and casing. Drilling mud descends through the pipe to cool the drilling bit as it bores into the rock and percolates back to the surface between the pipe and casing. On the sea, the rig is placed on an offshore oil platform. Modern drills can dig thousands of meters (or feet) deep.

Drilling is successful when the well strikes petroleum deposits below their seal rock. Natural pressure ejects gas and oil through the pipe. A wellhead is placed on the top of the well, and natural gas is separated from crude oil in a separator unit. Both products are collected and transported by intermediary pipelines connecting the many wells of a single field.

As long as natural pressure ejects petroleum, primary recovery continues. Once this pressure decreases, water or gas is pumped into the reservoir to create artificial pressure during secondary recovery. When maintaining artificial pressure is no longer economical, the well is considered exhausted.

Processing. Natural gas has to be cleaned in gas processing plants typically built close to the point of extraction. It is separated from its water and natural gas condensate (also called natural gasoline), and the latter is sent to a refinery. Then it is stripped of its acid gases (carbon dioxide, hydrogen sulfide, and other gases with sulfur content), which are desulfurized and often burned as tail gases. To gain as much pure methane, its main component, as is possible, natural gas undergoes a further series of processes to remove contaminants (such as mercury) and nitrogen. Natural liquid gases (NLGs) are removed for use as feedstock in petrochemical plants. Then, consumer-grade natural gas is ready for sale.

Crude oil is processed at a refinery. Refineries need not be close to oil fields as crude oil can be transported by pipeline or ocean tanker. A refinery, which is a highly integrated chemical plant, seeks to create the most desirable products out of crude oil. These products are hydrocarbon chains with few rather than many carbon atoms (called light instead of heavy).

First, crude oil is desalted, then it is separated into its various components, called fractions, by atmospheric distillation. As crude oil is heated, its different fractions vaporize at different temperatures and are removed separately. All fractions are processed further. The lightest fraction, distilled gas, is subjected to sulfur extraction. The heaviest fraction, called atmospheric bottom, undergoes vacuum distillation. Most distilled fractions undergo hydrotreatment, which is infusion of hydrogen to remove sulfur.

At the heart of a refinery are its fluid catalytic cracker and hydrocracker, which break down heavy (long) hydrocarbon chains into lighter (shorter) ones. A catalytic reformer is used to create higher octane reformate, a more valuable naphtha distillate. Isomerization and alkylation are two chemical processes designed to boost the distillate's octane rating. Refineries send their products to end users or the petrochemical industry.

Applications and Products

Gasoline. The most common and valuable petroleum product is gasoline. Typically, about 46 percent of crude oil is processed into gasoline of various qualities. As fuel for the internal combustion engine, gasoline is essential for any industrial society. It is used by most cars and many light trucks. Aviation gasoline fuels the piston engines of planes.

Refineries seek to optimize their gasoline output and strive to create blends that minimize damaging spontaneous combustion in engines (known as knocking). Lead was used until the 1970's, but its use has been prohibited in most industrialized nations because of its harmful effects. In its stead, chemists have developed a variety of alternative additives.

Diesel and Fuel Oils. The second most important petroleum products, which account for about 26 percent of crude oil, are diesel and other fuel oils. These fuel oils have longer hydrocarbon chains than gasoline and thus boil at higher temperatures. Diesel is used as fuel for trucks, buses, ships, locomotives, and automobiles. In Europe, diesel engines are very popular for automobiles as they are more fuel efficient. Environmentalists have expressed concern about burning these fuels because the process creates the pollutant sulfur dioxide.

Heavy fuel oils account for another 4 percent of petroleum products processed from crude oil. Among them, bunker oil is the heaviest. It is used to fuel large ship engines.

Natural Gas. Natural gas accounts for about 35 percent of the value of all petroleum products. It is used for generation of electric power, domestic and industrial heating, and increasingly transportation by bus or automobile. Europe has a dense networks of pipelines bringing natural gas, primarily from Russia, to industrial and domestic consumers.

Natural gas can be liquefied (as liquefied natural gas, LNG) for transport by gas tankers across oceans. This is the preferred method to bring natural gas from gas fields in the Middle East and North Africa to Europe. After reaching the destination, LNG is gasified and pumped into a pipeline at the port terminal.

Jet Fuel. Jet fuel powers gas-turbine engines of airplanes (jets). Its production amounts to about 9 percent of processed crude oil. Commercial jet fuel is produced according to international standards. It is either Jet A or Jet A-1, with Jet A freezing at -40 degrees Celsius (-40 degrees Fahrenheit) and A-1 at -47 degrees Celsius (-52.6 degrees Fahrenheit). Jet B is a lighter, more flammable jet fuel for cold regions. Military jet fuel is produced according to individual national standards.

Liquefied Petroleum Gas. The result of distilling the remaining gas from crude oil at a refinery, liquefied petroleum gas (LPG) is widely used for domestic heating and cooking in Asia, South America, and Eastern Europe, as well as a global fuel alternative for automobiles. LPG provides about 3 percent of worldwide fuel demand.

Petrochemicals. Petrochemicals are often by-products of gasoline production. They are grouped as olefins or aromatics. The value of petrochemicals, which can account for up to 8 percent of petroleum products made of crude oil, has been rising consistently as new chemical processes have found additional applications.

After they leave a refinery, most petrochemicals are used as feedstock for petrochemical plants. Among olefins, ethylene and propylene are building blocks for plastics and specialty chemicals. The olefin butadiene is processed to eventually form synthetic rubber. Among aromatics, benzene is used in the production of synthetic dyes, detergents, and such innovative products as polycarbonates that form light hard plastic shells in many electronic items such as mobile phones. Toluene and xylene are used for solvents or as building blocks for other chemicals, creating, for example, polyester fibers.

Petrochemicals are also used to create adhesives, food ingredients, and pharmaceuticals. They have become a source of much innovative chemical research.

Lubricating Oils. Although their volume accounts for only 1 percent of petroleum products leaving a refinery, lubricating oils and greases are important for the machines and engines of industrialized nations. Refineries create a great variety of lubricating oils through blending with different additives.

Asphalt. The heaviest product from crude oil, asphalt makes up about 3 percent of refinery products. Concrete asphalt for durable roads and runways accounts for about 85 percent of North American asphalt use; asphalt roof shingles are the next most common use.

Impact on Industry

Petroleum extraction and processing is the center of one of the world's most valuable industries. From 2005 to 2009, annual global production of crude oil and natural gas was remarkably constant at around 37 billion barrels of oil equivalent (BOE). This production is forecast to grow by 2 percent annually to reach 41 billion BOE by 2014.

However, the annual value of petroleum products is extremely volatile. This is because of the wildly swinging price for crude oil, one of the most heavily traded global commodities. In 2005, the global oil and gas market reached a value of $1.8 trillion. This value almost doubled, at constant production, to $3.1 trillion in 2008. In 2009, it collapsed by about one-third to $2.1 trillion, with almost the same amount of products created.

Government Agencies. Because of petroleum's essential role for any industrialized society, almost every developed and developing nation has set up government agencies with varying degrees of control over petroleum production, import, and taxation. By 2010, with the notable exception of the United States and Europe, almost all oil-producing countries had established national oil companies. These national companies may either extract and process crude oil and natural gas themselves, as Saudi Aramco does, or collaborate with private oil companies.

The Organization of Petroleum Exporting Countries (OPEC) is the most important international oil and gas agency. Founded in 1960 by five nations, including Saudi Arabia and Venezuela, OPEC sought to function as a cartel and stabilize the oil price. In 1973, OPEC triggered a global oil crisis by hiking prices and reducing production. By 2010, OPEC had twelve member states that accounted for one-third of global oil production and held two-thirds of global oil and gas reserves.

Universities and Research Institutes. Historically, university education and research into petroleum extraction and processing were done primarily by leading universities of the United States, Canada, Europe, Japan, and the former Soviet Union. With the key exception of the United States, most oil-producing nations were developing countries, particularly in the Middle East, southwest Asia, and Africa,

Fascinating Facts About Petroleum Extraction and Processing

- The term "petroleum" comes from Latin *petra* for rock and *oleum* for oil, and literally means rock oil. In China, since 1008 C.E., crude oil has been called *shiyou*, which means rock (*shi*) oil (*you*).
- The earliest known extraction of petroleum was before 2000 B.C.E. from oil pits near Babylon, in modern Iraq, for use as asphalt for city walls.
- By 347 C.E., the Chinese had drilled the first oil well and laid bamboo pipelines to salt springs. Crude oil was burned to evaporate brine and gain salt.
- Petroleum reservoirs were formed when ancient zooplankton and algae died and sank to ocean floors. There, the Earth's heat and pressure transformed them into oil and gas.
- The United States rose to power in part because of its unique combination of industrial assets and rich petroleum resources, being the world's leading petroleum producer until the mid-1970's.
- In 2010, the United States was still the third largest producer of petroleum, behind Saudi Arabia and Russia, producing 8.5 billion barrels per day.
- The relatively thinly populated oil-producing countries of the Middle East saw a massive jump in per capital income after the Organization of Petroleum Exporting Countries hiked prices in 1973.
- Power plants that use natural gas in the combined cycle mode, using both a gas and a steam turbine, are among the most fuel-efficient and least polluting means of electricity generation.
- Liquefied petroleum gas (LPG) is a primary domestic heating fuel in Asia. Trucks providing LPG canisters to households are a common sight even in highly industrialized Japan.
- The petroleum industry is cost intensive. Digging a deep offshore oil well costs more than $100 million, and a new world-class refinery costs up to $4 billion.

and lacked national petroleum know-how. By 2010, there was a strong tendency for universities in oil-producing nations to conduct research and provide education locally.

Major international and national oil companies allocate significant resources to corporate research and development in the oil and gas and petrochemical industry. Many of the technological advances in the extraction process, such as horizontal drilling, were developed by private companies. The same holds true for new seismic methods of petroleum reservoir exploration. Basic chemical research from universities is often applied to develop new petrochemical products at corporate plants.

Industry and Business Sectors. The oil and gas and petrochemical industry is commonly divided into different sectors known as streams. Upstream, often the exploration and production division of an oil company, involves the quest to find new petroleum reservoirs and access and exploit them. Midstream, the transport to processing plants, is usually included in downstream. Downstream involves natural gas processing, processing of crude oil at a refinery, and the petrochemical industry. Downstream also includes transport and marketing of finished petroleum products to customers. Some oil companies combine refining and marketing into one division.

Major Corporations. Because of the very high costs and risks involved, the oil and gas industry is dominated by large international or national companies with highly diversified but integrated operations. Saudi Aramco, Saudi Arabia's national oil company, is the world's largest oil and gas company. In 2007, it held 12 percent of the industry's global market share, producing 11 million barrels of crude oil equivalent every day.

In 2007, ExxonMobil was the world's largest private oil company with a market share of 10.7 percent. In 2009, ExxonMobil earned some $311 billion (down from a peak of $477 billion in 2008). With a profit margin of 6.2 percent, its 80,000 employees earned the company an average profit of $241,000 per person (which was half the 2008 value).

Together with Saudi Aramco and ExxonMobil, Royal Dutch Shell (9.8 percent) and BP (7.8 percent) accounted for 40 percent of the oil and gas market share in 2007. The capital-intensive activities of petroleum exploitation make for a very concentrated industry at its top end.

Careers and Course Work

Students interested in the petroleum industry should take science and mathematics courses, particularly chemistry and physics, in high school. In college, the aim should be a bachelor of science. The choice of science to study depends on whether the student wishes to pursue a career in exploration and production or processing and petrochemicals.

For the former, the geosciences, particularly geology, geophysics, mining, hydrology, and hydrogeology, as well as chemistry, are useful. Computer literacy and ability to work with specific applications in geology, particularly three-dimensional sensing and modeling, is important. Oil companies will welcome successful graduates who hold a bachelor of science in geology, hydrogeology and engineering, chemical engineering, geology and mining, or even physics. A master's or engineering degree in these fields will provide additional opportunities. For example, a master's of science in petroleum engineering or in environmental engineering geology can lead to an advanced entrance position as reservoir geologist. One of the most important criteria for top positions, however, is years of practical work experience. Because increasing responsibilities lead toward management, it would not be uncommon for a senior geologist to hold a master of business administration degree in addition to science degrees.

For a career in petroleum processing at a refinery or petrochemical plant, a bachelor of science in chemistry or a related field is a good foundation. A master's or doctoral degree in chemistry or chemical engineering, as well as in mechanical, environmental, or control engineering, will serve for an advanced position.

The oil and gas industry also employs many laboratory scientists and technicians. Corporate research and development departments are typically headed by scientists with doctorates in chemistry and related sciences.

Because of the cyclical nature of the oil and gas industry, there are periods in which skilled employees such as drilling engineers are in high demand, but these workers also experience layoffs during downturns. Work can be at remote locations, and global mobility is often expected.

Social Context and Future Prospects

The late-nineteenth-century discovery that crude

oil and natural gas could be turned into powerful fuels for industrialized society meant that those nations where oil and gas were found possessed a very valuable natural resource. However, use and distribution of wealth from petroleum was a huge social challenge.

The industrialized nations' dependence on petroleum products has significantly influenced global politics since the early twentieth century. Since the late 1980's, there has been increasing concern that the massive burning of hydrocarbon fuel causes global warming because of the accumulation of greenhouse gases in the atmosphere. As of 2010, no global consensus had been reached on how to address and mitigate this issue.

Oil spills, whether from wells or when crude oil tankers run aground, can be major environmental disasters. On April 20, 2010, an explosion destroyed the Deepwater Horizon oil drilling platform in the Gulf of Mexico. The ensuing spill of gas and crude oil from a well on the ocean floor, about 1,500 meters (about 5,000 feet) below the water, became the world's biggest spill, releasing between 35,000 to 60,000 barrels per day and proving very hard to contain.

There is ongoing scientific debate over when the Earth's petroleum resources will be depleted. In 1956, American geoscientist Marion King Hubbert published his peak oil theory, in which he tried to calculate when the volume of petroleum extracted would exceed the volume of remaining petroleum reserves. A concerted effort involving new exploration and extraction technologies seeks to discover even more remote reserves, particularly deep under the sea, and to develop oil shale and oil sands more efficiently and as economically as possible. Scientists have also predicted that future alternative fuels, including those derived from hydrogen or renewable resources, will one day take the role of petroleum.

R. C. Lutz, B.A., M.A., Ph.D.

FURTHER READING

Burdick, Donald. *Petrochemicals in Nontechnical Language.* 3d ed. Tulsa, Okla.: Penn Well, 2010. An accessible introduction to the field that relates serious science concepts in an informative fashion. Figures and tables.

Duffield, John. *Over a Barrel: The Costs of U.S. Foreign Oil Dependence.* Stanford, Calif.: Stanford University Press, 2008. Addresses economic and foreign policy in the United States as well as military responses to the problem. Concludes with a proposal to lower the costs of dependence. Tables, figures, maps.

Gary, James, et al. *Petroleum Refining: Technology and Economics.* 5th ed. New York: CRC Press, 2007. A well-written textbook that provides a good study of refining. The last two chapters cover economic issues. Five appendixes, index, photographs.

Hyne, Norman. *Nontechnical Guide to Petroleum Geology, Exploration, Drilling, and Production.* 2d ed. Tulsa, Okla.: Penn Well, 2001. A comprehensive description of all activities that take place before petroleum is processed. Tables, figures, index.

Raymond, Martin, and William Leffler. *Oil and Gas Production in Nontechnical Language.* Tulsa, Okla.: Penn Well, 2005. An accessible introduction that covers all major aspects. Lighthearted but very informative style.

Yeomans, Matthew. *Oil: Anatomy of an Industry.* New York: The New Press, 2004. A critical look at the oil industry, particularly in the United States. Provides a good historical overview and addresses issues of America's dependency on oil, world conflicts caused by oil, and the question of alternatives such as hydrogen fuel.

WEB SITES

American Association of Petroleum Geologists
http://www.aapg.org

American Petroleum Institute
http://www.api.org

Independent Petroleum Association of America
http://www.ipaa.org

National Petrochemical and Refiners Association
http://www.npra.org

Natural Gas Supply Association
http://www.ngsa.org

Petroleum Marketers Association of America
http://www.pmaa.org

See also: Plastics Engineering

PLANE SURVEYING

FIELDS OF STUDY

Mathematics; geometry; trigonometry; physics; civil engineering; geomatics engineering; geodesy; geodetic surveying.

SUMMARY

Surveying is the process of measuring distances, heights, distributions, and dimensions of features or areas that lie on, above, or below the Earth's surface. For mapping projects that cover small areas, such as the construction of roads and dams, or the surveys of mines and archaeological sites, plane surveying is used. This system is based on the assumption that the Earth's surface is a perfectly flat plane. For projects that cover larger areas, geodetic surveying is required. This system takes into consideration the curvature of the Earth's surface and is used primarily by governmental agencies in the mapping of states and countries.

KEY TERMS AND CONCEPTS

- **Alidade:** Telescope mounted parallel to a straightedge base.
- **Baseline:** Carefully measured, surveyed line that is the starting point for a map.
- **Elevation:** Point with known height above sea level.
- **Magnetic North:** Direction that a compass needle points.
- **Meridian of Longitude:** Line drawn on a globe that connects the poles.
- **Plane Table:** Surveyor's rectangular drawing board mounted on a tripod.
- **Rodman:** Surveyor's assistant who carries the stadia.
- **Stadia Rod:** Jointed, collapsible wooden rod graduated in feet and tenths of feet.
- **Theodolite:** Surveyor's telescope that allows two readings to be taken at one time.
- **Triangulation:** Surveying method that uses the measured sides and angles of a triangle.
- **Tripod:** Three-legged stand used for supporting a plane table or a theodolite.

Definition and Basic Principles

Plane surveying is a method used to describe the location of features on, above, or below the Earth's surface. These features are then shown on a map oriented with respect to north and with a uniform horizontal scale. One common surveying instrument is the plane table, a rectangular drawing board with a sheet of cloth-backed paper, mounted on a three-legged tripod. A telescopic alidade placed on the plane table is used for taking sights on distant objects. Another instrument is the theodolite, a telescope attached directly to the tripod head that can be used to take two readings at a time, either in a horizontal or vertical plane. Sights are taken on a ten- to fifteen-foot-long wooden stadia rod, jointed for ease in carrying and with numbers painted on it. The rod is carried by the surveyor's assistant, known as the rodman. His job is to hold the rod upright at the various "stations" that the surveyor wishes to locate on his map. Surveying has a long and distinguished history, used since ancient times to establish land boundaries and to make corners of buildings square. In the twenty-first century, surveying techniques are widely used by governments, corporations, engineers, builders and foundation contractors, geologists, archaeologists, topographers, hydrographers, and even underwater treasure hunters.

Background and History

Surveying is unquestionably one of the earliest skills utilized by humans as they began to adapt to life in communities and to construct buildings. Evidence for the ancient use of surveyors is found in the Egyptian pyramids, built around 3000 B.C.E., which have corners that are mathematically square, and at Stonehenge, the monument built on England's Salisbury Plain around 2800 B.C.E., which provided accurately surveyed lines of sight to help the Druids determine the starting dates for seasons and the dates that eclipses would occur. In 200 B.C.E., Greek scholar Eratosthenes was able to calculate the circumference of the Earth because he knew that the surveyed distance between two Egyptian cities (Alexandria and Syene) was 500 miles. The Romans employed surveyors to lay out their city streets and determine the proper placement for building foundations, and the

Incas in Peru built a mountaintop complex at Machu Picchu where the survey stones used for laying out the buildings can still be seen. In succeeding years, surveyors have been called on for the layout of city streets, the sites of bridges, the placement of building foundations, and the determination of property boundaries.

HOW IT WORKS

Plane surveying is the process of constructing a map based on measurements surveyed on the ground. It is prepared either on a level drawing board, known as the plane table, with the use of an alidade, or at the surveyor's office after the measurements have been taken in the field with the use of a theodolite. The measurements consist of lines drawn at angles to distant points, either at the same elevation as the instrument or higher and lower if the terrain is hilly. The plane table or the theodolite is mounted on a three-legged tripod, and a ten- to fifteen-foot-long stadia rod, graduated in feet and tenths of feet or in meters and tenths of meters, is held by a rodman at varying distances from the surveyor, where the sights are to be taken.

Equipment Required. When the alidade is used, the map is made on a sturdy drawing board that has a basal plate that allows it to be attached to the tripod. Cloth-mounted paper is preferred, and the sheets must be "seasoned" by exposing it to the climate it will be in for a few days prior to surveying so that shrinkage or expansion does not result from the local atmospheric conditions. The tripod head allows the drawing board to be rotated or tilted in any direction the operator wishes. The alidade consists of two parts: a base, often called the blade, which is a straightedge beveled on one or both sides, with a telescope above it, aligned with the base and mounted on a sturdy pedestal. The other common surveying instrument is a theodolite. This is basically a telescope that can measure two angles simultaneously. It is mounted directly on the tripod head, so there is no plane table board. This means that when a theodolite is used, the resulting map is not created in the field as sights are taken, but subsequently back at the surveyor's office based on the field notes.

Preparations for Mapping. When the plane table is used, the first step is for the operator to level the board using a bubble level. Next he must orient the board with respect to north, bearing in mind that two north-south lines will appear on his map. One will indicate true north and be parallel to the north-south meridians of longitude on a globe. The other line will indicate the direction that a compass needle points in the area where the map is being made. This direction is known as magnetic north. The operator must now select a scale for the map that will ensure that the entire area being surveyed will fall within the confines of the sheet of paper on the plane table drawing board. A baseline is also selected, between two points on the ground, and then drawn on the map. The locations to be surveyed should be visible from both ends of this baseline so that it becomes the starting point for the map. Extreme care must be taken in the measurement of the baseline and the plotting of it on the map, because any errors introduced here will result in corresponding errors throughout the map. Distances on the ground are measured with a steel tape or similar device.

Mapping Techniques. The operator begins by setting the tripod up at one end of the baseline. Ideally this should be a place of known elevation, such as a benchmark, which will make it possible to determine the comparative elevations of all the points surveyed on the map. Next, the surveyor sends out the rodman with the stadia rod, which will be held vertically at places where the surveyor wishes to take sights. Distances to the rod can be measured on the ground with a steel tape or determined by a process known as triangulation. Heights can be determined comparing the known height of the surveying instrument with the numbers painted on the stadia rod. After the surveyor has located enough points to provide the "control" needed for his map, topographic, hydrographic, and cultural features can then be sketched in. These features may include topographic contours, water features, property boundaries, roads, railroads, buildings of all kinds, or features of mining and archaeological interest.

APPLICATIONS AND PRODUCTS

Plane-surveying projects vary in size from those that map extensive areas in considerable detail (known as small-scale and intermediate-scale mapping projects), those that map smaller areas in considerable detail (known as large-scale mapping projects), and those that map either large or small areas in a sketchy fashion to meet temporary or immediate needs (known as reconnaissance mapping projects).

Fascinating Facts About Plane Surveying

- The straight sides and the perfect right-angle corners of the pyramids reveal that the Egyptians used surveying methods. The surveying instruments they used have not been found, but it is believed they measured distances with chains that had links all of the same size.
- When making a map using the plane table, the most important line drawn on the drawing board is the baseline. If the length of this line is incorrect by just the slightest fraction of an inch, corresponding errors will be introduced into every other line drawn on the map.
- While most surveying projects are conducted on the land surface, underground surveys are made in mines, and treasure hunters use surveying techniques under the ocean as they search for coins and other valuable objects in wrecks.
- The survey stones used by Inca surveyors as they laid out the mountaintop complex at Machu Picchu in Peru can still be seen in the ruins, indicating that surveying techniques were used in locating the building foundations.
- Maps in ancient Greece were considerably advanced, as it was known by the Greeks as early as 350 B.C.E. that the Earth was round.

The terms "large scale" and "small scale" when applied to maps can be confusing. The scale referred to here is the mileage scale on the base of the map. On a map of the United States, for example, the width of one mile on this scale would be tiny, and these maps are called small- scale maps. On a map of a town, on the other hand, the width of a mile on the mileage scale would be large, so these maps are referred to as large-scale maps.

Small- and Intermediate-Scale Mapping. The first step in the mapping of a large area, such as a county or state, is to determine the purpose of the map. Next, a search of the records for prior work in the area should be made. The surveyor needs to know what mapping has been done previously, and boundary lines and points of known elevation (benchmarks) should be located. Then a scale should be selected that will ensure the entire land area being studied will fall within the limits of the map, and if the map is to be contoured, a contour interval must be selected that will enable the draftsman to draw well-formed contours that are not too closely spaced. When the amount of funding is limited, a less detailed map might be considered, or the most important part of the project completed first. Comparison with the costs for similar projects in other areas will help in ascertaining the cost of the new project. If it is to be carried out in a tropical region, there may be only a certain number of months when the conditions for mapping are favorable, and prior to the start of fieldwork, the cloth-backed paper sheet on which the map is to be drawn should be seasoned. When the sights to be taken with the alidade or theodolite are long ones, the stadia rod must be durable, portable, not too heavy, and have numbers on it that will be legible from far away.

Large-Scale Mapping. Large-scale mapping projects are used for towns and cities, mining operations, archaeological sites, bridges, dams, and reservoir locations. Topographic contouring is frequently needed and, because of the large scale, details such as cultural features must be shown with more care than they are on the small- and intermediate-scale maps. Sight lines and angles have to be drawn with great accuracy, and the distances measured with pacing or even with a steel tape. Stadia rods marked in one-foot intervals are available for these projects, rather than the stadia rods with ten-foot intervals used for small- and intermediate-scale projects. Exact dimensions are required for details such as roads, buildings, and streams, and not just the picture symbols used on the small- and intermediate-scale maps.

Reconnaissance Mapping. The purpose of reconnaissance mapping, which may be on either a small, intermediate or large scale, is to provide information on a temporary or immediate basis and at as low a cost as possible. For this reason there can be a lack of uniformity in the amount of detail shown in different parts of the map, especially regarding the topographic or cultural features. Since these maps are generally made for a specific purpose, such as a geological survey or a highway location, features that have no bearing on the purpose for which the map is intended are often omitted entirely. Distances can be approximated using a measuring wheel rolled along the ground, or clocked with an automobile speedometer, and elevations in localities with extreme height can be approximated using a barometer calibrated in thousands of feet, such as the altimeters used in airplanes.

IMPACT ON INDUSTRY

Surveying is a profession that is found in every country where buildings have to be constructed or boundaries have to be surveyed to determine land ownership. Governments use surveyors where large tracts of land are concerned or where significant buildings are being constructed. Companies and private individuals utilize surveyors for smaller projects.

Government Agencies. One of the best-known governmental agencies utilizing surveyors is the U.S. Geological Survey (USGS). In 1879, the USGS undertook the responsibility for mapping the lower 48 states and has been the primary civilian mapping agency of the United States ever since. It has produced more than 55,000 1:24,000-scale topographic maps, also know as 7.5-minute quadrangle maps, covering the entire area of the lower 48 states in considerable detail. This mapping project was completed in 1992 and hard copies of the maps are available for sale. Subsequently the USGS has created a new series of maps called the National Map, and selections from the National Map can be displayed on a computer screen and downloaded. Another important governmental agency involved in surveying is the National Geodetic Survey (NGS), formerly known as the U.S. Coast and Geodetic Survey. This agency, which was founded by Thomas Jefferson in 1807, was originally responsible for charting the coastlines of the United States, but over time its duties gradually expanded to carrying out surveys of the interior of the country as well. In 1970, it became part of the National Oceanic and Atmospheric Administration (NOAA), but its surveying work has continued. As of 2011, it defines and manages anational coordinate system that provides the foundation for all the mapping and charting done in the United States. It does this by establishing the elevations of key points, benchmarks, throughout the country. These are labeled by metallic markers embedded in rock at the top of the highest point around, and they are the beginning points for most surveying. Other governmental agencies involved in mapping include federal agencies such as the U.S. Forest Service, the National Park Service, as well as numerous state, county, and city agencies

Universities. Several colleges and universities offer educational programs in surveying. Professors from these institutions and individuals publish their research in journals, such as *Surveying and Land Information Science*, which is published by the American Congress on Surveying and Mapping.

Industry, Business, and Corporations. The American Congress on Surveying and Mapping is a nonprofit organization with the goal of advancing the science of surveying, mapping, and related fields. It is composed of four independently incorporated organizations that include more than 5,000 surveyors and survey-related individuals. These individuals come from private industry, government, and the academic world. The congress hosts an annual conference, develops standards for the profession, distributes a bimonthly bulletin that updates readers on current events in surveying, publishes the journal *Surveying and Land Information Sciences*, and lists available surveying jobs in its Career Center. Another professional organization, which reaches an even wider audience, is the American Society of Civil Engineers. It is the world's largest publisher of civil engineering content, and has more than 400 branches and sections throughout the United States. In addition to publishing journals, running conferences, and providing career information, it also offers continuing-education programs for businesses and corporations.

CAREERS AND COURSE WORK

Traditionally, surveying was learned in an apprenticeship program. Interested young people would seek out a surveyor who needed assistants and was willing to teach them the skills involved. Years of fieldwork would be involved, and during this time, it was assumed that the apprentices would pick up the necessary background in mathematics, geometry, and trigonometry at a high school, in an adult education course, by attendance at a college or university, or even on their own. Eventual employment as a surveyor would result. The process has been speeded up. Colleges and universities offer surveying courses leading either to a two-year associate's degree in surveying, or a four-year bachelor of science degree in surveying. Numerous career paths are open to graduates of such programs. Employment by private surveying and engineering firms is one possibility. Another is working for the highway departments of cities, counties, or states. Other governmental agencies that employ surveyors are state lands commissions, the U.S. Forest Service, and the Bureau of Land Management, which is part of the U.S. Department of the Interior. A typical educational program is the one offered by the surveying education department at New Mexico State University. It leads to a bachelor of science in surveying engineering, and is accredited by the Engineering Accreditation Commission. Fifty students are enrolled in the program, and ten are graduated each year. They generally go on to jobs in construction surveying or boundary surveying, but those going into boundary surveying will have to take an exam to become licensed.

Social Context and Future Prospects

Surveying offers young people interesting careers with excellent future prospects. The working conditions are generally out-of-doors, and the job assignments can vary from setting up tripods in snake-infested swamps to lugging them up mountainsides to perch them on the crests of windswept peaks. Employment with private companies engaged in building construction is very common, as is the surveying of property boundary lines, including the locating of section corners and the determination of rights-of-way. For land surveyors, knowledge of the legal principles of boundary surveying is essential, and one must become licensed before work as a professional can begin. Surveyors are involved in many unusual tasks, such as mapping crime and accident scenes in forensic surveying or being called to testify as expert witnesses in court. Oil prospects, historic locations, mining claims, natural parks, and archaeological sites all require surveying, and a surveyor with business skills can form his own company. The demand for surveyors is not going to cease, and even when business downturns temporarily reduce the need for them in construction and boundary surveying, governmental agencies such as highway departments and environmental agencies will always need surveyors.

Donald W. Lovejoy, B.S., M.S., Ph.D.

Further Reading

Hill, Rosemary. *Stonehenge.* Cambridge, Mass.: Harvard University Press, 2008. A summary of findings at Stonehenge, with attention to the various surveying techniques needed to line up the massive stones with the proper orientation.

Ritchie, William, et al. *Surveying and Mapping for Field Scientists.* Harlow, England: Longman Scientific and Technical, 1994. Describes the various techniques and instruments used in plane surveying. Many helpful photographs and diagrams are included.

Tompkins, Peter. *Secrets of the Great Pyramid.* New York: Galahad Books, 1997. This lavishly illustrated book is a detailed treatment of the building of the Pyramid of Cheops, with numerous diagrams indicating how the monument was surveyed.

Wright, Kenneth R., and Alfredo Valencia Zegarra. *Machu Picchu: A Civil Engineering Marvel.* Reston, Va.: American Society of Civil Engineers, 2000. A profusely illustrated analysis of the mountaintop complex highlighting the surveying problems encountered and with photographs of the ancient survey stones.

Web Sites

American Congress on Surveying and Mapping
http://www.acsm.net

American Society of Civil Engineers
http://www.asce.org

National Council of Examiners for Engineering and Surveying
http://www.ncees.org

National Geodetic Survey
http://www.ngs.noaa.gov

See also: Civil Engineering; Trigonometry.

PLASTICS ENGINEERING

FIELDS OF STUDY

Organic chemistry; polymer chemistry; inorganic chemistry; physical chemistry; industrial chemistry; physics; environmental science; mechanical engineering; chemical engineering; fluid dynamics; machinist trades; mold-making trades; materials science; fiber and textile science; textile manufacturing; bioengineering; refinery operations; molding and manufacturing; process operations management.

SUMMARY

Plastics engineering functions on two principal fronts. The first is the application of the material properties of plastics to the solution of specific engineering problems. Examples in this aspect of the field include the application of Bakelite (a phenol-formaldehyde polymer) to the production of electric insulators and of epoxy-based fiber-reinforced polymers (FRPs) to the remediation of infrastructure. The second principal function of plastics engineering is the design and application of methods to produce the material objects desired. Typical examples of this aspect of plastics engineering include the design and manufacture of molding devices and of other machines to produce plastic materials and objects. These two fronts can be thought of as engineering with plastics and engineering of plastics.

KEY TERMS AND CONCEPTS

- **Block Copolymer:** Polymer whose molecules are composed of alternating segments of smaller polymeric molecules (oligomers).
- **Calender:** Production machine that uses a series of heated rollers to form thin plastic sheets while also producing a specific surface texture.
- **Curative:** Compound that must be included in a polymer formulation to bring about the desired polymerization reactions and final molecular structure of the polymer.
- **Engineering Plastics:** Synthetic polymeric materials that can be used in standard engineering applications in place of metals and other traditional materials.
- **Extrusion:** Forming method in which a mobile plastic material is forced under pressure through a die to take on the specific shape determined by the die profile.
- **Oligomer:** Small polymeric molecules generally containing ten or fewer monomer units.
- **Plasticity:** Extent to which a polymer will deform under the influence of pressure alone.
- **Pultrusion:** Forming method in which a bundle of fibers of glass or other suitable support material in a mobile plastic matrix is drawn through a die to form an fiber-reinforced polymer composite structure the shape of which is determined by the die profile.
- **Tacticity:** Inherent symmetry of the structure of a polymer molecule in respect to the relative positions of substituent groups along the polymer backbone.
- **Thermoplastic:** Property of softening to a state of higher deformability on heating; descriptive of plastics that become softer and more pliable with heating.
- **Thermosetting:** Property of becoming inflexible or brittle with heating; descriptive of any resin that polymerizes to form a solid mass that does not subsequently soften on heating

Definition and Basic Principles

The term "plastics engineering" can be taken to mean both engineering with plastics and engineering of plastics. In the first sense, plastics engineering refers to the use of plastics in engineering applications. In the second, plastics engineering is the development of plastic materials and applications or methods for their formation and manipulation.

Plastics in Engineering Applications. Traditional engineering materials such as metals and wood often provide structural strengths and weights that are excessive or have a negative impact on the product. They also pose problems in regard to production, energy consumption, and the degree to which they can be recycled. In many cases, traditional materials have been replaced by plastics because of their strength and lower manufacturing costs. The replacement of a metal gear by a corresponding plastic gear is a prime example. Metal gears require individual machining processes at each and every stage of production. Some of these processes include producing a die, boring center holes and setscrew holes, milling in keyways, and shaping each gear tooth. In many applications, only metal gears can provide the necessary physical strength the product needs. In many other applications, however, metal gears, which are more expensive and labor intensive to create, can be substituted with gears made from plastic. A plastic gear made from high-density nylon can be injection molded by the thousands from a single die at a fraction of both the cost and the time required to produce a single machined metal gear. The plastic gear is strong enough to withstand the strain placed on it in use, is much less massive—which means it requires fewer support mechanisms than metal—and is usually recyclable.

Plastic materials, especially thermoplastics, are also easily machined using standard woodworking and metalworking tools. They do not require the high-strength materials that are needed for machining metals, and they can be machined to close tolerances. Thermoplastics have the property of softening when heated. The heat generated by the cutting tool at the point of contact weakens their physical structure and makes them easier to machine. They are also much softer materials than the cutting tools used on them, so they can be cut easily in machining operations. Thermosetting plastics, however, are harder to work with as they do not soften with heating but become brittle and prone to fracture. The localized heating at the cutting edges in machining operations tends to make the surface of the material more brittle, requiring other steps such as sanding and polishing to produce a finished surface. Thermosets also tend to be considerably harder materials than thermoplastics, compounding the problems encountered in machining operations carried out on the materials. This is because thermosetting polymers tend to have highly cross-linked, three-dimensional structures at the molecular level. They are used primarily as matrix materials in fiber-reinforced polymers and composite structures for that reason.

The use of fiber-reinforced polymers allows the manufacture of structures that could not otherwise be made. A fiber-reinforced polymer consists of a fibrous material such as woven glass or carbon fibers, consolidated and bound into a solid polymer matrix. Fiber-reinforced polymer structures are typically laid up by hand, consolidated (compressed together) with the polymer resin matrix material, and cured with heat and pressure. This method is capable of producing uniquely shaped, strong, and lightweight structural pieces. Fiber-reinforced polymers can also be used in mass-production methods; thermoplastic materials can be employed to produce many relatively simple shapes that do not call for high strength. Variations on these methods, such as extrusion and pultrusion, represent combinations of these methodologies.

Development of Plastic Materials. Plastics engineering also relates to the development of the plastics themselves for specific properties and to the industrial methods of preparing, handling, and manipulating the materials. Plastics begin as monomers, which are compounds whose molecules can add to themselves repeatedly in a chain reaction to form very large polymer molecules. Each polymer molecule may be made up of several hundreds or thousands of monomer units. Thermoplastic polymers are almost always linear polymers, in which the monomers form a single head-to-tail chain from beginning to end. Thermosetting polymers are almost always made up of branched polymer molecules, in which monomer molecules have a structure that allows them to take part in more than one polymerization chain reaction. This results in a highly interconnected three-dimensional molecular structure. Plastics engineering at this level involves the molecular design and synthesis of the monomeric material and subsequent chemical and physical tests of the polymeric material. After the design of the molecule has been engineered, the method of producing the material in quantity must be devised. This involves the design, construction, and operation of processes from which the monomeric material is obtained in a stable state of sufficient purity that it will not undergo polymerization prematurely.

To get the material from the point of manufacture to the point of use, appropriate packaging and handling methods must be determined or devised. Once at the point of use, the materials must be manipulated to turn them into the desired products. Plastics engineering, at this point, deals with the design and preparation of the machinery and processes that are required for product formation. This includes any machining processes and an assortment of molding methods.

Background and History

The history of plastics engineering extends back several centuries. European explorers of the New World and Asia found the indigenous peoples using natural latexes and resins for such purposes as waterproofing clothing, bouncing balls, and other everyday objects. From the late eighteenth to the mid-nineteenth centuries, many efforts were made to adapt those materials to the purposes of European industry. This meant that methods to manipulate the materials had to be developed. At the time, chemistry did not include atomic theory, and practicing chemists were restricted to experimenting through trial and error. The first significant advance in the manipulation of natural plastics was the masticator, a device that could cut and recut a raw rubber mass. The process permitted the rubber to be calendered, or rolled out in thin sheets, without crumbling. Calendering could also be used to bind fabric and plastic together to form a waterproof material.

Industrial applications of plastic materials were limited to the use of natural resins until the accidental discovery of nitrocellulose. A semisynthetic resin formed by the nitration of cellulose (from cotton), nitrocellulose exhibited many of the properties of thermoplastic polymers and eventually became the basis of the celluloid industry. In 1884, George Eastman, founder of the Eastman Kodak Company, began to mechanically produce celluloid in thin flexible sheets through a process invented by Belgian chemist Leo Hendrik Baekeland. This celluloid film became the basis for the photography industry.

Further development of plastics was greatly assisted by the realization of a workable, predictive atomic theory. Rapidly developing understanding of molecular interactions led Baekeland to create the first fully synthetic polymer, Bakelite, a formulation of phenol and formaldehyde, in 1907. Bakelite was developed to fill a specific need: a material for electric insulators. It proved to be much more than that, as its properties made it an excellent replacement for many natural materials that were in ever-decreasing supply. One of its earliest applications was as a substitute for elephant ivory, used for making billiard balls. Bakelite was the first engineered thermosetting plastic and was developed simultaneously with the method of production. This method involved heating a mass of the component mixture, or prepolymer, to drive the polymerization reaction to completion while confining it under pressure within a shaped die or mold.

The many ways of modifying nitrocellulose, combined with an advancing comprehension of the polymerization process within the context of atomic theory, resulted in the development of numerous new materials. Between 1930 and 1939, for example, polymer research resulted in the production of an average of one new polymeric material per day; this necessitated the creation of new methods for turning those materials into products and of applying their properties to engineering problems. These factors were crucial to the outcome of World War II, and research and development in these areas has continued unabated.

The very properties that make plastics so useful also make them environmentally problematic, especially because the primary source of raw materials for plastics production has shifted from coal tar residues to petroleum. Many recyclable thermoplastics have been discarded in landfills and garbage heaps. By the early twenty-first century, plastics engineering had developed a more circumspect view of plastic applications and focused on developing methods of recycling discarded plastic goods, including thermoset plastics, which historically had not been recyclable.

How It Works

Polymerization. To understand the principles of plastics engineering, one should have an appreciation of the basic principles of polymerization. Polymers begin with specific molecules that are referred to as monomers. These contain a functional structure that allows them to form chemical bonds successively between individual molecules in a chain reaction. In a chain reaction, a molecule forms a bond to a second molecule, enabling the second to bond to a third, the third to bond to a fourth, and so on. Successive bond-forming reactions typically occur several thousand times before some condition is encountered that terminates the progress of the reaction chain. In linear polymers, this reaction process forms molecules that have a structure essentially as simple as that of their monomers but are much larger in size.

If the monomer molecules are selected in such a way that at least some of the molecules in the prepolymer can take part in more than one chain reaction at the same time, the reaction process produces a three-dimensional polymeric structure. In this structure, the linear molecular chains become bonded to each other, or cross-linked, as the monomer molecules become bonded to each other in the chain reactions. This results in a structure in which the polymers are intimately intertwined in a three-dimensional network that spans the entire mass. In principle, this can turn an entire multiton mass of resin into a single, very large molecule, although in practice this never happens because of the various and many ways in which polymerization chain reactions are terminated during the process.

Linear and Three-Dimensional Polymers. Linear polymers are generally thermoplastic in nature, becoming pliable when heated and eventually melting. Three-dimensional polymers, on the other hand, are generally thermosetting in nature, becoming brittle, breaking down, and decomposing when heated instead of becoming pliable and melting.

Thermosetting Materials. Thermosetting plastics, also known as thermoset, cannot be reformed once they have been polymerized, except through mechanical machining methods. As such, thermosetting materials are typically used in on-the-spot production methods such as injection molding and extrusion molding. Thermosetting plastics are stronger than thermoplastics because they are formed from rigid three-dimensional bonds.

Thermoplastic Materials. Thermoplastic materials can be reformed (and therefore recycled) when heated to a pliable state. They are capable of quick conversion from raw material to finished product and from waste material to new finished product. Plastics engineering at this stage begins with research and development of the molecular structure of the monomeric material to assess and characterize the properties of the polymeric material that it will produce. Such activity often includes chemically modifying and functionalizing the monomer molecules in specific ways to obtain desired properties in the resulting polymer. Once the molecular structure has been established, the engineering focus shifts from production of the material in the laboratory to production of the material in bulk. As plastics are typically produced on a scale of many millions of tons annually, continuous-flow methods of production are preferred, although some specialty plastics are best prepared by batch processes. The plastics engineer has the task of identifying and adapting or developing the most suitable process for the production of the material and of optimizing the process in practice. The processes used typically produce thermosetting resins for transshipment and use in polymerization reactions at other locations. Thermoplastics are produced by methods that make them easily transportable in solid form. A variety of process types are used in the subsequent processing and production of plastic objects.

Applications and Products

Fiber-Reinforced Polymers. Plastics engineering involving fiber-reinforced polymers, particularly their use in advanced composite materials, begins with engineering the design of the desired finished component. This includes the nature and form of the reinforcing materials, often polymeric materials, that will be employed in the composite structure and their distribution pattern within the finished structure. Because the shape and form of a thermosetting plastic cannot be altered after polymerization is complete, any particular object made from such materials must be prepared in that final shape. The plastics engineer uses molds and forms made specifically for that purpose. The active surfaces of the molds and forms correspond exactly in shape to the desired final shape of the object being made. The reinforcing material, impregnated with the thermosetting resin, is laid into the mold or form according to the design specifications, and the resulting stack is enclosed in special containment fabrics. Next, the contents are subjected to reduced pressure, usually by use of a vacuum pump. This permits an even distribution of atmospheric pressure that will compress and consolidate the stack and remove any extraneous gases that could interfere with the structural integrity of the product. The application of heat at this point completes the polymerization reaction while assisting with the elimination of gaseous wastes. The resulting composite structure, once cooled, is then removed from the mold, and if the part is of suitable quality, it is sent on to the next stage in the manufacturing process.

Pultrusion. Another method by which thermoplastic fiber-reinforced polymers are produced is pultrusion. In pultrusion, an appropriately designed bundle of continuous fiber strands is drawn through a die along with a molten thermoplastic matrix material. The die serves to consolidate the material combination, and as the matrix material solidifies on exiting the die, a continuous fiber-reinforced structure is produced. Examples of pultruded products include fiberglass rods and reinforced water hoses.

Injection Molding. Both thermosetting and thermoplastic polymers can be used in the injection-molding process, though the practical requirements are very different for the two types of polymers. In injection molding, a cavity in the shape of the desired object is filled with molten thermoplastic or a thermosetting prepolymer resin. When the solid polymer has formed, the object is removed from the mold and the process repeated. When a thermoplastic material is being used, the plastic material is fed into the mold under pressure from a heated hopper. If a thermosetting resin is being used, precautions must be taken to ensure that the polymerization reaction takes place only within the mold cavity and after the cavity has been filled. The process can rapidly produce a large quantity of objects of intricate design that would require many independent machining operations if the object were being made from traditional materials such as wood or metal. Successive molding steps enable the production of multicolored objects.

Thermoforming. The thermoforming process is applicable only to thermoplastic materials. In this process, a sheet of solid thermoplastic material, such as Plexiglas (polymethylmethacrylate) or polystyrene, is placed in a mold or form and heated to a temperature at which the material becomes highly pliable but does not melt. This allows the sheet to deform and adopt the shape of the mold, assisted by changing the pressure on one side of the sheet of material. The method is most suitable for designs that are relatively flat or low profile.

Blow Molding. The process of blow molding also applies only to thermoplastics. It is similar to thermoforming in that it uses a difference in pressure to force softened thermoplastic material to adopt the shape of a mold. Blow molding uses three-dimensional molds, however, and is typically used to produce open structures such as plastic bags, bottles, and other similar objects. In blow molding, a metered quantity of plastic material is set into a mold, where it is then blown outward by a sudden blast of heated air to the shape of the mold. The method is most suitable for open, three-dimensional shapes having little or no surface detail.

Extrusion. The process of extrusion is closely related to pultrusion. In extrusion, the plastic material is pushed through a die under pressure rather than drawn through with continuous fiber-reinforcement materials. Random-oriented fiber-reinforcement materials can be used in the extrusion process if they are blended with the molten plastic before entering the die. Extrusion can be used only to produce structures that have a constant cross-sectional profile along their entire length, as determined by the die profile. Quite complex cross-sectional designs can be produced in this way, but they are essentially only two-dimensional.

Compression and Transfer Molding. Both compression and transfer molding are used primarily for thermosetting resins and rely on heating a filled mold to drive the cross-linking reactions in the polymer body to completion. The processes use either a powdered resin or a preformed plug that will be set into a final desired form. Compression molding is the process that was developed by Baekeland for the production of Bakelite objects, and it uses heat and pressure to restrict the material to the final form of the product throughout the polymerization process. In transfer molding, the resinous material is first preheated and liquefied in a separate chamber before being transferred to the object mold and undergoing polymerization.

Rotational Molding. The process of rotational molding uses centrifugal force to distribute the plastic material within a spinning or rotating mold. The material may be injected as a liquid or added as a finely ground powder that is fused within the mold. The process is used exclusively for the production of hollow objects.

Foams. Various methods and treatments of resins, typically thermosetting resins, result in the formation of foams. The forceful addition or release of gas within the bulk polymer structure expands the material by pushing the component molecules apart. The gases may be produced as a by-product of the polymerization reaction or added as an extraneous component. Urea-formaldehyde foams are typically produced by the first method, as the polymerization reaction produces quantities of water vapor. Expanded foams such as Styrofoam are generally produced by adding a low-boiling liquid such as pentane to be absorbed by the solid plastic material. Subsequent rapid heating quickly vaporizes the liquid, and the resulting pressure of the vapor acts to expand the solid material. In other applications, combinations of materials such as urea and formaldehyde solutions react quickly to produce a solidified foam structure. Such materials are generally applied as sprays to provide an insulating and sealing layer on flat surfaces.

Casting and Encapsulation. These are basically simple molding procedures in which a liquefied plastic material is introduced into a mold without the application of increased or decreased pressure, forming a specific shape as the material solidifies.

Calendering. Thermoplastic materials can be formed into thin sheets by calendering, a process by which the material is pressed between heated rollers. The process can be used to laminate a cloth insert layer with a plastic material and to impart various surface textures to the resulting sheet.

Mechanical Machining Methods. The inherent pliability and shear properties of plastics, particularly of thermoplastics, makes them highly amenable to shaping with traditional tools such as saws, drills, planes, shapers, lathes, sanders, and millers. The materials are easily cut by steel tools, which are harder than any plastic. For this reason, mock-up designs are often initially constructed as plastic models rather than as metal constructs, allowing design engineers to test various physical properties, such as aerodynamic stability and wind resistance, that are directly related only to the shape of the structure and not to the materials used.

Impact on Industry

The influence of the plastics industry in the twentieth and early twenty-first centuries on world industries and the global economy cannot be overstated. For good or bad, plastics have become ubiquitous in the world. Internationally, billions of tons of plastics are produced per year and are used in a vast array of applications, from simple plastic tubes for drinking straws to skeletal joint replacements and advanced aircraft components. Virtually no aspect of modern life has not been affected or does not depend in some way on plastics engineering. North America has led the world in the consumption of plastic goods, although China, a major producer of plastics, has been increasing its usage.

Government and University Research. As an industry, plastics engineering has traditionally been driven by the large profits that result from patented processes and materials. As natural offshoots of organic chemistry, however, polymer science and plastics engineering have developed into major fields of study at the academic level. Research into new and improved polymerization reactions and methods has been conducted at many major universities and colleges, and some institutions specialize exclusively in polymer research.

In academic institutions, plastics engineering focuses on the study of the properties of plastics and the development of new, specialized materials and processes for their production. A very significant aspect of such research is the development of polymers with very specific stereochemical properties and molecular structures. Traditional engineering studies of plastic materials tend to focus on the application of those materials in various engineering roles. In the field of civil engineering, the application of fiber-reinforced polymers to the repair and maintenance of aging composite and other structures is a growing area of study that has shown a great deal of promise for the twenty-first century. Industrial engineering examines large-scale methods of producing polymeric materials with the intention of identifying better catalysts for polymerization reactions and of improving industrial methods for the production of plastics.

Government support of plastics engineering is primarily involved with military, infrastructure, and space science applications. Governmental agencies have expressed considerable interest in materials that provide structural advantages in aircraft, watercraft, armor, weapons, and personal protection for military, nonmilitary, and civil service applications such as fire, police, and emergency medical services.

Industry and Business. The specific production methods of many common polymers and plastics are patented and are closely guarded industry secrets. These patents have values of many tens of millions of dollars to their owners, who are generally large corporations with the resources to research and develop polymer science and are often suppliers of polymers and plastics. Other companies involved in plastic engineering work at an intermediate level between the plastics producers and the industries that turn these materials into consumer products. They include the various molding manufacturers and the ancillary companies and services that support their work.

Major Corporations. The largest corporations involved with plastics engineering are also those with

Fascinating Facts About Plastics Engineering

- In 1942, the U.S. Army Quartermaster Corps was trying to find a replacement for the standard issue brass U.S. Army bugle. Mills Plastics, of Chicago, worked closely with both a plastic toy trumpet manufacturer and a designer of woodwind instruments to design a bugle for testing.
- The invention of polyethylene improved the functionality of airborne radar, which was used by Allied Forces during World War II. The radar gave pilots a substantial advantage over German warplanes because Nazi Germany never was able to develop similar radar devices.
- During World War II, nylon, developed as a replacement for silk, was restricted to military uses, including the use of nylon stockings as bribery and leverage tools.
- Working at Rohm and Haas, Herman Bruson developed an ester of acrylic acid and found that it allowed oil to remain free-flowing even when very cold. The Soviet victory at Stalingrad in 1943 was made possible by the fact that the tanks, vehicles, and other equipment of the Soviet forces were equipped with this oil additive and continued to function properly in the bitter cold winter, while the equipment of the German forces that used only conventional oils and fluids froze up and ground to a halt.
- Teflon, discovered in 1938, does not burn. It does not melt below 620 degrees Fahrenheit but rather turns into a translucent gel. It does not conduct electricity at all, does not combine with oxygen, does not dissolve, is unaffected by acids, and is immune to molds, fungi, and bacteria. It was deemed completely useless until it was found to be the only material in existence that could enable the uranium enrichment process used in the Manhattan Project. When finally made available for public use, it was as nothing more exotic than a nonstick coating for cookware and steam irons.
- Leaching of bisphenol A (BPA) from the many commonly used plastic items used for food and beverage service has been linked to the earlier onset of puberty in Western civilizations. The effect may be caused by the molecular structure of bisphenol A, which resembles the structure of steroid hormones.
- As much as one-third of residential waste consists of packaging, as consumers have been given and have come to expect items to be individually wrapped.

the longest history of involvement in plastics. Some have grown from simple two-person operations into major international conglomerates, which speaks to the powerful growth and economic value of the plastics industry. Rohm and Haas, for example, originated as a partnership between two individuals who had worked together in the research department of a large chemical manufacturer. Less obvious major corporations entwined with the plastics industry include General Electric, which used Bakelite to manufacture electric insulators for telegraph equipment and used other resin materials for insulating electric wires and conductors.

Careers and Course Work

A career in plastics engineering requires a solid foundation in general and applied mathematics and in chemistry and physics, beginning in high school and continuing into postsecondary education. In college, students interested in a plastics engineering career should study organic chemistry, analytical chemistry, physical chemistry, physics, and applied mathematics, and take specialized courses in polymer chemistry, reaction kinetics, and industrial chemistry. This will guide the student into postgraduate academic programs and positions in industry. Those interested in careers related to the engineering of plastics materials for industrial applications should study chemistry, physics, applied mathematics, and engineering principles. Specialized course work will be determined by the chosen branch of engineering: mechanical, chemical, civil, industrial, or polymer.

The many applications of engineered plastics guarantee that many careers will involve training in the uses and applications of plastics. Civil engineering technologists and technicians can expect to encounter applications of fiber-reinforced polymers, forms, and molded plastics in the course of their work. Aircraft maintenance engineers will be required to undertake training in the use and repair of fiber-reinforced polymers and advanced composite structures as a regulated licensing requirement. Automotive technicians have the opportunity to specialize in repair techniques using engineered plastic components.

Social Context and Future Prospects

Plastics have become so entrenched in the workings of the modern world that careers involving plastics engineering seem destined to be as persistent as the materials themselves. The great strength of plastics, their chemical stability, is also their greatest failing. Plastic goods were never designed with the conservation of resources in mind. Rather, they have historically been produced as ultimately disposable items. The negative impact of this practice was realized only considerably after the use of plastics became widespread, at which time, the management of used plastics became an issue. The increasing price of plastics, tied to the ever-increasing price of the petroleum that is their source, has led to the consideration of mining landfill sites for their plastic content, particularly nylon. Scientists have also turned to developing methods of reusing and recycling plastics.

The plastics industry has continued to search for and devise new materials and products to maintain a constant supply of cheap, convenient, and ultimately disposable plastic products for consumer demand. At the same time, consumers increasingly demand that the plastics industry be more circumspect with and accountable for its use of plastics as a commodity. The long-term effects of the plastics industry include the accumulation of highly toxic and possibly carcinogenic materials in the environment, as well as problems resulting from the presence of leached monomers and other plastics-related chemicals.

The future of plastics engineering, therefore, appears to be developing along three distinct fronts, each offering broad opportunity as career choices. One is the traditional production and support base of the plastics industry. The second, and potentially the most significant, is the growing industry of plastics abatement, which involves the recovery of used plastics and the mitigation of plastic residues in the environment. The third front, only beginning to emerge, is the field of bioplastics, or plastics produced by living organisms through biological processes instead of through chemical alteration of petroleum sources. All three will require the services of plastics engineers in many different aspects.

Richard M. J. Renneboog, M.Sc.

Further Reading

Crawford, R. J. *Plastics Engineering.* 3d ed. Boston.: Butterworth-Heinemann, 1998. Details many aspects of plastics engineering, from basic principles of polymerization through industrial design and production practices. Also examines properties and processing behavior of plastics and composites.

Fenichell, Stephen. *Plastic: The Making of a Synthetic Century.* New York: HarperBusiness, 1996. A well-researched and readable account of the plastics industry from its beginnings to contemporary times, with as much focus on the social and historical contexts of plastics as on their technical development.

Margolis, James M., ed. *Plastics Engineering Handbook.* New York: McGraw-Hill, 2006. Examines the chemistry, properties, processes, products and designs, and applications.

Rosato, Dominick V., Donald V. Rosato, and Matthew V. Rosato. *Plastic Product Material and Process Selection Handbook.* New York: Elsevier, 2004. Presents properties and characteristics of 35,000 plastic formulations and discusses several fabrication processes, including blow molding, thermoforming, foaming, calendering, casting, reaction injection molding, rotational molding, compression molding, and fiber-reinforced polymers.

Strong, A. Brent. *Plastics: Materials and Processing.* 3d ed. Upper Saddle River, N.J.: Pearson Prentice Hall, 2006. A good fundamental overview with a lot of detail and examination of applications of plastics in many fields. Geared for nonspecialists.

Troughton, Michael J. *Handbook of Plastics Joining: A Practical Guide.* Norwich, N.Y.: William Andrew, 2008. A comprehensive review of plastics welding technologies that discusses heated tool, ultrasonic, vibration, spin, radio frequency, hot gas, extrusion, laser, infrared, resistive implant, induction, heat sealing, and flash-free plastic welding technologies.

Van der Zwaag, Sybrand, ed. *Self-Healing Materials: An Alternative Approach to Twenty Centuries of Materials Science.* Dordrecht, the Netherlands: Springer, 2007. A collection of articles from a conference on self-healing polymers and other plastics materials.

Web Sites

American Institute of Chemical Engineers
http://www.aiche.org

Society of Plastics Engineers
http://www.4spe.org

See also: Chemical Engineering; Engineering; Polymer Science; Structural Composites.

PNEUMATICS

FIELDS OF STUDY

Chemistry; physics; mathematics; electrical circuits; materials science; computer science; mechanics; biomedical engineering.

SUMMARY

Pneumatics technology generates and utilizes compressed gas, normally air, to provide energy to mechanical equipment performing work. Pneumatic technology can be used in large-scale industrial applications such as a drill for blasting rock to small actuators that move a prosthetic device such as an artificial leg. Pneumatics is based on the fact that all gases can be compressed, although the compression affects the volume and temperature. Air, the primary gaseous medium, is in endless supply, inexpensive, and environmentally safe.

KEY TERMS AND CONCEPTS

- **Actuator:** Mechanical portion of a system that performs the physical work desired.
- **Atmospheric Pressure:** Pressure at sea level that air exerts that will counterbalance a mercury column 29.9 inches high.
- **Bar:** Atmosphere of pressure.
- **Compressor:** Mechanical portion of a system that applies pressure on a gas reducing its volume and raising its temperature readying it to do work.
- **Filter:** Portion of a system that cleans gas of impurities.
- **Fluid Power:** Term used to describe the use of both liquids and gas for mechanical purposes.
- **Piston:** Cylinder of varying size that moves to perform mechanical work.
- **Receiver:** Container for holding gas under controlled temperature and pressure until it is needed.
- **Solenoid Valve:** Valve capable of converting electrical energy into mechanical energy, which in turn opens or closes the valve mechanically.
- **Valve:** Portion of a system designed to manipulate the gas quantity, distribution, and delivery regularly and precisely.

Definition and Basic Principles

Pneumatics is a technology resulting from the molecular and chemical composition of a gas, a medium that can move machinery. The machinery can be very large, such as a marine hovercraft, or very small, such as a portable ventilator for a human being. Pneumatics is half of a technology labeled as fluid power; the other half is hydraulics. Hydraulics uses many liquids because liquids are also dependent on pressure and temperature for their volume, although not as much as gases. In scientific terms, fluid is applied to both liquids and gases. Different laws govern the behavior of each medium.

Gases are molecular compounds. Matter in the gaseous state occupies a volume about one thousand times larger than the same amount of matter in the liquid state. The molecules that are in constant motion are also much farther apart. Because they have no inherent shape, they can be easily confined.

Air is the most common gas used in pneumatics. It is composed of nitrogen (78.09 percent), oxygen (20.95 percent), argon (0.93 percent), carbon dioxide (0.039 percent), and several other elements in trace amounts on the order of parts per million. Water vapor is also present in varying and unpredictable amounts, and water can be damaging to a system. Most pneumatic systems have a component for drying the air. An important value is the weight of a one-inch- square column of air from sea level to the beginning of outer space (thermosphere). Expressed in a variety of international measuring systems, the unit one atmosphere (atm) weighs is 101.3 kilopascal (kPa, named for French scientist Blaise Pascal), 14.7 pounds per square inch (psi), 760 torr (named for Italian physicist Evangelista Torricelli), or 29.9 inches of mercury because 1 atm can support a column of mercury that high.

Background and History

Pneumatics is derived from Latin meaning pertaining to air. The blacksmith's bellows pumping a fire is a primitive example. Windmills in twelfth century central Europe exhibited the technology's first large-scale possibility. Development was dependent on discoveries made in the seventeenth and eighteenth centuries about gas properties, especially how the volume of a gas is affected by pressure and temperature. The Italian physicist Evangelista Torricelli invented the mercury barometer in 1643 and then measured atmospheric pressure. He also formed a vacuum by inverting a mercury-filled tube. The first vacuum pump was invented by German scientist Otto von Guericke.

Robert Boyle, an Irish chemist, developed Boyle's law: For a gaseous substance at constant temperature, the volume is inversely proportional to the pressure. French physicist Jacques Charles's contribution was the volume of a given mass of gaseous substance increases or decreases $^1/_{273}$ for each degree Centigrade of temperature change, known as Charles's law. French chemist Joseph Louis Gay-Lussac discovered the relationship between volume and temperature. He devised a hydrometer and calculated the volume of a gaseous compound is equal to or less than the sum of the volumes of the combined gases. Amedeo Avogadro, an Italian physicist, learned that at the same temperature and pressure, equal volumes of different gases contain equal numbers of molecules. This fundamental concept is known as Avogadro's number.

Once the gaseous state was understood, gases could be manipulated to do work, and the field of pneumatics was possible.

How It Works

Pneumatics works by compressing a quantity of air (or some other gas), filtering it to make sure it is clean, drying it, and sending it through a system with one or more valves. The compressed air is delivered to a mechanical apparatus that can then move a force through a distance. The air is an energy-transfer medium.

Pneumatic System Components. The major components are: a compressor, an air storage tank (receiver), an air-cleaning apparatus composed of a main line filter, an air dryer, and a mist separator, all connected by a series of hoses. These sections are followed by an air-conditioning section composed of a filter, regulator, and lubricator, then a controlling section made up of a directional control valve and a speed controller, and, finally, the actuator or operating system for which the pneumatic system was designed to power.

This basic system is used in large factories where truck snow-plowing blades are spray painted. Such a system would be automated for mass production and include a timing apparatus—usually some sort of digital logic to control when and how much paint was sprayed. The paint is atomized and mixed with the pressurized air as it is released.

On a smaller scale, everyday handheld tools, such as a drill hammer for blasting slabs of concrete or air hammer for pounding nails, use the same principle but with less need or capacity for cleaning and removing moisture from the air sucked into the tool.

The compressed air pressure supplied for pneumatic tools is normally about 90 pounds per square inch gauge (psig).

Valves. Valves control the way the gas is used, stopped, or directed. They must function under a variety and range of temperatures. Control or proportional valves in a process system are power-operated mechanisms able to alter fluid flow. A pneumatic valve actuator adjusts valve position by making the air pressure either linear or rotary motion. Ball valves provide the shutoff capability. Gas valves are specialized to control the flow of another medium, such as natural gas. A pressure relief valve is a self-actuated safety valve that relieves pressure. A butterfly valve controls the flow of air or a gas through a circular disk or vane by turning the valve's pivot axis at right angles to the directing flow in the pipe.

Gaseous Medium. The gas or energy medium traveling through the system is unique in its behavior. Charles's law states a volume of a gas is directly proportional to the temperature. Boyle's law states the volume of a gas is inversely proportional to the pressure applied to it. If the amount of gas is calculated in moles (expressed as n), the temperature is in degrees Celsius (expressed as T), and the volume of the gas is in liters (expressed as V), and the pressure in atmospheres (expressed as p), then calculations can be made as to what happens to that gas changing one or more of those variables using the formula $pV = nRT$. R is known as the gas constant and has the same value for all gases. When changes are made to the conditions of a gas, the calculation can be made to find out what will happen to any one or two values of the gas in question. This is expressed $pV/T = pV/T$. This is one of the most important formulas in pneumatics.

Pneumatic System Categories. The type of actuator the air drives is one way of categorizing. These are rotor and reciprocating. Compressed air enters a rotating compressor and pushes on the vanes, rotating a central shaft or spindle. Tools such as drills or grinding wheels use this system. In a reciprocating piston compressor, the air enters a cylinder, expands, and forces the piston to move. In a reciprocating tool, the piston also has a return stroke, which is actuated by compressed air on the other end of the piston or a spring action. A riveting hammer is a tool using this technique.

Another way to categorize pneumatic devices is as either portable or rock drills. Portable devices include buffers, drills, screwdrivers, wrenches, and paint mixers. This category is powered by a rotary-vane type of air motor. Rock drills or percussion hammers are composed of high-carbon steels. In this type, the compressed air drives a piston down onto a loosely held drill inside a cylinder.

Applications and Products

Pneumatic systems have positive and negative features. Potentially, pneumatics can be used in any situation where mechanical work needs to be done. Its overarching desirable quality is a boundless supply of free air. The desirable features include stability under temperature changes, cleanliness (leaks do not cause contamination), and work at high speeds measured in feet per second (fps); the tools stall when they are overloaded with pressure and are therefore inherently safe. The undesirable features are: a constant actuator velocity is not possible; under normal conditions, usually 100 psi, the force is limited to between 4,500 and 6,700 pounds; and the gas must be prepared by removing dirt and humidity as these degrade a system. Since pneumatics is so versatile, it can be used across a broad spectrum of industries.

Heavy Loads. This is the traditional area. Factories often deal in products weighing tons and under extreme temperatures. Pneumatic systems can open heavy or hot doors. They can unload hoppers (containers usually with sloping floors used to carry bulk goods) in industries such as construction, steelmaking, and chemical. Slab molding machines are usually lifted and moved with pneumatic systems. Road or large flat-surface construction is usually rammed or tamped with them. The capability of moving heavy loads proves the efficiency of pneumatic systems.

Spraying. Pneumatic valves are capable of directing air and then atomizing liquids into fine droplets for this application. Large- and small-scale surfaces needing coverage include agriculture, where crops are sprayed with chemicals or water, and painting, as in automobile, motorcycle, or bicycle manufacturing.

Repetition. Repetitive motion has been used since the first mass production and has grown with production of ever-larger quantities in ever-shorter periods of time. Pneumatics is advantageous because of its low cost of operation and maintenance. Both large-scale and small-scale tools are also examples. Both handheld and robotic riveters can secure bolts and screws at very fast speeds. Pneumatic systems are ideal for holding maneuvers where pieces are clamped while other work is accomplished as in: woodworking, furniture making, gluing, heat sealing, or welding plastics. Guillotine blades for slicing precise quantities of a material by the millions are a classic example of this capability. Bottling and filling machines and household cleaners also use the systems.

Breaking and Smashing. The jackhammer, used for road repairs, is a pneumatic tool.

Fascinating Facts About Pneumatics

- The artificial hand developed at Germany's Aachen University of Technology uses pneumatic micro-actuators for finger movement.
- Audio-Animatronics, the form of robotics used by Walt Disney Imagineering, uses pneumatics in Disney theme park shows and rides such as Pirates of the Caribbean.
- The French scientist Guy Negre is experimenting with a minivan powered by four 100-liter steel bottles of compressed air at 300 times atmospheric pressure. He wants to use the compressed air to drive the minivan's motor.
- Pneumatic tubes, such as those used to deliver checks at the bank drive-through window, have stimulated ideas for transporting people inside pneumatic tubes. These would replace entire road systems, be fast, and nonpolluting.
- The artificial heart, made by SynCardia Systems in Tucson, Arizona, is driven entirely by a pneumatic system.
- Discoveries show that when cells are grown for biomedical research, they grow faster if they are placed on a pneumatically pulsed membrane in the petri dish. Researchers believe the pulsing stresses and pulls the cells, encouraging growth.

Medical Devices. Pneumatics is being used more in medical devices. When the dentist, drill or polisher in hand, steps on a small plate, a pneumatic system is activated that drives the instrument in a circular motion at a very high speed. Pneumatic systems are excellent for applications in patient ventilators. Ventilators can either take over breathing for an extremely sick patient or assist a patient with some breathing ability.

Robotics. Robotics for work as well as entertainment is another new application for the field. Most movable monsters or lifelike creatures in theme parks or museum are operated pneumatically.

Impact on Industry

Pneumatics is in every industrialized or semi-industrialized country in the world.

Societies and Organizations. Numerous societies and associations, such as the International Fluid Power Society, protect the integrity of the field with a variety of certification processes. Each country has its own division; in the United States there is also the National Fluid Power Association (NFPA) located in Milwaukee, Wisconsin.

Education. Because degrees are not awarded in the discipline, attracting and training new personnel is an ongoing effort. Not every engineering college has a fluid power department. One of the oldest is the Fluid Power Institute at Milwaukee School of Engineering in Milwaukee, Wisconsin, across town from the NFPA. Fluid power societies are increasingly conducting high school competitions with significant monetary awards to winning schools' teams as well as scholarships. Frequently, these are teamed with robotics competitions.

Most contemporary research is done in the United States, Germany, and France. Some British universities have certification programs.

Industry. Thousands of companies manufacture the component parts. Usually, they specialize in specific parts such as valves or compressors. One of the largest American companies is Parker Hannifin, based in Cleveland, Ohio, which specializes in motion control. Numatics Incorporated, in Novi, Michigan, is one of the leading providers of valves, cylinders, control systems, and air-preparation components.

Military. The military hires companies that can provide advances in weaponry and pneumatic tires for tactical vehicles. Pneumatics is also used in camp housing in the form of tents, sleeping bags, and decontamination showers. Medically, pneumatics is being used for anti-shock trousers, which halts severe blood loss on the battlefield.

Careers and Course Work

Pneumatics is a subset of the field of fluid power; fluid power is a subset of mechanical and electrical engineering. Students earn an engineering degree in one of the two then attend a university offering a master's-degree program in pneumatics. The ultimate ultimate designation in this field is fluid power engineer, which requires a bachelor's degree, eight years of work in the field, and successful completion of other certifications. Jobs are also attained by having a good aptitude for mechanical and electrical work then passing short courses followed by exams to earn a certification from the International Fluid Power Society. Examples of certifications are: fluid power connector and conductor, fluid power certified mechanic, fluid power certified technician, fluid power electronic controls, and fluid power specialist.

Certifications prepare people to be reliably expert in all stages of the pneumatic process from understanding theory, system design, schematics, efficiency improvement, controls, and safety.

Wherever tools are used, be they on an enormous scale as in mining or on a microscopic scale as in the biomedical field, opportunities exist for employment in the private sector and future growth is anticipated. Much of the growth will come about because of miniaturization of systems through computers and advances in materials. The military also makes great use of pneumatic systems in both weaponry and transportation. Teaching and research at universities is another career track.

Social Context and Future Prospects

Energy awareness, conservation, and environmental protection are forces that will allow pneumatics to remain a promising field into the future. Despite the commercial size of the field of pneumatics, few people know about it even though they use its technology daily.

From its inception, the industry has not changed much because the basic system remains the same. What does change is that component providers continue to improve on the efficiency and materials as well as the advances made by computerization and miniaturization of parts. Hospital ventilators at one time took up as much room as a hospital bed. Now, they can be placed in a belt pack and the patient sent home.

This field of nanotechnology works with unusual matter located between molecules in the size range of 0.000000001 meter. It allows pneumatic component parts such as pistons and seals to be manufactured with ultraprecision.

Judith L. Steininger, B.A., M.A.

Further Reading

Cundiff, John S. *Fluid Power Circuits and Controls: Fundamentals and Applications.* Boca Raton, Fla.: CRC Press, 2011. Both hydraulics and pneumatics are discussed, giving a good overview of fluid power field.

Daines, James R. *Fluid Power: Hydraulics and Pneumatics.* Tinley Park, Ill.: Goodheart-Willcox, 2009. Colorful, attractive presentation of the technology, including schematics.

Design News. "Score One for Pneumatics." (January 11, 2004). Cost efficiency of technology is discussed in this article, and *Design News* magazine provides year-round information about pneumatics.

Fleischer, Henry. *Manual of Pneumatic Systems Optimization.* New York: McGraw-Hill, 1995. The discussion of pneumatic systems efficiency is placed in social context of energy conservation.

Johnson, Olaf A. *Fluid Power: Pneumatics.* Chicago: American Technical Society, 1975. Though an older text, this is considered a bible in the industry.

Parr, Andrew. *Hydraulics and Pneumatics: A Technician's and Engineer's Guide.* 3d ed. Burlington, Mass.: Butterworth-Heinemann, 2011. An excellent update to classic text on pneumatics.

Rabie, M. Galal. *Fluid Power Engineering.* New York: McGraw-Hill, 2009. Pneumatic systems are explained with schematics and advanced math.

Web Sites

American Society for Mechanical Engineers
http://www.asme.org

International Fluid Power Society
http://www.ifps.org

National Fluid Power Association
http://www.nfpa.com

See also: Computer Science; Electrical Engineering; Mechanical Engineering.

POLYMER SCIENCE

FIELDS OF STUDY

Chemistry; physics; chemical engineering; materials science; organic chemistry; inorganic chemistry; physical chemistry; analytical chemistry; differential equations; computer science; structures and reactions of macromolecular compounds; kinetics and mechanisms of polymer synthesis; plastics engineering; synthetic rubber engineering; biopolymers; molecular biology; biochemistry; biophysics.

SUMMARY

Polymer science is a specialized field concerned with the structures, reactions, and applications of polymers. Polymer scientists generate basic knowledge that often leads to various industrial products such as plastics, synthetic fibers, elastomers, stabilizers, colorants, resins, adhesives, coatings, and many others. A mastery of this field is also essential for understanding the structures and functions of polymers found in living things, such as proteins and deoxyribonucleic acid (DNA).

KEY TERMS AND CONCEPTS

- **Addition Polymerization:** Chemical chain-reaction process in which monomers bond to each other in the step-by-step growth of a polymer, without generating any by-products.
- **Biopolymer:** Naturally occurring polymer.
- **Copolymer:** Polymer made from more than one kind of monomer, resulting in a macromolecule with more than one monomeric unit (or "mer") in the backbone.
- **Elastomer:** Substance composed of chemically (or physically) cross-linked linear polymers possessing the elastic properties of natural rubber.
- **Free Radical:** Short-lived and highly reactive molecular fragment with one or more unpaired electrons.
- **Inorganic Polymer:** Polymer made from inorganic monomers, such as silicates in nature and artificially made cements and lubricants.
- **Monomer (Monomeric Unit):** Largest atomic array (with possible projecting groups) constituting the basic building block of a polymer.
- **Nucleic Acid:** One of several nitrogenous compounds found in living cells and viruses, which, in their polymeric form as polynucleotides, contain the following principal parts: a base (either a purine or pyrimidine), a sugar, and a phosphoric-acid group.
- **Plastic:** Organic polymer, generally synthetic, often combined with additives such as colorants and curatives, capable of being molded or cast under heat and pressure into various shapes.
- **Polycondensation:** Chemical process in which nonidentical monomers react to form a polymer, with the concomitant formation of leaving molecules.
- **Polymerization:** Chemical reaction, usually catalyzed, in which a monomer or mixture of monomers form a chain-like macromolecule (or polymer).
- **Protein:** Complex polymer composed of chains of amino acids (groups that contain carbon, hydrogen, oxygen, nitrogen, and, sometimes, sulfur).

Definition and Basic Principles

Polymers are very large and often complex molecules constructed, either by nature or by humans, through the repetitive yoking of much smaller and simpler units. This results in linear chains in some cases and in branched or interconnected chains in others. The polymer can be built up of repetitions of a single monomer (homopolymer) or of different monomers (heteropolymer). The degree of polymerization is determined by the number of repeat units in a chain. No sharp boundary line exists between large molecules and the macromolecules characterized as polymers. Industrial polymers generally have molecular weights between ten thousand and one million, but some biopolymers extend into the billions.

Chemists usually synthesize polymers by condensation (or step-reaction polymerization) or addition (also known as chain-reaction polymerization). A good example of chain polymerization is the free-radical mechanism in which free radicals are created (initiation), facilitating the addition of monomers (propagation), and ending when two free radicals react with each other (termination). A general example of step-reaction polymerization is the reaction of two or more polyfunctional molecules to produce a larger grouping, with the elimination of a small molecule such as water, and the consequent repetition of the process until termination.

Besides free radicals, chemists have studied polymerizations utilizing charged atoms or groups of atoms (anions and cations). Physicists have been concerned with the thermal, electrical, and optical properties of polymers. Industrial scientists and engineers have devoted their efforts to creating such new polymers as plastics, elastomers, and synthetic fibers. These traditional applications have been expanded to include such advanced technologies as biotechnology, photonics, polymeric drugs, and dental plastics. Other scientists have found uses for polymers in such new fields as photochemistry and paleogenetics.

Background and History

Nature created the first polymers and, through chemical evolution, such complex and important macromolecules as proteins, DNA, and polysaccharides. These were pivotal in the development of increasingly multifaceted life-forms, including Homo sapiens, who, as this species evolved, made better and better use of such polymeric materials as pitch, woolen and linen fabrics, and leather. Pre-Columbian Native Americans used natural rubber, or *cachucha*, to waterproof fabrics, as did Scottish chemist Charles Macintosh in nineteenth century Britain.

The Swedish chemist Jöns Jakob Berzelius coined the term "polymer" in 1833, though his meaning was far from a modern chemist's understanding. Some scholars argue that the French natural historian Henri Braconnot was the first polymer scientist since, in investigating resins and other plant products, he created polymeric derivatives not found in nature. In 1836, the Swiss chemist Christian Friedrich Schönbein reacted natural cellulose with nitric and sulfuric acid to generate semisynthetic polymers. In 1843 in the United States, hardware merchant Charles Goodyear accidentally discovered "vulcanization" by heating natural rubber and sulfur, forming a new product that retained its beneficial properties in cold and hot weather. Vulcanized rubber won prizes at the London and Paris Expositions in 1850, helping to launch the first commercially successful product of polymer scientific research.

In the early twentieth century, the Belgian-American chemist Leo Baekeland made the first totally synthetic polymer when he reacted phenol and formaldehyde to create a plastic that was marketed under the name Bakelite. The nature of this and other synthetic polymers was not understood until the 1920's and 1930's, when the German chemist Hermann Staudinger proved that these plastics (and other polymeric materials) were extremely long molecules built up from a sequential catenation of basic units, later called monomers. This enhanced understanding led the American chemist Wallace Hume Carothers to develop a synthetic rubber, neoprene, that had numerous applications, and nylon, a synthetic substitute for silk. Synthetic polymers found wide use in World War II, and in the postwar period the Austrian-American chemist Herman Francis Mark founded the Polymer Research Institute at Brooklyn Polytechnic, the first such facility in the United States. It helped foster an explosive growth in polymer science and a flourishing commercial polymer industry in the second half of the twentieth century.

How It Works

After more than a century of development, scientists and engineers have discovered numerous techniques for making polymers, including a way to make them using ultrasound. Sometimes these techniques depend on whether the polymer to be synthesized is inorganic or organic, fibrous or solid, plastic or elastomeric, crystalline or amorphous. How various polymers function depends on a variety of properties, such as melting point, electrical conductivity, solubility, and interaction with light. Some polymers are fabricated to serve as coatings, adhesives, fibers, or thermoplastics. Scientists have also created specialized polymers to function as ion-exchange resins, piezoelectrical devices, and anaerobic adhesives. Certain new fields have required the creation of specialized polymers like heat-resistant plastics for the aerospace industry.

Condensation Polymerization. Linking monomers into polymers requires the basic molecular building blocks to have reaction sites. Carothers recognized that most polymerizations fall into two broad categories, condensation and addition. In condensation, which many scientists prefer to call step, step-growth, or stepwise polymerization, the polymeric chain grows from monomers with two or more reactive groups that interact (or condense) intermolecularly, accompanied by the elimination of small molecules, often water. For example, the formation of a polyester begins with a bifunctional monomer, containing a hydroxyl group

(OH, oxygen bonded to hydrogen) and a carboxylic acid group (COOH, carbon bonded to an oxygen and an OH group). When a pair of such monomers reacts, water is eliminated and a dimer formed. This dimer can now react with another monomer to form a trimer, and so on. The chain length increases steadily during the polymerization, necessitating long reaction times to get "high polymers" (those with large molecular weights).

Addition Polymerization. Many chemists prefer to call Carothers's addition polymerization chain, chain-growth, or chain-wise polymerization. In this process, the polymer is formed without the loss of molecules, and the chain grows by adding monomers repeatedly, one at a time. This means that monomer concentrations decline steadily throughout the polymerization, and high polymers appear quickly. Addition polymers are often derived from unsaturated monomers (those with a double bond), and in the polymerization process the monomer's double bond is rearranged in forming single bonds with other molecules. Many of these polymerizations also require the use of catalysts and solvents, both of which have to be carefully chosen to maximize yields. Important examples of polymers produced by this mechanism are polyurethane and polyethylene.

Applications and Products

Since the start of the twentieth century, the discoveries of polymer scientists have led to the formation of hundreds of thousands of companies worldwide that manufacture thousands of products. In the United States, about 10,000 companies are manufacturing plastic and rubber products. These and other products exhibit phenomenal variety, from acrylics to zeolites. Chemists in academia, industry, and governmental agencies have discovered many applications for traditional and new polymers, particularly in such modern fields as aerospace, biomedicine, and computer science.

Elastomers and Plastics. From its simple beginnings manufacturing Bakelite and neoprene, the plastic and elastomeric industries have grown rapidly in the quantity and variety of the polymers their scientists and engineers synthesize and market. Some scholars believe that the modern elastomeric industry began with the commercial production of vulcanized rubber by Goodyear in the nineteenth century. Such synthetic rubber polymers as styrene-butadiene, neoprene, polystyrene, polybutadiene, and butyl rubber (a copolymer of butylene and isoprene) began to be made in the first half of the twentieth century, and they found extensive applications in the automotive and other industries in the second half.

Although an early synthetic plastic derived from cellulose was introduced in Europe in the nineteenth century, it was not until the twentieth century that the modern plastics industry was born, with the introduction of Bakelite, which found applications in the manufacture of telephones, phonograph records, and a variety of varnishes and enamels. Thermoplastics, such as polyethylene, polystyrene, and polyester, can be heated and molded, and billions of pounds of them are produced in the United States annually. Polyethylene, a low-weight, flexible material, has many applications, including packaging, electrical insulation, housewares, and toys. Polystyrene has found uses as an electrical insulator and, because of its clarity, in plastic optical components. Polyethylene terephthalate (PET) is an important polyester, with applications in fibers and plastic bottles. Polyvinyl chloride (PVC) is one of the most massively manufactured synthetic polymers. Its early applications were for raincoats, umbrellas, and shower curtains, but it later found uses in pipe fittings, automotive parts, and shoe soles.

Carothers synthesized a fiber that was stronger than silk, and it became known as nylon and led to a proliferation of other artificial textiles. Polyester fibers, such as PET, have become the world's principal man-made materials for fabrics. Polyesters and nylons have many applications in the garment industry because they exceed natural fibers, including cotton and wool, in such qualities as strength and wrinkle resistance. Less in demand are acrylic fibers, but, because they are stronger than cotton, they have had numerous applications by manufacturers of clothing, blankets, and carpets.

Optoelectronic, Aerospace, Biomedical, and Computer Applications. As modern science and technology have expanded and diversified, so, too, have the applications of polymer science. For example, as researchers explored the electrical conductivity of various materials, they discovered polymers that have exhibited commercial potential as components in environmentally friendly battery systems. Transparent polymers have become essential to the fiber optics industry. Other polymers have had an important part in

Fascinating Facts About Polymer Science

- Life would not exist without polymers, since a lack of proteins would mean no enzymes for essential chemical reactions, and a lack of nucleic acids would mean that life forms could not replicate themselves.
- Many chemists initially rejected the idea that gargantuan polymers exist, and one eminent chemist stated that it was as if zoologists "were told that somewhere in Africa an elephant was found that was 1,500 feet long and 300 feet high."
- Lord Alexander Todd, the winner of the 1957 Nobel Prize in Chemistry, once stated that the development of polymerization was "the biggest thing that chemistry has done," since this has had "the biggest effect on everyday life."
- There are more chemists researching and developing synthetic polymers than in all other areas of chemistry combined.
- It takes about $1 billion to discover, develop, and introduce a new polymer into the marketplace.
- In the twenty-first century, employment figures show that more than half of American chemical industrial employment is in synthetic polymers.
- The plastic now known as Silly Putty was called Nutty Putty by the chemical engineer who invented it in the 1940's.
- In the new generation of commercial airplanes, such as the Boeing 787, about half of the construction materials come from polymers, polymer-derived fabrics, and composites.

the improvement of solar-energy devices through products such as flexible polymeric film reflectors and photovoltaic encapsulants. Newly developed polymers have properties that make them suitable for optical information storage. The need for heat-resistant polymers led the U.S. Air Force to fund the research and development of several such plastics, and one of them, polybenzimidazole, has achieved commercial success not only in aerospace but also in other industries as well.

Following the discovery of the double-helix structure of DNA in 1953, a multiple of applications followed, starting in biology and expanding into medicine and even to such fields as criminology. Nondegradable synthetic polymers have had multifarious medical applications as heart valves, catheters, prostheses, and contact lenses. Other polymeric materials show promise as blood-compatible linings for cardiovascular prostheses. Biodegradable synthetic polymers have found wide use in capsules that release drugs in carefully controlled ways. Dentists regularly take advantage of polymers for artificial teeth, composite restoratives, and various adhesives. Polymer scientists have also contributed to the acceleration of computer technology since the 1980's by developing electrically conductive polymers, and, in turn, computer science and technology have enabled polymer scientists to optimize and control various polymerization reactions.

Impact on Industry

According to the U.S. Department of Labor, polymer scientists, in the first decade of the twenty-first century, were the most extensively employed chemists, surpassing by four times their nearest competitors, chemists in the pharmaceutical industry. In terms of U.S. chemical exports, plastics and resins have been on a par with organic chemicals and pharmaceuticals. Because of the dramatic and increasing use of polymers in industries, some have called this period in technological history "the age of the macromolecule." Most important advances in chemistry and allied fields have involved synthetic polymers, a contemporary example being carbon nanotubes.

Government and University Research. In the history of polymer science, a major shift occurred, shared with several other fields, from the individual inventor of the nineteenth century to the collective creators of the twentieth. It is increasingly rare for an individual, working in an academic or industrial setting, to discover and develop new polymers, the way that Goodyear fashioned and exploited vulcanized rubber in the early nineteenth century. In response to intensifying international competition, American polymer scientists have fostered cooperation between academia and industry. An early example of this was DuPont's development of neoprene, which was rooted in Father Julius A. Nieuwland's work on the polymerization reactions of acetylene at Notre Dame University. Because of the escalating complexity of many new polymers, and the need to master a growing variety and intricacy of techniques, research in polymer science and engineering, whether academic, industrial, or collaborative, requires personnel from diverse backgrounds, ranging from polymer science and chemical engineering to computer science and technology. Cognizant of this, funders at government agencies have supported academic and industrial projects that involve large and diverse interdisciplinary research teams at centers around the country.

Small and Large Industries. Many new polymers have been developed and marketed by such major corporations as Dow in the United States and BASF in Germany. On the other hand, small and midsize companies involved with polymers outnumber the major corporations. Indeed, some observers have noticed a shift from the costly, high-risk and long-term research on new polymers undertaken by corporate research laboratories to cost-effective, short-term research on promising polymers by start-up companies. Because of the global nature of important large corporations, and because polymer-based products are marketed worldwide, international collaboration among research groups has become evermore significant. Furthermore, research teams have become increasingly internationalized, with American polymeric chemists working in Europe and their European counterparts working in the United States.

Careers and Course Work

Building on a base of advanced courses in chemistry, mathematics, and chemical engineering, undergraduates generally take an introductory course in polymer science. Graduate students in polymer science usually take courses in line with their chosen career goal. For example, students aspiring to positions in the plastics industry would need to take advanced courses in macromolecular synthesis and the chemical engineering of polymer syntheses. Students interested in biotechnology or bioengineering would need to take graduate courses in molecular biology, biomolecular syntheses, and so on.

Many opportunities are available for graduates with degrees in polymer science. The field is expanding, and careers in research can be forged in government agencies, academic institutions, and various industries. Chemical, pharmaceutical, biomedical, cosmetics, plastics, and petroleum companies hire polymer scientists and engineers. Because of concerns raised by the modern environmental movement, many companies are hiring graduates with expertise in biodegradable polymers. The rapid development of the computer industry has led to a need for graduates with an understanding of electrically charged polymeric systems. In sum, traditional and new careers are accessible to polymer scientists and engineers both in the United States and many foreign countries.

Social Context and Future Prospects

Barring a total global economic collapse or a cataclysmic environmental or nuclear-war disaster, the trend of expansion in polymer science and engineering, well-established in the twentieth century, should continue throughout the twenty-first. As polymer scientists created new materials that contributed to twentieth-century advances in such areas as transportation, communications, clothing, and health, so they are well-positioned to meet the challenges that will dominate the twenty-first century in such areas as energy, communications, and the health of humans and the environment. Many observers have noted the increasing use of plastics in automobiles, and polymer scientists will most likely help to create lightweight-plastic vehicles of the future. The role of polymer science in biotechnology will probably exceed its present influence, with synthesized polymers to monitor and induce gene expression or as components of nanobots to monitor and even improve the health of vital organs in the human body. Environmental scientists have made the makers of plastics aware that many of their products end up as persistent polluters of the land and water, thus fostering a search that will likely continue throughout the twenty-first century for biodegradable polymers that will serve both the needs of advanced industrialized societies and the desire for a sustainable world.

Robert J. Paradowski, M.S., Ph.D.

Further Reading

Carraher, Charles E., Jr. *Giant Molecules: Essential Materials for Everyday Living and Problem Solving.* 2d ed. Hoboken, N.J.: John Wiley & Sons, 2003. Called an "exemplary book" for general readers interested in polymers, *Giant Molecules* facilitates understanding with an apt use of illustrations, figures, and tables along with relevant historical materials.

_______. *Introduction to Polymer Chemistry.* 2d ed. Boca Raton, Fla.: CRC Press, 2010. This text, intended for undergraduates, explains polymer principles in the contexts of actual applications, while including material on such new areas as optical fibers and genomics. Summaries, glossaries, exercises, and further readings at the ends of chapters.

Ebewele, Robert O. *Polymer Science and Technology.* Boca Raton, Fla.: CRC Press, 2000. The author covers polymer fundamentals, the creation of polymers, and polymer applications, with an emphasis on polymer products.

Morawetz, Herbert. *Polymers: The Origins and Growth of a Science.* Mineola, N.Y.: Dover, 2002. This illustrated paperback reprint makes widely available an excellent history of the subject.

Painter, Paul C., and Michael M. Coleman. *Essentials of Polymer Science and Engineering.* Lancaster, Pa.: DEStech Publications, 2009. This extensively and beautifully illustrated book is intended for newcomers to the polymer field. Recommended readings at the ends of chapters.

Scott, Gerald. *Polymers and the Environment.* Cambridge, England: Royal Society of Chemistry, 1999. From an environmental viewpoint, this book introduces the general reader to the benefits and limitations of polymeric materials as compared with traditional materials.

Seymour, Raymond B., ed. *Pioneers in Polymer Science: Chemists and Chemistry.* Boston: Kluwer Academic Publishers, 1989. This survey of the history of polymer science emphasizes the scientists responsible for the innovations. Several of the chapters were written by scientists who were directly involved in these developments.

Web Sites

American Chemical Society
ACS Division of Polymeric Materials
http://pmse.sites.acs.org

American Institute of Chemical Engineers
http://www.aiche.org

Materials Research Society
http://www.mrs.org/home

See also: Computer Science; Plastics Engineering.

PROBABILITY AND STATISTICS

FIELDS OF STUDY

Mathematics; algebra; geometry; calculus; logic; topology; data collection and analysis; experimental design; survey design; physics; biology and biological mathematics; genetics; statistical modeling; distribution theory; game theory; stochastic processes; multivariate analysis; linear systems; time series analysis; actuarial science; information theory; signal processing; demography; quantitative finance and financial mathematics; meteorology; computer science; quantum computing; neurobiology; electrical engineering; mechanical engineering.

SUMMARY

Probability and statistics are two related fields covering the science of collecting, measuring, and analyzing information in the form of numbers. Both probability and statistics are branches of applied mathematics. Probability focuses on using numeric data to predict a future outcome. Statistics incorporates theory into the gathering of numerical data and the drawing of accurate conclusions. Because nearly all fields in applied science rely on the analysis of numbers in some way, probability and statistics are one of the most diverse areas in terms of subjects and career paths. Statisticians also practice in areas of the academic world outside of science and throughout industry.

KEY TERMS AND CONCEPTS

- **Bias:** Tendency to measure or draw conclusions from data in a specific way that influences the outcome.
- **Confidence Interval:** Likelihood that a conclusion based on statistical analysis will fit the population studied.
- **Critical Value:** Mathematical threshold used to determine whether a hypothesis is true or false.
- **Distribution:** Range within which a group of data falls and the way in which the data are arranged. Many statistical formulas measure distribution patterns.
- **Lurking Variable:** Variable outside the scope of a study that exerts an influence on the variables being studied.
- **Margin of Error:** Measurable amount by which a conclusion based on data analysis could be inaccurate.
- **Outlier:** Data value unusually far from the average when measured by standard deviation.
- **Population:** Group of items being studied. A statistical analysis might measure an entire population or use a random sample taken from the population.
- **Quantitative Value:** Value that can be measured and documented as a number; its opposite is categorical.
- **Random:** That which cannot be predicted in single instances but can be measured and predicted in multiple instances, such as a coin toss.
- **Standard Deviation:** Measure of the variance among data items in a set.

Definition and Basic Principles

Probability and statistics are two interconnected fields within applied mathematics. In both fields, principles of scientific theory are applied to the analysis of groups of data in the form of numbers. The main objective of probability and statistics is to ask and answer questions about data with as much accuracy as possible.

Defining "probability" can be a challenge, as multiple schools of thought exist. In one view, held by a group of scholars known as frequentists, probability is defined as the likelihood that a statement about a set of data will be true in the long run. Frequentists focus on the big picture, specifically at the collective outcome of multiple experiments conducted over time, rather than on specific data items or outcomes. In contrast, scholars known as Bayesians prefer to start with a probability-based assumption about a set of data, then test to see how close the actual data come to the initial assumption. On both sides of the debate, probabilists are seeking to understand patterns in data to predict how a population might behave in the future.

Statistics is a field with a broader scope than probability, but in some ways, it is easier to define. The academic discipline of statistics is based on the study of groups of numbers in three stages: collection, measurement, and analysis. At the collection stage, statistics involves issues such as the design of experiments and surveys. Statisticians must answer questions such as whether to examine an entire population or to work from a sample. Once the data are collected, statisticians must determine the level of measurement to be used and the types of questions that can be answered with validity based on the numbers.

No matter how rigorous an individual study might be, statistical findings are often met with doubt by scholars and the general public. A quote repeated often (and mistakenly attributed to former British prime minister Benjamin Disraeli) is "There are three kinds of lies: lies, damned lies, and statistics."

Background and History

Probability has been a subject of interest since dice and card games were first played for money. Gambling inspired the first scholarly discussions of probability in the sixteenth and early seventeenth centuries. The Italian mathematician Gerolamo Cardano wrote *Libellus de ratiociniis in ludo aleae* (*The Value of All Chances in Games of Fortune*, 1714) in about 1565, although the work was not published until 1663. In the mid-1600's, French mathematicians Blaise Pascal and Pierre de Fermat discussed principles of probability in a series of letters about the gambling habits of a mutual friend.

The earliest history of a statistical study is less clear, but it is generally thought to involve demographics. British scholar John Graunt studied causes of mortality among residents of London and published his findings in 1662. Graunt found that statistical data could be biased by social factors, such as relatives' reluctance to report deaths due to syphilis. In 1710, John Arbuthnot analyzed the male-female ratio of babies born in Britain since 1629. His findings—that there were more males than females—were used to support his argument in favor of the existence of a divine being.

A third branch of statistics, the design of experiments and the problem of observational error, has its roots in the eighteenth century work of German astronomer Tobias Mayer.

However, a paper by British theologian Thomas Bayes published in 1764 after his death is considered a turning point in the history of probability and statistics. Bayes dealt with the question of how much confidence could be placed in the predictions of a mathematical model based on probability. The convergence between probability and statistics has increased over time. The development of modern computers has led to major advances in both fields.

How It Works

In terms of scope, probability and statistics are some of the widest, most diverse fields in applied mathematics. If a research project involves items that must be counted or measured in some way, statistics will be part of the analysis. It is common to associate statistics with research in the sciences, but an art historian tracking changes in the use of color in painting from one century to another is just as likely to use a form of statistical analysis as a biologist working in a laboratory. Similarly, probability is used by anyone relying on numbers to make an educated guess about events in the future.

The word "statistic" can refer to almost any number connected to data. When statistics are discussed as a discipline, though, there is a multistep process that most projects follow: definition and design, collection, description and measurement, and analysis and interpretation.

Definition and Design. Much of the scholarship in the field of statistics focuses on data definition and the design of surveys and experiments. Statistical projects must begin with a question, such as "Does grade-point average in high school have an effect on income level after graduation?" In the data definition phase, the statistician chooses the items to be studied and measured, such as grades and annual earnings. The next step is to define other factors, such as the number of people to be studied, the areas in which they live, the years in which they graduated from high school, and the number of years in which income will be tracked. Good experimental design ensures that the rest of the project will gather enough data, and the right data, to answer the question.

Collection. Once the data factors have been defined, statisticians must collect them from the population being studied. Experimental design also plays a role in this step. Statistical data collection must be thorough and must follow the rules of the study. For example, if a survey is mailed to one thousand high school graduates and only three respond, more data must be collected before the survey's findings can be considered valid. Statisticians also must ensure that collected data are accurate by finding a way to check the reliability of answers.

Description and Measurement. Collected data must be stored, arranged, and presented in a way that can be used by statisticians to form statistics and draw conclusions. Grade-point averages, for example, might be compared easily if all the survey participants attended the same school. If different schools or grading systems were used, the statistician must develop rules about how to convert the averages into a form that would allow them to be compared. Once these conversions are made, the statistician would decide whether to present the data in a table, chart, or other form.

Analysis and Interpretation. In terms of statistical theory, the most complex step is the analysis and interpretation of data. When data have been collected, described, and measured, conclusions can be drawn in a number of ways—none of which is right in every case. It is this step in which statisticians must ask themselves a few questions: What is the relationship between the variables? Does a change in one automatically lead to a change in the other? Is there a third variable, or a lurking variable, not covered in the study that makes both data points change at the same time? Is further research needed?

One of the most common methods used for statistical analysis is known as modeling. A model allows the statistician to build a mathematical form, such as a formula, based on ideas. The data collected by the study can then be compared with the model. The results of the comparison tell a story that supports the study's conclusions. Some models have been found to be so innovative that they have earned their creators awards such as the Nobel Prize. However, even the best models can have flaws or can fail to explain actual data. Statistician George Box once said, "All models are wrong, but some models are useful."

Prediction. Probability deals with the application of statistical methods in a way that predicts future behavior. The goal of many statistical studies is to establish rules that can be used to make decisions. In the example, the study might find that students who achieve a grade-point average of 3 or higher earn twice as much, as a group, as their fellow students whose averages were 2.9 or lower. As an academic discipline, probability offers several tools, based on theory, that allow a statistician to ask questions such as: How likely is a student with a grade-point average of 3.5 to earn more than $40,000 per year?

For both statistics and probability, one of the primary objectives is to ask mathematical questions about a smaller population to find answers that apply to a larger population.

APPLICATIONS AND PRODUCTS

It would be nearly impossible to find a product or service that did not rely on probability or statistics in some way. In the case of a cup of coffee, for example, agricultural statistics guided where the coffee beans were grown and when they were harvested. Industrial statistics controlled the process by which the beans were roasted, packaged, and shipped. Statistics even influenced the strength to which the coffee was brewed—whether in a restaurant, coffeehouse, or a home kitchen. Probability played a role in each step as well. Forecasts in weather and crop yields, pricing on coffee bean futures contracts, and anticipated caffeine levels each had an effect on a single brewed cup.

One way to understand the applications and products of probability and statistics is to look at some general categories by function. These categories cover professional fields that draw some of the highest concentrations of professionals with a statistical background.

Process Automation. One of the broadest and most common ways in which statistical methods are applied is in process automation. Quality control is a leading example. When statistical methods are applied to quality control, measures of quality such as how closely products match manufacturing specifications can be translated into numbers and be tracked. This approach allows manufacturers to ensure that their products meet quality standards such as durability and reliability. It also verifies that products meet physical standards, including size and weight. Quality control can be used to evaluate service providers as well as manufacturers. If measures of quality such as customer satisfaction can be converted into numerical data, statistical methods can be applied to measure and increase it. One well-known quality control application, Six Sigma, was developed by the Motorola Corporation in the early 1980's to reduce product defects. Six Sigma relies on both probability and statistical processes to meet specific quality targets.

Another field in which process automation is supported by statistical analysis is transport and logistics. Transport makes up a significant amount of the cost of manufacturing a product. It also plays a major role in reliability and customer satisfaction. A manufacturer must ensure that its products will make the journey from one place to another in a timely way without being lost or damaged. To keep costs down, the method of transport must be as inexpensive as possible. Statistical methods allow manufacturers to calculate and choose the best transportation options (such as truck versus rail) and packaging options (such as cardboard versus plastic packaging) for their products. When fuel costs rise, the optimization of logistics becomes especially important for manufacturers. Probability gives manufacturers tools such as futures and options on fuel costs.

Biostatistics. Statistics are used in the biological sciences in a variety of ways. Epidemiology, or the study of disease within a population, uses statistical techniques to measure public health problems. Statistics allow epidemiologists to measure and document the presence of a specific illness or condition in a population and to see how its concentration changes over time. With this information, epidemiologists can use probability to predict the future behavior of the health problem and recommend possible solutions.

Other fields in the biological sciences that rely on statistics include genetics and pharmaceutical research. Statistical analysis has played a key role in the Human Genome Project. The amount of data generated by the effort of mapping human genes could be analyzed only with complex statistical processes, some of which are still being fine-tuned to meet the project's unique needs. Probability analyses allow geneticists to predict the influence of a gene on a trait in a living organism.

Pharmaceutical researchers use statistics to build clinical trials of new drugs and to analyze their effects. The use of statistical processes has become so widespread in pharmaceutical research that it is extremely difficult to obtain approvals from the U.S. Food and Drug Administration (FDA) without it. The FDA publishes extensive documentation guiding researchers through the process of complying with the agency's statistical standards for clinical trials. These standards set restrictions on trial factors ranging from the definition of a control group (the group against which the drug's effects are to be measured) to whether the drug effectively treats the targeted condition.

Spatial Statistics. Understanding areas of space requires the analysis of large amounts of data. Spatial statistics are used in fields such as climatology, agricultural science, and geology. Statistical methods provide climatologists with specialized tools to model the effects of factors such as changes in air temperature or atmospheric pollution. Meteorology also depends on statistical analysis because its data are time based and often involve documenting repeated events over time, such as daily precipitation over the course of several years. One of the best-known applications of probability to a field of science is weather forecasting.

In agricultural science, researchers use statistics and probability to evaluate crop yields and to predict success rates in future seasons. Statistics are also used to measure environmental impact, such as the depletion of nutrients from the soil after a certain kind of crop is grown. These findings guide recommendations, based on probabilistic techniques, about what kinds of crops to grow in seasons to come. Animal science relies on statistical analysis at every stage, from genetic decisions about the breeding of livestock to the environmental and health impacts of raising animals under certain kinds of farming conditions.

Geologists use statistics and probability in a wide range of ways. One way that draws a significant amount of interest and funding from industry is the discovery of natural resources within the Earth. These efforts include the mining of metals and the extraction and refining of products such as oil and gasoline. Mining and petroleum operations are leading employers of statisticians and probabilists. Statistical processes are critical in finding new geologic sites and in measuring the amount and quality of materials to be extracted. To be done profitably, mining and petroleum extraction are functions that must be carried out on a large scale, assisted by sizable amounts of capital and specialized equipment. These functions would not be possible without sophisticated statistical analysis. Statistics and probability are also used to measure environmental impact and to craft effective responses to disasters such as oil spills.

Risk Assessment. As a science, statistics and probability have their roots in risk assessment—specifically, the risk of losing money while gambling. Risk assessment remains one of the areas in which statistical analysis plays a chief role. A field in which statistical analysis and risk assessment are combined at an advanced level is security. Strategists are beginning to apply the tools of statistics and game theory to understanding problems such as terrorism. Although terrorism was once regarded as an area for study in the social sciences, people are developing ways to control and respond to terrorist events based on statistics. Probability helps strategists take lessons from terrorism in one political context or world region and apply them to another situation that, on the surface, might look very different.

Actuarial science is one of the largest and most thoroughly developed fields within risk assessment. In actuarial science, statistics and probability are used to help insurance and financial companies answer questions such as how to price an automobile insurance policy. Actuaries look at data such as birth rates, mortality, marriage, employment, income, and loss rates. They use this data to guide insurance companies and other providers of financial products in setting product prices and capital reserves.

Quantitative finance uses statistical models to predict the financial returns, or gains, of certain types of securities such as stocks and bonds. These models can help build more complex types of securities known as derivatives. Although derivatives are often not well understood outside of the finance industry, they have a powerful effect on the economy and can influence everyday situations such as a bank's ability to loan money to a person buying a new home.

Survey Design and Execution. Surveys use information gathered from populations to make statements about the population as a whole. Statistical methods ensure that the surveys gather enough high-quality data to support accurate conclusions. A survey of an entire population is known as a census. One prominent example is the United States Census, conducted every ten years to count the country's population and gather basic information about each person. Other surveys use data from selected individuals through a process known as sampling. In nearly all surveys, participants receive questions and provide information in the form of answers, which are turned into mathematical data and analyzed statistically.

Aside from government censuses, some of the most common applications of survey design are customer relationship management (CRM) and consumer product development. Through the gathering of survey data from customers, companies can use customer relationship management to increase the effectiveness of their services and identify frequent problems to be fixed. In creating and introducing new products, most companies rely on data gathered from consumers through Internet and telephone surveys and on the results of focus groups.

Impact on Industry

When statistical methods are introduced to a field, they create a system under which information can be gathered in new ways. The newly gathered information makes it possible for efficiency to be increased rapidly. Participants in the field, such as manufacturers, are forced to adapt to the new environment by taking advantage of these efficiencies. Failure to adapt to the new environment can make old processes expensive, difficult, and obsolete and can force participants out of the field. At the same time, participants can distinguish themselves and their offerings by focusing on the advantages of unique, handmade goods. An example of this process is the manufacturing of clothing, where the market is dominated by mass-produced goods but where couture apparel can be made by hand by specialists for a much higher price.

As disciplines, statistical analysis and probability went through explosive growth fueled by the Industrial Revolution in Europe and North America. This period began in the late 1700's and continued throughout the 1800's as industry, particularly manufacturing and transportation, adapted to the increased use of automation and machines. Before this period, functions such as quality control were overseen by merchants' guilds and the expertise of individuals in workshops because most goods were made or processed by hand.

William S. Gosset, an employee of Guinness's Brewery in Dublin, Ireland, at the beginning of the twentieth century, changed the field of statistics as a discipline when he documented a process known as the t-test. This test compares the results of a sample to a normally distributed population. Gosset's findings allowed manufacturers to measure and control quality within large batches of mass-produced goods.

Walter Shewhart, an engineer who worked at Bell Telephone Laboratories in Washington, D.C., from its founding in 1925, advanced the fields of statistics and probability with his research on statistical controls. Shewhart showed that data on manufactured goods can be measured in more effective ways than comparing the data to a normal curve, a discovery that led to better methods of quality control. W. Edwards Deming, one of the best-known

Fascinating Facts About Probability and Statistics

- In statistical terms, an unexpected event with extreme consequences that—in retrospect—seem obvious is known as a black swan.
- The butterfly effect is a large variance in possible outcomes in a situation that starts with changes on a much smaller scale. The phrase refers to a concept in probability that the flap of a butterfly's wings could, after a string of events, alter a tornado's path.
- A chain of numbers generated one at a time through a random process is known as a drunkard's walk. The name comes from the chain's appearance when plotted on a graph.
- The Monty Hall problem refers to a probability exercise in which a prize is hidden behind one of three closed doors. Two researchers at Whitman College in Washington found in an experiment that pigeons were more skilled at solving the Monty Hall problem than human subjects.
- Clinical trials for drugs often use statistical analysis to compare results to a placebo, such as a sugar pill. The placebo effect refers to statistically significant health improvements experienced by trial participants who are given a placebo. Researchers report that the placebo effect has grown over time, especially in trials of antidepressants, and therefore, direct marketing of drugs to consumers might increase patients' confidence in drug-based medical treatment.
- When statistics appear in the media, one of the most common mistakes is the confusion of "million" and "billion." Researchers believe that, aside from similar spelling, the two words refer to numbers so big that the human brain lacks an intuitive way to tell them apart easily.

names in modern statistical applications, continued Shewhart's work by applying ideas about process control and manufacturing quality to manufacturing in Japanese automakers.

The advance in computing technology revolutionized the fields of probability and statistics. Computers have made it possible for researchers to process and analyze large amounts of data much more rapidly. The science of data mining (extracting patterns from data) evolved from an intersection in the fields of computing, statistics, and marketing. Data mining came to be used in a wide variety of governmental activities as well as in nearly every branch of industry in the late 1990's. As with the Industrial Revolution of the eighteenth and nineteenth centuries, data mining and analysis have changed the world permanently.

The growth of the Internet, particularly the World Wide Web, has also increased the amount of information available for statistical and probability research. This information comes both in the form of raw data and in documented analysis. As of the end of December, 2009, there were more than 1.8 billion Internet users, nearly 27 percent of the world's population. As Internet access and use becomes more widespread, the amount of information generated by users will continue to increase. This increase will lead to new possibilities for statistical analysis based on usage and traffic patterns as well as the content contributed by users.

Probability and statistical analysis have become an integral part of most business and government operations. Many professionals who have a statistical component in their jobs also specialize in another field or function. If enough people hold jobs with a similar specialization, new fields of statistical study can develop. For example, the field of econometrics—the application of statistical theory to economics—has its own academic programs, research centers, associations, and scholarly journals. Econometricians work both in government agencies and in private industry. Biostatistics has become such a diverse area that its growth is more easily measured through the subfields it influences, such as epidemiology and genetics.

Careers and Course Work

The U.S. Bureau of Labor Statistics estimates that about 30 percent of the professionals who specialize in a branch of statistics work for federal, state, and local government agencies, with the rest in private industry and academic research. Of the statisticians employed by the federal government, the highest numbers were found in the Departments of Commerce, Agriculture, and Health and Human Services.

Statisticians and probabilists are likely to hold master's or doctoral degrees. Many hold a second degree in their fields of specialty. College-level course work for a statistician focuses on classes in mathematics, particularly calculus and differential equations. Statistics majors are likely to take courses on topics such as probability theory, statistical methods, mathematical modeling, experimental and survey design, and multivariate data analysis. Much of this course work depends on a skilled use of statistical computer software. Additional classes in computer programming or computer science are useful to many students in statistics and probability.

Although the Bureau of Labor Statistics counted about 23,000 statisticians in 2008, the number of professionals with a statistical component to their work is much higher. These professionals might include analysts, engineers, biologists, agricultural scientists, geologists, or bankers.

Actuarial science as a field is easier to track because of the qualifications needed to practice. There were almost 20,000 actuaries in 2008. More than half of the actuaries (55 percent) surveyed by the Bureau of Labor Statistics were employed by insurance companies, with another 16 percent working for consulting firms offering managerial, scientific, and technical advice. Very few actuaries are employed by government agencies. The earning power provided by an actuarial science degree is high: Median annual wages earned by actuaries in 2008 were $85,000.

To become an actuary in the United States, most professionals attend a four-year college and obtain a bachelor's degree in actuarial science from one of more than one hundred programs available. Course work in an actuarial science program includes mathematics (particularly calculus), probability, statistics, economics, and finance. An internship in the field while in school is encouraged and increases the chances of securing a job after graduation. Before practicing, an actuary must pass a series of licensing exams offered jointly by the two leading professional groups, the Society of Actuaries and the Casualty Actuarial Society. The initial exams can be taken before completion of the degree.

Social Context and Future Prospects

There is an increasing need for professionals with a knowledge of probability and statistics. The growth of the Internet has led to a rapid rise in the amount of information available, both in professional and in private contexts. This growth has created new opportunities for the sharing of knowledge. However, much of the information being shared has not been filtered or evaluated. Statistics, in particular, can be fabricated easily and often are disseminated without a full understanding of the context in which they were generated.

Statisticians are needed to design experiments and studies, to collect information on the basis of sound research principles, and to responsibly analyze and interpret the results of their work. Aside from a strong knowledge of this process, statisticians must be effective communicators. They must consider the ways in which their findings might be used and shared, especially by people without a mathematical background. Their results must be presented in a way that will ensure clear understanding of purpose and scope.

Although overall job growth for statisticians is predicted to remain steady, fields likely to see a higher rate of growth are those involving statistical modeling. An increase in computer software and other tools to support modeling has fueled higher demand for professionals with a familiarity in this area. Statistical modeling is useful in many contexts related to probability. These contexts range from the analysis of clinical trial data for pharmaceutical products to the forecasting of monetary gains and losses in a portfolio of complex financial instruments. Career growth prospects, particularly at the entry level, are most attractive for candidates with a background both in statistics and in their applied fields.

The number of jobs for actuaries is expected to grow more rapidly than for statisticians in general. However, the popularity of actuarial science as an area of study for college undergraduates may produce a surplus of qualified applicants. Insurers are expected to be the primary employers of actuaries for the foreseeable future, but demand for actuaries is growing among consulting firms. As the tools for modeling the impact of large-scale disasters become more sophisticated, companies in many industries will seek expertise in protecting themselves against risk—a trend that may heighten the need for actuarial expertise.

Julia A. Rosenthal, M.S.

Further Reading

Enders, Walter, and Todd Sandler. *The Political Economy of Terrorism.* New York: Cambridge University Press, 2005. A look at the effectiveness of terrorism through the lens of statistics, particularly game theory, and economics.

Fung, Kaiser. *Numbers Rule Your World: The Hidden Influence of Probabilities and Statistics on Everything You Do.* New York: McGraw-Hill, 2010. A discussion of introductory-level probability and statistics illustrated by everyday examples.

Mlodinow, Leonard. *The Drunkard's Walk: How Randomness Rules Our Lives.* New York: Pantheon, 2008. An explanation of the effects of randomness as a statistical concept and the ways in which it is often misunderstood in the media.

Nisbet, Robert, John Elder, and Gary Miner. *Handbook of Statistical Analysis and Data Mining Applications.* Burlington, Mass.: Elsevier, 2009. A thorough overview of the field of data mining, one of the largest applied areas in statistics, including several case studies.

Ott, R. Lyman, and Michael T. Longnecker. *An Introduction to Statistical Methods and Data Analysis.* Belmont, Calif.: Brooks/Cole, 2010. Provides a broad overview of statistical theory along with discussions of data inference in a variety of population types.

Takahashi, Shin. *The Manga Guide to Statistics.* San Francisco: No Starch Press, 2009. A set of introductory lessons in probability and statistics presented in the form of a graphic novel.

Web Sites

American Statistical Association
http://www.amstat.org

Casualty Actuarial Society
http://www.casact.org

Institute for Operations Research and the Management Sciences
Applied Probability Society
http://www.informs.org/Community/APS

Society of Actuaries
http://www.soa.org

See also: Applied Mathematics; Earthquake Prediction; Numerical Analysis; Risk Analysis and Management.

PROTEOMICS AND PROTEIN ENGINEERING

FIELDS OF STUDY

Protein structure and function; genomics; molecular biology.

SUMMARY

Proteomics is the study of an organism's complete set of proteins. The term "proteome" is commonly used to describe all the proteins that are made by an organism's cells; it can also be described as all the proteins that are synthesized by a particular cell at a particular time. Protein engineering is the process of developing useful or valuable proteins for practical use. Protein engineering uses two strategies for engineering proteins: rational design and directed evolution. Both techniques have been developed to synthesize and manipulate proteins. To study proteins, one needs to determine the sequence of each protein's amino acids and its corresponding three-dimensional structure.

KEY TERMS AND CONCEPTS

- **Affinity Chromatography:** Technique that identifies targets for proteins by attaching a protein to a solid matrix in a glass column that will either wash out after a solution has been added or stick to a target, depending on the nature of the protein.
- **Amino Acids:** Building blocks of proteins; there are twenty-one amino acids, each made of only seven to twenty-four atoms.
- **Directed Evolution:** Process of making random mutations to a protein and then selecting the variant proteins that have the qualities of interest.
- **Genome:** Complete set of genes in an organism, which is used as a coding template for protein synthesis.
- **Mass Spectrometry:** Technique of ionizing chemical compounds to generate charged molecules or molecule fragments and measuring their mass-to-charge ratios to determine the elemental composition of a protein.
- **Protein Chip:** Library of hundreds or thousands of different proteins for an organism individually placed on a known location of a "chip."
- **Protein Folding:** Physical process by which a polypeptide chain will fold into a three-dimensional shape to give rise to a functional protein.
- **Proteome:** Complete set of proteins synthesized in a cell.
- **Rational Protein Design:** Process of making desired changes to a protein based on the knowledge of the structure and function of that protein.

Definition and Basic Principles

Proteomics is the study of proteins: their structure and function and their interactions with other proteins. Proteins are made from the primary sequence of deoxyribonucleic acid (DNA), which is then transcribed into a messenger RNA (mRNA) molecule, which in turn is translated into polypeptide chains that will form a three-dimensional functional protein product. The study of proteins is complicated—more so than the study of genomics—because of the complexity of modifications that take place for the DNA sequence to become a protein product and because the proteome changes from cell to cell and at different time periods in a cell or organism's life.

Once the primary sequence of DNA forms an mRNA molecule, it is then translated into a polypeptide chain, in which posttranslational modifications (chemical modifications after translation) take place, which change the functional nature of that protein product. Specific posttranslational modifications include such processes as phosphorylation, ubiquitination, methylation, acetylation, glycosylation, oxidation, and nitrosylation. Although these processes may not be familiar to most people, all are processes that have the capability of modifying the organism's DNA to create various protein products. Once these changes in the protein take place, each cell may or may not express that protein because of variables such as the time in the organism's life, whether a functioning protein is needed in the type of cell in which the protein resides, or whether the specific conditions of the cell are conducive for a functioning protein. Also, proteins may have to communicate or interact with other proteins to become functional.

With all these possibilities to express certain proteins, the study of proteomics and the proteome is quite extensive and complex. However, understanding proteins and their functions will make it possible to design new drugs for the treatment of diseases and will provide researchers with a better knowledge of how cells work, how they interact with other cells, and how these interactions relate to a cell's ability to survive.

Background and History

The term "proteome" was coined by Marc Wilkins in 1994 as part of his doctoral thesis work on two-dimensional gel electrophoresis on proteins. Gel electrophoresis is a procedure that allows researchers to visualize the presence or absence of a protein in a specialized medium. Wilkins used this term to describe the entire complement of proteins expressed by the genome, which can be described as either all the proteins in a cell or an entire organism. The term "proteomics" was coined in 1997 to describe the study of proteins in the genome, thereby combining the two terms "protein" and "genomics" into one. When the Human Genome Project was completed in 2003, the entire human genome had been mapped. With the knowledge of the human DNA sequence, the next most logical step was to turn to the set of proteins in a cell or organism.

As the discipline of proteomics began, proteins were first studied by analyzing messenger RNA. Messenger RNA (mRNA) is a particular form of RNA that transcribes the genetic instructions of DNA to proteins for later expression. In early proteomic research, scientists took DNA, converted it into mRNA through transcription, and then studied the resulting mRNA to look for its protein product. It was found, however, that mRNA does not always correlate directly to protein content. First, mRNA may not always translate into a functional protein; second, the amount of protein produced from any mRNA differs based on the gene from which it is transcribed and how much protein that cell may need at any one time; third, the protein product may differ extensively from the original mRNA message based on posttranslational modifications; and fourth, proteins may need to interact with other proteins to be functional. All these complex features in studying the proteome have made new methods of study necessary, and new approaches to studying proteins have become available.

How It Works

With the advent of new molecular and cell biology technologies, which began with the Human Genome Project in the 1990's, the study of proteins in the early 2000's advanced significantly. Different technologies were developed, and the data generated from these studies grew so large that bioinformatic systems were developed to store all the data for interpretation and sharing with the scientific community at large. Protein studies have numerous steps and various technologies for isolating the proteins and analyzing them.

Isolation of Proteins. The first approach to the study of proteins is to disrupt cells and separate out the proteins from the other particles in the cell. Cells are first treated with a detergent or urea so that enzymes will not degrade the proteins. These steps can be accomplished using kits that will perform a whole-cell extraction, producing small pieces from which one can isolate the proteins of interest, for example, isolating the hydrophobic proteins that are involved as membrane proteins.

Separation and Purification. Once the proteins have been isolated, one can then begin to separate and purify them. A common method is to use affinity or liquid chromatography to separate out specific proteins or a family of related proteins through the binding of proteins to specific substrates, which will elute out specific proteins from a mixture. Certain technologies have made it possible to extract very small quantities of proteins from a mixture.

Proteins can also be labeled with an isotope to measure the quantity of protein separated by using mass spectrometry. Tags can also be peptide derived rather than isotopes; peptide-tagged protein can be isolated and purified directly when the proteins to be studied are already known. One can then gather information on the abundance of proteins in cells.

Two-Dimensional Gel Electrophoresis. Two-dimensional gel electrophoresis was developed in the 1970's and has enhanced the ability to separate out specific proteins. This method puts a mixture of proteins through two gels or dimensions. The first dimension separates proteins by isoelectric identification using a pH (acidity-alkalinity) gradient, and the second dimension is SDS-PAGE (sodium dodecylsulfate-polyacrylamide gel electrophoresis), which separates proteins by their molecular weight (larger molecules move slowly through a gel and smaller proteins move quickly and farther along the gel). A number of companies sell kits to perform two-dimensional gel electrophoresis, making these steps easier than in the past.

Protein Identification and Mass Spectrometry. Once the gels have been run, interpreting the data may be challenging. The data appear as spots for each of the protein products that have been separated. To identify these spots, certain dyes—fluorescent dyes or blue or silver stains—are used to stain specific proteins.

Analyzing different proteins is generally accomplished with mass spectrometry. Proteins are usually ionized by different processes. One then can use computer software to analyze the data to compare each of the proteins expressed.

Microarrays. Protein microarrays take a mixture of proteins and add them to antibodies, antigens, enzymes, substrates, membrane receptors, or ligands, which can recognize the protein with which they associate or interact. These specific proteins will light up on an array, or chip, that identifies specific proteins from the thousands of proteins in the mixture. This technique, which is being used but is still in development, is a powerful tool and has the potential of eliminating numerous laborious steps being used to identify proteins.

Applications and Products

Proteomics has immense potential for a broad number of practical applications in medicine and the pharmaceuticals industry.

Basic Research. One major area of proteomic research is to understand how the amino acid primary sequence specifies the stability and dynamics of protein conformation. This research would provide information on how to design novel functionalities of proteins and on disorders that work by changing a protein's three-dimensional structure, such as amyloidosis and prion diseases. Studying the folding and unfolding of proteins will help scientists understand the three-dimensional structures of proteins.

Disease Detection. One major application of proteomics is for disease detection. The National Cancer Institute is working on ways that proteomic technologies can be used for the detection of important proteins seen in disease. This technology would use biomarkers and target proteins that can be used to detect known diseases, including cancers, autoimmune disorders, and inflammatory diseases, as well as to screen for allergies.

Biomarkers that can be used for early detection of disease include gene mutations, gene transcription and translation modifications, and alterations in protein products. By looking at free DNA in serum, clinical testing has developed serum screens, and with the addition of biomarkers, testing has been expanded to use oncogene mutations, microsatellite instability regions in the DNA, and hypermethylation of promoter regions in DNA for the detection of cancer.

Clinical testing techniques using proteins as the basis for detection of disease include western blotting, immunohistochemical staining, enzyme-linked immunosorbent assay (ELISA), and mass spectrometry.

One such approach to disease detection is the use of proteomic pattern diagnostics to detect cancer. The first report of using this technique coupled with mass spectrometry was to identify ovarian cancer. More than two-thirds of ovarian cancer is not detected until it is in an advanced disease stage. Early detection is essential if this disease is to be treated and cured successfully.

One detection method is to identify discriminatory patterns of proteins that indicate ovarian cancer. Serum (obtained through a noninvasive procedure) from both normal women and those with ovarian cancer shows distinct patterns of proteins. This approach can be used for other cancers and diseases once the protein patterns are established.

Disease Treatment Though Novel Drugs. The development of novel drugs using proteomics for disease therapy holds tremendous potential. The major step forward is to identify proteins associated with disease, which can then be used as targets for drug development. First, a protein that is implicated in a disease must be identified. Then, drugs can be designed based on the protein's three-dimensional structure to interfere with the action of the protein. This can be achieved by developing molecules that fit into the protein at the active site to stop the enzymatic reaction that would normally occur. Inactivation of enzyme function may also be a means of developing personalized drugs based on different individuals' genetic, and thereby, protein makeup. With the advent of computer databases of protein structures, computer techniques can fit different molecules into a three-dimensional structure called virtual ligand screening.

For cancer diseases alone, drug development is a major area of scientific endeavor, and the industry continues to grow. For example, the development of novel proteins through protein engineering has led to new drugs. Top7, a fusion protein, was used to create an interleukin 1 blocker, Arcalyst (rilonacept), which was approved by the U.S. Food and Drug Administration in 2008 for the treatment of cryopyrin-associated periodic syndromes.

Protein Engineering. The field of protein engineering involves developing and creating useful or important proteins. Protein engineering encompasses two main strategies: rational design and directed evolution.

In rational design, mutations are induced to make changes to a protein to change the structure or function of that protein. Because mutagenesis techniques are well established, this technique has great potential; however, the difficulty in its success is that a protein's structure and function may not be known in detail. For proteins that are not well defined, it may be impossible to determine what mutations to incorporate.

Random mutations may be induced in a protein, and the variant proteins with the desired qualities sorted out from the protein mix. Once the variant proteins have been isolated, further mutations and selection of those variants are performed. This process of mutations and selection is called directed evolution because it simulates natural selection and may produce more fit or successful proteins. Another technique used in directed evolution is DNA shuffling, which mixes and matches pieces of variants to produce better protein products. This process is similar to the recombination process that occurs naturally in individual cells during sexual reproduction. One advantage of directed evolution is that it does not require prior knowledge of a protein or the knowledge of which mutation to induce. Rather, mutations are randomly induced, and those mutations are monitored to see what effect they have on the protein expression in a cell. The difficulty of directed evolution is that it requires high throughput, in which large amounts of DNA must be mutated and the protein products monitored for the desired qualities. This process may not be feasible for all proteins.

Impact on Industry

When the Human Genome Project was completed in 2003 and the entire human DNA sequence was decoded, the next project became the decoding of the human proteome. Out of the desire to be the first to decode the human proteome, many startup companies and established pharmaceutical and biotechnology companies worldwide, along with governmental agencies and universities, have begun spending billions of dollars in proteomic research. This field of research requires high-throughput technologies that still need to be developed and extensive data analysis to accomplish this goal. It also requires collaboration from all sectors, including government, universities, and industry.

Government and University Research. Governments and universities around the world are participating in the field of proteomic research, with major contributions from the United States, Canada, European countries, Japan, China and other industrial countries. In the United States alone, the funding for proteomic research from the National Institutes of Health exceeds $125 million, about 0.6 percent of all the awards granted by the institute in 2008. Canada has a National Research Council-sponsored proteomics laboratory at the Institute for Biological Sciences in Ottawa, where the proteomics team carries out proteomics and glycoproteomics research. Other organizations around the world include the Human Proteome Organisation, the British Society for Proteome Research, the Swiss Proteomic Society, the European Proteomics Association, the Italian Proteomics Association, and the Germany Society for Proteome Research. Worldwide, national, intergovernmental, and nonprofit agencies are recognizing the opportunity to increase the return on their research grants by requiring that the findings of research are made freely accessible on the Internet. This allows for free collaboration between agencies and the for-profit market. Universities that receive funding for proteomic research have been instrumental in leading the way toward major discoveries, and the collaborative research with commercial biotechnological firms has paved the way for quicker utilization of these discoveries.

Industry and Business Sectors. Companies worldwide are investigating new methods of proteomic research to improve the high throughput of protein identification and characterization. One major trend in the proteomics market is the development of laboratory equipment and biological reagents. Proteomic research has been expanded mainly through the use of highly coordinated laboratory equipment that isolates, separates, and visualizes proteins in the laboratory. Two-dimensional gel electrophoresis remains the most important technology because it is used to separate proteins from cells or tissues. Microarrays, which measure the expression or activity of proteins, may well advance this field further. The need to identify and characterize proteins gave rise to the creation of laboratory equipment that combines mass spectrometry and chromatography. These large, expensive, and complicated pieces of equipment are generally marketed through the business sector. The technology for this field must also be able to analyze the data stores in large databases open to the public. Such databases tend to be collaboration between university researchers and private industry. Proteomics is still emerging as a field of research, allowing an abundance of room for startup companies to establish new instrumentation and manufacture biologic products for the instruments in this growing world market.

Major Corporations. More than three hundred companies are involved in proteomics worldwide, and many play a large role in the advances of this field. The major companies are involved in instrumentation, biologics and biotechnology, pharmaceuticals, and bioinformatics. Some of the key companies are Amersham Biosciences together with Thermo Scientific, producing integrated systems for laboratory instruments; Applied Biosystems, producing various laboratory instruments; Micromass-Waters, involved in scanning electron microscopy and together with Bio-Rad Laboratories producing integrated systems for laboratory instruments and reagents; Proteome Systems, developing protein chips and a collaborative effort in bioinformatics with IBM; MDS Sciex, producing mass spectrometry and liquid chromatography systems; Takara Bio, making kits and reagents for molecular biology research; Roche Applied Science, producing high-throughput instrumentation and reagents for proteomic research; and PerkinElmer Life and Analytical Sciences, developing proteomic and array systems.

Fascinating Facts About Proteomics and Protein Engineering

- Human cells function through the concerted action of thousands of proteins that control their growth and differentiation, yet the specific function of most proteins is either unknown or poorly understood.
- Scientists estimate that once protein structure and function are understood, about 98 percent of disease therapies will target proteins in an individual's cells.
- In 1987, protein engineering enabled scientists to create two vaccines for hepatitis B: Engerix-B and Recombivax.
- Proteomics may be the method of choice for the early detection of disease by developing unique profiles for each individual genomic DNA and disease entity.
- Proteomics has demonstrated tremendous promise for providing early detection, disease risk assessment, and treatment protocols for chronic diseases such as cancer, chronic obstruction pulmonary disease, and asthma.
- The number of companies involved in proteomics has increased remarkably in the twenty-first century, with more than three hundred companies worldwide in the field.
- The surge in proteomic research is accompanied by a growing market value. In 2008, according to Front Line Strategic Consulting, the market value was $2.68 billion and growing rapidly.

Careers and Course Work

The field of proteomics is growing rapidly and offers exciting new careers in research for individuals with bachelor's through doctoral degrees. Careers in the field can be as diverse as medical doctors, doctorate-level researchers, and laboratory technicians. The number of laboratories that conduct protein research and therapeutics is growing rapidly, and funding for this type of research is expanding.

To enter the field of proteomics research, a person must first acquire a bachelor of science degree and conduct basic laboratory work. A master's degree in biology or a field of biology, such as genetics, microbiology, or molecular biology, may help in the competitive labor market. Research and laboratory technicians do much of the bench work. A doctorate is required to reach a position with decision-making power, such as becoming the director of a laboratory. Medical doctors often participate in research and are also involved in how the research is applied to patients in a clinical setting. The job titles most often seen in this field include medical doctors, principal investigators in research, laboratory directors, medical technologists, research technologists, laboratory technicians, and college professors involved in research.

Social Context and Future Prospects

With the coupling of advanced instrumentation and new technologies and informatics tools, the field of proteomics has great potential in disease detection, drug development, and basic research on proteins.

Another major aspect to the future of proteomics is whether a patent system that facilitates research and can effectively change with the dynamics of the field of genomics and proteomics can be developed. A definite conflict of interest exists between for-profit industries and nonprofit research institutes. This makes research studies difficult to finish and publish so that other researchers can freely access the results. This dilemma is worldwide and leads to higher costs of doing research and difficulties in converting research results to clinical useful applications. Intellectual property laws and patenting are difficult issues and becoming more complex and burdensome with time.

The issue of how to use systems biology, including proteomic research, to improve the health of individuals is a major priority. It is becoming increasingly apparent that proteomics will have a major role in creating a predictive, preventative, and personalized approach to medicine. This raises the question of how individualized medicine will be handled by insurance companies and whether there will be disparity between individuals in their access to individualized medical procedures and therapies. The idea that genomic and proteomic research may outpace societal change must be recognized and steps must be taken to address all issues.

Susan M. Zneimer, Ph.D.

Further Reading

Alberghina, Lilia. *Protein Engineering in Industrial Biotechnology.* Amsterdam: Hardwood Academics, 2000. An overview of the applications of proteins and how protein engineering can be used to solve problems in industrial biotechnology. Describes how protein engineering enhances purification of recombinant proteins and how it is applied to health care, including the development of new vaccines.

Hamacher, Michael, et al., eds. *Proteomics in Drug Research.* Weinham, Germany: Wiley-VCH, 2006. Contains information on technologies and applications, particularly in pharmaceuticals.

Liebler, Daniel. *Introduction to Proteomics.* Totowa, N.J.: Humana Press, 2002. Discusses key concepts in proteomics, including the methodology and instrumentation used. Describes the applications of mass spectrometry and provides an excellent introduction and overview of proteomics.

Mishra, Nawin C., and Günter Blobel. *Introduction to Proteomics: Principles and Applications.* Hoboken, N.J.: John Wiley & Sons, 2010. Introduces the field of proteomics and examines the role proteomics plays in the study of biological systems in general and disease in particular. Provides an understanding of the structure, function, and interactions of proteins and how they are used for identifying diseases and developing new drugs.

Pennington, Stephen, and Michael Dunn, eds. *Proteomics: From Protein Sequence to Function.* London: Garland, 2001. Gives an overview of how proteins can be identified, and used for clinical procedures and research. Discusses the technologies used in the field and protein structure and function.

Wulfkuhle, D. J., et al. "Proteomic Applications for the Early Detection of Cancer." *Nature Reviews* 3 (2003): 267-275. Describes cancer detection with a great review of techniques used in proteomics.

Web Sites

Human Proteome Organisation
http://www.hupo.org

National Cancer Institute
Clinical Proteomic Technologies for Cancer
http://proteomics.cancer.gov

National Institutes of Health
Proteomics Interest Group
http://proteome.nih.gov

See also: Bionics and Biomedical Engineering

PYROTECHNICS

FIELDS OF STUDY

Chemical engineering; chemistry; military science; physics; rocketry.

SUMMARY

Pyrotechnics is the science of controlling exothermic chemical reactions, using materials that are self-contained and self-sustained, to produce gas, heat, light, sound, and smoke. These types of exothermic chemical reactions have been developed in a variety of ways: to provide entertainment, promote safety, and generate compounds such as oxygen. Such reactions are used, for example, in fireworks, military ordnance, automotive air bags, chemical oxygen generators, pyrotechnic fasteners, and safety matches. Reactions can be carefully controlled and timed to automate remote actions or effect precisely timed sequences, such as those used in spaceflight. Those who study and practice in this field are called pyrotechnicians.

KEY TERMS AND CONCEPTS

- **Automotive Air Bag:** Passive automotive restraint consisting of a bag made of flexible material that inflates upon sensing a collision and prevents passengers from pitching forward or from side to side, depending on the type and position of the air bag.
- **Chemical Oxygen Generator:** Device that stores oxygen in a chemical form and then, after a chemical exothermic reaction is triggered, continuously releases oxygen. Often used in situations when oxygen in an area is depleted, such as a fire, mine, or depressurized airplane cabin, or where oxygen is not available, such as a submarine or spacecraft. Also called a chlorate or oxygen candle.
- **Endothermic Chemical Reaction:** Chemical reaction that needs heat to occur and absorb energy.
- **Exothermic Chemical Reaction:** Chemical reaction releasing heat because the energy generated by the reaction is greater than the energy needed to cause the reaction.
- **Explosive Bolt:** Bolt (of a nut-and-bolt combination) or other type of fastener containing an explosive charge that is detonated remotely and, when detonated, causes the bolt to shear and break in a certain way and at a certain time. Also called a pyrotechnic fastener.
- **Fireworks:** Variety of devices constructed for amusement containing explosives and combustibles that combine with other chemicals to generate colored lights, smoke, and noise, often deployed in series or in combination to produce aesthetically pleasing effects.
- **Safety Match:** Match specially designed to light only when struck against a specially prepared surface.

Definition and Basic Principles

The ever-changing field of pyrotechnics encompasses a broad range of exothermic chemical reactions that scientists and inventors have put to many uses. To the ordinary observer, the most visible use of an exothermic chemical reaction is that of fireworks for entertainment. However, this field has also been applied in many other areas, particularly to promote safety in air travel, spaceflight, mining, and firefighting.

This field of pyrotechnics does not include endothermic reactions, those that need energy to take place, such as an ice cube melting or a liquid evaporating. It also does not include anything related to the concept of "fire" in general, that is, in the sense of lighting things on fire.

Background and History

Fireworks are said to have originated in China more than 2,000 years ago. The story is that a cook mixed together charcoal, sulfur, and saltpeter (all materials commonly found in kitchens thousands of years ago) in a bamboo tube. When the tube filled with these substances got too close to a fire, it exploded. A Chinese monk, Li Tian, then took this explosive mixture a step further. He attempted to find ways of controlling the explosion, thus creating firecrackers. Li Tian lived in Liu Yang, in the Hunan province of China, which is still a main center for the production of fireworks that are shipped around the world. Lighting firecrackers became a common event around holidays such as Chinese New Year, as ghosts and evil spirits are thought to be frightened away by the loud noises of firecrackers. These exploding devices are still used at events marking births, deaths, and the New Year in Asia as well as other holidays celebrated around the world, including American Independence Day.

The explorer Marco Polo is believed to have brought the recipe for gunpowder—the combination of charcoal, sulfur, and saltpeter created by the Chinese cook—back to Europe after his visit to China in the thirteenth century. The Crusaders, who traveled from Europe to the Holy Land (modern Israel and Palestine) to fight Muslims and establish Christian rule, may also have been familiar with gunpowder after their trips to the Islamic world. Europeans began to use gunpowder for military purposes, but Italy was the first country where gunpowder was used to manufacture fireworks for display and entertainment. The Germans followed, and then the English, who developed fireworks and took them to new heights. Queen Elizabeth I created a highly regarded position in her royal court called the Fire Master of England, and King James II knighted his Fire Master for developing a brilliant fireworks display that celebrated James's coronation. William Shakespeare mentions fireworks in his writings, and George Frideric Handel is famous for composing music for the royal fireworks for King George II.

As scientists explored the chemical reactions occurring when certain substances were combined, they began to find uses for them beyond their entertainment value. For example, Swedish scientist Johan Edvard Lundström used this type of chemical reaction in his patent for safety matches in 1855. He put red phosphorus in sandpaper on the outside of a matchbox and the rest of the ingredients necessary to complete the chemical reaction he was seeking on the match heads inside the box. These special matches lit only when struck on the phosphorus-loaded sandpaper surface. These reactions were further studied and developed by scientists and inventors to bring about safety devices such as chemical oxygen generators and automotive air bags and controlled explosions such as those used in pyrotechnic fasteners.

How It Works

Pyrotechnics relies on exothermic chemical reactions, those that release heat or energy. A common example of an exothermic chemical reaction would be a fire. However, there are many other examples, such as the release of oxygen on demand, using a controlled chemical reaction other than fire. Controlled explosions that work underwater or in remote areas are another type of exothermic chemical reaction, representative of the field of pyrotechnics work.

Fireworks. The way pyrotechnics works can be detailed by the familiar use for fireworks. Fireworks consist of a shell composed of four parts: a container (a cylinder of paper and string), stars (balls made of a sparkler-type substance about the size of a pea), the bursting charge (similar to a firecracker), and a time-delayed fuse (to ensure the shell is at the correct height before exploding). The stars are packed into the container with, generally, a tube full of bursting charge through the center and more bursting charge (a gunpowder-like substance) sprinkled throughout. This shell is launched from a mortar (usually a short steel pipe) by a lifting charge exploding into the tube, where nitrogen and carbon dioxide expand and fling the firework into the sky. This lifting charge also lights the shell's fuse, which burns slowly until the shell reaches a safe altitude. It is also possible, but not as common, to use compressed air to launch fireworks and an electronic timer to detonate the charge.

These shells can be arranged as "multi-break shells," which explode in phases, similar to a two-stage rocket, and possibly contain stars of different types or colors. They may also contain noisemakers such as whistles. Each of these shells has a different time-delayed fuse that lights it so each shell can explode with, for example, brighter light or more sparks than the one before. These multi-break shells can be burst apart to explode farther away from each other and create designs by break charges (small explosive charges). The patterns created depend on the placement of the stars in the shell and the timing of the break charges that ignite them. For example, to create a heart pattern, the shell would be packed with stars in a heart shape, then a break charge would be tied to this group of stars to separate them all from the main shell at the same time.

Fireworks get color from basic chemical elements as they burn. Common colors produced are blue (copper), green (barium), red (strontium), white (magnesium), and yellow (sodium). Some of these elements are unstable and must be combined with other ingredients to maintain stability.

Chemical Oxygen Generators. A typical chemical oxygen generator contains a mix of sodium chlorate and iron powder. When this mixture is ignited, either by an automatic or manual switch, it smolders, releasing sodium chloride, iron oxide, and oxygen at a fixed rate through thermal decomposition. The most familiar type of chemical oxygen generator is probably that used on a commercial airliner. If the cabin of the airplane becomes decompressed, panels containing oxygen masks connected to a chemical oxygen generator open. When passengers pull on the mask, the retaining pins fall away and ignite the mixture, releasing oxygen.

Applications and Products

Fireworks. Fireworks, a class of low-explosive pyrotechnics, are an integral part of many cultural, religious, and patriotic celebrations around the world. They range from small devices that can be used by one person or a small group, such as firecrackers, to large displays for hundreds of people. They produce four primary effects: light, noise, smoke, and the dispersal of floating materials, such as confetti. They can be designed to burn in many colors, including red, orange, yellow, green, blue, purple, and silver. They are generally classified as either ground or aerial fireworks. Aerial fireworks may be skyrockets, which have their own propulsion system, or they may be designed using mortar shells and are shot into the air, where they explode. Fireworks are usually created using a pasteboard casing filled with some type of combustible material. Many casings are combined with different types of materials and colors to create shapes when exploded.

Safety Matches. Safety matches are used just about everywhere. Their ability to keep from spontaneously igniting makes them useful in safely lighting fires, such as those in fireplaces or barbecues. These matches are produced in such a way that they must be struck against a particular surface, setting off a chemical reaction that allows them to ignite. A safety match head contains sulfur and an oxidizing agent such as potassium chlorate. Other ingredients, such as powdered glass, color, and a binder made of glue and starch, keep the match head together until it is struck. The surface upon which the match must be struck may be powdered glass or sand and red phosphorus. The glass struck against glass generates heat that converts the red phosphorus to white phosphorus, a vapor that spontaneously ignites. This ignition liberates oxygen and causes the sulfur to burn and ignite the wood of the matchstick.

Chemical Oxygen Generators. Also known as oxygen candles, these generators hold a chemical reserve of oxygen that is released through an exothermic chemical reaction. They are used in places

Fascinating Facts About Pyrotechnics

- "Phossy jaw" was a disease caused by matches made in the 1830's with white phosphorus, which is poisonous. Those who made the matches or even used them frequently were crippled by deposits of phosphorus in the jaw. These deposits caused abscesses, swelling in the gums, and bones that glowed in the dark. This problem was solved with the introduction of red phosphorus, a safer compound that became common in matches.
- People who make and handle fireworks wear cotton clothes (down to their underwear) because the static electricity generated from synthetic fibers can ignite fireworks.
- For the 1996 Chinese New Year celebration in Hong Kong, a string of firecrackers was lit with the explosions lasting 22 hours.
- The Liu Yang region of the Chinese province of Hunan is the traditional birthplace of fireworks. It is still the main production area for fireworks worldwide.
- Self-contained rescue devices, a type of chemical oxygen generator, are used in mines (usually coal mines) to provide oxygen to miners after a fire or an explosion. However, they can be dangerous when stored because of their explosive capabilities and have been implicated in some mine disasters. They have also been targeted as a source of explosions on submarines.
- International Space Station crew members each use about one chemical oxygen generator per day to provide oxygen to them in space.
- The powder in an automotive air bag is talc or cornstarch, which keeps the bag flexible while it is stored in its case.
- A front-installed automotive air bag inflates and deflates in 0.04 seconds. A side bag must begin inflating in five milliseconds to protect a passenger effectively.

such as airplanes, mines, submarines, or a space station to hold a large amount of stored oxygen in a small space and in a lightweight form. For example, a chemical oxygen generator weighing about 40 pounds can provide about four days of oxygen. This type of generator usually contains a sodium chlorate pellet (but may contain inorganic superoxide or perchlorate) and an igniter, such as a firing pin, which is activated by a pull tab. This igniter can be triggered in a pyrotechnic reaction activated by friction or impact. Sodium chlorate gives up its oxygen as easily as almost pure oxygen when heated in a chemical reaction. As it decomposes (it does not truly burn), oxygen is released. This oxygen can then be used as a rescue inhaler, such as in a mining or firefighting situation, or be mixed with other gases to create an air-like environment in places such as a space station or submarine.

Explosive Bolts and Fasteners. Pyrotechnic fastener is another term for explosive bolt. This type of fastener may be a nut-and-bolt combination or any other type of fastener. It is made to be broken and incorporates a type of pyrotechnic charge that is detonated remotely. These kinds of fasteners are used in places where it is extremely important that the break occur at a certain predetermined time and place. An explosive bolt or nut is usually scored around the place where it needs to break. These kinds of breakable fasteners are used in the space shuttle, where a charge directed at the nuts around the bolts holding the space shuttle to the pad is detonated to release it at a certain point during the countdown. They are also used for separation during rocket stages. This type of fastener is more reliable than mechanical types of fasteners and are often lighter in weight and easier to control. Some typical chemical combinations used for this type of explosive are manganese, barium chromate, and lead chromate, or boron and potassium nitrate. The type of chemicals used depends on the amount of energy required to sever the connection and the burn rate desired. The detonator may be a blasting cap, though now it is possible to detonate with pulsed laser diodes and fiber-optic cables.

Automotive Air Bags. Engineers who developed air bags for automobiles had considered the idea for quite some time. However, it was not until small solid-propellant inflators were developed in the 1970's that these types of safety devices became a real possibility. These inflators allowed a chemical reaction (sodium azide reacting with potassium nitrate) that produces nitrogen gas to inflate the bag. The nitrogen gas explodes in hot blasts to inflate an air bag at up to 200 miles per hour, then quickly dissipates through small holes, deflating the bag. The whole process takes place in one-twenty-fifth of one second.

Impact on Industry

Exothermic chemical reactions are in use in many industries, including the automobile, safety, rocket, and mining industries. The government is interested in technicians and engineers who can provide these types of reactions for military and aerospace purposes. Theater and film are another possible industry for pyrotechnicians. Safe and exciting explosions are used in special effects for these industries.

Careers and Course Work

Two types of jobs in the pyrotechnic field are pyrotechnic engineers and pyrotechnicians. Pyrotechnic engineers design and create the explosions used in fireworks manufacture. They have a comprehensive understanding of the chemistry and physics used to craft a chemical reaction that is safe and aesthetically pleasing as well as training specific to creating fireworks. Pyrotechnic engineers may also work at developing pyrotechnic fasteners or automotive air bags.

A pyrotechnician is someone who organizes and sets off the fireworks at a performance. These individuals have a specific knowledge of safety rules and regulations as well as governmental requirements for this type of show. A Certified Display Operator certificate is often required for anyone performing in the fireworks field.

Many possibilities for employment in this field exist outside of the fireworks industry. Pyrotechnicians work in the field of rocketry, building and testing safe ways to detonate the type of explosions that are needed to launch rockets safely. Other possibilities include the safety field, which includes making and improving air bags for the auto industry and improving chemical oxygen generators for the auto, mining, and aircraft industries. Another possible field is that of special effects for television and movies, where one develops pyrotechnic displays or explosions that are safe for actors while the cameras roll. The military is another possible field for a pyrotechnician. Safe, controlled explosions that give military personnel an opportunity to experience them firsthand are necessary for training purposes.

Social Context and Future Prospects

Safety matches, fireworks, oxygen generators, and automotive air bags are part of everyday life; however, one does not often stop to think about the chemical reactions that must take place to make these possible.

Though safety matches have not changed for many years, new and more efficient ways to generate a safe chemical reaction are worth investigating.

Fireworks are a constantly changing field, and those who can engineer or deploy fireworks are always in demand. Engineers who can design interesting and beautiful displays would be challenged in this field. Deployments of pyrotechnics in the military field are changing as well, and those who can develop and improve safe ways to deploy ordnance are in demand.

Scientists, engineers, and researchers are constantly changing the way chemical oxygen generators are made and used to improve safety. New methods of igniting or beginning the chemical reaction in a safe way are being investigated, and new or different chemicals that combine to release oxygen easily can be found.

Auto air bags are constantly changing. In the beginning, most of the attention was focused on front and rear impact, even though 40 percent of serious injuries resulted from side impact. Many cars now offer side air bags, though the engineering is difficult. Head air bags are another type of air bag under development. Different ways of installing and triggering these types of bags are needed to improve safety in the future.

Though pyrotechnic fasteners are not generally part of one's everyday life, for those in the rocketry or underwater-construction field, engineering to make these chemical reactions safer and more effective is always welcome.

Marianne M. Madsen, M.S.

Further Reading

Agrawal, Jai Prakash. *High Energy Materials: Propellants, Explosives and Pyrotechnics.* Weinheim, Germany: Wiley-VCH, 2010. Covers high-energy materials from the 1950's to the present; discusses chemical and thermodynamic basics and focuses on safety.

Chan, Ching-Yao. *Fundamentals of Crash Sensing in Automotive Air Bag Systems.* Warrendale, Pa.: Society of Automotive Engineers, 2000. Includes discussions of how sensors and explosive elements must coordinate to produce safe and effective air bag deployment.

Conkling, John A., and Christopher J. Mocella. *Chemistry of Pyrotechnics: Basic Principles and Theory.* 2d ed. Boca Raton, Fla.: CRC Press, 2011. Covers basic chemical and safety principles, describes components of high-energy mixtures, and discusses production of smoke, sound, light, and color.

Donner, John. *A Professional's Guide to Pyrotechnics: Understanding and Making Exploding Fireworks.* Boulder, Colo.: Paladin Press, 1997. A "recipe" book of firecrackers with descriptions of how to assemble basic devices with simple diagrams.

Janssen, Thomas J. *Explosive Materials: Classification, Composition and Properties.* Hauppauge, N.Y.: Nova Science Publications, 2011. Discusses materials that store energy in amounts that can produce an explosion, such as chemical energy, pressurized compressed gas, and nuclear energy. Includes design of energetic materials, effects of temperature and humidity on explosive materials, and detection of post-blast materials.

Lancaster, Ronald. *Fireworks: Principles and Practice.* 4th ed. Gloucester, Mass.: Chemical Publishing Company, 2005. Considered the bible of pyrotechnics; contains overviews of principles with an extensive reference list.

Surhone, Lambert M., and Miriam T. Tennoe. Susan F. Henssonow, ed. *Pyrotechnic Fastener: Fastener, Explosive Material and Electricity.* Beau Bassin, Mauritius: Betascript Publishing, 2010. Includes descriptions of types of pyrotechnic fasteners, particularly those used in the aerospace industry.

Web Sites

American Pyrotechnics Association
http://www.americanpyro.com

Fireworks Foundation
http://www.fireworksfoundation.org

Pyrotechnics Guild International
http://www.pgi.org

See also: Chemical Engineering; Spacecraft Engineering.

Q

QUALITY CONTROL

FIELDS OF STUDY

Metrology; statistical process control; project management; manufacturing technology; quality assurance; pharmaceutical production; pharmaceutical testing; biochemical analysis; medical laboratory procedures; agricultural research and testing; agronomy.

SUMMARY

In its broadest sense, the concept of quality control refers to a process that ensures that the physical result of one's work matches the design concept of the work as closely and as consistently as possible. As goods began to be produced in large quantities, the importance of standardization, which creates easily interchangeable parts, and quality control, which ensures that goods are produced according to those standards, became apparent. The ideas, concepts, and practices for achieving quality control are readily transferable from one field to another. Thus, quality control as a working concept applies equally well to the preparation of drug compounds in a laboratory as to the manufacture of cast magnesium engine parts in a factory.

KEY TERMS AND CONCEPTS

- **Measurement Accuracy:** Degree of agreement between the true value and the measured value of a measurable quantity.
- **Precision:** Degree to which measurements of a quantity vary under the same conditions, sometimes referred to as the repeatability of a measurement.
- **Random Error:** Unavoidable errors in the measurement of a quantity.
- **Standard Deviation:** Value of the square root of the variance of a set of observations.
- **Statistical Process Control (SPC):** Use of statistical data from the ongoing evaluation of the output of a process to ensure process accuracy and precision.
- **Systematic Error:** Avoidable errors in the measurement of a quantity.
- **Total Quality Management (TQM):** Management program designed to minimize or eliminate all sources of error in a process through the implementation of high standards and accountability through extensive documentation.
- **True Quantity Value:** Actual and exact real value of some measurable quantity.
- **Variance:** Average value of the square of the difference between observed values and the mean of the observed values.

DEFINITION AND BASIC PRINCIPLES

Quality control can be defined as any process or procedure that has the purpose of maintaining an established or stated standard of quality for a product or process. The central aims of quality control are to ensure the consistent nature of the product or process of interest, to reduce costs and losses inherent in the process, and to maximize client satisfaction. These are generally embodied in the descriptive term "total quality management" (TQM).

TQM is an all-encompassing program designed to manage every aspect of the process of production of a good or service, from start to finish, and at all levels of a company. Generally, a TQM program includes thorough documentation of every stage through which the good or service passes and implementation of up-to-date processes and procedures designed to ensure quality. A TQM program always has a quality-control function. However, the converse is not always true, as quality-control activities and functions can be and often are employed on their own and not as an integral part of any TQM program.

Quality-control methods depend on the nature of the process involved. Methods employed, for example, in a manufacturing facility that turns out hundreds of identical machine parts per day will be very different from those used to ensure that teachers are performing to designated standards in their classrooms or that a complicated computer program performs to expectation without difficulty. All methods share the same purpose, however, which is to ensure that the output of the process meets the conditional criteria placed on its successful production.

In many instances, quality control consists of little more than visual inspection of a product to check for defects. This is a general and useful starting point for any quality-control function. Quality control is first and foremost a feedback process. When quality control is used to detect errors in the output stream, the information that it provides is used to adjust the manner in which the particular process treats the input stream. Determining the quality of the product (the output stream of the process) requires the creation of a product-specific system of quality-control testing and unequivocal methods of measuring designated features of the product, which can be compared with the ideal or standard features.

Metrology, the study and application of methods for the measurement of properties, plays an important role in quality control, which depends on measurement to obtain specific information about individual components of an output stream. Also critical in quality-control processes is the use of statistics, which determines and uses standard methods of obtaining generalized information relevant to specific aspects of the entire output stream. For example, metrology would be used to determine the achievement level of a particular student as a percentage mark on a specific test, while statistics would be used to determine the percentage mark typically achieved by a large number of students of the same age on the same test. In another setting, metrology would be used to accurately measure the distance between two features of individual machined parts in a factory, and statistics would be used to monitor how well, as a group, the parts being manufactured conform to the essential design criteria. The use of such statistical information as the basis for regulating and adjusting the manner in which the machine process is carried out is called statistical process control (SPC).

Background and History

People have always sought to create products that are satisfactory to others. Such workmanship generally attracts consumers and generates profits. In ancient times, most likely hunters preferred their arrowheads and spear points to come from the best flintknappers in the village because the best quality objects were the most effective; efficacy of the goods produced directly correlated with survival. Similarly, farmers who grew the best crops or craftspeople who made the best goods generally were able to sell or trade them more readily, thereby making a better living than those who produced goods of lesser quality.

With the development of skilled trades, the guild system developed. In a guild system, skilled tradespeople became represented in self-governing groups according to their specific type of trade. Each guild ensured that its members were able to perform the tasks associated with a specific trade to the guild standard. The guild system represents the first systematic approach to quality control. Quality control in the modern sense came about through the Industrial Revolution and the development of mass-production methods. In the late eighteenth century, the idea that machine components that were nominally the same should be interchangeable was put forward by American inventor Eli Whitney. Production methods, however, relied more on the machinist's art than anything else, and true interchangeability of parts was difficult to attain.

The number of standardized parts needed increased tremendously in the mid-twentieth century, largely because of the United States' involvement in World War II. Quality-control methods based on the statistical work of Walter Shewhart—often referred to as the father of statistical quality control—and American statistician W. Edwards Deming were developed as a means of ensuring absolute interchangeability of parts, eliminating waste, minimizing (or eliminating) liability issues, and maximizing return on investment in materials. Quality-control methods that were developed in the late twentieth century provided on-going feedback that allowed processes to be adjusted as needed to allow for essentially continuous production of components.

How It Works

Quality-control processes range from the exceedingly simple method of visual inspection and measurement typically used for parts produced in low quantities to the sophisticated automatic inspection and measurement systems employing advanced analytical techniques often used for parts produced in large quantities. All quality-control systems, however, refer to an ideal or design standard. The purpose of any quality-control technique is to determine how well the output item conforms to the ideal or design standard. Acceptable outputs are those that fall within a specific set of limiting values. Depending on the nature of the product, outputs may be subjected to several stages of examination before use and to ongoing inspection during use.

For example, in the production of cast magnesium rotors, the design standard for acceptable cast parts calls for uniform material distribution. Each part produced must pass a preliminary visual inspection. Those observed to have defects are rejected and recycled. The remaining parts proceed to the next stage, where they are checked for internal defects. Defective parts are sent to recycling, and acceptable parts are sent to the next stage. If a machining stage is required, the machined parts are checked for the correct dimensions. The quality checks continue until the parts are finished and acceptable.

Quality assessments are recorded as statistical data, which are used to maintain or control the specific process from which they were obtained. In the cast magnesium rotor example, if casting flaws frequently occurred in a specific location on the rotor, the casting process would be adjusted to eliminate the flaw or reduce the frequency of its occurrence. In other processes with high throughput (output over an extended period of time), it is not feasible to examine every single unit. In such cases, a random selection of individual outputs is tested and their conformance to the ideal is extrapolated to the entire output. This method relies on the output history as the basis for comparison, and variances in the output are tracked very closely to ensure that, overall, the individual components of the output stream remain within the parameters set in the design standard.

Applications and Products

The value of quality-control processes and procedures is widely recognized. The quality-control process has been integrated into virtually all aspects of human activity, from the simplest of mechanical production operations to the most insubstantial of services. Human activities themselves are subject to the application of quality-control measures.

The quality-control process is a feedback control system. The ultimate purpose of determining the quality of any output is to ascertain what aspect of the procedure is not performing adequately so that it can be corrected and the subsequent output of the process improved or at least maintained within the design standard. This applies equally to physical objects being assembled or manufactured in a factory and to services provided to customers in the retail environment. In essence, quality control ensures that work is being done in an optimal manner that produces the most output with the least waste.

Standardization. Standardization is key to achieving optimal quality control in the international economy. The International Organization for Standardization (ISO) and other organizations provide internationally recognized industrial and commercial standards. For example, the ISO provides standards regarding the magnetic strip on credit cards. The ISO also offers a family of quality management standards, which includes ISO 9000, 9001, and 14000. Other quality-control programs include the Motorola Corporation's Six Sigma and Lean Manufacturing (derived from the Toyota Production System). Such programs are generally not intended as stand-alone quality-control systems but instead define an overall management approach in which quality control is an integral component and tool of the project manager.

Quality Management Programs. Quality-control programs determine how standardization is achieved. To obtain certification from any program, a company must first commit to organizing its operations in accordance with the standards specified by the program. These standards govern most aspects of operations, including the documentation of supplies and other inputs, detailed procedures for each step of the manufacturing process, and the storage and internal delivery of materials. The accrediting body rigorously inspects the organization's operations and, if the operations are satisfactory, grants certification to the particular standard. The organization must continue to operate within the specific guidelines of the accreditation standard and undergo periodic checks and assessments to retain the accreditation. Note that certification to an ISO standard means that the company is following formalized business standards created by the ISO; it does not mean that the finished products have been tested. ISO certification represents a significant investment of resources for an organization and can cost tens of thousands of dollars per year to maintain. Much of this money goes to train key personnel and purchase equipment, especially if the certification

Fascinating Facts About Quality Control

- The Six Sigma program is designed to keep essentially all of a quality-controlled product within half of the variance allowed for each individual unit. This corresponds to one defect in a million possibilities.
- The January 28, 1986, *Challenger* space shuttle disaster was caused by defective O-ring seals on the outboard propellant tanks that were not detected during quality checks. Their failure resulted in an explosion that destroyed the vessel and killed all aboard.
- Quality-control standards developed from military specifications (mil-spec). Anecdotal histories suggest that these standards were established in an attempt to lessen the number of military fatalities caused by defective products.
- Quality control is a concept with deep historical roots. The Code of Hammurabi (c. 1800 B.C.E.) included such quality-control directives as "The mason who builds a house which falls down and kills the inmate shall be put to death."
- After World War II, Japanese industry looked to quality-control methods developed in the United States to optimize the manufacturing process and improve products. The Japanese adoption of the methods was so successful that "Made in Japan" became synonymous with quality.
- The devices used to measure variances have become increasingly precise and accurate. Measurement by eye, accomplished by trained inspectors in several minutes, has given way to near-instantaneous measurement to 0.0001 millimeters and less by automated machine processes.
- Quality control, quality inspection, and quality management are intimately related. Modern programs seek to ensure a thoroughly documented production path for each unit, from the source of the raw materials used for each part to its final disposition to its end user, including the specific procedures used in each stage.

program is focused on quality control itself. The value of such programs is in the return on this investment. Companies often report that using a quality management program such as Six Sigma has saved the organization hundreds of thousands of dollars per project.

Quality-Control Processes. A typical quality-control function consists of assessing the output of a process using a representative sample. This quality check can be qualitative (observing employee-customer interactions to gauge customer satisfaction) or based on physical measurements (comparing measurements to evaluate the consistency of a machining operation in a factory). Assessments based on actual physical measurements have spawned an entire industry and science based on metrology.

Simply, if the design standard of a particular piece of work calls for accuracy to within 0.0001 inch, then devices must be available that can be used by a trained individual to check the measurements of the product pieces. Precision mechanical devices such as micrometers, scales, and gauges have been largely replaced by programmable digital electronic devices that automatically carry out dimensional analyses on specific parts. These devices are capable of far more precise measurement than even the most skilled of human artisans and have much greater consistency and reliability.

IMPACT ON INDUSTRY

Without the concept of quality control—as it has been applied in the mass-production environment—the products of modern-day industry would be very different. In the 1970's, the Motorola Corporation, a U.S. company that was producing television sets, suffered from management practices that were costing the company millions of dollars annually in scrap units. The company hired a management team from Japan to apply its methodology to the workplace. After a short period, auditors of the procedures found that the incidence of scrap units and rejected parts had dropped to a level that was 70 percent lower than the best rate achieved by the traditional U.S. managers who had been replaced. Shortly after, Motorola developed its Six Sigma program.

Quality-control programs have demonstrated their value for waste reduction and optimization of materials, but they also provide less tangible benefits in the form of customer satisfaction. The products and services produced under the auspices of a quality-control program achieve a higher level of uniformity, and the end user can be assured that any individual piece will match up to any other piece for which it was designed. The same principle applies universally, whether the individual piece is a simple nut-and-bolt combination, a complex assembly of custom-designed parts, or a customer service program. Thus, quality-control programs enhance the profitability of the businesses that employ them.

The quality-control process has become an integral part of service and production industries in North America and around the world. It ensures both the quality and the dependability of the products, as well as the health, safety, and security of the consumer. Quality-control processes decrease the risk of failure and the liability of the provider while protecting the rights of the consumer through a high standard of accountability that is supported by thorough documentation.

Careers and Course Work

Specialization in quality-control procedures is a growing and very viable career option. Careers range from support and maintenance of quality-control programs to advanced project management in practical applications. Academic careers in quality control are available for those who wish to work on statistical procedures and models for the development of quality-control algorithms and procedures.

Because quality control is a universal concept, it has applications in careers in many fields. Quality-control procedures are used in both simple mechanical production processes and complex computer programming operations. The student who considers a career in quality control will be expected to acquire a sound grounding in mathematics and statistics in order to understand the basic principles of sampling, variance, and other particular features relevant to quality-control procedures. In postsecondary education, the student will be able to specialize in a particular field of study. Many community colleges offer two-year programs designed to prepare a student to become a quality-control specialist. Following completion of the program, the practicing quality assurance technician would be expected to maintain certification through an appropriate association such as the American Society for Quality and to master new and emerging quality-control procedures.

Quality control plays a significant role in many other postsecondary fields of study. Those pursuing a career in one of these fields must acquire a basic understanding of not only quality control but also the specific methodologies used in the particular field. For example, biochemical laboratory procedures require that the practitioner has intimate knowledge of biochemistry as well as the analytical procedures and sampling methods that are used in maintaining the quality standard of the tests being carried out. Similarly, a computer science specialist needs intimate knowledge of computer programming as well as the specific methods that are used to ensure the functional quality of software applications.

Social Context and Future Prospects

Quality control is more an application of knowledge than a tangible product. Sound quality-control practices are founded on the basic human desire to have access to the same goods and services that are available to others. The response of suppliers of those goods and services has been the development of methods to ensure that each unit of a good or service is as uniform as possible. These methods make it possible for sellers to offer virtually identical products to all customers.

The effects and value of quality control are far-reaching, and the absence or failure of quality control has equally far-reaching consequences. One can readily imagine the economic costs if the dependability of products were to decline. The cost per unit produced would rise as the proportion of unusable products increased. In the marketplace, more products would prove to be defective or fail, resulting in greater repair and replacement costs for the producer and consumer. If quality-control mechanisms affected service, resulting in less effective services, consumer satisfaction would also decline. Quality-control procedures most likely will become an increasingly prominent component of production as more effective methods are developed.

Richard M. J. Renneboog, M.Sc.

Further Reading

Allen, Theodore T. *Introduction to Engineering Statistics and Six Sigma: Statistical Quality Control and Design of Experiments and Systems.* London: Springer-Verlag London, 2006. Provides precise descriptions of many quality-control methods and details case studies in engineering and business to demonstrate the savings that can be achieved by a company through the application of an effective quality-control program. Designed to assist the reader in finding the appropriate method for his or her needs.

Clements, Richard Barrett. *Quality Manager's Complete Guide to ISO 9000: 1999 Cumulative Supplement.* Paramus, N.J.: Prentice Hall, 1999. An essential guide to the total quality management process known as ISO 9000. Includes a wealth of explanatory material regarding the goals and operation of the program, as well as step-by-step descriptive instructions on successfully accomplishing the mechanics of the program.

Deming, W. Edwards. *Out of the Crisis.* 1982. Reprint. Cambridge, Mass.: MIT Press, 2006. Deming sets forth his ideas about quality control in this classic work. He describes quality control in Japan and suggests how American managers can adopt his techniques.

Ott, Ellis R., Edward G. Schilling, and Dean V. Neubauer. *Process Quality Control: Troubleshooting and Interpretation of Data.* 4th ed. Milwaukee, Wis.: American Society for Quality, Quality Press, 2005. Uses case studies to illustrate the techniques and methods of collecting and analyzing data in regard to maintaining and troubleshooting working processes.

Wheeler, Donald J., and David S. Chambers. *Understanding Statistical Process Control.* 2d ed. Knoxville, Tenn.: Statistical Process Controls Press, 1992. A very practical guide to the fundamental statistics of statistical process control and their application in practice through the use of control charts and other charting methods.

Ziliak, Stephen T., and Deidre N. McClosky. *The Cult of Statistical Significance: How the Standard Error Costs Us Jobs, Justice, and Lives.* Ann Arbor: University of Michigan Press, 2008. Discusses the weakness of the concept of "statistical significance" as it is applied in many fields of research, including quality control.

WEB SITES

American Society for Quality
http://www.asq.org

General Electric
What Is Six Sigma?
http://www.ge.com/en/company/companyinfo/quality/whatis.htm

International Organization for Standardization
http://www.iso.org

Toyota
Toyota Production System
http://www2.toyota.co.jp/en/vision/production_system

See also: Engineering

R

RADIO

FIELDS OF STUDY

Broadcasting; communications; electronics; Global Positioning System; microwave technology; radar technology; radio technology; radio astronomy.

SUMMARY

Radio is a technology that involves the use of electromagnetic waves to transmit and receive electric impulses. Since its inception as a method of wirelessly transmitting Morse code, radio communications technology has had a tremendous impact on society. Although television has supplanted radio to a significant extent for public broadcasting, radio continues to play a significant role in this arena. As of 2011, radio broadcasts are delivered via technologies such as satellite, cable networks, and the Internet. Although the term "radio" brings to mind listening to news broadcasts and video, electromagnetic-wave transmission encompasses other fields, such as radar, radio astronomy, and microwave technology.

KEY TERMS AND CONCEPTS

- **Amplification:** Process of increasing the strength of an electronic transmission or a sound wave.
- **Amplitude:** Refers to the height of a radio wave.
- **Antenna:** Device that either converts an electric current into an electromagnetic radiation (transmitter) or converts electromagnetic radiation into an electric current (receiver).
- **Microphone:** Device that converts sound into an electrical signal.
- **Modulation:** Process of varying one or more properties of an electromagnetic wave; three parameters can be altered via modulation: amplitude (height), phase (timing), and frequency (pitch). Two common forms of radio modulation are amplitude modulation (AM) and frequency modulation (FM).
- **Radio Frequency:** Oscillation of a radio wave in the range of 3 kilohertz (3,000 cycles per second) to 300 gigahertz (300,000,000,000 cycles per second).
- **Radio Wave:** Electromagnetic radiation, which travels at the speed of light; radio waves are of a longer wavelength than infrared light.

Definition and Basic Principles

In contrast to sound waves, which require a medium such as air or water for propagation, electromagnetic waves can travel through a vacuum. In a vacuum such as outer space they travel at the speed of light (299,800 kilometers per second). In space, electromagnetic waves conform to the inverse-square law: the power density of an electromagnetic wave is proportional to the inverse of the square of the distance from a point source. Thus, all radio waves weaken as they travel a distance. When traveling through air the intensity of the waves is weakened. At some point, depending on the strength of the signal, the electromagnetic wave will no longer be discernible. Interference can weaken or destroy a radio signal. Other radio transmitters and accidental radiators (such as automobile ignition systems) produce interference and static. FM radios are much more resistant to interference and static than AM radios. Radio waves travel in a straight line; therefore, the curvature of the earth limits their range. However, radio waves can be reflected by the ionosphere, which extends their range.

The following steps occur in radio transmission: A transmitter modulates (converts) sound to a specific radio frequency; an antenna broadcasts the electromagnetic wave; the wave is received by an antenna; and a receiver tuned to the radio frequency demodulates the electromagnetic energy back into sound. Radio waves range from 3 kilohertz (kHz) to 300 gigahertz (GHz) and are categorized as: very low frequency (VLF; 3 to 30 kHz); low frequency (LF; 30 to 300 kHz); medium frequency (MF; 300 to 3,000 kHz); high frequency (HF; 3 to 30 megahertz [MHz]); very high frequency (VHF; 30 to 300 MHz); ultra high frequency (UHF; 300 to 3,000 MHz); super high frequency (SHF; 3 to 30 GHz); and extremely high frequency (EHF; 30 to 300 GHz). Each frequency range has unique characteristics and unique applications for which they are best suited.

Background and History

Electromagnetic waves were discovered in 1877 by the German physicist Heinrich Hertz, whose name is used to describe radio frequencies in cycles per second. Eight years later, American inventor Thomas Alva Edison obtained a patent for wireless telegraphy by discontinuous (intermittent) wave. A far superior system was developed in 1894 by the Italian inventor Guglielmo Marconi. Initially, he transmitted telegraph signals over a short distance on land. Subsequently, an improved system was capable of transmitting signals across the Atlantic Ocean. At the start of the twentieth century, Canadian inventor Reginald Aubrey Fessenden began experimenting with voice transmission via discontinuous waves while employed by the United States Weather Bureau. In 1902, he switched to continuous waves and successfully transmitted voice and music. In 1906, history was made when Fessenden transmitted voice and music from Massachusetts that was heard as far away as the West Indies. Lee de Forest developed the Audion tube (later known as the triode vacuum tube), which became a key component of radios. In 1920, the first radio news program was broadcast by station 8MK in Detroit; the station is still operational as of 2011 and broadcasts news as the CBS-owned WWJ. Also in 1920, the station WRUC in New York began broadcasting a series of Thursday night concerts, the initial range of which was 100 miles. However, it was soon expanded to 1,000 miles. WRUC also began sports broadcasts in the same year. Television first appeared in 1920; however, commercial television had its onset in the 1940's. In the early 1930's, frequency modulation (FM) and single sideband shortwave radio were invented by amateur radio operators. In 1954, Texas Instruments and Regency embarked on a joint venture to launch the Regency TR-1 transistor radio, which was powered by a 22.5-volt battery.

How It Works

Basic Example of a Radio Receiver. The humble crystal set, which first appeared at the close of the nineteenth century, is a radio receiver in its simplest form. It consists of an antenna, a tuned circuit, a crystal detector, and earphones. The tuned circuit consists of a tuning coil (a sequentially wound coil that can be tapped at any point) connected to a capacitor. This pair of components allows tuning of the receiver to a specific frequency, known as the resonant frequency. Only signals at the resonant frequency pass through the tuned circuit; other frequencies are blocked. The crystal is a semiconductor, which extracts the audio signal from the radio frequency carrier wave. This is accomplished by allowing current to pass in just one direction, blocking half the oscillations of the radio wave. This rectifies, or changes, the wave into a pulsing direct current, which varies with the audio signal. The earphones then convert the direct current into sound. The sound power is solely derived from the radio station that originated it. The electrical power and circuitry of more complex receivers serve to amplify this extremely weak signal to one that can power loudspeakers.

Amplitude Modulation (AM). With AM radio, the amplitude (height) of the transmitted signal is made proportional to the sound amplitude captured (transduced) by the microphone. The transmitted frequency remains constant. AM transmission is degraded by static and interference because sources of electromagnetic transmission such as lightning and automobile ignitions, which are at the same frequency, add their amplitudes to that of the transmitted signal. AM radio stations in the United States and Canada are limited to 50 kilowatts (kW). Early twentieth century, U.S. stations had powers up to 500 kW, some of which could be heard worldwide. Conventional AM transmission involves the use of a carrier signal, which is an inefficient use of power. The carrier can be removed (suppressed) from the AM signal. This produces a reduced-carrier transmission, which is termed a double-sideband suppressed-carrier (DSB-SC) signal. A sideband refers to one side of the mirror-image radio signal. DSB-SC has three times more power efficiency than an unsuppressed AM signal. Radio receivers for DSB-SC signals must reinsert the carrier. Another AM refinement is single-sideband modulation in which both one sideband and the carrier are stripped out. This modification doubles the effective power of the signal.

Frequency Modulation (FM). With FM, the variation in amplitude from the microphone causes fluctuations in the transmitter frequency. FM broadcasts have a higher fidelity than AM broadcasts and are resistant to static and interference. FM requires a wider bandwidth to operate and is transmitted in the VHF (30 to 300 MHz) range. These high frequencies travel in a straight line, and the reception range is generally limited to about 50 to 100 miles. FM signals broadcast from a satellite back to Earth do not have this distance limitation. An FM transmission can contain a subcarrier in which secondary signals are transmitted in a piggyback together with the main program. The subcarrier allows stereo broadcasts to be transmitted. The subcarrier can also transmit other information such as station identification and the title of the current song being played.

Digital Radio. Digital audio transmission consists of converting the analogue audio signal into a digital code of zeros and ones in a process known as digitizing. This technology allows for an increase in the number of radio programs in a given frequency range, improved fidelity, and reduction in fading (signal loss). Satellite radio is a digitized signal and can cover a distance in excess of 22,000 miles.

Applications and Products

Radio applications are widespread and include commercial radio, amateur radio, marine radio, aviation radio, radar, navigation, and microwave.

Commercial Radio. Radio broadcasts are available throughout the globe. They offer a variety of products, such as news, music, and political opinion. Many broadcasts are free of charge and available to receivers in range of the station. The station derives its revenue from advertising. An hour of broadcasting time typically contains ten to twenty minutes of advertising. Cable television and Internet services rebroadcast these stations either at no charge or with a fee. Cable television services and satellite radio broadcast a variety of commercial-free programs. In these cases, the subscription fee covers the cost of the broadcast.

Amateur Radio. Amateur radio (also known as ham radio) is the licensed use of designated radio bands for noncommercial exchange of messages, private recreation, emergency communication, and experimentation. Ham operators maintain and operate their own equipment. Through the years, they have made numerous contributions to radio technology, including FM and single sideband. They also perform a number of public services at no cost. For example, during the Vietnam War, the Military Auxiliary Radio System (MARS) allowed military personnel to call friends and relatives in the United States. A broadcast emanating from Vietnam was received by a U.S.-based amateur operator who patched the communication into the phone lines. MARS is still active; however, many of its services have been supplanted by the Internet and e-mail.

Shortwave Radio. Shortwave radio has many applications for long-range communication. The term "shortwave" refers to the wavelength of the frequency spectrum in which it operates: high frequency (3,000 to 30,000 kHz). The high-frequency wavelength is shorter than the ones first used for radio communications: medium frequency and low frequency. Shortwave broadcasts are readily transmitted over distances of several thousand kilometers, allowing intercontinental communication and ship-to-shore communication. Low-cost shortwave radios are available worldwide and facilitate the transfer of information to individuals where other forms of media are controlled for political reasons. A major disadvantage of shortwave radio is that it is subject to significant interference problems, such as atmospheric disturbances, electrical interference, and overcrowding of wave bands. The Internet and satellite radio have impacted shortwave radio, but it is still useful in areas where those services are unavailable or too expensive.

Marine Radio. All large ships and most small oceangoing vessels are equipped with a marine radio. Its purposes included calling for aid, communicating with other vessels, and communicating with shore-based facilities, such as harbors, marinas, bridges, and locks. Marine radio operates in VHF, from 156 to 174 MHz. Channels from 0 to 88 are assigned to specific frequencies. Channel 16 (156.8 MHz) is designated as the international calling and distress channel.

Aircraft Band. The aircraft band (or air band) operates in the VHF range. Different sections of the band are used for commercial and general aviation aircraft, air traffic control, radio-navigational aids, and telemetry (remote transmission of flight information). VHF omnidirectional range (VOR) or an instrument landing system operates in the aircraft band. A VOR is a ground-based station that transmits a magnetic bearing of the aircraft from the station. Unmanned aircraft (drones) are navigated by a radio system in which a pilot sits at a console on the ground and directs the flight.

Radio Telescope. A radio telescope is a large, directional, parabolic (dish) antenna that collects data from space probes and Earth-orbiting satellites. Many astronomical objects emit electromagnetic radiation in the radio frequency range, therefore radio telescopes can image astronomical objects such as galaxies. Some researchers are using radio telescopes to search for intelligent life forms in the universe.

Navigation. Navigation encompasses a number of radio applications including radio direction finding (RDF), radar, loran, and Global Positioning System (GPS).

An RDF system homes in on a radio transmission, which can be a specialized antenna that directs aircraft or commercial transmitters that have a known location.

Radar is an acronym for radio detecting and ranging. The device consists of a transmitter and receiver. The transmitter emits radio waves, which are deflected from a fixed or moving object. The receiver, which can be a dish or an antenna, receives the wave. Radar circuitry then displays an image of the object in real time. The screen displays the distance of the object from the radar. If the object is moving, consecutive readings can calculate the speed and direction of the object. If the object is airborne, and the radio is so equipped, the altitude is displayed. Radar is invaluable in foggy weather when visibility can be severely reduced.

Loran is an acronym for long-range navigation. The system relies on land-based low-frequency radio transmitters. The device calculates a ship's position by the time difference between the receipt of signals from two radio transmitters. The device can display a line of position, which can be plotted on a nautical chart. Most loran convert the data into longitude and latitude. Since GPS became available, the use of loran has markedly declined.

GPS is a space-based global navigation satellite system that provides accurate location and time information at any location on Earth where there is an unobstructed line of sight to four or more GPS satellites. GPS can function under any weather condition and anywhere on the planet. The technology depends on triangulation, just as a land-based system such as loran employs. GPS is composed of three segments: the space segment; the control segment; and the user segment. The U.S. Air Force operates and maintains both the space and control segments. The space segment is made up of satellites, which are in medium-space orbit. The satellites broadcast signals from space, and a GPS receiver (user segment) uses these signals to calculate a three-dimensional location (latitude, longitude, and altitude). The signal also transmits the current time, accurate within nanoseconds.

Fascinating Facts About Radio

- Radio pioneer Guglielmo Marconi initially transmitted radio signals over water. He was unable to transmit over land because prevailing laws in Europe gave government postal services a monopoly on message delivery.
- Another radio pioneer, Lee de Forest, who developed the Audion tube (triode vacuum tube) was indicted for using the mails to defraud because he was promoting "a worthless device." He was subsequently acquitted and his invention continued to be a key component of radios until vacuum tubes were supplanted by transistors.
- In this high-tech age, many deem Morse code to be an antiquated, inferior form of message transmission. However, on May 20, 2005, the *Tonight Show with Jay Leno* show featured two teams: a pair of amateur radio operators (hams) who were proficient at Morse code and another duo who were skilled at text messaging. The texters were out-messaged handily by the hams, who transmitted and received at a rate of twenty-nine words per minute.
- Radio gained acceptance more rapidly in the United States among amateur radio operators than the general public. In 1913, 322 amateurs were licensed; however, by 1917 there were 13,581, primarily boys and young men. Many older individuals considered radio to be a fad. They reasoned that listening to dots and dashes or occasional experimental broadcast of music or speech over earphones was a worthless endeavor.
- The microwave oven was a by-product of radar research. Raytheon engineer Percy Spencer was testing a new vacuum tube called a magnetron and found that a candy bar in his pocket had melted. Deducing that electromagnetic energy was responsible, he placed some popcorn kernels near the tube. The kernels sputtered and popped. His next trial was with an egg, which exploded in short order.

Impact on Industry

Radio has a significant impact on industry. The United States radio broadcasting and programming industry is composed of about 4,000 companies with combined annual revenue of about $16 billion. Major companies include CBS Radio, Citadel Broadcasting, Clear Channel Communications, Cumulus Media, and Sirius XM Radio. The fifty largest companies earn about 75 percent of the revenue. The United States TV broadcast network industry, which is interwoven with the radio industry, is even larger. It includes more than 1,350 networks and stations with a combined annual revenue of $80 billion. Beyond radio broadcasting are a host of wireless technologies that operate within radio frequencies. For example, the cell phone, which transmits and receives radio waves, is ubiquitous in developed nations. In fact, some individuals no longer possess a wired telephone service and rely entirely on their cell phone. The average home in the United States contains many devices that transmit or receive radio waves, including microwave ovens, garage-door openers, radio-controlled toys, and burglar-alarm systems.

Regulation. Radio transmission is regulated and closely monitored by government agencies. In the United States, it is regulated by the Federal Communications Commission (FCC). The FCC is an independent United States government agency charged with regulating interstate and international communications by radio, television, wire, satellite, and cable. The FCC's jurisdiction covers the fifty states, the District of Columbia, and U.S. possessions

Amateur radio licensing is governed by the FCC under strict federal regulations. Individuals of any age may be granted a license to operate amateur stations for personal use. To obtain a license, one must demonstrate an understanding of pertinent FCC regulations as well as knowledge of radio-station operation and safety considerations.

Industry and Business. Although the civilian population is a major purchaser of radio equipment, commercial and military applications make up a significant market share. In the commercial sector, the civilian aircraft and shipping industries are major purchasers of radio equipment. The National Aeronautics and Space Administration (NASA) requires extremely precise, complex, and expensive navigational and communications equipment for its missions. All branches of the military—Army, Navy, Air Force, Marines, and Coast Guard—purchase a great deal of radio equipment.

Government and University Research. One of the biggest sources of funding for radio research in the United States is the Defense Advanced Research Projects Agency (DARPA). It is also the biggest client for certain kinds of radio and navigation applications. The agency is primarily concerned with radio systems with a military focus, such as guidance systems. It also conducts research on satellite communication. Another source of funding for radio research is the U.S. Department of Energy (DOE).The DOE is currently focused on energy efficiency and renewable energy. The DARPA and the DOE supply funds to many universities in the United States for laser research and development. For example, as far back as 1959, DARPA began work with Johns Hopkins Applied Physics Laboratory to develop the first satellite positioning system.

Careers and Course Work

Many technical and nontechnical careers are available in radio. The technical fields require a minimum of a bachelor's degree in engineering or other scientific field. Many also require a postgraduate degree such as a master's or doctorate. Course work should include mathematics, engineering, computer science, and robotics. Positions are available in both the government and private sector. The ability to be a team player is often of value for these positions because ongoing research is often a collaborative effort. Less technical positions such as operation or maintenance of broadcasting equipment can be achieved with less training, which can be obtained at a technical college or trade school. A bachelor's degree in communications or related fields can qualify one for employment in radio or television broadcasting.

Social Context and Future Prospects

In view of the continuous advances in radio technology, further advances are extremely likely. Radio and related technologies are an integral component of everyday life in all but the most primitive societies. Radio has political implications, particularly in nations governed by tyrants and dictators, where radio is used as a propaganda tool and use of shortwave radio and the Internet is strongly discouraged. The Voice of America (VOA), which began broadcasting in 1942, is an international multimedia broadcasting service funded by the U.S. government. The VOA broadcasts about 1,500 hours of news, information, educational, and cultural programming each week to an estimated worldwide audience of 123 million people. In some repressed nations, the broadcast is jammed, which involves broadcasting noise at the same frequency to prevent reception. Even in the United States, political movements exist that wish to ban certain types of radio broadcasts (such as talk radio), which are deemed too conservative or too liberal.

Many associate microwave with a useful appliance to heat a frozen meal, but microwave is in the radio wavelength (0.3 GHz to 300 GHz) and is also used in communication. The microwave oven illustrates the fact that electromagnetic waves are a form of energy capable of heating—and damaging—tissue. For example, standing near or touching a powerful radio transmitter can result in severe burns. Microwave ovens are shielded to prevent exposure; other devices are not. The heating effect of an electromagnetic wave varies with its power and the frequency. The heating effect is measured by its specific absorption rate (SAR) in watts per kilogram. Many national governments as well as the Institute of Electrical and Electronics Engineers (IEEE) have established safety limits for exposure to various frequencies of electromagnetic energy based on their SAR.

Robin L. Wulffson, M.D., F.A.C.O.G.

Further Reading

Hallas, Joel. *Basic Radio: Understanding the Key Building Blocks.* Newington, Conn.: American Radio Relay League, 2005. An introduction to radio that reveals the key building blocks of radio: receivers, transmitters, antennas, propagation and their applications to telecommunications, radio navigation, and radio location. Includes simple, build-it-yourself projects to turn theory into practice, helping reinforce key subject matter.

Kaempfer, Rick, and John Swanson. *The Radio Producer's Handbook.* New York: Allworth Press, 2004. A definitive how-to guide for producing a radio show, explaining every duty a radio producer is expected to perform.

Keith, Michael C. *The Radio Station: Broadcast, Satellite and Internet.* 8th ed. Burlington, Mass.: Focal Press, 2010. The standard guide to radio as a medium. Discusses the various departments in all kinds of radio stations in detail.

Rudel, Anthony. *Hello, Everybody: The Dawn of American Radio.* Orlando, Fla.: Houghton Mifflin Harcourt, 2008. An overview of the birth of radio with an emphasis on the entrepreneurs, evangelists, hucksters, and opportunists who saw the medium's potential.

Silver, H. Ward. *ARRL Ham Radio License Manual: All You Need to Become an Amateur Radio Operator.* Newington, Conn.: American Radio Relay League, 2010. Informative text for one interested in becoming a ham radio operator.

Web Sites

American Radio Relay League
http://www.arrl.org

International Association for Radio, Telecommunications and Electromagnetics
http://www.narte.org

National Association of Broadcasters
http://www.nab.org

See also: Communications Satellite Technology; Computer Science; Engineering; Radio; Robotics.

RISK ANALYSIS AND MANAGEMENT

FIELDS OF STUDY

Actuarial science; applied mathematics; economics; project management; statistics.

SUMMARY

Risk analysis and management is a field that deals with the anticipation and management of risk. Relevant to nearly every industry in existence, risk analysis and management provide a comprehensive set of tools, techniques, and methodologies that enable risk practitioners to deal with risk as appropriate.

KEY TERMS AND CONCEPTS

- **Hazard:** Condition or situation that poses a threat to life, health, property, or environment and increases the probability or magnitude of a loss.
- **Probability:** Dimension of risk, the possibility that an event will occur; probability can be represented by a number from 0 (denoting that the event will never occur) to 1 (the event will certainly occur).
- **Residual Risk:** Risk that remains after risk-response strategies have been implemented.
- **Risk:** Combination of the probability of an event and its consequences.
- **Risk Indicator:** Event that acts as an early warning sign that a risk is about to occur, or an event that instigates a series of events that, if ignored, will lead to the risk occurring.
- **Risk Register:** List describing risks, their impact, and the strategy for managing each risk.
- **Threat:** Potential that a specific vulnerability may be exploited to cause harm to people, property, or the environment.
- **Vulnerability:** Degree of susceptibility of people, property, and the environment to the impact of a hazard.

Definition and Basic Principles

Risk analysis and management is the identification of possible risks, assessing their likelihood and impact, and following a structured approach to control, minimize, mitigate, or in some instances, exploit these identified risks. Strictly speaking, risk analysis and risk management are separate but closely related activities. The line between these two activities has become blurred and experts in one area or the other have gained skills that enable them to perform dual roles. Risk analysis and management is practiced in a wide range of fields, such as project management, information technology, security, insurance, and finance. Each field has customized existing risk analysis and management methods to suit its unique needs, and there are slightly different interpretations of what constitutes risk.

Background and History

Humans have been fascinated by risk for millennia. People of different cultures from different eras played games of chance in one form or the other, but they have always believed that their chances of winning or losing rested with the gods they worshipped. They had no method for calculating the probability of outcomes of their games or other life activities and adopted a passive attitude rooted in their belief that they could not affect their futures in any tangible manner.

This attitude began to evolve with the changes to the prevailing numbering system. Before the widespread use of the Hindu-Arabic numbering system, which uses the numbers zero to nine, many Western cultures used an alphabet-based numbering system. These systems were clumsy and made any type of complex mathematical calculations nearly impossible. With the adoption of the new numbering system at the beginning of the thirteenth century, complex calculations became possible and the calculation of probability became easier. However, it was not till the 1500's and 1600's, in the midst of the Renaissance, that prominent mathematicians of the time, including Galileo and Girolamo Cardano, began to study risk and how it applied to gambling. In the mid-1600's, French mathematicians Blaise Pascal and Pierre de Fermat invented the theory of probability as a solution to a 200-year-old math puzzle. As the years went by, mathematicians built on the foundation laid by Pascal and Fermat and developed multiple techniques for risk management that had far-reaching influence.

The oldest form of risk management was insurance. The earliest known record of risk management is found in Hammurabi's Code, an ancient Babylonian set of laws written around 1752 B.C.E. This form of insurance, known as bottomry, was a type of marine insurance in which the owner of a vessel could borrow money to buy cargo. If the ship was lost at sea, the vessel owner did not have to repay the debt. For centuries, insurance remained the primary method of risk management for individuals and businesses.

It was not until the late twentieth century, in the 1970's and 1980's, that preventive measures began to form a part of risk management. Insurance companies encouraged businesses to make their premises safer, businesses began to establish and adhere to quality standards for their products and services, and the government enacted legislation to make businesses consider the risks that existed for their employees. By the late 1990's and in the early years of the twenty-first century, multiple quality- and risk-management standards were created by members of various industries and different government regulatory agencies. Risk analysis and management is a full-fledged discipline with relevance in nearly every field of industry and business.

How It Works

Risk analysis and management is an adaptable, multistep process. Some of the steps have multiple parts. A general overview is provided below.

Risk Identification. This step involves determining and categorizing elements that may pose a risk. Techniques for identifying risks include brainstorming, interviewing, root-cause identification, reviewing historical data such as risk registers and failure histories from similar situations, and creating and analyzing risk-identification checklists. The information collected at this stage must be properly recorded to ensure it retains its usefulness in the future. Therefore, each risk should have a detailed description, a clearly defined category, and a risk indicator.

Risk Analysis. This step is split into two parts: qualitative risk analysis and quantitative risk analysis. Qualitative risk analysis involves prioritizing identified risks for further action. This helps identify high-priority risks and ensures that these risks are dealt with first. Factors used for prioritizing risks include their probability of occurring and the magnitude of their impact if they do occur, as well as when they are expected to occur and their impact on other interconnected areas such as cost and schedule.

The quantitative risk analysis process uses the prioritized list of risks produced by qualitative risk analysis to analyze the effect of the listed risks and assign a numerical rating to those risks. These ratings help in assessing the probability and impact of the identified risks. Quantitative risk analysis also creates an overall risk score that is applicable to the situation.

Risk-Response Planning. After the risks have been identified and prioritized, the next step is to develop options and determine the appropriate actions to take for the risks. This step addresses the risks by their priority and ensures that planned risk responses are cost effective, timely, realistic, appropriate to the significance of the risk, agreed on by all concerned parties, and owned by a responsible individual. Sometimes, it is necessary to choose the best risk response from a variety of options.

There are several risk-response strategies available to deal with different kinds of risks. For example, strategies for negative risks include avoidance, transfer, and mitigation, while strategies for positive risks include exploitation, sharing, and enhancement. Additional strategies include acceptance and contingency responses. Contingency responses are the possible actions that are identified and documented for dealing with a potential risk.

Risk Monitoring and Control. This is the process of keeping track of identified risks, identifying, analyzing, and planning for newly arising risks, monitoring risk indicators and the trigger conditions for contingency plans, reanalyzing existing risks, and monitoring residual risks. This step also involves reviewing the implementation of risk responses and evaluating their effectiveness. This step is an ongoing process and helps determine if the entire process has been successful.

Applications and Products

As mentioned earlier, risk analysis and management is a discipline that is used in various fields, some of which are described below.

Insurance. As the oldest form of risk management, the insurance industry has the most experience in developing and implementing risk analysis and management techniques. Many insurers have implemented solid risk-management strategies that ensure the accuracy of their data, no matter the format,

Fascinating Facts About Risk Analysis and Management

- The World Bank has determined that one U.S. dollar invested in disaster prevention saves seven U.S. dollars in disaster-related economic losses.
- Insurance companies began centuries ago in marketplaces, where individuals or corporations met to discuss, combine, and distribute risk. In 1688, individuals who provided shipping insurance often gathered at a coffee shop on Tower Street in London. To encourage their patronage, the owner of the shop provided them with reliable shipping information. The presence of the insurers attracted sailors, ship owners, and other merchants and before long the coffee shop became a commercial center for maritime trade. The name of the coffee shop owner was Edward Lloyd, and it was his "Edward Lloyd's Coffee House" that eventually became Lloyd's of London.
- Risk management has become so important that many risk-avoidance measures have been enacted into law to help ensure their effectiveness. For example, wearing a seatbelt while driving, wearing a helmet while riding a motorcycle, installing functional fire alarm systems in homes and office buildings, and providing lifeguards at public swimming pools are some of the risk-management activities that are mandated by law in many countries.
- Australia and Japan are two countries that are prone to severe natural disasters such as tsunamis, floods, wildfires, and earthquakes. As a result, both countries have developed and implemented effective disaster risk-management plans. Measures such as setting up and monitoring early-warning systems, limiting construction in disaster-prone areas, and educating the public on proper emergency procedures have helped reduce, and in some cases, eliminate, the loss of lives when disaster strikes.
- Immediately after the 9.0 magnitude earthquake that struck Japan on March 11, 2011, phone signals were shut off to provide more bandwidth for emergency services. The Internet remained accessible so people could contact friends and family and share news.

use stress testing to determine the effect of sudden market changes and ensure that the company is not adversely affected, and employ advanced risk-analytic tools to identify potential risks, apply effective risk-response strategies, and mitigate losses. These strategies also improve the accuracy, availability, reliability, and level of detail of the data that company executives require to make business decisions. Finally, these strategies reduce insurers' operating costs and save them thousands of work hours.

Environmental Risk Management. Since the mid-twentieth century, when the importance of environmental conservation came to the fore, leading environmental organizations have developed risk analysis and management methods that are unique to the field. The United States Environmental Protection Agency (EPA) has developed a risk-assessment process to provide information on potential health or ecological risks. The EPA uses this four-step process to characterize the nature and magnitude of health risks to humans and ecological receptors from chemical contaminants and other harmful physical or biological entities that may be present in the environment. Risk managers then use this information to help them decide how best to protect humans and the environment from contaminants or other harmful entities. Similar environmental-protection entities in other countries, such as the United Nations Environment Programme (UNEP), the European Environment Agency (EEA), and India's Ministry of Environment and Forests, as well as research organizations such as the Ecologic Institute and RTI International, also collect and compile data relevant to regions all over the world that assist managers in assessing the risks of operating in those regions.

Risk management in this field also deals with natural disasters such as earthquakes, floods, drought, wild fires, tornadoes, hurricanes, severe snowstorms, and tsunamis. Analyzing data from previous similar catastrophic events can help personnel in this field anticipate and plan for the recurrence of an event. For example, using radiocarbon-dated samples of soil collected from a trench dug across the San Andreas Fault in California, seismologist Kerry Sieh was able to predict with accuracy how often this particular fault would create large earthquakes in Southern California. Weather forecasting is another direct beneficiary of risk analysis and management. Meteorologists use historical data and apply probability calculations to the data they collect to predict weather patterns, both good and bad. Their efforts usually give local authorities and citizens of the affected area some time to prepare for the occurrence of the weather event.

Project Risk Management. Risk management is especially important in this area, mainly because of the various uncertainties associated with executing and managing projects. Many projects have insufficient or uncertain budgets, operate on a compressed schedule, and have goals and requirements that may change constantly over the life of the project. With these constraints in mind, it becomes clear that the management of project risks is a crucial part of successfully managing a project. Indeed, many new tools have been introduced to this field to help manage the inherent uncertainty of projects. Failure to manage the risks on a project can have multiple effects, including the complete failure of the project. Organizations that proactively and consistently manage risk on their projects have higher rates of success.

Information Technology (IT) Risk Management. In the light of all the risks associated with setting up and operating IT systems, it is clear that risk management is a vital part of this field. Hackers, viruses, accidental disclosure of sensitive information, natural disasters, and catastrophic system failures are only some of the risks that many IT systems face on a regular basis. A variety of risk-management methodologies and tools for managing risk in this field exists, including the National Institute of Standards and Technology (NIST) Special Publication (SP) 800-39, *Managing Information Security Risk: Organization, Mission and Information System View.* This details the United States' government standard. Another tool is the Operationally Critical, Threat, Asset and Vulnerability Evaluation (OCTAVE) process, which was developed by the Software Engineering Institute (SEI) at Carnegie Mellon University.

OCTAVE exists to help organizations improve their ability to manage and protect themselves from information-security risks. Other tools include the Facilitated Risk Assessment Process (FRAP), which was developed by Thomas R. Peltier, president of his own information security training firm, and the Consultative, Objective and Bi-Functional Risk Analysis (COBRA) process, which was originally created by C&A Systems Security. These and other techniques help improve risk management procedures in the IT field.

Impact on Industry

The impact of risk management on various industries cannot be truly quantified. From an analysis of gambling and games of chance, risk analysis and management has grown to affect nearly all industries in some shape or form. Research on the effectiveness of risk-management methods continues to highlight successes and areas for possible improvement. The evolution of new branches of risk management, such as enterprise risk management, ensures that government agencies and major corporations will continue to find better ways to reduce their risks and enhance their business operations.

Careers and Course Work

Risk-management professionals are in demand in virtually every industry. As more industries realize the importance of effective risk management, the demand for experts in this field can only continue to grow. The U.S. Bureau of Labor Statistics estimates that there are currently about 500,000 risk managers in the United States and that this number will increase by 13 percent, or 64,000, before 2016.

Pursuing a career in this field requires an affinity for numbers and calculations. High school course work should include intensive work in mathematics, especially in statistics and calculus.

Typical course work for a bachelor's degree in risk management includes actuarial science, applied mathematics, economics, statistics, and project management. Other course work includes algebra, finite math and calculus, decision theory, international business, insurance and risk, life and health risk management, and property and liability risk management. Currently, several universities in the United States offer bachelor's degrees in risk management.

An alternative to acquiring a bachelor's degree in risk management is to acquire a degree in a different but related field and then pursue a master's degree in risk management. Examples of related fields include insurance, actuarial science, applied mathematics, economics, international business, logistics management, statistics, and project management.

A master's degree in risk analysis and management or years of practical experience in the field lends an edge to job seekers, as employers prefer to hire professionals with an advanced degree or broader experience. Professional certifications are also highly valued and membership in professional organizations is encouraged.

Several universities in the United States offer master's degrees in risk management. A significant number of these degrees are also available online.

Because risk analysis and management is such a vital part of many businesses' operations, well-trained, experienced risk managers who display a strong grasp of the operations of various departments within their organization are prime candidates for promotion to top management positions. Some risk managers transfer to closely related positions within or outside their chosen field and those with extensive experience and sufficient funding may choose to start their own consulting firms.

Social Context and Future Prospects

There is still plenty of room for growth in this field. As more industries adopt risk-management procedures, more opportunities will arise. In the fields where risk analysis and management is firmly established, practitioners will continue to find ways to improve the tools and techniques currently in use.

For example, a study that analyzed the effectiveness of risk management in project risk planning across industries and countries was published by researchers Ofer Zwikael and Mark Ahn in 2011. They studied projects in three countries—Israel, Japan, and New Zealand—and drew several conclusions. They found risk management to be effective in reducing the impact of risk levels on project success rates. They also found that integrating risk management with other project management practices may improve the success of projects.

Ezinne Amaonwu, LL.B, M.A.P.W.

Further Reading

Crouhy, Michel, Dan Galai, and Robert Mark. *The Essentials of Risk Management.* New York: McGraw-Hill, 2006. The authors, corporate governance experts, provide a clear, well-written text about the field.

IBM Center for the Business of Government. *Managing Risk in Government: An Introduction to Enterprise Risk Management.* Washington, D.C.: Karen Hardy. *http://www.businessofgovernment.org/sites/default/files/RiskinGovernment.pdf.*

Institute of Risk Management. A Risk Management Standard. London. http://www.theirm.org/publications/PUstandard.html.

International Risk Management Institute. *Management: Why and How: An Illustrative Introduction to Risk Management for Business Executives.* Dallas: George L. Head. http://www.burnsagency.com/risk-management-why-and-how.pdf.

Vose, David. *Risk Analysis: A Quantitative Guide.* 3d ed. Hoboken, N.J.: John Wiley & Sons, 2008. This book offers a comprehensive and easy-to-understand overview of the field.

Web Sites

Global Association of Risk Professionals
http://www.garp.org

Professional Risk Managers' International Association
http://www.prmia.org

Public Risk Management Association
http://www.primacentral.org

Risk Management Association
http://www.rmahq.org/RMA

See also: Applied Mathematics; Earthquake Prediction.

ROBOTICS

FIELDS OF STUDY

Physics; mechanical engineering; electrical engineering; systems engineering; design; artificial intelligence; machine learning; computer science; mathematics; human-computer interaction; computer programming; field robotics; space robotics; medical robotics; underwater robotics; nanorobotics; molecular robotics; evolutionary robotics; neuroscience; kinematics; dynamics; feedback control; haptics; mechatronics; virtual reality; material science.

SUMMARY

Robotics is an interdisciplinary scientific field concerned with the design, development, operation, and assessment of electromechanical devices used to perform tasks that would otherwise require human action. Robotics applications can be found in almost every arena of modern life. Robots, for example, are widely used in industrial assembly lines to perform repetitive tasks. They have also been developed to help physicians perform difficult surgeries and are essential to the operation of many advanced military vehicles. Among the most promising robot technologies are those that draw on biological models to solve problems, such as robots whose limbs and joints are designed to mimic those of insects and other animals.

KEY TERMS AND CONCEPTS

- **Actuator:** Electromechanical device that enables a robot to perform physical motions.
- **Articulated:** Of a robot part, having joints that allow for movement.
- **Autonomous:** Robot operation that takes place without preprogrammed behaviors or human supervision.
- **Biomimetic Robot:** Robot whose design and engineering is based on a biological model that exists in nature.
- **Bottom-Up Control Structure:** Approach enabling robots to solve problems based on reasoning and observation, rather than preexisting programming. Also known as data-driven control structure; event-based control structure.
- **End Effector:** Device at the end of a manipulator that grasps and moves objects.
- **Manipulator:** Mechanical robot arm capable of moving materials, parts, and other objects.
- **Neural Network:** Computational framework, mimicking connections among neurons in human brains, that allows robots to learn from and adapt to their environments.
- **Servomechanism:** Feedback-control device used to control mechanical movements.
- **Servorobot:** Robot that follows a preprogrammed set of motions.
- **Tactile Sensing:** Ability of a robot to sense qualities of touch such as pressure, texture, temperature, and torque.
- **Task Environment:** Physical space in which robots operate, or the conditions of that space; also known as world space.
- **Teleoperation:** Management of an autonomous robot from a remote location.
- **Work Envelope:** Total physical space within which a manipulator and its end effector can move.

Definition and Basic Principles

Robotics is the science of robots—machines that can be programmed to carry out a variety of tasks independently, without direct human intervention. Although robots in science fiction tend to be androids or humanoids (robots with recognizable human forms), most real-life robots, especially those designed for industrial use, do not resemble humans physically. Robots typically consist of at least three parts: a mechanical structure (most commonly a robotic arm) that enables the robot to physically affect either itself or its task environment; sensors that gather information about physical properties such as sound, temperature, motion, and pressure; and some kind of processing system that transforms data from the robot's sensors into instructions about what actions to perform. Some devices, such as the search-engine bots that mine the Internet daily for data about links and online content, lack mechanical components. However, they are nevertheless considered robotic because they can perform repeated tasks without supervision.

Many robotics applications also involve the use of artificial intelligence. This is a complex concept with a shifting definition, but in its most basic sense, a robot with artificial intelligence possesses features or capabilities that mimic human thought or behavior. For example, one aspect of artificial intelligence involves creating parallels to the human senses of vision, hearing, or touch. The friendly voices at the other ends of customer-service lines, for example, are increasingly likely to be robotic speech-recognition devices capable not merely of hearing callers' words but also of interpreting their meanings and directing the customers' calls intelligently.

More advanced artificial intelligence applications give robots the ability to assess their environmental conditions, make decisions, and independently develop efficient plans of action for their situations—and then modify these plans as circumstances change. Chess-playing robots do this each time they assess the state of the chessboard and make a new move. The ultimate goal of artificial intelligence research is to create machines whose responses to questions or problems are so humanlike as to be indistinguishable from those of human operators. This standard is the so-called Turing test, named after the British mathematician and computing pioneer Alan Turing.

Background and History

The word "robot" comes from a Czech word for "forced labor" that the Czech writer Karel Čapek used in his 1921 play *R.U.R.* about a man who invents a humanlike automatic machine to do his work. During the 1940's, as computing power began to grow, the influential science-fiction writer Isaac Asimov began applying "robotics" to the technology behind robots. The 1950's saw the development of the first machines that could properly be called robots. These prototypes took advantage of such new technologies as transistors (compact, solid-state devices that control electrical flow in electronic equipment) and integrated circuits (complex systems of electronic connections stamped onto single chips) to enable more complicated mechanical actions. In 1959, an industrial robot was designed that could churn out ashtrays automatically. Over the ensuing decades, public fascination with robots expanded far beyond their actual capabilities. It was becoming clear that creating robots that could accomplish seemingly simple tasks—such as avoiding obstacles while walking—was a surprisingly complex problem.

During the late twentieth century, advances in computing, electronics, and mechanical engineering led to rapid progress in the science of robotics. These included the invention of microprocessors, single integrated circuits that perform all the functions of computers' central processing units; production of better sensors and actuators; and developments in artificial intelligence and machine learning, such as a more widespread use of neural networks. (Machine learning is the study of computer programs that improves their performance through experience.)

Cutting-edge robotics applications are being developed by an interdisciplinary research cohort of computer scientists, electrical engineers, neuroscientists, psychologists, and others, and combine a greater mechanical complexity with more subtle information processing systems than were once possible. Homes may not be populated with humanoid robots with whom one can hold conversations, but mechanical robots have become ubiquitous in industry. Also, unmanned robotic vehicles and planes are essential in warfare, search-engine robots crawl the World Wide Web every day collecting and analyzing data about Internet links and content, and robotic surgical tools are indispensable in health care. All this is evidence of the extraordinarily broad range of problems robotics addresses.

How It Works

Sensing. To move within and react to the conditions of task environments, robots must gather as much information as possible about the physical features of their environments. They do so through a large array of sensors designed to monitor different physical properties. Simple touch sensors consist of electric circuits that are completed when levers receive enough pressure to press down on switches. Robotic dogs designed as toys, for example, may have touch sensors in their backs or heads to detect when they are being petted and signal them to respond accordingly. More complex tactile sensors can detect properties such as torque (rotation) or texture. Such sensors may be used, for example, to help an assembly-line robot's end effector control the grip and force it uses to turn an object it is screwing into place.

Light sensors consist of one or more photocells that react to visible light with decreases in electrical resistance. They may serve as primitive eyes, allowing unmanned robotic vehicles, for example, to detect the bright white lines that demarcate parking spaces and maneuver between them. Reflectance sensors emit beams of infrared light, measuring the amounts of that light that reflect back from nearby surfaces. They can detect the presence of objects in front of robots and calculate the distances between the robots and the objects—allowing the robots either to follow or to avoid the objects. Temperature sensors rely on internal thermistors (resistors that react to high temperatures with decreases in electrical resistance). Robots used to rescue human beings trapped in fires may use temperature sensors to navigate away from areas of extreme heat. Similarly, altimeter sensors can detect changes in elevation, allowing robots to determine whether they are moving up or down slopes.

Other sensor types include magnetic sensors, sound sensors, accelerometers, and proprioceptive sensors that monitor the robots' internal systems and tell them where their own parts are located in space. After robots have collected information through their sensors, algorithms (mathematical processes based on predefined sets of rules) help them process that information intelligently and act on it. For example, a robot may use algorithms to help it determine its location, map its surroundings, and plan its next movements.

Motion and Manipulation. Robots can be made to move around spaces and manipulate objects in many different ways. At the most basic level, a moving robot needs to have one or more mechanisms consisting of connected moving parts, known as links. Links can be connected by prismatic or sliding joints, in which one part slides along the other, or by rotary or articulated joints, in which both parts rotate around the same fixed axis. Combinations of prismatic and rotary joints enable robotic manipulators to perform a host of complex actions, including lifting, turning, sliding, squeezing, pushing, and grasping. Actuators are required to move jointed segments or robot wheels. Actuators may be electric or electromagnetic motors, hydraulic gears or pumps (powered by compressed liquid), or pneumatic gears or pumps powered by pressurized gas. To coordinate the robots' movements, the actuators are controlled by electric circuits.

Motion-description languages are a type of computer programming language designed to formalize robot motions. They consist of sets of symbols that can be combined and manipulated in different ways to identify whole series of predefined motions in which robots of specified types can engage. Motion-description languages were developed to simplify the process of manipulating robot movements by allowing different engineers to reuse common sets of symbols to describe actions or groups of actions, rather than having to formulate new algorithms to describe every individual task they want robots to perform.

Control and Operation. A continuum of robotic control systems ranges from fully manual operation to fully autonomous operation. On the one hand, a human operator may be required to direct every movement a robot makes. For example, some bomb disposal robots are controlled by human operators working only a few feet away, using levers and buttons to guide the robots as they pick up and remove the bombs. On the other side of the spectrum are robots that operate with no human intervention at all, such as the KANTARO—a fully autonomous robot that slinks through sewer pipes inspecting them for damage and obstructions. Many robots have control mechanisms lying somewhere between these two extremes.

Robots can also be controlled from a distance. Teleoperated systems can be controlled by human operators situated either a few centimeters away, as in robotic surgeries, or millions of miles away, as in outer space applications. "Supervisory control" is a term given to teleoperation in which the robots themselves are capable of performing the vast majority of their tasks independently; human operators are present merely to monitor the robots' behavior and occasionally offer high-level instructions.

Artificial Intelligence. Three commonly accepted paradigms, or patterns, are used in artificial intelligence robotics: hierarchical, reactive, and hybrid. The hierarchical paradigm, also known as a top-down approach, organizes robotic tasks in sequence. For example, a robot takes stock of its task environment, creates a detailed model of the world, uses that model to plan a list of tasks it must carry out to achieve a goal, and proceeds to act on each task in turn. The performance of hierarchical robots tends to be slow and disjointed since every time a change occurs in the environment, the robot pauses to reformulate its plan. For example, if such a robot is moving forward to reach a destination and an obstacle is placed in its way, it must pause, rebuild its model of the world, and begin lurching around the object.

In the reactive (or behavioral) paradigm, also known as a bottom-up approach, no planning occurs. Instead, robotic tasks are carried out spontaneously in reaction to a changing environment. If an obstacle is placed in front of such a robot, sensors can quickly incorporate information about the obstacle into the robot's actions and alter its path, causing it to swerve momentarily.

The hybrid paradigm is the one most commonly used in artificial intelligence applications being developed during the twenty-first century. It combines elements of both the reactive and the hierarchical models.

Applications and Products

Industrial Robots. In the twenty-first century, almost no factory operates without at least one robot—more likely several—playing some part in its manufacturing processes. Welding robots, for example, consist of mechanical arms with several degrees of movement and end effectors in the shape of welding guns or grippers. They are used to join metal surfaces together by heating and then hammering them, and produce faster, more reliable, and more uniform results than human welders. They are also less vulnerable to injury than human workers. Another common industrial application of robotics is silicon-wafer manufacturing, which must be performed within meticulously clean rooms so as not to contaminate the semiconductors with dirt or oil. Humans are far more prone than robots to carry contaminants on them.

Six major types of industrial robots are defined by their different mechanical designs. Articulated robots are those whose manipulators (arms) have at least three rotary joints. They are often used for vehicle assembly, die casting (pouring molten metal into molds), welding, and spray painting. Cartesian robots, also known as gantry robots, have manipulators with three prismatic joints. They are often used for picking objects up and placing them in different locations, or for manipulating machine tools. Cylindrical robots have manipulators that rotate in a cylindrical shape around a central vertical axis. Parallel robots have both prismatic and rotary joints on their manipulators. Spherical robots have manipulators that can move in three-dimensional spaces shaped like spheres. SCARA (Selective Compliant Assembly Robot Arm) robots have two arms connected to vertical axes with rotary joints. One of their arms has another joint that serves as a wrist. SCARA robots are frequently used for palletizing (stacking goods on platforms for transportation or loading).

Service Robots. Unlike industrial robots, service robots are designed to cater to the needs of individual people. Robopets, such as animatronic dogs, provide companionship and entertainment for their human owners. The Sony Corporation's AIBO (Artificial Intelligence roBOt) robopets use complex systems of sensors to detect human touch on their heads, backs, chins, and paws, and can recognize the faces and voices of their owners. They can also maintain their balance while walking and running in response to human commands. AIBOs also function as home-security devices, as they can be set to sound alarms when their motion or sound detectors are triggered. Consumer appliances, such as iRobot Corporation's robotic vacuum cleaner, the Roomba, and the robotic lawn mover, the RoboMower, developed by Friendly Robotics, use artificial intelligence approaches to safely and effectively maneuver around their task environments while performing repetitive tasks to save their human users time.

Even appliances that do not much resemble public notions of what robots should look like often contain robotic components. For example, digital video recorders (DVRs) such as TiVos, contain sensors, microprocessors, and a basic form of artificial intelligence that enable them to seek out and record programs that conform to their owners' personal tastes. Some cars can assist their owners with driving tasks such as parallel parking. An example is the Toyota Prius, which offers an option known as Intelligent Parking Assist. Other cars have robotic seats that can lift elderly or disabled passengers inside.

Many companies or organizations rely on humanoid robots to provide services to the public. The Smithsonian National Museum of American History, for example, uses an interactive robot named Minerva to guide visitors around the museum's exhibits, answering questions and providing information about individual exhibits. Other professional roles filled by robots include those of receptionists, floor cleaners, librarians, bartenders, and secretaries. At least one primary school in Japan has even been experimenting with a robotic teacher developed by a scientist at the Tokyo University of Science. However, an important pitfall of humanoid robots is their susceptibility to the uncanny valley phenomenon. This is the theory that as a robot's appearance and behavior becomes more humanlike, people will respond to it more positively. The uncanny valley refers to a point at which this phenomenon reaches a peak and starts to reverse. In other words, an extraordinarily humanlike robot that is still identifiably a machine will cause people to feel revulsion and fear rather than empathy.

Medical Uses. Robotic surgery has become an increasingly important area of medical technology. In most robotic surgeries, a system known as a master-slave manipulator is used to control robot movements. Surgeons look down into electronic displays showing their patients' bodies and the robots' tool tips. The surgeons use controls attached to consoles to precisely guide the robots' manipulators within the patients' bodies. A major benefit of robotic surgeries is that they are less invasive—smaller incisions need to be made because robotic manipulators can be extremely narrow. These surgeries are also safer because robotic end effectors can compensate for small tremors or shakes in the surgeons' movements that could seriously damage their patients' tissues if the surgeons were making the incisions themselves. Teleoperated surgical robots can even allow surgeons to perform operations remotely, without the need to transport patients over long distances. Surgical robots such as the da Vinci system are used to conduct operations such as prostatectomy, cardiac surgery, bariatric surgery, and various forms of neurosurgery.

Humanoid robots are also widely used as artificial patients to help train medical students in diagnosis and procedures. These robots have changing vital signs such as heart rates, blood pressure, and pupil dilation. Many are designed to breathe realistically, express pain, urinate, and even speak about their conditions. With their help, physicians-in-training can practice drawing blood, performing cardiopulmonary resuscitation (CPR), and delivering babies without the risk of harm to real patients.

Other medical robots include robotic nurses that can monitor patients' vital signs and alert physicians to crises and smart wheelchairs that can automatically maneuver around obstacles. Scientists are also working on developing nanorobots the size of bacteria that can be swallowed and sent to perform various tasks within human bodies, such as removing plaque from the insides of clogged arteries.

Robot Exploration and Rescue. One of the most intuitive applications of robotic technology is the concept of sending robots to places too remote or too dangerous for human beings to work in—such as outer space, great ocean depths, and disaster zones. The six successful manned Moon landings of the Apollo program carried out during the late 1960's and early 1970's are dwarfed in number by the unmanned robot missions that have set foot not only on the Moon but also on other celestial bodies, such as planets in the solar system. The wheeled robots Spirit and Opportunity, for example, began analyzing material samples on Mars and sending photographs back to Earth in 2004. Roboticists have also designed biomimetic robots inspired by frogs that take advantage of lower gravitational fields, such as those found on smaller planets, to hop nimbly over rocks and other obstacles.

Robots are also used to explore the ocean floor. The Benthic Rover, for example, drags itself along the seabed at depths up to 2.5 miles below the surface. It measures oxygen and food levels, takes soil samples, and sends live streaming video up to the scientists above. The rover is operated by supervisory control and requires very little intervention on the part of its human operators.

Rescue robots seek out, pick up, and safely carry injured humans trapped in fires, under rubble, or in dangerous battle zones. For example, the U.S. Army's Bear (Battlefield Extraction-Assist Robot) is a bipedal robot that can climb stairs, wedge itself through narrow spaces, and clamber over bumpy terrain while carrying weights of up to three hundred pounds.

Impact on Industry

In 2009, the total value of the global robotics industry was estimated at $17.6 billion, with industrial robots and professional service robots making up the two largest segments of the market. According to a 2006 report by an international organization dedicated to assessing the state of technology research, Japan was leading the world in the development of humanoid robots, aspects of robot movement and manipulation, and entertainment and personal robots, while the United States was making the greatest strides toward improving robot navigation in difficult terrain, especially with regard to defense, space, and underwater applications. Significant robotics research in other areas has also been taking place in South Korea, Australia, Israel, and many European countries.

Government and University Research. One of the biggest sources of funding for robotics research in the United States is the Defense Advanced Research Projects Agency (DARPA). It is also the biggest client for certain kinds of robotics applications. The agency is primarily concerned with robots that have military uses, such as self-driven tanks capable of traversing rough terrain and maneuvering quickly around obstacles; robot medics that can treat soldiers on the battlefield; bipedal robots that can carry hundreds of pounds of supplies; and drone (pilotless) aircraft that can drop bombs on enemy targets. DARPA not only plans and executes its own research and development projects but also supports robotics research taking place in academic and corporate settings through direct grants and robotics competitions. For example, the DARPA Urban Challenge, which is backed by millions of dollars in prize money, invites robotics researchers to design and build autonomous ground vehicles capable of moving around independently in urban environments, performing tasks such as merging into traffic, driving around traffic circles, maneuvering through intersections, and avoiding obstacles—all while carrying out simulated military missions.

In 2008, academic researchers from the top robotics laboratories in the United States, including the Georgia Institute of Technology, Carnegie Mellon University, the University of California, Berkeley, Rensselaer Polytechnic Institute, Stanford University, and the Massachusetts Institute of Technology, formed an association known as the Computing Community Consortium. The consortium's goal is to identify major research opportunities in computing, such as robotics. It provides resources such as conferences, workshops, and planning support for projects and hopes to reduce what it sees as the gap between the level of robotics development investment in the United States and in countries such as Japan, South Korea, and the European Union.

Industry and Business. The manufacturing industry is the biggest user of robotics technologies, which help it cut costs and increase efficiency. Within manufacturing, the automotive industry is perhaps one of the biggest users of robots in its production methods. For example, welding robots are used heavily by car manufacturers such as General Motors, Toyota, Honda, and Ford, as are robotic manipulators that cut and shape parts, assemble the parts, inspect them for defects, and spray paint the finished cars. Some estimates hold that of every ten workers in automotive plants across the world, only eight are human—the other two are robotic arms. Robotics engineers are also in demand within the agricultural industry for such tasks as designing crop harvesters that can operate autonomously and improving the design of robotic milking machines.

The toy and entertainment industry is another sector that invests in the development of robotics applications. Companies such as Tamiya Corporation, Solarbotics, and OWI Robots sell build-your-own-robot kits designed for both children and adults. Complete with sensors, circuit boards, and mechanics, these kits help enthusiasts build robots that move independently and respond to light, sound, and touch. Additionally, aside from animatronic robopets such as AIBO, Sony produces a line of entertainment androids that can walk and dance. In 2007, Toyota launched a 5-foot-tall robot capable of using the seventeen different joints in its arms and hands to play musical compositions on a violin. Robots such as these are most often used in settings such as theme parks, casinos, and restaurants.

Careers and Course Work

Courses in advanced mathematics, physics, computer science, electrical engineering, and mechanical engineering make up the foundational requirements for students interested in pursuing careers as robotics engineers. Earning a bachelor of science degree in any of these fields would serve as an appropriate preparation for graduate work in a similar area. In most circumstances, either a master's degree or a doctorate is a necessary qualification for the most advanced future career opportunities in both academia and industry. However, an advanced degree is not generally required for a career as a robotics technician—someone who maintains and repairs robots rather than designs them.

Fascinating Facts About Robotics

- Each year since 1996, an international soccer competition has been held with robot players. The first soccer robots were designed with wheels or four legs, but later players have been, like humans, bipedal.
- NASA's Tumbleweed Rover, a large globe-shaped robot, was designed to explore ice caps on distant planets as well as the icy terrain of Antarctica, where it was first tested in 2004.
- The Japanese animatronic robot Paro, which resembles a furry seal, has been serving as a friendly and therapeutic companion to residents in nursing homes since 2005. Studies have shown spending time with Paro improves the mental and physical well-being of elderly subjects.
- When DARPA wanted a spy robot that could scale vertical surfaces, it developed Spinybot, which accomplishes this task using tiny clawlike spines and sticky adhesive—the same mechanisms that allow geckos to climb walls.
- In their quest to make robots more humanlike, engineers have created robots that, over time, evolve the capability to "lie" to one another about the location of scarce "food sources." The robots' artificial neural networks allow them to "learn" that their survival improves if they hide the real sources of "food" from their peers.
- Among the many human expressions some robots are able to express using a complex system of mechanical parts under realistic facial "skin" are anger, fear, sadness, happiness, surprise, curiosity, and disgust.
- Human beings have been imagining mechanical devices that operate autonomously since ancient times. One Greek myth, for example, tells of a massive bronze giant named Talos who protects the island of Crete from invaders. Talos was created by Hephaestus, the god of metalwork and technology.

Students should take substantial course work in more than one of the primary fields of study related to robotics (physics, mathematics, computer science, and engineering) because designing and testing robots requires skills drawn from multiple disciplines. In addition, anyone desiring to work in robotics should possess skills that go beyond the academic, including good physical coordination, excellent manual dexterity, and an aptitude for mechanical details. A collaborative mind-set is also an asset, as robotics work tends to be done by teams.

Careers in the field of robotics can take several different shapes. The manufacturing industry is the biggest employer of robotics engineers and technicians. Within this industry, robotics professionals might focus on developing, maintaining, or repairing production-line robots used in factory assembly. Other industries in which robotics engineers and technicians often find work include aviation, agriculture, nuclear energy, telecommunications, electronics, mining, health care and medicine, and education.

Many roboticists prefer employment within academic settings. Such professionals divide their time between teaching university classes on robotics and conducting their own research projects. Others find work in government agencies such as the National Aeronautics and Space Administration (NASA) and DARPA, focusing on large-scale robotics applications in such areas as space exploration, warfare, and disaster management.

Social Context and Future Prospects

In the twenty-first century, the presence of robots in factories all over the world is taken for granted. Meanwhile, robots are also increasingly entering daily life in the form of automated self-service kiosks at supermarkets, electronic lifeguards that detect when swimmers in pools are struggling, and cars whose robotic speech-recognition software enables them to respond to verbal commands. A science-fiction future in which ubiquitous robotic assistants perform domestic tasks such as cooking and cleaning may not be far away, but many technological limitations must be overcome for that to become a reality. For example, it can be difficult for robots to process multipart spoken commands that have not been preprogrammed—a problem for researchers in the artificial intelligence field of natural language processing.

Robots that provide nursing care or companionship to the infirm are not merely becoming important parts of the health care industry but may also provide a solution to the problem increasingly faced by countries in the developed world—a growing aging population who need more caretakers than can be found among younger adults. Robots are also likely to play a growing role in the American conversation about illegal immigration: As robotics technology improves and becomes less expensive, companies may well turn to cheap, efficient robots to do jobs that are typically performed by immigrant human labor, particularly in such areas as agriculture and manufacturing.

Meanwhile, some observers are concerned that the rise of industrial and professional service robots is already eliminating too many jobs held by American workers. Many of the jobs lost in the 2008-2009 recession within the struggling automotive industry, for example, are gone for good, because costly human workers were replaced by cheaper robotic arms. However, the issue is more complicated than that. In certain situations, the addition of robots to a factory's workforce can actually create more jobs for humans. Some companies, for example, have been able to increase production and hire additional workers with the help of robot palletizers that make stacking and loading their products much faster.

Safety concerns can sometimes hinder the acceptance of new robotic technologies, even when they have proven to be less likely than humans to make dangerous mistakes. Robotic sheep shearers in Australia, for example, have met with great resistance from farmers because of the small risk that the machines may nick a major artery as they work, causing the accidental death of a sheep.

M. Lee, B.A., M.A.

Further Reading

Faust, Russell A., ed. *Robotics in Surgery: History, Current and Future Applications.* New York: Nova Science, 2007. Includes chapters on telerobotics, applications for neurosurgery, urologic surgery, and cardiothoracic surgery.

Floreano, Dario, and Claudio Mattuissi. *Bio-Inspired Artificial Intelligence: Theories, Methods, and Technologies.* Cambridge, Mass.: MIT Press, 2008. Describes engineering methods and technologies based on biological systems. Includes chapters on evolutionary, cellular, neural, developmental, immune, behavioral, and collective systems.

Gutkind, Lee. *Almost Human: Making Robots Think.* 2006. Reprint. New York: W. W. Norton, 2009. An investigative journalist reports on his six years of observations of research at Carnegie Mellon's Robotics Institute.

Jones, Joseph L. *Robot Programming: A Practical Guide to Behavior-Based Robotics.* New York: McGraw-Hill, 2004. A guide to planning and programming a behavior-based robot. Designed for readers with no programming experience, but a mathematics background is helpful. Chapters contain programming exercises.

Siciliano, Bruno, et al. *Robotics: Modeling, Planning, and Control.* London: Springer, 2009. Comprehensive textbook on the technology behind robotic manipulators. Includes case studies, end-of-chapter problems, and an electronic solutions manual.

Siciliano, Bruno, and Oussama Khatib, eds. *The Springer Handbook of Robotics.* London: Springer, 2008. A comprehensive review of the history and existing technology in a huge number of subfields of robotics, including virtual simulation, sensor networks, and robots in agriculture.

Web Sites

Institute of Electrical and Electronics Engineers (IEEE)
Robotics and Automation Society
http://www.ieee-ras.org

National Aeronautics and Space Administration
Robotics Alliance Project
http://robotics.nasa.gov

National Science Foundation
A Special Report: Robotics
http://www.nsf.gov/news/special_reports/robotics/index.jsp

See also: Artificial Intelligence; Computer Engineering; Computer Science; Engineering; Software Engineering; Space Environments for Humans; Video Game Design and Programming; Virtual Reality.

S

SANITARY ENGINEERING

FIELDS OF STUDY

Algebra; geometry; drafting; surveying; chemistry; biology; mechanics; water supplies; bacteriology; microbiology; water transport and distribution; geography; topography.

SUMMARY

Sanitary engineering is the discipline of civil engineering that addresses matters of public health, including water quality (such as potable drinking water, wastewater treatment, storm-sewer drainage, and swimming pool cleanliness), sewage collection, recycling, and disposal. In some jurisdictions, sanitary engineering is now called environmental engineering and also covers issues of air pollution, water pollution, hazardous waste management, and land management. The importance of this discipline, which involves the practical application of engineering and life sciences, is its responsibility for providing a healthy environment for people.

KEY TERMS AND CONCEPTS

- **Aggregation:** When ionic repulsive forces are removed, small particles bind together to form larger particles.
- **Coagulation:** Smaller solid particles are induced to bind together to form larger particles.
- **Contamination:** Something in water that makes it unfit for drinking or cooking.
- **Desalination:** Process of removing salt from water.
- **Disinfection:** Killing of bacteria, viruses, fungi, and other disease-causing microbes.
- **Filtration:** Process of retaining particles based on size.
- **Hazardous Waste:** Discarded material that is potentially harmful to a person's health or the environment.
- **Pathogen:** Microbe that causes disease, especially in humans.
- **Pollution:** Fouling of water, air, or soil, typically with a substance produced by humans.
- **Potable:** Safe for drinking.
- **Sedimentation:** Process of using gravity to pull particles out of a solution.

DEFINITION AND BASIC PRINCIPLES

Sanitary engineering is the practical application of science and engineering to provide a safe natural environment in which people can live free from disease. It requires knowledge of organic and inorganic chemistry, biology, and bacteriology to recognize the health risks of various water impurities to humans and animals and the means to eliminate them. It also requires knowledge of mathematics (algebra, geometry, quantitative analysis, statistical significance), engineering (drafting, surveying, mechanics, materials, design, and construction), and environmental science (water supplies, topography, land management). Knowledge of the latter fields is necessary to develop and maintain the appropriate infrastructure for water transport, purification, and distribution.

Sanitary engineering works on very large scales (a municipal water-treatment tank can hold more than one million gallons of wastewater) and very small scales (lead in drinking water is measured in parts per billion per liter). Sanitary engineers are responsible for upholding public health regulations related to disease prevention in public venues that have toilets, drinking fountains, showers, swimming pools, Jacuzzis, and hot tubs. They may also review and approve plans for the installation of private septic systems and wells, taking into consideration the results of soil percolation tests and well water quality tests.

BACKGROUND AND HISTORY

Aqueducts, intentionally constructed channels for the transport of water for drinking and bathing, were created by the Romans in the third century B.C.E. Most of these were built in underground tunnels to protect them from contamination by enemies and diseases on the ground's surface. This was the archetype of the modern plumbing system.

Sanitary engineering arose from the contemporary scientific recognition that plumbing was related to cleanliness and thus had an impact on public health and disease prevention. Plumbing inspectors were trained to recognize and protect society from the use of inferior materials, reckless installation, and unsanitary water pathways. The American Society of Sanitary Engineering for Plumbing and Sanitary Research was formed on January 29, 1906. Henry B. Davis, the chief plumbing inspector for Washington, D.C., convened with twenty-five plumbing inspectors from other states to form this national association to standardize plumbing practices so as to prevent the transmission of diseases.

The 1970's brought significant legislation in the practice of sanitary engineering. In the Safe Drinking Water Act of 1974, the Environmental Protection Agency (EPA) established standards for drinking-water quality. In the Clean Water Act of 1977, the EPA established standards for the quality of surface waters into which foreign, polluting materials are discharged.

How It Works

Water Treatment. Water treatment is the overall process of removing contaminants from water to make it safe for drinking, bathing, cooking, and swimming. Without water treatment, waterborne diseases can cause illness and death, often from dehydration following diarrhea. Waterborne pathogens include *Cryptosporidium*, *Escherichia coli*, hepatitis A virus, and *Giardia intestinalis* parasite.

People expect clean water to be clear, colorless, odorless, and tasteless. This requires particulates be removed; microbes such as bacteria, viruses, and parasites be killed; and minerals such as iron, calcium, magnesium, manganese, and sulphur be bound and removed. To achieve this, a series of specific processes must be performed: physical separation of solids by settling and filtration, chemical reactions of coagulation and disinfection, and biological methods such as aeration, bacterial digestion of sludge, and filtration through natural materials.

The choice of processes depends on the nature and volume of the water to be purified. Analytical survey must be performed initially. There are two original sources of water: surface water and groundwater. Surface water encompasses rivers, lakes, streams, and ponds. Groundwater is accessible by digging wells and generally requires less water treatment than surface water, which contains more debris and pollutants.

Coagulation. When water is first received at the water-treatment plant, large pieces of solid material such as sewage are removed by a coarse screen and discarded. Then smaller solid particles are induced to bind together to form larger particles through coagulation. Ions with multiple charges (polyelectrolytes) change the pH of the water and trigger chemical reactions that cause aggregation. Alum is frequently added to attract dirt particles, which may contain herbicides and pesticides. Lime and soda ash cause calcium and magnesium to precipitate, thus "softening" the water.

Sedimentation. The material resulting from coagulation, called floc, has sufficient weight that it sinks to the bottom of the settling tank. This separation of solids by sedimentation is a time-consuming step. Algae rise to the top, where they may be skimmed. The clearer water on the top is then slowly siphoned off for filtration. Aerobic and anaerobic bacteria may be added to the withheld solids (sludge) to digest organic waste matter and neutralize pollutants. Carbon dioxide, ammonia, and methane gases are generated. The digested sludge may then be used as a fertilizer supplement in farming.

Filtration. Remaining particles in the water may be removed by filters made of artificial membranes, nets, or natural materials. Water may be filtered by passing it through beds of sand, gravel, or pulverized coal. Activated charcoal may be added to the water first to remove color, odor, taste, and radioactivity.

As another method of removing calcium and magnesium, water may be passed through ion-exchange columns, in which sodium ions compete with these cations for binding to porous material.

Aeration is used to remove dissolved elements such as iron, sulphur, and manganese. Air is forced into the water, and the oxygen removes carbon dioxide, hydrogen sulfide, and other gases. In diffused aeration, air is bubbled through the water. In spray aeration, water is sprayed through the air.

Desalination, the process of removing salt from the water, is often employed to make ocean water drinkable in places where freshwater is scarce. The salt is removed by microfiltration and reverse osmosis.

Disinfection. Disinfection is the general method of killing pathogens (bacteria, viruses, parasites). The most common method is chlorination of the water with sodium hypochlorite bleach. Less frequently used is ultraviolet light or ozone aeration. Water may be boiled at home for disinfection in cases of emergency.

Storage. Treated water must then be stored and delivered under clean conditions to prevent recontamination. It is stored in closed tanks or reservoirs; from there, it is piped to homes and businesses. Minimal chlorine may be added to maintain cleanliness. Fluoride may also be added to water to improve dental health by helping to prevent tooth decay.

Applications and Products

Water Treatment for Medical Purposes. Water used in medical, dental, and pharmaceutical procedures must meet exceptionally high quality standards. This ultrahigh quality of water is achieved with multiple technologies beyond chlorination, nanofiltration, and carbon adsorption. These technologies include reverse osmosis, deionization, ozonation, and ultraviolet irradiation.

Home Filtration Systems. The same filtration and purification methods used in large water treatment plants have been downscaled for home use. Faucet-mount filters use carbon filtration, ion-exchange filtration, and submicron filtration to reduce sediment, chlorine, lead, mercury, iron, herbicides, pesticides, insecticides, industrial solvents, volatile organic compounds, synthetic organic compounds, and trihalomethanes (THMs, chlorine and its by-products). These apparatuses rapidly provide filtered water that tastes and smells better with less cloudiness. Shower filters typically use copper-zinc oxidation media and carbon filtration to remove chlorine for softer skin and hair. Whole-house-use water filters are plumbed into the main water line and commonly include a sediment pre-filter, then copper-zinc oxidation media and crushed mineral stone or natural pumice to reduce chlorine, then activated carbon to remove other chemicals.

Home Water Softeners. Home water softeners are water-treatment systems that address hard water problems. Hard water contains high levels of dissolved magnesium, calcium, manganese, and iron. These minerals react with soap ingredients to create a filthy coating in sinks, bathtubs, and showers. They also react with detergent ingredients to make clothes, towels, and sheets feel abrasive. Drinking glasses washed in hard water may show spotting, streaking, or a cloudy coating. These minerals may accumulate to form crusty deposits inside pipes, showerheads, and teakettles, obstructing water flow. Home water softeners use ion exchange to replace the metal ions with sodium ions, which do not react with soap ingredients or accumulate in pipes.

Low-Flow Toilets. The Energy Policy Act of 1992 contained a mandate for low-flow toilets. This mandate went into effect in 1994 and stipulated that toilets be redesigned to reduce the water use from 5 to 7 gallons of water per flush to 1.6 gallons of water per flush. Design adjustments were made such as a wider flapper valve through which water flows from the tank into the bowl and a wider trap way through which water flows from the bowl to the sewage pipe. These modifications allowed the reduced amount of water to move with greater force and efficiency.

Low-Flow Showerheads. In keeping with the water conservation movement, showerheads were redesigned to reduce the water use by 30 to 40 percent with a flow of up to two gallons per minute. Design modifications were made such as enlarging the faceplate and increasing the water pressure locally so that the reduced amount of water is dispersed with increased spray power. Some heads are still adjustable to allow for a brisk massage or a gentle rinse.

Impact on Industry

Government Regulation. In this country, water quality is the responsibility of the United States Environmental Protection Agency (EPA). Specifically, the EPA's Office of Water is charged with ensuring the safety of drinking water, implementing legislation such as the Safe Drinking Water Act and the Clean Water Act, and maintaining and improving water ecosystems to protect the health of people, animals, fish, and plants. The Office of Water is further divided into the Office of Ground Water and Drinking Water, the Office of Science and Technology, the Office of Wastewater Management, and the Office of Wetlands, Oceans and Watersheds. While there are ten regional offices of the EPA, state governments are also responsible for drinking-water

Fascinating Facts About Sanitary Engineering

- Although 75 percent of the Earth's surface is covered with water, less than 1 percent can be used by people. The rest is saltwater, inaccessible underground water, or water frozen in polar ice caps and glaciers.
- In 1854, a cholera epidemic struck the Soho district of London. Physician John Snow traced the source of the disease to contaminated water being delivered through public pumps. His investigative methods led to the science of epidemiology.
- In 1945, fluoride was first added to the public water supply in Grand Rapids, Michigan, where the hypothesis that fluoride would prevent cavities was tested. A significant reduction in the incidence of cavities was seen, and by 1960, water fluoridation was a common practice in the United States, preventing dental disease in 50 million people.
- The national water infrastructure is aging and breaking. Every year in the United States, there are 75,000 cases of compromised sanitary sewers. They release annually from 3 billion to 10 billion gallons of untreated wastewater, contaminating recreational waters and causing 5,500 cases of illness as a result.
- The Centers for Disease Control and Prevention found in an inspection of 5,000 public hot tubs that 60 percent were capable of making people sick because they had not been properly cleaned. Although a chlorine-based disinfectant may have been used in routine cleaning, water temperatures above 84 degrees Fahrenheit reduce the disinfecting properties of chlorine.
- Scientists at the Environmental Protection Agency are working on the Virtual Embryo Project, computer-based models that predict the effect of environmental commercial chemicals, including those in the water supply, on embryonic development.

quality and the preservation of aquatic environments. Such state offices are usually associated with the state public health departments. Duties may then fall to county or municipal governments and their environmental and public health agencies. Wastewater treatment plants are operated locally. The water distribution system may be maintained by the public works department.

University Research. The Institute of Water Research at Michigan State University in East Lansing takes a multidisciplinary approach to water management in the Great Lakes region, with an emphasis on developing information systems and assessment tools.

The University of Texas at Austin houses the Center for Research in Water Resources. Graduate students and professors there conduct applied research with a focus on the complex water needs of the dry Southwest region.

The University of Arizona's Water Resources Research Center in Tucson is an extension of the College of Agriculture and Life Sciences. It works jointly with researchers in Mexico to assess and manage the aquifer shared by the two countries.

The Research Experience for Undergraduates program at the University of Florida provides students with eight weeks of research experience in various subjects, including water resources. Projects include detecting toxins produced by algae in water sources to keep it out of the drinking water and decentralized treatment of household gray water, wastewater that contains soap or detergent but does not include excretory waste.

Manufacturers. Culligan has become an acknowledged leader in water treatment since the company was founded in 1936. First known for its water softeners, its research and development team has made advancements in water treatment that have resulted in its holding more than 200 patents worldwide.

Brita, founded in 1966, specializes in water optimization. This company manufactures water-filtration products for professional and domestic purposes, including pitcher filters, faucet-mounted filters, and refrigerator filters. Brita products are positioned to improve the quality and taste of tap water and replace the cost of bottled water and its plastic containers.

PUR water-filtration products are manufactured as a specialty line of Procter & Gamble. Their technology claims to remove 99.9 percent of microbial *Cryptosporidium* and *Giardia intestinalis* and from 96 to 99 percent of trace levels of pharmaceuticals.

Careers and Course Work

Students interested in sanitary engineering typically pursue a bachelor of science degree in civil, chemical, or mechanical engineering, followed by a master of science degree in sanitary engineering. Doctorate degree programs are also available. To specialize in sanitary engineering, graduate students take classes in geographic information systems, hydrology, public health, urban water management and drainage, wastewater treatment processes, solid waste management, and water transport and distribution. Elective courses should include those in writing and public speaking, logic and problem-solving, computer modeling, societal governance, and regulatory compliance.

Sanitary engineers typically work for public health departments and government regulatory agencies. Some become professors and conduct academic research. Some work in private testing laboratories and research firms. Some operate water- treatment plants and pollution-control facilities. Others work as consultants to manufacturers, corporations, and private homeowners. Others may work abroad in countries where basic sanitation needs are not yet met. A professional engineering license is generally required for employment. This license distinguishes one's education and proficiency from that of a draftsman, machinist, mechanic, technician, plumber, or surveyor. Engineering is considered to be a safety-related practice, and licensure holds an engineer to legal liability.

Social Context and Future Prospects

Research projects in sanitary engineering include seeking processes and equipment for improved purification efficiency. One example is the development of large, portable water-treatment systems that are suitable for providing clean water to survivors of natural disasters and the bivouac medical units that treat them. Another example is a nanofiltration system that desalinates ocean water for use on naval ships, especially during times of conflict, and extended private offshore operations such as oil drilling. A related nanofiltration system is necessary for oil-spill cleanup. A third example is the specialized absorbent removal of microcontaminants that may be present in small yet detrimental amounts. These may include elements such as arsenic and lead, industrial solvents, and radioactive particles.

Most water treatment plants are not prepared to remove pharmaceuticals, including natural and synthetic hormones, that are flushed down the sink or toilet. Those that use chemical oxidative processes to remove estrogens and other medications generate disinfection by-products in the water supply that pose potential risks to human health. Some communities organize collections of unused or unwanted over-the-counter and prescription medications for disposal by authorized incineration.

Bethany Thivierge, B.S., M.P.H.

Further Reading

Binnie, Chris, and Martin Kimber. *Basic Water Treatment.* 4th ed. Cambridge, England: Royal Society of Chemistry, 2009. A comprehensive textbook on water-quality standards and practices in America and Europe.

Davis, Mackenzie L., and David A. Cornwell. *Introduction to Environmental Engineering.* 4th ed. New York: McGraw-Hill, 2006. A basic textbook for an easily understood overview of environmental engineering, including federal regulations and industry standards.

Morris, Robert D. *The Blue Death: Disease, Disaster, and the Water We Drink.* New York: HarperCollins, 2007. An epidemiologist who specializes in waterborne diseases discusses the history of water purification and the drinking-water industry.

Nathanson, Jerry A. *Basic Environmental Technology: Water Supply, Waste Management, and Pollution Control.* 5th ed. Upper Saddle River, N.J.: Prentice Hall, 2007. An introduction to the multifaceted field of environmental engineering, with illustrations for clarity and practice problems at the end of each chapter.

Sawyer, Clair N., Perry L. McCarty, and Gene F. Parkin. *Chemistry for Environmental Engineering and Science.* 5th ed. New York: McGraw-Hill, 2003. Focuses on the application of the principles and methods of chemistry in water purification and wastewater analysis.

Symons, James M. *Plain Talk About Drinking Water: Answers to Your Questions About the Water You Drink.* 5th ed. Denver: American Water Works Association, 2010. Consumers get clear information about drinking water in a question-and-answer format with easily understood language.

Web Sites

American Society of Sanitary Engineering
http://asse-plumbing.org

American Water Works Association
http://www.awwa.org

Centers for Disease Control and Prevention
Safe Water System
http://www.cdc.gov/safewater

Environmental Protection Agency
Water Treatment Process
http://water.epa.gov/learn/kids/drinkingwater/watertreatmentplant_index.cfm

See also: Civil Engineering; Environmental Engineering; Sewage Engineering.

SCANNING PROBE MICROSCOPY

FIELDS OF STUDY

Mathematics; physics; chemistry; biology; applied mathematics; surface science; material science; tribology

SUMMARY

Scanning probe microscopy is a methodology that allows direct observation of structures and properties at the atomic and molecular scales. The techniques are applicable to a wide variety of purposes, providing information that is otherwise inaccessible. Scanning probe microscopy is particularly appropriate to nanotechnology, permitting the direct construction of nanoscale objects in an atom-by-atom manner.

KEY TERMS AND CONCEPTS

- **NanoAmpere:** The typical unit of measurement of the tunneling current between probe tip and surface. One nanoAmpere equals 10^{-9} Ampere.
- **Piezo-Electric Element:** A component used to control the vertical movement of the scanning probe tip to maintain a constant force or current.
- **Probe Deconvolution:** Processing of image data to correct the resolution of data measured for steep gradients parallel to the probe tip.
- **Quantum Dot:** A small group of atoms that share an electron charge as though they were a single atom.
- **Van Der Waals Force:** A weak force acting between atoms that varies from weakly attractive to moderately attractive to intensely repulsive as the distance between the two atoms decreases.

Definition and Basic Principles

Scanning probe microscopy is a methodology that interfaces the macroscale of human observation (1–10^{-2} meters) to the atomic scale of the physical world, providing direct observation of features in the range of 100 micrometers to 10 picometers (10^{-4}–10^{-11} m). Though the upper range of resolution for scanning probe microscopy is well within the range of other methods of microscopy, it is the lower range that is of most interest, because it is at this range that direct observation of atomic and molecular scale properties is possible.

The basic principles of scanning probe microscopy are founded in the quantum mechanical properties of atoms. The methods use the measurement of electronic properties (current, voltage, or "atomic force") as the means of observing the nature of surfaces and surface phenomena.

Quantum mechanics describes and defines the behavior of electrons in atoms. One of the rules of quantum behavior is that electrons are constrained to specific locales known as orbitals within the structure of an atom and are not allowed to exist at the boundaries of those locales. An observable property called quantum mechanical tunneling occurs, however, which permits electrons to move from one locale to another across the orbital boundaries. In scanning probe microscopy the miniscule electrical current due to quantum mechanical tunneling between the atoms at a surface and the atoms at the tip of an atomic-scale probe is measured as a function of their relative positions. This provides a corresponding atomic scale map of the surface structure.

Background and History

The scanning tunneling microscope (STM) was invented in 1982 by German physicist Gerd Binnig and Swiss physicist Heinrich Rohrer. Their device could be conceived of as a sort of quantum mechanical phonograph, in which an exceedingly sharp metallic needle scans a surface in a manner similar to the way a phonograph needle scans the groove of a vinyl phonograph record. The needle would ideally taper down to a single atom at the point, enabling atom-to-atom interaction at the surface. Sensitive digital-electronic measurement devices would measure the electronic tunneling current between the tip and the surface. The devices would then relay that data directly to a computer that would then correlate the values according to the relative dimensions and spatial relationships of the probe tip and the surface. The result would be displayed as an image having resolution of atomic scale features.

In 1986, Binnig, along with American electrical engineer Calvin Quate and Swiss physicist Christoph Gerber, introduced the scanning force microscope

(SFM), which is also known as the atomic force microscope, or AFM, a variation on the STM that maintained a constant force between the scanning tip and the surface. This allowed any surface to be scanned, whereas the STM could only be used with electrically conductive surfaces. More recently, scanning near-field optical microscopy (SNOM) was developed, using measurement of short-range components of electromagnetic fields (a very small light source) between tip and surface to produce the equivalent of a photographic representation of the surface features.

How It Works

To appreciate the operation of scanning probe microscopy, it is necessary to understand the scale on which it operates. The unaided human eye can discern detail as small as approximately 0.1 millimeter. Optical microscopes can extend this to a resolution of about 0.0001 meter. Scanning electron microscopes can typically produce images with a resolution as fine as 10 micrometers (0.00001 meter). This is the range at which scanning probe microscopes only begin to work, and they typically provide information with a resolution of as little as 10 picometers (0.00000000001 meter).

At this scale, the operation is in the realm of quantum mechanical physics rather than classical physics, and the effects associated with that scale are very different from those that occur on a larger scale. The most important difference lies in what is meant by the word "surface."

Quantum Mechanical Physics. On scales that are significantly larger than atomic and molecular diameters, a "surface" is solid matter, analogous to a smooth tabletop. At the atomic scale of quantum mechanics, however, there is no such thing as a hard surface in that sense.

Quantum mechanics describes the structure of atoms as a very small, dense nucleus of massive protons and neutrons, surrounded by a cloud of electrons that is 100,000 times greater in diameter than the nucleus. The electron cloud is therefore very diffuse. The electrons in a neutral atom are equal in number to the protons contained in the nucleus, and are confined to specific three-dimensional regions, called orbitals, around the nucleus. and are allowed to have only very specific energies according to the orbitals they occupy. At this scale of operation, a scanning probe microscope measures the electromagnetic interaction of the electron clouds in the atoms of the probe tip and atoms of the surface being scanned.

Scanning Tunneling Microscopes (STMs). The STM operates by moving the atoms-wide point of the scanning tip across a metallic, and therefore electrically conducting, surface at a distance of less than one nanometer (10^{-9} meter). The device measures the magnitude of the "tunneling current" that arises between the probe tip and the surface atoms. This current is exceedingly small, and its measurement requires extremely sensitive digital sampling and amplification electronics, and computers to process the measured data according to the relative geometries of the tip and the surface. As the probe scans, the angle between the tip and the atomic surface changes, as does the distance between them. The tunneling current measurement thus has three-dimensional vector-field properties, and because the probe tip maintains a constant orientation, variations in the tunneling current are presumably caused by the three-dimensional shape of the atoms being scanned.

Atomic Force Microscopes (AFMs). The AFM operates in essentially the same manner as the STM, except that its function is to maintain a constant measured electrical force between the probe tip and the atomic surface being scanned. In this function, the probe tip follows the shape of the atomic surfaces directly, rather than measuring a property difference that changes according to the shape of the surface. Several different modes of operation are available within this context, such as constant contact, non-contact, intermittent contact, lateral force, magnetic force, and thermal scanning. Each mode provides a different type of information about the surface atoms.

Scanning Near-Field Optical Microscopes (SNFOMs). The SNFOMs use an extremely small-point light source as the probe, rather than a physical tip. Measurement of the effect that the surface has on the light provides the image data. The technique can employ a broad range of wavelengths to investigate different properties of the atoms being scanned.

Applications and Products

The field of scanning probe microscopy is a high technology research practice. Its direct applications are limited to the analytical study of surface phenomena and structures. The nature of the techniques of scanning probe microscopy makes it the method of choice for examination and study of surfaces that are not amenable to any other means of close examination, especially those of certain biological materials. Scanning probe microscopy, with its ability to probe and manipulate single atoms and so to form molecule-sized structures, is invaluable in chemistry and in the development of nanotechnology.

Surface Chemistry. Chemical reactions take place at the level of the outermost electronic orbitals of atoms, or at the electronic surface of the atoms involved. In catalyst-mediated reactions, the interaction among chemical species takes place on the surface of the catalyst material, where the atoms of the reactants have interacted electronically with the atoms of the catalytic material. This lowers the energy barriers that must be overcome for the reaction to occur, with the result that the desired reaction is facilitated.

Scanning probe microscopy has a spectroscopy mode that allows the direct measurement of electron energies in single atoms and of single atomic bonds between atoms in a molecule. The methodology provides a better understanding of the chemistry that takes place at surfaces, in turn contributing to the development of new and improved chemical processes. Perhaps the most important aspect of this is the enhanced knowledge of the mechanisms of surface phenomena such as oxidation and chemical corrosion.

Electronic Materials and Integrated Circuits. How small can a functional transistor be? Technology enables the construction of several million transistors on the surface of the small silicon chip that is the central processing unit (CPU) of a computer. Scanning probe microscopy enables the physical study of such structures in minute detail. It also enables the construction and study of transistor structures that are orders of magnitude smaller. The research in this field examines possible ways to construct integrated circuits and computers that exceed existing capabilities of production.

The available methods of production of integrated circuits have essentially reached the physical limit of their capabilities to construct viable semiconductor transistors. Atomic force microscopy is being used to investigate the construction of transistor-like structures based on quantum dots rather than on semiconductor junctions.

Another area of research is the construction of molecule-sized transistors made from graphene or carbon nanotubes and other materials. These technologies, when fully developed, will completely change the nature of computing by enabling the construction of a quantum computer, a device that could carry out in seconds calculations that existing computers would require possibly billions of years to complete.

Data storage would also be revolutionized by these innovations. Research indicates that data storage will soon reach capacities measured in terabits per square centimeter. The ultimate binary data storage density would have each bit stored in the space of one atom, a density that can be envisioned only with scanning probe microscope technology.

Nanostructures and Nanodesign. Nanotechnology works at the nanometer scale of 10^{-9} meters. To appreciate this scale, imagine the length of one millimeter divided into one million segments, each of which would be one nanometer. The concept of nanotechnology is to produce physical machines constructed to that scale. Because scanning probe microscopy can manipulate single atoms, it can be used to construct nanoscale, and even picoscale, physical mechanisms. The latter are essentially individual molecules whose physical structures imitate those of much larger devices and mechanisms, such as gears.

In September, 2011, researchers at Tufts University reported the successful use of low-temperature scanning tunneling microscopy to construct a working electric motor consisting of a single molecule. As can be imagined, this is a complex field of research, because quantum effects play a significant role in the interoperability of such small devices.

One application of scanning probe microscopy that is of immediate importance is the study of friction and abrasion at the atomic level, which is where those processes take place. Atomic force microscopy can be used to literally scratch the surface of a material, providing detailed information about how friction and abrasion actually work and about what might be done to lessen or prevent those effects.

Biological Studies. Scanning electron microscopy (SEM) has been the workhorse of biological research since its invention, providing detailed images of extremely small biological structures. The technology has some practical limits, however, because of the principles on which it functions. Many biological materials that are of interest cannot be studied in detail using SEM, but are amenable to study using scanning probe microscopy. The methods are useful in measuring the forces that exist among functional groups in biological and organic chemical structures.

Impact on Industry

The application of scanning probe microscopy in industrial research is expected to have at minimum one very important dividend: knowledge of abrasion and friction applicable to the field of tribology. This study, known as nanotribology, may provide the knowledge required to minimize, and in some cases eliminate, the effects of friction and abrasion. Ignorance of these effects is estimated to consume about 4 percent of a nation's gross national product annually. In the United States this corresponds to about $200 billion per year. It is also estimated that about one-third of all energy consumed is expended to overcome friction.

Scanning probe microscopy is widely applied in surface science. It is particularly useful in material science studies for controlling the nature of surfaces and to measure the roughness and hardness of surfaces. The methodology also permits the examination of magnetic structures at the atomic level.

Magnetism is a natural force that is still poorly understood and whose cause remains unknown. Quantum mechanics provides a mathematical basis for the phenomenon, but a definitive physical cause remains to be observed. The various modes of scanning probe microscopy may provide the direct observation of the properties responsible for magnetism on the atomic scale, leading to better understanding of how to manipulate and use that effect at the macroscale. Because magnetic interactions are also involved in frictional forces, understanding magnetism and magnetic structures on surfaces is an aspect of surface science that should not be overlooked.

Magnetic surface structures play an extremely important role in modern technology because they are the mechanisms whereby digital data is stored on computer drives and other magnetic media. Historically, this has been achieved in the relatively crude manner of coating an inert material with extremely fine particles of some magnetically susceptible material and imprinting the particles with a magnetic field according to the value of the data to be stored. The process has been remarkably effective, with the data so stored remaining intact almost indefinitely.

Fascinating Facts About Scanning Probe Microscopy

- The scanning tunneling microscope was invented in 1982 by Gerd Binnig and Heinrich Rohrer, for which they were awarded a Nobel Prize in Physics in 1986.
- Scanning probe microscopes can achieve resolution of structural surface details as small as 10 picometers.
- Scanning probe microscopes can move and assemble individual atoms to construct molecule-sized structures.
- Using a low-temperature scanning tunneling microscope, chemists at Tufts University successfully assembled and operated an electric motor consisting of a single molecule.
- Scanning probe microscopy can be used to examine biological molecules and materials that cannot be studied by any other method.
- The spectroscopy mode of scanning probe microscopes allows examination of the energy of single electrons in an atom and direct measurement of the energy of a bond between two atoms in a molecule.
- The tip of a scanning probe has a typical radius as small as 3 nanometers and can be as fine as a single atom.
- The probe tip of an atomic force microscope is attached to a cantilevered leaf spring. The up-and-down movements of the spring as the probe tip moves across the atoms of a surface are measured electronically.

However, the process has also been severely limited by the practical size and magnetic susceptibility of the magnetic particles, the physical size of the magnetic heads used to apply the magnetic imprint, and the ever-present possibility of accidentally eradicating the data by exposure of the medium to incidental magnetic fields. Intimate knowledge of magnetic surface structures obtained through scanning probe microscopy can lead to the development of ever better methods of reliably storing and retrieving data from magnetic media, by developing ever more effective magnetic structures.

Careers and Coursework

Career opportunities in scanning probe microscopy will arise as the methodology is developed for industrial applications. Students interested in a career in this field will require a recognized university degree as a minimum qualification. A bachelor's degree with specialization in surface chemistry or physics will require extensive study in mathematics; inorganic, analytical, and physical chemistry; physics; and electronics for a sound foundation in this field. Careers at an advanced level involving research and development will demand an advanced degree, either a master's or a doctorate, with specialization in a specific branch of scanning probe microscopy.

Social Context and Future Prospects

Scanning probe microscopy is a field that will have very little direct social context because of the extremely small scale of its subject matter. However, the secondary effects of the knowledge and technology derived from research and development in this field could have a large social impact, primarily because of the economic benefits from control of friction and from new technologies for integrated circuits and magnetic memory media. Any real predictions for the future prospects of the field of scanning probe microscopy are entirely conjectural.

Richard M. Renneboog, M.Sc.

Further Reading

Bhushan, Bharat, Harald Fuchs, and Masahiko Tomitori, eds. *Applied Scanning Probe Methods VIII: Scanning Probe Microscopy Techniques (NanoScience and Technology)*. Berlin: Springer, 2008. Provides the most up-to-date information on this rapidly evolving technology and its applications.

Howland, Rebecca, and Lisa Benatar. "A Practical Guide to Scanning Probe Microscopy." ThermoMicroscopes. March 2000. Web. Accessed September, 2011. This introductory guide to scanning probe microscopy describes several techniques and operating modes of the devices, as well as their structure and principles of operation, and discusses the occurrence of image artifacts in their use.

Meyer, Ernst, Hans Josef Hug, and Roland Bennewitz. *Scanning Probe Microscopy: The Lab on a Tip*. Berlin: Springer, 2004. This book provides an excellent overview of scanning probe microscopy before delving into more detailed discussions of the various techniques.

Mongillo, John. *Nanotechnology 110*. Westport, Conn.: Greenwood, 2007. Demonstrates the essential relationship between and value of scanning probe microscopy to nanotechnology.

See also: Applied Physics; Electronic Materials Production; Electronics and Electronic Engineering.

SEWAGE ENGINEERING

FIELDS OF STUDY

Geometry; algebra; bacteriology; microbiology; organic chemistry; inorganic chemistry; drafting; surveying; mechanics; hydrology.

SUMMARY

Sewage engineering is the discipline of civil engineering that addresses the separation, decontamination, and disposal of waste, especially human excrement, from water supplies. Domestic sewage contains solid and dissolved materials such as fecal matter, urine, toilet paper, vomitus, food waste, soaps, and other cleaning agents. Separation methods include filtration, adsorption, and precipitation. Decontamination methods include biocides, chemical neutralization, and phagocytosis. Disposal involves the collection, handling, and destruction of the final waste products in a manner that is safe for humans and the environment. The goal of these processes is to provide improved water quality for drinking, cooking, and bathing.

KEY TERMS AND CONCEPTS

- **Biogas:** Base product of the anaerobic bacterial breakdown of organic waste matter.
- **Eutrophication:** Ecological chain reaction in which the introduction of specific substances into a body of water causes increased plant growth and decay, which depletes the oxygen in the water, causing fish and other marine life to die.
- **Pathogen:** Microbe (bacteria, virus, fungus, prion, parasite) that causes disease, especially in humans.
- **Sedimentation:** Process of using gravity to pull particles out of a solution.
- **Sewage:** Wastewater carried away from homes and businesses.
- **Sewerage:** Infrastructure for the removal of sewage, including drains, pumping stations, and wastewater-treatment plants.
- **Sludge:** Untreated solid waste.
- **Toxic Waste:** Waste material that may cause illness or death to people and animals.
- **Vomitus:** Stomach contents expelled by vomiting.

Definition and Basic Principles

Sewage is more than 95 percent water. Sewage engineering addresses the remaining 5 percent, which contains a variety of materials and requires a variety of separation and disposal methods. Wastewater contains both nonpathogenic bacteria from the human digestion system and pathogenic microbes (bacteria, viruses, and parasites). It also contains both insoluble organic material (such as fecal matter, hair, food, vomitus, and paper) and soluble organic material (such as urine, sugars, proteins, and prescription medications). Similarly, it contains both insoluble inorganic matter (sand, metal particles) and soluble inorganic matter (such as ammonia, salt used to deice roads, salt from ocean water, and sulfur compounds). Wastewater also contains gases (carbon dioxide, methane, hydrogen sulfide), oily emulsions (hair dye, paint, glue, salad dressing), and toxins (pesticides, insecticides, herbicides). Finally, sewage contains large pieces of garbage (including condoms and tampons) and even dead animals (insects, fish, reptiles, and rodents).

The process of sewage treatment occurs in three phases. In the first stage, the largest solids are removed by mechanical means and then by sedimentation. The captured material is checked for toxicity before it is transported to landfills. In the second stage, the remaining solids are treated to become nonpathogenic; ironically, nonpathogenic bacteria are introduced to digest the organic matter. This material is often sold to farmers, nurseries, and garden centers for use as an organic soil enhancement. In the third stage, the remaining solids are treated to become nontoxic. This is a costly and difficult step, but some effort must be made to remove or neutralize minerals such as nitrates and phosphates. If these compounds get into a lake, they can cause eutrophication, a chain reaction of algae overgrowth, rotting, and oxygen depletion that destroys the aquatic ecosystem.

Background and History

Archaeological excavations have found that civilizations in Mesopotamia and Pakistan earlier than 2500 B.C.E. had sophisticated sewage drainage systems using clay pipes. Until the Roman Empire emerged,

Minoans on Crete had the most complex wastewater systems in that part of the world, including primitive toilets that flushed. The Roman Empire engineered a complex system of aqueducts, most of which were underground, with covered drains for storm water and household sewage. As the empire expanded, so did this technology. However, with the fall of the Roman Empire, the world reverted back to unsanitary conditions. Personal hygiene was not a priority, the cleanliness of homes and towns was similarly neglected, and water supplies were not protected from contamination. Plagues in the Middle Ages wiped out entire towns. Finally in the mid-1800's, after several cholera epidemics in Paris, the people there began to construct a sewer system. England, Germany, and the United States soon followed France's lead in developing networks of sewers, plumbing with pipes, and rudimentary toilets to create more sanitary living conditions. These facilities have continued to evolve as people's understanding of sanitation, public health, and sewage treatment has advanced.

How It Works

Wastewater is 99.94 percent water by weight. In homes, it goes down the sink, toilet, and drains from the bathtub, shower, dishwasher, and washing machine to either a septic tank or municipal sewage-treatment facility. Wastewater also comes from other sources such as schools, businesses, hospitals, commercial laundries, food-processing plants, and car washes. Urban runoff from rainstorms and melting snow goes down street drains to the treatment plant as well. This runoff carries pollution from roads, parking lots, and rooftops that would be harmful to natural waterways; therefore, this water gets treated as well.

In a sewage-treatment plant, large solids are retained by screens in preliminary treatment and then broken down by tumbling for further processing or allowed to settle by gravity for disposal as sludge. Material that floats is skimmed off the surface for disposal. Chemicals are added that cause small particles to form larger aggregates that can be more easily removed by straining and sedimentation. These processes, called primary treatment, remove about 60 percent of the solids. The remaining wastewater is subjected to further treatment.

In the next step, the wastewater is mixed with bacteria, which digest the organic material, and oxygen, which activates the bacteria. The bacteria may be free in solution or fixed to a medium matrix. The bacteria are collected for further digestion, and the undigested matter is collected as sludge. The remaining liquid is then treated to remove the fine particles and odors by chlorination, filtration, and reverse osmosis. These processes, called secondary treatment, remove about 90 percent of the remaining solids. The remaining water is subjected to further treatment.

In tertiary treatment, phosphorus is removed by binding it with aluminum-based compounds; these aggregates can then be filtered out. Soluble nitrogen is converted through a series to steps to nitrogen gas, which can safely be discharged into the air. Removal of these two elements prevents eutrophication when the clean water is discharged into local waterways. Any remaining microorganisms may be killed by ultraviolet radiation or ozone disinfection. The water is now safe for humans and the environment.

Sludge is untreated solid waste; biosolids have undergone waste treatment. Sludge is concentrated wastewater, but it is still only about 6 percent solids. To drive off the water, it may undergo conditioning, which is treatment with chemicals or heat, or it may undergo thickening, which is treatment with different chemicals or gravity. Stabilization involves subjecting the biosolids to anaerobic digestion, which produces methane gas. Biosolids may also be chemically digested with lime, which reduces odors. Finally, dewatering steps such as filtering, pressing, and centrifugation leave the solids with the consistency of dry dirt. These solids can be used for soil conditioning (bulking up soil to prevent erosion and retain more water), incinerated for energy, or disposed of in landfills. They may also be mixed with wood chips and leaves to create a natural compost product for parks and golf courses.

Applications and Products

Sewage Holding Tanks. Underground sewage holding tanks are necessary for places without access to a public sewer or on-site treatment facility. Such places include campgrounds, amusement parks, and commercial ventures. Each tank is built to hold from 4,000 to 10,000 gallons of sewage. These tanks were originally made of concrete, but they were vulnerable to acid erosion from human waste. Contemporary tanks are made of polyethylene or fiberglass, which are acid-resistant, and they are cast in one piece to be completely

Fascinating Facts About Sewage Engineering

- About 50 percent of the world's population does not have sewage treatment. Rivers downstream from large cities have become heavily polluted with fecal matter. The fecal coliform count in Asian rivers, the most contaminated in the world, is fifty times higher than the World Health Organization guidelines for appropriate sanitation.
- According to the International Conference on Water and the Environment, every year, worldwide, more than 1.2 billion people get sick from drinking, cooking with, and bathing in water fouled with human waste. Many die as a result, including an estimated 15 million children under the age of five.
- Every year, about 25 percent of U.S. beaches are posted as unsafe for swimming because of water-pollution conditions. This extends to freshwater as well—40 percent of America's rivers and 46 percent of its lakes are too polluted for fishing or swimming.
- The Mississippi River carries about 1.5 million metric tons of nitrogen from farmlands into the Gulf of Mexico each year. This creates eutrophication, resulting in an annual coastal dead zone the size of Massachusetts.
- In 1962, Chicago pollution-control engineers first used green vegetable dye to trace improper sewage discharge into the Chicago River. It has since become an annual tradition to dye the river green to celebrate St. Patrick's Day.
- Ancient Roman aqueducts made with clay pipes are well known. However, archaeologists have recovered a Roman child's toilet chair made of pottery dating back to the sixth century B.C.E. It is presently in the British Museum in London.
- Planet SKS, a sophisticated thirty-five-story community-housing high-rise with 184 three- and four-bedroom residential units in Mangalore, a major port of India, has its own on-site sewage-treatment plant.

watertight. These tanks require periodic emptying and cleaning.

Package Sewage Treatment Plants. For places without access to a public sewer, such as a small housing development or remote commercial site, where septic tanks do not meet existing regulations, a small on-site package sewage-treatment plant is appropriate. Depending on its size, one of these units may accommodate the needs of one household to as many as a community of 375 people. The unit employs sedimentation and aerobic bacterial digestion. The quality of the treated wastewater is such that it may be safety discharged into a nearby waterway or the ground with sufficient soil percolation. The unit requires periodic emptying and cleaning.

Onboard Sewage Treatment Systems. Sewage-treatment systems for use on workboats are becoming smaller, more efficient, and less objectionable to maintain. Such systems are practical for workboats such as fishing and Coast Guard vessels, fireboats, and harbor tugs. These onboard sewage-treatment systems typically consist of a tank packed with a medium on which bacteria are enmeshed. This medium has a high-surface-area-to-volume ratio for greater efficiency. The medium becomes immersed in sewage, and the bacteria digest it. Because the bacteria are fixed to the medium, the system can be flushed without manual cleaning.

Wastewater Lagoon Systems. Small rural communities that cannot afford to build or support a mechanical wastewater treatment plant often choose the less expensive, less labor-intensive option of wastewater lagoons. A new product developed by Wastewater Compliance Systems in Utah increases the efficiency of wastewater lagoons with modest expense. Bio-Domes, formerly called Poo-Gloos, are igloo-shaped plastic domes that work in clusters of at least two dozen sitting on the bottom of the lagoon. Each dome is really a set of nested domes with plastic packing between the layers to provide a large surface area on which bacteria are grown. Sewage is bubbled up through the bottom of the dome and out through a hole in the top. As it passes by the bacteria, they digest it in optimal dark, aerobic conditions.

Impact on Industry

Government Regulation. In this country, water quality is the responsibility of the U.S. Environmental Protection Agency (EPA). Specifically, the EPA's Office of Water is charged with ensuring the safety of drinking water, implementing legislation such as the Safe Drinking Water Act and the Clean Water Act and maintaining and improving water ecosystems to protect the health of people, animals, fish, and plants. The Office of Water is further divided into the Office of Ground Water and Drinking Water, the Office of Science and Technology, the Office of Wastewater Management, and the Office of Wetlands, Oceans and Watersheds. While there are ten regional offices of the EPA, state governments are also responsible for water quality and the preservation of aquatic environments. Such state offices are usually associated with the state public health departments. Duties may then fall to county or municipal governments and their environmental and public-health agencies. Sewage engineers provide services such as inspection of sewer design and construction, sewer cleaning and maintenance, wastewater treatment plant operation and maintenance, and urban-improvement planning.

University Research. Researchers from Oregon State University are developing the technology to convert municipal sewage into hydrogen gas for use in hydrogen fuel cells that may someday power electric cars. Their microbial- electrolysis system requires only 25 percent of the energy of water electrolysis to produce hydrogen gas. Equivalent hydrogen fuel energy is becoming comparable in price to a gallon of gasoline. However, more refinement is necessary for use on a commercial scale. This technology would be especially beneficial in remote areas and developing countries that do not have sewage-treatment facilities or the means to produce electricity because it would fill both needs.

University of Kansas researchers are using algae to digest sewage and absorb its environmentally harmful nitrogen, potassium, and even small amounts of pharmaceuticals. The algae then produce a clean-burning biofuel. Algae farms have advantages over those that raise corn or soybeans for biofuel: Because they are aquatic, they produce more biofuel on less land, as much as 5,000 gallons per acre. Corn produces 18 gallons per acre and soybeans 48 gallons per acre. Algae are less expensive to maintain because they do not require the irrigation and fertilization of food crops such as corn and soybeans. The research is still in the pilot stage and has not been scaled up for commercial use. A similar project is under way at Utah State University.

Business Ventures. British Gas has developed a method for the anaerobic digestion of fecal matter in a closed-loop system to recover biomethane gas for use in homes similar to that of natural gas. The treatment cycle takes 23 days and the biogas is being used in 130 homes in the United Kingdom county of Oxfordshire.

Capstone Turbine Corporation of California produces microturbines that run on different kinds of fuels, including biogas from landfills and wastewater-treatment plants. With the company's proprietary technology, the methane produced from human, agricultural, and food processing waste does not have to be free of hydrogen sulfide contaminants to burn cleanly and efficiently.

Ostara Nutrient Recovery Technologies, based in Vancouver, British Columbia, has found a way to extract nutrients from sludge and transform them into slow-release pellets of fertilizer. Their bed chemical reactor crystallizes phosphorus and ammonia and converts them into struvite, the same mineral that forms bladder stones in dogs and cats. When struvite forms uncontrolled on wastewater-treatment pipes, it creates clogs in the system. When it is formed intentionally, with the addition of other chemicals, it comes out of the system cleanly. In addition, pipes made of Kynar, a kind of Teflon, are used to discourage the struvite from sticking. The resulting fertilizer is then packaged and sold to golf courses and nurseries.

Careers and Course Work

Students interested in sewage engineering typically pursue a bachelor of science degree in civil, chemical, or mechanical engineering, followed by a master of science degree in sewage or waste-management engineering. Some schools also have a concentration program within a graduate environmental engineering degree. Doctorate degree programs are also available. To specialize in sewage engineering, graduate students take classes in geographic information systems, hydrology, public health, urban water management and drainage, wastewater treatment processes, solid waste management, and water transport and distribution. Elective courses should include those in writing and public speaking, logic and problem-solving, computer modeling, and societal governance, and regulatory compliance.

Sewage engineers typically work for public health departments and government regulatory agencies. They may be hired by large cities to oversee urban infrastructure expansion and improvement. Some become professors and conduct academic research. Some operate sewage-treatment plants and pollution-control facilities. Some work in agriculture, specializing in composting and organic-soil amendment. Others work as consultants to manufacturers, corporations, and private homeowners. Others may work in remote areas in the U.S. and abroad, where basic sanitation needs are not yet met. A professional engineering license is generally required for employment. This license distinguishes one's education and proficiency from that of a draftsman, machinist, mechanic, technician, plumber, or surveyor. Engineering is considered to be a safety-related practice, and licensure holds an engineer to legal liability.

Social Context and Future Prospects

Engineers are still attempting to find efficient and cost-effective ways to remove pharmaceutical compounds, including natural and synthetic hormones, from wastewater and sludge. These compounds are naturally excreted, but they are also typically flushed down toilets or rinsed down sinks as a convenient means of household disposal. Hormones that get into streams, rivers, and lakes have especially adverse effects on fish and subsequently on the animals and humans who eat them. Water-treatment plants that use chemical oxidative processes to remove estrogens and other medications generate disinfection by-products in the water supply that pose potential risks to human health. Some communities are organizing collections of unused and unwanted over-the-counter and prescription medications for disposal by authorized incineration.

Researchers have begun studies in sewage epidemiology, a new field in which information about the health status of a population may be gathered by studying its wastewater contents in real time. For example, a study on the consumption of illegal drugs in Brussels, Belgium, performed by daily sampling of the sewage coming into the wastewater treatment plant verified a significantly higher weekend use of cocaine and amphetamines, no daily variation in heroin and methadone use, and negligible consumption of methamphetamine, trends that were consistent with public health reports. A similar study was conducted in Barcelona, Spain, to evaluate drug abuse in prison systems. Although this methodology does not identify individual offenders, it provides needs-assessment data for a given population without subjects' active participation.

Bethany Thivierge, B.S., M.P.H.

Further Reading

Bagchi, Amalendu. *Design of Landfills and Integrated Solid Waste Management.* 3d ed. Hoboken, N.J.: John Wiley & Sons, 2004. Thorough coverage of all aspects of solid waste management including recycling, composting, and the microbiology and gas generation in landfills.

Davis, Mackenzie L. *Water and Wastewater Engineering: Design Principles and Practice.* New York: McGraw-Hill, 2010. A comprehensive illustrated textbook with practical applications presented as homework problems at the end of each chapter.

Hammer, Mark J., and Mark J. Hammer, Jr. *Water and Wastewater Technology.* 7th ed. Upper Saddle River, N.J.: Prentice Hall, 2011. Provides information on technical innovations, with illustrations of modern equipment and visual aids for mathematical concepts.

Russell, David L. *Practical Wastewater Treatment.* Hoboken, N.J.: John Wiley & Sons, 2006. A textbook for engineers with an emphasis on designing wastewater-treatment plants to handle burdens of specific industrial wastes.

Tchobanoglous, George, Franklin L. Burton, and H. David Stensel. *Wastewater Engineering: Treatment and Reuse.* 4th ed. New York: McGraw-Hill, 2003. A highly respected textbook on environmental regulations and technological advancements in wastewater treatment, including solid waste management.

Web Sites

American Public Works Association
http://apwa.net

American Society of Civil Engineers
http://www.asce.org

American Water Works Association
http://www.awwa.org

Solid Waste Association of North America
http://www.swana.org

See also: Civil Engineering; Environmental Engineering; Sanitary Engineering.

SOFTWARE ENGINEERING

FIELDS OF STUDY

Computer programming; systems engineering; Web engineering; computer science; engineering; business; project management; mathematics; physics.

SUMMARY

Software engineering is the science of software design, construction, and maintenance, including processes for managing the building of software. Software, the product of software engineering, is used in all facets of society. The most popular software is e-mail. In business, word processing and spreadsheet applications are omnipresent, and larger companies use enterprise resource planning software to reduce costs and optimize profits. Engineers and scientists have developed powerful software that runs on parallel processors for a variety of applications, from weather prediction to playing chess. Systems software, such as an operating system, provides the foundation on which all other software operates.

KEY TERMS AND CONCEPTS

- **Cloud Computing:** Software that is developed to execute on multiple servers and is delivered by a Web service.
- **Productivity Software:** Computer programs such as word processors, spreadsheets, databases, and presentation graphics packages.
- **Program Manager:** Person who at each phase of software development is responsible for the success of that phase.
- **Project Management:** Set of processes used to develop a piece of software.
- **Systems Analysis:** Set of procedures used to create a complete description of a software development project.
- **Systems Analyst:** Person who creates a detailed first description of a software project.
- **Systems Design:** Documentation produced from analyzing a software development project.
- **Systems Software:** Software used to make computer hardware available to application programs; includes operating systems, assemblers, and compilers.

Definition and Basic Principles

Software engineering is the science of software development. The software development process is characterized by a number of distinct phases. First, detailed specifications for the proposed software are created with the help of those who will be using the software. Then, an analysis is done to determine the resources needed for the project and the steps involved in its development. After the analysis is completed, a formal design is created. The design usually includes descriptions of the data needed for the system and the processes that need to be performed on the data, and some charts to help understand the proposed system's structure. Those developing business-oriented systems usually refer to this process as systems analysis and design rather than software engineering. After the analysis and design phase are completed, the software system must be implemented. Sometimes programmers must write computer code, but increasingly, software is built from components that someone else has developed. The final step in program development is testing to be sure the program performs as desired. After the system is finished, it needs to be modified to reflect new requirements, and ultimately, it must be retired or updated.

The amount and variety of software produced since the invention of the computer in the 1940's is truly phenomenal. World Wide Web browsers, e-mail, and personal productivity software (such as Microsoft Office) are used by people at home and at work. Operating systems, middleware, and integrated development environments (such as Microsoft's Visual Studio) are used by professional programmers to produce new computer applications every day. Web development systems are used to produce Web services that support applications on desktops and mobile devices.

Background and History

John von Neumann is credited with inventing the concept of a stored program for a computer. He recognized that it was possible to store a program in the memory of a computer along with the data. Von

Neumann's programs were written in a binary code, called machine language, that was hard for people to understand. A real breakthrough in the development of computer programs was the introduction of compilers, which allowed programmers to write code in high-level languages.

In the 1970's, Niklaus Wirth released Pascal, the first structured programming language. Pascal was designed to produce more reliable programs and had a great influence on the development of other programming languages. In the early 1970's, Dennis Ritchie introduced the C language, which became the programming language for applications development for the next few years. In the twenty-first century, program development uses many modern languages. The two most popular languages are Java and C++.

As more and more programs were developed, many recognized the need to develop a methodology for the program development process. In 1966, Corrado Böhm and Giuseppe Jacopini published "Flow Diagrams, Turing Machines, and Languages with Only Two Formation Rules" in the journal *Communications of the ACM,* thereby laying the foundation for structured programming. At the same time, Terry Baker at IBM applied the basics of structured programming to systems development, thereby providing the first steps in developing the science of software development, later called software engineering.

How It Works

The principal activity of software engineering is the production of software. There are a number of different approaches to developing software, and the best way to understand these approaches is to study the software development life cycle for each approach.

Waterfall Model. The waterfall model of software development is characterized by a sequential software development life cycle. The key characteristic of the waterfall model is that each step of the software development life cycle is completed before going to the next step. The first step is to perform an initial study of the problem to be sure it is feasible. In the second step, a detailed analysis of what the project entails is completed. This is one of the most important components of the waterfall model and often uses a specialist called a systems analyst. In the third step, the information contained in the systems analyst's report is transformed into a formal design. For the waterfall model, this often includes a dataflow diagram, a system chart, and structured flowcharts or pseudocode. In the forth step, the software is implemented. For most of the software being developed with the waterfall model, the implementation step requires that a program be written in a computer language such as COBOL (very popular in the mainframe era), Java, or C++. In the fifth step, the code is fully tested. Ideally, every conceivable set of inputs should be tried in the program to be sure that it always works. After the initial testing of the program, it is deployed and further testing is performed. The sixth, and final, step in the waterfall model is maintaining the system. This may include some small changes to the original program (no big changes for the waterfall model), correcting bugs discovered in the program, and determining when a new version of the system is needed.

Iterated Waterfall and Spiral Model. The iterated waterfall model is, as its name indicates, a modification of the waterfall model that allows programmers to redo a step of the waterfall model. When personal computers became a mainstay in the business world, programmers often had to return to earlier steps in the waterfall model and could not finish each step completely.

The spiral model of software development, introduced by Barry Boehm in 1987, was the next modification of the waterfall model. Boehm observed that rather than iterating the steps of the waterfall model, it made more sense to view the process of software development as a spiral of steps, many of which were the steps of the waterfall model. By placing the steps in a spiral, Boehm was able to select subsets of the waterfall model to repeat that more closely followed the actual software development process.

Object-Oriented Model. Object-oriented programming was first used with the Smalltalk language, developed at the Xerox PARC (later PARC) laboratories in the 1970's. The object-oriented model for software development was popularized by Grady Booch, James Rumbaugh, and Ivar Jacobson with their introduction of the Universal Modeling Language (UML) in 1997. Object-oriented software development uses the steps of the iterated waterfall model, but at all stages, it works with objects, their properties, and their methods. The key to software development with the UML is the development of a set of use cases during the analysis step that illustrate the threads of functionality of the software. During the design phase, developers create an object diagram and scenario diagrams for each use case to make coding in an object-oriented language such as Java easy.

Applications and Products

A vast amount of software has been developed for computers since the 1960's. Applications software, especially that designed for personal computers, dominates the software that has been developed. However, communications and systems software, designed to enable the operation of computers and networks, have also been important.

Applications Software. The personal computer was popularized by an article in the January, 1975, issue of *Popular Electronics* about kits used to build personal computers, such as the Altair 8800. These early computers had little software other than a simple command-line operating system. Programs were written in assembly language until 1975, when Bill Gates and Paul Allen ported a BASIC interpreter to the Altair. Simple word processors, based on a text-editing system for the Xerox Star (Xerox 8010 Information System) and IBM's Mag Card technology, were the first software made available for personal computers. In 1979, Dan Bricklin and Bob Frankston developed VisiCalc, the first spreadsheet for personal computers. Gary Durbin developed a simple database for personal computers in 1970, but it was the release of dBASE in 1980 that led to the popularity of personal computer databases. PowerPoint was released for the Macintosh in 1984, and with its release, the main components of Microsoft Office were all in place. Versions of Microsoft Office for Windows and the Macintosh were released in the 1990's, and they have dominated the personal productivity software market since then.

Gates and Allen founded Microsoft in 1975. When IBM entered into a contract with Microsoft to produce the software for the IBM personal computer in 1981, Microsoft began its transformation into the leading software company in the world. Its Windows operating system dominates the operating system market, Microsoft Office dominates the personal productivity software market, Internet Explorer dominates the browser market, and Outlook Express is the leader for e-mail use on personal computers.

In addition to Microsoft, Adobe is a leading producer of personal computer software. Adobe Acrobat is a powerful document production system. The Adobe reader is installed on most personal computers, and Adobe portable document format (PDF) files have become the standard for file exchange. Adobe's graphics and photo enhancement programs, such as Photoshop, are also widely used, and its Dreamweaver Web page editor also is popular.

Software for entertainment is also important. Hundreds of game programs are available for the personal computer, and numerous games can be played over the Internet. Many people listen to their favorite songs on their computers, watch a delayed broadcast of their favorite television shows, or enjoy the latest films after downloading them from one of the many online distributors. Mobile devices such as smartphones make use of many Web applications. Users can check their e-mail, get updates on the weather, and follow the latest developments in sports. Users of mobile devices also can use cloud computing to execute applications such as word processing on demand.

Many applications have been developed to support the storage of data. Initially, data storage software was a function of the operating systems. IBM's IMS hierarchal database, first released in 1970 and still in use, was the early leader in databases. In 1970, Edgar F. Codd published a paper describing a relational model for a database. It was used as the basis for Oracle in 1979 and IBM's DB2 in 1983. Powerful database-driven programs developed by SAP in 1972 and PeopleSoft in 1987 (PeopleSoft was acquired by Oracle Corporation in 2005) formed the basis of enterprise resource planning software, integrated systems that consolidate all business functions, including manufacturing, financials, human resources, supply chain management, project management, and customer relationship management.

Communications Software. One of the earliest uses of computer networks was to exchange files with messages. In the twenty-first century, e-mail has become the most popular program used on computers. Tim Berners-Lee developed a Web browser in 1990 while working at CERN (now the European Particle Physics Laboratory). Since then, Web browsers have become one of the world's main communications mediums. Much software has been developed for the Web, with the Google search engine being one of the most famous and most often used. In 1995, Ward Cunningham introduced the first Wiki, a set of online documents that are created and modified by those who use the documents. The online encyclopedia Wikipedia was created in 2001 and has since become an important source of information for almost all topics.

Early computer software used to play music was limited to WAV (binary waveform audio file format) and textual MIDI (musical instrument digital interface) files and, while interesting, was not widely used. Better file formats, in particular, MP3 (MPEG-1 audio layer 3) files, have produced high-quality reproductions that are small enough to download and store easily, and this has greatly increased audio player software. Early video software also was limited by file formats (such as Microsoft's audio video interleave, or AVI), low-speed networks, and slow processors. By 2000, network speed had increased, processors were faster, and the Windows media video (WMV) and QuickTime file formats made it practical to view streaming videos. As of 2010, Microsoft's Media Player and QuickTime's player are the most popular video display software, but many more players are available for download for those that are interested.

Software to support social networking by building a virtual community (such as Facebook) is increasingly popular. Although this can be seen as a simple extension of earlier instant messaging and message board software, the increased online communications represented by this software appear to be an important step forward in the evolution of communications.

Systems Software. Computers need software to function, and the most important software is the operating system. In the 1960's, IBM developed the first major operating system, the OS/360, for its System/360 line. In the 1970's, it introduced the MVS system, a descendant of the OS/360. It was reliable and provided solid support for the business and scientific applications of the 1970's. In addition, it supported multitasking and multiprogramming, and had good security. The UNIX operating system, originally developed at Bell Labs in 1969, provided a friendly interactive environment. It became very popular with systems programmers and scientists in the 1970's and 1980's but was not widely adopted for business.

The Xerox Star, released in 1981, was the first operating system to use a mouse and have a graphical user interface (GUI). In the early 1980's, Steve Jobs and Steve Wozniak produced the Macintosh computer for Apple. The Macintosh's operating system, which was based on the Xerox Star, was the first successful commercial Windows operating system. In 1992, Microsoft released Windows 3.1 and, in 1995, Windows 95. These operating systems made Windows the most popular operating system for both business and home computers.

Middleware is the software used to network computers. It includes network-enabled operating systems and the underlying communications software, such as implementations of TCP/IP (transmission control protocol/Internet protocol), and router software, such as CISCO's IOS (originally Internetwork Operating System) software. Mainframe and server software provide support for many network and Web applications. In addition to data storage, server-based applications provide support for audio, video, and images. An interesting Web server application is eBooks, which stores digital copies of books that can be rented or purchased and displayed on a wide range of communication devices.

The software development environment has improved considerably since the early days of computing. Most early computer programs were written in assembly (or machine) language. With the development of languages such as FORTRAN, BASIC, and COBOL, programs became easier to develop. In the 1980's, Borland introduced an Integrated Development Environment (IDE) that made it easy to develop programs, especially Pascal. Microsoft released its first IDE, Visual Studio, in 1997, and it has dominated the market since then. There are a large number of systems to support building Web applications. The leader is Microsoft's Visual Studio, but Macromedia's Dreamweaver and Microsoft's Expression Web are also used by many developers.

Impact on Industry

Almost every country in the world has a number of software engineers who produce software that is sold locally and worldwide. In the United States and Europe, where the computer industry started, application, system, and communication software are being developed at phenomenal rates. Asia has joined in the development of software. For example, although Google is the largest search engine company in the United States, the Chinese company that supports the Baidu search engine provides almost as many services as Google does.

In the United States, most of the software has been

developed by private companies such as Microsoft, IBM, and Oracle. Some very good software has been written by programmers who believe that most software should be made available to the public for little or no cost. For example, the GNU C compiler developed by the Free Software Foundation is a first-rate programming system free to anyone wishing to use it. The U.S. government and most other governments provide little direct support for software development, but they have considerable influence on the development of software by setting standards, such as requiring all computers purchased for government use to support TCP/IP.

Although the total value of software sold in a year is difficult to determine, a reasonable estimate is about $400 billion.

Government, Industry, and University Research. The U.S. government strongly influences the development of software by setting standards for software provided to the government. In the mainframe era, the requirement that a FORTRAN and COBOL compiler be available for each government computer encouraged all the early computer companies to develop these compilers. The National Science Foundation, the National Institutes of Health, and the Department of Defense have always provided support for university research into the development of computers and computer software, starting with the ENIAC (Electronic Numerical Integrator and Computer) constructed at the University of Pennsylvania in 1946. A great deal of research on the development of computer software has been done by industry in the United States and other countries. The IBM research laboratories, in addition to being a leader in computer hardware, have developed much advanced software. Microsoft has one of the largest industry research programs; it supports software development centers in Redmond, Washington; Cambridge, England; and at six other locations. In 2010, Microsoft announced Windows Azure, a cloud-computing operating system. As a part of Azure's development, Microsoft is moving to port Microsoft Office to the cloud, providing computing on demand to compete with Google's cloud office suite.

Software Companies. Many large and successful software companies exist, although there is some controversy as to exactly which companies are the largest in their fields. Microsoft is the largest software company in terms of volume of software produced and sold, with its operating system and office suite being very successful. The German company SAP is one of the largest producers of custom software with its supply chain management system software. IBM provides much of the server software (the new name for mainframe software) and has a large network management division, including Tivoli. Oracle is the largest and most successful database company, and with its acquisition of PeopleSoft and Sun, it is rated as the second- or third-largest software company.

The software industry contains many smaller but important companies. Symantec produces some important security software. Adobe Systems has a large number of graphics design programs. Apple still

Fascinating Facts About Software Engineering

- In 1983, Microsoft introduced its industry-leading word processor, Microsoft Word. Microsoft Word was based on an earlier WYSIWYG (what-you-see-is-what-you-get) text-editing system developed at the Xerox PARC laboratories.
- In 1964, John Kemeny and Thomas Kurtz co-developed BASIC, an interactive programming language. BASIC became immensely popular as a programming language for personal computers.
- In 1991, Linus Torvalds, a young student at the University of Helsinki in Finland, released the first version of Linux, an open-source operating system. As of 2010, only Microsoft Windows is more popular as an operating system.
- One of the great challenges in programming was to build a program that could play chess as well as a master. This was accomplished in 2006, when Vladimir Kramnik, the reigning world champion, was beaten by a computer program called Deep Fritz.
- In 1989, Tim Berners-Lee introduced the HTML language as a way to represent data on the Web. Later, Java and JavaScript were added to make programming on the Web almost as rich as programming on the desktop.
- Cloud computing, or providing software as a set of services to a computing device, is predicted to be the primary software model for the future. The origins of cloud computing go back to the early days of distributed computing, but it is not clear exactly who should get credit for its creation.

produces interesting operating system innovations. Quicken helps many small businesses bill their customers, and TurboTax is very popular at tax time.

Careers and Course Work

A major in computer science is the traditional way to prepare for a software engineering job. Courses in ethics, mathematics, and physics are required at the outset of a computer science degree program, followed by about thirty-six hours of computer hardware and software courses. Those with a computer science degree often take software engineering jobs in systems programming and scientific programming. Job tasks could include, for example, developing a new module for Microsoft Word or a new networking application for CISCO.

A major in information systems is another way to prepare for a software engineering job. First, students take courses in mathematics and business, then they take about thirty hours of courses on information systems development. Those with degrees in computer information systems often take software engineering jobs as systems analysts, business applications developers, or system managers. A typical job task includes helping implement a new enterprise resource management system for a university or local grocery chain.

Some schools offer a degree in software engineering. Students need a good background in mathematics and management for this degree. The degree emphasizes understanding how to build systems rather than building a strong programming background. Students taking this degree should choose a strong minor in science or business to prepare for a specific job after graduation. Students with a software engineering degree and an appropriate minor can get almost any software engineering job.

Social Context and Future Prospects

The future of software engineering is bright, with good jobs available in the computer industry, scientific laboratories, and businesses. All the traditional applications programs, such as payroll, supply chain management, or protein identification, need to be maintained and, in some cases, rewritten. However, a number of completely new applications are developing and show great promise as areas of employment. The development of an electronic health record (EHR) for each person is a huge software project, as is making the electronic health record portable. The creation of an electronic health record program will be one of the largest international software engineering projects ever conceived. The use of mobile devices to access computing from Web services is another important area of applications development. When accessing Web services in a browser or with an application on a computer is added to this mix, it becomes cloud computing, which will probably be the dominant form of software use in the future.

Although cloud computing offers an easy way to do computing, much work must be done to ensure that the data processed by these applications is secure and private. Standards for encrypting and signing data being transferred between mobile devices and cloud services do exist, but successfully implementing these standards will require much programming effort. Ensuring that individual and corporate data privacy issues are addressed in software implementations will require programmers to exercise great care in their program development and system management. In fact, many software engineering positions in the future are likely to be security and privacy analysts.

George M. Whitson III, B.S., M.S., Ph.D.

Further Reading

Hiltzik, Michael. *Dealers of Lightning: Xerox PARC and the Dawn of the Computer Age.* 1999. Reprint. New York: HarperBusiness, 2007. Describes the developments made at the Xerox PARC laboratories.

Pressman, Roger S. *Software Engineering: A Practitioner's Approach.* 7th ed. New York: McGraw-Hill Higher Education, 2010. Covers the entire field of software engineering. Includes excellent sections on Web engineering and project management.

Stair, Ralph, and George Reynolds. *Principles of Information Systems.* 9th ed. Boston: Course Technology, 2010. Contains a complete introduction to systems analysis and design and a good description of management information systems, decision support systems, and artificial intelligence software.

Whitten, Jeffrey L, and Lonnie D. Bentley. *Systems Analysis and Design Methods.* 7th ed. New York: McGraw-Hill/Irwin, 2007. Provides a comprehensive coverage of the analysis phase of software engineering, as well as a thorough introduction to the entire software development process from the business perspective.

Yourdon, Edward, ed. *Classics in Software Engineering.*

Upper Saddle River, N.J.: Yourdon Press, 1979. An outstanding compilation of articles written as software engineering developed, demonstrating how the field of software engineering developed.

Web Sites

Association for Computing Machinery
http://www.acm.org

IEEE Computer Society
http://sites.computer.org/ccse

Society of Software Engineers
http://sse.se.rit.edu

Software Engineering Institute, Carnegie Mellon
http://www.sei.cmu.edu

See also: Computer Engineering; Computer Graphics; Computer Languages, Compilers, and Tools; Computer Networks; Computer Science; Internet and Web Engineering; Parallel Computing; Video Game Design and Programming.

SPACECRAFT ENGINEERING

FIELDS OF STUDY

Physics; aerospace engineering; aerospace physiology and medicine; astronautics; astronautical engineering; mechanical engineering; electrical engineering; systems engineering; nuclear engineering; rocketry; computer science; computer programming; kinematics; thermodynamics; space physics; materials science; engineering management.

SUMMARY

Spacecraft engineering is an interdisciplinary engineering field concerned with the design, development, and operation of unmanned satellites, interplanetary probes, and manned spacecraft. Unmanned satellite missions include: commercial communications and remote sensing (including meteorological satellites), scientific research, military communications, navigation, and reconnaissance. Interplanetary probes are at present exclusively confined to scientific and exploratory missions. Manned spacecraft missions are currently confined to long-duration assignments at the International Space Station (ISS) and to short duration flights in support of the ISS on either the United States' Space Transportation System (STS) or the Russian Federation Soyuz spacecraft. As of 2011, manned missions returning to the Moon or first-time manned flights to Mars are likely decades away.

KEY TERMS AND CONCEPTS

- **Bus:** Portion of a spacecraft that houses and supports the payload.
- **Clarke Belt:** Collective term for geosynchronous orbits, named for Arthur C. Clarke, noted science fiction author, who was the first to propose the use of satellites in these orbits for communications.
- **Deorbit:** Disposal of a spacecraft at end of mission either by sending it to burn up in the atmosphere during reentry or by moving it into a little-used orbit, where it will not be a threat to other spacecraft.
- **Eclipse:** Passage through the Earth's shadow, where no sunlight is available for the spacecraft to use for heat or to warm solar panels.
- **Geosynchronous Orbit:** Circular orbit directly over the equator with an orbital period of precisely twenty-four hours; satellites in these orbits appear to be stationary in the sky as seen from the Earth's surface.
- **Hypoxia:** Deficient amounts of oxygen in the blood or body tissues.`
- **Outgassing:** Spontaneous release of gas or volatile liquid in a vacuum.
- **Payload:** Portion of a spacecraft tasked with performing the flight mission.
- **Photovoltaic:** Of or related to the generation of voltage from the absorption of light.
- **Plasma:** Ionized gas consisting of free electrons mixed with atoms having too many or too few electrons.
- **Reentry:** Return to Earth of a spacecraft, accompanied by extremely high atmospheric temperatures and pressures that are capable of destroying any unprotected spacecraft.

Definition and Basic Principles

Spacecraft engineering is the process of designing, constructing, and testing vehicles for deployment and operation in the full expanse of space above the Earth's atmosphere, generally regarded as beginning at an altitude of fifty miles. Launch vehicles and rocket propulsion are considered part of the related but separate field of rocketry.

Spacecraft have to be robust enough to survive several harsh environments. The vibrational loads generated by the launch vehicle will damage or destroy weak or poorly designed structures. The electronic components must function reliably in a high-radiation environment. All parts of the spacecraft cycle from extreme heat to extreme cold each orbit as the spacecraft moves into and out of the Earth's shadow. Outgassing in a vacuum can contaminate solar panels and camera optics. Motion through the plasma of the ionosphere creates potentially damaging static-electric charges.

Energy is expensive in space: It must either be gathered from sunlight from solar panels of limited area or generated from onboard fuel supplies, the exhaustion of which will end the mission. Batteries wear out through repeated charging and discharging

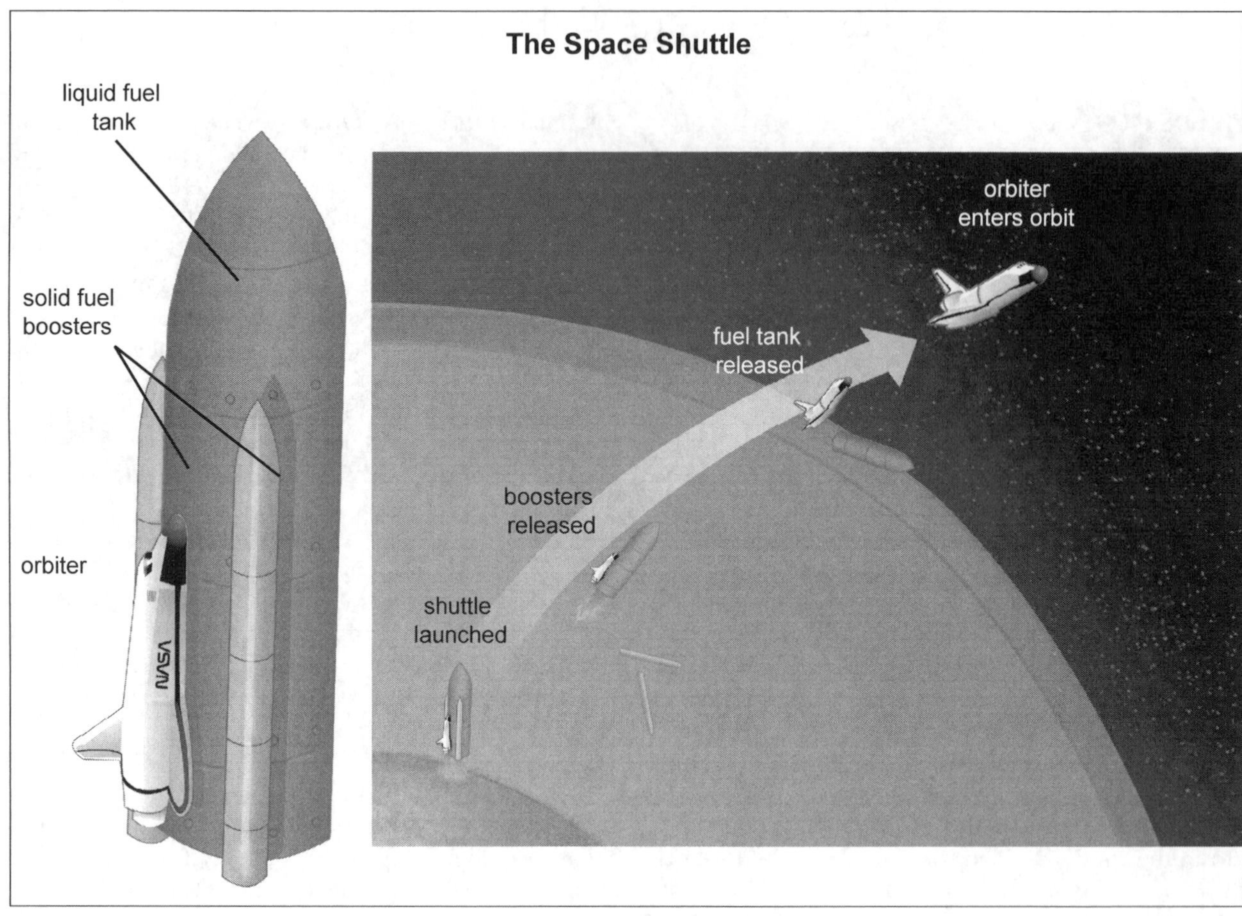

and must be managed carefully to last as long as possible. Electrical components must operate with high efficiency and draw as little power as possible; inactive components have to be kept off or in a low-power standby state. Everything must be designed with the knowledge that repairs or replacements will be impossible at worst and difficult, dangerous, and expensive at best.

Batteries, fuel cells, pressure vessels, propulsion systems, and nuclear power modules are all inherently hazardous devices; a launch failure can be catastrophic. All launch ranges operate under rigidly enforced safety standards for the protection of the spacecraft, the launch vehicle, and the personnel working around them. Reliability, survivability, and safety have to be designed into the spacecraft from conception. Rigorous testing throughout the development and manufacturing process plays a major role in spacecraft engineering.

Background and History

Spacecraft engineering did not exist before the development of the A-4 (more popularly known as the V-2) rocket by the German army during World War II. Both the Untied States and the U.S.S.R. integrated technology gotten from captured German scientists, engineers, and rocket hardware into their domestic long-range ballistic-missile development efforts. The primary goal was the development of intercontinental-range ballistic missiles for the delivery of nuclear warheads in case of war. The A-4, however, flew very poorly without the ton of high explosives it was originally designed to carry. The ordnance payload was replaced with scientific instruments for exploration of atmospheric conditions at high altitude and astronomical observations unimpeded by atmospheric interference. The extreme conditions of rocket flight and the harsh environment of space posed new challenges to the instrument developers.

Techniques that evolved to meet those challenges laid the foundation for the new field of spacecraft engineering.

In the early years, spacecraft were of necessity small and lightweight. The main challenges were miniaturizing components and operating on small amounts of electrical power. The newly invented transistor and its packaging into integrated circuits were adopted by spacecraft engineers immediately in spite of their initially high cost. Power came from compact high-performance batteries supplemented by recently developed silicon solar cells to produce electricity from sunlight. The initiation of manned spaceflight added the challenges of providing a livable environment for the crew and returning them safely to Earth. Spacecraft engineering had to confront the biological issues of providing air, water, and food while disposing of waste products. Long-duration manned missions raised quality-of-life issues such as comfort, privacy, personal hygiene, and physical fitness.

In the 1970's, interplanetary spacecraft ranged far from the Sun and strained the capabilities of solar panels to provide sufficient electrical power. Spacecraft engineers harnessed the energy of radioactive decay to power deep-space missions and to keep the spacecraft warm so far from the Sun. Missions to Mercury and Venus, so near the Sun, posed the opposite challenge of keeping the spacecraft from overheating.

Command, control, and data acquisition from remote platforms with limited power available for broadcasting has always been and continues to be a major communications challenge for the spacecraft engineer.

How It Works

Spacecraft. Spacecraft are composed of a payload and a bus. The bus is designed as a major system composed of seven or more subsystems. The electrical power system (EPS) provides the power to operate the active components. The communications system (Comms) maintains contact with ground control. The command and data handling system (C&DH) issues electronic commands to all onboard units and collects data from each unit for transmission to the ground. The thermal control system (TCS) regulates the temperatures of all onboard units to keep them within acceptable operating ranges. The attitude control system (ACS) controls the rotational dynamics of the spacecraft to achieve and maintain the required orientation in space. The propulsion system (PS) makes necessary changes in the trajectory of the spacecraft to keep it on course. The structure holds all of the spacecraft components and provides the mechanical support necessary during manufacture, transport, and launch. Manned spacecraft include additional environmental control and life-support systems (ECLSS). Redundant (duplicate) components and subsystems are used as much as possible to maximize reliability.

Electrical Power. The electrical power system is responsible for power generation, capture, or storage, plus delivery of conditioned electrical power to all parts of the spacecraft. Power may be generated by onboard reactors such as fuel cells, or captured from sunlight by solar panels. Power from sunlight is stored in rechargeable batteries for later use when the spacecraft is in eclipse, or for times when power demand temporarily exceeds the total available from solar panels alone. Power-conditioning circuits are necessary to provide electricity at the voltage, current, and stability required by individual components.

Communications (Comms). Communications consists of the antennas, transmitters, receivers, amplifiers, modulators, and demodulators necessary for communications with the ground. Comms may also include equipment for encrypting and decrypting the signal to prevent interception and to block attempts at illegally seizing control of the spacecraft with false messages. Communication must be maintained across distances that may stretch billions of miles in the case of interplanetary probes using signals of modest strength because of the limited amount of electrical power available. Static-free frequency modulation (FM) signals are preferred to minimize errors in transmission. High frequencies, on the order of billions of cycles per second, allow large amounts of data to be moved quickly. Spacecraft engineers are currently experimenting with Internet-type communications protocols.

Command and Data Handling. The command and data handling system centers on the flight computer. The computer monitors the status of all components: turns them on and off in accordance with schedules transmitted up from the ground, collects housekeeping data from all units, and science data from any science packages onboard.

Attitude Control System. High-gain antennas must

be accurately pointed toward the Earth to maintain communications with ground control. Remote-sensing instruments and science packages need to be pointed at their study targets. Manned spacecraft must maintain proper attitude for safe reentry. To achieve all of these, the altitude control system senses the orientation of the spacecraft relative to the fixed stars and reorients the spacecraft as necessary to fulfill the mission.

Spacecraft orientation is determined by reference to the Sun, the Earth, and the brightest stars. The Sun and the bright stars can be located optically; the Earth can be sensed even during eclipse by the infrared radiation emitted by its warm surface. Interplanetary probes can locate the Earth by homing in on the radio signal coming from ground control.

Some spacecraft require three-axis stability where rotation about any axis must be rigorously suppressed. If the spacecraft begins to rotate in an undesired manner, onboard gyroscopes are spun up to absorb the additional rotational momentum and, by reaction, leave the spacecraft as a whole stationary. Many other spacecraft maintain stability by rotation about a fixed axis. Control of the altitude control system is the responsibility of the command and data handling system. When total rotational momentum of the spacecraft gets too large, the excess is eliminated by firing the attitude thrusters in the propulsion system.

Propulsion System. The propulsion system performs occasional maneuvers required to keep Earth-orbiting satellites on station or interplanetary probes on course. The propulsion system consists of rocket thrusters, propellant, pumps, valves, and pressure vessels. Attitude control thrusters control the rotational dynamics of the spacecraft. Course correction thrusters change the speed or direction of motion of the spacecraft. The propellant must be storable for long periods of time under the harsh conditions of space. Special pressurization techniques are necessary to move liquid propellants from tanks to thrusters in zero gravity. The spacecraft must carry enough propellant for the planned mission lifetime plus a reserve necessary for deorbit at end of life.

Structure. Spacecraft structures must satisfy the competing requirements of strength and low weight. Spacecraft structures must not bend or sag under the acceleration loads experienced during launch. Nuts and bolts must contain a locking feature to prevent them from loosening under vibrational loads. The mass of the structure plays a passive role in thermal control by conducting heat from warmer to colder parts of the spacecraft.

Thermal Control System. The thermal control system uses active and passive methods of moving heat to maintain normal operating temperatures for critical components. Active methods include the use of heaters to warm cold objects and pump-driven fluids to move heat from hot to cold areas. Passive devices include reflective coatings and insulation.

Environmental Control and Life-Support Systems (ECLSS). Life-support systems manage air, water, food, and waste. Humans require a pressurized atmosphere that provides oxygen for respiration and humidity for comfort, while removing carbon dioxide. Too little oxygen leads to hypoxia, while too much leads to oxygen toxicity. Too much carbon dioxide leads to carbon dioxide poisoning. Too little humidity leads to extreme crew discomfort and possibly dangerous electrostatic discharges; too much humidity leads to condensation, which can interfere with electrical systems and nurture the growth of bacteria and fungi. Humans need about seven pounds of water a day for proper hydration plus an additional seven pounds for hygiene and housekeeping. Active adults require 2,500 to 5,000 calories a day to function without losing body mass. Corresponding amounts of waste are generated.

Captured waste products must either be recycled, returned to Earth, or dumped overboard. Short-duration flights can be open-loop, that is to say, all the required consumables can be onboard at launch and the waste products dumped overboard or disposed of after landing. The mass involved is prohibitive for long-duration flights. Open-loop systems can continue to be used if regular resupply flights are possible, as is done for the ISS. For manned missions to Mars, by contrast, resupply will be difficult if not impossible, driving the ECLSS design toward full regenerative recycling, where carbon dioxide is taken up by plants that provide food and oxygen and wastewater is purified and reused. The ISS has an important role as a test bed for these emerging technologies.

Applications and Products

Commercial Spacecraft. To date, all commercial

Fascinating Facts About Spacecraft Engineering

- Sputnik I, the world's first spacecraft, was launched by the U.S.S.R. on October 4, 1957. It deorbited three months later, on January 3, 1958.
- The United States launched its first spacecraft, Explorer I, on January 31, 1958. It deorbited on March 31, 1970.
- The oldest spacecraft still in orbit is Vanguard I, launched by the United States on March 17, 1958.
- As of November 11, 2010, there were 3,476 spacecraft in Earth orbit.
- The fastest spacecraft ever flown are the twin Helios probes in orbit around the Sun. At closest approach to the Sun, they travel at 150,000 miles per hour.
- Currently, four spacecraft are leaving the solar system and are headed into interstellar space: Voyagers 1 and 2, and Pioneers 11 and 12.
- Voyager 1 has been traveling for 33 years and is currently 10,900,000,000 miles from Earth.
- CubeSats are miniature satellites 2.5 inches on a side and are simple and inexpensive enough for student design teams to engineer and fabricate successfully.
- Twenty satellites of the Orbiting Satellite Carrying Amateur Radio (OSCAR) series allow licensed amateur radio operators to experiment with communications via satellite.
- More than 250,000 people worked in the U.S. space sector in 2008. The average annual wage was $90,000.

spacecraft have been communications satellites or remote-sensing satellites. Geosynchronous satellites appear stationary in the sky to an observer on Earth and make radio contact through fixed antennas possible. Television and radio programming sent up from the ground are amplified and rebroadcast to anyone with a line of sight to the spacecraft. Because the satellite never sets, the signal is uninterrupted. These satellites also provide radio and telephone communications for ships at sea and people living in remote locations. The market for these satellites supports a number of spacecraft manufacturers worldwide.

Remote-sensing satellites observe the surface of the Earth across a broad range of the electromagnetic spectrum that stretches from radio waves, through the infrared and optical bands, and into the ultraviolet. The oldest and most mature application of remote sensing is in the field of weather forecasting. Infrared and optical photography from space is used in land-use planning, mapping, crop surveys, and pollution monitoring.

Passenger-carrying suborbital spacecraft are under development and may soon be carrying tourists to the threshold of space.

Scientific Spacecraft. The first U.S. spacecraft, Explorer I, made the first major scientific discovery of the space age when it discovered belts of protons and electrons trapped by the Earth's geomagnetic field. Discoveries by scientific spacecraft of all nations have profoundly changed mankind's knowledge of the Earth and its place in the universe. Interplanetary probes have mapped almost every significant body in the solar system and discovered dozens of planetary moons undetectable from Earthbound telescopes. The Hubble Space Telescope has photographed stars at birth, stars at death, and a black hole at the center of the Milky Way. Energy that does not penetrate the Earth's atmosphere such as X rays, gamma rays, and high-energy ultraviolet rays can be studied only from space platforms.

Military Spacecraft. Military spacecraft include communications satellites, reconnaissance and surveillance satellites, missile-attack early-warning satellites, and navigation-support satellites such as the Global Positioning System (GPS) constellation. Military spacecraft are considered force-enhancement or force-support assets. They do not directly engage in hostilities.

Impact on Industry

Government and Military Sector. The U.S. government budgeted $62.6 billion for military and civilian space activities in 2007. Although space missions are conceived and managed by government space agencies or military departments, the actual design, development, and manufacture of the spacecraft are done by private industry under government contract. Typically, a prime contract is negotiated with one company, and it assumes total responsibility for delivery of the finished spacecraft. This prime contractor then divides the work into segments that are awarded to subcontractors. Either the prime contractor or one of the subcontractors assumes the role of integrating contractor and bears responsibility of receiving each subcontractor's finished product and assembling the actual spacecraft. For military spacecraft, the prime contractor is a company from the defense sector. For scientific spacecraft, the prime contractor is usually the university or research institution employing the principal investigator of the research project. An individual spacecraft project can easily involve dozens of subcontractors and hundreds of vendors supplying parts and materials to the subcontractors.

All parts and materials used in spacecraft construction must meet rigid and demanding flight-certification standards. The standards applied to manned spacecraft are the most demanding of all. Formal and well-documented quality control and testing procedures are mandatory at all stages from raw material to finished product. Government oversight and review of these procedures is keen and penetrating. Many government projects are also classified in part or in total. Key positions can be filled only by personnel cleared by the government for access to classified information, work areas must be secured with access limited to authorized personnel, and sensitive documents generated by the project must be protected and controlled. Quality control and security teams frequently outnumber the engineers. A specialized segment of the defense industry has evolved to satisfy these secondary needs.

Business Sector. Sales of GPS equipment and GPS-capable chipsets totaled $56.19 billion in 2007. Direct-to-home television and satellite radio generated $65.47 billion in sales that same year. Insurance companies collected $530 million in premiums for policies covering spacecraft during launch and on orbit. Global space activity revenues exceeded $251 billion.

Careers and Course Work

Undergraduate study in almost any field of engineering is sufficient entry-level training, but this should be followed up with graduate study at an institution with a strong spacecraft-design program—preferably one with faculty involved in an ongoing spacecraft project. Entry-level jobs are available with spacecraft manufacturers, government space agencies, intelligence agencies, and military departments. Entry into the field is also possible through appointment as a commissioned officer in the armed forces that have assigned space missions and responsibilities.

Upper- and mid-level-management jobs require administrative and budgeting skills that can be acquired on the job but are significantly enhanced by business or financial management courses. These courses can be completed as part of a secondary master's degree, through professional development training, or continuing education. Strong written and oral communication skills are essential.

Most spacecraft development work involves trade secrets and access to classified information. Individuals who cannot qualify for a security clearance, have a bad credit history or a criminal record, or who are not a citizen of the country in which they reside will not readily find employment in this field.

Social Context and Future Prospects

Spacecraft and the services they provide are now part of everyday life. Airliners navigating across oceans and pedestrians navigating city streets rely on GPS devices to find their destinations, and embedded GPS chips in cell phones allow parents to track their children anywhere. Satellites bring television straight to the home and radio straight to the automobile. Public databases allow anyone connected to the Internet to acquire a satellite photo of almost any spot on Earth. Accurate long-range weather forecasting exerts a daily influence on almost all business and personal planning. Spacecraft are now an indispensable part of the global economic and social infrastructure. This infrastructure must be replaced as it ages and wears out, creating a continuing demand for the services of the spacecraft engineer.

Spacecraft have become so numerous in low-Earth orbit that disposal of spacecraft at end of life is a major design challenge for the modern spacecraft engineer. On February 10, 2009, Iridium-33, a U.S. communications satellite, collided with Cosmos-2251, a defunct Russian military communications satellite. The debris generated by the collision now threatens other satellites at the same orbital altitude. Future collisions are certain to happen more often as the population of spacecraft increases: They are best prevented by deliberately deorbiting nonoperational spacecraft so that they burn up in the atmosphere in reentry or by moving them to orbits at seldom-used altitudes. Tougher regulation by United States and international agencies is highly probable.

Current trends indicate that the National

Aeronautics and Space Administration (NASA) will withdraw from manned spaceflight with the anticipation that commercial firms will take over this function. Commercial firms must first develop the necessary expertise in rocketry and then learn to produce safe and reliable manned spacecraft.

Billy R. Smith, Jr., Ph.D.

Further Reading

Edwards, Bradley C., and Eric A. Westling. *The Space Elevator: A Revolutionary Earth-to-Space Transportation System.* Houston, Tex.: B.C. Edwards, 2003. A glimpse at truly exotic and daring spacecraft engineering. Excellent example of the way requirements drive design decisions in an engineering project.

Jenkins, Dennis R. *Space Shuttle: The History of the National Space Transportation System—The First One Hundred Missions.* 3d ed. Dennis R. Jenkins, 2001. A history of the space shuttle from a spacecraft-design perspective. Detailed schematic diagrams and voluminous photos of construction, maintenance, and operation of the shuttle reveal much about the art of manned spacecraft engineering.

Mishkin, Andrew. *Sojourner: An Insider's View of the Mars Pathfinder Mission.* New York: Berkeley Books, 2003. The author is a senior systems engineer for NASA. His book is a personal glimpse of spacecraft engineering and operations.

Seedhouse, Erik. *Tourists in Space: A Practical Guide.* Chichester, England: Praxis, 2008. A glimpse of manned spacecraft engineering as it is being done in what may be the start of a new era in manned spaceflight.

Sellers, Jerry Jon. *Understanding Space: An Introduction to Astronautics.* Edited by Douglas K. Kirkpatrick. 3d ed. New York: McGraw-Hill, 2005. Chapter 1 outlines the elements of a space mission; Chapter 3 discusses the space environment; Chapter 11 discusses spacecraft design overall; Chapters 12 and 13 discuss spacecraft subsystems. This is an introductory college text but has a great deal of material accessible to intelligent high school students.

Swinerd, Graham. *How Spacecraft Fly: Spaceflight Without Formulae.* New York: Copernicus Books, 2008. Chapters 7 through 9 give an informative discussion of spacecraft engineering without mathematics. They begin with a discussion of spacecraft at the system level and follow up with examinations of spacecraft subsystems. Chapter 10 discusses what the twenty-first century may hold for spaceflight.

Web Sites

Aerospace Industries Association
http://www.aia-aerospace.org

American Institute of Aeronautics and Astronautics
http://www.aiaa.org

Jet Propulsion Laboratory
http://www.jpl.nasa.gov

Satellite Industry Association
http://www.sia.org

See also: Electrical Engineering; Engineering; Mechanical Engineering.

SPACE ENVIRONMENTS FOR HUMANS

FIELDS OF STUDY

Aeronautics; astrobiology; space engineering; life sciences; astronomy; computer science; environmental science; nanotechnology; robotics and embedded systems.

SUMMARY

Space environments for humans concerns the potential development of habitats, spacecraft, and techniques that will allow humans to create and live in sustainable environments in space, such as space stations, or on planets and satellites, such as the Moon or Mars. Advances in space science, engineering, and technology since the mid-twentieth century resulted in plausible scenarios for human colonization of space, as well as means for harvesting and using extraplanetary resources for purposes of human settlement. Although no definitive off-Earth scenario that addresses how humans will live in space yet exists, it does appear possible that human environments in space will remain a subject of concern for scientists, engineers, space agencies, and technologists into the future.

KEY TERMS AND CONCEPTS

- **Extraplanetary Resources:** Resources that are available in off-Earth environments; these could include natural resources used for powering spacecraft and settlements.
- **Extravehicular Activity (EVA):** Activity that occurs outside the protective environments of spacecraft, such as a spacewalk.
- **Life-Support System:** System developed to allow human and other biological life-forms to live in space environments.
- **Space Engineering:** Science of engineering as applied to off-Earth environments. By their nature, these environments have different physical properties and therefore require knowledge specifically related to space science and physics.
- **Space Habitat:** Semipermanent or permanent human living environment on off-Earth planetary bodies (such as satellites or planets) or aboard orbiting spacecraft.
- **Space Radiation:** Cosmic and solar energy or matter that could have detrimental effects on space travelers, as well as their spacecraft, equipment, and environments.
- **Space Tourism:** Use of space environments for leisure activities such as short- and long-term trips in orbit around Earth, as well as vacations aboard spacecraft or in human settlements on planetary bodies.

Definition and Basic Principles

Although human settlement in space has not become a scientific reality, three major space initiatives allowed scientists and technologists to realize the first stages of creating long-term human living environments in space. The first was a series of National Aeronautics and Space Administration (NASA) programs—Mercury, Gemini, and Apollo—that culminated in a successful lunar landing in 1969 and subsequent return trips to the Moon. The second and third developments were the successful deployment of NASA's space shuttle program and several Earth-orbiting space stations—including, most notably, the Russian Mir and the International Space Station (ISS). The data collected from each of these projects broadened understanding of the possibilities of long-term survival of humans in various space environments and opened up possibilities for continued ventures in space.

Areas of continuing research related to human environments in space include the effects of solar and other types of radiation on spacefaring humans and their spacecraft, the availability of extraplanetary resources that could sustain human life in space and power the spacecraft, the creation of materials to protect humans in space or on other planets, realistic colonization prospects for humans on other planets or satellites, funding and management of space projects related to human environments in space, and the determination of whether living in space is a realistic option for human beings.

As scientists, space agencies, and technologists grapple with such questions, scientific research continues. In large part, this is the result of scientific curiosity, and the knowledge gained from living and working in space adds to understanding of how people interact with their home planet. However, because of environmental issues such as global warming and possible energy shortages, there are also major concerns about the continued sustainability of Earth and its resources for human habitation.

Background and History

A long-established subject for science-fiction and fantasy writers, the dream of living and traveling widely in space and on other planets precedes the scientific reality of being able to do so. Much of what has been discovered about the needs of humans in space environments was learned during the space race in the 1950's and 1960's, when the Soviet Union and the United States competed to see who would be the first nation to put a man in space and return him safely to Earth. Also adding to knowledge of the effects of spaceflight on humans were the subsequent missions in NASA's space shuttle program and the many space stations launched since the 1970's. By this time, scientists had recognized that space habitats for humans could become viable options for space programs.

In the years before space travel, however, many scientific visionaries considered the special needs of environments for humans in space. Russian scientist Konstantin Tsiolkovsky, for example, was a pioneer in theorizing about spaceflight. Among Tsiolkovsky's numerous influential ideas were the space elevator (a contraption anchored to Earth that would allow travel high into Earth's orbit without the use of rockets) and the notion of colonizing space as an answer to the potential dangers faced by humans on Earth, such as catastrophic impacts from solar bodies. Similar concerns are still used as justification by some parties in insisting that space colonization is an important next step for humankind.

In the twenty-first century, the idea of building space environments for humans is still alive and, more than ever before, has taken on an international tone. Numerous nations, including the United States, Russia, China, Japan, and India, have been considering flights to the Moon and Mars, and some have made plans for the creation of permanent or semipermanent lunar and Martian bases, as well as long-term journeys within the solar system.

How It Works

Because of the harsh conditions awaiting humans in off-Earth environments, special consideration must be given to the protective clothing worn by spacefarers, as well as the protective materials that will be used to build their stationary and floating habitats. Just as important are the design of habitats in which humans are to live and work, as well as the resources that will power their habitats and vehicles and possibly even feed entire colonies.

For space environments for humans to become feasible, further research and development must take place in three main areas: life-support systems and shielding, types of space-bound habitats, and resources to sustain habitat populations. Given the vast number of possibilities within these scenarios, there are many considerations that will take precedence over others as humankind progresses toward a more developed future in space. For example, the operating conditions, power resources needs, and amenities aboard a habitat built for long-term, near-Earth orbit will be very different from those for a roundtrip to the Moon or Mars. Despite these differences, there are many other areas in which research and development could uncover certain universally applicable materials and techniques.

Space Habitats. Since it was first understood that it is possible to live in an off-Earth environment, many designs have been proposed to accommodate humans in diverse space environments. The only living and working environments ever used by humans in space have been the capsules used to transport astronauts to and from the Moon and space shuttles and space stations launched into orbit from Earth. In the case of the lunar capsules, each was designed for a short voyage of less than a month's duration. Similarly, space shuttles, while offering a wider range of movement than capsules, as well as more storage capacity, were not designed for human habitation out of Earth's atmosphere. Space stations, however, because of their size and the ability to add modules (modular compartments, designed for many purposes, whether personal, scientific, or operational), offer a glimpse into what orbiting habitats will probably be like during the initial stages if spacefaring becomes more common. They are also durable and can be built to the discreet specifications of the sponsor of the vessel.

Several types of habitats may be needed as

housing, storage, and work areas for humans in space. NASA and other space agencies have been considering the establishment of base camps on the Moon and perhaps even Mars. One plan uses torus-shaped designs that can be inflated and connected together in a manner similar to inflated inner tubes connected to each other by portals. The tori would be constructed of highly durable protective fabric that is light enough to transport from Earth to the Moon. Other possible designs include the use of bunkers dug into the surface of the Moon; the advantage of this design is that, instead of being exposed to the extremely cold air on the Moon, the bunkers could be temperature-regulated and would have a natural barrier—the surface regolith (soil)—against cosmic rays. Other proposed habitats include designs for mobile vehicles that can be joined together for periods of time to form base camps and separated for movement when appropriate.

Life-Support Systems and Shielding. The physiological effects of the weather and atmospherics awaiting astronauts in space and on extraplanetary bodies are different from those found on Earth. For example, long-term exposure to solar radiation can cause major health problems for astronauts. In addition, continuous living and working in zero-gravity environments can lead to cardiovascular problems, loss of bone density, muscle atrophy, and many other issues that can lessen the productivity of astronauts and even cause life-threatening conditions. A further consideration is protection from meteoroids that can damage machinery and habitats. With these things in mind, designs for special protective suits and gear are being created that will lower exposure to solar particles and energy. The protective suits will also encourage efficient oxygen circulation and regulate temperature and air pressure, flexibility and mobility, and weight and comfort.

Life-Sustaining Space Resources. Because of the very high cost of supplying life-sustaining resources such as oxygen, water, food, and power from Earth, it is very likely that vessels and habitats will need to become self-reliant. Gathering, storing, and manufacturing these resources once initial launch supplies are depleted remains a significant issue. Although orbiting space vessels will remain reliant on terrestrial sources, lunar (and possibly even Martian) colonies may be able to extract some of these things from the natural resources available at hand. Scientists think that both water and oxygen can be processed from lunar regolith, giving those settlements a nearly endless source of these resources. Power generation could easily be solved by using solar units, but only during the lunar day, which lasts for fourteen days. During the other fourteen days, lunar settlers would probably have to use a different source of power, such as battery cells and possibly even power resources beamed in from satellites.

Applications and Products

Human-inhabited environments in space have many potential uses, and like many of humankind's earliest journeys on Earth, they will be expensive and potentially hazardous. They may also be extremely rewarding and alter the course of human endeavor for centuries to come. However, as of the early twenty-first century, space programs were directed toward exploration rather than settlement.

Lunar and Martian Bases. The space shuttles and manned space station traversing the sky are used for scientific research. Lunar or Martian bases would most likely be used in the same way but could offer much more data on a wider variety of subjects. They would be constructed in environments that allow study of such topics as the capabilities of natural resource development in space, the long-term effects of non-Earth atmospheres and conditions on humans (also plants and animals), the social and psychological effects of spacefaring, and the technological capabilities of the human race. All of these are very broad subjects, but they are vastly important to an understanding of the requirements of continuing to voyage into space.

Mining and Harvesting of Space Resources. Human settlements in space could help solve specific resource needs faced by humans on Earth, such as environmental and power needs. One proposed endeavor would be to use lunar regolith as a source of energy that could be beamed back to Earth through microwaves. Other possibilities include using the helium-3 found in the lunar soil as a fusion-power source to fuel rockets. If proper transport was arranged, helium-3 could also be used for Earth's power needs. Perhaps the most important use for resources found in space would be to sustain human settlements and help them become self-sufficient.

Space Tourism. Since 2001, a handful of wealthy, private individuals have traveled into space as tourists, all of them aboard Russian spacecraft. Willing to pay tens of millions of dollars, these so-called space tourists are the first clients of a potentially lucrative—although extremely dangerous—international industry that may someday offer trips to low-Earth-orbit hotels, space stations, and lunar or Martian bases. The cost of such trips is likely to decrease over time until, for example, short-term trips into orbit cost tens of thousands of dollars. Before space tourism becomes common, it is likely that this industry will be heavily regulated.

Living and Traveling in Space. Each new space venture helps determine whether people can live in orbit or on other planetary bodies. Once the feasibility is determined, it will open the door for many other considerations that have remained merely speculation. Questions and areas for research include whether it will be possible to create environments that can grow into self-sustaining human settlements, perhaps with farms and industry as a source of income; whether human settlements need governance in the same way colonies once did; and whether space colonies might seek independence from or become hostile to governments on Earth. Though these questions seem to be the stuff of science fiction, they could become real issues affecting people on Earth.

Impact on Industry

There are many types of environments that could be built for humans in space, each based on the conditions that are specific to the environment. For example, low-orbiting space stations are not at all like space shuttles in their design needs, while a Martian colony would differ greatly from a lunar colony. In this sense, the design and study of space environments for humans is highly theoretical. The nontheoretical work is being done in suborbital environments, designs for next-generation space shuttles and rockets, and whatever modules are being planned for the International Space Station or other space stations. This field remains mainly the focus of national and international space agencies using private contractors as partners, but the potential impact of a growing private industry should not be disregarded. Of major importance, too, are the universities around the world that are contributing much-needed research to the field.

National Space Programs. By far, the most important and far-ranging research into space environments for humans is conducted by, or at the behest of, national and international space agencies. NASA, the European Space Agency, the Japan Aerospace Exploration Agency, the China National Space Administration, and the Russian Federal Space Agency are among the largest and most powerful space agencies. Each of these agencies has engaged in or planned major space projects designed to further knowledge. For example, much of what is known about operating and managing space stations comes from the Russian agency, as well as those parties involved in the International Space Station. The United States, China, and Japan all have been developing plans for manned lunar missions.

Private Space Initiatives. Although the for-profit aerospace industry can claim a large share of the credit for successfully and safely sending humans into space, it has done so primarily as a partner with government agencies. However, as the prospects for humankind in space grow, it is likely that nongovernmental, private enterprises will eventually engage in profitable space ventures. Several private entities have been operating in space, but none has used manned vessels, although at least one, Bigelow Aerospace, has been experimenting with orbital designs that have the potential to carry humans. In 2004, the X Prize Foundation awarded a multimillion-dollar prize in the field of suborbital spaceflight, in part to attract the kind of outside-the-box thinking that tends to occur in the private sector. It is very likely that private industry—as it has done historically—will develop products and services in the wake of the research and efforts of national and international space agencies.

Public-Private Initiatives. The network of space industries involves the use of private contractors as partners in the design, manufacture, and distribution of goods (for example, launch vehicles and propulsion units) and services (for example, engineering and telecommunications). Although some of this work involves technologies and techniques common in nonspace uses, such as the manufacture of ocean liners or airplanes, the special needs of in-space environments may require highly specialized workers in such areas as nanotechnology and robotics. For example, space environments could use nanomaterials to solve problems of electrostatic lunar dust or even hygiene in cabins, and the environments could be controlled partially by robotic systems that are independent of direct human control.

Fascinating Facts About Space Environments for Humans

- In zero-gravity environments, human height increases by a few centimeters. This is because of the effects of reduced spinal curvature and the expansion of intervertebral discs in the back. The downside of this is lower back pain for astronauts.
- Lunar dust can pose pesky problems for astronauts working on the Moon. At sunrise and sunset, lunar dust tends to levitate and stick to surfaces. If lunar dust seeps into machinery, it can cause system failures, making this an important engineering concern for future lunar base builders.
- Being a space tourist is extremely expensive. Space visitor Guy Laliberté paid about $35 million to travel aboard a Russian shuttle to the International Space Station. It is estimated that, someday, such trips will be less expensive and will accommodate more tourists.
- If large-scale space habitats for humans are ever built, it is likely that they will be based at least somewhat on designs such as the O'Neill cylinder and the Bernal sphere (similar to ships featured in the 1968 film *2001: A Space Odyssey* and 1994-1998 television series *Babylon 5*, respectively). The designs for these habitats promise living space for thousands of people and enough resources to have an agricultural industry in space.
- Becoming an astronaut is not easy. Aside from the professional and academic credentials needed for the job, prospective astronauts must also have perfect eyesight, stand between 62 and 75 inches high, have excellent blood pressure, and be able to tread water for ten minutes while wearing a flight suit.

University Research. Some of the most important pioneering research in space environments for humans has come from universities. For example, work done under the auspices of the Universities Space Research Association (which includes the Massachusetts Institute of Technology, Harvard University, Stanford University, and more than one hundred other universities) in medical and biotechnology fields has studied the use of microwave devices for closing wounds and the development of a means for measuring the effects of radiation on astronauts.

Typically, national space agencies award money for research into areas of vital interest concerning future space projects, including studies in specific areas, such as on-board system controls, environmental designs for living and working, and the creation of devices that will make life easier in human habitats.

Careers and Course Work

Many possible career paths are available in the area of space environments for humans. For those interested in creating and designing these habitats, focus should be given to areas of space biology, space engineering, computer science, robotics and embedded systems, nanotechnology, and aeronautics, because space environments will draw on the knowledge of workers from all these areas. To become involved in the use of these environments, some of these same areas of focus will be needed, but the quickest path to doing so is to become an astronaut. To be a NASA astronaut, one needs to meet basic science and math requirements as well as to hold a bachelor's degree or higher in an area of biological or life science, engineering, or math. Professional experience in the military and aviation could be of use to students seeking a career in piloting spacecraft.

The industry that has grown up around the needs related to space environments for humans, although relatively small, is a highly technical field. There are many very specialized areas of focus for technicians, project managers, and other workers, many of whom will have to acquire advanced degrees in their fields of study to be considered for even entry-level jobs. Some careers, such as robotics and embedded systems engineers, require degrees in computer science, while others, such as environment designers, need a degree in engineering or physics.

Social Context and Future Prospects

Humans may someday use environments in space as regularly as they use terrestrial spaces. Space environments may be used as private living spaces for entire families and public spaces for commerce. Of course, these are merely dreams, but by identifying what humans need to survive in space—whether it is clothing, habitats, or means of transportation—and by building these things, humans will move one step closer to making those dreams a reality.

The first phases of learning about the needs of humankind in space have come and gone. The days of the first space race between the United States and the Soviet Union taught space scientists and astronauts the fundamentals of taking humans into space and returning them safely to Earth. However, research is allowing people to envision far more than short-term journeys. For example, the space suits designed for NASA's Constellation program are made for a variety of purposes and time frames. Similarly, the types of habitats that have been envisioned appear to be small-scale versions of what could someday become cities in space (either floating or stationary). In short, scientists no longer worry about whether humans can survive in space environments but instead about which designs will be used first and for what purposes.

However, humankind is not yet living in space. Unlike science-fiction stories in which entire civilizations are packed into starships and shipped out across the galaxy and beyond, plausible plans for living in space are limited to the concerns of the small groups of astronauts who will be involved in the initial test runs of these plans. Growth in this industry is most likely to occur as a result not only of national and international interests but also of corporate concerns and the desires of people to make the next small steps toward human-friendly space environments.

Craig Belanger, M.S.T., B.A.

Further Reading

Harris, Philip. *Space Enterprise: Living and Working Offworld in the Twenty-First Century.* New York: Springer Praxis, 2009. Overview of commercial, political, social, environmental, legal, and other practical aspects of human development of space. Features extensive illustrations of people and concepts related to the themes of each chapter, as well as citations for researchers seeking further information.

Howe, A. Scott, and Brent Sherwood, eds. *Out of This World: The New Field of Space Architecture.* Reston, Va.: American Institute of Aeronautics and Astronautics, 2009. These essays address the requirements of space architecture—the creation of habitats for humans in space—and look at possible structures on the Moon and Mars.

Seedhouse, Erik. *Tourists in Space: A Practical Guide.* New York: Springer Praxis, 2008. Presents an overview of information related to space tourism, including physiological effects of spaceflight on the human body, companies involved in human spaceflight, and astronaut training and instructional material.

Thirsk, Robert, et al. "The Space Flight Environment: The International Space Station and Beyond." *Canadian Medical Association Journal* 180, no. 12 (June, 2009): 1216-1220. Discusses the basic science behind the effects of space travel and extended periods of time in zero-gravity on the human body and details specific risks to astronauts aboard the International Space Station.

Zubrin, Robert. *Entering Space: Creating a Spacefaring Civilization.* New York: Jeremy P. Tarcher/Putnam, 2000. Presents a scientist's perspective on human spaceflight beyond what has been possible, with a deep focus on natural resources gathering and usage, as well as the author's speculation about the next steps for human space travel.

Web Sites

China National Space Administration
http://www.cnsa.gov.cn/n615709/cindex.html

European Space Agency
http://www.esa.int/esaCP/index.html

Japan Aerospace Exploration Agency
http://www.jaxa.jp/index_e.html

National Aeronautics and Space Administration
http://www.nasa.gov

Russian Federal Space Agency
http://www.federalspace.ru/main.php?lang=en

Virgin Galactic
http://www.virgingalactic.com

See also: Spacecraft Engineering

SPORTS ENGINEERING

FIELDS OF STUDY

Sports; biomechanics; ergonomics; kinesiology; engineering; mathematics; physics; physiology.

SUMMARY

Sports engineering is the study of how sports equipment affects performance and safety. Sports engineers use technologies such as three-dimensional imaging and computer modeling in combination with engineering analysis to optimize the overall performance of athletes and their equipment. The scope of sports engineering includes optimizing safety equipment such as helmets and pads, fields and buildings such as hockey arenas, as well as equipment required for sports such as skis and racquets. Sports engineering is a relatively new discipline, founded as a degree program in 1998 at the University of Sheffield in the United Kingdom. The International Sports Engineering Association (ISEA) was subsequently formed to promote the field of sports engineering and to provide a forum for sports engineers to discuss and collaborate.

KEY TERMS AND CONCEPTS

- **Aerodynamics:** Study of air motion as air interacts with matter.
- **Biomechanics:** Application of mechanical principles for biologic organisms.
- **Kinesiology:** Study of movement.
- **Neural Network Computing:** Computers that simulate the processing of neurons in the human brain. These systems are used in solving complex problems.
- **Physiology:** Study of biologic systems.

Definition and Basic Principles

Sports engineering is the application of engineering principles to the study of sports equipment and sports venues. This field most closely resembles mechanical engineering. Mechanical engineering uses physics and mathematics to study and design physical processes and mechanical systems. Sports engineering uses the same techniques to study and design sport-specific equipment and venues. Since the 1990's, sports engineers have been involved in sports as varied as golf, hockey, speed skating, and tennis.

Background and History

Sports engineering is a relatively new field that was formed as a subdiscipline of mechanical engineering. The best-known sports engineering program is at the University of Sheffield in the United Kingdom. Founded in 1998, the program divides the discipline into the areas of aerodynamics, sports surfaces, impact, and friction. Although the specific field of sports engineering is new, the fundamentals that make up this subdiscipline are based on centuries of study.

The earliest recorded Olympic competition dates back to ancient Greece. As technology improves it has been applied to virtually all sports. Even clothing has advanced to provide sports specific advantages such as improved aerodynamics, breathable fabric, thinner insulating fabrics, and superior waterproofing. Sports engineers also work with the governing bodies of sports organizations to meet the regulations of those specific sports.

The fields of engineering evolved from physics and mathematics. The first societies were formed in the nineteenth century. As the specific field of mechanical engineering evolved, the principles of physics and mathematics were already being applied to sports equipment. Biomechanics and kinesiology developed as specialized fields in the subject of human movement; however these fields did not specifically address sports equipment. The University of Sheffield provided a focused area of study in sports equipment and venues by founding the sports engineering program along with the Sports Engineering Research Group (SERG). The International Sports Engineering Association (ISEA) was then founded to provide a forum for engineers interested in this subject. There are now a handful of universities around the world that offer degrees in sports engineering and other mechanical engineering programs that offer classes in sports engineering.

How It Works

Mechanical engineering applies mathematics and physics to the design and study of physical and mechanical processes. Sports engineering uses mechanical engineering knowledge to focus on the study and design of sports equipment and venues. Refinements in equipment lead to improved performance by athletes as well as improved safety for participants. Regulations of various sports have been modified in order to accommodate new forms of equipment while maintaining fairness between competitors and improving sports safety. It is useful to use the University of Sheffield's divisions of aerodynamics, sports surfaces, impact, and friction to understand the areas of study that make up sports engineering.

Aerodynamics. Aerodynamics uses mathematics and physics to describe and model the airflow around objects. For any sport that uses balls, it is helpful to understand the trajectories that may result under different circumstances to develop techniques for improved accuracy and speed. Aerodynamics also applies to any sports that involve speed. Skeleton, speed skating, bobsledding, sprinting, downhill skiing, and many other sports use specialized equipment and clothing to reduce the effects of air flow on performance.

To study the aerodynamics of sports equipment, there are a variety of techniques that can be employed. Wind-tunnel tests and laser scanners are used in the study of sports aerodynamics along with sophisticated mathematics techniques such as computational fluid dynamics (CFD). Forces such as lift and drag can be calculated using these methods. Variations in surface roughness of a ball or of clothing, spin on a ball, or other factors will change the way an object moves through space. The most familiar example of this is the variation in the behavior of a baseball depending on the pitch.

Sports Surfaces. The interactions of athletic shoes or equipment and the sports surface play a large role in performance and can result in injuries. Surfaces may vary even within a sport, such in as tennis and soccer. Information about these interactions is used by shoe manufacturers and other manufacturers to design equipment that will reduce injuries and optimize performance. Traction-test devices are used to simulate conditions in some cases. A traction-test device is simply a shoe surface that is mounted on a plate that can be tested on various surfaces.

There are numerous variables involved in the interactions of surfaces. Each sport has different materials and different requirements for movements on the surface, and there are also inter-athlete differences. The mathematic models needed to study these interactions are so complex that SERG researchers have turned to neural-network modeling to generate information that can be used to design better equipment.

Impacts. One of the most-studied sports has been tennis. The variables of racquet design, swing technique, ball design, and area of the strings used will all change the trajectory and speed of the tennis ball in play. Numerous other sports can be studied in terms of impacts, including baseball, squash, hockey, field hockey, and cricket.

Three-dimensional ideography and high-speed photography can be used to study the effects of different variables in the impact stage and resulting trajectory of the ball. This information can then be used to refine the athlete's technique and to design more effective equipment.

Friction. Surface characteristics of balls, fields, and other sports equipment will affect sports performance. The interactions between the athlete's skin and grip must be considered in these circumstances in addition to environmental conditions such as moisture. The effects of sports surfaces on human skin is also studied to reduce injuries.

This area of sports engineering has overlap with biomechanics since it involves the interaction of the skin and other tissues with the sports equipment or sports surface.

Special Considerations. There are areas where these SERG divisions will overlap or where specialized knowledge is required. For speed skaters, the interaction of the skate blade and the ice surface creates a melting effect that must be understood in order to design faster skates. There are experts in this specific interaction that are working to improve skate-blade design. For changing conditions such as outdoor skiing or biking, there are different types of equipment that are designed for specific conditions. One example of this is the powder ski, which has different characteristics in terms of shape and design as compared with slalom or downhill skis.

The varied requirements for each sport results in a wide array of specialization for sports engineers.

Applications and Products

Engineering techniques have been used to

improve sports equipment of all types even before there was a designation called sports engineering. There are many examples of this in sporting history. The pole vault record has increased to 6.14 meters as a result of improvements in materials and design during the twentieth century. This is a 53 percent increase from the first official competition. Sports engineering has resulted in improvements in areas as diverse as fly-fishing rods and Frisbees. Amateur, professional, and Olympic competitors have all benefited from improvements in equipment and venues.

Olympic Sports. Sports engineering has increasingly been used to improve Olympic-athlete performance. Researchers at the University of Sheffield were credited with helping the British cycling team win four gold metals at the 2004 summer Olympic Games in Athens. Researchers then turned to studying skeleton and used complex modeling and digital shape sampling and processing to improve performances in the 2006 winter Olympic Games in Turin, Italy. Sports engineering can be particularly helpful in those sports where one-hundredth of a second might be the different between a gold and silver medal.

Amateur Sports. Sporting-equipment manufacturers use designers and engineers to produce better-performing shoes, skis, racquets, and other products intended for the general public. Well-known sports manufacturers hire sports engineers to improve on existing products continually. These improvements have benefited the casual sportsman as well as more serious amateurs.

For example, lighter materials and the development of the oversize tennis racquet in the 1970's made it easier for beginners. Improved materials have increased racquet stiffness, which allows for a more efficient transfer of force to the ball, which leads to faster speeds.

In mountain biking, the addition of front and rear shock absorbers, lightweight frames, improved gearing, and step-in toe clips have made it easier for the less experienced riders to navigate difficult terrain. Similarly, skis and snowboards have evolved to provide increased ease of turning, which helps beginner and intermediate skiers and snowboarders.

Lightweight helmets are now worn by most bikers and skiers, which improves safety. Many of the ski helmets have improved features such as removable inserts for warmth and vents that can be opened or closed for cooling. Wearable water backpacks are now commonplace among many athletes and are more convenient than the standard water bottle.

As of 2011, virtually every sport has benefited from sports engineering.

Professional Sports. Improvements in sports equipment is driven by the demand for increased performance by Olympians and professional athletes. Advances in technology, such as the larger tennis racquet head and increased racquet stiffness, that have benefited amateur tennis players have also benefited professional tennis players. The larger head allows highly skilled professionals to achieve more topspin and higher speeds.

Advanced golf club and ball designs have been used by professional golfers. Differences in weighting and shaft construction influence performance. Golf balls are also evolving with the introduction of solid cores, which lead to longer drives. The dimpling on the golf balls are designed to give maximum flight. These dimples are modified in some newer balls to optimize flights. Equipment improvements have benefited professional sports of all types, and professional organizations continue to push for the technologic edge that will make them more competitive.

Sports Venues. Sports engineers are also involved in designing and studying sports venues. An example of this is the recent development by SERG engineers of a trueness meter for greens evaluation. This is a device that helps greens keepers evaluate the smoothness of turf. This improves the ability of the greenkeeper to achieve optimum conditions for best performance.

In 2011, the National Hockey League (NHL) discussed the need for improved safety to reduce concussions in players. Discussions include possible changes to the glass surrounding the rinks. This type of design change to improve the safety of a sporting venue is an area where sports engineers may get involved.

Sports Safety. Sports-equipment engineering has had an impact on injury prevention at all levels. Helmets have become commonplace for sports such as downhill skiing and bicycling, and in sports such as hockey, a helmet is now required at both the amateur and professional levels. Baseball is another sport in which a helmet is required for certain positions. In some jurisdictions, bicycle helmets are required by law to be worn by riders under the age of eighteen. Many amateur participants will wear helmets to reduce the chance of injury even if they are not mandatory. There is evidence in the peer-reviewed medical literature that helmet use for sports such as bicycling and hockey does reduce injury. Full facial protection has also been shown to reduce injury in hockey players.

For venues, sports engineering can be used to modify existing venues or to design newer safer venues. In the 2010 winter Olympic Games in Vancouver, the death of twenty-one-year-old Georgian athlete Nodar Kumaritashvili during a skeleton run sparked a public debate about the track safety. Some modifications were made shortly after the accident to improve safety. There continues to be debate about the role of inexperience versus track design as a cause for the accident. This accident is likely to influence future skeleton track designs, which is an area where sports engineers will be involved.

Safety considerations include the durability of equipment and risk of sudden failure of equipment and impact the choice of materials that sports engineers consider when they design sports equipment. These considerations affect athletes at all levels, sporting organizations, governing bodies, and manufacturers.

Clothing. Clothing for high-speed sports has been improved to reduce the drag from air flow. High-tech clothing has been credited with improvements in sprinters, speed skaters, and downhill skiers. Shoes are another area of innovation with improvements in material durability and performance characteristics. Weight, durability, cushioning, spikes, and flexibility are engineered to provide the best attributes for each sport.

Fascinating Facts About Sports Engineering

- Marathon runners can have measurably increased cardiac enzymes in the blood following a race. These are the same enzymes that are elevated after a heart attack. This elevation is temporary and of unknown significance. Studies are under way to investigate the long-term effects of repeated marathon running.
- The first snowboards were made in the early 1900's from boards or from skies attached together.
- Ed Hedrick, who is considered a founder of Frisbee golf, was so passionate about the sport that he had his cremated ashes made into Frisbees after he died. The Frisbees were given to family and friends to use for playing Frisbee golf.
- The modern marathon distance is based on the story of Phidippides of Greece, who ran 26 miles to Athens during the Battle of Marathon in 490 B.C.E. to warn everyone about the Persian invaders who had landed at Sparta.
- The roof of the Minneapolis Metrodome football stadium collapsed under the weight of snow in December, 2010. The design of this stadium uses pressure inside the building to hold up the domed roof, which is made of a material similar to a trampoline and weighs 265 metric tons. The Minnesota Vikings had to play the remainder of their season at other stadiums out of state.
- Pigeon shooting was an Olympic sport in the 1900 Paris games.
- There are reported to be more than three billion soccer fans worldwide, making it the most popular sport in the world.

Impact on Industry

Sports engineering continues to grow as an academic field and continues to impact the development of sports at all levels. Manufacturers use sports engineers in the design and modification of sports products. Sports organizations and governing bodies continue to work with sports engineers to keep up with technologic advances.

Academics. The first sports engineering program was implemented by the University of Sheffield in the United Kingdom in 1998. Since that time, programs have been developed in Australia and the United States. The United States' first program was started by the University of Colorado Denver in 2010. There are other universities around the world that have sports engineering research and courses as part of a mechanical engineering program. Some links to university research departments are listed on the ISEA Web site.

Areas of research in sports engineering are wide ranging. An important part of research involves improved safety. Helmets, pads, face guards, and shoe design are all areas where injury prevention can be considered.

Manufacturing. Manufacturers have been a driving force for innovations in sports technology. They employ sports engineers for design and modification of products. Olympics and professional sports are two areas where a logo can have a big marketing impact.

Governing Organizations. It has been a challenge for sports organizations to keep up with changing technology. There are two main considerations for creating regulations for sports competitions: athlete safety and fairness. Regardless of the sport, the goal of athletic competitions is to award the winning medal to the best athlete. If an athlete has an unfair technologic advantage, then the advantage is due to technology and not physical skill.

In some cases, safety issues may not be obvious when technology is first introduced. For example, composite baseball bats, which were prone to breakage from wear, were used in youth baseball for a while before safety concerns were raised. An example of fairness can be found in golf. The shaft characteristics of the club include elasticity. There are regulations that are now in place for professional competitions to make sure that no athlete has an unfair advantage due to technology. It is challenging for governing bodies to adapt to the constantly changing technology. Consultations with sports engineers can be helpful in making decisions regarding rule changes and safety considerations.

In professional hockey, the NHL has been faced with a push to reduce the risk of concussions in players. There are several considerations that have been debated as factors in the rate of injury, and there is some discussion about the rink design in the prevention of concussion. The solution may be a combination of rule changes and technology changes. This is an area where sports engineering can have a positive impact.

Careers and Course Work

Sports engineering is a newer field compared with other engineering disciplines. There are some university programs that offer a sports engineering degree and others that have courses in sports engineering within their mechanical engineering degree program. Bachelor of engineering degrees will usually require three to five years to complete. A strong background in mathematics and physics is needed. Advanced degrees such as a master's or doctorate may be required for some career paths, such as teaching. Specific jobs may require some knowledge of biomechanics, kinesiology, or anatomy.

As of 2011, the University of Sheffield in the United Kingdom continues to be one of the leading programs in sports engineering. There are programs in Australia and a program at the University of Colorado Denver in the United States. For prospective students interested in considering a career in sports engineering, it may be helpful to contact universities with mechanical engineering programs as well to see what courses may be offered in sports engineering.

An interest in sports or a particular skill in sports can be helpful in the field of sports engineering. In some cases, athletes will go on to become sports engineers. For example, British aerospace engineer and Olympic skeleton competitor Kristan Bromley has done research on sled design to optimize performance.

For some applications, additional expertise or training in materials, architecture, physiology, kinesiology, or biomechanics may be needed. This type of specialization might come with an advanced degree or through work experience. The applications of sports engineering are wide ranging, so there are a number of paths an individual career might take.

Social Context and Future Prospects

As sports engineers become more commonplace and sports engineering degree programs are created, this discipline is likely to be more widely recognized as a field separate from mechanical engineering. It is similar to the field of geomatics engineering, which was a branch of civil engineering and is now recognized as a separate field of engineering.

Since the middle of the twentieth century, new sports have evolved through material and design changes. The Frisbee and in-line skates were invented, and snowboarding was created. In Nordic skiing, a new technique called skate skiing has become popular as a result of technique and equipment changes. This in turn has changed the way many Nordic trails are prepared such that the classic ski tracks are to one side and a larger flat area of packed snow is created for the skate skiers. Skateboarding and skateboard parks are another example of a sport and venue that have become popular as a result of improvements in equipment. These types of innovations will continue to occur in the future with modifications to existing sports and inventions of new sports.

The benefits to amateur and leisure sports will be enhanced comfort, performance, and safety as manufacturers continually improve their products.

Entrepreneurs will create new products with the help of sports engineers, who will likely play an important role in the evolution of sports going forward. In fact, a 2011 symposium organized by Loughborough University, Tokyo Institute of Technology, and the Japan Society for the Promotion of Science London was titled "Sports Engineering: Engineering the Future of Sport."

Ellen E. Anderson Penno, M.D., M.S., F.R.C.S.C., Dip. A.B.O.

Further Reading

Baine, Celeste. *High Tech Hot Shots: Careers in Sports Engineering.* Alexandria, Va.: National Society of Professional Engineers, 2004. This easy-to-understand book is written with the high school or pre-engineering student in mind. Includes sports engineers' stories and a compilation of resources for students considering sports engineering as a career.

Cho, Adrian. "Engineering Peak Performance." *Science* 305 (July 30, 2004): 643-644. Profile of mechanical engineer Mont Hubbard and his sports engineering lab in Davis, California.

Estivalet, Margaret, and Pierre Brisson, eds. *The Engineering of Sport 7.* Vol. 2. Paris: Springer, 2009. Excellent volume of essays on a wide variety of sports engineering sectors, including design of balls, running shoes, sports venues, surfaces, tennis racquets.

Farish, Mike. "Bottoms Up." *Engineering* 248 no. 3 (2007): 45-47. This article describes the work of Kristan Bromley in the engineering of skeleton sleds.

James, David. "Design Engineering—Managing Technology: Slight Advantage." *The Engineer* (May 5, 2008): 34. David James, a sports engineer from the University of Sheffield, describes advancements in the engineering of the tennis racquet.

Moritz, Eckehard Fozzy, and Steve Haake, eds. *The Engineering of Sport 6: Developments for Disciplines.* Vol. 2. New York: Springer, 2006. The variety of contributors to this volume detail the foundations for a career in sports engineering—biomechanics, measurement techniques, modeling equipment and systems, testing, protoyping, and benchmarking.

Web Sites

International Sports Engineering Association (ISEA)
http://www.sportsengineering.co.uk

Sports Engineering Research Group (SERG)
http://www.serg.group.shef.ac.uk

University of Sheffield
http://www.shef.ac.uk

See also: Computer Graphics; Engineering; Mechanical Engineering.

STRUCTURAL COMPOSITES

FIELDS OF STUDY

Mechanical engineering; chemical engineering; civil engineering technology; construction trades; machinist trades; mold-making trades; aircraft design and manufacturing; aircraft maintenance engineering; aeronautical engineering; transportation design and manufacturing; sporting goods design and manufacturing; watercraft design and manufacturing; shipbuilding; construction materials; engineered wood products; electronic design; surface chemistry; hydraulic engineering; dentistry; orthopedic surgery; nanotechnology.

SUMMARY

A composite material is formed when two or more separate materials are combined to make a new material that has its own characteristic properties. Structural composites are used to fabricate structures and other objects for which that material's properties are advantageous. Examples of structural composites range from simple or low-tech materials such as adobe brick, plywood, and reinforced concrete to more developed or high-tech materials such as the advanced composite materials used in modern aircraft and spacecraft. The range and variety of structural composites and their applications, even within a single type, is nearly limitless, with each new material resulting in the development of even newer materials and applications.

KEY TERMS AND CONCEPTS

- **Autoclave:** Programmable device that provides controlled temperatures and pressures in a sealed chamber.
- **Consolidation:** Final stage in the preparation of a structural composite when the ultimate distribution of components is achieved.
- **Creep:** Change in dimensions over time of a structure under load.
- **Damage Tolerance:** Measure of the growth rate of cracks in a structure; the ability of a composite structure to withstand damage and to continue performing as designed.
- **Delamination:** Separation of components in a structural composite.
- **Galvanic Corrosion:** Corrosion associated with an electric current produced by contact between two dissimilar substances, such as aluminum and carbon.
- **Hygrothermal Effect:** Change in the properties of a (composite) material caused by absorption of water and change of temperature.
- **Isotropic:** Exhibiting uniform properties in all directions.
- **Lamination:** Ordering of layers of materials in a structural composite.
- **Matrix:** Material that contains or binds together the components of a structural composite.
- **Prepreg:** Common term to describe a fiber material that has been previously impregnated with a matrix resin for use in structural composite fabrication processes.
- **Quasi-Isotropic:** Simulating isotropic properties.
- **Ramping:** Programmed raising and lowering of temperatures and pressures during the curing stage of resin-based composite structures.
- **Scarfing:** Method of preparing the damaged section of a laminated composite structure for repair.
- **Ultrasonic Welding:** Using the heat of friction produced by ultrasonic vibrations to fuse thermoplastic materials.
- **Warp Clock:** Guide for the orientation of fiber strands in structural composite fabrication.

Definition and Basic Principles

Structural composites, or composite materials, are specialized materials made by combining two or more simpler materials. In the resulting composite material, the properties of the component materials combine to produce new and unique properties. In composite materials, the whole is truly greater than the sum of the parts. The vast majority of structural composite materials can be described as being made from two types of materials: a matrix or binding material and a structural reinforcement material. The matrix material serves to contain and bind together the reinforcement material and to prevent it from bending, deforming, or breaking under load. The reinforcement material provides the load-bearing strength and other properties that are the foundation of the composite material's intrinsic value.

Structural composites are designed to respond to forces only in the direction of the forces placed on them. They may be constructed to have strength on one axis or in one direction only, or to have strength in several directions. A structural composite such as concrete exhibits the same strength in all directions and is described as being isotropic. Structural composites such as fiberglass-reinforced polymer (FRP) sheets have varying strengths in different directions and are described as anisotropic. If the layers of reinforcement material in a laminated structural composite are designed to complement each other, the resulting material can be made to exhibit essentially equal strengths in all directions. This condition is called quasi-isotropic and is an extremely important consideration in designing and constructing structural composites. The essential differences between anisotropic, isotropic, and quasi-isotropic composite materials are best appreciated by examining the properties of some exemplary structural composites.

Basic concrete consists of a mixture of sand, gravel, Portland cement, and water. Individually, these materials have practically no structural value, but when mixed together, they form the composite material known as concrete. A chemical reaction between the water and the anhydrous silicate compounds of the Portland cement results in the formation of mass quantities of hydrated silicate crystals. These crystal blooms grow into and mesh with one another to bind the sand and gravel components into a solid mass of stonelike consistency. Because its components have no directionality of their own and are uniformly blended, concrete displays identical properties in all directions and is thus isotropic.

In a fiberglass-reinforced polymer structure, fibers of glass, carbon, or other materials are employed as the reinforcement material. Each fiber has compressive or tensile strength only in the direction of its long axis; lateral pressure can break and destroy the fibers. Stacking numerous fibers together and enveloping them in a polymer resin matrix material can greatly increase the force required to break them, but their strength nevertheless remains primarily along their main axes. The resulting structure thus displays varying strength properties in different directions, making it anisotropic.

If the reinforcement fibers are woven together like cloth, a significant strength component is added orthogonally (at right angles) to the length of the structure as the fiber weaves back and forth across the material. Arranging successive layers of woven fibers—so that their main axes point in different directions—results in the formation of a structure whose major strengths are almost uniformly distributed in many different directions. Because this is only an apparent distribution and not an actual distribution of behavior, the material is called quasi-isotropic.

Background and History

Structural composites and the principles of their construction have been known and used for literally thousands of years. Adobe brick, concrete, and laminated wood are the earliest known examples. The Assyrians and Babylonians are known to have used a type of concrete based on clay. A more robust type of concrete made from heated powdered clam and oyster shells that used lime (calcium oxide) as the binding agent was developed in ancient Egypt. Laminated structures such as wood pieces that have been mechanically fastened together or bonded together with an adhesive material of some kind have a similarly long history.

The development and growth of chemical knowledge, especially since the Industrial Revolution of the eighteenth and nineteenth centuries, brought about tremendous growth in the variety and quality of composite materials that could be made. Portland cement was developed (or perhaps redeveloped) by Joseph Aspdin in 1824, and it has been the cornerstone of the structural composites industry ever since. All other types of concrete used in the twenty-first century are derived from the basic Portland cement variety.

Modern chemistry and materials, especially since the development of polymers, have resulted in a veritable explosion of the structural composites field that continues unabated. The vast majority of composite materials in use or being developed are based on polymeric matrices such as epoxies and resins and on mineral or metal fibers such as glass, aluminum, steel, basalt, graphite, and boron. The use of low-density, high-strength metals as the matrix material is also being explored. One example of this approach is the structural composite known as glass-reinforced fiber metal laminate (GLARE), a composite of glass fibers in a matrix of aluminum metal that is used in the construction of the Airbus A-380 aircraft.

The field of structural composites, or composite materials, continues to grow, especially as materials developed for one purpose find other uses, usually in applications in which high strength and low weight as well as unique shapes and formability are desirable properties.

How It Works

The basic principle behind structural composites is that of additive strength. A composite material is made by combining a material with a particular set of properties with another material with a different set of properties. Ideally, the properties of the composite material surpass or greatly differ from those of the individual materials used to create it. The principle is the same regardless of whether it is applied at the atomic or very large scale. The results and usefulness of the particular application are generally seen in the bulk properties of the composite material.

The additive strength principle can be demonstrated using sticks. First, take a single stick that has the thickness of a pencil and bend and break it. Then, tie several pencil-size sticks together and attempt to bend and break them as a single unit. Logically, it should require only a force equal to that required to break one stick, multiplied by the number of sticks. In practice, however, a great deal more force is required. This is because the individual units support one another against an applied force. A single unit is easily deformed to the point of failure, but the addition of units makes it much more difficult to achieve the amount of deformation needed to break any of them. This classic demonstration shows that the strength of composite structures is enhanced beyond the sum of the parts.

Individual long structures such as fibers, sticks, pipes, and rods are anisotropic, meaning they have varying strengths in different directions because of their three-dimensional shapes, even though the actual materials from which they are made may be completely isotropic. Iron atoms are isotropic, and a steel sphere, for example, is equally isotropic. If a steel ball is manipulated into a long, thin rod, the shape confers vastly different properties from those that the steel had as a sphere. Laterally, the rod will have an equally high compressive strength as the sphere because of its circular cross-sectional shape. Longitudinally, however, the rod will not bear compressive loads as well and will easily undergo lateral deformation such as bending. Theoretically, because of the isotropic nature of the iron atoms in the steel rod, it must be equally resistant to compression as it is to tension, and if prevented from deforming or bending, the rod should bear a tremendous load. This is the function of the matrix material in a structural composite. Matrix materials are invariably isotropic in nature; they must bind the other materials equally well in all directions. A matrix material is also required to be chemically and physically inert in regard to the nonmatrix materials that it contains.

The principle of additive strength can be applied equally, regardless of the type of composite material. Composite materials can be broadly classified as laminates, reinforced plastics, cermets, fabrics, and filled composites. Each class of composite materials has its own particular strengths and weaknesses and its own methods of production, although there is some crossover in the areas of application.

Laminates. Laminates include plywood, corrugated cardboard, and similar products. In these structural composites, an adhesive binder is used to unite layers of material. In plywood, sheets of wood veneer—oriented at right angles to each other according to the direction of the wood grain—are alternately stacked with an interposing layer of a strong adhesive, and the assembly is compressed by high-pressure rollers. The alternating orientation of the wood fibers gives a product many of the qualities of a regular board cut in the traditional manner but provides much higher resistance to inherent problems such as splitting or warping and enhances flexibility. The manufacture of plywood also allows a much higher percentage of a harvested log to be used instead of being turned into sawdust.

Corrugated cardboard is constructed in a similar manner, using sheets of heavy manila paper. Typically a sheet of the same material is formed into a corrugated pattern instead of a flat sheet. The corrugated sheet is then glued and sandwiched between the two flat sheets. The corrugations have strength properties that mimic those of an array of separate tubes, and they impart that strength to the composite structure. Additional layers of corrugated and plain sheet material can be applied either in parallel or orthogonal orientations to produce larger cardboard structures with strength rivaling that of wooden boards. In addition, larger laminated structures of alternating layers of wood or other anisotropic material and adhesive are commonly used. The most apparent of these is the laminated beam composed of strips of wood and adhesive. Such beams provide a strength and versatility that simply cannot be achieved by traditional one-piece beams.

Reinforced Plastics. Reinforced plastics represent the most technologically advanced field of structural composites. They also have the broadest and most diverse range of applications. Fiber-reinforced plastics (FRPs) are divided into normal and advanced composites. The methods of working with each are very similar in many respects. The designation relates to the nature of the matrix and reinforcement materials and the tolerances required of the finished products. To create a fiber-reinforced plastic, an array of selected fiber reinforcement material is impregnated with a polymer resin and formed into a desired shape. The mass is then compacted to consolidate the fibers together as tightly as possible, and the matrix material polymerizes to form a solid mass about the reinforcement fibers.

A very broad variety of matrix materials, as polymer resins, are available for use with a similarly broad assortment of reinforcement materials. The fiber materials come in many forms, from single strands to thick woven sheets. Fiber materials typically include various grades of glass; natural fibers such as silk and hemp; various metals such as steel, tungsten, aluminum, and boron; mineral fibers such as basalt; high-strength polymers such as Kevlar and other polyaramides; and carbon fibers that provide a versatile combination of high strength and low weight.

Cermets. Cermets are ceramic and metal powders that are consolidated with heat and pressure into specific shapes. They are then processed into a final form. The resulting forms can have machinability like ordinary metal parts but with reduced mass and improved wear resistance or superior hardness and strength properties for use as machining tools.

Fabrics. Fabrics qualify as structural composites when they are produced as a combination of different fiber materials such as cotton and polyester. The particular value of fabrics as structural composites is in their ability to drape to a given shape or to provide a protective covering with specific properties. The apparent simplicity of fabric structure belies the amount and complexity of research undertaken in the fabric industry, primarily for the development of new materials for fabric use and applications for those fabrics.

Filled Composites. Filled composites are the single largest area of the structural composites field, primarily because of the volume of concrete and highway paving blends that are used around the world. A filled composite consists of a mass of fine particulate inert matter blended into a matrix material. Filled composites include concrete, asphalt, and various materials such as linoleum sheets and floor tiles, particle board, and composite flooring.

Applications and Products

On the Atomic Scale. The most fundamental of structural composites are metal alloys. These materials do not appear to have the material interfaces that normally characterize structural composites. One could argue that these interfaces are provided by the atomic boundaries themselves, but the point is academic. The important feature is that the principle of additive strength begins at the atomic and molecular level and carries over to the bulk properties of the composite material. In alloys, the combination of two or more distinctly different materials at the atomic level produces a new material with distinctive properties of its own.

Although alloys do not typify the concept of structural composites or composite materials, they do demonstrate that the principle of their functionality applies on both very small and very large scales. The application of those principles on that small scale is increasingly relevant in the development of new and useful materials and methods and is intimately bound to the manipulation of atoms and molecules to produce specific atomic and molecular structures. Two examples illustrate the formation of structural composites from the atomic scale: boron fiber and the combination of materials to produce specific electronic properties or physical structures.

Boron fiber was first produced for commercial testing and use in 1964. The process by which it is produced makes boron fiber a structural composite in its own right. To make boron fiber, a very fine filament of the element tungsten is passed slowly through a sealed reaction chamber that contains an atmosphere of vaporized boron. Boron atoms deposit on the tungsten filament and build up in metallic crystal form to produce an enhanced fiber of desired dimensions. The filament then passes out of the reaction chamber to be wound on a spool. The boron fibers are very stiff in comparison with other fibers normally used in structural composites, and this property lends itself to purposes for which other fibers are neither adequate nor suitable. For example, boron fiber composites are extremely well suited to applications calling for extreme stiffness or rigidity coupled to very low weight, making it a major application in aircraft and spacecraft structural technology.

An even more important example of atomic scale structural composites is the combination of materials to produce specific electronic properties or physical structures, which has increasing significance in microtechnology and nanotechnology as well as in electronics. For example, silicon and germanium are combined to construct silicon-germanium (Si-Ge) diodes, transistors, and other semiconductor devices that are the basis of modern electronics. Modern integrated circuits and computer processing chips can contain literally hundreds of thousands of such miniature structures on a single unit. With the development of devices such as the scanning tunneling microscope (STM) and the atomic force microscope (AFM), specific composite structures can be constructed on an atomic scale, literally atom by atom. This remains a developing technology, the value of which increases as the field of nanotechnology evolves.

Structural vs. Composite Structures. The similar terms of "structural composites" and "composite structures" have very different meanings. A composite structure can be defined as any construct of more than one material in which the function of each material remains separate from the functions of the other materials. A structural composite, on the other hand, is a combination of different materials in which the separate materials combine in their functionality so that the structural composite behaves as though it were just a single material. It is this unity of functionality that makes structural composites so useful and valuable

Building and Construction Materials. In terms of volume usage, filled composites make up the single largest category of composite materials. Filled composites consist of a mixture of small particles in a solidified matrix. The matrix material or binder can be any mobile material into which a quantity of particulate matter can be blended and which becomes solid—or at least highly resistant to structural change—after emplacement. Examples of filled composite building materials include linoleum in sheet and tile form, roofing shingles, composite flooring (simulated hardwoods), particle board, medium-density fiberboard (MDF), Aspenite (waferboard), synthetic wood (typically, a blend of recovered thermoplastics and sawdust extruded into standard lumber shapes), sealing putties, concretes, plasters, patching compounds, paints, and paving mixtures. Other composite building materials include plywood, laminated beams, reinforced glass and mirrors, fiber-reinforced plastics, structural components, and paper or fabric surfaces covering materials. Of these examples, some have a limited number of forms or compositions, while others can be produced in limitless variety.

Fiber-Reinforced Plastics. The term "structural composites" has come to refer generally to materials constructed from polymeric resins and fiber reinforcements, called fiber-reinforced plastics (FRP). One major advantage of FRP structures is that the methodology can be employed for the production of a single item as well as multiple items; the "product" and the "application" are often the same thing. The second major advantage of FRP structures is the complexity of the shape that can be produced in one step, with strengths and properties relative to weight that cannot be matched by other structural technologies. FRP structural composites are thus widely used in the broadest variety of applications, from the simplest fiberglass patching compounds to the most advanced structural components of existing aircraft and spacecraft.

Common uses of FRP structures include simple fabrications such as tub and shower enclosures, spas and hot tubs, automobile body and marine patching compounds, customized and regular replacement automobile body components, and low-tolerance or low-performance shrouds and housings. Production methods for such structural composites in low-tech applications are relatively simple; they do not call for critical placement of reinforcing materials or extensive consolidation of the structure before the curing process. Similar, but far more demanding methods are used for the production of advanced FRP structures that must meet close tolerances and high performance specifications.

Molds and forms used for advanced FRP fabrications are constructed from specialized materials that closely match the thermal responses of the materials being molded and are typically custom-machined with exacting precision. Highly specialized thermosetting resins and advanced reinforcement materials such as

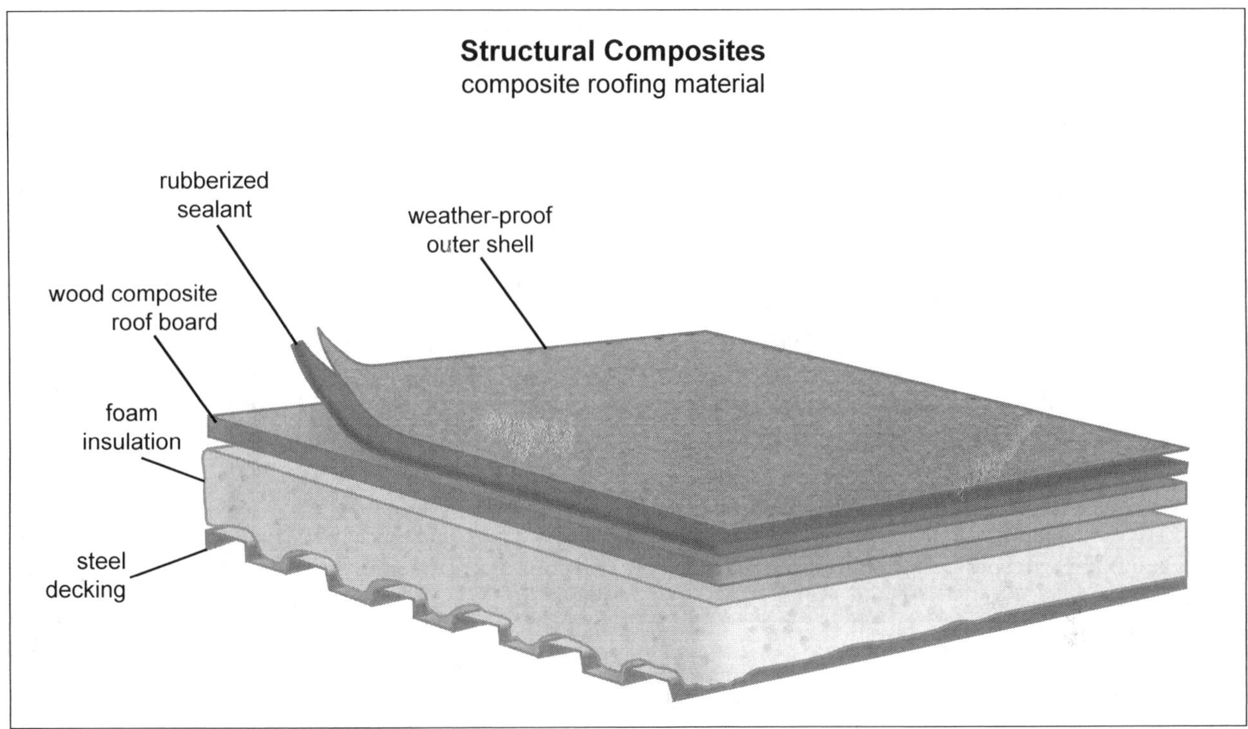

carbon, aramid, boron, and titanium fibers are used in very specific patterns. The reinforcing materials are preimpregnated with the matrix resin on site (alternatively, commercially available prepregs are used) and laid into the mold or form by hand. The resulting stack is then consolidated or compacted under reduced pressure and heated to cure the matrix resin to its solid state. This stage of the process often requires the use of an autoclave to provide a controlled sequence of temperature and pressure changes for the consolidation and curing of the structure.

Advanced composites are used primarily in the aerospace industries, but advanced fiber-reinforced plastics are increasingly being applied in other areas because of their versatility, low cost, ease of preparation and repair, and the strength and durability of FRP structures. For example, FRP methods and materials have been used in the construction of sections of multilane highways that pass through problematic terrain and of bridges for both pedestrian and vehicular traffic. The inherent strength of reinforcement materials such as carbon fiber have also been applied to the repair of traditional structures such as masonry walls, concrete pillars, and wooden beams to extend their working lives and, in many cases, to eliminate the need for replacement of those structures. Since the development of waterproof polymers and fiberglass combinations, simple FRP construction has also been used in the manufacture of boats. Advanced FRP composites and methods are being applied to the construction and modification of watercraft on an increasing scale, from the simplest computer designs to the most advanced military warships.

Cermets. Cermets are composite materials made from ceramic and metal powders. The material combination is formed and pressed into the desired shape and consolidated by sintering (making objects from powder). It is then processed to produce the finished product. Like FRP composites, cermets are also amenable to use in the production of a single unit or of multiple components. Cermets are used in two ways. In one mode of use, the final product is intended to be primarily ceramic in nature, and the metal component is either chemically altered or removed after forming, leaving the ceramic component. This method is used primarily for the production of reduced weight components with high wear resistance and strength properties. In the other mode of use, the presence of the metal is required to maintain and support the ceramic component. This method is used primarily for the production of high-strength, high-durability metal-cutting tools. The specific properties of the cermet material, especially variations that include carbonitride ceramic materials, offer enhanced working lives and performance characteristics at a lower cost than specialty metals alloys.

Fabrics. Structural composites in fabrics are used in an entirely different manner from other structural composites, because fabrics are typically used on their own strengths and merit rather than in combination with a binding or matrix material. Fabric composites may employ combinations of very specialized materials, especially if they are to be used in FRP applications. The vast majority of fabric structural composites, such as rayon-cotton or cotton-polyester, are used in the garment and furniture industries for clothing and for protective enclosures and coverings. Specialized fabrics combining such materials as carbon and Kevlar fibers are used in FRP structural composites.

Impact on Industry

Structural composites of all types are used throughout the world, largely in the international trade and transportation industries. The quantity and technological level of research and development being carried out in the field of structural composites generally corresponds directly to the level of affluence and technological development of nations. Poorer nations tend to be users of the technologies, while affluent nations lead the way in the development of new materials and applications.

Government and University Research. Often, the research and development of structural composites such as advanced FRPs is associated with the aerospace and transportation industries. National governments are deeply involved in this area because of their obligation to maintain effective military services and to provide other publicly funded services. Similarly, major manufacturers in the aerospace, marine, and transportation industries are involved in the active research and development of new materials and applications, both for their own purposes and in their status as outsourced or contracted suppliers for government agencies.

The intimate relationship between materials and the disciplines of chemistry and engineering also ensures that many academic institutions are involved in fundamental research and development in the field of structural composites. The field is so broad that institutions often focus their efforts on very specific aspects and applications of structural composites, rather than on the field as a whole. It is also sufficiently broad that trained individuals and small to medium-sized businesses are able to make significant contributions and be profitable as well, although these businesses tend to focus on small-scale applications and the service side rather than the research and development side of structural composites.

The primary factor in the continuing development and use of structural composites is their innate versatility. The materials and methods offer unparalleled opportunities for providing new solutions to problems and, at the same time, challenge scientists to develop materials that provide even better solutions. This adaptability has given rise to a large industry based solely on the application of structural composites that continues to grow and develop. The areas of application and the various businesses that have developed within them are relevant to essentially every facet of society—from basic infrastructure services such as road construction to the most advanced structural technologies.

Primary business activities are based on the applications of structural composites as the business product. Secondary business activities serve the needs of the primary industries, often as suppliers of raw materials. A third level of business activity involves providing basic and specialized training for the use of structural composites in specific applications and markets. As an example, the primary industry might be an aircraft manufacturer or a construction engineering firm, the secondary industry could then be a cement supplier or a resin and reinforcement fiber manufacturer, and the tertiary industry could be an organization that provides training in the use of advanced composite materials for aircraft maintenance workers or in the use of concrete and paving blends. In a very real sense, the opportunity for an individual's success in this field is as limitless as the opportunities for new developments and applications of the materials.

Careers and Course Work

Because structural composites is a discipline based on engineering and materials, those who undertake a career in the field will be required to acquire mathematical skills and a basic knowledge of physics as a minimum standard body of technical knowledge.

Fascinating Facts About Structural Composites

- The historical use of structural composites spans several thousand years, probably dating from the first time early builders found that adding fibrous materials such as straw or dry grasses produced a stronger and more durable adobe brick.
- Fiber-reinforced structural composites are widely found in nature and include such diverse materials as bamboo, flax, and hemp (which contain long cellulose-based fibers) and hooves, fingernails, and rhinocerous horn, which contain long fibers of the protein keratin.
- Filled composites also occur in nature, the most obvious examples being the composite rock type known as breccias and fossil-containing shale and limestone. Sintered limestone is used to produce Portland cement for the production of the filled composite known as concrete.
- The Portland cement formulation was developed in 1824, but concrete has been in use for thousands of years in various forms. Examples have been found that date back to ancient Egypt, Assyria, and Babylon. The most recognizable example of an ancient concrete structure is the Roman Coliseum, completed in 80 C.E.
- Vehicles are being tested that contain no metal components whatsoever. Their driveshafts, axles, and other high-stress components have been made entirely from advanced carbon-fiber composites. Modern advanced aircraft such as tactical and stealth fighters and bombers as well as many passenger craft could not be built without the use of advanced structural composites.
- Cermets have been known and used since the 1920's. Since 2003, innovations in technology have made cermet metal-working cutters the preferred tool for many metal machining operations.
- A carbon-fiber bicycle frame is lighter and yet considerably stronger than a comparable steel frame. A custom-made Calfee Dragonfly frame weighs in at 2.1 pounds and sells for $4,495 (carbon fiber costs about $5 per pound). A fully equipped bike using this frame could weigh as little as 12 pounds. A similar frame constructed from boron fiber would have much greater rigidity but would cost a bit more, as boron fiber costs about $700 per pound.

This will allow the individual to perform basic work with structural composite materials. Specific techniques and methods can be learned through subsequent hands-on training provided on the job by experienced and qualified practitioners. Completion of a secondary school technical program that includes a mathematics component of geometry and trigonometry, as well as physical sciences, is the necessary foundation for work in this field.

Individuals who wish to pursue more advanced practical training at an applied level should undertake a postsecondary program of study. Practical fields related to structural composites include the construction and machinist trades, civil engineering technology, aircraft maintenance, engineering technology, and chemical engineering technology. These areas of study will involve design and practical applications of filled composites, fiber-reinforced plastics, laminates, fabrics, and the use of cermets in machining operations.

To pursue research and development of advanced applications and materials, undergraduate and graduate training in a university or college program is typically required. The exact nature of the program will most likely depend on the type of materials and design applications being pursued. At a minimum, one should expect to undertake training in organic and polymer chemistry, inorganic chemistry, analytical chemistry, physical chemistry, industrial chemistry, chemical engineering, mechanical engineering, strengths of materials, physics, and applied mathematics. Depending on the individual's particular area of specialization, a selection from these courses of study will provide the graduate with an understanding of the theoretical principles necessary for the design and development of new materials as well as the applications of structural composites.

Social Context and Future Prospects

The broad applicability of structural composites in society, along with the equally broad applications that already exist, make it very difficult to predict any specific future developments in the field. It is certain that an ongoing need exists for the maintenance and repair of the present composites infrastructure. Concrete structures age and require maintenance or replacement. Similarly, FRP materials break down over time through environmental action, eventually requiring repair or replacement. In addition, each area in which structural composites are used continues to produce new products required by the increasing population. At the same time, new materials and methods are continually being developed and applied.

Undoubtedly, the value and importance of structural composites will increase with the growth of nanotechnology and its applications. This will be particularly true for the design and production of nanostructures and microstructures that carry out very specific tasks. An example is the work being carried out in the production of the so-called laboratory on a chip, a device designed to perform very specific analytical tests on numerous extremely small samples. Test protocols in use are for atmospheric, aquatic, and biological sample analysis to detect the presence of specific compounds, environmental contaminants, and pathogens. Future developments in this area are expected to provide hundreds of analyses from nearly microscopic amounts of sample materials and emplaced reagents. This in turn will call for the development and application of very specific materials and physical structures that will be best provided by structural composites.

Certainly, as the interest in space exploration continues, the applicability of present materials must inevitably be examined. The usefulness of carbon-fiber structures in space has been well proven by the application of that technology in the National Aeronautics and Space Administration's space shuttle program with the Canadarm remote manipulators. The suitability of other materials and methods has yet to be determined, and questions in this area of application are many. These questions include, for example, the feasibility of preparing concrete on the surface of the Moon or on Mars using indigenous materials and the properties of this concrete as compared with terrestrial concrete. Additionally, scientists are questioning whether FRP production processes can take place in an orbiting facility, with orders-of-magnitude lower pressures than can be readily produced on Earth. The answers to these questions, and others like them, will have very significant effects on the global economy as they are applied to nonspace applications, in much the same way that developments from the aerospace industry have historically contributed to private industry and the well-being of society.

Richard M. J. Renneboog, M.Sc.

Further Reading

Baker, Alan, Stuart Dutton, and Donald Kelly, eds. *Composite Materials for Aircraft Structures.* 2d ed. Reston, Va.: American Institute of Aeronautics and Astronautics, 2004. A comprehensive look at structural composites and their uses in air transport industries.

Bank, Lawrence C. *Composites for Construction. Structural Design with FRP Materials.* Hoboken, N.J.: John Wiley & Sons, 2006. A textbook for the application of FRP structural composites in the construction industry for new construction, retrofitting, and repairs.

Chung, Deborah D. L. *Composite Materials: Functional Materials for Modern Technologies.* London: Springer-Verlag, 2004. A comprehensive guide to the various applications and many types of structural composite materials available.

Forbes, Peter. "Self-Cleaning Materials." *Scientific American* 299 (August, 2008): 88-95. Describes structural amendments that impart a variety of useful properties; the amendments described are nanoscale and molecular scale in dimension.

Gay, Daniel, and Suong V. Hoa. *Composite Materials: Design and Applications.* 2d ed. Boca Raton, Fla.: CRC Press, 2007. This comprehensive work covers the methods and techniques of structural composites, particularly fiber-reinforced plastics.

Jaworska, L., et al. "Functionally Graded Cermets." *Journal of Achievements in Materials and Manufacturing Engineering* 17, nos. 1-2 (July/August, 2006): 73-76. Describes the microstructures and properties of various cermet materials.

National Research Council. *Materials Science and Technology.* Washington, D.C.: National Academics Press, 2003. Offers a historical overview of structural composites and predictions of future applications.

Web Sites

American Society for Composites
http://www.asc-composites.org

Federation of European Materials Societies
http://www.fems.org

Society for the Advancement of Material and Process Engineering
http://www.sampe.org

See also: Plastics Engineering; Polymer Science.

SUBMARINE ENGINEERING

FIELDS OF STUDY

Metallurgy; chemistry; engineering; computer science; computer-aided design; material science; mathematics.

SUMMARY

Submarines are naval vessels capable of operating beneath the surface of the water. The early models were powered by simple diesel engines, but most modern submarines are nuclear powered. The world's navies operate most of the submarines, which are used as military vessels during wars. The submarine's ability to submerge beneath the waves allows it to hide from adversaries and avoid detection while carrying out its mission. Submarines are also used in civilian scientific capacities to study the deep portions of the ocean inaccessible by traditional diving methods.

KEY TERMS AND CONCEPTS

- **Attack Submarine:** General-purpose military submarine designed to hunt and destroy enemy submarines and surface ships.
- **Ballistic Missile Submarine:** Submarine armed with strategic nuclear-armed missiles; often called a "boomer" by U.S. Navy crews.
- **Bathyscaphe:** Civilian research submarine specifically designed for voyages to the deepest part of the ocean.
- **Cavitation:** Bubbles caused by the propeller spinning through the water; these bubbles produce noise, which can reveal the submarine's location.
- **Conning Tower:** Structure atop the submarine hull that gives the crew increased visibility.
- **Diving Plane:** Small "wing" that projects from either the side of the submarine hull or the conning tower and provides directional control when the submarine is submerged.
- **Hydrophone:** Listening device mounted on the submarine hull that allows the crew to listen for external threats.

Definition and Basic Principles

Submarine engineering is the field of designing, constructing, and improving submarines to work in their unique environment. Submarines are the only moving human-made objects designed to operate underwater, a hostile environment with many hazards and potential dangers to both vessel and crew. In addition, most submarines are weapons of war and must deal with the threat of combat in addition to the perils of the sea. Early submarines were more accurately submersibles, craft that could submerge beneath the surface but that spent most of their time on the surface. Early submarines possessed only limited electric battery power to drive the engines while submerged and had to return to the surface after a relatively short period of time to recharge their batteries. Modern submarines, especially those with nuclear power, are true submarines, spending most of their time submerged and little time on the surface.

As tools of war, the weapons found on submarines have also changed over time to deal with different threats. Early submarines were armed with simple explosive charges on the end of a pole, but the invention of the torpedo in the late nineteenth century gave submarines a long-distance weapon that allowed them to take full advantage of their ability to hide underwater, sneak up on an enemy vessel, and destroy it. By the late twentieth century, submarines came equipped with modern offensive and defensive systems.

Submarines are also used in peaceful pursuits. Small research submarines are ideally suited to explore the depths of the ocean because they can go much deeper than divers equipped with portable sources of oxygen. Because of the intense water pressure found in deep waters, these research submarines must be built very strong. Despite the dangers of operating at depth, submarines have visited even the deepest portions of the ocean, thousands of feet beneath the surface.

Background and History

Although several inventors tried to construct submarines before the Industrial Revolution, none succeeded because they lacked an effective power source. These early submersibles, such as David Bushnell's *Turtle*, built in 1776, failed because they had to rely on human power. Later, when steam power became available, inventors such as Robert Fulton designed powered submarines, but the steam engines could not work underwater because they needed a source of oxygen for the fires in their boilers.

In 1897, the American inventor John Holland solved most of these problems. His submarines, albeit small, featured two power sources. The submarine had an electric motor that ran on large batteries for operating underwater and a diesel engine for cruising on the surface. The diesel engine also ran a dynamo to recharge the batteries so that the submarine could spend more time underwater. The U.S. Navy, along with agencies from several other countries, purchased Holland's submarine design. By the twentieth century, submarines had matured into advanced weapons of war, and in both World War I and World War II, submarines sank thousands of enemy ships on both sides of the conflict.

The biggest weakness of the submarine was its periodic need to resurface to recharge its batteries. That problem was solved in 1954, when the U.S. Navy commissioned the USS *Nautilus*, the first nuclear-powered vessel in the world. With an unlimited amount of power that did not need an oxygen supply, nuclear-powered submarines could stay submerged indefinitely, and because they did not need to take on diesel fuel, they could also operate at full power indefinitely, allowing them to range across the ocean with little fear of discovery.

How It Works

Diving and Maneuvering. Submarines, with their ability to operate beneath the surface, require special equipment to disappear below the surface (and come back up) and to maneuver while in the water. On the surface, the submarine operates like a traditional ship, relying on buoyancy to remain afloat and a rudder to maneuver. To submerge, however, submarines need to lose their buoyancy, which is accomplished using ballast tanks. When the submarine is on the surface, the ballast tanks are filled with air. When the air is compressed into tanks and the ballast tanks fill with water, the submarine becomes heavy enough to submerge. Once submerged, the submarine maneuvers using diving planes, small "wings" attached to the hull that direct the water flow past the submarine, to maneuver the submarine up or down. To surface, the submarine blows the stored air back into the ballast tanks, and the submarine rises.

Propulsion. Early submarines required two things that limited their range and ability to stay underwater: air and battery power. Large battery arrays allowed submarines to remain submerged but only for a limited time. Submarines had to use their diesel engines to recharge the batteries, but the diesel engines needed air so the recharging had to occur on the surface. The advent of nuclear power in the 1950's, however, removed the major limitations on propulsion. Nuclear power plants did not need air to operate, and the nuclear submarine's steam turbine did not need batteries. Because of these features, some modern submarines are employed on missions during which they may not surface for six months or more. By the 1990's, new air-independent propulsion systems permitted the operation of diesel engines underwater without the expense of a nuclear power plant.

Sensors. Early submarines relied on human sight to sense obstacles and targets around them. This led to the creation of the conning tower, a structure built atop the submarine hull that allowed a lookout to view the area around the submarine while the bulk of the vessel remained just below the surface of the water. Once submerged, early submarines were blind and needed other means to find their way. After World War I, submarines used hydrophones, sensitive microphones mounted on the submarine hull, to listen for other ships and submarines around them. By the 1950's, sonar allowed submarines to actively detect other vessels.

Navigation. When early submarines submerged, they lost their ability to navigate because they lacked external reference points, such as the stars, by which to navigate. Submarines had to navigate into the general area of a target, then submerge to conduct their attack. When nuclear submarines began to spend more time submerged than surfaced, however, crews needed a more accurate means of establishing their location, especially after submarines were armed with nuclear missiles that required precise targeting. The solution was inertial navigation, whereby the submarine crew took extremely precise measurements from motion sensors mounted throughout the submarine and used the data relative to the amount of time and direction the submarine had traveled to calculate their location.

Stealth. Submarines can hide beneath the ocean, but that does not mean they are not detectable. Sonar allows aircraft, surface ships, and other submarines to detect the presence of submarines because of the noise they generate. To counter this, submarines strive to be extremely quiet, and all efforts are made to eliminate or suppress noise. Submarine propellers move very slowly to limit cavitation, the creation of air bubbles as the propeller blade moves through the water. A submarine's machinery is placed on racks fitted with rubber cushions so that the machinery vibration is not transmitted into the hull and then into the water. Many submarines are coated on the outside with anechoic tiles, material that absorbs sonar waves instead of bouncing them back.

Shape. Early submarines were shaped like surface craft. They were long and narrow and had a pointed bow to pierce the waves. Although this shape worked well on the surface, it was not suitable for underwater speed, as the hull shape created drag that was not hydrodynamic. When nuclear power permitted submarines to spend virtually all of their time submerged, hull forms began to reflect the change. Early nuclear submarines had hulls with a tubular cigar shape, but by the 1960's, U.S. nuclear submarines had adopted a teardrop shape that emulated the shape of whales, with a bulbous bow and a stern that tapered to a point. This proved to be the most hydrodynamic shape, and all subsequent submarines have copied it.

Armament. The invention of the torpedo in the late nineteenth century made the submarine an effective ship-destroying weapon, but submarines had other weapon systems as well. Most submarines carried a deck gun in the 4- to 6-inch caliber range to deal with unarmed targets. When aircraft became a viable weapon, submarines began to carry light antiaircraft guns, in the 20- to 40-millimeter caliber range, for self-defense. After nuclear power permitted submarines to stay underwater most of the time, these deck-mounted weapons became unnecessary. Instead, submarines added missiles to their armament. Submarines still used increasingly advanced torpedoes, but by the late twentieth century, they could fire either ballistic missiles armed with nuclear weapons or long-range cruise and antiship missiles.

Applications and Products

Military Use of Submarines. The technological breakthroughs that led to the modern submarine make it the most lethal naval weapons system available to any nation that can afford it. As military craft, submarines fulfill three functions: defending shores, attacking enemy ships and submarines in the open ocean, and launching ballistic missiles.

Small diesel-electric submarines, usually equipped with air-independent propulsion systems, are used for coastal defense and short-range local operations, where their limited diving depth and relatively short operational range is not a limitation. As diesel-electric submarines are less expensive than nuclear-powered submarines, they are most common in the navies of smaller countries with less money to spend on defense.

Larger attack submarines, usually nuclear powered, are used to patrol the open ocean during wartime, searching for and sinking enemy surface ships and submarines. Submarines can sneak up close to possible targets and destroy them with either modern torpedoes or long-range cruise missiles. Because of their stealth features, attack submarines are also often used for intelligence-gathering missions against possible enemies.

The last military use of submarines is as missile carriers. These submarines, known as ballistic missile submarines, carry a number of long-range ballistic missiles armed with nuclear warheads and act as a reliably safe deterrent to enemy nuclear attack. Land-based nuclear forces are vulnerable because they are visible and therefore vulnerable to enemy attack, but submarine-based missiles hiding and moving beneath the ocean are virtually untraceable. The first ballistic missile submarine, the USS *George Washington*, was commissioned into the U.S. Navy in 1959. A number of countries other than the United States also operate ballistic missile submarines.

Civilian Use of Submarines. Not all submarine engineering is for military purposes. Technological advances allowed for expanded civilian use of submarines. A number of oceanographic and scientific institutes operate deep diving research submarines, known as bathyscaphes or deep submergence research vessels (DSRVs), to explore the depths of the ocean. These scientific submarines are very different from military submarines. Although military submarines must be large in order to range across the oceans, research submarines are typically small, have a limited range, and must be conveyed to their research area by a supporting surface ship. Although military submarines have a relatively limited maximum diving depth, generally about 1,000 feet, research submarines are constructed with thick walls to survive depths significantly deeper. In 1960, a U.S.-Swiss research submarine, the *Trieste*, reached the bottom of Challenge Deep, the deepest point in the world, at a depth of more than 36,000 feet.

The other primary civilian use of submarines is as entertainment and tourism. A number of resorts, especially in the Caribbean, use small submarines to take fee-paying tourists on a short and shallow ride into the ocean. The submarines are popular because they allow the rider to view the ocean and its wildlife in a comfortable, safe environment. The submarine also eliminates the need for bulky scuba gear and being trained in its use. Instead, tourists can simply climb aboard and see much of what a scuba diver would without the expense and inconvenience.

Impact on Industry

Submarine development and construction involves a number of industries. Besides shipbuilding, submarines, especially military ones, require specialized equipment and materials to operate in their harsh underwater environment. Submarines need special hardened steel to survive the crushing depths, sophisticated communications and electronics gear, and modified weapons systems that no other naval vessel uses. None of this is cheap. The standard ballistic missile submarine of the U.S. Navy, the Ohio class, each cost more than $1 billion, and each of their twenty-four Trident D-5 ballistic missiles costs $29 million. The most modern U.S. attack submarine, the Virginia class, costs $1.8 billion. That high cost limits the number of submarines that a navy can acquire. Consequently, only a few countries can afford submarines, especially nuclear-powered ones.

Only a handful of companies can make nuclear-powered submarines. In the United States, for instance, the Electric Boat Company in Groton, Connecticut, is the only company that manufactures submarines for the U.S. Navy. There were other companies, but the increasing specialization and rising costs have left Electric Boat as the sole supplier. The rising costs of nuclear-powered submarines triggered an increasing interest in diesel-electric power, the system that drove earlier submarines, among nations that want submarines but cannot afford nuclear power. Sweden and Germany have taken the lead in developing air-independent propulsion systems that give small coastal submarines at least some capability approaching nuclear-powered submarines but at a fraction of the cost. As a result, Germany and Sweden have exported submarines to many smaller nations around the world or licensed the production of their air-independent propulsion system for installation in locally built submarines.

Civilian submarines, on the other hand, are a growing industry. In addition to a growing number of research submarines and tourist submarines, private manufacturers are now producing submarines for private owners. Although small and of limited range and depth, an increasing number of submarines are privately owned. These submarines range from small (20-foot) personal craft to large luxury yachts capable of submerging for extended time periods. Although most personal craft are used for personal enjoyment, many political experts fear that submersibles and submarines in private hands might be used

Fascinating Facts About Submarine Engineering

- With few exceptions, the U.S. Navy traditionally names its submarines after fish and marine life.
- In 1864, during the American Civil War, the Confederate submarine CSS *Hunley*, a 40-foot-long submarine powered by a hand crank turned by its eight-man crew, became the first submarine to destroy an enemy warship when it sank the USS *Housatonic* off the coast of South Carolina. Although it completed its mission successfully, the *Hunley* sank, killing its entire crew.
- In 1915, the German submarine *U-20* sank the British passenger ship *Lusitania*, killing 1,200 passengers, an event that is often cited as the cause of the United States' entry into World War I.
- In 1944, the American submarine *Tang* was accidentally sunk by one of its own torpedoes, which ran in a circular path and came around to hit the submarine.
- In 1958, the nuclear-powered submarine *Nautilus* was the first vessel to reach the North Pole by passing under the ice.
- In 1960, the nuclear-powered submarine *Triton* circumnavigated the world, tracing the voyage of explorer Ferdinand Magellan, in only sixty-eight days. Magellan's trip in the sixteenth century took nearly three years.
- On its 1960 dive to the Challenger Deep, the *Trieste* took nearly five hours to descend 36,000 feet, moving at about 1 mile per hour.

for illegal purposes or terrorism. In 2008, antidrug authorities in Colombia intercepted a major shipment of cocaine bound for the United States concealed in a small submarine operating just below the ocean surface.

Careers and Course Work

Submarine engineering requires very specialized and advanced training, usually a doctorate, in addition to a number of years experience in the form of training or apprenticeship. In addition to basic shipbuilding, engineers in the submarine field require advanced training in the particular attributes and challenges of constructing ships designed to go underwater. Materials engineering, especially involving high-tensile steels, is a prime example of a specialized skill required to construct a submarine. Engineering of nuclear power plants is a process separate from ship construction and requires an equal amount of specialized training in the design, use, and handling of nuclear fuels. Because of the myriad subsystems in every submarine, engineering related to submarine construction includes fields ranging from electrical/electronic engineering to environmental engineering and waste management. Course work required for such a level of engineering involves advanced engineering in a range of subdisciplines, including chemistry, metallurgy, and computer-aided design.

Social Context and Future Prospects

The submarine will be around for a long time to come. As a military weapon, submarines, especially nuclear ones, are stealthy, capable, and flexible in the missions they can accomplish. They can dominate large areas of ocean, and their ability to carry a variety of weapons systems makes them useful to navies that have them and attractive to navies that want them. As the backbone of a nation's nuclear deterrent force, missile-firing submarines, although expensive, are less vulnerable than land-based systems, guaranteeing their presence in the world's major navies for the foreseeable future.

Civilian submarines are also an area of growing interest. Although scientific submarines have mapped significant portions of the oceans, the majority of the ocean floor remains unexplored. As natural resources on the land become scarcer, humanity will rely more and more on the ocean to fill the need for materials, and the exploration of the ocean will take on even more significance. Civilian submarines also fill the need to educate the public about environmental protection. By providing a view of the ocean and demonstrating its importance to human survival, civilian submarines promote the cause of environmentalism and the preservation of natural resources.

Steven J. Ramold, Ph.D.

Further Reading

Burcher, Roy, and Louis Rydill. *Concepts in Submarine Design.* New York: Cambridge University Press, 1995. A good general study on submarine design.

Clancy, Tom, and John Gresham. *Submarine: A Guided Tour Inside a Nuclear Warship.* New York: Berkeley Trade, 2003. A layman's perspective of the development, use, and future of submarine warfare, including a good study of life aboard a modern submarine and hypothetical future uses of submarines.

Friedman, Thomas. *U.S. Submarines Through 1945: An Illustrated Design History.* Annapolis, Md.: Naval Institute Press, 1995. An exhaustive study of American and comparative foreign submarine design by one of the leading experts on naval construction.

Friedman, Thomas, and James L. Christley. *U.S. Submarines Since 1945: An Illustrated Design History.* Annapolis, Md.: Naval Institute Press, 1994. Covers the development of nuclear-powered submarines in great detail.

National Research Council, Committee on Future Needs in Deep Submergence Science. *Future Needs in Deep Submergence Science: Occupied and Unoccupied Vehicles in Basic Ocean Research.* New York: National Academies Press, 2004. Features an excellent history of the use of submarines in oceanic research and charts the path for the future use of deep-water submarines.

Parrish, Thomas. *The Submarine: A History.* New York: Penguin 2004. The best single-volume history of submarines, from their origins to modern usage.

Web Sites

Federation of American Scientists
Deep Submergence Vehicles
http://www.fas.org/programs/ssp/man/uswpns/navy/intel/dsv.html

Marine Technology Society
Manned Underwater Vehicles Committee
http://www.mtsmuv.org

Naval Submarine League
http://www.navalsubleague.com/NSL/default.aspx

Submarine Industrial Base Council
http://www.submarinesuppliers.org

U.S. Navy
The Submarine
http://www.navy.mil/navydata/ships/subs/subs.asp

See also: Antiballistic Missile Defense Systems; Naval Architecture and Marine Engineering.

T

THERMAL POLLUTION CONTROL

FIELDS OF STUDY

Biology; chemistry; civil engineering; ecology; environmental science; hydrology; mechanical engineering.

SUMMARY

Thermal pollution is heated water discharged into lakes and rivers. By raising the ambient water temperature, aquatic plant and wildlife are often threatened. The industry that dumps the greatest amount of heat into lakes and rivers is thermoelectric power, which consists of coal, oil, and natural gas-combusting plants, and nuclear reactors. Because of the laws of thermodynamics, this waste heat cannot be eliminated, but it can be kept from waterways by means of cooling towers or cooling ponds or used in cogeneration industrial parks.

KEY TERMS AND CONCEPTS

- **British Thermal Unit (Btu):** Amount of heat required to raise the temperature of one pound of water by one degree Fahrenheit.
- **Carnot Engine:** Heat engine performing at the maximum possible theoretical efficiency when transforming heat into mechanical energy. The efficiency depends upon the temperature difference between the input and the output heat.
- **Cogeneration:** Process whereby waste heat from a thermoelectric power plant is used as space heating for contiguous industries.
- **Condenser:** Device used to condense steam from the power plant turbine by transferring steam's vaporization heat into water or air passing through the condenser.
- **Heat Engine:** Thermodynamic device that transforms heat into mechanical energy.
- **Watt:** Unit of power equal to 1 joule per second.

Definition and Basic Principles

Thermal pollution is any increase of water temperature due to the discharge of warm or hot water into lakes or rivers. Most thermal pollution results from thermoelectric power plants using a local supply of water to transfer heat from a condenser to the environment, raising the ambient water temperature, which often adversely impacts aquatic life.

No machine is completely efficient, but heat engines, constrained by the second law of thermodynamics (which states that since not all energy within a heat engine can be transformed into useful mechanical energy, there will always be waste heat), are particularly inefficient. Since efficiency depends on the temperature difference between the input and output heat reservoirs, one would like the input temperature to be as high as possible and the output temperature to be as low as possible. The maximum temperature is constrained by the highest temperature and steam pressure the boiler can withstand; the output temperature is determined by the environment. Typically, a fossil fuel plant operates with an input temperature of about 1,000 degrees Fahrenheit, the practical maximum. The output temperature is typically 100 degrees Fahrenheit, giving a theoretical efficiency of about 60 percent. In practice this is impossible to achieve for several reasons. It is difficult to maintain the steam at 1,000 degrees Fahrenheit, turbines are less than 90 percent efficient, and the conversion of stored energy (such as coal) into heat is only about 88 percent efficient. Consequently, the actual efficiency of a fossil-fueled power plant is typically about 40 percent. Nuclear plants (operating at lower temperatures) are only 33 percent efficient; two-thirds of the energy consumed is released as waste heat.

Background and History

Although federal legislation governing waterways began in the nineteenth century, it was not until 1948 that the first law governing water quality, the Federal Water Pollution Control Act, was enacted.

Enforcement provisions were strengthened in 1956, expanded in 1965, and restructured under the Environmental Protection Agency (EPA) in 1972. With increased awareness of environmental issues, the Clean Water Act (CWA), was passed in 1972, with major amendments in 1977. This legislation enforced minimizing environmental impacts to water caused by contaminants. The EPA standards, recently upgraded, ensure that the best available technology is used to design cooling water structures and that large thermoelectric power plants upgrade existing facilities to be in compliance. Since this affects more than 400 fossil-fuel plants as well as thirty-eight nuclear power plants, the requirements are continuously being challenged by the electric power industry as being too stringent. Retrofitting or adding cooling towers to existing power plants would incur such high costs that the industry estimates about 10 percent of its generating capacity would need to be retired.

How It Works

The Problem. Thermoelectric power plants (coal, oil, natural gas, and nuclear) use heat from combustion or a nuclear reactor to transform water into steam in a boiler. The steam impinges on the blades of a large turbine and then is converted back to liquid by a condenser on the output side of the turbine. Because of the huge pressure drop when the steam condenses, considerable force is exerted on the turbine blades, causing them to rotate. The rotational energy powers a generator that transforms the mechanical energy into electrical energy. The condenser, a large heat exchanger, removes heat by transferring it to the environment. The condensed steam is then pumped back into the boiler, where it is again vaporized. The heat removed from the condenser and dumped into the environment is termed thermal pollution. The least expensive means of heat disposal is to draw water from a river or lake; by passing this water through the condenser, the heat required to condense the steam back to liquid is transferred to the water, raising its temperature by 8 degrees to 25 degrees Fahrenheit.

A 1,000-megawatt nuclear power plant operating at 33 percent efficiency generates 10.3 billion British thermal units (Btu's) of heat energy every hour, of which 6.9 billion Btu's are discharged as waste heat. Limiting the coolant temperature increase to 20 degrees Fahrenheit requires 700,000 gallons per minute of water to flow through the condenser. This is equivalent to a stream seventy-six feet wide and ten feet deep flowing at two feet per second. In contrast, a 1,000-megawatt coal-fired plant operating at 40 percent requires 5.1 billion Btu's to be discharged every hour, considerably less than the nuclear plant.

Environmental Effects. Although the temperature is not raised substantially, dumping waste heat into rivers or lakes can profoundly affect aquatic life. Every fish species has a lethal water temperature, which if encountered will kill the fish within an hour. Salmon, for example, prefer water with a temperature in the sixties; their lethal temperature is 77 degrees Fahrenheit. Even a modest water temperature increase of 15 degrees Fahrenheit could render a river uninhabitable for salmon. Also, an increased temperature in a river or near a lake shore could disrupt a species' spawning grounds. A power plant on the Columbia River in Washington State once raised the water temperature in its surroundings to such a degree that salmon could not swim upstream to lay their eggs. Additionally, as water temperature increases, the dissolved oxygen content is modified, which adversely affects both fish and plant life, such as algae. Furthermore, when a power plant goes offline during winter for repairs, the sudden drop in water temperature has a biologically detrimental effect on fish. Finally, there are subtle effects such as a higher water temperature causing a small minnow to disappear, which deprives the larger species of fish that was relying on that minnow as a food source.

The extent to which a power plant affects the aquatic environment depends on how much water is used for cooling. For relatively narrow and shallow rivers the entire flow of the river may be heated to a higher temperature, requiring a considerable downstream distance to be dissipated. Even a relatively modest increase in water temperature may ruin a river or lake by destroying natural habitats, particularly for fish that prefer cooler water. Higher water temperatures also can lower the dissolved oxygen content to the point that fish cannot experience respiration. Reduced oxygen content also encourages the unrestrained growth of aquatic plants such as algae, which clogs the water and further lowers the oxygen content. Finally, warmer water may concentrate the lethal effect of any toxic chemicals present.

Traditional Solutions. Given the increasing demand for electricity requiring more power plants, most of the water used in the United States passes through a power plant and is heated by about 20 degrees Fahrenheit, placing a severe demand on the aquatic environment. Although it is cost-effective to use a local body of water to dispose of waste heat, it is generally not an environmentally sound practice. Two other methods of disposing of the heat are cooling towers, which transfer heat directly to the atmosphere, and cooling ponds, man-made lakes that absorb the heat to create warm bodies of water at about 100 degrees Fahrenheit. The cooling pond slowly releases its heat into the atmosphere. There are advantages and disadvantages to each of these systems.

Applications and Products

Cooling Towers. There are two types of cooling towers, wet towers and dry towers. Wet towers cool by evaporation and require constant water replacement. Warm water from the condenser is sprayed over baffles; unevaporated water drips to the foundation, where it is collected and recirculated. Evaporation requires about 3 percent of the water to be replaced continuously.

Two different types of dry-cooling methods are possible: direct dry cooling using an air-cooled condenser or a water-cooled condenser, which transfers its heat to the atmosphere through finned tubes in the tower. Direct dry cooling functions like an automobile radiator: Air flowing past the condenser tubes absorbs heat by conduction and radiation. The generating plant uses less than 10 percent of the water required for a wet-cooled system, but the enormous fans employed consume up to 1.5 percent of the station's electrical output. Although the fans allow greater control over cooling, these systems are much less efficient than circulating water systems, thus a larger and more complex cooling plant is required. Closed water loop systems have no evaporative loss, but such systems are seldom employed because of their high initial price, increased operating costs, and reduced efficiency. Hybrid wet and dry towers are used occasionally; a dry section is positioned above the wet section. This reduces the dry tower cost, lessens the evaporative water loss of a wet tower, and moderates the inefficiency of dry cooling.

Cooling towers have both beneficial and detrimental aspects. On the positive side, towers do not affect fish or land animal life, and they occupy a relatively small land area compared with cooling ponds. On the negative side, these structures are huge (typically at least 400 feet high with a base diameter exceeding 300 feet) and costly to construct (the cost of a power plant incorporating a cooling tower would be increased by about 7 percent). Typically, power plants require more than one cooling tower, which increases the cost even more. The water lost to the atmosphere creates higher humidity, increasing fog, and winter icing. Economic considerations have established wet cooling towers as the preferred method of condenser cooling for thermoelectric plants in the United States constructed since 1970.

Cooling Ponds. Cooling ponds, because of their warm temperature, can be stocked with fish having a high lethal temperature, such as catfish, which can be harvested. Additional positive aspects of these ponds compared with cooling towers are that evaporative losses are significantly reduced, ponds are less costly to construct, and the environmental impact is considerably diminished. The unfortunate aspect of these ponds is that they require a substantial land area, about one to two acres per megawatt of power. A typical 1,000-megawatt plant would require up to 2,000 acres (about three square miles) for the cooling pond. The warm water could also cause more regional fog. Cooling ponds are most often used in the Southwest, where land is available and humidity is low. Sometimes the water is mechanically sprayed into the air to encourage evaporative cooling, which reduces the pond size. About 15 percent of thermoelectric plants in the United States use cooling ponds as their primary condenser cooling method.

Repurposing of Waste Heat. Although waste heat can be disposed of through cooling towers or cooling ponds, the heat ultimately still enters the environment. Rather than merely dumping the heat, several methods may be employed to use the heat in more environmentally benign manners. Waste heat could be used for desalination plants, which evaporate saltwater to obtain distilled water. Heated water could also be used to irrigate crops, helping to extend growing seasons. Waste-treatment plants could use the heated water to evaporate water from sewage more rapidly, and certain rivers, such as the St. Lawrence, could be kept ice-free all winter from the heat emanating from a series of riverside power plants. Finally, if the power plant were associated with an industrial park, the waste heat could be piped to nearby buildings as a source of winter heat—a process termed cogeneration.

Impact on Industry

Complying with EPA regulations on thermal pollution forces utility companies to divert capital from expanding demand-driven capacity to replacing once-through cooling systems with more expensive cooling towers. Because of this, the EPA now requires economic analyses of the impact on industry of new pollution-control regulations in an attempt to balance environmental controls with economic considerations.

Due to dwindling water resources, the Department of Energy's (DOE) National Energy Technology Laboratory (NETL) released a study in August, 2010, forewarning utility companies that nearly 350 coal-based power plants will be vulnerable to potential water supply and demand conflicts over the next twenty years—particularly in the Southeast. Utilities are concerned that impending EPA regulations could require all new thermoelectric plants to install cooling towers, raising construction costs and substantially increasing water usage. Considering that 40 percent of the water used in the United States is consumed by thermoelectric power plants, and water shortages are becoming increasingly problematic, the scope of the problem becomes apparent. If carbon dioxide stack emissions are also to be reduced, water consumption could increase by an additional 30 to 40 percent for fossil fuel plants.

The NETL study analyzed data from more than 500 coal plants, looking at their hydrologic environment and the regional projected population growth. Of the analyzed plants, 347 were deemed vulnerable to water shortages; one-third of these are in the southeastern United States, where population growth and recurrent long-term droughts exacerbate the problem. More than half of the vulnerable plants use once-through cooling systems; the others employ cooling towers. Of the vulnerable plants using once-through systems, 80 percent use freshwater, 10 percent use cooling ponds or canals, and about 10 percent use saltwater. In response to this impending crisis, utilities are planning to modernize and enlarge some power plants while eliminating a number of the smaller and less efficient generating units. Consideration is also being given to substituting "nontraditional" waters for some portion of the cooling water currently used. Nontraditional waters include water from deep saline aquifers, coal-bed methane fields, mine pools, oil and gas fields, and water from shale gas extraction. Although about half of the vulnerable plants are located near a non-traditional source of water, few are located in the Southeast. Although the once-through cooling method makes relatively modest demands on water usage, a sizable power plant can require more water than is accessible from a river during the summer if its temperature is not to be raised excessively. For example, the 3,000-megawatt Civaux Nuclear Power Plant in France keeps almost 5.3 million gallons of water stored by upstream dams to guarantee a sufficient cool water supply during droughts.

Due to concerns of efficiency and economics, dry cooling towers have been removed from consideration for new power plants. A 2009 U.S. DOE study states they are three to four times more expensive to implement than wet cooling systems, and their operating costs are higher due to lower efficiency. Dry systems are also unreliable during hot weather because

Fascinating Facts About Thermal Pollution Control

- World's largest cooling tower is 656 feet tall and is located in Niederaussem, Germany.
- Fossil fuel plant requirements for once-through cooling: For a 15-degree Fahrenheit rise in cooling water temperature, about 1 cubic foot per second of water is required for every megawatt of power of generated. A 1,000-megawatt plant requires a 15-foot-diameter pipe with water flowing at about 10 feet per second.
- Why is it impossible for the steam just to be circulated without being condensed back into water? The condenser is necessary to create the large pressure drop across the turbine in order to rotate the blades that drive the generator.
- In the United States there are sixty-five nuclear power plants with 104 nuclear-generating units and 1,493 coal-fired plants.
- Based on the Clean Water Act, in 2010 new rules defining large power-plant cooling standards were proposed. These requirements will affect 422 fossil fuel plants and 38 nuclear plants. The law is being challenged by industry as too expensive to implement.
- In the United States, 85 percent of electrical energy comes from thermoelectric power plants, the vast majority of which are coal fired; 40 percent of the water used nationwide is consumed by these thermoelectric plants.

cooling occurs by heat transfer through metal fins. Since the rate of transfer is proportional to temperature difference, efficiency is considerably diminished when the ambient temperature is high. There are also safety considerations concerning the removal of heat in the event of an emergency shutdown. It is therefore improbable that new or retrofitted power plants will implement dry cooling towers in the foreseeable future.

Careers and Course Work

Controlling thermal pollution from electric utility plants while containing costs will require new, more efficient designs for cooling towers as well as new technologies to alleviate the problem by more cost-effective means. Those interested in working in these areas will need a technical background best achieved by a college major in physics or engineering. If one is more interested in the environmental effects of thermal pollution, the obvious college major would be biology or environmental studies. The EPA also needs people with legal training, such as lawyers and paralegals, because of the complicated legislation that needs to be interpreted and applied to industries that pollute the aquatic environment.

There is an increasing need to provide cooling for thermoelectric power plants without compromising the environment. Creative people with technological training will be necessary to fulfill future demands to develop new water-treatment technologies for purifying the cooling water needed for thermoelectric plants from contaminated sources such as reclaimed wastewater, agricultural runoff, or salty groundwater.

Social Context and Future Prospects

Coal-fired power plants, requiring more than 3 million tons of coal per year for each 1,000 megawatts of generating capacity, must be located relatively close to a fuel supply, typically inland. Since nuclear power plants are not subject to this constraint, they can be situated in coastal regions and employ once-through seawater cooling. Coastal nuclear power plants also have the advantage of being able to use the waste heat for desalination. Countries in North Africa and the Middle East already employ waste heat from fossil fuel plants for this purpose.

A 2008 DOE/NETL study estimating future power plant water consumption based on a projected increase of generated power and anticipated technological changes indicates that future nuclear plants will have higher thermal efficiency, thus requiring less water. Although future coal plants will also employ new technologies to reduce water usage, the mandated removal of sulfur compounds (which create sulfuric acid) will increase water consumption. If greenhouse-gas emissions are also to be limited, 50 to 90 percent more water will be required, making fossil fuel plants considerably more water intensive than nuclear plants.

Since oceans are an immense heat sink, it has been proposed that power plants be built several miles offshore—they would scarcely be visible, and detrimental biological effects should be minimal. Offshore plants would also allow power to be generated in the vicinity of large coastal energy users, thus minimizing transmission line losses.

George R. Plitnik, B.A., B.S., M.A., Ph.D.

Further Reading

Farber, Daniel A., Jody Freeman, and Ann E. Carlson. *Cases and Materials on Environmental Law.* 8th ed. St. Paul, Minn.: West, *2010.* Comprehensive treatment of major federal environmental legislation including the influence of constitutional decisions on thermal pollution regulation.

Hinrichs, Roger A., and Merlin Kleinbach. 4th ed. *Energy: Its Use and the Environment.* Belmont, Calif.: Brooks/Cole, 2006. Comprehensive overview of energy issues—radioactive waste, municipal solid waste, global warming—and their impact on society and the environment.

Houck, Oliver A. *The Clean Water Act TMDL Program: Law, Policy, and Implementation.* 2d ed. Washington, D.C.: Environmental Law Institute, 2002. Complete guide to the total maximum daily load (TMDL) requirements for industry on clean water, including thermal pollution regulations.

Langford, T. E. L. *Ecological Effects of Thermal Discharges.* New York: Elsevier Applied Science, 1990. Comprehensive treatment of all thermal effects including water chemistry, bacteria, fungi, algae, invertebrates, and fish.

Laws, Edward A. *Aquatic Pollution: An Introductory Text.* 3d ed. Hoboken, N.J.: John Wiley & Sons, 2000. Chapter 11 covers power-plant design, details toxic effects on biota, and presents a case study and its correctives.

Web Sites

Association for Environmental Studies and Sciences
http://www.aess.info

National Resources Defense Council
http://www.nrdc.org

Society for Environmental Toxicology and Chemistry
http://www.setac.org

See also: Civil Engineering; Mechanical Engineering.

TIME MEASUREMENT

FIELDS OF STUDY

Horology; metrology; mathematics; physics; particle physics; classical mechanics; quantum mechanics; geophysics; geology; nuclear physics; astronomy; cosmology; navigation; psychology; physiology; neurobiology; telecommunications; computer science; mechanical engineering; physical engineering.

SUMMARY

Time measurement is the science and practice of counting the repetitions of recurring phenomena and subdividing the intervals between each repetition into smaller units that are capable of being measured by a variety of devices, including mechanical and atomic clocks. Time measurement is an important part of everyday life. It enables people to schedule activities, measure distances and speed, and navigate from place to place. The accurate measurement of immensely small units of time is essential to a huge number of fields of basic and applied science, including physics, computing, and medicine.

KEY TERMS AND CONCEPTS

- **Accuracy:** Degree to which the pulse or signal measured by a particular clock conforms to a reference pulse or signal, such as coordinated universal time. The opposite of accuracy is drift.
- **Atomic Clock:** Clock that either contains an internal atomic oscillator or uses an external atomic oscillator as a reference.
- **Atomic Oscillator:** Oscillator that uses changes in the energy level of atoms to determine the intervals between the signals it produces.
- **Coordinated Universal Time (UTC):** Twenty-four-hour international reference time scale, based on the atomic definition of a second that is the standard timekeeping reference for the world; measures the hour, minute, and second at the prime meridian (0 degrees longitude) located near Greenwich, England; and is sometimes called Greenwich Mean Time, in casual, nonscientific usage.
- **Frequency Standard:** Oscillator whose signals are used by other devices as a reference for measuring time intervals.
- **Oscillator:** Device that generates a constant, periodic signal that varies in magnitude around a central point.
- **Quality Factor (Q):** Number of vibrations an oscillator can make before it requires an additional burst of energy to continue working.
- **Resonance:** Vibration or oscillation of an atom as it emits electromagnetic radiation. Any atom of a given element will resonate at precisely the same frequency, or rate.
- **Second:** Time required for a cesium-133 atom to oscillate 9,192,631,770 times; can be further subdivided into units such as nanoseconds (one billionth of a second), picoseconds (one trillionth of a second), and femtoseconds (one quadrillionth of a second).
- **Stability:** Degree to which a particular clock can produce a repeating pulse or signal at exactly the same frequency over a given period of time.

Definition and Basic Principles

The science of time measurement involves devising methods to count the number of iterations of phenomena that repeat themselves at regular intervals. For example, the rotation of the Earth on its axis, the shifts in the phases of the Moon, and the changing of the seasons are all familiar units of time based on observable changes in material objects.

Time itself is a slippery concept that has no simple scientific definition, though physicists consider it one of the four fundamental dimensions of the universe, along with length, width, and depth. Although scientists have not managed to satisfactorily define time, they have developed extraordinarily accurate ways of quantifying the passage of time so that individual events can be referred to on a consistent scale. The goal of time measurement is to supply information about time of day, or the instant at which an event takes place; time interval, or the duration between two events; and frequency, or the rate at which a repeated event takes place.

One of the basic principles of time measurement is the notion that time is not a physical constant but can shrink or expand in response to other forces. According to Albert Einstein's general theory of relativity, strong gravitational fields can stretch the interval between the signals given by a clock. This phenomenon can affect both observations of the simultaneity of events and measurements of the duration of events. Clocks located in space, such as the ones on the International Space Station or on satellites orbiting the Earth, must be adjusted to correct for the effects of relativity. The special theory of relativity shows that two observers moving in relation to one another will observe duration and simultaneity differently; the differences between their observations, however, are negligible except when the speeds involved approach the speed of light.

Background and History

Attempts to mark the passage of days, months, and seasons using the movements of the Sun, Moon, and stars long predate attempts to measure shorter periods of time. Calendars, in other words, are older than clocks. Early clocks such as sundials and obelisks made use of the changing length of shadows over the course of the day to mark intervals that roughly corresponded to modern hours. Other primitive means of keeping track of the passage of time include water clocks, which were designed to drip at a relatively constant rate; hourglasses, which used the same principle but incorporated sand instead of water; and clocks that used the burning of candles or incense to measure time.

In the fourteenth century, mechanical (machine-powered) clocks began to appear. Some relied on a device known as an escapement, which controlled an unwinding spring that rotated a series of gears that, in turn, caused the hands of the clock to tick forward steadily. Others used a pendulum (a weight on a string), the back-and-forth motion of which served as a natural oscillator. Over the next several hundred years, inventors and engineers labored to reduce inaccuracies in timekeeping by compensating for such factors as friction, changes in temperature, and interference from other moving parts within the clocks.

The twentieth century saw two major advances in time measurement: the development of quartz clocks, which used electric circuits to generate constant electrical vibrations in quartz crystals, and the invention of atomic clocks, which take advantage of the natural resonance frequency of atoms to create an immensely stable oscillator. The atomic clock has become the standard tool for modern scientific timekeeping.

How It Works

Physical time scales span a dazzlingly wide range. The smallest scale that physicists are able to work with mathematically is Planck time, a single unit of which (about 10^{-43} second) is defined as the length of time it takes for a photon traveling at the speed of light to cross a distance of one Planck length (about 10^{-33} centimeter). In contrast, the cosmological time scale—on which events such as the beginning of the universe and the formation of stars and planets is marked—consists of periods as vast as tens of billions of years. Only a portion of the time scales that exist within this range can be accurately measured using the technology described in this essay.

Household Clocks. The majority of clocks, watches, and small electronic circuits used in everyday life are built on oscillators that use quartz crystals to generate a consistent pulse. Quartz crystals (whether natural or synthetic) have the property known as piezoelectricity, which means that the crystals expand and contract—or vibrate—when they receive an electric force. The combination of a quartz crystal with a battery that applies and then reverses an electric voltage produces a regular oscillation, the frequency of which depends on the size of the crystal and the form into which it is cut. A quartz oscillator found in an ordinary household clock will probably have a quality factor (Q) of about 10^4 to 10^6 and be accurate to about a few seconds per month—perfectly adequate for everyday use.

Scientific Clocks. Scientific and industrial purposes demand atomic clocks with far more accuracy and stability than that of household clocks or watches. Atomic clocks make use of the fact that all atoms are capable of existing at a number of different discrete (noncontinuous) levels of energy. As an atom jumps back and forth between a higher and a lower energy level, it resonates at a particular frequency, and this frequency is exactly the same for every atom of a given element. For example, cesium-133 atoms (one of the two types of atoms most commonly used in atomic clocks; rubidium is the other) resonate between two particular energy levels at 9,192,631,770 cycles per second. (In fact, the time it takes for this number of oscillations of a cesium-133 atom to take place serves as the definition of a second in the International System of Units, or SI.) This stable vibration can be thought of as paralleling the swinging back and forth of the pendulum in an old-fashioned mechanical clock.

There are many different forms of atomic clocks, but the basic mechanics involved are relatively consistent. A laser beam is shone into a cloud of atoms, tossing them high into the air. The frequency of light at which the laser shines is adjusted until the vast majority of the atoms experience a change in energy state. This process tunes the laser's frequency to match the resonance frequency of the atoms themselves, and it can then be used to mark the passage of time with great accuracy and reliability. Most atomic clocks also make use of a technique known as laser cooling, in which the movement of atoms is slowed by dropping the temperature within the clock to something very close to absolute zero. This lengthens the period of time during which the atoms can be properly observed.

Standards in Time Measurement. Within the field of time measurement, standards refer to devices or signals that serve as benchmarks for particular measurements, such as time intervals or frequencies. Standards allow other clocks to be precisely adjusted so that they all keep the same time and can be recalibrated according to the same measure if they should happen to gain or lose time. For example, the National Institute of Standards and Technology's cesium fountain atomic clock (known as the NIST-F1), located in Boulder, Colorado, is the standard atomic cesium clock on which all other clocks in the United States are calibrated. NIST-F1 is one of hundreds of highly accurate atomic clocks around the world that together define the interval between each second in UTC, the official global time of day. NIST-F1 will gain or lose a single second only once every 60 million years and has a Q factor of about 10^{14}.

Applications and Products

Navigation. The ability to measure time accurately and to precisely synchronize more than one clock is essential for many forms of navigation, including the United States-based Global Positioning System (known as GPS, or NAVSTAR) used in many cars, boats, airplanes, missile guidance systems, cell phones, watches, computer clocks, network clocks, and other devices. To determine a location, a GPS receiver calculates the time it takes for signals from four separate satellites to reach it, multiplies that time by the speed of a radio wave, and uses this figure to determine the exact distance between the receiver and each satellite. Then the receiver creates four imaginary spheres, each of which has a radius that is the same length as the distance between the receiver and one of the satellites. The latitude, longitude, and altitude at the point where the spheres cut across each other is the latitude, longitude, and altitude of the device. This process is known as triangulation. The United States is not the only country with a global navigation satellite system. Russia, China, and the European Union are among those working on developing similar applications.

Even the tiniest inaccuracies in the measurement of the time interval between when the signal is sent and when it is received can cause substantial errors—differences of several feet or more—in the triangulation calculation that determines the user's physical location. As a result, GPS receivers rely heavily on standard atomic frequency references. Each satellite has a small handful of local atomic clocks onboard (multiple clocks are used to provide a system of redundancies, a way to ensure reliability and accuracy). Each clock is calibrated to a master atomic clock measuring UTC and managed by the U.S. Naval Observatory. In general, each clock on a GPS receiver, no matter where on Earth it is located, will be no more than 500 nanoseconds to 1 millisecond out of sync with UTC. Time measurement not only facilitates the movement of cars, boats, and airplanes on the earth and sea and in the sky but also is fundamental to interplanetary space travel. In fact, every National Aeronautics and Space Administration (NASA) spacecraft is equipped with standards-referenced atomic clocks that enable accurate and reliable navigation.

Telecommunications. Every telecommunications system in use—including standard telephone systems, wireless communication networks, radio and television broadcasting systems, cable networks, satellite telephones, and the Internet—relies on the ability to synchronize exactly the signals sent between transmitters and receivers. For example, wireless telephones transmit information to each other in the form of small chunks of data called packets. The signal between the transmitter and receiver must be perfectly synchronized to be able to identify where each packet begins and ends. In addition, the transmitter and receiver must process data at the same speed so as to prevent lags or data loss. Finally, for multiple telephones to send data using the same frequency channel, each telephone is assigned a particular time slot. Individual transmitters send their signals in series, never interfering with one another's data even though they share the same medium of transmission. (Radios broadcast their signals using exactly the same system, which is known as time division multiple access, or TDMA.) Because cell phones move from location to location as their users travel, they also require a precise time measurement system—usually pegged to the atomic clocks on GPS satellites. These systems help make minute changes to the timing of the cell phone signals to compensate for the fact that the distance between the telephone and the base station is changing.

Electronics and Computers. Every computer contains a small, built-in quartz-based oscillator that serves as an internal marker of time intervals for the machine. The central processing unit uses this clock to determine the intervals at which its microprocessor is directed to complete instructions, as well as for purposes such as scheduling automatic processes and time-stamping events. It is also important for computers that are sending and receiving information over a network or the Internet to be highly synchronized with each other for data to be transmitted accurately. Since the piezoelectric qualities of quartz crystals change with temperature, however, computer clocks tend to drift as the machinery inside them heats up with use. For this reason, most computer networks are equipped with a Network Time Protocol (NTP) server that uses an atomic frequency standard such as the signal produced by NIST-F1 to synchronize the internal clocks of the computers to a more accurate and stable time signal.

Impact on Industry

Academic Research. The ability to accurately measure time of day, duration, and frequency is fundamental to any number of scientific fields. Geophysicists, for instance, use time measurement to synchronize the data they collect from different places about lightning, earthquakes, seismic events, and other geological phenomena. Astronomers use it to triangulate the radio waves transmitted by stars to determine the location of individual celestial objects and put together maps of the sky. Atomic clocks that use hydrogen atoms to measure frequency have enabled physicists to test both the general and the special theories of relativity proposed by Einstein. Researchers have shot such clocks tens of thousands of kilometers into the air and observed how the speed and altitude of the rocket affected the pulse the clocks produced. In addition, because time has become one of the measurements that people are able to make with the greatest possible accuracy, many other scientific measurements are based on units of time. For example, length is defined by the distance that light is capable of traveling within a certain period of time.

Industry and Business. The ability to accurately and consistently measure frequency, duration, and time of day is also essential for the fundamental activities of a vast number of global industries. Accurate time measurement is particularly important for the transportation industry. The aviation industry requires accurate time-interval measurements to efficiently schedule a vast array of daily flights, as well as to enable air traffic control systems to safely direct pilots amid the crowded skies. In addition, every communication between airplanes and ground control centers is stamped with the time of day. These records help investigators to establish the exact origin of a malfunction or other unforeseen event. Train and bus transportation companies demand highly synchronized, accurate clocks for similar reasons. Power companies are another industry in which time plays an important role. They rely on time and frequency information to measure the usage of electricity by their customers, maintain a balance of supply and demand across their networks, and prevent sudden outages.

Other businesses that make use of time and frequency measurements include hospitals, which collect biomedical readings from patients and require their monitoring instruments to be precisely calibrated; camera and audiovisual equipment manufacturers, who must engineer their products to time-stamp the recordings they make; manufacturers of home appliances such as microwave ovens, cooking stoves, dishwashers, and televisions, all of which require timers; and, of course, watch and clock manufacturers and repairers.

Time measurement has some surprising applications in other industries as well. Accurate frequency

Fascinating Facts About Time Measurement

- As the Moon slowly spirals away from the Earth, its gravitational pull on the planet lessens, making each day last a tiny fraction longer and, over time, causing each year to contain fewer days.
- In 1761, a horologist named John Harrison won $20,000 from the British government for building a chronometer that remained accurate and stable even on the turbulent waters of the sea, thus enabling sailors to calculate longitude to within one-third of a degree.
- The body's internal molecular timekeeper, known as the circadian clock, regulates cycles of sleep and wakefulness. When it is out of sync, it can cause disorders such as jet lag and depression.
- Conventional time-of-day measurements are based on the changing position of the Sun in the sky, but shifts in the positions of stars can also be used to define a kind of time known as sidereal time. One sidereal day lasts a few minutes less than one solar day.
- Highly accurate time-of-day information, based on atomic frequency standards, is available through National Institute of Science and Technology radio broadcasts twenty-four hours a day.
- Station WWVB in Colorado broadcasts on 60 kilohertz; stations WWV and WWVH are located in Colorado and Hawaii, respectively, and broadcast on five frequencies from 2.5 to 20 megahertz.
- A leap second is added to UTC every so often (usually less than once per year) to bring it into sync with the more stable astronomical time, which is based not on atomic clocks but on the rate at which the Earth rotates.

signals are required, for instance, to properly tune musical instruments such as guitars, violins, and pianos to the correct tone. For example, 440 hertz has become a standard pitch for orchestras tuning up. The correct frequency signal can be obtained not only from a tuning fork but also from an oscillator or even a national radio service such as the ones operated by the National Institute of Standards and Technology.

Major Corporations. Agilent Technologies, which grew out of technology giant Hewlett Packard, is one of the world's leading scientific/industrial measurement corporations. Agilent produces a variety of time measurement products, including a time-to-digital converter that uses a system of six stopwatches to time the duration between a single start event and several ending events. Other important companies in the electronic measurement industry include National Instruments and Tektronix.

Careers and Course Work

A student contemplating a career as an academic or industry-based researcher in time measurement or metrology (the study of measurement itself) or planning to enter a related field such as astronomy, geophysics, or engineering should begin by pursuing a rigorous course of study in science, technology, and mathematics, preferably leading to a bachelor of science degree in physics. Of prime importance within the field of physics are subjects such as atomic structure, special and general relativity, electromagnetics, cosmology, the physics of the solar system, and quantum mechanics. In mathematics, geometry and trigonometry are especially instructive to the science of time measurement.

An interest in time measurement might lead down a number of career paths. For instance, one might enter the field of horology and become a designer of timekeeping apparatuses such as clocks, watches, timers, and marine chronometers. One might become a staff researcher investigating the physics of time at a government or university laboratory or observatory such as the National Institute of Standards and Technology, the U.S. Naval Observatory, or the NMi Van Swinden Laboratorium in the Netherlands. Alternatively, one might pursue a job developing, maintaining, or repairing time measurement instrumentation for scientific or industrial purposes, in which case, course work in mechanical and electrical engineering would also be required.

Social Context and Future Prospects

Scientists are working on a new generation of optical frequency atomic clocks that use atoms such as strontium-87 and mercury to produce oscillations at optical frequencies about 100,000 times faster than conventional cesium clocks, which oscillate at microwave frequencies between 0.3 and 300 gigahertz. These clocks are still experimental but hold the potential to be even more accurate than conventional cesium clocks (gaining or losing one second every 400 million years) and capable of measuring time at higher resolutions (subdividing it into smaller units). Time measurement research has the potential to revolutionize a host of fields. Such devices could allow military navigators to bypass jammed GPS signals, enable e-mails to be encrypted at a far safer and more complex level than ever before, or be used in inexpensive portable medical imaging devices to scan patients' hearts, brains, and other organs at high resolution even outside hospitals.

M. Lee, B.A., M.A.

FURTHER READING

Audoin, Claude, and Bernard Guinot. *The Measurement of Time: Time, Frequency, and the Atomic Clock.* Translated by Stephen Lyle. New York: Cambridge University Press, 2001. Offers information on the physics behind time measurement, with sections on atomic frequency standards, types of oscillators and atomic clocks, and scientific and industrial applications of time measurement. Includes appendixes listing global time laboratories and definitions of SI units.

Dunlap, Jay C., Jennifer J. Loros, and Patricia J. DeCoursey, eds. *Chronobiology: Biological Timekeeping.* Sunderland, Mass.: Sinauer Associates, 2004. Discusses the anatomy, physiology, genetics, and molecular biology of living organisms with circadian clocks.

Eidson, John C. *Measurement, Control, and Communication Using IEEE 1588.* London: Springer, 2006. An overview of the Institute of Electrical and Electronics Engineers (IEEE) standard 1588, the standard for a precision clock synchronization protocol for networked measurement and control systems, an engineering technology designed to synchronize devices operating on real-world time.

Jones, Tony. *Splitting the Second: The Story of Atomic Time.* Philadelphia: Institute of Physics Publishing, 2000. An accessible account of the history of modern atomic timekeeping, beginning in the 1950's. Contains numerous figures, tables, and graphs illustrating technical points.

Kaplan, Elliot D., and Christopher Hegarty, eds. *Understanding GPS: Principles and Applications.* 2d ed. Boston: Artech House, 2006. A comprehensive technical overview of how the Global Positioning System works, including chapters on time and GPS and the effects of clock errors.

Lombardi, Michael A. "Radio Controlled Wristwatches." *Horological Journal* (May, 2006): 187-190. Describes in detail various types of radio-controlled wristwatches and discusses solutions to the independence problem.

Lombardi, Michael A., Thomas P. Heavner, and Steven R. Jefferts. "NIST Primary Frequency Standards and the Realization of the SI Second." *Measure: The Journal of Measurement Science* 2, no. 4 (December, 2007): 74-89. Explains the history and principles behind the NIST frequency standards. Includes references, figures, tables, and color photographs of NIST cesium clocks.

WEB SITES

National Institute of Standards and Technology
Physics Laboratory, Time and Frequency Division
http://www.nist.gov/physlab/div847/index.cfm

The Official U.S. Time
http://time.gov

U.S. Naval Observatory
Precise Time
http://www.usno.navy.mil/USNO/time

See also: Computer Languages, Compilers, and Tools

TOPOLOGY

FIELDS OF STUDY

Mathematics; geometry; graph theory; set theory; abstract algebra; physics; dimension theory; knot theory.

SUMMARY

Topology is the mathematical study of the relationships and properties of objects in any number of dimensions, independent of size. Topology plays a crucial role in many branches of mathematics, since arguments that can be formulated in topological terms are more general than those that depend on geometrical properties. Topology has also served to provide models for cosmologists seeking to describe the structure of the universe. Predictions about the outcomes of interactions between objects frequently use topological language rather than the exact calculations of classical mechanics. Even biology has found it helpful to use topology to describe the relationships between individual living things and species.

KEY TERMS AND CONCEPTS

- **Chaos:** Ability of two nearby points to end up arbitrarily farther apart after the repeated application of a function.
- **Continuous Function:** Function that can be graphed without lifting pencil from paper.
- **Dimension:** Number of variables that could be used to describe a relationship, extended to fractional dimensions for objects characterized by infinite processes.
- **Disk:** Sphere in any number of dimensions together with its interior.
- **Function:** Description of how one set of points can be changed into another set of points.
- **Graph:** Collection of points and line segments used to represent a collection of objects and the relations between them.
- **Homeomorphism:** Equivalence of two objects used to serve as a characterization for having the same properties.
- **Knot:** Generalization of a physical knot, thought of without thickness and in an arbitrary number of dimensions.
- **Manifold:** Subset of space used as the basis for the action of a function.
- **Metric:** Means for measuring distances that generalizes the Pythagorean theorem in familiar space.
- **Orientable:** Of a surface that can be described in familiar terms as having two distinct sides.
- **Sphere:** Object topologically equivalent to the surface of a ball in any number of dimensions.
- **Stability:** Characteristic of a system that does not change under further application of a function.
- **Topology:** Collection of subsets of a space that describes how its pieces fit together in general.

Definition and Basic Principles

Topology is a generalization of geometry, which historically served as a description of the structure of space. In topology, issues of size and measurement are not considered important, and questions about what kinds of overlap two sets of points must have depend on more general properties of the sets. Topology emerged out of the traditional fields of calculus, geometry, and algebra, and there are subdivisions of the field closely related to each of them. The language of topology looks like that of sets, but there is also room for carrying over ideas from the study of geometrical objects in familiar spaces. The terminology is also based on imagery familiar from geometry.

The applications of topology cut across branches of science and mathematics that can use the language of relationships to describe the objects they study. Topology is abstract enough that it can sometimes seem as though it is not about any particular kind of object, but that makes it even more widely applicable. One of the common characterizations of topology is that it is "rubber sheet geometry," where one studies the properties of objects that are preserved under the kind of stretching and squeezing that rubber can undergo. Mathematicians have come up with more abstract ways of describing these properties, which draw on abstract algebra.

Background and History

The word "topology" is a relatively new coinage, but other terms such as "analysis situs" have a longer history. The mathematician Gottfried Wilhelm

Leibniz in the seventeenth century first proposed the idea of a geometry that would depend on the properties of space more generally than the Euclidean geometry universally studied at the time. In the eighteenth century, the Swiss mathematician Leonhard Euler came up with the first concrete example of an application of topological ideas when he solved a problem about whether one could traverse all the bridges in a certain town without repeating any part of one's path.

The subject remained dormant until the late nineteenth century, when the French mathematician Jules Henri Poincaré pursued a couple of lines of study that led to different branches of topology. He looked at the relationship between the algebraic properties of an object and its geometrical properties, which gave rise to geometrical topology. He also studied physical processes in which classical mechanics seemed to be unable to describe the results, giving rise to the field of nonlinear dynamics, one of the branches that contributed to the rise of chaos theory a hundred years later.

In the twentieth century, the rise of set theory provided a language for talking about collections of points in a more fundamental way than had been possible previously, and an entire school of Polish mathematicians produced fundamental results in what is called point-set topology. As abstract algebra became more sophisticated, its language became crucial for talking about topological spaces, and the discipline of algebraic topology was created to use algebraic techniques for understanding topology. Later in the century, the rise of computers gave empirical evidence for the way in which the repetition of functions produced certain kinds of patterns surprisingly independent of the set from which one started and sometimes even the kind of function. Finally, Albert Einstein's general theory of relativity suggested that traditional geometry would not be capable of describing the behavior of light, which brought physicists to appreciate the need to generalize geometry. The rise of various models for the structure of the universe (especially string theory) has put topology at the center of the concern of theoretical astrophysics.

How It Works

Point-Set Topology. The study of sets starts from the notion of a set as a collection of elements. In the simplest cases, one can think of this collection as a list, but the mathematically interesting cases usually require an infinite collection, for which the notion of list is unhelpful. Instead, one uses the notion of a property to define the members of a set as part of a larger collection. The ideas of intersection and union as ways of combining sets help to define the notion of a closed set and an open set. A closed set is one that includes its boundary, while an open set does not include any of its boundary points. For example, a circle with its interior is a closed set, while the interior by itself is an open set. Point-set topology concerns the results of taking intersections and unions of closed and open sets, frequently to the extent of allowing an infinite number of sets in the operation. Such interactions (when combined with the notion of complementation, taking everything that is not in a given set) give rise to sequences of possible outcomes from repeated processes. The foundations of point-set topology lie in the axioms used for the theory of sets, so different results can come from a selection of different axioms.

The interest of point-set topology outside the area of set theory comes from considering what happens when functions are applied to various kinds of sets of points. In particular, one is frequently interested in continuous functions, which do not allow for jumps in the values of the function. Traditionally, a physical model for a function suggested that it ought to be continuous, but the laws of quantum mechanics have made the assumption of continuity a little less useful and have produced the notion of a distribution as a more general sort of function than the continuous case.

The kind of set of points on which functions act in topology is often a manifold. This is a piece of space that is locally like Euclidean space (the familiar version). In other words, one can find some kind of function that connects the manifold's behavior with that of ordinary space. Knowing how the function behaves on Euclidean space and knowing how Euclidean space is related to the manifold provides a description for how the function applies on the manifold. This process of going from the manifold to Euclidean space, operating on Euclidean space, and then returning to the manifold is characteristic of mathematics in the twentieth century and beyond.

Algebraic Topology. Algebra became a much more abstract discipline as the nineteenth century went along, and by the arrival of the twentieth century, it was far from its roots of solving polynomial equations using the standard arithmetic operations. First, it had been discovered that there were such equations that could not be solved by those operations. Then there was the challenge of trying to identify which equations could be solved by the old-fashioned means. Finally, there was the question of how to get useful information about the solutions of an equation even when they could not be identified explicitly. The result was the creation of theories of algebraic objects such as groups, rings, and fields. These had laws corresponding to those for various sets of real numbers, in some cases imitating those of the whole numbers and in other cases those of larger sets. In addition, the use of the language of algebra extended beyond sets of numbers to functions and other kinds of mathematical objects.

With all these algebraic structures under consideration, those studying geometrical objects could try to come up with algebraic characteristics for geometry. Just as René Descartes in the seventeenth century had explained how to use coordinate axes to turn questions about geometry into questions that could be handled by the algebra of the time, so the higher-powered algebra of the twentieth century could be brought to bear on identifying and characterizing geometric objects. One of the issues that arose, once algebraic characteristics could be attached to geometric objects, was whether one could tell that two geometric objects were homeomorphic from enough algebraic information. Different kinds of algebraic objects gave rise to different levels of classification, and the algebraic information offered paths for exploring the geometry.

Applications and Products

Graph Theory. One of the areas created by Euler and not followed up on for a century or more was the study of graphs. These are not the objects of algebra that represent the relationship between two variables on a pair of coordinate axes but instead offer a more general way of describing relationships between objects. One uses vertices (points) to stand for the objects themselves and edges (line segments) between the vertices to represent their having a certain kind of property. For example, one can use vertices to stand for numbers and put edges between them if they are within a certain range of one another (less than one unit apart, for example). One can also make the graph directed by putting arrows on the edges if the relationship between the objects represented by the vertices is not symmetric. For example, if the points represent numbers, one could use an edge with an arrow pointing from one point to another if the number corresponding to the first point is a factor of the number corresponding to the second.

There are many issues connected with a graph when it has been used to model the relationship between objects. For example, one can ask what the shortest path is between two points in the graph, where distance is defined not by the usual length from geometry but the number of edges required to get between the points. There may be points that cannot be connected by any path, especially when the edges are directed. Results going back to Euler help resolve some of these questions, but others depend on algorithms that take a long time to run.

Applications of graph theory in other sciences are countless. In the middle of the nineteenth century, graphs were already being used in chemistry to represent the processes of transforming one substance into another. In modern times, graphs have been used to represent the proximity of different species to one another in the natural world. Because molecular biology can be used to describe a species, the genetic code is used to characterize the species in mathematical terms. Then the distance between two codes can be measured by the number of changes required to get from one code to another. For example, the word "lick" can be changed to the word "luck" by altering only one letter, but "lick" requires two changes to get to "risk." This has become a battleground for those concerned about the theory of evolution because there are questions about whether certain kinds of paths would be possible in whatever time is given for the existence of the Earth. It should also be clear that topology does not answer every question in the disciplines to which it can be applied. The question of how likely an alteration in the genetic code is to take place is not part of the domain of graph theory. In general, graph theory has offered a language of representing information and asking questions about it that is independent of geometry, and topology has been invaluable as a means of describing situations, but biologists have not thereby been shown to be dispensable.

Chaos Theory and Fractals. In the nineteenth century, much science fiction was written about the notion of molecules' being small worlds with their own inhabitants. The idea was also extended to the possibility that perhaps the galaxies of which the stars and planets are a part are only particles in a yet larger system. Although these speculations have not been borne out in investigating the universe, the notion that there could be similarity across a difference of scale has been one of the most fruitful ideas in the area of mathematics known as chaos theory. It arose from the empirical evidence that one of the assumptions of classical mathematics was not well-founded: If one takes two numbers that are reasonably close together and applies a continuous function to them, they will stay close together. That is almost the definition of continuity, but problems arose when the application of the function was repeated not just once or twice, but thousands or millions of times. What became evident was that points that started as close together as one could imagine would end up unimaginably far apart. What was even more noticeable was that after a sufficiently large number of repetitions of applying a function, the patterns that resulted looked almost as though they had been designed by an artist.

The search for an explanation for these results led to the notion of a fractal. The word is short for "fractional dimension" and refers to the ability of some objects to have a dimension somewhere between two whole numbers. For example, a curve is usually thought of as a one-dimensional object, since it is the result of applying a function to a straight line. There are, however, curves that can be constructed by an infinite process that are called space filling because they seem to be able to fill a two-dimensional region (or at least a large part of it). Coming up with a notion of dimension for such objects requires a little more mathematics than the whole numbers. This approach to fractional dimension turns out to be useful in studying, for example, the coastline of a country, where magnification frequently leads to a similar shape to the one visible to the naked eye or on a map. There is a fair amount of visual beauty in fractals, which accounts for their prevalence as a subject for books designed to appeal to the general reader, but their importance in studying physical phenomena does not depend on their aesthetic characteristic.

If a function is applied sufficiently often, the resulting pattern may turn out to be stable. Perhaps the most striking conclusion of chaos theory is the observation that the stability of a configuration is independent of the original configuration from which it arose. There are models for the evolution of a system (such as the game of Life, devised by the British-American mathematician John Horton Conway), which offer insights into population changes and dynamics, and the same is true of the study of cellular automata, which are grids with patterns determined by simple rules. Simple rules, if allowed to govern the evolution of a system through enough generations, produce wildly improbable results. Topology offers the best explanation, precisely because it works with the smallest number of initial conditions.

Impact on Industry

Government and University Research. The study of topology started as one of the main ingredients in the Polish school of mathematicians in Warsaw and elsewhere between the two World Wars. After the destruction of the Polish educational system during World War II, many of the Polish mathematicians who survived migrated to other parts of the world, including the United States. They proceeded to put their topological set-theoretical experience to work by the side of the developments in abstract algebra, and the subject became a part of the curriculum for graduate students in mathematics in the United States. The abstractness of the language used to present the results sometimes led to the subject's being parodied as a kind of abstract nonsense. The continued ability of topological results to be applied broadly in mathematics overcame the doubts. The usefulness of topology in describing the structure of the universe has also guaranteed that physicists are consumers of the subject. Scientists recognized that traditional geometry does not describe the universe and that the theories of Einstein and others called for something more general than Euclidean space, and topological manifolds fit the bill.

Business. The structure of chemicals and how to manufacture them more efficiently is a concern of a good deal of industry. Some chemical processes involve the crossing and uncrossing of the constituents, which is a process like that involved in knots. As a result, knot theory plays a role in biochemistry and chemistry, trying to figure out which chemicals might serve as catalysts for making a given crossing more likely as needed. Again, it is not the topologist dictating to the chemist but the combination of

Fascinating Facts About Topology

- The Möbius strip is a one-sided surface. It is made by taking a strip of paper and twisting it before taping the ends together. Cutting a Möbius strip in half or in thirds produces highly unusual effects.
- The Klein bottle is a higher-dimensional version of the Möbius strip. There is no separation between its inside and its outside. Both the strip and the bottle are examples of nonorientable surfaces.
- The ability to take off a sweater from under a jacket without taking off the jacket is a demonstration that the sweater is not really "inside" the jacket.
- It takes four colors to color any map in such a way that adjacent regions have different colors. If one has the map on a torus instead, it takes seven colors.
- There are algorithms for deciding whether a point is on the inside or the outside of a curve. This may not seem necessary, but when the curve is complicated enough, there may be no easy visual way of telling.
- If one adds up the number of faces and vertices on a solid with polygonal faces, the number will be two more than the number of edges.

topology and chemistry that makes it possible to synthesize compounds more efficiently.

In many situations, businesses need to worry about allocating resources in different locations. It can be crucial to decide whether there is an efficient way of connecting all the locations to determine what needs to be left at each site. Graph theory has a role to play in determining whether paths can be found that will enable materials to be supplied in the right order and even the right quantity. As a result, the methods of graph theory will be called on to come up with improvements over arrangements that were originally produced by accident rather than design.

Careers and Course Work

The student interested in pursuing a career in topology would benefit from exposure to as many branches of mathematics as possible. Even though topology is in some ways a generalization of geometry, the geometrical foundation for topology does not become lost. Upper-level algebra courses provide part of the language for doing topology. Doctoral work in mathematics or computer science is the best path to a career strictly in topology. However, students with an eye on applications of topology in areas such as physics, chemistry, biology, or even business will need to learn about those fields rather than expecting topology to supply all the answers.

Most of those pursuing research in topology are involved in teaching mathematics at the university level. Some employment of topologists could come with an organization such as the National Security Agency, with a strong interest in cryptanalysis. The same kind of techniques that apply to questions about the genetic code apply to code breaking in general. Although physicists and cosmologists put a good deal of topology to use, it is generally in an academic setting.

Social Context and Future Prospects

Topology has always suffered from hostility on the part of those who are inclined to see classical mathematics such as geometry as sufficient for understanding the universe. It has also been a victim of some of the battles fought within other areas of mathematics. Set theory was involved with challenges as a result of paradoxes coming out of certain axiomatic foundations. Abstract algebra was viewed as a discipline without content. Catastrophe theory came into existence as a way of applying topology to many disciplines but never lived up to the claims made by its creators. Nevertheless, topology has managed to earn a place within the curriculum at all universities offering a modern picture of mathematics. It was the most exciting discipline in mathematics in the United States for much of the second half of the twentieth century.

Topology should be able to maintain its central role in mathematical education at the university level as a result of its alliance with so many other disciplines. Books on topology in chemistry, physics, and other sciences bear witness to the usefulness of topology as a language and a set of ideas. Graduate students no longer need to fear being held up to scorn for studying a discipline like topology. It has survived the occasional overambitious claim and found a lasting role in mathematics and the sciences.

Thomas Drucker, B.S., M.S.

Further Reading

Adams, Colin. *The Knot Book.* Providence, R.I.: American Mathematical Society, 2001. Uses pictures to explain the importance of knots without expecting much mathematical background.

Adams, Colin, and Robert Franzosa. *Introduction to Topology: Pure and Applied.* Upper Saddle River, N.J.: Pearson Prentice Hall, 2008. Almost the only book to provide both the theoretical background for topology and the variety of applications for the discipline.

Flapan, Erica. *When Topology Meets Chemistry: A Topological Look at Molecular Chirality.* New York: Cambridge University Press, 2000. A mathematician's detailed look at how topology helps distinguish compounds with the same molecular composition.

Gilmore, Robert, and Marc Lefranc. *The Topology of Chaos.* New York: John Wiley & Sons, 2002. Illustration of the role that topology plays in understanding dynamical systems.

Mandelbrot, Benoit. *The Fractal Geometry of Nature.* 1983. Reprint. New York: W. H. Freeman, 2006. A pioneering study of how topology serves as a guide to various phenomena in Nature.

Richeson, David S. *Euler's Gem: The Polyhedron Formula and the Birth of Topology.* Princeton, N.J.: Princeton University Press, 2008. Introduction for the nonmathematician to basic ideas in topology and its historical development.

Web Sites

American Mathematical Society
http://www.ams.org

International Linear Algebra Society
http://www.ilasic.math.uregina.ca/iic

Mathematical Association of America
http://www.maa.org

Society for Industrial and Applied Mathematics
http://www.siam.org

See also: Algebra; Applied Mathematics; Chaotic Systems; Fractal Geometry; Geometry.

TRANSPORTATION ENGINEERING

FIELDS OF STUDY

Physics; thermodynamics; fluid dynamics; materials science; probability and statistics; calculus; civil engineering; urban planning; land use; environmental science and policy; community and economic development; aerospace engineering; geographic information systems; decision science.

SUMMARY

Transportation engineering is an area of specialty within the field of civil engineering. Transportation engineers are concerned with the design and building of large-scale systems that move people from one place to another. These systems include roads, railway and subway networks, and airports. The greatest demand for transportation engineers comes from cities, where the concentration of people living together is highest and where traffic is most likely to occur. The field of transportation engineering is growing due to a rise in urban populations overall during the first decade of the twenty-first century. Increased awareness of fuel costs and environmental-impact issues are also driving the growth of transportation systems in cities.

KEY TERMS AND CONCEPTS

- **Geographic Information Systems (GIS):** Systems that integrate and analyze data to create detailed maps.
- **Infrastructure:** Urban features that support growth, such as roads, railroads, bridges, and sewers.
- **Intelligent Systems:** Transportation systems that adapt to changing conditions, such as tolls that increase during peak-use hours.
- **Light Rail:** Form of public transportation in which rail cars are fueled by electricity and operated on a rail network at ground level, often next to a highway or road.
- **Mass Transit:** Transportation modes in areas with high-density populations that provide alternatives to automobile use; examples include trains, subways, light rail systems, and buses.
- **Mode:** Within transportation, a category such as rail, car, bus, or airplane.
- **Toll:** Charge for use of a specific transportation system, often on a per-vehicle or per-trip basis; refers most commonly to roads.
- **Zoning:** Laws defining and restricting the ways in which land can be used.

Definition and Basic Principles

Transportation engineering is the design, development, and building of systems that move large numbers of people from one place to another. The need for systems like these is greatest in cities. Transportation engineers work closely with urban planners and engineers to make sure that the systems they design are a good fit with a city's overall needs, geography, and budget. Other stakeholders in transportation systems are local businesses, state and federal government agencies, and environmental groups. Transportation engineers also take into account the requirements of these stakeholders when a new system is planned.

Transportation systems can be categorized into modes defined by the type of infrastructure needed to run them. Buses, for example, need a fleet of vehicles and a network of roads and streets. Systems that need rails include trains, subways, light rail and monorail (considered separate categories from standard train systems such as commuter rail), and streetcars and trams. Ferries are used in many cities with extensive waterfront areas. Airports also fall within the domain of transportation engineering. While most transportation systems are designed with the needs of a geographically focused population in mind, airports are unique in that they serve passengers from all over the country—or the world.

Background and History

Transportation engineering before the nineteenth century was limited. Problems with traffic on city roads had existed since the days of ancient Rome, but there were few mechanical alternatives to traveling by individual carriage, on horseback, or on foot.

The first public transportation systems are believed to be networks of stagecoaches that took paying passengers from a designated point within a city to one of several destinations outside of it. These networks appeared in major cities such as Paris, London, and New York at the beginning of the 1800's. They were followed by horse-drawn omnibuses that ran regularly on routes within city limits. By the 1860's, the first elevated trains were being designed and launched in American cities. Engines ran on steam at first but were quickly replaced by electric systems by the 1880's. Electric railways made it possible to shift track systems from elevated lines to underground tunnels. A boom in the building of subway systems followed in the United States and Europe with the turn of the century.

Public-transportation ridership peaked in the United States in 1944 and 1945. Rapid advancements in automobile technology, combined with the passage of the Federal Aid Highway Act in 1956, led to a decline in use of mass transit. Streetcar systems in many major U.S. cities ceased in the mid-1950's. However, sharp increases in gasoline prices in the 1970's renewed interest in public transportation.

How It Works

Transportation engineers are involved at every stage of the process in developing and building transit systems. Some projects are large and may take many years, such as the construction of a new subway line. Others are as small as changes in an individual bus route. All transportation engineering projects go through a similar set of steps from start to finish.

Identification of Transit Needs. Before a new transit system can be designed, transportation engineers must have a clear understanding of the needs of the passengers it will serve. Questions asked at this stage are: How many passengers will use the system each day? How will these patterns vary on weekdays versus weekends? When during the day will usage be highest (also known as peak time), and how much will this vary from off-peak times? Transit engineers also consider issues such as the choice of transportation mode, such as light rail versus bus and the ways in which ridership would be affected. To answer these questions, transportation engineers work with urban planners and other professionals to gather information. The information comes from sources such as population maps, road traffic data, and surveys of residents living and working in the targeted areas.

Identification of Project Limits. All transit projects need resources to be built. Transit engineers must consider the budgets of city, state, and federal authorities in planning new systems. Funding from the private sector, such as large corporations or associations of businesses in an area, is used often, especially when companies are likely to benefit from the new system. Geography is another major factor in planning. The mode of transportation chosen and the budget needed to build the system depends heavily on the amount and type of land (or, in some cases, water) being covered. The environmental impact of a new transit system must also be studied. While public transportation is nearly always better for the environment on the whole than the same number of passengers driving individual cars, different modes of transportation affect the environment in different ways.

Model Creation and System Design. Once information has been gathered and the questions above have been answered, transportation engineers begin to create models of potential transit solutions, including the mode or modes to be used. These models are based on mathematical and engineering principles. They show the ways in which a new transit system would provide benefits for the costs involved. The models also track factors such as the geographic area covered, the density of the network (the number of routes and the distance passengers would need to travel to reach each one), the frequency and scheduling of routes, and the prices of fares. Transportation engineers must build models that show how the proposed system will meet both the current and the future needs of a community.

System Selection. Transportation engineers present their models and solutions to the stakeholders in a new transit system. These stakeholders include community residents, government agencies such as city councils and budget committees, and environmental groups. Once stakeholders have provided feedback, engineers incorporate their input into a revised version of the plan and seek final approval. In most cases, multiple meetings and rounds of feedback are needed before approval is granted by all stakeholders.

System Implementation. The approved transit system is ready to be brought into action. If the system is a small-scale change such as a shift in bus routes, transportation engineers may set a specific period of time during which the change is tested. The engineers will establish goals and gather data during the test period to make sure the change is effective. If new equipment or infrastructure is needed, engineers will seek bids from manufacturers to get the highest-quality products for a competitive price. Vendors are selected, orders are placed, and the system is built.

Followup. Once the new system is in place, transportation engineers ensure that it runs properly and receives the ongoing maintenance and supplies it needs.

Applications and Products

Transportation engineers hold jobs that range from the preserving of historic railroads to the designing of new technologies. One of the easiest ways to understand the field of transportation engineering is to look at it by infrastructure type.

Highways, Roads, and Streets. The design and building of roads and the vehicles that use them are some of the largest fields employing transportation engineers. Roads must meet the needs of the greatest number of drivers and passengers with a minimum of traffic delays. They must follow federal, state, and local requirements governing the safety of their design and the quality of their construction materials. Roads must provide enough capacity for the needs of businesses while ensuring quality of life for consumers, especially in terms of traffic, noise, and pollution. Transportation engineers find ways to combine networks of highways, roads, and streets in order to gain the greatest benefit overall.

Bus systems are the primary form of mass transit found on roads. Buses reduce the amount of traffic and pollution generated by road travel as well as the cumulative wear and tear of tires on road surfaces. An advantage of bus systems is that they can be used in areas with lower population density, such as suburbs or small towns, for a smaller budget than would be needed to build a train or subway. Bus routes and frequency can also be changed easily, either in response to short-term demands such as construction or long-term shifts such as neighborhood growth. A disadvantage of bus transit is that, unless buses use dedicated lanes, bus routes can be subject to the same traffic problems faced by cars.

Heavy Rail. Transportation networks known as "heavy rail" include subway systems, commuter or regional rail lines, and intercity train systems such as Amtrak in the United States. In densely populated areas, particularly in the stretch of major cities known as the Northeast corridor (ranging approximately from Boston to Washington, D.C.), there is a great deal of overlap between heavy rail networks, and it is often easy for passengers to transfer from one type of rail network to another simply by crossing a train platform.

Subways, also known as metros or rapid transit systems in some urban areas, are networks of trains that serve the needs of one city. Subway tracks often run through underground tunnels or on rails that are separated physically from passengers, pedestrians, and motor and bicycle traffic. This design allows many subway networks to use a third rail, charged with electricity, for power—a design that would not be possible or safe otherwise. Subways may also use rails elevated above the roadway or at grade (or ground) level. A subway system is distinguished from other types of heavy rail in that its trains often run more frequently, serve stations that are located closer together, and can sometimes (though not always) be ridden by passengers for a flat fare rather than a price determined by distance. However, not all subway systems worldwide follow this approach to fares. The London Underground, also known as the "tube," because of the cylindrical shape of many of its trains and tunnels, is a subway system that prices fares in part based on distance traveled.

At the other end of the spectrum of heavy rail networks are interurban, or inter-city, train systems. These systems connect large cities to each other via a network that can cover an entire country or continent. In some countries, these networks are managed by a governmental agency or operated by a legally permitted and closely regulated monopoly. In other countries, a network of regional rail authorities works together to maintain the track networks and train schedules. Interurban trains have the lowest frequency in terms of the number of trains per day. They also serve a small number of train stations within a metropolitan area, often one or a few stations located in a central or downtown area. Fare pricing is based on distance and route traveled, though passes for unlimited travel by distance within a given time period—such as a month or a year—are often available.

Commuter rail can be seen as a hybrid between subway systems and interurban train systems. Like subways, commuter rail systems serve the needs of a single city or metropolitan area. Commuter trains use a network of tracks that provide service between a city's downtown and its suburbs. Stations are designed to serve the needs of passengers traveling a longer distance, so stops are spaced farther apart than on a subway line. Fares for individual rides are priced by distance traveled, but most passengers rely on commuter rail to travel regularly between work and home and are more likely to buy monthly or yearly passes.

Light Rail, Streetcars, and Trams. In contrast to heavy rail, many light rail and tram systems operate on a network specifically designed for the carrying of a limited number of passengers. Light rail trains often run on rails separated from roadways, while streetcars share space and right-of-way on streets with motor vehicles. Trams may or may not share space on roads in the way that streetcars do, depending on the city and system. This category generally does not use an electrified third rail for power. Instead, trains are more likely to be powered by a network of overhead wires to which each train maintains a direct connection. Diesel engines are also used to power trams. These systems usually require less capital to build and maintain than heavy rail.

Ferries and Water Taxis. Ferries, or ferryboats, are the primary form of mass transportation across bodies of water. Cities next to harbors, rivers, or coastal waters often establish ferry services as an integrated part of their mass transit systems. The Staten Island Ferry in New York City and the Star Ferry service in Hong Kong are two examples of ferries of this type. These ferries essentially serve as shuttles between two points, one on either side of a harbor or river. Water taxis, also called water buses, are used in cities with canals such as Venice and Amsterdam and follow routes with multiple stops. Other ferries, such as the commercially operated services between islands in Greece or Slovenia, carry passengers for longer trips and are closer in function to interurban trains.

Airports. Airports fall into a unique category in transportation engineering. Airports are managed by a local authority, often within the government agency that also oversees other forms of transportation infrastructure such as shipping ports. Airports do not need a physical network of rails or roads to connect them to each other. However, they serve the needs of commercial and government-run airline and cargo services using a communications network that manages air traffic control. The design, building, and maintenance of airports is a highly specialized field of engineering with close ties to aerospace.

Impact on Industry

The demand for professionals in the field of transportation engineering comes from many different sources. Most urban transit systems are managed and regulated by government agencies. Government funding also provides the financial support for new transportation systems and upgrades to existing ones. Many transportation engineers are employed directly by government agencies. However, these agencies also hire commercial engineering firms for design and building work. Professional opportunities for transportation engineers can be found at these firms as well.

The Institute of Transportation Engineers (ITE) is the leading professional association globally. As of 2011, ITE reported a total membership of almost 17,000 engineers in more than ninety countries. Because transportation engineering falls within the larger field of civil engineering, the American Society of Civil Engineers (ASCE) also includes a number of specialists among its members. Many of these members belong to the ASCE's Transportation and Development Institute, which hosts conferences and publishes books, journals, and research papers on topics of interest.

One of the leading areas of interest is intelligent transportation systems (ITS). ITS is a broad category that includes any system with the technology to respond to information collected and analyzed in real time. The possibilities for innovation with ITS are so vast that the U.S. Department of Transportation has launched a special agency, the Intelligent Transportation Systems Joint Program Office, within its Research and Innovative Technology Administration. The program is looking at ways to improve areas such as highway safety and the effectiveness of mass transit with the integration of real-time data.

One proposal focuses on a system known as Applications for the Environment: Real-Time Information Synthesis (AERIS) that would help people make decisions about the best and most efficient mode of transportation to use for a particular trip. Under the proposal, AERIS would provide information comparing current travel times from one point to another depending on the mode of transportation. The information would be drawn from

Fascinating Facts About Transportation Engineering

- The ancient Greeks believed that the dead took public transportation— in the form of a ferry-boat—from the living world to Hades.
- The term "omnibus" came from the city of Nantes, France, in the 1820's. A passenger carriage for hire that followed a standard route would pull up in front of a shop sign that read *Omnes Omnibus*, or All for All.
- In 1890, streetcar ridership in the United States had reached 2 billion passengers per year. By 1960 nearly all major cities had closed down their streetcar systems.
- Passengers took about 10.2 billion trips on mass transit systems in the United States in 2010.
- On an average weekday, about 35 million passengers ride U.S. public transportation.
- Mass transit ridership has reduced fuel usage in the United States by 640 million gallons each year as of 2010. It has also reduced travel time by 785 million hours.
- At more than 840 miles of track, the network of the New York City subway is the longest in the world. About 660 miles of track are used to carry passengers, while the rest are in train yards or on maintenance lines.
- A renovation of the Athens Metro in Greece begun in 1992 led to the discovery of about 30,000 artifacts. More than fifty archaeologists assisted with the project.

traffic data on highways and roads, public transit schedules, up-to-the-minute information on delays, and weather data. The system's interface might list a traveler's options by transportation mode, ranking the options by travel time, cost, or carbon footprint. These options could be viewed on a mobile device such as a cell phone.

The program is also looking at ways to increase safety for subway and train passengers. The ITS Rail Exploratory Initiative extends a proposal already in place for highway traffic in which vehicles could communicate directly with one another about hazardous conditions ahead on the road. Under the Rail Exploratory Initiative, transportation engineers are considering whether this technology could also be used by heavy and light rail and even by freight transportation systems. Applications of this technology might allow trains to respond in real time to problems such as blocked roadway crossings or congestion at heavily used track junctions. A significant number of accidents involving passenger injuries and deaths stem from communication problems between train operators. Innovations under the Rail Exploratory Initiative, if successful, could dramatically improve train safety.

A second major area of interest for the field of transportation engineering is the system's dependence on fossil fuels. Steady increases in the price of petroleum-based fuels over the last several years have made the operation of many transit systems more expensive. Increased costs are passed on, in part, to passengers through higher ticket and fare prices. Petroleum-based fuels also can have a detrimental effect on the environment. ITE, the U.S. Department of Transportation, and other professional groups have identified a multipronged approach to the situation. Solutions involve increasing the efficiency of existing systems so less fuel is needed; finding alternative energy sources and developing cost-effective ways to use them; and changing public awareness and travel habits when it comes to making environmentally sound choices about modes of transportation.

Careers and Course Work

Most transportation engineers hold a bachelor's or graduate degree. As a field, transportation engineering is too highly specialized to warrant its own major, so many students pursue majors such as civil engineering. Degrees in areas such as mechanical engineering, urban planning, geography, and probability and statistics would also have a number of applications within transportation engineering, though some additional courses dealing with transit systems might be needed.

Students begin with course work providing a general foundation in the principles of engineering. These courses would include physics, mathematics, materials science, general engineering, engineering technology, and design. Courses in civil engineering would examine the systems and structures needed to support large numbers of people, such as buildings, water systems, roads, railway networks, and bridges.

Opportunities to specialize further in transportation

engineering vary by school. The ITE has compiled a list of college-level courses taught in the United States, Canada, and Australia on topics within transportation engineering. This list provides insight into the types of advanced education available. The list of courses also indicates which colleges and universities offer the widest range of course work options. The courses cover topics such as the design of traffic systems, intelligent systems and transportation, transit system modeling, and environmental issues in transportation design. The ITE also offers extensive continuing-education opportunities to professionals, including a Transportation Leadership Graduate Certificate.

Demand for new hires in the field of transportation engineering rises and falls with the availability of funding, especially from government agencies, for transit projects. However, technological developments and a changing awareness of the importance of environmental issues gives an advantage to students with recent educations. There is also demand for engineering technicians, a career path that generally requires an associate's degree and involves more training on the job.

Social Context and Future Prospects

Public transit systems have been a critical part of major cities for the last few centuries. With the development and mass production of the automobile, particularly since the 1950's, there were predictions that mass transit would be sharply reduced or would disappear completely. Cars, it was believed, would become the preferred way for people to get around for the foreseeable future.

Public opinion began to change in the late 1980's. Consumers were becoming more aware of the negative environmental impact of burning vast quantities of petroleum-based fuel. One response to this situation was the development of car engines that use energy more effectively, such as hybrid engines. Many consumers, particularly those living in large urban areas, also changed their habits and began to use public transportation or to ride bicycles more frequently. In the 2000's, large and sudden increases in gasoline prices gave consumers another incentive to choose a bus, train, subway, or bike ride over a car trip.

High fuel prices and environmental concerns are two forces that are likely to keep demand for public transit, and transportation engineering, steady in the twenty-first century. Transportation engineers will be asked to find new solutions for moving large numbers of passengers in cities where growth is taking place both downtown and in the far suburbs.

Julia A. Rosenthal, B.A., M.S.

Further Reading

Altshuler, Alan, and David Luberoff. *Mega-Projects: The Changing Politics of Urban Public Investment.* Washington, D.C.: Brookings Institution, 2003. Examines the building of some of the largest urban transportation projects in the twentieth and twenty-first centuries and the funding, and politics, that supported them.

Gifford, Jonathan L. *Flexible Urban Transportation.* Oxford, England: Elsevier Science, 2003. Describes the impact of the U.S. highway system on urban mass transit and provides recommendations for future approaches to transit planning.

Goddard, Stephen B. *Getting There: The Epic Struggle Between Road and Rail in the American Century.* Chicago: University of Chicago Press, 1996. Offers a well-researched discussion and opinion on the rise of the highway system and the decline of rail networks.

Hanson, Susan, and Genevieve Giuliano, eds. *The Geography of Urban Transportation.* 3d ed. New York: Guilford Press, 2004. Provides a thorough analysis of mass transit systems; used frequently as a textbook in transportation engineering courses.

Moore, Terry, and Paul Thorsnes, with Bruce Appleyard. *The Transportation/Land Use Connection.* Chicago: American Planning Association, 2007. Analyzes the role of transportation engineering within the context of urban planning.

Post, Robert C. *Urban Mass Transit: The Life Story of a Technology.* Baltimore: The Johns Hopkins University Press, 2009. Provides a readable and engaging overview of the relationship between transit systems as infrastructure and social issues in cities; also contains a time line of transit history.

Web Sites

American Public Transportation Association
http://www.publictransportation.org/news/transit-facts.asp

American Society of Civil Engineers
http://www.asce.org

Institute of Transportation Engineers
http://www.ite.org/councils/Education/index.asp

U.S. Department of Transportation
Intelligent Transportation Systems Joint Program Office
http://www.its.dot.gov/strategic_plan2010_2014/index.htm#two-a-5

See also: Bridge Design and Barodynamics; Civil Engineering; Engineering; Geoinformatics; Probability and Statistics.

TRIGONOMETRY

FIELDS OF STUDY

Algebra; geometry.

SUMMARY

Trigonometry is the study of triangles, namely the relationship between their three angles and three sides. It is concerned with both planar (two-dimensional) and spherical (three-dimensional) triangles. Any situation that involves angles or triangles can be seen as an application of trigonometry. For example, surveyors often measure one distance and two angles of a triangle as aids in marking boundaries. Trigonometry provides the other two distances and the other angle. In fields such as optics or acoustics, light and sound waves can be expressed in terms of sine, one of the six trigonometric functions.

KEY TERMS AND CONCEPTS

- **Cosecant:** Reciprocal of the sine value.
- **Cosine:** In a right triangle, the ratio of the length of the side adjacent to an acute angle (not the hypotenuse) and the length of the hypotenuse.
- **Cotangent:** Reciprocal of the tangent value.
- **Degree Measure:** System for measuring angles in which one degree is equal to $^1/_{360}$ of a circle.
- **Hypotenuse:** Longest side of a right triangle.
- **Plane Triangle:** Triangle whose three vertices lie in a plane.
- **Radian Measure:** System for measuring angles in which one radian is the measure of a central angle of a circle that intercepts an arc equal in length to the radius of the circle.
- **Secant:** Reciprocal of the cosine value.
- **Sine:** In a right triangle, the ratio of the length of the side opposite to an acute angle and the length of the hypotenuse.
- **Spherical Triangle:** Triangle whose three vertices are located on the surface of a sphere.
- **Tangent:** In a right triangle, the ratio of the length of the side opposite to an acute angle and the length of the adjacent side (not the hypotenuse).

Definition and Basic Principles

Trigonometry is the study of triangles, namely the relationships that exist between the measures of the three interior angles and the lengths of the three sides. To solve a triangle, one must determine the measures of each of its angles and the lengths of each of its sides. This is frequently done by means of the trigonometric ratios, especially the two primary ones, sine and cosine.

In Euclidean geometry, the sum of the measures of the interior angles of any plane triangle is 180 degrees. (However, for spherical triangles that total is always more than 180 degrees and is not always the same.) Thus for any plane triangle, if two angle measures are known, the third is immediately deduced. By means of this fact and others, principally the Pythagorean theorem, the law of sines, and the law of cosines, any triangle can be solved or proven not to exist as long as three of the six measurements of the triangle are known.

There are two exceptions. First, if only the three angle measures are known, the triangle cannot be solved, since similar triangles have the same angle measures but different side lengths. Second, if two sides and an angle are known, but the angle is not between the two known sides, this gives rise to the so-called ambiguous case. Two different triangles can satisfy the three known facts. In this event, both triangles are given as solutions, unless additional information rules out one of these possibilities.

Finally, the trigonometric relationships are commonly viewed not simply as ratios but as functions. They are then studied and used in calculus and higher mathematics, as any other function would be.

Background and History

The Egyptian Rhind papyrus dates to about 1650 B.C.E. It contains several problems regarding the slopes of pyramids. The solutions to these problems are equivalent to finding the cotangent ratios, although there is no evidence that the Egyptians ever thought in terms of angles. In the second century B.C.E., Hipparchus of Nicaea developed a table of chord lengths in a circle depending on the central angle. Bisecting the chord produced a triangle inside the circle; the chord values led directly to the sine ratio. Because his table was essentially the world's first table of trigonometric values, Hipparchus is often called the inventor of trigonometry.

The six trigonometric functions had all been studied by Arab scholars by the tenth century, most commonly in the service of astronomy. It was not until the fifteenth century that trigonometry was finally seen as an independent discipline within mathematics and not simply as a tool for solving astronomical problems. This came as a result of the publication of *De triangulis omnimodis* (wr. 1464, pb. 1533; *On Triangles*, 1967) by Viennese astronomer Regiomontanus (Johannes Müeller). The word "trigonometry" was first applied to the subject in 1595 by Bartholomäus Pitiscus. Soon after, trigonometry was subjected to new methods of analysis and took its modern form.

How It Works

Triangles can be constructed either on planes or on spheres. In addition, the six trigonometric functions are used apart from any direct reference to triangles.

Spherical Trigonometry. A spherical triangle has its vertices on the surface of a sphere. The total measure of its three interior angles is always greater than 180 degrees but less than 540 degrees; the exact amount depends on the particular triangle. The vertices of the triangle are connected by three great arcs, where a great arc is a portion of a great circle, the circle whose circumference is the largest possible on that sphere. Like the line segment, its counterpart in two dimensions, a great arc is the shortest distance between two points on a sphere.

In the system of radian measure, lengths of arcs correspond directly to the measures of angles. Thus, in a spherical triangle whose angles are represented by A, B, and C, and whose corresponding opposite sides are represented by a, b, and c, one may din relationships among the angles and sides by thinking always in terms of angles. For example, $(\sin a)/(\sin A) = (\sin b)/(\sin B) = (\sin c)/(\sin C)$, and $\cos a = (\cos b)(\cos c)+(\sin b)(\sin c)(\cos A)$. These equations (or related forms of the cosine rule) can be used to solve spherical triangles.

Plane Trigonometry. A plane triangle is two-dimensional. Its three vertices are connected by line segments and the sum of the measures of its three interior angles is 180 degrees.

The six trigonometric ratios apply to right triangles, that is, triangles containing a 90 degree angle. In this case, the famous Pythagorean theorem also applies, namely that the sum of the squares of the two legs of the triangle equals the square of the hypotenuse. If any two sides or any two angles are known, the third can then be deduced. If all sides, or all angles and one side, are known, the trigonometric ratios are used to find any unknown values.

By extension, these ratios can be used to analyze any triangle, because a perpendicular line segment can be dropped from any vertex to a base of the triangle, thus introducing one or two right triangles into the original diagram. These oblique triangles are usually solved by means of two identities, or laws. The law of sines states that $(\sin A)/a = (\sin B)/b = (\sin C)/c$, where the notation is the same as in the case of spherical triangles. The law of cosines asserts that $c^2 = a^2+b^2-2ab(\cos C)$. Similar forms of this law involve the cosine of angles A and B.

Trigonometric Functions. By the seventeenth century, the trigonometric ratios were considered as functions whose input was an angle expressed in radian measure. As functions, they can be graphed. All six graphs are periodic, meaning that they repeat themselves indefinitely. The sine and cosine functions have the so-called sine wave or sinusoid for their graphs. The other four functions have vertical asymptotes that divide the graphs into infinitely many pieces.

In calculus, derivative and antiderivative relationships among the trigonometric functions are established. Of primary importance are the ones involving sine and cosine: Cosine is the derivative function of sine, while the negative sine is the derivative function of cosine.

Applications and Products

Navigation. Historically navigators would measure angles to landmarks on the Earth or in the sky and consult tables with known positions or distances to evaluate the position of the vessel. Beginning in the mid-twentieth century, the technology that formed the basis of the Global Positioning System (GPS) was developed. The United States deploys between twenty-four and thirty-two satellites that continually broadcast signals to Earth, including the precise time

Fascinating Facts About Trigonometry

- One prominent scientist of antiquity is Eratosthenes of Cyrene, who lived about 200 B.C.E. He identified the tilt of the Earth's axis with respect to its plane of orbit as about 23.5 degrees, a number still considered accurate. He also suggested adding a leap day to the calendar every four years.
- Eratosthenes noticed that the Sun's rays seemed to be directly overhead in Syene, Egypt, at noon on the summer solstice, while at the same time on that day, in Alexandria, Egypt, the rays were about 7 degrees from the vertical. Eratosthenes used this information to estimate the circumference of the spherical Earth to be about 29,000 miles. The correct figure is closer to 25,000 miles.
- Trigonometry was considered simply a branch of astronomy from ancient times until the fifteenth century, and spherical trigonometry was the important area of study. Only later was plane trigonometry regarded as the focus.
- Every point on a plane has coordinates that can be expressed in terms of the cosine and sine of angle. This concept is connected to Sir Isaac Newton's definition of the polar coordinate system. Newton, though, was not the first to adopt this idea. Honeybees use essentially the same system to send other bees to sources of nectar.
- It is impossible to map a sphere onto a flat surface without distortion. Greenland is actually much smaller than South America, despite its appearance to the contrary on many maps. Cartographers have developed various projections, using trigonometric functions, that maintain accuracy in maps of the Earth in at least some respects, depending on the intended application.
- In 1852, Sir George Everest and his survey team used trigonometry to measure the height of Peak XV in the Himalayas from a distance of 100 miles. They estimated the peak at exactly 29,000 feet above sea level but arbitrarily added 2 feet for fear that others would think it was an estimate. The height of Peak XV, now known as Mount Everest, is 29,028 feet.

at which the signal was sent. A position on Earth can be thought of as a function of four variables: latitude, longitude, altitude, and time. If signals are received from four different satellites, the receiver is programmed to solve a system of equations and return the value of the four variables. The solution often involves the method of trilateration, in which a coordinate system is rotated through two different angles and the new coordinates are expressed in terms of sines and cosines of the angles.

Trigonometry is also useful when a navigator is attempting to steer a vessel on the sea or in the air in the presence of wind or sea currents. First, the navigator begins with the desired path, both its angular direction and its velocity. This becomes one side of a triangle. At the end of that side, the navigator attaches a second side representing the current, according to its direction. The lengths of these sides represent the velocities. Connecting the ends of these segments with a third side produces a triangle in which two side lengths and one angle are known. The length of the third side represents the velocity of the vessel after the effect of the current has been accounted for, a value called the ground speed. Its angular position determines the true course of the vessel. The ground speed is typically discovered by the law of cosines and the true course by the law of sines.

Surveying. Triangulation has been used for surveying since the sixteenth century. The surveyor marks two positions and precisely measures the distance between them. Angles are measured between each of these two positions and a third landmark. The surveyor now knows two angles of the triangle and the length of one of the sides. The remaining angle is then computed, while the lengths of the other two sides are determined by means of the law of sines. The surveyor uses one side of this triangle as a basis and repeats the process using a fourth landmark to complete a second triangle. By repeating this process, the surveyor is able to cover the region in question with a series of triangles, all of whose measurements are known. These measurements are typically used to determine property boundaries.

Harmonic Motion. Simple harmonic motion is one in which an object repeatedly cycles through the same set of positions. Examples (ignoring friction) include the vibration of a string, the motion of a pendulum, and the voltage of household electricity. All of these can be represented by the equation $y = a \sin(bt)$, where t represents the input (usually in units of time), a represents the amplitude of the motion, and b is a factor derived from the frequency of the motion. Knowing this formula, specific types of harmonic motion can be attained by solving for the variables a and b.

Force. Forces are represented by vectors because they have both magnitude and direction. Trigonometry can be used to solve for forces by considering the lengths of the vectors to be the magnitude of the associated force and the angles to come from the direction in which the forces are being applied. One example is finding the tensions in two wires that are holding an object. The weight of the object is itself a force and can be represented as a vector whose direction is vertical (meaning, pointing toward the center of the earth). The typical solution procedure is to break a vector into its horizontal and vertical components, so that the vector is now the hypotenuse of a right triangle whose legs are these components. Because the total tension in the wires must equal the opposite of the weight vector (so that the object is actually suspended in the air), a set of equations is obtained in the horizontal and vertical directions. The sine and cosine functions are always involved in these solutions, the results of which are needed by the engineers, who must ensure that the structure they are building will support the weight as required.

Impact on Industry

In research or industrial endeavors, trigonometry typically plays a subservient role. It is a tool used to achieve a desired end and not the focus of the research. However, trigonometry plays multiple roles in research and industry.

Government and University Research. A prime example of government research connected with trigonometry is the area of geodesy. This field concerns accurate positioning on the Earth: on land, on sea, or even in the atmosphere. It can be a difficult problem because the Earth is not truly spherical. Calculations must account for its equatorial bulge. Trigonometry is used to provide triangulation of a region; in particular, spherical trigonometry is employed in determining three-dimensional coordinates. Agencies that conduct research in this area include the U.S. Geological Survey, part of the U.S. Department of the Interior, and the National Geodetic Survey, a branch of the National Oceanic and Atmospheric Administration.

Another agency making use of trigonometry is the National Aeronautics and Space Administration (NASA). In addition to using trigonometry for navigational needs, NASA continues to investigate the shape of the universe using the Wilkinson Microwave Anisotropy Probe (WMAP). The idea is to measure very large triangles in the universe. If the total angle measure is 180 degrees, the universe is flat. If it is more than 180 degrees, the universe is spherical, and if it is less than 180 degrees, the universe is shaped like a horse's saddle.

Industry and Business Sectors. Industry conducts research that makes use of trigonometry and not research into trigonometry itself.

In robotics, machines are designed to accomplish a wide variety of tasks. Trigonometry is used to help plot the path that the robot must take. One example of the usefulness of this technology is in the area of physical rehabilitation. Patients in weakened conditions often need to undergo many weeks of rehabilitation. However, there can be a large number of patients and a relatively small number of therapists. Robots are designed to manipulate the body in ways appropriate for the needs of the patient so that all patients can be served in a timely manner.

Laser scanning systems use trigonometry to derive the coordinates needed to pinpoint a position. These technologies can be used to create highly accurate digital models of three-dimensional objects. Analysis can then be performed on the digital model.

Lock-in amplifiers (LIA) use trigonometry to detect and measure small alternating current (AC) signals. They are becoming increasingly common and are most helpful in the presence of much noise. When the frequency of the desired signal is known, a sinusoidal wave at that frequency can be given as input to the LIA. The LIA then locks in this frequency and excludes all others so that the desired AC signal can be received while other signals are rejected.

A Fourier series is used to express a periodic function as a combination of sines and cosines. It is related to the Fourier transform and the discrete Fourier transform. Among many uses of this procedure is file compression, such as that used in

JPEG (Joint Photographic Experts Group) images. Important research in forensic science is centered on ways of detecting image tampering. One avenue of investigation is to study the Fourier transforms of images and manipulated images so that disturbances in the original image can be more easily observed.

Careers and Course Work

Trigonometry is a branch of elementary mathematics. As such, many jobs for which trigonometry is important will also require higher levels of mathematics or physics. An interested college student will go beyond trigonometry and major in mathematics, physics, or engineering. Many of these students will go on to earn a master's or doctoral degree before beginning their careers. Those who work as mathematicians take courses in calculus, differential equations, probability and statistics, linear and abstract algebra, analysis, and other courses. They will teach, work as researchers in pure or applied mathematics, or apply mathematics to solve problems in government or industry. Most physicists, in addition to taking many courses in mathematics, will also study fields, waves, heat, mechanics, quantum physics, and electricity and magnetism. Most physicists work in research or in developing new technologies.

Surveyors will typically take classes in drafting and mechanical drawing in addition to trigonometry. They must pass exams to become licensed. They may work as geodetic surveyors, working on relatively large areas on the surface of the Earth, or perhaps as marine surveyors whose interest is a body of water. They also serve as cartographers and work in other fields.

Other professions making use of trigonometry include machinists, sheet metal workers, and forensic scientists. Most of these professions are projected to have good to above-average job growth, with the exception of machinists, where the number of jobs is expected to decline slowly.

Social Context and Future Prospects

The importance of trigonometry is that it lies at the foundation of the work of mathematics, engineering, and physics. For example, through their experimentation, physicists learn the principles at work in the universe, from the astronomical to the subatomic scales. One example of this is the high-profile experimentation being done using particle accelerators. Physicists hope to use these to recreate conditions that existed immediately following the big bang. They expect to gain insight into the forces and energies that first shaped the universe and continue to do so.

In itself, trigonometry is socially neutral. It does not generate difficult social or ethical dilemmas. For this reason, there are no controversies concerning trigonometry or its applications.

Michael J. Caulfield, B.S., M.S., Ph.D.

Further Reading

Klintberg, Bo. "Hipparchus' 3600'-Based Chord Table and Its Place in the History of Ancient Greek and Indian Trigonometry." *Indian Journal of History of Science* 40, no. 2 (2005): 169-203. The article corrects the mistaken conclusion of a 1973 paper and shows that Indian mathematicians may have produced sine tables without referencing the work of Hipparchus.

Maor, Eli. *Trigonometric Delights.* Princeton, N.J.: Princeton University Press, 1998. Maor isolates the most interesting developments and characters found in the history of trigonometry and shares those stories.

McKeague, Charles P., and Mark D. Turner. *Trigonometry.* Belmont, Calif.: Thomson Brooks/Cole, 2008. Introduces readers to all the topics covered in an introductory trigonometry course.

Sultan, Alan. "CORDIC: How Hand Calculators Calculate." *The College Mathematics Journal* 40, no. 2 (March, 2009): 87-92. Sultan explains how an ingenious use of trigonometric rotation allows calculators to compute very quickly.

Van Brummelen, Glen. *The Mathematics of Heaven and Earth: The Early History of Trigonometry.* Princeton, N.J.: Princeton University Press, 2009. Presents a comprehensive review of the history of trigonometry, from its early days as a tool for astronomers through the heliocentric theory of Copernicus in the sixteenth century.

Web Sites

American Association for Geodetic Surveying
http://www.aagsmo.org

American Congress on Surveying and Mapping
http://www.acsm.net

American Mathematical Society
http://e-math.ams.org

National Aeronautics and Space Administration
Wilkinson Microwave Anisotropy Probe
http://map.gsfc.nasa.gov

National Oceanic and Atmospheric Administration
National Geodetic Survey
http://www.ngs.noaa.gov

U.S. Geological Survey
http://www.usgs.gov

See also: Applied Mathematics; Calculus; Geoinformatics; Geometry.

TYPOGRAPHY

FIELDS OF STUDY

Graphology; cartography; computer science; computer graphics; software engineering; graphic design; industrial design; ergonomics; information science; information architecture; psychology; social psychology; linguistics; neurobiology; psychophysics; social neuroscience; anthropology; data visualization; mathematics; human factors; anthropometrics.

SUMMARY

Typography is the art of designing proportionally spaced lettering arranged in the form of words, sentences, and blocks of text—also known as type. The central aim of typography is to create legible, readable, and aesthetically appealing textual material that clearly communicates its central message and is designed appropriately for its audience. The typographer is concerned, for instance, with issues having to do with the shape, size, and spacing of individual letterforms, as well as with the character and effect of different typefaces. The applications of typography are ubiquitous in daily life and can be found in every form of written communication, from books, newspapers, and advertising billboards to street signs and e-mail messages.

KEY TERMS AND CONCEPTS

- **Character:** Basic typographic element.
- **Family:** Group of related but slightly varied fonts.
- **Font:** Complete set of characters (including upper and lowercase letter forms, numbers, symbols, and punctuation) sharing the same basic size and style; also known as typeface.
- **Font Format:** Standard specification for how fonts are designed. The three major font formats used in digital type are PostScript, TrueType, and OpenType.
- **Justified:** Text whose lines are all the same length; achieved by small adjustments to the spaces between words; its opposite is unjustified.
- **Kerning:** Adjustments made to the space between characters within words.
- **Leading:** Space between lines of text.
- **Ligature:** Character that consists of two or more letterforms joined together.
- **Majascules:** Uppercase or capital letterforms; minuscule letterforms are lowercase.
- **Pica:** Unit of measurement for typefaces; one pica is 12 points, or about $^1/_6$ of an inch.
- **Roman:** Upright rather than italic or cursive letterforms.
- **Serif:** Slight projection at the end of the vertical stroke of a letterform; styles without such projections are called sans serif.
- **Weight:** Thickness of the vertical lines, or stems, in individual letterforms.
- **X-Height:** Height of the main body of lowercase letterforms in a given font.

Definition and Basic Principles

Typography is the art and practice of designing, selecting, and placing letterforms into words, sentences, and blocks of type so that they may either be printed or electronically displayed in a digital format, such as a Web page. Typography is a functional art, meaning that if practiced well, it does not call attention to itself. Instead, it is a means of facilitating the expression of an idea or ideas. For example, in environmental print such as a sign indicating a traffic rule about where to turn, the task of ensuring that the rule itself is clearly conveyed is obviously more important than leading the viewer to admire the font in which it is displayed. Nevertheless, whether a written communication is attractive and easy to read are important factors in determining how well its message is understood and whether a viewer will choose to continue reading a given piece of writing through to the end.

The two fundamental principles of typography—the basic qualities that a typographer is always striving to achieve—are readability and legibility. Readability can be defined as how easy a block of text is to read, while legibility can be defined as how quickly and easily individual words or phrases can be recognized by the reader. Myriad factors influence how easy a block of text is to read and how recognizable words and letters are. Among these are x-height, weight, font size, leading, kerning, margin size, the color of type, the style of the lettering (italics, bold, and so on), and the choice of font or typeface, including whether a serif or sans serif font is used. In addition, factors having to do with the physiology of human vision—such as whether a reader has impaired eyesight or is colorblind—can affect readability and legibility.

Background and History

Typography has its origins in a very specific technological innovation, and therefore, its history, unlike that of many fields of science, has a definite start date. The birth of typography was 1450, the year in or around the invention of the first printing press by German printer Johann Gutenberg. Gutenberg's press depended on a technology called movable type—a method of printing in which each letter was cast on a separate piece of metal and could be moved around to arrange, or set, words and sentences. Gutenberg used his movable-type printing press to create the first printed book, the Gutenberg Bible.

Less than a hundred years after the invention of the printing press, printed books were reaching large numbers of readers. Specialized presses had also been developed to enable the printing of music scores. As a result, before long, the efficiency of the printing press rendered the art of calligraphy obsolete in most of Western Europe, including Germany, Italy, France, and England. Still, the roots of typography are clear: The three families of type that came to dominate Western printing (gothic, roman, and italic) were all adapted from the writing styles of calligraphers.

In the nineteenth century, mechanical type—in which blocks of text for printing could be produced by manipulating an early version of a keyboard rather than by individually setting each letter of type—made the process of printing even faster and more practical. Later, photo typesetting transformed printing yet again. This technology involved projecting characters onto photographic film using large rotating drums.

Undoubtedly, the great advancement in typography that took place during the twentieth century was the development of digital type. With this technology, characters do not have a physical form but are stored as bitmaps (maps of colored and blank pixels) or converted into mathematical formulae that describe how the lines and curves of each figure should be displayed (known as outline or vector fonts). Digital type has made desktop publishing—producing electronic versions of typographic documents such as newspapers, posters, and even entire books—the easiest and most inexpensive way to distribute written communications.

How It Works

Typeface Anatomy. The standard typographic features associated with any given typeface are known collectively as anatomy (because the names of body parts are often used to describe the parts of letters). Typeface anatomy is useful because it enables typographers to break down a character into its individual parts to analyze, manipulate, or modify each component in isolation. The terms baseline, meanline, x-height, and cap height are all used to help define the range of physical space occupied by characters. The baseline is an imaginary line denoting where characters seem to sit. A few letters, such as *g* and *q*, have parts that extend far below the baseline. The meanline is the invisible line that rests just on top of most lowercase letters, such as the tips of the arms in a lowercase *y*. It is important for the baseline and meanline to have a consistent relationship to the letters within a font. The x-height is the distance between the baseline and the meanline and corresponds to the height of the lowercase letter *x*. The cap height is the distance between the baseline and the tops of uppercase letters such as *H*.

Many other elements of typeface anatomy pertain to the characters themselves. Ascenders and descenders, for instance, are any letter or symbol parts that extend above or below the meanline or baseline. The stem of a letter is its main body; bars are any horizontal or diagonal lines that are open on at least one side, such as the three parallel lines in an uppercase *E*. If a bar connects two lines, as in the middle stroke of an uppercase *H*, it is known as a crossbar. If it seems to serve as a base for it, as in the bottom stroke of an uppercase *L*, it is known as a leg. Bowls are circular lines that form a defined space, such as in the letters *b* and *d*. The inside of a bowl is known as a counter.

Ultimately, the anatomy of a typeface has a profound influence and effect on the reader, including how readable and legible the text produced is. For example, typefaces with prominent ascenders and descenders and large, open bowls and counters, as well as serif fonts (fonts in which letter strokes are finished with a slight projection), are generally considered to be more readable than others. In addition, a typographer designing a font to be as legible as possible must attend to the frequent confusion that arises between certain characters, including *i* and *j* and the letter *o* and the number *0*.

Page Layout. Besides the qualities of the typeface

itself, the way in which words and sentences are arranged on the page is an important component of typographical design. Some conventions that seem intuitive were in fact the result of careful thought on the part of typographers. For instance, the widespread use of the indent to indicate a new paragraph dates to the seventeenth century. Earlier, various other techniques had been employed to mark a new paragraph, including the placement of the first letter of the new paragraph in the margin rather than in the main body of the text.

Typographers are greatly concerned with using white space to proper effect. For example, when laying out a page of text, it is important not to use leading (the vertical spaces between lines of text) that is too wide, because this can make it harder for the reader's eye to travel from one line to the next. However, leading that is too narrow can make text less readable, as the ascenders in one line may become confused with the descenders of the previous line.

Kerning is another important aspect of spacing and layout. To kern is to selectively adjust the amount of space between particular letters or words in a block of type to achieve a desired effect. For example, a typographer may decrease the spaces between letters as the size of type on a page increases in order to keep the overall shape of the words uniform. Increasing the spaces between words can also help slow the reader down as he or she moves through the text.

Digital Type. Digital type is the creation of electronic documents and other pieces of text (as opposed to the traditional production of print in physical form). Digitized fonts can be stored in two ways: One is as bitmaps, or rasters. These are coded maps that deliver instructions to a computer program, telling it to turn on a specific set of pixels on the screen to form a particular character. The major drawback of bitmap fonts is that when they are scaled up, or displayed in a larger size, the jagged edges of the individual pixels become more and more evident, making typefaces look uneven rather than smooth. Most digital type is produced using outline, or vector, fonts. These are sets of data in mathematical form that deliver instructions to a computer program, describing particular characters as points on an outline. Because the lines between the points are not fixed, vector fonts can be scaled up or down without causing distortions.

APPLICATIONS AND PRODUCTS

Print Publishing. Print publishing—the production of materials such as newspapers, magazines, and books—would not be possible without the effective use of typography. For example, the text found in newspapers and magazines is commonly set in a series of justified vertical columns, a layout that allows publishers to squeeze in as many words of copy as possible without the page seeming cramped or cluttered. Because text in a newspaper or magazine serves so many purposes, parts of the copy are formatted in different ways so as to draw attention to their functions. Headlines are set in larger, bold type, for instance, while captions may be set in smaller type or italicized. The demands of book publishing are slightly different. A reader generally holds a book at arm's length, looking at it straight on without moving his or her head from side to side. This enables the typographer to space letters more closely together without compromising readability.

Desktop Publishing. Copy designed to be read on the upright, brightly backlit, sometimes rather small screen of a computer requires different typographical techniques than that designed to be printed on paper. For example, one difference between the two formats is the relatively low resolution of most contemporary computer screens—somewhere between 72 and 300 pixels per inch—compared with the resolution of printed materials—somewhere around 2,500 dots per inch. This constraint led typographers to the realization that although serif fonts are easier to read on the printed page, sans serif fonts are much easier to read on a screen.

Advertising. Typography plays a pivotal role in every form of advertising, from billboards to subway posters to brochures, logos, and television tag lines. The way a designer chooses to handle the type, or copy, in an advertisement can have an enormous impact on the bottom line of the business placing the advertisement. Far more than in a book or a newspaper (where the content is usually more important to consumers than they way it is presented), in an advertisement, typographical choices can either drive away or draw in a viewer. One of the biggest typographical decisions a company can make is how it presents its brand name in a logo. If done successfully and used consistently, the color, weight, spacing, and font used to display a brand name can become so recognizable that the typeface itself, even if seen in a different context, is enough of a signature to create instant brand recognition on the part of the viewer.

For example, the connection consumers made between the brand identity of Swedish furniture giant Ikea and the typeface the company used (a customized version of the sans serif font Futura) was so powerful that in 2009, when Ikea changed the typeface it used on posters and in its catalog to Verdana, a font with a taller x-height and more rounded ascenders, many consumers complained vigorously.

Signage. Way-finding signage, which is any form of visual communication intended to direct people to where they need to go, is one of the most functional applications of typography. In this context, unlike in advertising, the message conveyed by the sign rather than the mood or emotion suggested by the chosen typeface is obviously of paramount importance. Still, typographical choices are just as crucial. The words on a way-finding sign need to be absolutely readable, and if possible, they should be legible to the viewer as quickly as possible, because many such signs are read by someone passing by on foot or in a vehicle. Typography was used to improve the effectiveness of signage when the U.S. Federal Highway Administration (FHA) undertook a redesign of highway road signs. The old typeface, known as E-Modified, was found to be too difficult to comprehend, especially when upper and lowercase letters were mixed and when viewed by older drivers with poor eyesight. The FHA hired a design firm to create an entirely new typeface, one that had letters with about the same weight but with interior shapes—bowls and counters—that were larger and more open. In addition, lowercase letters in the new font, known as Clearview, were heavier, distinguishing them more clearly from uppercase letters. In 2004, several states began replacing their old signs with Clearview signs. Studies of the readability and legibility of the signs have found that drivers can determine their meanings from up to 200 feet away.

Fascinating Facts About Typography

- Although many early printing technologies have become obsolete, the vocabulary associated with them has survived into the digital age. Movable Type, a popular Web blog publishing platform founded in 2001, is named after the very first printing press technology.
- Typographers often say that "good typography is invisible," but the feature-length documentary film *Helvetica* (2007), which follows the history and cultural impact of an iconic modern font, brought the field squarely into the public eye.
- Amazing things can happen when typography meets art. One trend involves artists using intricate typographic designs to create portraits. These artists create shapes and faces by using letters and words instead of drawing lines.
- The most frequently used font worldwide is Times New Roman, originally designed for the British newspaper *The Times* and later picked up by Microsoft as its default font in the word-processing program Microsoft Word.
- Urban graffiti is an informal application of typography. Many street artists become known for the distinctive fonts they use in their lettering, or tags.
- Typography is sometimes used as a component of visual design in printed poetry. The twentieth-century American poet E. E. Cummings made frequent use of typographic elements such as odd spacing between words and lines and unexpected punctuation marks.
- In typographic jargon, a "widow" is a line that ends a paragraph but appears by itself at the beginning of the next page or column. An "orphan" is a word, part of a word, or very short line that appears by itself at the end of a paragraph.

Impact on Industry

Industry and Business. Besides the publishing and advertising industries (in both print and online media), whose primary business functions are intimately connected with the world of typography, virtually every business sector that makes use of written communication is affected in some way by typographical technologies and trends or seeks to make use of typography to further its goals. Every public relations department, for example, makes careful choices about the typefaces it uses to represent the person, product, or service it is trying to market. The Gotham font used by Barack Obama's presidential campaign team in 2008 on all its posters and other graphics is a good example. The typeface's clean, modern lines helped reinforce the candidate's message of fresh ideas and change. Another industry that has come to rely on sound principles of digital typography is the personal electronic device sector, companies that produce products such as smartphones and personal music players. Because the screens on such devices are so small, clever typographical techniques are required to make the text they display readable.

Major Corporations. In the age of digital type, software companies are by far the biggest producers of typographical tools and applications. Some of the biggest commercial font companies (known as foundries just like the metal-casting factories that make physical type) include Adobe, Linotype (part of the Monotype corporation), Bitstream, and Hoefler and Frere-Jones. Computer technology companies such as Apple and Microsoft are among the biggest licensers of the font families produced by these and other foundries.

Careers and Course Work

In the past, typography was a specialized expertise on which a professional could base his or her entire career. It has become one of a suite of skills associated with a career in fields such as graphic design, advertising, communications, packaging design, and software design. Preparation for such a career should begin with any courses that are available in computer science at the high school level, such as computer graphics, software engineering, and computer-aided-design. Traditional fine arts courses in drawing, sketching, printmaking, and the graphic arts are also appropriate. A number of schools of the arts offer typography certificate, diploma, or associate degree programs; however, students who intend to complete a bachelor's degree should consider concentrating in areas such as computer science, computer graphics, graphic design, or visual communications. Perhaps the most important skills required for any professional working with typography are those that pertain to specific pieces of software commonly used in the industry. These include Adobe products such as InCopy and InDesign and the desktop publishing software XML Professional Publisher.

Social Context and Future Prospects

Typography is a powerful instrument of communication. As such, it has the potential to influence the beliefs, behaviors, and ideas of its audience in both positive and negative ways. Choices about fonts, spacing, and layout on a poster urging residents of a city to use less water, for instance, could have an important impact on the effectiveness of an environmental intervention. By the same token, however, smart typographical techniques used to create propaganda materials or pamphlets containing hate speech could help spread incendiary ideas more effectively. Every typographical artifact, from an e-mail exchange between friends to a billboard by the side of the highway—no matter the message—creates a medium of expression that conveys far more than just the meanings of the words it contains. With the advent of desktop publishing and software tools that enable anyone to design a customized font and publish his or her own writings, typography is swiftly becoming a democratized field that is virtually without barriers to entry.

M. Lee, B.A., M.A.

Further Reading

Baines, Phil, and Andrew Haslam. *Type and Typography.* 2d ed. London: Laurence King, 2005. A basic textbook that covers the history of typography and addresses issues of form, function, structure, and common conventions. Includes a time line of type designers from 1400 to the 2000's.

Bringhurst, Robert. *The Elements of Typographic Style.* 3d ed. Point Roberts, Wash.: Hartley and Marks, 2005. A guide to choosing type, using elements such as subheads and tables, adjusting legibility using kerning, and other issues of typographical design. Includes a glossary, a list of unusual characters, and a guide to further reading.

Graham, Lisa. *Basics of Design: Layout and Typography for Beginners.* 2d ed. Clifton Park, N.Y.: Thomson/Delmar Learning, 2005. An approachable, heavily illustrated handbook for typographical design. Uses before and after images to demonstrate techniques and walks the reader through a series of common design projects, such as business cards and graphical reports.

Jury, David. *What Is Typography?* Hove, England: Rotovision, 2006. This practical guide is full of photographs and is especially strong on the anatomy of typography. The "Portfolio" section includes examples of the work of sixteen different designers or design firms.

Web Sites

Association Typographique Internationale
http://www.atypi.org

The Society of Typographic Aficianados
http://www.typesociety.org\

See also: Communication; Computer Graphics; Computer Science; Internet and Web Engineering.

URBAN PLANNING AND ENGINEERING

FIELDS OF STUDY

Urban design; land use; transportation design; environmental policy; community and economic development; geographic information systems.

SUMMARY

Urban planning and engineering is an interdisciplinary field focusing on the study of land within cities, towns, and metropolitan areas and the ways in which it is used. Urban planners conduct thorough assessments of specific areas of land, often with the support of statistical tools and mapping software such as geographic information systems. They analyze historical trends affecting the land's physical, economic, and social environments. Based on these findings, planners make recommendations about ways in which land can be used more effectively in the future. Planners advise lawmakers, government agencies, commercial developers, businesses, and residents. The rapid growth of urban areas worldwide is fueling a high demand for professionals in the field.

KEY TERMS AND CONCEPTS

- **Brownfield:** Vacant land contaminated by previous use.
- **Geographic Information System (GIS):** System that integrates and analyzes data to create detailed maps.
- **Green Building:** Construction using materials from environmentally sustainable sources and minimizing energy waste.
- **Infill:** Building on vacant land between areas already developed.
- **Infrastructure:** Urban features that support growth, such as roads, railroads, bridges, and sewers.
- **Mixed-Use Construction:** Construction blending retail and residential use.
- **New Urbanism:** Land use that places housing close to workplaces, retail, and transit.
- **Urban Sprawl:** Low-density land use.
- **Zoning:** Laws defining and restricting the ways land can be used.

Definition and Basic Principles

Urban planning and engineering is the design of cities and towns according to a plan. Such plans provide guidance on questions ranging from where to place new roads to how many supermarkets are needed in a neighborhood. Urban planning, also known as urban studies or city planning, ensures that the land in a city or town will be used in a way that will benefit the largest number of people living and working there. It helps officials in local government and land developers avoid problems such as traffic, overcrowding, and urban sprawl.

Geography is a major factor affecting a city's urban planning efforts. The types of businesses that make up the local economy are another factor. Cities that rely on heavy industry such as oil refining, for example, have different urban planning needs than those for which technology or agriculture is more important. The topic of urban planning is often associated with newer high-growth cities where vacant land is being developed for the first time. Older urban areas also must depend on planning when renewal is needed.

Urban engineering is a specific field within urban planning and is more commonly known as civil engineering or municipal engineering. It includes the design of transportation and its infrastructure.

Background and History

Urban planning and engineering in North America began with the development of the first large cities on the East Coast. These cities started as small settlements. As they grew, particularly in the seventeenth and eighteenth centuries, their roads and buildings were influenced by the design of European cities.

The use of a grid-based system for laying out streets and neighborhoods became popular in the 1800's, led by innovations in cities such as New York and Philadelphia. As railroads crossed the continent, many cities followed standard plans developed by the railroad companies. The Industrial Revolution led many cities to place the needs of manufacturing operations over those of citizens, creating pollution, overcrowding, and slums.

In the twentieth century, the U.S. government

passed several major laws to guide the development of cities and towns. The U.S. Department of Housing and Urban Development, created in 1965, addressed common problems of urban growth on a national level. The rise of suburbs, influenced by the spread of automobiles and highway systems, created new challenges for city planners. As metropolitan areas have sought ways to fight urban sprawl and to redevelop land left underutilized by a general movement toward the suburbs, the demand has grown for urban planners.

How It Works

Land-Use Study. When developing a new plan for a city or region, urban planners start by assessing the way in which the land is being used. Planners identify who is using the land and how. In urban areas, land use is often classified into business use (such as retail or manufacturing) and residential use (such as houses and parks). Even if land is vacant, it is likely to have been business or residential property at some point in the city's history. Once planners have defined who the land's users are, they create a detailed report and a map. The report and the map are known as the land-use study. The report contains information on the types of businesses in the area, the population living there, and the ways in which both have changed over time. The map covers geographic features such as mountains and lakes, but its primary focus is on the features added by businesses and people using the land. It includes information about transportation infrastructure, utilities such as power and sewer lines, and the location and density of public resources such as fire stations, hospitals, and libraries.

Urban Plan Development. Once the land-use study is complete, urban planners evaluate the ways in which the land's present uses meet or fail to meet the needs of its businesses and inhabitants. Most urban planners are hired by officials from city and regional governments. These officials are likely to bring in urban planners to help solve problems such as traffic congestion or economic stagnation. Because of this, an urban plan often focuses on the ways in which land use could be improved. Urban planners apply their professional knowledge about the best ways in which an area of a city can be designed. Planners also spend time with business owners and residents to learn more about their specific needs. Planners consult with government officials on budget matters, as land-use changes are often financed by tax revenues. Finally, urban planners research zoning and building codes and environmental regulations. Planners must ensure that any recommended changes to land use are legal and will receive all necessary approvals.

Conflict Management. Urban planners are most frequently hired to help resolve conflicts about land use. One of the most important roles played by many urban planners is that of mediator. Planners, as outside parties, gather information from all sides and incorporate this information into their recommended plans. This process can involve participating in community meetings and defending findings in front of businesses, residents, and government committees that may be critical of the planners' advice.

Technology. To create plans, urban planners rely on tools such as mapping software and geographic information systems. These systems help planners assess existing land use and forecast the impact of recommended changes. Planners must also be skilled in the use of databases on population statistics and trends, laws and regulations, and geographic and environmental data.

Applications and Products

Most professionals in the urban planning field have an area of specialty. Some of the most common areas are residential planning, transportation engineering and design, business and economic planning, historic preservation, environmental planning, and zoning and code enforcement.

Residential Planning. Planners who focus on residential issues answer questions such as whether a community offers enough housing to meet its existing needs and how these needs will change in the next ten, twenty, or fifty years. They make recommendations that guide the building of single-family homes, townhouses, and apartment buildings within a community. Residential planners also study community income levels and develop plans for housing that will be affordable as well as comfortable. A subfield within residential planning is public housing. Many cities have moved away from public housing models in which residents were concentrated within single, large-scale communities, where problems such as crime and economic blight developed. Residential planners in this subfield are creating new approaches to public housing that improve the quality of life for all community members.

Transportation Engineering and Design. If a main road is choked with traffic, does the road need to be rebuilt or is a new highway needed? What is the best route for a new train line between the suburbs and downtown? Transportation engineers help cities find the best ways for people and commercial goods to get from one point to another. They evaluate whether existing structures are working and develop plans for the building of new roads, bridges, tunnels, rail lines, and public transit systems. Transportation engineers also measure the impact that the building of new structures will have on communities.

Business and Economic Planning. Planners who focus on business and economic issues work in one of the most diverse areas of urban planning. They address questions such as how a city that depended on heavy manufacturing can adapt when companies close their factories and whether a neighborhood should build a large shopping mall or several small groups of stores. Some planners specialize in guiding the long-term economic health of a community. Others specialize in the design, creation, and refurbishing of business and retail buildings. When economic planners succeed, communities can support a range of businesses that, in turn, create jobs. Some subfields of economic planning extend beyond the businesses themselves. For example, economic planners might recommend that a city build more schools to equip residents with new skills. The workforce created by these schools might attract industries to the area, which would boost economic growth in the long run.

Historic Preservation. Historic preservation tries to determine issues such as whether decrepit older houses that make up the bulk of homes in a neighborhood should be protected or torn down and what elements give a city's downtown its unique character. This field has been on the rise since the 1960's and 1970's, when residents and businesses of many cities began to migrate from the urban core to the newly built suburbs. Many urban planners in this field are involved with the creation of urban historic districts, neighborhoods that receive economic support from the city in exchange for preserving the architectural styles of their older buildings. Other specialists in historic preservation advise city governments on ways to reuse buildings that have historical value but have outlived their original purpose. Urban planners in this field help cities retain their distinct personalities while meeting the needs those who live and work in these cities.

Environmental Planning. What is the most cost-effective way to clean up a river full of industrial pollutants? How can a city encourage more of its residents to take public transit or bike to work? Environmental planners help urban areas address questions such as these. Like the movement toward historic preservation, the importance of environmental protection has been on the rise since the 1960's and 1970's. Businesses and city governments must comply with increasingly strict laws about the cleanliness of air, water, and land. There is also greater public interest in protecting the environment, which has led to changes such as the increase in citywide recycling programs. Green buildings, or those that use materials with a minimal environmental impact and seek to conserve as much energy as possible, have become a priority for many cities and are becoming a larger part of their urban plans. Planners who specialize in this field help cities benefit as much as possible from new approaches to environmental protection.

Zoning and Code Enforcement. What kinds of businesses should be built around a new sports stadium? How far from a power plant should homes be built? Zoning and code enforcement is a specialty within urban planning that focuses on the laws and regulations that govern the ways land can be used. It includes both the writing of new codes and the process of ensuring that governments, businesses, and residents comply with them. Zoning specialists protect an urban plan's long-term vision by defining the ways in which specific projects must follow the plan.

Impact on Industry

Urban planning and engineering is a field that is highly fragmented by country. The majority of professionals who identify themselves as urban planners and engineers are based in the United States. Information on the profession outside of a few leading countries is limited.

Professional Associations and Membership. The Urban Land Institute is one of the largest organizations of urban planners with a global geographic mandate. In 2009, the institute reported 35,000 members in sixteen countries. Roughly 90 percent of its members were based in the Americas, the vast majority of whom were in the United States and Canada.

Fascinating Facts About Urban Planning and Engineering

- Planned communities are cities built from an urban plan since their founding. Some of the oldest are located in the Indus Valley of what is now Pakistan. These cities, planned on street grids with sewage and drainage systems, date from before 2500 B.C.E.
- One of the oldest known planned communities in the United States is St. Augustine, Florida. It was founded in 1565 by colonists from Spain.
- Navi Mumbai, India, is the largest planned community in the world. Located next to Mumbai, it covers 344 square kilometers and has a population of about 2.6 million.
- Theme park developers often hire urban planners when new parks are being designed. Planners advise developers on matters such as site selection, park layout, transportation within the park, landscaping, and crowd control.
- Urban planning was a central component of the 1893 World's Columbian Exposition in Chicago. The White City designed by architect Daniel Burnham went on to influence the city beautiful movement in the United States and elsewhere.
- The design of urban supermarkets has begun to change in response to the new urbanism movement. Because shoppers are more likely to visit on foot than by car and to shop more frequently, many grocery stores in downtown areas have smaller parking lots and offer a wider range of fresh foods.
- In an editorial in the *Evening Post* in 1844, William Cullen Bryant first called for the building of a large public park in New York. Bryant's idea went on to become Central Park.
- The video game series *Sim City* teaches players about the importance of urban planning. A player is evaluated by a team of advisers, which includes a city planner and a transportation specialist, within the game.

The remaining members were divided between Europe, the Middle East, Africa, and India (8 percent) and Asia (2 percent).

The American Planning Association, the largest professional group of urban planners in the United States, has more than 43,000 members. Nearly all of its members work in the United States. A member survey conducted by the association in 2008 found that 67 percent of respondents worked for public agencies, primarily organizations within local governments. Private sector consulting firms employed another 25 percent, with the remaining 8 percent of respondents working in other settings. Most of the urban planners working in the private sector were employed by law firms and real estate developers.

Geographic Information Systems. The use of Geographic Information Systems (GIS) by urban planners is increasing. Planners use GIS technology to generate maps containing information drawn from environmental, topographic, social, and economic sources. These maps help planners document the ways in which land is being used. GIS tools are particularly useful in tracking the relationships among factors such as local geography and industry. With the assistance of GIS software, urban planners can forecast the effects of their proposed plans and ensure that their recommendations meet the needs of the greatest number of people.

One of the most significant uses of GIS worldwide is the 1,000 Cities Project. The project, a joint venture between the United Nations Human Settlements Programme and California-based software firm ESRI, was established in 2003 as one of the first uses of GIS for urban planning purposes on a global scale. ESRI has distributed copies of its GIS software to 1,000 applicants from cities in developing countries. The data collected by the software will be used to advance urban planning in these cities. Data gathered will also support the work of the Global Urban Observatory, a worldwide network of urban planning research centers that have pledged to help implement the United Nation's habitat agenda. The 1,000 Cities Project is unusual in that it brings together urban planners from around the globe and that it focuses on areas where urban planning efforts have lagged.

In the United States, one of the leading applications of GIS in the field of urban planning is the Urban Dynamics Research Program at the U.S. Geological Survey's Land Cover Institute. The program has gathered information on changes in land use over time in major American cities. Although the emphasis is on changes taking place in modern times, the program offers data for some cities going back as far as 1800. Information is gathered from historic maps, photographs, and satellite images and analyzed by the program's GIS software. The program's objectives are to support long-term urban planning that is environmentally sustainable and to allow cities to project the future impact of their existing land-use patterns. One of the program's benefits is that it allows urban planners and city governments to track whether antipollution measures placed on natural resources are meeting their goals.

Careers and Course Work

The magazine *U.S. News & World Report* named urban and regional planning one of its best careers of 2009. Although governmental agencies employ the largest number of urban planners and engineers, consulting firms in the private sector are increasingly hiring urban planners and engineers. Planners many increase their earning power and promotion opportunities by developing areas of specialty. Fields such as environmental sustainability and transportation engineering are areas of strong growth.

In 2009, the Planning Accreditation Board had approved fifteen bachelor's degree programs and sixty-seven master's degree programs in urban planning at U.S. colleges and universities. Many students who enter the field at the graduate level hold bachelor's degrees in fields such as environmental science, geography, or political science. Course work covers urban design, economic and environmental planning, community development, land use, and zoning and code enforcement. Related course work might include architecture, public administration, economics, and law. A specialization in GIS would require a student to take classes in mathematics, statistics, and computer science. At the graduate level, most students take internships or part-time jobs in applied settings such as government agencies.

About half of all urban planners hold a bachelor's or graduate degree in the field. However, most entry-level jobs in urban planning offered by government agencies require a master's degree. In addition, urban planners may seek certification from the American Institute of Certified Planners to increase their skills and earning potential. Planners working in Michigan must be registered, a process that involves passing examinations and verifying professional experience. A similar licensing requirement exists in New Jersey.

Social Context and Future Prospects

Demand for urban planners and engineers is on the rise. The Bureau of Labor Statistics forecast that jobs in the urban and regional planning sector would grow at a rate of 19 percent from 2008 to 2018. Planning jobs in the private sector that involve scientific and technical services are expected to grow by a remarkable 40 percent. Positions in government agencies are predicted to increase by 14 percent.

Many of the new jobs in the private sector can be found with consulting firms that specialize in architecture and engineering services. An increase in demand for environmentally sustainable development is fueling the need for urban planners with up-to-date knowledge. This type of development includes everything from green buildings to the preservation of natural resources. The rise of new urbanism in the redevelopment of older communities and the building of new ones will continue to foster a need for urban planners familiar with its principles. For many cities, the most rapid growth is taking place in the outermost suburbs. Urban planners will be needed to manage the expansion of transportation and other infrastructure in a way that will avoid problems such as traffic congestion and pollution.

As reliance on GIS and other technologies increases, urban planners and engineers will be expected to have strong backgrounds in statistics and computer science. Professionals who focus on these areas are likely to see the strongest possibilities for long-term career growth.

Julia A. Rosenthal, B.A., M.S.

Further Reading

Ascher, Kate. *The Works: Anatomy of a City.* 2005. Reprint. New York: Penguin Press, 2007. An illustrated manual on New York's infrastructure.

Bayer, Michael, Nancy Frank, and Jason Valerius. *Becoming an Urban Planner: A Guide to Careers in Planning and Urban Design.* Hoboken, N.J.: John Wiley & Sons, 2010. A comprehensive handbook about urban planning as a profession.

Chavan, Abhijeet, Christian Peralta, and Christopher Steins. *Planetizen's Contemporary Debates in Urban Planning.* Washington, D.C.: Island Press, 2007. Contains essays on urban planning from more than twenty-five contributors.

Duany, Andres, Elizabeth Plater-Zyberk, and Jeff Speck. 10th anniversary ed. *Suburban Nation: The*

Rise of Sprawl and the Decline of the American Dream. New York: North Point Press, 2010. The widely cited perspective of three urban planners on new urbanism and city growth.

Jacobs, Jane. *The Death and Life of Great American Cities.* 1961. Reprint. New York: Random House, 2002. A classic work, still regarded as one of the most influential books on modern urban planning.

Levy, John M. *Contemporary Urban Planning.* 9th ed. White Plains, N.Y.: Longman, 2011. Covers many aspects of urban planning, including developments such as the economic downturn of 2009 and the resulting policies of the Barack Obama administration.

Maantay, Juliana, John Ziegler, and John Pickles. *GIS for the Urban Environment.* Redlands, Calif.: ESRI Press, 2006. An overview of the use of geographic information systems in urban planning and related fields.

Web Sites

American Planning Association
http://www.planning.org

United Nations
U.N.-Habitat for a Better Urban Future
http://www.unhabitat.org/categories.asp?catid=254

Urban Land Institute
http://www.uli.org

U.S. Geological Survey, Land Cover Institute
Urban Dynamics Research Program
http://landcover.usgs.gov/urban/intro.php

See also: Architecture and Architectural Engineering; Civil Engineering; Transportation Engineering.

V

VIDEO GAME DESIGN AND PROGRAMMING

FIELDS OF STUDY

Animation; computer animation; computer programming; computer graphics.

SUMMARY

Video game design and programming combines artistic and computer programming skills to create products for a variety of gaming platforms, such as the Xbox 360, iPhone, and personal computers (PC). Whereas the products of video game design and programming were once primarily electronic amusements for children and adolescents, a number of factors have resulted in video games being used for other purposes, such as medical or military training, fitness, and physical rehabilitation. The popularity of video games has also resulted in the growth of game design and programming curricula in education.

KEY TERMS AND CONCEPTS

- **End User:** Person, also known as a gamer, who purchases and plays a video game.
- **Game Artificial Intelligence (AI):** Function within a game program that gives players an experience that resembles that of interacting with an intelligent player.
- **Game Engine:** Software used to create or modify a video game; many different games can be created from a single engine.
- **Level Design:** Design and creation of different levels, or fields of play, within a video game; levels often can be distinguished from one another by differing locales, characters, or game goals.
- **Mod:** Modification of a video game using computer software, often for the purpose of adding new elements to a game or creating a new version of a game.
- **Moore's Law:** Observation or principle in the area of computing hardware that states that the number of transistors on a chip will double every two years.
- **Platform:** Combination of hardware and software used to play a video game; examples include PC games, console games, arcade games, and handheld games.
- **Quality Assurance:** Process by which a video game is tested for software defects and other issues to achieve maximum quality before product launch and release.
- **Serious Games:** Genre of video games most commonly used in training and rehabilitation, the opposite of entertaining games.

DEFINITION AND BASIC PRINCIPLES

Video game design and programming is the field of computer science primarily responsible for the creation of video games to be played across a variety of electronic and digital platforms. This field combines artistic skills such as computer graphics and traditional drawing with computer programming skills to produce games ranging from casual games, a term denoting simple, Web-based games designed for quick diversions, to elaborate games featuring immersive worlds, multiple levels, and a fair amount of skill development for play.

Since their early history, video games have grown from being simple, entertaining diversions that could be experienced on occasion (such as those found in restaurants, bars, and arcades in the 1970's) to lifestyle-enhancing entertainment systems that are playable from almost any type of consumer electronics device with audio-visual capabilities. The mass adoption of video games in contemporary culture has resulted in the creation of a highly influential industry that rivals the sound-recording and motion-picture industries and may even surpass those industries in terms of profitability and user bases.

Since the mid-1990's, the widespread popularity of video games has given rise to game design and programming curricula in higher education. There are also game design programs in some high schools, but these are not as common. Game design and programming degrees are offered at a number of universities and colleges in the United States, including DigiPen Institute of Technology, Full Sail University, and the University of Advancing Technology.

Background and History

The first video games were created by computer scientists and programmers after World War II. However, it was not until the 1960's and 1970's that the combination of graphics and audio capabilities produced what the modern user would recognize as a standard video game. For many years, the earliest games, such as *Spacewar!*, *Pong*, and *Tank*, provided a generic model for many copycat games. However, by the mid-1980's, many genres of video games had been created, including strategic life-simulation games (such as *The Sims*), first-person shooters, adventure games, role-playing games (RPGs), and even the earliest versions of online games, all of which were popular genres of board or pencil-and-paper games.

The massive growth of this industry since the 1980's is in part because of the continuous improvements made to the hardware and software with which games are created and on which they are played. PCs and communications devices (cell phones, for example) follow the rule of Moore's law. According to the law, which is named after Gordon Moore, the cofounder of Intel, the power and speed of computer chips doubles about every eighteen to twenty-four months. Thus, a computer that sold for a thousand dollars in the 1990's has far less processing power and speed than a computer selling for the same price in the 2010's. This increase in processing power allows games to be created using far more elaborate programming and design tools than were used in the past. Thus, video games have become far more realistic than before, although the cost has remained roughly the same.

How It Works

The development of video games can be divided into several phases, from initial concept to design and programming to final product. Each of these phases has a number of processes.

Concept. The initial idea for a video game is known as the concept. The design document, whether a single sheet of paper or a complicated, multipart document, will spell out all the important information that the entire design and programming team will follow for the duration of the game's development. The simplicity of the initial concept will not necessarily have any impact on how complex the final product will be, but it can guide certain decisions along the way. These include the size of the design team, the tools to be used, the platform for which the game is being designed, and who the target audience will be. Early phases of game development involve decisions about the intended plot and style of the game, the language of the game, the design and character of the game world and environments, and the characters that will populate the game world. Much of this is guided by designers and programmers as they oversee game development during each of its phases.

Design and Programming. Once a concept has been envisioned, the next phases involve the production of the game in the desired video game format and platform, such as a video game console or PC. This involves many different steps with many levels of expertise: programming of game elements for the chosen platform, selection of the game engine or software used to program the game and the modeling and other tools to create pleasing characters and environments, and creation of the earliest models of the game so it may be tested. The roles of each development team member at this point are fairly fixed, and they work together to make sure that each element of the game matches the other elements to ensure proper game play (all experiences during a player's interaction with game systems). Major considerations during this phase include determining which game engine or software and audio and graphics tools to use. The programming languages commonly used in game design include C, C++, Java, and Python. Some companies make development kits available to budding game designers. At some point in this process, a version of the game is tested to ensure that software defects, also known as bugs, are eliminated and that the game works as intended.

Crunch Time and Postproduction. Once a video game is near its final, or completion, phase, it is adjusted by designers and programmers and undergoes further testing and quality assurance. Although the game is continuously tested and adjusted by programmers and designers throughout the development phases, it is up to professional game testers to see if the game works from a user's perspective. A marketing team is usually brought in at this level to prepare for the launch and release of the game. As the video game industry is a multibillion-dollar industry, these later phases can be just as important as the early phases of design and development.

Applications and Products

The games that existed before the boom in popularity of video games in the late 1970's—the era of video arcades and home systems—were relatively simple programs that had little compatibility beyond the specific gaming systems for which they were designed, such as the television console or the large, boxy coin-operated machines found at arcades. Therefore, these games were considered novelty pastimes that were enjoyed in specific places at specific times and by a fairly small percentage of the population. However, the evolution of video games, like that of motion pictures, is based on continuous adjustments to the technology used to create them. The consequence of these improvements over time is a shift in the way that video games are perceived and an expansion of the market for these games to include consumers other than traditional gamers, who were typically male adolescents.

Video Games for All Ages. The primary purpose of video game design and programming is the creation of consumer goods for entertainment. Video games, once primarily entertainment for male adolescents and young men, are increasingly played by people of all ages and are being adapted to train and educate people as well as to entertain them. In the 1990's and the early twenty-first century, many new applications were created that not only elevated the gaming experience but also enabled video games to be used for purposes other than pure entertainment. Video games have become valuable tools in diverse areas such as user interfaces, storytelling, training and education, the arts, marketing, medicine, physical rehabilitation, and the military.

Beginning in the late 1980's and early 1990's, with the advent of video game production houses—and video game franchises such as *Doom, Final Fantasy,* and *Super Mario Bros.*, which generated millions of dollars—game designers and programmers have constantly pushed the boundaries of computer science and graphic arts. Innovations in video game design and programming include the use of cell phones and personal electronic devices as important new platforms for gaming and the introduction of new systems designed to appeal to nontraditional gamers, such as the Nintendo Wii. These innovations reflect the creativity of game makers in carving out new markets for themselves.

Innovative Platforms. Video game programmers and designers have increasingly developed games to run on different platforms and use novel controllers, thereby drawing in consumers who historically have not been regularly associated with video gaming. In the case of handheld devices such as the Apple iPhone, Android smartphones, and BlackBerries, consumers recognize the utility of having a device that can be used not only for telecommunications but also for entertainment. Extended battery life, ease of purchasing games, and user-friendly interface designs are also factors in drawing new users.

Fascinating Facts About Video Game Design and Programming

- In 1997, the total sales of video games in the United States was $5.1 billion. In 2008, sales of video games, game systems, and accessories reached $21.3 billion.
- Video games continue to innovate based on new technologies. For example, the game *Spore* uses the Internet for a feature that enables creatures and buildings created by one player to appear in another player's game. This means that the environment of each player's game is generated in part by the imaginations of other players.
- Several conferences specifically for game designers and programmers are held each year. One of the most popular is the Game Developers Conference (GDC). Often, game companies reveal new titles and features at these conferences, which receive worldwide attention from Internet and press journalists. These conferences are excellent places for student designers and programmers to meet prospective employers.
- During the arcade game era of the 1970's and 1980's, video games often were based on popular motion pictures. By the early twenty-first century, that trend had begun to reverse itself, with many motion pictures based on popular video games.
- As video games evolved, the job of a game designer became very similar to that of a film director. As video games become more like motion pictures and films continue to use more special effects, there may come a time when the roles of game designer and film director are interchangeable.

Similarly, the Nintendo Wii game console is responsible for bringing game technology to users who would not traditionally be part of the game market. The simple addition of a wireless, motion-sensitive game controller called the Wii Remote resulted in a new gaming environment for users. Users can control the game with physical gestures, which means that they can play sports in a virtual environment that feels much like the actual activity. The Wii Fit, introduced in 2008, contains a balance board that allows users to engage in yoga, aerobics, strength training, and balance games. The interactive nature of these fitness games has been noted by health and physical rehabilitation professionals, resulting in a burgeoning research field in which more and more innovative gaming technologies are being applied to very serious health issues.

Games as Educational Tools. An important new field for video game designers and programmers is serious games, or those with a primary purpose other than pure entertainment. The idea of a video game environment being used to train pilots and soldiers (two of the more common occupations long associated with gaming) is not new, but the increasing realism, control, and complexity of the tools available for training make video games more and more attractive for a variety of applications. In the medical and health fields, for example, motion-sensitive controllers are useful for training physicians in surgical techniques and aiding in the rehabilitation of physical mobility in elderly patients, as well as those with limited physical movement because of congenital or age-related conditions.

Serious games have been used to teach people about such social issues as the Arab-Israeli conflict and the guerrilla war in Darfur (started in 2003). For example, in the game *Darfur Is Dying*, players take on the identities of members of a refugee family to learn more about the hazards of everyday life in that region. Other serious games deal with peak oil scenarios, climate change, and a host of other situations that could benefit from more public awareness. In particular, the relationship between game design and military recruitment highlights a possible trend in serious games and occupational fields, in that in addition to scenario simulations, games may increasingly become a part of recruitment and training.

Impact on Industry

Developments in video game design and programming span international borders and are overwhelmingly related to profitability. As applications for video games continue to be developed for the medical, military, and other fields, video game design and programming is likely to become more international and involve more cultures and nations.

Government and University Research. The federal government has begun to exploit the growth in video game programs at colleges and universities; however, government-university research in these fields has been somewhat limited. At the University of Advancing Technology, for instance, students may earn internship credit while working on educational games for the U.S. Department of Defense. Much of the governmental research, however, appears to be conducted in areas such as traditional computer science and hardware development rather than video game design and programming.

Industry and Business. Most of the research and progress in video game design and programming occurs within for-profit corporations because the financial health and viability of these companies depend on their making products that are competitive and will keep gamers actively interested in their franchises and stand-alone titles.

Both large and small companies drive innovation in the field of video game design and programming. As of the early twenty-first century, the largest corporations involved in the creation of video games were Electronic Arts (EA), known for *The Sims*, *Madden NFL*, and the *Rock Band* series; Sony, maker of the PlayStation game console and many popular game titles; and Nintendo, maker of the Wii and DS (dual screen or developer's system) platforms. These companies are long-time innovators within the field of video game design and programming and are well placed to continue to grow into the twenty-first century, particularly as the games market becomes even more popular with all ages and numerous types of households.

Careers and Course Work

One important result of the surge in video gaming since the 1980's and 1990's is the creation of video game design and programming curricula at universities and colleges. The overall framework for a degree in a video-game-related field is based largely on the student's focus. Most programs will include courses designed to create a general understanding of video games, such as the history of video games and an introduction to game design and game programming. After these lower-division courses, the focus narrows to more specific courses in aspects of design, such as level design, production and documentation, and methods for working with artists and programmers. Game programmers, on the other hand, continue course work in computer science as applied to video game development, as well as specific programming tools for game development. A number of colleges offer game design and programming degrees, with some even offering master's degrees in management and production.

The career track for a video game designer or programmer could start with a job at a small company with only a handful of employees who create just a few games a year. In this environment, a designer should expect to work very closely with other team members and to have occasional duties in other areas, such as programming. The alternative is to be hired by a much larger company, such as Sony Online Entertainment, in which there may be many departments simultaneously working on different game titles. In this environment, one should expect to maintain a high degree of focus on just one skill set. In either environment, the same levels of skill and workmanship are highly valued and can be quite rewarding for a talented game developer.

Social Context and Future Prospects

To understand why the video game industry has grown as rapidly as it has, one must look at the far-ranging applications that each new generation of game design and programming technology makes possible. Video games have a wider range of uses than ever before and additional practical applications are likely to be developed as gaming increasingly becomes an important part of everyday life for millions of people around the world. Future prospects include immersive environments that expand game playing into levels of interaction and sensory experience that previously could be found only in science fiction. Other possibilities include very powerful, yet relatively inexpensive displays and controllers, as well as tools that allow almost anyone to create his or her own video games on a typical home computer.

Video gaming, once considered a pastime for mostly male adolescents, is becoming popular among people of all ages and women. According to the Entertainment Software Association, as of 2007, most households in the United States have some type of gaming equipment. The average age for a video game buyer is over thirty-five, and an estimated 25 percent of gamers are over the age of fifty. Women make up almost half of the gamer population, and women over the age of eighteen play more games than teenage boys do. Video games are increasingly being used for nontraditional activities, such as health and fitness regimens. These trends point to a broad change in the acceptance level of video gaming across all age groups.

Craig Belanger, M.S.T., B.A.

Further Reading

Adams, Ernest. *Break into the Game Industry: How to Get a Job Making Video Games.* Emeryville, Calif.: McGraw-Hill/Osborne, 2003. This book acts as a guide for those interested in finding work in the video games industry, offering a comprehensive breakdown of specific jobs in the industry, as well as tips and tricks for preparing to become a developer.

_______. *Fundamentals of Game Design.* 2d ed. Berkeley, Calif.: New Riders, 2010. An up-to-date guide to principles, tools, and techniques of game design, with examples drawn from published games. Includes valuable exercises for development.

Koster, Raph. *A Theory of Fun for Game Design.* Scottsdale, Ariz.: Paraglyph Press, 2004. An interestingly laid-out work on the virtue of fun in game design written by a Sony Online Entertainment executive. Offers visual as well as narrative explanations of the fundamentals of game play.

Lecky-Thompson, Guy W. *Video Game Design Revealed.* Boston: Charles River Media, 2007. Examines the programming techniques needed to create a video game.

Marcovitz, Hal. *Video Games.* Detroit: Lucent Books, 2010. Examines the history of video games and their place in society.

Thompson, Jim, Barnaby Berbank-Green, and Nic Cusworth. *Game Design: Principles, Practice, and Techniques.* Hoboken, N.J.: Quarto, 2007. A visual reference guide to the history of video game design and techniques. Also offers a discussion of tools used by designers.

Web Sites

Entertainment Software Association
Games in Daily Life
http://www.theesa.com/gamesindailylife/art.asp

Gamasutra: The Art and Business of Making Games
http://www.gamasutra.com

Gamedev.net
http://www.gamedev.net

International Game Developers Association
http://www.igda.org

See also: Artificial Intelligence; Computer Engineering; Computer Graphics; Computer Science; Internet and Web Engineering; Software Engineering; Virtual Reality.

VIRTUAL REALITY

FIELDS OF STUDY

Physics; mathematics; mechanical engineering; electronics; electrical engineering; systems engineering; robotics; artificial intelligence; machine learning; computer science; computer systems architecture; human-computer interaction; software programming; new media design; dynamics; mechatronics; feedback control; kinematics; haptics; optics; acoustics; experimental psychology; neuroscience; human biology; media ethics.

SUMMARY

The applied science of virtual reality engages in the design and engineering of and research related to special immersive interactive computer systems. These virtual reality systems synthesize environments, or worlds, which are simulations of reality that are usually rendered using three-dimensional computer images, sounds, and force feedbacks. Virtual reality applications are used for pilot and astronaut training, entertainment, communication, teleoperation, manufacturing, medical and surgery training, experimental psychology, psychotherapy, education, science, architecture, and the arts. This technology, which submerges humans into altered environments and processes, intensifies experience and imagination, thereby augmenting research and education. Virtual reality training systems can simplify and improve manufacturing and maintenance, while simultaneously reducing risk exposure.

KEY TERMS AND CONCEPTS

- **Augmented Reality (Display):** Head-mounted display with transparent glasses that show semitransparent data or images superimposed over the wearer's field of vision to augment perception of or knowledge about reality.
- **Cave Automatic Virtual Environment (CAVE):** Form of virtual reality in which images are projected on screens (walls, ceiling, or floor).
- **Equilibrioception:** Sense of balance.
- **Force/Tactile Feedback Devices:** Devices that transmit and stimulate haptic and kineasthetic sensations such as pressure, force, vibration, weight, torque, and resistance.
- **Head-Mounted Display (HMD):** Device that inhibits visual sensations from the real environment and replaces them with three-dimensional computer simulations; also known as goggles.
- **Immersion:** Impression of one's presence in a virtual reality environment; the more senses involved, and the more holistic and realistic the sensations, the more thorough the immersion.
- **Sensing Glove:** Hand-covering input device, wired with sensors, for gesture recognition. It enables three-dimensional interaction and navigation in virtual reality environments. Also known as a wired glove, DataGlove, or CyberGlove.
- **Six Degrees Of Freedom (6-DOF):** Ability to perform six different movements (directions and rotations) and combinations thereof: forward-back, left-right, up-down, roll, pitch, and yaw.
- **Telepresence:** Feeling of being present in a mediated environment through virtual reality technology involving teleconferencing.
- **Tracking System:** System that locates an object's or subject's position and orientation through optic, ultrasonic, magnetic, electric, and mechanical sensors.

Definition and Basic Principles

A virtual reality system is an interactive technology setup (software, hardware, peripheral devices, and other items) that acts as a human-to-computer interface and immerses its user into a computer-generated three-dimensional environment. Virtual reality is the environment or world that the user experiences while using such a system. Although the term "virtual" implies that this simulated world does not actually exist, the term "reality" refers to the user's experience of the simulated environment as being real. The more that the senses are involved in a compelling fashion, the more genuine the perceived experience will be, and the more intense the imagination. Most virtual reality systems stimulate the senses of sight, hearing, touch, and other tactile-kinesthetic sense perceptions, such as equilibrioception, torque, and even temperature. Less often they include smell, and with existing technology, they exclude taste. Virtual reality must be almost indistinguishable from reality in some applications such as pilot training, but it can often differ significantly from the real world in, for example, games.

Virtual reality in a narrow sense (a computer-generated simulation that exists virtually but not materially) is not the same as augmented reality (enhanced reality) or telepresence (in the sense of teleconferencing). Augmented reality technology improves the perception of and supplements the knowledge about existing entities or processes (highlighting data of interest while abstracting the less important information). Telepresence (as teleconferencing) refers to a remote virtual re-creation of a real situation (for example, to enable audio and visual interactions between people at diverse places). However, a wider notion of virtual reality includes notions of both augmented reality and telepresence.

Background and History

The idea of simulated reality is often traced back to the ancient Greek philosopher Plato's allegory of the cave in *Politeia* (fourth century B.C.E.; *Republic*, 1701). In the allegory, spectators observe images (shadows) of objects on a cave wall that they take for real objects. The term "virtual" is derived from the Latin word *virtue* (which means goodness or manliness). "Virtual" then means existing in effect or in essence but not in actuality. The notion of virtual reality can be traced back to the French theater director, actor, playwright, and illustrator Antonin Artaud who described theater as *la réalite virtuelle* in his influential *Théâtre et son double* (1938; *The Theatre and Its Double*, 1958). Computer scientist and artist Myron W. Krueger coined the technical term "artificial reality" in his book of the same title, published in 1983. Computer scientist and artist Jaron Lanier popularized the notion of virtual reality as a technical term in the 1980's. Many artists, science-fiction authors, and directors incorporated concepts of computer-generated, simulated, or augmented reality in their creations. One of the most prominent examples is the holodeck, an advanced form of virtual reality featured on the television program *Star Trek: The Next Generation* (1987-1994).

Technology. In the late 1960's, computer scientist Ivan Sutherland created the first head-mounted device, which was capable of tracking the user's viewing direction. In the 1970's, Sutherland and David Evans developed a computer graphic scene generator. In the 1970's and 1980's, force feedback was incorporated into tactile input devices such as gloves and interactive wands. Lanier and Thomas Zimmerman developed sensing gloves, which recognized finger and hand movements.

This kind of virtual reality technology was intended to improve flight simulators and applications for astronaut training. In 1981, the National Aeronautics and Space Administration (NASA) combined commercially available Sony liquid crystal display (LCD) portable television displays with special optics for a prototype stereo-vision head-mounted device called the Virtual Visual Environment Display (VIVED). NASA then created the first virtual reality system, which combined a host computer, a graphics computer, a noncontact user position-tracking system, and VIVED.

Scott Fisher and Elizabeth Wenzel developed hardware for three-dimensional virtual sound sources in 1988. Also in the late 1980's, Lanier built a virtual reality system for two simultaneous users that he named RB2 (Reality Built for Two). Fisher incorporated sound systems, head-mounted device technology, and sensor gloves into one system called the Virtual Interactive Environment Workstation (VIEW), also used by NASA. The first conference on virtual reality, "Interface for Real and Virtual Worlds," was held in Montpellier, France, in March, 1992. That same year, scientists, engineers, and medical practitioners assembled in San Diego, California, for the "Medicine Meets Virtual Reality Conference." In September, 1993, the Institute for Electrical and Electronics Engineers (IEEE) organized its first virtual reality conference in Seattle.

How It Works

To create a realistic computer-generated world, several high-end technologies must be integrated into a single virtual reality system. This kind of high-end system is used at university, military, governmental, and private research laboratories. Adequate computing speed and power, fast image and data processors, broad bandwidth, and sophisticated software are essential. Other requirements include high-tech input-output devices or effectors (such as head-mounted devices), three-dimensional screens, surround-sound systems, and tactile devices (such as wired gloves and suits, tracking systems, and force feedback devices, including motion chairs and multidirectional treadmills).

Input. The input devices used for virtual reality systems—sensing gloves, trackballs, joysticks, wands, treadmills, motion sensors, position trackers, voice recognizers, and biosensors—are typically more complex than those used for personal computers. Biosensors recognize eye movement, muscle activity, body temperature, pulse, and blood pressure, all of which are vital to surgical applications. Position trackers and motion sensors identify and monitor the user's position and movements. The tracking systems used in virtual reality systems are mechanical, optical, ultrasonic, or magnetic devices. Steering wheels, joysticks, or wands are used for pilot training and games. Sophisticated devices for research and experiments (such as those for molecular modeling) offer six-degrees-of-freedom input. Such input devices allow the computer to adjust the virtual environment according to the data received from the user. When motion sensors and position trackers detect the user's movement, the data are processed by the computer in real time, and the display, also in real time, has to accurately render the image (such as the interior of a building). If slow data processing creates a time lag, the user may experience simulator sickness or motion sickness (nausea or dizziness), especially if the user's senses register conflicting data. In a virtual reality parachute training session, for example, equilibrioception might be in conflict with visual perception if movement is represented faster by the display than by the force feedback system.

Output. Output devices are intended to stimulate as many senses as possible for a high degree of immersion. The important visual output devices are head-mounted devices, LCDs, and projectors and screens. Developers compete to make these displays the most immersive. All displays must be able to render three-dimensional images. Other output devices are for sound and touch. Sound systems can assist in conveying the impression of three-dimensional space. Force feedback systems give the user the sense of physical resistance (essential in surgical virtual reality systems), torque, tilt, and vibration (appreciated for games and essential for pilot training).

The CAVE (Cave Automatic Virtual Environment) is a surround-sound, surround-screen, projection-based room-sized virtual reality system developed by the Electronic Visualization Laboratory at the University of Illinois at Chicago (the name is trademarked by the University of Illinois Board of Regents). Users put on lightweight stereo glasses and walk around inside the room, interacting with virtual objects. One user is an active viewer, controlling the projection, and the others are passive viewers, but all users can communicate while in the CAVE. The system was designed to help with visualization of scientific concepts.

Hardware and Software. Standard personal computers have both limited memory capacity and limited performance capability for running professional virtual reality applications. Therefore, high-end hardware and software have been developed for specific purposes such as games, research programs, and pilot, combat infantry, and medical training. Computer languages such as C++, C#, Java, and Quest3D are used for programming virtual reality software. Virtual reality computers handle tasks such as data input and output, and the interaction, integration, and recomposition of all the data required in virtual environment management. Because a virtual reality system needs high computing, processing, and display speeds, the virtual reality computer architecture can at times use several computers or multiple processors.

Applications and Products

Aerospace and Military. One of the early applications of virtual reality was in pilot training. Modern flight simulators are convincingly close to real flight experiences, although the simulation of acceleration and zero gravity are still challenges. In the military, virtual reality applications are used not only by pilots, paratroopers, and tank drivers but also by battle strategists and combat tacticians. These military applications make targeting more precise, thereby reducing human casualties and collateral damage. Virtual reality systems are also used to evaluate new weapons systems. In space exploration, virtual reality systems help astronauts prepare for zero-gravity activities such as the repair of solar panels on the outside of a spaceship. Virtual reality training systems give the user the opportunity to review and evaluate specific sequences in a training session or the entire session.

Entertainment and Games. The military took commercial games and adapted them to create flight simulators and other professional virtual reality applications, which, in turn, were adapted to use as games. Virtual reality game applications are not common in gaming arcades or in home entertainment systems, but some companies lease virtual reality game equipment (such as training applications for golf, racing, and shooting) to customers for entertainment. Some virtual reality applications enable users to journey through fantastic and futuristic worlds. The common equipment for virtual reality games—depending on the quality and level of sophistication of the games—includes head-mounted devices, several LCD screens, tracking systems, omnidirectional treadmills for walking, force feedback or rumbling seats (or platforms) for flight and race simulations, batons for tennis, and guns, wands, or sticks for shooting. The more sensory feedback provided, the more immersive the virtual reality game application.

Education. The more senses that are involved in the process of learning and training, the better the educational impact, especially when it comes to learning skills or practical content. The employment of a virtual reality system enables data and images to become interactive, colorful, three-dimensional, and accompanied by sound. For example, a visitor to a virtual museum is able to virtually touch the artifacts, taking them from their shelves, turning them around, and viewing them from different perspectives. In an immersive experience, a user can witness historical events such as famous battles as they take place. Virtual reality applications exist in almost every area of education, including sports (such as golfing), sciences (such as astronomy, physics, and chemistry), the humanities (such as history), and vocational training (such as medicine procedures and mechanical engineering techniques).

Art. Although classic artworks can be immersive, few of them are interactive. People are not allowed to touch exhibits in most art galleries and museums. In 1993, the Solomon R. Guggenheim Museum became the first major museum to dedicate an entire exhibition to virtual reality art. The exhibition featured virtual reality installations from Jenny Holzer, Thomas Dolby, and Maxus Systems International. Virtual reality technology enables artists to blur or combine genres (such as music, graphic arts, and video) and involves the viewer in the creation of art. Every viewer or user of the artwork perceives the work in a different way depending on his or her input into the virtual reality work of art. A virtual reality art exhibit can be programmed to produce sound or visual feedback according to, for example, a visitor's footsteps or voice input from an audience. Artworks thus become interactive, and viewers become cocreators of the art.

Science, Engineering, and Design. Virtual reality systems can advance the creative processes involved in science, engineering, and design. Results can be tested, evaluated, shared, and discussed with other virtual reality users. Chemists at the University of North Carolina, Chapel Hill, used virtual reality systems for modeling molecules, and similar systems have been used to "observe" atoms. Buildings, automobiles, and mechanical parts can be designed on computers and partially evaluated in a virtual reality environment. An architect can take a client on a virtual walk through a building before it is constructed, or the aerodynamics of a new automobile design can be evaluated before manufacturing a model or prototype. Virtual reality applications also can simulate crash tests. The use of virtual reality applications reduces cost, waste, and risk.

Business. Business applications include stores on the Internet featuring virtual showrooms and 360-degree, three-dimensional views of products. Another important application is teleconferencing. In contrast to the traditional conference call, virtual reality applications may permit multisensory evaluations of new products. Virtual reality systems also help network, combine, and display data from diverse sources to analyze financial markets and stock exchanges. In these situations, virtual reality applications serve as decision support systems.

Medicine, Therapy, and Rehabilitation. Experienced surgeons as well as physicians in training use virtual reality systems to practice surgery. The images for such training programs are taken from X rays, computed tomography (CT) scans, and magnetic resonance imaging (MRI). Virtual operations can be recorded and repeated as many times as desired, in contrast to practice operations on animals or human corpses, which cannot be repeated. Force feedback gloves give practitioners the realistic feel and touch needed, for example, to determine how much force is needed for certain incisions.

In psychotherapy, virtual reality applications help treat clients with mental disorders such as phobias (anxiety disorders), who often undergo desensitization treatment. Virtual reality allows clients who fear enclosed spaces (claustrophobia), dirt and germs (mysophobia), or snakes (ophidiophobia) to gradually confront their fears without being exposed to the

Fascinating Facts About Virtual Reality

- Some extremely useful virtual reality anatomy programs owe their existence to convicted murderer Joseph Paul Jernigan, who donated his body to medical research in 1993.
- An early head-mounted device, developed in 1968 by Ivan Sutherland and Bob Sproull, displayed three-dimensional images by using two cathode-ray tubes. This helmet was so heavy that it had to be suspended from the ceiling. It was dubbed the Sword of Damocles, after the sword that hung on a horse hair over the head of Damocles, a courtier in ancient Greece.
- The first commercial head-mounted devices were called EyePhones. They were introduced by Jaron Lanier in 1989, sold for the price of $11,000 per set, and weighed 2.4 kilograms. The liquid crystal display (LCD) had a very low resolution–only 360 × 240 pixels.
- *The NewScientist* reported in April, 2005, that the Japanese manufacturer Sony had patented a description of a device that stimulates the brain directly in a nonintrusive way. The device uses ultrasonic sound to stimulate the five senses.
- Deere & Company, manufacturer of agricultural and construction equipment, entered into a partnership with Virtual Reality Applications Center in 1994 to use virtual reality to simulate vehicle operations, examine ergonomics, and improve its manufacturing process.

actual condition or object. Virtual reality applications can also help clinicians understand certain psychological problems better by enabling them to experience what the client or the patient experiences. For example, a psychiatrist, psychologist, or counselor may take a virtual bus ride in the role of a schizophrenic client, during which they experience some simulated symptoms of this disorder, such as distorted images viewed through the windows of the bus or strange voices that seem to appear from nowhere.

Impact on Industry

Virtual reality research is undertaken in many countries, including the United States, members of the European Union, Japan, Canada, Switzerland, South Korea, and Israel. Virtual reality products, or parts for them, are manufactured in these countries and in China, Hong Kong, Taiwan, and several Southeast Asian nations. Traditionally, the United States has led research and development in the area of virtual reality, primarily because of its work in military and aerospace applications. Both the United States and the European Union are activity engaged in research and development in the entertainment, medical, educational, and engineering sectors. The Research Center for Virtual Environments and Behavior at the University of California, Santa Barbara, conducts research related to education, social psychology, the societal impact of virtual reality, and the human visual perception of space. The German institute Frauenhofer IGD leads the world in applied research in visual computing. Its researchers develop virtual reality and augmented reality visual components for medicine, training, education, business, and multimedia applications. Researchers at the Virtual Reality Applications Center at Iowa State University are working on such diverse projects as virtual teleoperation of unmanned vehicles, virtual power plants, and virtual and immersive interactive learning modules for cell biology. The University of Nottingham has established the AS Interactive Project to use virtual reality applications to support people with Asperger's syndrome, a form of autism.

Industry. High-end virtual reality systems are used by research laboratories, large manufacturers, and banks. Mid-range virtual reality applications are often used for educational purposes, and low-end applications are mainly associated with arcade games and home entertainment. Virtual reality research and the development of high-end applications often intertwine when large corporations sponsor research at world-renowned research institutes. The production of virtual reality hardware, software, and peripheral devices is highly specialized, with manufacturers often specializing in only one narrow segment of virtual reality. For example, Evans & Sutherland specializes in the development and production of virtual reality specifically for immersive planetarium theaters. Leica Geosystems produces laser scanners for capturing topographic data; these are used to recreate images of environments. Similar laser scanners and sensors are produced by SICK, a company based in Germany. Some of its scanners and sensors were used by Google to capture three-dimensional images for its Street View project.

Careers and Course Work

Because the development of virtual reality systems and applications involves many areas of applied science, students pursuing careers related to virtual reality systems should be versatile. A bachelor of science in a field such as information technology or mechanical or electrical engineering is an appropriate foundation for a career in virtual reality, but most positions in the field require multidisciplinary skills and experience. Therefore, the best approach is to take additional courses in other fields.

For high-level positions or to work as a researcher in industry or academia, a master of science degree or a doctorate is recommended. Work can also be found with governmental entities (such as NASA) and the military as well as manufacturers of virtual reality systems and products (such as head-mounted devices, screens, force feedback systems, and tracking systems). The producers of virtual reality software and hardware need skilled professionals such as electrical and mechanical engineers, roboticists, software developers, and information technologists. As yet, job opportunities related to virtual reality technology and research are still somewhat limited because of the field's highly technical and specialized nature and also because virtual reality applications are not yet in widespread use.

Social Context and Future Prospects

Some experts believe that virtual reality, like the computer and the Internet, soon will become a commonplace and indispensable part of everyday life, but others do not see the potential for such a wide implementation. Most agree, however, that once computing speed and power, broad bandwidth, and peripheral systems become more affordable, virtual reality will be more widely used. Beyond science, education, and other professional applications, the entertainment industry is thought most likely to want affordable virtual reality innovations. Critics of virtual reality applications point to personal and societal risks such as isolation, desocialization, and alienation, but advocates emphasize the technology's proven potential for augmenting people's lives. Both critics and advocates agree that experiences in virtual reality can alter people's perceptions of and responses to the real world. These alterations are intentional and welcome in most cases but sometimes they take place in an unintended and potentially dangerous manner. Airplane pilots have reportedly made mistakes that could be linked to the limitations of training with a flight simulator, which, for example, is incapable of realistically simulating acceleration. However, virtual reality applications are valuable in highly technical and precise areas of industry, business, and research and are likely to gain more users, despite the potential risks and the expense.

Roman Meinhold, M.A., Ph.D.

Further Reading

Burdea, Grigore C., and Philippe Coiffet. *Virtual Reality Technology*. 2d ed. Hoboken, N.J.: Wiley-Interscience, 2003. Provides broad coverage of virtual reality technology. Includes illustrations and index.

Craig, Alan B., William R. Sherman, and Jeffrey D. Will. *Developing Virtual Reality Applications. Foundations of Effective Design*. Burlington, Mass.: Morgan Kaufmann, 2009. In-depth treatment of various virtual reality applications with illustrations, extensive bibliography, and index.

Heim, Michael. *Virtual Realism*. New York: Oxford University Press, 1998. Critical appraisal from a metaperspective on virtual reality technology by a virtual reality philosopher.

Sherman, William R., and Alan B. Craig. *Understanding Virtual Reality: Interface, Application and Design*. San Francisco: Morgan Kaufmann, 2003. Substantive introduction to VR technology and its applications, with case studies, illustrations, and index.

Vince, John. *Introduction to Virtual Reality*. New York: Springer, 2004. Good overview of virtual reality graphics, hardware, software, and applications with glossary and index.

Web Sites

International Virtual Reality Association
http://www.vr.org.au/wordpress/?page_id=2

Society for Modeling and Simulation International
http://www.scs.org

Virtual Reality Site
http://www.vrs.org.uk

See also: Aeronautics and Aviation; Computer Science; Information Technology; Video Game Design and Programming.

Appendixes

BIOGRAPHICAL DICTIONARY OF SCIENTISTS

Alvarez, Luis W. (1911-1988): A physicist and inventor born in San Francisco, Alvarez was associated with the University of California, Berkeley, for many years. He explored cosmic rays, fusion, and other aspects of nuclear reaction. He invented time-of-flight techniques and conducted research into nuclear magnetic resonance for which he was awarded the 1968 Nobel Prize in Physics. He contributed to radar research and particle accelerators, worked on the Manhattan Project, developed the ground-controlled approach for landing airplanes, and proposed the theory that dinosaurs were rendered extinct by a massive meteor impacting Earth.

Archimedes (c. 287-c. 212 B.C.E.): A Greek born at Syracuse, Sicily, Archimedes is considered a genius of antiquity, with interests in astronomy, physics, engineering, and mathematics. He is credited with the discovery of fluid displacement (Archimedes' principle) and a number of mathematical advancements. He also developed numerous inventions, including the Archimedes screw to lift water for irrigation (still in use), the block-and-tackle pulley system, a practical odometer, a planetarium using differential gearing, and several weapons of war. He was killed during the Roman siege of Syracuse.

Babbage, Charles (1791-1871): An English-born mathematician and mechanical engineer, Babbage designed several machines that were precursors to the modern computer. He developed a difference engine to carry out polynomial functions and calculate astronomical tables mechanically(which was not completed) as well as an analytical engine using punched cards, sequential control, branching and looping, all of which contributed to computer science. He also made advancements in cryptography, devised the cowcatcher to clear obstacles from railway locomotives, and invented an ophthalmoscope.

Bacon, Sir Francis (1561-1626): A philosopher, statesman, author, and scientist born in England, Bacon was a precocious youth who at the age of thirteen began attending Trinity College, Cambridge. Later a member of Parliament, a lawyer, and attorney general, he rejected Aristotelian logic and advocated for inductive reasoning—collecting data, interpreting information, and carrying out experiments—in his major work, *Novum Organum* (*New Instrument*), published in 1620, which greatly influenced science from the seventeenth century onward. A victim of his own research, he experimented with snow as a way to preserve meat, caught a cold that became bronchitis, and died.

Baird, John Logie (1888-1946): A Scottish electrical engineer and inventor, Baird successfully transmitted black-and-white (in 1925) and color (in 1928) moving television images, and the BBC used his transmitters to broadcast television from 1929 to 1937. He had more than 175 patents for such far-ranging and forward-thinking concepts as big-screen and stereo TV sets, pay television, fiber optics, radar, video recording, and thermal socks. Plagued with ill health and a chronic lack of financial backing, Baird was unable to develop his innovative ideas, which others later perfected and profited from.

Bardeen, John (1908-1991): A Wisconsin-born electrical engineer and physicist, Bardeen worked for Gulf Oil, researching magnetism and gravity, and later studied mathematics and physics at Princeton University, where he earned a doctoral degree. While working at Bell Laboratories after World War II he, Walter Brattain (1902-1987), and William Shockley (1910-1989) invented the transistor, for which they shared the 1956 Nobel Prize in Physics. In 1972, Bardeen shared a second Nobel Prize in Physics for a jointly developed theory of superconductivity; he is the only person to win the same award twice.

Barnard, Christiaan (1922-2001): A heart-transplant pioneer born in South Africa, Barnard was a cardiac surgeon and university professor. He performed the first successful human heart transplant in 1967, extending a patient's life by eighteen days, and subsequent transplants—using innovative operational techniques he devised—allowed new heart recipients to survive for more than twenty years. He was one of the first surgeons to employ living tissues and organs from other species to prolong human life and was a contributor to the effective design of artificial heart valves.

Bates, Henry Walter (1825-1892): A self-taught

naturalist and explorer born in England, Bates accompanied anthropologist-biologist Alfred Russel Wallace (1823-1913) on a scientific expedition to South America between 1848 and 1852, which he described in his 1864 work, *The Naturalist on the River Amazons*. He collected thousands of plant and animal species, most of them unknown to science, and was the first to study the survival phenomenon of insect mimicry. For nearly thirty years he was secretary of the Royal Geographical Society and also served as president of the Entomological Society of London.

Becquerel, Antoine-Henri (1852-1908): A French physicist and engineer born into a family boasting several generations of scientists, Becquerel taught applied physics at the National Museum of Natural History and at the Polytechnic University, both in Paris, and also served as primary engineer overseeing French bridges and highways. He served as president of the French Academy of Sciences and received numerous awards for his work investigating polarization of light, magnetism, and the properties of radioactivity, including the 1903 Nobel Prize in Physics, which he shared with Pierre and Marie Curie.

Bell, Alexander Graham (1847-1922): A Scottish engineer and inventor whose mother and wife were deaf, Bell researched hearing and speech throughout his life. He began inventing practical solutions to problems as a child. His experiments with acoustics led to his creation of the harmonic telegraph, which eventually resulted in the first practical telephone in 1876 and spawned Bell Telephone Company. Bell became a naturalized American citizen and also invented prototypes of flying vehicles, hydrofoils, air conditioners, metal detectors, and magnetic sound and video recording devices.

Benz, Carl (1844-1929): A German engineer and designer born illegitimately as Karl Vaillant, Benz designed bridges before setting up his own foundry and mechanical workshop. In 1888, he invented, built and patented a gas-powered, engine-driven, three-wheeled horseless carriage named the Benz Motorwagen, which was the first automobile available for purchase. In 1895, he built the first trucks and buses and introduced many technical innovations still found in modern automobiles. The Benz Company merged with Daimler in the 1920's and introduced the famous Mercedes-Benz in 1926.

Berzelius, Jons Jakob (1779-1848): A physician and chemist born in Sweden, Berzelius was secretary of the Royal Swedish Academy of Sciences for thirty years. He is credited with discovering the law of constant proportions for inorganic substances and was the first to distinguish organic from inorganic compounds. He developed a system of chemical symbols and a table of relative atomic weights that are still in use. In addition to coining such chemical terms as "protein, "catalysis," "polymer," and "isomer," he identified the elements cerium, selenium, silicon, and thorium.

Bessemer, Henry (1813-1898): The English engineer and inventor is chiefly known for development of the Bessemer process, which eliminated impurities from molten pig iron and lowered costs in the production of steel. Holder of more than one hundred patents, Bessemer also invented items to improve the manufacture of glass, sugar, military ordnance, and postage stamps, and built a test model of a gimballed, hydraulic-controlled steamship to eliminate seasickness. His steel-industry creations led to the development of the modern continuous casting process of metals.

Birdseye, Clarence (1886-1956): The naturalist, inventor, and entrepreneur was born in Brooklyn, New York. He began experimenting in the early 1920's with flash-freezing fish. Using a patented process, he was eventually successful in freezing meats, poultry, vegetables, and fruits, and in so doing changed consumers' eating habits. Birdseye sold his process to the company that later became General Foods Corporation, for whom he continued to work in developing frozen-food technology. His surname—split in two for easy recognition—became a major brand name that is still familiar.

Bohr, Niels (1885-1962): A Danish theoretical physicist, Bohr introduced the concept of atomic structure, in which electrons orbit the nucleus of an atom, and laid the foundations of quantum theory, for which he was awarded the 1922 Nobel Prize in Physics. He later identified U-235, an isotope of uranium that produces slow fission. During World War II, after escaping from Nazi-occupied Denmark, he worked as consultant to the Manhattan Project. Following the war, he returned to Denmark and became a staunch advocate for the nondestructive uses of atomic energy.

Bosch, Carl (1874-1940): The German-born chemist,

metallurgist, and engineer devised a high-pressure chemical technique (the Haber-Bosch process) to fix nitrogen, used in mass-producing ammonia for fertilizers, explosives, and synthetic fuels. He was awarded (along with Friedrich Bergius, 1884-1949) the 1931 Nobel Prize in Chemistry for his work. He was a founder and chairman of the board of IG Farben, for a time the largest chemical company in the world, but was ousted in the late 1930's for criticizing the Nazis.

Brahe, Tycho (1546-1601): A nobleman born of Danish heritage in what is modern-day Sweden, Brahe became interested in astronomy while studying at the University of Copenhagen. He made improvements to the primitive observational instruments of the day but never had access to the telescope. Nonetheless, he was able to study the positions of stars and planets accurately and produced useful catalogs of celestial bodies, particularly for the planet Mars, which helped Johannes Kepler (1571-1630) to formulate the laws of planetary motion. Craters on the Moon and on Mars are named in Brahe's memory.

Brunel, Isambard Kingdom (1806-1859): A British-born civil engineer and inventor, Brunel designed and built tunnels, bridges, and docks—many still in use—often devising ingenious solutions to problems in the process. He is best remembered for developing the SS *Great Britain*, the largest and most modern ship of its time and the first ocean-going iron ship driven by a propeller. Brunel was also a railroad pioneer, serving as chief engineer for Great Western Railway, for which he specified a broad-gauge track to allow higher speeds, improved freight capacity, and greater passenger comfort.

Burbank, Luther (1849-1926): Despite having only an elementary-school education, the Massachusetts-born botanist and horticulturist was a pioneer in the field of agricultural science. Working from a greenhouse and experimental fields in Santa Rosa, California, Burbank developed more than 800 varieties of plants, including new strains of flowers, peaches, plums, nectarines, cherries, peaches, berries, nuts, and vegetables, as well as new crossbred products such as the plumcot. One of his most useful creations, the Russet Burbank, became the potato of choice in food processing, particularly for French fries.

Calvin, Melvin (1911-1997): A Minnesota-born chemist of Russian heritage, Calvin taught molecular biology for nearly fifty years at the University of California, Berkeley, where he founded and directed the Laboratory of Chemical Biodynamics (later the Structural Biology Division) and served as associate director of the Lawrence Berkeley National Laboratory. He and his research team traced the path of carbon-14 through plants during photosynthesis, greatly enhancing understanding of how sunlight stimulates chlorophyll to create organic compounds. He was awarded the 1961 Nobel Prize in Chemistry for his work.

Carnot, Sadi (1796-1832): A French physicist and military engineer, Carnot was an army officer before becoming a scientific researcher, specializing in the theory of heat as produced by the steam engine. His *Reflections on the Motive Power of Fire* focused on the relationship between heat and mechanical energy and provided the foundation for the second law of thermodynamics. His work greatly influenced scientists such as James Prescott Joule (1818-1889), William Thomson (Lord Kelvin, 1824-1907), and Rudolf Diesel (1858-1913) and made possible more practically and efficiently designed engines later in the nineteenth century. Carnot's career was cut short by his death from cholera.

Carson, Rachel (1907-1964): A marine biologist and author born in Springdale, Pennsylvania, Carson worked for the U.S. Bureau of Fisheries before turning full-time to writing about nature. Her popular and highly influential articles, radio scripts and books, including *The Sea Around Us, The Edge of the Sea, Under the Sea-Wind,* and *Silent Spring* enlightened the public about the wonders of nature and the dangers of pesticides such as DDT, which was eventually banned in the United States. Carson is credited with spurring the modern environmental movement.

Celsius, Anders (1701-1774): A Swedish astronomer, Celsius studied the aurora borealis and was the first to link the phenomena to the Earth's magnetic field. He also participated in several expeditions designed to measure the size and shape of the Earth. Founder of the Uppsala Astronomical Observatory, he explored star magnitude, observed eclipses, and compiled star catalogs. He is perhaps best known for the Celsius international temperature scale, which accounts for atmospheric pressure in measuring the boiling and freezing points of water.

Clausewitz, Carl von (1780-1831): As a Prussian-born soldier and military scientist, Clausewitz participated in numerous campaigns, beginning in the early 1790's, and fought in the Napoleonic Wars. After his appointment in 1818 to major general, he taught at the Prussian military academy and helped reform the state army. His principal written work, *On War*, unfinished at the time of his death from cholera, is still considered relevant and continues to influence military thinking via its practical approach to command policies, instruction for soldiers, and methods of planning for strategists.

Colt, Samuel (1814-1862): An inventor born in Hartford, Connecticut, Colt designed a workable multishot pistol while working in his father's textile factory. In the mid-1830's he patented a revolver and set up an assembly line to produce machine-made weapons featuring interchangeable parts. The perfected product, the Colt Peacemaker, was used in the Seminole and Mexican-American wars and became popular during America's western expansion, and Colt became a millionaire. Colt's Manufacturing Company continues to produce a wide variety of firearms for civilian, military, and law-enforcement purposes.

Copernicus, Nicolaus (1473-1543): The Polish mathematician, physician, statesman, artist, linguist, and astronomer is credited with beginning the scientific revolution. His major work, published the year of his death, *De revolutionibus orbium coelestium* (*On the Revolutions of the Heavenly Spheres*), was the first to propose a heliocentric model of the solar system. The book inspired further research by Tycho Brahe (1546-1601), Galileo Galilei (1564-1642), and Johannes Kepler (1571-1630) and stimulated the birth of modern astronomy.

Cori, Gerty Radnitz (1896-1957): A biochemist born in Prague (now the Czech Republic), Cori came to the United States in 1922 and became a naturalized American citizen in 1928. She worked with her husband Carl at what is now Roswell Park Cancer Institute in Buffalo, New York, researching carbohydrate metabolism and discovered how glycogen is broken down into lactic acid to be stored as energy, a process now called the Cori cycle. She was awarded the 1947 Nobel Prize in Physiology or Medicine, the first American woman so honored.

Cousteau, Jacques (1910-1997): A French oceanographer, explorer, filmmaker, ecologist, and author, Cousteau began underwater diving in the 1930's, and it became a lifelong obsession. He coinvented the Aqua-Lung in the 1940's—the precursor to modern scuba gear—and began making nature films during the same decade. He founded the French Oceanographic Campaigns in 1950 and aboard his ship *Calypso* explored and researched the world's oceans for forty years. In the 1970's he created the Cousteau Society, which remains a strong ecological advocacy organization.

Crick, Francis (1916-2004): An English molecular biologist and physicist, Crick designed magnetic and acoustic mines during World War II. He was later part of a biological research team at the Cavendish Laboratory. Focusing on the X-ray crystallography of proteins, he identified the structure of deoxyribonucleic acid (DNA) as a double helix, a discovery that greatly advanced the study of genetics. He and his colleagues, American James D. Watson (b. 1928) and New Zealander Maurice Wilkins (1916-2004), shared the 1962 Nobel Prize in Physiology or Medicine for their groundbreaking work.

Curie, Marie Sklodowska (1867-1934) and Pierre Curie (1859-1906): Polish-born chemist-physicist Marie was the first woman to teach at the University of Paris. She married French physicist-chemist Pierre Curie in 1895, and the couple collaborated on research into radioactivity, discovering the elements polonium and radium. She and her husband shared the 1903 Nobel Prize in Physics for their work; she was the first woman so honored. After her husband died, she continued her research and received the 1911 Nobel Prize in Chemistry, the first person to receive the award in two different disciplines. She founded the Radium (later Curie) Institute.

Daimler, Gottlieb (1834-1900): The German-born mechanical engineer, designer, inventor, and industrial magnate was an early developer of the gasoline-powered internal combustion engine and the automobile. He and fellow industrial designer Wilhelm Maybach (1846-1929) began a partnership in the 1880's to build small, high-speed engines incorporating numerous devices they patented—flywheels, carburetors, and cylinders—still found in modern engines. After creating the first motorcycle they founded Daimler Motors and began selling automobiles in the early 1890's. Their Phoenix model won history's first auto race.

Darwin, Charles Robert (1809-1882): An English naturalist and geologist, Darwin participated in the five-year-long worldwide surveying expedition of the HMS *Beagle* during the 1830's, observing and collecting specimens of animals, plants, and minerals. The voyage inspired numerous written works, particularly *On Natural Selection, On the Origin of the Species,* and *The Descent of Man,* which collectively supported his theory that all species have evolved from common ancestors. Though modern science has virtually unanimously accepted Darwin's findings, his theory of evolution remains a controversial topic among various political, cultural, and religious groups.

Davy, Sir Humphry (1778-1829): A chemist, teacher, and inventor born in England, Davy began conducting scientific experiments as a child. As a teen he worked as a surgeon's apprentice and became addicted to nitrous oxide. He later researched galvanism and electrolysis, discovered the elements sodium, chlorine, and potassium, and contributed to the discovery of iodine. He invented the Davy safety lamp for use in coal mines, was a founder of the Zoological Society of London, and served as president of the Royal Society.

Diesel, Rudolf (1858-1913): A French-born mechanical engineer and inventor of German heritage, Diesel designed an innovative refrigeration system for an ice plant in Paris and improved the efficiency of steam engines. His self-named, patented diesel engine introduced the concept of fuel injection. The efficient diesel engine later became commonplace in trucks, locomotives, ships, and submarines and, after redesign to reduce weight, in modern automobiles. Diesel disappeared while on a ship. His body was later discovered floating in the sea, but it is still unknown whether he fell overboard, committed suicide, or was murdered.

Edison, Thomas Alva (1847-1931): A scientist, inventor and entrepreneur born in Milan, Ohio, Edison worked out of New Jersey in the fields of electricity and communication and profoundly influenced the world. Credited with more than 1,000 patents, he is best known for creating the first practical incandescent light bulb, which has illuminated the lives of humans since 1879. Other inventions include the stock ticker, a telephone transmitter, electricity meters, the mimeograph, an efficient storage battery, and the phonograph and the kinetoscope, which he combined to produce the first talking moving picture in 1913.

Einstein, Albert (1879-1955): A German-born theoretical physicist and author of Jewish heritage, Einstein came to the United States before World War II. Regarded as a genius, and one of the world's most recognized scientists, he was awarded the 1921 Nobel Prize in Physics. He developed general and special theories of relativity, particle and quantum theories, and proposed ideas that continue to influence numerous fields of study, including energy, nuclear power, heat, light, electronics, celestial mechanics, astronomy, and cosmology. Late in life, he was offered the presidency of Israel but declined the honor.

Euclid (c. 330-c. 270 B.C.E.): A Greek mathematician who taught in Alexandria, Egypt, Euclid is considered the father of plane and solid geometry. He is remembered principally for his major extant work, *The Elements*—a treatise containing definitions, postulates, geometric proofs, number theories, discussions of prime numbers, arithmetic theorems, algebra, and algorithms—which has served as the basis for the teaching of mathematics for two thousand years. He also explored astronomy, mechanics, gravity, moving bodies, and music and was one of the first scientists to write about optics and perspective.

Everest, Sir George (1790-1866): A geographer born in Wales, Everest participated for 25 years in the Great Trigonometrical Survey of the Indian subcontinent—which surveyed an area encompassing millions of square miles while locating, measuring, and naming the Himalayan Mountains—and served as superintendent of the project from 1823 to 1843. Later knighted, he served as vice president of the Royal Geographical Society. The world's tallest peak, Mount Everest in Nepal (known locally as Chomolungma), was named in his honor.

Faraday, Michael (1791-1867): A self-educated British chemist and physicist, Faraday served an apprenticeship with chemist Sir Humphry Davy (1778-1829), during which time he experimented with liquefied gases, alloys, and optical glasses. He invented a prototype of what became the Bunsen burner, discovered benzene, and performed experiments that led to his discovery of electromagnetic induction. He built the first electric dynamo, the precursor to the power generator, researched the relationship between magnetism and light, and made numerous other contributions to the studies of electromagnetism and electrochemistry.

Farnsworth, Philo (1906-1971): An inventor born in Utah, Farnsworth became interested in electronics and mechanics as a child. He experimented with television during the 1920's, and late in the decade he demonstrated an electronic, nonmechanical scanning system for image transmissions. During the early 1930's, he worked for Philco but left to carry out his own research. In addition to significant contributions to television, Farnsworth held more than 300 patents and devised a milk-sterilizing process, developed fog lights, an infrared telescope, a prototype of an air traffic control system, and a fusion reaction tube.

Fermi, Enrico (1901-1954): An Italian-born experimental and theoretical physicist and teacher who became an American citizen in 1944, Fermi studied mechanics and was instrumental in the advancement of thermodynamics and quantum, nuclear, and particle physics. Awarded the 1938 Nobel Prize in Physics for his research into radioactivity, he was a member of the team that developed the first nuclear reactor in Chicago in the early 1940's and served as a consultant to the Manhattan Project, which produced the first atomic bomb. He died of cancer from sustained exposure to radioactivity.

Feynman, Richard (1918-1988): A physicist, author, and teacher born in New York, Feynman participated in the Manhattan Project and made numerous contributions to a diverse field of specialized scientific disciplines including quantum mechanics, supercooling, genetics, and nanotechnology. He shared the 1965 Nobel Prize in Physics—with Julian Schwinger (1918-1994) and Sin-Itiro Tomonaga (1906-1979)—for work in quantum electrodynamics, particularly for his lucid explanation of the behavior of subatomic particles. He was a popular and influential professor for many years at California Institute of Technology.

Fleming, Alexander (1881-1955): A Scottish-born biologist, Fleming served in the Royal Army Medical Corps during World War I and witnessed the deaths of many wounded soldiers from infection. He was a professor of bacteriology at a teaching hospital, and he specialized in immunology and chemotherapy research. In 1928 he discovered an antibacterial mold, which over the next decade was purified and mass-produced as the drug penicillin, which played a large part in suppressing infections during World War II. A major contributor to the development of antibiotics, he shared the 1945 Nobel Prize in Physiology or Medicine.

Forrester, Jay Wright (b. 1918): An engineer, teacher, and computer scientist born in Nebraska, Forrester built a wind-powered electrical system while in his teens. Associated with the Massachusetts Institute of Technology as a researcher and professor for many years, he developed servomechanisms for military use, designed aircraft flight simulators, and air defense systems. He founded the field of system dynamics to produce computer-generated mathematical models for such tasks as determining water flow, fluid turbulence, and a variety of mechanical movements.

Fourier, Joseph (1768-1830): A French-born physicist, mathematician, and teacher Fourier accompanied Napoleon's expedition to Egypt, where he served as secretary of the Egyptian Institute and was a major contributor to *Description of Egypt,* a massive work describing the scientific findings that resulted from the French military campaign. After returning to France, Fourier explored numerous scientific fields but is best known for his extensive research on the conductive properties of heat and for his theories of equations, which influenced later physicists and mathematicians.

Franklin, Benjamin (1706-1790): A Boston-born author, statesman, scientist, and inventor, Franklin worked as a printer in his youth and from 1733 to 1758 published *Poor Richard's Almanack.* A key figure during the American Revolution and a founding father of the United States, he established America's first lending library and Pennsylvania's first fire department, served as first U.S. postmaster general and as minister to France and Sweden. He experimented with electricity and is credited with inventing the lightning rod, bifocal glasses, an odometer, the Franklin stove, and a musical instrument made of glass.

Freud, Sigmund (1856-1939): An Austrian of Jewish heritage, Freud studied neurology before specializing in psychopathology and conducted extensive research into hypnosis and dream analysis to treat hysteria. Considered the father of psychoanalysis and a powerful influence on the field, he originated such psychological concepts as repression, psychosomatic illness, the unconscious mind, and the division of the human psyche into the id, ego, and superego. He fled from the Nazis and went to London. Riddled with cancer from years of cigar smoking, he took morphine to relieve his suffering and hasten his death.

Frisch, Karl von (1886-1982): The Vienna-born son of a surgeon-university professor, von Frisch initially studied medicine before switching to zoology and comparative anatomy. Working as a teacher and researcher out of Munich, Rostock, Breslau, and Graz universities, he focused his research on the European honeybee. He made many discoveries about the insect's sense of smell, optical perception, flight patterns, and methods of communication that have since proved invaluable in the fields of apiology and botany. He was awarded the 1973 Nobel Prize in Physiology of Medicine in recognition of his pioneering work.

Fuller, R. Buckminster (1895-1983): The Massachusetts-born architect, philosopher, engineer, author, and inventor developed systems for lightweight, weatherproof, and fireproof housing while in his twenties. Teaching at Black Mountain College in the late 1940's, Fuller perfected the geodesic dome, built of aluminum tubing and plastic skin, and afterward developed numerous designs for inventions aimed at providing practical and affordable shelter and transportation. He coined the term "synergy" and advocated exploiting renewable sources of energy such as solar power and wind-generated electricity. He was awarded the Presidential Medal of Freedom in 1983.

Galen (129-c. 199): An ancient Roman surgeon, scientist, and philosopher, Galen traveled and studied widely before serving as physician to Roman emperors Marcus Aurelius (121-180), Commodus (161-192), Septimus Severus (146-211), and Caracalla (188-217). In the course of his education he explored human and animal anatomy, became an advocate of proper diet and hygiene, and advanced the practice of surgery by treating the wounds of gladiators and ministering to plague victims. His medical discoveries and healing methods, detailed in numerous written works, influenced medicine for more than 1,500 years.

Galilei, Galileo (1564-1642): A physicist, astronomer, mathematician, and philosopher born in Pisa, Italy, Galileo is known as the father of astronomy and the father of modern science. A keen astronomical observer, he made significant improvements to the telescope, through which he studied the phases of Venus, sunspots, and the Milky Way, and he discovered Jupiter's four largest moons. He risked excommunication and death championing the heretical Copernican heliocentric view of the solar system. He also invented a military compass and a practical thermometer and experimented with pendulums and falling bodies.

Galton, Francis (1822-1911): An anthropologist, geographer, meteorologist, and inventor born in England, Galton was a child prodigy. He traveled widely, exploring the Middle East and Africa, and wrote about his expeditions. Fascinated by numbers, he devised the first practical weather maps for use in newspapers. He also contributed to the science of statistics, studied heredity—coining the term "eugenics" and the phrase "nature or nurture"—and was an early advocate of using fingerprints in criminology. Galton is responsible for inventing a high-frequency whistle used in training dogs and cats.

Gates, Bill (b. 1955): An entrepreneur, philanthropist, and author born in Seattle, Gates became interested in computers as a teenager. He left Harvard in 1975 to cofound and to serve as chairman (until 2006) of Microsoft, which developed software for IBM and other systems before launching its own system in 1985. The result, Microsoft Windows, became the dominant software product in the worldwide personal computer market. Profits from his enterprise made Gates one of the world's richest people, and he has used his vast wealth to assist a wide variety of charitable causes.

Goddard, Robert H. (1892-1945): A physicist, engineer, teacher, and inventor born in Massachusetts, Goddard became interested in science as a child and experimented with kites, balloons, and rockets. He received the first of more than 200 patents in 1914 for multistage and liquid-fuel rockets, and during the 1920's he conducted successful test flights using liquid fuel. Goddard experimented with solid fuels and ion thrusters and is credited with developing tail fins, gyroscopic guidance systems, and many other basics of rocketry that greatly influenced the designs of rocket scientists who came after him.

Grandin, Temple (b. 1947): An animal scientist born in Massachusetts, Grandin was diagnosed with autism as a child. As an adult she earned advanced degrees before receiving a doctorate from the University of Illinois in 1989. A professor at Colorado State University, an author, and an autism advocate, she has made numerous humane improvements to the design of livestock-handling facilities that have been incorporated into meat-processing plants worldwide to reduce or eliminate animal stress, pain, and fear.

Haber, Fritz (1868-1934): A chemist and teacher of Jewish heritage (later a convert to Christianity) born in Germany, Haber developed the Haber process to produce ammonia used in fertilizers, animal feed, and explosives, for which he was awarded the 1918 Nobel Prize in Chemistry. At Berlin's Kaiser Wilhelm Institute (later the Haber Institute) between 1911 and 1933, he developed chlorine gas used in World War I, experimented with the extraction of gold from seawater, and oversaw production of Zyklon B, the cyanide-based pesticide that was employed at extermination camps during World War II.

Halley, Edmond (1656-1742): An English astronomer, mathematician, physicist, and meteorologist, Halley wrote about sunspots and the solar system while a student at Oxford. In the 1670's, he cataloged the stars of the Southern Hemisphere and charted winds and monsoons. Inventor of an early diving bell and a liquid-damped magnetic compass, a colleague of Sir Isaac Newton (1642-1727), and leader of the first English scientific expedition, Halley is best remembered for predicting the regular return of the comet that bears his name.

Heisenberg, Werner (1901-1976): A theoretical physicist and teacher born in Germany, Heisenberg conducted research in quantum mechanics with Niels Bohr (1885-1962) at the University of Copenhagen. There he developed the uncertainty principle, which proves it is impossible to determine the position and momentum of subatomic particles at the same time. Awarded the 1932 Nobel Prize in Physics for his work, he also contributed research on positrons, cosmic rays, spectral frequencies, matrix mechanics, nuclear fission, superconductivity, and plasma physics to the continuing study of atomic theory.

Herschel, William (1738-1822) and Caroline Herschel (1750-1848): A German-born astronomer, composer, and telescope maker who moved to England in his teens, William spent the early part of his career as a musician, playing cello, oboe, harpsichord, and organ and wrote numerous symphonies and concerti. In the 1770's he began building his own large reflecting telescopes and with his diminutive (4 feet, 3 inches) but devoted sister Caroline spent countless hours observing the sky while cataloging nebulae and binary stars. The Herschels are credited with discovering two of Saturn's moons, the planet Uranus and two of its moons, and coining the word "asteroid."

Hersey, Mayo D. (1886-1978): A mechanical engineer born in Rhode Island, Hersey was a preeminent expert on tribology, the study of the relationship between interacting solid surfaces in motion, the adverse effects of wear, and the ameliorating effects of lubrication. He worked as a physicist at the National Institute of Standards and Technology (1910-1920) and the U.S. Bureau of Mines (1922-1926) and taught at the Massachusetts Institute of Technology (1910-1922). He was a consultant to the Manhattan Project and won numerous awards for his contributions to lubrication science.

Hippocrates (c. 460-c. 377 B.C.E.): An ancient Greek physician born on the island of Kos, Hippocrates was the first of his time to separate the art of healing from philosophy and magical ritual. Called the father of Western medicine, he originated the belief that diseases were not the result of superstition but of natural causes, such as environment and diet. Though his concept that illness was the result of an imbalance in the body's fluids (called humors) was later discredited, he pioneered such common modern clinical practices as observation and documentation of patient care. He originated the Hippocratic Oath, which for many centuries served as the guiding principle governing the behavior of doctors.

Hooke, Robert (1635-1703): A brilliant, multitalented British experimental scientist with interests in physics, astronomy, chemistry, biology, geology, paleontology, mechanics, and architecture, Hooke was instrumental as chief surveyor in rebuilding the city of London following the Great Fire of

1666. Among many accomplishments in diverse fields he is credited with inventing the compound microscope—via which he discovered the cells of plants and formulated a theory of fossilization—devised a balance spring to improve the accuracy of timepieces, and either created or refined such instruments as the barometer, anemometer, and hygrometer.

Howlett, Freeman S. (1900-1970): A horticulturist born in New York, Howlett was associated with the Ohio State University as teacher, administrator, and researcher for more than forty-five years and was considered an expert on the history of horticulture. His investigations focused on plant hormones, embryology, fruit setting, reproductive physiology, and foliation for a variety of crops, including fruits, vegetables, and nuts. He created five new varieties of apples popular among consumers. A horticulture and food science building at Ohio State is named in his honor.

Hubble, Edwin Powell (1889-1953): An astronomer born in Missouri, Hubble was associated with the Mount Wilson Observatory in California for more than thirty years. Using what was then the world's largest telescope, he was the first to discover galaxies beyond the Milky Way, which greatly expanded science's concept of the universe. He studied red shifts in formulating Hubble's law, which confirmed the big bang or expanding universe theory. The American space telescope launched in 1990 was named for him, and he was honored in 2008 with a commemorative postage stamp.

Huygens, Christiaan (1629-1695): Born in the Netherlands, Huygens was an astronomer, physicist, mathematician, and prolific author. He made early telescopic observations of Saturn and its moons and was the first to suggest that light is made up of waves. He discovered centrifugal force, proposed a formula for centripetal force, and developed laws governing the collision of celestial bodies. An inveterate inventor, he patented the pendulum clock and the pocket watch and designed an early internal combustion engine. He is also considered a pioneer of science fiction for writing about the possibility of extraterrestrial life.

Jacquard, Joseph Marie (1752-1834): A French inventor, Jacquard created a series of mechanical looms in the early nineteenth century. His experiments culminated in the Jacquard loom attachment, which could be programmed, via punch cards, to weave silk in various patterns, colors, and textures automatically. The labor-saving device became highly popular in the silk-weaving industry, and its inventor received royalties on each unit sold and became wealthy in the process. The loom inspired scientists to incorporate the concept of punch cards for computer information storage.

Jenner, Edward (1749-1823): A surgeon and anatomist born in England, Jenner experimented with cowpox inoculations in an attempt to prevent smallpox, a virulent infectious disease of ancient origin with a high rate of mortality that killed millions of people. In the early nineteenth century, Jenner successfully developed a method of vaccination that provided immunity from smallpox and late in life became personal physician to King George IV. The smallpox vaccination was made compulsory in England and elsewhere, and the disease was declared eradicated worldwide in 1979.

Jobs, Steven (b. 1955-2011): An inventor and entrepreneur of Syrian and American heritage born in San Francisco, Jobs worked at Hewlett-Packard as a teenager and was later employed at Atari designing circuit boards. In 1976, he and coworker Steve Wozniak (b. 1950) and others founded Apple, which designed, built, and sold a popular and highly successful line of personal computers. A multibillionaire and holder of more than 200 patents, Jobs continued to make innovations in interfacing, speakers, keyboards, power adaptation, and myriad other components related to modern computer science until his death in late 2011

Kepler, Johannes (1571-1630): A German mathematician, author and astronomer, he became interested in the cosmos after witnessing the Great Comet of 1577. He worked for Tycho Brahe (1546-1601) for a time and after Brahe's death became imperial mathematician in Prague. A major contributor to the scientific revolution, Kepler studied optics, observed many celestial phenomena, provided the foundation for Sir Issac Newton's theory of gravitation, and developed a set of laws governing planetary motion around the Sun—including the discovery that the orbits of planets are elliptical—that were confirmed by later astronomers.

Krebs, Sir Hans Adolf (1900-1981): A biochemist and

physician born the son of a Jewish surgeon in Germany, Krebs was a clinician and researcher before moving to England after the rise of the Nazis. As a professor at Cambridge University, he explored metabolism, discovering the urea and citric acid cycles—biochemical reactions that promote understanding of organ functions in the body and explain the cellular production of energy—for which he shared the 1953 Nobel Prize in Medicine or Physiology. He was also knighted in 1958 for his work.

Lawrence, Ernest O. (1901-1958): A South Dakota-born physicist and teacher, Lawrence researched the photoelectric effect of electrons at Yale. In 1928, he became a professor at the University of California, Berkeley, where he invented the cyclotron particle accelerator, for which he was awarded the 1939 Nobel Prize in Physics. During World War II, he was involved in the Manhattan Project. Lawrence popularized science and was a staunch advocate for government funding of significant scientific projects. After his death, laboratories at the University of California and the chemical element lawrencium were named in his honor.

Leakey, Louis B. (1903-1972) and Mary Nicol Leakey (1913-1996): Louis, born in Kenya, was an archaeologist, paleontologist, and naturalist who married London-born anthropologist and archaeologist Mary Nicol. Together and often with their sons, Jonathan, Richard, and Philip, they excavated at Olduvai Gorge in East Africa, where they unearthed the tools and fossils of ancient hominids. Their discoveries of the remains of Proconsul africanus, Australopithecus boisei, Homo habilis, Homo erectus, and other large-brained, bipedal primates effectively proved Darwin's theory of evolution and extended human history by several million years.

Leonardo da Vinci (1452-1519): An Italian genius considered the epitome of the Renaissance man, da Vinci was a superb artist, architect, engineer, mathematician, geologist, musician, mapmaker, inventor, and writer. Creator of such famous paintings as the *Mona Lisa* and *The Last Supper*, he is credited with imagining the helicopter, solar power, and the calculator centuries before their invention. His far-ranging mind explored such subjects as anatomy, optics, vegetarianism, and hydraulics, and his journals, written in mirror-image script, are filled with drawings, ideas, and scientific observations that are still closely studied.

Linnaeus, Carolus (1707-1778): A Swedish botanist, zoologist, physician, and teacher, Linnaeus began studying plants as a child. As an adult, he embarked on expeditions throughout Europe observing and collecting specimens of plants and animals and wrote numerous works about his findings. He devised the binomial nomenclature system of classification for living and fossil organisms—called taxonomy—still used in modern science, which provides concise Latin names of genus and species for each example. Linnaeus also cofounded the Royal Swedish Academy of Science.

Lippershey, Hans (c. 1570-c. 1619): A master lens grinder and spectacle maker born in Germany who later became a citizen of the Netherlands, Lippershey is credited with designing the first practical refracting telescope (which he called "perspective glass"). After fruitlessly attempting to patent the device, he built several prototypes for sale to the Dutch government, which distributed information about the telescope across Europe. Other scientists, such as Galileo, soon duplicated and improved upon Lippershey's invention, which became a primary instrument in the science of astronomy.

Lumière, Auguste (1862-1954) and Louis Lumière (1864-1948): The French-born brothers worked at their father's photographic business and devised the dry-plate process for still photographs. From the early 1890's, they patented several techniques—including perforations to guide film through a camera and a color photography process—that greatly advanced the development of moving pictures. From 1895 to 1896, they publicly screened a series of short films to enthusiastic audiences in Asia, Europe, and North and South America, demonstrating the commercial potential of the new medium and launching what would become the multibillion-dollar film industry.

Maathai, Wangari Muta (b. 1940-2011): An environmental and political activist of Kikuyu heritage born in Kenya, Maathai studied biology in the United States before becoming a research assistant and anatomy teacher at the University of Nairobi, where she was the first East African woman to earn a Ph.D. She founded the Green Belt Movement, an organization that plants trees, supports environmental conservation, and advocates for women's rights. A former member of the Kenyan Parliament and former Minister of Environment, she was awarded the 2004 Nobel Peace Prize for her work and is the first African woman to receive the award.

McAdam, John Loudon (1756-1836): A Scottish engineer, McAdam became a surveyor in Great Britain and specialized in road building. He devised an effective method—called "macadam" after its inventor—of creating long-lasting roads using gravel on a foundation of larger stones, with a camber to drain away rainwater, which was adopted around the world. He also introduced hot tar as a binding agent (dubbed "tarmac," an abbreviation of tarmacadam) to produce smoother road surfaces. Modern road builders still use many of the techniques he innovated.

Mantell, Gideon (1790-1852): A British surgeon, geologist, and paleontologist, Mantell began collecting fossil specimens from quarries as a child. As an adult, he was a practicing physician and pursued geology in his spare time. He discovered fossils that were eventually identified as belonging to the Iguanodon and Hylaeosaurus—which he named Megalosaurus and Pelorosaurus—and he became a recognized authority on dinosaurs. His major works were *The Fossils of South Downs: Or, Illustrations of the Geology of Sussex* (1822) and *Notice on the Iguanodon: A Newly Discovered Fossil Reptile* (1825).

Marconi, Guglielmo (1874-1937): An Italian-born electrical engineer and inventor, Marconi experimented with electricity and electromagnetic radiation. He developed a system for transmitting telegraphic messages without the use of connecting wires and by the early twentieth century was sending transmissions across the Atlantic Ocean. His devices eventually evolved into radio, and the transmitter at his factory in England was the first in 1920 to broadcast entertainment to the United Kingdom; he shared the 1909 Nobel Prize in Physics with German physicist Ferdinand Braun (1850-1918).

Maxwell, James Clerk (1831-1879): A Scottish-born mathematician, theoretical physicist, and teacher, Maxwell had an insatiable curiosity from an early age and as a teenager began presenting papers to the Royal Society of Edinburgh. He experimented with color, examined hydrostatics and optics, and wrote about Saturn's rings. His most significant work, however, was performed in the field of electromagnetism, in which he showed that electricity, magnetism, and light are all results of the electromagnetic field, a concept that profoundly affected modern physics.

Mendel, Gregor Johann (1822-1884): Born in Silesia (now part of the Czech Republic), Mendel became interested in plants as a child. In the 1840's he entered an Augustinian monastery, where he studied astronomy, meteorology, apiology, and botany. Called the father of modern genetics, he is best known for his experiments in hybridizing pea plants, which evolved into what later were called Mendel's laws of inheritance. Though his work exerted little influence during his lifetime, his concepts were rediscovered early in the twentieth century and have since proven invaluable to the study of heredity.

Mendeleyev, Dmitri Ivanovich (1834-1907): A Russian chemist, teacher, and inventor, Mendeleyev studied the properties of liquids and the spectroscope before becoming a professor in Saint Petersburg and later serving as director of weights and measures. He created a periodic table of the sixty-three elements then known arranged by atomic mass and the similarity of properties (a revised form of which is still employed in modern science) and used the table to correctly predict the characteristics of elements and isotopes not yet found. Element 101, mendelevium, discovered in 1955, was named in his honor.

Meng Tian (259-210 B.C.E.): A general serving under Qin Shi Huang, first emperor of the Qin Dynasty (221-207 B.C.E.), Meng Tian led an army of 100,000 to drive warlike nomadic tribes north out of China. Descended from architects, he oversaw building of the Great Wall to prevent invasions, cleverly incorporating topographical features and natural barriers into the defensive barricade, which he extended for more than 2,000 miles along the Yellow River. After a coup following Emperor Qin's death, Meng Tian was forced to commit suicide. The Qin Dynasty fell just three years later.

Montgolfier, Joseph Michel (1740-1810) and Jacques-Etienne Montgolfier (1745-1799): Born in France to a prosperous paper manufacturer, the Montgolfier brothers designed and built a hot-air balloon, and in 1783 Jacques-Etienne piloted the first manned ascent in a lighter-than-air craft. The French Academy of Science honored the brothers for their exploits, which inspired further developments in ballooning. The Montgolfier brothers subsequently wrote books on aeronautics and continued experimenting. Joseph is credited with designing a calorimeter and a hydraulic ram, and Jacques-Etienne invented a method for the manufacture of vellum.

Morse, Samuel F. B. (1791-1872): An artist and inventor born in Massachusetts, Morse painted portraits and taught art at the City University of New York before experimenting with electricity. In the mid-1830's, he designed the components of a practical telegraph—a sender, receiver, and a code to translate signals into numbers and words—and in 1844 sent the first message via wire. Within a decade, the telegraph had spread across America and subsequently around the world. The invention would inspire such later advancements in communication as radio, the Teletype, and the fax machine.

Nernst, Walther (1864-1941): A German physical chemist, physicist, and inventor, Nernst discovered the Third Law of Thermodynamics—defining the chemical reactions affecting matter as temperatures drop toward absolute zero—for which he was awarded the 1920 Nobel Prize in Chemistry. He also invented an electric lamp, and developed an electric piano and a device using rare-earth filaments that significantly advanced infrared spectroscopy. He made numerous contributions to the specialized fields of electrochemistry, solid-state chemistry, and photochemistry.

Newton, Sir Isaac (1642-1727): The English physicist, mathematician, astronomer, and philosopher is considered one of the most gifted and scientifically influential individuals of all time. He developed theories of color and light from studying prisms, was instrumental in creating differential and integral calculus, and formulated still-valid laws of celestial motion and gravitation. He was knighted in 1705, the first British scientist so honored. From 1699 until his death he served as master of the Royal Mint and during his tenure devised anticounterfeiting measures and moved England from the silver to the gold standard.

Nobel, Alfred (1833-1896): A Swedish chemist and chemical engineer, Nobel invented dynamite while studying how to manufacture and use nitroglycerin safely. In the course of building a manufacturing empire based on the production of cannons and other armaments, he experimented with combinations of explosive components, also producing gelignite and a form of smokeless powder, which led to the development of rocket propellants. Late in his life, he earmarked the bulk of his vast estate for the establishment of the Nobel Prizes, annual monetary awards given in recognition of outstanding achievements in science, literature, and peace.

Oppenheimer, J. Robert (1904-1967): A brilliant theoretical physicist, researcher, and teacher born to German immigrants in New York City, Oppenheimer was the scientific director of the Manhattan Project, which developed the atomic bombs dropped on Japan during World War II. Following the war, he was primary adviser to the U.S. Atomic Energy Commission and director of the Institute for Advanced Study in Princeton, New Jersey. He contributed widely to the study of electrons and positrons, neutron stars, relativity, gravitation, black holes, quantum mechanics, and cosmic rays.

Owen, Richard (1804-1892): An English biologist, taxonomist, anti-Darwinist, and comparative anatomist, Owen founded and directed the natural history department at the British Museum. He originated the concept of homology, a similarity of structures in different species that have the same function, such as the human hand, the wing of a bat, and the paw of an animal. He also cataloged many living and fossil specimens, contributed numerous discoveries to zoology, and coined the term "dinosaur." Owen advanced the theory that giant flightless birds once inhabited New Zealand long before their remains were found there.

Paré, Ambroise (c. 1510-1590): A French royal surgeon, Paré revolutionized battlefield medicine, developing techniques and instruments for the treatment of gunshot wounds and for performing amputations. He greatly advanced knowledge of human anatomy by studying the effects of violent death on internal organs. He pioneered the lifesaving practices of vascular ligating and herniotomies, designed prosthetics to replace amputated limbs, and was the first to create realistic artificial eyes from such substances as glass, porcelain, silver, and gold.

Pasteur, Louis (1822-1895): A chemist, microbiologist, and teacher born in France, Pasteur focused on researching the causes of diseases and methods for preventing them after three of his children died from typhoid. He proposed a germ theory, demonstrating that microorganisms affect foodstuffs. This ultimately led to his invention of pasteurization—a method of killing bacteria in milk, which was later applied to other substances. A pioneer in immunology, he also developed vaccines to combat anthrax, rabies, and puerperal fever.

Pauli, Wolfgang (1900-1958): An Austrian theoretical physicist of Jewish heritage who converted to Catholicism, Pauli earned a Ph.D. at the age of twenty-one. While lecturing at the Niels Bohr Institute for Theoretical Physics, he researched relativity and quantum physics. He discovered a new law governing the behavior of atomic particles and the characteristics of matter, called the Pauli exclusion principle, for which he was awarded the 1945 Nobel Prize in Physics. During World War II, he moved to the United States and became an American citizen but later relocated to Zurich.

Pauling, Linus (1901-1994): Born in Portland, Oregon, Pauling earned advanced degrees in chemical engineering, physical chemistry, and mathematical physics. A Guggenheim Fellow, he studied quantum mechanics in Munich, Copenhagen, and Zurich before teaching at the California Institute of Technology. He specialized in theoretical chemistry and molecular biology and greatly advanced understanding of the nature of chemical bonds. A political activist who warned of the dangers of nuclear weapons, he became one of a handful of scientists to receive Nobel Prizes in two fields: the 1954 prize in chemistry and the 1982 peace prize.

Pavlov, Ivan (1849-1936): A Russian physiologist and psychologist, Pavlov began investigating the digestive system, which led to experiments with the effects of behavior on the nervous system and the body's automatic functions. He used animals in researching conditioned reflex actions to a variety of visual, tactile, and sound stimuli—including bells, whistles, and electric shocks—to discover the relationship between salivation and digestion and was able to make dogs drool in anticipation of receiving food. He was awarded the 1904 Nobel Prize in Physiology or Medicine for his work.

Planck, Max (1858-1947): A German theoretical physicist credited with founding quantum theory—which affects all matter in the universe—Planck earned a doctoral degree at the age of twenty-one before becoming a professor at the universities of Kiel and Berlin. He explored electromagnetic radiation, quantum mechanics, thermodynamics, blackbodies, and entropy. He formulated the Planck constant, which describes the proportions between the energy and frequency of a photon and provides understanding of atomic structure. He was awarded the 1918 Nobel Prize in Physics for his discoveries.

Ptolemy (c. 100-c. 178): A mathematician, astronomer, and geographer of Greek heritage who worked in Roman-ruled Alexandria, Egypt, Ptolemy wrote several treatises that influenced science for centuries afterward. His *Almagest*, written in about 150, contains star catalogs, constellation lists, Sun and Moon eclipse data, and planetary tables. Ptolemy's eight-volume *Geographia* (*Geography*) followed and incorporates all known information about the geography of the Earth at the time and helped introduce the concept of latitudes and longitudes. His work on astrology influenced Islamic and medieval Latin worlds, and his writings on music theory and optics pioneered study in those fields.

Pythagoras (c. 580-c. 500 B.C.E.): An ancient Greek philosopher and mathematician from Samos, Pythagoras traveled widely seeking wisdom and established a religious-scientific ascetic community in Italy around 530 B.C.E. He had interests in music, astronomy, medicine, and mathematics, and though none of his writings survived, he is credited with the discovery of the Pythagorean theorem governing right triangles (the square of the hypotenuse is equal to the sum of the squares of the other two sides). His life and philosophy exerted considerable influence on Plato (c. 427-347 B.C.E.) and through Plato greatly affected Western thought.

Reiss, Archibald Rodolphe(1875-1929): A chemist, photographer, teacher, and natural scientist born in Germany, Reiss founded the world's first school of forensic science at the University of Lausanne, Switzerland, in 1909. He published numerous works that greatly influenced the new discipline, including *La photographie judiciaire* (*Forensic photography*, 1903) and *Manuel de police scientifique. I Vols et homicides* (*Handbook of Forensic Science: Thefts and Homicides*, 1911). During World War I he investigated alleged atrocities in Serbia and lived there for the rest of his life. The institute he founded more than a century ago has become a major school offering numerous courses in various forensic sciences, criminology, and criminal law.

Röntgen, Wilhelm Conrad (1845-1923): A German physicist, Röntgen studied mechanical engineering before teaching physics at the universities of Strassburg, Giessen, Würzburg, and Munich. He experimented with fluorescence and electrostatic charges. In the process of his work he discovered X rays—and also discovered that lead could effectively block the rays—meanwhile laying the foundations of what would become radiology: the medical specialty that uses radioactive imaging to diagnose disease. He was awarded the first Nobel Prize in Physics in 1901. Element 111, roentgenium, was named in his honor in 2004.

Rutherford, Ernest (1871-1937): A chemist and physicist born in New Zealand, Rutherford studied at the University of Cambridge before teaching physics at McGill University in Montreal and at the University of Manchester. He made some of the most significant discoveries in the field of atomic science, including the relative penetrating power of alpha, beta, and gamma rays, the transmutation of elements via radioactivity, and the concept of radioactive half-life. His work, for which he received the 1908 Nobel Prize in Chemistry, was instrumental in the development of nuclear energy and carbon dating.

Sabin, Albert Bruce (1906-1993): A microbiologist born of Jewish heritage as Albert Saperstein in Russia, Sabin later became an American citizen and changed his name. Trained in internal medicine, he conducted research into infectious diseases and assisted in the development of a vaccine to combat encephalitis. His major contribution to medicine was an effective oral polio vaccine, which was administered in mass immunizations during the 1950's and 1960's and eventually led to the eradication of the disease worldwide. Among other honors, he received the Presidential Medal of Freedom in 1986.

Sachs, Julius von (1832-1897): A German botanist, writer, and teacher, Sachs made great strides in the investigation of plant physiology, morphology, heliotropism, and germination while professor of botany at the University of Würzburg. In addition to numerous written works on photosynthesis, water absorption, and chloroplasts that significantly advanced the science of botany, he also invented a number of devices useful to research, including an auxanometer to measure growth rates, and the clinostat, a device that rotates plants to compensate for the effects of gravitation on botanical growth.

Sakharov, Andrei (1921-1989): A Russian nuclear physicist and human rights activist, Sakharov researched cosmic rays, particle physics, and cosmology. He was a major contributor to the development of the hydrogen bomb but later campaigned against nuclear proliferation and for the peaceful use of nuclear power. He received the 1975 Nobel Peace Prize, and though he received several international honors in recognition of his humanitarian efforts, he spent most of the last decade of his life in exile within the Soviet Union. A human rights center and a scientific prize are named in his honor.

Scheele, Carl Wilhelm (1742-1786): A chemist born in a Swedish-controlled area of Germany, Scheele became a pharmacist at an early age. Though he discovered oxygen through experimentation, he did not publish his findings immediately, and the discovery was credited to Antoine-Laurent Lavoisier (1743-1794) and Joseph Priestly (1733-1804), though science later gave the Scheele recognition he deserved. Scheele also discovered the elements barium, manganese, and tungsten, identified such chemical compounds as citric acid, glycerol, and hydrogen cyanide, experimented with heavy metals, and devised a method of producing phosphorus in quantity for the manufacture of matches.

Shockley, William (1910-1989): A physicist and inventor born to American parents in England, Shockley was raised in California. After earning a doctoral degree, he conducted solid-state physics research at Bell Laboratories. During World War II, he researched radar and anti-submarine devices. Following the war, he was part of the team that invented the first practical solid-state transistor, for which he shared the 1956 Nobel Prize in Physics with John Bardeen (1908-1991) and Walter Brattain (1902-1987). He later set up a semiconductor business that was a precursor to Silicon Valley. His major work, *Electrons and Holes in Semiconductors* (1950), greatly influenced many scientists.

Sikorsky, Igor (1889-1972): A Ukrainian engineer and test pilot who immigrated to the United States and became a naturalized American citizen, Sikorsky was a groundbreaking designer of both airplanes and helicopters. Inspired as a child by the drawings of Leonardo da Vinci (1452-1519), he created and flew the first multi-engine fixed-wing aircraft and the first airliner in the 1910's. He built the first flying boats in the 1930's—the famous Pan Am Clippers—and in 1939 designed the first practical helicopter, which introduced the system of rotors still used in modern helicopters.

Spilsbury, Sir Bernard Henry (1877-1947): The first British forensic pathologist, Spilsbury began performing postmortems in 1905. He investigated cause of death in many spectacular homicide cases—including those of Dr. Crippen and the Brighton trunk murders—that resulted in convictions and enhanced the science of forensics. He was a consultant to Operation Mincemeat, a successful World War II ruse (dramatized in the 1956 film *The Man Who Never Was*) involving the corpse of an alleged Allied courier, which deceived the Axis powers about the invasion of Sicily. Spilsbury was found dead in his laboratory—a victim of suicide.

Stephenson, George (1781-1848): A British mechanical and civil engineer, Stephenson invented a safety lamp for coal mines that provided illumination without the risk of explosions from firedamp. He designed a steam-powered locomotive for hauling coal, which evolved into the first public railway line in the mid-1820's, running on his specified track width of 4 feet, 8.5 inches. This measurement became the worldwide standard railroad gauge. He worked on numerous rail lines, in the process making many innovations in the design and construction of locomotives, tracks, viaducts, and bridges that greatly advanced railroad transport.

Teller, Edward (1908-2003): An outspoken theoretical physicist born in Hungary, Teller came to the United States in the 1930's and taught at George Washington University while researching quantum, molecular, and nuclear physics. A naturalized American citizen, he was a member of the atomic-bomb-building Manhattan Project. A strong supporter for nuclear energy development and testing for both wartime and peacetime purposes, he cofounded and directed Lawrence Livermore National Laboratory and founded the department of applied science at the University of California, Davis.

Tesla, Nikola (1856-1943): Born in modern-day Croatia, the brilliant if eccentric Tesla came to the United States in 1884 to work for Thomas Edison's company and later for Edison's rival George Westinghouse (1846-1914). In 1891, Tesla became a naturalized American citizen. A physicist, mechanical and electrical engineer, and an inventor specializing in electromagnetism, he created fluorescent lighting, pioneered wireless communication, built an alternating- current induction motor, and developed the Tesla coil, variations of which have provided the basis for many modern electrical and electronic devices.

Vavilov, Nikolai Ivanovich (1887-1943): A Russian botanist and plant geneticist, Vavilov served for two decades as director of the Institute of Agricultural Sciences (now the N. I. Vavilov Research Institute of Plant Industry) in Leningrad (now Saint Petersburg). During his tenure, he collected seeds from around the world, establishing the world's largest seed bank—with more than 200,000 samples—and conducted extensive research on genetically improving grain, cereal, and other food crops to produce greater yields to better feed the world. Arrested during World War II for disagreeing with Soviet methods of agronomy, he died of complications from starvation and malnutrition.

Vesalius, Andreas (1514-1564): A physician and anatomist born as Andries van Wesel in the Habsburg Netherlands (now Belgium), Vesalius taught surgery and anatomy at the universities of Padua, Bologna, and Pisa in Italy. Dissatisfied at the inaccuracies in the standard texts of the day—based solely on the 1,400-year-old work of ancient physician Galen, since Rome had long discouraged performing autopsies—he dissected a human corpse in the presence of artists from Titian's studio. This resulted in the seven-volume illustrated work, *De humani corporis fabrica libri septem* (*On the Fabric of the Human Body*, 1543), which served as the foundation for modern anatomy.

Vitruvius (c. 80-c. 15 B.C.E.): A Roman architect and engineer, Vitruvius served in many campaigns under Julius Caesar (100-44 B.C.E.), for whom he designed and built mechanical military weapons, such as the ballista (a projectile launcher) and siege machines. His major written work, *De architectura* (*On Architecture*, c. 27 B.C.E.), set the standard for building structures solidly, usefully, and attractively. The book covers the construction of machines—including cranes, pulleys, sundials, and water clocks. It discusses construction materials and describes ancient Roman building innovations that greatly influenced later architects, particularly during the Renaissance.

Watt, James (1736-1819): A Scottish mechanical and civil engineer, Watt designed a steam engine to pump water out of mines. Refinements of his engine were used in grinding, milling, and weaving, and further improvements—including gauges, throttles, gears, and governors—enhanced the engine's efficiency and safety, making it the prime mover of the Industrial Revolution and the power source of choice for early trains and ships. Watt also devised an early copying machine and discovered a method for producing chlorine for bleaching. The unit of electrical power is named for him.

Wegener, Alfred (1880-1930): A German meteorologist, climatologist, and geophysicist, Wegener was one of the first to employ weather balloons. He was first to advance the theory of continental drift, proposing that the Earth's continents were once a single mass that he called Pangaea; his ideas, however, were not accepted until long after his death. From 1912, he worked in remote areas of Greenland examining polar airflows and drilling into the ice to study past weather patterns. He died in Greenland during his last ill-fated expedition.

Westinghouse, George (1846-1914): Born in Central Bridge, New York, Westinghouse was an engineer, inventor, entrepreneur, and a rival of Thomas Edison (1847-1931). He built a rotary steam engine while still a teenager and in his youth patented several devices—including a fail-safe compressed-air braking system—to improve railway safety. He developed an alternating-current power distribution network that proved superior to Edison's direct-current scheme, invented a power meter still in use, built several successful hydroelectric generating plants, and devised shock absorbers for automobiles.

Whittle, Sir Frank (1907-1996): Born the son of an engineer in England, Whittle joined the Royal Air Force as an aircraft mechanic and advanced to flying officer and test pilot before eventually rising to group captain. While in the Royal Air Force, he began designing aircraft engines that used turbines rather than pistons. In the mid-1930's, he formed a partnership, Power Jets, which produced the first effective turbojet design before the company was nationalized. He later developed a self-powered drill for Shell Oil and wrote a text on gas turbine engines.

Wiener, Norbert (1894-1964): A mathematician born in Missouri, Wiener was a child prodigy. He began college at the age of eleven, earned a bachelor's degree in math at fourteen, and a doctorate in philosophy from Harvard at the age of eighteen. During World War I, he researched ballistics at Aberdeen Proving Ground and afterward spent his career teaching mathematics at Massachusetts Institute of Technology. A pioneer of communication theory, he is credited with the development of theories of cybernetics, robotics, automation, and computer systems, and his work greatly influenced later scientists.

Woodward, John (1665-1728): An English naturalist, physician, paleontologist, and geologist, Woodward was an early collector of fossils, which served as the basis for his *Classification of English Minerals and Fossils* (1729), a work that influenced geology for many years. He also conducted pioneering research into the science of hydroponics. His collection of specimens formed the foundation of Cambridge University's Sedgwick Museum, and his estate was sold to provide a post in natural history, now the Woodwardian Chair of Geology at Cambridge.

Wright, Orville (1871-1948) and Wilbur Wright (1867-1912): The Wright brothers were American aviation pioneers who began experimenting with flight in their teens. In the early 1890's they opened a bicycle sales and repair shop, which financed their research into manned gliders. They soon progressed to designing powered aircraft. They eventually invented and built the first practical fixed-wing aircraft and piloted the world's first sustained powered flight—a distance of more than 850 feet over nearly a minute—in 1903 at Kitty Hawk, North Carolina. The Wright Company later became part of Curtiss-Wright Corporation, a modern high-tech aerospace component manufacturer.

Zeppelin, Ferdinand von (1838-1917): Born in Germany, Zeppelin served in the Prussian army and made a balloon flight while serving as a military observer in the American Civil War. After returning to Europe, he designed and constructed airships and devised a transportation system using lighter-than-air craft. He created a rigid, streamlined, engine-powered dirigible in 1900 and was instrumental in the creation of duralumin, which later led to lightweight all-metal airframes. By 1908, he was providing commercial air service to passengers and mail, which had an enviable record for safety until the *Hindenburg* disaster in 1937.

Zworykin, Vladimir (1889-1982): A Russian who emigrated to the United States after World War I, Zworykin worked at the Westinghouse laboratories in Pittsburgh. An engineer and inventor who patented a cathode ray tube television transmitting and receiving system in 1923, he later worked in development for the Radio Corporation of America (RCA) in New Jersey, where his inventions were perfected in time to be used to telecast the 1936 Olympic Games in Berlin. He also contributed to the development of the electron microscope.

Jack Ewing

GLOSSARY

absolute zero: The complete absence of thermal energy, resulting in a temperature of -273.15 degrees Celsius. This temperature is the basis for the Kelvin scale (starting at 0 Kelvin) developed by the British physicist, Lord Kelvin, in 1848. What living organisms feel as heat or warmth is a difference in temperature between two objects, which results in a transfer of thermal energy. Molecules at absolute zero have no thermal energy to transfer but can receive thermal energy from contact with a warmer object. *See also* cold, heat, temperature.

acid: A compound containing hydrogen ions (with a positive charge) in its molecules, which are released when the acid is dissolved in water. Acids include such familiar hazardous substances as sulphuric, nitric, and hydrochloric acid, essential nutrients such as ascorbic acid (vitamin C), and common flavorings or preservatives such as acetic or ethanoic acid (vinegar). Acids react chemically with substances known as bases. The balance of acids and bases in a solution is measured by the pH scale, from 0 (strongly acidic) to 7 (neutral) to 14 (strongly alkaline). *See also* alkali, basic chemical.

alkali: A base that is dissolved in water. Alkaline substances are identified by a measurement from 8 to 14 on the pH scale. *See also* basic chemical.

alpha particle: One of three common forms of radiation from the nuclei of unstable radioactive elements, consisting of two protons and two neutrons, identical to the nucleus of a helium atom, without its electron shell. It has a velocity in air of one-twentieth the speed of light. *See also* beta particle, gamma ray.

amino acids: Biological molecules that serve as the building blocks of proteins and enzymes. Amino acids are incorporated into proteins by transfer RNA, according to the genetic code contained in DNA. The majority of amino acids have names ending with -ine, and are complex arrangements of atoms of carbon, nitrogen, hydrogen, and oxygen. *See also* enzyme, protein.

animal husbandry: The art and science of breeding, raising, and caring for domesticated animals, primarily in small- or large-scale agriculture, as sources of food, leather, wool, and other products useful to humans. Husbandry skills are not only required for many jobs in agriculture but for zookeepers, maintaining rodent and amphibian populations in laboratories, and for large-scale veterinary and animal-vaccination practices.

antiseptic: Any chemical substance that kills or inhibits the growth of microorganisms causing sepsis—putrefaction, decay, or other infection—generally applied to surface tissues of human or other living organisms or to nonliving surfaces that may harbor microorganisms.

atmosphere: The layers of gas surrounding the solid or liquid surfaces of a planet. The atmosphere of the Earth is 78.08 percent nitrogen, 20.95 percent oxygen, less than one percent argon, and hundredths or thousandths of a percent neon, helium, and hydrogen. The amounts of water vapor, carbon dioxide, methane, nitrous oxide, and ozone vary with biological (and more recently industrial) processes. Water can rise to as high as four percent. The atmosphere has been divided by different studies into five to six distinct layers: the troposphere, tropopause, stratosphere, mesosphere, and ionosphere (or thermosphere), plus the very thin exosphere fading into interplanetary space. The ozone layer is in the upper level of the stratosphere.

atom: The smallest particle of matter that has the characteristics of an element, such as oxygen, iron, calcium, or uranium. Three subatomic particles are common to all atoms: protons, neutrons, and electrons. The characteristics of any atom are determined by the number of these particles, particularly the negatively charged electrons in the outer shell. *See also* compound, element, molecule, periodic table of the elements.

atomic number: The number of protons (positively charged particles) in the nucleus of an atom, also the number of electrons (negative charge) in the atom in its standard form. Ions of an atom have larger or smaller levels of electron charge. *See also* electron, ion, periodic table of the elements, proton.

atomic weight: The total mass of the protons and neutrons in an atomic nucleus, with a tiny addition for the weight of electrons. Uranium has the atomic number 92, for 92 protons and 92 electrons, but different isotopes such as U-235 (atomic weight 235, adding 143 neutrons to 92 protons) or U-238 (atomic weight 238, adding 146 neutrons to 92 protons).

ballistics: The science of propelling objects, from rocks and spears to spacecraft. Mastery of this field requires mathematical precision in determining the energy required to put a stationary object into motion in a desired direction, and adjust its course, considering friction from wind or water, or the absence of friction in relatively empty vacuum, and the effect of any body powerful enough to exert gravitational pull, such as the Earth, Moon, or Sun.

basic chemical: Any substance that reacts with an acid. Bases include some metals, such as sodium, calcium, zinc, and aluminum when not protected by an aluminum oxide coating. Other bases include carbonates, hydroxides, and metal oxides (compounds formed by burning metals in oxygen). When a base reacts with an acid, the result is a metal salt and water. *See also* acid, alkali.

battery: In electricity, any device for storing an electrical charge so that it can be used later to power a machine, heater, or light source. Common types of batteries include lead-acid batteries, used for internal combustion automobiles and backup power for industries and military bases; solid alkaline and carbon-zinc batteries, used for flashlights and portable radios; mercury oxide batteries, used in small electronic equipment such as hearing aids, rechargeable nickel-cadmium and nickel hydride batteries; and lithium-ion batteries, an advanced rechargeable type used in portable computers, iPods, and hybrid or electric motor vehicles. Every battery relies on an oxidation-reduction chemical reaction induced by passing a current through its component materials and a reverse reaction that gives off an electric current when plugged into a circuit.

beta particle: One of three common forms of radiation from the nuclei of unstable radioactive elements, carrying a negative charge, similar to an electron, but moving at a high rate of speed, formed when a neutron (neutral electrical charge) transforms into a proton (positive electrical charge) by ejecting a negative charge. *See also* alpha particle, gamma ray.

binary number system: A mathematical system having only two numerals, 0 and 1, most commonly used in computer hardware and software, because at its most basic, a computer can turn a series of switches on (value = 1) or off (value = 0). *See also* computer.

biosphere: First defined by Austrian geologist Edward Seuss in 1875 as "the place on Earth where life dwells." The concept has been expanded to include all living organisms on Earth, dead organic matter, and the biological component of dynamic processes such as the cycling of carbon, nitrogen, oxygen, phosphorous, and other elements.

British thermal unit (Btu): The heat required to raise one pound of water one degree Fahrenheit, equal to 252 calories. Also, the heat required to produce 779.9 foot-pounds of energy in a mechanical system. This is a common measure of the energy potential in fuels. *See also* heat.

calorie: The heat required to raise the temperature of one gram of water one degree Celsius. This is a common measure of the energy potential in food but can be applied to fuels and mechanical processes. *See also* heat.

carboniferous fuels: Any source of energy obtained from carbon-based compounds, particularly coal and oil. These two common fuel sources accumulated during the Carboniferous period of the Paleozoic era.

catalyst: A substance that makes a chemical reaction between two other substances proceed at a significantly faster rate, without being consumed in the reaction. (See also enzyme.)

ceramics: Inorganic, nonmetallic solids, particularly made from clay, that are processed at high levels of heat, then cooled. In addition to common use for pottery and tableware, ceramics have many industrial applications, such as ceramic-based thermocouples to measure high temperatures, ceramic insulators, laser components, heat storage and diffusion, and capacitors.

chromosome: A basic unit of heredity in living cells. Each chromosome is composed of proteins and DNA, which carry thousands of genes. In a healthy, normal, human cell, there are twenty-three pairs of chromosomes. In sexual reproduction, one chromosome in each pair comes from the father, the other from the mother. *See also* DNA, gene.

climate: Prevailing weather conditions in a specific region or area over the course of a year, characterized by a typical range of temperatures, seasonal or year-round humidity, precipitation, prevailing winds, and extremes of seasonal variation.

cold: The sensation felt when a living organism is in contact with a substance of a lower temperature. Thermal energy naturally flows from a warmer object to a colder object when they are in physical contact. *See also* absolute zero, heat, temperature.

compound: A chemical substance composed of molecules, containing atoms of two or more different elements, forming a single particle. For example, water is a compound, in which each water molecule contains two atoms of the element hydrogen and one atom of the element oxygen. *See also* atom, molecule, element, periodic table of the elements.

computer: Any device that can be programmed to perform mechanical or electrical computation, processing numbers. Since the middle of the twentieth century, the term is commonly used for equipment that can be programmed using a binary number system to perform a variety of work. For a computer to process letters, words, graphic images, or maps, human programmers have to encode nonnumerical data in a numeric form. *See also* binary number system.

cryptology: The science of encrypting or decrypting information, including creating codes for privacy or security and finding means to break a code by working from a message in an unknown code to learn the pattern. Any code in which a symbol is substituted for each letter of the alphabet is particularly vulnerable to decryption, because in a large sample, there is a probability for how often each letter will appear, with "e" being used more often than any other. *See also* encoding.

demography: Variations in human population that can be studied statistically, in defined groups, rather than individual behavior. Almost any characteristic can form the basis for demographic research: ethnicity, diet, religion, wealth, language, education, urbanization, occupation, marriage customs, class or caste distinctions.

desalination: Removing salt from water for use in drinking or agriculture. Ninety-seven percent of the water on earth is salt water, mostly in the oceans. The dissolved salt content is harmful to freshwater land plants and land animals. Removing the salt is energy intensive and therefore has a high cost, but in places where freshwater is in short supply, it is sometimes considered worth the expense.

detergent: A type of surfactant that has the property of removing stains or particles from a surface, keeping it suspended in water, and allowing the suspended solids to be rinsed away. *See also* surfactant.

distillation: Isolating and purifying a liquid substance by heating a solution to the exact boiling point of the desired end product, producing a vapor, then capturing and condensing the vapor in a separate container. Water, perfumes, and alcohols are all common examples of liquids that can be distilled.

DNA (deoxyribonucleic acid): The complex molecule making up genes, encoding inheritance in all living species. It is known for a unique double-helix structure, with the code in an "alphabet" of four types of molecule: cytosine pairs with guanine, while adenine pairs with thymine. DNA is increasingly used for identification, particularly in crime scenes, to determine paternity of a child, and to study inheritance of both individuals and demographic groups. Study of DNA is also leading to new treatments for genetically inherited diseases. *See also* chromosome, gene.

ecology: A branch of biology that studies the complex relationships between living organisms, their environment, the manner in which a variety of plant and animal life forms a mutually interdependent ecosystem, and the competition between life forms within and between ecosystems.

electrical storage: See battery.

electrolysis: Applying an electric current to take apart (decompose) the molecules of a substance in solution. If an electric current is run through a container of water, the hydrogen and oxygen atoms in the water molecule will separate and form diatomic molecules (two atoms in each molecule) of oxygen and hydrogen gas. Electrolysis can be applied to many compounds. Industrial uses include separating chlorine and caustic soda from brine and refining metals such as sodium, calcium, magnesium, and aluminum from common ore compounds.

electromagnetism: A fundamental force of nature acting on all electrically charged particles. The previously separate studies of electricity, magnetism, and optics (initially visible light, which is only one spectrum of electromagnetic radiation) were unified by the work of English scientists Michael

Faraday and Robert Maxwell in the mid-nineteenth century.

electron: A negative charge in the outer shell of every atom. Electrons are sometimes described as particles, but within an atom they act more as electrical charges in shells, rather than particles in orbit. In a stable atom, the number of electrons exactly equals the number of protons in the nucleus. Ions have fewer or additional electric charges. *See also* atom, ion, neutron, proton.

element: A substance formed by atoms of a single type, found in the periodic table of the elements. Hydrogen, helium, and oxygen are all examples of elements. *See also* atom, compound, molecule, periodic table of the elements.

encoding: Representing information by a system of symbols or characters. Encoding processes exist in human memory, heredity, computer programming, and in written or verbal communication, including military or business communications intended to be secret. The most common use is to take a message in a plain language and convert it into a sequence of characters that can be read only by a person instructed in the code—or by a cryptologist who can break the code by mathematically analyzing the pattern. Natural encoding processes include the genetic code, which stores in long molecules of DNA the structure, physiology, and metabolism of a complete living organism. *See also* binary number system, computer, cryptology, DNA.

engineering: Practical application of the knowledge of pure science, and sometimes of art as well, not only to construct buildings, bridges, infrastructure, and engines, but to plan and organize industrial and community processes. There are many branches of engineering, including electrical, industrial, mechanical, civil, aeronautical, geotechnical, transportation, water management, disaster preparedness and management, and telecommunications.

entropy: The spontaneous direction of any natural process, tending to lose energy or to become more chaotic. All things naturally tend toward equilibrium: Two objects of different temperatures, placed in contact, will equalize to a common temperature as heat is transferred from the warmer object to the colder object. A solid dissolved in a solution spreads evenly throughout the solution, gas at two different pressures will equalize, but in none of these examples will objects spontaneously develop uneven temperatures, pressures, or concentrations in solution.

enzyme: A biological catalyst, any protein molecule within a living organism that speeds up biochemical reactions to a rate that will sustain life. The effect may speed up metabolic reactions by a factor of one million, compared with what would occur chemically outside the body. Names and classification of enzymes are regulated by the International Commission on Enzymes. Most enzymes are named by adding -ase to the root of a corresponding substrate, the molecule an enzyme acts upon. Sucrase catalyzes the hydrolysis of sucrose into glucose and fructose. A living cell has a unique set of 3,000 enzymes, each defined by the cell's DNA.

erosion: The process of wind or water wearing away soil or rock. The existence of soil is due in part to erosion of stone surfaces from the time solid rock first formed on Earth. Whole mountain chains have been worn down by erosion—the Appalachian Mountains were once as high as the Rockies. Sandstone is formed by compression of eroded rock under the oceans. Soil erosion can destroy cultivated land. Sheet erosion removes soil in a uniform layer, while rill erosion cuts small channels into the soil, and gully erosion forms deeper channels carrying away a large volume of soil.

eukaryotic cell: A complex cell on which all life more complex than a bacterium or yeast is based, characterized by a nucleus containing the cell's DNA and a number of specialized organelles, supplied with energy by mitochondria, which individually resemble the more primitive prokaryotic cell. *See also* prokaryotic cell.

exosphere: A layer of Earth's atmosphere from 500 kilometers above the surface to between 10,000 and 190,000 kilometers—or halfway to the Moon. At 190,000 kilometers, the force of solar radiation is more powerful than the force of Earth's gravity on the thinly distributed atmospheric molecules, but many scientists consider 10,000 kilometers to be the boundary with interplanetary space. Within the exosphere, a gas molecule can travel hundreds of kilometers before bumping into another gas molecule.

fermentation: A biological process for breaking down complex organic compounds into simpler compounds. One of the most familiar in human history is the conversion by yeast of sugar to carbon dioxide, alcohol, and water. Fermentation also occurs in cells, including animal muscle cells, breaking down glucose to produce lactic acid, lactate, carbon dioxide, and water, as well as adenosine triphosphate, a source of energy. It is less efficient than cellular respiration but occurs when muscles are short of oxygen. Many anaerobic bacteria ferment sugars: Lactobacillus ferment milk to produce yogurt. Fermentation also produces lactic acid in a variety of foods, such as sauerkraut and sourdough bread.

forensic: The application of science to legal concerns. The analysis of crime scenes, firearms, DNA, and the pathology of dead bodies are common subjects of forensic investigation, but dentists, toxicologists, psychiatrists, engineers, and practitioners in many other fields can also be called upon.

fuel cell: A source of electric current that operates in a manner similar to a battery, generating electricity as a by-product of a chemical reaction. The difference is that a fuel cell continues generating power as long as it is supplied with fuel. The hydrogen fuel cell, one of the most commonly known, generates electricity, heat, and water.

gamma ray: One of three common forms of radiation from the nuclei of unstable radioactive elements, which has no mass and no electrical charge. It is made up of photons, the fundamental particle of light, moving at the speed of light, emitted at a high energy level. X rays are similar but originate in electron fields rather than in the nucleus of atoms. *See also* alpha particle, beta particle.

gene: A unit of hereditary information found within a chromosome that determines the characteristics of an organism. Each gene is an ordered series of nucleotides, which are subunits of DNA, composed of a base molecule containing nitrogen, a phosphate molecule, and a pentose sugar molecule. *See also* chromosome, DNA.

Global Positioning System (GPS): A system owned by the United States government and operated by the Air Force for determining the exact position on the surface of the Earth of any user or defined landmark. By receiving signals from any of 24 orbiting satellites that are in direct line of site to a user's position, GPS devices can calculate latitude, longitude, altitude, and time.

gravity: As defined by Sir Isaac Newton's law of universal gravitation, a force that attracts any two objects in the universe. The strength of gravitational attraction depends on the mass of the two objects and decreases according to the distance between them, squared.

halogens: Elements fluorine, chlorine, bromine, iodine, and astatine in the periodic table of the elements. Being one electron short of filling the outermost shell of each atom, halogens form chemical bonds easily with other elements. They often form salts, including common table salt (sodium chloride), and calcium chloride—the salt applied to roads in winter to melt ice.

heat: The transfer of thermal energy, the kinetic energy of molecules, between two objects as the result of a temperature difference. Thermal energy is a vibration at the molecular level, which increases the volume of a substance, or increases the pressure of a gas or liquid in a closed space. Heat can be used to accomplish work mechanically; it is also produced by friction of moving parts, which wastes energy applied to accomplish work in a mechanical system.

husbandry: The cultivation of land to raise edible plant crops (and textile crops) or breed and raise domestic animals for food. *See also* animal husbandry.

hydrocarbons: Among the simplest of organic molecules, made up of a number of carbon and hydrogen atoms. Compounds with a benzene ring in the molecular structure are called aromatic hydrocarbons. Those without a benzene ring are called aliphatic hydrocarbons, which include alkanes (single carbon bond), alkenes (one double bond), and alkynes (one triple bond). There are a nearly unlimited number of derivative carbon compounds that can be formed by adding oxygen atoms to hydrocarbons.

hydrology: The study of water and, more specifically, of the way water cycles through lakes, streams, ponds, rivers, oceans, the atmosphere, and underground water flows and reservoirs, including evaporation, rain-, and snowfall. This includes study of contamination in water, as well as movement and distribution.

hydroponics: Growing plants in a solution of water and selected nutrients, without need for soil. Hydroponics can rely on sunlight or can be established in a closed, indoor environment using grow lights.

inflammable: An object or substance that catches fire easily, from the Latin *inflammare*, meaning to kindle a flame. Often confused, because "flammable" has the same meaning, and for English words, the prefix in- often means opposite, such as invisible. Safety officials have encouraged use of "flammable" to avoid misunderstanding and "nonflammable" to mean a substance that will not burn easily. Inflammability of any substance increases with higher concentration of oxygen in the surrounding atmosphere and decreases with lower oxygen concentration. Burning is an oxidation process—combining oxygen with the molecules of the burning substance—which can begin at lower temperatures with a higher concentration of oxygen. Once started, sufficient heat is given off to make the fire self-sustaining.

ion: An atom that has more or fewer electrons in its outermost shell than protons in its nucleus. *See also* atom, compound, element, molecule, periodic table of the elements.

ionosphere: The outermost layer of Earth's atmosphere, so named because solar radiation causes many atoms at this level to ionize, gaining or losing an electron. *See also* ion.

irradiation: Exposing any substance, or living organism, to radiation most commonly used to destroy bacteria, viruses, fungi (mold), and insects in food or on surfaces used for food preparation. The level of radiation used is not strong enough to disintegrate the nucleus of any atom making up the substance of any food item, so the food itself does not become radioactive. Irradiation is also used in treating cancer, checking luggage at airports, and sterilizing many items.

joule: A measure of heat or energy, used more commonly in scientific research than in industry; the energy required to accelerate a body with a mass of one kilogram using one newton of force over a distance of one meter. Equal to 0.2390 calories or 0.738 foot-pounds.

laser: A light source that emanates from a well-defined wavelength; originally an acronym for light amplified by stimulated emission of radiation.

light-emitting diode (LED): A diode constructed to provide illumination from the movement of electrons through a semiconductor, which is housed in a bulb that concentrates the light in a desired direction. Because there is no filament to warm up, as in a conventional electric light bulb, LEDs use ten percent or less electrical current to provide the same brightness of light and last up to twenty times longer. LEDs can be constructed to provide almost any desired color or hue of light.

lithography: A printing process that is the most common method of printing and publishing. The earliest process, invented by Alois Senefelder in Germany in 1798, is still used by hand-applying a greasy ink to a specially prepared block of limestone, which is moistened with water. The water is repelled by the ink, and in turn repels an oil-based ink, applied with a roller, from moistened areas, creating the desired image when paper is applied against the plate. Modern offset lithography burns a photosensitive metal plate through a negative film image to create a pattern of roughened areas—to which oil- or rubber-based inks will adhere—while a thin layer of water repels the ink from unexposed smooth metal surfaces.

magnetism: One aspect of electromagnetism, involving fields either generated by a current moving through a wire or a magnetized object, in which the molecules are aligned with magnetic north in one uniform direction and magnetic south in the opposite direction.

mean: Sometimes called the arithmetic mean, or average, it is the sum of a series of figures divided by the total number of figures.

median: The middle number in a group of numbers arranged in order from lowest to highest. Comparing the mean and the median can help to correct for the distortion of extreme highs or lows.

mesosphere: A layer of Earth's atmosphere, between the stratosphere and the ionosphere (or thermosphere), variously defined as beginning at an altitude from 30 to 55 kilometers above sea level and continuing to an altitude of about 80 to 90 kilometers.

metabolism: Physical processes and chemical reactions within a living body that convert or use energy, including those associated with digestion, excretion, breathing, blood circulation, growing and using muscle tissues, communication through the nervous system, and body temperature. *See also* enzyme, protein.

metals: A majority of known elements, generally

shiny in appearance and good conductors of heat and electricity. In ionic compounds, metals usually provide the positive ion. Many metals react with acids and therefore act as a base.

metric system: Scales of measurement in which each unit is one tenth of the next largest and ten times the next smallest, simplifying conversion, recording, and mathematical operations. The system is built around the unit of the meter (equivalent to 39.39 inches), which the system's inventors defined as one-ten millionth the distance from the North Pole to the equator.

microbe: Any microscopic form of life, also called a microorganism, particularly bacteria, protozoa, fungi, or virus. Most commonly, this term refers to pathogenic microscopic life—those that cause infection, disease, decay, sepsis, or gangrene. However, biologists are identifying an increasing number of microbes that are beneficial, even essential to life, including a variety of those found in the human intestine.

mitochondria: A type of organelle within eukaryotic cells, where oxygen and nutrients are converted into adenosine triphosphate, the molecule that stores chemical energy for the cell. This process, called aerobic respiration, is possible only in the presence of oxygen. Mitochondria are rod-shaped, have their own DNA, and reproduce independently within the cell, resembling some primitive prokaryotic cells. This suggests that prokaryotic cells were absorbed within the cell walls of evolving eukaryotic cells in a symbiotic relationship. Mitochondria enable cells to produce adenosine triphosphate fifteen times more efficiently than is possible by anaerobic respiration.

mode: The most frequent value in a series of numbers. For example, if the students in a class are all age sixteen, seventeen, or eighteen, and there are more sixteen-year-olds than either seventeen- or eighteen-year-olds, then sixteen would be the mode.

molecule: A particle formed by two or more atoms of the same element or of different elements. A water molecule is formed by one oxygen atom and two hydrogen atoms. Bonds holding molecules together are formed by the electrons in each atom's outer shell. *See also* atom, compound, element, periodic table of the elements.

navigation: The art and science of plotting a course from a starting point to a desired destination, most commonly piloting a boat on water or travel in outer space. The term is sometimes used for travel on land as well, particularly in a desert or flatland without significant landmarks, or in recent years, using a GPS navigation device for ordinary driving. Historically, navigation has been accomplished by using certain stars as reference points for latitude. Longitude was often guesswork, based on the time of day at the starting point compared with the current location, until invention of the seagoing chronometer in 1764. Radio beacons, radar, the gyroscopic compass, accurate maps, and the satellite-based GPS have all provided increased precision in navigation.

neuron: A nerve cell, the basic unit of the spinal cord, nervous system, and brain of human beings and other mammals. Neurons communicate information in chemical and electrical forms throughout an organism. Sensory neurons provide information to the brain from receptors in every part of the body. Motor neurons transmit direction from the brain to muscles. Shortly after birth, neurons stop reproducing, while other cell types continue to do so. Neurons have specialized structures called axons and dendrites to send and receive signals at connections known as synapses.

neutron: A neutral particle within the nucleus of an atom, having neither a positive or negative electrical charge and a weight slightly more than that of a proton. *See also* atom, electron, proton.

noble gases: Elements helium, neon, argon, krypton, xenon, and radon in the periodic table of the elements. They are called "noble" because having their outer electron shell filled to capacity, they do not react with other elements or form compounds. Radon, however, is a radioactive element. All of these elements exist in the form of gases, forming liquids or solids only at extremely low temperatures, approaching absolute zero.

nucleus (atomic): The tightly packed protons and neutrons at the core of every atom, which account for nearly all of an atom's mass. The total size of an atom is defined by its electron shells.

nucleus (cell): An organelle within each living eukaryotic cell that acts as a control center, storing genes on chromosomes, producing messenger RNA molecules (which transfer code for essential proteins from genes in the chromosomes), producing ribosomes, and organizing replication of DNA, including complete copies for cell division.

optics: The study of light, including all systems for gathering, concentrating, and manipulating light, such as mirrors, spectacles, telescopes, microscopes, cameras, spectroscopes, lasers, fiber optic communications, and optical data storage and retrieval.

organic compounds: Compounds that are created in or by living cells, rather than as a result of spontaneous physical processes. The simplest organic compounds are hydrocarbons, composed of carbon and hydrogen. Sugars and some other compounds are composed of carbon, hydrogen, and oxygen. The most complex organic compounds are composed of carbon, hydrogen, oxygen, nitrogen, occasionally sulfur, and sometimes small traces of metals.

oxidation: The process of any other element forming chemical bonds with atoms of oxygen. Common examples include the formation of rust on metal, and the burning of wood or any other inflammable substance. Oxidation occurs at many points in human and animal metabolism, including oxygen binding to hemoglobin in the blood stream and many chemical reactions within each cell in the body.

oxygenation: Infusion of oxygen gas into a solution or organic process, such as the transfer of oxygen through the membranes in the lung to enter the bloodstream.

ozone layer: An outer layer of Earth's atmosphere made up of ozone, a molecule of oxygen containing three atoms, instead of the two atoms of the oxygen breathed by living things. While ozone is toxic to life, causing chemical burns at relatively low concentrations, the ozone layer absorbs much of the ultraviolet radiation in sunlight, protecting life at the planet's surface.

pasteurization: A process discovered by French microbiologist Louis Pasteur for rapidly heating milk to destroy disease-causing bacteria, while leaving the nutritional content of the milk unaffected. It is also used to destroy bacteria in wine and beer manufacturing.

periodic table of the elements: The table of chemical elements arranged according to their atomic number. *See also* atom, atomic number, element, valence.

photon: The particle of light. Since light is a wave of electromagnetic radiation, it is something of a paradox that it also travels in particles. Quantum mechanics is based on the observation that electromagnetic radiation travels in discrete quantities of energy, called quanta. The photon is also the particle that, moving at a high volume of energy, is called a gamma ray.

photosynthesis: A chemical process in which carbon dioxide and water are converted into carbohydrates and oxygen by the energy from a light source, generally the Sun. This reaction, which all plants and many bacteria rely on, is the source of the unnatural presence of oxygen in Earth's atmosphere and supplies the entire food chain upon which animal life depends for existence.

photovoltaic: Electricity from light—any process using the energy of light to generate electricity directly. The most common practical application is the photovoltaic cell, made of a thin semiconductor wafer, treated to form an electric field, with electrical conductors attached to its positive and negative sides. When these poles are connected by a circuit, an electrical current is generated by photons from incoming sunlight, knocking loose electrons in the semiconductor.

physiology: Study of the entire system of physical functions in a living body: its mechanical, biochemical, and bioelectrical processes, the purpose and operation of each organ, and the interaction of different organs and parts.

polarity: The existence of two opposite characteristics, such as the north and south poles of a magnet or the positive and negative poles of a battery.

prokaryotic cell: The earliest and most primitive type of cell, probably the first life form on Earth, lacking a cell nucleus. Most bacteria are prokaryotic. Most one-celled animals, such as paramecium, and simple plants such as algae have the more complex eukaryotic cell.

protein: A long chain of amino acids. There are thousands of different proteins in each cell of the human body, and since each species has slightly different proteins in its cells, there are millions of different proteins in the biosphere. A balance of all necessary proteins is essential to the continued life of any organism. Food consumption must either supply each complete protein that the human body cannot manufacture for itself or a wide variety of incomplete proteins that can be assembled into complete proteins.

proton: The positively charged particles in the nucleus of every atom. Along with neutrons, protons account for most of the weight in an atom, because electrons have very little weight. *See also* atom, electron, neutron.

protoplasm: The living substance of a cell, including the content of the cell membrane and the substance within the cell—a transparent gelatinous material composed of inorganic substances (90 percent water with mineral salts and gases such as oxygen and carbon dioxide), and organic substances (proteins, carbohydrates, lipids, nucleic acids, and enzymes). Protoplasm outside the cell nucleus is called cytoplasm.

radiation: Emission of subatomic particles or of photons at high energy levels from radioactive atoms. *See also* alpha particles, beta particles, gamma rays.

radio waves: A band of wavelengths in the electromagnetic spectrum that can be generated by a spark gap in an electrical circuit and are commonly used for broadcast communication. *See also* electromagnetism, ionosphere.

refrigerant: A compound that transfers thermal energy in cooling systems, including air conditioners, freezers, refrigerators, and low-temperature manufacturing processes. Releasing the refrigerant at low pressure, in tubes that are in physical contact with the area to be cooled, transfers heat to the refrigerant, which is then transferred to radiator coils outside the area to be cooled, by compressing the refrigerant to a liquid state. Common refrigerants include ammonia, dichlorodifluoromethane, propane, hydrochlorofluorocarbon (HCFC), and hydrofluorocarbon (HFC).

RNA (ribonucleic acid): A complex molecule similar to, but less complex than, DNA. RNA molecules transfer genetic information from genes in longer DNA molecules forming the chromosomes of living cells to the active metabolic proteins in a living cell. *See also* DNA.

scientific method: A process for investigating nature by observation and experiment, creating hypotheses to make sense of observations.

seismology: The study of earth movement, particularly the mechanism and causes of earthquakes.

semiconductor: A material that conducts electricity more efficiently than an insulator, but less efficiently than a conductor, useful in constructing diodes, which conduct electricity in only one direction. Common semiconductor materials include silicon, germanium, and selenium. Semiconductors are essential to construction and design of computers and most electronic equipment.

sewage: Waste, usually carried in water, which is either chemical industrial waste or organic waste. Industrial sewage often includes metals, other toxins, and complex molecules that are not naturally metabolized. In most industrialized economies, sewage goes through several stages of treatment to remove solid waste for disposal, and the water is returned to lakes and rivers in relatively clean condition.

spectroscopy: Study of the spectrum of wavelengths of electromagnetic radiation, particularly the wavelengths of light visible to the human eye. When light reflected off any substance is viewed or projected through a spectroscope, each element casts a unique pattern of lines on the visible spectrum. This is useful in astronomy for identifying the chemical composition of stars and planets, since physical samples cannot be obtained, as well as in analytical chemistry.

statistics: Methods of obtaining, organizing, analyzing, interpreting, and presenting numerical data, used in many areas of science and industry, as well as in demographic studies of human populations. *See also* mean, mode, median.

sterilization (of microbes and pathogens): Killing all or most microbes on a working surface, or on the surface of instruments to be used in medical care or food preparation, by means of heat (including pasteurization), irradiation, or chemical antiseptics.

sterilization (pertaining to reproduction): Surgically or chemically preventing an organism from producing offspring. This includes neutering or spaying of pets and domesticated farm or service animals, castration of male animals, irradiating male or female insects as a pest-control measure, and, in humans, severing and tying the Fallopian tubes to prevent pregnancy in a woman or vasectomy to prevent a man inseminating a woman.

stratosphere: A layer of Earth's atmosphere from 18 kilometers to 50 kilometers. Unlike the troposphere, airflow is mostly horizontal with no weather patterns. The ozone layer is in the upper level of the stratosphere.

surfactant: A chemical substance that reduces the surface tension of water. A common use of surfactants is in manufacture of detergents and soaps. Surfactants are complex molecules with one component attracted to water molecules (hydrophilic) and the other component repelled by water molecules (hydrophobic).

temperature: In theory, a measure of the average kinetic energy of molecules: The larger or denser an object is, the more heat is required to raise its temperature by one degree. The most common scales for measuring temperature select arbitrary fixed points and assign arbitrary numerical values. The Fahrenheit scale assigns a value of 32 degrees to the temperature at which water freezes and 212 degrees to the temperature at which water boils. The Celsius scale assigns zero to the freezing point and 100 degrees to the boiling point. The Kelvin scale begins with absolute zero, the temperature at which a substance lacks any thermal energy. *See also* absolute zero, cold, heat.

thermosphere: *See* ionosphere.

toxin: A poison; any substance that will have a toxic effect on organic life, particularly proteins produced by bacteria, plants, or in specialized glands of certain animals. Toxins may damage or paralyze without killing or have a caustic effect, but in high enough concentrations, exposure to most toxins will cause death. Some toxins, in low doses, can be used in medical treatment: Two of the seven types of botulinum toxin are used to inhibit muscle spasms, smooth wrinkles of the upper face (Botox), and treat cervical dystonia.

troposphere: The layer of Earth's atmosphere closest to the surface, up to 14 kilometers, where all weather takes place, with rising and falling air currents. Air pressure at the upper limit of the troposphere is about 10 percent of the pressure at sea level.

ultrasound: Vibrations or sound waves at a higher frequency than the human ear can detect.

vacuum: The absence of matter, including air or other gases, inside an enclosed space or in remote areas of outer space. No perfect vacuum is known, since interstellar space is estimated to contain at least one hydrogen atom per cubic meter.

valence: The capacity of the atoms of an element to combine with other atoms of the same element or another element. *See also* atom, molecule, periodic table of the elements.

Charles Rosenberg

TIMELINE

The Time Line below lists milestones in the history of applied science: major inventions and their approximate dates of emergence, along with key events in the history of science. The developments appear in boldface, followed by the name or names of the person(s) responsible in parentheses. A brief description of the milestone follows.

2,500,000 B.C.E.	**Stone tools:** Stone tools, used by Homo habilis and perhaps other hominids, first appear in the Lower Paleolithic age (Old Stone Age).
400,000 B.C.E.	**Controlled use of fire:** The earliest controlled use of fire by humans may have been about this time.
200,000 B.C.E.	**Stone tools using the prepared-core technique:** Stone tools made by chipping away flakes from the stones from which they were made appear in the Middle Paleolithic age.
100,000-50,000 B.C.E.	**Widespread use of fire by humans:** Fire is used for heat, light, food preparation, and driving off nocturnal predators. It is later used to fire pottery and smelt metals.
100,000-50,000 B.C.E.	**Language:** At some point, language became abstract, enabling the speaker to discuss intangible concepts such as the future.
16,000 B.C.E.	**Earliest pottery:** The earliest pottery was fired by putting it in a bonfire. Later it was placed in a trench kiln. The earliest ceramic is a female figure from about 29,000 to 25,000 B.C.E., fired in a bonfire.
10,000 B.C.E.	**Domesticated dogs:** Dogs seem to have been domesticated first in East Asia.
10,000 B.C.E.	**Agriculture:** Agriculture allows people to produce more food than is needed by their families, freeing humans from the need to lead nomadic lives and giving them free time to develop astronomy, art, philosophy, and other pursuits.
10,000 B.C.E.	**Archery:** Archery allows human hunters to strike a target from a distance while remaining relatively safe.
10,000 B.C.E.	**Domesticated sheep:** Sheep seem to have been domesticated first in Southwest Asia.
9000 B.C.E.	**Domesticated pigs:** Pigs seem to have been domesticated first in the Near East and in China.
8000 B.C.E.	**Domesticated cows:** Cows seem to have been domesticated first in India, the Middle East, and sub-Saharan Africa.
7500 B.C.E.	**Mud bricks:** Mud-brick buildings appear in desert regions, offering durable shelter. The citadel in Bam, Iran, the largest mud-brick building in the world, was built before 500 B.C.E. and was largely destroyed by an earthquake in 2003.
7500 B.C.E.	**Domesticated cats:** Cats seem to have been domesticated first in the Near East.
6000 B.C.E.	**Domesticated chickens:** Chickens seem to have been domesticated first in India and Southeast Asia.
6000 B.C.E.	**Scratch plow:** The earliest plow, a stick held upright by a frame and pulled through the topsoil by oxen, is in use.
6000 B.C.E.	**Electrum:** The substance is a natural blend of gold and silver and is pale yellow in color like amber. The name "electrum" comes from the Greek word for amber.

6000 B.C.E.	**Gold:** Gold is discovered—possibly the first metal to be recognized as such.
6000-4000 B.C.E.	**Potter's wheel:** The potter's wheel is developed, allowing for the relatively rapid formation of radially symmetric items, such as pots and plates, from clay.
5000 B.C.E.	**Wheel:** The chariot wheel and the wagon wheel evolve—possibly from the potter's wheel. One of humankind's oldest and most important inventions, the wheel leads to the invention of the axle and a bearing surface.
4200 B.C.E.	**Copper:** Egyptians mine and smelt copper.
4000 B.C.E.	**Moldboard plow:** The moldboard plow cut a furrow and simultaneously lifted the soil and turned it over, bringing new nutrients to the surface.
4000 B.C.E.	**Domesticated horses:** Horses seem to have been domesticated first on the Eurasian steppes.
4000 B.C.E.	**Silver:** Silver can be found as a metal in nature, but this is rare. It is harder than gold but softer than copper.
4000 B.C.E.	**Domesticated honeybees:** The keeping of bee hives for honey arises in many different regions.
4000 B.C.E.	**Glue:** Ancient Egyptian burial sites contain clay pots that have been glued together with tree sap.
3500 B.C.E.	**Lead:** Lead is first extruded from galena (lead sulfide), which can be made to release its lead simply by placing it in a hot campfire.
c. 3100 B.C.E.	**Numerals:** Numerals appeared in Sumerian, Proto-Elamite, and Egyptian hieroglyphics.
3000 B.C.E.	**Bronze:** Bronze, an alloy of copper and tin, is developed. Harder than copper and stronger than wrought iron, it resists corrosion better than iron.
3000 B.C.E.	**Cuneiform:** The method of writing now known as cuneiform began as pictographs but evolved into more abstract patterns of wedge-shaped (cuneiform) marks, usually impressed into wet clay. This system of marks made complex civilization possible, since it allowed record keeping to develop.
3000 B.C.E.	**Fired bricks:** Humans begin to fire bricks, creating more durable building materials that (because of their regular size and shape) are easier to lay than stones.
3000 B.C.E.	**Pewter:** The alloy pewter is developed. It is 85 to 99 percent tin, with the remainder being copper, antimony, and lead; copper and antimony make the pewter harder. Pewter's low melting point, around 200 degrees Celsius, makes it a valuable material for crafting vessels that hold hot substances.
2700 B.C.E.	**Plumbing:** Earthenware pipes sealed together with asphalt first appear in the Indus Valley civilization. Greeks, Romans, and others provided cities with fresh water and a way to carry off sewage.
2650 B.C.E.	**Horse-drawn chariot (Huangdi):** Huangdi—a legendary patriarch of China—is possibly a combination of many men. He is said to have invented—in addition to the chariot—military armor, ceramics, boats, and crop rotation.

2600 B.C.E.	**Inclined plane:** Inclined planes are simple machines and were used in building Egypt's pyramids. Pushing an object up a ramp requires less force than lifting it directly, although the use of a ramp requires that the load be pushed a longer distance.
c. 2575-c. 2465 B.C.E.	**Pyramids:** Pyramids of Giza are built in Egypt.
1750 B.C.E.	**Tin:** Tin is alloyed with copper to form bronze.
1730 B.C.E.	**Glass beads:** Red-brown glass beads found in South Asia are the oldest known human-formed glass objects.
1600 B.C.E.	**Mercury:** Mercury can easily be released from its ore (such as cinnabar) by simply heating it.
1500 B.C.E.	**Iron:** Iron, stronger and more plentiful than bronze, is first worked in West Asia, probably by the Hittites. It could hold a sharper edge, but it had to be smelted at higher temperatures, making it more difficult to produce than bronze.
1500 B.C.E.	**Zinc:** Zinc is alloyed with copper to form brass, but it will not be recognized as a separate metal until 1746.
1000 B.C.E.	**Concrete:** The ancient Romans build arches, vaults, and walls out of concrete.
1000 B.C.E.	**Crossbow:** The crossbow seems to come from ancient China. Crossbows can be made to be much more powerful than a normal bow.
1000 B.C.E.	**Iron Age:** Iron Age begins. Iron is used for making tools and weapons
700 B.C.E.	**Magnifying glass:** An Egyptian hieroglyph seems to show a magnifying glass.
350 B.C.E.	**Compass:** Ancient Chinese used lodestones and later magnetized needles mostly to harmonize their environments with the principles of feng shui. Not until the eleventh century are these devices used primarily for navigation.
350-100 B.C.E.	**Scientific method (Aristotle):** Aristotle develops the first useful set of rules attempting to explain how scientists practice science.
300 B.C.E.	**Screw:** Described by Archimedes, the screw is a simple machine that appears to be a ramp wound around a shaft. It converts a smaller turning force to a larger vertical force, as in a screw jack.
300 B.C.E.	**Lever:** Described by Archimedes, the lever is a simple machine that allows one to deliver a larger force to a load than the force with which one pushes on the lever.
300 B.C.E.	**Pulley:** Described by Archimedes, the pulley is a simple machine that allows one to change the direction of the force delivered to the load.
221-206 B.C.E.	**Compass:** The magnetic compass is invented in China using lodestones, a mineral containing iron oxide.
215 B.C.E.	**Archimedes' principle (Archimedes of Syracuse):** Archimedes describes his law of displacement: A floating body displaces an amount of fluid the weight of which is equal to the weight of the body.
200 B.C.E.	**Astrolabe:** A set of engraved disks and indicators becomes known as the astrolabe. When aligned with the stars, the astrolabe can be used to determine the rising and setting times of the Sun and certain stars, establish compass directions, and determine local latitude.

40 C.E.	**Ptolemy's geocentric system (Ptolemy):** A world system with the Earth in the center, and the Moon, Venus, Mercury, Sun, Mars, Jupiter, Saturn, and fixed stars surrounding it. The geocentric Ptolemaic system would remain the most widely accepted cosmology for the next fifteen hundred years.
90 C.E.	**Aeolipile (Hero of Alexandria):** The aeolipile—a steam engine that escaping steam causes to rotate like a lawn sprinkler—is developed.
105 C.E.	**Paper and papermaking (Cai Lun):** Although papyrus paper already existed, Cai Lun creates paper from a mixture of fibrous materials softened into a wet pulp that is spread flat and dried. The material is strong and can be cheaply mass-produced.
250 C.E.	**Force pump (Ctesibius of Alexandria):** Ctesibius develops a device that shoots a jet of water, like a fire extinguisher.
815 C.E.	**Algebra (al-Khwārizmī):** al-Khw{amacr}rizm{imacr} develops the mathematics that solves problems by using letters for unknowns (variables) and expressing their relationships with equations.
877 C.E.	**Maneuverable glider (Abbas ibn Firnas):** A ten-minute controlled glider flight is first achieved.
9th century	**Gunpowder:** Gunpowder is invented in China.
1034	**Movable type:** Movable type made of baked clay is invented in China.
1170	**Water-raising machines (al-Jazari):** In addition to developing machines that can transport water to higher levels, al-Jazari invents water clocks and automatons.
1260	**Scientific method (Roger Bacon):** Bacon develops rules for explaining how scientists practice science that emphasize empiricism and experimentation over accepted authority.
1284	**Eyeglasses for presbyopia (Salvino d'Armate):** D'Armate is credited with making the first wearable eyeglasses in Italy with convex lenses. These spectacles assist those with farsightedness, such as the elderly.
1439	**Printing press (Johann Gutenberg):** Gutenberg combined a press, oil-based ink, and movable type made from an alloy of lead, zinc, and antimony to create a revolution in printing, allowing mass-produced publications that could be made relatively cheaply and disseminated to people other than the wealthy.
1450	**Eyeglasses for the nearsighted (Nicholas of Cusa):** Correcting nearsightedness requires diverging lenses, which are more difficult to make than convex lenses.
1485	**Dream of flight (Leonardo da Vinci):** On paper, Leonardo designed a parachute, great wings flapped by levers, and also a person-carrying machine with wings to be flapped by the person. Although these flying devices were never successfully realized, the designs introduced the modern quest for aeronautical engineering.
1543	**Copernican (heliocentric) universe:** Copernicus publishes *De revolutionibus* (*On the Revolutions of the Heavenly Spheres*), in which he refutes geocentric Ptolemaic cosmology and proposes that the Sun, not Earth, lies at the center of the then-known universe (the solar system).

1569	**Mercator projection (Gerardus Mercator):** The Mercator projection maps the Earth's surface onto a series of north/south cylinders.
1594	**Logarithms (John Napier):** Napier's logarithms allow the simplification of complex multiplication and division problems.
1595	**Parachute (Faust Veranzio):** Veranzio publishes a book describing sixty new machines, one of which is a design for a parachute that might have worked.
1596	**Flush toilet (Sir John Harington):** Harington's invention is a great boon to those previously assigned to empty the chamber pots.
1604	**Compound microscope (Zacharias Janssen):** Janssen, a lens crafter, experiments with lenses, leading to both the microscope and the telescope.
1607	**Air and clinical thermometers (Santorio Santorio):** Santorio develops a small glass bulb that can be placed in a person's mouth, with a long, thin neck that is placed in a beaker of water. The water rises or falls as the person's temperature changes.
1608	**Refracting telescope (Hans Lippershey):** Lippershey is one of several who can lay claim on developing the early telescope.
1609	**Improved telescope (Galileo Galilei):** Galileo grinds and polishes his own lenses to make a superior telescope. Galileo will come to be known as the father of modern science.
1622	**Slide rule (William Oughtred):** English mathematician and Anglican minister Oughtred invents the slide rule.
1629	**Steam turbine (Giovanni Branca):** Branca publishes a design for a steam turbine, but it requires machining that is too advanced to be built in his day.
1642	**Mechanical calculator (Blaise Pascal):** Eighteen-year-old Pascal invents the first mechanical calculator, which helps his father, a tax collector, count taxes.
1644	**Barometer (Evangelista Torricelli):** Torricelli develops a mercury-filled barometer, in which the height of the mercury in the tube is a measure of atmospheric pressure.
1650	**Vacuum pump (Otto von Guericke):** After demonstrating the existence of a vacuum, von Guericke explores its properties with other experiments.
1651	**Hydraulic press (Blaise Pascal):** Pascal determines that hydraulics can multiply force. For example, a 50-pound force applied to the hydraulic press might exert 500 pounds of force on an object in the press.
1656	**Pendulum clock (Christiaan Huygens):** Huygens discovers that, for small oscillations, a pendulum's period is independent of the size of the pendulum's swing, so it can be used to regulate the speed of a clock.
1662	**Demography (John Graunt):** Englishman Graunt develops the first system of demography and publishes *Natural and Political Observations Mentioned in the Following Index and Made Upon the Bills of Mortality*, which laid the groundwork for census taking.
1663	**Gregorian telescope (James Gregory):** The Gregorian telescope produces upright images and therefore becomes useful as a terrestrial telescope.

1666	**The calculus (Sir Isaac Newton):** Newton (and independently Gottfried Wilhelm Leibniz) develop the calculus in order to calculate the gravitational effect of all of the particles of the Earth on another object such as a person.
1670	**Spiral spring balance watch (Robert Hooke):** Hooke is also credited as the author of the principle that describes the general behavior of springs, known as Hooke's law.
1672	**Leibniz's calculator (Gottfried Wilhelm Leibniz):** Leibniz develops a calculator that can add, subtract, multiply, and divide, as well as the binary system of numbers used by computers today.
1674	**Improvements to the simple microscope (Antoni van Leeuwenhoek):** Leeuwenhoek, a lens grinder, applies his lenses to the simple microscope and uses his microscope to observe tiny protozoa in pond water.
1681	**Canal du Midi opens:** The 150-mile Canal du Midi links Toulouse, France, with the Mediterranean Sea.
1698	**Savery pump (Thomas Savery):** Savery's pump was impractical to build, but it served as a prototype for Thomas Newcomen's steam engine.
1699	**Eddystone Lighthouse (Henry Winstanley):** English merchant Winstanley designs the first lighthouse in England, located in the English Channel fourteen miles off the Plymouth coast. Winstanley is moved to create the lighthouse after two of his ships are wrecked on the Eddystone rocks.
1700	**Piano (Bartolomeo Cristofori):** Cristofori, a harpsichord maker, constructs an instrument with keys that can be used to control the force with which hammers strike the instrument's strings, producing sound that ranges from piano (soft) to forte (loud)—hence the name "pianoforte," later shortened to "piano."
1701	**Tull seed drill (Jethro Tull):** Before the seed drill, seeds were still broadcast by hand.
1709	**Iron ore smelting with coke (Abraham Darby):** Darby develops a method of smelting iron ore by using coke, rather than charcoal, which at the time was becoming scarce. Coke is made by heating coal and driving off the volatiles (which can be captured and used).
1712	**Atmospheric steam engine (Thomas Newcomen):** Newcomen's engine is developed to pump water out of coal mines.
1714	**Mercury thermometer, Fahrenheit temperature scale (Daniel Gabriel Fahrenheit):** Fahrenheit uses mercury in a glass thermometer to measure temperature over the entire range for liquid water.
1718	**Silk preparation:** John Lombe, owner of the Derby Silk Mill in England, patents the machinery that prepared raw silk for the loom.
1729	**Flying shuttle (John Kay):** On a loom, the shuttle carries the horizontal thread (weft or woof) and weaves it between the vertical threads (warp). Kay develops a shuttle that is named "flying" because it is so much faster than previous shuttles.
1738	**Flute Player and Digesting Duck automatons (Jacques de Vaucanson):** De Vaucanson builds cunning, self-operating devices, or automatons (robots) to charm viewers.

1740	**Steelmaking:** Benjamin Huntsman invents the crucible process of making steel.
1742	**Celsius scale (Anders Celsius):** Celsius creates a new scale for his thermometer.
1745-1746	**Leiden jar (Pieter van Musschenbroek and Ewald Georg von Kleist):** Von Kleist (1745) and Musschenbroek (1746) independently develop the Leiden jar, an early type of capacitor used for storing electric charge.
1746	**Clinical trials prove that citrus fruit cures scurvy (James Lind):** Others had suggested citrus fruit as a cure for scurvy, but Lind gives scientific proof. It still will be another fifty years before preventive doses of foods containing vitamin C are routinely provided for British sailors.
1752	**Franklin stove (Benjamin Franklin):** Franklin develops a stove that allows more heat to radiate into a room than go up the chimney.
1752	**Lightning rod (Benjamin Franklin):** Franklin devises a iron-rod apparatus to attach to houses and other structures in order to ground them, preventing damage during lightning storms.
1756	**Wooden striking clock (Benjamin Banneker):** Banneker's all-wood striking clock operates for the next fifty years. Banneker also prints a series of successful scientific almanacs during 1790's.
1757	**Nautical sextant (John Campbell):** When used with celestial tables, Campbell's sextant allows ships to navigate to within sight of their destinations.
1762	**Marine chronometers (John Harrison):** An accurate chronometer was necessary to determine a ship's position at sea, solving the pressing quest for longitude.
1764	**Spinning jenny (James Hargreaves):** Hargreaves develops a machine for spinning several threads at a time, transforming the textile industry and laying a foundation for the Industrial Revolution.
1765	**Improved steam engine (James Watt):** A steam condenser separate from the working pistons make Watt's engine significantly more efficient than Newcomen's engine of 1712.
1767	**Spinning machine (Sir Richard Arkwright):** Arkwright develops a device to spin fibers quickly into consistent, uniform thread.
1767	**Dividing engine (Jesse Ramsden):** Ramsden develops a machine that automatically and accurately marks calibrated scales.
1770	**Steam dray (Nicolas-Joseph Cugnot):** Cugnot builds his three-wheeled fardier à vapeur to move artillery; the prototype pulls 2.5 metric tons at 2 kilometers per hour.
1772	**Soda water (Joseph Priestley):** Priestley creates the first soda water, water charged with carbon dioxide gas. The following year he develops an apparatus for collecting gases by mercury displacement that would otherwise dissolve in water.
1775	**Boring machine (John Wilkinson):** Wilkinson builds the first modern boring machine used for boring holes into cannon, which made cannon manufacture safer. It was later adapted to bore cylinders in steam engines.

1776	**Bushnell's submarine (David Bushnell):** Bushnell builds the first attack submarine; used unsuccessfully against British ships in the Revolutionary War, it nevertheless advances submarine technology.
1779	**Cast-iron bridge:** Abraham Darby III and John Wilkinson build the first cast-iron bridge in England.
1779	**Spinning mule (Samuel Crompton):** Crompton devises the spinning mule, which allows the textile industry to manufacture high-quality thread on a large scale.
1781	**Uranus discovered (Sir William Herschel):** Herschel observes what he first believes to be a comet; further observation establishes it as a planet eighteen times farther from the Sun than the Earth is.
1782	**Hot-air balloon (Étienne-Jacques and Joseph-Michel Montgolfier):** Shaped like an onion dome and carrying people aloft, the Montgolfiers' hot-air balloon fulfills the fantasy of human flight.
1782	**Oil lamp (Aimé Argand):** Argand's oil lamp revolutionizes lighthouse illumination.
1783	**Parachutes (Louis-Sébastien Lenormand):** Lenormand jumps from an observatory tower using his parachute and lands safely.
1783	**Wrought iron (Henry Cort):** Cort converts crude iron into tough malleable wrought iron.
1784	**Improved steam engine (William Murdock):** In an age when much focus was on steam technology, Murdock works to improve steam pumps that remove water from mines. He will go on to invent coal-gas lighting in 1794.
1784	**Bifocals (Benjamin Franklin):** Tired of changing his spectacles to see things at close range as opposed to objects farther away, Franklin designs eyeglasses that incorporate both myopia-correcting and presbyopia-correcting lenses.
1784	**Power loom (Edmund Cartwright):** Cartwright's power loom forms a major advance in the Industrial Revolution.
1785	**Automated flour mill (Oliver Evans):** Evans's flour mill lays the foundation for continuous production lines. In 1801, he will also invent a high-pressure steam engine.
1790	**Steamboat (John Fitch):** Fitch not only invents the steamboat but also proves its practicality by running a steamboat service along the Delaware River.
1792	**Great clock (Thomas Jefferson):** Jefferson's clock, visible and audible both within Monticello and outside, across his plantation, is designed to maintain efficiency. He also invented an improved portable copying press (1785) and will go on to invent an improved ox plow (1794).
1792	**Coal gas (William Murdock):** Murdock develops methods for manufacturing, storing, and purifying coal gas and using it for lighting.
1793	**Cotton gin (Eli Whitney):** Whitney's engine to separate cotton seed from the fiber transformed the American South, both bolstering the institution of slavery and growing the "cotton is king" economy of the Southern states. Five years later, Whitney develops an assembly line for muskets using interchangeable parts.

1793	**Semaphore (Claude Chappe):** Chappe invents the semaphore.
1796	**Smallpox vaccination (Edward Jenner):** Jenner's vaccine will save millions from death, culminating in the eradication of smallpox in 1979.
1796	**Rumford stove (Benjamin Thompson):** The Rumford stove—a large, institutional stove—uses several small fires to heat the stove top uniformly.
1796	**Hydraulic press (Joseph Bramah):** Bramah builds a practical hydraulic press that operates by a high-pressure plunger pump.
1796	**Lithography (Aloys Senefelder):** Senefelder invents lithography and a process for color lithography in 1826.
1799	**Voltaic pile/electric battery (Alessandro Volta):** Volta creates a pile—a stack of alternating copper and zinc disks separated by brine-soaked felt—that supplies a continuous current and sets the stage for the modern electric battery.
1800	**Iron printing press (Charles Stanhope):** Stanhope invents the first printing press made of iron.
1801	**Pattern-weaving loom (Joseph M. Jacquard):** Jacquard invents a loom for pattern weaving.
1804	**Monoplane glider (George Cayley):** Cayley develops a heavier-than-air fixed-wing glider that inaugurates the modern field of aeronautics. Later models carry a man and lead directly to the Wright brothers' airplane.
1804	**Amphibious vehicle (Oliver Evans):** Evans builds the first amphibious vehicle, which is used in Philadelphia to dredge and clean the city's dockyards.
1805	**Electroplating (Luigi Brugnatelli):** Brugnatelli develops the method of electroplating by connecting something to be plated to one pole of a battery (voltaic pile) and a bit of the plating metal to the other pole of the battery, placing both in a suitable solution.
1805	**Morphine (Friedrich Setürner):** Setürner, a German pharmacist, isolates morphine from opium, but it is not widely used for another ten years.
1806	**Steam locomotive (Richard Trevithick):** After James Watt's patent for the steam engine expires in 1800, Trevithick develops a working steam locomotive. By 1806 he has developed his improved steam engine, named the Cornish engine, which sees worldwide dissemination.
1807	**Internal combustion engine (François Isaac de Rivaz):** De Rivaz builds the first vehicle powered by an internal combustion engine.
1807	**Paddle-wheel steamer (Robert Fulton):** Fulton's steamboat becomes far more commercially successful than those of his competitors.
1808	**Law of combining volumes for gases (Joseph-Louis Gay-Lussac):** Gay-Lussac discovers that, when gaseous elements combine to make a compound, the volumes involved are always simple whole-number ratios.
1810	**Preserving food in sealed glass bottles (Nicolas Appert):** Appert answers Napoleon's call to preserve food in a way that allows his soldiers to carry it with them: He processes food in sealed, air-tight glass bottles.

1810	**Preserving food in tin cans (Peter Durand):** Durand follows Nicolas Appert in preserving food for the French army, but he uses tin-coated steel cans in place of breakable bottles.
1815	**Miner's safety lamp (Sir Humphry Davy):** Davy devises a miner's safety lamp in which the flame is surrounded by wire gauze to cool combustion gases so that the mine's methane-air mixture will not be ignited.
1816	**Macadamization (John Loudon McAdam):** McAdam designs a method of paving roads with crushed stone bound with gravel on a base of large stones. The roadway is slightly convex, to shed water.
1816	**Kaleidoscope (Sir David Brewster):** The name for Brewster's kaleidoscope comes from the Greek words *kalos* (beautiful), *eidos* (form), and *scopos* (watcher). "Kaleidoscope," therefore, literally means "beautiful form watcher."
1816	**Stirling engine (Robert Stirling):** The Stirling engine proves to be an efficient engine that uses hot air as a working fluid.
1818	**First photographic images (Joseph Nicéphore Niépce):** Niépce creates the first lasting photographic images.
1819	**Stethoscope (René-Théophile-Hyacinthe Laënnec):** Laënnec invents the stethoscope to avoid the impropriety of placing his ear to the chest of a female heart patient.
1820	**Dry "scouring" (Thomas L. Jennings):** Jennings discovered that turpentine would remove most stains from clothes without the wear associated with washing them in hot water. His method becomes the basis for modern dry cleaning.
1821	**Diffraction grating (Joseph von Fraunhofer):** Von Fraunhofer's diffraction grating separates incident light by color into a rainbow pattern. The various discrete patterns reveal the structure of specific atomic nuclei, making it possible to identify the chemical compositions of various substances.
1821	**Braille alphabet (Louis Braille):** Braille develops a tactile alphabet—a system of raised dots on a surface—that allows the blind to read by touch.
1821	**Electromagnetic rotation (Michael Faraday):** Faraday publishes his work on electromagnetic rotation, which is the principle behind the electric motor.
1822	**Difference engine (Charles Babbage):** Babbage's "engine" was a programmable mechanical device used to calculate the value of a polynomial—a precursor to today's modern computers.
1823	**Waterproof fabric is used in raincoats (Charles Macintosh):** Macintosh patents a waterproof fabric consisting of soluble rubber between two pieces of cloth. Raincoats made of the fabric are still often called mackintoshes (macs), especially in England.
1824	**Astigmatism-correcting lenses (George Biddell Airy):** Airy develops cylindrical lenses that correct astigmatism. An astronomer, Airy will go on to design a method of correcting compasses used in ship navigation and the altazimuth telescope. He becomes England's astronomer royal in 1835.

1825	**Electromagnet (William Sturgeon):** Sturgeon builds a U-shaped, soft iron bar with a coil of varnished copper wire wrapped around it. When a voltaic current is passed through wire, the bar becomes magnetic—the world's first electromagnet.
1825	**Bivalve vaginal speculum (Marie Anne Victoire Boivin):** Boivin develops the tool now widely used by gynecologists in the examination of the vagina and cervix.
1825	**"Steam waggon" (John Stevens):** Stevens builds the first steam locomotive to be manufactured in the United States.
1826	**Color lithography (Aloys Senefelder):** Senefelder invents color lithography.
1827	**Matches (John Walker):** Walker coats the ends of sticks with a mixture of antimony sulfide, potassium chlorate, gum, and starch to produce "strike anywhere" matches.
1827	**Water turbine (Benoît Fourneyron):** Fourneyron builds the first water turbine; it has six horsepower. His larger, more efficient turbines powered many factories during the Industrial Revolution.
1828	**Combine harvester (Samuel Lane):** Patent is granted to Lane for the combine harvester, which combines cutting and threshing.
1829	**Rocket steam locomotive (George Stephenson):** Stephenson builds the world's first railway line to use a steam locomotive.
1829	**Boiler (Marc Seguin):** Seguin improves the steam engine with a multiple fire-tube boiler.
1829	**Polarizing microscope (William Nicol):** Nicol invents the polarizing microscope, an important forensic tool.
1830	**Steam locomotive (Peter Cooper):** Cooper's four-wheel locomotive with a vertical steam boiler, the *Tom Thumb*, demonstrates the possibilities of steam locomotives and brings Cooper national fame. His other inventions and good management enable Cooper to become a leading industrialist and philanthropist.
1830	**Lawn mower (Edwin B. Budding):** Budding, an English engineer, invents the lawn mower.
1830	**Paraffin (Karl von Reichenbach):** Von Reichenbach, a German chemist, discovers paraffin.
1830	**Creosote (Karl von Reichenbach):** Von Reichenbach distills creosote from beachwood tar. It is used as an insecticide, germicide, and disinfectant.
1831	**Alternating current (AC) generator (Michael Faraday):** Faraday constructs the world's first electric generator.
1831	**Mechanical reaper (Cyrus Hall McCormick):** McCormick's reaper can harvest a field five times faster than earlier methods.
1831	**Staple Bend Tunnel:** The first railroad tunnel in the United States is built in Mineral Point, Pennsylvania.
1832	**Electromagnetic induction (Joseph Henry):** Henry discovers that changing magnetic fields induce voltages in nearby conductors.

1832	**Codeine (Pierre-Jean Robiquet):** French chemist Robiquet isolates codeine from opium. Because of the small amount found in nature, most codeine is synthesized from morphine.
1834	**Hansom cab (Joseph Aloysius Hansom):** English architect Hansom builds the carriage bearing his name.
1835	**Colt revolver (Samuel Colt):** The Colt revolver becomes known as "one of the greatest advances of self-defense in all of human history."
1835	**Photography (Joseph Nicéphore Niépce):** Niépce codevelops photography with Louis-Jacques-Mandé Daguerre.
1836	**Daniell cell (John Frederic Daniell):** Daniell invents the electric battery bearing his name, which is much improved over the voltaic pile.
1836	**Acetylene (Edmund Davy):** Davy creates acetylene by heating potassium carbonate to high temperatures and letting it react to water.
1837	**Electric telegraph (William Fothergill Cooke and Charles Wheatstone):** Wheatstone and Cooke devise a system that uses five pointing needles to indicate alphabetic letters.
1837	**Steam hammer (James Hall Nasmyth):** Nasmyth develops the steam hammer, which he will use to build a pile driver in 1843.
1837	**Steel plow (John Deere):** Previously, plows were made of cast iron and required frequent cleaning. Deere's machine is effective in reducing the amount of clogging farmers experienced when plowing the rich prairie soil.
1837	**Threshing machine (Hiram A. and John A. Pitts):** The Pitts, brothers, develop the first efficient threshing machine.
1838	**Fuel cell (Christian Friedrich Schönbein):** Schönbein's fuel cell might use hydrogen and oxygen and allow them to react, producing water and electricity. There are no moving parts, but the reactants must be continuously supplied.
1838	**Propelling steam vessel (John Ericsson):** Swedish engineer Ericsson invents the double screw propeller for ships allowing them to move much faster than those relying on sails.
1839	**Nitric acid battery (Sir William Robert Grove):** The Grove cell delivered twice the voltage of its more expensive rival, the Daniell cell.
1839	**Daguerreotype (Jacques Daguerre):** Improving on the discoveries of Joseph Nicéphore Niépce, Daguerre develops the first practical photographic process, the Daguerreotype.
1839	**Vulcanized rubber (Charles Goodyear):** Adding sulfur and lead monoxide to rubber, Goodyear processes the batch at a high temperature. The process, later called vulcanization, yields a stable material that does not melt in hot weather or crack in cold.
1840	**Electrical telegraph (Samuel F. B. Morse):** Others had already built telegraph systems, but Morse's system was superior and soon replaced all others.

1841	**Improved electric clock (Alexander Bain):** With John Barwise, Bain develops an electric clock with a pendulum driven by electric impulses to regulate the clock's accuracy.
1841	**First negatives in photography (William Henry Fox Talbot):** Talbot, an English polymath, invents the calotype process, which produces the first photographic negative.
1842	**Commercial fertilizer (John B. Lawes):** Lawes develops superphosphate, the first commercial fertilizer.
1843	**Rotary printing press (Richard March Hoe):** Patented in 1847, the steam-powered rotary press is far faster than the flatbed press.
1843	**Multiple-effect vacuum evaporator (Norbert Rillieux):** Rillieux develops an efficient method for refining sugar using a stack of several pans of sugar syrup in a vacuum chamber, which allows boiling at a lower temperature.
1845	**Suspension bridges (John Augustus Roebling):** A manufacturer of wire cable, Roebling wins a competition for an aqueduct over the Allegheny River and goes on to design other aqueducts and suspension bridges, culminating in the Brooklyn Bridge, which his son, Washington Augustus Roebling, completes in 1883.
1845	**Sewing machine (Elias Howe):** Howe develops a machine that can stitch straight, strong seams faster than those sewn by hand.
1846	**Neptune discovered (John Galle):** German astronomer Galle observes a new planet, based on irregularities in the orbit of Uranus calculated the previous year by England's John Couch Adams and France's Urbain Le Verrier.
1847	**Nitroglycerin (Ascanio Sobrero):** Italian chemist Sobrero creates nitroglycerin.
1847	**Telegraphy applications (Werner Siemens):** Siemens refines a telegraph in which a needle points to the alphabetic letter being sent.
1849	**Laryngoscope (Manuel P. R. Garcia):** Spanish singer and voice teacher, Garcia, known as the father of laryngology, devises the first laryngoscope.
1851	**Foucault's pendulum (Léon Foucault):** Foucault's pendulum proves that Earth rotates.
1851	**Sewing machine (Isaac Merritt Singer):** Singer improves the sewing machine and successfully markets it to women for home use.
1851	**Ophthalmoscope (Hermann von Helmholtz):** Helmholtz invents a device that can be used to examine the retina and the vitreous humor. In 1855, he will invent an ophthalmometer, an instrument that measures the curvature of the eye's lens.
1854	**Kerosene (Abraham Gesner):** Canadian geologist Gesner distills kerosene from petroleum.
1852	**Hypodermic needle (Charles G. Pravaz):** French surgeon Pravaz devises the hypodermic syringe.

1855	**Bunsen burner (Robert Wilhelm Bunsen):** Bunsen—along with Peter Desaga, an instrument maker, and Henry Roscoe, a student—develops a high-temperature laboratory burner, which he and Gustav Kirchhoff use to develop the spectroscope (1859).
1855	**Bessemer process (Sir Henry Bessemer):** Bessemer creates a converter that leads to a process for inexpensively mass-producing steel.
1856	**Synthetic dye (William H. Perkin):** British chemist Perkin produces the first synthetic dye. The color is mauve, which triggers a mauve fashion revolution.
1857	**Safety elevator (Elisha Graves Otis):** Otis's safety elevator automatically stops if the supporting cable breaks.
1858	**Internal combustion engine (Étienne Lenoir):** Lenoir's engine, along with his invention of the spark plug, sets the stage for the modern automobile.
1858	**Transatlantic cable (Lord Kelvin):** Kelvin helps design and install the underocean cables for telegraphy between North America and Europe, serving as a chief motivating force in getting the cable completed.
1859	**Signal flares (Martha J. Coston):** Coston's brilliant and long-lasting white, red, and green flares will be adopted by the navies of several nations.
1859	**Lead-acid battery (Gaston Planté):** French physicist Planté invents the lead-acid battery, which led to the invention of the first electric, rechargeable battery.
1860	**Refrigerant (Ferdinand Carré):** French inventor Carré introduces a refrigerator that uses ammonia as a refrigerant.
1860	**Electric incandescent lamp (Joseph Wilson Swan):** Swan produces and patents an incandescent electric bulb; in 1880, two years after Edison's light bulb, Swan will produce a more practical bulb.
1860	**Web rotary printing press (William Bullock):** Bullock's press has an automatic paper feeder, can print on both sides of the paper, cut the paper into sheets, and fold them.
1860	**Henry rifle (Tyler Henry):** American gunsmith Henry designs the Henry rifle, a repeating rifle, the year before the Civil War begins.
1860	**First mail service:** Pony Express opens overland mail service. The service eventually expands to include more than 100 stations, 80 riders, and more than 400 horses.
1861	**Machine gun (Richard Gatling):** Gatling develops the first machine gun, called the Gatling gun. It has six barrels that rotate into place as the operator turns a hand crank; the shells were automatically chambered and fired.
1861	**First color photograph:** Thomas Sutton develops the first color photo based on Scottish physicist James Clerk Maxwell's three-color process.
1861-1862	**USS Monitor (John Ericsson):** Ericsson develops the first practical ironclad ship, which will be use during the Civil War. He goes on to develop a torpedo boat that can fire a cannon from an underwater port.

1862	**Pasteurization (Louis Pasteur):** Pasteur's germ theory of disease leads him to develop a method of applying heat to milk products in order to kill harmful bacteria. He goes on to develop vaccines for rabies, anthrax, and chicken cholera (1867-1885).
1863	**Subway:** The first subway opens in London; it uses steam locomotives. It does not go electric until 1890.
1865	Pioneer **(Pullman) sleeping car (George Mortimer Pullman):** Pullman began working on sleeping cars in 1858, but the *Pioneer* is a luxury car with an innovative folding upper birth to allow the passenger to sleep while traveling.
1866	**Self-propelled torpedo (Robert Whitehead):** English engineer Whitehead develops the modern torpedo.
1866	**Transatlantic telegraph cable:** The first successful transatlantic telegraph cable is laid; it spans 1,686 nautical miles.
1867	**Dynamite (Alfred Nobel):** Nobel mixes clay with nitroglycerin in a one-to-three ratio to create dynamite (Nobel's Safety Powder), an explosive the ignition of which can be controlled using Nobel's own blasting cap. He goes on to patent more than three hundred other inventions and devotes part of the fortune he gained from dynamite to establish and fund the Nobel Prizes.
1867	**Baby formula (Henri Nestlé):** Nestlé combines cow's milk with wheat flour and sugar to produce a substitute for infants whose mothers cannot breast-feed.
1867	**Steam velocipede motorcycle (Sylvester Roper):** Roper spent his lifetime making steam engines lighter and more powerful in order to make his motorized bicycles faster. His velocipede eventually reaches 60 miles per hour.
1867	**Flat-bottom paper bag machine (Margaret E. Knight):** Knight designs a machine that can manufacture flat-bottom paper bags, which can stand open for easy loading.
1867	**Dry-cell battery (Georges Leclanché):** French engineer Lelanche invents the dry-cell battery.
1868	**Typewriter (Christopher Latham Sholes):** American printer Sholes produces the first commercially successful typewriter.
1869	**Periodic table of elements (Dmitry Ivanovich Mendeleyev):** The periodic table, which links chemical properties to atomic structure, will prove to be one of the great achievements of the human race.
1869	**Air brakes for trains (George Westinghouse):** In 1867, Westinghouse developed a signaling system for trains. The air brake makes it easier and safer to stop large, heavy, high-speed trains.
1869	**Transcontinental railroad:** The United States transcontinental railroad is completed.
1869	**Celluloid (John Wesley Hyatt):** American inventor Hyatt produces celluloid, the first commercially successful plastic, by mixing solid pyroxylin and camphor.
1869	**Suez Canal opens:** The canal, 101 miles long, took a decade to build and connects the Red Sea with the eastern Mediterranean Sea.

1871	**Fireman's respirator (John Tyndall):** The respirator grows from Tyndall's studies of air pollution.
1871	**Commercial generator (Zénobe T. Gramme):** Belgian electrical engineer Gramme builds the Gramme machine, the first practical commercial generator for producing alternating current.
1872	**Blue jeans (Levi Strauss):** Miners tore their pockets when they stuffed too many ore samples in them. Strauss makes pants using heavy-duty material with riveted pocket corners so they will not tear out.
1872	**Burbank russet potato (Luther Burbank):** Burbank breeds all types of plants, using natural selection and grafting techniques to achieve new varieties. His Burbank potato, developed from a rare russet potato seed pod, grows better than other varieties.
1872	**Automatic lubricator (Elijah McCoy):** McCoy uses steam pressure to force oil to lubricate the pistons of steam engines.
1872	**Vaseline (Robert A. Chesebrough):** Chesebrough, an American chemist, patents his process for making petroleum jelly and calls it Vaseline.
1873	**QWERTY keyboard (Christopher Latham Sholes):** After patenting the first practical typewriter, Sholes develops the QWERTY keyboard, designed to slow the fastest typists, who otherwise jammed the keys. The basic QWERTY design remains the standard on most computer keyboards.
1874	**Barbed wire (Joseph Farwell Glidden):** An American farmer, Glidden invents and patents barbed wire. Barbed-wire fences make farming and ranching of the Great Plains practical. Without effective fences, animals wandered off and crops were destroyed. At the time of his death in 1906, Glidden is one of the richest men in the country.
1874	**Medical nuclear magnetic resonance imaging (Raymond Damadian):** Damadian and others develop magnetic resonance imaging (MRI) for use in medicine.
1876	**Four-stroke internal combustion engine (Nikolaus August Otto):** In order to deliver more horsepower, Otto's engine compresses the air-fuel mixture. His previous engines operated near atmospheric pressure.
1876	**Ammonia-compressor refrigeration machine (Carl von Linde):** Breweries need refrigeration so they can brew year-round. Linde refines his ammonia-cycle refrigerator to make this possible.
1876	**Telephone (Elisha Gray):** Gray files for a patent for the telephone the same day that Alexander Graham Bell does so. While the case is not clear-cut, and Gray fought with Bell for years over the patent rights, Bell is generally credited with the telephone's invention.
1877	**Phonograph (Thomas Alva Edison):** Edison invents the phonograph—an unexpected outcome of his telephone research.
1878	**First practical lightbulb (Thomas Alva Edison):** Twenty-two people have invented lightbulbs before Edison and Joseph Swan, but they are impractical. Edison's is the first to be commercially viable. Eventually, Swan's company merges with Edison's.

1878	**Loose-contact carbon microphone (David Edward Hughes):** Hughes's carbon microphone advances telephone technology. In 1879, he will invent the induction balance, which will be used in metal detectors.
1878	**Color photography (Frederic Eugene Ives):** American inventor Ives develops the halftone process for printing photographs.
1879	**Saccharin (Ira Remsen):** Remsen synthesizes a compound that is up to three hundred times sweeter than sugar; he also establishes the important *American Chemical Journal*, serving as its editor until 1915.
1880	**Milne seismograph (John Milne):** Milne invents the first modern seismograph for measuring earth tremors. He will come to be called the father of modern seismology.
1881	**Improved incandescent lightbulb (Lewis Howard Latimer):** Latimer develops an improved way to manufacture and to attach carbon filaments in lightbulbs.
1881	**Sphygmomanometer (Karl Samuel Ritter von Basch):** Von Basch invents the first blood pressure gauge.
1882	**Induction motor (Nikola Tesla):** Tesla's theories and inventions make alternating current (AC) practical.
1882	**Two-cycle gasoline engine (Gottlieb Daimler):** Daimler builds a small, high-speed two-cycle gasoline engine. He will also build a successful motorcycle in 1885 and (with Wilhelm Maybach) an automobile in 1889.
1883	**Solar cell (Charles Fritts):** American scientist Fritts designs the first solar cell.
1883	**Shoe-lasting machine (Jan Ernst Matzeliger):** The machine sews the upper part of the shoe to the sole and reduces the cost of shoes by 50 percent.
1884	**Fountain pen (Lewis Waterman):** The commonly told story is that Waterman was selling insurance and lost a large contract when his pen leaked all over it, prompting him to invent the leak-proof fountain pen.
1884	**Vector calculus (Oliver Heaviside):** Heaviside develops vector calculus to represent James Clerk Maxwell's electromagnetic theory with only four equations instead of the usual twenty.
1884	**Roll film (George Eastman):** Roll film will replace heavy plates, making photography both more accessible and more convenient. In 1888, Eastman and William Hall invent the Kodak camera. These developments open photography to the masses.
1884	**Roll film (George Eastman):** Roll film will replace heavy plates, making photography both more accessible and more convenient. In 1888, Eastman and William Hall invent the Kodak camera. These developments open photography to the masses.
1884	**Steam turbine (Charles Parsons):** Designed for ships, Parsons's steam turbine is smaller, more efficient, and more durable than the steam engines in use.
1884	**Census tabulating machine (Herman Hollerith):** Hollerith's machine uses punch cards to tabulate 1890 census data. He goes on to found the company that later becomes International Business Machines (IBM).

1885	**Machine gun (Hiram Stevens Maxim):** Maxim patents a machine gun that can fire up to six hundred bullets per minute.
1885	**Bicycle (John Kemp Starley):** English inventor Starley is responsible for producing the first modern bicycle, called the Rover.
1885	**First gasoline-powered automobile (Carl Benz):** Benz not only manufactures the first gas-powered car but also is first to mass-produce automobiles.
1885	**Incandescent gas mantle (Carl Auer von Welsbach):** The Austrian scientist invents the incandescent gas mantle.
1886	**Dictaphone (Charles Sumner Tainter):** Tainter, an American engineer who frequently worked with Alexander Graham Bell, designs the Dictaphone.
1886	**Dishwasher (Josephine Garis Cochran):** Like modern washers, Cochran's dishwasher cleans dishes with sprays of hot, soapy water and then air-dries them.
1886	**Electric transformer (William Stanley):** Stanley, working at Westinghouse, builds the first practical electric transformer.
1886	**Gramophone (Emile Berliner):** A major contribution to the music recording industry, Berliner's gramophone uses flat record discs for recording sound. Berliner goes on to produce a helicopter prototype (1906-1923).
1886	**Linotype machine (Ottmar Mergenthaler):** Pressing keys on the machine's keyboard releases letter molds that drop into the current line. The lines are assembled into a page and then filled with molten lead.
1886	**Electric-traction system (Frank J. Sprague):** Sprague's motor can propel a tram up a steep hill without its slipping.
1886	**Hall-Héroult electrolytic process (Charles Martin Hall and Paul Héroult):** The industrial production of aluminum from bauxite ore made aluminum widely available. Prior to the electrolytic process, aluminum was a precious metal with a value about equal to that of silver.
1886	**Coca-Cola (John Stith Pemberton):** Developed as pain reliever less addictive than available opiates, the original Coca-Cola contains cocaine from cola leaves and caffeine from kola nuts. It achieves greater success as a beverage marketed where alcohol is prohibited.
1886	**Yellow pages (Reuben H. Donnelly):** Yellow paper was used in 1883 when the printer ran out of white paper. Donnelly now purposely uses yellow paper for business listings.
1886	**Fluorine (Henri Moissan):** French chemist Moissan isolates fluorine and is awarded the Nobel Prize in Chemistry in 1906. Compounds of fluorine are used in toothpaste and in public water supplies to help prevent tooth decay.
1887	**Radio transmitter and receiver (Heinrich Hertz):** Hertz will use these devices to discover radio waves and confirm that they are electromagnetic waves that travel at the speed of light; he also discovers the photoelectric effect.
1887	**Distortionless transmission lines (Oliver Heaviside):** Heaviside recommends that induction coils be added to telephone and telegraph lines to correct for distortion.

1887	**Olds horseless carriage (Ransom Eli Olds):** Olds develops a three-wheel horseless carriage using a steam engine powered by a gasoline burner.
1887	**Synchronous multiplex railway telegraph (Granville T. Woods):** Woods patents a variation of the induction telegraph that allows messages to be sent between moving trains and between trains and railway stations. He will eventually obtain sixty patents on electrical and electromechanical devices, most of them related to railroads and communications.
1888	**Cordite (Sir James Dewar):** Dewar, with Sir Frederick Abel, invents cordite, a smokeless gunpowder that is widely adopted for munitions.
1888	**Pneumatic rubber tire (John Boyd Dunlop):** Dunlop's pneumatic tires revolutionize the ride for cyclists and motorists.
1888	**Kodak camera:** George Eastman, founder of Eastman Kodak, introduces the first Kodak camera.
1889	**Electric drill (Arthur James Arnot):** Arnot's drill is used to cut holes in rock and coal.
1889	**Bromine extraction (Herbert Henry Dow):** Dow's method for extracting bromine from brine enables bromine to be widely used in medicines and in photography.
1889	**Rayon (Louis-Marie-Hilaire Bernigaud de Chardonnet):** Bernigaud de Chardonnet, a French chemist, invents rayon, the first artificial fiber, as an alternative to silk.
1889	**Celluloid film:** George Eastman replaces paper film with celluloid.
1890	**Improved carbon electric arc (Hertha Marks Ayrton):** The carbon arc produces an intense light that is used in streetlights.
1890	**Pneumatic (air) hammer (Charles B. King):** A worker with a pneumatic hammer can break up a concrete slab many times faster than can a worker armed with only a sledgehammer.
1890	**Smokeless gunpowder (Hudson Maxim):** Maxim (perhaps with brother Hiram) develops a version of smokeless gunpowder that is adopted for modern firearms; he goes on to develop a smokeless cannon powder that will be used during World War I.
1890	**Rubber gloves in the operating room:** American surgeon William Stewart Halsted introduces the use of sterile rubber gloves in the operating room.
1891	**Rubber automobile tires (André and Édouard Michelin):** The Michelin brothers manufacture air-inflated tires for bicycles and later automobiles, which leads to a successful ad campaign, featuring the Michelin Man (Bibendum).
1891	**Carborundum (Edward Goodrich Acheson):** Attempting to create artificial diamonds, Acheson instead synthesizes silicon carbide, the second hardest substance known. He will develop an improved graphite-making process in 1896.
1892	**Kinetoscope (Thomas Alva Edison):** Edison completes Kinetoscope; the first demonstration is held a year later.
1892	**Calculator (William Seward Burroughs):** Burroughs builds the first practical key-operated calculator; it prints entries and results.

1892	**Dewar flask (Sir James Dewar):** Dewar invents the vacuum bottle, a vacuum-jacketed vessel for storing and maintaining the temperature of hot or cold liquids.
1892	**Artificial silk (Charles F. Cross and Edward J. Bevan):** British chemists Cross and Bevan create viscose artificial silk (cellulose acetate).
1893	**Color photography plate (Gabriel Jonas Lippmann):** Also known as the Lippmann plate for its inventor, the color photography plate uses interference patterns, rather than various colored dyes, to reproduce authentic color.
1893	**Alternating current calculations (Charles Proteus Steinmetz):** Steinmetz's calculations make it possible for engineers to determine alternating current reliably, without depending on trial and error, when designing a new motor.
1894	**Cereal flakes (John Harvey Kellogg):** Kellogg, a health reformer who advocates a diet of fruit, nuts, and whole grains, invents flaked breakfast cereal with the help of his brother, Will Keith Kellogg. In 1906 Kellogg established a company in Battle Creek, Michigan, to manufacture his breakfast cereal.
1894	**Automatic loom (James Henry Northrop):** Northrop builds the first automatic loom.
1895	**Streamline Aerocycle bicycle (Ignaz Schwinn):** Through hard work and dedication, Schwinn develops a bicycle that eventually makes his name synonymous with best of bicycles.
1895	**Victrola phonographs (Eldridge R. Johnson):** Johnson develops a spring-driven motor for phonographs that provides the constant record speed necessary for good sound reproduction.
1895	**Cinématographe (Auguste and Louis Lumière):** The Lumière brothers' combined motion-picture camera, printer, and projector helps establish the movie business. Using a very fine-grained silver-halide gelatin emulsion, they cut photographic exposure time down to about one minute.
1895	**Antenna:** Aleksandr Stepanovich Popov demonstrated radio reception with a coherer, which he also used as a lightning detector.
1896	**Wireless telegraph system (Guglielmo Marconi):** Marconi is the first to send wireless signals across the Atlantic Ocean, inaugurating a new era of telecommunications.
1896	**Aerodromes (Samuel Pierpont Langley):** Langley's "Aerodrome number 6," using a small gasoline engine, makes an unmanned flight of forty-eight hundred feet.
1896	**Four-wheel horseless carriage (Ransom Eli Olds):** Oldsmobile patents Olds's internal combustion engine and applies it to his four-wheel horseless carriage, naming it the "automobile."
1896	**High-frequency generator and transformer (Elihu Thomson):** Thomson produces an electric air drill, which advances welding to improve the construction of new appliances and vehicles. He will also invent other electrical devices, including an improved X-ray tube.

1896	**X-ray tube (Wilhelm Conrad Röntgen):** After discovering X radiation, Röntgen mails an X-ray image of a hand wearing a ring and paving the way for the medical use of X-ray imaging—one of the most important discoveries ever made for medical science.
1896	**Better sphygmomanometer (Scipione Riva-Rocci):** Italian pediatrician Riva-Rocci develops the most successful and easy-to-use blood-pressure gauge.
1897	**Modern submarine (John Philip Holland):** Holland's submarine is the first to use a gasoline engine on the surface and an electric engine when submerged.
1897	**Oscilloscope (Karl Ferdinand Braun):** The oscilloscope is an invaluable device used to measure and display electronic waveforms.
1897	**Jenny coupler (Andrew Jackson Beard):** Beard's automatic coupler connects the cars in a train without risking human life. The introduction of automatic couplings reduces coupling-related injuries by a factor of five.
1897	**Escalator (Charles Seeberger):** Before Seeberger built the escalator in its now-familiar form, it was a novelty ride at the Coney Island amusement park.
1897	**Automobile components (Alexander Winton):** The Winton Motor Carriage Company is incorporated, and Winton begins manufacturing automobiles. His popular "reliability runs" helps advertise automobiles to the American market. He will produce the first American diesel engine in 1913.
1897	**Diesel engine (Rudolf Diesel):** Diesel's internal combustion engine rivals the efficiency of the steam engine.
1897	**Electron discovered (J. J. Thomson):** Thomson uses an evacuated tube with a high voltage across electrodes sealed in the ends. Invisible particles (later named electrons) stream from one of the electrodes, and Thomson establishes the particles' properties.
1898	**Flashlight (Conrad Hubert):** Hubert combines three parts—a battery, a lightbulb, and a metal tube—to produce a flashlight.
1898	**Mercury vapor lamp (Peter Cooper Hewitt):** Hewitt's mercury vapor lamp proves to be more efficient than incandescent lamps.
1899	**Alpha particle discovered (Ernest Rutherford):** Rutherford detects the emission of helium 4 nuclei (alpha particles) in the natural radiation from uranium.
1900	**Aspirin:** Aspirin is patented by Bayer and sold as a powder. In 1915 it is sold in tablets.
1900	**Dirigibles (Ferdinand von Zeppelin):** Von Zeppelin flies his airship three years before the Wright brothers' airplane.
1900	**Gamma ray discovered (Paul Villard):** Villard discovers gamma rays in the natural radiation from uranium. They resemble very high-energy X rays.
1900	**Brownie camera:** George Eastman introduces the Kodak Brownie camera. It is sold for $1 and the film it uses costs 15 cents. The Brownie made photography an accessible hobby to almost everyone.
1901	**Acousticon hearing aid (Miller Reese Hutchison):** Hutchison invents a battery-powered hearing aid in the hopes of helping a mute friend speak.

1901	**Vacuum cleaner (H. Cecil Booth):** Booth patents his vacuum cleaner, a machine that sucks in and traps dirt. Previous devices, less effective, had attempted to blow the dirt away.
1901	**String galvanometer (electrocardiograph) (Willem Einthoven):** Einthoven's device passes tiny currents from the heart through a silver-coated silicon fiber, causing the fiber to move. Recordings of this movement can show the heart's condition.
1901	**Silicone (Frederick Stanley Kipping):** English chemist Kipping studies the organic compounds of silicon and coins the term "silicone."
1902	**Airplane engine (Charles E. Taylor):** Taylor begins building engines for the Wright brothers' airplanes.
1902	**Lionel electric toy trains (Joshua Lionel Cowen):** Cowen publishes the first Lionel toy train catalog. Lionel miniature trains and train sets become favorite toys for many years and are prized by collectors to this day.
1902	**Air conditioner (Willis Carrier):** Whole-house air-conditioning becomes possible.
1903	**Windshield wipers (Mary Anderson):** At first, the driver operated the wiper with a lever from inside the car.
1903	**Wright Flyer (Wilbur and Orville Wright):** The Wright Flyer is the first heavier-than-air machine to solve the problems of lift, propulsion, and steering for controlled flight.
1903	**Safety razor with disposable blade (King Camp Gillette):** Gillette's razor used a disposable and relatively cheap blade, so there was no need to sharpen it.
1903	**Space-traveling projectiles (Konstantin Tsiolkovsky):** Tsiolkovsky publishes "The Exploration of Cosmic Space by Means of Reaction-Propelled Apparatus," in which he includes an equation for calculating escape velocity (the speed required to propel an object beyond Earth's field of gravity). He is also recognized for the concept of rocket propulsion and for the wind tunnel.
1903	**Ultramicroscope (Richard Zsigmondy):** Zsigmondy builds the ultramicroscope to study colloids, mixtures in which particles of a substance are dispersed throughout another substance.
1903	**Spinthariscope (Sir William Crookes):** Crookes invents a device that sparkles when it detects radiation. He also develops and experiments with the vacuum tube, allowing later physicists to identify alpha and beta particles and X rays in the radiation from uranium.
1903	**Crayola crayons (Edwin Binney):** With his cousin C. Harold Smith, Binney invents dustless chalk and crayons marketed under the trade name Crayolas.
1903	**Motorcycle:** Harley-Davidson produces the first motorcycle, built to be a racer.
1903	**Electric iron:** Earl Richardson introduces the lightweight electric iron.
1904	**Glass bottle machine:** American inventor Michael Joseph Owens designs a machine that produces glass bottles automatically.

1905	**Novocaine (Alfred Einkorn):** While researching a safe local anesthetic to use on soldiers, German chemist Einkorn develops novocaine, which becomes a popular dental anesthetic.
1905	**Special relativity (Albert Einstein):** At the age of twenty-six, Einstein uses the constancy of the speed of light to explain motion, time, and space beyond Newtonian principles. During the same year, he publishes papers describing the photoelectric effect and Brownian motion.
1905	**Intelligence testing:** French psychologist Alfred Binet devises the first of a series of tests to measure an individual's innate ability to think and reason.
1906	**Hair-care products (Madam C. J. Walker):** Walker trains a successful sales force to go door-to-door and sell directly to women. Her saleswomen, beautifully dressed and coiffed, are instructed to pamper their clients.
1906	**Broadcast radio (Reginald Aubrey Fessenden):** In broadcast radio, sound wave forms are added to a carrier wave and then broadcast. The carrier wave is subtracted at the receiver leaving only the sound.
1906	**Klaxon horn (Miller Reese Hutchison):** Hutchison files a patent application for the electric automobile horn.
1906	**Chromatography (Mikhail Semenovich Tswett):** Tswett, a Russian botanist, invents chromatography.
1906	**Freeze-drying (Jacques Arsène d'Arsonval and George Bordas):** D'Arsonval and Bordas invent freeze-drying, but the practice is not commercially developed until after World War II.
1907	**Sun valve (Nils Gustaf Dalén):** Dalén's device uses sunlight to activate a lighthouse beacon. His other inventions make automated acetylene beacons in lighthouses possible.
1907	**Mantoux tuberculin skin test (Charles Mantoux):** French physician Mantoux develops a skin-reaction test to diagnose tuberculosis. He builds on the work of Robert Koch and Clemens von Pirquet.
1908	**Helium liquefaction (Heike Kamerlingh Onnes):** Kamerlingh Onnes produces liquid helium at a temperature of about 4 kelvins. He will also discover superconductivity in several materials cooled to liquid helium temperature.
1908	**"Tin Lizzie" (Model T) automobile (Henry Ford):** Ford's development of an affordable automobile, manufactured using his assembly-line production methods, revolutionize the U.S. car industry.
1908	**Electrostatic precipitator (Frederick Gardner Cottrell):** The electrostatic precipitator is invaluable for cleaning stack emissions.
1908	**Geiger-Müller tube (Hans Geiger):** Geiger invents a device, popularly called the Geiger counter, that is a reliable, portable radiation detector. Later his student Walther Müller helps improve the instrument.
1908	**Vacuum cleaner (James Murray Spangler):** Spangler receives a patent on his electric sweeper, and his Electric Suction Sweeper Company eventually becomes the Hoover Company, the largest such company in the world.

1908	**Cellophane (Jacques Edwin Brandenberger):** Brandenberger builds a machine to mass-produce cellophane, which he has earlier synthesized while unsuccessfully attempting to develop a stain-resistant cloth.
1908	**Water treatment:** Chlorine is used to purify water for the first time in the United States, in New Jersey, helping to reduce waterborne illnesses such as cholera, typhoid, and dysentery.
1908	**Audion (Lee De Forest):** De Forest invents a vacuum tube used in sound amplification. In 1922, he will develop talking motion pictures, in which the sound track is imprinted on the film with the pictures, instead of on a record to be played with the film, leading to exact synchronization of sound and image.
1909	**Synthetic fertilizers (Fritz Haber):** Haber also invents the Haber process to synthesize ammonia on a small scale.
1909	**Maxim silencer (Hiram Percy Maxim):** The silencer reduces the noise from firing the Maxim machine gun.
1909	**pH scale:** Danish chemist Søren Sørensen introduces the pH scale as a standard measure of alkalinity and acidity.
1910	**Chlorinator (Carl Rogers Darnall):** Major Darnall builds a machine to add liquid chlorine to water to purify it for his troops. His method is still widely used today.
1910	**Bakelite (Leo Hendrik Baekeland):** Bakelite is the first tough, durable plastic.
1910	**Neon lighting (Georges Claude):** Brightly glowing neon tubes revolutionize advertising displays.
1910	**Syphilis treatment:** German physician Paul Ehrlich and Japanese physician Hata Sahachiró discover the effective treatment of arsphenamine (named Salvarsan by Ehrlich) for syphilis.
1911	**Colt .45 automatic pistol (John Moses Browning):** Commonly called the Colt Model 1911, an improved version of the Colt Model 1900, the Colt .45 is the first autoloading pistol produced in America. Among Browning's other inventions are the Winchester 94 lever-action rifle and the gas-operated Colt-Browning machine gun.
1911	**Gyrocompass (Elmer Ambrose Sperry):** Sperry receives a patent for a nonmagnetic compass that indicates true north.
1911	**Atomic nucleus identified (Ernest Rutherford):** Rutherford discovered the nucleus by bombarding a thin gold foil with alpha particles. Some were deflected through large angles showing that something small and hard was present.
1911	**Ductile tungsten (William David Coolidge):** Coolidge also invented the Coolidge tube, an improved X-ray producing tube.
1911	**Ochoaplane (Victor Leaton Ochoa):** In addition to inventing this plane with collapsible wings, Ochoa also developed an electricity-generating windmill.
1911	**Automobile electric ignition system (Charles F. Kettering):** Kettering invents the first electric ignition system for cars.
1912	**Automatic traffic signal system (Garrett Augustus Morgan):** Morgan also invents a safety hood that served as a rudimentary gas mask.

1913	**Gyrostabilizer (Elmer Ambrose Sperry):** Sperry develops the gyrostabilizer, a device to control the roll, pitch, and yaw of a moving ship. He will go on to invent the flying bomb, which is guided by a gyrostabilizer and by radio control.
1913	**Erector set (Alfred C. Gilbert):** Erector sets provide hands-on engineering experience for countless children.
1913	**Zipper (Gideon Sundback):** While others had made zipper-like devices but had never successfully marketed them, Sundback designs a zipper in approximately its present form. He also invents a machine to make zippers.
1913	**Improved electric lightbulb (Irving Langmuir):** Langmuir fills his lightbulb with a low-pressure inert gas to retard evaporation from the tungsten filament.
1913	**Industrialization of the Haber process (Carl Bosch):** Bosch scales up Haber's process for making ammonia to an industrial capacity. The process comes to be known as the Haber-Bosch process.
1913	**Bergius process (Friedrich Bergius):** Bergius develops high-pressure, high-temperature process to produce liquid fuel from coal.
1913	**Electric dishwasher:** The Walker brothers of Philadelphia produce the first electric dishwasher.
1913	**Stainless steel (Harry Brearley):** Brearley invents stainless steel.
1913	**Thermal cracking (William Burton and Robert Humphreys):** Standard Oil chemical engineers Burton and Humphreys discover thermal cracking, a method of oil refining that significantly increases gasoline yields.
1914	**Backless brassiere (Caresse Crosby):** The design of a new women's undergarment leads to the expansion of the U.S. brassiere industry. Caresse was originally a marketing name that Mary Phelps Jacob eventually adopted as her own.
1915	**Panama Canal opens:** The passageway between the Atlantic and Pacific oceans creates a boon for the shipping industry.
1915	**General relativity (Albert Einstein):** Einstein refines his 1905 theory of relativity (now called special relativity) to describe the theory that states that uniform accelerations are almost indistinguishable from gravity. Einstein's theory provides the basis for physicists' best understanding of gravity and of the framework of the universe.
1915	**Jenny (Glenn H. Curtiss):** The Jenny becomes a widely used World War I biplane, and Curtis becomes a general manufacturer of airplanes and airplane engines.
1915	**Pyrex:** Corning's brand name for glassware is introduced.
1915	**Warfare:** Depth-charge bombs are first used by the Allies against German submarines.
1916	**By-products of sweet potatoes and peanuts (George Washington Carver):** Carver publishes his famous bulletin on 105 ways to prepare peanuts.
1919	**Proton discovered (Ernest Rutherford):** After bombarding nitrogen gas with alpha particles (helium 4 nuclei), Rutherford observes that positive particles with a single charge are knocked loose. They are protons.
1919	**Toaster (Charles Strite):** Strite invents the first pop-up toaster.

1920	**Microelectrode (Ida H. Hyde):** Hyde's electrode is small enough to pierce a single cell. Chemicals can also be very accurately deposited by the microprobe.
1921	**Ready-made bandages:** Johnson & Johnson puts Band-Aids on the market.
1921	**Antiknock solution (Thomas Midgley, Jr.):** While working at a General Motors subsidiary, American mechanical engineer Midgley develops an antiknock solution for gasoline.
1921	**Insulin:** University of Toronto researchers Frederick Banting, J. J. R. Macleod, and Charles Best first extract insulin from a dog, and the first diabetic patient is treated with purified insulin the following year. Banting and Macleod win the 1923 Nobel Prize in Physiology or Medicine for their discovery of insulin.
1923	**Improved telephone speaker (Georg von Békésy):** Békésy's studies of the human ear lead to an improved telephone earpiece. He will also construct a working model of the inner ear.
1923	**Quick freezing (Clarence Birdseye):** Birdseye's quick-freezing process preserves food's flavor and texture better than previously used processes.
1924	**Coincidence method of particle detection (Walther Bothe):** Bothe's method proves invaluable in the use of gamma rays to discover nuclear energy levels.
1924	**Ultracentrifuge (Theodor Svedberg):** Svedberg's ultracentrifuge can separate isotopes, such as uranium 235 from uranium 238, from each other—a critical step in building the simplest kind of atomic bomb.
1924	**EEG:** German scientist Hans Berger records the first human electroencephalogram (EEG), which shows electrical patterns in the brain.
1925	**Leica I camera:** Leitz introduces the first 35-millimeter Leica camera at the Leipzig Spring Fair.
1925	**First U.S. television broadcast:** Charles Francis Jenkins transmits the silhouette image of a toy windmill.
1926	**Automatic power loom (Sakichi Toyoda):** Toyoda's loom helps Japan catch up with the western Industrial Revolution.
1926	**Liquid-fueled rocket (Robert H. Goddard):** A solid-fueled rocket is either on or off, but a liquid-fueled rocket can be throttled up or back and can be shut off before all the fuel is expended.
1927	**Aerosol can (Erik Rotheim):** Norwegian engineer Rotheim patents the aerosol can and valve.
1927	**Adiabatic demagnetization (William Francis Giauque):** Adiabatic demagnetization is part of a refrigeration cycle that, when used enough times, can chill a small sample to within a fraction of a kelvin above absolute zero.
1927	**All-electronic television (Philo T. Farnsworth):** Farnsworth transmits the first all-electronic television image using his newly developed camera vacuum tube, known as the image dissector. Previous systems combined electronics with mechanical scanners.
1927	**First flight across the Atlantic:** Charles Lindbergh flies the Spirit of St. Louis across the Atlantic. He is the first to make a solo, nonstop flight across the ocean.

1927	**Iron lung (Philip Drinker):** Drinker, a Harvard medical researcher, assisted by Louis Agassiz Shaw, devises the first modern practical respirator using an iron box and two vacuum cleaners. Drinker calls the device the iron lung.
1927	**Garbage disposal (John W. Hammes):** American architect Hammes develops the first garbage disposal to make cleaning up the kitchen easier for his wife. It is nicknamed the "electric pig" when it first goes on the market.
1927	**Adjustable-temperature iron:** The Silex Company begins to sell the first iron with an adjustable temperature control.
1927	**Analogue computer (Vannevar Bush):** Bush builds the first analogue computer. He is also the first person to describe the idea of hypertext.
1928	**Sliced bread (Otto F. Rohweddeer):** Bread that came presliced was advertised as "the greatest forward step in the baking industry since bread was wrapped." Today the phrase "the greatest thing since sliced bread" is used to describe any innovation that has a broad, positive impact on daily life.
1928	**First television programs:** First regularly scheduled television programs in the United States air. They are produced out of a small, experimental station in Wheaton, Maryland.
1928	**Link Trainer (Edwin Albert Link):** Link's flight simulator created realistic conditions in which to train pilots without the expense or risk of an actual air flight. Link also developed a submersible decompression chamber.
1928	**New punch card:** IBM introduces a new punch card that has rectangular holes and eight columns.
1928	**Radio network:** NBC establishes the first coast-to-coast radio network in the United States.
1928	**Pap smear (George N. Panpanicolaou):** Greek cytopathologist Panpanicolaou patents the pap smear, a test that helps detect uterine cancer.
1928	**Portable offshore drilling (Louis Giliasso):** Giliasso creates an efficient portable method of offshore drilling by mounting a derrick and drilling outfit onto a submersible barge.
1929	**Iconoscope (Vladimir Zworykin):** Zworykin claims that he, not Philo T. Farnsworth, should be credited with the invention of television.
1929	**Strobe light (Harold E. Edgerton):** Edgerton's strobe is used as a flash bulb. He pioneers the development of high-speed photography.
1929	**Dymaxion products (R. Buckminster Fuller):** Fuller's "Dymaxion" products feature an energy-efficient house using prefabricated, easily shipped parts.
1929	**Van de Graaff generator (Robert Jemison van de Graaff):** Van de Graaff invents the Van de Graaff generator, which accumulates electric charge on a moving belt and deposits it in a hollow glass sphere at the top.

1929-1936	**Cyclotron (Ernest Orlando Lawrence and M. Stanley Livingston):** Lawrence and Livingston are studying particle accelerators and develop the cyclotron, which consists of a vacuum tank between the poles of a large magnet. Alternating electric fields inside the tank can accelerate charged particles to high speeds. The cyclotron is used to probe the atomic nucleus or to make new isotopes of an element, including those used in medicine.
1930	**Schmidt telescope (Bernhard Voldemar Schmidt):** Schmidt's telescope uses a spherical main mirror and a correcting lens at the front of the scope. It can photograph large fields with little distortion.
1930	**Pluto discovered (Clyde Tombaugh):** Tombaugh observes a body one-fifth the mass of Earth's moon. Pluto comes to be regarded as the ninth planet of the solar system, but in 2006 it is reclassified as one of the largest-known Kuiper Belt objects, a dwarf planet.
1930	**Freon refrigeration and air-conditioning (Charles F. Kettering):** After inventing an electric starter in 1912 and the Kettering Aerial Torpedo in 1918 (the world's first cruise missile), Kettering and Thomas Midgley, Jr., use Freon gas in their cooling technology. (Freon will later be banned because of the effects of chlorofluorocarbons on Earth's ozone layer.)
1930	**Synthetic rubber (Wallace Hume Carothers):** Carothers synthesizes rubber and goes on to develop nylon in 1935. His work professionalizes polymer chemistry as a scientific field.
1930	**Scotch tape (Richard G. Drew):** After inventing masking tape, Drew invents the first waterproof, see-through, pressure-sensitive tape that also acted as a barrier to moisture.
1930	**Military and commercial aircraft (Andrei Nikolayevich Tupolev):** Tupolev emerges as one of the world's leading designers of military and civilian aircraft. His aircraft set nearly eighty world records.
1930's	**Washing machine (John W. Chamberlain):** Chamberlain invents a washing machine that enables clothes to be washed, rinsed, and have the water extracted from them in a single operation.
1931	**Electric razor (Jacob Schick):** Schick introduces his first electric razor, which allows dry shaving. It has a magazine of blades held in the handle.
1931	**Radio astronomy (Karl G. Jansky):** One of the founders of the field of radio astronomy, Janksy detects radio static coming from the Milky Way's center.
1932	**Positron discovered (Carl D. Anderson):** Anderson discovers the positron, a positive electron and an element of antimatter.
1932	**Neoprene (Julius Nieuwland):** The first synthetic rubber is marketed.
1932	**Neutron discovered (James Chadwick):** Chadwick detects the neutron, an atomic particle with no charge and a mass only slightly greater than that of a proton. Except for hydrogen 1, the atomic nuclei of all elements consist of neutrons and protons.

1932	**Phillips-head screw (Henry M. Phillips):** The Phillips-head screw has an X-shaped slot in the head and can withstand the torque of a machine-driven screwdriver, which is greater than the torque that can be withstood by the conventional screw.
1932	**Duplicating device for typewriters (Beulah Louise Henry):** Henry's invention uses three sheets of paper and three ribbons to produce copies of a document as it is typewritten. Henry also develops children's toys—for example, a doll the eye color of which can be changed.
1932	**Cockroft-Walton accelerator (John Douglas Cockcroft and Ernest Thomas Sinton Walton):** The Cockcroft-Walton accelerator is used to fling charged particles at atomic nuclei in order to investigate their properties.
1932	**Richter scale (Charles Francis Richter):** Richter develops a scale to describe the magnitude of earthquakes; it is still used today.
1932	**Neutron (Sir James Chadwick):** Chadwick proves the existence of neutrons; he is awarded the 1935 Nobel Prize in Physics for his work.
1933	**Nuclear chain reaction (Leo Szilard):** Szilard conceives the idea of a nuclear chain reaction. He becomes a key figure in the Manhattan Project, which eventually builds the atomic bomb.
1933	**Magnetic tape recorder (Semi Joseph Begun):** Begun builds the first tape recorder, a dictating machine using wire for magnetic recording. He also develops the first steel tape recorder for mobile radio broadcasting and leads research into telecommunications and underwater acoustics.
1933	**Electron microscope (Ernst Ruska):** Ruska makes use of the wavelengths of electrons—shorter than those of visible light—to build a microscope that can image details at the subatomic level.
1933	**Recording:** Alan Dower Blumlein's patent for stereophonic recording is granted.
1933	**Polyethylene (Eric Fawcett and Reginald Gibson):** Fawcett and Gibson of Imperial Chemical Industries in London accidentally discover polyethylene. Hula hoops and Tupperware are just two of the products made with the substance.
1933	**Modern airliner:** Boeing 247 becomes the first modern airliner.
1933	**Solo flight:** Wiley Post makes the first around-the-world solo flight.
1934	**First bathysphere dive:** Charles William Beebe and Otis Barton make the first deep-sea dive in the Beebe-designed bathysphere off the Bermuda coast.
1934	**Langmuir-Blodgett films (Katharine Burr Blodgett):** A thin Langmuir-Blodgett film deposited on glass can make it nearly nonreflective.
1934	**Passenger train:** The Burlington Zephyr, America's first diesel-powered streamlined passenger train, is revealed at the World's Fair in Chicago.
1935	**Frequency modulation (Edwin H. Armstrong):** Armstrong exploits the fact that, since there are no natural sources of frequency modulation (FM), FM broadcasts are static-free.

1935	**Diatometer (Ruth Patrick):** Patrick's diatometer is a device placed in the water to collect diatoms and allow them to grow. The number of diatoms is sensitive to water pollution.
1935	**Kodachrome color film (Leopold Mannes and Leopold Godowsky, Jr.):** Mannes and Godowsky invent Kodachrome, a color film that is easy to use and produces vibrant colors. (With the digital revolution of the late twentieth century, production of Kodachrome is finally retired in 2009.)
1935	**Physostigmine and cortisone (Percy Lavon Julian):** Julian synthesizes physostigmine, used to treat glaucoma, and cortisone, used for arthritis. He will hold more than 130 patents and will become the first African American chemist inducted into the National Academy of Sciences.
1935	**Mobile refrigeration (Frederick McKinley Jones):** Mobile refrigeration enables the shipping of heat-sensitive products and compounds, from blood to frozen food.
1935	**Radar-based air defense system (Sir Robert Alexander Watson-Watt):** Watson-Watt's technical developments and his efforts as an administrator will be so important to the development of radar that he will be called the "father of radar."
1935	**Fallingwater (Frank Lloyd Wright):** Wright designs and builds a showcase house blending its form with its surroundings. One of the greatest architects of the twentieth century, he will produce many architectural innovations in structure, materials, and design.
1936	**Field-emission microscope (Erwin Wilhelm Müller):** Müller completes his dissertation, "The Dependence of Field Electron Emission on Work Function," and goes on to develop the field-emission microscope, which can resolve surface features as small as 2 nanometers.
1936	**Pentothal (Ernest Volwiler and Donalee Tabern):** Pentothal is a fast-acting intravenous anesthetic.
1937	**Muon discovered (Seth Neddermeyer):** Neddermeyer, working with Carl Anderson, J. C. Street, and E. C. Stevenson discover the muon (a particle similar to a heavy electron) while examining cosmic-ray tracks in a cloud chamber.
1937	**Concepts of digital circuits and information theory (Claude Elwood Shannon):** Shannon's most important contributions were electronic switching and using information theory to discover the basic requirements for data transmission.
1937	**X-ray crystallography (Dorothy Crowfoot Hodgkin):** Hodgkin uses X-ray crystallography to reveal the structure of molecules. She goes on to win the 1964 Nobel Prize in Chemistry.
1937	**Model K computer (George Stibitz):** The model K, an early electronic computer, employs Boolean logic.
1937	**Artificial sweetener:** American chemist Michael Sveda invents cyclamates, which is used as a noncaloric artificial sweetener until it is banned by the U.S. government in 1970 because of possible carcinogenic effects.
1937	**First pressurized airplane cabin:** The first pressurized airplane cabin is achieved in the United States with Lockheed's XC-35.

1937	**Antihistamines (Daniel Bovet) :** Swiss-born Italian pharmacologist Bovet discovers antihistamines. He is awarded the 1957 Nobel Prize in Physiology or Medicine for his work.
1938	**Teflon (Roy J. Plunkett):** Plunkett accidentally synthesizes polytetrafluoroethylene (PTFE), now commonly known as Teflon, while researching chlorofluorocarbon refrigerants.
1938	**Electron microscope (James Hillier and Albert Prebus):** Adapting the work of German physicists, Hillier and Prebus develop a prototype of the electron microscope; and in 1940 Hillier produces the first commercial electron microscope available in the United States.
1938	**Xerography (Chester F. Carlson):** Xerography uses electrostatic charges to attract toner particles to make an image on plain paper. A hot wire then fuses the toner in place.
1938	**Walkie-talkie (Alfred J. Gross):** Gross's portable, two-way radio allows the user to move around while sending messages without remaining tied to a bulky transmitter. Gross invents a pager in 1949 and a radio tuner in 1950 that automatically follows the drift in carrier frequency due to movement of a sender or receiver.
1937-1938	**Analogue computer (George Philbrick):** Philbrick builds the Automatic Control Analyzer, which is an electronic analogue computer.
1939	**Helicopter (Igor Sikorsky):** Sikorsky, formerly the chief construction engineer and test pilot for the first four-engine aircraft, tests his helicopter, the Vought-Sikorsky 300, which after improvements will emerge as the world's first working helicopter.
1939	**Jet engine (Hans Joachim Pabst von Ohain):** The first jet-powered aircraft flies in 1939, while the first jet fighter will fly in 1941.
1939	**Atanasoff-Berry Computer (John Vincent Atanasoff and Clifford Berry):** The ABC, the world's first electronic digital computer, uses binary numbers and electronic switching, but it is not programmable.
1939	**DDT (Paul Hermann Müller):** Müller discovers the insect-repelling properties of DDT. He is awarded the 1948 Nobel Prize in Physiology or Medicine.
1940's	**Solar technology (Maria Telkes):** Telkes develops the solar oven and solar stills to produce drinking water from ocean water.
1940	**Cavity magnetron (Henry Boot and John Randall):** Boot and Randall develop the cavity magnetron, which advances radar technology.
1940	**Penicillin:** Sir Howard Walter Florey and Ernst Boris Chain isolate and purify penicillin. They are awarded, with Sir Alexander Fleming, the 1945 Nobel Prize in Physiology or Medicine.
1940	**Blood bank (Charles Richard Drew):** Drew establishes blood banks for World War II soldiers.
1940	**Color television (Peter Carl Goldmark):** Goldmark produces a system for transmitting and receiving color-television images using synchronized rotating filter wheels on the camera and on the receiver set.

1940	**Paintball gun (Charles and Evan Nelson):** The gun and paint capsules, invented to mark hard-to-reach trees in the forest, are eventually used for the game of paintball (1981), in which people shoot each other with paint.
1940	**Audio oscillator (William Redington Hewlett):** Hewlett invents the audio oscillator, a device that creates one frequency (pure tone) at a time. It is the first successful product of his Hewlett-Packard Company.
1940	**Antibiotics (Selman Abraham Waksman):** Waksman, through study of soil organisms, finds sources for the world's first antibiotics, including streptomycin and actinomycin.
1940	**Plutonium (Glenn Theodore Seaborg):** Seaborg synthesizes one of the first transuranium elements, plutonium. He becomes one of the leading figures on the Manhattan Project, which will build the atomic bomb. While he and others urged the demonstration of the bomb as a deterrent, rather than its use on the Japanese civilian population, the latter course was taken.
1940	**Thompson submachine gun (John T. Thompson):** Thompson works with Theodore Eickhoff and Oscar Payne to invent the American version of the submachine gun.
1940	**Automatic auto transmission:** General Motors offers the first modern automatic automobile transmission.
1941	**Jet engine (Sir Frank Whittle):** Whittle develops the jet engine independent of Hans Joachim Pabst von Ohain in Germany. After World War II, they meet and become good friends.
1941	**Solid-body electric guitar (Les Paul):** Paul's guitar lays the foundation for rock music. He also develops multitrack recording in 1948.
1941	**Z3 programmable computer (Konrad Zuse):** Zuse and his colleagues complete the first general-purpose, programmable computer, the Z3, in December. In 1950, Zuse will sell a Z4 computer—the only working computer in Europe.
1941	**Velcro (Georges de Mestral):** Burrs sticking to his dog's fur give de Mestral the idea for Velcro, which he perfects in 1948.
1941	**Dicoumarol:** The anticoagulant drug dicoumarol is identified and synthesized.
1941	**RDAs:** The first Recommended Dietary Allowances (RDAs), nutritional guidelines, are accepted.
1942	**Superglue (Harry Coover and Fred Joyner):** After developing superglue (cyanoacrylate), Coover rejects it as too sticky for a 1942 project. Coover and Joyner rediscover superglue in 1951, when Coover recognizes it as a marketable product.
1942	**Aqua-Lung (Jacques-Yves Cousteau and Émile Gagnon):** The Aqua-Lung delivers air at ambient pressure and vents used air to the surroundings.
1942	**Controlled nuclear chain reaction (Enrico Fermi):** In 1926 Fermi helped develop Fermi-Dirac statistics, which describe the quantum behavior of groups of electrons, protons, or neutrons. He now produces the first sustained nuclear chain reaction.

1942	**Synthetic vitamins (Max Tishler):** After synthesizing several vitamins during the 1930's, Tishler and his team develop the antibiotic sulfaquinoxaline to treat coccidiosis. He also develops fermentation processes to produce streptomycin and penicillin.
1942	**Bazooka:** The United States military first uses the bazooka during the North African campaign in World War II.
1943	**Meteorology:** Radar is first used to detect storms.
1944	**Electromechanical computer (Howard Aiken and Grace Hopper):** The Mark series of computers is built, designed by Aiken and Hopper. The U.S. Navy uses it to calculate trajectories for projectiles.
1944	**Colossus:** Colossus, the world's first vacuum-tube programmable logic calculator, is built in Britain for the purpose of breaking Nazi codes.
1944	**Phased array radar antennas (Luis W. Alvarez):** Alvarez's phased array sweeps a beam across the sky by turning hundreds of small antennas on and off and not by moving a radar dish.
1944	**V-2 rocket (Wernher von Braun):** Working for the German government during World War II, von Braun and other rocket scientists develop the V-2 rocket, the first long-range military missile and first suborbital missile. Arrested for making anti-Nazi comments, he later emigrates to the United States, where he leads the team that produces the Jupiter-C missile and launches vehicles such as the Saturn V, which help make the U.S. space program possible.
1944	**Quinine:** Robert B. Woodward and William von Eggers Doering synthesize quinine, which is used as an antimalarial.
1945	**Automatic Computing Engine (Alan Mathison Turing):** While the Automatic Computing Engine (ACE) was never fully built, it was one of the first stored-program computers.
1945	**Atomic bomb (J. Robert Oppenheimer):** Oppenheimer, the scientific leader of the Manhattan Project, heads the team that builds the atomic bomb. On the side of military use of the bomb to end World War II quickly, Oppenheimer saw this come to pass on August 6, 1945, when the bomb was dropped over Hiroshima, Japan, killing and maiming 150,000 people; a similar number of casualties ensued in Nagasaki on August 9, when the second bomb was dropped. Japan surrendered on August 14.
1945	**Dialysis machine (Willem Johan Kolff):** Kolff designs the first artificial kidney, a machine that cleans the blood of patients in renal failure, and refuses to patent it. He will construct the artificial lung in 1955.
1945	**Radioimmunoassay (RIA) (Rosalyn Yalow):** RIA required only a drop of blood (rather than the tens of milliliters previously required) to find trace amounts of substances.
1945	**Electronic Sackbut (Hugh Le Caine):** Le Caine builds the first music synthesizer, joined by the Special Purpose Tape Recorder in 1954, which could simultaneously change the playback speed of several recording tracks.

1945	**ENIAC computer (John William Mauchly and John Presper Eckert):** The Electronic Numerical Integrator and Computer, ENIAC, is the first general-purpose, programmable, electronic computer. (The Z3, developed independently by Konrad Zuse from 1939 to 1941 in Nazi Germany, did not fully exploit electronic components.) Built to calculate artillery firing tables, ENIAC is used in calculations for the hydrogen bomb.
1945	**Microwave oven (Percy L. Spencer):** The microwave oven grew out of the microwave generator, the magnetron tube, becoming more affordable.
1946	**Tupperware (Earl S. Tupper):** Tupper exploits plastics technology to develop a line of plastic containers that he markets at home parties starting in 1948.
1946	**Carbon-14 dating (Willard F. Libby):** Libby uses the half-life of carbon 14 to develop a reliable means of dating ancient remains. Radiocarbon dating has proven to be invaluable to archaeologists.
1946	**Magnetic tape recording (Marvin Camras):** Camras develops a magnetic tape recording process that will be adapted for use in electronic media, including music and motion-picture sound recording, audio and videocassettes, floppy disks, and credit card magnetic strips. For many years his method is the primary way to record and store sound, video, and digital data.
1946-1947	**Audiometer (Georg von Békésy):** Békésy invents a pure-tone audiometer that patients themselves can control to measure the sensitivity of their own hearing.
1946	**Radioisotopes for cancer treatment:** The first nuclear-reactor-produced radioisotopes for civilian use are sent from the U.S. Army's Oak Ridge facility in Tennessee to Brainard Cancer Hospital in St. Louis.
1947	**Transistor (John Bardeen, Walter H. Brattain, and William Shockley):** Hoping to build a solid-state amplifier, the team of Bardeen, Brattain, and Shockley discover the transistor, which replaces the vacuum tube in electronics. Bardeen is later part of the group that develops theory of superconductivity.
1947	**Platforming (Vladimir Haensel):** American chemical engineer Haensel invents platforming, a process that uses a platinum catalyst to produce cleaner-burning high-octane fuels.
1947	**Tubeless tire:** B.F. Goodrich announces development of the tubeless tire.
1948	**Holography (Dennis Gabor):** Gabor publishes his initial results working with holograms in Nature. Holograms became much more spectacular after the invention of the laser.
1948	**Long-playing record (LP) (Peter Carl Goldmark):** Goldmark demonstrates the LP playing the cello with CBS musicians. The musical South Pacific is recorded in LP format and boosts sales, making the LP the dominant form of recorded sound for the next four decades.
1948	**Gamma-ray pinhole camera (Roscoe Koontz):** Working to make nuclear reactors safer, Koontz invents the gamma-ray pinhole camera. The pinhole should act like a lens and form an image of the gamma source.
1948	**Instant photography (Edwin Herbert Land):** Land develops the simple process to make sheets of polarizing material. He perfects the Polaroid camera in 1972.

1948	**Synthetic penicillin (John C. Sheehan):** Sheehan develops the first total synthesis of penicillin, making this important antibiotic widely available.
1949	**First peacetime nuclear reactor:** Construction on the Brookhaven Graphite Research Reactor at Brookhaven Laboratory on Long Island, New York, is completed.
1949	**Magnetic core memory (Jay Wright Forrester):** Core memory is used from the early 1950's to the early 1970's.
1950's	**Fortran (John Warner Backus):** Backus develops the computer language Fortran, which is an acronym for "formula translation." Fortran allows direct entry of commands into computers with Englishlike words and algebraic symbols.
1950	**Planotron (Pyotr Leonidovich Kapitsa):** Kapitsa invents a magnetron tube for generating microwaves. He becomes a corecipient of the Nobel Prize for Physics in 1978 for discovering superfluidity in liquid helium.
1950	**Purinethol (Gertrude Belle Elion):** Elion develops the first effective treatment for childhood leukemia, 6-mercaptopurine (Purinethol). Elion later discovers azathioprine (Imuran), an immunosuppressive agent used for organ transplants.
1950	**Artificial pacemaker (John Alexander Hopps):** Hopps develops a device to regulate the beating of the heart to treat patients with erratic heartbeats. By 1957, the device is small enough to be implanted.
1950	**Contact lenses (George Butterfield):** Oregon optometrist Butterfield develops a lens that is molded to fit the contours of the cornea.
1951	**Fiber-optic endoscope (fibroscope) (Harold Hopkins):** Hopkins fastened together a flexible bundle of optical fibers that could convey an image. One end of the bundle could be inserted into a patient's throat, and the physician could inspect the esophagus.
1951	**The Pill (Carl Djerassi):** The birth-control pill, which becomes the world's most popular and is possibly most widely used contraceptive, revolutionizes not only medicine but also gender relations and women's status in society. Its prolonged use is later revealed to have health consequences.
1951	**Field-emission microscope (Erwin Wilhelm Müller):** Müller develops the field-ion microscope, followed by an atom-probe field-ion microscope in 1963, which can detect individual atoms.
1951	**Maser (Charles Hard Townes):** The maser (microwave amplification by stimulated emission of radiation) is a "laser" for microwaves. Discovered later, the "laser" patterned its name the acronym "maser."
1951	**Artificial heart valve (Charles Hufnagel):** Hufnagel develops an artificial heart valve and performs the first heart-valve implantation surgery in a human patient the following year.
1951	**UNIVAC (John Mauchly and John Presper Eckert):** Mauchly and Eckert invent the Universal Automatic Computer (UNIVAC). UNIVAC is competitor of IBM's products.
1952	**Bubble chamber (Donald A. Glaser):** In a bubble chamber, bubbles form along paths taken by subatomic particles as they interact, and the bubble trails allow scientists to deduce what happened.

1952	**Photovoltaic cell (Gerald Pearson):** The photovoltaic cell converts sunlight into electricity.
1952	**Improved electrical resistor (Otis Boykin):** Boykin's resistor had improved precision, and its high-frequency characteristics were better than those of previous resistors.
1952	**Language compiler (Grace Murray Hopper):** Hopper invents the compiler, an intermediate program that translates English-language instructions into computer language, followed in 1959 by Common Business Oriented Language (COBOL), the first computer programming language to translate commands used by programmers into the machine language the computer understands.
1952	**Amniocentesis (Douglas Bevis):** British physician Bevis develops amniocentesis.
1952	**Gamma camera (Hal Anger):** Nuclear medicine pioneer Anger creates the first prototype for the gamma camera. This leads to the inventions of other medical imaging devices, which detect and diagnose disease.
1953	**Medical ultrasonography (Inge Edler and Carl H. Hertz):** Edler and Hertz adapt an ultrasound probe used in materials testing in a shipyard for use on a patient. Their technology makes possible echograms of the heart and brain.
1953	**Inertial navigation systems (Charles Stark Draper):** Draper's inertial navigation system (INS) is designed to determine the current position of a ship or plane based on the initial location and acceleration.
1953	**Heart-lung machine (John H. Gibbon, Jr.):** American surgeon Gibbon conducts the first successful heart surgery using a heart-lung machine that he constructed with the help of his wife, Mary.
1953	**First frozen meals:** Swanson develops individual prepackaged frozen meals. The first-ever meal consists of turkey, cornbread stuffing, peas, and sweet potatoes.
1954	**Geodesic dome ® (Buckminster Fuller):** After developing the geodesic dome, Fuller patents the structure, an energy-efficient house using prefabricated, easily shipped parts.
1954	**Atomic absorption spectroscopy (Sir Alan Walsh):** Atomic absorption spectroscopy is used to identify and quantify the presence of elements in a sample.
1954	**Synthetic diamond (H. Tracy Hall):** Hall synthesizes diamonds using a high-pressure, high-temperature belt apparatus that can generate 120,000 atmospheres of pressure and sustain a temperature of 1,800 degrees Celsius in a working volume of about 0.1 cubic centimeter.
1954	**Machine vision (Jerome H. Lemelson):** Machine vision allows a computer to move and measure products and to inspect them for quality control.
1954	**Hydrogen bomb (Edward Teller):** The first hydrogen bomb, designed by Teller, is tested at the Bikini Atoll in the Pacific Ocean.
1954	**Silicon solar cells (Calvin Fuller):** Silicon solar cells have proven to be among the most efficient and least expensive solar cells.

1954	**First successful kidney transplant (Joseph Edward Murray):** American surgeon Murray performs the first successful kidney transplant, inserting one of Ronald Herrick's kidneys into his twin brother, Richard. Murray shares the 1990 Nobel Prize for Physiology or Medicine with E. Donnall Thomas, who developed bone marrow transplantation.
1954	**Transistor radio:** The first transistor radio is introduced by Texas Instruments.
1954	**IBM 650:** The IBM 650 computer becomes available. It is considered by IBM to be its first business computer, and it is the first computer installed at Columbia University in New York.
1954	**First nuclear submarine:** The United States launches the first nuclear- powered submarine, the USS *Nautilus.*
1955	**Color television's RGB system (Ernst Alexanderson):** The RGB system uses three image tubes to scan scenes through colored filters and three electron guns in the picture tube to reconstruct scenes.
1955	**Floppy disk and floppy disk drive (Alan Shugart):** Working at the San Jose, California, offices of International Business Machines (IBM), Shugart develops the disk drive, followed by floppy disks to provide a relatively fast way to store programs and data permanently.
1955	**Hovercraft (Sir Christopher Cockerell):** Cockerell files a patent for his hovercraft, an amphibious vehicle. He earlier invented several important electronic devices, including a radio direction finder for bombers in World War II.
1955	**Pulse transfer controlling device (An Wang):** The device allows magnetic core memory to be written or read without mechanical motion and is therefore very rapid.
1955	**Polio vaccine (Jonas Salk):** Salk's polio vaccine, which uses the killed virus, saves lives and improves the quality of life for millions afflicted by polio.
1956	**Fiber optics (Narinder S. Kapany):** Kapany, known as the father of fiber optics, coins the term "fiber optics." In high school, he was told by a teacher that light moves only in a straight line; he wanted to prove the teacher wrong and wound up inventing fiber optics.
1956	**Scotchgard (Patsy O'Connell Sherman):** Sherman develops a stain repellent for fabrics that is trademarked as Scotchgard.
1956	**Ovonic switch (Stanford Ovshinsky):** Ovshinsky invents a solid-state, thin film switch meant to mimic the actions of neurons.
1956	**Videotape recorder (Charles P. Ginsburg):** The video recorder allows programs to be shown later, to provide instant replays in sports, and to make a permanent record of a program.
1956	**Liquid Paper (Bette Nesmith Graham):** Graham markets her "Mistake Out" fluid for concealing typographical errors.
1956	**Dipstick blood sugar test (Helen M. Free):** Free and her husband Alfred co-invent a self-administered urinalysis test that allows diabetics to monitor their sugar levels and to adjust their medications accordingly.

1956	**350 RAMAC:** IBM produces the first computer disk storage system, the 350 RAMAC, which retrieves data from any of fifty spinning disks.
1957	**Wankel rotary engine (Felix Wankel):** Having fewer moving parts, the Wankel rotary engine ought to be sturdier and perhaps more efficient than the common reciprocating engine.
1957	**Laser (Gordon Gould, Charles Hard Townes, Arthur L. Schawlow, Theodore Harold Maiman):** Having conducted research on using light to excite thallium atoms, Gould tries to get funds and approval to build the first laser, but he fails. Townes (inventor of the maser) and Schawlow of Bell Laboratories will first describe the laser, and Maiman will first succeed in building a small optical maser. Gould coins the term "laser," which stands for light amplification by stimulated emission of radiation.
1957	**Intercontinental ballistic missile (ICBM):** The Soviet Union develops the ICBM.
1957	**First satellite:** The Soviet Union launches Sputnik, the first man-made satellite.
1958	**CorningWare:** CorningWare cookware is introduced. It is based on S. Donald Stookey's 1953 discovery that a heat-treatment process can transform glass into fine-grained ceramics.
1958	**Integrated circuit (Robert Norton Noyce and Jack St. Clair Kilby):** The microchip, independently discovered by Noyce and Kilby, proves to be the breakthrough that allows the miniaturization of electronic circuits and paves the way for the digital revolution.
1958	**Ultrasound:** Ultrasound becomes the most common method for examining a fetus.
1958	**Planar process (Jean Hoerni):** Hoerni develops the first planar process, which improves the integrated circuit.
1960's	**Lithography:** Optical lithography, a process that places intricate patterns onto silicon chips, is used in semiconductor manufacturing.
1960	**Measles vaccine (John F. Enders):** Enders, an American physician, develops the first measles vaccine. It is tested the following year and is hailed a success.
1960	**Echo satellite (John R. Pierce):** The first passive-relay telecommunications satellite, Echo, reflected signals. The signals, received from one point on Earth, "bounce" off the spherical satellite and are reflected back down to another, far distant, point on Earth.
1960	**Automatic letter-sorting machine (Jacob Rabinow):** Rabinow's machine greatly increased the speed and efficiency of mail delivery in the United States. He also invented an optical character recognition (OCR) scanner.
1960	**Ruby laser (Theodore Harold Maiman):** Maiman produces a ruby laser, the world's first visible light laser.
1960	**Helium-neon gas laser (Ali Javan):** Javan produces the world's second visible light laser.
1960	**Chardack-Greatbatch pacemaker (Wilson Greatbatch and William Chardack):** Greatbatch and Chardack create the first implantable pacemaker.

1960	**Radionuclide generator:** Powell Richards and Walter Tucker and their colleagues at Brookhaven Laboratory in New York invent a short half-life radionuclide generator for use in nuclear medicine diagnostic imaging procedures.
1961	**Audio-animatronics (Walt Disney):** Disney established WED, a research and development unit that developed the inventions he needed for his various enterprises. WED produced the audio-animatronic robotic figures that populated Disneyland, the 1964-1965 New York World's Fair, films, and other attractions. Audio-animatronics enabled robotic characters to speak or sing as well as move.
1961	**Ruby laser:** The ruby laser is first used medically by Charles Campbell and Charles Koester to excise a patient's retinal tumor.
1961	**First person in space:** Soviet astronaut Yuri Gagarin becomes the first person in space when he orbits the Earth on April 12.
1962	**Soft contact lenses (Otto Wichterle):** Wichterle's soft contacts can be worn longer with less discomfort than can hard contact lenses.
1962	**Continuously operating ruby laser (Willard S. Boyle and Don Nelson):** The invention relies on an arc lamp shining continuously (rather than the flash lamp used by Theodore Maiman in 1960).
1962	**Light-emitting diode (Nick Holonyak, Jr.):** Holonyak makes the first visible-spectrum diode laser, which produces red laser light but also stops lasing yet remains a useful light source. Holonyak has invented the red light-emitting diode (LED), the first operating alloy device—the "ultimate lamp."
1962	**Telstar satellite (John R. Pierce):** The first satellite to rebroadcast signals goes into operation, revolutionizing telecommunications.
1962	**Quasar 3C 273 (Maarten Schmidt):** Schmidt shows that this quasar is very distant and hence very bright. Further research shows quasars to be young galaxies with active, supermassive black holes at their centers.
1962	**First audiocassette:** The Philips company of the Netherlands releases the audiocassette tape.
1962	**Artificial hip (Sir John Charnley):** British surgeon Charnley invents the low-friction artificial hip and develops the surgical techniques for emplacing it.
1963	**Learjet (Bill Lear):** The Learjet, a small eight-passenger jet with a top speed of 560 miles (900 kilometers) per hour, can shuttle VIPs to meetings and other engagements.
1963	**Self-cleaning oven:** General Electric introduces the self-cleaning electric oven.
1963	**Artificial heart (Paul Winchell):** Winchell receives a patent (later donated to the University of Utah's Institute for Biomedical Engineering) for an artificial heart that purportedly became the model for the successful Jarvick-7.
1963	**6600 computer (Seymour Cray):** The 6600 was the first of a long line of Cray supercomputers.
1963	**Carbon fiber (Leslie Philips):** British engineer Philips develops carbon fiber, which is much stronger than steel.

1964	**Three-dimensional holography (Emmett Leith):** Leith and Juris Upatnieks present the first three-dimensional hologram at the Optical Society of America conference. The hologram must be viewed with a reference laser. The hologram of an object can then be viewed from different angles, as if the object were really present.
1964	**Moog synthesizer (Robert Moog):** The Moog synthesizer uses electronics to create and combine musical sounds.
1964	**Cosmic background radiation (Arno Penzias and Robert Wilson):** Penzias and Wilson detect the cosmic background radiation, which corresponds to that which would be radiated by a body at 2.725 kelvins. It is thought to be greatly redshifted primordial fireball radiation left over from the big bang.
1964	**BASIC programming language (John Kemeny and Thomas Kurtz):** Kemeny and Kurtz develop the BASIC computer programming language. BASIC is an acronym for Beginner's All-purpose Symbolic Instruction Code.
1965	**Minicomputer (Ken Olsen):** Perhaps the first true minicomputer, the PDP-8 is released by Digital Equipment Corporation. Founder Olsen makes computers affordable for small businesses.
1965	**Aspartame (James M. Schlatter):** Schlatter discovers aspartame, an artificial sweetener, while trying to come up with an antiulcer medication.
1965	**First space walk:** Soviet astronaut Aleksei Leonov is the first person to walk in space.
1966	**Gamma-electric cell (Henry Thomas Sampson):** Sampson works with George H. Miley to produce the gamma-electric cell, which converts the energy of gamma rays into electrical energy.
1966	**Handheld calculator (Jack St. Clair Kilby):** While working for Texas Instruments, Kilby does for the adding machine what the transistor had done for the radio, inventing a handheld calculator that retails at $150 and becomes an instant commercial success.
1966	**First unmanned moon landing:** Soviet spacecraft Luna 9 lands on the moon.
1967	**Electrogasdynamic method and apparatus (Meredith C. Gourdine):** Gourdine develops electrogasdynamics, which involves the production of electricity from the conversion of kinetic energy in a moving, ionized gas.
1967	**Pulsars (Jocelyn Bell and Antony Hewish):** Pulsars, rapidly rotating neutron stars, are discovered.
1968	**Practical liquid crystal displays (James Fergason):** Fergason develops an liquid crystal display (LCD) screen that has good visual contrast, is durable, and uses little electricity.
1968	**Lasers in medicine:** Francis L'Esperance begins using the argon-ion laser to treat patients with diabetic retinopathy.
1968	**Computer mouse (Douglas Engelbart):** Engelbart presents the computer mouse, which he had been working on since 1964.
1968	**Apollo 7:** Astronauts on Apollo 7, the first piloted Apollo mission, take photographs and transmit them to the American public on television.

1968	**Interface message processors:** Bolt Beranek and Newman Incorporated win a Defense Advanced Research Projects Agency (DARPA) contract to develop the packet switches called interface message processors (IMPs).
1969	**Rubella vaccine:** The rubella vaccine is available.
1969	**First person walks on the moon:** Neil Armstrong, a member of the U.S. Apollo 11 spacecraft, is the first person to walk on the moon.
1969	**Boeing 747:** The Boeing 747 makes its first flight, piloted by Jack Waddell.
1969	**Concorde:** The Concorde makes its first flight, piloted by André Turcat.
1969	**Charge-coupled device (Willard S. Boyle and George E. Smith):** Boyle and Smith develop the charge-coupled device, the basis for digital imaging.
1969	**ARPANET launches:** The Advanced Research Projects Agency starts ARPANET, which is the precursor to the Internet. UCLA and Stanford University are the first institutions to become networked.
1970's	**Digital seismology:** Digital seismology is used in oil exploration and increases accuracy in finding underground pools.
1970's	**Mud pulse telemetry:** Mud pulse telemetry becomes an oil-industry standard; pressure pulses are relayed through drilling mud to convey the location of the drill bit.
1970	**Optical fiber (Robert Maurer and others):** Maurer, joined by Donald Keck, Peter Schultz, and Frank Zimar, produces an optical fiber that can be used for communication.
1970	**Compact disc (James Russell):** The compact disc (CD) revolutionizes the way digital media is stored.
1970	**UNIX (Dennis Ritchie and Kenneth Thompson):** Bell Laboratories employees Ritchie and Thompson complete the UNIX operating system, which becomes popular among scientists.
1970	**Network Control Protocol:** The Network Working Group deploys the initial ARPANET host-to-host protocol, called the Network Control Protocol (NCP), establishing connections, break connections, switch connections, and control flow over the ARPANET.
1971	**Computerized axial tomography (Godfrey Newbold Hounsfield):** In London, doctors performed the first CAT scan of a living patient and detected a brain tumor. In a CAT (or CT) scan, X rays are taken of a body like slices in a loaf of bread. A computer then assembles these slices into a detail-laden three-dimensional image.
1971	**First videocassette recorder:** Sony begins selling the first videocassette recorder (VCR) to the public.
1971	**Microprocessor (Ted Hoff):** The computer's central processing unit (CPU) is reduced to the size of a postage stamp.
1971	**Electronic switching system for telecommunications (Erna Schneider Hoover):** Hoover's system prioritizes telephone calls and fixes an efficient order to answer them.
1971	**Intel microprocessors:** Intel builds the world's first microprocessor chip.

1971	**Touch screen (Sam Hurst):** Hurst's touch screen can detect if it has been touched and where it was touched.
1972	**First recombinant DNA organism (Stanley Norman Cohen, Paul Berg, and Herbert Boyer):** The methods to combine and transplant genes are discovered when this team successfully clones and expresses the human insulin gene in the Escherichia coli.
1972	**Far-Ultraviolet Camera (George R. Carruthers):** The Carruthers-designed camera is used on the Apollo 16 mission.
1972	**Cell encapsulation (Taylor Gunjin Wang):** Wang develops ways to encapsulate beneficial cells and introduce them into a body without triggering the immune system.
1972	**Pioneer 10:** The U.S. probe Pioneer 10 is launched to get information about the outer solar system.
1972	**Networking goes public:** ARPANET system designer Robert Kahn organizes the first public demonstration of the new network technology at the International Conference on Computer Communications in Washington, D.C.
1972	**Pong video game (Nolan K. Bushnell and Ted Dabney):** Bushnell and Dabney register the name of their new computer company, Atari, and issue Pong shortly thereafter, marking the rise of the video game industry.
1973	**Automatic computerized transverse axial (ACTA) whole-body CT scanner (Robert Steven Ledley):** The first whole-body CT scanner is operational. Ledley goes on to spend much of his career promoting the use of electronics and computers in biomedical research.
1973	**Packet network interconnection protocols TCP/IP (Vinton Gray Cerf and Robert Kahn):** Cerf and Kahn develop transmission control protocol/Internet protocol (TCP/IP), protocols that enable computers to communicate with one another.
1973	**Automated teller machine (Don Wetzel):** Wetzel receives a patent for his ATM. To make it a success, he shows banks how to generate a group of clients who would use the ATM.
1973	**Food processor:** The Cuisinart food processor is introduced in the United States.
1973	**Air bags in automobiles:** The Oldsmobile Tornado is the first American car sold equipped with air bags.
1973	**Space photography:** Astronauts aboard Skylab, the first U.S. space station, take high-resolution photographs of Earth using photographic remote-sensing systems. The astronauts also take photographs with handheld cameras.
1974	**Kevlar (Stephanie Kwolek):** Kwolek receives a patent for the fiber Kevlar. Bullet-resistant Kevlar vests go on sale only one year later.
1975	**Ethernet (Robert Metcalfe and David Boggs):** Metcalfe and Boggs invent the Ethernet, a system of software, protocols, and hardware allowing instantaneous communication between computer terminals in a local area.

1975	**Semiconductor laser:** Scientists working at Diode Labs develop the first commercial semiconductor laser that will operate continuously at room temperature.
1976	**First laser printer:** IBM's 3800 Printing System is the first laser printer. The ink jet is invented in the same year, but it is not prevalent in homes until 1988.
1976	**Apple computer (Steve Jobs):** Jobs cofounds Apple Computer with Steve Wozniak.
1976	**Jarvik-7 artificial heart (Robert Jarvik):** The Jarvik-7 allows a calf to live 268 days with the artificial heart. Jarvik combined ideas from several other workers to produce the Jarvik-7.
1976	**Apple II (Steve Wozniak):** Wozniak develops the Apple II, the best-selling personal computer of the 1970's and early 1980's.
1976	**First Mars probes:** The National Aeronautics and Space Administration (NASA) launches Viking 1 and Viking 2, which land on obtain images of Mars.
1976	**Kurzweil Reading Machine (Ray Kurzweil):** Kurzweil develops an optical character reader (OCR) able to read most fonts.
1976	**Microsoft Corporation (Bill Gates):** Gates, along with Paul Allen, found Microsoft, a software company. Gates will remain head of Microsoft for twenty-five years.
1976	**Conductive polymers:** Hideki Shirakawa, Alan G. MacDiarmid, and Alan J. Heeger discover conductive polymers. They are awarded the 2000 Nobel Prize in Chemistry.
1977	**Global Positioning System (GPS) (Ivan A. Getting):** The first GPS satellite is launched, designed to support a navigational system that uses satellites to pinpoint the location of a radio receiver on Earth's surface.
1977	**Fiber-optic telephone cable:** The first fiber-optic telephone cables are tested.
1977	**Echo-planar imaging (Peter Mansfield):** British physicist Mansfield first develops the echo-planar imaging (EPI).
1977	**Gossamer Condor (Paul MacCready):** MacCready designs the Gossamer Albatross, which enables human-powered flight.
1978	**Smart gels (Toyoichi Tanaka):** Tanaka discovers and works with "smart gels," polymer gels that can expand a thousandfold, change color, or contract when stimulated by minor changes in temperature, magnetism, light, or electricity. This capacity makes them useful in a broad range of applications.
1978	**Charon discovered (James Christy):** Charon is discovered as an apparent bulge on a fuzzy picture of Pluto. Its mass is about 12 percent that of Pluto.
1978	**First cochlear implant surgery:** Graeme Clark performs the first cochlear implant surgery in Australia.
1978	**The first test-tube baby:** Louise Brown is born in England.
1978	**First MRI:** The first magnetic resonance image (MRI) of the human head is taken in England.
1979	**First laptop (William Moggridge):** Moggridge, of Grid Systems in England, designs the first laptop computer.

1979	**First commercially successful application:** The VisiCalc spreadsheet for Apple II, designed by Daniel Bricklin and Bob Frankston, helps drive sales of the personal computer and becomes its first successful business application.
1979	**USENET (Tom Truscott, Jim Ellis and Steve Belovin):** Truscott, Ellis, and Belovin create USENET, a "poor man's ARPANET," to share information via e-mail and message boards between Duke University and the University of North Carolina, using dial-up telephone lines.
1979	**In-line roller skates (Scott Olson and Brennan Olson):** After finding some antique in-line skates, the Olson brothers begin experimenting with modern materials, creating Rollerblades.
1980's	**Controlled drug delivery (Robert S. Langer):** Langer develops the foundation of controlled drug delivery technology used in cancer treatment.
1980	**Alkaline battery (Lewis Urry):** Eveready markets alkaline batteries under the trade name Energizer. Urry's alkaline battery lasts longer than its predecessor, the carbon-zinc battery.
1980	**Interferon (Charles Weissmann):** Weissmann produces the first genetically engineered human interferon, which is used in cancer treatment.
1980	**TCP/IP:** The U.S. Department of Defense adopts the TCP/IP suite as a standard.
1981	**Ablative photodecomposition (Rangaswamy Srinivasan):** Srinivasan's research on ablative photodecomposition leads to multiple applications, including laser-assisted in situ keratomileusis (LASIK) surgery, which shapes the cornea to correct vision problems.
1981	**Scanning tunneling microscope (Heinrich Rohrer and Gerd Binnig):** The scanning tunneling microscope shows surfaces at the atomic level.
1981	**Improvements in laser spectroscopy (Arthur L. Schawlow and Nicolaas Bloembergen):** Schawlow shares the Nobel Prize in Physics with Nicolaas Bloembergen for their work on laser spectroscopy. While most of Schawlow's inventions involved lasers, he also did research in superconductivity and nuclear resonance.
1981	**First IBM personal computer:** The first IBM PC, the IBM 5100, goes on the market with a $1,565 price tag.
1982	**Compact discs appear:** Compact discs are now sold and will start replacing vinyl records.
1982	**First artificial heart:** Seattle dentist Barney Clark receives the first permanent artificial heart, and he survives for 112 days.
1983	**Cell phone (Martin Cooper):** The first mobile (wireless) phone, the DynaTAC 8000X, receives approval by the Federal Communications Commission (FCC), heralding an age of wireless communication.
1983	**Internet:** ARPANET, and networks attached to it, adopt the TCP/IP networking protocol. All networks that use the protocol are known as the Internet.
1983	**Cyclosporine:** Immunosuppressant cyclosporine is approved for use in transplant operations in the United States.

1983	**Polymerase chain reaction (Kary B. Mullis):** While driving to his cottage in Mendocino, California, Mullis develops the idea for the polymerase chain reaction (PCR). PCR will be used to amplify a DNA segment many times, leading to a revolution in recombinant DNA technology and a 1993 Nobel Prize in Chemistry for Mullis.
1984	**Domain name service is created:** Paul Mockapetris and Craig Partridge develop domain name service, which links unique Internet protocol (IP) numerical addresses to names with suffixes such as .mil, .com, .org, and .edu.
1984	**Mac is released:** Apple introduces the Macintosh, a low-cost, plug-and-play personal computer with a user-friendly graphic interface.
1984	**CD-ROM:** Philips and Sony introduce the CD-ROM (compact disc read-only memory), which has the capacity to store data of more than 450 floppy disks.
1984	**Surgery in utero:** William A. Clewall performs the first successful surgery on a fetus.
1984	**Cloning:** Danish veterinarian Steen M. Willadsen clones a lamb from a developing sheep embryo cell.
1984	**AIDS blood test (Robert Charles Gallo):** Gallo and his colleagues identify the virus HTLV-3/LAV (later renamed human immunodeficiency virus, or HIV) as the cause of acquired immunodeficiency syndrome, or AIDS. Gallo creates a blood test that can identify antibodies specific to HIV. This blood test is essential to keeping the supply in blood banks pure.
1984	**Imaging X-ray spectrometer (George Edward Alcorn):** Alcorn patents his device, which makes images of the source using X rays of specific energies, similar to making images with a specific wavelength (color) of light. It is used in acquiring data on the composition of distant planets and stars.
1984	**DNA profiling (Alec Jeffreys):** Noticing similarities and differences in DNA samples from his lab technician's family, Jeffreys discovers the principles that lead to DNA profiling, which has become an essential tool in forensics and the prosecution of criminal cases.
1985	**Windows operating system (Bill Gates):** The first version of Windows is released.
1985	**Implantable cardioverter defibrillator:** The U.S. Food and Drug Administration (FDA) approves Polish physician Michel Mirowski's implantable cardioverter defibrillator (ICD), which monitors and corrects abnormal heart rhythms.
1985	**Industry Standard Architecture (ISA) bus (Mark Dean and Dennis Moeller):** Dean and Moeller design the standard way of organizing the central part of a computer and its peripherals, the ISA bus, which is patented in this year.
1985	**Atomic force microscope:** Calvin Quate, Christoph Gerber, and Gerd Binnig invent the atomic force microscope, which becomes one of the foremost tools for imaging, measuring, and manipulating matter at the nano scale.
1986	**Mir:** The Soviet Union launches the Mir space station, the first permanent space station.
1986	**Burt Rutan's Voyager:** Dick Rutan (Burt's brother) and Jeana Yeager make the first around-the-world, nonstop flight without refueling in the Burt Rutan-designed Voyager. The Voyager is the first aircraft to accomplish this feat.

1986	**High-temperature superconductor (J. Georg Bednorz and Karl Alexander Müller):** Bednorz and Müller show that a ceramic compound of lanthanum, barium, copper, and oxygen becomes superconducting at 35 kelvins, a new high- temperature record.
1987	**Azidothymidine:** The FDA approves azidothymidine (AZT), a potent antiviral, for AIDS patients.
1987	**Echo-planar imaging:** Echo-planar imaging is used to perform real-time movie imaging of a single cardiac cycle.
1987	**Parkinson's treatment:** French neurosurgeon Alim-Louis Benabid implants a deep-brain electrical-stimulation system into a patient with advanced Parkinson's disease.
1987	**First corneal laser surgery:** New York ophthalmologist Steven Trokel performs the first laser surgery on a human cornea. He had refined his technique on a cow's eye. Trokel was granted a patent for the Excimer laser to be used for vision correction.
1987	**UUNET and PSINet:** Rick Adams forms UUNET and Bill Schrader forms PSINet to provide commercial Internet access.
1988	**Transatlantic fiber-optic cable:** The first transatlantic fiber-optic cable is installed, linking North America and France.
1988	**Laserphaco probe (Patricia Bath):** Bath's probe is used to break up and remove cataracts.
1989	**Method for tracking oil flow underground using a supercomputer (Philip Emeagwali):** Emeagwali receives the Gordon Bell Prize, considered the Nobel Prize for computing, for his method, which demonstrates the possibilities of computer networking.
1989	**World Wide Web (Tim Berners-Lee and Robert Cailau):** Berners-Lee finds a way to join the idea of hypertext and the young Internet, leading to the Web, coinvented with Cailau.
1989	**First dial-up access:** The World debuts as the first provider of dial-up Internet access for consumers.
1990's	**Environmentally friendly appliances:** Water-saving and energy-conserving washing machines and dryers are introduced.
1990	**Hubble Space Telescope:** The Hubble Space Telescope is launched and changes the way scientists look at the universe.
1990	**Human Genome Project begins:** The U.S. Department of Energy and the National Institutes of Health coordinate the Human Genome Project with the goal of identifying all 30,000 genes in human DNA and determining the sequences of the three billion chemical base pairs that make up human DNA.
1990	**BRCA1 gene discovered (Mary-Claire King):** King finds the cancer-associated gene on chromosome 17. She demonstrates that humans and chimpanzees are 99 percent genetically identical.

1991	**Nakao Snare (Naomi L. Nakao):** The Snare is a device that captures polyps that have been cut from the walls of the intestine, solving the problem of "lost polyp syndrome."
1991	**America Online (AOL):** Quantum Computer Services changes its name to America Online; Steve Case is named president. AOL offers e-mail, electronic bulletin boards, news, and other information.
1991	**Carbon nanotubes (Sumio Iijima):** Although carbon nanotubes have been seen before, Iijima's 1991 paper establishes some basic properties and prompts other scientists' interest in studying them.
1991	**The first hot-air balloon crosses the Pacific (Richard Branson and Per Lindstrad):** Branson and Lindstrad, who teamed up in 1987 to cross the Atlantic, make the 6,700-mile flight in 47 hours and break the world distance record.
1992	**Newton:** Apple introduces Newton, one of the first handheld computers, or personal digital assistants, which has a liquid crystal display operated with a stylus.
1993	**Mosaic (Marc Andreessen):** Andreessen launches Mosaic, followed by Netscape Navigator in 1995—the first Internet browsers. Both Mosaic and Netscape allow novices to browse the World Wide Web.
1993	**Flexible tailored elastic airfoil section (Sheila Widnall):** Widnall applies for a patent for this device, which addresses the problem of being able to measure fluctuations in pressure under unsteady conditions. She serves as secretary of the Air Force (the first woman to lead a branch of the military) and also serves on the board investigating the space shuttle Columbia accident of 2003.
1993	**Light-emitting diode (LED) blue and UV (Shuji Nakamura):** Nakamura's blue LED makes white LED light possible (a combination of red, blue, and green).
1994	**Genetically modified (GM) food:** The Flavr Savr tomato, the first GM food, is approved by the FDA.
1994	**Channel Tunnel:** Channel Tunnel, or Chunnel, opens, connecting France and Britain by a railway constructed beneath the English Channel.
1995	**51 Pegasi (Michel Mayor and Didier Queloz):** Mayor and Queloz detect a planet orbiting another normal star, the first extrasolar planet (exoplanet) to be found. As of June, 2009, 353 exoplanets were known.
1995	**Saquinavir:** The FDA approves Saquinavir for the treatment of AIDS. It is the first protease inhibitor, which reduces the ability of the AIDS virus to spread to new cells.
1995	**iBot (Dean Kamen):** Kamen invents iBOT, a super wheelchair that climbs stairs and helps its passenger to stand.
1995	**Global Positioning System (Ivan A. Getting):** The GPS becomes fully operational.
1995	**Illusion transmitter (Valerie L. Thomas):** A concave mirror can produce a real image that appears to be three-dimensional. Thomas's system uses a concave mirror at the camera and another one at the television receiver.
1996	**LASIK:** The first computerized excimer laser (LASIK), designed to correct the refractive error myopia, is approved for use in the United States.

1996	**First sheep is cloned:** Scottish scientist Ian Wilmut clones the first mammal, a Finn Dorset ewe named Dolly, from differentiated adult mammary cells.
1997	**Robotic vacuum:** Swedish appliance company Electrolux is the first to create a prototype of a robotic vacuum cleaner.
1998	**PageRank (Larry Page):** The cofounder of Google with Sergey Brin, Page devises PageRank, the count of Web pages linked to a given page and a measure how valuable people find that page.
1998	**UV Waterworks (Ashok Gadgil):** The device uses UV from a mercury lamp to kill waterborne pathogens.
1998	**Napster:** College dropout Shawn Fanning creates Napster, an extremely popular peer-to-peer file-sharing platform that allowed users to download music for free. In 2001 the free site was shut down because it encouraged illegal sharing of copyrighted properties. The site then became available by paid subscription.
1999	**Palm VII:** The Palm VII organizer is on the market. It is a handheld computer with 2 megabytes of RAM and a port for a wireless phone.
1999	**BlackBerry (Research in Motion of Canada):** A wireless handheld device that began as a two-way pager, the BlackBerry is also a cell phone that supports Web browsing, e-mail, text messaging, and faxing—it is the first smart phone.
2000	**Hoover-Diana production platform:** A joint venture by Exxon and British Petroleum (BP), the Hoover-Diana production platform goes into operation in the Gulf of Mexico. Within six months it is producing 20,000 barrels of oil a day.
2000	**Clone of a clone:** Japanese scientists clone a bull from a cloned bull.
2000	**Minerva:** The Library of Congress initiates a prototype system called Minerva (Mapping the Internet Electronic Resources Virtual Archives) to collect and preserve open-access Web resources.
2000	**Supercomputer:** The ASCI White supercomputer at the Lawrence Livermore National Laboratory in California is operational. It can hold six times the information stored in the 29 million books in the Library of Congress.
2001	**XM Radio:** XM Radio initiates the first U.S. digital satellite radio service in Dallas-Ft. Worth and San Diego.
2001	**Human cloning:** Scientists at Advanced Cell Technology in Massachusetts clone human embryos for the first time.
2001	**iPod (Tony Fadell):** Fadell introduces the iPod, a portable hard drive-based MP3 player with an Internet-based electronic music catalog, for Apple.
2001	**Segway PT (Dean Kamen):** Kamen introduces his personal transport device, a self-balancing, electric-powered pedestrian scooter.
2003	**First digital books:** Lofti Belkhir introduces the Kirtas BookScan 1200, the first automatic, page-turning scanner for the conversion of bound volumes to digital files.
2003	**Aqwon (Josef Zeitler):** The hydrogen-powered scooter Aqwon can reach 30 miles (50 kilometers) per hour. Its combustion product is water.

2003	**Human Genome Project is completed:** After thirteen years, the 25,000 genes of the human genome are identified and the sequences of the 3 million chemical base pairs that make up human DNA are determined.
2004	**Stem cell bank:** The world's first embryonic stem cell bank opens in England.
2004	**SpaceShipOne and SpaceShipTwo (Burt Rutan):** Rutan receives the U.S. Department of Transportation's first license issued for suborbital flight for SpaceShipOne, which shortly thereafter reaches an altitude of 328,491 feet. Rutan's rockets are the first privately funded manned rockets to reach space (higher than 100 kilometers above Earth's surface).
2004	**Columbia supercomputer:** The NASA supercomputer Columbia, built by Silicon Graphics and Intel, achieves sustained performance of 42.7 trillion calculations per second and is named the fastest supercomputer in the world. It is named for those who lost their lives in the explosion of the space shuttle Columbia in 2003. Because technology evolves so quickly, the Columbia will not be the fastest for very long.
2005	**Blue Gene/L supercomputer:** The National Nuclear Security Administration's BlueGene/L supercomputer, built by IBM, performs at 280.6 trillion operations per second and is now the world's fastest supercomputer.
2005	**Eris (Mike Brown):** Working with C. A. Trujillo and D. L. Rabinowitz, Brown discovers Eris, the largest known dwarf planet and a Kuiper Belt object. It is 27 percent more massive than Pluto, another large Kuiper Belt object.
2005	**Nix and Hydra discovered (Pluto companion team):** The Hubble research team—composed of Hal Weaver, S. Alan Stern, Max Mutchler, Andrew Steffl, Marc Buie, William Merline, John Spencer, Eliot Young, and Leslie Young—finds these small moons of Pluto.
2006	**Digital versus film:** Digital cameras have almost wholly replaced film cameras. *The New York Times* reports that 92 percent of cameras sold are digital.
2007	**First terabyte drive:** Hitachi Global Storage Technologies announces that it has created the first one-terabyte (TB) hard disk drive.
2007	**iPhone (Apple):** Apple introduces its smart phone, a combined cell phone, portable media player (equal to a video iPod), camera phone, Internet client (supporting e-mail and Web browsing), and text messaging device, to an enthusiastic market.
2008	**Roadrunner:** The Roadrunner supercomputer, built by IBM and Los Alamos National Laboratory, can process more than 1.026 quadrillion calculations per second. It works more than twice as fast as the Blue Gene/L supercomputer and is housed at Los Alamos in New Mexico.
2008	**Mammoth Genome Project:** Scientists sequence woolly mammoth genome, the first of an extinct animal.
2008	**Columbus lands:** The space shuttle Atlantis delivers the Columbus science laboratory to the International Space Station. The twenty-three-foot long laboratory is able to conduct experiments both inside and outside the space station.

2008	**Retail DNA test (Anne Wojcicki):** Wojcicki (wife of Google founder Sergey Brin) offers an affordable DNA saliva test, 23andMe, to determine one's genetic markers for ninety traits. The product heralds what *Time* magazine dubs a "personal-genomics revolution."
2009	**Large Hadron Collider:** The Large Hadron Collider (LHC) becomes the world's highest energy particle accelerator.
2009	**Hubble Space Telescope repairs (NASA):** STS-125 astronauts conducted five space walks from the space shuttle Atlantis to upgrade the Hubble Space Telescope, extending its life to at least 2014.
2009	**AIDS vaccine:** Scientists in Thailand create a vaccine that seems to reduce the risk of contracting the AIDS virus by more than 31 percent.
2010	**Jaguar supercomputer:** The Oak Ridge National Laboratory in Tennessee is home to Jaguar, the world's fastest supercomputer, the peak speed of which is 2.33 quadrillion floating point operations per second.

Charles W. Rogers, Southwestern Oklahoma State University, Department of Physics; updated by the editors of Salem Press

GENERAL BIBLIOGRAPHY

Aaboe, Asger. *Episodes from the Early History of Astronomy.* New York: Springer-Verlag, 2001.

Abbate, Janet. *Inventing the Internet.* Cambridge, Mass.: MIT Press, 2000.

Abell, George O., David Morrison, and Sidney C. Wolff. *Exploration of the Universe.* 5th ed. Philadelphia: Saunders College Publishing, 1987.

Achilladelis, Basil, and Mary Ellen Bowden. *Structures of Life.* Philadelphia: The Center, 1989.

Ackerknecht, Erwin H. *A Short History of Medicine.* Rev. ed. Baltimore: The Johns Hopkins University Press, 1982.

Aczel, Amir D. *Fermat's Last Theorem: Unlocking the Secret of an Ancient Mathematical Problem.* Reprint. New York: Four Walls Eight Windows, 1996.

Adler, Robert E. *Science Firsts: From the Creation of Science to the Science of Creation.* Hoboken, N.J.: John Wiley & Sons, 2002.

Alberts, Bruce, et al. *Molecular Biology of the Cell.* 2d ed. New York: Garland, 1989.

Alcamo, I. Edward. *AIDS: The Biological Basis.* 3d ed. Boston: Jones and Bartlett, 2003.

Aldersey-Willliams, Hugh. *The Most Beautiful Molecule: An Adventure in Chemistry.* London: Aurum Press, 1995.

Alexander, Arthur F. O'Donel. *The Planet Saturn: A History of Observation, Theory, and Discovery.* 1962. Reprint. New York: Dover, 1980.

Alioto, Anthony M. *A History of Western Science.* 2d ed. Upper Saddle River, N.J.: Prentice Hall, 1993.

Allen, Oliver E., and the editors of Time-Life Books. *Atmosphere.* Alexandria, Va.: Time-Life Books, 1983.

Ames, W. F., and C. Rogers, eds. *Nonlinear Equations in the Applied Sciences.* San Diego: Academic Press, 1992.

Andriesse, Cornelis D. *Christian Huygens.* Paris: Albin Michel, 2000.

Angier, Natalie. *Natural Obsessions: Striving to Unlock the Deepest Secrets of the Cancer Cell.* Boston: Mariner Books/Houghton Mifflin, 1999.

Annaratone, Donnatello. *Transient Heat Transfer.* New York: Springer, 2011.

Anstey, Peter R. *The Philosophy of Robert Boyle.* London: Routledge, 2000.

Anton, Sebastian. *A Dictionary of the History of Science.* Pearl River, N.Y.: Parthenon Publishing, 2001.

Archimedes. *The Works of Archimedes.* Translated by Sir Thomas Heath. 1897. Reprint. New York: Dover, 2002.

Arms, Karen, and Pamela S. Camp. *Biology: A Journey into Life.* 3d ed. Philadelphia: Saunders College Publishing, 1987.

Armstrong, Neil, Michael Collins, and Edwin E. Aldrin. *First on the Moon.* New York: Williams Konecky Associates, 2002.

Arrizabalaga, Jon, John Henderson, and Roger French. *The Great Pox: The French Disease in Renaissance Europe.* New Haven, Conn.: Yale University Press, 1997.

Arsuaga, Juan Luis. *The Neanderthal's Necklace: In Search of the First Thinkers.* Translated by Andy Klatt. New York: Four Walls Eight Windows, 2002.

Artmann, Benno. *Euclid: The Creation of Mathematics.* New York: Springer- Verlag, 1999.

Asimov, Isaac. *Exploring the Earth and the Cosmos.* New York: Crown, 1982.

_______. *The History of Physics.* New York: Walker, 1984.

_______. *Jupiter, the Largest Planet.* New York: Ace, 1980.

Aspray, William. *John von Neumann and the Origins of Modern Computing.* Boston: MIT Press, 1990.

Astronomical Society of the Pacific. *The Discovery of Pulsars.* San Francisco: Author, 1989.

Audesirk, Gerald J., and Teresa E. Audesirk. *Biology: Life on Earth.* 2d ed. New York: Macmillan, 1989.

Aughton, Peter. *Newton's Apple: Isaac Newton and the English Scientific Revolution.* London: Weidenfeld & Nicolson, 2003.

Aujoulat, Norbert. *Lascaux: Movement, Space, and Time.* New York: Harry N. Abrams, 2005.

Aveni, Anthony F., ed. *Skywatchers.* Rev. ed. Austin: University of Texas Press, 2001.

Baggott, Jim. *Perfect Symmetry: The Accidental Discovery of Buckminsterfullerene.* New York: Oxford University Press, 1994.

Baine, Celeste. *Is There an Engineer Inside You? A Comprehensive Guide to Career Decisions in Engineering.* 2d ed. Belmont, Calif.: Professional Publications, 2004.

Baker, John. *The Cell Theory: A Restatement, History and Critique.* New York: Garland, 1988.

Baldwin, Joyce. *To Heal the Heart of a Child: Helen Taussig, M.D.* New York: Walker, 1992.

Barbieri, Cesare, et al., eds. *The Three Galileos: The Man, the Spacecraft, the Telescope: Proceedings of the Conference Held in Padova, Italy on January 7-10, 1997.* Boston: Kluwer Academic, 1997.

Barkan, Diana Kormos. *Walther Nernst and the Transition to Modern Physical Science.* New York: Cambridge University Press, 1999.

Barrett, Peter. *Science and Theology Since Copernicus: The Search for Understanding.* Reprint. Dorset, England: T&T Clark, 2003.

Bartusiak, Marcia. *Thursday's Universe.* New York: Times Books, 1986.

Basta, Nicholas. *Opportunities in Engineering Careers.* New York: McGraw-Hill, 2003.

Bates, Charles C., and John F. Fuller. *America's Weather Warriors, 1814-1985.* College Station: Texas A&M Press, 1986.

Bazin, Hervé. *The Eradication of Smallpox: Edward Jenner and the First and Only Eradication of a Human Infectious Disease.* Translated by Andrew Morgan and Glenise Morgan. San Diego: Academic Press, 2000.

Beatty, J. Kelly, and Andrew Chaikin, eds. *The New Solar System.* 3d rev. ed. New York: Cambridge University Press, 1990.

Becker, Wayne, Lewis Kleinsmith, and Jeff Hardin. *The World of the Cell.* New York: Pearson/Benjamin Cummings, 2006.

Berlinski, David. *A Tour of the Calculus.* New York: Vintage Books, 1997.

Bernstein, Jeremy. *Three Degrees Above Zero: Bell Labs in the Information Age.* New York: Charles Scribner's Sons, 1984.

Bernstein, Peter L. *Against the Gods: The Remarkable Story of Risk.* New York: John Wiley & Sons, 1996.

Bertolotti, M. *Masers and Lasers: An Historical Approach.* Bristol, England: Adam Hilger, 1983.

Bickel, Lennard. *Florey: The Man Who Made Penicillin.* Carlton South, Victoria, Australia: Melbourne University Press, 1995.

Bizony, Piers. *Island in the Sky: Building the International Space Station.* London: Aurum Press Limited, 1996.

Blackwell, Richard J. *Galileo, Bellarmine, and the Bible.* London: University of Notre Dame Press, 1991.

Bliss, Michael. *The Discovery of Insulin.* Chicago: University of Chicago Press, 1987.

Blumenberg, Hans. *The Genesis of the Copernican World.* Translated by Robert M. Wallace. Cambridge, Mass.: MIT Press, 1987.

Blunt, Wilfrid. *Linnaeus: The Compleat Naturalist.* Princeton, N.J.: Princeton University Press, 2001.

Bodanis, David. *Electric Universe: The Shocking True Story of Electricity.* New York: Crown Publishers, 2005.

Bohm, David. *Causality and Chance in Modern Physics.* London: Routledge & Kegan Paul, 1984.

Bohren, Craig F. *Clouds in a Glass of Beer: Simple Experiments in Atmospheric Physics.* New York: John Wiley & Sons.

Boljanovic, Vukota. *Applied Mathematics and Physical Formulas: A Pocket Reference Guide for Students, Mechanical Engineers, Electrical Engineers, Manufacturing Engineers, Maintenance Technicians, Toolmakers, and Machinists.* New York: Industrial Press, 2007.

Bolt, Bruce A. *Inside the Earth: Evidence from Earthquakes.* New York: W. H. Freeman, 1982.

Bond, Peter. *The Continuing Story of the International Space Station.* Chichester, England: Springer-Praxis, 2002.

Boorstin, Daniel J. *The Discoverers.* New York: Random House, 1983.

Bottazzini, Umberto. *The Higher Calculus: A History of Real and Complex Analysis from Euler to Weierstrass.* New York: Springer-Verlag, 1986.

Bourbaki, Nicolas. *Elements of the History of Mathematics.* Translated by John Meldrum. New York: Springer, 1994.

Bowler, Peter J. *Charles Darwin: The Man and His Influence.* Cambridge, England: Cambridge University Press, 1996.

_______. *Evolution: The History of an Idea.* Rev. ed. Berkeley: University of California Press, 1989.

_______. *The Mendelian Revolution: The Emergence of Hereditarian Concepts in Modern Science and Society.* Baltimore: The Johns Hopkins University Press, 1989.

Boyer, Carl B. *A History of Mathematics.* 2d ed., revised by Uta C. Merzbach. New York: John Wiley & Sons, 1991.

Bracewell, Ronald N. *The Fourier Transform and Its Applications.* 3d rev. ed. New York: McGraw-Hill, 1987.

Brachman, Arnold. *A Delicate Arrangement: The Strange Case of Charles Darwin and Alfred Russel Wallace.* New

York: Times Books, 1980.

Bredeson, Carmen. *John Glenn Returns to Orbit: Life on the Space Shuttle.* Berkeley Heights, N.J.: Enslow, 2000.

Brock, Thomas, ed. *Milestones in Microbiology, 1546-1940.* Washington, D.C.: American Society for Microbiology, 1999.

Brock, William H. *The Chemical Tree: A History of Chemistry.* New York: W. W. Norton, 2000.

Brooks, Paul. *The House of Life: Rachel Carson at Work.* 2d ed. Boston: Houghton Mifflin, 1989.

Browne, Janet. *Charles Darwin: The Power of Place.* New York: Knopf, 2002.

Brush, Stephen G. *Cautious Revolutionaries: Maxwell, Planck, Hubble.* College Park, Md.: American Association of Physics Teachers, 2002.

Brush, Stephen G., and Nancy S. Hall. *Kinetic Theory of Gases: An Anthology of Classic Papers With Historical Commentary.* London: Imperial College Press, 2003.

Bryant, Stephen. *The Story of the Internet.* London: Pearson Education, 2000.

Buffon, Georges-Louis Leclerc. *Natural History: General and Particular.* Translated by William Smellie. Avon, England: Thoemmes Press, 2001.

Burger, Edward B., and Michael Starbird. *Coincidences, Chaos, and All That Math Jazz: Making Light of Weighty Ideas.* New York: W. W. Norton, 2005.

Burke, Terry, et al., eds. *DNA Fingerprinting: Approaches and Applications.* Boston: Birkhauser, 2001.

Byrne, Patrick H. *Analysis and Science in Aristotle.* Albany: State University of New York Press, 1997.

Calder, William M., III, and David A. Traill, eds. *Myth, Scandal, and History: The Heinrich Schliemann Controversy.* Detroit: Wayne State University Press, 1986.

Calinger, Ronald. *A Contextual History of Mathematics.* Upper Saddle River, N.J.: Prentice Hall, 1999.

Canning, Thomas N. *Galileo Probe Parachute Test Program: Wake Properties of the Galileo.* Washington, D.C.: National Aeronautics and Space Administration, Scientific and Technical Information Division, 1988.

Cantor, Geoffrey. *Michael Faraday: Sandemanian and Scientist: A Study of Science and Religion in the Nineteenth Century.* New York: St. Martin's Press, 1991.

Carlisle, Rodney. *Inventions and Discoveries: All the Milestones in Ingenuity—from the Discovery of Fire to the Invention of the Microwave.* Hoboken, N.J.: John Wiley & Sons, 2004.

Carlson, Elof Axel. *Mendel's Legacy: The Origin of Classical Genetics.* Woodbury, N.Y.: Cold Spring Harbor Laboratory Press, 2004.

Carola, Robert, John P. Harley, and Charles R. Noback. *Human Anatomy and Physiology.* New York: McGraw-Hill, 1990.

Carpenter, B. S., and R. W. Doran, eds. *A. M. Turing's ACE Report of 1946 and Other Papers.* Cambridge, Mass.: MIT Press, 1986.

Carpenter, Kenneth J. *The History of Scurvy and Vitamin C.* Cambridge England: Cambridge University Press, 1986.

Carrigan, Richard A., and W. Peter Trower, eds. *Particle Physics in the Cosmos.* New York: W. H. Freeman, 1989.

_______, eds. *Particles and Forces: At the Heart of the Matter.* New York: W. H. Freeman, 1990.

Cassanelli, Roberto, et al. *Houses and Monuments of Pompeii: The Works of Fausto and Felice Niccolini.* Los Angeles: J. Paul Getty Museum, 2002.

Caton, Jerald A. *A Review of Investigations Using the Second Law of Thermodynamics to Study Internal-Combustion Engines.* London: Society of Automotive Engineers, 2000.

Chaikin, Andrew. *A Man on the Moon: The Voyages of the Apollo Astronauts.* New York: Penguin Group, 1998.

Chaisson, Eric J. *The Hubble Wars.* New York: HarperCollins, 1994.

Chaisson, Eric J., and Steve McMillan. *Astronomy Today.* 5th ed. Upper Saddle River, N.J.: Pearson Prentice Hall, 2004.

Chandrasekhar, Subrahmanyan. *Eddington: The Most Distinguished Astrophysicist of His Time.* Cambridge, England: Cambridge University Press, 1983.

Chang, Hasok. *Inventing Temperature: Measurement and Scientific Progress.* Oxford, England: Oxford University Press, 2004.

Chang, Laura, ed. *Scientists at Work: Profiles of Today's Groundbreaking Scientists from "Science Times."* New York: McGraw-Hill, 2000.

Chant, Christopher. *Space Shuttle.* New York: Exeter Books, 1984.

Chapman, Allan. *Astronomical Instruments and Their Users: Tycho Brahe to William Lassell.* Brookfield, Vt.: Variorum, 1996.

Chase, Allan. *Magic Shots.* New York: William Morrow, 1982.

Check, William A. *AIDS.* New York: Chelsea House, 1988.

Cheng, K. S., and G. V. Romero. *Cosmic Gamma-Ray Sources.* New York: Springer-Verlag, 2004.

Christianson, John Robert. *On Tycho's Island: Tycho Brahe and His Assistants, 1570-1601.* New York: Cambridge University Press, 2000.

Chung, Deborah D. L. *Applied Materials Science: Applications of Engineering Materials in Structural, Electronics, Thermal and Other Industries.* Boca Raton, Fla.: CRC Press, 2001.

Clark, Ronald W. *The Life of Ernst Chain: Penicillin and Beyond.* New York: St. Martin's Press, 1985.

_______. *The Survival of Charles Darwin: A Biography of a Man and an Idea.* New York: Random House, 1984.

Cline, Barbara Lovett. *Men Who Made a New Physics.* Chicago: University of Chicago Press, 1987.

Clos, Lynne. *Field Adventures in Paleontology.* Boulder, Colo.: Fossil News, 2003.

Clugston, M. J., ed. *The New Penguin Dictionary of Science.* 2d ed. New York: Penguin Books, 2004.

Coffey, Patrick. *Cathedrals of Science: The Personalities and Rivalries That Made Modern Chemistry.* New York: Oxford University Press, 2008.

Cohen, I. Bernard. *Benjamin Franklin's Science.* Cambridge, Mass.: Harvard University Press, 1990.

_______. *The Newtonian Revolution.* New York: Cambridge University Press, 1980.

Cohen, I. Bernard, and George E. Smith, eds. *The Cambridge Companion to Newton.* New York: Cambridge University Press, 2002.

Cole, K. C. *The Universe and the Teacup: The Mathematics of Truth and Beauty.* Fort Washington, Pa.: Harvest Books, 1999.

Cole, Michael D. *Galileo Spacecraft: Mission to Jupiter: Countdown to Space.* New York: Enslow, 1999.

Collin, S. M. H. *Dictionary of Science and Technology.* London: Bloomsbury Publishing, 2003.

Connor, James A. *Kepler's Witch: An Astronomer's Discovery of Cosmic Order Amid Religious War, Political Intrigue, and the Heresy Trial of His Mother.* San Francisco: HarperSanFrancisco, 2004.

Conrad, Lawrence, et al., eds. *The Western Medical Tradition, 800 B.C. to A.D. 1800.* New York: Cambridge University Press, 1995.

Cook, Alan. *Edmond Halley: Charting the Heavens and the Seas.* New York: Oxford University Press, 1998.

Cooke, Donald A. *The Life and Death of Stars.* New York: Crown, 1985.

Cooper, Geoffrey M. *Oncogenes.* 2d ed. Boston: Jones and Bartlett, 1995.

Cooper, Henry S. F., Jr. *Imaging Saturn: The Voyager Flights to Saturn.* New York: H. Holt, 1985.

Corsi, Pietro. *The Age of Lamarck: Evolutionary Theories in France, 1790-1830.* Berkeley: University of California Press, 1988.

Coulthard, Malcolm, and Alison Johnson, eds. *The Routledge Handbook of Forensic Linguistics.* New York: Routledge, 2010.

Craven, B. O. *The Lebesgue Measure and Integral.* Boston: Pitman Press, 1981.

Crawford, Deborah. *King's Astronomer William Herschel.* New York: Julian Messner, 2000.

Crease, Robert P., and Charles C. Mann. *The Second Creation: Makers of the Revolution in Twentieth Century Physics.* New York: Macmillan, 1985.

Crewdson, John. *Science Fictions: A Scientific Mystery, A Massive Cover-Up, and the Dark Legacy of Robert Gallo.* Boston: Little, Brown, 2002.

Crick, Francis. *What Mad Pursuit: A Personal View of Scientific Discovery.* New York: Basic Books, 1988.

Crump, Thomas. *A Brief History of Science as Seen Through the Development of Scientific Instruments.* New York: Carroll & Graf, 2001.

Cunningham, Andrew. *The Anatomical Renaissance: The Resurrection of the Anatomical Projects of the Ancients.* Brookfield, Vt.: Ashgate, 1997.

Cutler, Alan. *The Seashell on the Mountaintop: A Story of Science, Sainthood, and the Humble Genius Who Discovered a New History of the Earth.* New York: Dutton/Penguin, 2003.

Dalrymple, G. Brent. *The Age of the Earth.* Stanford, Calif.: Stanford University Press, 1991.

Darrigol, Oliver. *Electrodynamics from Ampère to Einstein.* Oxford, England: Oxford University Press, 2000.

Dash, Joan. *The Longitude Prize.* New York: Farrar, Straus and Giroux, 2000.

Daston, Lorraine. *Classical Probability in the Enlightenment.* Princeton, N.J.: Princeton University Press, 1988.

Davies, John K. *Astronomy from Space: The Design and Operation of Orbiting Observatories.* New York: John Wiley & Sons, 1997.

Davies, Paul. *The Edge of Infinity: Where the Universe Came from and How It Will End.* New York: Simon & Schuster, 1981.

Davis, Joel. *Flyby: The Interplanetary Odyssey of Voyager 2.* New York: Atheneum, 1987.

Davis, Martin. *Engines of Logic: Mathematicians and the Origin of the Computer.* New York: W. W. Norton, 2000.

Davis, Morton D. *Game Theory: A Nontechnical Introduction.* New York: Dover, 1997.

Davis, William Morris. *Elementary Meteorology.* Boston: Ginn, 1894.

Dawkins, Richard. *The Ancestor's Tale: A Pilgrimage to the Dawn of Evolution.* New York: Houghton Mifflin, 2004.

_______. *River Out of Eden: A Darwinian View of Life.* New York: Basic Books, 1995.

Day, Michael H. *Guide to Fossil Man.* 4th ed. Chicago: University of Chicago Press, 1986.

Day, William. *Genesis on Planet Earth.* 2d ed. New Haven, Conn.: Yale University Press, 1984.

Dean, Dennis R. *James Hutton and the History of Geology.* Ithaca, N.Y.: Cornell University Press, 1992.

Debré, Patrice. *Louis Pasteur.* Translated by Elborg Forster. Baltimore: The Johns Hopkins University Press, 1998.

DeJauregui, Ruth. *100 Medical Milestones That Shaped World History.* San Mateo, Calif.: Bluewood Books, 1998.

De Jonge, Christopher J., and Christopher L. R. Barratt, eds. *Assisted Reproductive Technologies: Current Accomplishments and New Horizons.* New York: Cambridge University Press, 2002.

Delaporte, François. *The History of Yellow Fever: An Essay on the Birth of Tropical Medicine.* Cambridge, Mass.: MIT Press, 1991.

Dennett, Daniel C. *Darwin's Dangerous Idea: Evolution and the Meanings of Life.* New York: Simon & Schuster, 1995.

Dennis, Carina, and Richard Gallagher. *The Human Genome.* London: Palgrave Macmillan, 2002.

DeVorkin, David H. *Race to the Stratosphere: Manned Scientific Ballooning in America.* New York: Springer-Verlag, 1989.

Dewdney, A. K. *The Turing Omnibus.* Rockville, Md.: Computer Science Press, 1989.

Diamond, Jared. *The Third Chimpanzee: The Evolution and Future of the Human Animal.* New York: HarperCollins, 1992.

DiCanzio, Albert. *Galileo: His Science and His Significance for the Future of Man.* Portsmouth, N.H.: ADASI, 1996.

Dijksterhuis, Eduard Jan. *Archimedes.* Translated by C. Dikshoorn, with a new bibliographic essay by Wilbur R. Knorr. Princeton, N.J.: Princeton University Press, 1987.

Dijksterhuis, Fokko Jan. *Lenses and Waves: Christiaan Huygens and the Mathematical Science of Optics in the Seventeenth Century.* Dordrecht, the Netherlands: Kluwer Academic, 2004.

Dimmock, N. J., A. J. Easton, and K. N. Leppard. *Introduction to Modern Virology.* 5th ed. Malden, Mass.: Blackwell Science, 2001.

Dore, Mohammed, Sukhamoy Chakravarty, and Richard Goodwin, eds. *John Von Neumann and Modern Economics.* New York: Oxford University Press, 1989.

Drake, Stillman. *Galileo: Pioneer Scientist.* Toronto: University of Toronto Press, 1990.

_______. *Galileo: A Very Short Introduction.* New York: Oxford University Press, 2001.

Dreyer, John Louis Emil, ed. *The Scientific Papers of Sir William Herschel.* Dorset, England: Thoemmes Continuum, 2003.

Duck, Ian. *One Hundred Years of Planck's Quantum.* River Edge, N.J.: World Scientific, 2000.

Dudgeon, Dan E., and Russell M. Mersereau. *Multidimensional Digital Signal Processing.* Englewood Cliffs, N.J.: Prentice Hall, 1984.

Dunham, William. *The Calculus Gallery: Masterpieces from Newton to Lebesgue.* Princeton, N.J.: Princeton University Press, 2005.

_______. *Euler: The Master of Us All.* Washington, D.C.: Mathematical Association of America, 1999.

_______. *Journey Through Genius.* New York: John Wiley & Sons, 1990.

Durham, Frank, and Robert D. Purrington. *Frame of the Universe.* New York: Cambridge University Press, 1983.

Easton, Thomas A. *Careers in Science.* 4th ed. Chicago: VGM Career Books, 2004.

Edelson, Edward. *Gregor Mendel: And the Roots of Genetics.* New York: Oxford University Press, 2001.

Edey, Maitland A., and Donald C. Johanson. *Blueprints: Solving the Mystery of Evolution.* Boston: Little, Brown, 1989.

Edwards, Robert G., and Patrick Steptoe. *A Matter of Life.* New York: William Morrow, 1980.

Ehrenfest, Paul, and Tatiana Ehrenfest. *The Conceptual Foundations of the Statistical Approach in Mechanics.* Mineola, N.Y.: Dover, 2002.

Ehrlich, Melanie, ed. *DNA Alterations in Cancer: Genetic and Epigenetic Changes.* Natick, Mass.: Eaton, 2000.

Eisen, Herman N. *Immunology: An Introduction to Molecular and Cellular Principles of the Immune Responses.* 2d ed. Philadelphia: J. B. Lippincott, 1980.

Espejo, Roman, ed. *Biomedical Ethics: Opposing Viewpoints.* San Diego: Greenhaven Press, 2003.

Evans, James. *The History and Practice of Ancient Astronomy.* New York: Oxford University Press, 1998.

Fabian, A. C., K. A. Pounds, and R. D. Blandford. *Frontiers of X-Ray Astronomy.* London: Cambridge University Press, 2004.

Fara, Patricia. *An Entertainment for Angels: Electricity in the Enlightenment.* New York: Columbia University Press, 2002.

_______. *Newton: The Making of a Genius.* New York: Columbia University Press, 2002.

_______. *Sex, Botany, and the Empire: The Story of Carl Linnaeus and Joseph Banks.* New York: Columbia University Press, 2003.

Farber, Paul Lawrence. *Finding Order in Nature: The Naturalist Tradition from Linnaeus to E. O. Wilson.* Baltimore: The Johns Hopkins University Press, 2000.

Fauvel, John, and Jeremy Grey, eds. *The History of Mathematics: A Reader.* 1987. Reprint. Washington, D.C.: The Mathematical Association of America, 1997.

Feferman, S., J. W. Dawson, and S. C. Kleene, eds. *Kurt Gödel: Collected Works.* 2 vols. New York: Oxford University Press, 1986-1990.

Feldman, David. *How Does Aspirin Find a Headache?* New York: HarperCollins, 2005.

Ferejohn, Michael. *The Origins of Aristotelian Science.* New Haven, Conn.: Yale University Press, 1991.

Ferguson, Kitty. *The Nobleman and His Housedog: Tycho Brahe and Johannes Kepler—The Strange Partnership That Revolutionized Science.* London: Headline, 2002.

Ferris, T. *Coming of Age in the Milky Way.* New York: Doubleday, 1989.

Ferris, Timothy. *Galaxies.* New York: Harrison House, 1987.

Field, George, and Donald Goldsmith. *The Space Telescope.* Chicago: Contemporary Books, 1989.

Field, J. V. *The Invention of Infinity: Mathematics and Art in the Renaissance.* New York: Oxford University Press, 1997.

Fincher, Jack. *The Brain: Mystery of Matter and Mind.* Washington, D.C.: U.S. News Books, 1981.

Finlayson, Clive. *Neanderthals and Modern Humans: An Ecological and Evolutionary Perspective.* New York: Cambridge University Press, 2004.

Finocchiaro, Maurice A., ed. *The Galileo Affair: A Documentary History.* Berkeley: University of California Press, 1989.

Fischer, Daniel. *Mission Jupiter: The Spectacular Journey of the Galileo Spacecraft.* New York: Copernicus Books, 2001.

Fischer, Daniel, and Hilmar W. Duerbeck. *Hubble Revisited: New Images from the Discovery Machine.* New York: Copernicus Books, 1998.

Fisher, Richard B. *Edward Jenner, 1741-1823.* London: Andre Deutsch, 1991.

Flowers, Lawrence O., ed. *Science Careers: Personal Accounts from the Experts.* Lanham, Md.: Scarecrow Press, 2003.

Ford, Brian J. *The Leeuwenhoek Legacy.* London: Farrand, 1991.

_______. *Single Lens: The Story of the Simple Microscope.* New York: Harper & Row, 1985.

Fournier, Marian. *The Fabric of Life: Microscopy in the Seventeenth Century.* Baltimore: The Johns Hopkins University Press, 1996.

Fowler, A. C. *Mathematical Models in the Applied Sciences.* New York: Cambridge University Press, 1997.

Foyer, Christine H. *Photosynthesis.* New York: Wiley-Interscience, 1984.

Frängsmyr, Tore, ed. *Linnaeus: The Man and His Work.* Canton, Mass.: Science History Publications, 1994.

Franklin, Benjamin. *Autobiography of Benjamin Franklin.* New York: Buccaneer Books, 1984.

French, A. P., and P. J. Kennedy, eds. *Niels Bohr: A Centenary Volume.* Cambridge, Mass.: Harvard University Press, 1985.

French, Roger. *William Harvey's Natural Philosophy.* New York: Cambridge University Press, 1994.

Fridell, Ron. *DNA Fingerprinting: The Ultimate Identity.* New York: Scholastic, 2001.

Friedlander, Michael W. *Cosmic Rays.* Cambridge, Mass.: Harvard University Press, 1989.

Friedman, Meyer, and Gerald W. Friedland. *Medicine's Ten Greatest Discoveries.* New Haven, Conn.: Yale University Press, 2000.

Friedman, Robert Marc. *Appropriating the Weather: Vilhelm Bjerknes and the Construction of a Modern Meteorology.* Ithaca, N.Y.: Cornell University Press, 1989.

Friedrich, Wilhelm. *Vitamins.* New York: Walter de Gruyter, 1988.

Friedrichs, Günter, and Adam Schaff. *Microelectronics*

and Society: For Better or for Worse, a Report to the Club of Rome. New York: Pergamon Press, 1982.

Frist, William. *Transplant.* New York: Atlantic Monthly Press, 1989.

Fuchs, Thomas. *The Mechanization of the Heart: Harvey and Descartes.* Rochester, N.Y.: University of Rochester Press, 2001.

Gallo, Robert C. *Virus Hunting: AIDS, Cancer, and the Human Retrovirus: A Story of Scientific Discovery.* New York: Basic Books, 1993.

Galston, Arthur W. *Life Processes of Plants.* New York: Scientific American Library, 1994.

Gamow, George. *The New World of Mr. Tompkins.* Cambridge, England: Cambridge University Press, 1999.

Gani, Joseph M., ed. *The Craft of Probabilistic Modeling.* New York: Springer-Verlag, 1986.

García-Ballester, Luis. *Galen and Galenism: Theory and Medical Practice from Antiquity to the European Renaissance.* Burlington, Vt.: Ashgate, 2002.

Gardner, Eldon J., and D. Peter Snustad. *Principles of Genetics.* 7th ed. New York: John Wiley & Sons, 1984.

Gardner, Robert, and Eric Kemer. *Science Projects About Temperature and Heat.* Berkeley Heights, N.J.: Enslow Publishers, 1994.

Garner, Geraldine. *Careers in Engineering.* 3d ed. New York: McGraw-Hill, 2009.

Gartner, Carol B. *Rachel Carson.* New York: Frederick Ungar, 1983.

Gasser, James, ed. *A Boole Anthology: Recent and Classical Studies in the Logic of George Boole.* Dordrecht, the Netherlands: Kluwer, 2000.

Gay, Peter. *The Enlightenment: The Science of Freedom.* New York: W. W. Norton, 1996.

_______. *Freud: A Life for Our Time.* New York: W. W. Norton, 1988.

Gazzaniga, Michael S. *The Social Brain: Discovering the Networks of the Mind.* New York: Basic Books, 1985.

Geison, Gerald. *The Private Science of Louis Pasteur.* Princeton, N.J.: Princeton University Press, 1995.

Gell-Mann, Murray. *The Quark and the Jaguar: Adventures in the Simple and the Complex.* New York: W. H. Freeman, 1994.

Georgotas, Anastasios, and Robert Cancro, eds. *Depression and Mania.* New York: Elsevier, 1988.

Gerock, Robert. *Mathematical Physics.* Chicago: University of Chicago Press, 1985.

Gest, Howard. *Microbes: An Invisible Universe.* Washington, D.C.: ASM Press, 2003.

Gesteland, Raymond F., Thomas R. Cech, and John F. Atkins, eds. *The RNA World: The Nature of Modern RNA Suggests a Prebiotic RNA.* 2d ed. Cold Spring Harbor, N.Y.: Cold Spring Harbor Laboratory Press, 1999.

Gigerenzer, Gerd, et al. *The Empire of Chance: How Probability Theory Changed Science and Everyday Life.* New York: Cambridge University Press, 1989.

Gilder, Joshua, and Anne-Lee Gilder. *Heavenly Intrigue: Johannes Kepler, Tycho Brahe, and the Murder Behind One of History's Greatest Scientific Discoveries.* New York: Doubleday, 2004.

Gillispie, Charles Coulston, Robert Fox, and Ivor Grattan-Guinness. *Pierre-Simon Laplace, 1749-1827: A Life in Exact Science.* Princeton, N.J.: Princeton University Press, 2000.

Gingerich, Owen. *The Book Nobody Read: Chasing the Revolutions of Nicolaus Copernicus.* New York: Walker, 2004.

_______. *The Eye of Heaven: Ptolemy, Copernicus, Kepler.* New York: Springer-Verlag, 1993.

Glashow, Sheldon, with Ben Bova. *Interactions: A Journey Through the Mind of a Particle Physicist and the Matter of This World.* New York: Warner Books, 1988.

Glass, Billy. *Introduction to Planetary Geology.* New York: Cambridge University Press, 1982.

Gleick, James. *Chaos: Making a New Science.* New York: Penguin Books, 1987.

_______. *Isaac Newton.* New York: Pantheon Books, 2003.

Glen, William. *The Road to Jaramillo: Critical Years of the Revolution in Earth Science.* Stanford, Calif.: Stanford University Press, 1982.

Glickman, Todd S., ed. *Glossary of Meteorology.* 2d ed. Boston: American Meteorological Society, 2000.

Goddard, Jolyon, ed. *National Geographic Concise History of Science and Invention: An Illustrated Time Line.* Washington, D.C.: National Geographic, 2010.

Goding, James W. *Monoclonal Antibodies: Principles and Practice.* New York: Academic Press, 1986.

Godwin, Robert, ed. *Mars: The NASA Mission Reports.* Burlington, Ont.: Apogee Books, 2000.

_______. *Mars: The NASA Mission Reports.* Vol. 2. Burlington, Ont.: Apogee Books, 2004.

_______. *Space Shuttle STS Flights 1-5: The NASA Mission Reports.* Burlington, Ont.: Apogee Books, 2001.

Goetsch, David L. *Building a Winning Career in*

Engineering: 20 Strategies for Success After College. Upper Saddle River, N.J.: Pearson/Prentice Hall, 2007.

Gohlke, Mary, with Max Jennings. *I'll Take Tomorrow.* New York: M. Evans, 1985.

Gold, Rebecca. *Steve Wozniak: A Wizard Called Woz.* Minneapolis: Lerner, 1994.

Goldsmith, Donald. *Nemesis: The Death Star and Other Theories of Mass Extinction.* New York: Berkley Publishing Group, 1985.

Goldsmith, Maurice, Alan Mackay, and James Woudhuysen, eds. *Einstein: The First Hundred Years.* Elmsford, N.Y.: Pergamon Press, 1980.

Golinski, Jan. *Science as Public Culture: Chemistry and Enlightenment in Britain, 1760-1820.* Cambridge, England: Cambridge University Press, 1992.

Golthelf, Allan, and James G. Lennox, eds. *Philosophical Issues in Aristotle's Biology.* Cambridge, England: Cambridge University Press, 1987.

Gooding, David, and Frank A. J. L. James, eds. *Faraday Rediscovered: Essays on the Life and Work of Michael Faraday, 1791-1867.* New York: Macmillan, 1985.

Gordin, Michael D. *A Well-Ordered Thing: Dmitrii Mendeleev and the Shadow of the Periodic Table.* New York: Basic Books, 2004.

Gornick, Vivian. *Women in Science: Then and Now.* New York: Feminist Press at the City University of New York, 2009.

Gould, James L., and Carol Grant Gould. *The Honey Bee.* New York: Scientific American Library, 1988.

Gould, Stephen Jay. *Time's Arrow, Time's Cycle: Myth and Metaphor in the Discovery of Geological Time.* Cambridge, Mass.: Harvard University Press, 1987.

_______. *Wonderful Life: The Burgess Shale and the Nature of History.* New York: W. W. Norton, 1989.

Govindjee, J. T. Beatty, H. Gest, and J.F. Allen, eds. *Discoveries in Photosynthesis.* Berlin: Springer, 2005.

Gow, Mary. *Tycho Brahe: Astronomer.* Berkeley Heights, N.J.: Enslow, 2002.

Graham, Loren R. *Science, Philosophy, and Human Behavior in the Soviet Union.* New York: Columbia University Press, 1987.

Grattan-Guinness, Ivor. *The Norton History of the Mathematical Sciences.* New York: W. W. Norton, 1999.

Gray, Robert M., and Lee D. Davisson. *Random Processes: A Mathematical Approach for Engineers.* Englewood Cliffs, N.J.: Prentice-Hall, 1986.

Greene, Mott T. *Geology in the Nineteenth Century: Changing Views of a Changing World.* Ithaca, N.Y.: Cornell University Press, 1982.

Gregory, Andrew. *Harvey's Heart: The Discovery of Blood Circulation.* London: Totem Books, 2001.

Gribbin, John. *Deep Simplicity: Bringing Order to Chaos and Complexity.* New York: Random House, 2005.

_______. *Future Weather and the Greenhouse Effect.* New York: Delacorte Press/Eleanor Friede, 1982.

_______. *The Hole in the Sky: Man's Threat to the Ozone Layer.* New York: Bantam Books, 1988.

_______. *In Search of Schrödinger's Cat: Quantum Physics and Reality.* New York: Bantam Books, 1984.

_______. *In Search of the Big Bang.* New York: Bantam Books, 1986.

_______. *The Omega Point: The Search for the Missing Mass and the Ultimate Fate of the Universe.* New York: Bantam Books, 1988.

_______. *The Scientists: A History of Science Told Through the Lives of Its Greatest Inventors.* New York: Random House, 2002.

Gribbin, John, ed. *The Breathing Planet.* New York: Basil Blackwell, 1986.

Gutkind, Lee. *Many Sleepless Nights: The World of Organ Transplantation.* New York: W. W. Norton, 1988.

Hackett, Edward J., et al., eds. *The Handbook of Science and Technology Studies.* 3d ed. Cambridge, Mass.: MIT Press, 2008.

Hald, Anders. *A History of Mathematical Statistics from 1750 to 1950.* New York: John Wiley & Sons, 1998.

_______. *A History of Probability and Statistics and Their Applications Before 1750.* New York: John Wiley & Sons, 1990.

Hall, A. Rupert. *The Scientific Revolution, 1500-1750.* 3d ed. New York: Longman, 1983.

Halliday, David, and Robert Resnick. *Fundamentals of Physics: Extended Version.* New York: John Wiley & Sons, 1988.

Halliday, David, Robert Resnick, and Jearl Walker. *Fundamentals of Physics.* 7th ed. New York: John Wiley & Sons, 2004.

Hankins, Thomas L. *Science and the Enlightenment.* Reprint. New York: Cambridge University Press, 1991.

Hanlon, Michael, and Arthur C. Clarke. *The Worlds of Galileo: The Inside Story of NASA's Mission to Jupiter.* New York: St. Martin's Press, 2001.

Hanson, Earl D. *Understanding Evolution.* New York: Oxford University Press, 1981.

Hargittai, István. *Martians of Science: Five Physicists Who Changed the Twentieth Century.* New York: Oxford

University Press, 2006.

Harland, David M. *Jupiter Odyssey: The Story of NASA's Galileo Mission.* London: Springer-Praxis, 2000.

_______. *Mission to Saturn: Cassini and the Huygens Probe.* London: Springer-Praxis, 2002.

_______. *The Space Shuttle: Roles, Missions, and Accomplishments.* New York: John Wiley & Sons, 1998.

Harland, David M., and John E. Catchpole. *Creating the International Space Station.* London: Springer-Verlag, 2002.

Harrington, J. W. *Dance of the Continents:* New York: V. P. Tarher, 1983.

Harrington, Philip S. *The Space Shuttle: A Photographic History.* San Francisco: Brown Trout, 2003.

Harris, Henry. *The Birth of the Cell.* New Haven, Conn.: Yale University Press, 1999.

Harrison, Edward R. *Cosmology: The Science of the Universe.* Cambridge England: Cambridge University Press, 1981.

Hart, Michael H. *The 100: A Ranking of the Most Influential Persons in History.* New York: Galahad Books, 1982.

Hart-Davis, Adam. *Chain Reactions: Pioneers of British Science and Technology and the Stories That Link Them.* London: National Portrait Gallery, 2000.

Hartmann, William K. *The Cosmic Voyage: Through Time and Space.* Belmont, Calif: Wadsworth, 1990.

_______. *Moons and Planets.* 5th ed. Belmont, Calif.: Brooks-Cole Publishing, 2005.

Hartwell, L. H., et al. *Genetics: From Genes to Genomes.* 2d ed. New York: McGraw-Hill, 2004.

Harvey, William. *The Circulation of the Blood and Other Writings.* New York: Everyman's Library, 1990.

Haskell, G., and Michael Rycroft. *International Space Station: The Next Space Marketplace.* Boston: Kluwer Academic, 2000.

Hathaway, N. *The Friendly Guide to the Universe.* New York: Penguin Books, 1994.

Havil, Julian. *Gamma: Exploring Euler's Constant.* Princeton, N.J.: Princeton University Press, 2003.

Hawking, Stephen W. *A Brief History of Time.* New York: Bantam Books, 1988.

Haycock, David. *William Stukeley: Science, Religion, and Archeology in Eighteenth-Century England.* Woodbridge, England: Boydell Press, 2002.

Hazen, Robert. *The Breakthrough: The Race for the Superconductor.* New York: Summit Books, 1988.

Headrick, Daniel R. *Technology: A World History.* New York: Oxford University Press, 2009.

Heath, Sir Thomas L. *A History of Greek Mathematics: From Thales to Euclid.* 1921. Reprint. New York: Dover Publications, 1981.

Heilbron, J. L. *The Dilemmas of an Upright Man: Max Planck As a Spokesman for German Science.* Berkeley: University of California Press, 1986.

_______. *Electricity in the Seventeenth and Eighteenth Centuries: A Study in Early Modern Physics.* Mineola, N.Y.: Dover Publications, 1999.

_______. *Elements of Early Modern Physics.* Berkeley: University of California Press, 1982.

_______. *Geometry Civilized: History, Culture, and Technique.* Oxford, England: Clarendon Press, 1998.

Heilbron, J. L., and Robert W. Seidel. *Lawrence and His Laboratory: A History of the Lawrence Berkeley Laboratory.* Berkeley: University of California Press, 1989.

Heisenberg, Elisabeth. *Inner Exile: Recollections of a Life with Werner Heisenberg.* Translated by S. Cappelari and C. Morris. Boston: Birkhäuser, 1984.

Hellegouarch, Yves. *Invitation to the Mathematics of Fermat-Wiles.* San Diego: Academic Press, 2001.

Henig, Robin Marantz. *The Monk in the Garden: The Lost and Found Genius of Gregor Mendel, the Father of Genetics.* New York: Mariner Books, 2001.

Henry, Helen L., and Anthony W. Norman, eds. *Encyclopedia of Hormones.* 3 vols. San Diego: Academic Press, 2003.

Henry, John. *Moving Heaven and Earth: Copernicus and the Solar System.* Cambridge, England: Icon, 2001.

Herrmann, Bernd, and Susanne Hummel, eds. *Ancient DNA: Recovery and Analysis of Genetic Material from Paleographic, Archaeological, Museum, Medical, and Forensic Specimens.* New York: Springer-Verlag, 1994.

Hershel, Sir John Frederic William, and Pierre-Simon Laplace. *Essays in Astronomy.* University Press of the Pacific, 2002.

Hillar, Marian, and Claire S. Allen. *Michael Servetus: Intellectual Giant, Humanist, and Martyr.* New York: University Press of America, 2002.

Hobson, J. Allan. *The Dreaming Brain.* New York: Basic Books, 1988.

Hodge, Paul. *Galaxies.* Cambridge, Mass.: Harvard University Press, 1986.

Hodges, Andrew. *Alan Turing: The Enigma.* 1983. Reprint. New York: Walker, 2000.

Hofmann, James R., David Knight, and Sally Gregory Kohlstedt, eds. *André-Marie Ampère: Enlightenment and Electrodynamics.* Cambridge, England: Cambridge University Press, 1996.

Holland, Suzanne, Karen Lebacqz, and Laurie Zoloth, eds. *The Human Embryonic Stem Cell Debate: Science, Ethics, and Public Policy.* Cambridge, Mass.: MIT Press, 2001.

Holmes, Frederic Lawrence. *Antoine Lavoisier, the Next Crucial Year: Or, the Sources of His Quantitative Method in Chemistry.* Princeton, N.J.: Princeton University Press, 1998.

Horne, James. *Why We Sleep.* New York: Oxford University Press, 1988.

Hoskin, Michael A. *The Herschel Partnership: As Viewed by Caroline.* Cambridge, England: Science History, 2003.

_______. *William Herschel and the Construction of the Heavens.* New York: Norton; 1964.

Howse, Derek. *Greenwich Time and the Discovery of Longitude.* New York: Oxford University Press, 1980.

Hoyt, William G. *Planet X and Pluto.* Tucson: University of Arizona Press, 1980.

Hsü, Kenneth J. *The Great Dying.* San Diego: Harcourt Brace Jovanovich, 1986.

Huerta, Robert D. *Giants of Delft, Johannes Vermeer and the Natural Philosophers: The Parallel Search for Knowledge During the Age of Discovery.* Lewisburg, Pa.: Bucknell University Press, 2003.

Hummel, Susanne. *Fingerprinting the Past: Research on Highly Degraded DNA and Its Applications.* New York: Springer-Verlag, 2002.

Hunter, Michael, ed. *Robert Boyle Reconsidered.* New York: Cambridge University Press, 1994.

Hynes, H. Patricia. *The Recurring Silent Spring.* New York: Pergamon Press, 1989.

Ihde, Aaron J. *The Development of Modern Chemistry.* New York: Dover, 1984.

Irwin, Patrick G. J. *Giant Planets of Our Solar System: Atmospheres, Composition, and Structure.* London: Springer-Praxis, 2003.

Isaacson, Walter. *Benjamin Franklin: An American Life.* New York: Simon & Schuster, 2003.

Jackson, Myles. *Spectrum of Belief: Joseph Fraunhofer and the Craft of Precision Optics.* Cambridge, Mass.: MIT Press, 2000.

Jacobsen, Theodor S. *Planetary Systems from the Ancient Greeks to Kepler.* Seattle: University of Washington Press, 1999.

Jacquette, Dale. *On Boole.* Belmont, Calif.: Wadsworth, 2002.

Jaffe, Bernard. *Crucibles: The Story of Chemistry.* New York: Dover, 1998.

James, Ioan. *Remarkable Mathematicians: From Euler to Von Neumann.* Cambridge, England: Cambridge University Press, 2002.

Janowsky, David S., Dominick Addario, and S. Craig Risch. *Psychopharmacology Case Studies.* 2d ed. New York: Guilford Press, 1987.

Jeffreys, Diarmuid. *Aspirin: The Remarkable Story of a Wonder Drug.* London: Bloomsbury Publishing, 2004.

Jenkins, Dennis R. *Space Shuttle: The History of the National Space Transportation System: The First 100 Missions.* Stillwater, Minn.: Voyageur Press, 2001.

Johanson, Donald, and B. Edgar. *From Lucy to Language.* New York: Simon and Schuster, 1996.

Johanson, Donald C., and Maitland A. Edey. *Lucy: The Beginnings of Humankind.* New York: Simon & Schuster, 1981.

Johanson, Donald C., and James Shreeve. *Lucy's Child: The Discovery of a Human Ancestor.* New York: William Morrow, 1989.

Johnson, George. *Strange Beauty: Murray Gell-Mann and the Revolution in Twentieth-Century Physics.* New York: Alfred A. Knopf, 1999.

Jones, Henry Bence. *Life and Letters of Faraday.* 2 vols. London: Longmans, Green and Co., 1870.

Jones, Meredith L., ed. *Hydrothermal Vents of the Eastern Pacific: An Overview.* Vienna, Va.: INFAX, 1985.

Jones, Sheilla. *The Quantum Ten: A Story of Passion, Tragedy, and Science.* Toronto: Thomas Allen, 2008.

Jones, W. H. S., trans. *Hippocrates.* 4 vols. 1923-1931. Reprint. New York: Putnam, 1995.

Jordan, Paul. *Neanderthal: Neanderthal Man and the Story of Human Origins.* Gloucestershire, England: Sutton, 2001.

Joseph, George Gheverghese. *The Crest of the Peacock: The Non-European Roots of Mathematics.* London: Tauris, 1991.

Jungnickel, Christa, and Russell McCormmach. *Cavendish: The Experimental Life.* Lewisburg, Pa.: Bucknell University Press, 1999.

Kaplan, Robert. *The Nothing That Is: A Natural History of Zero.* New York: Oxford University Press, 2000.

Kargon, Robert H. *The Rise of Robert Millikan: Portrait of a Life in American Science.* Ithaca, N.Y.: Cornell University Press, 1982.

Katz, Jonathan. *The Biggest Bang: The Mystery of Gamma-Ray Bursts.* London: Oxford University Press, 2002.

Kellogg, William W., and Robert Schware. *Climate Change and Society: Consequences of Increasing Atmospheric Carbon Dioxide.* Boulder, Colo.: Westview Press, 1981.

Kelly, Thomas J. *Moon Lander: How We Developed the Apollo Lunar Module.* Washington, D.C.: Smithsonian Books, 2001.

Kemper, John D., and Billy R. Sanders. *Engineers and Their Profession.* 5th ed. New York: Oxford University Press, 2001.

Kepler, Johannes. *New Astronomy.* Translated by William H. Donahue. New York: Cambridge University Press, 1992.

Kermit, Hans. *Niels Stensen: The Scientist Who Was Beatified.* Translated by Michael Drake. Herefordshire, England: Gracewing 2003.

Kerns, Thomas A. *Jenner on Trial: An Ethical Examination of Vaccine Research in the Age of Smallpox and the Age of AIDS.* Lanham, Md.: University Press of America, 1997.

Kerrod, Robin. *Hubble: The Mirror on the Universe.* Richmond Hill, Ont.: Firefly Books, 2003.

_______. *Space Shuttle.* New York: Gallery Books, 1984.

Kevles, Bettyann. *Naked to the Bones: Medical Imaging in the Twentieth Century.* Reading, Mass.: Addison Wesley, 1998.

Kiessling, Ann, and Scott C. Anderson. *Human Embryonic Stem Cells: An Introduction to the Science and Therapeutic Potential.* Boston: Jones and Bartlett, 2003.

King, Helen. *Greek and Roman Medicine.* London: Bristol Classical, 2001.

_______. *Hippocrates' Woman: Reading the Female Body in Ancient Greece.* New York: Routledge, 1998.

Kirkham, M. B. *Principles of Soil and Plant Water Relations.* St. Louis: Elsevier, 2005.

Kline, Morris. *Mathematical Thought from Ancient to Modern Times.* New York: Oxford University Press, 1990.

Klotzko, Arlene Judith, ed. *The Cloning Sourcebook.* New York: Oxford University Press, 2001.

Knipe, David, Peter Howley, and Diane Griffin. *Field's Virology.* 2 vols. New York: Lippincott Williams and Wilkens, 2001.

Knowles, Richard V. *Genetics, Society, and Decisions.* Columbus, Ohio: Charles E. Merrill, 1985.

Koestler, Arthur. *The Sleepwalkers.* New York: Penguin Books, 1989.

Kolata, Gina Bari. *Clone: The Road to Dolly, and the Path Ahead.* New York: William Morrow, 1998.

Komszik, Louis. *Applied Calculus of Variations for Engineers.* Boca Raton, Fla.: CRC Press, 2009.

Kramer, Barbara. *Neil Armstrong: The First Man on the Moon.* Springfield, N.J.: Enslow, 1997.

Krane, Kenneth S. *Modern Physics.* New York: John Wiley & Sons, 1983.

Lagerkvist, Ulf. *Pioneers of Microbiology and the Nobel Prize.* River Edge, N.J.: World Scientific Publishing, 2003.

Lamarck, Jean-Baptiste. *Lamarck's Open Mind: The Lectures.* Gold Beach, Ore.: High Sierra Books, 2004.

_______. *Zoological Philosophy: An Exposition with Regard to the Natural History of Animals.* Translated by Hugh Elliot with introductory essay by David L. Hull and Richard W. Burckhardt, Jr. Chicago: University of Chicago Press, 1984.

Landes, Davis S. *Revolution in Time: Clocks and the Making of the Modern World.* Rev. ed. Cambridge, Mass.: Belknap Press, 2000.

Langone, John. *Superconductivity: The New Alchemy.* Chicago: Contemporary Books, 1989.

Lappé, Marc. *Broken Code: The Exploitation of DNA.* San Francisco: Sierra Club Books, 1984.

_______. *Germs That Won't Die.* Garden City, N.Y.: Doubleday, 1982.

La Thangue, Nicholas B., and Lasantha R. Bandara, eds. *Targets for Cancer Chemotherapy: Transcription Factors and Other Nuclear Proteins.* Totowa, N.J.: Humana Press, 2002.

Laudan, R. *From Mineralogy to Geology: The Foundations of a Science, 1650-1830.* Chicago: University of Chicago Press, 1987.

Lauritzen, Paul, ed. *Cloning and the Future of Human Embryo Research.* New York: Oxford University Press, 2001.

Leakey, Mary D. *Disclosing the Past.* New York: Doubleday, 1984.

Le Grand, Homer E. *Drifting Continents and Shifting Theories.* New York: Cambridge University Press, 1988.

Levy, David H. *Clyde Tombaugh: Discoverer of Planet Pluto.* Tucson: University of Arizona Press, 1991.

Lewin, Benjamin. *Genes III.* 3d ed. New York: John Wiley & Sons, 1987.

_______. *Genes IV.* New York: Oxford University Press, 1990.

Lewin, Roger. *Bones of Contention: Controversies in the Search for Human Origins.* New York: Simon & Schuster, 1987.

Lewis, Richard S. *The Voyages of Columbia: The First True Spaceship.* New York: Columbia University Press, 1984.

Lindberg, David C. *The Beginnings of Western Science.* Chicago: University of Chicago Press, 1992.

Lindley, David. *Degrees Kelvin: A Tale of Genius, Invention, and Tragedy.* Washington, D.C.: Joseph Henry Press, 2004.

Linzmayer, Owen W. *Apple Confidential: The Real Story of Apple Computer, Inc.* San Francisco: No Starch Press, 1999.

Lloyd, G. E. R., and Nathan Sivin. *The Way and the Word: Science and Medicine in Early China and Greece.* New Haven, Conn.: Yale University Press, 2002.

Logan, J. David. *Applied Mathematics.* 3d ed. Hoboken, N.J.: John Wiley & Sons, 2006.

Logsdon, John M. *Together in Orbit: The Origins of International Participation in the Space Station.* Washington, D.C.: National Aeronautics and Space Administration, 1998.

Longrigg, James. *Greek Medicine: From the Heroic to the Hellenistic Age.* New York: Routledge, 1998.

_______. *Greek Rational Medicine: Philosophy and Medicine from Alcmaeon to the Alexandrians.* New York: Routledge, 1993.

Lorenz, Edward. *The Essence of Chaos.* Reprint. St. Louis: University of Washington Press, 1996.

Lorenz, Ralph, and Jacqueline Mitton. *Lifting Titan's Veil: Exploring the Giant Moon of Saturn.* London: Cambridge University Press, 2002.

Loudon, Irvine. *The Tragedy of Childbed Fever.* New York: Oxford University Press, 2000.

Luck, Steve, ed. *International Encyclopedia of Science and Technology.* New York: Oxford University Press, 1999.

Lutgens, Frederick K., and Edward J. Tarbuck. *The Atmosphere: An Introduction to Meteorology.* 2d ed. Englewood Cliffs, N.J.: Prentice-Hall, 1982.

Lyell, Charles. *Elements of Geology.* London: John Murray, 1838.

_______. *The Geological Evidences of the Antiquity of Man with Remarks on Theories of the Origin of Species by Variation.* London: John Murray, 1863.

_______. *Principles of Geology, Being an Attempt to Explain the Former Changes of the Earth's Surface by Reference to Causes Now in Operation.* 3 vols. London: John Murray, 1830-1833.

Lynch, William T. *Solomon's Child: Method in the Early Royal Society of London.* Stanford, Calif.: Stanford University Press, 2000.

Ma, Pearl, and Donald Armstrong, eds. *AIDS and Infections of Homosexual Men.* Stoneham, Mass: Butterworths, 1989.

MacDonald, Allan H., ed. *Quantum Hall Effect: A Perspective.* Boston: Kluwer Academic Publishers, 1989.

MacHale, Desmond. *George Boole: His Life and Work.* Dublin: Boole Press, 1985.

Machamer, Peter, ed. *The Cambridge Companion to Galileo.* New York: Cambridge University Press, 1998.

Mactavish, Douglas. *Joseph Lister.* New York: Franklin Watts, 1992.

Mader, Sylvia S. *Biology.* 3d ed. Dubuque, Iowa: Win. C. Brown, 1990.

Magner, Lois. *A History of Medicine.* New York: Marcel Dekker, 1992.

Mahoney, Michael Sean. *The Mathematical Career of Pierre de Fermat, 1601-1665.* 2d rev. ed. Princeton, N.J.: Princeton University Press, 1994.

Mammana, Dennis L., and Donald W. McCarthy, Jr. *Other Suns, Other Worlds? The Search for Extrasolar Planetary Systems.* New York: St. Martin's Press, 1996.

Mandelbrot, B. B. *Fractals and Multifractals: Noise, Turbulence, and Galaxies.* New York: Springer-Verlag, 1990.

Mann, Charles, and Mark Plummer. *The Aspirin Wars: Money, Medicine and 100 Years of Rampant Competition.* New York: Knopf, 1991.

Marco, Gino J., Robert M. Hollingworth, and William Durham, eds. *Silent Spring Revisited.* Washington, D.C.: American Chemical Society, 1987.

Margolis, Howard. *It Started with Copernicus.* New York: McGraw-Hill, 2002.

Marshak, Daniel R., Richard L. Gardner, and David Gottlieb, eds. *Stem Cell Biology.* Woodbury, N.Y.: Cold Spring Harbor Laboratory Press, 2002.

Martzloff, Jean-Claude. *History of Chinese Mathematics.* Translated by Stephen S. Wilson. Berlin: Springer, 1987.

Massey, Harrie Stewart Wilson. *The Middle Atmosphere as Observed by Balloons, Rockets, and Satellites.* London: Royal Society, 1980.

Masson, Jeffrey M. *The Assault on Truth: Freud's Suppression of the Seduction Theory.* New York: Farrar,

Straus and Giroux, 1984.

Mateles, Richard I. *Penicillin: A Paradigm for Biotechnology.* Chicago: Canadida Corporation, 1998.

Mayo, Jonathan L. *Superconductivity: The Threshold of a New Technology.* Blue Ridge Summit, Pa.: TAB Books, 1988.

McCarthy, Shawn P. *Engineer Your Way to Success: America's Top Engineers Share Their Personal Advice on What They Look for in Hiring and Promoting.* Alexandria, Va.: National Society of Professional Engineers, 2002.

McCay, Mary A. *Rachel Carson.* New York: Twayne, 1993.

McDonnell, John James. *The Concept of an Atom from Democritus to John Dalton.* Lewiston, N.Y.: Edwin Mellen Press, 1991.

McEliece, Robert. *Finite Fields for Computer Scientists and Engineers.* Boston: Kluwer Academic, 1987.

McGraw-Hill Concise Encyclopedia of Science and Technology. 6th ed. New York: McGraw-Hill, 2009.

McGraw-Hill Dictionary of Scientific and Technical Terms. 6th ed. New York: McGraw-Hill, 2002.

McGrayne, Sharon Bertsch. *Nobel Prize Women in Science: Their Lives, Struggles and Momentous Discoveries.* 2d ed. Washington, D.C.: Joseph Henry Press, 1998.

McIntyre, Donald B., and Alan McKirdy. *James Hutton: The Founder of Modern Geology.* Edinburgh: Stationery Office, 1997.

McLester, John, and Peter St. Pierre. *Applied Biomechanics: Concepts and Connections.* Belmont, Calif.: Thomson Wadsworth, 2008.

McMullen, Emerson Thomas. *William Harvey and the Use of Purpose in the Scientific Revolution: Cosmos by Chance or Universe by Design?* Lanhan, Md.: University Press of America, 1998.

McQuarrie, Donald A. *Quantum Chemistry.* Mill Valley, Calif: University Science Books, 1983.

Menard, H. W. *The Ocean of Truth: A Personal History of Global Tectonics.* Princeton, N.J.: Princeton University Press, 1986.

Menzel, Donald H., and Jay M. Pasachoff. *Stars and Planets.* Boston: Houghton Mifflin Company, 1983.

Merrell, David J. *Ecological Genetics.* Minneapolis: University of Minnesota Press, 1981.

Mettler, Lawrence E., Thomas G. Gregg, and Henry E. Schaffer. *Population Genetics and Evolution.* 2d ed. Englewood Cliffs, N.J.: Prentice-Hall, 1988.

Meyers, Robert A., ed. *Encyclopedia of Physical Science and Technology.* 18 vols. San Diego: Academic Press, 2005.

Meyerson, Daniel. *The Linguist and the Emperor: Napoleon and Champollion's Quest to Decipher the Rosetta Stone.* New York: Ballantine Books, 2004.

Middleton, W. E. Knowles. *A History of the Thermometer and Its Use in Meteorology.* Ann Arbor, Mich.: UMI Books on Demand, 1996.

Miller, Ron. *Extrasolar Planets.* Brookfield, Conn.: Twenty-First Century Books, 2002.

Miller, Stanley L. *From the Primitive Atmosphere to the Prebiotic Soup to the Pre-RNA World.* Washington, D.C.: National Aeronautics and Space Administration, 1996.

Mishkin, Andrew. *Sojourner: An Insider's View of the Mars Pathfinder Mission.* New York: Berkeley Books, 2003.

Mlodinow, Leonard. *Euclid's Window: A History of Geometry from Parallel Lines to Hyperspace.* New York: Touchstone, 2002.

Monmonier, Mark. *Air Apparent: How Meteorologists Learned to Map, Predict, and Dramatize Weather.* Chicago: University of Chicago Press, 2000.

Moore, Keith L. *The Developing Human.* Philadelphia: W. B. Saunders, 1988.

Moore, Patrick. *Eyes on the University: The Story of the Telescope.* New York: Springer-Verlag, 1997.

_______. *Patrick Moore's History of Astronomy.* 6th rev. ed. London: Macdonald, 1983.

Morell, V. *Ancestral Passions: The Leakey Family and the Quest for Humankind's Beginnings.* New York: Simon & Schuster, 1995.

Morgan, Kathryn A. *Myth and Philosophy from the Presocratics to Plato.* New York: Cambridge University Press, 2000.

Moritz, Michael. *The Little Kingdom: The Private Story of Apple Computer.* New York: Morrow, 1984.

Morris, Peter, ed. *Making the Modern World: Milestones of Science and Technology.* 2d ed. Chicago: KWS Publishers, 2011.

Morris, Richard. *The Last Sorcerers: The Path from Alchemy to the Periodic Table.* Washington, D.C.: Joseph Henry Press, 2003.

Morrison, David. *Voyages to Saturn.* NASA SP-451. Washington, D.C.: National Aeronautics and Space Administration, 1982.

Morrison, David, and Tobias Owen. *The Planetary System.* 3d ed. San Francisco: Addison Wesley, 2003.

Morrison, David, and Jane Samz. *Voyage to Jupiter.* NASA SP-439. Washington, D.C.: Government

Printing Office, 1980.

Moss, Ralph W. *Free Radical: Albert Szent-Györgyi and the Battle over Vitamin C.* New York: Paragon House, 1988.

Muirden, James. *The Amateur Astronomer's Handbook.* 3d ed. New York: Harper & Row, 1987.

Mullis, Kary. *Dancing Naked in the Mind Field.* New York: Pantheon Books, 1998.

Mulvihill, John J. *Catalog of Human Cancer Genes: McKusick's Mendelian Inheritance in Man for Clinical and Research Oncologists.* Foreword by Victor A. McKusick. Baltimore: The Johns Hopkins University Press, 1999.

Nahin, Paul J. *Oliver Heaviside: Sage in Solitude.* New York: IEEE Press, 1987.

Ne'eman, Yuval, and Yoram Kirsh. *The Particle Hunters.* New York: Cambridge University Press, 1986.

Netz, Reviel. *The Shaping of Deduction in Greek Mathematics: A Study in Cognitive History.* New York: Cambridge University Press, 2003.

Neu, Jerome, ed. *The Cambridge Companion to Freud.* New York: Cambridge University Press, 1991.

North, John. *The Norton History of Astronomy and Cosmology.* New York: W. W. Norton, 1995.

Nutton, Vivian, ed. *The Unknown Galen.* London: Institute of Classical Studies, University of London, 2002.

Nye, Mary Jo. *Before Big Science: The Pursuit of Modern Chemistry and Physics, 1800-1940.* New York: Twayne, 1996.

Nye, Robert D. *Three Psychologies: Perspectives from Freud, Skinner, and Rogers.* Pacific Grove, Calif.: Brooks-Cole, 1992.

Oakes, Elizabeth H. *Encyclopedia of World Scientists.* Rev ed. New York: Facts on File, 2007.

Olson, James S., and Robert L. Shadle, ed. *Encyclopedia of the Industrial Revolution in America.* Westport, Conn.: Greenwood Press, 2002.

Olson, Steve. *Mapping Human History: Genes, Races and Our Common Origins.* New York: Houghton Mifflin, 2002.

Ozima, Minoru. *The Earth: Its Birth and Growth.* Translated by Judy Wakabayashi. Cambridge, England: Cambridge University Press, 1981.

Pagels, Heinz R. *The Cosmic Code.* New York: Simon & Schuster, 1982.

_______. *The Cosmic Code: Quantum Physics As the Law of Nature.* New York: Bantam Books, 1984.

_______. *Perfect Symmetry: The Search for the Beginning of Time.* New York: Simon & Schuster, 1985.

Pai, Anna C. *Foundations for Genetics: A Science for Society.* 2d ed. New York: McGraw-Hill, 1984.

Pais, Abraham. *The Genius of Science: A Portrait Gallery of Twentieth-Century Physicists.* New York: Oxford University Press, 2000.

Palmer, Douglas. *Neanderthal.* London: Channel 4 Books, 2000.

Parker, Barry R. *The Vindication of the Big Bang: Breakthroughs and Barriers.* New York: Plenum Press, 1993.

Parslow, Christopher Charles. *Rediscovering Antiquity: Karl Weber and the Excavation of Herculaneum, Pompeii, and Stabiae.* New York: Cambridge University Press, 1998.

Parson, Ann B. *The Proteus Effect: Stem Cells and Their Promise.* Washington, D.C.: National Academies Press, 2004.

Pedrotti, L., and F. Pedrotti. *Optics and Vision.* Upper Saddle River, N.J.: Prentice Hall, 1998.

Peitgen, Heinz-Otto, and Dietmar Saupe, eds. *The Science of Fractal Images.* New York: Springer-Verlag, 1988.

Peltonen, Markku, ed. *The Cambridge Companion to Bacon.* New York: Cambridge University Press, 1996.

Penrose, Roger. *The Emperor's New Mind: Concerning Computers, Minds, and the Laws of Physics.* New York: Oxford University Press, 1989.

Persaud, T. V. N. *A History of Anatomy: The Post-Vesalian Era.* Springfield, Ill.: Charles C. Thomas, 1997.

Peterson, Carolyn C., and John C. Brant. *Hubble Vision: Astronomy with the Hubble Space Telescope.* London: Cambridge University Press, 1995.

_______. *Hubble Vision: Further Adventures with the Hubble Space Telescope.* 2d ed. New York: Cambridge University Press, 1998.

Pfeiffer, John E. *The Emergence of Humankind.* 4th ed. New York: Harper & Row, 1985.

Piggott, Stuart. *William Stukeley: An Eighteenth-Century Antiquary.* New York: Thames and Hudson, 1985.

Pike, J. Wesley, Francis H. Glorieux, David Feldman. *Vitamin D.* 2d ed. Academic Press, 2004.

Plionis, Manolis, ed. *Multiwavelength Cosmology.* New York: Springer, 2004.

Plotkin, Stanley A., and Edward A. Mortimer. *Vaccines.* 2d ed. Philadelphia: W. B. Saunders, 1994.

Polter, Paul. *Hippocrates.* Cambridge, Mass.: Harvard University Press, 1995.

Popper, Karl R. *The World of Parmenides: Essays on the Presocratic Enlightenment.* Edited by Arne F. Petersen and Jørgen Mejer. New York: Routledge, 1998.

Porter, Roy. *The Greatest Benefit to Mankind: A Medical History of Humanity, from Antiquity to the Present.* New York: W. W. Norton, 1997.

Porter, Roy, ed. *Eighteenth Century Science.* Vol. 4 in *The Cambridge History of Science.* New York: Cambridge University Press, 2003.

Poundstone, William. *Prisoner's Dilemma.* New York: Doubleday, 1992.

Poynter, Margaret, and Arthur L. Lane. *Voyager: The Story of a Space Mission.* New York: Macmillan, 1981.

Principe, Lawrence. *The Aspiring Adept: Robert Boyle and His Alchemical Quest.* Princeton, N.J.: Princeton University Press, 1998.

Prochnow, Dave. *Superconductivity: Experimenting in a New Technology.* Blue Ridge Summit, Pa.: TAB Books, 1989.

Pullman, Bernard. *The Atom in the History of Human Thought.* New York: Oxford University Press, 1998.

Pycior, Helena M. *Symbols, Impossible Numbers, and Geometric Entanglements: British Algebra Through the Commentaries on Newton's Universal Arithmetick.* New York: Cambridge University Press, 1997.

The Rand McNally New Concise Atlas of the Universe. New York: Rand McNally, 1989.

Rao, Mahendra S., ed. *Stem Cells and CNS Development.* Totowa, N.J.: Humana Press, 2001.

Raup, David M. *The Nemesis Affair: A Story of the Death of Dinosaurs and the Ways of Science.* New York: W. W. Norton, 1986.

Raven, Peter H., and George B. Johnson. *Biology.* 2d ed. St. Louis: Times-Mirror/Mosby, 1989.

Raven, Peter H., Ray F. Evert, and Susan E. Eichhorn. *Biology of Plants.* 6th ed. New York: W. H. Freeman, 1999.

Reader, John. *Missing Links: The Hunt for Earliest Man.* Boston: Little, Brown, 1981.

Reichhardt, Tony. *Proving the Space Transportation System: The Orbital Flight Test Program.* NASA NF-137-83. Washington, D.C.: Government Printing Office, 1983.

Remick, Pat, and Frank Cook. *21 Things Every Future Engineer Should Know: A Practical Guide for Students and Parents.* Chicago: Kaplan AEC Education, 2007.

Repcheck, Jack. *The Man Who Found Time: James Hutton and the Discovery of the Earth's Antiquity.* Reading, Mass.: Perseus Books, 2003.

Rescher, Nicholas. *On Leibniz.* Pittsburgh, Pa.: University of Pittsburgh Press, 2003.

Reston, James. *Galileo: A Life.* New York: HarperCollins, 1994.

Rhodes, Richard. *The Making of the Atomic Bomb.* New York: Simon & Schuster, 1986.

Rigutti, Mario. *A Hundred Billion Stars.* Translated by Mirella Giacconi. Cambridge, Mass.: MIT Press, 1984.

Ring, Merrill. *Beginning with the Presocratics.* 2d ed. New York: McGraw-Hill, 1999.

Riordan, Michael. *The Hunting of the Quark.* New York: Simon & Schuster, 1987.

Roan, Sharon. *Ozone Crisis: The Fifteen-Year Evolution of a Sudden Global Emergency.* New York: John Wiley & Sons, 1989.

Robinson, Daniel N. *An Intellectual History of Psychology.* 3d ed. Madison: University of Wisconsin Press, 1995.

Rogers, J. H. *The Giant Planet Jupiter.* New York: Cambridge University Press, 1995.

Rose, Frank. *West of Eden: The End of Innocence at Apple Computer.* New York: Viking, 1989.

Rosenthal-Schneider, Ilse. *Reality and Scientific Truth: Discussions with Einstein, von Laue, and Planck.* Detroit: Wayne State University Press, 1980.

Rossi, Paoli. *The Birth of Modern Science.* Translated by Cynthia De Nardi Ipsen. Oxford, England: Blackwell, 2001.

Rowan-Robinson, Michael. *Cosmology.* London: Oxford University Press, 2003.

Rowland, Wade. *Galileo's Mistake: A New Look at the Epic Confrontation Between Galileo and the Church.* New York: Arcade, 2003.

Rudin, Norah, and Keith Inman. *An Introduction to Forensic DNA Analysis.* Boca Raton, Fla.: CRC Press, 2002.

Rudwick, Martin, J. S. *The Great Devonian Controversy: The Shaping of Scientific Knowledge Among Gentlemanly Specialists.* Chicago: University of Chicago Press, 1985.

Rudwick, M. J. S. *The Meaning of Fossils: Episodes in the History of Paleontology.* Chicago: University of Chicago Press, 1985.

Ruestow, Edward Grant. *The Microscope in the Dutch Republic: The Shaping of Discovery.* New York:

Cambridge University Press, 1996.

Ruse, Michael. *The Darwinian Revolution: Science Red in Tooth and Claw.* 2d ed. Chicago: University of Chicago Press, 1999.

Ruspoli, Mario. *Cave of Lascaux.* New York: Harry N. Abrams, 1987.

Sagan, Carl. *Cosmos.* New York: Random House, 1980.

Sagan, Carl, and Ann Druyan. *Comet.* New York: Random House, 1985.

Sandler, Stanley I. *Chemical and Engineering Thermodynamics.* New York: John Wiley & Sons, 1998.

Sang, James H. *Genetics and Development.* London: Longman, 1984.

Sargent, Frederick. *Hippocratic Heritage: A History of Ideas About Weather and Human Health.* New York: Pergamon Press, 1982.

Sargent, Rose-Mary. *The Diffident Naturalist: Robert Boyle and the Philosophy of Experiment.* Chicago: University of Chicago Press, 1995.

Sauer, Mark V. *Principles of Oocyte and Embryo Donation.* New York: Springer, 1998.

Schaaf, Fred. *Comet of the Century: From Halley to Hale-Bopp.* New York: Springer-Verlag, 1997.

Schatzkin, Paul. *The Boy Who Invented Television: A Story of Inspiration, Persistence, and Quiet Passion.* Silver Spring, Md.: TeamCon Books, 2002.

Schiffer, Michael Brian. *Draw the Lightning Down: Benjamin Franklin and Electrical Technology in the Age of the Enlightenment.* Berkeley: University of California Press, 2003.

Schlagel, Richard H. *From Myth to Modern Mind: A Study of the Origins and Growth of Scientific Thought.* New York: Peter Lang Publishing, 1996.

Schlegel, Eric M. *The Restless Universe: Understanding X-Ray Astronomy in the Age of Chandra and Newton.* London: Oxford University Press, 2002.

Schliemann, Heinrich. *Troy and Its Remains: A Narrative of Researches and Discoveries Made on the Site of Ilium and in the Trojan Plain.* London: J. Murray, 1875.

Schneider, Stephen H. *Global Warming: Are We Entering the Greenhouse Century?* San Francisco: Sierra Club Books, 1989.

Schofield, Robert E. *The Enlightened Joseph Priestley: A Study of His Life and Work from 1773 to 1804.* University Park: Pennsylvania State University Pres, 2004.

Scholz, Christopher, and Benoit B. Mandelbrot. *Fractals in Geophysics.* Boston: Kirkäuser, 1989.

Schonfelder, V. *The Universe in Gamma Rays.* New York: Springer-Verlag, 2001.

Schopf, J. William, ed. *The Earth's Earliest Biosphere.* Princeton, N.J.: Princeton University Press, 1983.

Schorn, Ronald A. *Planetary Astronomy: From Ancient Times to the Third Millennium.* College Station: Texas A&M University Press, 1999.

Schwinger, Julian. *Einstein's Legacy.* New York: W. H. Freeman, 1986.

Sears, M., and D. Merriman, eds. *Oceanography: The Past.* New York: Springer-Verlag, 1980.

Seavey, Nina Gilden, Jane S. Smith, and Paul Wagner. *A Paralyzing Fear: The Triumph over Polio in America.* New York: TV Books, 1998.

Segalowitz, Sid J. *Two Sides of the Brain: Brain Lateralization Explored.* Englewood Cliffs, N.J.: Prentice-Hall, 1983.

Segrè, Emilio. *From X-Rays to Quarks.* San Francisco: W. H. Freeman, 1980.

Sekido, Yataro, and Harry Elliot. *Early History of Cosmic Ray Studies: Personal Reminiscences with Old Photographs.* Boston: D. Reidel, 1985.

Sfendoni-Mentzou, Demetra, et al., eds. *Aristotle and Contemporary Science.* 2 vols. New York: P. Lang, 2000-2001.

Shank, Michael H. *The Scientific Enterprise in Antiquity and the Middle Ages.* Chicago: University of Chicago Press, 2000.

Sharratt, Michael. *Galileo: Decisive Innovator.* Cambridge, Mass.: Blackwell, 1994.

Shectman, Jonathan. *Groundbreaking Scientific Experiments, Investigations, and Discoveries of the Eighteenth Century.* Westport, Conn.: Greenwood Press, 2003.

Sheehan, John. *The Enchanted Ring: The Untold Story of Penicillin.* Cambridge, Mass.: MIT Press, 1982.

Shilts, Randy. *And the Band Played On: Politics, People, and the AIDS Epidemic.* New York: St. Martin's Press, 1987.

Silk, Joseph. *The Big Bang.* Rev. ed. New York: W. H. Freeman, 1989.

Silverstein, Arthur M. *A History of Immunology.* San Diego: Academic Press, 1989.

Simmons, John. *The Scientific Hundred: A Ranking of the Most Influential Scientists, Past and Present.* Secaucus, N.J.: Carol, 1996.

Simon, Randy, and Andrew Smith. *Superconductors: Conquering Technology's New Frontier.* New York: Plenum Press, 1988.

Simpson, A. D. C., ed. *Joseph Black, 1728-1799: A Commemorative Symposium.* Edinburgh: Royal Scottish

Museum, 1982.

Singh, Simon. *Fermat's Enigma: The Epic Quest to Solve the World's Greatest Mathematical Problem.* New York: Anchor, 1998.

Slayton, Donald K., with Michael Cassutt. *Deke! U.S. Manned Space: From Mercury to the Shuttle.* New York: Forge, 1995.

Smith, A. Mark. *Ptolemy and the Foundations of Ancient Mathematical Optics.* Philadelphia: American Philosophical Society, 1999.

Smith, G. C. *The Boole-De Morgan Correspondence, 1842-1864.* New York: Oxford University Press, 1982.

Smith, Jane S. *Patenting the Sun: Polio and the Salk Vaccine.* New York: Anchor/Doubleday, 1991.

Smith, Robert W. *The Space Telescope: A Study of NASA, Science, Technology and Politics.* New York: Cambridge University Press, 1989.

Smyth, Albert Leslie. *John Dalton, 1766-1844.* Aldershot, England: Ashgate, 1998.

Snider, Alvin. *Origin and Authority in Seventeenth-Century England: Bacon, Milton, Butler.* Toronto: University of Toronto Press, 1994.

Sobel, Dava. *Galileo's Daughter: A Historical Memoir of Science, Faith, and Love.* New York: Penguin Books, 2000.

_______. *Longitude: The True Story of a Lone Genius Who Solved the Greatest Scientific Problem of His Time.* New York: Penguin Books, 1995.

Spangenburg, Ray, and Diane Kit Moser. *Modern Science: 1896-1945.* Rev ed. New York: Facts on File, 2004.

Spilker, Linda J., ed. *Passage to a Ringed World: The Cassini-Huygens Mission to Saturn and Titan.* Washington, D.C.: National Aeronautics and Space Administration, 1997.

Stanley, H. Eugue, and Nicole Ostrowsky, eds. *On Growth and Form: Fractal and Non-Fractal Patterns in Physics.* Dordrecht, the Netherlands: Martinus Nijhoff, 1986.

Starr, Cecie, and Ralph Taggart. *Biology.* 5th ed. Belmont, Calif.: Wadsworth, 1989.

Stefik, Mark J., and Vinton Cerf. *Internet Dreams: Archetypes, Myths, and Metaphors.* Cambridge, Mass.: MIT Press, 1997.

Stein, Sherman. *Archimedes: What Did He Do Besides Cry Eureka?* Washington, D.C.: Mathematical Association of America, 1999.

Steiner, Robert F., and Seymour Pomerantz. *The Chemistry of Living Systems.* New York: D. Van Nostrand, 1981.

Stewart, Ian, and David Tall. *Algebraic Number Theory and Fermat's Last Theorem.* 3d ed. Natick, Mass.: AK Peters, 2002.

Stigler, Stephen M. *The History of Statistics.* Cambridge, Mass.: Harvard University Press, 1986.

Stine, Gerald. *AIDS 2005 Update.* New York: Benjamin Cummings, 2005.

Strathern, Paul. *Mendeleyev's Dream: The Quest for the Elements.* New York: Berkeley Books, 2000.

Streissguth, Thomas. *John Glenn.* Minneapolis, Minn.: Lerner, 1999.

Strick, James. *Sparks of Life: Darwinism and the Victorian Debates over Spontaneous Generation.* Cambridge, Mass.: Harvard University Press, 2000.

Strogatz, Steven H. *Nonlinear Dynamics and Chaos: With Applications to Physics, Biology, Chemistry and Engineering.* Reading, Mass.: Perseus, 2001.

Struik, Dirk J. *The Land of Stevin and Huygens: A Sketch of Science and Technology in the Dutch Republic During the Golden Century.* Boston: Kluwer, 1981.

Stryer, Lubert. *Biochemistry.* 2d ed. San Francisco: W. H. Freeman, 1981.

Stukeley, William. *The Commentarys, Diary, & Common-Place Book & Selected Letters of William Stukeley.* London: Doppler Press, 1980.

Sturtevant, A. H. *A History of Genetics.* 1965. Reprint. Woodbury, N.Y.: Cold Spring Harbor Laboratory Press, 2001.

Sullivan, Woodruff T., ed. *Classics in Radio Astronomy.* Boston: D. Reidel, 1982.

_______. *The Early Years of Radio Astronomy. Reflections Fifty Years After Jansky's Discovery.* New York: Cambridge University Press, 1984.

Sulston, John, and Georgina Ferry. *The Common Thread: A Story of Science, Politics, Ethics, and the Human Genome.* Washington, D.C.: Joseph Henry Press, 2002.

Sutton, Christine. *The Particle Connection.* New York: Simon & Schuster, 1984.

Suzuki, David T., and Peter Knudtson. *Genethics.* Cambridge, Mass.: Harvard University Press, 1989.

Swanson, Carl P., Timothy Merz, and William J. Young. *Cytogenetics: The Chromosome in Division, Inheritance, and Evolution.* 2d ed. Englewood Cliffs, N.J.: Prentice-Hall, 1980.

Swetz, Frank, et al., eds. *Learn from the Masters.* Washington, D.C.: Mathematical Association of America, 1995.

Tanford, Charles. *Franklin Stilled the Waves.* Durham, N.C.: Duke University Press, 1989.

Tarbuck, Edward J., and Frederick K. Lutgens. *The Earth: An Introduction to Physical Geology*. Columbus, Ohio: Charles E. Merrill, 1984.

Tattersall, Ian. *The Last Neanderthal: The Rise, Success, and Mysterious Extinction of Our Closest Human Relatives*. New York: Macmillan, 1995.

Taub, Liba Chaia. *Ptolemy's Universe: The Natural Philosophical and Ethical Foundations of Ptolemy's Astronomy*. Chicago: Open Court, 1993.

Tauber, Alfred I. *Metchnikoff and the Origins of Immunology: From Metaphor to Theory*. New York: Oxford University Press, 1991.

Taubes, Gary. *Nobel Dreams: Power, Deceit and the Ultimate Experiment*. New York: Random House, 1986.

Taylor, Michael E. *Partial Differential Equations I: Basic Theory*. 2d ed. New York: Springer, 2011.

Taylor, Peter Lane. *Science at the Extreme: Scientists on the Cutting Edge of Discovery*. New York: McGraw-Hill, 2001.

Thomas, John M. *Michael Faraday and the Royal Institution: The Genius of Man and Place*. New York: A. Hilger, 1991.

Thompson, A. R., James M. Moran, and George W. Swenson, Jr. *Interferometry and Synthesis in Radio Astronomy*. New York: John Wiley & Sons, 1986.

Thompson, D'Arcy Wentworth. *On Growth and Form*. Mineola, N.Y.: Dover, 1992.

Thoren, Victor E., with John R. Christianson. *The Lord of Uraniborg: A Biography of Tycho Brahe*. New York: Cambridge University Press, 1990.

Thrower, Norman J. W., ed. *Standing on the Shoulders of Giants: A Longer View of Newton and Halley*. Berkeley: University of California Press, 1990.

Thurman, Harold V. *Introductory Oceanography*. 4th ed. Westerville, Ohio: Charles E. Merrill, 1985.

Tietjen, Jill S., et al. *Keys to Engineering Success*. Upper Saddle River, N.J.: Prentice-Hall, 2001.

Tillery, Bill W., Eldon D. Enger, and Frederick C. Ross. *Integrated Science*. New York: McGraw-Hill, 2001.

Tiner, John Hudson. *Louis Pasteur: Founder of Modern Medicine*. Milford, Mich.: Mott Media, 1990.

Todhunter, Isaac. *A History of the Mathematical Theory of Probability: From the Time of Pascal to that of Laplace*. Sterling, Va.: Thoemmes Press, 2001.

Tombaugh, Clyde W., and Patrick Moore. *Out of Darkness: The Planet Pluto*. Harrisburg, Pa.: Stackpole Books, 1980.

Toulmin, Stephen, and June Goodfield. *The Fabric of the Heavens: The Development of Astronomy and Dynamics*. Chicago: University of Chicago Press, 1999.

Townes, Charles H. *How the Laser Happened: Adventures of a Scientist*. New York: Oxford University Press, 1999.

Traill, David A. *Schliemann of Troy: Treasure and Deceit*. London: J. Murray, 1995.

Trefil, James S. *The Dark Side of the Universe. Searching for the Outer Limits of the Cosmos*. New York: Charles Scribner's Sons, 1988.

_______. *From Atoms to Quarks: An Introduction to the Strange World of Particle Physics*. New York: Charles Scribner's Sons, 1980.

_______. *Space, Time, Infinity: The Smithsonian Views the Universe*. New York: Pantheon Books, 1985.

_______. *The Unexpected Vista*. New York: Charles Scribner's Sons, 1983.

Trefil, James, and Robert M. Hazen. *The Sciences: An Integrated Approach*. New York: John Wiley & Sons, 2003.

Trefil, James, ed. *The Encyclopedia of Science and Technology*. New York: Routledge, 2001.

Trento, Joseph J. *Prescription for Disaster: From the Glory of Apollo to the Betrayal of the Shuttle*. New York: Crown, 1987.

Trinkhaus, Eric, ed. *The Emergence of Modern Humans: Biocultural Adaptations in the Later Pleistocene*. Cambridge, England: Cambridge University Press, 1989.

Trounson, Alan O., and David K. Gardner, eds. *Handbook of In Vitro Fertilization*. 2d ed. Boca Raton, Fla.: CRC Press, 1999.

Tucker, Tom. *Bolt of Fire: Benjamin Franklin and His Electrical Kite Hoax*. New York: Public Affairs Press, 2003.

Tucker, Wallace H., and Karen Tucker. *Revealing the Universe: The Making of the Chandra X-Ray Observatory*. Cambridge, Mass.: Harvard University Press, 2001.

Tunbridge, Paul. *Lord Kelvin: His Influence on Electrical Measurements and Units*. London, U.K.: P. Peregrinus, 1992.

Tuplin, C. J., and T. E. Rihll, eds. *Science and Mathematics in Ancient Greek Culture*. New York: Oxford University Press, 2002.

Turnill, Reginald. *The Moonlandings: An Eyewitness Account*. New York: Cambridge University Press, 2003.

United States Office of the Assistant Secretary for

Nuclear Energy. *The First Reactor.* Springfield, Va.: National Technical Information Service, 1982.

University of Chicago Press. *Science and Technology Encyclopedia.* Chicago: Author, 2000.

Van Allen, James A. *Origins of Magnetospheric Physics.* Expanded ed. 1983. Reprint. Washington, D.C.: Smithsonian Institution Press, 2004.

Van Dulken, Stephen. *Inventing the Nineteenth Century: One Hundred Inventions That Shaped the Victorian Age.* New York: New York University Press, 2001.

Van Heijenoort, Jean. *From Frege to Gödel: A Source Book in Mathematical Logic, 1879-1931.* Cambridge, Mass.: Harvard University Press, 2002.

Verschuur, Gerrit L. *Hidden Attraction: The History and Mystery of Magnetism.* New York: Oxford University Press, 1993.

_______. *The Invisible Universe Revealed: The Story of Radio Astronomy.* New York: Springer-Verlag, 1987.

Villard, Ray, and Lynette R. Cook. *Infinite Worlds: An Illustrated Voyage to Planets Beyond Our Sun.* Foreword by Geoffrey W. Marcy and afterword by Frank Drake. Berkeley: University of California Press, 2005.

Viney, Wayne. *A History of Psychology: Ideas and Context.* Boston: Allyn & Bacon, 1993.

Vogt, Gregory L. *John Glenn's Return to Space.* Brookfield, Conn.: Millbrook Press, 2000.

Von Bencke, Matthew J. *The Politics of Space: A History of U.S.-Soviet/Russian Competition and Cooperation in Space.* Boulder, Colo.: Westview Press, 1996.

Wagener, Leon. *One Giant Leap: Neil Armstrong's Stellar American Journey.* New York: Forge Books, 2004.

Wakefield, Robin, ed. *The First Philosophers: The Presocratics and the Sophists.* New York: Oxford University Press, 2000.

Waldman, G. *Introduction to Light.* Englewood Cliffs, N.J.: Prentice Hall, 1983.

Walker, James S. *Physics.* 2d ed. Upper Saddle River, N.J.: Pearson Prentice Hall, 2004.

Wallace, Robert A., Jack L. King, and Gerald P. Sanders. *Biosphere: The Realm of Life.* 2d ed. Glenview, Ill.: Scott, Foresman, 1988.

Waller, John. *Einstein's Luck: The Truth Behind Some of the Greatest Scientific Discoveries.* New York: Oxford University Press, 2002.

_______. *Fabulous Science: Fact and Fiction in the History of Science Discovery.* Oxford, England: Oxford University Press, 2004.

Walt, Martin. *Introduction to Geomagnetically Trapped Radiation.* New York: Cambridge University Press, 1994.

Wambaugh, Joseph. *The Blooding.* New York: Bantam Books, 1989.

Wang, Hao. *Reflections on Kurt Gödel.* Cambridge, Mass.: MIT Press, 1985.

Watson, James D. *The Double Helix: A Personal Account of the Discovery of the Structure of DNA.* Reprint. New York: W. W. Horton, 1980.

Watson, James D., and John Tooze. *The DNA Story.* San Francisco: W. H. Freeman, 1981.

Watson, James D., et al. *Molecular Biology of the Gene.* 4th ed. Menlo Park, Calif.: Benjamin/Cummings, 1987.

Weber, Robert L. *Pioneers of Science: Nobel Prize Winners in Physics.* 2d ed. Philadelphia: A. Hilger, 1988.

Weedman, Daniel W. *Quasar Astrophysics.* Cambridge, England: Cambridge University Press, 1986.

Wells, Spencer. *The Journey of Man: A Genetic Odyssey.* Princeton, N.J.: Princeton University Press, 2002.

Westfall, Richard S. *Never at Rest: A Biography of Isaac Newton.* New York: Cambridge University Press, 1980.

Wheeler, J. Craig. *Cosmic Catastrophe: Supernovae and Gamma-Ray Bursts.* London: Cambridge University Press, 2000.

Whiting, Jim, and Marylou Morano Kjelle. *John Dalton and the Atomic Theory.* Hockessin, Del.: Mitchell Lane, 2004.

Whitney, Charles. *Francis Bacon and Modernity.* New Haven, Conn.: Yale University Press, 1986.

Whyte, A. J. *The Planet Pluto.* New York: Pergamon Press, 1980.

Wilford, John Noble. *The Mapmakers.* New York: Alfred A. Knopf, 1981.

_______. *The Riddle of the Dinosaur.* New York: Alfred A. Knopf, 1986.

Wilkie, Tom, and Mark Rosselli. *Visions of Heaven: The Mysteries of the Universe Revealed by the Hubble Space Telescope.* London: Hodder & Stoughton, 1999.

Will, Clifford M. *Was Einstein Right?* New York: Basic Books, 1986.

Williams, F. Mary, and Carolyn J. Emerson. *Becoming Leaders: A Practical Handbook for Women in Engineering, Science, and Technology.* Reston, Va.: American Society of Civil Engineers, 2008.

Williams, Garnett P. *Chaos Theory Tamed.* Washington,

D.C.: National Academies Press, 1997.

Williams, James Thaxter. *The History of Weather.* Commack, N.Y.: Nova Science, 1999.

Williams, Trevor I. *Howard Florey: Penicillin and After.* London: Oxford University Press, 1984.

Wilmut, Ian, Keith Campbell, and Colin Tudge. *The Second Creation: The Age of Biological Control by the Scientists That Cloned Dolly.* London: Headline, 2000.

Wilson, Andrew. *Space Shuttle Story.* New York: Crescent Books, 1986.

Wilson, Colin. *Starseekers.* Garden City, N.Y.: Doubleday, 1980.

Wilson, David B. *Kelvin and Stokes: A Comparative Study in Victorian Physics.* Bristol, England: Adam Hilger, 1987.

Wilson, Jean D. *Wilson's Textbook of Endocrinology.* 10th ed. New York: Elsevier, 2003.

Windley, Brian F. *The Evolving Continents.* 2d ed. New York: John Wiley & Sons.

Wojcik, Jan W. *Robert Boyle and the Limits of Reason.* New York: Cambridge University Press, 1997.

Wolf, Fred Alan. *Taking the Quantum Leap.* San Francisco: Harper & Row, 1981.

Wollinsky, Art. *The History of the Internet and the World Wide Web.* Berkeley Heights, N.J.: Enslow, 1999.

Wolpoff, M. *Paleoanthropology.* 2d ed. Boston: McGraw-Hill, 1999.

Wood, Michael. *In Search of the Trojan War.* Berkeley: University of California Press, 1988.

Wormald, B. H. G. *Francis Bacon: History, Politics, and Science, 1561-1626.* New York: Cambridge University Press, 1993.

Yen, W. M., Marc D. Levenson, and Arthur L. Schawlow. *Lasers, Spectroscopy, and New Ideas: A Tribute to Arthur L. Schawlow.* New York: Springer-Verlag, 1987.

Yoder, Joella G. *Unrolling Time: Huygens and the Mathematization of Nature.* New York: Cambridge University Press, 2004.

Yolton, John W. ed. *Philosophy, Religion, and Science in the Seventeenth and Eighteenth Centuries.* Rochester, N.Y.: University of Rochester Press, 1990.

Zeilik, Michael. *Astronomy: The Evolving Universe.* 4th ed. New York: Harper & Row, 1985.

Index

SUBJECT INDEX

Note: Page numbers in **bold** indicate main discussion

C

Q

R

S

T

U

X

Y

Z